Ecology and Management of the North American Moose

A Wildlife Management Institute Book

Ecology and Management of the North American Moose

Second Edition

Compiled and edited by
Albert W. Franzmann and Charles C. Schwartz

Technical editor Richard E. McCabe

Field sketch illustrations by William D. Berry

UNIVERSITY PRESS OF COLORADO

The Wildlife Management Institute (WMI) is a private, nonprofit, scientific and educational organization, based in Washington, D.C. WMI's sole objective, since its founding in 1911, has been to help advance restoration and sound management of North America's wildlife resources. As part of the Institute's program, scientific information generated through research and management experiences is consolidated, published, and used to strengthen resource decision making, management opportunities and methodologies, and general understanding of and appreciation for wildlife and their habitats. *Ecology and Management of the North American Moose* is one of thirty books produced by WMI since 1942, including the award-winning *Ducks, Geese and Swans of North America; Big Game of North America; Mule and Black-tailed Deer of North America; Elk of North America; White-tailed Deer: Ecology and Management; Ecology and Management of the Mourning Dove, Ecology and Management of the Wood Duck; North American Elk: Ecology and Management;* and *Pronghorn: Ecology and Management.* For additional information about WMI, its publications, programs and membership, go to *www.wildlifemanagementinstitute.org.*

Published by the University Press of Colorado
5589 Arapahoe Avenue, Suite 206C
Boulder, Colorado 80303

Previously published by Smithsonian Institution Press, Washington D.C.

Printed in the United States of America

The University Press of Colorado is a proud member of the Association of American University Presses.

The University Press of Colorado is a cooperative publishing enterprise supported, in part, by Adams State College, Colorado State University, Fort Lewis College, Mesa State College, Metropolitan State College of Denver, University of Colorado, University of Northern Colorado, and Western State College of Colorado.

The paper used in this publication meets the minimum requirements of the American National Standard for Information Sciences—Permanence of Paper for Printed Library Materials. ANSI Z39.48-1992

Library of Congress Cataloging-in-Publication Data

Ecology and management of the North American moose / compiled and edited by Albert W. Franzmann and Charles C. Schwartz ; technical editor, Richard W. McCabe ; field sketch illustrations by William D. Berry. — 2nd ed.
p. cm.
"A Wildlife Management Institute book."
Includes bibliographical references and index.
ISBN 978-0-87081-895-0 (hardcover : alk. paper) 1. Moose—North America. 2. Wildlife management—North America. I. Franzmann, Albert W. II. Schwartz, Charles C.
QL737.U55E336 2007
639.97'9657—dc22

2007039015

16 10 9 8 7 6 5 4 3 2

Dust jacket photograph by Victor Van Ballenberghe
Endsheet illustrations by H. R. Timmerman

Copy editor: Danielle Ponsolle
Design by Janice Wheeler
Typesetting and layout by Blue Heron

Dedication

We originally decided to dedicate this book to pioneer moose biologists, by listing those who paved the way for sound management of the species. As we began compiling the names of those responsible, we realized that there were many and each, in his or her own way, was an important contributor. We also realized that we were likely to omit some inadvertently, including persons who contributed significantly without being active "moose" biologists. The art of managing a species depends on information and insights from a variety of sources, and it would be impossible to include all the pioneers. Therefore, we dedicate this book to all those who helped build the information base of moose science and management. Without them, this project could not have been conceived or started. We include, too, their spouses and family members, whose support was necessary for them to accomplish their work.

Contributors

Warren B. Ballard is professor and Bricker Chair in Wildlife Management, Department of Natural Resources Management, Texas Tech University, Lubbock, Texas.

Arnold H. Boer retired as director of the Fish and Wildlife Branch of the Department of Natural Resources and Energy in Fredericton, New Brunswick, and currently is a consultant.

Anthony B. Bubenik, deceased, was a research scientist for the Wildlife Branch of the Ontario Department of Lands and Forests in Maple, Ontario.

M. E. "Mike" Buss retired as regional wildlife analyst for the Ontario Ministry of Natural Resources in Bracebridge, Ontario.

Kenneth N. Child retired as regional wildlife biologist for the British Columbia Ministry of Environment, Lands and Parks and as senior environmental coordinator for the Northern Region of British Columbia Hydro in Prince George.

Vincent F.J. Crichton is manager of the Game, Fur and Problem Wildlife Section of the Wildlife and Ecosystem Protection Branch of Manitoba Conservation in Winnipeg, Manitoba.

Albert W. Franzmann retired as wildlife research veterinarian and is a former director of and consultant for the International Wildlife Veterinary Service, Inc., in Soldotna, Alaska.

Kris J. Hundertmark is assistant professor of wildlife ecology at the Institute of Arctic Biology and at the Department of Biology and Wildlife, University of Alaska, Fairbanks.

Patrick D. Karns, retired as a wildlife biologist for the Minnesota Department of Fish and Wildlife and worked extensively in the boreal forest ecosystem.

Murray W. Lankester retired as professor of biology at Lakehead University in Thunder Bay, Ontario, after thirty-two years and as co-editor of the journal *Aces* for the past twelve years.

Richard E. McCabe is executive vice president of the Wildlife Management Institute in Washington, D.C.

James M. Peek is professor emeritus of wildlife resources at the University of Idaho in Moscow.

Henry M. "Milt" Reeves retired as a biologist for the U.S. Fish and Wildlife Service.

Wayne L. Regelin retired as deputy commissioner of the Alaska Department of Fish and Game in Juneau, Alaska.

Lyle A. Renecker is a wildlife biologist and consultant in Stratford, Ontario, and an adjunct professor in the Department of Biology, Laurentian University, Sudbury, Ontario.

William M. Samuel is professor emeritus in the Department of Biological Sciences, University of Alberta, Edmonton.

Charles C. Schwartz is leader of the Interagency Grizzly Bear Study Team at the U.S. Geological Survey's Northern Rocky Mountain Science Center in Bozeman, Montana.

Robert R. Stewart is principal BASSA, Tropical Wood Production in Managua, Nicaragua.

Ian D. Thompson is a wildlife research biologist with the Canadian Forest Service in Sault Ste. Marie, Ontario.

H. R. "Tim" Timmermann retired as regional wildlife biologist for the Ontario Ministry of Natural Resources and as co-editor of the journal *Aces.*

Victor Van Ballenberghe, a former wildlife research biologist for the Research Branch of the U.S. Forest Service in Anchorage, Alaska, currently is an affiliate professor of biology and wildlife at the University of Alaska, Fairbanks.

Contents

Figures

Tables

In Memoriam

Anthony (Tony) Bubenik, author of two chapters in this book, passed away on February 4, 1995. He was born in 1913 in Jevicko, Moravia, now part of the Czech Republic. Tony obtained a Ph.D. in organic chemistry from the University of Brno, but applied himself professionally to wildlife research. He became head of the Department of Wildlife Research in the Czechoslovak Academy of Sciences. With the communist takeover after World War II, Tony was forced to work as a laborer. He later resumed his research in Yugoslavia and Switzerland before emigrating to Canada in 1970 to work for the Ontario Department of Lands and Forests at the Southern Research Station in Maple, Ontario.

Tony quickly developed an interest in moose and attended his first North American Moose Conference in 1972. Over the years, he regularly attended the annual North American Moose Conferences and not only contributed papers, but became a sounding board and challenger of ideas and theories in the workshop sessions. Tony's interests and knowledge were very diverse, and he was involved in a wide variety of studies on moose. Fluent in six languages, he corresponded regularly with colleagues from all over the world. His research activities resulted in nearly 300 publications and articles and several books. He received the Distinguished Moose Biologist Award from his peers in 1982.

One of Tony's great interests and strengths was in the field of hunter education. In the 1970s, he started moose hunter seminars in Ontario, based on his European experiences. He covered moose biology, sex and age identification, field care and handling, hunting techniques, ethics and hunting law. He challenged hunters to improve their shooting skills using life-sized moose silhouettes. He had a special rapport with hunters and easily gained their attention and respect.

Tony was instrumental in introducing new management strategies for moose in many North American jurisdictions. He stimulated the concept of selective harvest strategies designed to protect mature breeding stock and increase moose populations. He introduced biologists to the concepts of moose population social order, disorder, balance and social classes. His terms "teenager," "prime" and "senior" soon were being used to describe moose population structures. In the field of moose biology, Tony was a pioneer, an innovator, a leader.

Above all, Tony Bubenik was an achiever. He was a man of great resourcefulness conviction, integrity, artistry and a passionate scientist. He gave advice and encouragement freely and had profound impact on all who were fortunate to know and work with him. He challenged us all with his enthusiasm, lust for knowledge and life. Tony was an inspiration to many, and leaves a rich, diverse and enduring legacy.

Preface

Why a book about moose?

During happy hour one evening at the 1991 North American Moose Conference in Anchorage, Alaska, a group of biologists discussed this very subject. More than 35 years had passed since anything substantial on the moose had appeared in the professional literature. Randolph Peterson's 1955 tome *North American Moose* was the last. Of course, a great deal of information had been generated and published on moose in the interim, but not in a synthesized and composite source. Everyone in the happy hour group—a biased assemblage, to be sure—agreed that a single and definitive book on the moose was timely. More so it was needed because of the increasing complexity of managing the species and its habitats.

The conviction that a new, up-to-date and comprehensive monograph on moose was needed and the group's enthusiasm to initiate the work were sufficient to propel the notion beyond happy hour consensus. A tentative outline and schedule emerged, and prospective contributors were queried about participating.

We enlisted authors who had extensive experience with and knowledge of the various outline topics. All were accomplished biologists and dedicated natural resource conservationists. Most had devoted a significant portion of their professional careers to managing and/or researching moose. Each eagerly committed to the monumental labor that the book represented.

Certain, then, that a new book *could* be accomplished, we approached the Wildlife Management Institute to determine its interest in handling our planned opus. WMI, with a tradition of producing high-quality and enduring books on wildlife species and topics, agreed to review the prospectus, but also gave us some forewarning about publishing in general and wildlife publications in particular. We were advised that numerous wildlife topics deserve book treatment, but very few are undertaken because of matters of readability and marketability. A WMI book, we learned, must be, besides state-of-the-science, written so that persons other than just moose biologists and other scientists can understand and appreciate the significance of the information included. It also must appeal to a broad audience, because a book narrow in readership is a losing proposition financially. A book that costs too much or has too few buyers for whatever other reason has a short shelf life. And a short shelf life in the wildlife literature translates to little use or application of the information and messages. These considerations, WMI informed us, were as practically determining as was need for establishing a base in the literature. Furthermore, to our heightened consternation, we learned that WMI had approved and taken on only a very small percentage of the book proposals it received in the past score of years.

We were elated when WMI agreed to work with us, and the 6-year process of book production—from generation to reviews and multiple edits to publication—has been extraordinarily intense, detail-oriented, time-consuming, energizing, enervating and sometimes frustrating for everyone. The outcome is entirely gratifying.

We believe this book is highly readable, interesting, informative and useful. We did *not* intend and do *not* consider it to be the final word on all things moose. To the contrary, we see it mainly as a "new and improved" informational base, from which and against which research and data can be compared.

The foremost expectation of the authors of *Ecology and Management of the North American Moose,* although perhaps not that of WMI, is that this first edition will be rapidly outdated. That will mean significant advances have been made in wildlife science and technology to the betterment of North American moose.

Every happy hour should conclude this satisfactorily.

Albert W. Franzmann and Charles C. Schwartz

Acknowledgments

Any work of the type and scope of this volume requires the cooperation and assistance of a great many people. And in the hope that no one has been overlooked, the editors and authors of this book gratefully acknowledge the significant contributions of those individuals listed below alphabetically.

In addition, the editors recognize that this volume would not have been envisioned were it not for the persons, organizations and businesses that have actively supported progressive wildlife management throughout North America. The editors and authors recognize that this volume could not have been reasonably undertaken were it not for the efforts of many citizens past and present who detailed by word, artwork or photograph the ecology of moose. The editors and authors also recognize that, in the absence of the knowledge and experiences of virtually every resource specialist who has studied or managed moose, this volume would not have been completed.

Special thanks are due to David W. Johnston of Fairfax, Virginia, for valuable editorial insights, and the Research Librarian Staff, William Jasper Kerr Library, Oregon State University, Corvallis, for literature research services. Finally, enormous credit is owed Kelly G. Wadsworth, Wildlife Management Institute, for coordinating the manuscript through its multiple stages of preparation.

K. Absolon, *Masaryk University, Brno*
L. Adams, *U.S. National Park Service, Anchorage, Alaska*
D.A. Aitken, *Department of Biology, College of New Caledonia, Prince George, British Columbia*
C. Alexander, *Vermont Fish and Wildlife Department, St. Johnsbury*
A. Allen, *Thunder Bay, Ontario*
J. Andrews, *Department of Forest Sciences, University of Alaska-Fairbanks*
J. Augusta, *Charles University, Prague, Switzerland*
A.A. Azarolli, *University of Florence, Italy*
L. Balciauskas, *Vilnius, Lithuania*
A.G. Bannikov, *Russian Academy of Sciences, Moscow*
S.P. Barry, *British Columbia Ministry of the Environment, Fish and Wildlife Branch, Prince George*
T.D. Beach, *University of New Brunswick, Fredericton*
J.H. Beattie, *Hudson's Bay Company Archives, Winnipeg, Manitoba*
R. Beaulieu, *Saskatchewan Parks and Renewable Resources, Hudson Bay*
P.D. Beaudette, *New Brunswick Department of Natural Resources and Energy, Fish and Wildlife Branch, Fredericton*
T.J. Bellhouse, *Port Loring, Ontario*
E. Berry, *Fairbanks, Alaska*
P.H. Bewick, *Ontario Ministry of Natural Resources, Sudbury*
H. Bicker-Sturzenegger, *Switzerland*
H. and U. Bicker, *Perdell-Grabs, Switzerland*
A.R. Bisset, *Ontario Ministry of Natural Resources, Kenora*
A.R. Bjorn, *Kluane National Park, Haines Junction, Yukon*
M.K. Black, *Thomas Jefferson Memorial Foundation, Inc., Charlottesville, Virginia*
D. Blower, *British Columbia Ministry of the Envirinment, Victoria*

E.M. Bogomlova, *Moscow, Russia*
R. Bonar, *Weldwood of Canada, Ltd., Hinton, Alberta*
K. Bontaites, *New Hampshire Fish and Game Department, New Hampton*
G.N. Bos, *Alaska Department of Fish and Game, Anchorage*
T. Bouin, *La Maurice National Park, Shawinigan, Quebec*
R.T. Bowyer, *Institute of Arctic Biology, University of Alaska-Fairbanks*
W. Bradford, *Jasper National Park, Jasper, Alberta*
W. Brock, *Voyageurs National Park, International Falls, Minnesota*
M. Bryant, *Seedshedee National Wildlife Refuge, Green River, Wyoming*
G.A. Bubenik, *University of Guelph, Guelph, Ontario*
M. Bubenik, *Thornhill, Ontario*
P.C. Calkin, *University of Buffalo, Department of Geology, Buffalo, New York*
D. Carney, *Blackfeet Fish and Game, Browning, Montana*
G. Carroll, *Alaska Department of Fish and Game, Anchorage*
R. Case, *Department of Renewable Resources, Northwest Territories*
S. Cedarleaf, *Minnesota Department of Natural Resources, St. Paul*
A. Chalk, *Wildlife Research Institute, Ontario Ministry of Natural Resources, Maple*
J-G. Chavarie, *Forillon National Park, Gaspé, Quebec*
C.S. Churcher, *University of Toronto*
B. Collins, *Newfoundland Department of Environment and Lands, Wildlife Division, St. John's.*
W. Collins, *Alaska Department of Fish and Game, Palmer*
D. Connor, *Red Lake Department of Natural Resources, Red Lake, Minnesota*
J. Cromer, *West Virginia Division of Natural Resources, Elkins*
A.F. Cunning, *Alaska Department of Fish and Game, Anchorage*
W. Dailey, *Harvard College Library, Cambridge, Massachusetts*
D.A. Dast, *Ontario Ministry of Natural Resources, Gogama*
J.L. Davis, *Fairbanks, Alaska*
L. Deede, *Tamarac National Wildlife Refuge, Rochert, Minnesota*
P. Deering, *Fundy National Park, Alma, New Brunswick*
G. Del Frate, *Alaska Department of Fish and Game, Homer*
D. Demare, *Ontario Ministry of Natural Resources, Sault Ste. Marie*
J. Dempsey, *New Brunswick Department of Natural Resources and Energy, Fish and Wildlife Branch, Fredericton*
L.J. De Vos, *Dubois Collection, Leyden, The Netherlands*
S. Dudinski, *Kakabeka Falls Game Farm, Kakabeka Falls, Ontario*
J. Dunder, *U.S. Bureau of Land Management, Rock Springs, Wyoming*
G. Eason, *Ontario Ministry of Natural Resources, Wawa*
D.S. Eastman, *British Columbia Ministry of the Environment, Victoria*
M. Eliiuk, *Ontario Ministry of Natural Resources, Sioux Lookout*
M. Ellingwood, *North Franklin, Connecticut*
G. Erickson, *Montana Department of Fish, Wildlife and Parks, Helena*
D. Euler, *School of Forestry, Lakehead University, Thunder Bay, Ontario*
K. Evenden, *Glenbow Museum, Calgary, Alberta*
P. Fargey, *Prince Albert National Park, Waskesiu Lakes, Saskatchewan*
J. Faro, *Alaska Department of Fish and Game, Soldotna*
G. Fenton, *Pukaskwa National Park, Marathon, Ontario*
S. Ferguson, *Newfoundland Department of the Environment and Lands, Wildlife Division, St. John's*
C. Fiest, *Minnesota Department of Natural Resources, St. Paul*
C.P. Filonov, *Russian Academy of Sciences, Moscow*
M. Fitzsimmons, *Prince Albert National Park, Waskesiu Lakes, Saskatchewan*
J. Flag, *Mt. Revelstoke and Glacier National Parks, Revelstoke, British Columbia*
B. Flahey, *Manotick, Ontario*
K.K. Flerow, *University of Moscow, Russia*
Z. Frankenberger, *Medical Faculty, Charles University of Prague*
D.M. Franzmann, *Soldotna, Alaska*
C. Gagnon, *Newburg, Oregon*
B. Gamblit, *Minnesota Department of Natural Resources, St. Paul*
J. Gardner, *U.S. Bureau of Land Management, Salt Lake City, Utah*
A.W. Gentry, *British Museum of Natural History, London*
L. Gerrard, *U.S. Bureau of Land Management, Buffalo, Wyoming*
L. Ginsburg, *Museum of Natural History, Paris, France*
R. Graf, *Northwest Territories Department of Renewable Resources, Fort Smith*
M. Graves, *Voyageurs National Park, International Falls, Minnesota*
H. Griese, *Alaska Department of Fish and Game, Anchorage*
K.Gustafson, *New Hampshire Fish and Game Department, Concord*
D. Guthrie, *University of Alaska-Fairbanks*
B. Gutierrez, *Wildlife Management Institute, Washington, D.C.*
J. Haarala, *U.S. Bureau of Indian Affairs, Saulte Ste. Marie, Michigan*
O. Hadley, *U.S. Bureau of Land Management, Dillon, Montana*
M.J. Harrison, *American Philosophical Society, Philadelphia, Pennsylvania*

B. Hawkinson, *Minnesota Department of Natural Resources, St. Paul*
I. Hatter, *British Columbia Ministry of the Environment, Victoria*
M. Hedden, *Maine Historical Commission, Augusta*
M. Hedrick, *National Elk Range, Jackson, Wyoming*
E. Helgren, *Caesar Kleberg Research Institute, Kingsville, Texas*
J. Henley, *Parker River National Wildlife Refuge, Plum Island, Massachusetts*
V.G. Heptner, *Moscow, Russia*
K. Heuer, *Banff National Park, Banff, Alberta*
C. Heusser, *New York University, Department of Biology, Tuxedo*
A. Hicks, *New York Department of Environmental Conservation, Delmar*
P. Hill, *Thomas Gilcrease Institute of American History and Art, Tulsa, Oklahoma*
Z.N. Hnezl, *Masaryk University, Brno*
W. Hobgood, *U.S. Bureau of Land Management, Fairbanks, Alaska*
R.J. Hudson, *Department of Renewable Resources, University of Alberta, Edmonton*
M. Huff, *Kluane National Park, Haines Junction, Yukon*
G.L. Hundertmark, *Accurate Secretarial, Sterling, Alaska*
D. Hunt, *Minnesota Department of Natural Resources, St. Paul*
B. Husson, *Environment Canada, Ottawa*
R.S. Irwin, *Polson, Montana*
C. Jane, *Ontario Ministry of Natural Resources, Wildlife Policy Branch, Toronto*
G. Jense, *Utah Division of Wildlife Resources, Salt Lake City*
D. Jessup, *California Department of Fish and Game, Rancho Cordova*
M. Johnson, *Camas National Wildlife Refuge, Pocatello, Idaho*
R. Kangas, *Thunder Bay, Ontario*
N. Karns, *Atlanta, Michigan*
R. Kacyon, *Alaska Department of Fish and Game, Anchorage*
D.M. Keppie, *University of New Brunswick, Fredericton*
C. Keqink, *Shanghai University, China*
R.N. Kervin, *Ontario Ministry of Natural Resources, Kirkland Lake*
V.D. Kheruvimov, *Tambov, Russia*
J.G. Kie, *Forestry Sciences Laboratory, USDA Forest Service, Fresno, California*
J. Kimball, *Utah Division of Wildlife Resources, Ogden*
D.R. Klein, *Alaska Cooperative Wildlife Research Unit, Fairbanks*
R. Knight, *Reindeer Research Program, Agricultural and Forestry Experiment Station, University of Alaska-Fairbanks*
C. Kogod, *National Geographic Society, Washington, D.C.*
J. Kotok, *Agassiz National Wildlife Refuge, Middle River, Minnesota*
M.V. Kozhukhov, *Yaksha, Siberia*
K. Kraemer, *Portland, Oregon*
P.R. Krausman, *University of Arizona, Tuscon*
A. Krzywinski, *Popielno, Poland*
R. Kufeld, *Colorado Division of Wildlife, Denver*
J. Kunsall, *Utah Division of Wildlife Resources, Ogden*
C. Lafferty, *Nahanni National Park Reserve, Fort Simpson, Northwest Territories*
G. LaJeuness, *Shosone and Arapaho Tribal Fish and Game Department, Fort Washakie, Wyoming*
G. Lamontagne, *Quebec Ministère du Loisir de la Chasse et de la Pêche, Quebec City*
E. Langineau, *Michigan Department of Natural Resources, Lansing*
C. Lapp, *Rice Lake National Wildlife Refuge, McGregor, Minnesota*
D. Larsen, *Yukon Department of Renewable Resources, Fish and Wildlife Branch, Whitehorse*
J. Leinders, *University of Utrecht, The Netherlands*
M. Lenarz, *Minnesota Department of Natural Resources, Wildlife Branch, Grand Rapids*
P.C. Lent, *Boulder, Colorado*
K. Lorenz, *Lorenz Institute of Animal Behavior, Altenberg, Austria*
A. Lovass, *U.S. National Park Service, Anchorage, Alaska*
J. Lykke, *Forestry Director, Vuku, Norway*
T. Lytle, *Colorado Division of Wildlife, Denver*
J.G. MacCracken, *College of Forest, Wildlife, and Range Sciences, University of Idaho, Moscow*
D. MacDonald, *Canadian National Railway, Smithers, British Columbia*
S. Madsen, *U.S. Bureau of Land Management, Vernal, Utah*
K. Martin, *Ontario Ministry of Natural Resources, Huntsville*
V. Masse, *La Mauricie National Park, Shawinigan, Quebec*
M. Masteller, *Alaska Department of Fish and Game, Anchorage*
A.G. Mathews, *Ontario Ministry of Natural Resources, Sioux Lookout*
M.S. McCabe, *Madison, Wisconsin*
T.R. McCabe, *U.S. National Biological Service, Anchorage, Alaska*
K. McCafferty, *Wisconsin Department of Natural Resources, Rhinelander*
D. McCleery, *U.S. Bureau of Land Management, Missoula, Montana*
R.G. McColm, *Quetico Provincial Park, Atikokan, Ontario*
C. McComb, *Little Pend Orielle National Wildlife Refuge, Colville, Washington*
M. McDonald, *Alaska Department of Fish and Game, Anchorage*
H. McIntyre, *Reindeer Research Program, Agricultural and Forestry Experiment Station, University of Alaska-Fairbanks*

J. McKenzie, *North Dakota Game and Fish Department, Bismarck*
B. Meisner, *Prince George, British Columbia*
F. Messier, *Department of Biology, University of Saskatchewan, Saskatoon*
P. Meyer, *White House Historical Association, Washington, D.C.*
V. Miller, *Quetico Provincial Park, Atikokan, Ontario*
D. Miquelle, *Smithsonian Institution, Washington, D.C.*
R.D. Modafferi, *Alaska Department of Fish and Game, Palmer*
D. Moore, *Hudson's Bay Company Archives, Winnipeg, Manitoba*
K. Morris, *Maine Department of Inland Fisheries and Wildlife, Bangor*
K. Morrison, *Ontario Ministry of Natural Resources, Huntsville*
D. Moyles, *Alberta Fish and Wildlife, Edmonton*
D. Mullen, *Moosehorn National Wildlife Refuge, Calais, Maine*
A.A. Nasimovic, *Moscow, Russia*
R. Nelson, *U.S. Forest Service, Washington, D.C.*
T. Nette, *Nova Scotia Department of Lands and Forests, Kentville*
K. Niethammer, *Red Rock Lakes National Wildlife Refuge, Lima, Montana*
K. Nygren, *Ilomantsi, Finland*
M. Obradovich, *U.S. Bureau of Land Management, Huntsville, Utah*
J. Oelfke, *Isle Royale National Park, Houghton, Michigan*
L. Oldenburg, *Idaho Department of Fish and Game, Boise*
J. Olterman, *Colorado Division of Wildlife, Denver*
E. Packee, *Agricultural and Forestry Experiment Station, University of Alaska-Fairbanks*
S. Parker, *Pukaskwa National Park, Marathon, Ontario*
T. Paul, *Alaska Department of Fish and Game, Juneau*
P. Pavlansky, *Charles University, Prague*
M. Payne, *Ontario Ministry of Natural Resources, Huntsville*
C.J. Peddicord, *Wildlife Management Institute, Washington, D.C.*
W.D. Peteete, *Institute for Space Studies, New York, New York*
G.R. Peters, *Superior National Forest, Duluth, Minnesota*
W. Peterson, *Minnesota Department of Natural Resources, Grand Marais*
M. Peterson, *Wood Buffalo National Park, Fort Smith, Northwest Territories*
R. Peterson, *School of Forestry and Wood Products, Michigan Technological University, Houghton*
S. Peterson, *Alaska Department of Fish and Game, Juneau*
L. Pfanmuller, *Minnesota Department of Natural Resources, St. Paul*
F. Phillips, *Newfoundland Wildlife Division, Goose Bay*
M. Plante, *LaMauricie National Park, Shawinigan,*
A. Post, *U.S. Geological Survey, Vashon, Washington*
R. Pratt, *National Geographic Society, Washington, D.C.*
G. Redmond, *New Brunswick Department of Natural Resources and Energy, Fish and Wildlife Branch, Fredericton*
C.S. "Paddy" Reid, *Ontario Ministry of Culture and Communications, Kenora*
J. Reynolds, *Kootenai National Wildlife Refuge, Bonners Ferry, Idaho*
B. Richard, *Kouchibouguac National Park, Kent County, New Brunswick*
J.Rieck, *Washington Department of Wildlife, Olympia*
C.S. Reid, *Ontario Ministry of Culture and Communications, Kenora*
T. Rinkes, *Great Divide Resource Area, Rawlings, Wyoming*
K. Risenhoover, *Department of Fisheries and Wildlife, Texas A & M University, College Station*
P. Rizor, *Arapaho National Wildlife Refuge, Walden, Colorado*
L. Roach, *Ontario Ministry of Natural Resources, Woodview*
C.T. Robbins, *Department of Zoology, Washington State University, Pullman*
D. Roberts, *U.S. Bureau of Land Management, Cheyenne, Wyoming*
K. Robinson, *Terra Nova National Park, Glovertown, Newfoundland*
R. Rothwell, *Wyoming Game and Fish Department, Cheyenne*
D.L. Sabine, *University of New Brunswick, Fredericton*
D. Salinas, *Minnesota Department of Natural Resources, St. Paul*
A. Sands, *U.S. Bureau of Land Management, Boise, Idaho*
P. Schladweiler, *Montana Department of Fish, Wildlife and Parks, Bozeman*
S.M. Schmitt, *Rose Lake Wildlife Pathology Laboratory, East Lansing, Michigan*
G. Schoonveld, *Colorado Division of Wildlife, Fort Collins*
K.J. Schwartz, *Soldotna, Alaska*
M.A. Schwartz, *Soldotna, Alaska*
G. Sedlacek, *Pickering, Ontario*
W. Sharpe, *Trappers International Marketing Service, Prince George, British Columbia*
S.G. Shetler, *Smithsonian Institution, Washington, D.C.*
B. Shoonan, *Kootenay National Park, Radium Hot Springs, British Columbia*
K. Siman, *Czechoslovakia Forests, Prague*
R. Sjostrom, *Bear Lake National Wildlife Refuge, Montpelier, Idaho*
F. Skuncke, *Stockholm, Sweden*
H. Smith, *Ontario Ministry of Natural Resources, Wildlife Policy Branch, Toronto*
P. Smith, *Wildlife Research Institute, Ontario Ministry of Natural Resources, Maple*
S. Smith, *Wyoming Game and Fish Department, Cheyenne*
R.D. Sparrowe, *Wildlife Management Institute, Washington, D.C.*
A. Spiess, *Maine Historical Commission, Augusta*
T. Spraker, *Alaska Department of Fish and Game, Soldotna*

B. Stenquist, *Minnesota Department of Natural Resources, St. Paul*
J. Summers, *Minnesota Department of Natural Resources, Grand Rapids*
M. Sweeny, *Sunkhaze Meadows National Wildlife Refuge, Old Town, Maine*
Z. Tachezy, *Charles University of Prague*
M. Tansy, *Seney National Wildlife Refuge, Seney, Michigan*
P. Tarleton, *Prince Albert National Park, Prince Albert, Saskatchewan*
W. Taylor, *Alaska Department of Fish and Game, Anchorage*
E.S. Telfer, *Canadian Wildlife Service, Edmonton, Alberta*
R. Thompson, *Cape Breton Highlands National Park, Nova Scotia*
A. Todd, *Alberta Department of Environmental Protection, Natural Resources Service, Whitecourt*
C. Todesco, *Ontario Ministry of Natural Resources, Chapleau*
J. Toepfer, *North Dakota*
L. Turs, *U.S. Bureau of Land Management, Spokane, Washington*
L. Upham, *U.S. Bureau of Land Management, Lakewood, Colorado*
R. Vernimen, *U.S. Bureau of Land Management, Anchorage, Alaska*
A. Verone, *Sagamore Hill National Historic Site, Oyster Bay, New York*
G. Vevellio, *Massachusetts Division of Fisheries and Wildlife, Westborough*
M. Vukelich, *Ontario Ministry of Natural Resources, Timmins*
K.D. Wade, *Pukaskwa National Park, Marathon, Ontario*
W. Wagner, *Smithsonian Institution, Washington, D.C.*
L. Walton, *Glenbow Museum, Calgary, Alberta*
R. Ward, *Yukon Department of Renewable Resources, Whitehorse*
F.W. Waters, *Royal Air Force, Irvinestown, Northern Ireland*
R. Watt, *Watertown Lakes National Park, Watertown Park, Alberta*
G.S. Watts, *British Columbia Ministry of the Environment, Fish and Wildlife Branch, Prince George*
K.N. Wehrmeister, *Moose International, Mooseheart, Illinois*
A.J. Welch, *U.S. Bureau of Land Management, Lander, Wyoming*
J. Wentzell, *Kejimikujik National Park, Maitland Bridge, Nova Scotia*
C. Wheaton, *Oregon Department of Fish and Wildlife, Portland*
R. Wheeler, *Ontario Ministry of Natural Resources, Kapuskasing*
B. Wilkinson, *Wildlife Research Institute, Ontario Ministry of Natural Resources, Maple*
O. Williams, *Paris, Ontario*
S. Williamson, *Wildlife Management Institute, North Stratford, New Hampshire*
M. Wilton, *Ontario Ministry of Natural Resources, Algonquin Park District, Whitney*
A. Wolterson, *British Columbia Ministry of the Environment, Fish and Wildlife Branch, Cranbrook*
G. Wright, *University of Idaho, Moscow*
O.E. Yegorov, *Yakutsk, Russia*
P.B. Yurgenson, *Moscow, Russia*
R. Zarnke, *Alaska Department of Fish and Game, Fairbanks*
L. Zieman, *Chicago Historical Society, Chicago, Illinois*

JUNE 15, '71 - HOGAN CREEK, McKINLEY PARK, ALASKA - (EVENING) - FROM LIFE + MEMORY (COMP. JUNE 16)
A SHARP HOLLOW DEFINES BASE OF EAR
EAR WITH BUFFY RIM - A PATCHY REMNANT OF OLD HAIR - BUT BACK OF EAR GRAY-BROWN, LEATHERY LOOKING -
SHORT, NEW DARK HAIR AROUND EYE, ACCENTUATED BY (OR EMPHASIZING) HOLLOW UNDER "BROW" TO GIVE "MELANCHOLY" EFFECT - ONE REASON A COW MOOSE HAS WORN OR WEARY LOOK TO US -
COW LESS PATCHY-LOOKING THAN SOME - ONLY SMALL STREAKS OF WHITE OR WHITISH SHOWING - BUT PLANES OF HIP + SHOULDER LESS EVIDENT THAN ON YEARLING - LOST UNDER SURFACE TEXTURE -
YEARLING APPEARED TO HAVE MORE TILT TO PELVIS THAN COW (?)
GALLOPING DOWN SLOPE TO CATCH UP WITH MAMA -
SHARP BREAKS IN HAIR TONE ON MID-TIBIA, LOWER SIDE, NECK DUE TO OPPOSING HAIR DIRECTION -
YEARLING: OLD COAT EVIDENT CHIEFLY ON TOP OF RUMP + MANE - "SMOOTHER" THAN COW, BUT HEAVY, PLUSHY COAT - GENERALLY, AN EFFECT OF DARK HAIR WITH PALER, BUFFIER TIPS, SO THAT COMPACTED HAIR (AS ON NECK) FORMS LIGHT "RIDGES" - VAREST, ON LEGS & BELLY, LOOKING INTO THE HAIR - SO THIS CHANGES WITH ANGLE -
COW + YEARLING BULL BROWSING UP N SIDE OF HOGAN CRK - YEARLING STAYED FAIRLY CLOSE TO COW, BUT HEADED UP INTO ONE LARGE WILLOW PATCH QUITE INDEPENDENTLY - COW CROSSED CREEK AT BRIDGE, + YEARLING, MISSING HER, TROTTED + THEN GALLOPED DOWN TO STREAM.
THE TWO BROWSED ON UP A DRAW S OF ROAD, WANDERING UP OVER THE RIDGE + DOWN ON SANCTUARY RIVER SIDE -
COW APP. PREGNANT - ROUND-BELLIED, + AGGRESSIVE TOWARD YEARLING WHENEVER HE CAME TOO CLOSE TO HER, RUSHING HIM WITH UPTHRUST MUZZLE - COW'S UDDER DEVELOPED -

1

HENRY M. REEVES AND RICHARD E. McCABE

Of Moose and Man

What the Buffalo was to the Plains, the Whitetail Deer to the Southern woods, and the Caribou to the Barrens, the Moose is to this great Northern belt of swamp and timberland. . . . It is the creature that enables the natives to live.—Ernest Thompson Seton 1929[III, 2]: 189

The moose is a Creature, not only proper, but is thought peculiar, to North America, and one of the Noblest Creatures of the Forest.—Paul Dudley 1721: 165

. . . from the Natives who are intimately acquainted with them, and make them their peculiar study . . . first in order is the Moose Deer, the pride of the forest, and the largest of all the Deer. . . .—David Thompson in Tyrrell 1916: 95

Long before humans and moose first encountered one another in the New World, they apparently had shared a semblance of harmony, if not always equilibrium, in the boreal forest landscape of Eurasia. Despite arrival in North America at separate times, they completed a circumpolar, boreal region cohabitation and established a relationship eventually similar to the one that evolved in the Old World.

The Beringian Connection

At various times during the highly active glaciation of the Pleistocene epoch in the northern hemisphere, expanding continental ice sheets lowered sea levels, periodically exposing a narrow "bridge" between Asia and North America. Approximately 56 miles (90 km) long, the Bering Strait land bridge, or Beringia, joined today's Siberian Chukchi Peninsula with Alaska's Seward Peninsula. When not blocked by glacial shelf, Beringia was literally an avenue of migration for land mammals, facilitating exchange between the two continents. Most emigration, including of moose and humans, was from Eurasia.

Compared with most of the emigrant species, the moose probably was a late arrival in North America, perhaps toward the middle of the fourth and last period of major glaciation—the Wisconsinian stage—approximately 70,000 to 10,000 years B.P. (Hibbard et al. 1965). Moose fossils from the Wisconsinian stage, and perhaps the earlier Illinoian stage, have been found in Alaska (Péwé and Hopkins 1967). Distinction between these fossilized moose and an earlier, related form—the stag-moose (*Cervalces scotti*)—is uncertain (Grant 1902, Kurtén and Anderson 1980). Despite its relatively late appearance in the New World, moose very likely preceded human presence on the continent.

Most authorities believe that the first human inhabitants arrived—also via the Bering Strait (either by land bridge or boat)—somewhere between 40,000 to 10,000 years B.P.

The newcomers, Beringians, were nomadic Paleo-hunters, adapted to the similar taiga (coniferous, evergreen forestland of the subarctic region) of Ice Age Siberia, who had a few rudimentary weapons (clubs, spears and the atlatl, but probably not the bow and arrow). These hunters made use of fire and may have been accompanied by dogs.

The America that moose and then humans first occupied was mainly interred in glacial ice. Undoubtedly prompted by weather, social competition and food scarcity (Franzmann 1981), moose dispersed southward along the present-day McKenzie River basin in an ice-free corridor that had been created by retreats and divisions of the continental icecap into two ice sheets (Wright 1892, Bliss 1939, MacNeish 1959a, 1959b, 1963, see also Johnston 1933, Roberts 1940). To the east was the Laurentian ice sheet, which blanketed nearly two-thirds of North America from the Arctic to south of the Great Lakes and from the St. Lawrence River to the eastern face of the Rockies. To the west was the Cordilleran ice sheet, which covered what is now Oregon to the Aleutian archipelago (Clark 1981). In small groups (several families), some of the Paleo-hunters necessarily pursued their food sources southward.

Global warming beginning about 14,000 years B.P. prompted a gradual, pendulous unveiling of new and barren landscape from beneath the massive ice sheets. Scraped to bedrock or soils only marginally fertile, the land first produced tundra, then coniferous, then deciduous vegetation. Moose acclimated to such habitat, probably populating the northern hemisphere by moving northward in the wake of glacial melt. It appears that the species' northeasternmost range (farthest from Berengia) was in place about 9,000 years B.P. (Hughes 1988, see also Rowe 1972). By about 4,500 years B.P., when the ice sheets receded beyond the current tundra ecoregion, a vast boreal forest established across the continent. It emerged in the postglacial compromise of climate, moisture and soils.

At various densities, temporally and spatially, moose apparently pioneered into all of the continuous boreal forest, most of the tundra, much of the Pacific coastal/subalpine forest and some of the northern intermontane. Also inhabited were portions of the arctic alpine tundra and Canadian parkland that were transitional to tundra and boreal forest, respectively (Canada Surveys and Mapping Branch 1974). It appears that moose did not extensively establish home ranges in habitats where the mean summer (July) temperature exceeded about 60°F (15.5°C) or where mean winter (January) temperature was below about –5°F (–20°C). With the post- Pleistocene shifts to climatic homeostasis, the species' subarctic range likely oscillated somewhat above and particularly below the prevailing zone of discontinuous permafrost (see Hunt 1967). At its prehistoric maximum, the moose range appears to have encompassed about 3.05 million square miles (7.9 million km^2) (see Seton 1909, Peterson 1955, Olsen 1964).

Only Ernest Thompson Seton's (1929: 175) calculation of "around 1,000,000" provides any sense of magnitude of continental moose population size before 1900. Seton based his approximation at least in part on reported densities ranging from 6 per square mile (2.3/km^2) in a section of Ontario, to 2 per square mile (0.9/km^2) for most of the Kenai Peninsula, Alaska, to 1 per square mile (0.39/km^2) in Manitoba, to 0.5 per square mile (0.19/km^2) in the Caughnawanna Reserve of Quebec. He acknowledged a crude, total range estimate of 3.5 million square miles (9.07 million km^2). We have no data with which to argue as too liberal a rangewide density of 0.29 moose per square mile (0.11/km^2) extrapolated from Seton's estimations, or an even higher density of 0.33 moose per square mile (0.13/km^2) factored from his moose population number and our refinement of the species' range.

Walcott (1939: 266) deemed Seton's assessment of 1 million moose as "rather high." However, given the recent reliable estimates of North America's moose population (see Chapter 2) closely approximate 1 million, that the species' pre-Columbian range is not yet fully reoccupied, and considering the relative impacts of former natural predation versus modern predation, harvest and incidental mortality rates, we believe Seton's number is reasonable and perhaps even conservative.

Less constrained by temperature, more catholic in food habits and certainly more gregarious than moose, humans eventually occupied and habituated to all but the most extreme low desert and high arctic portions of North America. The atmospheric conditions and physiographic circumstances that mediated the southern limit of moose range served chiefly to demarcate a northern demographic boundary of human habitation below which essential resources were more easily, readily and seasonally available, thereby permitting relatively dense, economically stable and permanent human populations (Figure 1).

It was mostly in North America's Subarctic that Paleo-hunters, their descendants of the Shield Archaic, Northern Plano and Woodland traditions, and moose developed rhythms and relationships that were to endure for the most part intact to historic time (see Clark 1981, Noble 1981, Wright 1981). The ancients who colonized the boreal region did not develop a written record per se. However, their rudimentary pictographs and symbols etched on stone, hides, bark and wood, plus a thin and discontinuous vein of artifacts from prehistory, document little in the way of cultural or economic change for the ±3,500 years predating the arrival of Europeans. As reported in Newton and Hys-

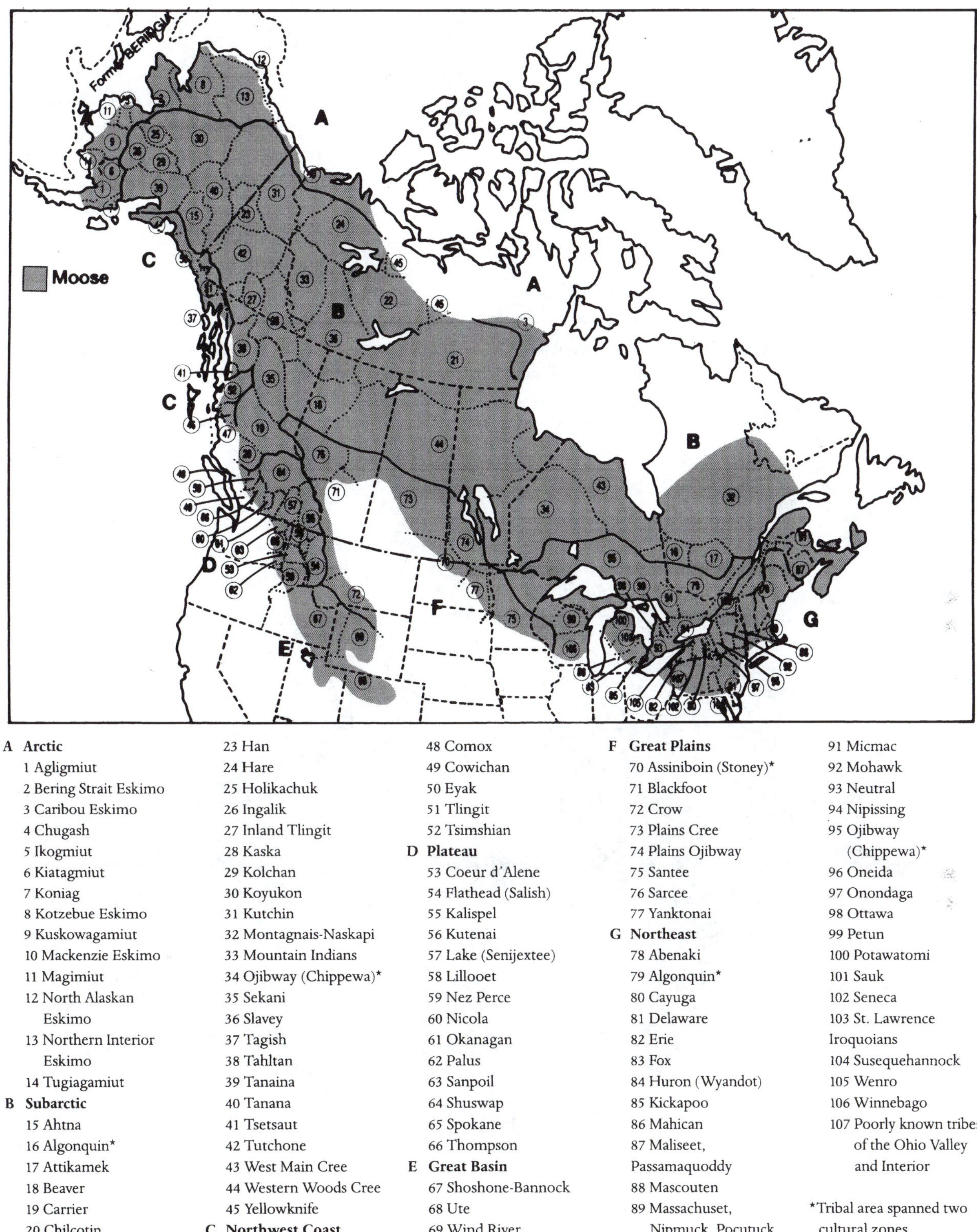

A Arctic
1 Agligmiut
2 Bering Strait Eskimo
3 Caribou Eskimo
4 Chugash
5 Ikogmiut
6 Kiatagmiut
7 Koniag
8 Kotzebue Eskimo
9 Kuskowagamiut
10 Mackenzie Eskimo
11 Magimiut
12 North Alaskan Eskimo
13 Northern Interior Eskimo
14 Tugiagamiut

B Subarctic
15 Ahtna
16 Algonquin*
17 Attikamek
18 Beaver
19 Carrier
20 Chilcotin
21 Chipewyan
22 Dogrib
23 Han
24 Hare
25 Holikachuk
26 Ingalik
27 Inland Tlingit
28 Kaska
29 Kolchan
30 Koyukon
31 Kutchin
32 Montagnais-Naskapi
33 Mountain Indians
34 Ojibway (Chippewa)*
35 Sekani
36 Slavey
37 Tagish
38 Tahltan
39 Tanaina
40 Tanana
41 Tsetsaut
42 Tutchone
43 West Main Cree
44 Western Woods Cree
45 Yellowknife

C Northwest Coast
46 Bella Bella
47 Bella Coola
48 Comox
49 Cowichan
50 Eyak
51 Tlingit
52 Tsimshian

D Plateau
53 Coeur d'Alene
54 Flathead (Salish)
55 Kalispel
56 Kutenai
57 Lake (Senijextee)
58 Lillooet
59 Nez Perce
60 Nicola
61 Okanagan
62 Palus
63 Sanpoil
64 Shuswap
65 Spokane
66 Thompson

E Great Basin
67 Shoshone-Bannock
68 Ute
69 Wind River Shoshone

F Great Plains
70 Assiniboin (Stoney)*
71 Blackfoot
72 Crow
73 Plains Cree
74 Plains Ojibway
75 Santee
76 Sarcee
77 Yanktonai

G Northeast
78 Abenaki
79 Algonquin*
80 Cayuga
81 Delaware
82 Erie
83 Fox
84 Huron (Wyandot)
85 Kickapoo
86 Mahican
87 Maliseet, Passamaquoddy
88 Mascouten
89 Massachuset, Nipmuck, Pocutuck, Wampanoag
90 Menominee
91 Micmac
92 Mohawk
93 Neutral
94 Nipissing
95 Ojibway (Chippewa)*
96 Oneida
97 Onondaga
98 Ottawa
99 Petun
100 Potawatomi
101 Sauk
102 Seneca
103 St. Lawrence Iroquoians
104 Susequehannock
105 Wenro
106 Winnebago
107 Poorly known tribes of the Ohio Valley and Interior

*Tribal area spanned two cultural zones.

Figure 1. Approximate cultural zones and tribal areas of North American Indians within the range of moose at the time of European contact, 1500 to 1650 (adapted from Hall 1981, National Geographic Society 1982). Through intertribal contact and trade, other tribes in other zones undoubtedly knew of moose. Also tribal boundaries were rarely static; instead, they shifted or fluctuated in response to warfare, alliances, weather, disease and other factors.

lop (1995: 32): "Little changed in those millennia; a hard winter in 2000 B.C. was much like a hard winter in 1500 A.D."

At the time of the first substantive European contact, in the sixteenth century, the Native population of the extant subarctic moose range (representing at least 65 percent of the total moose range) was thought to number 60,000 (Kroeber 1939), or about 0.03 person per square mile (0.012/km²). Within the vast range of moose, the population density of Native people is thought to have been less than 0.01 per square mile (<0.004/km²) in the Arctic to 1.0 to 4.0 per square mile (0.4–1.5/km²) in the relatively arable Great Lakes/St. Lawrence River region (Mooney 1928, Driver 1969, Garrett 1988, see also Leacock 1969, VanStone 1974, Dobyns 1976, Smith 1981b, Ridington 1981).

Most of the people of that time span and geographic region—deemed by sociologists as the "Subarctic cultural area" (see Driver and Massey 1957, Murdock and O'Leary 1975)—became seminomadic hunters, trappers and fishermen. "Flesh foods," according to Gillespie (1981a: 15), were "the only significant form of sustenance" for the entire region (Figure 2). The Natives subsisted secondarily on fish, small mammals, marine mammals, fowl, eggs and, to a very minor extent, plant foods. Big game ungulates were of primary importance, because they represented the greatest return (food volume and nutrition) for the peoples' foraging investments. The animals of widespread, primary subsistence importance were caribou and moose, usually in that order. Variously within the region, white-tailed and

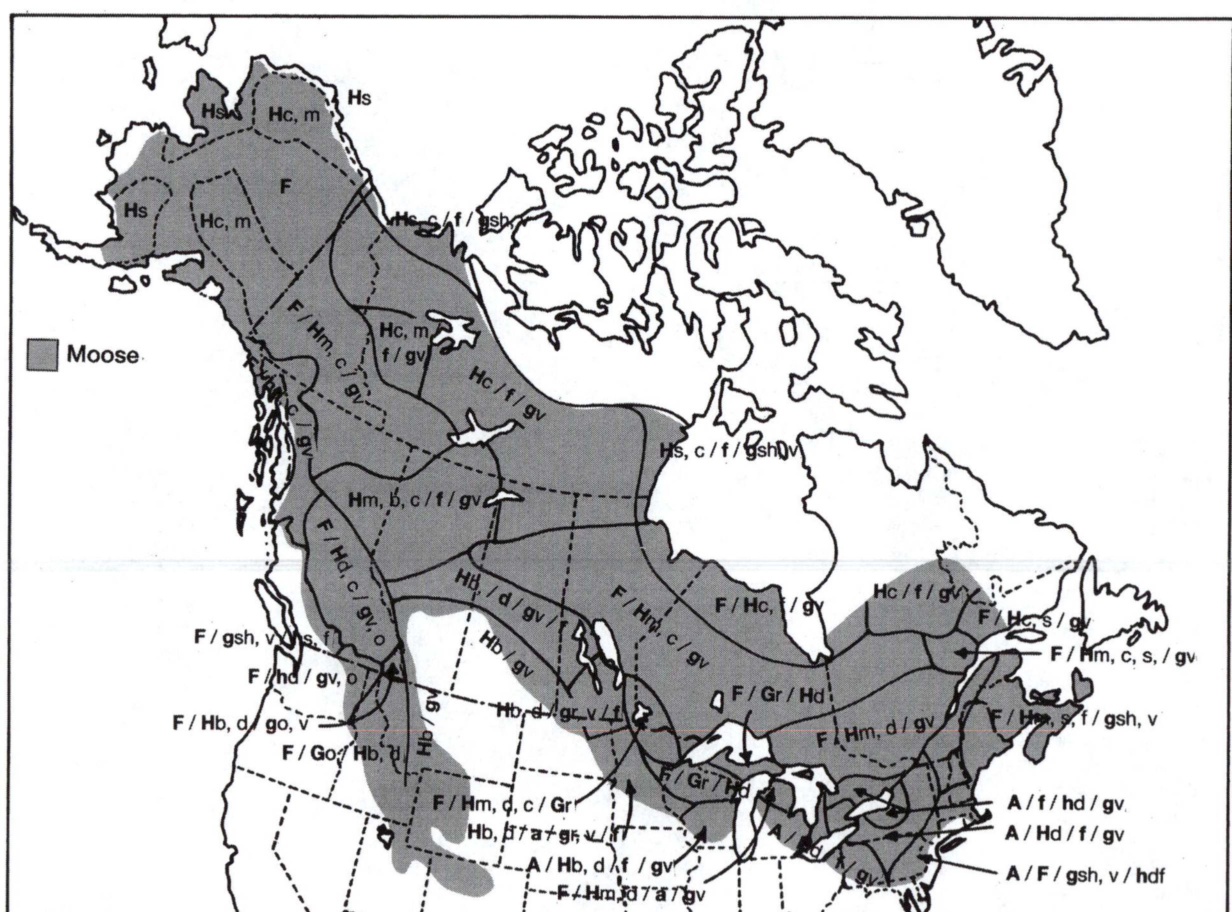

Codes: Dominant activity = F (fishing), H (hunting), G (gathering) and A (agriculture)
Secondary activity = f (fishing), h (hunting), g (gathering) and a (agriculture)

Major food sources* = b (bison), c (caribou), d (deer, elk, mountain goat or mountain sheep), m (moose), s (sea mammals), f (fowl), v (vegetables, nuts, berries), r (wild rice), sh (shellfish, seaweed) and o (roots)
*No major fish specialization, except salmon along West Coast; agriculture crops = corn, beans and cucurbits

Figure 2. Types of subsistence used by protohistoric North American Natives within the postcontact range of moose, based on archeological data (adapted from Harris and Matthews 1987). The analysis of discarded bones and shells from an archeological site can indicate when the site was occupied, what animals were eaten and the relative importance of the various foods. However, where soils are acidic, as in the Canadian Shield, such remains are rarely preserved, and few prehistoric hunting and fishing artifacts survive. Overall, archeological data permit only an incomplete picture of patterns of subsistence in the late prehistoric period.

mule deer, wood and Plains bison, elk, Dall's sheep, mountain goat, and musk-oxen also were exploited, mostly providentially (Figure 3).

Of moose, Marc Lescarbot noted in 1609, "It is the most abundant game which the Savages have after the Fish" (Grant 1914: 893, see also Bethune 1937). Caton (1877: 73) wrote that the "aborigines . . . depended upon the flesh of the Moose for their support." Maxwell (1978: 112) noted that "moose probably was the most important component in the diet of these Indians [Northeast Woodland semi-nomads]" (see also Thwaites 1897b, 1897e). Rogers (1973) added that no other food item represented as provident a food item for certain Woodland Indians as did the moose. And Seton (1909: 180) summarized: "In all the vast region

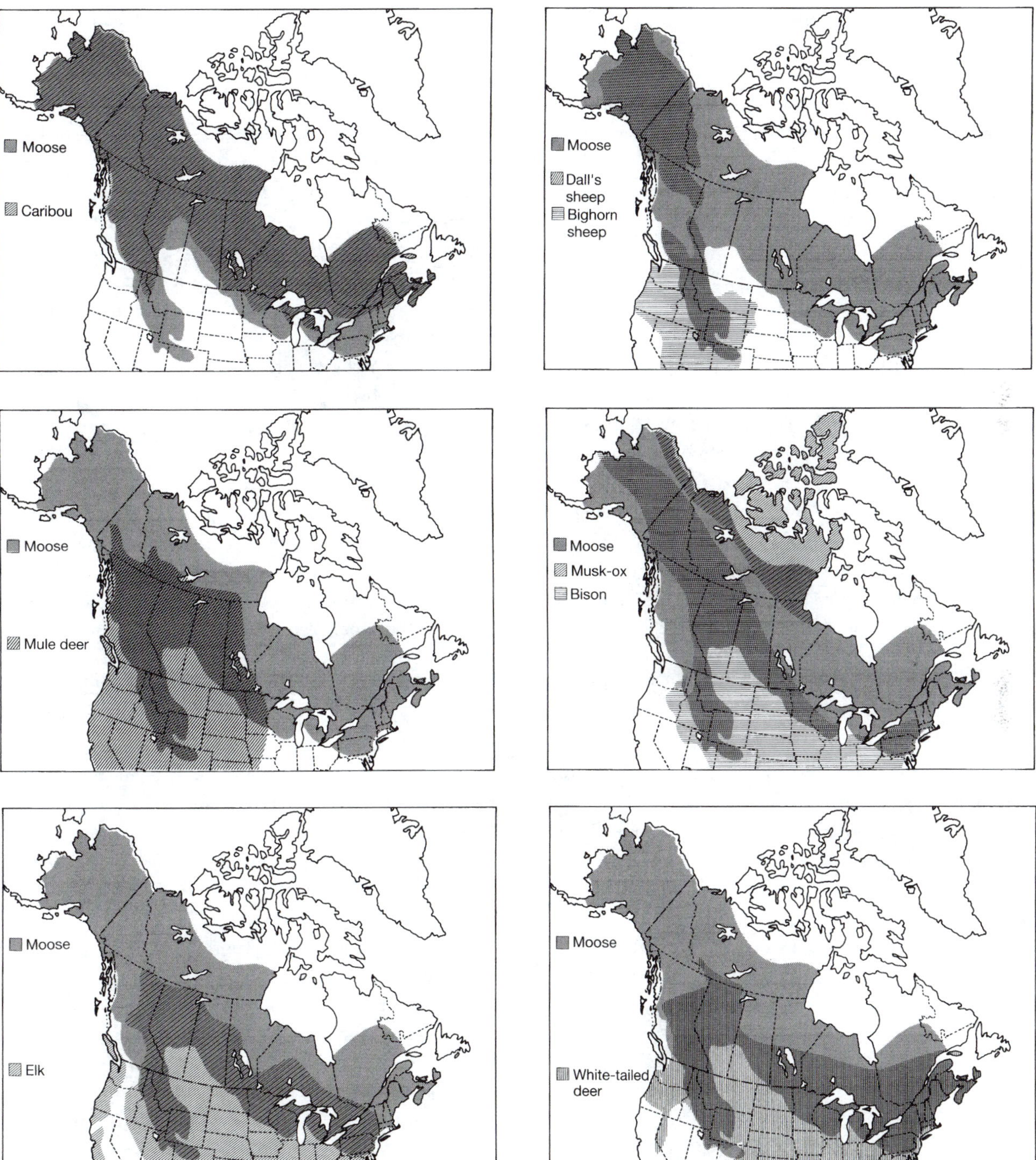

Figure 3. Approximate ranges of moose and certain other big game species at the time of European contact, 1500–1650 (adapted primarily from Schmidt and Gilbert 1978, Hall 1981, Wallmo 1981, Thomas and Toweill 1982, Halls 1984).

that is dotted on my map the Moose is, or was, the Indians' staff of life" (see also Macoun 1882). Nevertheless, moose were less abundant than caribou, but in most areas, the perpetually migratory caribou were available only seasonally, sometimes only for a few days or weeks per year. This region was (and is) as Rostlund (1952: 153) declared, "a land of feast or famine. The game is afoot with the seasons. . . ."

The Americana Traditions

In pristine moose range, where human survival was a fairly constant food quest, the cultural and spiritual identities of the people were linked closely to the resources whose acquisition represented their foremost occupation. And this was the case concerning moose and its hunters. The primary relationship of Native Americans (also and variously known as First People, First Nations, Amerinds, Natives and Indians) to moose was that of predator and prey.

Hunting

The boreal forest hunters of yesteryear, as noted, were chiefly meat eaters, therefore their existence depended on regular procurement of "game," with moose being a preferred prey. Moose represented an extraordinarily high return per time and energy investments and, depending on hunting technique, involved an acceptable safety risk. Despite and perhaps because of their crude weaponry, the people of the Subarctic were exceptionally skilled and tenacious hunters (see e.g., Milton and Cheadle 1865). Their livelihood made no allowance for incompetence or inefficiency. "No where was the Indian's keenness of observation more displayed than in fishing and hunting," wrote Canadian ethnologist Jenness (1932: 53–54), "Generally speaking . . . the average Indian, whatever his tribe, possessed more [hunting] ability than the average white man. . . ."

As quarry, the moose was considered by Natives a formidable challenge. John Tanner, who lived among the Ojibway in the early 1800s, indicated that those Indians and other Algonkians acknowledged that moose were the most difficult animals to kill (James 1956). Samuel Hearne, explorer for the Hudson's Bay Company in the 1700s, wrote that, "of all the large beasts in those parts [essentially, central Canada] the buffalo is easiest to kill and the moose are the most difficult" (Tyrrell 1911: 254). Explorer David Thompson (Tyrrell 1916: 95), who traveled with and among the Indians of the Missouri River and much of interior Canada from 1799 to 1814, reported that the moose "is of a most watchful nature; it's [*sic*] long, large, capacious ears enables it to catch and discriminate every sound; his sagacity for self preservation is almost incredible. . . ." Big game hunter *cum* U.S. President Theodore Roosevelt (1893) opined that, on par with white-tailed deer, moose were somewhat warier than elk and blacktails (mule deer). Ward (1878: 451) contributed the view that, "When alarmed, this ponderous animal moves away with the silence of death, carefully avoiding all obstructions, and selecting the moss-carpeted bogs and swales, through which he threads his way with a persistence that often sets at defiance all the arts and endurance of even the practiced Indian hunter." Somewhat contrarily, Hollingsworth (1787: 85–86) reported that, "when chased by Indians, its horns are laid back upon its shoulders, and in this posture its strength and velocity are so great as to break down and destroy small trees and branches of considerable size." In any case, the Indian hunter surely knew of the dangers attendant to confronting a cornered or wounded moose. The early literature is replete with tragic examples of hunters who misjudged or underestimated the aggressiveness of harassed, angered moose (e.g., Audubon and Bachman 1845–1848, Gillmore n.d., Roosevelt 1916).

In sum, as Arctic explorer and naturalist John Richardson (1829: 234) observed, ". . . the art of moose-hunting is looked upon as the greatest of an Indian's acquirements."

TIMING

The availability or unavailability of other food sources often was a consideration in whether, when and to what extent moose were hunted. However, records indicate that Natives took the prized moose opportunistically in all seasons (e.g., Hosley 1981a), but planned hunts for these animals typically occurred in autumn when the animals could be lured during the rut (Denys 1672, Grant 1894, Osgood 1936, 1971, Nelson 1973, Cronon 1983). Also, in winter, tracking was possible (Richardson 1929) when moose tended to "yard" and their movements were restricted and deep snows impeded escape (Hoffman 1839, Hardy 1855, Peterson 1955, Davies and Johnson 1963, Feit 1973). Later, in spring, hunters on snowshoes, often hauling toboggans (see McKennan 1981), could easily overtake moose foundering in crusted snow (Wallace 1932, Rogers 1962, Slobodin 1981). De Laguna and McClellan (1981) reported that, after guns were obtained, Ahtna Indians of Alaska hunted moose most successfully in April. Mason (1946: 16) noted that moose "forms quite an item in the native [Slave] dietary, particularly in the summer months."

In the 73-volume documentation of Jesuit missionary efforts, travels and explorations in New France from 1610 to 1791, moose hunting by Indians in southern and eastern Canada generally was undertaken "in the season of snow" (Thwaites 1897c: 313). Various Jesuits identified such hunting somewhat more specifically, including by Micmacs

Hunting moose, particularly mature bulls during the rut, could be a risky endeavor, and injuries and narrow escapes for the hunters are mentioned frequently in the early literature. Above is the apparent doubly fatal result of an encounter. Despite efforts to identify the scene, its location, date and scenario, no specific information was found. Dr. F.E. Anderson of Red Wing, Minnesota, took the photograph, probably in Canada in the late 1920s. The scene, showing a nearly complete moose skeleton and scattered human bones, also reveals heel-less shoes or moccasins (one on each side of the moose skull). In all likelihood, the encounter occurred in autumn or winter before heavy snowfall that hid the remains subsequently bleached by summer sun. *Photo courtesy of the Wildlife Management Institute.*

"when the heavy snows come" (Thwaites 1897c: 277) and from February to mid-March (Thwaites 1897a), by Iroquois in winter (Thwaites 1899c) and during the deep, crusted snows of April (Thwaites 1899e), and by Ottawas in spring (Thwaites 1899h) until June 1 (Thwaites 1899d).

Alexander Henry (the younger), a fur trader with the Northwest Company who traveled among the Indians of the Red, Saskatchewan, Missouri and Columbia rivers from 1799 to 1814, wrote on September 30, 1810, from a point north of current Edmonton, Alberta: "It requires little precaution to kill them ['buck moose'] at this season, when they are rutting, and frequently mistake a horse for a doe [*sic*: cow moose]" (Coues 1897: 634).

Richardson (1829: 234) pointed out that "The skill of a moose-hunter is most tried in the early part of the winter; for during the summer the moose, as well as other animals are so much tormented by musquitoes, that they become regardless of the approach of man." Snow (1981: 604) wrote, concerning the Ingalik of westcentral Alaska, "Moose were . . . hunted during the winter, although there were easier times of the year to do it." Across the continent, the seminomadic peoples of the northeastern woodlands found moose hunting easiest during periods of deep, soft snow in winter (Maxwell 1978) or later toward spring when thawing and refreezing crusted the snow surface.

Samuel Hearne, the first Caucasian to reach the Arctic Ocean by mainland North America, in 1717 mentioned that, in summer, moose swimming lakes or rivers commonly were killed by Indians in canoes (Tyrrell 1911).

TOOLS OF THE CHASE

The arsenal of moose-hunting Indians was limited but effective. Their primary weapons, before acquiring muzzle-loading firearms, were the so-called "hand weapons" (knives, clubs and thrusting spears) and "projectile weapons" (including bows and arrows, casting spears or lances, and perhaps atlatls).

The bows typically were made of wood (primarily juniper, birch, and willow, but also yew, ash, oak, hickory and witchhazel), some sinew-backed. Arrow (willow, dogwood) and spear shafts also were wood. Most blades and weapon points, often serrated, were chipped of flint, chert or obsidian, shaved antler or bone, or native copper.

Because moose were not easily or safely approached within effective killing range of the primitive weapons, and because of the animals' solitary nature, communal hunting was difficult and infrequent. Most "hunting" actually involved some form of entrapment or immobilization (discussed later). As reported in Thwaites (1897c), only by a great stroke of luck could a hunter get close enough to a moose to kill it with a bow and arrow. Weapons were used more to administer the coup de grace than in the traditional sense or notion of bringing a free-moving animal to ground by wounding. A.E. Speiss (personal communication: 1996) conveyed that arrow-shots at stalked moose were taken at less than 40 paces, and rarely was the animal brought down with a single arrow. The wounded animal usually ran off after being hit and had to be stalked and shot one or more additional times.

There were exceptions, of course, such as described by Jesuit missionary Paul Le Jeune, who lived among the Montagnais in 1633–1634 in southern Quebec. Fr. Le Jeune reported that Indians pursued moose on foot and killed the animals "with thrusts from javelins which are fastened on long poles for this purpose, and which they hurl when they dare not or cannot approach the beast" (Thwaites 1897c: 295). Nicolas Denys (1672) observed that, for hunting moose, the effective range of the Micmac (Acadia) bow and arrow was 45 to 50 paces. Another exception was the pursuit by Indians in canoes of swimming moose.

Caton (1877: 71, 73) summarized that the "rude weapons of the natives seemed not to have any abiding or fatal terror for the Moose," but "Even before the introduction of firearms among them the aborigines were successful in their [moose] capture." On the other hand, Hosley (1981b) wrote that, because of the solitary nature of moose, they did not lend themselves to communal hunting techniques of Natives of the Alaska Plateau, so did not become an important source of food for those Indians until the advent of rifles.

TECHNIQUES

Indians of the Subarctic, who lived year-round primarily on big game tended to hunt individually or in pairs for moose. Northern Woodland Indians of eastern Canada and the New England states, who subsisted mainly by fishing and/or gathering, relegated purposeful hunts for moose to the late-autumn and winter months. And these hunts typically were the enterprise of pairs or groups of hunters (see Trigger 1978, cf. Maxwell 1978). This generalization emphasizes dietary and geographic distinctions, but the foremost explanations rest with relative resource availability (economic opportunity), cultural modes and subsistence skills.

The most common practices of Indian hunting for moose, as reported in the literature, involved some manner of immobilizing, disabling or otherwise disadvantaging the quarry so it could not escape or counterattack before being dispatched.

Snaring. Snaring was a widely used technique for taking moose. Made of strong, woven or braided cords of moose or caribou hide twisted, stretched and dried, the snares then were worked sufficiently to make them pliable (Merrill 1916). Mackenzie (1801: 185) wrote: "Those snares are . . . made of the Green Skin of the Rein or Moos Deer cut so that it requires from 10 to 30 Strand to make this Cord which is no thicker than a Cod Line & strong enough to resist the Strength of any Animals in the Country." Honigmann (1946) reported that 0.5-inch (1.3-cm) wide moose hide babiche (cordage cut from wet partially tanned skin stretched and dried) dried in cold weather was the best snare line. Goodchild (1984: 86) indicated that "A snare [loop] for moose was anywhere between three and six feet [0.9–1.8 m] in diameter and about three feet [0.9 m] from the ground. Often, a double slip knot was used so that the noose would stay tight when the animal began to struggle."

As Merrill (1916: 136–137) described, suspended weights sometimes were attached to the distal ends of snare cords: "A slip-noose . . . was suspended where moose would be likely to pass—over a runway or near a spring. The line was run over a strong upper limb of a tree, and a heavy clog was attached to the end farthest from the moose. The animal's head once in the noose, the strain would release the clog, which would fall, and the noose would be drawn taunt. The animal would struggle . . . but the end was never greatly in doubt. The moose's indifferent vision made this method hunting easy, and many moose have been taken by Indians in this way."

John Martyn of Quebec (in Audubon and Bachman 1845–1848: 236) also gave a good account of clog or drag-pole snaring: "A rope is passed over a horizontal branch of a tree, with a large noose and slip-knot at one end, whilst a heavy log is attached to the other, hanging across the limb or branch, and touching the ground. The Moose, as it walks along, passes its head through the noose, and the farther it advances, the tighter it finds itself fastened, and whilst it plunges terrified onwards, the log is raised from the ground until it reaches the branch, when it sticks, so that no matter in what manner the Moose moves, the log keeps a continued strain, rising and falling, but not giving the animal the least chance to escape, and at last the poor creature dies miserably" (see also Cooper 1938, Honigmann 1946).

Snares were placed along trails moose habitually used, and vegetation or other impediments were positioned to guide the animal's head into the noose. Warren (1884) reported that Ojibways around Lake Superior caught moose

Unlike most other big game hunted by North American Indians, moose rarely were driven to enclosures or waiting shooters. However, some Natives artfully constructed wing fences of wood or brush that extended as much as a quarter mile (0.4 km) in length narrowing to gaps where snares were set. The moose were not driven, but left to follow the course of least resistance to the "trap" end, which was checked periodically by the Indians. The above scene from 1921 shows an abandoned Indian "pole-fence" for routing moose in the Northwest Territories. *Photo by Olaus J. Murie; courtesy of the U.S. National Archives.*

in spring-pole snares, with hide nooses hung over paths the animals commonly used to reach food and water. Sometimes, drift fences of downed trees and brush were constructed in openings to guide moose to snare locations set 200 to 300 feet (61–91 m) apart (De Laguna and McClellan 1981). Honigmann (1946) indicated that wing fences in open areas sometimes extended 0.25 mile (0.4 km), but were shorter in dense cover. In parts of Alaska, Natives constructed fences so sturdy that moose and caribou were forced to gaps where snares were set. A gold miner of undetermined veracity asserted that, while prospecting between the Yukon and Tanana rivers, he located one such "game fence" that was 30 miles (48 km) in length (U.S. Senate 1900: 477).

Nicholas Perrot (1864: 106–107), commandant of the French Royal Commission in the Great Lakes region, in the late seventeenth and early eighteenth century, described an Indian snaring technique that involved driving: "Hunting the caribous is usually practiced on the great flat plains [savanes]; and at the outset they surround the game with trees and poles planted at intervals, in which they stretch snares of rawhide, which enclose a narrow passage purposely left. When all these snares have been prepared, they go far away, marching abreast and uttering loud yells; this unusual noise frightens the animals and drives them to flight on every side; no longer knowing which way to go, they encounter this obstruction which has been made ready in their course. Not being able to clear it, they are compelled to follow it until they reach the passage in which the snares are laid with running knots, which seize them by the neck. It is in vain that they strive to escape; rather, they tear up the stakes [of the snare] and drag these with them as far as the larger trees; in short, their utmost efforts to extricate themselves only serve to strangle them more quickly. The moose are hunted in about the same manner. . . ." If, in fact, "in about the same manner" included communal, upland drive, it was a fairly unique strategy for taking moose and likely was practiced regularly only by northern Woodland groups of the Laurentian Highlands.

On the other hand, Peet (1890: 6), writing on effigy mounds of prehistoric Indians in the Great Lakes region, noted: "It is remarkable that effigies of buffaloes, moose and elk are more frequently associated with game drives than any other animal." In boreal/tundra regions, driving moose was considered virtually impossible (McClellan 1981a, 1981b), although N. M. Simmons (in Gillespie 1981b: 332), reporting on the Mountain Indians of western Northwest Territories, indicated that "I saw one [drift fence] set-up that probably was used to snare moose or at least guide them close to hunters, but this was not a common situation."

Snares had the advantages of simplicity, relative ease of manufacture and replacement, and self-functioning operation; further, one "hunter" could set and monitor many. Goodchild (1984) supposed that snaring in aboriginal times likely was a principal method of taking moose. He noted, however, that the meat of snared moose might be tainted (presumably if an animal was not recovered in reasonable time) and that butchering carcasses found frozen was a difficult proposition.

McKennan (1959: 48) wrote: "I believe that the importance of the big-game snare in the Athapaskan cultural pattern is not fully appreciated. In the days of the bow and arrow it was the most effective method for securing game. According to Upper Tanana natives [eastcentral and southcentral Alaska], the bow and arrow was practically never used during extremely cold weather, for should the first arrow miss the mark, the twang of the bowstring was so loud as to be sure to frighten the animal. Among the Kutchin [northwestern Yukon/northeastern Alaska], moose were likewise more generally snared than shot."

Perrot (1864) reported that Indians snared more than 2,400 moose on Manitoulin Island in Lake Huron in the winter of 1670–1671. Peterson (1955), however, questioned the claim and suggested that the animals were caribou.

Trapping. Although Indians in moose range extensively used nets, enclosure traps and pitfalls to capture and kill big game, there is only scant evidence that these techniques were used for moose, because the animals' size, wariness, relatively low population density and, again, solitary nature made them poor candidates for capture, which made investment in the efforts impractical. However, part-Ojibway William Warren (1884: 95) wrote of his Indian ancestors in northern Wisconsin: "To catch the moose . . . they built long and gradually narrowing inclosures of branches, wherein they would first drive and then kill them, one after another, with their barbed arrows." Honigmann (1946: 36) noted that Slave Indians sometimes attempted to drive moose to rocky terrain, "into what might be regarded as a natural surround—a gully or ravine, and there kill the trapped animal." And in southern New England, Ninnimissinuok "commonly" used "traps" of unidentified type to capture moose (Vaughan 1977, Bragdon 1996). Also, Goodchild (1984) reported that pitfalls never were used, but Martyn (in Audubon and Bachman 1845–1848: 236) wrote that "They [moose] are . . . 'pitted' at times, but their legs are so long that this method of securing them seldom succeeds, as they generally manage to get out."

Waterborne Pursuit. Another widespread moose hunting practice was the taking of swimming animals. Although excellent swimmers, moose could be overtaken by canoers and either fatally wounded at close range or "herded" to concealed shooters on shore. Natives of the Subarctic and other physiographic regions of moose range traveled extensively by boat or canoe and undoubtedly frequently encountered moose feeding in water or swimming or wading across waterways. When the animals were found in or enticed to deep water, they were readily pursued. This occurred throughout the ice-free months, but particularly at times when moose foraged extensively on aquatic vegetation or when tormenting insects drove the animals from dense vegetation into water. As Samuel Hearne reported of moose: "In Summer they are generally found to frequent banks of rivers and lakes . . . to avoid the innumerable multitudes of muskettos and other flies that pester them exceedingly during that season" (Tyrrell 1911: 259). Hunting from canoes and boats had two other significant advantages. First, the hunters, too, were relatively relieved of the insect swarms, as mentioned earlier (Richardson 1829, see also Honigmann 1981). Second, kills on or near shore greatly eased and expedited the considerable burden of transporting the bulky carcass because boats and canoes could be used. Because a moose killed in water will sink (not float, contrary to general belief) and be difficult if not impossible to recover fresh, Indians attempted to inflict wounds that would cause the moose to expire *after* reaching shore.

British artist, mountaineer and journalist, Frederick Whymper (1869: 214–216), who ascended the Yukon River in 1866 and 1867, wrote: "At this season [June] the musquitoes in the woods are a terrible scourge, and even the moose cannot stand them. He plunges into the water, and wades or swims as the case may be, often making for the islands. . . . The natives do not always waste powder and shot over them, but get near the moose, manoevring round in their birch-bark canoes till the animal is fatigued, and then stealthily approach and stab it in the heart or loins."

In the 1770s, Hearne witnessed Chipewayan and Cree Indians take swimming moose, and later wrote of the prey's docility: "When pursued in this manner, they are the most inoffensive of all animals, never making any resistance; and the younger ones are so simple, that I remember to have seen an Indian paddle his canoe up to one of them, and take it by the poll without the least opposition: the poor harmless animal seeming at the same time as contented alongside the canoe, as if swimming by the side of its dam, and looking up in our faces with the same fearless innocence that a house-lamb would, making use of its fore-foot almost every instance to clear its eyes of muskettos, which at that time were remarkably numerous . . ." (Tyrrell 1911: 259–260).

Another early account of hunting moose from a canoe appeared in the *Jesuit Relations* (Thwaites 1899d: 163): "In this Lake we encountered a Moose swimming across. . . . We straightway dispatched a little bark canoe in pursuit, manned by two Frenchmen and two Algonquin Savages. . . . These men, being still more dexterous in the water than the animal, made it turn and double back many times in that great Lake, where its actions were like those of a Stag chased by Hunters in the open country. It was a pleasure to see how, by dint of bursts of speed and convulsive movements, he tried to gain the land, and how the Hunters at the same time, tossing on the water in their Canoe, blocked its way and guided him despite himself toward the Bark [barkentine] where men were waiting to despatch it—which they finally did."

Asche (1981) noted that swimming moose were killed by bow and arrow, spears and clubs. Jones (1906), writing of the modes of killing game by central Algonquins, reported that moose overtaken by canoes were killed with knives used to cut the animals' throats. Jones indicated, as well, that moose were killed by drowning after a paddle was used to punch a hole between the ribs. Among Indians of the

Native hunting of moose frequently involved finding the animals in water or driving or luring them to water where they were forced to swim to escape. Essentially defenseless and no match for experienced canoers, moose were quickly overtaken and killed by club, arrow, lance or knife wounds that punctured the rib cage. *Top scene "Man Against Moose" was painted by Frank B. Hoffman; photo courtesy of Kramer Gallery, Inc., Minneapolis. Bottom scene "The Moose Chase" was painted in 1888 by George DeForest Brush; photo courtesy of the Smithsonian Institution National Museum of American Art.*

Alaska Plateau, "the lance or spear was widely used . . . to stab moose . . . from a canoe" (Hosley 1981a: 535, see also Osgood 1958).

Fortuitous opportunities to take swimming moose probably were more commonplace but somewhat less successful than planned "water hunts," which had numerous variations. For example, jack-lighting (a.k.a. firelighting, nightlighting and shining) was an effective way to approach moose along shorelines or in shallow water at night. Of Pemigewassetts and Nipmuck Indians of New England, Little (1870: 37) wrote: "In early autumn, when moose and deer fed at night on the grassy shores of the lakes and rivers, the Indian hunter, with rude lantern brightly flashing in front, placed in the prow of the canoe, would paddle noiselessly in the dark shadow behind, and when sufficiently near his spell-bound victim would send his feathered shaft on its silent but fatal mission. Every dark night of autumn these spectral fires might be seen gliding like will-o'-the-wisps over the rivers, ponds, and lakes in the Pemigewassett country." Powell (1856: 344) reported that jack-lighting could involve more risk than was normally invited: ". . . in hunting on a dark night, in a canoe, on the water, when in pursuit of deer, &c, a flambeau, or torch, or candle, can be used to great advantage, the animals being apparently bewildered or fascinated by the bright, steady light which approaches them so noiselessly and still; but the moose, as soon as he perceives it, approaches it, quickening his pace as he comes nearer, till . . . he charges full upon it, destroying the canoe, and frequently injuring the occupant. However, with the extinction of the torch his fury ceases."

Few records exist of Natives killing swimming moose by leaping on the animal's back to administer a fatal wound (left). However, because moose in deep water cannot easily defend against "bulldogging," the technique was plausible, as further evidenced by the two photos, circa 1912, at right. *Left scene "Killing a Moose in the Water" by Henry Sandham; photo from Stumer (1978). Top and right photos by Warwick S. Carpenter; courtesy of the Wildlife Management Institute.*

Driving. When efforts were made to drive moose, they usually involved large groups of people, sometimes entire villages, including dogs (the only domestic animal of the aborigines). Moose were chased to water where hunters hidden nearby in canoes undertook pursuit (Merrill 1916). Nicolas Perrot (1864) told that Cree Indians of the Lake Superior region loosed dogs to drive moose to water and hunters waiting in canoes. Little (1870: 37) wrote of Indians forming grand hunting parties annually to perform semicircle drives over New England hills, with skillful spearers and archers positioned near the apex of a V-shaped abattis of trees: "Then with shouts, and yells, and wild whoops, the moose and deer, bears and wolves, were roused with the smaller game." In the vicinity of Lake Champlain in 1759, J.C.B. (1941: 109) wrote that: "The northern savages have another way of hunting this animal without any risk. They divide into two bands. One band gets in canoes and, joining in line, they form a half circle, each end touching the shore. The other band on shore make another large circle and loose their dogs to start up moose enclosed in this area. The dogs chase the moose before them and force them to plunge into the water. No sooner are they in the water, than fire [shooting arrows] is opened upon them from all the canoes. They very rarely escape from such an attack." Whymper (1869: 214) documented that, "In some cases the Indians in numbers surround an island known to have moose . . . and a regular *battue* ensues."

Fire also was used to drive moose, as Purchas (1625[XIX]: 282) noted: ". . . in a great Lland upon the Coast, called by our people Mount Mansell, whither the savages goe at certaine seasons to hunt them; the manner whereof is by making of severall fires, and setting the Countrey with people, to force them into the Sea to which they are naturally addicted, and then there are others that...in their Boates with Bowes and weapons of severall Kindes . . . slay [them]."

Enticing. Enticing or luring also was a popular technique used by moose- hunting Indians, but apparently was used more by Natives of the Northeast than elsewhere. There were several variations of "calling" moose. The most com-

Birch- or cedar-bark horns used to call bull moose were about 16 to 20 inches (41–51 cm) long, and usually 0.75 inch (1.9 cm) in diameter at the small end and 3.5 to 4 inches (9–10 cm) in diameter at the other. Hardy (1855) noted that Native moose-callers far excelled white callers because the formers' efforts sounded more like they originated from moose lungs rather than from human vocal chords (see also Gillmore no date). *Left photo (C24379) courtesy of the National Archives of Canada. Right photo from Ward (1878), artist unknown.*

mon was a vocalization that imitated the mating call of an adult bull moose in the rut, which sounded to Selous (1907: 227) like "a human being in the throes of sea-sickness." Merrill (1916: 101) advised that "Calling is effective for a limited season . . . only when the moose is thrown off his guard by the violence of his passions." According to Merrill, the value of the moon as an aid to hunting in the calling season was not to be overestimated. Also, calling in the morning, best begun half an hour or more before sunrise, was more likely to produce results than evening calling. Indians were well aware that moose typically fed in early morning before sunrise and, when sated, then found a place to ruminate and sleep. It was believed that the morning feeding was voracious enough and the postfeeding sleep sound enough to mask the approach of a hunter more effectively than at other times of the day or night.

To magnify the imitated grunt of a cow, described as a sound "halfway between a 'bark' and a 'moo'" (Goodchild 1984: 53), some Indians used a cedar- or birch-bark horn to entice amorous bulls. Erickson (1978: 129) reported that, among the Maliseet-Passamaquoddy of northern Maine, southern New Brunswick and southern Quebec, "The birchbark moose call was an indispensable part of the hunter's equipment." With such a call, moose hunters would "lure an animal into the open where tribesmen lay in wait" (Maxwell 1978: 348).

Hibbs (1890: 31) wrote that "native hunters, in the northeastern woods, are said to use the birch-bark horn with such terrible results to unsuspecting game. The horn has never been used in the Rocky Mountains, to my knowledge, and I have never heard any such noise here as is attributed to the Moose in the woods of Maine and Canada. The cow Moose, I have reason to believe, never utters a cry of any kind, here, and the bull of our region simply whistles, like the Elk and Deer." On the other hand and without geographic reference, Seton (1909[I]: 171) reported that, "By softly modulated squeals, whines, and other sounds suggestive of a female Moose, a skilful caller can decoy the great beast within a few yards. . . ." Hardy (1855) related that Indians seldom called during windy conditions, because the moose invariably were relatively cautious at those times.

To attract bull moose, some Indians simulated the sound of a cow urinating, by pouring water from a container into still water. As Denys (1672: 427) described: "At that time the hunting was done at night upon the rivers in a canoe. Counterfeiting the cry of the female, the Indians [Micmacs] with a dish of bark would take up some water, and let it fall into the water from a height. The noise brought the male, who thought it was a female making water [see also Wallis and Wallis 1955]. For this object they let themselves go softly along the stream; if they were ascending, they paddled very softly, and from time to time they made water fall, counterfeiting always the female. . . . Those who were in the canoe would hear him coming, because of the noise the beast made in the woods, and they kept on constantly imitating the cry of the female, which made him come close to them. They were all ready to draw upon him, and never missed him."

Indian hunters also imitated the noises of combatant bull moose. The Tahltan or Kaska Indians of northwestern British Columbia, for example, pounded on trees or thrashed willows with antlers or a cow's dry shoulder blade with the central ridge removed (Honigmann 1946). Jones (1872: 322) wrote that, "in the fall, when the moose are rutting, the hunter provides himself with a shoulder blade of the same animal; he then approaches the male as close as possible, and rubs the bone against the trees. The moose charges at once, mistaking the sound for that made by

Indians lured bull moose into archery range during the rut by pouring water into a lake or river, imitative of a urinating cow moose. *Illustration by J.B. Clemens; photo from Irwin (1984: 272).*

another male rubbing his horns against the trees." Other Indians clashed antlers together to simulate fights between prime bulls. Bulls were attracted to the excitation of a dominance fight, also apparently signaling the proximity of one or more prospective mates (Merrill 1916).

Stalking. Stalking moose, curiously called "still-hunting" or "creeping" (Hardy 1855: 185) in the early literature, was a frequent technique in autumn, when insect activity abated, and in winter, when snowfall eased tracking or retarded moose escape from hunters on snowshoes. It was only infrequently attempted in summer, because of the difficulty in following sign and the animals' relatively "wretched" condition at that time of year (Slobodin 1981, see also Osgood 1936, Nelson 1973). Caton (1877) reported, however, that moose were stalked on islands in summer where they sometimes swam to escape insects. And in reference to the hunting practices of Indians of the Great Slave Lake region, Mason (1946: 16) wrote that, because of the solitary nature of moose, it "must be hunted by stalking." Mason further noted that moose formed "quite an item in the native dietary, particularly in the summer months," which suggests an active season of stalking by these people. Baron Lahontan wrote that stalking moose "is compass'd by crauling [*sic*] like Snakes among the Trees and Thickets, and approaching 'em upon the Leeward side, so that they may be shot . . ." (Ganong 1910: 59).

According to David Thompson, Natives (Algonkians) of the Stoney region (essentially, the eastern half of Canada [see Glover 1962]) distinguished between "tracking"—which involved following an animal's footprints—and "tracing"—which involved "following the marks of feeding, rubbing itself against the ground, and against trees, and lying down" (Tyrrell 1916: 95).

James Hector, medical officer and naturalist with the Palliser Expedition of western Canada in 1857–1860, characterized: "The Indians in hunting are very observant of the cropping of the willow tops, and there was something quite exciting . . . as he pointed out where the willow tops were yet wet with the saliva of the animal [moose], or when, in walking rapidly through the wood, he would stop suddenly and pick up a morsel of half chewed leaf which it had dropped, and when he found that it had stopped to take several bites from one bush. . . . At last he sees him, but all the trouble and fatigue is not to be wasted on a miss, so he patiently works up close to the animal, who, if at rest, is generally buried in a thicket, perhaps only visible to the keen Indian's eye by the lazy flap of his large ear. They generally get within 20 yards [18 m] before they fire" (Spry 1968: 314–315).

Generally, stalking was acknowledged as the most difficult of hunting methods because moose tended to be too secretive and wary to permit hunters to get within killing range of their primitive weapons. During the rut, however, as previously described by Alexander Henry (in Coues 1897), the mating urges of bulls in particular, but cow moose as well, overrode their innate alertness or concern for imminent danger. Wrote Grant (1894: 353): "Indian hunters say that when they hear in the twilight the breaking of the undergrowth and the crash of antlers in one of these mighty [dominance] battles, they slip up close and shoot the cow as she stands placidly at one side watching the result with languid interest. When she falls the bulls fight on with redoubled fury, and so intent are they on the duel that both can be killed with ease." Wallis and Wallis (1955) and Maxwell (1978) indicated that Micmacs of New Bruswick wore disguises to stalk moose. In 1674, John Josselyn indicated that Native hunters overcame the alertness of moose by shooting when the animals were feeding with their heads submerged in water (Lindholdt 1988).

As a rule, stalking was the challenge of lone hunters or, preferably, pairs (Honigmann 1946, see also Rogers 1962, Slobodin 1981). Also, the ideal conditions for stalking were days of high wind, and damp or lightly snow-covered ground. Martyn (in Audubon and Bachman 1845–1848) identified moonlit nights with snow-covered ground as rewarding stalking time. A light snowfall overnight resulted in moose tracks to be followed and muffled the hunter's footfall (Goodchild 1984). Robert Hood, a member of Sir John Franklin's first expedition into the Northwest Territories, wrote that, "In stormy weather, the uproar of the wind in the forest renders the quickness of its [the moose's] ears useless . . ." (Houston 1974: 60). David Thompson told of a "Doe Moose" that a Western Woods Cree hunter left undisturbed from May until early October 1804, when gale winds finally permitted a successful stalk (Tyrrell 1916: 97).

Intimately familiar with characteristic behaviors of moose, experienced Indian hunters use "semicircling" stalking (Spry 1968, see also Osgood 1936, Honigmann 1946), a technique referred to as *onatawa·hike·win* by Round Lake Ojibway of northwestern Ontario (Rogers 1962). Locating the track of a feeding moose and identifying its general meandering direction, the hunter would follow the trail to the downwind side, looping in at about half-mile (0.8-km) intervals and, as cover permitted, to relocate the track. When it could not be found, the stalker knew that, as is typical of moose, his quarry had doubled back to bed down (see Milton and Cheadle 1865). Still to the windward side, the hunter turned back and made smaller semicircles or loops back to the moose's reversed trail (Figure 4). "Then," according to Hector (Spry 1968: 315), "if the ground is hard and the branches of the trees dry and likely to crackle, he

Figure 4. The "semicircling" moose stalking technique (adapted from Nelson 1973 and Merrill 1916).

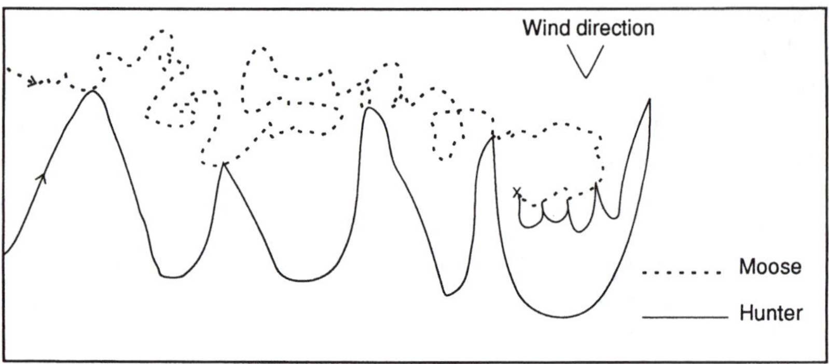

strips off all his clothes, and glides like a snake through the forest, peering and prying to catch the first glimpse of the animal." A variation on the hunters' concluding preparation was provided by Rogers (1962: 42): "When the man knows he is close, he removes his snowshoes, which are noisy, and proceeds on foot. The hunter may also reverse his parka since the soft inner side makes less noise . . . when it rubs against a branch."

Pursuing. Openly pursuing or chasing down moose was another widely practiced hunting method having multiple variations. Gillespie (1981b) identified this as a method of the Mountain Indians, but done with greater difficulty than for caribou and mountain sheep (see also Jenness 1937).

Even during summer in a land of "feast or famine," the procurement of meat such as from a moose might necessitate chases that were nothing less than exhausting marathons. Dudley (1721: 167) noted: "After you unharbour a Moose, he will run a Course of 20 or 30 miles [32.2–51.5 km], before he turns about, or comes by a Bay. . . ." Lahontan related that, "The Savages assure us, that in Summer [fleeing moose] 'twill trot three Days and three Nights without intermission" (Ganong 1910: 57). And as David Thompson wrote: ". . . when chased it [a moose] can trot (it's [*sic*] favorite pace) about twenty five to thirty miles [40–48 km] an hour. . . . If not molested it travels no farther than to find it's food, and is strongly attached to it's first haunts" (Tyrrell 1916: 96). Hector (Spry 1968: 315) pointed out that, after an Indian has shot at a moose, he will race to head off the animal, which, hit or not and unless dealt an immediately fatal wound, invariably halts at about 60 yards (55 m) from the point of disturbance. If a wounded moose runs off, Hector explained, the Indian goes to the animal's "lair" to ascertain how telling the shot was, for "the animal will run a distance proportional to the wound it received and will only stop short of a 5 or 6 hours' flight from mere weakness. He [the Indian hunter] follows the track, and in nine out of ten cases the moose is found dead or dying within half a mile (0.8 km)."

U.S. Army officer Randolph B. Marcy (1866) related that, on a hunting excursion in the vicinity of Saginaw Bay, Michigan, in 1844, he engaged an old Chippewa Indian, Petowanquad, as guide. Once, while hunting on snowshoes, Petowanquad reportedly struck the tracks of seven moose. That day, he followed two of them and killed both with his rifle. The five other moose were followed and killed over a 3-day period.

Goodchild (1984) wrote that a moose could be run down over a period of hours, but the chase was considered to render the animal's meat unpalatable. Goodchild added that if a moose was unintentionally frightened, it was likely to bolt downwind of the disturbance, probably to catch the scent of its pursuers. Indian hunters knew to search to their downwind side for a retreating moose. And if a bull and cow pair was spooked and separated in autumn, the two animals likely circled back to the same area to find one another. Hunters merely lay in ambush for their return.

Goodchild also informed that Indians occasionally used dogs to track and chase moose, chiefly in spring (see also Thwaites 1896, McClellan and Denniston 1981). Nicolas Denys (1672: 430–431) provided a graphic, yet representative account of chase-hunting with dogs: "When they took their Dogs to hunt the Moose in spring, summer, and autumn, the Dogs would run about for some time, some in one direction and some in another. The one which first met some track followed it without giving tongue. If he overtook the beast, he got in front of it, jumping for the nose. Then he howled. The Moose amused himself, and wished to kick the Dog in front. All the other Dogs which heard it came running up and attacked it from all sides. It defended itself with its feet in front; the Dogs tried to seize its nose or ears. In the meantime the Indian arrives, and tries without being seen to approach within shot below the wind. For if the animal perceives him or his smell, the Moose takes to flight and scorns the Dogs, unless the hunter gives it an arrow-shot. Being injured, it has difficulty saving itself from the Dogs, which follow it incessantly, as does also the Indian, who overtakes it and shoots again. But sometimes the

Marc Lescarbot gave an early testimony on hounding for moose: "They [Natives] have dogs much like foxes in form and size, and with hair of all colours, which follow them; and although these do not give tongue, yet they know very well how to find the haunt of the beast which they are seeking, on finding which they pursue it courageously, and never give over until they have it down. Finally, having wounded her mortally, they worry her with their hounds till she is forced to fall down. Then they rip up her belly, give the spoil to the dogs, and also take their own share" (Grant 1911–1914: 221). Jesuit missionary and artist Nicolas Point, who lived among Native tribes in the Intermountain region of Idaho and Montana in the 1840s and painted the scene above, was astounded by the Indian dogs' ferocity and fearlessness (Donnelly 1967). *Photo courtesy of the Smithsonian Institution Anthropological Archives.*

Dogs, which have seized the ears or the muzzle, drag it to earth before the Indian has come up. They are not inclined to abandon it, for very often they have had nothing to eat for seven to eight days. The Indian arrives, completes the kill, splits open the belly, and gives all the entrails to his Dogs, which have a great junket. It is this which makes them so keen in the chase."

Traveling with Koyukon Indians on the Yukon River in 1866, Whymper and William Dall related that the Indians' dogs followed along the shore and foraged for themselves. In scouring the woods for food, the dogs proved useful by starting game that the Indians sometimes intercepted. "In the evening [June 10] they [the dogs] found a young moose, which they surrounded till the Indians were enabled to kill it" (Whymper 1869: 212).

Crusting. "Crusting"—the chase of moose in deep, ice-encrusted snow by Indians on snowshoes—is by far the most frequently mentioned aboriginal hunting technique in early journals (e.g., Ganong 1910, Grant 1911–1914). In 1603, Samuel de Champlain was perhaps the first to document pursuit of moose foundering in deep snow by Indians on snowshoes with "filling" (webbing) of moose hide strips (Grant 1907). Documentation by Jesuit missionaries reveal that winters of scant snowfall were seasons of misery for Natives, because moose did not yard where anticipated and the few that could be found were not easily approached or killed (see Thwaites 1897c, 1898e, 1899c, 1899e).

A combination of deep snow and an icy surface was needed for good crusting. Maxwell (1978: 348) noted that "Windy, sunny days right after a snowfall were best for hunting. The wind deadened any sound the hunter ['wearing snowshoes'] made, while the fresh snow, crusted by the sun, outlined the animals' tracks." As E.A. Kendall stated (in Audubon and Bachman 1845–1848: 233–234): "In March, when the sun melts the snow on the surface and the nights

Native "crust-hunting" on "rackets" (or "racquettes") for moose ("Orignalds") was illustrated in Lahontan (1703: plate 55). The drawing is particularly fanciful not only because of the inaccurate size comparisons of the Indian hunter and his antlered quarry, but also because crusting was not known to have been attempted in snow less than 2 to 3 feet (0.6–0.9 m) deep, which was not the case depicted. *Photo courtesy of the Library of Congress.*

are frosty, a crust is formed, which greatly impedes the animal's progress, as it has to lift its feet perpendicularly out of the snow or cut the skin from its shanks by coming in contact with the icy surface. It would be useless to follow them when the snow is soft, as their great strength enables them to wade through it without any difficulty."

Hearne also commented on crusting in late March and early April: ". . . yet in the middle of the day it [the thaw] was very considerable: it commonly froze hard in the nights; and the young men took the advantage of the mornings, when the snow was hard crusted over, and ran down many moose; for in those situations a man with a good pair of snow-shoes will scarcely make any impression on the snow, while the moose . . . will break through at every step up to the belly. . . . The moose are so tender-footed and so short-winded [?], that a good runner will generally tire

"Crusting" was undertaken when snow conditions permitted in winter and where moose yarded, but the usual time for such hunting was spring, when snowpack was exposed to occasional thawing and refreezing. Observed Jesuit missionary Fr. Paul Le Jeune: "Sometimes . . . [Algonkian hunters] chase one of these animals for two or three days, the snow being neither hard nor deep enough; while at other times a child could almost kill them, for, the snow being frozen after a slight thaw or rain, these poor moose are hurt by it, and cannot go far without being slaughtered" (Thwaites 1897c: 295). *Scene entitled "Hunting Moose Deer in Winter" painted by George Catlin; photo courtesy of The Royal Ontario Museum.*

them in less than a day, and very frequently in six or eight hours; though I have known some of the Indians continue the chace [*sic*] for two days, before they could come up with, and kill the game. On those occasions the Indians, in general, only take with them a knife or bayonet, and a little bag containing a set of fire-tackle, and are as lightly clothed as possible; some of them carry a bow and two or three arrows. . . . the Indians who have neither a bow or arrows . . . are generally obligated to lash their knives or bayonets to the end of a long stick, and stab the moose at a distance" (Tyrrell 1911: 279–280).

Richardson (1829: 235), commenting on the crusting of moose, noted: "An instance is recorded in the narrative of Captain Franklin's second journey, where three hunters pursued a moose-deer for four consecutive days, until the footsteps of the chace [*sic*] were marked with blood, although they had not yet got a view of it. At this period of the pursuit the principal hunter had the misfortune to sprain his ankle, and the two others were tired out; but one of them, having rested for twelve hours, set out again, and succeeded in killing the animal, after a further pursuit of two days' continuance."

Crusting sometimes was so easy that even boys and women, normally relegated to the tasks of butchering and transporting meat, were allowed to partake in this manner of hunt (Houston 1974). And at least some Indians—notably the Hare (Richardson 1829, Wentzel 1889–1890) and Mountain (Gillespie 1968–1971) Indians of the Northwest Territories, Kutchin of the Yukon (Jones 1872), Western Woods Cree of central Canada (Coues 1897), Micmac of New Brunswick (Thwaites 1896), and Massachusett and Eastern Abenaki of southern, coastal Maine (Lindholdt 1988)—used dogs in crusting moose. "If the snow is less than two feet [0.6 m] deep, dogs, as many as three, are employed. When the tracks of moose are found the dogs are let loose. Although the moose is too heavy to be supported by the crust, he can run rather fast because of the shallow depth of the snow, but the dogs, supported by the crust, can overtake the animal. When they do, they circle the moose and hold it at bay until the hunter has a chance to arrive. If the snow is deep, dogs are not used" (Rogers 1962: 43).

Despite the technique's popularity, crusting was strenuous labor. "Even this 'easy' hunt required speed and stamina" (Slobodin 1981: 517). And a consequence of this and other chases was diminished meat quality. David Thompson wrote: ". . . when run by Men, Dogs, or Wolves for any distance, it's [*sic*] flesh is alltogether [*sic*] changed, becomes weak and watery . . . the change is so great, one can hardly be persuaded it is the meat of a Moose Deer" (Tyrrell 1916: 96). However, as will be discussed, meat palatability did not always affect consumption.

Subsistence

As previously noted, moose was a staple food for many Subarctic Indians and only somewhat less so for Northern Woodland Natives (Figure 5). For virtually all, moose flesh was, as represented in the lexicon of the Naskapi, *notcimi'ʼumi'tcə'm*—"pure food" (Speck 1935: 78).

Noting the impossibility of determining how vital a role moose played in traditional Native economy, Nelson (1973: 85) nevertheless wrote that, to the Chalkyitsik Kutchin, "'meat' is almost synonymous with moose. Whereas other animals may be considered delicacies or treats, moose is probably the one meat they could least think of doing without."

DISTRIBUTION AND TRANSPORTATION

Practices of handling a dead moose varied widely among tribes. Traditionally, hunting was the vocation and responsibility of men, whereas the laborious, time-consuming tasks of processing a carcass and hauling the meat, hide and other parts to camp fell to women and older children. At a kill site, Beaver Indians, like others, usually divided a moose among members of the band, with each family providing someone to help pack in the parts (Ridington 1981). Sometimes during times of game scarcity or poor hunting conditions, a kill might be made a distance of several days' travel from the main Indian camp. If hunger was great, a small group of hunters might simply encamp by the carcass and, over a week or more, fully consume it. Among Western Woods Cree during the winter of 1804–1805, David Thompson reported that, if the distance to transport a moose carcass to a campsite or to a canoe was considerable, "the meat is split and dried by smoke, in which no resinous wood must be used; this reduces the meat to less than one third of its weight" (Glover 1962: 83).

After a hunter killed a moose and needed either to continue hunting or return to camp for assistance, the carcass was cached. Caching for a day or two necessitated only a thick layer of willows to ward off jays (gray or Canada jays were colloquially known as "moose-birds" [see Seton 1909, Merrill 1916]) and ravens. If the meat was to be cached for a longer period, up to a week, it was hung in trees as protection against wolverines, wolves and bears. Caching for more than a week often involved constructing a sturdy log structure, because ravens could eat too much of a carcass cached in a tree in that time (Nelson 1973).

In warm weather, brush was gathered and placed near the carcass so that the meat removed could be kept off the ground. In winter, the meat was simply placed on the hide spread on compacted snow. Some moose-hunting Indians took special precautions not to spill or smear blood on

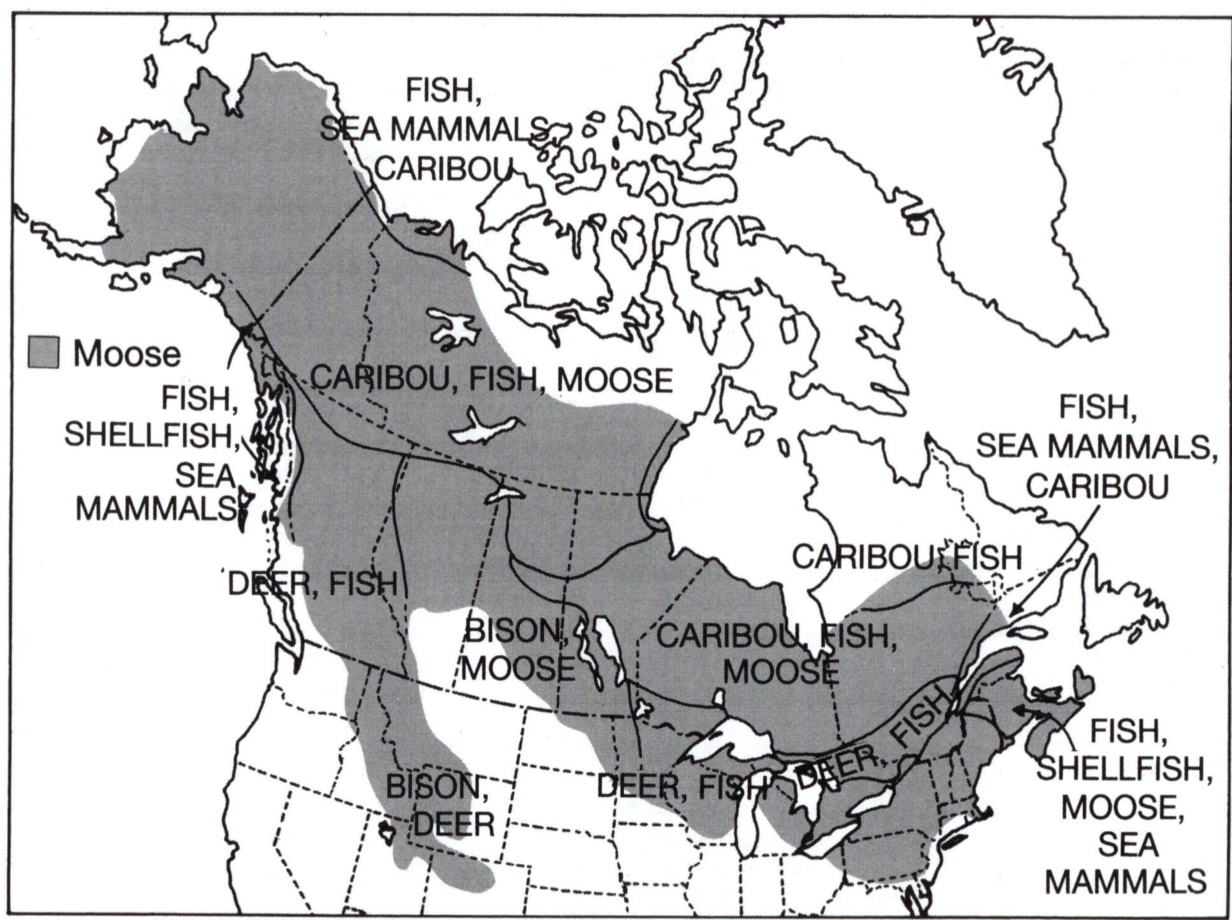

Figure 5. Types of subsistence used by Indian tribes within the postcontact range of the moose, based on ethnohistoric data (adapted from Driver and Massey 1957, Trigger 1978, Helm 1981, Harris and Matthews 1987).

snow, which they felt was a sign of disrespect to the animal, which could bring on a blizzard.

Butchering commenced with skinning, which was initiated by a cut from the throat down the center of the belly and up the insides of the legs. Then, the skin was carefully pulled or cut free, usually as a single piece (see Maxwell 1978). Only rarely was the moose quartered with the hide on. The carcass usually was cut into 14 pieces: hind legs (2), forelegs (2) including scapulae, head, neck, rib sections (2), brisket, backbone sections (3) and backstraps (2). Except for use of a maul or sharp axe to sever large bones or major joints, knives were the only other butchering tools. Some of the pieces might be further divided for easier transport, depending on the available assistance.

Although there was wide disparity among tribes in salvaging internal organs, the heart, brains, colon, intestines, mesentery, adrenal glands and reticulum invariably were saved. The lungs were the only organ fairly universally discarded. If transport necessitated greater efficiency (lightening of load), the head (except the nose meat and tongue), most of the stomach, lower vertebrae, liver or hide were abandoned. Large antlers usually were not transported.

Some Cree hunters reportedly left antlers in trees so the animals' spirits could see the sunrise and reassure other moose not to fear yielding themselves to hunters. Montagnais hunters left the skulls of moose, with antlers attached, elevated in trees (Speck 1935). Thompson wrote that "The bones of the head of a Moose must be put into water or covered with earth or snow" (Tyrrell 1916: 97). In reference to other Nahathaway (Cree) Indians, Thompson also related: "At the death of a Moose Deer, the hunter in a low voice cries 'wut, wut, wut'; cuts a narrow strip of skin [the bell] from off the throat and hangs it up to the Manito [an abstract and impersonal supernatural power]" (Glover 1962: 75). Sir John Franklin (1823: 64) wrote of the Cumberland House band of Western Woods Cree: "Many of the . . . hunters are careful to prevent a woman from partaking of the head of a moose-deer, less it should spoil their future hunts, and for the same reason they avoid bringing it up to a fort, fearing lest the white people should give the bones to dogs."

Among most tribes, processing a moose carcass was the task of women, although men, children and dogs usually assisted in the transport to camp of meat and other salvaged parts. The division of labor, however, varied greatly among the Natives, often as much a product of circumstance (e.g., kill location) as tradition. By whomever the butchering was done, it was a practiced undertaking. Skinner's (1912: 135) observation of the process by a team of Saulteaux Indians indicated an efficiency that was more norm than exception: "The animal was first drawn, skinned and quartered, the neck severed, the head cut off and the ribs separated from the spine, the entire process not taking more than fifteen minutes." Wrote David Thompson: "The Natives are very dextrous in cutting up, and separating the joints of a [moose] Deer" (Tyrrell 1916: 97). *Scene entitled "The Passion of* Ah-We-Ah*" painted by Frederic Remington; photo from Remington (1899: 10).*

Mountain Indians were known to suspend pieces of viscera of a slain moose from tree branches at the butchering site "as an offering to the spirit of the dead moose" (Gillespie 1981b: 337). Hodge (1910) noted that sacrifices of moose were homages paid by Iroquois and Algonquin tribes. The same was true of Cree (Milton and Cheadle 1865). Skinner (1912: 164) revealed that Saulteaux Indians left moose "beards" and other hide parts elevated at the kill site to mark the locations where the animals were slain. Skinner also observed moose skulls and bones and other animal parts hung in trees at a deserted campsite, which may belie his impression that the displays were merely hunting landmarks. Furthermore, Slavey Indian children were taught that unused or discarded bones simply tossed away rather than piled up would eventually prevent the killing of moose (Honigmann 1946).

Other simple observances of respect for game hunted by Subarctic Indians generally included avoidance of unwanton killing, prescribed and careful butchering so as not to break vertebrae or other bones in the process, and careful disposal of the skull and other bones so they would not be ravaged by dogs (see Speck 1935, Honigmann 1946, 1956, Rousseau 1952, Hallowell 1960, Helm 1961, Preston 1964). Other taboos related to Natives' spiritual associations with moose are discussed later.

Savishinsky (1970) described three patterns of sharing game, depending on the species or number of animals killed, by Hare Indians in the Northwest Territories. For a moose, the kill was given away by the successful hunter to his father or a close friend. All the men and some young women of a settlement traveled to the kill site, where the gift recipient supervised the butchering and designated its distribution among the people. Nelson (1973) indicated that moose meat tended to be distributed among tribal members (and families) according to need. Honigmann (1946) wrote that, among Fort Nelson Slave Indians, those invited to retrieve meat from a moose carcass "owned" the meat they packed home.

Transporting camp meat was most difficult in summer because of heat, humidity, impeding brush, bothersome insects and the likelihood of spoilage. Lacking horses, boreal Indians either carried the meat on their backs and loaded pack dogs with small quantities, or loaded the bounty into boats to be paddled or towed. Winter snows, particularly if hard-surfaced, enabled use of toboggans.

Huge celebrations customarily were held when a meat transport party returned. Moose meat figured prominently in Indian feasts, as related by Fr. Paul Le Jeune (Thwaites 1897c: 281): "In all their feasts, as well as in their ordinary repasts, each one is given his part, from which it happens that only two or three have the best pieces, for they do not divide them. For example, they will give the tongue of a Moose and all the giblets to a single person . . . these are the best pieces, which they call *Mascanou*, 'the Captain's part.'"

PROCESSING AND PRESERVING

Depending on their hunger, Indians ate moose parts both cooked and raw. According to Hearne, as soon as a moose was killed by a "Southern" Indian, "the hunter rips up its belly, thrusts in his arm, snatches out the kidneys, and eats them warm, before the animal is quite dead. They also at times put their mouths to the wound . . . and suck the blood; which they say quenches thirst, and is very nourishing" (Tyrrell 1911: 308).

Price (1945) reported that Indians removed the adrenal glands from a moose, cut each into two pieces, and willfully ate them and the walls of the reticulum to prevent the effects of scurvy, at a time when whites were unaware of the antiscorbutic benefits of vitamin C in this regard. And, according to Price, no other plant or animal tissue was a richer source of this vitamin than the adrenal glands. (We were unable to verify that, as implied, Indians were familiar with scurvy and this particular means of its prevention.)

Also eaten raw were round lumps of suety fat stripped of the "last entrail," and marrow extracted from cracked shank bones (Kendall *in* Audubon and Bachman 1845–1848).

Nelson (1973) wrote that, if shank bones were not discarded at the kill site, they were cooked there to secure marrow. And from antlered bulls taken in late summer, antler velvet was stripped off, roasted at the kill site and consumed (Smith 1984).

The easiest and quickest way to cook meat was simply to impale it on a green stick and broil it near or over the fire. Besides roasting, the most common and widespread method of cooking moose meat was by stone-boiling, first and aptly described by Lescarbot: ". . . after the roast we had boiled meat, and broth abundantly, made ready in an instant by a savage, who framed with his hatchet a tub or trough of the trunk of a tree, in which he boiled the flesh . . . [with] stones made red hot in the fire and [reheated] until the meat is boiled" (Grant 1911–1914[III]: 221–222). De Laguna and McClellan (1981) noted that Ahtna Indians of southern Alaska stone-boiled moose meat in spruce or willow bark baskets (not birch [Mason 1946]) or in hardened moose or caribou rumens (see also Honigmann 1981).

Fairly universal delicacies from moose included tongues, meat from the upper lips and nose hump, intestines, fetuses and heads. Moose tongue was smoked, roasted or boiled, then sliced. The upper lip or muzzle and meat from the hump of a moose nose was called "moufles" (Thwaites 1898b), "moufflon" (Audubon and Bachman 1845–1848), "moofle" (Spry 1968), "moosel" (Murray 1908) and "muffle" (Seton 1909, Merrill 1916), and greatly favored by Natives (Fraser 1881), being rich and greasy when cooked, much like beaver tail (see Whymper 1869, Hibbs 1890, Roosevelt 1893, Glover 1962). "The Nose is look'd upon [by Natives] as great Dainty's; I have eat [*sic*] several of them my self; they are perfect Marrow" (Dudley 1721: 167). Muffle usually was boiled, often along with the tongue. Ralph (1892) noted that muffle was the meat most prized, even more so than gastronomically renowned bison humps, by Hudson's Bay Company trappers (see also Donnelly 1967). "As to the fat intestines of the Moose, which are their greatest delicacies," Fr. Le Jeune (Thwaites 1897c: 281) reported, "they usually roast them and let everyone taste them, as they do another dish, which they hold in high esteem,—namely, the large intestine of the beast filled with grease, and roasted, fastened to a cord, hanging and turning before the fire." A member of the Palliser expeditionary party in western Canada in 1858 reported on moose "gut sausage which our hunters and half-breeds prepared with great skill" (Spry 1968: 244). Fetuses were a "great dainty," either roasted or boiled (Denys 1672: 433), but usually were given to the tribal elders who relished the tender, flavorful meat. Likewise, moose veal was highly esteemed. Moose heads taken to camp, or when camps were moved to kill sites (which was not usual), were stripped of hair and, with the tongue, muffle and brains removed, boiled in a large container to which vegetables, including digesta from the moose rumen (see Savishinsky and Hara 1981), were added. This produced a "moosehead soup" that was "a great delicacy" (Nelson 1973: 111). Miller (1986) told that Micmacs particularly favored roasted heads of young moose.

According to Milton and Cheadle (1865: 132), Western Woods Cree hunters immediately consumed the breast, liver, kidneys and tongue "at a single meal" because "moose is a sacred animal." At James Hector's camp in September 1858, near Lake Louise, Alberta, Assiniboin women prepared a feast of "moose nose and entrails, boiled blood and roast kidneys" (Spry 1968: 314, see also Tyrrell 1911).

Clearly, the favorite part of moose was bone marrow. High in fat and essential fatty acids, marrow also was a very important source of nutrition. It was sucked from cracked bones, particularly long bones of the legs and sometimes from ribs. In his *New Relation of Gaspesia,* Recollect missionary Crestien Le Clercq wrote that the Natives considered moose "grease . . . so delicious that they drink it wholly pure, with as much gusto as if it were the most pleasing liquour in the world. They eat it also quite raw as something exquisite" (Ganong 1910: 118). The leg bones were not discarded, even after the marrow was removed. (In the Subarctic, dogs were mainly fed dried fish meat and rarely meat or bones from large terrestrial mammals [see Helm and Lurie 1961, Helm 1981, Denys 1672].) Instead, bones were broken into small pieces or pounded into powder and boiled to produce fat. This process of rendering fat from the bones of a moose (excluding marrow) yielded 5 to 6 pounds (2.3–2.7 kg) of grease "as white as snow, and firm as wax" (Denys 1672: 423). Called *camaco* or *kumoo* by Micmac Indians and "moose butter" by whites, the rendered fat often was put up in cakes of 9 to 10 pounds (4.0–4.5 kg) and transported in birchbark baskets or "rogans" (Skinner 1912), and sometimes in moose bladders (Honigmann 1946). Furthermore, in anticipation of periods of food depletion and hunger, some people stockpiled bones that ultimately were crushed up and boiled into a fatty soup (Nelson 1973, McKennan 1981). "White as milk," moose grease soup reportedly was believed by Indians to be "as good for the chest as a large glass of brandy, or as the best of our meat broths" (Ganong 1910: 119).

In winter, meat was kept frozen in caches placed in trees or otherwise elevated well above the reach of marauding carnivores, particularly wolves (Mayer 1989). Salt was reportedly used by Abenakis as a means of preserving moose meat (Thwaites 1899a), but how widespread this practice was is not known (see Merrill 1916, Mason 1946), nor whether salt was used in this manner prehistorically.

Along the Churchill River in Saskatchewan, a Chipewyan woman pounds dried moose meat in the process of making pemmican—a staple for most northern Natives. Mixed with hot grease and berries, and packed firmly in air-tight containers, pemmican was a mainstay during times of travel and periods of food shortage. *Photo (PA20387) by C.A. MacDonald; courtesy of the National Archives of Canada.*

Moose meat not readily consumed was preserved frozen, air-dried, smoked or made into pemmican—a nearly universal food prepared for times when fresh meat was unavailable. "When heavy snows came," wrote Fr. Le Jeune, "they eat fresh Moose meat; they dry it to live upon the rest of the time until September" (Thwaites 1897c: 277). Pemmican was made by first drying the meat, pounding it finely, and mixing it with hot fat and, if available, dried berries or fruits. It was packed firmly into skin bags for efficient transportation and storage. Highly concentrated and nutritious, pemmican was a valuable trade commodity with other tribes and, later, with white explorers, trappers and traders.

Fr. Le Jeune provided a vivid account of how Indians dried moose meat: "When they are engaged in drying meat, they will throw down upon the ground a whole side of the Moose, beat it with stones, walk over it, trample upon it with their dirty feet; the hairs of men and of animals, the feathers of birds, if they have killed any, dirt and ashes,—all these are ground into the meat, which they make almost as hard as wood with the smoke. Then when they come to eat this dried meat, all goes together into the stomach, for they have not washed it" (Thwaites 1897c: 265).

Early English and French explorers and traders usually were pleased, if not delighted, to obtain moose meat, whether "green" (fresh) or dried, from Indians. In the latter half of the 1600s, Governor of New France Louis de Buade, Count de Frontenac (1867: 88) wrote: "The generosity of our savages merited the most lively gratitude on our part; already for some time not having been able to find suitable hunting grounds, we had been compelled to eat nothing

Moose meat not consumed fresh typically was smoked. Cut into strips of 2 to 3 inches (5.1–7.6 cm) wide and 12 to 24 inches (30.5–61 cm) long, the meat was hung on racks over green wood fires and out of the reach of camp dogs. Smudge pots often were used to keep insects away. When acceptably jerked—"hard as a stick of wood," according to Italian Jesuit Fr. Bressani (Thwaites 1899h: 115)—the strips of meat were lightly pounded to soften them somewhat for compact storage and eventual chewability. *Photo (PA20094) by R.D. Davidson; courtesy of the National Archives of Canada.*

but bacon; the moose and elk which they gave us removed the disgust we began to have for our ordinary fare."

QUALITY AND QUANTITY

Indians relished wild meat and, among many Subarctic Natives, "moose is the one that really counts" (Nelson 1973: 112). Le Clercq recorded that "The meat of the moose is that which our Gaspesians esteem the most" (Ganong 1910: 118). And as Jesuit Fr. Cramoisy wrote in 1672: "God has given . . . to these cold regions the bear, moose, beaver, and porcupine; they constitute a food supply which, for bracing the stomach . . . are well worth the figs and oranges [of the tropics]" (Thwaites 1899i: 205). At least among Montagnais-Naskapi, according to Speck (1935: 79), an important deduction was that, by eating animals—such as caribou, bear, beaver and moose—that fed on vegetal matter, which was the original source of medicinal pharmacology, the Indians were "taking-medicine." In essence, a diet of wild game was prophylactic. Famed Arctic explorer Vilhjalmur Stefansson (1944: 97) indicated that during starvation periods, when the Native diet was almost exclusively fresh meat boiled or roasted, pemmican was nibbled "to supply from its high fat content the lacking fat of the buffalo, caribou, moose or other game." Stefansson also pointed out that a constraint of the Hudson's Bay Company logistical operations was its insistence on European foodstuffs (although not entirely) and customary European style of food preparation. The Northwest Company became a serious rival of the Hudson's Bay Company, as noted later, in part because its trappers and traders readily adopted Native foods, especially summer pemmican.

Numerous early accounts tell of the extraordinarily rapacious appetites of boreal Indians and of many Europeans who lived or visited among them. "The quantity and variety of food which some Indians are capable of consuming is beyond the comprehension of a white man" (Hoffman 1896: 287). Hector related one such incident, after the killing of a 3-year-old moose: "We at once made a fire by the carcass, which lay among fallen timber where the snow was about four feet [1.2 m] deep. Our appetite was tremendous, so that, although the flesh of the animal was so lean that at other times we would not have eaten it, we continued cooking, eating, and sleeping the remainder of the day, and the whole of the next, by which time there was little left of the moose but the coarser parts of the meat. . . . indeed both the dogs and masters conducted themselves more like wolves than was altogether seemly, excepting under such circumstances" (Spry 1968: 385).

Hunters of the Shield Subarctic required great quantities of food for themselves and their families. During the winter, daily per capita requirements of 4,500 to 5,000 calories were needed, with the summer need being somewhat less (Burton and Edholm 1955, Feit 1973). To meet these needs, at least 4 pounds (1.8 kg) of flesh food had to be consumed daily (Rogers 1963), and part of this had to be fat to fulfill caloric and essential fatty acids requirements. Generally, big game yielded the most food per labor unit input (Rogers 1973). But when the daily yield from big game dropped below about 3,500 calories, hunters turned to small game and fish for dietary diversity and taste variety, despite the greater labor input required. If alternative food supplies were unavailable, starvation soon commenced (see Thwaites 1898e, 1899c, 1899d, 1899h, Cronon 1983).

In particular, for example, Beaver Indians sought to organize their activities so that *at least* 2 pounds (0.9 kg) of meat per capita a day was regularly available (Ridington 1981). Thus, a group of about 10 people needed about 140 pounds (64 kg) of meat a week, or a moose about every three weeks (see also Feit 1973). On the other hand, a group of 100 people would need 1,400 pounds (635 kg) a week, or nearly four moose per week. Calculating a food yield of 759 pounds (344.6 kg) per adult moose averaging 1,105 pounds (501.7 kg) live weight, McCabe (1982) approximated that 2.7 moose were needed weekly to support a band of 100 Indians (Table 1). Because of these requirements, plus the availability of moose in their foraging range, the Beaver Indians tended to form optimal-sized groups consisting of about 30 individuals. If the group was too large, a short period of meat scarcity would quickly bring the group to starvation. Conversely, a very small group placed too much reliance on the well-being and skills of only one or two hunters.

Inasmuch as fresh meat was not continually available, Indians tended to gorge themselves whenever it was—sometimes eating to the point of sickness, of which Alexander Mackenzie (1801: 66) wrote: "I should imagine these complaints ['fluxes'] must frequently proceed from their immoderate indulgence in fat meat . . . particularly when . . . preceded by long fastings."

In his ethnozoological examination of elk, McCabe (1982) analyzed the "approximate" importance of various nonvegetal foods in the tribal diets of Indians. He ranked moose and fish both second behind caribou in the subarctic region; in the northeast region, moose was ranked third, behind deer (first), and small mammals and birds (tied for second); and moose did not appear among the four "importance" ranks in the Great Lakes region. In light of further analysis for this work, moose would have to be considered closely behind only deer and fish for many northern Great Lakes tribes.

Other than numerous reports that Indians generally esteemed the flavor of fresh moose meat, well-cooked and clean (see Nelson 1973), most comments on its quality came from whites. Paul Dudley (1721: 167), for example, noted that "The Meat of Moose is excellent Food, and tho' it be not so delicate as the common Venison, yet it is more substantial, and will bear salting. . . . The Indians have told me, that they can travel three times as far after a Meal of Moose, as after any other Flesh of the Forest." David Thompson wrote that "The flesh of a Moose in good condition, contains more nourishment than that of any other Deer; five pounds [2.3 kg] of this meat being held equal in nourishment to seven pounds [3.2 kg] of any other meat even of the Bison, but for this, it must be killed where it is quietly feeding . . ." (Tyrrell 1916: 96). Among the Flathead, Coeur d'Alene and Blackfoot of the Intermountain West, 1840–1847, Jesuit Nicolas Point observed that moose, "in the quality of its meat, is almost the equal of buffalo" (Donnelly 1967: 167). "It does not inconvenience the stomach," wrote Denys (1672: 383). "One can eat all his fill of it, and then an hour later can eat as much again. . . . As to the taste it suggests venison a little, and is at least as pleasing to eat as the Stag [North American elk or Eurasian red deer]." Morton (1632: 74–75) noted the flesh of moose was "very good foode, and much better than our redd Deare of England."

From the vicinity of Lake Champlain in 1759, J.C.B. (1941: 109) confirmed that moose "flesh has a very good flavor, and is nourishing but not heavy." British military officer Captain Jonathan Carver (1838: 277) traveled extensively through the upper Great Lakes region from 1766 to 1768 and commented that moose meat "is exceeding good food, easy of digestion, and very nourishing." George Nelson, a fur trader with the Northwest Company, who spent the 1802–1803 winter in the St. Croix Valley of northwestern

Table 1. Approximate maximum number of adult moose and other big game animals of various species hypothetically necessary to support a band of 100 North American Indians (modified and expanded from McCabe 1982: 89)

Species[a]	Approximate mean live weight in pounds (kg)[b]	Approximate food yield in pounds (kg)[c]	Number of animals needed per period of time[d]			
			Day	Week	Month	Year
Moose	1,105 (501.7)	759 (344.6)	0.39	2.7	11.9	142.4
Caribou	295 (133.8)	212 (96.2)	1.41	9.9	42.4	516.5
Bison	1,382 (627.4)	937 (425.4)	0.32	2.2	9.7	116.9
Elk (Rocky Mountain)	600 (272.4)	407 (184.8)	0.73	5.1	22.2	266.5
Deer	165 (74.9)	115 (52.2)	2.60	18.2	79.0	949.0

[a] Only major ungulate species within the range of the moose are included.

[b] Computed annual average weights of mature males and females of the various subspecies, recognizing that Natives historically did not select exclusively for prime age animals.

[c] Calculated on the basis of 90 percent of dressed carcass weight (minus most bones), plus 60 percent of visceral weight. This calculation also assumes a 7 percent higher take of meat and fat and a 95 percent higher take of viscera by pre-1850 Natives than by modern hunters.

[d] Based on an estimated year-round meat consumption rate of 3.0 pounds (1.36 kg) per person per day.

Wisconsin, reminisced that a meal of "moose dear" was "excellent though it had only been *passed* thro' the water [undercooked]" (Barton and Nute 1948: 17). Winthrop (1861) attested that moose meat combined the flavors of beef and turtle. Theodore Roosevelt (1893: 214) opined that "The flesh of moose is very good; although some deem it coarse." Seton (1909[I]: 181) wrote that moose steaks were "delicious." Moose flesh was "considered very good" by Audubon and Bachman (1845–1848: 187), "as good as beef" by William Wood in 1634 (Vaughan 1977, cf. Gillmore n.d.), and "far above deer or even reindeer [caribou] meat" by Whymper (1869: 215, see also Russell 1898). Shields (1890: 20) observed that "The venison of the moose is good, winter or summer. It is coarse-grained—even more so than that of the elk. . . . the flesh is dark and uninviting to the eye, but sweet and juicy to the palate." Merrill (1916: 206) indicated that "even an old moose, if in good condition in other respects, is excellent in flavor." A rare dissenter was Pierre Esprit Radisson, who worked to establish fur trade with Indians in the northern Great Lakes region during the 1650s. Radisson confessed, ". . . I killed an oriniak. I could have killed more, but we liked the fowls better" (Adams 1961: 127).

These favorable assessments of moose meat quality notwithstanding, it should be recognized that Native gastronomy was markedly different than that of whites. Nutritional impulses and limited dietary variation appear to have influenced a wider range of food palatability than was the case for relatively discriminating whites, as evidenced by the latters' amazement and poorly disguised repugnance with certain Indian eating habits and choices (see Denys 1672, Hibbs 1890, Grant 1907, Ganong 1910, Merrill 1916, cf. Nelson 1973). The forementioned Indian preference for fat-laden intestines is one example.

Also, Indians sometimes salvaged decayed moose carcasses found in their travels, and ate them, particularly in times of starvation. In December 1821, Indians with John Richardson's expedition retrieved for food a moose that had drowned in a rapid part of the river partially covered with ice (Houston 1984). In spring 1855, Henry T. Allen's exploration party along the Chettyna River, Alaska, came upon an Indian who had salvaged some morsels from a moose that wolves had killed during the winter (U.S. Senate 1900: 429). Midshipman Robert Hood, with the Franklin expedition in 1819–1821, reported that an unborn moose calf possibly weighing 30 pounds (13.6 kg) was eaten readily by the Indians, although it had been partially consumed by wolves the night before (Houston 1974: 57). Among pregnant Indian women of the Kenai Peninsula, the calf moose meat was thought to facilitate their own delivery (Palmer *in* Hosley 1949: 38). Conversely, some Indians had constraints about eating embryos, but not because of aversion to the taste: "They do not eat the little embryos of Moose, which they take from the wombs of the mothers, except at the end of the chase [hunting season?] for this animal. The reason is that their mothers love them, and they would become angry and difficult to capture, if their offspring were eaten so young" (Thwaites 1897c: 34).

We were unable to identify either generally or specifically whether there existed an age or sex selection of moose by Indian hunters. Except for mention in Denys (1672) that, when a cow and a bull were started together, the cow likely was pursued because it might have the prize of a single or twin fetuses. Otherwise, and in the absence of sufficient archeological data or convincing ethnological documentation, we assume that the Natives' subsistence patterns would have led them to select for the larger animals (most meat)—bull versus cow, adult versus calf—when a choice availed itself.

Skin and Bones

Truly, as the bison was to the Great Plains Indians, the elk was to many Indians of the northern Plains, Plateau and Intermountain regions, the white-tailed deer was to the Great Lakes, midwestern prairies, and eastern and southeastern forests, and the caribou was to many nearctic aboriginals, the moose was a veritable commissary for most Indians within the species' range. Wrote Gillespie (1981a: 15) "For the Indians living in the Northwestern Transition section of the boreal forest . . . of Canada and in the plateau of Alaska and the Barren Ground caribou are the subsistence core, with the moose an important subsidiary resource. For the rest of the Subarctic the woodland caribou and moose are the essential large game. These animals have also provided the raw materials for much of the material culture of the natives." Reporting on Slavey and Dogrib Indians he met in July 1789, Alexander Mackenzie (1801: 36) noted, "Their clothing is made of the dressed skins of the rein or moose-deer, though more commonly of the former." For provisioning those and other Denes—Athapaskan-speaking groups of western Northwest Territories—Thompson (1994: 5) confirmed that "moose and caribou were the most important sources." Regarding many Indians of the northeast coastal region and the St. Lawrence lowlands, "The moose was not only the main staple of the . . . diet, but the primary source of their clothing" (Maxwell 1978: 112, see also Thwaites 1897b). Added Seton (1909[I]: 181): "Assisted in warm weather by various fish, it [moose] bears practically the burden of their support." And although the literature suggests at least equally diverse uses of caribou by Subarctic and Woodland Indians, Merrill (1916: 202) believed that "moose furnished more than any of the others [animals]."

Bernard R. Ross (1861: 433, 437), a long-time employee of the Hudson's Bay Company in the Mackenzie River District of the Canadian Northwest, presented a partial overview of Indian uses of the moose other than for food: "The hide supplies parchment, leather, lines, and cords; the sinews yield thread and glue; the horns serve for handles to knives and awls, as well as to make spoons of; the shank bones are employed as tools to dress leather with; and with a particular portion of the hair, when dyed, the Indian women embroider garments . . . The capotes, gowns, fire-bags, mittens, moccasins, and trousers made of it are often richly ornamented with quills and beads, and when new look very neat and becoming. . . ."

SKIN

Other than meat, the moose's skin probably was its most valuable product. Clothing made of moose hides was preferred by many Indian tribes and, among some, there was no substitute. Wrote Morton (1632: 74) of moose: "Their hids [*sic*] are by the Salvages converted into very good lether, and dressed as white as milke." Dudley (1721: 168) noted that "The Skin of the Moose, when well dress'd, makes excellent Buff [thick leather]. . . ."

Moose hide preparation was the work of Indian women, and it was labor intensive (De Laguna and McClellan 1981) and required skill and patience (Rachel Robert *in* Thompson 1994). Various degrees of treatment were involved depending on the intended use of the skin—from mere rawhide to hair-on hide, to fully tanned products. In all cases, the process was complex (McClellan and Denniston 1981, cf. Russell 1898), but particularly the tanning to soft and supple suede. Improper treatment could ruin a skin, at least for its intended use. Also, although the process generally was similar from tribe to tribe, there were some considerable variations in treatment tools and concoctions.

Generally, skins taken from moose in late summer and autumn were considered the best (McClellan 1981a, Rogers and Smith 1981). And despite the widespread use of moose hide, especially for clothing, it was deemed by some as inferior to that of other big game, such as from deer and caribou (Merrill 1916) and bison (Bethune 1937, Morrow 1975), because it was thicker, more porous, less fibrous and relatively easily stretched yet less elastic. Finished moose hide, wrote Russell (1898: 185), "is soft and pliable, nearly as thick though far inferior in wearing quality to cowhide." Honigmann (1946) indicated that bearskin babiche was much superior in its tensile strength to moose hide babiche. On the other hand, J.C.B. (1941: 109) related that moose "skin is very soft and flexible, is used for chamois and makes good buff leather" (see also Carver 1838). Mason (1946: 16) noted that, among Dogrib, moose hide was preferred "for many purposes" to that of the caribou. And regarding moose, David Thompson offered: "His skin makes the best leather" (Glover 1962: 84), an opinion shared by Seton (1909). In any case, it equated closely with elk skin in terms of usefulness and durability. Also, Tatsi Wright, a 99-year-old Mountain Indian woman reported in 1971 that, compared with other big game hides, those of moose were difficult to dress and sew because of their thickness (see Gillespie 1981b).

In 1795, Samuel Hearne wrote: "To dress those [moose] skins according to the Indian method, a lather is made of the brains and some of the softest fat or marrow of the animal, in which the skin is well soaked, when it is taken out, and not only dried by the heat of a fire, but hung up in smoke for several days; it is then taken down and well

A Penobscot Indian (Eastern Abenaki) wears a deer-scalp mask preliminary to participation in the Clown or Trade Dance—a nighttime gaming ceremony. Also being worn is a hair-on moose hide coat. *Photo from 1912 by Frank Speck; courtesy of the Smithsonian Institution National Museum of the American Indian.*

soaked and washed in warm water, till the grain of the skin is perfectly open, and has imbibed a sufficient quantity of water, after which it is taken out and wrung as dry as possible, and then dried by the heat of a slow fire; care being taken to rub and stretch it as long as any moisture remains in the skin. By this simple method, and by scraping them afterwards, some of the moose skins are very delicate both to the eye and the touch" (Tyrrell 1911: 263).

Tro-chu-tin (Klondike band of Han) Indians of Alaska processed moose skins somewhat differently, as Adney (1900: 502) related: "The hides are brought in-doors and women at once set to work dressing them. The hair was shaved off; then the skin was turned over, and all the sinew and meat adhering was removed by means of a sort of chisel made from a moose's shin bone; and finally scraped, a work requiring a whole day of incessant and tiresome labor. The skin was now washed in a pan of hot water, and then wrung dry with the help of a stick as a tourniquet. After which the edges were incised for subsequent lacing into a frame, and then hung out-doors over a pole. The tanning, with a 'soup' of liver and brains, is done the next summer."

Denys (1672: 411) detailed the Micmac moose skin-dressing process: ". . . these are soaked and stretched in the sun, and are well-heated on the skin side for pulling out the hair. Then they stretch them and pull out the hair with bone instruments made on purpose. . . . Then they rub it with bird's liver and a little oil. Next, having rubbed it well between the hands, they dress it over a piece of polished wood made shelving on both sides. . . . They rub it until it becomes supple and manageable. Then they wash it and rub it with sticks many times, until it leaves the water clean. Then they spread it to dry."

Jones (1906: 114–115) elaborated on the Ojibway technique of processing moose hide: "To dress a skin it was first soaked in water alone or in a preparation of brain boiled in water; it was then stretched on a rectangular frame of four poles fastened at the corners with thongs. The frame was leaned against a solid support, and the hair was then scraped off by means of a short, round, thick-handled tool with a short blade attached to the bent neck of the handle; it worked like a hoe. . . ." Besides "hoeing"—scraping against the grain—to remove hair, many Indians soaked the hide first in a lye mixture of water and wood ashes for 3 to 6 days, then buried it in a bundle. After some months, the

Moose hide preparation invariably began with use of a moose leg bone "flesher" or "punch" with a serrated end, to scrape off adhering meat, fat and other tissue (left). To tan a moose hide, a skin was subjected to a series of treatments, but ultimately stretched on a frame after being soaked, then rubbed vigorously and repeatedly with smooth sticks to soften and dry the leather (right). "With no other tools than a knife and the leg bones of the animal killed, and no other tanning agent than its brain, an 'old wife' can convert a green moose-skin, weighing fifty pounds [22.7 kg], into light serviceable leather in five days," wrote Russell (1898: 185). Honigmann (1946) estimated that, because of the enormous amount of time and labor required, an Indian woman could tan only about 10 moose hides per year, besides performing other chores. *Left photo courtesy of the Library of Congress. Right photo (PA42114) courtesy of the National Archives of Canada.*

hair was softened and relatively easily scraped away. Parker (1954: 76) stated that, to remove hair from moose hide, the hide had to be soaked in ashes and water or have ashes sprinkled on and rubbed in, then moistened with water. In both cases, the skins were rolled up until "the lye of the ashes weakens the texture of the skin." Slavey Indians held that hairs scraped from a moose hide had to be discarded in such a manner that people would not step on them, for fear of showing disrespect to moose and jeopardize hunting of the species (see Honigmann 1946).

Essentially, moose hide was softened by soaking in a pasty concoction of water and fermented brain (McClellan and Denniston 1981), boiled moose brains and moose fat (Honigmann 1946), or moose brains, grease and spinal fluid (Thompson 1994), and the fibers subsequently broken down further by scraping and smoking. After a hide was dehaired, Dogribs kneaded in a mixture of hair and grease to clean and soften it again, then dried and smoked the hide to let the grease permeate (Mason 1946). Then the hide was soaked in hot water overnight, and stretched and smoke-dried daily for 4 successive days. Moose hides dressed in the hair reportedly were treated only with liver rubbed in by hand on the skin side and "sawed" over sticks until sufficiently pliable (Denys 1672).

Jones (1906) noted that "tanned" coloration was given to skins by smoking them in smudges, with sumac and rotted spruce being among the best fuels. A young recruit of the Oblates of Mary Immaculate, Father Émile Petitot (1888: 374) noted in 1862 that Slavey clothing worn by men and women were made of closely fitted moose hide smoked to a "beautiful saffron" hue. Thompson (1994) indicated that smoking with rotten spruce or willow fuel made moose hide more resistant to moisture.

Moose hide was used primarily for clothing, and virtually all articles of apparel worn by boreal and woodlands Indians were fashioned from moose hide by one or more Indian groups within historic time (Table 2). The most widespread use was for footwear. "Of this lether [*sic*]," wrote Morton (1632: 74), "the Salvages make the best shooes. . . ." Added Russell (1898: 172): "Dressed moose leather is the best material attainable by the Northern Indians for the manufacture of moccasins." Among the many tribes known to have manufactured moccasins and/or mukluks from moose skin were the Dogrib, Slavey, Western Woods Cree, Hare, Saulteaux, Beaver, Carrier, Kaska, Island Tlingit, Tagish, Han, Ahtna, Micmac, Western Abenaki, Delaware, Sekani, Huron, Chippewa, Ottawa, St. Lawrence Iroquois and Northern Ojibway.

As best as can be determined, moose hide moccasins were essentially of six types, either as a continuous part of pants or leggings (see Allen 1887, Powell 1909), mukluks or in patterns known as "beavertail," "rabbit nose," "Athapascan" and "Tlingit" (see Hatt 1916, Mason 1896). Traditionally, the latter two were sewn with calf-high wraps among northern tribes, and with ankle flaps by New England and Great Lakes Indians. A few tribes, such as the Chippewa (Maxwell 1978), Dogrib (Mason 1946) and Koyukon (Clark 1981), featured winter moccasins with moose skin soles and deer hide or caribou hide uppers. Denys (1672: 412) noted the Indian practice of moccasins made from "old robes of Moose skin, which are greasy and better than new [freshly tanned hide]," although a Fort Nelson Slavey woman noted that moccasins of this make and age easily burned if the wearer got too close to a fire. Depending on the style and size, 10 to 15 pairs of moccasins could be made from a single moose hide (Honigmann 1946).

Generally, fully tanned moose hide moccasins were worn in summer and hair-on moccasins (with the hair side in) for winter footwear. Skin from the moose hock was choice for winter footwear because it was "nearly waterproof" (Day 1978: 154). It also has a natural bend, con-

Table 2. Some articles of clothing made of moose hide worn by North American Indians

Garment	Tribe or region[a]	Source
Apron (shaman's)	Tlingit	Newton and Hyslop 1992
Belt	Slavey	Thompson 1994
Bracelet	Chipewayan	Tyrrell 1911
Cap	Western Abenaki	Day 1978
Cape	Penobscot	Snow 1981
Capote	Mackenzie River	Ross 1861
Coverlets	Micmac	Denys 1672
Diaper	Slavey	Asche 1981
Dress	Dogrib	Rogers and Smith 1981
Fringe (trim)	Tagish	McClellan 1981a
Gloves	Ahtna	De Laguna and McClellan 1981
Gown	Mackenzie River	Ross 1861
Hair roach	Southwestern Chippewa	Ritzenthaler 1978
Hat ("stick")	Kaska	Honigmann 1946
Hood	Hare	Savishinsky and Hara 1981
Hunting frock	Kutchin	Russell 1898
Jacket	Hare	Savishinsky and Hara 1981
Leggings	Sekani	Denniston 1981
Mantle	New England	Morton 1632
Mitten gauntlets	Saulteaux	Steinbring 1981
Mittens	Western Woods Cree	Smith 1981a
Moccasins	Montagnais	Hatt 1916
Mukluks	Northern Ojibwa	Rogers and Taylor 1981
Pants	Tagish	McClellan 1981a
Robe	Ojibwa	Jones 1906
Shirt	Kutchin	Slobodin 1981
Tunic	Carrier	Tobey 1981

[a] The clothing identified was not necessarily exclusive to any tribe or group.

forming to the human ankle and foot. Indians of some tribes, such as the Dogrib and Western Abenaki, wore footwraps of muskrat or snowshoe hare skins inside moose hide moccasins.

Among the many other uses of moose hide by various Native groups at one time or another within historic time and perhaps earlier was a soft, tanned strap worn by expectant Koyukon women below the abdomen to help support the weight of the unborn child (Clark 1981). Han women gave birth by squatting against or next to a horizontal support pole padded with moose hide (Crow and Obley 1981). Slavey mothers carried and nursed their infants in a moss-lined moose hide bag (Asche 1981). They also fashioned diapers from moose calfskin, lined with moss.

Some Delaware warriors carried into battle heat-toughened moose hide shields that covered the body to the shoulders (Goddard 1978). Carrier (Tobey 1981) and Tlingit (De Laguna 1990) wore moose hide "armor" during warfare. Slavey, Montagnais, Eastern Cree, Chippewa, Ojibway and Eastern Abenaki used moose rawhide knife sheaths, and some Slaveys had quivers of hide taken from the lower hind legs of moose (Honigmann 1946). McClellan (1981c) identified postcontact ammunition pouches carried by Tagish men.

Western Abenaki reportedly wore caps of moose hide taken from the animals' shoulders—"the long white hairs of the moose hump forming a natural crest" (Day 1978: 154). Fort Nelson Slavey men wore "stick" hats that featured moose hide covering and tassels (Honigmann 1946). Kutchin wore their hair long, bound in three "queues" held together with a cord "neatly worked with [caribou or moose] hair artificially colored" (Mackenzie 1801: 49). Chipewyan men and women sometimes wore headbands of moose hide (Tyrrell 1911). And fashionable among single Micmac women were ornamental hair decorations featuring "threads of leather from unborn moose" (Denys 1672: 414).

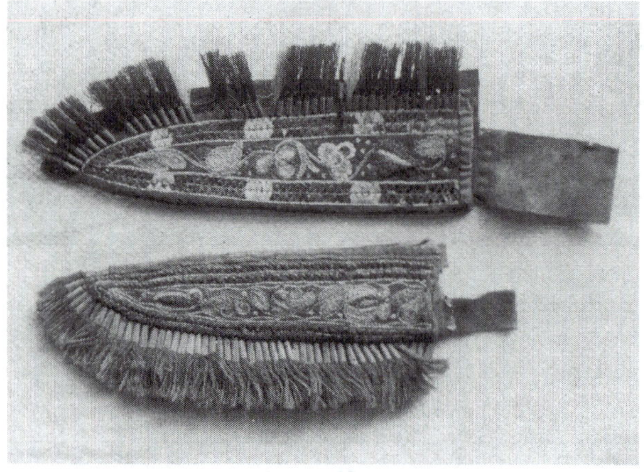

Finely beaded Micmac knife sheaths of moose hide with moose hair fringe. *Photo courtesy of the Smithsonian Institution Anthropological Archives.*

Western Woods Cree, Hare, Beaver, Slavey and others used tanned moose hides as blankets (see Mason 1946, Smith 1981a), and hair-on moose hide served as bedding for Athapaskans of the Alaska Plateau (Hosley 1981a) and Algonkians of the northeast (Denys 1672, Maxwell 1978).

To transport light packs, Ojibways used tumplines of moose skin passed over the arms and across the chest; for heavy loads, a moose hide tumpline was placed over the forehead (Jones 1906). Moose hide straps also were used to secure packs to Kaska (Honigmann 1946) and Ahtna (Allen 1887) dogs, which were said to be capable of carrying loads of 35 and 25 to 40 pounds (15.9 and 11.3–18.1 kg), respectively, as far as 20 miles (32.2 km) in a day. Southern Tutchone Indians fit their dogs with moose hide "shoes" to protect the animals' feet from shard ice (McClellan 1981d). For harnessing of dog teams, moose hide was preferable to caribou skin by Dogribs, because of the former's thickness and strength (Helm et al. 1981).

A number of Indian tribes—including Carrier (McClellan and Denniston 1981), Inland Tlingit (McClellan 1981a), Ahtna (De Laguna and McClellan 1981) and Dogrib (Mason 1946)—lined the runners of sleds with hair-on skin from the forelegs of moose. The skins were sewn together so that the pelage grain ran to the back, reducing drag and minimizing ice build-up on the runners.

Despite the generally low regard with which moose hide was held by some Indian tribes, because of its weight and porosity, quite a few Natives used moose skin for boats and canoes. Ahtna Indian boats featuring moose hide were up to 30 feet (9.1 m) in length and held 10 to 30 paddlers (Abercrombie 1900). Moose hide boats used by Nalani Indians were approximately 35 feet (10.7 m) long and 10 to 12 feet (3.0–3.7 m) wide (Honigmann 1946). The frame and supports were made from spruce. Twenty moose skins comprised the hull, supported by spruce ribs bent around the hull on the outside of the skins. For some interior travel, Tlingit Indians temporarily covered light cottonwood-frame boats with raw moose hides (McClellan 1981a). Kutchin and Tanaina Indians occasionally used mooseskin boats for summer travel (Osgood 1937). Maliseet (see Gyles 1851, Adney and Chapelle 1964), Hare (Savishinsky and Hara 1981) and various Eastern Abenaki (Pennant 1784, Snow 1978) tribes built moose hide-covered canoes. Usually after hunts, and to facilitate the transport of game, Penobscots stretched wet, dehaired moose hide over elm-pole gunwhales to construct canoes, using cedar staves as ribs (Speck 1940). Such craft typically were dismantled when the hunting party returned home, and the hides were put to other uses. To some large canoes, Micmac Indians added

Mountain Indians' boats made of moose rawhide typically required 8 to 10 skins sewn together and stretched (with hair on the outside and the grain toward the stern) over a roughly made frame with large, pliable, green willow ribs, and the seams sealed with "hard grease" (Camsell 1954: 49). The skin boat above, near Fort Norman in the Northwest Territories, was about 25 feet (7.6 m) in length, with a beam of approximately 5 feet (1.5 m), was suitable for river travel and slid easily over submerged rock, but was too heavy for navigating shallow streams. *Photo (PA18653) by F.V. Seibert; courtesy of the National Archives of Canada.*

sails of bark or the "well-dressed skin of a young moose" (Denys 1672: 422) and, with as many as 8 to 10 passengers and a favorable wind, the canoe "went as swiftly as the throw of a stone."

Also for travel, moose hide received widespread use for snowshoes. Greatly diverse in frame form and size, snowshoes were used by nearly all Natives for traversing deep snows. The preferred webbing leather apparently was bearskin or moose hide babiche (e.g., see Townsend 1981). Nicolas Denys (1672: 420) described snowshoes "corded with Moose skin dressed to parchment [rawhide] both thick and thin. The thick were placed in the middle part of the snow-shoe, where the foot rests between two sticks, while the thin were used at the two ends."

Also concerning transportation, there exist museum specimens of Ojibway and Plains Cree pad saddles of moose hide from the 1800s (Hail 1980, see also J.C.B. 1941). (Incidentally, when Iroquois were first exposed to horses, they called them the "Moose of France," and were astonished that the equine "were so tractable and so obedient to man's every wish" [Thwaites 1899f: 81]).

Moose hide served as lodge covering by a few tribes, including Western Woods Cree (Smith 1981a) and Beaver Indians (Ridington 1981), but most Natives seem to have selected lighter and relatively portable hides, such as caribou or elk, for this purpose. Hibbs (1890) indicated that some Indians of the Rocky Mountains region made tents of moose hide. Some Slavey Indians, whose tents were of caribou skins, used moose hide entry flaps (Honigmann 1946).

Another commonplace use for partially tanned moose hide was as packsacks and containers for clothing and dried foodstuffs (Denys 1672, Honigmann 1946, McClellan and Denniston 1981). Personal items typically were stored in tanned calf hide bags often elaborately decorated. Athapaskans reportedly made bags from the skins of fetal moose (Rogers and Smith 1981). Some Blackfoot and Plains Cree used tanned moosehide for pipe bags (Hail 1980). Partially tanned moose hide or rawhide generally sufficed for storage of pemmican, dried berries and smoked food. Some Tahltan "firebags"—leather pouches for carrying materials to start fire—were made of moose hide (Ross 1861, Newton and Hyslop 1995). Also widespread was the use of moose hide clothing ties and belts, bundle straps, snare lines, cordage and handwraps for such tools as beamers, fleshers, flakers, sewing awls and knives. Eastman (1916) referenced a moose hide drum head stretched on a basswood cylinder.

During dire periods of severe food deprivation, moose hide was boiled for soup or eaten outright. Jesuit missionaries among Ottawa Indians told of sharing moose hide meals with starving Indians. When foods became scarce during the winter of 1670–1671, for example, an Ottawa woman offered Fr. Cramoisy "an excellent dish . . . I was told that a piece of the door had been put in the pot. . . . It was an old Moose-skin . . . and it last me twenty-four hours. . . . I had little hope of prolonging my life therewith . . . after eating such unaccustomed diet" (Thwaites 1899h: 145, 151, 153). Weeks later, and suffering further from malnutrition, Cramoisy wrote: "It is true, I was occasionally given a moose-skin but that was a feast by no means common." And just before spring ice melt, the famished missionary admitted that "Acorns, rock tripe, and moose-skins were then delicious to me" (see also Thwaites 1899a).

BONES

Comparatively few material uses were made of moose bones by North American Natives, but most of those reported were associated with various tools. The foremost use, such as by Montagnais and Naskapi (Rogers and Leacock 1981), Eastern Cree (Rogers 1967), Piegan (Russell 1898) and Tanana (McKennan 1981), was that of hide beamers and fleshers from the cannon bone. Wilbur (1978) indicated that New England Indians sometimes chose moose metacarpal and metatarsal bones for awls and spear, harpoon and fish hook points. These were created by splitting bone lengthwise, and bone slivers broken off and ground to shape. Warren (1884) reported that the Ojibway made knives from rib bones, and moose bone occasionally served as pipe stem reamers for Micmacs (Denys 1672). Gaming dice cut from moose scapulae were used by Micmacs and Penobscots (Culin 1907). Sekani Indians once used moose jaw clubs in warfare (Morice 1895, Denniston 1981). Seton (1929) indicated that Natives used moose bone to cut ice, but identified neither the Indians nor the particular bones. And, as previously noted, moose shoulder blades were used in hunting, to simulate the sound of antler thrashing. Indians of the Northwest coast reportedly made fishhooks from the naturally block-shaped maxillary bone of the moose nose (Goodchild 1984). Fire-induced cracks in charred moose scapulae supposedly enabled conjurers in certain tribes, such as the Misstassini Eastern Cree, to divine the future (Newton and Hyslop 1995).

Denys (1672: 383) wrote that within the moose heart "there occurs a little bone of which the Indian women make use to aid in childbirth; they reduce it to powder, and swallow it in water, or in the soup made from the animal." Le Clercq indicated that the heart bone was called *Oagando hi guidanne*, or "bone for giving birth," and was "a sovereign remedy for easing the confinement, and for relieving the spasms and suffering of childbirth" (Ganong 1910: 275). Merrill (1916: 268) reported that the bone was not imaginary, but actually the *os cordis*—"a local ossification of the septum between the ventricles of the heart, and is found in a number . . . of ruminants . . . after they pass a certain age. Its medicinal value is nil."

USE OF OTHER MOOSE PARTS

Sinew from tendons along the moose backbone was a popular thread for sewing clothing, wrapping arrowheads and fletching, manufacturing glue or "size" (by boiling down), fashioning rabbit snares and weaving fishing nets. Strong and somewhat elastic, bundles of large tendons were scraped clean, dried and preserved. When sinew was needed, a strand or two were extracted and moistened in the mouth before use (Honigmann 1946). When softened, the fibers were separated (by pounding, according to Denys [1672]) as needed, then stretched, twisted together (usually three or four strands rolled along the seamstress' leg) and finally sewn through or wrapped around the article of attachment (see Thompson 1994). When the thread or threads dried, they contracted, tightly closing seams or securing the item. "The moose furnishes the best quality of this article," wrote Ross (1861: 437). Bowstrings used by Indians of the New England woodlands "were often three twists of moose sinew" (Wilbur 1978: 55, see also Hutchinson 1764).

Moose hair was extensively used by Natives, mainly for embroidering decorative motifs on moccasins, other clothing and leather and birchbark containers; in fact, we were unable to identify any tribe within the former moose range that did not embroider design on some articles. Wrote Carver (1838: 277): "The hair of the moose is light grey, mixed with a blackish red. It is very elastic, for though it be beaten ever so long, it will retain its original shape." The long, black-tipped hairs of the moose mane, bell, rump, shoulders and cheeks were sewn by women into often elaborate geometric and circulinear patterns. Before the widespread availability of trade beads, porcupine quills sometimes supplemented moose hair for decoration. Many tribes, including Slavey (Helm et al. 1981), Micmac (Snow 1978), Iroquois (Fenton 1978), Ottawa and Potawatomi (Quimby 1960) and Huron (Speck 1911, Maxwell 1978), used vegetable dyes to color moose hair. Rose Tsetso (*in* Thompson 1994) indicated that Denes dyed moose hair black using charcoal, yellow from alder bark, red from currants or red ochre, and blue from blueberries. After being washed, dyed and dried, the moose hairs usually were stored in pouches according to color. When needed, hairs were moistened in water or the mouth, and three or four were sewn together by a diagonal stitch with sinew to make a single strand. Just before the stitch was pulled tight, it was twisted to increase the strand's tensile strength and tighten the pattern. According to Wilbur (1978), moose hair for decoration was wrapped three times around each woven cord twist or line stitch (Figure 6).

Other uses of moose hair included woven burdenstraps and tumplines by Iroquois (Morgan 1954) and hair roaches worn by Chippewa Indians (Ritzenthaler 1978). Some Slavey Indians put moose hair in gloves for extra warmth (Honigmann 1946). Blackfeet sometimes stuffed moose hair into bisonhide pad saddles (Ewers 1958). Fr. Le Jeune wrote of a shaman who used moose hair in charming ceremonies (Thwaites 1897c).

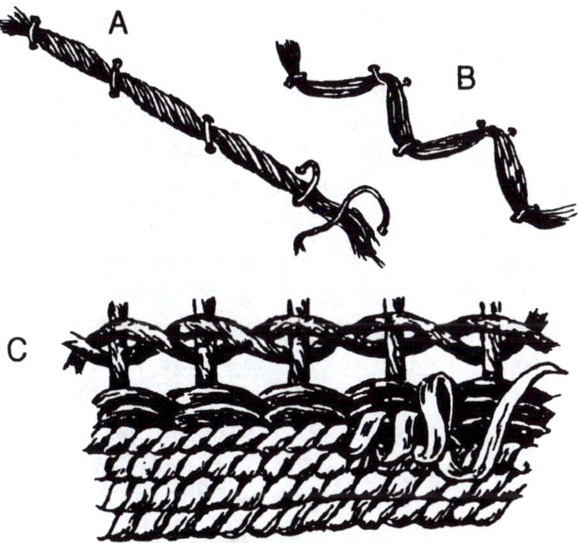

Figure 6. Moose hair was used extensively in decoration of hides for Subarctic and Woodlands Natives, especially before the widespread availability of beads. Moose hair for such purpose, primarily from the animal's bell, mane, cheeks and rump, was 4 to 6 inches (10.2–15.2 cm) in length and whitish with black tips, usually was dyed. When ready for use, hairs were moistened, and three or four strands were combined. Line stitches (A) were sewn onto leather with diagonal stitches of sinew. Before the stitches were tightened, the hair "bundles" were twisted slightly to stretch the strands and give a bead-like effect. Zig-zag stitches (B) also involved three or four hairs bundled together and sewn into pattern by parallel lines of sinew thread. For embroidery (C), moose hair typically was wrapped three times around each stitch. *Illustrations from Wilbur (1978).*

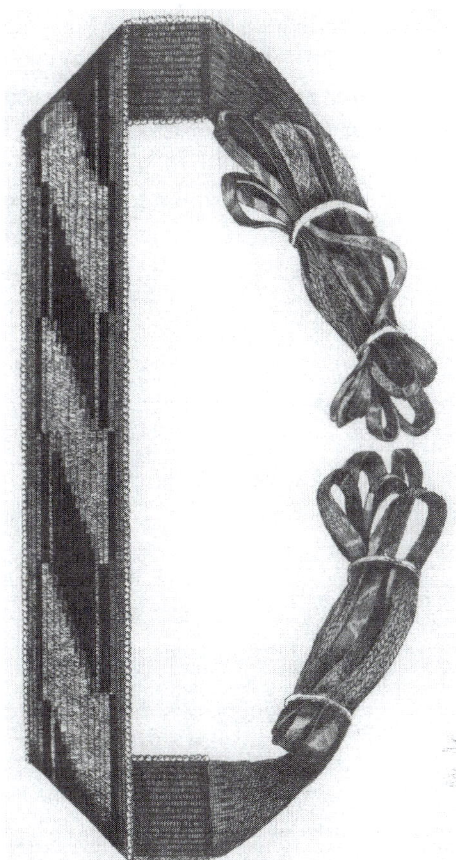

Beaded burden strap of moose hair used by Iroquois. *Photo from Morgan (1954).*

Day (1978) noted that the chief materials of every Western Abenaki woman's "tool kit" were tanned hide, fur pieces, birch bark, spruce root, paint, porcupine quills and moose hair.

Unlike with elk and certain other wildlife, extraction, collection and use of moose teeth apparently were not common aboriginal practices. Interestingly, the few references found concerning Indian uses of moose teeth differ significantly and evince little overlap among tribes. The Eastern Cree, for example, reportedly extracted teeth from slain moose as trophies, talismanic symbols or even some sort of "offering" of respect (Skinner 1912: 68). Culin (1907) documented moose teeth dice among Nova Scotia Micmacs. Tobey (1981) cited moose molars used as flaking devices by Carrier women. And John Josselyn wrote: "The Indian Webbes [women weavers] make use of the broad Teeth of the Fawns [moose calves] to hang about their Childrens Neck when they are breeding of their Teeth [teething]" (Lindholdt 1988: 20).

Similarly, the moose's massive antlers had limited application, perhaps in part because they were cumbersome to carry and difficult to cut or split compared with caribou or deer antler. Nevertheless, Hurons used moose antlers for hoes (Heidenreich 1978). Mason (1946: 24) indicated that Dogribs used portions of moose antlers for ice chisels, axes, arrowpoints, knife handles and "various heavy implements." Ross (1861) related that some Chippewayan Indians also fashioned knife handles from moose antler, and awl handles and spoons as well. Honigmann (1946) reported that Kaska used this antler to make cutting tool handles. Seton (1929) wrote that Natives (unidentified) made rattles from moose antler. Dudley (1721: 166) indicated that, of antler palms, "Indians make good Ladles, that will hold a Pint." And Josselyn reported that "moose horns are better for physick [headaches] than [is] harts horn [pasque flower]" (Lindholdt 1988: 19–20). That assessment might be explained by Roy's (1974) comment that moose antler was prized for its abundant ammonia.

Some moose supposedly experienced a malady that it could cure itself, as described by Denys (1672: 382–383): "They are subject to fall from epilepsy. The Indians say that when the Moose feels it coming on it stops, and then with the left hind foot it scratches itself behind the ear so much that the blood flows, and this cures it." As early as about 1612, Fr. Joseph Jouvency noted: "The savages . . . are cured by the hoof of its left hind leg. In this hoof there is certain marvelous and manifold virtue, as is affirmed by the testmony of the most famous physicians [of Europe]. It avails especially against the epilepsy. . . . Nor does it have less power in the cure of pleurisy, dizziness, and, if we may believe those familiar with it, six hundred other diseases" (Thwaites 1896: 247, 249, see also Boucher 1664).

Le Clercq affirmed the belief of the epilepsy cure, "but it is necessary to secure it [left hind moose hoof], the savages say, at a time when the animal is itself suffering from this malady . . ." (Ganong 1910: 472). In 1686, Baron Lahontan (1703) wrote that the left hind foot of female moose was believed—by Frenchmen, if not Indians—to cure "falling sickness." Added Pierre François Xavier de Charlevoix in 1721 of the hoof-rubbing cure: "a circumstance which has made his hoof be taken for a specific against the falling sickness. This is applied over the heart of the patient, which is also done for palpitation of the heart; they place it in the left hand, and rub the ear with it. . . This horny substance is also believed to be good in the pleurisy, in cholic pains, in Fluxes, vertigoes, and purples, when pulverized and taken in water . . ." (Kellogg 1923[1]: 183–184). However, Charlevoix is thought to have ascribed Abenaki belief in the diseases and cure to European influence (Merrill 1916).

A possible source of the belief about the magical powers of the moose hoof may be found in a statement by a Major [Hamilton?] Smith: "During its [the moose's] progress it holds the nose up, so as to lay the horns horizontally back. This attitude prevents it seeing the ground distinctly; and as the weight is carried very high upon its elevated legs, it is said sometimes to trip by treading on its fore-heels, or otherwise, and occasionally to give itself a heavy fall. It is probably owing to this occurrence that the elk [moose] was believed by the ancients to have frequent attacks of epilepsy, and to be obliged to smell its hoof before it could recover; hence the . . . reputation, especially of the fore-hoofs, as a specific against the disease" (Richardson 1829: 237).

In any case, the belief or superstition of moose hoof as a remedy for any ailment apparently waned as it was removed from European pharmacopeia in the 1760s (Howell 1930). Otherwise, the only other identified Native use of moose hooves was as rattles (Seton 1929). Curiously, hooves of other large ungulates (including bison, elk and deer) were extensively used by Indians across North America in the production of glue, by shaving and boiling. We were unable to locate any reliable mention of this use of moose hoof.

Yet another supposed Native medicinal by-product of moose was its dung, which was applied to relieve the effects of sprains (Mackenzie 1801).

In 1637, Fr. Le Jeune revealed that Indians sometimes found a stone "in the heart, or the shoulder, or some other part of an Elk [moose], or of another animal" (Thwaites 1898a: 13). The Indians believed that the possessor of such a stone would have fortune in hunts. In 1901, Ernest Thompson Seton (1929: 158) also learned of the mysterious object that was "great medicine" to Indians (presumably Western Cree), who called it *Peeto-mong-sons* or "Little Moose in the Big Moose." Seton was informed that only 1 in 500 moose had a "little medicine Moose" in its skin. By cross-examination and inquiry, Seton determined that such "a mascot of eternal good luck in Moose hunting" was, in fact, a dermoid cyst—a cyst or sac of developmental origin, containing hair follicles, sweat glands and sebaceous glands (A.W. Franzmann personal communication: 1996). Given that Fr. Le Jeune, Jesuit superior of the "Residence of Kebec," remained principally among Montagnais-Naskapi in the St. Lawrence River Valley, and Seton's source was from Manitoba, and because the separate reports spanned more than 260 years, we conclude that Indian belief in the charm of dermoid cysts in moose was both temporally and ethnogeographically widespread.

Fr. Le Jeune also noted that Iroquoian Indians rubbed moose fat on their children's skin (Thwaites 1897d). The missionary did not mention whether this was done to protect the youngsters from the sun rays or wind or insects, but he indicated that it seemed to have the effect of darkening the complexion. Kaska Indians reportedly stored grease and blood in moose stomachs, intestines ("gut") and bladders (Honigmann 1946: 54). Fr. Pierre Biard noted during

1611 that Passamaquoddy Indians made oil from seal fat and stored it in moose bladders "which are two or three times as large and as strong as our pig-bladders" (Thwaites 1897a: 79). After the arrival of traders in the Subarctic, Indians learned to make lamps of tin cups in which moose grease was burned by means of rag wicks (Honigmann 1946).

Aspects of Culture

COMMERCE

Protohistoric and early contact exchange of animal meat and parts by boreal and woodlands aboriginals seems to have been predicated nearly as much on matters of socialization and confederation as on the utility of provisioning with scarce resources. Thereafter, it was mainly an enterprise of material competition (e.g., see Ray 1974, Fitzhugh 1985). For many, if not most Natives, moose meat and hides were important commodities of such transactions among themselves and eventually, fatefully, with whites.

Indicative of intertribal commerce was that of the Kutchin in Alaska, who exchanged dressed moose hide and wolverine pelts with Eskimos for walrus hide, seal skins, whale bone and ivory (McClellan 1981b). However, the trade was sporadic because of environmental, linguistic and cultural constraints (Burch 1979). Canadian Cordillera Natives who traveled westward in summer to fish for salmon took along tanned moose and caribou hides and ground-squirrel skins to trade with coastal tribes (e.g., Chilkat, Chilkoot and Taku) for shells, seal skins, eulachon oil, dried seaweed and other marine delicacies (McClellan and Denniston 1981, Cruikshank 1986). Among tribes of the northern Pacific coast region, "where dentalia were not so much valued, elk and moose skins seem formerly [before the availability of trade blankets] to have constituted one of the standards of value" (Hodge 1907: 448). Across the continent, Montagnais exchanged moose hides for Huron corn, cereals and tobacco (Thwaites 1897c, see also Thomas 1985). Wrote Thomas Morton (1632: 74): ". . . the Salvages [*sic*]. . . use to barter away the skinnes [of moose] to other Salvages, that have none of that kinds of bests in the parts where thay live." Also, mention is made in the literature of trade among various Iroquois and Algonquin Indians of moose skins "somewhat worn" (e.g., Kenton 1956: 251), which helps to validate the aforementioned Native preference for such hide in the manufacture of moccasins. Also likely is that well-worn moose hide was easier or better to decorate with quills, hair and beads than was relatively fresh hide.

From the times of earliest exploration by whites in moose range across the continent, Natives often provisioned the newcomers with moose meat and fat, either by trade, purchase, commission or gift. In the early 1600s, Samuel de Champlain advised that Indians provided the French with game (moose and elk) in exchange for bread and other items (Grant 1907). Morton (1632: 75) observed: "There is such abundance of them [moose] that the Salvages [*sic*], at hunting time, have killed of them so many, that they have bestowed six or seaven at a time, upon one English man whome they have borne affection to." And near Sault Ste. Marie, Ontario, in 1669, Sulpican priests François Dollier de Casson and René de Bréhaut de Galinee noted that "Meat is so cheap here that for a pound [0.45 kg] of glass beads I had four minots [1 minot = 1.3 bushels] of fat entrails of moose, which is the best morsel of the animal. This shows how many these people [probably Nipissing] kill" (Kellogg 1917: 207).

David Thompson reported that, while wintering during 1798–1799 at Lake Biche, 105 miles (169 km) northeast of Edmonton, his expeditionary party received during 5 months "forty nine Moose all within twenty miles [32 km] of the [trading] House and a few Bull Bisons" (Tyrrell 1916: 305). There, moose accounted for about two-thirds of the party's meat, secured by their own hunting efforts, by hiring Indian hunters and by trade for meat from unengaged Indians. Thompson's associate, Alexander Henry, wrote of that experience: "In 1799, at my winter-quarters on the Terre Blanche, animals were so scarce as to oblige me to hire my hunters upon extravagant terms: For every moose . . . , six skins [credits in terms of "made beaver" pelts]; for every red deer [elk], five skins; to be paid for in whatever article of dry goods they might think proper to take . . ." (Coues 1897: 2–3).

Moose skins were an early trade item between the Lower Ahtna of Alaska and Russians who, since 1819, had a small post on the Copper River near Taral (De Laguna and McClellan 1981). Moose hides and lynx and beaver pelts were obtained from the Upper Ahtna.

Among most Natives, sharing and gift-giving were social imperatives, based on universally periodic need. Besides the survival benefits of generosity, the custom bestowed prestige on the giver, without discrediting recipients wanting or not. As missionaries learned during starvation periods (discussed earlier), Indians readily shared what they had. And as one Jesuit reported: "When they give anything as a present, it is well not to refuse it, and to reward them for it. They nearly always give more than is given them in return. For instance, I have seen Some of them give ten moose-Tongues, etc., and be well content with a platter of peas, some Indian corn, 2 biscuits, a chunk of tobacco, and a drink of brandy" (Thwaites 1900a: 259).

Leacock (1995: 170) suggested that, as items of trade with white entrepreneurs, hides of moose and caribou "were of

More so in the moose range of the north than in the intermountain region of the Louisiana Purchase, the fur and hide business was a matter of trade with Natives who procured and prepared the animal skins. Certainly, British and French trappers also "worked" the rivers, streams, woods and parklands of Canada, but the mainstay of the commerce was trade with Indians (see Chittenden 1935, Phillips 1961, Heidenreich and Ray 1976). Similarly, the northern companies engaged and relied heavily on Native hunters to provision their fort and post network personnel with food. The illustration above, by William Rogers, appearing in the *Harper's Weekly* for January 25, 1859, shows the Hudson's Bay Company's Fort Garry (Winnipeg, Manitoba) trading post. *Photo courtesy of the Glenbow Museum Archives, Calgary, Alberta.*

little significance," and the leather was needed more by Indians than that to be gained by its barter. Nevertheless, as reported in the early 1700s by French surgeon/botanist Sieur de Dièreville (1708: 131), "The skin [of moose] is traded, for its advantages are known, & and it sells for a fair price." As important to some Natives as the fairness of exchange was the evidence of transaction: . . . "trade goods represented success in hunting and trading, skills highly admired . . ." (Thompson 1994: 53).

At Tadousac, a trading post at the mouth of the Saguenay River in Quebec, 151 skins had been traded by Indians by June 1646 (Thwaites 1898d) and more than 500 in 1648 (Thwaites 1898e). Furthermore, thousands of skins were accepted in trade at the many Hudson's Bay Company factories and posts on Hudson Bay, with York Factory (Fort Nelson) being the major conduit for exported skins. Also, Nicholas Denys (1672) related that, during his time (1635 to 1650) among the Micmac and Maliseet, in the region from the head of the Bay of Fundy to the St. John's River, the Natives traded more than 3,000 moose skins per year. Miller (1986: 362) wrote that "hundreds of thousands of beaver, moose, and other skins were taken out of the Maritime area before the fur trade came to an end in the late eighteenth century." In central and northern Canada at that time, "From 2 to 5 dollars is paid for a dressed [moose] skin and about half as much for the meat" (Russell 1898: 230).

Ultimately, trade or sale of animal resources to whites was the manner in which Natives unwittingly bargained away their health, welfare and self-reliance (see Skinner 1912, Bishop 1969, 1981, Brasser 1978, Cronon 1983, Hamell 1987). The lure of white traders' metals, guns, kettles, fabrics, beads, needles, mirrors, alcohol and other novelties subtly induced Indians to ignore or abandon traditional subsistence patterns and ecological associations. It also ex-

Native acquisition of firearms is widely credited as the cause of moose population declines in North America in the 1700s and 1800s. Of course, the killing power of rifles exceeded that of traditional Indian weapons, but Natives efficiently took moose before guns were introduced. The real effects of firearms on both Indians and moose rest primarily with the fact that "trade guns" were one of a number of attractive goods available from whites in exchange for certain animal hides and meat. The lure of these goods prompted the Natives to abandon their traditional lifestyles and beliefs to supply themselves with trade products. This included hunting moose year-round, instead of seasonally, and often taking many animals solely for their hides, as opposed to killing only enough to supply their immediate needs. Besides the collapse of Native economy as a result, Indians also forsook their spiritual ties to wildlife, which were touchstones of their cultures. In this seemingly imaginative and unlikely scene, a bull moose is driven by a hunter on snowshoes into an Indian camp where hunters lie in ambush. In fact, Denys (1672) noted that, after a moose was stalked and arrow-shot for a second time, Natives made an effort to turn the animal toward camp. *Illustration from the Dankoler Collection; photo courtesy of the State Historical Society of Wisconsin.*

posed them catastrophically to Old World pathogens. In mere decades, the Natives were dispossessed of lifestyles delicately balanced over millennia with nature, including its fauna. To Indians, the price of commerce in wildlife, regardless of the value of trade goods received, was cultural bankruptcy. When the economics of moose and other animals preempted the perceived spiritual kinship with those wildlife, the Native identity was undermined and forever distorted or lost (e.g., see Martin 1978a, 1978b, Krech 1981).

SOCIAL ORGANIZATION

Throughout North America, aboriginal social organization was highly variable and predicated on such factors as intertribal conflict, custom, religion and resource availability. In moose country, available resources were the predominant factor, primarily by relegating social unit size to small numbers, because hunting-dependent groups required mobility and could ill-afford too many members to feed in a region of sparsely distributed and frequently scarce food sources. Except during summer gatherings, most Indians of the moose range formed bands of 10 to 30, usually consisting of extended family.

Within tribes, groups or bands, organization and socialization were also achieved by clans and societies, typically identified totemically with an animal, both materially and spiritually important to the unit. There were tribal, territorial and linguistic linkages as well. In many cases, the spiritual ideology and subsistence resources were not mutually exclusive, except that the animals chosen for representation by a particular group of people invariably were venerated for noble traits (usually of survival, health or productivity), *in addition* to their practical values as food and fiber. Accordingly, such affiliations gave their members both spiritual and cultural identity.

Although the moose apparently was not a common clan or society totem or symbol, perhaps because its behavioral characteristics did not lend themselves as readily as those of other wildlife species to mystical or supernatural association, some tribes did have moose units. For example, the Southwestern Chippewa's moose clan was one of three clans that comprised the marten (Monsone) phratry (Ritzenthaler 1978). A Menominee phratry was divided into moose, elk, marten and fisher totems (clans) (Hoffman 1896). Moose was one of at least 18 Ottawa clans (Quimby 1960) and one of 5 "original" Ojibway clans (Rajnovich 1994: 17). And the West Main Cree residing in the vicinity of where the Moose River flows into James Bay were known both as the Moose and Moose River Indians (Richardson 1852, Skinner 1912), and their totemic symbol was the moose.

WARFARE

Grant (1902: 228) indicated that many Canadian Indian tribes entered the "North Woods" (Adirondacks) hunting grounds of the Six Nations to secure winter supplies of moose meat and hides. Warfare resulted "much in the same manner as the northern and southern Indians warred for control of Kentucky."

SPIRITUALISM

Any attempt to generalize or encapsulate the cosmology or spiritual associations of a past society carries the risks of gross simplification and misrepresentation. To do so for many groups of culturally diverse people and for a period spanning at least several centuries is foolhardy. Nevertheless, because moose was a significant feature in the spirituality of many Native tribes, an effort is made here to relate, with only limited interpretation and analysis, at least some of what is known about Indian belief system as it pertained to the species.

Spirituality, or religion, appears not to have been just an adjunct to the daily life of boreal and northeastern woodlands Indians. Instead, it intrinsically circumscribed and accounted for virtually all of it. Culture and religion were indistinct and literal. What Natives believed about animals—although the beliefs varied significantly among them and especially over time—was conceived of as reality, not merely fables, moralizations or object lessons. Apparently, most precontact Indians saw themselves as unremarkable animals, but perceived other animals as in possession of mystical traits or powers, either good or bad, benevolent or malevolent. And how Indians treated and entreated those other animals logically influenced or dictated their own well-being. This is not to say that any or many aboriginals were zootheists; rather, their belief systems seem to have been firmly grounded in complex forms of pantheism.

To outsiders, North America's moose range very much seemed to be a land of feast or famine. But it was not similarly envisioned by the indigenous peoples, who did not equate resource plenty or scarcity with the limitations of the land. Elements of nature, including wildlife, were viewed as charismatic, and their harmony (cooperativeness) was predicated by human conduct and propitiation or by vague spirit beings—evil, prankish or both. Accordingly, human well-being—much less survival—in the boreal/woodland North was not an ecological teeter between feast and famine, but instead a matter of spiritual integrity and luck. Those who kept faith with nature and avoided disturbance by the execrable spirit being(s) retained access to essential resources—mainly food.

We found little evidence to suggest that the moose was regarded as a truly mystical animal—possessor of certain ascribable or teachable powers. On the other hand, many Natives embraced moose as knowledgeable, sentient and autonomous beings, with whom they had to establish both trust and understanding (see Ridington 1986). Among Slavey Indians, for example, moose engendered great respect, unlike wolf and otter, for which the prevailing attitude was fear (Honigmann 1946). Naskapi held both bears and moose in especially high religious regard (Chamberlain 1906), whereas Iroquois reportedly believed that moose, unlike bears, brought good luck (J.C.B. 1941). Furthermore, the Iroquois contended that one who dreamed of moose could anticipate a long life (J.C.B. 1941). For the Cree, a dream about moose portended a successful moose hunt. And a dream of moose by Montagnais translated as a dream of food (Newton and Hyslop 1992, 1995). The spiritual quality of moose, at least so far as we were able to determine, almost always bore on its role as provender.

Moose apparently were not deified, even among those individuals and groups (including clans) that embraced the species totemically, but many Natives subscribed to the belief that the species' preternatural benevolence hinged on respect shown the animals in life and death by those who used and were sustained by them. Most Indians apparently held that moose ultimately gave themselves to respectful hunters, who had a reciprocal obligation "to provide the conditions in which animals can grow and survive" (Feit 1986: 189). In essence, hunters could search for moose but might not find or kill them unless the moose allowed itself to be located and dispatched by hunters who were competent, solicitous and reverent. A hunter's spirituality was affirmed at each outing.

Algonkians generally accepted that every bird, mammal, fish and reptile was represented in the spirit (manitou) world by a "master" animal (Rajnovich 1994: 36). It was to this master animal, such as the Moose Manitou, that consideration and respect was owed.

Evidence of ability to kill big game such as moose was prerequisite to a Micmac boy's transition to manhood (Bock 1978); moose hunting success acknowledged a spiritual maturity and attainment. Tanana River Indians and Northern Saulteaux honored their young men's first moose with a feast. With the former, the first moose could not be eaten by the young hunter or his family, but instead was gifted to cross-relatives (Guédon 1981). With the latter, the youngster was expected to sit up all night, drumming and singing prayers for future hunting success (Skinner 1912).

Many of the accommodations of respect for moose pertaining to treatment of freshly killed or butchered carcasses were noted earlier. In addition, and apparently universally, was the taboo of exposing the moose in life or death and its hunters to menstrual blood. Among Mountain Indians, for example, menstruating women could not walk on moose hunters' trails, touch or walk over hunting weapons, or use main entrances to shelters (Gillespie 1981b). Pregnant or menstrual Koyukon women were not allowed to eat fresh moose meat (Clark 1981). And Micmac menstruants had to show particular care for rules of conduct, such as careful disposal of sticks and bones used in cooking (Bock 1978). In nearly all of the continent's aboriginal communities, young women were isolated in some degree or manner during their menarche, to prevent their "sickness" from infecting the purity of daily life and spiritual relationships. We identified no cases in which such taboo was associated only with moose.

Other rituals or taboos pertaining to moose or moose hunting (although not necessarily exclusively) included belief that moose bones burned on a fire would drive those animals away (Kutchin); speaking badly of a moose made its kind less willing to be killed (Kutchin); if a hunter was selfish with his quarry, other moose would be offended and refuse to be found or killed by that hunter (Slavey); and bone slivers must not be used to extract marrow from moose bones (Northern Athapaskans). Also, women were not allowed in shelters where a moose carcass was lain (Montagnais); hunters could not speak in boastful terms of moose they had killed (Cree); heads or bodies of moose could not be left where dogs could eat them (Kutchin); women could not express excitement at the return of moose hunters and children could not rush to greet them (Cree); and fresh moose meat had to be presented to group elders, and dogs and children could not impede that contact (Cree). In addition, moose meat and hides could be brought into a shelter only through smoke holes, not by normal passageways (Ahtna); moose bones were burned to prevent dogs from getting them (Ahtna); and, when eating moose, an elder had to place a piece of meat in the fire for each family member to appease the spirit of the slain animal (Montagnais).

In other regards, Ahtna believed that the spirit of a slain moose stayed with the animal for 3 days after death, and any contact of the body parts with human infants would cause those youngsters to become ill (De Laguna and McClellan 1981). Father Le Jeune indicated that the Montagnais apparently believed that hail was somehow helpful in killing moose (Thwaites 1897c). Also, souls of deceased Montagnais were thought to hunt for the souls of beaver, porcupine and moose, to sustain the former in afterlife. And when Montagnais children emerged from their housing each morning, they shouted "Come Porcupines, Come Beaver, Come Moose," which, according to Le Jeune, was "all of their prayers" (Thwaites 1897c: 203).

MYTHOLOGY

Native American mythology was (and is) a complex interrelationship of cultural traditions and spiritual values. It provided a sense of order, logic and continuity to the universe and, nearly as important, served as both substance and process of history. Perpetuated orally through stories, songs, placenames and genealogies, and visually through a variety of pictographic forms, mythology defined and explained for Natives their individual and social identities, cultural distinctions, and both secular and metaphysical orientations. More than a body of narratives that chronicled the past, Native myths were the mechanism of correlating perceived recurrent and essential truths to prevailing circumstances and future probabilities.

Virtually all physical entities and elements factored into aboriginal mythology, especially those that bore directly on the peoples' diverse material economies. Accordingly, the moose was significantly and widely represented in northern Natives' mythological panoply, as indicated by the following legends.

The Micmacs of Nova Scotia believed that moose originally came from the sea, which explained to them why moose sought refuge in water when hunted too persistently (Hardy 1855, Wallis and Wallis 1955). Governor Neptune, an 89-year-old Penobscot living in Old Town, Maine, told Henry David Thoreau (1864) the story of how moose (and deer) originated from whales. According to Abenaki mythology, *Glooskap,* the central figure and giant guardian of the Indian race, created the moose (Leland *in* Merrill 1916). He first made moose huge, but finding them too large and powerful for Indians to kill, decided to shrink them. *Glooskap's* squeezing them down unevenly into a smaller figure accounted for the species' short body, humped back and bulging nose.

Curtis (1923) related a Native legend that explained how the moose became its present size. Originally, the moose was so large that it browsed on the tops of trees and was dangerous to people. Therefore, the Great Spirit sent messenger *Ksiwhambeh* to tell the people that he was going to change the animal. From a strip of birchbark, *Ksiwhambeh* made a horn to call the moose, and when the animal was attracted, it was told that it was going to be made smaller. The moose obediently lowered its head, and *Ksiwhambeh* grabbed the animal between the horns, pushed it down to its present size, and told the moose he should never come close unless called.

A Menomini legend told of the demigod *Manabush* encountering the "Moose people"—four-footed hoofed carnivores. *Manabush* eventually killed all but two of the Moose people, and these he condemned to living in a cedar swamp and feeding on the "moismiu" (willows), which was proclaimed their "food for all time" (Young *in* Merrill 1916: 257).

A Dogrib myth related how *Hottah,* "the two-year-old moose, cleverest of all northern animals" aided in the creation of the northern environment (Bell 1903: 81). In this legend, *Hottah* told *Ithenhiela,* an enslaved Woods Cree boy, who, with *Hottah's* help, was fleeing from his captor *Naba-Chan*—the Big Man, who consumed a whole moose, or 2 caribou, or 50 partridges each day—to throw out a rock to slow the pursuing evil giant. *Ithenhiela* did so, and the rock grew in size to form the Rocky Mountains. Other actions created separate landscape features of and social harmony in the region.

Other Native legends involving moose included "Moose Demands a Wife" and "The Fairey Wives" (Spence 1994), "The Moose and the Hare" and "The Moose and the Jackfish" (Russell 1898), "The Hunter and the Elk People" and "How the Moose Were Defeated" (Hoffman 1896), "Manabush and the Moose" and "The Catfish and the Moose" (Judson 1914), "Team" (Merrill 1916) and "The Moose and Woodpecker" (Williams 1956).

Accounts of gigantic moose are common in Native mythology. One such account, based on Indian narrative, told that "All the largest Moose are only little dwarfs compared with this one; he has legs so long that, however deep the snow may be, he is never inconvenienced by it, while the others are almost buried in it. . . . He has a skin that is arrow-proof and bullet-proof and he seems invulnerable. They add that he carries a fifth leg, which grows out from his shoulders and which he uses like a hand in preparing his bed. He never goes alone, and does not appear without being escorted by a greater number of other Moose . . ." (Thwaites 1899g: 273). Marshall (in Merrill 1916) believed that the mythical giant moose with a fifth appendage may have originated from ancient accounts of the mammoth and its proboscis. Also, the origin of the giant moose may be traced to Native belief that it arose from the whale. *Illustration by Henry Sandham; photo from Grant (1894: 352).*

SYMBOLISM

Natives often named months, moons or suns after naturally recurring events, including those pertaining to animals. For example, Dogrib Indians called their tenth moon (in autumn) *ĕk-olă-chin-co-să,* meaning "paddle shoulder sun"—or the time when moose could be lured by striking scapula bones against trees (Cope 1919). Similarly, Slavey Indians called their eighth moon (in early autumn) *colo-ye-kĕn-ak-e-ne-i-a-să,* or the "moose rutting sun" (Russell 1898). Among Eastern Cree, *wâ-wâs-kis o pes-im,* or "when the moose cast their 'horns'," was the ninth month. And the Montagnais called their tenth month *béni-tsí' éli*—the month of the "moose deer."

SEMIOTICS

Over time, North American Indians devised an elaborate sign language by which linguistically unrelated people could communicate at least fundamentally through universal gestures. For moose, instructed Clark (1885: 261), "Generally the sign for ELK [bring the hands alongside the head, palms in, and fingers and thumbs extended, separated and pointed up; move the hands by wrist action back and forth two or three times] is made . . . and then holding left hand still in its position, carry right in front of and touching it; move right to front, and left to rear, separating hands a few inches; this is to denote the great width of the horns." Fronval and Dubois (1978: 45) described a refined and simplified sign for moose: "Put the tips of both thumbs at a level with the ears. Then spread the fingers upwards. Finish by touching the front teeth with the index fingers."

Native American picture-writings that feature moose have been found on stone, wood, bark, pottery, pipe bowls and hides (Mallery 1893). They are thought to be primarily of Algonkian origin and date roughly from 3,000 years before present to the current century, and mostly from the Late Woodland period (M. Hedden, A. Watchman and P. Reid personal communications: 1996).

Perhaps most impressive and intriguing of these moose images are rock paintings and petroglyphs. They have been found from British Columbia (Bell 1979) to Maine (Mallery 1893, Hedden 1996), but the majority are among more than 300 pictograph sites of the Canadian Shield in the northern Great Lakes region, especially in northwestern Ontario (see Dewdney and Kidd 1973). Some of the paintings are crude representations of animals, figures mythical or real, and inanimate objects; some are quite abstract; others are highly stylized and artistic. Moose are occasionally depicted in the paintings, but not particularly emphasized among the animals represented. When evident, however, most are clearly representative of the species (M. Graves personal communication: 1996).

These pictographs typically are reddish (the color red being powerful "medicine," reportedly symbolic of life, virtue and good fortune), having been painted with red ochre from finely ground hematite mixed with vegetable oils, blood, egg white or oil from sturgeon, forming a substance more durable than modern paints (Rajnovich 1994).

Rock paintings, like other picture-writings from antiquity, are sign language—thoughts, events, customs and histories conveyed symbolically and conceptually as substitutes for or complements of intrinsically significant gestures and words. Wrote Mallery (1893: 26): "Whether remaining purely ideographic, or having become conventional, picture-writing is the direct and durable expression of ideas of which gesture language gives the transient expression." The semiotics of picture-writing are accomplished by signs, symbols and metaphors (Rajnovich 1994).

The paintings are believed to be idiosyncratic messages to other Natives (i.e., didactic and/or directional [see Schoolcraft 1851]), to the spirits or manitous (i.e., appeasements or propitiations [see Hoffman 1888, 1891]), to memorials (Mallery 1893), to dream visions (e.g., totems [Rajnovich 1987]), to medicine quests (Rajnovich 1994), or to benchmarks of religious sites (M. Graves personal communication: 1996) or to sociological principles (V. Miller personal communication: 1996). Whether of secular or sacred

Moose are featured in Native American petroglyphs (images or symbols incised in rock) and pictographs (record of pictoral symbols drawn or painted) dating from 2,000 to 3,000 years ago to the twentieth century. Rock paintings (pictographs) are believed to be of both sacred and secular types (see Mallery 1893, Dewdney and Kidd 1973, Rajnovich 1994). Above is a moose cow and calf from Darky Lake, Quetico Provincial Park, Ontario. The circular and dash lines below the animals have been interpreted as hoofprints, perhaps signifying a journey or metaphorical quest of some kind. The hash marks above the animals may indicate time periods, and may represent the nonconsecutive dream occurrences of moose by the artist/spiritualist. Also, on the front inside cover of this book is a bull moose rock painting from Lac la Croix, Quetico Provincial Park. On the back inside cover is a pictograph of a spirit moose, perhaps a Moose Manitou, from Crooked Lake, Minnesota. The "pipe" emanating from the figure is a "power line," indicating spirituality. Emanating from the mouth, the power line indicates speech or message delivery. That the animal's bell is curved may mean that the message is in song or further spiritual connection of the moose to the serpent—a very powerful animal master. *Photos by H.R. Timmermann; courtesy of the Royal Ontario Museum, Toronto.*

intent, there is little evidence that the artistic nature of the paintings was a consideration beyond clarity of the figures or symbols.

Anthropologists have given much attention to aboriginal picture-writing in North America, yet much remains a mystery. At the very least, the expressions make clear that the popular notion that recorded history dates from the documentations of post-Columbian scribes is indeed myopic.

Newcomers

"From the beginning of the white man's exploration and settlement of North America, the majestic moose has been a familiar figure in the northern wooded part of the continent" (Peterson 1955: 4).

Within the millenary span of second wave arrival and colonization of the continent by Europeans, Asians and Africans, hundreds of thousands, if not millions, of the newcomers have encountered moose. Hardly can it be imagined that each first witness and most subsequent observations of the species have not been unique and no less than a marvel. The following are a few such early accounts and reflections (Figure 7, Table 3).

Some Notable Chroniclers

VANGUARDS

Jacques Cartier explored the St. Lawrence Valley in 1535 and overwintered there. Among the various wild animals he observed the Natives hunt were "dains" and "cerfz." One of Cartier's translators, H.B. Stephens, interpreted "dains" as being moose, but he was somewhat uncertain. Almost surely, Cartier saw or learned of moose that winter, so it is probable that either "dains" or "cerfz" was his name for moose (Merrill 1916: 5). If so, Cartier may have been the first to write of the North American species. Later investigators believed that moose was one of the animals that provided food for Cartier's men through barter with the Indians (cf. Seton 1929).

With 200 followers, Jean-François de La Roque, Sieur de Roberval, attempted to colonize Montreal Island in 1541–1542, then the site of the Iroquoian village Hochelaga. In documenting that effort, Roberval noted the various species of animals in the vicinity, including "Bugles," which Biggar (1924: 268) interpreted as "no doubt the moose."

The first clear mention of moose seems to be that of Gorges, in Purchas' *Pilgrimage* (1625: 755): "Captain Hanham sayled to the River of Sagadahoc [in] 1606. He relateth of their beasts . . . redde Deare, and beast bigger, called the Mus."

Another early certain account of moose in North America is that of Samuel de Champlain, the "Father of New France," who wrote in the early 1600s: "The savages who dwell here are few in number. During the winter, in the deepest snows, they hunt elks [the common English name for the European moose] and other animals, on which they live most of the time. . . . When they go a hunting, they use a kind of snow-shoe twice as large as those hereabouts, which they attach to the soles of their feet, and walk thus over the snow without sinking in. . . . They search for the track of animals, which, having found, they follow until they get sight of the creature, when they shoot at it with their bows, or kill it by means of dagger attached to the end of a short pike, which is very easily done, as the animals cannot walk on the snow without sinking in" (Grant 1907: 55). Champlain listed 12 species of game upon which the aborigines of the St. Lawrence Valley subsisted in 1603. The first mentioned were the "orignacs," or moose.

The Captains Meriwether Lewis and William Clark's expedition across the Louisiana Purchase in 1804 to 1806 was, in essence, a military expedition. "The Corps of Discovery" subsisted largely off the land, but failed to bag a single moose, even though its route took it twice through the range of the Shiras moose, an "undiscovered" subspecies inhabiting the central and northern Rocky Mountains of the U.S. (see Moulton 1986–1996, Cutright 1989). However, the expedition's members may have been the first white men to have seen this subspecies (although a case might be made for possible first observation by Alexander Mackenzie). On July 7, 1806, Lewis reported that Reuben Field had wounded a "moose-deer" near their camp, then located near the Continental Divide. Also, Sgt. John Ordway reported seeing several near the mouth of the Milk River in Montana, on May 10, 1805. Finally, in their concurrent journal entries for June 2, 1806, both captains reported being told by Nez Perce Indians of plenty of "Moos" on the East Branch (Salmon River) of Lewis' (Snake) River (Thwaites 1904–1905[5]: 99, 102). Not ones to be deterred by consistent spellings, Lewis and Clark variously recorded moose as "mooce," "mose," "mooce deer" and "moose-deer."

Captain John Palliser, a widely traveled soldier and sportsman, is best known for his surveys of the western prairies and mountains of Canada. In his journal entry for January 22, 1858, Palliser wrote of an Indian who reportedly had killed 57 "moose-deer" the previous year (Spry 1968), explaining at least in part why Palliser decided to engage him as a guide the following summer. Palliser and his associates frequently recorded seeing or killing moose. On July 7, 1858, for example, Palliser's contingent wounded a large bull, and member John Sullivan's account of its demise follows: "At last he turned to bay, terrifying the horses (some

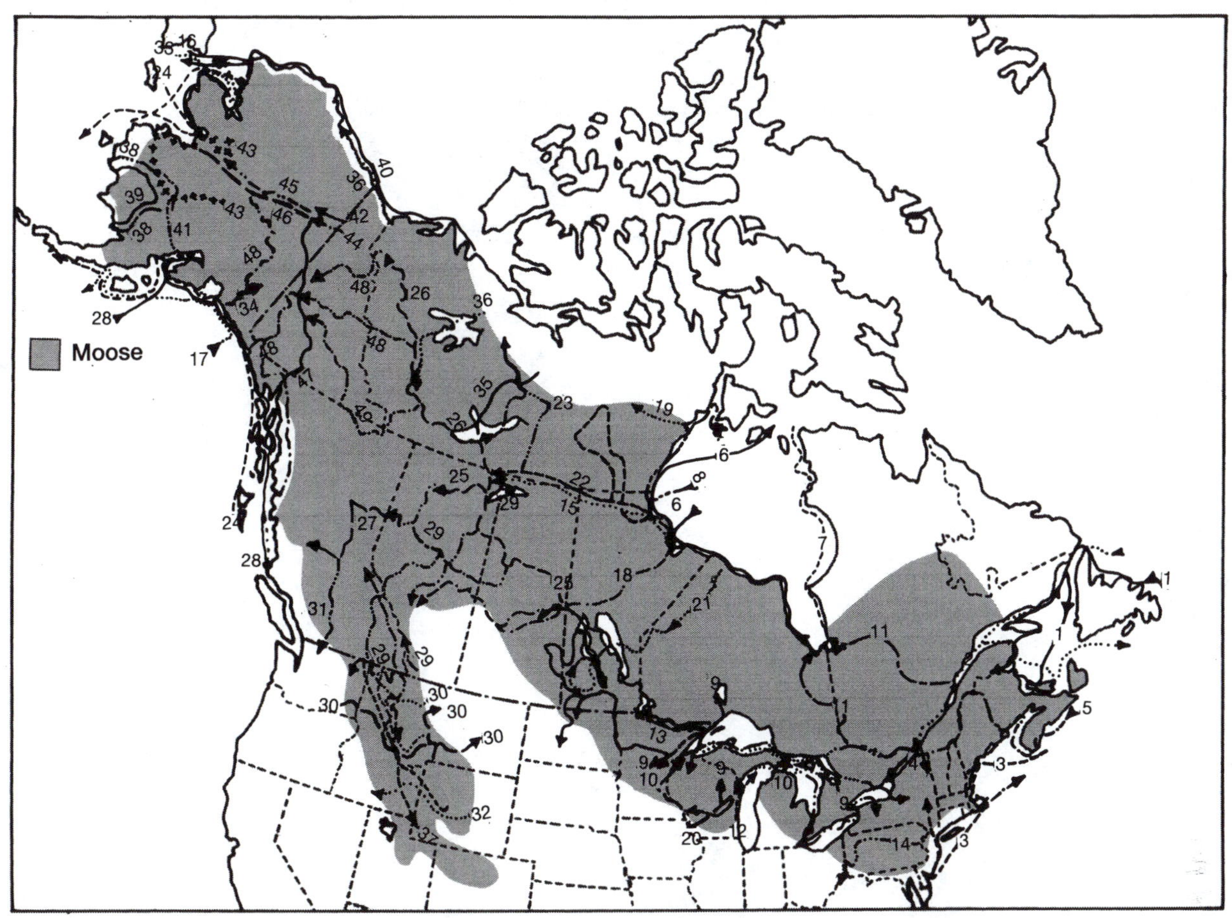

Explorer	Nationality	Exploration years
1 Jacques Cartier	French	1534–1536
2		1535–1536
3 Samuel de Champlain	French	1603–1607
4		1603–1615
5		1604–1607
6 Henry Hudson	British	1609
7		1610
8 Thomas Button and William Baffin	British	1612–1613
9 Jesuit and other missionaries	French	1641–1670
10 Sieur des Groseilliers	French	1659–1660
11 French journeys to Hudson Bay	French	1671–1686
12 Louis Jolliet and Pere Marquette	French	1673–1674
13 Jaques de Noyon	French	1686
14 Arnout Viele	American	1692–1694
15 William Stuart	British	1715–1716
16 Vitus Bering (Danish)	Russian	1728, 1741
17 Sieur de La Verendrye	French	1732–1743
18 Anthony Henday	British	1754–1755
19 Capts. Christopher and M. Norton	British	1761–1762
20 Jonathan Carver	American	1766–1767
21 William Tomison	British	1767–1770
22 Samuel Hearne	British	1769
23		1770–1772
24 James Cook	British	1776–1780
25 Peter Pond	American	1778–1780
26 Alexander Mackenzie	British	1789
27		1793
28 George Vancouver	British	1793–1794
29 David Thompson	British	1797–1812
30 Meriwether Lewis and William Clark	American	1804–1805
31 Simon Fraser	British	1808
32 The Astorians	American	1811–1812
33 Otto von Kotzebue	Russian	1815
34 A. Klimovskiy	Russian	1819
35 John Franklin, George Back and John Richardson	British	1819–1822
36		1825–1827
37 Jedediah Smith	American	1824
38 I. Vasilev and S. Lukeen	Russian	1829
39 A. Kolmakov	Russian	1832–1839
40 Thomas Simpson	British	1833
41 A. Glazanov and S. Lukeen	Russian	1834
42 V. Malakov	Russian	1838–1843
43 L.A. Zagoskin	Russian	1842–1844
44 Alexander Murray	British	1847
45 I. Lukeen	Russian	1863
46 Telegraph survey	American	1865–1866
47 Skagway/Klondike/Yukon route	Various	1896–1898
48 Copper River/Klondike route	Various	1896–1898
49 Canadian Klondike routes	Various	1896–1898

Figure 7. Major, early European and Asian explorations into the range of North American moose (screened area) (adapted from Goetzmann 1959, 1966, Ferris 1968, Gerlach 1970, Waldman 1985, Harris and Matthews 1987, Newman 1989, Fernández-Armesto 1991, Gentilcore 1993). Exploration was the initial stage of foreign intrusion into the New World. It familiarized the white trailblazers with the continent's Native inhabitants, including their level of development, beliefs and means of survival. Exploration also indicated where settlement was feasible and the natural produce and products of trade value. To the church and clerics, it revealed a trove of souls to be saved. To sovereigns, it was the vehicle for imperialism. Generally within moose range, France dominated early explorations of eastern Canada, the great Lakes region and the upper Mississippi River valley; the English pressed into the regions of Hudson Bay and western Canada, the Maritimes, New England and the mid-Atlantic coast; and Russians first penetrated Alaska.

Table 3. Selected historic references to the moose, arranged chronologically and by subject category

Authority	Approximate time	Type of information[a]													Source
		De	T/N	B	Di	Ab	Nh	F	O	Nr	Co	Sh	Cu	Il	
Cartier, Jacques	1535		[b]												Merrill 1916
Champlain, Samuel de	1604–1618						3	2							Grant 1907
Hanham, Capt. T.	1606		1												Purchas 1625
Lescarbot, Marc	1609	3	2			1	4	2	1					1	Grant 1911–1914
Jesuit Relations	1610–1791	1	1	1	1	2	5	5	3	2	3				Thwaites 1896–1901
Gorges, Sir Ferdinando	1606–1620	2	1				1								Purchas 1625
Smith, Capt. John	1616	1													Smith 1616
Sagard, Fr. G.	1632	2	1			1					1				Sagard 1632
Wood, William	1634	1					2	1	1						Vaughan 1977
Morton, Thomas	1632	2	1	1	1	1	1	1	1		1				Morton 1632
Le Clercq, Fr. C.	1641	1		2			3	5		2					Ganong 1910
Williams, Roger	1643		1												Trumbull 1963
Boucher, Pierre	1664	2	1	1	1	2	2	2	1	1				2	Boucher 1664
Radisson and Groseilliers	1668–1669	2	2	1	2	1	2	2	2						Adams 1961
Dollier, Fr. F. and Fr. R. Galineé	1669–1670										2				Kellogg 1917
Hudson's Bay Company	1670–1885				1	1	1	3	2		5				Hudson's Bay Record Society 1938–1969
Denys, Nicolas	1672	2			2	1	5	3	3	2					Denys 1672
Josselyn, John	1674	2					3	1		1	1				Lindholdt 1988
Perrot, Nicolas	1680–1718		[b]	2			3	1							Perrot 1864
Lahontan, Baron	1703						2				1			2	Lahontan 1703
Diéreville, Sieur de	1708			1			2	2		1	1				Dièreville 1708
Oldmixon, John	1708						1	1	1	1					Oldmixon 1741
Charlevoix, Pierre de	1721	2	1	2	1		3	1	2	2					Charlevoix 1761
Catesby, Marc	1743	2	2		2										Catesby 1743
Isham, James	1743–1747	1	1			1		2			1				Rich and Johnson 1949
Kalm, Peter	1747–1751	2	2								1				Kalm 1770
La Potherie, B. de	1753							1		2					Tyrrell 1931
Graham, Alexander	1767–1791	2	1	1	1			1	1		1				Williams 1969
Long, John	1768–1791		1					1							Thwaites 1904
Hearne, Samuel	1769–1772	3		3			4		3						Tyrrell 1911
Henry, Alexander	1779–1814		2					2			1				Coues 1897
Pennant, Thomas	1784	3	2	3	2		3	2	2	2				1	Pennant 1784
Thompson, David	1784–1812			2		2	3	2	2	1					Tyrrell 1916
Belknap, Jeremy	1784, 1791	2	1	2				1	1	1					Belknap 1812
Jefferson, Thomas	1787	3	2	2	1										Jefferson 1787
Mackenzie, Alexander	1789, 1792				1	1		2	2		2				Mackenzie 1801
Barton, Benjamin Smith	1792	1		1	2			2							Sterling 1974
Harris, Thaddeus Mason	1803				1	1									Thwaites 1904–1907
Lewis M. and W. Clark	1804–1806				2	1					1				Thwaites 1904–1905, Moulton 1986–1996
Pike, Zebulon M.	1805–1806				1			1							Jackson 1966
Franchère, Gabriel	1811–1841, 1854							2							Thwaites 1904–1907
Hood, Robert	1819–1821	2					3	2	2					2	Houston 1974
Franklin, Sir John	1819–1822		2												Franklin 1823
Simpson, Sir George	1820–1821							1							Rich 1938
Richardson, Sir John	1820–1822	3	3	2	2		2	2	2						Houston 1984
Harlan, Richard	1825	4	2												Harlan 1825
Ogden, Peter Skene	1825				1			1							Rich and Johnson 1950
Godman, John D.	1826	4	3	3	2	2		2	1					1	Godman 1826
Maximilian, Prince	1832–1834	1	1		1										Maximilian 1843
De Kay, James E.	1842	4	3	3	2	2		2	2			1		2	De Kay 1842
Zagoskin, L.A.	1842–1844				2	2	1	2							Michael 1967
Farnham, Thomas J.	1843							1	1						Thwaites 1904–1907

Table 3 continued

Authority	Approximate time	Type of information[a]													Source
		De	T/N	B	Di	Ab	Nh	F	O	Nr	Co	Sh	Cu	Il	
De Smet, Fr. Pierre J.	1845–1846		1					2	1						De Smet 1847
Audubon, J.J. and J. Bachman	1846–1853	5	4	4	3	2	4	3	1		1	2		5	Audubon and Bachman 1845–1848
Thoreau, Henry David	1846–1857	2	1	2			2	1		3		4	2		Thoreau 1864
U.S. Army, Alaska	1849–1899		1		3	2	2	3	2		1			5	U.S. Senate 1900
U.S. Railroad surveys	1853–1854				2	2									U.S. Senate 1855–1860
Hardy, Campbell	1855			2			1					4			Hardy 1855
Palliser, John and James Hector	1857–1859	3					4	2							Spry 1968
Baird, Spencer F.	1859	2	4												Baird 1859
Kane, Paul	1859				1		1	2	2		2				Garvin 1925
Whymper, Frederick	1862–1866		2			3		2							Whymper 1869
Dunraven, Earl of	1870–1876											2			Dunraven 1876
Caton, John D.	1877	5	5	4	2	2	2								Caton 1877
Ward, Charles C.	1878	2			2	2	3		1			5			Ward 1878
Roosevelt, Theodore	1885–1900	2		4	1	1	1	1	1			5		1	Roosevelt 1908, 1909, 1916
Webb, John S.	1897							2	1						Webb 1898
Lyddeker, Richard	1898	5	5	4	3	2								2	Lyddeker 1898
Preble, Edward A.	1902				2										Preble 1902
Shiras, George III	1906–1932	2	4	3	3	2						2		5	Shiras 1906, 1908, 1912, 1913, 1921, 1932, 1935
Selous, F.C.	1907			2				1				3			Selous 1907
Preble, Edward A.	1908			3	4	2	1								Preble 1908
Hornaday, William T.	1910	4	2	4	2										Hornaday 1900, 1910

[a] Key to subject categories: De = physical description; T/N = taxonomy, names; B = behavior; Di = distribution; Ab = abundance; Nh = Native hunting; F = food and subsistence; O = other use; Nr = Native religion and mythology; Co = gifts, trade and commerce; Sh = sport hunting (post-Columbian); Cu = cultural (post-Columbian); Il = illustrations only (no text). Value ratings 1 (low) through 5 (high) are subjective.

[b] Species identification uncertain.

of which threw their riders and ran off), but surrounded on all sides he at length fell, gallantly facing his enemies and riddled with balls and arrows. We halted in the neighborhood to enjoy a feast of the moose meat . . . the animal was in the prime of life, seven years old, and in splendid condition. His proportions, measured with a tape by Dr. James Hector, were as follows: length, 8 feet 6 inches [2.59 m] . . . height of shoulder, 6 feet 4 inches [1.93 m] . . . girth behind the shoulders, 7 feet 4 inches [2.24 m] . . . antlers, palmated with four prongs, 1 foot 8 inches long [0.51 m], but as yet not quite developed, being in the velvet, and quite soft" (Spry 1968: 244).

BLACK ROBES

Both Jesuit and Recollect missionaries (called "Black Robes" by Natives) in New France wrote much about the moose and how it was used by the Indians and themselves. The 73 volumes of the *Jesuit Relations* (Thwaites 1896–1901), written and published annually during 1610–1791, are remarkable in their scope, detail and veracity. Some of the references to moose in the collection have been cited earlier in this chapter.

From New England in 1630, Reverend Francis Higginson (1929: 310) wrote of "a great Beast called a Molke as bigge as an Oxe."

In 1642–1643, missionaries told how moose indirectly advanced the adoption of Christianity among Indians. A party of Atticameges and Algonquins were traveling together when they encountered the trails of two moose. While the Algonquins delayed pursuit to pray for hunting success, the Atticameges hurriedly followed one trail. The Algonquins set out only near midday, but soon found and killed their moose. They then backtracked to where they first encountered the trails, and eventually found and killed the moose

Possibly the earliest European illustrations of North American moose. From published and public accounts of explorations in New France, particularly Jaques Cartier's second voyage to the New World, Italian geographer, historian and statesman Gian Battista Ramusio (1556: 3) produced a map and diagram of an Iroquoian village at Hochelaga, the present site of Montreal. The map appeared in volume three of Ramusio's *Navigationi e Viaggi,* a year before his death in 1557. Ramusio never visited North America, and his depiction of the Native village is entirely too orderly and idyllic. However, at the left side of the "plan," beyond the village borders, four animals of perhaps three species are shown. Two are bears; one is a deer or elk; and one—with palmated antlers—may represent a moose (top left). If the animal, shown as a detail, actually is a moose, it may be the first-published European illustration of North American moose. Conversely, Merrill (1916) probably incorrectly speculated that a 1609 map of Port Royal, New France—now Annapolis, Nova Scotia—drafted by Marc Lescarbot and featuring a bull moose (top right in detail) at the bottom and to the right of a short river named "R. de l'Orignac" (now Moose River) was the earliest historic illustration of the North American moose (Grant 1911–1914). And like Ramusio, Flemish goldsmith Theodor de Bry (1528–1598) made engravings of the New World based on accounts and sketches of others. In the bottom scene, by either de Bry or his contemporary Matthaeus Merianus (1573–1650), Indians were shown idyllically hunting, fishing and berry-picking in the relatively carefree time before the intrusion of Europeans. The large, antlered, horse-like creature in the foreground very possibly was intended as a moose, for the hunting methods being used against it in the background are precisely those described in the literature. "Moose" are being driven into water by fires, and swimming "moose" are being clubbed and pursued by archers in canoes (Vachon 1982: 160). On the other hand, the same techniques were used by eastern Natives to take white-tailed deer and North American elk—species with which de Bry (1590) also was unfamiliar. But the fact that the animal in the foreground has palmated antlers and a dewlap strongly suggests moose. *Top photos courtesy of the Library of Congress. Bottom photo courtesy of the Smithsonian Institution Anthropological Archives.*

that the Atticameges had been tracking unsuccessfully, thus leaving the latter with "an excellent opinion of our holy faith, and a desire for Baptism" (Thwaites 1898c: 53).

Insight also was given about what made for happiness in Indian converts: "As they live only from Day to Day, they do not desire much; and all their wishes end in having something to eat. It is a savage's supreme good to have fresh meat; he then considers himself the happiest person in the world. . . . I am often asked [by them] if they eat the meat of moose, bear, etc. in paradise; And I Answer them that, if they Desire to eat it, their desires will be satisfied" (Thwaites 1899g: 129).

In the Intermountain West during the 1840s, a royal "cuisine à la Sauvage" was served Belgian Fr. Pierre-Jean De Smet (1847: 210–211) by his companion, Morigeau, in appreciation for De Smet performing the marriage ceremony for him and his wife, followed by the baptism of the couple's six small Indian children. The first dish consisted of two bear paws, followed by a roast porcupine and a moose's muzzle, topped off with a substantial soup containing remains of beef, buffalo, venison, beavers' tails, hare and partridge.

The moose depicted in this early illustration were erroneously identified as caribou by Fr. Louis Nicolas in his *Codex Canadensis* (Vachon 1982: 273). The sketch, in sepia on parchment, was drawn by Nicolas during 1664–1674 when he served as a Jesuit priest among the Ottawa and Mohawk Indians (S. Erwin personal communication: 1992). His *Codex Canadensis* contained 180 watercolor and ink sketches of the Indians and their activities and manufactures, and the flora and fauna of New France (Hall 1991: 30). *Photo courtesy of the Thomas Gilcrease Institute of American History and Art, Tulsa, Oklahoma.*

COUREURS DE BOIS

The most important early commercial enterprise within the range of the moose was that of the fur trapping and trading Hudson's Bay Company (HBC—sometimes humorously or derisively interpreted as "Here Before Christ"). Figuratively speaking, the moose was fuel for the Hudson's Bay Company's far-flung endeavors. It fed company employees and officials during much of the year, as well as the Natives who traded furs and other commodities (including moose skins and meat) at Company factories and outposts (Figure 8). Indians and Metís hunters were normally employed at the larger Hudson's Bay Company posts and some forts to ensure a steady supply of meat. Similarly, voyageurs of the rival Northwest Fur Company, traveling and trading via the St. Lawrence River and Great Lakes, depended heavily on moose for food.

Many Hudson's Bay Company officers and explorers wrote of the moose in a noncommercial context. For example, Governor George Simpson, while on an extensive tour of Hudson's Bay Company posts recorded that, on arrival at Fort Vermilion along the Peace River in northern Alberta on August 21, 1828, he and his party were treated to "a sumptuous supper of hot moose steaks and potatoes" (Rich 1947[XXIV]: 14). Under Chief Factor John Rowand's guidance, Edmonton House (site of present Edmonton, Alberta) became the Company's most important and profitable trading post in the Northwest. At his "Big House," Rowand hosted many prestigious visitors, including the itinerant prairie artist Paul Kane in 1847. Their meal that evening included "boiled premature buffalo calves, dried moose noses, wild goose, and whitefish delicately browned in buffalo marrow" (Newman 1987: 238).

CURIOUS OTHER CHRONICLERS

Pierre Boucher was one of the very earliest (and least recognized) to describe systematically the natural history of any region of North America. His *Histoire Veritable et Naturelle* . . ., written about the flora and fauna of New France, preceded by decades far better-known works by the Bartrams, Catesby, Audubon, Wilson, Lawson and others. Regarding the moose, Boucher (1664: 54–55) wrote: "Let us

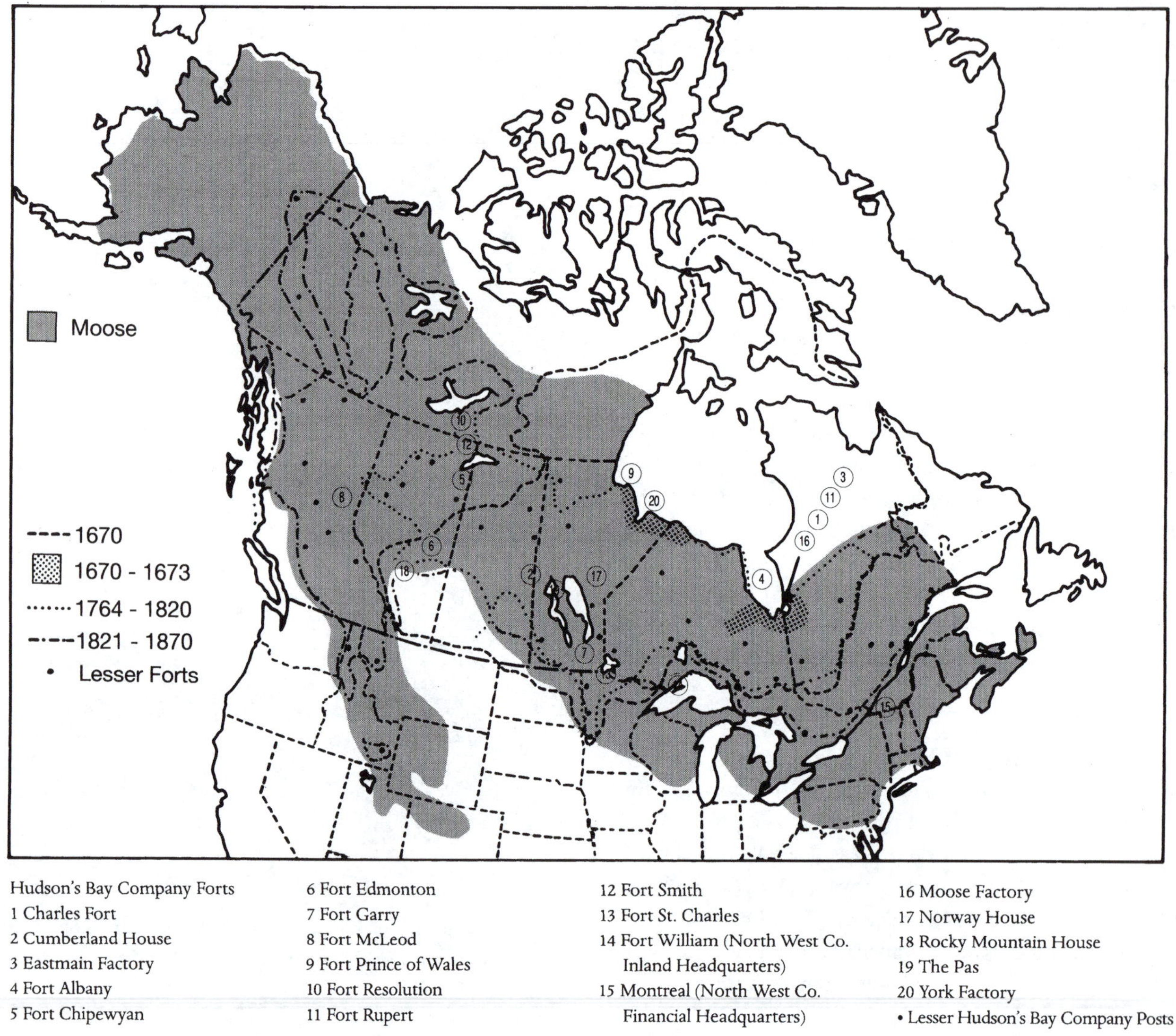

Figure 8. Hudson's Bay Company forts, posts and territory, 1670–1870 (adapted from Ferris 1968, Goetzmann 1959, 1966, Gerlach 1970, Waldman 1985, Harris and Matthews 1987, Newman 1989, Gentilcore 1993). The exploitation of furs, notably the beaver, was the most important single impetus for foreigners to probe the uncharted domain of the moose. On May 2, 1670, King Charles II granted the Hudson's Bay Company a trade monopoly over the vast geographical area whose waters flowed into Hudson Bay, otherwise known as "Rupert's Land." In a practical sense, the corporation also exercised all administrative responsibilities for the region. Initially, Hudson's Bay Company traders clung to a half-dozen or so posts ("factories") along the shores of Hudson Bay—"asleep by the frozen sea"—depending on Indians to bring furs to them (Newman 1989: 18). In contrast, the competing North West Fur Company, operating since about 1760 but essentially formed in 1775 and headquartered in Montreal, relied on French voyageurs to undertake annual canoe trips into remote western and northern Canada to conduct trade and return with furs. Forced by this competition and frequent incursion into their lands, the Hudson's Bay Company dispatched traders and trappers westward, and established new trading posts. In 1821, the two companies merged. A second invasionary force was the missionary zeal of Jesuits and other chiefly Roman Catholic religious sects. The "Black Robes" often accompanied fur traders, but sometimes mounted explorations of their own in search of souls needing salvation.

begin with the most common and most natural of all animals in this country, the 'Elan', which is called in these parts 'Original.' They are normally bigger than large mules with a head shaped more or less the same. The difference is that the males have antlers similar to the deer, but flat. These fall off every year and each year add one new fork. The flesh is good and light and never does any harm. The skin is worn in France and passes for buffalo, the bone marrow has medicinal value for nerve ailments. It is said that the hoof of the left foot is good for the illness 'caduc.' It is an animal with long legs and a good disposition. It has a split hoof, and no tail. It defends itself with its front hoofs like a deer."

Hudson's Bay Company personnel on a scow on Dease River, British Columbia, about 1909. For much of Hudson's Bay Company's nearly 330-year history, and that of the North West Company, employees subsisted to a great extent on foods provided by the Natives with whom they traded. Moose meat was a regular and generally preferred fare. *Photo by E. Thomson; courtesy of the Glenbow Museum Archives, Calgary, Alberta.*

Nicolas Denys' *Description Géographique et Historique des Costes de L'Amerique Septentrionale* is remarkably insightful documentation on the natural history and ethnology of New France, also in the mid-seventeenth century. Denys (1672: 382) described the moose as being "as powerful as a mule. The head is almost the same; the neck is longer, the whole much more lightly built, the legs long and very lean, the hoof cleft, and a little stump of tail. Some have the hair grey-white, others red and black, and when they grow old the hair is hollow. . . . It carries great antlers on its head, flat and forked in the form of a hand. Some are seen which have about a fathom [6 feet: 1.8 m] of breadth across the top, and which weigh as much as a hundred [45 kg] or a hundred and fifty pounds [68 kg]. They are shed like those of deer. . . . All of it is good to eat. The male is fat in the summer, and the female in winter. When one is pregnant, she bears one and sometimes two young."

Pehr (Peter) Kalm, a Swede and colleague of the great Linnaeus, visited America from 1747 to 1751. Aware that very long and branched horns (antlers) of a mysterious beast (perhaps the extinct Giant Deer or Irish Elk) had been excavated in Ireland, Kalm inquired about the possible presence of the species in North America. Renowned Philadelphia naturalist John Bartram told Kalm that he could find no reliable information on such an animal. However, following a discussion with Benjamin Franklin, Kalm (1770: 156) related "that he [Franklin] had, when a boy, seen two of the animals which they call Moose-deer, but he well remembered that they were not of such size as they must have been, if the horns found in Ireland were to fit them. The two animals which he saw, were brought to Boston to be sent to England to Queen Anne. Anyone who wanted to see them had to pay twopence. A merchant paid for a number of school-boys who wanted to see them, among whom was Franklin. . . . The elks are called 'Originals' by the French in Canada."

Another such chronicler was Alexander Philipp Maximilian, Prince of Wied-Neuwied, a German aristocrat who, with artist Karl Bodmer in the early 1830s was among the first white men after Lewis and Clark to ascend the Missouri River. Although Maximilian's journey did not carry him into moose range, he learned of its presence upstream on today's Milk River in Montana. A guide brought the German adventurer part of the skin and head of an "original" that he had killed there (Maximilian 1843: 203). As with Boucher (1664), Kalm (1770) and others, Maximilian or his editors, typographers or translators referred to moose as "originals"—likely a corruption of the French word "orignals" for moose, but also possibly because many features of the moose (large size, large and palmated antlers, massive nose, humped shoulders, long legs, pendulous bell, etc.) were so unusual or unique as to invite the "original" characterization.

Thomas Jefferson, third United States President, once confided that "Natural history is my passion" (Foley 1900: 294). Little known as a naturalist and amateur biologist, at least during his own time (1743–1826), Jefferson disputed George Louis Leclerc, Comte de Buffon, of Louis XIV's court, who was then perhaps the world's leading zoologist. In an early edition of *Histoire Naturelle Générale et Particulière*, Buffon (1749–1769) had declared that when an animal species was found in both the Old and New Worlds, the latter form was smaller and inferior. Jefferson took strong exception to the assertion and undertook to set the record straight (Cutright 1989). He wisely or fortuitously chose the North American moose with which to make a counter argument, and proceeded to gather information and specimens of the animal (which did not occur in his native Vir-

By the mid-1600s, voyageurs and French officials (both civil and military), usually with Indian guides, hunted moose for sport as well as subsistence. "Moose are hunted on the snow; and many of our Frenchmen have killed thirty or forty apiece. . . . Formerly [before adaptation of the Indian technique of tracking or chasing moose to deep snow], the hunting of them appeared to our Frenchmen an impossibility, and now it seems to them as recreation" (Thwaites 1899c: 193). *Top photo from Waterman (1973), artist unknown. Center scene "Hunting Moose" (circa 1860), by Dennis Gale; photo (C40184) courtesy of the National Archives of Canada. Bottom scene "Moose Tracking" (circa 1867), by Henry Buxton Lawrence (ca. 1870); photo (C40800) courtesy of the National Archives of Canada.*

ginia). In his manuscript, *Notes on the State of Virginia* (completed in December 1781), Jefferson (1787: 53) wrote: "Kalm tells us that the Moose . . . of America is as high as a tall horse; and Catesby, that it is about the biggness of a middle sized ox. (I have seen a skeleton 7 feet [2.1 m] high, and from good information believe that they are often considerably higher. The elk of Europe is not two-thirds of his height.)" By virtue of supporting information (Figure 9) and a specimen Jefferson had acquired and its skin and skeleton sent to Buffon in Paris (although it arrived in poor condition after a 4-month voyage), Buffon concluded: "I should have consulted you, sir, before publishing my natural history, and then I should have been sure of my facts" (Parton *in* Foley 1900: 68).

The Viviparous Quadrupeds of North America, by John James Audubon and the Reverend John Bachman, assisted by Audubon sons John W. and Victor G., was the most ambitious effort through the mid-1800s to illustrate and describe North American mammals. A lengthy text, based on the authors' personal observations and testimony from others, described the moose and its habits. The lordly, majestic moose depicted by J. W. Audubon appeared quite different from Major Hamilton Smith's characterization of the animal (*in* Audubon and Bachman 1845–1848: 232): ". . . when we had the gratification of bringing one down with a rifle and of examining him in detail as he lay before us, we confess he appeared awkward in his gait, clumsy and disproportionate in limbs, uncouth and inelegant in form, and possessing less symetry and beauty than any other species of the deer family."

Ernest Thompson Seton is widely known for his popular books and excellent wildlife field sketches and illustrations, and his scientific contributions were astute and progressive. They culminated in the two-volume *Life-Histories of Northern Animals* (1909) and its revised and expanded sequel, *Lives of Game Animals* (1929). Chapter IV in volume I of the former is devoted to "The Moose, or Flat-horned Elk."

Seton's (1940: 274) first and only successful moose hunt, in October 1884, occurred in the vicinity of Carberry, Manitoba, and entailed more than 300 miles (483 km) of arduous stalking in snow over the course of 19 days, but he had fulfilled his dream to bag "the grandest beast that roams

John McDuffee to John Sullivan, with Answers to Queries concerning the Moose
Rochester March 5th. 1784

1. *Is not the Caribou and the Black Moose one and the same Animal? The Carabou Calevan or Indian Shovler is an Animal very different from the Moose. His hoofs are like a Horses his Horns are short and have no prongs, is about the bigness of a small Horse, lives in Heaths and Swamps, feeds chiefly on Roots which he digs up with his feet, is seldom seen in this quarter of the Country.*
2. *Is not the grey Moose and the Elk one and the same Animal and quite different from the former? The Elk is Deer of a large size and is known by the name of the Newfoundland Deer. The Black and Grey Moose are one and the same Animal. The Black are mostly found to the Eastward, the Grey to the Southward.*
3. *What is the heighth of the grey Moose at the weathers, its length from the Ears in the root of the Tail, and its circumference where largest? One of the largest is about 8½ feet high at the weathers, 8½ or 9 feet long, behind his fore legs is about 7 feet in circumference.*
4. *Has it a Sollid or Cloven Hoof? Cloven.*
5. *Do their feet make a loud ratling as they run? They do.*
6. *Is the under part of the Hoof covered with Hair? It is not.*
7. *Are they a Swift Animal? They trott exceeding fast and have no other gait.*
8. *Do they sweat when hard run or only drip at the tongue? Only drip at the Tongue.*
9. *At what season do they shed their Horns, and when recover them? They shed their Horns in January, push'd off by the succeeding Horns and recover them gradually 'till push'd off again by Succession.*
10. *Has the Doe Horns as well as the Buck? They have not.*
11. *How many young does She produce at a time? She has but one the first time, ever after two.*
12. *What is their Food? Brouse and Bark of Trees chiefly of Maple.*
13. *How far southward are they known? No farther than Hudsons River, to the southward of that the Moose are grey.*
14. *Have they ever been tamed and used to any purpose? They are Easily tamed, but can bear no labour.*
15. *Are the Horns of the Elk Palmated, or are they round and pointed? The Horns are the same as a Deers only flatted at the topp.*
16. *Has the Elk always or ever a white spot a foot in Diameter round the root of the Tail? I never saw but one; that had none.*

Sir

The above answers are the best I am able to give to your Queries, would only observe that there is a Moose's Horn in one of the Pigwackett Towns so large that is Used for a Cradle to rock the Children in. Am Sir with due Respects your most Humble Servant,

John McDuffee

Figure 9. To gather information to support his contention that the North American moose was equal to if not larger than its Old World counterpart (the great French naturalist Buffon had asserted the contrary), Thomas Jefferson sent questionnaires to individuals knowledgeable about the moose (Cullen 1983[7]: 23–24). He distributed 16 questions, and these, plus replies from John McDuffee of Rochester, New York, on March 5, 1784, are reproduced. Although some of McDuffee's responses were fairly accurate, others were vague if not incorrect. Nevertheless, they ultimately served in support of Jefferson's proposition.

When Theodore Roosevelt campaigned (unsuccessfully) for a third term, in 1912, on behalf of the Bull Moose (Progressive) party, a campaign photo showed him astride the party symbol (right). Although the photo was a poorly disguised composite, it served to reveal the candidate as a man of courage, initiative and vitality. *Photo courtesy of the Theodore Roosevelt Collection, Harvard College Library.*

America's woods." But as he gazed on the superb animal, a large bull "turned into a pile of butcher's meat, for the sake of a passing thrill of triumph. . . . [I] then and there made a vow, which I have lived up to ever since—that so long as they are threatened with early extinction, I will never again lift my rifle against any of America's big game."

Theodore Roosevelt may be remembered by many almost as much for being a hunter and preeminent conservationist as for his role as statesman. In his book, *The Wilderness Hunter*, the twenty-sixth United States President described his first successful moose hunt, in 1887, high in the Bitterroot Mountains forming the Montana/Idaho boundary: "At last we reached the hummock, and I got into position for a shot, taking a final look at my faithful 45-90 Winchester to see that all was in order. Peering cautiously through the shielding evergreens, I at first could not make out where the moose was lying, until my eye was caught by the motion of his big ears, as he occasionally flapped them lazily forward. Even then I could not see his outline; but I knew where he was, and having pushed my rifle forward on the moss, I snapped a dry twig to make him rise. My veins were thrilling and my heart beating with that eager, fierce excitement, known only to the hunter of big game, and forming one of the keenest and strongest of the many pleasures which with him go make up 'the wild joy of living.' As the sound of the snapping twig smote his ears the moose rose nimbly to his feet, with a lightness on which one would not have reckoned in a beast so heavy in body. He stood broadside to me for a moment, his ungainly head slightly turned, while his ears twitched and his nostrils sniffed the air. Drawing a fine bead against his black hide, behind his shoulder and two thirds his body's depth below his shaggy withers, I pressed the trigger . . ." (Roosevelt 1909: 209).

Roosevelt (1916) later told how he inadvertently violated the game laws of Quebec while hunting moose in that province in 1915. On a September morning, he shot a bull with an antler spread of 52 inches (132 cm). That afternoon, while paddling their canoe, he and a companion encountered a large bull that went ashore and, for more than an hour, charged the hunters whenever they tried disembarking. The moose finally disappeared, but returned and charged the men when they had landed. Roosevelt tried to dissuade the moose with a shot fired over its head, but it came on, forcing Roosevelt to shoot the animal at a distance of 30 feet (9.1 m). Because only one moose could legally be taken in a day, Roosevelt filed an affidavit attesting to the circumstances of his technical violation. No official notice was taken against him.

After serving two terms as President (1901–1909), Roosevelt initially supported William Howard Taft as his Republican party successor. But Roosevelt's dissatisfaction with Taft caused him to bolt the party and pursue a third term as nominee of the new Progressive Party, popularly known as the "Bull Moose party" (Bishop 1920: 337). While campaigning in Milwaukee, Wisconsin, on October 14, 1912, Roosevelt was shot point blank in the chest as he was bowing to his audience from an automobile. The would-be

Colonel Roosevelt, as he liked to be known, posed with his guide Arthur Lirette, and the antlers of one of two moose he shot on September 19, 1915 (Roosevelt 1916: 165). Presumably, the 52-inch (132-cm) "rack" is of the bull he killed legally that morning, and not the bull he shot in self-defense that afternoon (see text). At the time, Roosevelt was the guest of Dr. Alexander Lambert, an old hunting companion, at the Tourilli Club in Quebec. Roosevelt's rifle is a 1903 Springfield, model 6000, which was his chief weapon during the previous dozen years when he killed more than 300 head of game on three continents. However, the moose was among Roosevelt's favorite game species. He shot his first moose in the Bitterroot Range, which straddles the western Montana/Idaho border, in September 1887 (see Roosevelt 1909: 203). The twenty-sixth U.S. president was a prestigious writer of outdoor experiences and natural history. Of his 32 books, at least 7 dealt with big game animals and the hunting of them. *Photo courtesy of the U.S. National Archives.*

John Woodhouse Audubon did nearly half the animal paintings and Victor G. Audubon did many of the backgrounds in *The Viviparous Quadrupeds of North America* (Audubon and Bachman 1845–1848), begun by their father, John James Audubon, who died in 1851. The younger Audubons' "drawn from nature" rendition of "Moose Deer" (plate LXXVI) shows a standing bull and recumbent cow and calf in an unnaturally open setting. The three Audubons seemed more facile painting birds than mammals, although at its time, *The Quadrupeds* . . . was far the best available book on American mammals. *Photo courtesy of the U.S. Library of Congress.*

assassin was immediately seized, and Roosevelt, untreated, insisted that he was not seriously hurt and would fulfill a nearby scheduled appearance.

Stepping to the platform where he was to speak, the eventually unsuccessful candidate reached inside his coat for the folded manuscript of his speech. He discovered that the pistol bullet had penetrated it and his metal spectacle case. Although startled at first, Roosevelt held up the latter to the audience and declared "It takes more than that to kill a Bull Moose!" (Bishop 1920: 338). He was examined later that evening by doctors who determined that removing the projectile was too risky. Consequently, Roosevelt carried the bullet with him to his death 7 years later.

George Shiras III of Pennsylvania was a politician, conservationist and pioneer wildlife photographer. As a U.S. congressman, Shiras had drafted a bill that was the first recognition by the federal government of the need to preserve and protect migratory birds. It served as foundation for the Weeks–McLean bill enacted in 1913, the migratory bird treaty signed in 1916 by the United States and Canada and the enabling U.S. legislation passed in 1918. As a private conservationist, Shiras publicized the plight of American wildlife, and helped arouse public interest and support for badly needed federal conservation laws.

Much of Shiras' effectiveness as a conservationist arose from his photographs of wildlife taken throughout much of North America. Dozens appeared in *National Geographic Magazine*, and later in *Hunting Wildlife with Camera and Flashlight* (1935). Shiras was the first to perfect night-flash photography, and took the first such picture of a moose.

The moose always held a special fascination for Shiras; it was the subject of many of his day and night photographs, taken in Michigan, Wyoming, New Brunswick and Alaska. The moose appeared in 41 pictures (including 11 taken with flash) in six magazine articles and in 65 photos in his book. To countless Americans, these were their first images of the moose. Fittingly, Edward W. Nelson (1914) of the U.S. Bureau of Biological Survey named the Rocky Mountain subspecies for his friend, calling it the Shira's moose, now also known as the Yellowstone moose (Peterson 1955).

Subsistence

Remote from the trappings and predilections of European civilization, explorers, visitors and settlers readily adopted the North American Natives' subsistence patterns, although decidedly more avariciously at times. With them, the newcomers brought the Judeo-Christian tenet of human dominion over wildlife, an "ethic" that was antithetical to the Natives' relatively pure view of ecology. The whites also were possessors of firearms and had inclinations to extract more than a lifestyle from the New World wilderness. But foremost, at first, they needed nourishment.

Explorer Alexander Henry the younger (Coues 1897: 3–4) remarked when food was in short supply in western Canada: "Animals were so scarce that we suffered much from hunger. On Dec. 19th, 20th, and 21st [1799] we ate nothing until the evening of the last day, when I received a moose's head, which was boiled and divided among 17 persons."

Only rarely were specific quantities of foods eaten per period of time mentioned in explorers' journals and, even then, the size of the party was seldom given. One of the earliest quantified records of moose harvest was given by Baron Lahontan: "This, Sir, was our Diversion for three Months in the Woods. . . . We took fifty six Elks [moose], and might have kill'd twice as many, if we had hunted for the benefit of skins" (Ganong 1910: 59). While his party

George Shiras III, besides being a champion of wildlife conservation as a U.S. congressman, was a pioneering photographer. He is credited with perfecting night-flash photography and wild animals were his subjects. He took the first such photo of a moose—a cow feeding along a shoreline, which was so startled by the flash that it sought escape in water. Shiras had taken the picture from a canoe and the cow that nearly capsized the craft did manage to knock the camera overboard. Nevertheless, the glass plate was saved. Shiras took numerous other diurnal and nocturnal photos of moose, including of the subspecies named for him. The porcupine above was captured on film while gnawing on a moose antler during Shiras' first visit to Alaska in 1911. *Photo by George Shiras III; courtesy of the National Geographic Society.*

spent 26 days in May 1845, ". . . at the source of the Athabaska," Fr. De Smet (1847: 257) recorded that they killed and ate "twelve moose deer; two reindeer [caribou]; thirty large mountain sheep or bighorn; two porcupines; two hundred and ten hares; one beaver; two muskrats; twenty-four bustards [geese]; one hundred and fifteen ducks; twenty-one pheasants [grouse]; one snipe; one eagle; and one owl; plus thirty to fifty whitefish everyday and twenty trout daily."

The Hudson's Bay Company occasionally documented the quantities of Native commodities used at the various posts. For example, York Factory manifests show that 125 dressed moose skins, equivalent to 250 beaver pelts, were used there in 1801 (Johnson 1967). However, at inland posts, such as Cumberland House (present day extreme eastcentral Saskatchewan), moose provided a larger proportion of the food than at posts on Hudson Bay, where concentrations of migrating geese usually supplied an abundance of food. Cumberland House journals regularly showed that Indians brought for trade "green" (unprocessed) moose flesh, dried moose, "beat" (pulverized dried) moose meat and moose fat.

Discussing the value of moose to early settlers of New England, Grant (1894: 355–356) wrote: "The early settlers in Vermont and New Hampshire found their meat a most welcome source of food; in fact, the numbers of moose alone enabled the colonists at first to keep from starvation during the long winters." Similarly, Ward (1878: 450) noted: "To the early settlers in the states of Maine, Vermont, and New Hampshire, and the provinces of Nova Scotia and New Brunswick, the flesh of the moose was the main-stay, and his hide furnished them with serviceable clothing."

Nathan Moore, a market hunter for logging camps and settlements in the Maine woods, kept a record of the number of moose he had killed (Merrill 1916). When he died in 1906 at the age of 88, his tally stood at 276. Thoreau (1864) confirmed that Maine lumbermen often ate moose meat.

Merrill (1916) stated that moose meat is the only kind of venison suitable for preserving in brine. In Nova Scotia, farmers near moose country preserved its flesh for winter use by a solution of salt, saltpeter, and sometimes molasses, cloves and other spices.

Moose supplied meat to members of the Western Union Telegraph Expedition, which began surveying a telegraph route across Alaska in 1865 to connect with one planned for eastern Siberia via the Bering Strait. William Dall (1870), of the expedition's scientific corps, reported that moose frequently were shot for food or purchased from local Indians.

During times of extreme food shortages in Alaska, soldiers even resorted to eating spoiled moose. One of H.T. Allen's men reported that, in the Copper River Valley in 1885, "This is Lieutenant Allen's birthday, and he celebrated it by eating rotten moose meat. . . . Rotten moose meat would be a delicacy now. . . . All so weak that we were

In July 1872, the photographic division of the Ferdinand V. Hayden Expedition—first of the "Great Surveys" authorized by the U.S. Department of the Interior to map the Western Territory and promote and publicize the Yellowstone region—encamped in the shadow of the Grand Teton Mountain. "The game was plentiful in this region; bighorned mountain sheep, bears and elk ranged over the high plateau, and one of the men got a young moose" (Jackson 1947: 164). Also killed was a young elk (on the ground) and an adult cow moose (head only hanging next to the moose calf). *Photo by William H. Jackson; courtesy of the U.S. Geological Survey.*

dizzy, and would stagger like drunken men" (U.S. Senate 1900: 430).

Moose also was an important food for many of the thousands of argonauts who rushed to the Yukon during the late 1890s. Many miners took the steamboat up the Yukon River to the Klondike gold fields. The sighting of a moose by a steamer pilot and the shout "Moose! Moose!" brought invariably a rush of prospectors and Indians to the rail with loaded rifles. On at least one such occasion, a hapless moose was riddled with bullets from 60 or 70 blazing rifles, and its carcass retrieved and immediately converted to food (Webb 1898).

Moose frequently were shot in the vicinity of mining camps and towns, and sold directly to miners or through butcher shops. In Dawson, one shop advertised the availability of moose meat by setting out antlers. Although few Indians worked the placers, many made a business of market hunting by providing fresh meat and fish (McKennan 1981). Adney (1900: 507) reported that "The Indians killed in all about eighty moose and sixty-five caribou, much of which they sold to the miners in Dawson, . . . and invested the proceeds in finery and repeating rifles."

In his annual report on the administration of the Alaska Game Law dated November 1, 1915, Governor Strong recounted the effect of the gold stampede in the late 1890s in the Cook Inlet country on big game (Merrill 1916). At the peak of the frenzy, 1,000 people were in the two settlements of Hope and Sunrise. They depended chiefly on moose and mountain sheep for their meat supply, and these animals often were wantonly and wastefully killed. Some old-timers boasted they killed moose merely for its tongue—the choicest part—and left the carcass to go to waste. Probably as a result, moose became scarce near new human habitations.

Some Other Uses

HIDES

Early colonists in New England and farther north used moose skin for clothing. Alexander Bradford of Dorchester, Massachusetts, bequeathed in 1645, a "Moose Suite & a musket & Sworde & bandilieres & vest" (Merrill 1916: 17). Moose skins were used in making "toggies" or "banians" (originally, a jacket or shirt worn in India), worn by Hudson's Bay Company men in winter (Rich and Johnson 1952, Graham *in* Williams 1969). This loose-fitting outer garment, reaching almost to the feet, had cuffs and a cape of beaver or otter, but in very cold weather, it was insufficiently warm

Market hunting for moose during the Klondike gold rush era served both the gunners (right), many of whom were Natives, and hungry miners. Above left is the California Market in Dawson, Yukon, Northwest Territories, in 1901, at a time when moose meat sold in the territory for $0.30 per pound ($0.66/kg), compared with $0.45 to $0.75 per pound ($0.99 to $1.65/kg) for beef (Mayer 1989). At the peak of the gold rush, moose meat sold for as much as $1.00 to $1.50 per pound ($0.45 to $0.68/kg) (Bolotin 1980). *Left photo courtesy of the Selid Collection, Archives, Alaska and Polar Regions Department, University of Alaska, Fairbanks. Right photo courtesy of the U.S. National Archives.*

and was replaced by a toggy of beaver or supplemented with other clothing. Dehaired moose skins also were used for covering windows of Hudson's Bay Company forts and posts. Their thickness withstood severe winds yet transmitted some precious light during the long, dark winters.

MERCANTILISM

Fur trappers and traders who worked the moose range were interested primarily in beaver and fox (particularly "black" or blue-phase arctic fox). Their attraction to moose as a marketable commodity was limited at first, although the animal's meat was in demand for subsistence, as previously noted. In the first decade of the 1700s, "great loads of skins from deer, elk [moose], and bear" were exported to England from New York ports alone (Norton 1974: 86). With escalating foreign competition, principally English and French, and erosion of European beaver market in the 1830s, trade emphasis increased on big game hides for leather (see Chittenden 1935, Rich 1960, Phillips 1961, Clayton 1967, Ray 1978, 1987).

A Dogrib "moose suit," featuring moose hide shirt and leggings. The pouch, knife sheath, gun case and moccasins likely also were fashioned from smoked moose hide. The apparel was modeled by C.H. Keefer in 1937 at Fort Wrigley in the Northwest Territories. *Photo (PA73924) by C.A. Keefer; courtesy of the National Archives of Canada.*

In his *New Voyages to North-America,* Baron Lahontan reported the commercial values of certain Native "products" of New France (Ganong 1910: 259). Among them were:

	Livres	Sous
Of dry or common beavers, *per pound* . . .	3	0
The Skins of Elks before they are dress'd, are worth *per pound* about	0	12
The Skins of Stags are worth *per pound* about	0	8

The "skins of elks" actually were moose skins, as Lahontan wrote elsewhere of "Orignalds or Elks" (plate 55); the "stags" no doubt were elk. Interestingly, moose skins were more highly valued than elk skins, but still well below the value of beaver pelts.

The best source of information on the quantities and commercial value of moose hides and other commodities is the voluminous records of the Hudson's Bay Company. A review of some of those trade, sale, cargo and auction documents at the Hudson's Bay Company Archives in Winnipeg, Manitoba, revealed that, although the Hudson's Bay Company was a tightly regimented company, considerable differences exist among factors and clerks at the various posts and over time. Thus, the records are not standardized. We found that moose sometimes were listed as elk, sometimes not at all and sometimes skins and hides of all "deer-like" species were lumped together.

Although commerce in beaver skins was the enduring mainstay of its trade, the Hudson's Bay Company, since its establishment in 1670, always made great effort to find other commodities that might find a profitable market in England and on the continent (Leechman 1974). Some moose skins were received from Indians in the very early days of the company. On December 12, 1671, Hudson's Bay Company Committee (the governing body in London) authorized the sale of "ottar & mouse [moose] Skins as they shall see fit" (Rich 1942: 16); apparently this was the parcel of "Otter and Moose Skins sold" on April 30, 1672 for £10 (Rich 1942: 39). Many other lots of moose skins were sold in subsequent years.

On June 6, 1689, the Committee instructed its officials at Hudson Bay: "Wee would have you *send home, what Buffeloes, Moose Skins, . . . or any New Commodity you Can meete with, at the Cheapest prizes* you Can which is left to your Discretion . . ." (Rich and Johnson 1957: 58). And on May 30, 1694, Hudson's Bay Company Governor James Knight was instructed: "*Moose skins* or *Buffelo hydes* are a great commodity here this Warr time, therefore if any are to be had pray send them home & suffer them not to be used in the Factory Upon any accott. whatsoever haveing sent Leather

&ca. to Supply in stead thereof. . . . Wee must alsoe Perticulerly recommend to you the Improvement of . . . procureing all the Moose Skins you can & not suffer any of them to be cutt or otherwise used in the Factory . . ." (Rich and Johnson 1957: 231, 236).

After only a few years, London instructed that no more moose skins be sent. A company official posted on Hudson Bay questioned the directive: "I am doubtful by your letter that moose skins are no commodity, because you write to let the men take up what they will. . . . Here is a great many to go home, and I wonder if they are not a commodity in England. The French used to encourage the trade of them more than any other sort of commodity" (Davies and Johnson 1965: 62).

From September 8, 1714, to August 1, 1715, 11,120 whole parchment beaver, 2,080 half parchment beaver, 3,250 coat beaver; and 2,390 moose, 20 small moose, and 5 moose "fawn" skins were received in trade at Hudson's Bay Company's York Fort (Gilbert 1973). The moose then was valued as equivalent to two beaver; the small moose equaled one beaver; and a moose calf hide was worth half of a beaver pelt. At that time, the trade value of moose was about 22.8 percent that of beaver. In 1716, cargo being prepared at another Hudson's Bay Company post for shipment to London included 18,602 whole parchment beaver, 3,220 half parchment beaver, "4000 [made-beaver equivalent] in 2000 moose," and "639 in 424 dressed moose" (Davies and Johnson 1965: 74).

The London Committee that micromanaged Hudson's Bay Company's trading policies in British North America included few members who had ever been to America and had little insight into the practicality of the guidance they were providing. They failed to appreciate that the fur catch and moose hide trade could not be turned on and off like a faucet or the great value and use being made of the moose products at far-flung company posts.

Emphasis gradually shifted from the efficacy of trade in moose skins to the exchange value of moose in terms of "made-beaver," the universal trading unit (a skin from a beaver taken during its prime and in good condition). To standardize trade with the natives, the Hudson's Bay Company developed two schedules: the *Comparative* of values among commodities received from the Indians (e.g., castoreum, goose quills, moose hides), and a *Standard of Trade* setting the number of beaver skins required for certain items of European manufacture offered to the Indians for trade (e.g., guns, powder, cloth, beads). To illustrate, the *Comparative* in effect in the late eighteenth century set the following as equivalent to one made-beaver: 1 moose skin "in parchment" (pr 1.5 moose skins, dressed); 3 skins of old bears, 4 black fox skins; 1 skin of an old otter, 2 prime marten skins, 2 pounds (0.91 kg) of castoreum or 2,000 goose quills (Graham *in* Williams 1969: 274). One made beaver (or one moose skin "in parchment") could be exchanged for 0.75 pound (0.34 kg) of Brazil tobacco or 1.0 pound (0.45 kg) of powder. On the other hand, 14 beavers or moose skins in parchment were required for one gun, 4 for 1.0 gallon (3.81 l) of British brandy, 2 for 1.0 pound (0.45 kg) of English beads, 1.5 for a brass kettle (a very popular item) and 2 for a pair of shoes. The intended stability sought by the standard was upset when it was variously set two moose equivalent to one beaver, changed back to the original rate and reset at a half moose per beaver.

Hudson's Bay Company Governor George Simpson, unlike his predecessors, traveled widely throughout the empire. While in the Western Caledonia Columbia River District (now British Columbia) in 1829, he reported that a dressed moose skin was monetarily valued at 3s 9d.; at the same time, a plain "three-point" blanket was worth 8s 6d., more than twice as much (Rich 1947).

From review of Hudson's Bay Company archival documents and the fur trade literature, we suspect that moose hides occasionally were omitted from London sales leaflets. Hudson's Bay Company was royally chartered, but still a privately owned enterprise. Individual stockholders were not supposed to obtain high-grade imports for private sale, but apparently sometimes this occurred. Also, highly sought commodities, such as moose hide, might have been withheld from public auction and sold privately by the company—a dubious but fairly common practice among importers. Also fairly clear is that, at some posts and times, employees were reluctant to ship precious moose hide, opting instead to outfit themselves with the skins. Wrote Russell (1898: 230): "Mooseskins are supplied to all [Hudson's Bay] Company posts where moose are not found, from which to make moccasins, etc., for officers and servants."

TROPHIES

Moose antlers and mounted heads have long been among the most prized of all game trophies worldwide. Queen Elizabeth I (reigned 1558–1603) assembled a great collection of game trophies from all parts of her domain and areas explored by her voyagers (Merrill 1916). These were displayed in the "Horn Room," near the Great Hall of Hampton Court Palace, just west of London on the Thames. Officials of her trans-Atlantic colonies supplied five sets of moose antlers (these may be seen today by visitors to Hampton Court). After Henry VIII's great palace ceased to be a royal residence, the trophies were scattered in various halls and apartments of the vast structure, but eventually were given places of honor in the Great Hall itself. There, in the distinguished company of the aristocracy

During Theodore Roosevelt's two terms as U.S. president, 1901 through 1909, the State Dining Room of the White House in Washington, D.C. was decorated with mounts of various big game animals Roosevelt had shot, including a bull moose over the main fireplace (left). Political wags and cartoonists of the time wrote that the White House's "trophy room" decor was a result of "teddyfication." *Photo courtesy of the Library of Congress.*

of English game animals from 350 years ago, they have presided over state banquets of a succession of British sovereigns (indeed, much as Theodore Roosevelt's moose heads presided over gatherings in the White House's State Dining Room during his presidency). The greatest spans of the royal racks measured 59, 48.5, 47, 42.5 and 38 inches (150, 123, 119, 107 and 97 cm) and, with the exception of the first one, well short of recent-year trophies. They may be, however, the oldest extant antlers of North American moose (Merrill 1916).

A larger pair of moose antlers from the Canaan River area of New Brunswick, measuring 62 inches (157 cm) in spread, was given to Edward VII, who, as Prince of Wales, visited Canada in 1860 (Merrill 1916).

Moose antlers long have been used for display or decorative purposes. A set of large moose antlers (having a spread of 64.5 inches [164 cm]) was displayed over the fireplace in the studio of the great western artist Albert Bierstadt, who had shot the animal in New Brunswick in 1880. In 1889, Lord Stanley, the Canadian governor general, traveled from Quebec to Vancouver, then the longest stretch of railroad in the world, to promote western settlement (Samuels and Samuels 1985). His special train was pulled by a wood-

Monticello, south of Charlottesville, Virginia, features nearly all the original furniture, instruments, artwork and scientific specimens brought to the estate by third United States President Thomas Jefferson, who designed, built and lived there for his last 56 years (Wernick 1993). The entrance hall was a mini-museum, including a mastodon jaw, a bison skull, and elk and moose antlers. Various investigators have suggested that the moose antlers were from the Lewis and Clark expedition (see e.g., Ambrose 1996). The journals of Meriwether Lewis do not make mention of the kill of a single moose, although one was wounded by expeditioner Reubin Field on July 7, 1805, east of what now is Lincoln, Montana, and a few others were encountered. On the other hand, the journal entries account for only 442 of the expedition's 863 days. Ambrose (1996) indicated that Indian artifacts and both the elk and moose antlers were shipped from Fort Mandan in South Dakota in spring 1805. Ultimately, 25 boxes of artifacts that reached Washington, D.C. were shipped to Richmond, presumably enroute to Monticello. The ship was stranded and everything except some "horns" (perhaps the moose and elk antlers) were lost. Equally as likely is that the moose antlers may have been among those collected by Jefferson in his attempt to gather evidence to convince preeminent French zoologist Buffon that the North American moose was not a stunted version of the Scandinavian form. *Photo by Robert Lautman; courtesy of the Thomas Jefferson Memorial Foundation, Inc.*

burning locomotive decorated with a large pair of moose antlers mounted over the lantern.

CAPTIVITY AND DOMESTICATION

Early French officials suggested the possibility of domesticating moose to facilitate travel to distant outposts. In 1636, Fr. Le Jeune wrote that the governor of New France had two bull moose and one cow in captivity that he was hoping to domesticate (Thwaites 1897e).

Numerous passages from other early journals describe efforts to keep moose in captivity. For example, Fr. Gabriel Sagard (1632: 225) reported that, when he was in New France during 1623–1624, a young moose was being held in the Fort at Quebec, with the intention of taking it to France. Unfortunately, the animal died of wounds inflicted by dogs. Also, according to Andrew Graham, a live moose was sent from Churchill Fort to his King George III in 1767, and was kept in Richmond Park in southwestern metropolitan London (Williams 1969). The moose was the female survivor of a pair. "It cost the Governor and Committee [of the ever frugal Hudson's Bay Company] £9 10s. 11d to feed this moose from October 1767 to February 1768 . . . and on 25 May 1768 they wrote to tell Moses Norton not to send any more livestock home. Yet more moose did come, and at least two of them from Hudson Bay. A marginal note that Joseph Colen wrote in his copy of Hearne's Journey records that 'I had two of these animals Tame and followed me the same as any domestic animal—They were sent to England and are now in the Park of the Duke of Norfolk.' A later note continues, 'Since the above note was written, I have heard that one is Dead.' The Duke of Norfolk offered [£?] 1,220 for 6 brace. . . . This pair of moose must have been shipped between 1785 and 1798. . . ."

Samuel Hearne's editor and annotator, J.B. Tyrrell (1911) added that the moose formerly sent to the King was from Churchill; however, it (apparently the male of the pair) died enroute to England. Tyrrell also reported that the same Indian who provided this moose brought two others to the Factory in 1777. These followed him along shore while he paddled downstream in a canoe. At night, when the Indian landed, the moose reportedly came up to him to be fondled like domestic stock.

"The moose are also the easiest to tame and domesticate of any of the deer kind," wrote Samuel Hearne (in Tyrrell 1911: 260). "I have repeatedly seen them at Churchill as tame as sheep, and even more so; for they would follow their keeper any distance from home, and at his call return with him, without the least trouble, or ever offering to deviate from the path." And Lantz (1910: 18) summarized: "Perhaps no other American deer is naturally so well adapted to domestication as the moose. Professor [Spencer F.] Baird relates that a pair of animals were kept by a man living near Holton, Me. These had been trained to draw a sleigh, which they did with great steadiness and swiftness, subject, however, to the inconvenience that, when they once took it into their heads to cool themselves in a neighboring river or lake, no effort could prevent them."

Martyn (*in* Audubon and Bachman 1845–1848: 235) reported that "A Mr. Bell, residing at Three Rivers, has a Moose which has been taught to draw water in a cart or sleigh during the winter, but there is no possibility of working it during the rutting season. We never heard of any attempt to ride on the Moose deer." According to Ganong, Governor Sir Edmund Head, who lived in New Brunswick from 1848 to 1854, owned a moose that often was driven in harness in races on the ice against horses ridden by Army officers (Merrill 1916).

Seton (1909[I]: 180–181) related a communication from George H. Mesham about a domesticated moose: "A neighbor of mine, Henry Stoggett, of Shoal Lake [Manitoba/Saskatchewan border], had a tame moose for a considerable time. It was very affectionate, and when called would come like a dog; also it was as playful as a kitten, and would, like a kitten, play with a round pebble or croquet ball, striking it with its front feet and running after it. It would also gambol with the children, dogs, or young cattle."

Moose in captivity or under domestication have shown a fondness for many human foods. For example, mashed, boiled potatoes were relished by one young moose (Powell 1856), another showed a preference for ship-biscuits, and yet another made a nuisance of himself begging for chewing tobacco (Shiras 1935).

Perhaps the greatest North American effort to restrain a group of moose within a large enclosure is that of Austin Corbin, who established the Corbin Game Preserve in New Hampshire, toward the end of the nineteenth century (Ferris 1897). At one point, the 40.6-square mile (105-km^2) preserve reportedly contained some 4,000 animals, including moose, bison, elk, big-horned sheep and various deer. At the time, it was said to be the largest private game park in the world, except for one owned by the Duke of Sutherland in Scotland, and one or two royal demesnes on the continent. The dozen introduced moose eventually increased to 150.

LANGUAGE AND LITERATURE

In his *Dictionary of the English Language*, English lexicographer Samuel Johnson (1755) described the elk (moose) as "a large stately animal of the stag kind. The neck is short and slender; the ears nine inches in length, and four in breadth. The colour of its coat in Winter is greyish, in Summer it is paler; generally three inches in length, and equalling horsehair in thickness. The upper lip of the *elk* is large. The artic-

Hardy (1855: 202) wrote that "it is a singular circumstance, that the moose when taken young can be easily domesticated, and becomes at once perfectly tame and audaciously familiar." Powell (1856: 344) concurred: "The moose is easily tamed, and when domesticated, exhibits much sagacity, and, if well treated, a very affectionate disposition. I kept a young one (one year old) a short time, which manifested as much docility as a lamb. But when insulted or injured they are very revengeful and unforgiving." Despite the expressed ease of moose domestication, it apparently was not a widespread or frequent practice, for the obvious reason that the animals' subsistence and material values far outweighed those of pet or beast of burden. *Top left photo (C8186) by Ernest Brown and top right photo (PA149811) by Ernest F. Keir; both courtesy of the National Archives of Canada. Bottom photo by George Clark; courtesy of the Glenbow Museum Archives, Calgary, Alberta.*

ulations of its legs are close, and the ligaments hard, so that its joints are less pliable than those of other animals. The horns of the male *elk* are short and thick near the head, where it by degrees expands into a great breadth, with prominences in its edges. Elks live in herds, and are very timorous. The hoof of the left hinder foot only, has been famous for the cure of epilepsies; but it is probable that the hoof of any other animal will do as well."

Given the size and significance of North American moose, surprisingly few books have been written entirely about the animal. As early as 1855, a two-volume treatise on moose hunting in New Brunswick and Nova Scotia was written by a British officer/sportsman (Hardy 1855). Samuel Merrill's *The Moose Book* (1916) is an insightful blend of historical, cultural and biological information about the species. A relatively scientific treatment, *North American Moose* by Randolf Peterson (1955), is based on the author's field and library research in Canada, and remains an important reference. Adolph Murie's *The Moose of Isle Royale* (1934) reported on his field studies of moose inhabiting that island in Lake Superior. Technical bulletins, such as by Hosley (1949) and Houston (1968), have provided management-oriented updates of the species or subspecies. Joe Van Wormer's *The World of the Moose* (1972) is a popularly written account of the species.

On the other hand, many authors have featured the moose in the plots, themes and narratives of fiction and poetry. For example, the moose is the linchpin of Henry David Thoreau's classic, *The Maine Woods* (1864), in which Thoreau brings together his personal experiences and feelings gained during three canoe trips into the Maine wilderness. The values and meanings of moose to the local Indians are blended with the author's own insights into the relationships between contemporary man and his environment. In his autobiography *Trails of an Artist-Naturalist,* Ernest Thompson Seton (1940) devoted an entire chapter to his single, successful moose hunt. Adolph Murie (1961) wrote exclusively of the moose in Chapter 9, "Picturesque Moose," in *A Naturalist in Alaska*. Conservationist Sigurd Olson frequently mentioned the moose in his *Runes of the North* (1963), a story set in the vast wilderness that sweeps from the Quetico/Superior and Hudson Bay to the Yukon and Alaska.

Many shorter works on the moose appeared in literary magazines and sporting journals of the late nineteenth century. Examples of the former include: *Moose Hunting* by Henry P. Wells (1887), a well-illustrated moose-hunting story; *Antoine's Moose-Yard* by Julian Ralph (1890), an account of Frederic Remington's moose hunt near Mattawa, on the Ottawa River, Canada; and *The Vanishing Moose, and Their Extermination in the Adirondacks* by Madison Grant (1894), a scholarly account of the demise of moose in New York, which did much to arouse public interest in the species and concern about its disappearance in the northeast.

William Wood was among the early authors of North America who interjected poetry among his prose. In listing the "Beasts that Live on the Land" in New England in 1634, Wood (in Vaughan 1977: 59) penned:

> "The kingly lion [panther?] and the strong-armed bear,
> The large-limbed mooses, with the tripping deer,
> Quill-darting porcupines, and racoons be
> Castled in the hollow of an aged tree. . . ."

Part of a poem by Arthur Wentworth Eaton (in Merrill 1916: 247–248) reads:

> "Glooskap it was who taught the use
> Of the bow and spear, and sent the moose
> Into the Indian hunter's hands;
> Glooskap who strewed the shining sands. . . ."

In *The Wilderness Hunter*, Theodore Roosevelt (1909: Preface) included a passage from Walt Whitman about a moose hunt:

> "Where winter wolves bark amid wastes of snow and
> ice-clad tree . . .
> The moose, large as an oxe, cornerd by hunters, plunging
> with his
> forefeet, the hooves as sharp as knives . . .
> The blazing fire at night, the sweet taste of supper, the talk,
> the bed
> of hemlock boughs, and the bear skin."

The moose also has figured in children's literature. Three excellent nonfiction books that accurately describe and illustrate the moose are *Moose* (Scott and Sweet 1981), *Denaki, an Alaskan Moose* (Berry 1965) and *Meet the Moose* (Rue and Owen 1985). Many fictional works featuring the moose for the preadolescent audience include *Blue Moose* and *Return of the Moose* (Pinkwater 1993), *Moose for Jessica* (Wakefield and Carrarra 1990), *Moose, Goose & Little Nobody* (Raskin 1974), *A Moose is Not a Mouse* (Berson 1975) and *Moose in the Garden* (Carlstrom 1990).

The moose—like ducks, mice, bears and coyotes—has been a boon to the entertainment industry and allied enterprises. The seeming ungainliness of moose, its peculiar physical features and its mishaps make it a popular caricature for youngsters' television shows and videos, perhaps best exemplified by "Bullwinkle," developed by Jay Ward and Bill Scott. Marty Stouffer's 1992 feature, "Magnificent as a Moose," focused on the species in Alaska and its interactions with Athabascan Indians.

Table 4. Some North American terms and colloquialisms derived from or associated with moose

Term or colloquialism	Definition	Source[a]
Bull Moose Party	Popular name for U.S. Progressive Party, 1912–1916	Adams 1940(I): 250
Moose beat	Moose winter area	*Knickerbocker* 1938(XII): 293
Moose berry	Squashberry	Franklin 1823: 88
Moose bird	Gray jay	Williamson 1832: 150
Moosebush	Hobblebush viburnum	Murray 1908(10): 684
Moose butter	Food made from boiled, pulverized moose bones and marrow	Denys 1672: 423
Moose call	Birch bark horn used to lure bull moose	Hodge 1907: 940
Moose cat	Anyone with great ability, strength or charisma (slang)	Rickaby 1926: 212
Moose disease or sickness	Parelaphostrongylosis	See Chapter 15
Moose dog	Domestic dog trained to hunt moose and grab them by the nose	Howells 1873: 52
Moose elm	Slippery elm	Hodge 1907: 940
Moose-face	A rich, ugly man	Matsell 1859: 56
Moose flower	Trillium	Cooper 1968: 73
Moose fly	*Lyperosiops alcis*	Audubon 1834(II): 437
Moosegrass	Beargrass	Thompson 1938: 86
Moose-ground	Moose habitat	*New York Mirror* October 28, 1837
Moosehead plant	"some river-side plant"	Murray 1908(10): 648
Moose horn	(see Moose call)	Hodge 1907: 940
Mooseling	Calf moose	Winthrop 1861: 55
Moose man	A moose hunter	Murray 1908(10): 648
Moose maple	Striped maple	Hoffman 1839: 38
Moose-misse	American mountain-ash	*American Folk-Lore* 1902(XV): 249
Moose pen	(see Moose beat)	Hammond 1854: 71
Moose plum	Various types of plum	Greenleaf 1829: 113
Moose suit	Moose skin outfit	Murray 1908(10): 648
Moose tick	Wood or winter tick	Peterson 1955: 32
Moose tree	Leatherwood	Greatrex 1854: 119
Moose warden	Conservation officer	Goode 1883: 83
Mooseweed	Pickerelweed	*Harper's Magazine* October 1881
Moosewood	(see Moose tree)	Bartlett 1859: 278
Moose woods	(see Moose-ground)	*Spirit of the Times* September 6, 1856
Moosey	Suggestive of the presence of moose	Thoreau 1864: 377
Moose yard	(see Moose beat)	Hoffman 1839: 58

[a] Sources are select, and neither the first nor only use of the terms identified.

Oddly, the moose does not appear in two important sagas of the American "north woods." It is absent from *The Song of Hiawatha* by Henry Wadsworth Longfellow (1855), and the fanciful tales of folk hero Paul Bunyan (Hoffman 1952, Stevens 1948).

The term "moose" has been widely represented in the North American lexicon, primarily as a nonpossesive noun modifier (Table 4). The neologistic constructions—many colloquial and antiquated—relate to characteristic behavioral or morphological association with the animal. The most prevalent contemporary idiomatic use of "moose" is as a descriptor for something or someone large and powerful. Slang among the U.S. military, "moose" refers to a young Japanese or Korean wife or mistress of a serviceman stationed in Japan or Korea. And "moose" has a place in the final frontier, in the form of Manned Orbital Operations Safety Equipment (MOOSE)—a space lifeboat for astronauts.

ARTWORK

It is only natural that the wild moose has been a favorite subject of many notable North American artists past and present (see McIntosh 1993). A selected list of moose artwork appears in Table 5.

On June 23 and 24, 1989, a complete set of illustrations from Audubon's *Quadrupeds . . .* was put up for auction in New York by Sotheby's (1989). Item 465, a sheet having both the moose and bighorn sheep illustrations had a pre-

Table 5. Sampling of artwork featuring North American moose

Artist	Title	Source
Douglas Allen	Untitled	O'Connor 1961: 135
John Woodhouse Audubon	"Moose Deer"	Audubon and Bachman 1845–1848: plate lxxvi
Robert Bateman	"Autumn Overture—Bull Moose"	Bateman and Derry 1981: 95
Carter Beard	"Head of Moose"	Roosevelt 1909: 214
Joe Beard	"Moose-birds"	Ward 1878: 459
Gisèle Benoît	"Rivalry"	McIntosh 1993: 124
Monique Benoît	"Twins—Moose calves"	McIntosh 1993: 126
Thomas Bewick	"The Elk"	Bewick 1807: 120
Albert Bierstadt	"Moose Hunters' Camp, Nova Scotia"	Elman 1972: plate 17
Jim Biesinger	"Challenged"	McIntosh 1993: 125
Allan Brooks	"Moose"	Banfield 1974: plate 38
George De Forest Brush	"Killing the Moose"	Brush 1892: 600
A.B. Frost	"The Moose at Bay"	Farnham 1884: 395
Louis Agassiz Fuertes	"Moose"	Nelson 1916: 462
Philip R. Goodwin	"Unexpected Game"	Elman 1972: plate 44
Campbell Hardy	"Moose Hunting"	Hardy 1855: frontispiece

continued on next page

Table 5 continued

Artist	Title	Source
Robert Hood	"Doe Moose Pasquia Hills 3 April 1820"	Houston 1974: plate 9
Francis Lee Jaques	"Moose in Swamp"	Jaques 1973: 175
Paul Krapf	"Moose Country"	*Wildlife Art News* 1993(6): 181
Bob Kuhn	"Bull Moose"	Bashline and Saults 1976: 114
Thomas Landreen	"Bull Moose"	Kastner 1991: 63
Robert McNamara	"Bristling Intensity—Cow and Calf Moose"	McIntosh 1993: 126
Wayne Meineke	"Moose"	McIntosh 1993: 122
John Frederick Miller	"Moose"	Dance 1978: 180
Daniel P. Metz	"Unchallenged"	Johnston 1995: 69
R.A. Muller	"Head of a Bull Moose"	Wells 1887: 449
Ron Parker	"Autumn Foraging"	Hume 1987: 66
Titian Ramsay Peale	Untitled sketches	Files of American Philosophical Society, Philadelphia
Frederic Remington	"A Moose Bull Fight"	Ralph 1890: 650
Judi E. Rideout	"Early Snow"	McIntosh 1993: 125
Carl Rungius	"The American Moose, in New Brunswick"	Hornaday 1910: frontispiece
Charles Russell	"Caught Napping"	Murray 1984: 24
Henry Sandham	"A Moose Fight"	Ward 1878: 449
Frank Schoonover	"Moose on Track"	Schoonover 1976: 94
Charles W. Schwartz	Untitled bull and cow	Schmidt and Gilbert 1978: 66
Alexander Seidel	"Bull Moose"	O'Connor 1961: 142
Ernest Thompson Seton	"Moose Family in Early Winter"	Seton 1909(I): 145
Ray Sexton	"Eyeing Up—Moose"	McIntosh 1993: 121
J. Smit	"Moose"	Ferris 1897: 930
Sharon Sommers	"Crown Royal"	McIntosh 1993: 123
Walter A. Weber	"Bull Moose"	Cahalane and Weber 1946: 20
K. Wolf	"A Moose Family"	Ward 1878: 451
Fredrick Arthur Verner	"Lake, North of Lake Superior"	Murray 1984: 46
N.C. Wyeth	"The Moose Hunter"	Wyeth 1912: 48

auction estimated value of $1,400 to $1,800. The print actually fetched $4,400.

Superb photography also can be superb art, as exemplified by Michio Hoshinso's photographic essay *Moose* (1988). *Booklist* highly praised the work resulting from this Japanese photographer's pursuit of the moose in Alaska over 8 years. Moose photos by such professional photographers as William Ruth, Johnny Johnson, Leonard Lee Rue III, Tom Walker, Judd Cooney, Michael Francis and others have graced many a magazine cover and feature.

Perhaps the most widely viewed sculptures of moose are the pair designed by Alexander Phiminster Proctor (1862–1950), born in Ontario and raised in the American West (Armstrong et al. 1976). As a little-known artist then, he sculpted 37 life-sized denizens of the American wilderness in bronze for the World's Columbian Exposition held in Chicago during 1893 to mark the quadricentennial of Columbus' "discovery" of America (Buel 1894). Proctor's pair of moose (one animal is visible in the lower right) welcomed visitors onto the east steps of the bridge across the South Lagoon leading to the Exposition's ornate Administration Building. A pair of Proctor's bison guarded the bridge's west end, and to the distant right is a pair of elk. The World's Columbian Exposition was America's bold, ostentatious notice that the nation was an emerging world power. Nearly 20.5 million people visited the 686-acre (278 ha), $28 million "City of Palaces" situated on the shore of Lake Michigan (Badger 1979). For most, crossing the bridge would be their closest encounter with a "moose." *Photo courtesy of the Smithsonian Institution Museum of American History.*

German-born Carl Rungius (1869–1959) is widely acknowledged as the preeminent portrayer of western big game, but also as an artistic master of the continent's wilderness. When Rungius first visited America in 1894, he failed to bag a moose while hunting in Maine that autumn. "If Carl had taken a moose, he might have returned happily to Germany," wrote Whyte and Hart (1985: 13). Instead, an uncle persuaded him to remain for another chance the following year. Within a year, the budding wildlife artist's fascination with and affinity for the West ultimately provided focus and dedication to his life's work in North America and to his remarkable legacy. Moose frequently were the subject of Rungius' artwork, perhaps because they "symbolized for Europeans the spirit of wilderness" (Whyte and Hart 1985: 48). Although greatly admired and appreciated, including by such contemporary notables as Theodore Roosevelt, William T. Hornaday and Frederic Remington for his illustrations and paintings of all big game, Rungius was deemed by some as "the Rembrandt of the moose." When asked of his favorite subject, he playfully responded, "Mooses in spruces." Of Rungius, preeminent wildlife artist Robert Bateman averred: "In his chosen genre, his artistic accomplishments have never been surpassed" (Whyte and Hart 1985: 1). Top left is Rungius in his Greenpoint (Brooklyn, New York) studio about 1900. Top right is the artist on a moose hunting trip in New Brunswick in 1905. Bottom scene is Rungius' "Zero Weather," a 1908 oil on canvas 24 by 32 inches (61 by 81 cm). *Photos courtesy of the Glenbow Museum Archives, Calgary, Alberta.*

For 50 years, 1857 to 1907, the firm of Currier and Ives of New York City produced unknown numbers of colored stone lithographic prints of more than 7,000 images. The inexpensive prints typically were sentimental, idealistic, idyllic or heroic views of North America and its evolving history. Some prints featured wildlife, usually dramatic action scenes in "sporting" adventure or predation. The date of "Moose and Wolves: A Narrow Escape" is unknown, as is the name of the original artist. For certain, the artist was not one of the naturalist–illustrators occasionally employed or copied, because the moose has elk-type antlers and a cow bison head and body. According to Kipp (1991), the estimated retail value of this particular print in 1991 was $305. *Photo courtesy of the Library of Congress.*

"Moose Deer of North America (17 Hands High)" is an aquatint painting by George Heriot (1759–1839). It is especially remarkable for featuring a cow moose, as opposed to a heavy-antlered bull generally selected as the principal subject by illustrators and fine artists of the time and since. *Photo (C12782) courtesy of the National Archives of Canada.*

SYMBOLISM

The moose as a feature on the Hudson's Bay Company's official seal is an enigma of royal significance. The beaver and moose were lifeblood of the Hudson's Bay Company in its exploitative heyday; the former provided profits, while the latter made its very existence possible in much of boreal British America. Although the beaver (and fox) fittingly appear on the Hudson's Bay Company coat-of-arms, the two animals flanking and supporting the shield and crest were, for nearly three centuries, elk, not moose.

Furthermore, a provision of the Hudson's Bay Company charter called for the payment "of two Elkes and two Black Beavers whensoever and as often as wee our heires and Successors shall happen to enter upon the said Countryes Territoryes and Regions hereby granted" (Anonymous 1970). Although elk did occur in parts of the Hudson's Bay Company's domain, it was an unimportant trade item and a minor food source.

The enigma must be examined in the context of May 20, 1670, when King Charles II granted a royal charter to the "Governor and Company of Adventurers of England Trading into Hudson Bay." Then, and even today, the European counterpart of the North American moose was called "elk." Of course, the drafting of the charter armorial design was done in England. For nearly three centuries, this problem of "elk or moose" was studiously and discretely ignored, and British sovereigns perfunctorily received their "rents" of two "elkes" whenever they visited Canada.

However, as early as 1927, two obvious moose adorned the sculpted company seal over the entrance to Beaver House, the Company's London headquarters (Traquair 1945). And the Hudson's Bay Company seal shown on the cover of *The Beaver* issue featuring the 275th anniversary of the company in 1945 also featured two bull moose. Furthermore, the coat-of-arms of the Hudson Bay Company was redesigned in 1962, and bull moose were installed as the new crest and shield supporters. Rechartering of the Hudson's Bay Company in 1970 under Canadian law removed the "rental" provision and the persistent moose/elk identity problem.

While admiring the Battle of Britain window in the Royal Air Force Memorial Chapel, Westminster Abbey in June 1992, the senior author was astonished to see the head of a bull moose portrayed on one of the small stained glass panes commemorating each of the various participating squadrons. An on-duty warden advised that the particular badge honored a Royal Canadian Air Force squadron that participated in the monumental air battle. A further inquiry to the Royal Air Force Museum, Herndon, brought a response from Flight Lt. F.W. Waters, giving details about that squadron, No. 242; and another RCAF squadron, No. 419. Each used a moose as the central feature of its insignia, and both squadrons were staffed initially with Canadian pilots.

The initial coat-of-arms of the Hudson's Bay Company, chartered on May 2, 1670 by King Charles II, had two rampant (standing) elk supporting a shield featuring four beavers, and a fox atop a crown. This seal was in use for three centuries. The beaver and fox represented commodities of trade, but the elk was relatively unimportant in the company's history. Apparently, the moose was intended in place of elk, because their hides were a significant trade item and moose meat was an invaluable food for company employees. However, the counterpart of the North American moose in Europe was called "elk," and through confusion unrecorded, the North American elk rather than the North American moose was designated by London armorists. The error was finally corrected, and moose (shown above) now serve as the shield supporters (Hobusch 1980: 172). *Pro pelle cutem* translates to "A Skin for a Skin"—an apt descriptor of the company's enterprise and indicator of the price that its employees paid in the harsh northlands of the New World. *Photo courtesy of the Hudson's Bay Company Archives.*

Rather than being a conservation organization for the welfare of the moose, Moose International, together with its auxiliary, Women of the Moose, is one of the world's largest international benevolent and social organizations (Anonymous 1988). Its 1.8 million members are located in the United States, Canada and Great Britain, and organized into 2,300 Moose Lodges, directed from its headquarters in Mooseheart, Illinois. Appropriately, its symbol is the silhouette of a bull moose. The rationale for selection of moose as the organization's figurehead was given by K.N. Wehrmeis-

The Hudson's Bay Company's Royal Charter of May 2, 1670, granted by Charles II, specified that whenever an English sovereign visited its domain, the "Governor and Company of Adventurers Trading into Hudson's Bay," would pay rent of "two Elkes and two Black Beavers." King George VI dutifully accepted payment in 1939 (top), as did Queen Elizabeth II in 1959 (bottom left) and again in 1970 (bottom right). The payment of "elkes" undoubtedly was drafted with the intent of moose, per the European species, but the terms were abided by literally, as evidenced by the North American elk antlers on display at each ceremony. The confusing and potentially embarrassing rental provision was eliminated with the rechartering of the Hudson's Bay Company in Canada in 1970 on the tricentennial of its founding. *Photos courtesy of the Hudson's Bay Company Archives.*

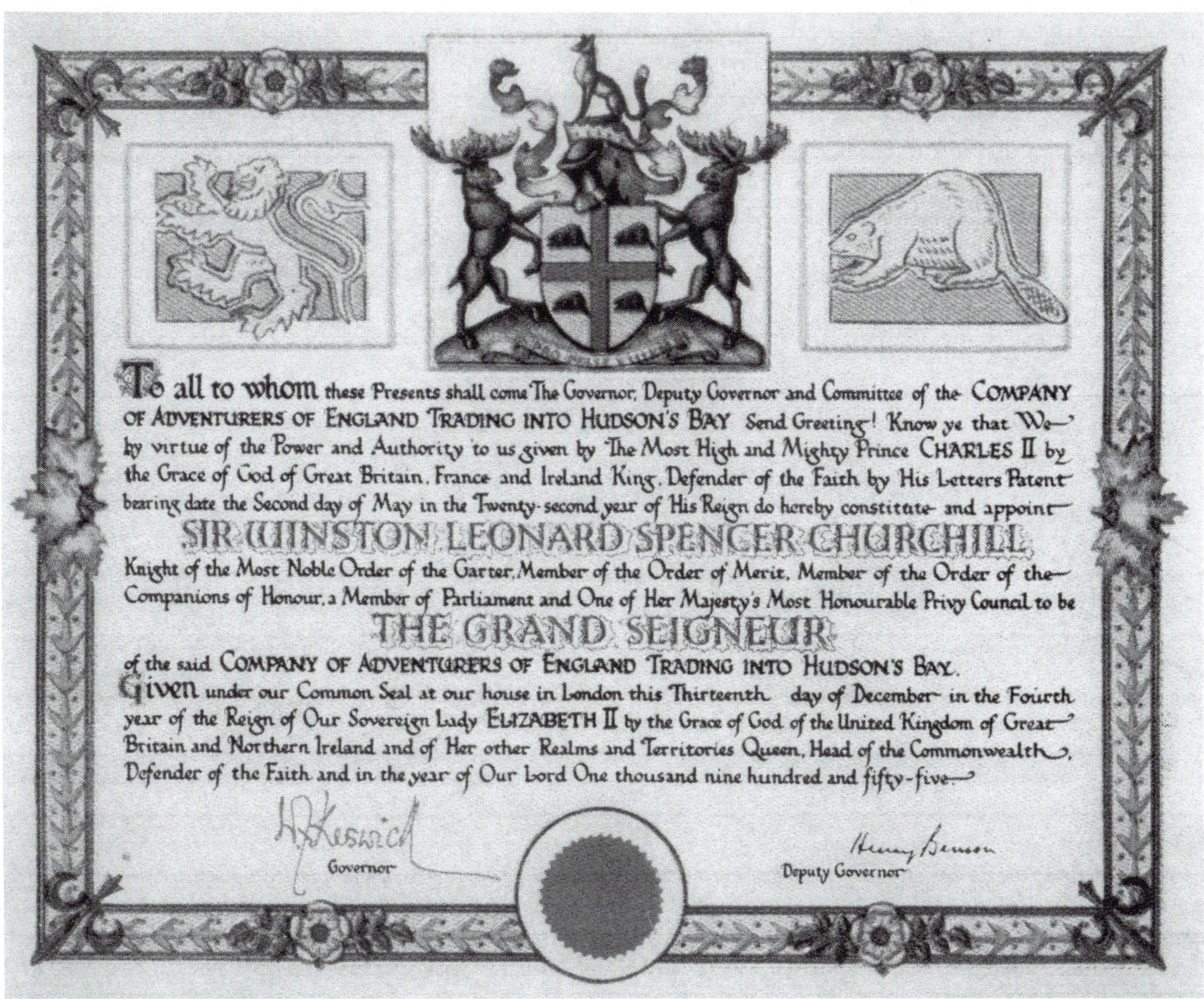

To all to whom these Presents shall come The Governor, Deputy Governor and Committee of the COMPANY OF ADVENTURERS OF ENGLAND TRADING INTO HUDSON'S BAY Send Greeting! Know ye that We by virtue of the Power and Authority to us given by The Most High and Mighty Prince CHARLES II by the Grace of God of Great Britain, France and Ireland King, Defender of the Faith by His Letters Patent bearing date the Second day of May in the Twenty-second year of His Reign do hereby constitute and appoint

SIR WINSTON LEONARD SPENCER CHURCHILL

Knight of the Most Noble Order of the Garter, Member of the Order of Merit, Member of the Order of the Companions of Honour, a Member of Parliament and One of Her Majesty's Most Honourable Privy Council to be

THE GRAND SEIGNEUR

of the said COMPANY OF ADVENTURERS OF ENGLAND TRADING INTO HUDSON'S BAY.

Given under our Common Seal at our house in London this Thirteenth day of December in the Fourth year of the Reign of Our Sovereign Lady ELIZABETH II by the Grace of God of the United Kingdom of Great Britain and Northern Ireland and of Her other Realms and Territories Queen, Head of the Commonwealth, Defender of the Faith and in the year of Our Lord One thousand nine hundred and fifty-five

Governor

Deputy Governor

In 1955, Sir Winston Churchill was appointed the Hudson's Bay Company's first Grand Seigneur—the highest honor the Company could bestow—thereby continuing a long and celebrated association with the Churchill family. Lord John Churchill, First Duke of Marlborough (often known simply as Marlborough), succeeded the Duke of York as Governor of the Hudson's Bay Company in 1685 when the latter became King James II. Lord John's tenure was marked by prosperity for the Company, which presented him in 1690 with a gold plate "for his great care and troubles in their concernes" (Anonymous 1956). Largely through his resounding victories on the North American continent was Britain able to regain control of Hudson Bay in 1714. Fort Churchill and the Churchill River were named for Lord John. Just before VE Day in 1945, Sir Winston wrote the governor of the Hudson's Bay Company on the occasion of the Company's 275th anniversary: "Its life has been filled with achievements of which you may be justly proud. Yours is a fine record of enterprise in opening the vast territories of North America and in serving their people." The deed appointing Sir Winston as Grand Seigneur is marked with the rampant lion and fleur-de-lys from the Spencer/Churchill coat-of-arms, whereas the Hudson's Bay Company seal—marked by two rampant moose, four beaver and a fox—top the document. *Photo courtesy of the Hudson's Bay Company Archives.*

ter (personal communication: 1993): "Strong and majestic, the moose is unique in all the animal kingdom. . . . He takes only what he needs, nothing more. He loves freedom and is the master of his domain. Yet for his great size and strength he lives in peace with other creatures. The moose uses his size and power not to dominate but to protect, not to spoil but to preserve. He is a fierce protector, a loyal companion, and a generous provider who brings comfort and security to those within his vision. . . ."

The moose has served as a symbol in many other ways. For example, it was featured for many years on the masthead of *Forest and Stream* magazine, an influential American sporting journal now defunct. Several ships have been named for the moose. An Hudson's Bay Company sloop

Insignias of the two Royal Canadian Air Force "Moose squadrons" that participated in World War II. Hawker Hurricanes of Squadron 242 (left) helped protect the astonishingly successful British evacuation at Dunkirk in June 1940 and participated in the Battle of Britain. Its Spitfires fought in North Africa. The squadron's last assignment was the Berlin airlift in 1949. The nickname of Squadron 419 (right) originated with its first commanding officer, Wing Commander John "Moose" Fulton, a native of Kamloops, British Columbia. Over more than 3 years, "Moose" bomber squadron logged 4,325 sorties, mostly over Germany. It suffered huge casualties, losing 310 airmen killed, 867 missing, 187 taken prisoner and 129 aircraft lost. *Photos reproduced courtesy of the Directorate of History, Canadian National Defence Headquarters.*

carried the name (Davies and Johnson 1963), as have two U.S. Navy vessels. The first, a wooden, sternwheel gunboat served illustriously on the Mississippi River during the Civil War; the second, *Moose* (IX-124), was used in the South Pacific Ocean during World War II as a fuel-storage vessel (U.S. Navy Department 1969).

Moose frequently have been used symbolically in advertising. The head of a bull moose also is the logo for a Canadian beer, Moosehead, bottled in New Brunswick and Nova Scotia by Moosehead Breweries, "Canada's Oldest Independent Brewery." A bull moose appears in the center of an advertisement by Komatsu, a Japanese manufacturer of earthmoving equipment, with the rhetorical question, "What's your view of the world? . . . Do you see life as it is, or as it can be?" (*Wall Street Journal*, March 19, 1993). The moose, presumably representative of undeveloped wilderness, is ominously surrounded by symbols of civilization—well-manicured farms, houses, highways and a city. More positively, moose antlers appear in another advertisement for airline flights bound for Anchorage, Alaska.

A popular television show in the mid-1990s was Columbia Broadcasting System's "Northern Exposure," about the people and life in fictitious Cicely, Alaska. The opening and credit segments of the hour-long show featured a young moose wandering through town, offering the audience comic and intrigue and, symbolically, a sense of the town's remote setting and wilderness ambiance.

NOMENCLATURE

The North American moose has been given many names by various peoples over the years (Table 6). The name "moose" supposedly originates from one or more Indian names: *moos* (Narragansett), *mus* (Abenaki) and/or *muns* (Penobscot) (Murray 1908: 648). Barnhart (1988: 676) indi-

The former logo (left) of the Loyal Order of Moose, and the current logo of the International Order of Moose, an international fraternal and benevolent organization. *Photos courtesy of the International Order of Moose.*

Table 6. Some historic vernacular and aboriginal names for moose in North America[a]

Language	Names	Source
English	American black elk	Smith *in* Audubon and Bachman 1845–1848: 232
	Black elk	De Kay 1842(1): 116
	Black moose	Dudley 1721: 165, Franklin 1823: 665, Griffith 1827(4): 73, Pennant 1784: 18, Seton 1929(3): 153
	Bugle[b]	Roberval *in* Biggar 1924: 268
	Elk	Audubon and Bachman 1845–1848: 232, Charlevoix 1761: 199, Denys 1672: 109, Dièreville 1708: 129, Graham *in* Williams 1969: 16, Lahontan 1703(1): 1705, Grant 1911–1914(III): 220
	Elk deer	Isham *in* Rich and Johnson 1949: 154
	Elke	Josselyn *in* Lindholdt 1988: 88, Morton 1632: 74
	Flat-horned elk	De Kay 1842(1): 116, Merrill 1916: 243, Seton 1909(1): 144, Seton 1929(3): 153
	Forest oxen	Vanderdonck *in* DeKay 1842: 116
	Molke	Higginson 1929: 310
	Moos	Hudson's Bay Company records *in* Rich 1946: 35, Lewis *in* Moulton 1986–1996(3): 432, Smith 1616: 342
	Moose	Catesby 1743(2): xxvii, Denys 1672: 109, Grant 1911–1914(III): 220, Merrill 1916: 243, Sagard 1632: 224, Seton 1909(I): 144, Rich 1945: 68, Rich and Johnson 1957: 58
	Moose deer	Dudley *in* Audubon and Bachman 1845–1848: 232, Franklin 1823: 665, Harlan 1825: 229, Hood *in* Houston 1974: 44, Lewis *in* Moulton 1986–1996(3): 432, Spry 1968: 181, Pennant 1784: 16, Richardson 1829(1): 232
	Moose-deer	Graham *in* Williams 1969: 16, Merrill 1916: 243, Spry 1968: 180
	Mose	Lewis *in* Moulton 1986–1996(3): 433, Morton 1632: 74, Oldmixon 1741: 187
	Mosse	Gorges *in* Purchas 1625(XIX): 281
	Mouse	Hudson's Bay Company records *in* Rich 1942: 16, 1945: 67
	Stag	DeMonts *in* Audubon and Bachman 1845–1848: 232
	Stagg	De Monts *in* Richardson 1829(1): 232
	Tree topper	Caton 1877: 72
French	Elan[c]	DeMonts *in* Audubon and Bachman 1845–1848: 232, Dièreville 1708: 129
	Ellan	De Monts *in* Richardson 1829(1): 232
	Eslan ou orinal	Sagard-Theodat *in* Audubon and Bachman 1845–1848: 232, Richardson 1829(1): 232, Sagard 1632: 749
	Horiniack	Kellogg 1917: 58
	l'elan	Catesby 1743(II): xxiv, Ganong 1910: 274
	l'orignac	Seton 1929(3): 153

continued on next page

The provincial coat-of-arms of Ontario (top) and the official seals of the states of Maine (center) and Michigan (bottom) feature moose. In the Maine seal, a moose rests at the base of a pine tree. *Photos courtesy of the respective jurisdictions.*

Table 6 continued

Language	Names	Source
	L'Orignal	Pennant 1784: 18, Seton 1909(I): 144, 1929(3): 153
	l'orignat	Seton 1909(I): 147
	L'Orinal	Martyn *in* Audubon and Bachman 1845–1848: 236
	Original[d]	Merrill 1916: 238, Harlan 1825: 229, Kalm 1770: 157, Maximilian 1843: 203
	Orignac	Denys 1672: 382, Dièreville 1708: 129, Ganong 1910: 274, Grant 1911–1914(III): 220, Merrill 1916: 243, Richardson 1829(1): 232
	Orignal	Charlevoix 1761: 197, Dièreville 1708: 129, Lahontan 1703(1): 104, Grant 1911-1914(III): 220, Merrill 1916: 243, Richardson 1829(1): 232, Sagard 1632: 224, Seton 1929(3): 160
	Orignald	Lahontan 1703(1): 55
	Orignas	Dièreville 1708: 129
	Orignat	Dièreville 1708: 129, Merrill 1916: 243, Seton 1929(2): 160
	Orignaux	Rasles *in* Thwaites 1900b: 211
	Origniac	Ganong 1910: 275
	Orinac	Dièreville 1708: 129
	Orinal	La Hontan [*sic*], Charlevoix *in* Audubon and Bachman 1845–1848: 232
	Oriniack	Radisson *in* Adams 1961: 95
	Oriniak	Radisson *in* Draper and Thwaites 1903–1915: 71
Dutch	Eelenden	Montanus *in* O'Callaghan 1849–1851(IV): 75
Italian	Gran bestie	Bressani *in* Thwaites 1899b: 112
Russian	Los	Zagoskin *in* Michael 1967: 311
Aboriginal		
Abenaki	*Aianbe* (male)	Rasles *in* Thoreau 1864: 110
"Minnitarris (Grosventres)"	*Apatapa*	Maximilian 1843: 274
Micmac	*Aptaptou*	DeMonts *in* Audubon and Bachman 1845–1848: 232, Grant 1911–1914(III): 220, Richardson 1829(1): 232
Slave	*Cikittisso*	Henry and Thompson *in* Coues 1897: 535
Slavey	*Co-lö*^{n}	Russell 1898: 229
Kutchin	*Dinjik*	Mueller 1964: 26
Chipeweyan	*Dinyai*	Mackenzie 1801: 157
Abenaki	*Hèrar* (female)	Rasles *in* Thoreau 1864: 110
Slave	*Ko'lg*	Honigmann 1946: 155
Unidentified	*Mong-soa*	Seton 1909(I): 147, 1929(3): 160
Unidentified	*Mongswa*	Seton 1909(I): 147, 1929(3): 160
"Ojibuas, Ojibeuas, Chipewas, or Algonkins"	*Mons*	Maximilian 1843: 227, Merrill 1916: 237, J. Long *in* Thwaites 1904–1907 (II): 243

Table 6 continued

Language	Names	Source
Chippewa	*Mo^{n}s*	Hodge 1907: 940
Menominee	*Mō^{n}s*	Hoffman 1896: 42
Algonquin	*Monse*	Mackenzie 1801: 142
Algonquin	*Monsoll*	Harlan 1825: 232
Milecite	*Moosu*	Hardy 1855: 194
Cree	*Monswa*	Hodge 1907: 940, Merrill 1916: 237
Abenaki	*Monz*	Hodge 1907: 940, Merrill 1916: 237
Algonquin, Massachusetts and Narraganset	*Moos*	Hodge 1907: 940, Merrill 1916: 237, Murray 1908(10): 648
Cree	*Moose*	Seton 1909(I): 144, 1929(3): 153
"Ojib"	*Moose*	Seton 1909(I): 144
Ojibway	*Moose*	Seton 1929(3): 153
Montagnais	*Moosh*	Merrill 1916: 237
Cree	*Moosöă*	Richardson 1829(1): 232
Unidentified	*Moòs-sóog*	Trumbull 1963: 120
Unidentified	*Moosu*	Graham *in* Williams 1969: 16
Cree	*Mooswa*	Hood *in* Houston 1974: 172, Russell 1898: 13
Cree	*Moos-wa*	Seton 1909(I): 144
Kinistineau (Cree)	*Mooswah*	Henry and Thompson *in* Coues 1897: 535
Cree	*Mooswu*	Hood *in* Houston 1974: 172
Mohegan	*Mooth*	Sterling 1974
Chippewa	*Moouse*	J. Long *in* Thwaites 1904–1907(II): 243
Unidentified	*Mooze*	Murray 1908(10): 648
Delaware	*Mos*	Hodge 1907: 940, Merrill 1916: 237
Unidentified	*Mose*	Merrill 1916: 14, Morton 1632: 74, Murray 1908(10): 648
Cree	*Moswa*	Hood *in* Houston 1974: 172
Unidentified	*Mosse*	Murray 1908(10): 648
Ojibwa	*Mous*	Warren 1884: 45
Ojibway	*Mouse*	Henry and Thompson *in* Coues 1897: 535
Knisteneaux (Cree)	*Mouswah*	Mackenzie 1801: 142
Ojibway	*Mouze*	Henry and Thompson *in* Coues 1897: 535
Penobscot	*Muns*	Murray 1908(10): 648
Abenaki	*Mus*	Murray 1908(10): 648
Passamaquoddy	*Mus.*	Hodge 1907: 940, Merrill 1916: 237
Algonkin	*Musu*	Pennant 1784: 18
Cree	*Mus-wa*	Russell 1898: 229
Mandan	*Páhchub-ptptá*	Maximilian 1843(3): 248
Blackfoot	*Sikitisuh*	Maximilian 1843(3): 220
Huron	*Sondareinta*	Sagard 1632: 225
Sioux	*Ta*	Seton 1929(3): 153
Assiniboine	*Tah*	Henry and Thompson *in* Coues 1897: 535
Ogallala Sioux	*Tah*	Seton 1909(I): 144
Yankton Sioux	*Tahg-chah*	Seton 1909(I): 144
Koyukon	*Tanaiger*	Whymper 1869: 321
Micmac	*Teäm* (male)	Grant 1911–1914(III): 220
	Teeam	Hardy 1855: 194
Midnoósky (Ahtna)	*Tenayga*	U.S. Senate 1900
Dog Rib	*Tĕn-di*	Russell 1898: 229
Chipewyan	*Ten-neé*	Seton 1909(I): 144
Loucheux	*Tĭ^n-gik*	Russell 1898: 229
Kotch-a-kutchin	*Tĭn-jī-yuk'*	Whymper 1869: 326
Inkilik (Ingalik)	*Ttanika*	Zagoskin *in* Michael 1967
Slave	*Ulon*	Honigmann 1946: 155
Arikara	*Wah-suchárut*	Maximilian 1843(3): 211

[a] Likely some names originate from misspellings, particularly for aboriginal names within the same tribes that reflect differing phonetic spellings, e.g., "The differences in spelling in the various dialects are partially explainable perhaps by the fact that the Indians employed a sound which cannot be closely indicated by the letters of the English alphabet" (Merrill 1916: 237). Furthermore, some English-speaking authors, such as Lewis and Clark, did not spell consistently. Translators and editors may have introduced other errors. Tribal names are as given by author.

[b] Moose presumed.

[c] French name for the European elk, but also used to designate the North American moose.

[d] Supposedly from "Orenac," a Basque word for deer.

cated that "moose" apparently was from the Algonquian *moosu*, meaning "he strips off [young tree bark]." Also Merrill (1916) devoted an entire chapter (XI) to names given the moose (or European elk) in the New and Old Worlds (see also Dudley 1721, Dale 1736).

PLACENAMES

Early conservationist Madison Grant (1894: 345) noted that moose and caribou "shrink back before the most advanced outpost of civilization, and soon vanish altogether, leaving behind the names of lakes, rivers, and mountains as the only evidence of their existence." Indeed, hundreds of natural geological features and human geographical sites scattered throughout much of forested boreal North America bear the stamp "moose."

The popularity and inspiration of "moose" as a placename can be demonstrated in such widely separated places as Maine and Alaska. The official Geographic Names Information System (GNIS) maintained by the U.S. Geological Survey (1985) showed 79 Maine locational names based on "moose," "moosehead" and "moosehorn," including: Moose Brook (10); Moose Cove (4); Moose Hill (4); Moose

Island (6); Moose Mountain (5); Moose Pond (16); Moose Pond Brook (2); Moose River (3); and Moosehorn Stream (2). Mooselookmeguntic is from an Indian word for "where the hunters watch the moose by night" (Douglas-Lithgow 1909). Cape Rosier, in Penobscot Bay of coastal Maine, formerly was called Moosecajik, from an "ancient Indian name" meaning "the moose's rump" (Merrill 1916: 55). Others, such as Mooseduk and Mooseleuk, are of less certain origin. The GNIS list for Alaska (U.S. Geological Survey 1979) has 81 such names including: Moose Creek (47); Moose Island (2); Moose Lake (7); Moose Pass (2); Moose Point (2); and Moosehorn Lake (2).

Cities in other provinces and states include Moosonee, L'Orignal and Moosehill, Ontario; Moosehorn, Manitoba; Moose, Wyoming; Moosomin, Saskatchewan; and Moosup, Connecticut. Moose Jaw, in southwestern Saskatchewan and situated on Moosejaw Creek, supposedly got its name because it is "where white man mended cart with jawbone of moose." Also, Moosehide was the name of a Klondike gold camp along the Yukon.

The moose was the centerpiece of Daniel Beard's illustration "Evicted Tenants of the Adirondacks" featured in an 1885 issue of *Harper's Weekly*. Even earlier, in his journal for March 23, 1856, Henry David Thoreau lamented the "sweeping" landscape alterations in slightly more than 200 years of European settlement of North America: "When I consider that the nobler animals have been eliminated here,—the cougar, panther, lynx, wolverine, wolf, bear, moose, deer, the beaver, the turkey, etc., etc.,—I cannot feel as if I lived in a tamed, and, as it were, emasculated country" (Torrey 1906: 220–221). *Photo courtesy of the Library of Congress.*

Postscript

In much of northern forested America, human existence essentially depended on the moose or the caribou. The vastness and diversity of the northern forests ensured that the annual productivity of moose far exceeded the inroads made on it by Natives seeking food and other uses. The wary moose was readily vulnerable only during certain seasons or conditions. In particular, the unavailability of moose because of light snowfall periodically threatened aboriginal survival from one year to the next. During most winters, however, it was the moose that often tided tribes over to times of food abundance.

Quite understandably, the moose assumed an extremely important, if not paramount role in the cultures of many northern Natives. Its influence extended into almost every aspect of Indian life and society. At least until 400 years ago, Indians apparently used the gift of moose wisely and respectfully.

This changed when Europeans intruded upon the North American scene. These were people of various national, religious and social origins, who had no such compunction against exploiting moose for personal, imperial and commercial purposes. Their ambition and technology greatly altered the moose's environment—only rarely with long-term benefit to either the moose or its environment. Within the niggling span of less than three centuries, moose vanished or became scarce within the reach and foresight of the tide of new civilization.

Wrote Englishman, the Earl of Dunraven (1876: 27), who visited and toured North America in the 1870s: "Poor *Cervus Alces*! your ungainly form has an old-time look about it; your very appearance seems out of keeping with the present day. The smoke of the chimney, the sound of the axe, are surely though slowly encroaching on your wild domains. The atmosphere of civilization is death to you, and in spite of your exquisitely keen senses of smell and hearing you too will soon have to be placed in the category of things that have been."

The Earl's epitaph for moose narrated similar forecasts by others for the species before and briefly after the turn of the century. Thoreau, Grant, George Bird Grinnell, Roosevelt, and William T. Hornaday all agreed that the moose was a vanishing resource. By the early 1900s, the lament was loud and sufficiently summary that it found manifestation in exploitation restraint. It also contributed to the interest and enthusiasm of certain sportsmen and a few others to conceive, formulate and implement the new science of wildlife management, an achievement of which has been the recovery of moose throughout most of its prehistoric North American range.

OCT. 9
YOUNG BULL -
DEEP CHEST, FAIRLY STRAIGHT LINE TO BELLY - ADDS TO HEAVY-SHOULDERED LOOK

SNOWING LIGHTLY - BULL HAD LAYER OF SNOW ON ANTLERS & SNOW CLINGING TO TOP PLANES OF BACK

(H) - DARK AROUND EYE, EXTENDING BELOW EYE; DARK LINE EXTENDING DIAGONALLY BACK OF EYE - LOWER PART OF HEAD & MUZZLE DARK. LIGHTER AREA IMMEDIATELY AROUND & IN FRONT OF EYE - LIGHT NOSTRIL
DARK MANE -

OCT. 9, MCKINLEY PARK - SAME GENERAL AREA AS BULL AT 6 -

INSIDE OF E[ARS] LIGHT, GRE[...] CONSPICUOUS PATTERN -
EARS EXTEND ABOVE BEAM OF ANTLERS -

(MEMORY)

SOLITARY BULL FEEDING IN WILLOWS - O[...] ONLY PALMS OF ANTLERS VISIBLE, M[...] ALONG TH[E] BRUSH -

WILLOW FOLIAGE S[...] CLINGING T[...] MANY BRA[...] NOW FAIN[...] BUSH, O[...]

SCATTERED SPRUCE - DWARF BI[...] WITH FW [...]

(MEMORY)

BIG BULL - FAIRLY LIGHT COLORATION - LIGHT BROWN WASH OVER SHOULDERS & BACK, DARK (BLACK APPEARING) UNDERPARTS - LIGHT LEG - ANTLERS CHIPPED - ONE TINE SNAPPED OFF - HAIR ON RUMP BUNCHED - WET (SNOWING). SLOW, PONDEROUS GAIT - HEAD LOW.

(LIFE & MEMORY)
COW MOOSE STRADDLING FOR FOOD UNDER SNOW -
JACKSON HOLE, WYO. APRIL 6, '56

LIGHT GRAY "STOCKINGS", LIGHT NOSTRIL & UPPER LIP - DARKEST ON UNDERPARTS - DARK BROWN ABOVE

2

ANTHONY B. BUBENIK

Evolution, Taxonomy and Morphophysiology

Because adequate fossil records are lacking in both time and space, tracing the evolution of a species such as the moose up through its relatives and ultimately to its recent ancestors is difficult. About 40 million years ago, the early ancestors of today's ruminants existed in both North America and Eurasia. The North American forms known as pseudodeer (*Merycodontinae*) (A.B. Bubenik 1990) possessed live bony appendages that, after some unknown period, shed their covering of presumably keratinized skin. These appendages probably were the closest approximations to antlers as they are presently known. However, the pseudodeer did not take the next evolutionary step of actively casting the dead bony parts or of regrowing replacements (Frick 1937, Goss 1983, A.B. Bubenik 1990).

All extinct and living deer probably evolved from Eurasiatic stocks known as protodeer (Dicroceridae). The first true Cervidae appeared in the early Miocene of Eurasia, about 20 million years before present (B.P.) (Clutton-Brock et al. 1982). From then and there, their descendants migrated to Europe and the Americas.

Fossil Records

Almost 2 million years ago, during the late Pliocene to middle Pleistocene, the ancestors of the current *Alces* evolved. During this period, they grew larger, their necks shortened and the vertebrae enlarged. Their hypsodont (high crown) dentition changed to brachyodont (low crown) form, but the skulls of the ancient and extinct broad-fronted moose *Libralces* and the tripalmated stag moose *Cervalces* still had long nasal bones as did other deer that articulated with the intermaxillae. Their skulls were broad and high compared with those of today's moose. These early ancestors culminated in body and antler size during the glacial Pleistocene (Frenzel 1967), but only *Alces* survived.

During the Pleistocene, there was a period of accelerated evolution when many bizarre giants occurred in various mammalian families (Reynolds 1929, Jaczewski 1980, Sher 1987). Among them was the giant stag moose *Cervalces scotti*, which lived in the late Pleistocene and probably migrated to North America over the Bering Strait between 100,000 and 200,000 years ago (Figure 10). These moose remained in the periglacial tundra of northwestern Alaska until the appearance of the corridor between the Cordilleran and Laurentide ice shields about 40,000 years B.P. (Bryson and Wendland 1967, Rutter 1980, Catto and Mandryk 1990), which enabled southward movement into lower North America. Remains have been discovered in New Jersey (Scott 1885). The stag moose had massive tripalmated antlers 4.9 feet (1.5 m) in spread, which grew laterally from the skull in a fashion similar to that of the modern moose (Frick 1937). It probably represented a transitional stage between *Libralces* and *Alces*. These cervalces moose went extinct toward the end of the Holocene about 11,000 years B.P. (Churcher and Pinsof 1987).

Another early and possibly transitional form of *Alces* from the late Villafranchian through the Pleistocene was the broad-fronted moose *Libralces*. According to Vislobo-

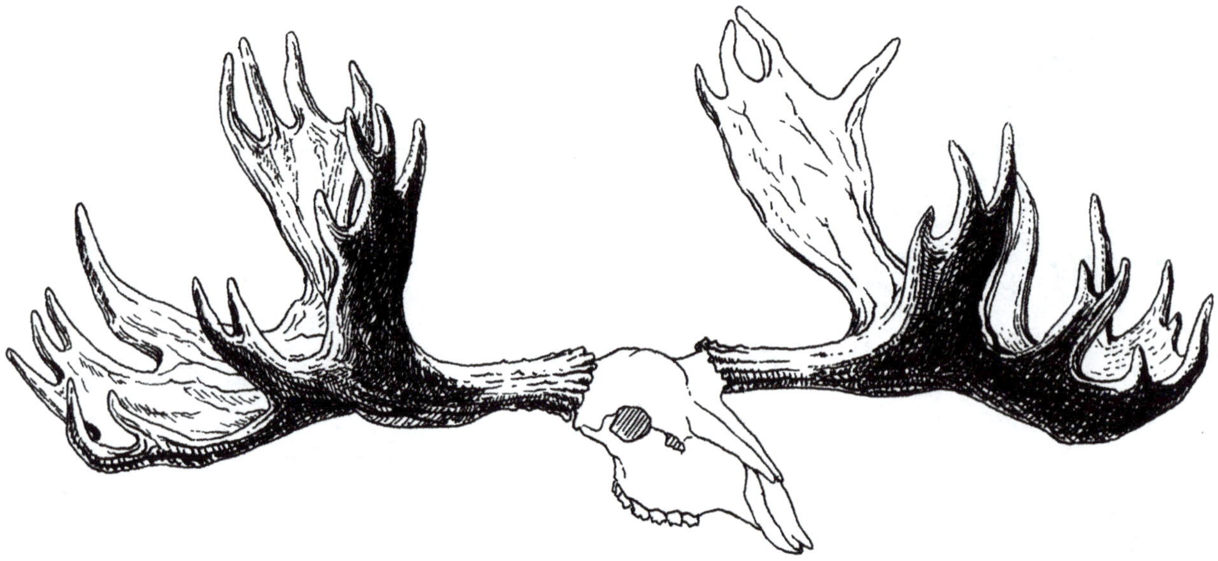

Figure 10. Probable shape of the stag moose *Cervalces scotti*. A giant moose of the late Pleistocene, it probably migrated to North America over the Beringia 100,000 to 200,000 years ago. *Reconstructed from different sources by A.B. Bubenik.*

kova (1986, 1990), it possibly was the first *Alces*, although more specimens are needed to prove this.

The earliest known *Alces* was the savannah-dwelling *Alces gallicus* (Heintz and Poplin 1981), also known as *Libralces gallicus* (Azzaroli 1981), which is known from the late Villafranchian (late Pleistocene) and middle Pleistocene deposits in Europe. It is the oldest and possibly smallest broad-fronted moose from England and Europe. Its skeletal structure suggests that it was an endurance runner. Its antlers, spanning about 10 feet (3 m), were characterized by a long beam and small semicircular palms that were cupped upward and rimmed with tines (Kurten 1968). The anterior points of *Libralces gallicus*' antlers faced downward as shown in Figure 11.

The largest known ancestor of today's moose was *Alces latifrons* (Heintz and Poplin 1981), also known as *Libralces latifrons* (Johnson 1874) (Figure 12). This libralces emerged in the middle Pleistocene and spread eastward through Siberia and over the Beringia into North America. It is unknown whether this crossing occurred during the Illinoian or Sangamonian period (Geist 1987). This species died out before the last Wisconsinian glaciation.

Except for its larger skeleton, *Libralces latifrons* had all the typical features of the genus *Alces*. The antlers varied between specimens, but were relatively symmetrical, with slightly convex surfaces. The spread varied up to 8.5 feet (2.5 m) (Kahlke 1956). They grew radially, and the longest tines were on the dorsal and frontal edges.

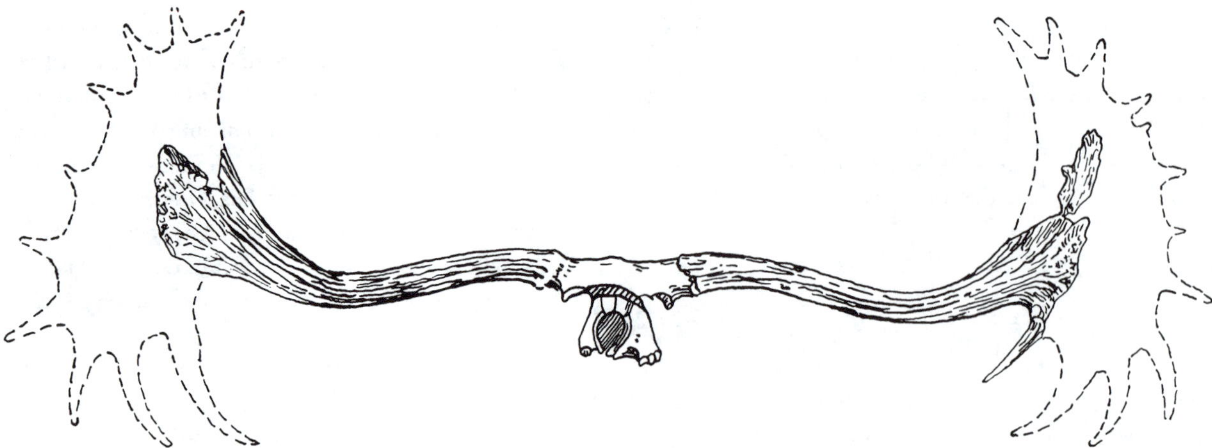

Figure 11. The presumed shape of antlers of *Libralces gallicus*, the earliest known *Alces* and whose remains were found in Europe (from Azzaroli 1950 and Bubenik 1994). Antler span was about 10 feet (3 m).

Figure 12. Reconstructed head of *Libralces latifrons,* the largest known ancestor of today's moose. It emerged during the middle Pleistocene and spread eastward through Siberia and into North America through the Bering land bridge. *Illustration by A.B. Bubenik, from A.B. Bubenik (1992b).*

North American Invasion

All North American moose seem to have originated from Siberia. However, views of when they arrived in the taiga around the Great Lakes differ. The first invasion probably did not occur before the end of the last glaciation (Wisconsin), 10,000 to 14,000 years B.P. However, there is disagreement whether the invasion occurred during the second or third interglacial period (Hopkins 1967, Pewe and Hopkins 1967, Khan 1970, McFarland et al. 1985, Catto and Mandryk 1990; see also figures 1, 3 and 4 *in* Rutter 1988). The absence of fossil moose in northwestern Alaska during the Upper Pleistocene seems inexplicable. This is especially true during the Illinoian glacial stage (≥500,000 years ago) or the Sangamonian interglacial period (≤120,000 years ago) (McFarland et al. 1985) when other large mammals, such as the mammoths, bears and even elk crossed the Beringia (Guthrie 1966, 1990a), and the libralcin and alcin moose were present in northeastern Siberia (Sher 1987). Both appear in northwestern Alaska much later, during the Wisconsinian period, when glaciation prevented any southward migration (Guthrie 1990a).

Peterson (1955) hypothesized that moose reached the taiga before development of the Wisconsinian ice shield (Figure 13). As a result of the expansion of the ice cap during the Wisconsinian glaciation, the southern populations were pushed into three different refugia, and the northern population into one refugia. Peterson speculated that these four distinct refugia became the centers of origin of the four modern subspecies of moose in North America. The three southern races were separated by extensive grasslands, whereas the northern race was separated by the extensive ice shield itself (see also Chapter 1). However, Peterson did not indicate whether one or two invasions of moose from the north penetrated south, or whether the southerly invading moose were different from those re-

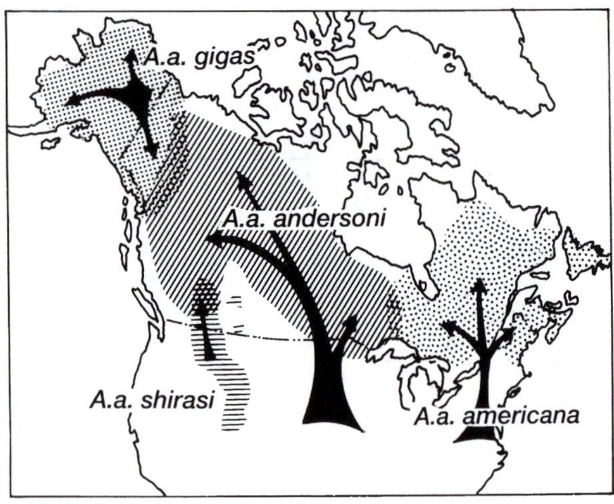

Figure 13. Postglacial dispersion of moose in North America (from Gasaway et al. 1987, after Peterson 1955 and Kelsall and Telfer 1974). Peterson (1955) theorized that moose reached the taiga before the Wisconsinian glacial period. With expansion of the ice shield, southern populations were forced into three refugia that became the centers of origin for the four subspecies of moose in North America. Geist (1987) offered the theory that all current moose originated from a single stock of *Alces alces*.

maining in Alaska. He even speculated about a reverse invasion from the south to the north when the corridor between the Cordilleran and Laurentian ice sheet opened.

Geist (1987) provided another hypothesis regarding the southerly invasion of moose. He assumed that the first wave of moose during the Illinoian period (>500,000 years ago) was actually *Alces latifrons*. The next wave arrived during the Sangamonian and Wisconsinian periods as the stag-moose, *Cervalces scotti*. Geist speculated that the recent moose, *Alces alces*, emerged in Alaska during the Wisconsinian period, and appeared in lower North America during the megafaunal extinction, when *Cervalces scotti* went extinct. He felt that all current moose originated from this single stock of *Alces alces*, and their southward expansion was limited by climatic and edaphic factors. Geist (1987: 21) also considered Peterson's (1955) hypothesis "... a bold step toward embracing Ice Age history. This hypothesis assumes the coexistence of *Alces alces* and *Cervalces scotti* south of the ice sheath during the Sangamonian and Wisconsinian periods. Of this there is no evidence." However, Churcher (1983) provided evidence, from fossil remnants of *Alces alces* from the Peace River country of Alberta (±11,000 years B.P.), for this coexistence toward the end of the Wisconsinian period.

I hypothesized that two invasions occurred over the Beringia (Bubenik 1986). First, the taiga moose of the Far East entered North America and migrated south to become geographically isolated from Alaska by the Wisconsin ice shield. The tundra moose of Siberia then invaded Alaska. However, this hypothesis must be rejected because Guthrie (1990b) established that the first fossil record of moose from Alaska was only 8,740 years old. These fossils should belong to the forest/tundra subspecies *Alces a. gigas*, whereas the fossils of the taiga moose from the Peace River country are about 3,000 years older and could hardly belong to the *A. a. gigas* lineage.

Another, although yet unproved, possibility is that an alternate migration route existed directly from Siberia (Chukotsk Peninsula) to the Canadian taiga. This route, in my opinion, likely led from the Kamchatka Peninsula (Kristchinsky 1974) and coastal British Columbia (Figure 14). At peak of the Wisconsinian glacial period (about 18,000 years B.P.), moose likely had about 2,000 to 8,000 years to traverse this eastward and then southward migration route during the retreat of sea ice (Sancetta and Robinson 1983, T. Hughes personal communication: 1991). During this time, the coast frequently was covered by glaciers. However, it was a time influenced by a relatively warm and humid (oceanic) climate. These weather patterns made it possible for trees and possibly small forests to grow on the surface of the Aleutian and Alaskan glaciers, such as the Yakataga, Bering, Martin River or Malaspina glaciers (Heusser 1960, Post and Streveler 1976, Peteet 1986). Regardless of the presence of forested glaciers, such migrations appear possible if shrub vegetation was present on islands connected by sea ice and shorelines were lower than today by approximately 100 yards (91 m).

Hence, during the mid-Wisconsinian glacial period, sea ice in the southern Bering Sea possibly provided a migration corridor for taiga moose from the Far East over the Aleutian chain without intermixing with the more northern forest/tundra moose. During this same period, sea ice from the Bering Sea enabled mammals of Kamchatka to invade Japan over the Kurril Islands (Flint 1948).

That hypothetical migration route would help explain why the dark taiga (Eastern) moose appeared in lower North America earlier than did moose appear in Alaska. The possibility of two chronologically different invasions is supported by distinct physical characteristics of the forest/tundra and taiga moose on both sides of the Pacific Ocean. A lack of mitochondrial DNA variability within North American moose (Cronin 1989) supports Geist's (1987) theory (see Bowyer et al. 1991), but does not prove it. The lack of genetic variability found in moose by Cronin (1989) is an exception among deer (Ryman et al. 1980). In addition, comparative data from both the taiga and forest/tundra moose of Siberia are lacking.

If my assumption about two different subspecies of moose invading North America is correct, it helps explain

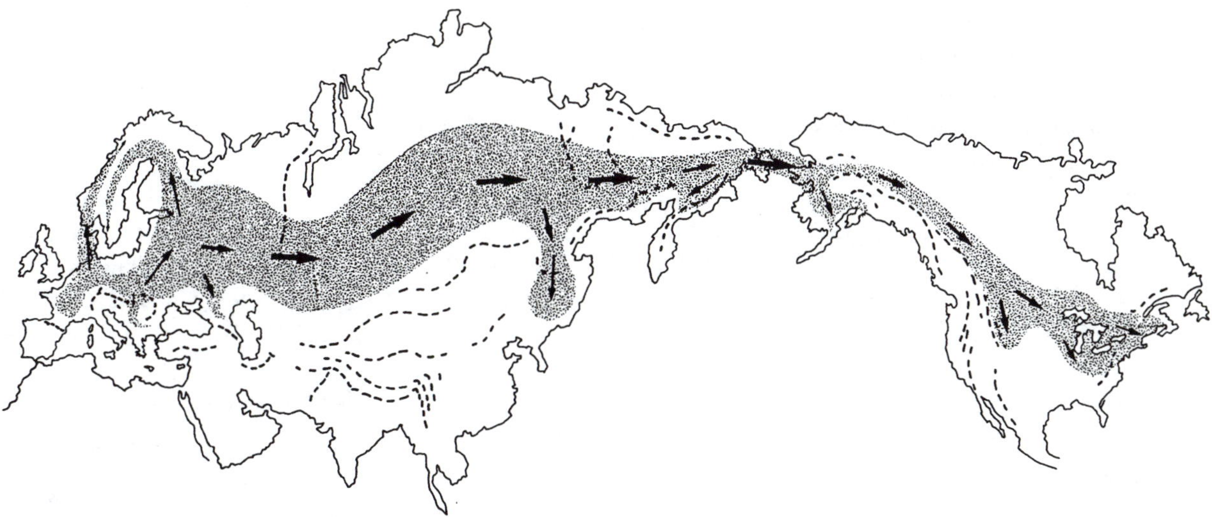

Figure 14. Possible Pleistocene migration routes of the taiga and tundra moose. *Illustration by A.B. Bubenik.*

their rapid adaptation to the two vastly different ecosystems. It also explains why neither of these subspecies expanded into the habitat of the other (Peterson 1955, Gauthier and Larsen 1985, Kelsall 1987).

Obviously, there are several viable theories about the prehistoric invasion of moose into North America. Hopefully, as genetic technologies improve (see Cronin 1989), we can unravel some of the mysteries about the ancestral evolution of deer in general and moose in particular.

Taxonomy

The ruminant Artiodactyls are a diverse and complex group, and their phylogenetic position, including the Cervidae (deer family) within this assemblage, has been subject to controversy (Webb and Taylor 1980, Groves and Grubb 1987, 1990, A.B. Bubenik 1990). It is not within the scope of this chapter to cover the various views on ruminant classification, but I feel it is important to mention this fact, should one of these new taxonomic classifications become commonly accepted. Until then, I choose to use the classification of Simpson (1945), who listed moose and other deer species in the family Cervidae in the order Artiodactyla (even-toed), infraorder Pecora, suborder Ruminantia. The more advanced Pecora are divided into three superfamilies, the Cervoidae, Giraffoidea and Bovoidae. Simpson (1946) classified the Alcins as follows:

Suborder Ruminantia
Infraorder Pecora
Family Cervinae
Subfamily Odocoileinae
Tribe Alcini = Alcinae Jerdon 1974 (Pleistocene to recent)
Genera Cervalces Scott (extinct since Holocene)
Alce = Alces (Pleistocene to recent)

The most obvious features of the Cervidae are the deciduous antlers of all genera except for the antlerless water deer (*Hydropotes*) and musk deer (*Moschus*). The antlers contrast with the Bovidae that possess unbranched, nondeciduous supraorbital or postorbital horns, consisting of a perennial bony core covered by a hollow, keratinized sheath. Thus, it is a misnomer to consider deer and bovids as close relatives. It is also incorrect to think of antlers as another sort of horn. Antlers are unique in that they are replaced each year and regrow if forcibly removed (Bubenik 1966a).

The systematics of the living deer are based on the anatomical features of their ancestors, including retention of certain bones of the lower foreleg and characteristics of the skull, including the nasal bones, vomer and bone pedicles (antler bases) with their distal parts (A.B. Bubenik 1990), which are periodically shed. Moose and other related deer possess the following skull characteristics: (1) upper jaw without incisors; (2) upper canine teeth virtually always absent; (3) lower canines modified as a quasi fourth incisor, the first premolar lost; (4) a gap between the incisor row and the second premolar, known as the "diastema"; (5) second and third premolars tightly spaced; (6) maxillary bone articulates with the nasals in moose but in other deer also with the premaxillary; (7) pedicles short and the first antlers of Odocoileidae appear 4 to 6 months after birth; and (8) two orifices to the lacrimal duct situated on or inside of the orbit.

Moose and other deer also have the following postcranial skeletal characteristics: (1) during the evolution of deer, the third and fourth metacarpal bones fused to form the "cannon bone"; (2) the third and fourth digits formed the hooves; (3) dewclaws were formed from the second and fifth digits, and the hooves have long soles; and (4) the second and fifth metacarpal bones underwent regression, with only their distal or proximal parts retained. Brooke (1878) used this criteria to classify deer into two different groups. The plesiometacarpal (with distal remnants) were referred to as Old World deer, whereas the telemetacarpal (with proximal remnants) were considered the New World lineage (Geist 1986, Sher 1987). However, this New World/Old World delineation must be considered questionable (Vislobokova 1986, A.B. Bubenik 1990) because the roe deer (*Capreolus*) is telemetacarpal but has features of Old World species, and all New World deer actually evolved in Eurasia and migrated to the Americas.

Subfamily Odocoileinae, Tribe *Alcini*

Here again, I follow the classification provided by Simpson (1945) who placed the moose in the subfamily Odocoileinae (Pocock 1923), which was originally labeled Alcinae by Jerdon (1874). There is no doubt that moose should be classified with reindeer/caribou and other American deer, excluding elk. The common origin of these deer species is supported by the following common features: (1) the living members telemetacarpal; (2) nasal cavity normal, except short in moose; (3) vomer short and, in living moose, the premaxillae not connected to nasals; (4) permanent dentition completed within 18 months after birth; (5) the distal part of pedicles grows and hardens into "first antlers" during the first 6 months of life (Frick 1937, Simpson 1945, Flerov 1950, 1952, A.B. Bubenik 1966a, 1990); (6) rumen with only two blind chambers (A.B. Bubenik 1959, 1990); (7) paired tarsal glands (except in few equatorial species) well developed, and both sexes urinate on them (Haltenorth 1963, Hershkowitz 1969); (8) all species multiparous (i.e., can bear more than one young), except the pudu of South America; and (9) intermediary type penis (Slijper 1938, Walton 1960). When semierect, the penis hangs down, enabling urination on the heels when the bull assumes a "horseshoe" posture (hind legs put forward and kept together).

The taxonomy of *Alcini* still is not firmly established, and different views exist on the relationships among extinct and extant clans or species. There is general agreement that all living moose are subspecies of *Alces alces* Linnaeus (1758) of Fennoscandia, where they spread from northern Eurasia into North America. There is disagreement whether all genera or subgenera are offshoots of one lineage of *Alces* Gray (1821), Heintz and Poplin 1981, whether *Libralces* or *Cervalces* should have the subgeneric rank (Azzaroli 1981, 1985, Heintz and Poplin 1981, Churcher and Pinsof 1987, Sher 1987, Vislobokova 1990), or whether *Alces* had an independent evolution.

Genus *Alces*

In English, the Latin name *Alces* means "elk," a term erroneously used by the first American settlers for the North American elk (*Cervus elaphus*) or "wapiti." Historically, in Eurasia, the moose is referred to as the elk, and red deer has been synonymous with wapiti. Presently in Eurasia, the terms elk and moose are synonymous. The present moose *Alces alces* appeared in Europe about 0.5 million years ago (Flerov 1931, 1950, Vereshchagin 1967). The accepted genus *Alces* was originally named *Cervus* by Linnaeus (1758), but also included *Alce* (Hamilton-Smith 1827) and *Paralces* (Allen 1902).

Systematics of North American Moose

The systematics of North American moose are not definitive and some changes may occur. Presently, the following subspecies are accepted.

Shira's or Yellowstone Moose

Taxonomy. Alces alces shirasi Nelson (1914).
Type specimen. U.S. National Museum, no. 202975.
Type location. Snake River, Lincoln County, Wyoming.
Body. Total length and height over shoulders are unknown. Bulls weigh less than or equal to 816 pounds (370 kg).
Skull. Total length is 22.4 to 23.4 inches (57–59 cm), and the nasal aperture is relatively wide (Peterson 1952).
Coat coloration. Coat color differs strongly in early winter from typical *Alces americana* from Maine, Nova Scotia and New Brunswick. The entire top of the back, including the upper side of the neck, is rather pale, rusty or yellowish-brown, with slightly dusky hair tips. According to Peterson (1955), the ears are paler, grayer and closer to the coloration pattern of the Alaskan moose than to that of the taiga moose. However, J.M. Peek (personal communication: 1986) indicated that he had seen *A. a. shirasi* in winter that generally looked more like *A. a. andersoni* than *A. a gigas*. There are some very black specimens of *A. a. shirasi* in winter. The coat does fade in the sunlight and the animals' sex and age influence coat coloration.
Antlers. On the basis of moose that I observed in Wyoming, *A. a. shirasi* has the smallest antlers of all North American moose. This was confirmed by Gasaway et al. (1987), who concluded that the *A. a. shirasi* subspecies had the smallest mean antler size for prime bulls in North America.
Breeding behavior. According to the description of Altmann

(1959), both serial and harem mating have been recorded (see Chapter 4).

Geographic distribution. Occurs in western Wyoming, eastern and northern Idaho, western Montana and northward into southern Alberta, southeastern British Columbia, and northeastern Utah; also successfully reintroduced to Colorado (see Chapter 3).

Eastern or Taiga Moose

Taxonomy. Alces alces americana Clinton (1822).

Type specimen. Unknown.

Type location. Unknown.

Body. Live weight is between 794 to 1,323 pounds (360–600 kg); carcass weight up to 838 pounds (380 kg). Total length is equal to or greater than 9.2 feet (280 cm); shoulder height is 6.1 to 6.4 feet (185–195 cm). Before mating season, the front hoof width of a prime bull is maximum and measures about 5.1 inches (13 cm). The hoof width of a cow is 3.7 inches (9.5 cm). Hoof width of calves in autumn is around 2.8 inches (7 cm).

Skull. Total length is 22.4 to 26.0 inches (57–66 cm) (Peterson 1952). Craniometric parameters have a wide range (Youngman 1975). The septomaxillary bones are wedged between the end of the upper edge of the premaxillae, nasals and maxillae, about 60 percent of its length. Brain of both sexes is fully developed around 36 months of age, but the range in volume is great—between 25.6 to 33.6 cubic inches (420–550 cm^3). Calves are born with an eye angle of 36 $\pm$ 6 degrees, which widens at maturity to 44 $\pm$ 6 degrees; this may be an adaptive process for enhancing the stereoscopic vision, which is necessary in the relatively dense taiga (Bubenik and Bellhouse 1980).

Coat coloration. Early winter coloration pattern is age-, social rank- and sex-dependent. Before the rut, submature or low-ranking prime bulls have dark brown or partially black faces; cows of breeding age have brown faces and almost black winter coats; postprime bulls have brown "cow faces." The face of the postprime cow is dark. The carpal and tarsal parts of the legs are ivory yellow. Over their torsos, bulls in full prime have "stripwise" longer hair, which create parallel, downward-running, darker lines. A vulva patch (Mitchell 1970) is present in cows. Antorbital glands are shallow, and the tarsal glands are covered by strong white hairs. During the rut, owing to frequent urination on the heels, tarsal hairs of adult cows

Taiga moose from Quetico Park in Ontario inhabit an area in the moose range that is considered an integrate between the eastern *(Alces alces americana)* and northwestern *(A. a. andersoni)* subspecies. It is questionable if this subspeciation is valid for taiga moose. Nevertheless, descriptions of typical specimens have provided a basis for such subspeciation. *Photo by H.R. Timmermann.*

and bulls become dark brown. On its surface, the interdigital gland has long, green-colored hairs.

Antlers. Antler spread rarely exceeds 65 inches (165 cm); a very good spread is 55 inches (140 cm). The maximum dry weight of antlers rarely exceeds 40 pounds (18.1 kg). The specific gravity is great, and the largest range of variation occurs in immature males and decreases toward prime age. Most prime antlers are of the butterfly-type, and well-developed front palms are less frequent than in the Alaskan/Yukon moose. Generally, prime bulls have a brow-tine ramified into three or four points. The typical shell-type antlers are rare. First antlers of calves are small buds. Yearlings carry second antlers as massive spikes, but around 12 percent of 1- and 2-year-olds have a unilaterally or bilaterally forward-oriented prong above each coronet. Small, two- or three-pointed cervicorn antlers in yearlings are rare. Palmated antlers in yearlings are an exception. Normally, the palmation begins to develop with the third antler set. Maximum antler development is achieved between the tenth and twelfth years of life (Timmermann 1971, Bubenik et al. 1978a).

Breeding behavior. In the taiga moose, breeding occurs as pairs, with the rut controlled by the cow (Bubenik 1985, 1987). The estrus of cows is not synchronized. Each cow has her own mating arena in which she lures the bull. Calves remain with their mothers when the bull is present (see Chapters 4 and 5).

Other characteristics. The main odoriferous component of urinary pheromones is *o*- cresol (Dombalagian 1979).

Geographic distribution. The taiga moose occurs from Maine and Nova Scotia westward through Quebec to central northern Ontario, where it apparently intergrades with *A. a. andersoni*. It was introduced into Newfoundland where it now is established (see Chapter 3).

Northwestern Moose

Taxonomy. Alces alces andersoni Peterson (1952). In my view, there is some question about the validity of this subspecies. Youngman (1975) considered *A. a. andersoni* an intergrade population, connecting *A. a. gigas* with the other forms and, therefore, not worthy of subspecific rank. However, the Northwestern moose currently is accepted as a valid subspecies, therefore it is included in these discussions.

Type specimen. Adult male (skin and complete skeleton), number 20068, Royal Ontario Museum of Zoology and Paleontology.

Type location. Section 27, Township 10, Range 16, Sprucewood forest reserve 15 miles (24 km) east of Brandon, Manitoba.

Body. The body size is within the variance of *A. a. americana*. According to Peterson (1952, 1955), it is a medium-size form, different chiefly in cranial details.

Skull. A medium-size moose differing from other North American forms chiefly in cranial details, especially with respect to the shape of the palate (relatively wider than *A. a. americana* and narrower than *A. a. shirasi* and *A. a. gigas*) (Peterson 1955).

Coat coloration. Similar to *A. a. americana*.

Antlers. Similar to *A. a. americana*.

Breeding behavior. Similar to *A. a. americana* (see Chapters 4 and 5).

Geographic distribution. Northern Michigan and Minnesota, western Ontario, westward to central British Columbia, north to eastern Yukon Territory and Mackenzie River Delta, and Northwest Territories. A more precise delimitation is lacking. The impact of the oceanic climate is clearly visible between southern and northern Alberta and northern British Columbia populations, and in the Rockies where hybridization with either *A. a. shirasi* or *A. a. gigas* moose is quite possible (see Chapter 3).

Alaskan/Yukon or Tundra Moose

Taxonomy. Alces alces gigas Miller (1899).

Type specimen. U. S. National Museum, number 86166.

Type location. From the north side of Tustumena Lake, Kenai Peninsula, Alaska.

Body. Subspecies as large as the Siberian moose *A. a. buturlini*. Mean length of adult bulls from different regions of Alaska varied from 9.5 to 10 feet (290–306 cm) and adult cows varied from 9.3 to 9.9 feet (283–302 cm) (Franzmann et al. 1978). Withers height of bulls is 6.2 to 7.0 feet (190–212 cm), possibly more. Live weight of cows up to 1,100 pounds (500 kg); bulls weigh 1,566 pounds (710 kg) and possibly more (Schwartz et al. 1987a). Frontal hoof width of prime cows is ±4.3 inches (11 cm), and ±5.5 inches (14 cm) for prime bulls. Tarsal and interdigital glands are not described, but they probably are of the same structure as in taiga moose.

Skull. Total length of the skull is 24.2 to 27 inches (61–69 cm) long; maximum width is 18.7 to 20.5 inches (47–52 cm). The palate is relatively wide in relation to the length of tooth row. The septomaxillary bone begins at the upper edge of the maxillary bones, which ossify and fuse with the maxillae much earlier than in any other subspecies. The nose is convex, the occiput is high, and the horizontal angle of the orbits in mature individuals is 36 ± 6 degrees (Bubenik and Bellhouse 1980).

Coat coloration. Summer color is unisex brown, but brighter in cows than in bulls. The winter coat is sex, age and rank

The taxonomic designation for Alaskan moose is *Alces alces gigas*. The subspecies often is referred to as the tundra moose, as many moose in Alaska and the Yukon live in and near tundra areas. However, it also inhabits boreal forest areas and intergrades with the northwestern subspecies in the Yukon. The Alaskan moose generally is larger than the other subspecies in body size and antler development, and often is referred to as the tundra moose, whereas the eastern and northwestern subspecies frequently are called taiga moose. *Photo by Charles C. Schwartz; courtesy of the Alaska Department of Fish and Game.*

dimorphic, as is the taiga moose. However, the torso coloration pattern is different: the upper part of the neck and torso is ocherous; the dark brown, sometimes almost black hairs on the hump are up to 9.8 inches (25 cm) long and proceed as a strip toward the rump; the shoulders are bright ocherous to almost yellow, visibly contrasting with the upper body parts; in some individuals, a brighter saddle appears over the torso, and the lower part of the neck, torso and upper part of the forelegs and the hindquarters are practically black, with bluish glitter over the biceps and buttocks. The carpal and tarsal parts of the legs are ivory white. Bulls in full prime have parallel, downward-running strips of hairs over the torso as in taiga moose.

Antlers. Alaskan/Yukon moose antlers generally are bipalmated, with well-expressed brow-palms. Compared with the taiga moose, the antler ontogeny is faster. Yearlings frequently carry small palms with a brow-tine. Antler spread of 82.7 inches (210 cm) has been recorded (Gasaway et al. 1987). Antler weight (maximum, 64 pounds [29 kg]; average, 44.1–55.2 pounds [20–25 kg]) is substantially higher than in taiga moose.

Breeding behavior. The Alaskan/Yukon moose breeds in harems of cows that voluntarily aggregate on the rutting arena of the bull, but remain agonistic toward each other (see Chapters 4 and 5).

Other characteristics. The main odoriferous component of the urinary pheromones is *o*- cresol, as in the taiga moose (Dombalagian 1979). Potent sex pheromones—the unsaturated C_{19} steroids—with a musk scent are in the saliva of bulls (Schwartz et al. 1990).

Geographic distribution. The Alaskan/Yukon subspecies presently is found from the Kivalina River in northwestern Alaska (Hall 1973) down to the Yukon Territory and the most northern part of British Columbia—approximately north of the 60 degree parallel. The limiting factor southward may be the taiga and humid climate (Youngman 1975, Gauthier and Larsen 1985, Geist 1987; see also Chapter 3).

Morphophysiology

Skeletal Development

The skeletal bones of moose grow more allometrically than do those in any other deer (Petrov 1964). Because skeletal growth (especially ankylosis of the pelvic vertebrae) continues from conception until roughly 4.5 to 5.5 years of life, the gestalt of the moose changes many times.

SKULL

The skull consists of two portions: the cranial part surrounds the brain (Figure 15) and the remainder is the facial component. The brain case is formed by a perforated bony wall on the interior nasal cavity and the olfactory bulbs. The roof of the brain case is built by the parietal bone and posterior sides of the temporal bones. The base of the brain case is closed by a conical parasphenoid bone (Romer and Parsons 1978) with the cavity for the pituitary gland.

In deer, as a rule, growth of the skull slows down at about 14 months of age, when cartilage between the occipital bone and the sphenoid begins to fuse. The anterior edge fuses later and more slowly with the adjacent presphenoid bone. When the anterior end of the sphenoid begins to amalgamate with the presphenoid brain case growth is complete (Schuhmacher 1939). Hence, brain case and brain volume are fully developed at the age of 36 to 40 months.

Fusion between sphenoid and presphenoid bones is a relatively slow process, occurring around the fourth year of life in cows and the fifth year in bulls (Bubenik and Bellhouse 1980). The rate of disappearance of this grove can serve as a reliable indicator of age between 3.5 and 5.5 years, agreeing with the estimate of Peterson (1955).

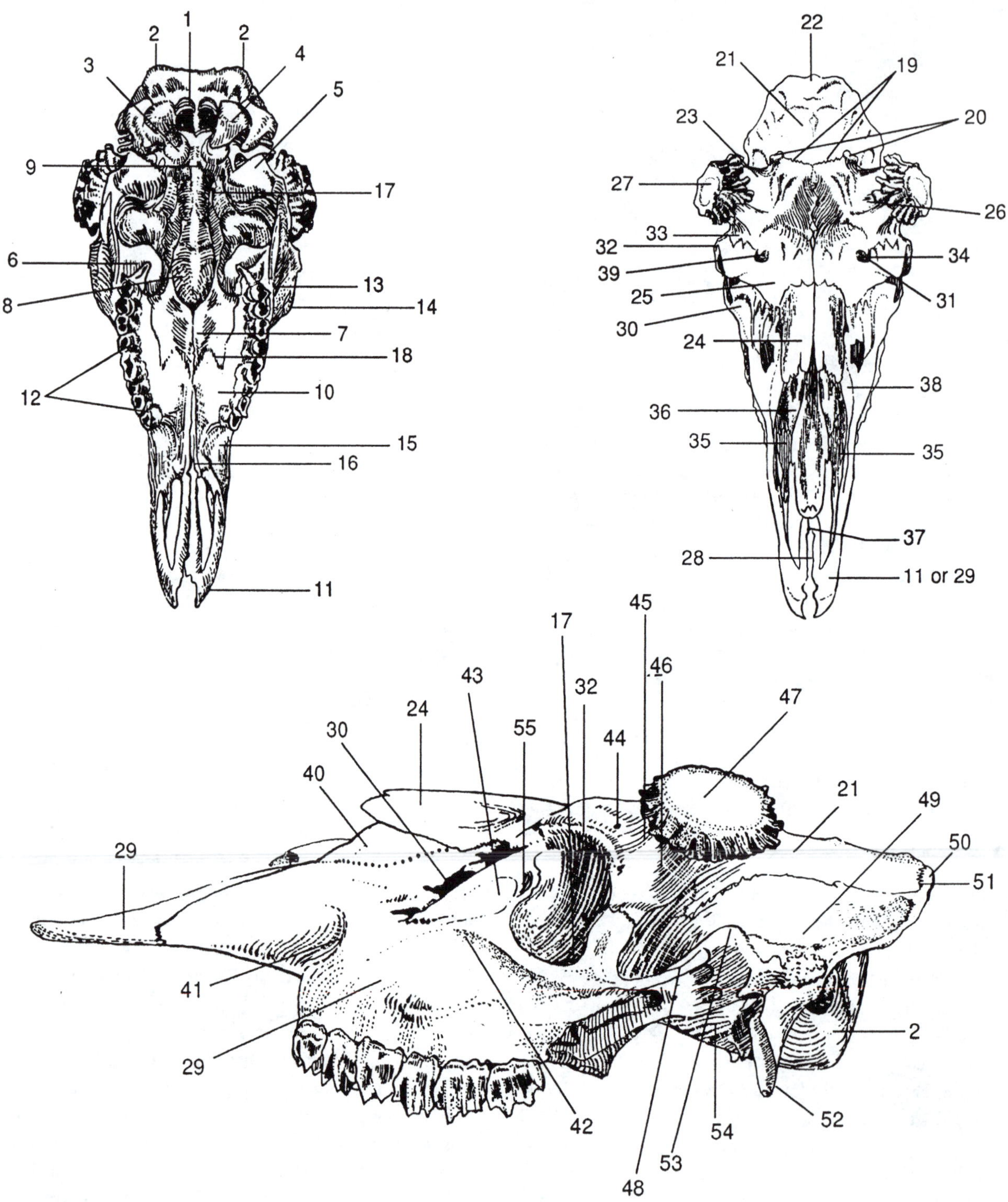

Figure 15. Cranial bones in North American moose, in ventral (top left), dorsal (top right) and lateral (bottom) views (see Kolda 1951): 1 = occipital foramen; 2 = occipital condyle; 3 = tuberum musculae (roughened area of bone where muscle attaches); 4 = jugular process; 5 = tympanic bulla; 6 = lacrimal bulla; 7 = chaonae or turbinal bones; 8 = nasopharynx; 9 = sphenoid bone; 10 = palate bones; 11 = maxillary bone; 12 = premolars and molars; 13 = zygomatic bone; 14 = palatine process; 15 = molar enlargement; 16 = interalveolar space; 17 = optical foramen; 18 = transverse palatine suture; 19 = intercornual protuberances (in mature bulls, reinforced by prominent ridge with two small pedicle-like protuberances = 20); 21 = parietal bone; 22 = occipital bone; 23 = cornual process (called pedicle in cervids); 24 = nasal bone; 25 = frontal bone; 26 = cornual fossa; 27 = cornual bone (in cervids, it is the pedicle from which antler grows); 28 = intermaxillary space; 29 = maxillary bone; 30 = lacrimal depression; 31 = supraorbital foramen; 32 = supraorbital rim; 33 = supra-orbital depression; 34 = supraorbital trench; 35 = ventral turbinate; 36 = dorsal turbinate; 37 = palatine fissure; 38 = septonasal bone

FACIAL BONES

The face of the moose represents 69 to 70 percent of the total head length, longest among deer. In comparison, faces of the Giant Irish deer and North American elk embody only 63 to 64 percent and 52 to 58 percent of total head length, respectively. Peterson (1950, 1952) considered shape and length of both nasals and intermaxillae, and minimal width of the frontal aperture in relation to the upper tooth row, important parameters to determine subspecies.

FRONTAL BONES

A unique feature found in the moose and no other deer is the suture formed where frontal bones join the nasal bones. The profile at this junction is concave, but the sutures rise sharply with the center of the parietal bone, forming a small crest. The suture consists of two perpendicular interlocked meanders that cannot be physically separated. It is a masterpiece of bioengineering that probably is a morphological adaptation to the enormous pressure imposed by both the mass of the antlers and collision with hard objects.

The growth rate of frontal bones is sexually dimorphic, being exponential in bulls but nearly linear in cows (Figure 16). These differences in growth rate can be interpreted as a biological necessity in males, to develop strong and pressure-resistant frontal bones on which the weight and leverage of antlers rely.

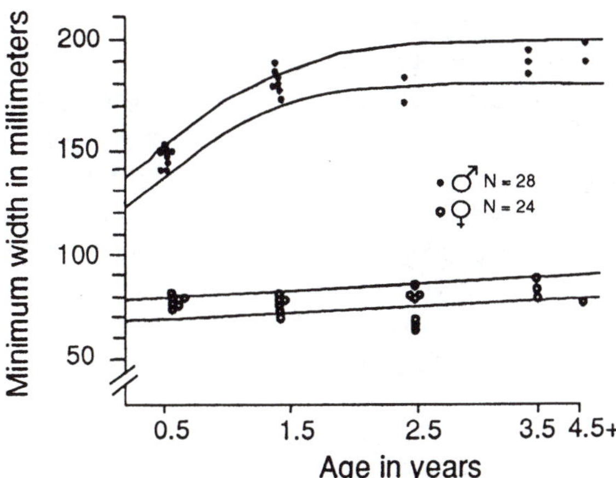

Figure 16. Development of frontal bones in cow and bull moose (from Bubenik and Bellhouse 1980). The width of frontal bones is a perfect sexually dimorphic parameter. The relationship in females is almost linear, and exponential in males.

PEDICLES

Pedicles are bony protuberances where the antlers develop on the skull. In well-conditioned male calves, pedicles begin to develop by the end of the summer, and their apices are ossified by late autumn. Pedicles are slightly divergent in calves, but with each subsequent antler cycle, they orient more horizontally and lose length. They can be partially rebuilt in the subsequent antler development. Despite their very solid cortex, pedicles can be broken, resulting in a lifelong handicap, because the subsequent regenerating antler is mostly malformed and may be reoriented.

ORBITS

Orbits are the protruding bony rims around the eye sockets. The overlapping angles of their horizontal axis indirectly form the angle of the eye and show a low degree of stereoscopy. The eye angle of neonates of taiga moose changes from 36 ± 6 degrees in calves to 44 ± 6 degrees in mature animals. This contrasts with an eye angle of 34 ± 4 degrees in four specimens of tundra moose from interior Alaska, ranging in age from 6 to 14 years of age. This may be the reason why the taiga moose has limited stereoscopic vision when compared with the almost monoscopic vision of the tundra moose (see Chapter 5).

NASAL BONES

Nasal bones of the moose are substantially shorter, whereas the maxillae have pushed the premaxillae forward as their connection with the nasals was lost. According to Slaby (1990), only the maxilloturbinal, but not nasoturbinal, is present in moose (cf. Pinus 1928). This resulted in development of a large amount of room between the nasals and anterior edge of premaxillae. This space provided the room necessary for development of a muscular prehensile nose with widely spaced nostrils.

CHEEK BONES

Cheek bones in the moose skull are built not only by maxillary, but also by the septomaxillary and maxillary bones in front of the nose. The septomaxillae are bones wedged between the nasals and the maxillae, and are found in primi-

←

(its suture in moose more than 2 to 4 years old is visible only in transparent light as a faint white line, otherwise it is fused completely with the maxillary bones); 39 = supraorbital canal; 40 = maxillary septum (not visible in moose more than 3 or 4 years old because of complete fusion with the maxillary bones—the former suture is indicated by the dotted line); 41 = infraorbital foramen; 42 = facial crest; 43 = lacrimal fossa; 44 = optical foramen; 45 = antler base; 46 = antler column (pedicle in cervids); 47 = antler shaft (cut off here); 48 = zygomatic arch; 49 = squamous portion of the temporal bone; 50 = external occipital protuberance; 51 = lamboidea suture; 52 = fungularal process; 53 = pterigoid process; 54 = sphenoid bone; and 55 = lacrimal canaliculi. *Illustrations by A.B. Bubenik.*

tive amphibians (Romer 1956). Their upper edge begins to descend before it reaches the nasals. In fetuses and newborn calves, the septomaxillae are composed of cartilaginous tissue that ossifies very fast. By the time a calf is 6 months old, the ossification process is complete, and the lower corner fuses with the upper edge of the maxilla. Traces of the suture disappear from the surface between 12 and 36 months of age, but there are a few specimens in which either the anterior and posterior ends of the suture remain visible longer. The suture also can be traced in transparent light (Bubenik 1986).

VOMER

The vomer is the vertical V-shaped bone separating the nasal cavity into two chambers. In moose, compared with other *Odocoileinae* (except the roe deer), the vomer is high and short. It descends posteriorly and does not divide the aperture of posterior nares. The anterior portion is long and shallow, proceeding with a large vertical cartilage wedged into the anterior end of the vomer.

CHOANAE OR TURBINAL BONES

These paired scrolls of bone develop in the oral roof of the palate and are the interior openings of the nasal passage. They increase substantially the surface area of the nasal passages and fuse with adjacent elements of the maxillae and nasal bones. According to Slaby (1990), in contrast to Pinus (1928), only the maxilloturbinal, but not the basoturbinals, are developed in moose. The choanae of moose cover a large surface area, which enhances olfactory discrimination.

MANDIBLE

The lower jaw is very specialized. It narrows according to the shape of the premaxillae and is equipped with many cristae for attachment of the massive chewing muscles. The length and growing period of the diastema correlates with the total length of the skull, as has been shown by Peterson (1955). The mandible anatomy is typical for ruminant browsers (Likhachev 1956). However, the *processus angularis*—the outer angle between the mandibular ramus and mandible—has a sexually dimorphic shape. In cows, it does not protrude as sharply or as much as in bulls.

DENTITION

The normal dental formula for the moose is Incisors 0/3 + canines 0/1 + premolars 3/3 + molars 3/3 (× 2 sides) = 32.

The number of teeth found in all deer species follows the same evolutionary trend. Upper incisors were lost sometime during the Middle Oligocene, about 30 million years ago. The canines regressed from long tusk-like teeth to mere stumps or were lost entirely. In the lower jaw, as a consequence of rumination, larger molars evolved with more columns and a longer tooth row (Obergfell 1957, Sablina 1970). In species that prefer browse forage, the oral cavity enlarged, with a greater distance between the incisors and premolars (diastema). The canines shifted and adapted functionally as a fourth incisor.

There is little published information on the timing of tooth eruption in North American moose. Based on a small sample of mandibles from Ontario moose, Peterson (1955) provided the only description. From more recent studies, it now is known that tooth development varies considerably between individuals (Bubenik and Bellhouse 1980, Peterson et al. 1983). In a sample collected by Bubenik and Bellhouse (1980) in northwestern Ontario, tooth replacement was delayed in 69 percent of male calves and 16 percent of female calves. Markgren (1964) also observed such deviation from the "norm" in Swedish moose. Variation in tooth eruption can result from differences in date of birth, stress as a result of crowding in roe deer (Passagre 1971) or, more likely, from nutritional condition during growth (Peterson et al. 1983). To compensate for loss in tooth height, the angle of incisors is permanently raised to meet the gum. To compensate for the wear of the conical crowns, the molars and premolars are pushed forward to prevent the loss of mutual contact. Hence, in older individuals, the tooth row can be as much as 0.6 inch (1.5 cm) shorter (Schuhmacher 1939, A.B. Bubenik 1982a).

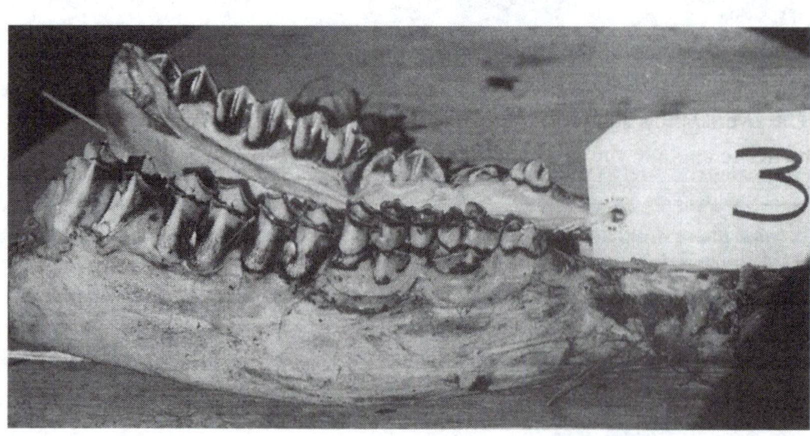

A jawbone with teeth from a yearling moose shows premolars 2 and 3 (deciduous) in the process of shedding on the right (near) side of the jaw. Permanent premolars 2 and 3 on the opposite side are erupting. *Photo by H. R. Timmermann.*

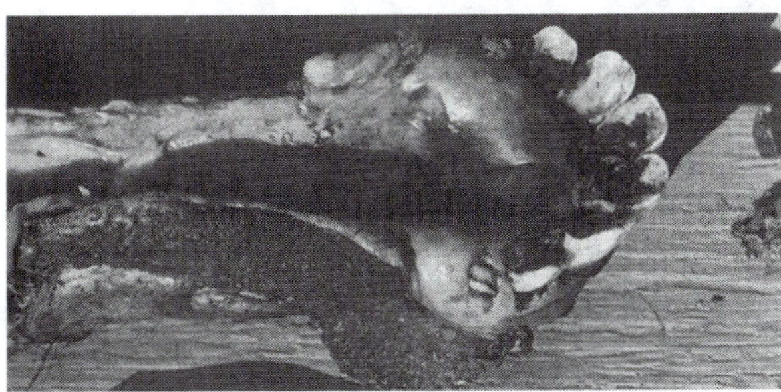

A view of the lower front teeth (incisors) from a yearling moose shows erupting third and fourth incisors. The dental formula for a mature moose is 6 incisors (lower only), 2 canines (lower only), 12 premolars (6 upper, 6 lower), and 12 molars (6 upper, 6 lower) for a total of 32 permanent teeth. *Photo by H. R. Timmermann.*

TOOTH HISTOLOGY

The body of any tooth is built by dentine, which surrounds the pulp that contains the tooth nerve. The crown dentine is shielded by a thin layer of enamel, whereas the roots are protected by cementum, which protects the dentine from bacterial decay.

Biting and chewing forage exert grinding pressure on the tooth surface and wear away the sharp edges of the crescent-shaped tooth cusps (Peterson 1955, Riney 1955, Romer and Parsons 1978). The physical pressure exerted on the teeth stimulates development of new protective layers of cementum around the roots. Layers that develop during summer are thicker and more dense than those developed in winter. When sectioned and stained, these summer and winter layers appear as bands or lines that, when counted, provide an accurate estimation of the animal's age (Sergeant and Pimlott 1959). The method can be confounded by (1) quality of the cut annuli zone, (2) variation in deposition of the first cementum layer, and (3) variation among methods used (Dalton and Francis 1988). False annuli can also complicate interpretation (Gasaway et al. 1978); thus, caution in cementum census is necessary (Simkin 1968, Cumming and Evans 1978, Dalton and Francis 1988). In moose, the cementum layers, on the roots of the second molar are so thick that the lines are readable with a 10-power magnifying glass, and it is unnecessary to decalcify, stain and cut them (Figure 17).

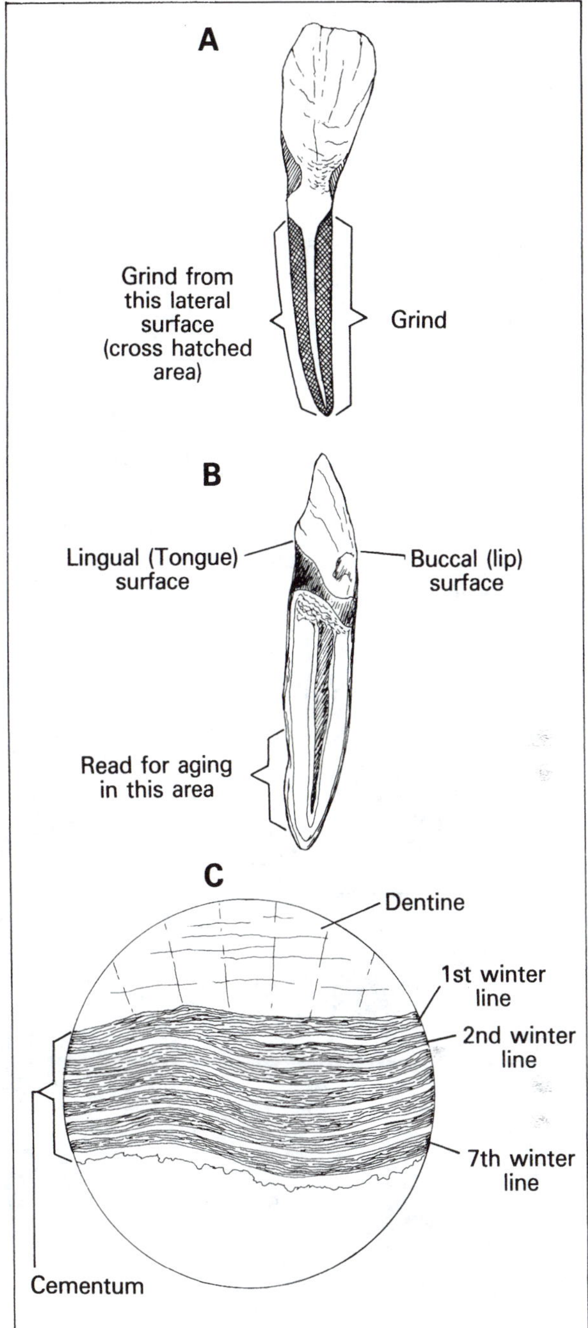

Figure 17. Frontal (A) view of moose tooth, indicating lateral surfaces (in cross-hatching) to be ground away for age determination; lateral (B) view of ground down tooth; and (C) 30-power magnification of cross-sectioned tooth of a 7-year-old moose killed in autumn. *Illustrations by Celia Carl Anderson; redrawn from original by Dave Harkness, Alaska Department of Fish and Game.*

With the approach of prime age, tooth wear of moose begins to deviate from the norm identified by Passmore et al. (1955). Errors in aging up to 3 years are possible and depend on hardness of the dentine. Soft dentine is attributed to certain nutritional conditions, including a lack of fluorine, the presence of abrasive substances on or within plant tissues and possibly genetics and overpopulation (Schuh 1982).

The continuous chewing of twigs and bark during winter is the major reason why moose teeth wear in a progressive manner. As moose age past prime, the crowns of their molars wear and tooth rows shorten. Gaps between teeth are filled by molar teeth pushing forward as has been shown in elk (Bubenik 1982a). Eventually, crowns on the teeth are worn away, resulting in only the tips of the roots showing. This eventually impairs browsing and chewing, lowers digestive capability and probably is the main reason why moose seldom live beyond 16 to 18 years of age.

ABNORMAL DENTITION

Peterson (1955) discussed supernumerary incisiform teeth on either side of the incisor row. These teeth protruded right or left, and in only three cases were extra teeth present on both sides. In this situation, one of them had the incisor rotated 180 degrees, with the occlusal surface facing down instead of up. Radiographs of mandibles with missing teeth display incisors embedded horizontally and failing to erupt on the lingual side of the jaw. First premolars and regressed upper canines are rare. Canines rarely appear as sharp stumps in milk dentition. In permanent teeth, they are rare (Kozhukhov 1990) and probably without enamel, as in red deer (Frankenberger 1951).

Postcranial Skeleton

The postcranial skeleton is composed of two main parts—the vertebral column (atlas/axial complex) and the bones of the appendages (appendicular skeleton) (Figure 18). The appendicular skeleton has 13 pairs of ribs, paired shoulder blades, a pelvic bone at the end of the trunk and the limbs as supportive structures (Garrod 1877, Kolda 1951, Severtsov 1951, Likhachev 1955, 1956, Petrov 1957, 1958a, 1958b, 1964, Pilarski and Roskosz 1959, Essen and Sparre 1975, Bubenik et al. 1978a, Timmermann and Lankester 1978, Bartosiewicz 1987).

The gait of an animal can be classified as a slipper, duiker, galloper or runner, depending on the ratio between withers and rump height and the neck/trunk ratio. The gait of the moose is that of a typical endurance runner, well adapted to soft ground (Mangel 1991) and high wind velocity. The scapular spine and the axis of the humerus in the moose

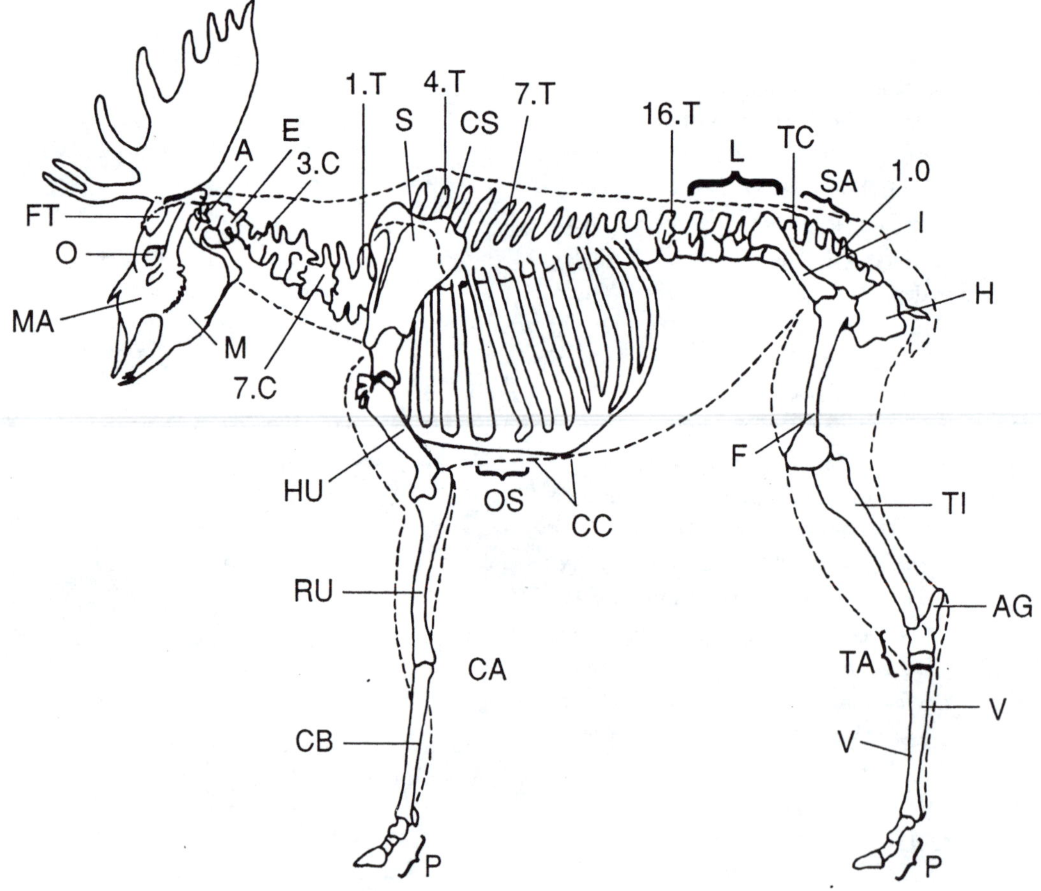

Figure 18. Skeletal structure of a moose and terminology of the main bones: O = eye orbits; MA = maxilla; M = mandible; A = atlas; E = epistropheus; FT = temporal fossa; 3.C = cervical vertebrae; 7.C = seventh cervical vertebrae; 1.T = first thoracic vertebrae; 4.T = fourth thoracic vertebrae; 16.T = sixteenth thoracic vertebrae; 7.T = seventh thoracic vertebrae; L = lumbar vertebrae; SA = sacral bones; 1.0 = first tailbone; S = scapula; CS = cartilaginous part of the scapula; CC = costal or rib bone cartilage; OS = costal or rib bones (1–13); HU = humerus; RU = radius and ulna (fused); CA = carpal bones; CB = metacarpus (cannon bone); P = phalanges or digital bones; TC = tuber coxae; I = illium; H = ischium; F = femur; TI = tibia; AG = tuber calcis or astragalus (cervids); TA = tarsal bones; and V = metatarsus. *Redrawn from Likhachev (1956) and Kolda (1951) by A.B. Bubenik.*

form an angle of about 140 degrees, compared with 105 to 117 degrees in elk and 115 in the horse. This increased angle enables the moose to take a greater stride than does the horse because the scapula can be shifted farther backward. Similarly, the pelvic bone has a relatively sharp slope of about 45 degrees with the ground, compared with only 22 degrees in elk. Because of the 65-degree angle between the pelvis and femoral axis, the moose stride is greater than that of the horse.

VERTEBRAE OF THE SKELETON

The postcranial skeleton has 33 vertebrae, with a total length in mature individuals ranging from 109 to 122 inches (276–310 cm). Neck length is almost equal to skull length, which is unique among cervids. The relatively short neck of the moose aids in carrying the large head and heavy antlers (Kolda 1951, Petrov 1958a, Richardson 1990).

LIMBS

Relationships between the long bones of moose forelimbs and hind limbs provide evidence for an endurance runner with a long stride. The bones' length seems to be adapted to allow easy movement through deep snow and/or blowdown within forest.

On the forelimbs, the humerus is short and massive. About 60 percent of the body mass rests on this part of the skeleton, compared with only 54 percent in cattle (Petrov 1958a). The ulna is fused with the radius, and the angle between the two measures about 145 degrees, nearly identical to that of both the elk and horse (140 degrees). the length of the radius represents about 56 percent of the whole forelimb length without hooves, in contrast to 47 to 49 percent in other deer species (Petrov 1957, 1958a, 1964, 1985b, Petrov et al. 1992).

Two pairs of flat carpal bones build a flexible connection to the metacarpus. In moose, only the distal remnants of the two lateral metacarpals are present, which puts the alcins among the telemetacarpal deer. The cannon bone of the moose still is separated into two compartments by a septum, the remnant of the fused metacarpal, but it does not reach the proximal end of the cannon bone. In contrast, white-tailed deer have only distal and proximal remnants of this septum. In caribou, the uninterrupted septum does not reach the opposite (anterior) side of the cannon.

The dewclaws, as well as the digits, are connected by cartilage and controlled by strong tendons (Likhachev 1956). They enable the hooves to spread on soft ground or squeeze together before lifting the foot. The exceptional angle between the shin bone and the metatarsus also permits a very special gait, observed only in moose in deep snow (described in Chapter 5).

Owing to the considerable load on the forelegs, the dewclaws and digits are long, frontally narrow and wide laterally (Figure 19). Thus, with a wither height of 69 to 75 inches (175–190 cm), the chest and abdomen are about 38.6 inches (98 cm) above the ground (Franzmann et al. 1978). This height allows the moose to move easily through powder snow up to 3 feet (91 cm) deep.

Hooves are built by keratinized skin covering the last

Dewclaws on the back of the feet of moose and the digits or hooves (left) are connected by cartilage and controlled by strong tendons that enable the hooves to splay outward for stability and traction on soft ground. Moose spend a great amount of time in bogs and along lakes and streams and this aspect of their anatomy helps keep this large animal from sinking and enables freer movement in those areas. The track of a moose, with dewclaw imprint (right), shows how weight is distributed. *Photos by Charles C. Schwartz; courtesy of the Alaska Department of Fish and Game.*

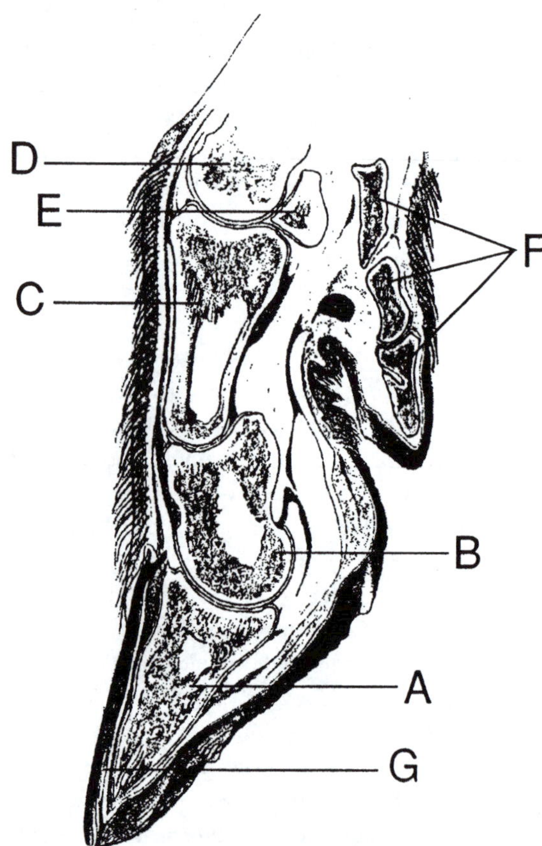

Figure 19. Cross section of the foot of a moose: A = third phalanx; B = second phalanx; C = first phalanx; D = metacarpus (front leg) or metatarsus (back leg); E = proximal sesamoid bone; F = phalanges of the dewclaw; and G = hoof. *Illustration by A.B. Bubenik.*

phalanges or toe. The hoof has a long sole, and the external toe is longer than the internal toe.

Normal wear of the moose hoof is compensated by fast growth (Marma 1972, Kühme 1974, Bubenik et al. 1978b). However, when a bull begins rutting, the increased level of activity results in additional hoof wear, which first appears on the tips of the toes, then the lateral edges and finally the sole. Sole wear may result in cracks, bleeding and infection (A.B. Bubenik et al. 1975).

Antlerogenesis and Antlers

Fossil findings suggest that natural selection of antlers progressed toward deciduous protuberances with great morphogenic but species-specific diversity. Ritualized forms of behavior transformed antlers into structures of considerable social significance (A.B. Bubenik 1990). Their unique neural control, the lineage-specific peculiarities of endocrine control and their multipotent effect on conspecifics contributed to the view that antlers are "bones of contention" (Henshaw 1971, Chapman 1975). We have progressed in our knowledge of antlerogenesis of many deer species, but the mechanisms in moose remain virtually unknown. Therefore, it is necessary when discussing moose antlerogenesis to rely on the antlerogenic mechanisms of other deer subfamilies (Timmermann 1971, Goss 1983, A.B. Bubenik 1990, G.A. Bubenik 1990a, 1990b, Suttie 1990, Suttie and Fenessy 1991, 1992, Suttie et al. 1991, Lincoln 1992).

Evolution

Antlers first developed as perennial, but later as "disposable" protuberances of the skull. To understand the evolutionary history of antlerogenesis, we must go back some 24 million years. The first appearance of antler-like growth showed velericorn (skin-covered) perennial protuberances occurring in both sexes of some species. We do not know their primary function, but perhaps it was to protect exposed head edges, such as eye orbits, from thorny vegetation or injuries during sparring. They had to be very efficient because, through natural selection, they developed into morphologically large, socially effective and environmentally acceptable organs. However, the fragility and sensitivity of these perennial bones that lacked a covering of muscles likely limited even longer growth. Nevertheless, these small protuberances visually accentuated the head, and glands of the skin provided a scent-dissipating mechanism at the nose horizon of peers (A.B. Bubenik 1990, G.A. Bubenik 1993a).

The fragility and tactile sensitivity of these protuberances might be why there is no evidence that they were used as weapons. Instead, species with both long tusks and antlers retained the former as weapons, like the antlered muntjacs (Barette 1977). Those deer, whose tusks regressed to stumps (Frankenberger 1951, 1954) or disappeared (Frick 1937, Obergfell 1957), fought with their front legs, by neck fight, or powered and lifted their opponents, as roe deer do (A.B. Bubenik 1971, 1982c). To minimize the fragility of long porous protuberances, their apices began to mineralize. Once the bone petrified, the skin either dried and peeled off or remained adhered as mummified tissue. It took approximately 12 million years to achieve the stage where the first solid-bone protoantlers appeared (A.B. Bubenik 1990).

How long these protoantlers remained attached to their live bases (protopedicles) is unknown. It is assumed that they detached and were lost as soon as their base (pedicle) regenerated. Fossil evidence indicates that, if accidentally lost, the apex and protopedicle regenerated, and that each pedicle exerted morphogenetic processes independent of the other. This genetic blueprint remains present today in all deer species. The danger that the long protuberances, as levers could, break out the supraorbital rims likely resulted in their ramification (dividing and spreading) instead of growth in length. This resulted in very sophisticated lineage- and species-specific patterns of construction and shape (A.B. Bubenik 1990).

Similar to perennial horns, protoantlers could only change size and shape very slowly. Consequently, only class membership but not individual recognition was possible. This handicap was overcome when the centrifugal ossification (inside to outside) reversed into centripetal ossification, and a compact cortex with a porous core developed. This process shortened the growth and hard periods of antlers from years to months (A.B. Bubenik 1990, 1993). Such change made it possible for the shape and size of antlers to reflect annual fitness and social status of the animal. In addition, insensitive apices could now be used as weapons.

This evolutionary event was achieved about 6 million to 8 million years ago and profoundly enhanced the importance of cranial protuberances. Nevertheless, the supraorbital location of their bases was still a handicap for progressive growth and tool diversity. This disadvantage was solved by translocation of the pedicle bases from the orbital rims to the outer edges of the frontal bones. This change permitted selection of antlers that were large, socially and environmentally acceptable, and advantageous to the species (Geist 1986, A.B. Bubenik 1990). Near the end of the Miocene and the beginning of the Pliocene (about 5 million years ago), deer with pedicles on the frontal bone crests began to dominate the quickly vanishing protodeer.

Antlered ruminants, the deer, evolved more than once from different lineages. Therefore, not all deer are close relatives and some are very distant. Ancestors of moose developed in a seasonally changing climate that influenced their evolution. Thus, the antler cycle became species specific. The cycle was timed by adaptations to hormone secretion, which resulted in a concise reproductive period so that calving coincided with the beginning of the vegetative period.

Histology

Antlers develop as apices (extensions) of rod-shaped, perennial pedicles by apposition of cartilaginous bone matrix. The cells of embryonic connective tissue grow and transform into amorphous cartilage and then into fibrous elements that ultimately ossify. The immature forms later change into beam-like lamellar (plate-like) trabeculae (Wheather et al. 1979). As solid, organized trabeculae, they form the compact bone or may form a spongy mass called "cancellous bone." Generally the transformation occurs by endochondral ossification, except in those deer species where intramembranous ossification occurs (Lojda 1956, Banks and Newbrey 1982, Goss 1983, Suttie and Fenessy 1990). Depending on the species, this successive process develops at various depths below the growing apex under a special skin called "velvet."

Calcium and Phosphorus

Hydroxyapatite is the main mineral in antlers and is responsible for antler strength. Chemically, it is a crystalline calcium phosphate with a calcium/phosphorus ratio of 2.15:1 (Chaddock 1940, Babicka et al. 1944, Kay et al. 1981, Wuthier 1984, Muir et al. 1987a, 1987b, Muir and Sykes 1988).

In hard antlers, calcium in the ratio is lower in comparison with that of hydroxyapatite, because phosphorus also is a constituent of proteins, phospholipids and alkaline phosphatase that comprise about 34 percent of the antler (Bernhard et al. 1953, Kuhlmann et al. 1962, Kay et al. 1981, Muir et al. 1987a, 1987b, Ivankina et al. 1990, Labetskaya et al. 1990). The calcium/phosphorus ratio changes during growth depending on the progress of mineralization and remains stable after antler death (Rerabek and Bubenik 1963, Kay et al. 1981).

Mineralization

Mineralization proceeds slowly during the first two-thirds of the antler growth. Both the antler cortex and core remain relatively porous and fragile. During the last third of growth, mineralization is dynamic, and the solid cortex develops with a rapid rise in testosterone secretion (Muir et al. 1987a). Formation of the hydroxyapatite in antlers is unknown, but changes of the calcium/phosphorus ratio points to a gradual buildup (possibly from the simple dicalcium phosphate hydrate) over the amorphous calcium phosphate, to the final crystalline form. Toward the end of the mineralization phase, growth stops and the antlers become pointed.

The antler surface experiences intense subperiosteal ossification during the last phase of dynamic mineralization when growth declines. The coronets (burrs), ridges and pearls develop between and above vessels, and long prongs eventually grow from the cortex (Bubenik 1956, 1959, 1982b, Bubenik and Munkačević 1967, Bubenik et al. 1978a). The nature of this process is poorly understood. Perhaps, it is during this phase that the hydroxyapatite crystals become oriented to provide antler strength and resistance to the blows and pressures of sparring and fighting (Kitchener 1992).

Generally, with maturity, the density and specific gravity of antlers decline from 1.9 to 1.2. Miller et al. (1985) suggested that only small antlers (spikes and forks) are nearly totally compact, thus are not strong enough to resist the

Mineralization of antlers, primarily by calcium and phosphorus, requires that during their rapid growth moose must derive these minerals from food intake or depletion of body reserves, primarily bone. When antlers are shed, moose lose that store of minerals. However, other animals benefit, particularly rodents, that chew on shed antlers to obtain the minerals therein for themselves. *Photo by Vince Crichton.*

Shed moose antler leaves a scar or pedicle. The skin rim seals the cleft between the bone and skin of the pedicle, which protects against infection. The rim is where blood supply develops after casting the coronary circuits, and where new vessels grow to supply the growing antler. Mature bulls with large antlers generally cast them in early winter, and immature bulls with small antlers generally retain them until late winter and some even into spring. *Photo by Charles C. Schwartz; courtesy of the Alaska Department of Fish and Game.*

torque imposed on their flexibility during fights. In contrast, palmated antlers develop not only large surfaces, but also a spongy core that strengthens the antler and absorbs the impact of blows and twists during fighting.

The Pedicle and Antler Death

The pedicle is covered by a skin containing many hair follicles with many sebaceous glands, and very rarely, sudoriferous glands. The skin rim seals the potential cleft between the bone and the skin of the pedicle. It protects against infection and prepares itself for future proliferation of the bone-forming tissue after the antler sheds. It is at this location that thick arteries and veins develop after casting the coronary circuits, and where new branches of vessels later grow to deliver nutrients to the growing antler (Rorig 1900, Suttie and Fennessy 1990, Goss et al. 1992).

The circumference of the antler base grows with age, but remains identical with the pedicle. Its outline represents an individual seal whose shape can be recognized on any subsequent antler cast. The seal profile can change from concave to convex depending on the progress of osteoclastic activity at the top of the pedicle. Dead antlers remain connected to the pedicle as long as the core and cortex of the pedicle do not begin to regenerate.

Velvet

According to Goss (1983), velvet begins to develop when the pedicle epidermis overgrows the bare pedicle stump. Hence, all antler growth and mineralization develop under the velvet (Vacek 1955, Goss 1983, 1985, Goss et al. 1992, Bubenik 1993b). Velvet thickness, size, density of sebaceous glands, color and orientation of its hairs vary with the evolutionary age of the species. According to Vacek (1954), velvet is the most sensitive skin among mammals.

The length of the velvet period in species is age dependent and may last from 60 to 160 days. In prime moose, it lasts about 140 days. It is believed that (1) the direct cause of velvet death is impairment of blood flow to the cortex due to narrowing of sinuses by deep mineralization (Wislocki and Waldo 1954) or (2) high testosterone levels are the impetus for shedding (Waldo and Wislocki 1951, Wislocki and Waldo 1953, Bubenik 1966a, Goss 1983). Events such as spontaneous peeling of velvet, velvet shedding before complete ossification, velvet regeneration under dying velvet, and casting of antlers with adherent and mummified velvet (Bubenik and Bubenik 1978, A.B. Bubenik 1990, A.B.

A prime Kenai Peninsula (Alaskan) bull moose in full "velvet." All antler growth and mineralization develops under the velvet. Velvet is the most sensitive skin among mammals and is highly vascular. Damage to velvet during antler development affects normal antler growth, and extensive damage can result in extreme deformities. During the velvet stage, bulls are very protective of their growing antlers when moving through forest. They have a great sense of the antlers' size and thereby avoid injuring them. The "bump" on the skull seen occasionally between the antlers in older bulls is proliferated bone tissue, the result of sparing and fighting during the rut. *Photo by Albert W. Franzmann; courtesy of the Alaska Department of Fish and Game, Soldotna.*

Bubenik et al. 1990a) speak instead for a photoperiodically controlled velvet-vessels anoxia or infarction (Bubenik and Bubenik 1986). A predisposition for this may be the high sensitivity of velvet arteries to vasomotor constriction (Jaczewski et al. 1965, Wika and Edvinsson 1978). The hypothesis of photoperiodically induced anoxia is explained by velvet shedding independent to the progress of antler ossification and the relatively narrow range of shedding dates in contrast to casting dates (Goss 1983). The loss of velvet is the reason why Wika and Krog (1980: 422) call it "a disposable vascular bed."

Vascularization of Velvet and Antler Growth

To keep any tissue alive, oxygen and nutrients must be supplied and waste must be removed. Velvet antlers that transform from mesenchyme to fully ossified bone of 0.5 to 0.75 inch (1.3–1.9 cm) in length per day need a very dense and sophisticated pattern of vessels.

The pedicle is supplied by lateral and medial divisions of the supraorbital branch of the temporal artery (Figure 20). However, the first antlers are supplied only by the lateral branch (Suttie and Fenessy 1990). Several parallel arteries and veins proceed over the shaft. Some of them diverge and develop tines or points; those that remain parallel proceed as beams. Both sides of palmated antlers show different patterns of branching and crisscrossing. At the tip of any point, the arteries converge and narrow into capillaries that submerge like a whirlpool into the core of the growing tip.

Arteries adhere tightly to the posterium, whereas veins subsist more outward in the stratum corneum. The periosteum is attached to the antler bone by Sharpey's fibers. The velvet surface is more or less keratinized, with numerous embedded sebaceous glands and few sudoriferous glands

Normal velvet death is hypothesized to be photoperiodically controlled velvet vessel anoxia or infarction. The length of the velvet period in prime moose lasts about 140 days, and shedding of velvet generally begins in late August or early September. Bulls rub their antlers against trees and shrubs to help the shedding process. Such rubbing also colors antlers to brownish tones. Just after the velvet is shed, the antlers are white. *Photo by Charles C. Schwartz; courtesy of the Alaska Department of Fish and Game.*

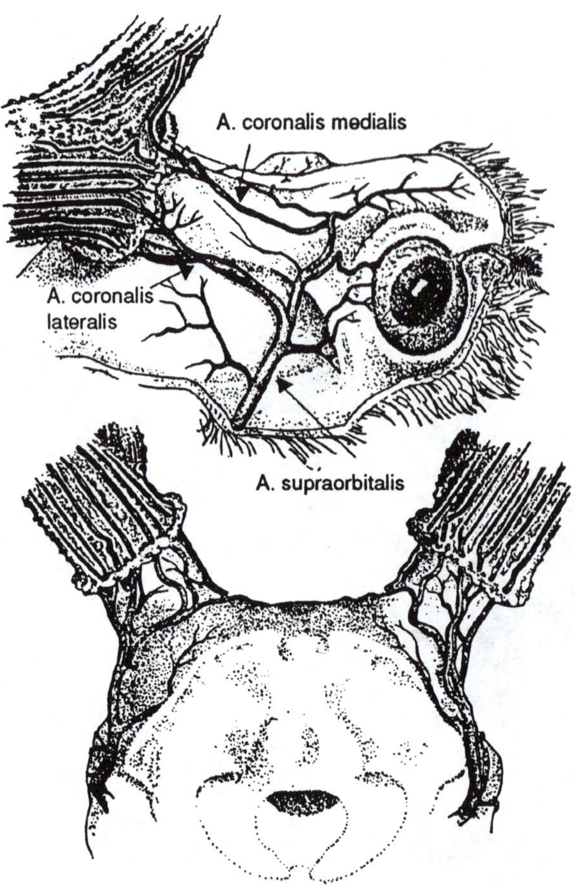

Figure 20. Frontal (top) and dorsal (bottom) views of the circulation system for blood supply to moose antler pedicle and beam from the temporal artery and its branches (from A.B. Bubenik 1966b). A very dense and sophisticated pattern of vessels is needed to transform the mesenchyme to fully ossified bone at the rate of 0.5 to 0.75 inches (1.3–1.9 cm) in length per day.

(Frankenberger 1954, Vacek 1955, Goss 1985, G.A. Bubenik 1993a). Due to intense subperiosteal ossification, the veins and arteries are overgrown by coronet and become embedded within it.

Antler color depends partly on the amount of oxidized blood on the cortex surface and partly on chemical reaction between the blood and juices from plants on which the antlers are rubbed (Bubenik 1966a). Only in well-mineralized antlers do the tips and ridges remain ivory white, because of complete petrification of their osteon. In moose, the frontal area of the palm is less pigmented because this area is more difficult to rub on plants.

Whether velvet antlers have thermoregulatory qualities is still controversial. Ohtaishi and Too (1974) recorded thermoregulatory response against changes of ambient temperature in the velvet of sika deer, but others detected none (Krog et al. 1969, Wika and Krog 1971, Wika et al. 1975, Rayner and Ewen 1981). Hence, it seems that Goss (1983: 163) was correct in stating that thermoregulatory reactions of velvet may be species specific, but "the luxurious blood flow in deer antlers is logically interpreted as a mechanism to ensure elevated temperature, conducive to the rapid proliferation of cells in a structure which must grow fast enough to reach full dimensions in just a few months."

Organic Factors and Antlerogenesis

The timing of velvet shedding and antler cast, the strategy of antler growth and mineralization, as well as the development of antler "bauplans" (innate direction that antler growth and form [architecture] are destined to take) are complicated processes (see Gould and Levontin 1979) that evoke endocrine (hormonal) impacts, enzyme influence

Shed or partially shed velvet of moose antlers reveals extensive grooves and tracts of blood vessels that fed the developing antler. Antlers may grow from 0.5 to 0.75 inch (1.3–1.9 cm) in length per day. *Photo by Charles C. Schwartz; courtesy of the Alaska Department of Fish and Game.*

and neural control. I will not outline all the information available on this subject as it is complex and most comes from deer species other than moose. However, some of the main effects and concepts need to be emphasized.

Hormones

Nearly every hormone-producing tissue and hormone in the body has an influence on antlerogenesis. Effects may be indirect, direct or combined. In the adult, testosterone is important in direct control of antler growth cycles (Blauel 1935, 1936, Wislocki et al. 1947) and is adjusted by photoperiod (Goss 1983). The following hormones influence antlerogenesis.

MELATONIN

Melatonin secretion from the pineal gland positively correlates with length of night and is considered the primary messenger of photoperiodically mediated seasonality in boreal deer (G.A. Bubenik 1990a).

PROLACTIN

Prolactin produced by the anterior lobe of the pituitary gland suppresses testosterone production. Melatonin directly affects secretion of prolactin. Thus, prolactin is considered the secondary messenger of photoperiodically mediated seasonality in boreal deer.

GROWTH HORMONE

This hormone is chemically close to prolactin and suspected of being involved with antler growth (G.A. Bubenik 1990a). Its level rises abruptly with the renewal of antler growth, but drops before dynamic growth begins.

Insulin-Like Growth Factor (IGF_1)

This factor is mainly produced in the liver and to a lesser extent in other tissues (Suttie and Fennessy 1990). It is affected by growth hormone but not under its control, and it peaks with the peak of velvet antler growth.

LUTEINIZING HORMONE

This gonadotropin stimulates the production and secretion of testosterone, which attains its maximum levels during prime years and correlates with the maturation progress (Bubenik and Schams 1986, Bubenik 1991), and social rank in red deer (Freeman et al. 1992), possibly in moose (Schwartz et al. 1990), antelopes (Alados and Escos 1992) and humans (Barlow 1992).

TESTOSTERONE

As the primary androgen (masculine hormone), it plays a triple role in the antler cycle (Bubenik et al. 1975, G.A. Bubenik 1990a, 1993a, Suttie and Fennessy 1991): (1) As a stimulator or inductor of antler growth. This role is still not clear and, thus, controversial. It seems to be essential for induction of pedicle development, but its small peak at the beginning of new antler growth has been recorded many times. The question is whether this level is involved directly or indirectly. (2) As a stimulant of vigorous mineralization of the ossifying trabeculae of antler bone directly or indirectly through corresponding neural pathways in the velvet and bone matrix in the third phase of velvet growth (Bubenik et al. 1975, Li et al. 1988, Barrell and Staples 1991). However, its maximum level is achieved just before peak of the rut. Casting of antlers occurs when the declining testosterone rate reaches a critical level below which osteoclasts of the pedicle are activated. Testosterone level also has a dramatic impact on the speed and direction of mineralization of the pedicle and the profile of the antler base seal (A.B. Bubenik 1990, G.A. Bubenik 1990b). (3) As a probable vector or transmitter of a neural engram, which predestines the quality of subsequent antlers (G.A. Bubenik 1982, 1990a, G.A. Bubenik et al. 1978a, A.B. Bubenik 1990). The importance of testosterone in breeding behavior, social rank and maturity is discussed in Chapter 5.

The absence or deficiency of testosterone achieved by castration or noted in cryptorchids (males with undescended testicles) has a tremendous effect on antlerogenesis. Castration during the hard-antler stage induces casting within 2 weeks and new but uncoordinated antler growth. The antler tissue of castrates proliferates sometimes into bizarre features called freak antlers, perukes, cactus or velveted antlers (Millais 1915, Murie 1928, Seton 1953, Goss 1983, Baber 1987, G.A. Bubenik 1990a, A.B. Bubenik et al. 1990).

Velveted antlers are perennial and frequently retain the species-specific shape (in contrast to perukes or freak antlers) but have long and numerous pearls on the surface, developed by subperiosteal ossification. Their fragile tips can be easily broken, or they are often sequestered and regenerated after years of slow mineralization. Velveted antlers have been recorded in both sexes of moose (Kapherr 1924, Murie 1928, Skuncke 1949, Seton 1953, Wishart 1980, 1990, Hohle and Lykke 1986). A.B. Bubenik et al. (1990) first obtained information on the histology of velveted antlers (Figure 21). This was a prime bull, and the velveted antlers were partly broken and regenerated many times. The cross section and radiographs revealed centrifugal mineralization with simultaneous subperiosteal ossification. Consequently, the antlers were continually gaining volume and could regenerate lost parts. The superficial blood vessels were bedded deeper and deeper into the antler core by enveloping them in new compact osteon.

Abnormal moose antler resembling coral in structure and texture is referred to as a "peruke head." In all likelihood, this bull had low testosterone output attributable to advanced age (senescence) or was an accidental castrate. Generally, the normal calcification or hardening process is reduced and such antlers continue to grow. Blood supply to these structures is reduced, however, and eventually, through freezing, the "antlers" fall off. Female hermaphrodites (female having both male and female sex organs but with primary female characteristics) also may develop peruke antlers. *Photo by Rolf Peterson.*

TESTOSTERONE DERIVATIVES

G.A. Bubenik (1990a) indicated that testosterone derivatives cannot be excluded as possible factors involved in the antler cycle, but further studies are necessary.

GLUCOCORTICOIDS

Cortisol is the major glucocorticoid produced in the adrenal gland. It inhibits growth by slowing cartilage development and thus reducing the length of antlers, as observed in roe deer (Bubenik et al. 1976). Cortisol secretion is regulated by adrenocorticotrophic hormone (ACTH). White-tailed deer bucks that responded poorly to intravenous ACTH administration produced generally inferior antlers, but bucks that responded vigorously to ACTH produced high cortisol levels and grew excellent antlers (Bubenik 1991).

MINERALOCORTICOIDS

Mineralocorticoids are a product of the adrenal glands and have not been studied in deer in general or in antlerogenesis in particular, but they may be involved indirectly in antler mineralization.

ADRENAL ANDROGENS

Adrenal androgens may substitute for testicular androgens (testosterone) in cryptorchids or in females with velvet antlers (A.B. Bubenik 1990, G.A. Bubenik 1990a).

ANDROSTENEDIONE (A_4)

A_4 is produced by the adrenals and testes, especially during puberty. It may be a precursor or metabolite of testosterone. In white-tailed deer, it begins to increase in blood with the initiation of new antler growth, and peaks first between June and August and then again in November. The November peak coincides with the acme of testosterone during the rut. It was postulated that A_4, possibly in synergism with estradiol, is responsible for the subperiosteal ossification, which leads to the development of coronets, pearls and ridges on the surface of nearly completed antlers (G.A. Bubenik et al. 1990).

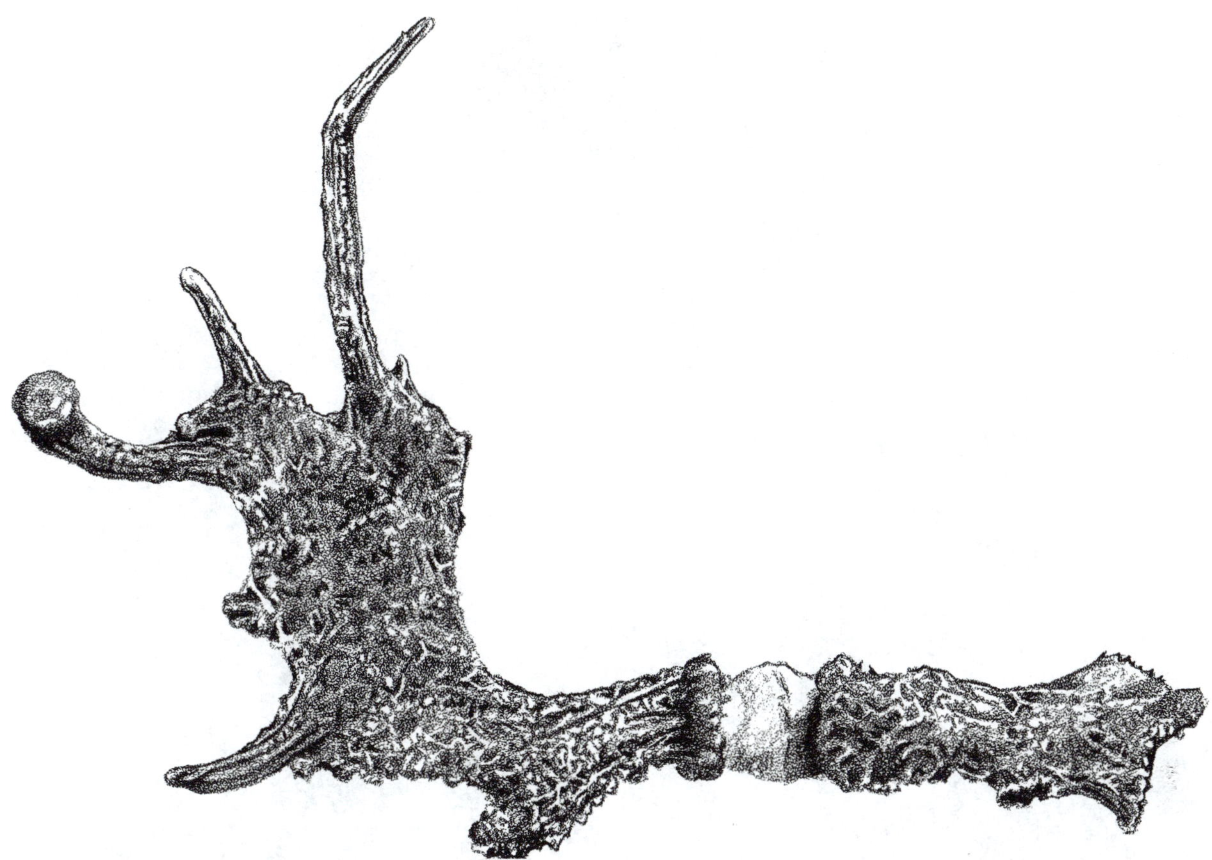

Figure 21. Frontal view of velveted antlers of a prime, cryptorchid moose. *Illustration by A.B. Bubenik.*

THYROXINE (T_4) AND TRIIODOTHYRONINE (T_3)

T_4 and T_3 are both anabolic hormones. T_3 is produced in the thyroid, and T_4 is its tissue active metabolite. During the most intensive antler growth period of white-tailed deer (June–July), T_4 was significantly higher and T_3 significantly lower in blood within antler veins compared with levels in the jugular and saphenous veins (Bubenik et al. 1987b). These differences disappeared during the mineralization phase in August. According to Fischer (1987), T_4 shows antagonistic trends toward testosterone. The importance of T_3 is evident from the study of Hasling et al. (1987), who reported that this hormone is involved in bone remodeling.

CALCITONIN AND PARATHYROID HORMONE (PTH)

Calcitonin produced in the thyroid and PTH in the parathyroid are heavily involved in calcium turnover of the skeleton and antlers. As will be discussed later, calcium metabolism relies on calcitonin and PTH supplied from forage and also that resorbed from the skeleton. It is postulated that both sources of calcium are involved indirectly in the growth and mineralization of antlers.

OSTEOCALCIN

Osteocalcin likely stimulates the development of osteoblasts—the building elements of antler tissue—with alkaline phosphates and T_3 (Hasling et al. 1987).

Vitamins, Enzymes and Amino Acids

Other organic compounds are involved directly or indirectly in antlerogenesis. Some are products of, or control calcium metabolism. Two vitamins (1,25 dihydroxyvitamin D; vitamin D_3) and one enzyme (alkaline phosphatase) have synergism during the intense phase of antler growth (G.A. Bubenik 1990a).

Serum hydroxyproline, an amino acid, is an indicator of collagen (the connective tissue of bone) degradation and,

These abnormal (peruke) antlers have adorned the home of Henry Woodsworth of Biscotasing, Ontario, since 1909. While working with the Hudson's Bay Factory at Bisco, Woodsworth was informed by a terrified Indian that the Indian had just shot a moose he called *weetogo* (meaning "devil" in Cree) and left it in the woods. He wanted no part of it. Woodsworth went to the site and found these antlers. They weighed 18 pounds (8.2 kg) and had a 41-inch (104-cm) spread. The color was normal but the antlers resembled coral in structure and roughness. Local Indians had great fear of the antlers and would not touch or go near them. Although this peruke was in view in the Chapleau office of the Ontario Department of Lands and Forests, an Indian woman who saw them covered her face and rushed from the building. Reportedly, the Cree who killed the moose was himself killed a year later when run over by a train. *Photo courtesy of the Ontario Department of Lands and Forests.*

thus, of calcium flux from bone reserves to other organs. Flynn et al. (1980) found that the magnitude of calcium flux in male moose was greater than in females (Figure 22). They attributed this to greater utilization of skeletal reserves in the last phase of antler growth. This was later confirmed in several reports (Kay et al. 1981, Muir et al. 1987a, Muir and Sykes 1988, Van der Eems et al. 1988, Burr and Martin 1989). Flynn et al. (1980) reported that, in both sexes, the calcium flux was markedly higher in May through August than in February. However, the magnitude of the male response was greater and proceeded until July. According to Rasmussen (1977), a lactating rat can absorb as much as 50 percent of skeletal ash. How much skeletal absorption moose experience to grow antlers is unknown, but apparently it does occur, as discussed later.

Van der Eems et al. (1991) found two components detectable in urine that are indicators of enhanced antler mineralization. They were urinary cyclic AMP (a messenger that mediates parathyroid hormone action) and gamma-carboxyglutamic acid.

Where they have a direct impact, local growth factors are produced by certain tissues or they may be transported to special target organs. Those suspected of being involved in antlerogenesis are retinoic acid (Thaller and Eichelle 1987,

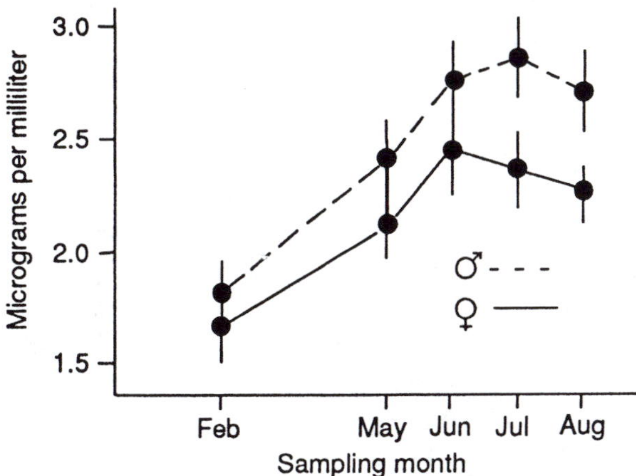

Figure 22. Seasonal curve of serum hydroxyproline levels from female and male moose (from Flynn et al. 1980). Hydroxyproline is an indicator of the degradation of connective tissue in bone, and its measurement reveals calcium flux from bone to other tissues. The amount in the blood of males is greater attributable to demands of antler growth.

G.A. Bubenik 1990b, Gianni et al. 1991, Noji et al. 1991), platelet-derived growth factor, fibroblast growth factor, cartilage-derived growth factor, bone morphogenic protein (Suttie and Fennessy 1991), epidermal growth factor (Ko et al. 1986, Yuncheung et al. 1987) and nerve-growth factor (Davies et al. 1987, Suttie and Fennessy 1991).

Neural Control

At least one important, yet puzzling characteristic of antler growth is the independence of each pedicle and antler (Gaskoin 1856, Brandt 1901). Bubenik and Pavlansky (1965) concluded that the antler bauplan was controlled in the brain, and Bubenik (Bubenik and Pavlansky 1965) hypothesized that, despite being a disposable part of the skeleton, antlers were retained in the phantom memory. They hypothesized that two independent centers for antler growth may be responsible for the phantom memory for velvet antler injuries and hormone levels. This hypothesis was later supported by results of electrical stimulation of the trigeminal nerve (Bubenik et al. 1982) and pedicle transplants (Jaczewski 1990).

Other support for neural control comes from studies of the asynchronous antler cycle. Distinct antler phases may be switched on or off under neural command but not necessarily at the same time in both pedicles. Studying the antler cycle in red deer, Topinski (1975) reported that stags under stress have the most delayed asynchrony. When they limited the time of synchrony to 24 hours, Bartos and Perner (1991) did not find any difference between red deer with synchronous or asynchronous casting. However, the asynchronous interval correlated negatively with individual rank.

Regardless of where the antler bauplan is controlled, the neural impact on size, shape, symmetry and asynchrony of antler growth cannot be neglected. Memory for velvet injury in early stages of antler growth and levels of those hormones that are involved in growth are not anecdotal. This must be considered a unique neural phenomenon.

Antlers of Moose

Antler Terminology

Because there is no official terminology for antlers, and for moose antlers in particular (A.B. Bubenik 1982b), I use those depicted in Figure 23. The right palm shows the antler in a normal view, the left is a replica obtained by tracing the outline on paper in which the antler was wrapped.

CORONET OR BURR

The coronet or burr is the elevated bony rim around the antler base just above the skin of the pedicle.

SEAL

Seal is the term for the mostly ivory-looking bone at the antler base. Because its shape corresponds to that of the pedicle, it is individually formed and can be used as a "fingerprint" of each specimen.

SHAFT

The shaft is the perennial and annually generated part of the antler base before branching occurs. In moose, it forms an ellipse in cross section.

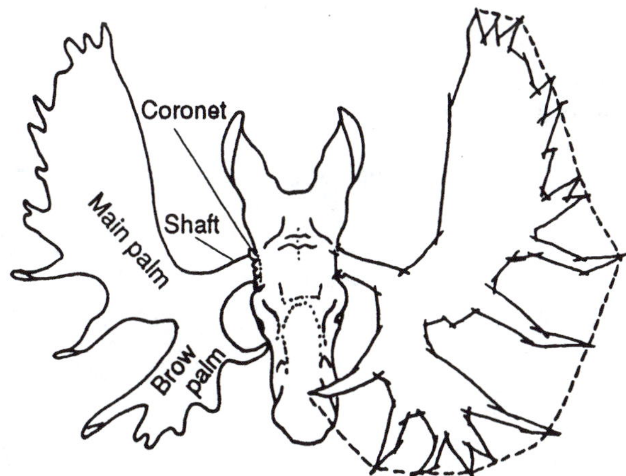

Figure 23. Frontal view of actual (left) and replica (right) moose antler. *Illustration by A.B. Bubenik.*

The coronet or burr of moose antler is an elevated bony rim around the antler base just above the skin of the pedicle. Bone carvers and artisans seek shed antlers for their trade, and especially those from prime bulls that have the large coronet, which is crafted into belt buckles. *Photo by Charles C. Schwartz; courtesy of the Alaska Department of Fish and Game.*

Basic Shapes of Moose Antlers

The long historic development of recent moose antlers from the libralcin design progressed so slowly that it is difficult to decide if alcins are an independent subfamily of the Middle Pleistocene or just the end of libralcin evolution (Figure 24). This slow trend and genetic adaptation may be why there are different varieties of moose antlers.

SIZE AND SHAPE OF ANTLER SURFACE

The spherical shape of moose antlers make them ideal for tracing their outline on paper. The resulting traced shape, as a flat replica, can be either a single or deep, engulfed ellipse. Because of the curved plane of the antler's surface, only the paper replica allows for a reliable measurement of the actual surfaces and an estimate of the degree of asymmetry by overlapping the right and left replicas.

CERVICORN ANTLERS

This antler form (Figure 25) was reported as common in yearlings and only in adult moose in Maine and adjacent to New Brunswick. This appears to be a localized phenomenon because other Eastern moose seldom produce cervicorn antlers at or after 2 years of age (Gasaway et al. 1987).

PALMICORN ANTLERS

Palmicorn antlers are the form seen most often in North American moose 2 years of age and older (Gasaway et al. 1987). In my view, the development of dorsally elongated palms (Figure 26) resulted from (1) natural selection for greater body size, (2) social effect that the large surface had as a visual and scent releaser and (3) female choice for the largest socially tolerable species-specific antlers. From the Early to Middle Pleistocene, alcine palms experienced change in shape and size. This alternating change was influenced by the variability of habitat and climatic condi-

Figure 24. Depiction of the process of transformation of libralcin antlers to *Alces* palms. The dashed line represents the presumed shape, intermediate to *Alces alces*. This process progressed very slowly and, in the absence of examples of clear progression, some questions remain. The slow adaptation may explain the present different varieties of moose antlers. *Illustration by A.B. Bubenik, based on a photo at the Museum of Natural History in Zagreb, Croatia.*

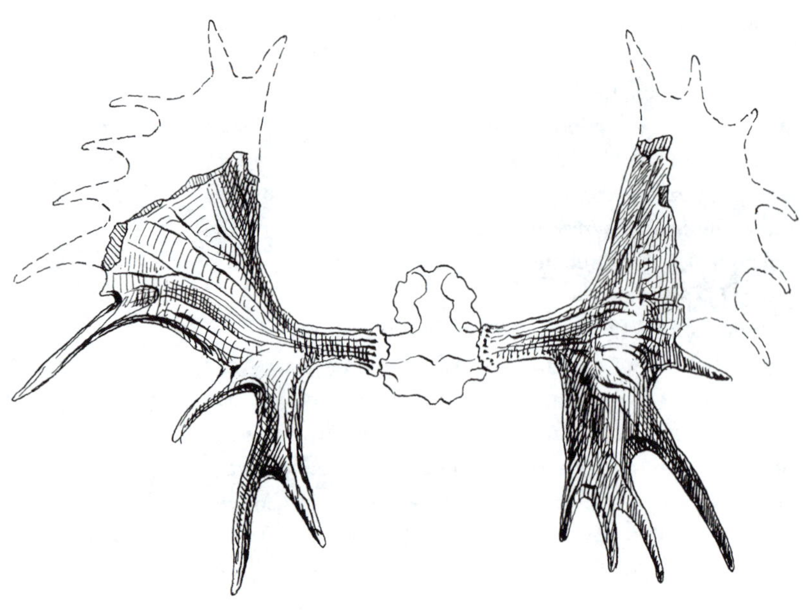

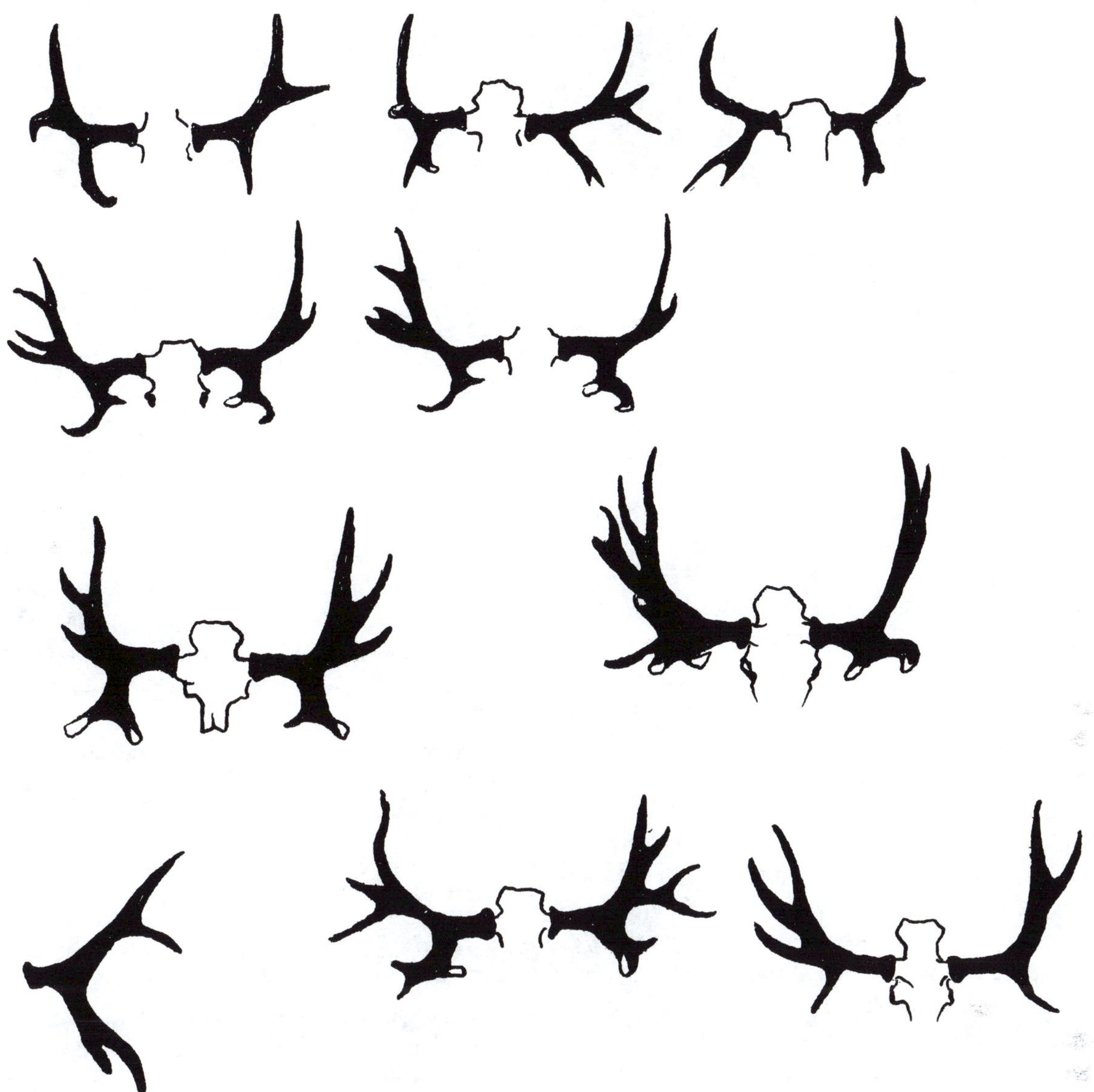

Figure 25. Cervicorn moose antlers grow radially, not subsequently as do true cervicorn antlers, such as those of red deer. *Illustration by A.B. Bubenik.*

tions (Sher 1987). From this, the symmetrical ellipsoidal or shell-form developed.

Further development led to the dorsally elongated and frontally shortened ellipsoid. This architectural change made it possible to maintain one solid palm that shielded the forehead. Eventually, development of an engulfment between the medium and frontal arteries permitted the accrual of a curved antler surface and a bipalmated or "butterfly" antler with the frontal part lifted and tilted above the forehead, producing an ideal shield for the face.

It is interesting to note that the wide brow-palm with numerous points is more frequent in the relatively mild tempered tundra moose, whereas the more aggressive taiga moose has a narrower and less pointed brow-palm. Generally, a multipointed brow-tine is characteristic of mature bulls.

SHAFT CIRCUMFERENCE

Shaft circumference in taiga moose increases gradually—to approximately 7 inches (18 cm)—until 6 years of age. Thereafter, the increase is much slower, reaching maximum circumference by 9.5 years (Timmerman 1971, Bubenik et al. 1978a).

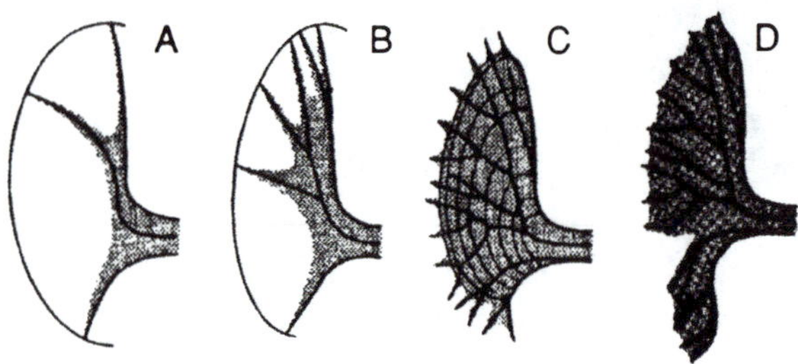

Figure 26. Palmicorn antlers evolved from the cervicorn type by elongation of the dorsal palm. A = cervicorn antler; B = developing palmation; C = palmicorn antler of the "shell type"; D = palmicorn antler of the butterfly type. *Illustration by A.B. Bubenik.*

ANTLER SPREAD

Antler spread is the largest distance between the outermost points of the antlers. This measurement fascinates naturalists and hunters, but there is no sociobiological evidence to justify that it is the most important feature of antlers. Any pair of antlers with the same surface volume and weight can have either a narrow or wide spread, depending on how close to parallel the inner edges of the main palm are and the orientation of the outer points.

The antlers of Alaskan moose are larger, heavier and have greater spread than those of taiga or Shira's moose (Gasaway et al. 1987). The spread increases with age for the first 5 to 8 years and then little through age 12. Alaskan (tundra) moose attain maximum spreads of more than 80 inches (203 cm), taiga moose more than 70 inches (178 cm), and Shira's moose more than 60 inches (152 cm) (Timmermann 1971, Gasaway et al. 1987).

The general north-to-south decline of antler weight may follow the rule that antler weight in large deer positively correlates with body weight (Huxley 1931). It also complies with Bergmann's rule of body size and Geist's (1986) theory of dispersal.

MORPHOMETRIC RELATIONSHIPS

Bubenik et al. (1978a) analyzed morphological characteristics of 273 small- to medium-sized antlers, with fresh weights up to 26.4 pounds (12 kg) and surface areas of as much as 496 square inches (3,200 cm^2). They found highly significant linear correlations between (1) surface area to antler volume, (2) surface area to air-dried weight and (3) main palm surface to whole surface area. The specific gravity of medium- to optimal-sized antlers was 1.3 ± 0.12.

Morphological Evolution

As discussed earlier, there is reasonable suspicion that the libralcins may represent an ancestral lineage of deer from the Middle Pliocene. If this is correct, I assume that cervoalcin antlers had slender tines of the "pseudocervicorn" bauplan, because the almost crescent-shaped antlers of the first libralcins seemed to grow radially and not successively as do cervicorn antlers of red and white-tailed deer. Under radial growth, almost all points stop growth at once. In cervicorn construction, the lowest tines develop earlier than do the uppermost tines (Rorig 1906, Muir et al. 1987b, Fennessy et al. 1991; Figure 27).

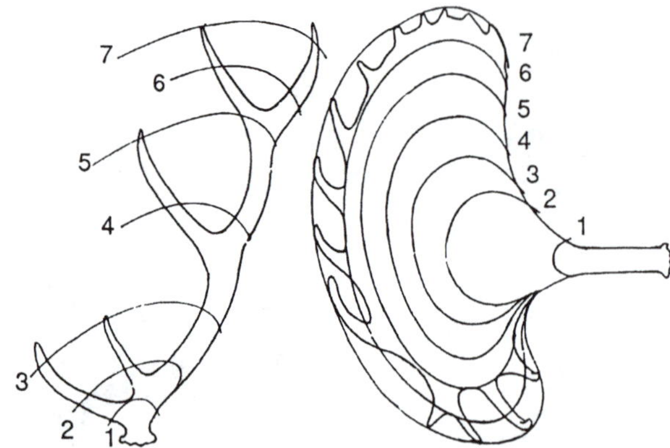

Figure 27. Schematic representations of cervicorn antler growth in red deer (left) and radial antler growth in moose (right). *Illustration by A.B. Bubenik.*

Patterns of Antler Growth

There is evidence of parallel waves visible on the corrugated surface of the moose palm. On a radiograph, these waves appear as discrete bands of denser and thicker bone (Kay et al. 1981). On the growing antler, they appear as strips of white (pigmentless) hair on velvet, creating a puzzling pattern. Sometimes, the white strips run parallel to the expanding edge and points. Other times, their course is erratic, making it impossible to correlate them with palm surface growth.

Kay et al. (1981) and Goss (1983) assumed the bands on the palm, about 0.8 inches (2 cm) apart, developed in 24-hour periods. However, based on prime antler development of 50 bands in 140 days (Van Ballenberghe 1982), I calculate that almost 48 hours are required for one wave. My recent studies indicate that the discrete band formation is much more complex, and more intensive work is needed.

Antlerogeny: Development of Subsequent Antlers

In moose, antler form develops successively from the second, mostly cervicorn set of antlers. It generally takes 4 or 5 years for the final antler shape and form to develop. Final

Some patterns of white (pigmentless) hair on the antler velvet of moose run parallel with the expanding edge and points, but on others, the course is erratic, making it impossible to correlate them to palm surface growth. *Photo by Charles C. Schwartz; courtesy of the Alaska Department of Fish and Game.*

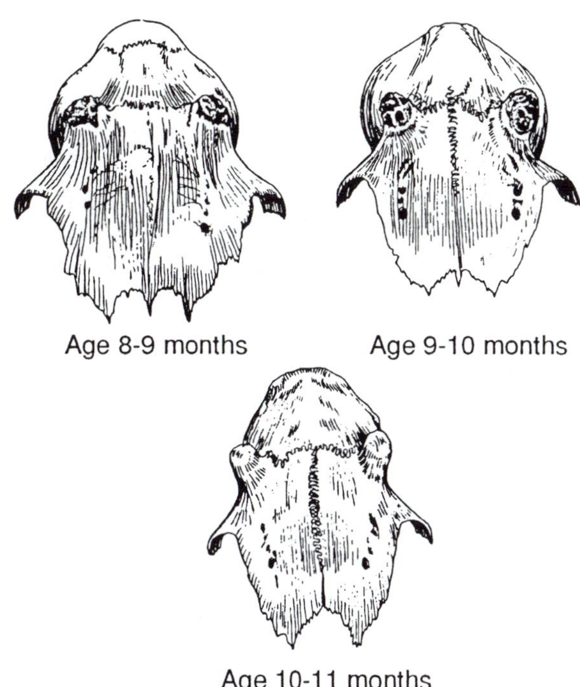

Figure 28. First antlers as they appear on the skull of a moose calf. *Illustration by A.B. Bubenik.*

size and weight appear near the end of sexual maturity. It can be concluded that, as a rule, the rate of antlerogeny is an indicator of population fitness.

FIRST ANTLERS OF CALVES

In well-developed male calves, pedicles appear during September or October. Their apices ossify at different lengths between November and December when the calf is 6 to 7 months old (Figure 28). If 0.5 inch (1.3 cm) or less of apex becomes mineralized, it remains covered by skin, and shiny almost hairless epidermis indicates that the skin was rubbed. If 1.0 inch (2.5 cm) or more of the apex mineralizes, the skin is shed and minute thumb-like antlers appear (Lunberg 1956, Bubenik 1963, Sugden 1964, Bubenik et al. 1978a, Filonov 1983).

Characteristic of first antlers is the absence of the coronet. A swollen, occasionally pearled base indicates the zone between the pedicle and the antler. If the first antlers develop, they generally are cast in March or April. Those that remain under the skin begin to grow after their osteon presumably demineralizes and the cartilaginous trabeculae is incorporated into the tip of the second antlers.

YEARLING ANTLERS

The second set of antlers develops during late spring and summer, when the moose becomes 1 year old. Each antler

Well-developed male moose calves have antler pedicles that appear during September or October. Their apices ossify in November or December, at 6 months of age. *Photo by Charles C. Schwartz; courtesy of the Alaska Department of Fish and Game.*

is about 10 to 13 inches (25–33 cm) long, with three or more points, a visible coronet and occasional palmation (Figure 29).

Prongs are very common in the second antlers and infrequent in the third antlers of bulls in good fitness (A.B. Bubenik 1982c). Prongs grow directly from the cortex, not from bifurcation of the beam as in other deer (A.B. Bubenik 1975a, 1982c, Bubenik and Munkačević 1967). The prongs of moose grow right above the coronet as small quasi-brow-tines parallel to the eyes. They also may be a relict from cervoalcin ancestry, but their significance and lack of appearance after 3 years are unknown.

Another interesting feature of second antlers is the elevated position of the pedicles and antlers. This probably is a homoplasic reminder of the cervoalcin ancestry. Why the shafts of some antlers converge is unknown. Generally, it takes 3 years for the pedicles to assume an almost horizontal orientation, possibly attributable to the weight of the palms. Second antlers generally are cast during February or March.

THIRD TO FIFTH ANTLERS

Bull moose from 2 to 4 years of age do not carry fully developed antlers. These submature bulls have incomplete head pigmentation, small palms, sometimes only brow forks, and main palms that seldom diverge more than 65 degrees.

During antlerogeny, the number of points increases to a maximum of around 15 per antler. Generally, the fifth antler is close to the shape of a prime antler, but the surface area and the shielding effect for the forehead still are not fully developed.

ANTLERS OF PRIME BULLS

Prime antlers are carried by moose aged 5 to 12 years. These antlers show a great range of variance in both taiga and tundra moose (Gasaway et al. 1987). They grow steadily in size until the bulls' tenth year, when they reach optimal size and form. Between the eleventh and thirteenth

Figure 29. Spike and forked-type antlers from immature bull moose. Some harvest programs are focused on harvesting immature bulls, and the spike or forked antler designation is used to describe legal bulls (see Chapter 17). *Illustration by A.B. Bubenik.*

years, a plateau in antler growth seems to occur (Timmermann 1971, Bubenik et al. 1978).

The architecture of the brow-palms and their points appear to have behavioral significance. In my opinion, only brow-points provide optimum defense of the forehead. Most prime bulls have this characteristic. Divergent tines gradually narrow from the third to fifth year (Bubenik et al. 1977, 1978b). The offensive and defensive shaping of the brow-points in prime class taiga moose shows great variability (Bubenik et al. 1977, 1978b), but the reasons are unknown.

Antler sets were taken from the same bull in consecutive years at the Moose Research Center, Alaska. The lower set, from age 1.4 years, had a 40.5-inch (103-cm) spread. The next set up, from 2.4 years, had a spread of 50.8 inches (129 cm). The next, from 3.4 years, featured a spread of 58 inches (147 cm). And the top antlers, when the moose was 4.4 years old, had a spread of 61 inches (155 cm). This growth rate is from an animal that had an unlimited food supply and had maximized his growth and development. *Photo by Charles C. Schwartz; courtesy of the Alaska Department of Fish and Game.*

ANTLERS OF SENIOR AND SENESCENT BULLS

On the basis of present knowledge of moose antlerogeny, it is unknown if bulls 12 or 14 years old achieve seniority or if senescence begins before or after 16 years of age. We know that bulls whose faces turn brownish, whose antler palms become narrow or twisted and whose brow-points become reduced to a single fork or one relatively long tine are close to the end of seniority (Figures 30 and 31).

Antler Growth and Calcium Requirements

Knowledge of antler chemistry of moose is lacking, especially concerning the calcium/phosphorus ratio and dry matter content during the growing phases. According to Kay et al. (1981), the calcium/phosphorus ratio in antlers begins to increase from 0.4 to 1.5, and when this ratio reaches 2.0, the antlers become hard. In red deer, Muir and Sykes (1988) reported that dynamic mineralization occurs during the last 20 days before velvet shedding. This is likely the time frame for moose as well. It occurs with the cessation of growth and rapid development of the matrix throughout the antler and deposition of large amounts of hydroxyapatite (Fennessy et al. 1988). Studies have shown that, in deer, osteons of the skeleton are in a permanent rebuilding process and, during the mineralization phase, the skeleton undergoes temporary osteoporosis (softening and fragility by decalcification) (Meister 1956, Banks et al. 1968a, 1968b, Muir et al. 1987a, 1987b, Muir and Sykes 1988, Fennessy et al. 1991). Peterson (1977) and Hindelang et al. (1992) detected osteoporosis in moose from skeletal remains on Isle Royale National Park, Michigan.

If moose develop a calcium deficiency as deer do, then calcium must be resorbed, probably first from costal skeleton (Banks et al. 1968b, Brown 1990). Ribs and shoulder blades, which may lose as much as 40 percent of their calcium during this process (Banks et al. 1968b), become osteoporotic, perhaps explaining why so many moose have broken ribs and shoulder blades (Skunke 1949, Knorre 1959) or why they have a tendency to become arthritic (Peterson 1974, Timmermann and Lankester 1978, Peterson 1988). Calcium and phosphorus deposited in antlers is irreversibly lost. Deer can remineralize their costal skeleton by September (Banks et al. 1968b), but it is not known if the same holds for moose. Bull moose enter the rut shortly after mineralization of the antlers and do not eat during this period. Consequently, bone maladies of moose may result from the impact of rutting activities and prolonged osteoporosis. Of course, it is natural that bulls with smaller antlers may have a smaller calcium deficit, which may explain why small bulls tend to become dominant toward the end of the rut.

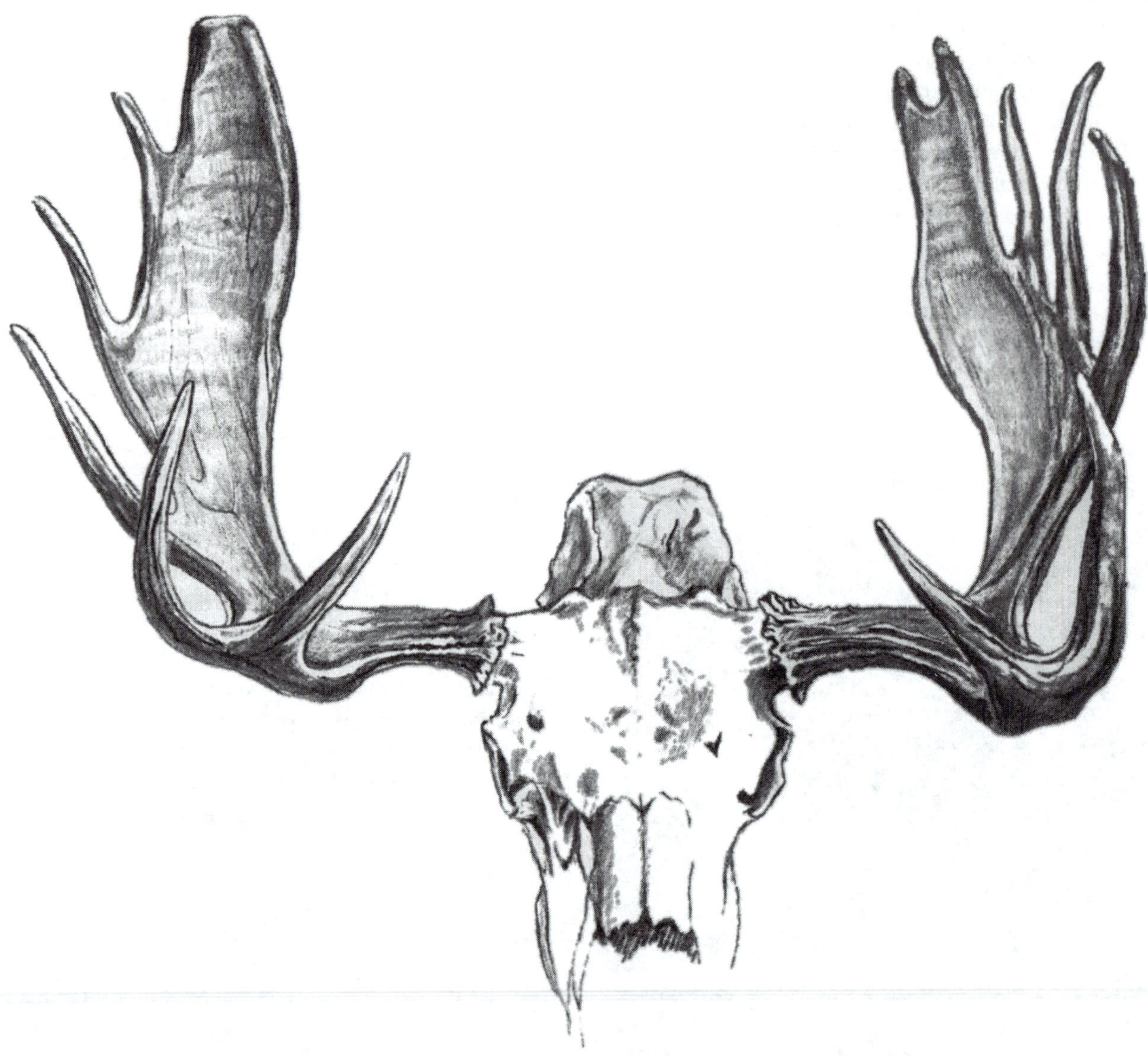

Figure 30. Antlerogeny—the development of subsequent antler sets during aging—of senior bull moose seems to be quite variable, probably depending on the rate of decline in sexual activity, reduction of social rank and diminished nutritional intake or digestion due to tooth wear. By virtue of a dearth of senior bulls in many moose populations during the past 30 to 50 years, knowledge about seniority is incomplete. Antler mass of senior bulls is increasingly concentrated toward the lower portions of the antlers, which is characteristic of virtually all senior male deer (A.B. Bubenik 1966, 1990). *Illustration by A.B. Bubenik.*

Other causes of osteoporosis include lowered calcium absorption, storage and stress (Albon et al. 1992, Bubenik and Pond 1992).

Antler Anomalies

Antler anomalies are almost as common in moose as in other deer. Only a lack of testicular testosterone in cryptorchids or physical castrates has a profound effect on antlerogenesis. Otherwise, the common anomalies are as follows.

Abnormal Asymmetry

Antlers, in contrast to horns, generally show some degree of asymmetry, which is considered quite normal (A.B. Bubenik 1990). Abnormal asymmetry generally is caused by injury at the beginning of antler growth (Figure 32). These trophic responses disappear in the next or subsequent antlers (Bubenik 1966a). It is unknown if such asymmetries are hereditary.

My collection corroborates that, in moose, transspecific

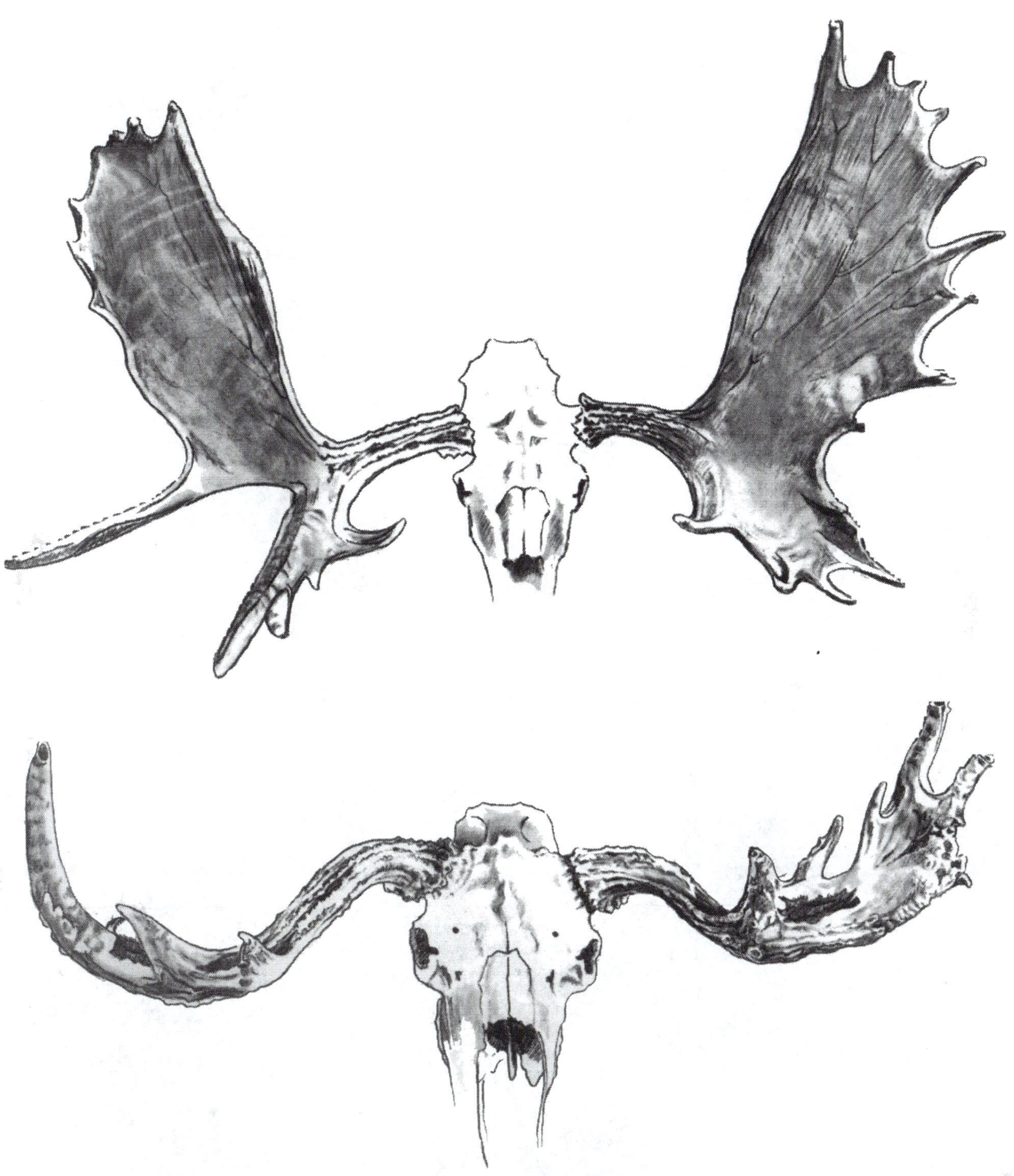

Figure 31. As with antlerogeny of senior bull moose, information about antler development of senescent bulls is inadequate. Nevertheless, it is known that antler mass of senescents shrinks substantially, palms can become notably asymmetrical (top), and shafts may appear undulated and are unable to develop or maintain palms (bottom). Shaft, beam and tips tend to be malformed and fragile, and often have pieces of mummified velvet adhering until casting, evidence of lost drive to shed velvet and lost sex drive. The top antler set is from a bull 16 years old, and bottom set is that of a typical senescent specimen that was about age 20. *Illustrations by A.B. Bubenik.*

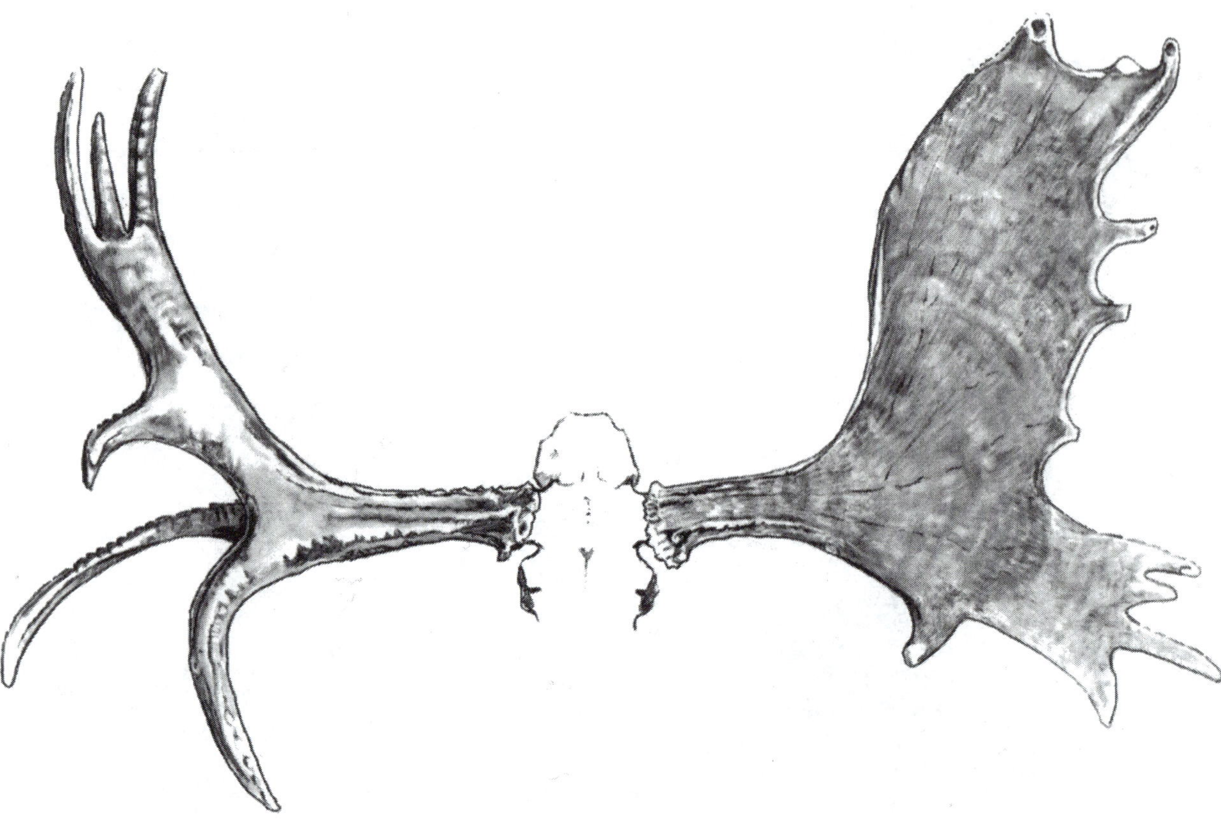

Figure 32. Antler asymmetry is the rule rather than the exception, and is most commonly explained in moose by injury to early growth antler. The antlers featured above were cast synchronously. *Illustration by A.B. Bubenik.*

Antlers of tundra (Alaskan) bull moose are about 40 percent heavier than those of taiga (Eastern) bulls. It is speculated that the difference is in the thickness, assuming that a density of 1.2 g/cm^3 is the same. The reason for the thicker antlers in tundra bulls, such as the one above that is attaining maximum antler development, is unknown. *Photo by Charles C. Schwartz; courtesy of the Alaska Department of Fish and Game.*

When developing and poorly mineralized, moose antler palms may be pierced by sharp branches, resulting in partially healed holes. *Photo by Vince Crichton.*

Moose, like other deer, occasionally develop antlers with multiple palms. These occur when the developing antler bud splits vertically. *Photo by Karen Morris; courtesy of the Maine Department of Inland Fisheries and Wildlife.*

bauplans appear perhaps as the result of some genetic mishap (A.B. Bubenik 1966a, 1990).

Split or Double Antlers

Like other deer, moose occasionally develop antlers with multiple palms. These double or multiple palms occur when the developing antler bud splits vertically in response to a longitudinal or transversal incision (Bubenik and Pavlansky 1965).

Velericorn or Antlered Cow Moose

Antler protuberances and antlers have been reported from cow moose that, for some reason (e.g., tumors, cystic or senescent ovaries), produce testosterone (Alston 1879, Goss 1983, Hofmann 1987, Wishart 1990). Some antlered cows with external female characteristics (Wishart 1990) may be hermaphrodites, as have been found in red deer (Stewart-Scott et al. 1990). Also, fertile, lactating, velericorn cows have been reported (Wishart 1990).

Genetics and Antlerogenesis

Very little is known about moose genetics and nothing about genetics as they relate to moose antler size and shape. This knowledge gap, coupled with the controversy surrounding selective culling programs, social well-being and trophy hunting, make imperative that more be learned about the role of genetics and the genetic links to antlers of moose.

Sociobiological Versus Formula Scoring of Antlers

A curious phenomenon of human conceptualism is the desire to measure and compare. In science, measurement is purposeful and necessary. In the case of antlers and other trophies, measurement and comparison have significance only if they relate to parameters important for the species' ecological well-being.

Scoring formulas for harvested trophies are biased because they were developed by hunters according to subjective criteria of imagination and aesthetics (Webb 1984). None is based on parameters of sociobehavioral importance for the species, and I wonder why biologists also use them for morphometric comparisons when the leading parameters are the backbone of the scoring formula (Bubenik 1968b). Inasmuch as antlers are one of the best indicators of fitness, there is both need and opportunity for development of a scoring formula that suits the species and allows for comparisons among biologically fit and unfit antlers and for assessment of population status.

Moose managers and hunters should first be aware that antlers are a product of neurally controlled metabolism and its rules. Hence, the produced antler bone mass deserves primary attention and should not be evaluated with the one-dimensional measurement systems currently used. Second, the formula must respect architectural importance, including offensive and defensive capabilities of the whole structure and the range of their limitations. Third, ranges of antler quality must be significantly dependent to show sociobiological value. Finally, although antlers are a disposable part of the skull, craniometric parameters also must be considered for qualitative and taxonomic assessment and indexing.

From a naturalistic viewpoint, moose display antlers best when viewed in a vertical position in or below our horizon. This also is the perspective used by moose that best shows the grandeur and beauty of these magnificent products of nature (Figure 33).

Musculature of Moose

Because of the peculiarities of the musculature and limbs, moose can rest and even sleep standing. Similar to those of other deer and cattle, the bright red muscles of moose are interwoven with many strong tendons (Figure 34). Unlike in domestic cattle, there is very little fat deposited within and between the fibers of moose muscle; thus, the meat is very lean. The largest muscles are on the neck, shoulders and hind legs.

Knorre (1959) estimated that a mature, well-conditioned bull moose produced 62 percent meat. Peterson (1955) estimated a 59 percent yield for Ontario moose. According to

Figure 33. Best positions to display moose antlers. The head-down presentation (left) gives an optimal perspective on the form of moose antlers. *Illustration by A.B. Bubenik.*

Peterson's data from a 1,300 pound bull (590 kg), one forequarter (apparently with the neck muscle) weighed 218 pounds (99 kg), and the hindquarter weighed 167 pounds (76 kg) (see Chapter 18). For comparisons with other deer, see Komarek (1958), Hegerova and Štěrba (1959), Thomas and Toweill (1982) and Halls (1984).

Muscle Swelling

The neck and shoulder muscles of male deer swell during the rut (Knorre 1959). Schnare and Fischer (1987) found in fallow deer that the swelling is caused by raised and accumulated creatinine, whose concentrations correlate well with blood plasma testosterone. This likely explains the swollen neck in rutting bull moose.

Central Nervous System

Brain

When the intelligence of an animal species is ranked, often it is based on the size of their brain and expanded gray matter (Wirz 1950). Comparing the cortical surface of the moose brain with brain size of other mammals (Jerison 1982), the moose ranks high in the mammalian hierarchy. It may be close to that of the horse and higher than red deer

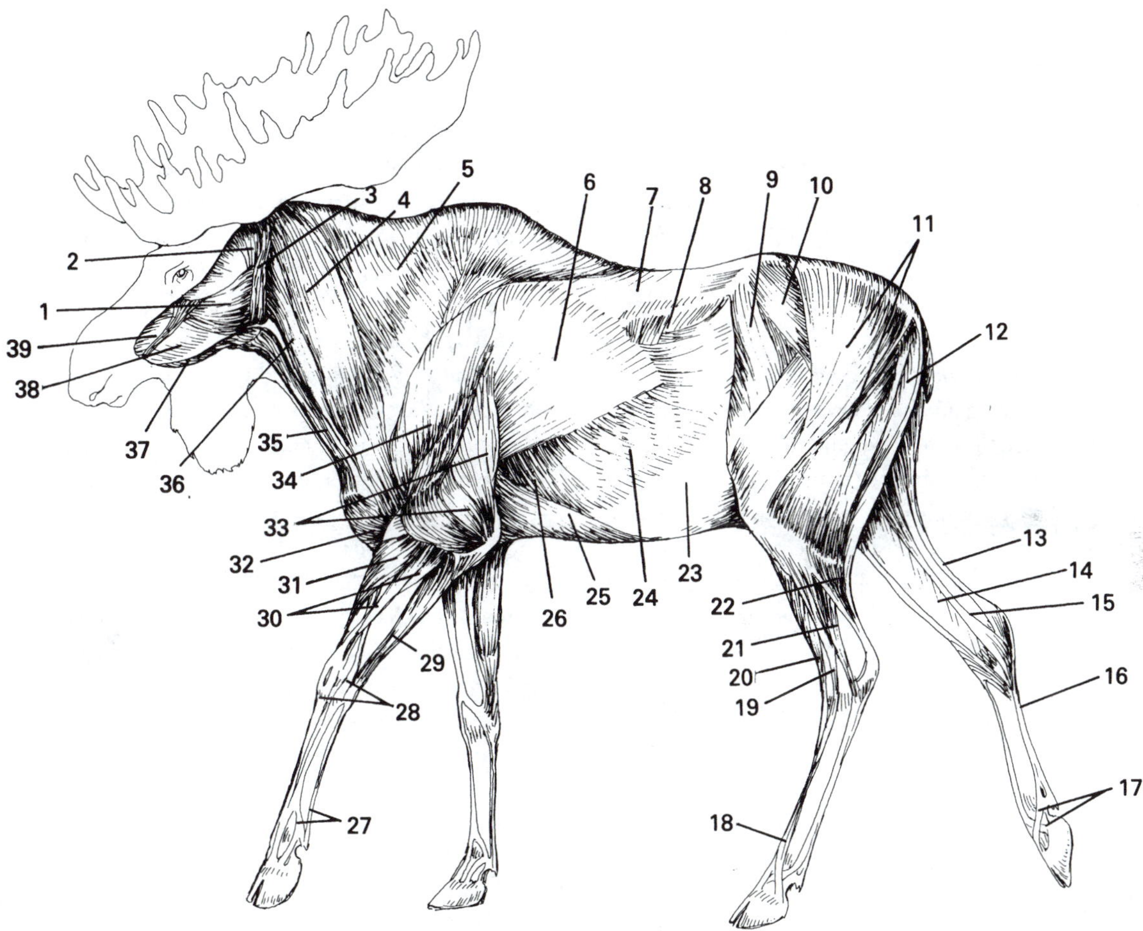

Figure 34. Lateral view of the major superficial musculature and tendons of a moose. 1 = masseter; 2 = arcus zygomaticus; 3 = parotido-auricularis; 4 = brachiocephalicus; 5 = trapezius; 6 = latissimus dorsi; 7 = lumbo-dorsal facis; 8 = serratus ventralis; 9 = tensor faciae latae; 10 = gluteus medius; 11 = biceps femoris; 12 = semitendinosis; 13 = Achilles tendon; 14 = flexor digitorum; 15 = tarsal tendon of biceps femoris; 16 = superficial and deep flexor tendons; 17 = suspensory ligaments; 18 = anterior tendons; 19 = lateral extensor; 20 = long extensor; 21 = deep flexor; 22 = gastrocnemius; 23 = aponeurosis; 24 = obliquus abdominus externus; 25 = deep pectoral; 26 = serratis ventralis; 27 = flexor tendons of metatarsus; 28 = extensor tendons; 29 = ulnarus lateralis; 30 = extensor digiti; 31 = extensor carpi radialis; 32 = superficial pectoral; 33 = triceps; 34 = deltoid; 35 = sterno-cephalicus; 36 = brachiocephalicus; 37 = mylohyoideus; 38 = buccinator; and 39 = zygomaticus. *Illustration by Celia Carl Anderson.*

(Wirz 1950). This could partially explain why moose become relatively tame when raised by humans (Knorre 1961, 1974, Renecker et al. 1987, Dudin and Mikhailov 1990).

The gross subdivisions of the brain are the cerebrum, cerebellum and brain stem. Occupying the upper skull, the cerebrum is the main part of the brain, and represents the largest part of the central nervous system in the moose. The outermost gray matter, or cortex, of the cerebrum is the seat of all conscious sensation and actions, memory and intelligence.

The anterior part of the cerebrum (telencephalon) branches into two outgrowths called the "olfactory bulbs," the receptors of smell. Farther back in the cerebrum (diencephalon) are found the thalamus and hypothalamus. The thalamus is a relay center for nerve fibers connecting the two halves (hemispheres) of the cerebrum with the brain stem and spinal cord. The hypothalamus and pituitary gland are the most important endocrine organs in the brain case. The nuclei of the hypothalamus exert control over visceral activities, water balance, thermoregulation, sleep, hunger, thirst, rage, sexual drive and other behaviors. This part of the brain has changed little throughout vertebrate evolution and controls many forms of complex instinctive behavior.

PINEAL ORGAN

Embedded over the brain stem between the hypothalamus and thalamus is the glandular representative of the evolutionarily regressed third eye—a light-sensitive pineal organ. It secretes melatonin, a substance that controls responses to photoperiodicity. Consequently, the pineal organ is partially responsible for controlling onset of the rut, antler growth, changes in metabolic rates and other physiological phenomena that cycle seasonally or rhythmically.

LIMBIC SYSTEM

The limbic system is an important component of the midbrain. It is an ancient brain area concerned with moods, sexual urges and strong emotions. It mediates physiological activity between the hippocampus and cortex.

Brain Size

In moose, individual brain size varies considerably. In taiga moose of northwestern Ontario, the brains of 6-month-old calves are between 19.0 to 26.2 cubic inches (312–430 cm^3), and increases to 22.9 to 32.3 cubic inches (375–530 cm^3) in full- grown adults (Bubenik and Bellhouse 1980). There are great variations in yearlings as well. The large ranges within an age class are enigmatic because they seem to be typical not only for individuals but possibly of generations. The size variations would seem to indicate corresponding variation in intelligence of individual moose, but this is not easily measured for this species.

There are no published data on moose brain anatomy or the degree of cerebralization. Originally, Bubenik and Bellhouse (1980) presumed, as is known for other mammals (Hahn et al. 1979), that the ratio for convolutedness correlated with survival of moose. However, my statistical analysis of a large sample of brain sizes from Isle Royale provided to me by Dr. Rolf Peterson did not support the notion that brain size was positively related to survival. However, this does not imply that, among moose, there may not be individual differences in intelligence, resulting from fetal or neonatal malnutrition (Winick 1976, Diamond 1990a, 1990b).

Endocrine Organs

The endocrine and nervous systems of the moose and other mammals are intimately related and responsible for the integrative organization of bodily functions. Metabolism, growth, sexual differentiation, reproduction, maintenance of the internal environment and adaptation to the external environment are affected largely by hormones secreted by the endocrine system. Endocrine glands produce hormones that pass directly into the circulatory system, whereas other glands secrete their products through duct networks (McEwan et al. 1979, Gall and Isakson 1989).

Chemically, hormones can be roughly divided into two classes—steroids and protein derivatives. Hormones secreted by the adrenal cortex and gonads characterize steroids. The pituitary, thyroid, parathyroid, pancreas and adrenal medulla secrete protein derivatives (Dickson 1970). This is a simplified classification but provides a beginning point for discussing the major endocrine organs. The following descriptions of hormone-producing endocrine glands and functions were derived from a variety of sources (Turner 1948, Dickson 1970, Schmidt-Nielsen 1975, Bubenik 1982b, Fraser 1991).

PITUITARY GLAND

The pituitary is a small gland attached to the brain by a short stalk. The gland is extremely important because its secretions regulate the function of other endocrine glands. Its primary actions are to stimulate the gonads, thyroid, adrenal cortex, parathyroid and mammary glands. It is responsible for stimulating growth, uterine contraction and blood vessel contraction. It also influences carbohydrate metabolism.

ADRENAL GLANDS

These paired glands are located at the anterior poles of the kidneys. In mammals, the glands consist of the cortex surrounding the internal medullary tissue. Secretions of the adrenal cortex are complex and produce hormones whose actions affect carbohydrate metabolism, fluid balance and renal function. The medulla produces epinephrine, which has a similar effect as the sympathetic nervous system.

THYROID GLAND

The thyroid gland is located on the anterior surface of the larynx. It consists of two lateral lobes and a narrow connecting isthmus. The gland produces thyroxin, which regulates the rate of oxidation of all tissues of the body, which means it regulates body metabolism.

PARATHYROID GLANDS

These glands are small bodies embedded in the thyroid gland, which is primarily responsible for regulating calcium/phosphorus metabolism.

PANCREAS

The pancreas is located in the abdominal cavity. It is a compound gland whose exocrine portion produces digestive juices that enter the small intestine through the pancreatic duct and whose endocrine cells produce insulin and lipocaic. These hormones regulate the storage and utilization of carbohydrates.

OVARIES

The ovaries are egg-shaped organs at the ends of the oviducts of the female reproductive system. They develop follicles that produce estrogen, which stimulates female sex accessory organs, regulates secondary sex characteristics and influences sexual behavior. The corpus luteum of the ovary produces progesterone, which prepares the female for pregnancy and lactation (see Chapters 4 and 5).

TESTES

These male sex glands develop in the abdomen and descend through the inguinal canal into the scrotum. The testes produce testosterone, which maintains the functional status of sex accessories and secondary sex characteristics. Testosterone also influences sexual behavior (see Chapters 4 and 5).

Other Glands and Tissues

Other glands and tissues associated with hormone production include (1) the pineal body, which is an outgrowth from the roof of the diencephalon; (2) thymus gland, which has right and left lobes located on either side near the junction of the thorax and neck; (3) placenta; (4) spleen; (5) liver; (6) kidney cortex; and (7), as discussed, the brain.

Sensory Organs and Perception

Vision, hearing, smell, taste and touch are sensory impulses transmitted to the brain by the eyes, ears, nose, palate, tongue, body surfaces from hooves to skin and possibly antlers. It is important to understand that an animal's response to external cues within the environment can range from irresistibility to one of very low importance. Among the former are the so-called super stimuli, such as urine scent of an estrous female or the head of a large antlered deer, which can suppress the perception of lower-ranking stimuli (Lorenz 1954, 1965, 1966, Tinbergen 1955, 1990, Lorenz and Leyhausen 1968).

Eyes and Vision

Little is known of the acuity of vision in moose, but it is recognized that moose depend less on sight than on hearing and smell to detect activity in their proximity. Moose have nearly monoscopic vision and must move their eyes to assess objects at close range. They can move their eyes independent of one another. Details of the morphology of the eye and the role of vision are discussed in Chapter 5.

Ears and Hearing

Compared with humans, moose have very acute hearing. Both large ears, called pinae, and possibly the antlers play an important role in hearing. In Chapter 5, more detail is provided on hearing sensitivity and why both ears and possibly antlers play an important role in silent language among moose. However, I am uncertain whether ear movement serves more as a means of echolocation of sound or as a visual cue of behavioral expression, the common behavioral interpretation.

Chemical Reception

As in other members of the deer family, chemical communication is of paramount importance to the moose. There are three organs that serve as receptors of chemical signals—the nose or rhinarium, the vomeronasal or Jacobson's organ, and the tongue (Figure 35).

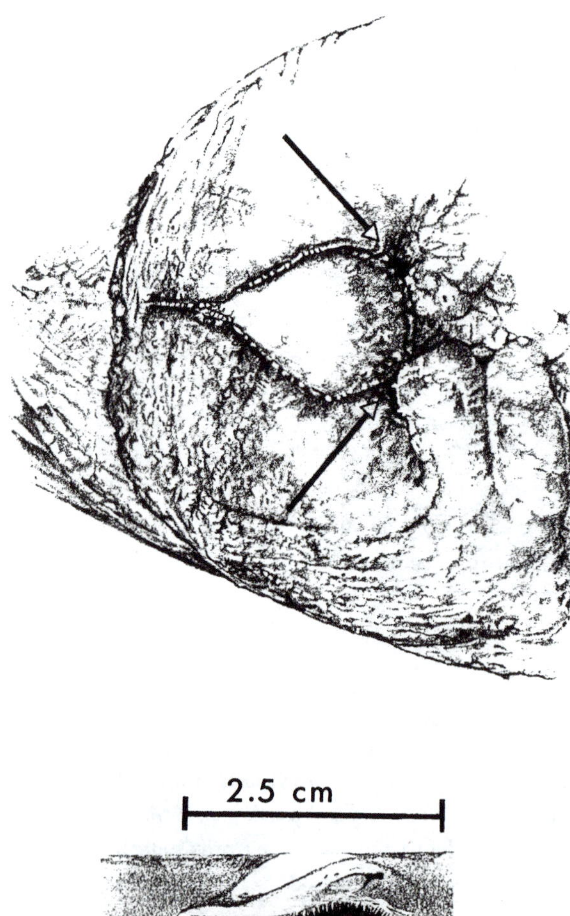

Figure 35. The palate of a moose (top), with openings (arrows) of the vomeronasal or Jacobson's organ, and a longitudinal section through the cavern of the vomeronasal organ (bottom). The acute sense of smell in moose depends on this organ. *Illustration by A.B. Bubenik, from Bubenik (1987).*

The nose of the moose is one of its unique characteristics. There is a large space between the nasals and protruding premaxillae that allows for the development of a large rhinarium with huge nostrils and a prehensile muzzle. A characteristic nasolabial spot on the end of the nose is hairless and may be shaped differently among subspecies. The spot (an inverted triangle on the moose) is used as the beginning point in measuring total body length. The sense of smell in moose is extremely well developed. *Photo by Charles C. Schwartz; courtesy of the Alaska Department of Fish and Game.*

NOSE

Compared again with humans, moose have a very acute sense of smell. The coating of the nasal cavity has a large surface area lined with a layer of moist, mucus-bathed tissue containing millions of smell-sensing cells. These cells or sensory receptors perceive and respond to distinct scents. When an odor molecule is dissolved in the mucus of the nose, and its receptor sensitivity is properly matched, the neuron becomes agitated and fires a signal to the brain. However, this great and fine responsiveness to scents assumes that either the receptors are sensitive to many chemicals or that, on each cilia, there are many different receptors. It is not know which is correct for the moose.

In the moose, there is a large space between the short nasals and forward protruding premaxillae. This has allowed for development of a large nose with huge nostrils and a prehensile muzzle. The distance between the nostrils is almost as great as is the distance between human ears.

JACOBSON'S ORGAN

Some chemicals do not evaporate easily and are not detected by the nose. These compounds reach the olfactory receptors as airborne droplets or aerosols and are detected by the vomeronasal or Jacobson's organ (Jacobson 1811, Meredith 1980, Bubenik 1987, Nagpal et al. 1988). Many of the chemical compounds detected by the Jacobson's organ are associated with sexual activity. Consequently, because of the location of the Jacobson's organ and its function in sexual activity, Cooper and Burghardt (1990) recommended these chemical substances be called vomerones.

The Jacobson's organ begins as paired caverns at the anterior part of the palate (Figure 35). In moose, these caverns are about 1 inch (2.5 cm) long, lined with mucosal tissue and olfactory ciliates. Around the cavities are muscles. Rhythmical contractions of these muscles suck in the inhaled air and concentrate airborne substances, making them detectable. To activate Jacobson's organ, a moose must raise the upper lip, open the mouth and perform the so-called flehmen (see Chapter 5). Amazingly, there is no reference to flehmen in cow moose or other female deer, despite being a gesture quite common in females of horned ungulates (Estes 1972, Briedermann 1993). The regressed Jacobson's organ in cow moose probably explains why they do not perform flehmen.

While the nose can detect and identify many different smells, additional odors are detected by the retronasal route. Vomerones entering the mouth with the air or food are dissolved in the saliva and perceived by clusters of spindly cells of taste buds of the tongue.

TONGUE AND TASTE

The moose tongue has the longest movable part (36 percent of total) among tongues of all deer (Hofmann and Nygren 1992). It is not known whether moose can differentiate between more than the four basic tastes of sweet, salty, sour and bitter, as is known for humans (Freedman 1993). Ends of the taste buds project up into the tiny pores of the tongue's surface. Although taste cells are not neurons, as are the olfactory ciliate, they can send electric impulses to associated neurons, which then convey the taste message to the brain. Although the number of taste papillae in ruminants is much lower than in humans (Freedman 1993), their density in moose is highest among deer (Sokolov 1964, Hofmann and Nygren 1992). According to the size and function, three different papillae are present in the tongue.

Some of the chemical constituents of saliva act as potent sexual pheromones. These are transferred to the opposite sex by exhalation as aerosols, on antlers splattered with urine-soaked mud from the pitholes (see Chapter 5) or by licking genitals. It is possible that they directly activate corresponding neurons or enter the bloodstream through the capillary system of the vulva skin by licking. In this way, they can penetrate the skin and enter the bloodstream, changing the mood of the partner.

Skin

Moose skin is similar to that of other deer, particularly caribou. Skin thickness varies from 0.53 inch (1.34 cm) on the withers and 0.09 inch (0.23 cm) on the legs (Sokolov and Chernova 1987). The skin is thickest during winter. The epidermis varies between 39 and 100 m in thickness, of which the dense stratum corneum can make up as much as 93 percent of this thickness in winter, but only 14 percent in summer. According to Kheruvimov (1969), the skin of a mature moose constitutes 7.3 to 7.4 percent of the body mass. The skin of the belly is pigmented (Sokolov 1964, Sokolov and Chernova 1987).

The skin of the moose contains numerous glands. Around each guard hair are paired sebaceous glands, whereas only one such gland is associated with each underfur hair. The gland ducts are large and secretions also are believed to be large (Sokolov and Chernova 1987), and it is suspected that their primary function is lubrication of hairs. In winter, the secretory activity is depressed. Sweat glands are large and present the entire year (Sokolov 1964). In summer, they contribute to thermoregulation. It is unknown whether sebaceous and/or sweat glands contribute to the specific scent of moose pelage. The scent of moose hairs is easily detected by humans and may have some pheromonal function.

Skin receptors called *neuromas* convey the messages of touch, temperature and pain through the spinal cord to the brain. Each skin receptor seems to have its own center of memory. The density of skin receptors in moose for pleasant and unpleasant sensation is unknown.

Besides the skin-sensing tactile stimulation, hairs or bundle of hairs—including the vibrissae around the mouth and eyelashes—are important sensors of touch. They are scattered over the body, but not in a uniform manner. This explains the different responsiveness to touch and pain at different sites.

Hair

HAIR TYPE AND STRUCTURE

Moose have four types of hair—guard hair, wool hair or underfur, eyelashes, and vibrissae. Guard hairs of varying length cover the entire body. Wool hairs also cover the body, but are absent on the legs and face. Vibrissae are concentrated around the eyes and mouth. A small nasolabial

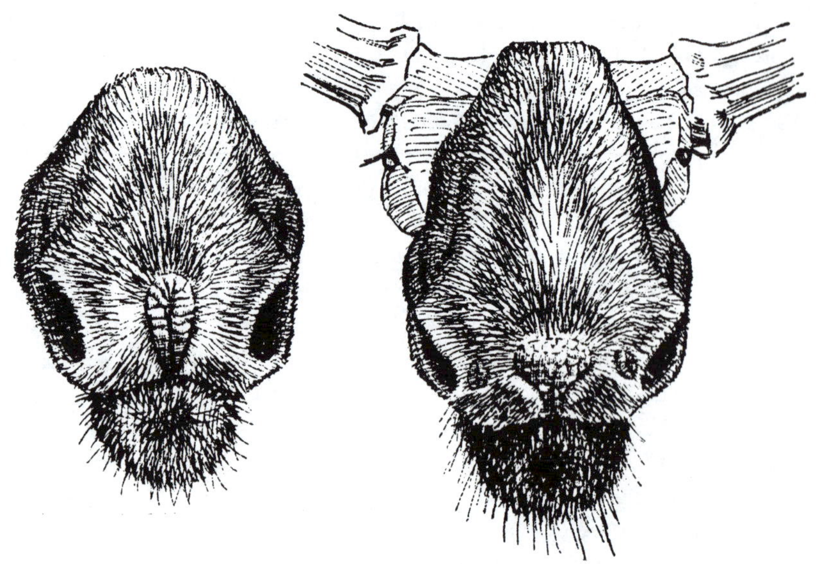

Figure 36. Frontal views of the noses of two moose feature different sizes and shapes of nasolabial spots. *Illustration by A.B. Bubenik, from Flerov (1952).*

spot (Figure 36), located on the end of the nose, is hairless and may be shaped differently among moose subspecies (Flerov 1952).

Guard hairs have a species-specific surface with a thin cortical layer (Williamson 1961) that is seasonally different. The center of the hair, the medulla, is both spacious and air filled. The medulla represents 83 percent of the winter chest hair surface area and 52 percent of summer chest hairs (Heptner and Nasimovich 1974). The medulla serves both for insulation and as an aid in flotation when moose must navigate rivers or lakes.

The guard hairs of moose are less brittle than those of caribou. In winter, on the moose neck and torso, they can be about 4 inches (10.2 cm) long. Longer guard hairs grow over the spine near the hump and can be 10 inches (25.4 cm) or longer. Their erector muscles (arrectores pili) are so powerful that these hairs can be raised to enlarge the hump impression substantially. However, in late winter, guard hairs become so long that they partially separate along the spine and hang to both sides of the neck, displaying their white bases.

The wooly hairs of the underfur are twisted and can reach 1.0 inch (2.54 cm) in length. They probably only grow during summer. The underfur is the most important insulative sheet over the body.

The stiff eyelashes, 0.6 to 0.8 inch (1.5–2.0 cm) long, and the vibrissae around them are sensory organs. They stimulate the eyelids to close to avoid injury from foreign objects. Vibrissae around the muzzle serve the same purpose, in addition to facilitating search for underwater plants. The vibrissae are sparse and about 1.6 to 3.0 inches (4.0–7.5 cm) long (Sokolov 1964, Kheruvimov 1969, Sokolov and Chernova 1987).

HAIR COLORATION

There is a basic difference in moose pelage coloration in Fennoscandia, the North American taiga, and forest/tundra from Siberia to North America, as discussed earlier. Hair coloration varies with body site, season, sex and age. In well-conditioned individuals, the hairs are shiny. They reflect a blue or ocherous light, depending on their basic tone and color.

HAIR CHEMISTRY

Flynn and Franzmann (1987) reviewed studies investigating the chemistry of hairs. Hair can serve as an indicator of local and seasonal abundance or shortage of essential elements such as sodium, potassium, calcium, magnesium, copper, iron, manganese and zinc. It also may be used to detect the presence of such toxic elements as cadmium and lead (Figure 37).

The hormone thyroxine seems to be responsible for the shining luster of hairs and hence is indicative of thyroxine fitness (Bubenik and Smith 1986).

HAIR MOLT

In moose calves, the juvenile wooly hairs are exchanged for the winter coat with guard hairs and underfur. According to Kaletskyi (1965), this process has an individual pattern and is complete between August and the second half of September. It begins with the back and head and ends with posterior part of the thighs. Metabolically, development of winter hair is strenuous, and is associated with a sudden rise of food intake and a stagnation of weight gain (Kheruvimov 1969).

Bulls in good condition and barren cows are the first to

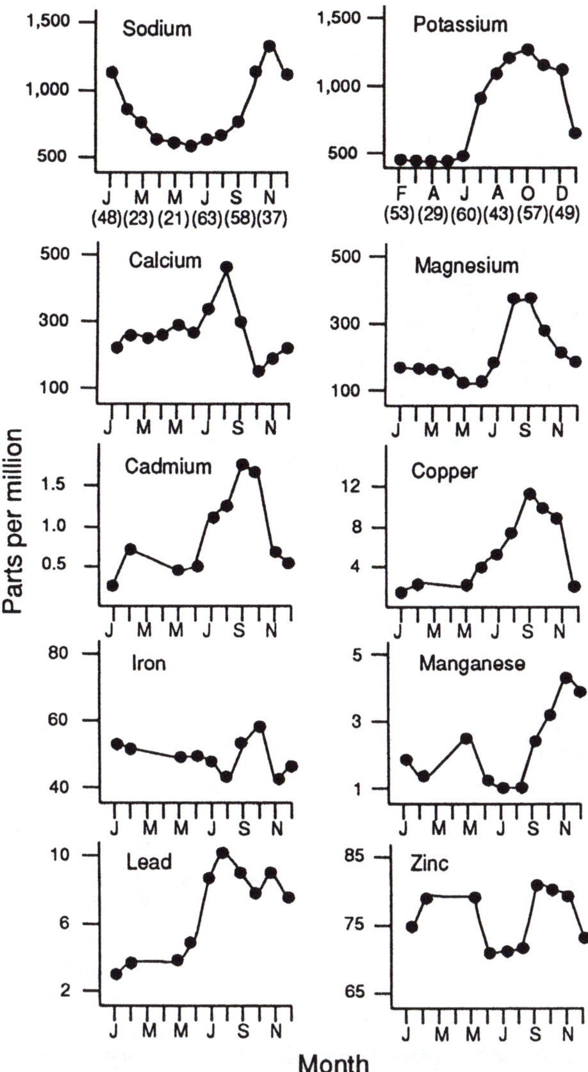

Figure 37. Fluctuation of hair element levels in Alaskan moose (from Flynn and Franzmann 1987). The mean levels of hair elements from moose at the Moose Research Center (1972–1974) reflect the seasonal uptake of these elements. Hair acts as a recording filament, which makes it possible to use it as an indicator of element intake. A copper deficiency in a subpopulation of Alaskan moose was detected by applying this principle. Numbers in parentheses are sample sizes.

grow both the summer and winter coats. Lactating cows and juveniles molt last. As in all deer, the spring molt is conspicuous; the annual molt is hardly noticed.

HAIR LOSS

Total hair loss, which leaves the skin bare, is known only in the taiga moose seriously infested with winter ticks (see Chapter 15).

Bell

The bell of the moose is a sexually dimorphic structure that functions as a visual communicator. Details are discussed in Chapter 5.

Timmermann (1979) did not find any special secretory activity in the bell skin or inside the bell itself. Timmermann's histological and blood vascularization (Figure 38) investigations provided evidence that the bell is an evagination of the hair-covered skin united with that of the lower cheeks. Its sebaceous and sweat glands do not differ from those of the neighboring skin area. A single bell artery branches either from the left or right lingual artery to supply blood. Also, a single bell vein returns the blood to the jugular vein. Dense mesh arterioles that anastomose into dense capillaries do not always protect the bell from freezing. Occasionally, part of a bell may freeze and fall off.

Skin Glands

In moose, certain parts of skin are expanded or invaginated as reservoirs or beds that secrete special compounds. These are called skin glands.

UDDER

The udder, with four hairless nipples, weighing 2.6 to 3.3 pounds (1.2–1.5 kg), is the largest skin gland (Sokolov and Chernova 1987). It swells with milk during lactation. The udder skin is slightly pigmented and sparsely covered by 2- to 5-inch (5.1–12.7-cm) guard hairs and 0.8-inch (2-cm) wool hairs. The base is covered by adipose (fat) tissue (Sokolov 1964). The udder cannot be seen when viewing the cow from the side, and is not highly conspicuous from behind unless full of milk (Stringham 1974).

FOREHEAD GLANDS

These seasonally active apocrine glands are present in Capreolidae (Schuhmacher 1939) and *Odocoileus* (Volkman et al. 1978). Their presence in moose seems highly probable, based on observations of moose rubbing their foreheads on "signal" trees and afterward testing the smell (Miller et al. 1991).

ANTORBITAL (LACRIMAL) GLANDS

The paired antorbital glands are located at the anterior edge of the eyes. They are small in the moose, as in other subfamilies of the Odocoileidae, possibly due to the shallow lacrimal cavity. The skin of the gland has a thickness of about 0.02 inch (0.051 cm). The sebaceous and sweat glands form a continuous layer, and about 0.25 inch (0.64 cm) be-

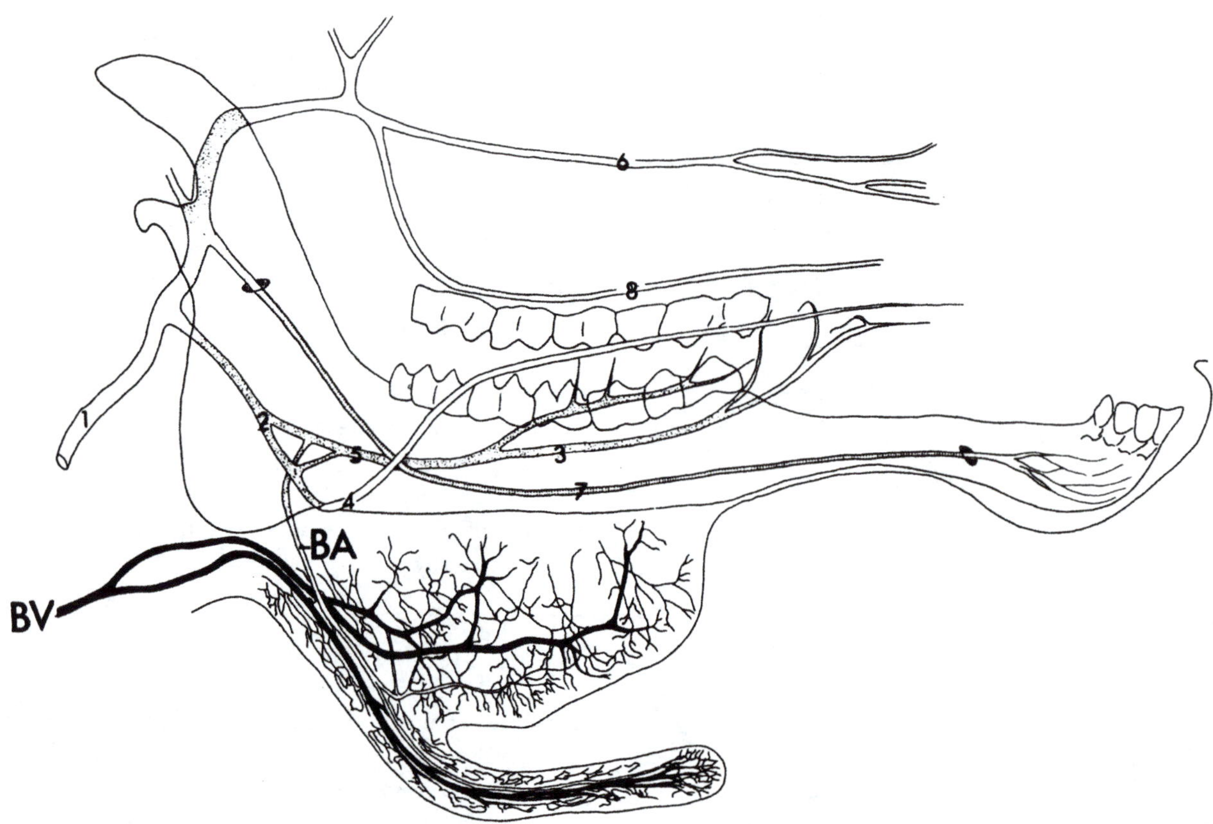

Figure 38. Vascularization of the moose bell. 1 = common carotid artery; 2 = lingual/facial branch; 3 = sublingual artery; 4 = facial artery; 5 = lingual artery; 6 = infraorbital artery; 7 = mandibular/alveolar artery; 8 = maxillary artery; 9 = bell tail; BA = bell artery; and BV = bell vein (from Timmermann 1979). Despite this vascular network, the lower part of a moose bell may freeze and drop off.

low them are tubular glands that enter the hair follicles. The role of these tubular gland secretions is unknown (Sokolov and Chernova 1987).

INTRANASAL AND METATARSAL GLANDS

These glands commonly occur in many Odocoileidae, but are absent in moose (Atkeson et al. 1988).

TARSAL GLANDS

The tarsal glands are located on the inner sides of the heels or hocks, and identified by bushy, whitish hairs. They are equipped with huge arrectores pili of 0.02-inch (0.051-cm) diameter. In an agitated moose, these hairs are erected, the glands become visually enlarged and their scents are more exposed to the air current. The skin in this area is 0.24-inch (0.62-cm) thick, with large sebaceous glands. These sebaceous glands are similar on both bulls and cows and do not change seasonally. There are data to suggest that the bushy hairs of the center of tarsal glands may have special "scent hairs," termed osmetrichia, like those found in mule deer (Müller-Schwarze et al. 1977). On their surfaces, osmetrichia have large chambers or "pockets" in which urine and lipids are trapped and later smeared on vegetation. If osmetrichia are present in moose, they may serve as scent communication.

CAUDAL GLAND

The caudal gland has not been described in moose, although it probably is present because it occurs in both Odocoileidae and Cervidae (Lewin and Stelfox 1967, Hofmann and Thomé 1986).

CIRCUMVAGINAL GLANDS

The circumvaginal gland area, which is only active during the estrus period, has been described by Frankenberger (1957) for red, fallow and roe deer. The presence of these glands in moose is indicated by the olfactory checking of cow's vulva area by the bull.

INTERDIGITAL GLANDS

In both sexes of moose, the interdigital glands are developed on all legs, as pockets of huge sebaceous and sweat glands embedded in narrow epidermis with long, greenish hairs protruding from the hoof surface. The glandular area of these interdigital pockets is shallow on the forelegs and deep on the hind legs. The apocrine–sebaceous cells show a low secretory rate in late spring. Secretions peak in late September and early October. In contrast, the sweat glands are active year-round (Chapman 1985). The ultrastructure has been described by Sokolov and Stepanova (1980).

Soft Organs of Moose

I am unaware of published data detailing mass or volume of soft organs of North American moose (Figure 39).

LUNGS

The mass of lungs may indicate the health status of moose; it should be weighed minus the trachea. On the basis of my data for red and roe deer, the lungs do not fluctuate seasonally as do the other soft organs described below.

HEART

The main function of the heart is to pump oxidized blood through the body, but it also functions as an endocrine gland. In moose, as in red deer, heart weight shows great individual variability (Bubenik 1984) and seasonal fluctuation.

SPLEEN

The spleen of the moose serves two major purposes: it breaks down red blood cells and acts as a storage organ for blood cells. Stored red blood cells are released into the bloodstream when needed. Based on histological studies, Blumenthal (1952) concluded that the moose spleen has a large storage capacity, more similar to the horse than to other ruminants. White pulp, which is substantially reduced in favor of red pulp, consists of leukocytes, which are immunobiologically important (Van Rooijen 1977, Weber and Pert 1989, Weiss 1990). Red pulp consists mainly of red blood cells. Whether this is species specific or an environmental adaptation is unknown (Hartwig and Hartwig 1975).

The spleen's relative mass is related to the animal's circulating blood volume. It must be viewed cautiously. My investigations of spleen mass of red deer, roe deer and chamois have shown that there is strong correlation between spleen mass and how the animal dies. With instant death, the spleen has a normal weight representative of a live animal. Animals that are stressed before death, and those that may remain unconscious before death (i.e., immobilized animals) likely pump a large volume of blood cells from the spleen into the circulatory system, therefore their spleen mass undoubtedly is diminished. Nonetheless, the relative mass of the moose spleen must be considered as the largest among deer (Hartwig and Hartwig 1975).

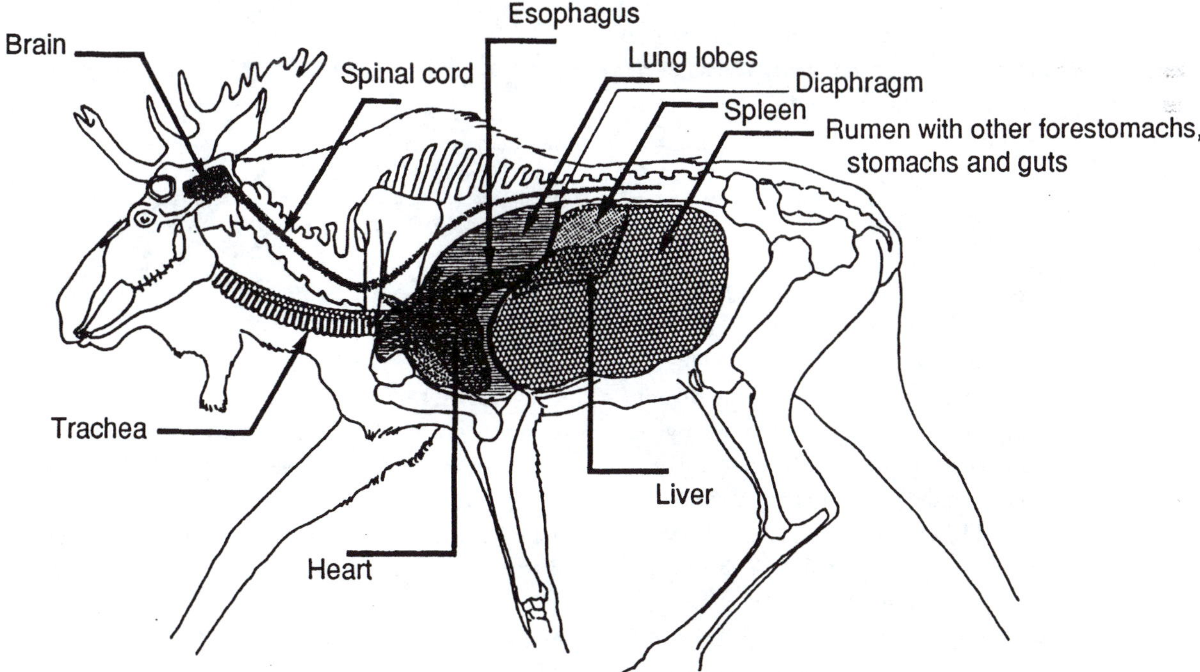

Figure 39. Diagrammatic longitudinal section of a moose, featuring the principal soft organs in relation to skeleton. *Illustration by A.B. Bubenik.*

KIDNEYS

Kidneys form the upper part of the urinary system. Functionally, they filter arterial blood, removing catabolic wastes. Moose kidneys are smooth, not lobulated. The left and right kidney may differ slightly in mass. Animals in good-to-excellent body condition have a thick layer of fat covering the kidneys. Kidney fat indices—the ratio of kidney fat mass to kidney mass (see Riney 1955, Flook 1970, McGillis 1972)—are a fitness indicator.

LIVER

The liver is a dark red gland located on the right side of the moose, just behind the diaphragm. The liver filters venous blood returning from the digestive system. It produces and stores glycogen—the sugar that fuels the body. The liver also acts as an endocrine gland, producing hormones involved in growth and blood-coagulating substances. It is the source of the bile with cholesterol necessary for digestion of fat. The moose, like other deer species, does not have a gall bladder. Bile is secreted directly from the liver into the gut.

The liver stores large amounts of vitamin A (Drescher-Kadren and Walser-Karst 1970). In males, it has the capability of temporarily storing up to 20 percent of its mass in fat before the rut, similar to North American elk (Flook 1970). This fat reserve is used as a direct source of energy in a process termed *reversible steatosis*. During this process, the liver changes from dark red to a more-or-less pinkish-white color. The normally granulated texture also changes to a softer more amorphous mass.

Bull moose do not eat for up to 3 weeks during the rut and rely on stored liver fats during this time. The pinkish-colored liver encountered by hunters during this time is quite normal.

Table 7. Adult Alaskan moose blood values (from Franzmann and Schwartz 1983)[a]

Blood value	Measure[b]	Mean	Standard deviation	Sample size
Glucose	mg/dL	132.1	40.2	1,164
Cholesterol	mg/dL	83.7	17.4	1,165
Triglyceride	mg/dL	16.8	13.4	210
Lactic dehydrogenase (LDH)	U/L	306.1	96.7	1,164
Glutamic oxaloacetic transaminase (GOT)	U/L	153.3	69.6	1,163
Glutamic pyruvic transaminase (GPT)	U/L	66.4	108.2	160
Alkaline phosphatase (APT)	U/L	71.4	78.0	1,161
Phosphorus (P)	mg/dL	4.9	1.4	1,165
Calcium (Ca)	mg/dL	10.4	1.0	1,165
Ca/P	ratio	2.12		1,165
Iron (Fe)	mg/dL	150.2	40.3	85
Sodium (Na)	mEq/L	137.8	6.2	212
Potassium (K)	mEq/L	5.3	1.4	212
Chlorine (Cl)	mEq/L	96.2	5.3	212
Carbon dioxide (CO_2)	mEq/L	16.8	9.0	210
Blood urea nitrogen (BUN)	mg/dL	11.5	10.6	1,163
Creatinine (C)	mg/dL	2.5	0.7	212
C/BUN	ratio	0.75		173
Bilirubin	mg/dL	0.44	0.27	1,165
Uric acid	mg/dL	0.33	0.22	1,131
Total protein (TP)	g/dL	7.3	0.9	1,183
Albumin (A)	g/dL	4.4	0.8	1,170
Globulin (G)	g/dL	2.9	0.8	1,155
A/G	ratio	1.65		1,155
$Alpha_1$ globulin	g/dL	0.35	0.28	1,168
$Alpha_2$ globulin	g/dL	0.56	0.27	1,155
Beta globulin	g/dL	0.71	0.31	1,169
Gamma globulin	g/dL	1.34	0.46	1,170
Hemoglobin (Hb)	g/dL	18.0	2.5	857
Packed cell volume (PCV)	%	46.9	7.1	775
Mean corpuscular hemoglobin concentration (MCHC)	%	38.7	2.9	489

[a] Thirty-seven months of age or older.

[b] mg/dL = milligrams per deciliter; U/L = units per liter; mEq/L = milliequivalents per liter; g/dL = grams per deciliter.

Blood and Blood Parameters

Blood makes up approximately 10 percent of total live body weight or mass. It has the following functions (Dukes 1947): (1) transports nutrients from the alimentary canal to the tissues; (2) transports oxygen from the lungs to the tissues; (3) removes waste products from the tissues to the organs of excretion; (4) transports secretions of the endocrine organs; (5) aids in equalizing the water content of the body; (6) aids in regulating body temperature; (7) regulates various ion concentrations in the body; and (8) assists in the body defenses against microorganisms. Blood is an opaque, rather viscid fluid composed of plasma in which are suspended red corpuscles (erythrocytes), white corpuscles (leukocytes) and platelets. Many of the functions of blood are directed toward maintaining a constant internal environment—a condition know as *homeostasis* (Dukes 1947).

There are many constituents in blood that can be chemically analyzed. On the basis of laboratory findings, normal or expected blood values or parameters are established. Variation from these normal values indicates a potential disease state. The important point in this regard is that the body is constantly working to maintain homeostasis and alterations in blood values often are delayed.

The normal values established for a species should be

based on extensive sampling under controlled conditions. This is rarely possible for wild animals. Generally, the best that can be done is to identify the possible sources of variation and classify the samples by sex, age, season, excitability, restraint (drugs used), gestation, estrus or rut, and fat determinations (probes, condition classifications) (Franzmann 1985).

Blood parameter alterations in moose have been tested and applied to general condition assessment of moose, but the number of parameters are limited (Franzmann 1985, Franzmann et al. 1987). Extensive blood analyses of moose were done in Alaska (Franzmann and Schwartz 1983); they provide the largest collection of baseline values available for moose (Table 7). The application of blood parameters to condition in moose is discussed in Chapter 15.

(SKETCHED IN FROM LIFE- COMPLETED FROM MEMORY)
JULY 28, '56
IGLOO CREEK

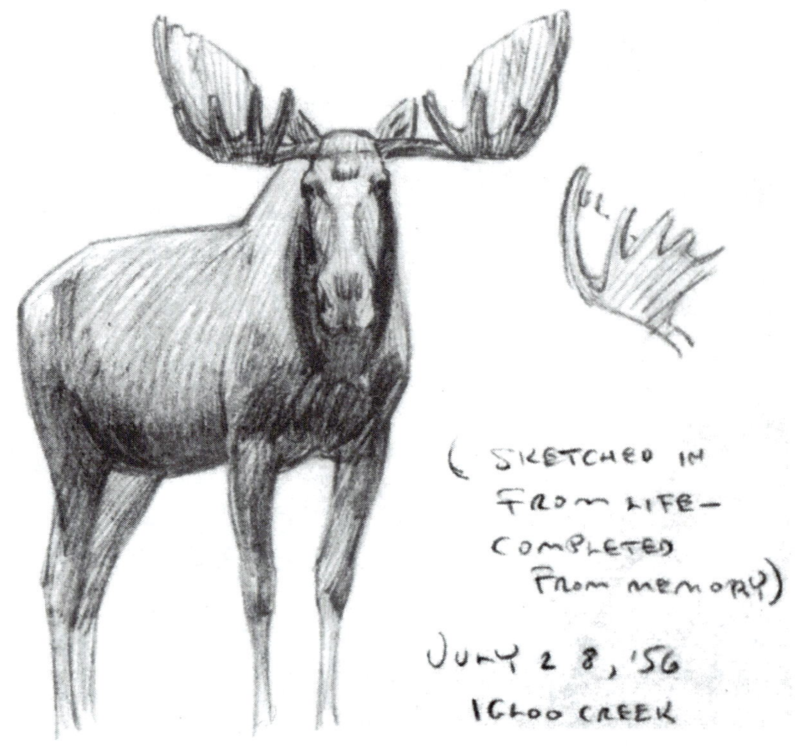

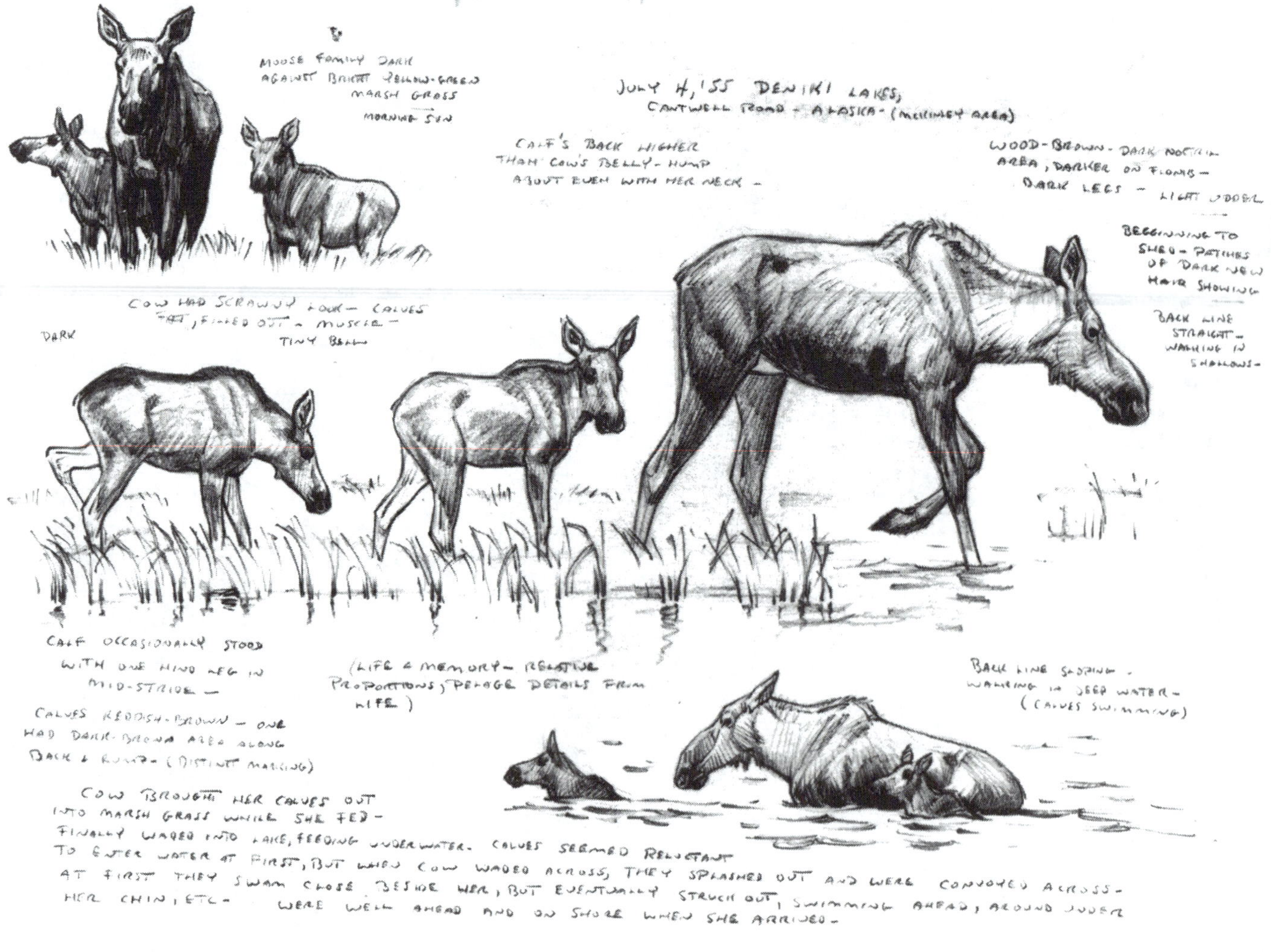

MOOSE FAMILY DARK AGAINST BRIGHT YELLOW-GREEN MARSH GRASS
MORNING SUN
JULY 4, '55 DENIKI LAKES,
CANTWELL ROAD - ALASKA - (MCKINLEY AREA)
CALF'S BACK HIGHER THAN COW'S BELLY - HUMP ABOUT EVEN WITH HER NECK -
WOOD-BROWN - DARK NOSTRIL AREA, DARKER ON FLANKS - DARK LEGS - LIGHT UDDER
BEGINNING TO SHED - PATCHES OF DARK NEW HAIR SHOWING
BACK LINE STRAIGHT - WALKING IN SHALLOWS -
COW HAD SCRAWNY LOOK - CALVES FAT, FILLED OUT - MUSCLE - TINY BELLY
DARK
CALF OCCASIONALLY STOOD WITH ONE HIND LEG IN MID-STRIDE -
(LIFE & MEMORY - RELATIVE PROPORTIONS; PELAGE DETAILS FROM LIFE)
BACK LINE SLOPING - WALKING IN DEEP WATER - (CALVES SWIMMING)
CALVES REDDISH-BROWN - ONE HAD DARK-BROWN AREA ALONG BACK & RUMP - (DISTINCT MARKING)
COW BROUGHT HER CALVES OUT INTO MARSH GRASS WHILE SHE FED - FINALLY WADED INTO LAKE, FEEDING UNDERWATER. CALVES SEEMED RELUCTANT TO ENTER WATER AT FIRST, BUT WHEN COW WADED ACROSS, THEY SPLASHED OUT AND WERE CONVOYED ACROSS - AT FIRST THEY SWAM CLOSE BESIDE HER, BUT EVENTUALLY STRUCK OUT, SWIMMING AHEAD, AROUND UNDER HER CHIN, ETC - WERE WELL AHEAD AND ON SHORE WHEN SHE ARRIVED -

3

PATRICK D. KARNS

Population Distribution, Density and Trends

Although moose have been present on the North American continent for thousands of years, their distribution has been very dynamic in recent times. In many parts of their present range, moose are relative newcomers. Only within the last century and a half or so have they occupied extensively the Rocky Mountain chain in the western United States, established themselves north of Lake Superior and extended their range into Labrador. Moose continue to occupy new territory.

Distribution

Moose are found only in the northern hemisphere, in a broad, global band of boreal forest dominated by spruce, fir and pine trees, where fire is a major factor in shaping of the vegetative communities (Odum 1983, Telfer 1984; Figure 40). Moose occur in Canada, United States, Russia, Finland, Sweden, Denmark, Norway, Lithuania, Estonia, Poland, Czechoslovakia, Manchuria and China. The primary difference in moose habitat between Eurasia and North America is that the former is more affected by agriculture and forestry (Gill 1990). The association of moose with boreal forest of the northern hemisphere is unique; there are no counterparts of boreal forest, or moose, in the southern hemisphere (Shelford 1963).

Rowe (1982) described the boreal forest region in North America as a continuous belt from Newfoundland to the Rocky Mountains and northwestward into Alaska, with white and black spruces being the dominant trees, and a mixture of white birch, trembling aspen and balsam poplar (Figure 41). In eastern Canada and the United States, moose also occur in the Great Lakes/St. Lawrence vegetative region, which consists of eastern white and red pines, eastern hemlock and yellow birch. Other tree species include sugar and red maples, basswood, white elm and, to a lesser extent, white cedar, large-toothed aspen, beech, white oak, butternut and white ash intermixed with more northern boreal species. Moose also are found in the Acadian forest of eastern portion of the continent, which is dominated by red spruce associated with balsam fir, yellow birch and sugar maple. In the middle of the continent, extending from northwestern Minnesota, through southern Manitoba, central Saskatchewan and Alberta, moose also are found in the transition zone between prairie and northern forest (Berg and Phillips 1974). This area is on fertile alluvial soils, often covered with peat, where slight changes in elevation greatly influence the vegetation, and large areas of marsh and willows are interspersed with stands of coniferous and deciduous trees.

The primary factors limiting the geographic distribution of moose are food and cover to the north (Kelsall and Telfer 1974) and climate to the south (Renecker and Hudson 1986b). The most critical factor, especially to the southern distribution of moose, is temperature—in particular, heat. In winter coats, moose become stressed by temperatures more than 23°F (5.1°C). In summer coats, they experience stress at temperatures of 59°F (14°C) or higher. The lower critical temperature for moose is unknown; moose have

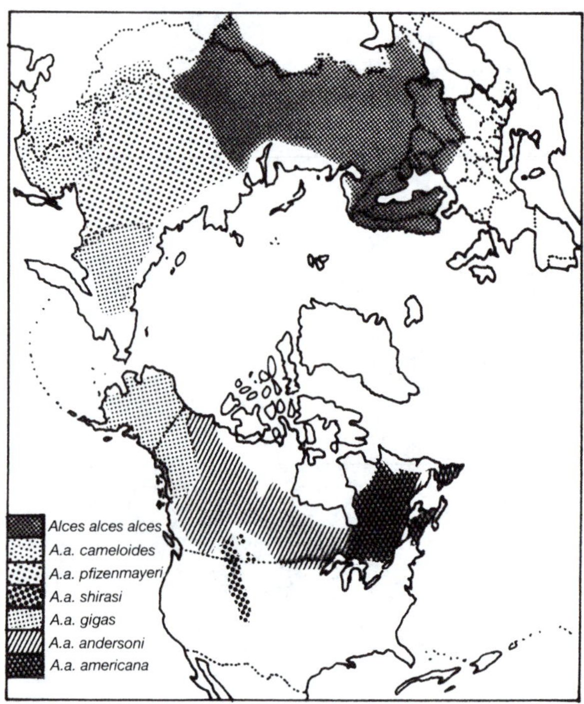

Figure 40. Circumpolar distribution of moose (from Telfer 1984).

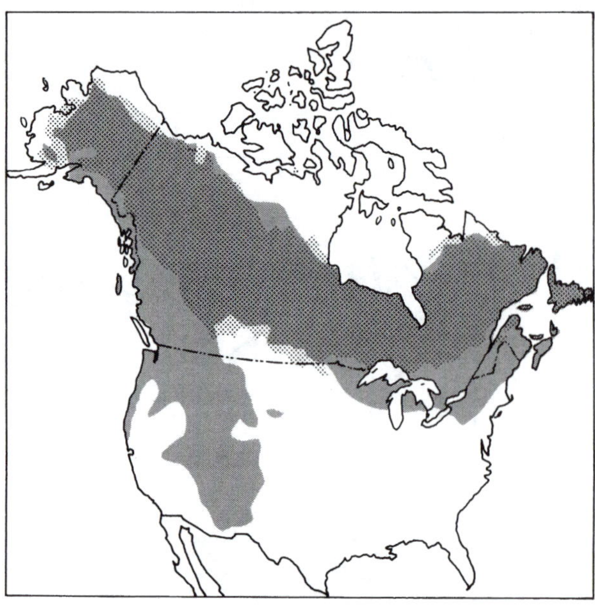

Figure 41. Distribution of spruce (solid screen) and boreal forest (dot screen) in North America (from Halfpenny and Ozanne 1989).

been observed, apparently unaffected, at −40°F (−40°C). In late winter (mid-January through mid-April) and summer (June through mid-September), when ambient temperatures are high enough to elicit panting to dissipate the excess heat, moose actively seek sites for shade and cooling (Schwab and Pitt 1991). In late winter, they commonly are found in association with the protective cover of conifer trees; in the summer, they generally are near water. For the most part, the thermal patterns that regulate moose distribution are those that circumscribe North America's boreal region.

The distribution of cervids, including moose, is determined largely by physiological adaptation to climate and vegetation (Schwartz 1992a; Figure 42). The lower critical temperatures for various cervid species in winter coat are: Rocky Mountain mule deer, 9°F (–23°C); Columbian black-tailed deer, 45°F (7°C); and northern white-tailed deer, 0°F (–18°C) (Mautz et al. 1985). Parker and Robbins (1984) found thermally critical temperature extremes for mule deer to be −4°F (–20°C) and above 41°F (5°C) in winter, and above 77°F (25°C) in summer. For elk, Parker and Robbins determined winter temperature extremes of below −4°F (–20°C) and above 68°F (20°C). Holter et al. (1975) found thermal panting in white-tailed deer occurred at 95°F (35°C), and shivering started at 59° and −4°F (15° and −20°C) in summer and winter coats, respectively.

Within the northern hemisphere, moose still are expanding their range into areas that have not been occupied by them at least since retreat of glacial ice some 10,000 to 14,000 years ago (Peterson 1955, Kelsall 1972, Geist 1987, Cronin 1992, Hundertmark et al. 1992b, Zheleznov 1993; see also Chapters 1 and 2). During interglacial periods, moose migrated from Asia by way of a land bridge connecting Siberia and Alaska (see Chapter 1). After retreat of the last continental ice sheet, moose gradually colonized North America's northern hemisphere, adapting to conditions as they occupied new areas to the south and east of the interglacial refugium. With few exceptions, the present moose range is area occupied by the species for the past 2,000 years, although not necessarily continuously (Figures 43 and 44). The exceptions, particularly in recent decades, are significant expansions into new areas (Telfer 1984, Boer 1992b). Moose are not faring well in areas where human in-

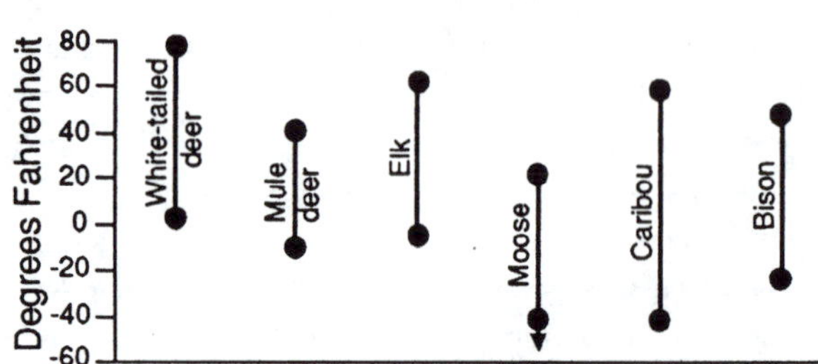

Figure 42. Zones of thermal neutrality for several North American ungulate species.

tolerance for moose is high, such as in developed areas of Massachusetts and Connecticut (Vecellio et al. 1994), and in certain agricultural areas, such as northwestern Minnesota.

Beginning in Alaska, moose are found in most suitable habitat throughout the mainland portion of that state, except on the northernmost coastal plain (LeResche et al. 1974a, S.R. Peterson personal communication: 1992). Moose moved into southeastern Alaska in the 1920s or 1930s. Those animals were augmented by translocations in the late 1950s and 1960s. Moose also were translocated to the Copper River Delta in the 1950s. They began to immigrate to the Seward Peninsula in northwestern Alaska in the mid- to late 1930s. Moose are found throughout the Yukon Territory (R. Ward personal communication: 1992), across the MacKenzie Delta and south of the treeline (Graf 1992) in the Subarctic Ecoclimatic Province (Ecoregions Working Group 1989) to northwestern Hudson Bay in Manitoba (Crichton 1992b). The Subarctic Ecoclimatic Province is described as the transition between the boreal forest to the south and the tundra to the north. Vegetation is characterized by stunted and open stands of black spruce and tamarack with some white spruce, dwarf birch, willow, northern Labrador-tea and lichens; moose is one of its characteristic life forms (Ecoregions Working Group 1989). Working in northern Alaska, Coady (1980) reported that, although suitable moose habitat existed along the major streams in the area, moose were conspicuous targets and were hunted by the Nuamiut Indians until the Natives were removed from the area in 1920, after which moose populations increased. Coady saw the same scenario on the Seward Peninsula, where moose were extremely scarce before 1950, but rapidly expanded their range and numbers in the 1960s and 1970s after the demise of mining in the area and desertion by most humans. Wolves occurred in both areas, but fed mainly on caribou. The northern limits of moose distribution follows the Hudson Bay coast through Ontario (McNicol 1990) and into Quebec, extending to the Leaf River, and just beyond the George River to the northwest, just to Ungava Bay (Brassard et al. 1974, Kelsall 1987). Moose still are spreading into Labrador from the south—a natural postglacial expansion of their range (Mercer and Kitchen 1968). This northern extent of moose range is wholly in the Subarctic Ecoclimatic Province. Moose were introduced into Labrador in 1953 (Northcott 1974) from Newfoundland. Moose also occur in the riparian areas along the major rivers to the north of this line (Kelsall 1972, Graf 1992), but are limited by a lack of woody food plants from inhabiting the tundra (Kelsall and Telfer 1974).

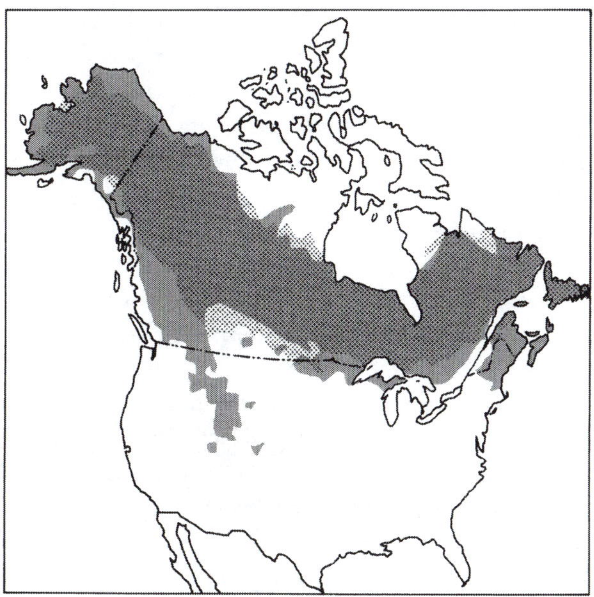

Figure 44. Boreal forest region (dot screen) and recent distribution (solid screen) of moose in North America (from Kelsall 1987).

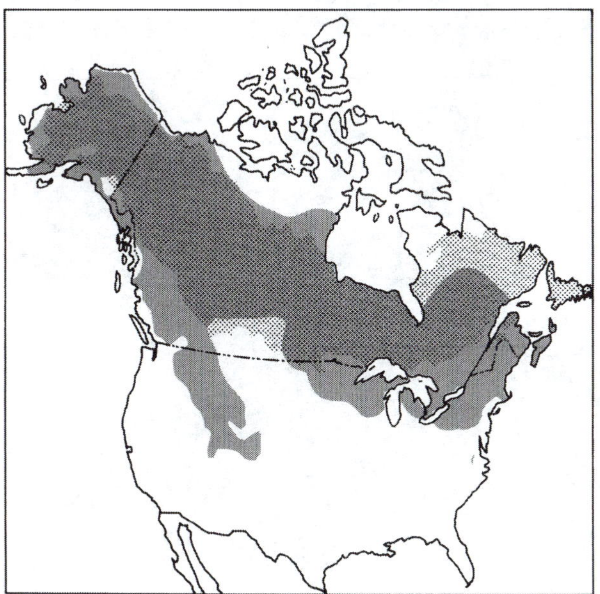

Figure 43. Boreal forest region (dot screen) and known former range of moose distribution (solid screen) in North America (from Hall and Kelson 1959).

Before introductions of moose in 1895 and 1904, caribou were the only member of the deer family present on Newfoundland (Peterson 1955). By 1920, moose had spread 50 miles (80.5 km) from their release site at Howley and, by 1945, occupied the entire island (Northcott 1974). They are now well-established throughout the island and are the major big game species in Newfoundland in terms of hunter demand and meat value. The meat value of a single moose has been estimated to be $1,320, and the annual harvest was recently valued at $41 million (Oosenbrug et al. 1991). Before the introduction of moose, Newfoundland's southern forest had an abundant shrub layer that now is essentially gone due to heavy browsing by moose (Mercer and Manuel 1974). This is a particular challenge to managing Newfoundland's national parks in as nearly a natural condition as possible without human intervention (Corbett in press).

Moose have long occurred in northern British Columbia, but are relative newcomers to western and southern British Columbia (Kelsall 1987). Before 1860, there were no moose in the interior area of British Columbia, and the Indians living in this region reportedly had no name for them in native language (MacGregor 1987). Moose occupied central and southern British Columbia in the early 1900s (D. Blower personal communication: 1992). Presently, they are found throughout British Columbia, except in the rugged Coastal Mountains, coastal islands and grasslands of the southern interior (Eastman and Ritcey 1987). Along the southern borders of British Columbia moose have expanded their range into Washington and Idaho. In Idaho, moose population size and distribution has increased most dramatically since the 1970s, and Compton and Oldenburg (1994) predicted that most of the available range would soon be occupied. Moose began to show up in the Selkirk Mountains in northeastern Washington in the 1950s, and continue to spread into new areas of that state (Zender 1991). A few moose now occur in Oregon, where, before 1960, several attempts to introduce moose failed (C. Wheaton personal communication: 1992).

Moose are relative newcomers to the western states, being there for only the past 150 years or so. Moose were not reported by early travelers or trappers in the area (Smith 1982, Chittendon 1986, Leege 1990, Compton and Oldenberg 1994). Residents of Wind River Indian Reservation in central Wyoming reported that moose were scarce before the 1930s (Smith 1982, 1985). During the 1930s, moose came to occupy their present range in the Wind River area,

Much of the boreal forest is a mosaic of cover types created by many natural forces and commercial logging. Moose populations can be a significant factor in affecting the vegetation. In some areas, such as Newfoundland (above), where moose were introduced in 1895 and 1904, moose have altered the plants in many areas by heavy browsing. *Photo by Albert W. Franzmann; courtesy of the Alaska Department of Fish and Game, Soldotna.*

and became quite numerous by the 1960s, when they declined because of overhunting.

Moose became established in Montana and Idaho during the early to mid-1800s, and during the next 100 years they spread through Wyoming and into northern Utah. With animals captured near Moran in 1948 and 1950, a moose population was established in the Big Horn Mountains of Wyoming (Pimlott and Carberry 1958). Natural distribution has been augmented by translocations (Duvall and Porter 1987, G.K. Jense personal communication: 1992). Since 1950, moose have extended their range in Utah to the Snowy Range and upper Platte River Valley from animals translocated to Colorado in 1978. A few moose have shown up in the Laramie Peak area in the 1970s, and now are in the Bighorn Mountains (R. Rothwell personal communication: 1992).

The historic range of moose did not extend as far south as Colorado, although individual moose were sighted there in the past 100 years (Peterson 1955). Duvall and Porter (1987) indicated that moose slowly moved their range southward, and predicted that they would colonize Colorado from southern Wyoming and northern Utah. Several habitat and suitability studies indicated that good moose habitat existed in Colorado and plans were made to introduce them to the state. Thirty-six moose were captured in Utah and Wyoming and moved to North Park and Laramie River in 1978 and 1979. By 1988, the moose population was estimated to number 250 (Duvall and Schoonveld 1988). By 1990, this population increased to approximately 700 animals (T. Lytle personal communication: 1992). During December 1991, January 1992 and January 1993, 106 moose were captured in Colorado, Utah and Wyoming and released in the upper Rio Grande River basin in southwestern Colorado. These moose have established themselves on an approximately 50-square mile (130-km^2) area around the release site (Olterman et al. 1994).

The spread of moose down the Rocky Mountains in western United States has occurred mostly in the twentieth century, as a continuation of the postglacial distribution to available habitats.

Moose are well established across the northern two-thirds of the prairie provinces of Alberta, Saskatchewan and Manitoba (Peterson 1955), where they have been for hundreds of years. Alexander Henry, in his trip from Pembina to Ft. Vermillion in 1808, following the route through Lake Winnipeg, to Grand Rapids, Cumberland House, and on the Saskatchewan River, mentioned encountering moose and buying moose meat on many occasions (Coues 1965). Here, moose occur in the large expanses of northern boreal forest and the parkland-to-prairie ecotone (Todd 1991). Moose distribution in Manitoba has changed in recent years, with moose extending their range northward, occupying all of the province with the exception of the Hudson Bay Lowlands (Crichton 1992b).

Moose in the prairie provinces also occur in "islands" of isolated habitat. Most notable are Elk Island National Park and adjacent Blackfoot and Ministik areas east of Edmonton, the Cypress Hills area on the Alberta/Saskatchewan border (Todd 1991), Moose Mountain Provincial Park in Saskatchewan, and Turtle Mountain Provincial Park, Riding Mountain National Park and Spruce Woods Provincial Forest in Manitoba. Moose are found in similar habitats in adjacent areas of North Dakota and Minnesota, including in the Turtle Mountains, Pembina Hills, escarpment and beach areas bordering prehistoric Lake Aggasiz, and along major rivers in eastern North Dakota and northwestern Minnesota. This was traditional moose range, but moose were absent, or very rare, from the 1900s to the early 1950s, when moose began to reappear in the area. Since the 1950s, moose sightings have become more frequent and widespread in these areas, and moose are now seen in areas that are far from traditional habitat on the prairie edge. Moose populations in these areas have increased dramatically since the 1950s, and hunting seasons for moose were instituted in 1971 in Minnesota, and 1977 in North Dakota.

In 1982, the Minnesota Department of Natural Resources recognized the uniqueness of the prairie moose population and established a separate population survey unit, the Northwest Prairie Unit, separate from the Northwest Forest Unit—typical boreal forest. The Northwest Prairie Unit consists of a zone of transition between prairie to the south and coniferous forest to the north. It is a narrow ecotone of agricultural land, with large expanses of willow brush, marsh and scattered deciduous forest (Berg and Phillips 1974). Moose and associated recreation reportedly have a bright future in North Dakota (Knue 1991) and adjacent areas of similar habitat in northwestern Minnesota (Anonymous 1990). Moose in the prairie area are smaller bodied than those in the boreal forest (Karns 1976), and have smaller antlers (Gasaway et al. 1984).

The moose range extends from Manitoba into adjoining Ontario, continuing in the boreal forest, to the north and east of the prairie moose. Absent from much of the area north of Lake Superior until the twentieth century, moose have only recently colonized much of northern Ontario, again, as a part of the postglacial dispersal that is still in progress (Peterson 1955, deVos 1964). Today, moose are found across northern Ontario in the boreal and Great Lakes forest areas of the province, extending from the international border north to Hudson Bay and from Manitoba to Quebec. There are no moose in southern Ontario (McNicol 1990).

To the south of Ontario and Quebec, moose occur in a

Shira's moose have expanded their range during the past century primarily as a result of natural progression, protection and translocations. They are the smallest of North American moose, and occur in Colorado, Utah, Wyoming, Idaho, Montana, southwestern Alberta and southeastern British Columbia. *Top photos courtesy of the Wyoming Game and Fish Department. Bottom photo by Judd Cooney.*

belt of boreal forest and in coniferous/deciduous ecotone forests through Minnesota, Wisconsin and the upper peninsula of Michigan (deVos 1964, Karns et al. 1974, Krefting 1974b, Peterson 1955). They are common in the boreal and transitional forests in northeastern Minnesota, with well-established populations and hunting seasons. Moose were abundant in northern Wisconsin until the 1850s, but essentially nonexistent by 1900 (Krefting 1974b). Today, they are rare across northern Wisconsin, with only 30 to 40 animals in the state. Some calves are seen every year, suggesting a resident population likely augmented by emigrants from Minnesota and Michigan. Proposals in the early 1990s to translocate moose, elk and/or caribou to Wisconsin have been decided in favor of elk (K. McCafferty personal communication: 1992).

Historically, moose were common throughout Michi-

Moose are found in the aspen parklands south of the boreal forest in northwestern Minnesota, North Dakota, Manitoba, Saskatchewan (above) and Alberta. They have been expanding their range into these areas in recent decades. Conflicts with agriculture are common throughout the region. *Photo by Albert W. Franzmann; courtesy of the Alaska Department of Fish and Game, Soldotna.*

gan's upper peninsula and the northern and southeastern portions of the lower peninsula (Hall and Kelson 1959). The last moose sighting in the lower peninsula was in 1883, near Black Lake in Presque Isle County (Baker 1983). Moose fared better in the upper peninsula, probably because of better habitat, and the fact that it was not logged until after the Civil War (Burt 1954, Baker 1983). A reintroduction from Isle Royale in the late 1930s failed because the moose were in poor physical shape and poached heavily after the institution of meat rationing during World War II (Anonymous 1991). A population of 25 to 50 moose has lived on the eastern end of the upper peninsula for the past few decades, periodically augmented by animals moving across the St. Mary's River from Ontario (Baker 1983). Sixty-one moose were translocated from Algonquin Park in Ontario to northern Michigan in the winters of 1985 and 1987 (Aho and Hendrickson 1989). The resulting population in the upper peninsula appears to be self-sustaining, and public acceptance is very high, with very low poaching losses (Aho and Hendrickson 1989). The current moose population is

The Pembina River Valley in northwestern North Dakota is the state's primary moose range. Moose immigrated there in the 1960s from similar habitat in northwestern Minnesota. *Photo courtesy of the North Dakota Game and Fish Department.*

estimated at 300 to 400 animals (E. Langenau personal communication: 1992).

Moose were reported in the vicinity of Detroit in the early eighteenth century, and were found elsewhere in the lower peninsula until they were extirpated in the late 1800s (Baker 1983). Suitable habitat exists, but is not occupied (Kelsall 1987). The lack of moose probably is attributable to increased pressures exerted by human development, and perhaps by increasing ambient temperatures (Baker 1983).

Isle Royale National Park—a 210-square mile (544-km^2) island in Lake Superior, established in 1940—is home to a moose herd that has been the subject of long-term ecological studies of predator/prey relations for many decades. Caribou originally occupied the island, but disappeared in the early 1900s, about the time that moose appeared, crossing on ice from the mainland. Gray wolves came to the island in the late 1940s (Peterson 1977).

Historically, moose are known to have occurred in southern Ontario and much of Pennsylvania, New York and the New England states (Hall and Kelson 1959; see also Chapter 1).

Moose were found in southern Quebec when the European settlers arrived in the 1600s, and still occur throughout the province except in the southern forests and northern tundra (Brassard et al. 1974, Joyal 1987). The northern limit for moose in Quebec is the Leaf River in the northwest, and just beyond the George River in the northeast, just touching Ungava Bay (Kelsall 1987).

Moose extended their range into Labrador from Quebec, in a northeasterly direction along riparian habitats (Mercer and Kitchen 1968). The expansion probably was limited by the presence of trappers and migratory Indians in the river valleys, who subsisted from the 1880s through the 1940s on moose when it was available. With the demise of the fur industry, the trappers moved out of the river valleys and moose increased. In 1953, seven bulls and five cows were translocated from Newfoundland to Labrador and released in the St. Lewis River region (Pimlott and Carberry 1958). Today, moose are found in most of the available habitat in Labrador (Northcott 1974, F. Phillips personal communication: 1992).

Moose reached New Brunswick at least 2,500 years before present (see Boer 1992b). Currently they are found throughout the province, except on coastal islands. The fire frequency is approximately 340 years in the Maritime lowland region, and 625 years in the hardwood and coniferous forests found at the higher elevations (Wein and Moore 1977). These compare with a 100-year fire frequency throughout the boreal forest (Heinselman 1973). Therefore, with the advent of logging in the 1800s, and the fires it produced, moose habitat in New Brunswick probably was enhanced considerably. Fire suppression and improved road access since the 1920s has reduced the frequency of fires and the size of burns (Boer 1992b).

Although moose are found throughout Nova Scotia (Benson and Dodds 1977), they are at low densities throughout the western portion of the province and at relatively high density on Cape Breton (T. Nette personal communication: 1992).

In the New England states, moose disappeared from Pennsylvania in the 1790s, Massachusetts early in the 1800s, New York in the 1860s, and Vermont and New Hampshire about 1900 (Peterson 1955). In 1950, the New England distribution of moose was limited to northern and central portions of Maine, and only a few animals in extreme northern New Hampshire. Subsequently, the range expanded in both of those states. Moose reappeared in Vermont by the 1970s, and population growth accelerated in the 1980s and early 1990, with the range expanding from a small area in the northern part of the state to encompass nearly the whole state by 1990 (C. Alexander personal communication: 1992). From Vermont, moose have moved into northeastern New York and established a small but viable population (Hicks 1986, Garner and Porter 1990). Moose sightings have increased in recent years in Massachusetts, where a small population is established. Moose are seen occasionally in Connecticut. Most of the area of the states is unsuitable for moose from a human compatibility standpoint, and moose often are hazed, immobilized and moved, or euthanized, depending on the threat they pose to human safety (Vecellio et al. 1994).

Since retreat of the last continental glaciers, moose have occupied essentially all of the available habitat in North America by the latter part of the twentieth century. Within historic time, their numbers and range have fluctuated in response to direct and indirect human exploitations and conservation. Within the past century, and especially since 1950, habitat and harvest management and translocation practices have enabled moose populations to grow, pioneer new areas and repopulate areas from which moose have long been absent.

Density

Moose population size and density usually are determined from some sort of aerial survey conducted under rigid conditions during winter months (see Chapters 6 and 17). Although moose are big animals, they are very difficult to count, and accurate results are impossible if surveys are not conducted under fairly ideal conditions. The many jurisdictions that census moose use different techniques, and the results are not always directly comparable. However, results can give indications of relative abundance and population trends for specific areas.

Farmland abandonment, forest-cutting practices and harvest protection have allowed moose populations to increase close to many urban/suburban areas across North America. Conflicts have arisen between moose and humans, and the presence of moose in many of these areas has become socially unacceptable. *Photo by Ken Love; courtesy of the* Sunday Journal, *Auburn, Maine.*

Moose population densities primarily are products of the habitat quality, which is generally associated with seral forest succession, as a result of fire, windstorms, insects, or alluvial depositions (Krefting 1974b, Haggstrom 1994). In most parks and game reserves, moose are not hunted and, in some cases, predators are absent. Consequently, high densities are attained on relatively small areas (Corbett in press). On the two large burn areas on the Kenai National Moose Range in Alaska, moose reached their highest densities of four moose per square mile (1.5/km^2) 17 to 26 years after the fires, and then declined at about 9 percent per year thereafter (Regelin et al. 1987a, Loranger et al. 1991). Telfer (in press) evaluated the role of fire on the production of woody browse in boreal forest, and showed browse yield to be maximum with a 60- to 80-year burn rotation. This approximates presettlement (pristine) conditions, and is the successional pattern needed to maintain moose populations best. Telfer also demonstrated that fire control effectively instituted in the 1950s resulted in reduced moose forage, which contributed to reduced moose populations in Alberta, Saskatchewan and Manitoba. In fact, the moose populations reported by Elliot (1988) for fire-suppressed northern Manitoba were 2.5 percent the size reported for moose in other areas of the boreal forest (Telfer in press).

Table 8 represents, for the most part, crude density estimates for moose populations occurring over large areas, ascertained or extrapolated from various studies. Only a few of the estimates reflect ecological densities—those in a specific habitat type at a specific season—and those are the highest figures. The densities range from less than 0.1 to more than 24 moose per square mile (0.04–9.3/km^2). The higher densities generally occur in park situations where hunting is not allowed and predation is absent or minimal. The moose populations of the Chapleau and Quetico Crown game preserves in northwestern Ontario, for example, have gray wolves and black bears as predators, but are not subject to hunting harvest. The population densities in these two areas are about twice that of the surrounding areas that have the same predators and are hunted.

Trends

North American Population

The North American moose population has shown a positive growth at least since the middle of the twentieth century (Figure 45). Peterson (1955) estimated the population of moose in North America at 341,700 moose in 1948.

Table 8. Moose population densities for selected areas of North America

Area	Moose per square mile (km²)	Source	Area	Moose per square mile (km²)	Source
Chapleau Crown Game Preserve, Ontario[a]	0.8–0.9 (0.31–0.35)	Thompson and Euler 1987	British Columbia (continued)		
Quetico Crown Game Preserve, Ontario[a]	0.8–0.9 (0.31–0.35)		Wet interior	0.3 (0.12)	
Ontario[b]	0.4 (0.15)		Subboreal	2.0 (0.77)	
Gaspesie Park, Quebec[a]	5.2 (2.0)	Crête 1989	Maine	0.1–1.0 (0.04–0.40)	Kelsall and Telfer 1974
Quebec (province-wide)[b]	0.3 (0.12)		Minnesota	0.3–3.7 (0.12–5.29)	Kelsall and Telfer 1974, R. Beaulieu personal communication: 1992
Isle Royale, Michigan[c]	5.2 (2.0)		Saskatchewan		
Northwest Territories	0.1–0.4 (0.04–0.15)	Graf 1992	Northern	0.3 (0.12)	
Northern Manitoba		Elliot 1988	Northcentral	0.5 (0.19)	
Area 1			Central	1.0 (0.40)	
Closed conifer	0.1 (0.04)		Eastcentral	1.8 (0.69)	
Open conifer	<0.1 (0.04)		Southeastern	1.8 (0.69)	
Mature mixed wood	<0.1 (0.04)		Fundy National Park, Nova Scotia[a]	0.5 (0.19)	Corbett in press
Young mixed wood	0.5 (0.19)		Kouchibouguac National Park, New Brunswick[a]	1.0 (0.40)	
Burn	0.1 (0.04)		Cape Breton Highlands National Park, Nova Scotia[a]	13.6 (5.25)	
Area 2			Terra Nova National Park, Newfoundland[a]	1.0 (0.40)	
Closed conifer	0.1 (0.04)		Gros Morne National Park, Newfoundland[a]	2.9–17.8 (1.12–6.87)	
Open conifer	<0.1 (0.04)		General		Telfer 1984
Mature mixed wood	<0.1 (0.04)		Subarctic areas	<0.3 (0.12)	
Young mixed wood	0.4 (0.15)		Better ranges	0.3–0.8 (0.12–0.31)	
Burn	0.3 (0.12)		Excellent ranges	1.0–2.6 (0.40–1.00)	
Quebec hunting zones		Courtois and Crête 1993	Concentrations in Ontario, Alberta and Minnesota	10.5–24.0 (4.05–9.27)	
South	2.1 (0.81)		Kenai National Moose Range, Alaska		Loranger et al. 1991
North center	2.4 (0.93)		1947 burn	9.4 (3.63)	
West	5.5 (2.12)		1969 burn	11.3 (4.36)	
Center	4.7 (1.81)				
North	1.0 (0.40)				
Minnesota		Karns 1982			
Northeast	1.0 (0.40)				
Northwest	0.5 (0.19)				
British Columbia		Eastman and Ritcey 1987			
Boreal lowland	1.6 (0.62)				
Boreal upland	1.8 (0.69)				
Coastal	0.2 (0.08)				
Dry interior	0.6 (0.23)				

[a] Not hunted.

[b] Hunted.

[c] Nonhunted, island population.

Kelsall (1987) placed it at 888,000 in 1985, and Gill (1990) estimated there to be 1 million moose in North America in 1990. The latter two estimates are reasonably consistent with estimates compiled by the various provinces and states (Table 9). Peterson's (1955) estimate for moose in the late 1940s probably was too low, or the population would have had to experience about a 275 percent increase between 1948 and 1960. This would have represented an annual rate of increase of 7.6 percent for that period, which is highly unlikely in a moose population subject to losses from natural causes. Four to 5 percent is what can be achieved under

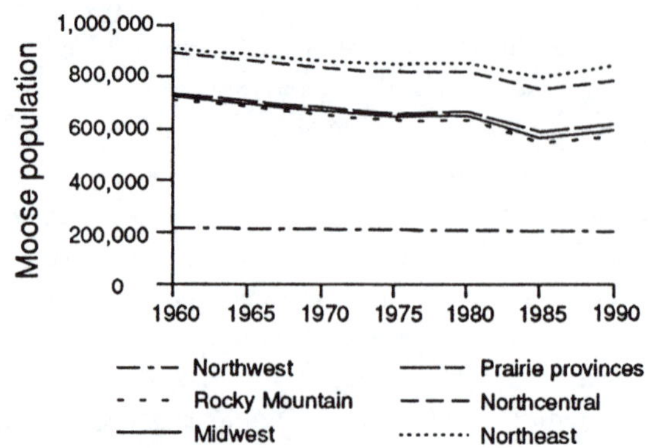

Figure 45. Trends of cumulative regional moose populations in North America, 1960–1990.

Forest fires occur naturally throughout the boreal forest region and perpetuate moose habitat. Fires occur naturally about 1 in every 100 years throughout most of this region. The burned areas provide good habitat for moose until the woody vegetation grows out of their reach. *Left photo by Albert W. Franzmann; courtesy of the Alaska Department of Fish and Game, Soldotna. Right photo by Charles C. Schwartz; courtesy of the Alaska Department of Fish and Game.*

natural conditions in Ontario (Bisset 1991) and 4 percent in northeastern Minnesota (Karns 1982). Applying a more conservative rate of increase, say 4.5 percent, would place the population in the late 1940s at 500,000 moose—some 158,300 more moose in North America in the late 1940s than estimated by Peterson (1955). This does not seem unreasonable, recognizing that the state of knowledge of moose population and distribution when Peterson made his estimates was really sparse. Beginning in the late 1950s, more emphasis was being placed on moose biology across the continent, enabling better population estimates to be made. Beginning in the 1960s, moose populations have increased in many areas since Peterson's work, and they moved into new areas where they were unknown in recent times (Kelsall 1987).

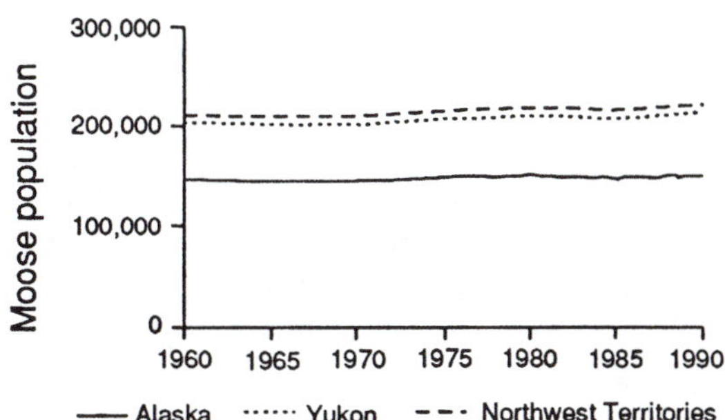

Figure 46. Trends of cumulative regional moose populations in northwestern North America, 1960–1990.

Regional Populations

On the basis of available population data, regional moose population trends from 1960 to 1990 were variable (Table 9, Figure 45). Moose populations in Alaska, Yukon and Northwest Territories experienced slight population increases during the period from 1960 through 1990 (Figure 46). The moose populations in this vast area have been maintained by a series of severe winters in the late 1960s and early 1970s, human harvests and predation by gray wolves, brown bear and black bear (Franzmann 1993, G. Carroll personal communication: 1992, R. Case personal communication: 1992, R. Ward personal communication: 1992).

Moose populations declined in British Columbia, Alberta, Saskatchewan and Manitoba (Figure 47). Telfer (in press) explained that reduced burning of the boreal forest since the 1950s permitted plant succession to age beyond optimal browse quantity and quality for moose. This habitat circumstance, in conjunction with predation and hunter harvest pressure and subsistence hunting are responsible for the moose population declines in this large region.

Moose populations in the western United States in-

Table 9. Moose population trends across North America, 1960 through 1990

State or province	1960	1965	1970	1975	1980	1985	1990
Alaska	145,000	145,000	145,000	150,000	152,000	150,000	155,000
Alberta	120,000	116,500	113,500	110,000	107,000	104,000	100,500
British Columbia	300,000	280,000	250,000	240,000	240,000	170,000	175,000
Colorado	0	0	0	0	36	61	700
Idaho	4,100	4,400	4,600	4,700	4,900	5,100	5,100
Labrador	750	1,000	1,500	2,000	3,500	4,000	5,000
Maine	4,000	4,700	6,700	9,200	11,250	16,900	22,500
Manitoba	38,000	40,000	45,000	32,500	21,000	23,000	27,000
Michigan	40	40	40	40	40	47	350
Isle Royale	575	725	1,325	1,275	875	1,075	1,596
Minnesota	1,500	2,800	4,600	4,800	8,050	10,200	12,000
Montana	4,500	4,500	4,500	4,500	4,500	4,500	6,000
Newfoundland	35,000	35,000	45,000	70,000	67,000	107,000	111,000
New Brunswick	3,500	4,200	6,000	8,200	10,000	15,000	20,000
New Hampshire	500	600	900	1,600	2,000	2,500	4,000
New York	0	0	0	0	6	11	20
Northwest Territories	8,500	8,500	8,500	8,500	8,800	8,700	9,000
North Dakota	1	10	40	75	150	500	700
Nova Scotia	3,000	3,000	3,000	3,000	3,000	3,000	4,000
Ontario	78,000	80,000	82,000	91,000	83,000	99,000	103,000
Quebec	92,000	90,000	87,000	83,000	75,000	75,000	69,000
Saskatchewan	40,000	40,000	40,000	40,000	60,000	50,000	65,000
Utah	200	500	750	1,000	1,500	2,000	3,750
Vermont	15	25	50	100	200	500	1,300
Washington	<10	<10	<10	<10	35	150	185
Wisconsin	40	40	40	40	40	40	40
Wyoming	3,750	4,500	5,900	7,500	7,600	8,700	13,500
Yukon	56,000	56,000	56,000	58,000	59,000	59,500	60,000
Total	938,981	922,050	911,955	931,040	930,482	920,484	975,241

The boreal forest in North America reaches its northern- and westernmost distribution in northwestern Alaska. Moose populations in these areas, such as the Susitna River Valley (above), have been stable or increasing in recent years. *Photo by Albert W. Franzmann; courtesy of the Alaska Department of Fish and Game, Soldotna.*

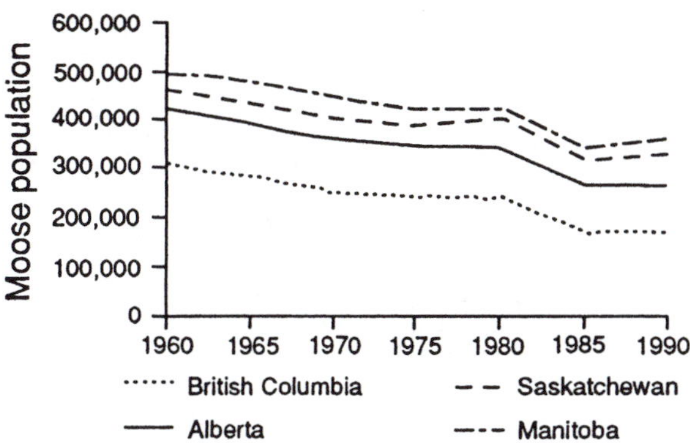

Figure 47. Trends of cumulative regional moose populations in Canada's western provinces, 1960–1990.

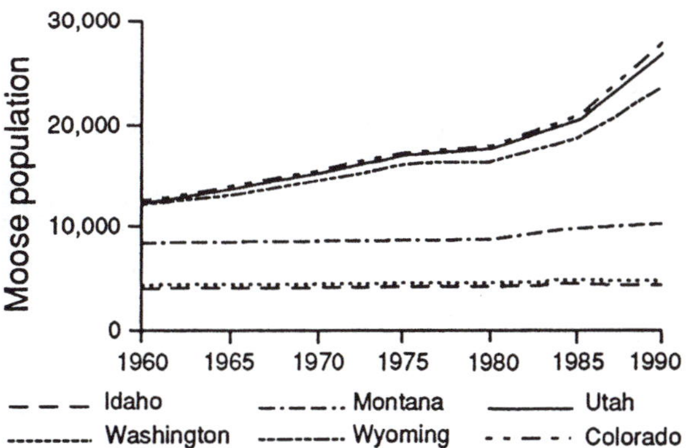

Figure 48. Trends of cumulative regional moose populations in the western United States, 1960–1990.

creased and expanded geographically by means of translocations and emigration (Figure 48). Both the population and area occupied by moose have increased dramatically in the Selkirk Mountain area of Washington State in the past 20 years (Zender 1991). The populations also have increased in areas of the western states of Idaho, Montana, Wyoming and Utah through immigration and occupation of new habitats (Peek 1974, Compton and Oldenburg 1994, L. Oldenburg personal communication: 1992). Wilson (1971) reported that moose had extended their range in the Unita Mountains of Utah and increased in population by 500 percent from 1965 to 1971. Continuing pressures from timber companies, road building, urban sprawl and increasing human populations will have to be mitigated to preserve ade-

All across North America, including the intermountain region of the central (above) and northern Rockies, moose are reestablishing themselves where they were once extirpated or becoming established on new range. *Photo by Joe Van Wormer; courtesy of the Denver Museum of Natural History.*

Effective forest fire suppression began across northern North America in the 1950s and has reduced both the frequency and size of forest fires. This has reduced the amount and quality of moose habitat across broad areas, and has generally reduced moose populations from Manitoba (above) to British Columbia. *Photo by Vince Crichton.*

quate moose habitat (Compton and Oldenburg 1994), as well as unregulated harvest (John et al. 1985) in western states.

Midwestern moose populations also increased (Figure 49). Wisconsin's population was stable at 30 to 50 animals, but did not significantly expand its range. Wisconsin is at the southern edge of moose range, where heat may be a limiting factor, as is poaching (K. McCafferty personal communication: 1992). For years, moose in Michigan's upper peninsula varied from 30 to 50 animals, and depended on animals crossing the St. Mary's River from Ontario to maintain within that range. In 1985 and 1987, the population was supplemented by 61 animals translocated from Algonquin Park, Ontario (Aho and Hendrickson 1989). No moose occurred in their former range in the lower peninsula (Baker 1983).

Isle Royale National Park is a 210-square mile island in Lake Superior. Caribou were at one time abundant on the island, but disappeared in the early 1900s. Moose arrived on Isle Royale sometime around 1905. They found abundant forage and probably exceeded 3,000 by the mid-1930s, when there was a winter die-off (Mech 1966, Hickie no date). Gray wolves came to the island in the mid-1940s, and have since fed there almost exclusively on moose (Mech 1966). The moose population has fluctuated (Figure 50) mainly in response to severe winters (Peterson 1977).

Moose populations in Ontario and Quebec declined from 1960 to 1990 (Figure 51), as the result of additive mortality of hunting and predation (Bergerud 1981, Crête 1987, Thompson and Euler 1987). In Labrador, moose are expanding their range, and their population is increasing (Mercer and Kitchen 1968, F. Phillips personal communication: 1992).

Since they were introduced in the late 1800s, moose populations of Newfoundland have shown a steady growth from just a few animals to an estimated 111,000 in 1990 (Figure 52) (Northcott 1974, S. Ferguson personal communication: 1992). There are no effective predators on the island, and the few declines since the 1960s were attributed to overhunting (Mercer and Manuel 1974, S. Ferguson personal communication: 1992).

In the maritime provinces and New England states,

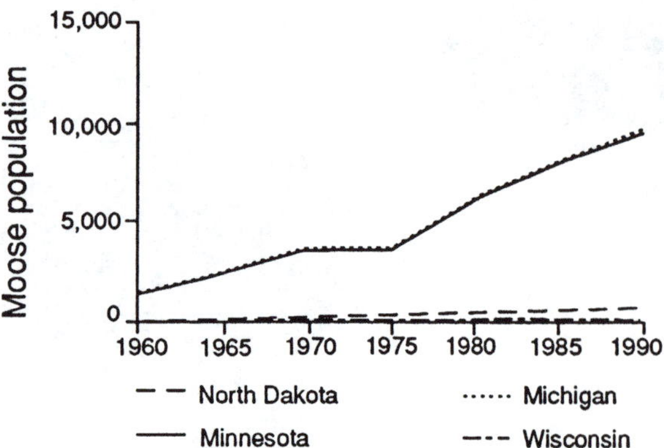

Figure 49. Trends of cumulative regional moose populations in the midwestern United States, 1960–1990.

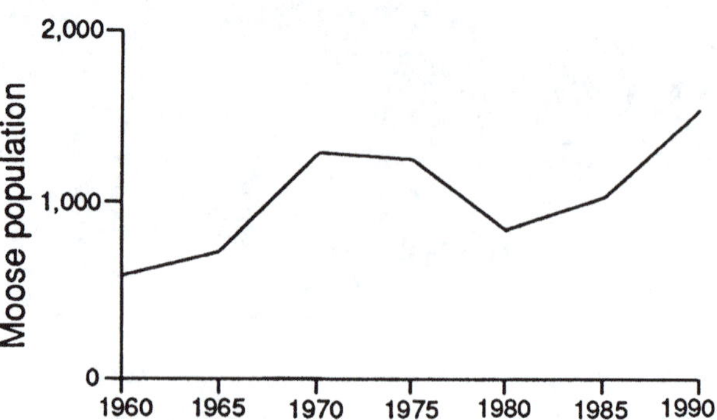

Figure 50. Trend of moose population for Isle Royale National Park, Michigan, 1960–1990.

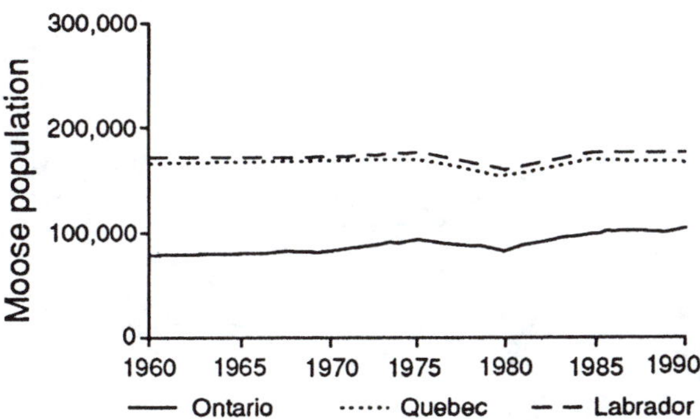

Figure 51. Trends of cumulative regional moose populations in Ontario, Quebec and Labrador, 1960–1990.

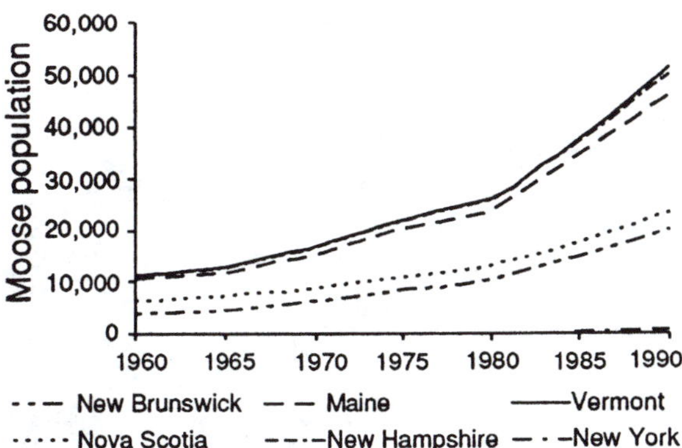

Figure 53. Trends of cumulative regional moose populations in northeastern North America, 1960–1990.

moose increased substantially in number (Figure 53) and range in the past few decades (Karns et al. 1974). Moose hunting was curtailed in New Brunswick in 1937 due to low population, and remained closed until 1960, by which time the population had recovered sufficiently to warrant a limited season (Boer 1992). Moose populations in Nova Scotia have remained at a relatively low level throughout most of the province, except Cape Breton Island, where poor road access limits moose vulnerability to hunters (T. Nette personal communication: 1992). The moose population in Maine has increased from 2,000 in the early 1900s to 22,500 and still growing in the 1990s, and hunting seasons that had been closed since 1935 were reopened in 1980 (Karns et al. 1974, Morris and Elowe 1993). Low moose populations prompted the closure of moose seasons in New Hampshire in 1901. Population recovery was slow and, after 80 years, it had reached 500 (Bontaites and Guftason 1993). The population reached about 4,000 in 1990. The recent history of moose in Vermont is similar; moose populations increased from 200 in 1980 to 1,500 in 1993. After population recoveries, moose hunting seasons were reopened in 1980 in New Hampshire (Bontaites and Guftason 1993) and 1993 in Vermont (Alexander 1993).

From the Vermont population, moose have begun to reestablish themselves in Massachusetts and Connecticut,

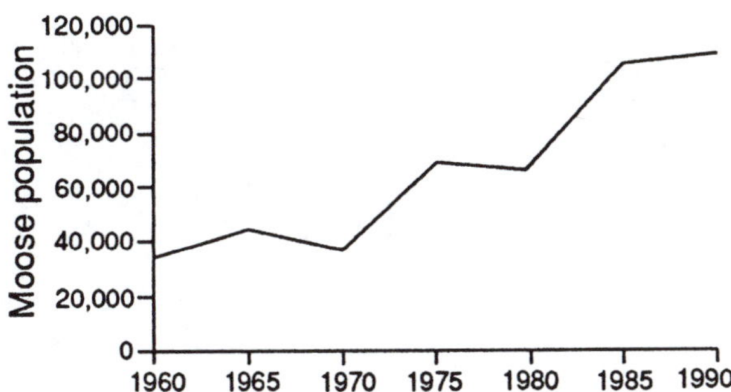

Figure 52. Trend of moose population in Newfoundland, 1960–1990.

where their presence has brought them into conflict with human activities in these densely settled areas (Vecellio et al. 1994, M. Ellingwood personal communication: 1992).

Reasons given for the moose population increases include the absence of predators, reduced deer populations and a consequent reduction of the brain worm *Parelaphostrongylus tenuis* in moose, reversion of farmland to forest, larger clearcuts, increased wetland habitats created by higher beaver populations and legal protection (Karns et al. 1974, Hicks 1986, Boer 1992b, Alexander 1993, Bontaites and Guftason 1993, Morris and Elowe 1993).

Summary

Despite reduced predation in most areas and increased spatial competition with human activity, North American moose have prospered in the twentieth century. Improved understanding of moose ecology, careful management of hunter harvests, and habitat conservation and management—including forestry practices—sensitive to the species' needs have enabled overall population growth and expansion. That is the continuing trend. Nevertheless, the obstacles faced by moose populations now include the efficiency and effectiveness with which forest fires are dealt, even in remote areas. This is exacerbated by a growing legion of people moving into "the bush," and their demand for protection from forest fires. Dams in western areas also have reduced flooding on streams—another important phenomenon of rejuvenation of moose habitat. In urbanized areas, moose pose a threat to human life and property, and are killed annually in vehicle and train collisions (see Chapter 8). And, again, the growth and spread of moose, concurrent with human population growth and sprawl, potentially jeopardizes their continuing prosperity by virtue of public intolerance of such "imposition."

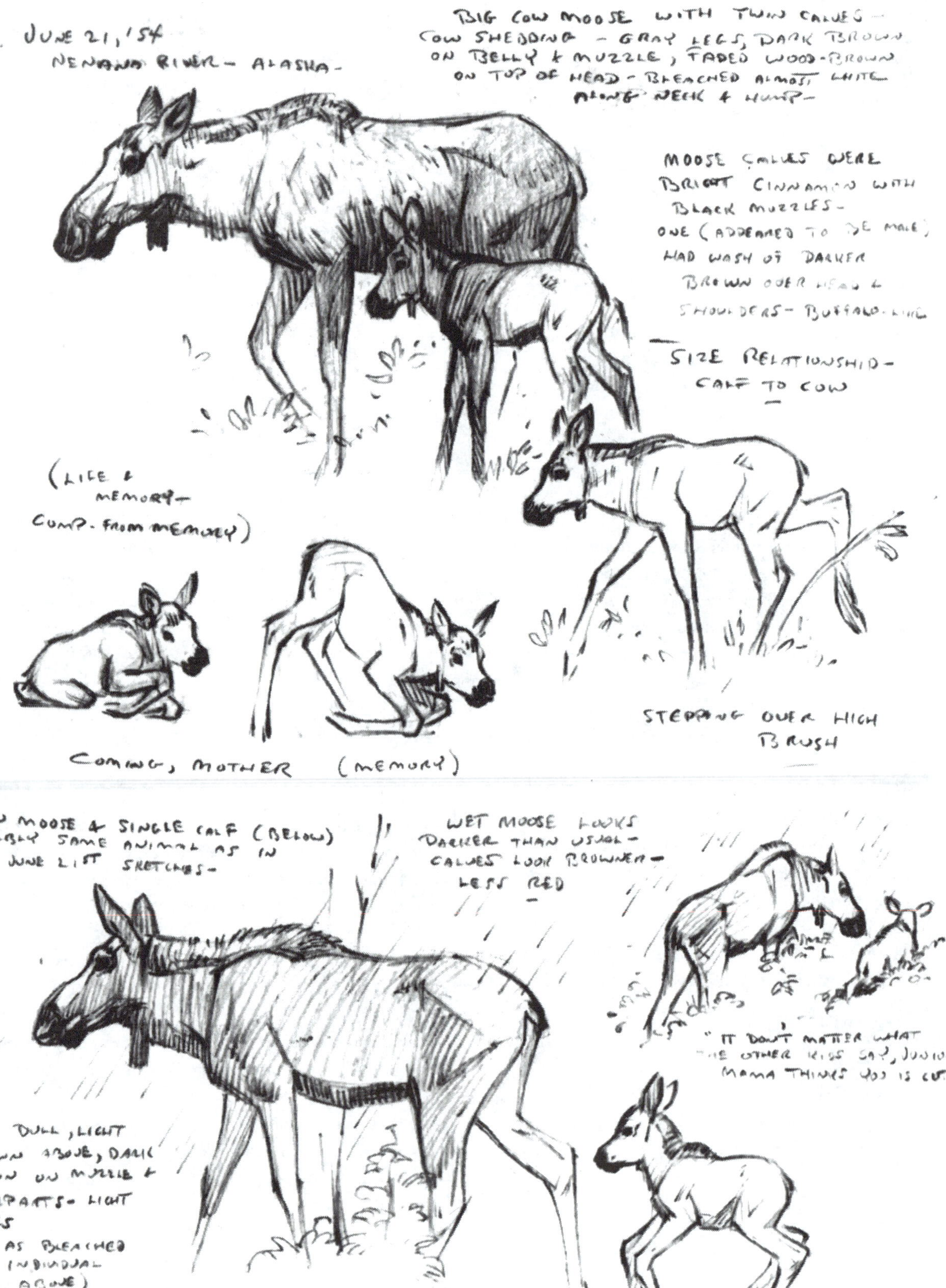

JUNE 21, '54
NENANA RIVER – ALASKA –
BIG COW MOOSE WITH TWIN CALVES – COW SHEDDING – GRAY LEGS, DARK BROWN ON BELLY & MUZZLE, FADED WOOD-BROWN ON TOP OF HEAD – BLEACHED ALMOST WHITE ALONG NECK & HUMP –
MOOSE CALVES WERE BRIGHT CINNAMON WITH BLACK MUZZLES – ONE (APPEARED TO BE MALE) HAD WASH OF DARKER BROWN OVER HEAD & SHOULDERS – BUFFALO-LIKE
SIZE RELATIONSHIP – CALF TO COW
(LIFE & MEMORY – COMP. FROM MEMORY)
STEPPING OVER HIGH BRUSH
COMING, MOTHER (MEMORY)
COW MOOSE & SINGLE CALF (BELOW) PROBABLY SAME ANIMAL AS IN 1ST JUNE 21ST SKETCHES –
WET MOOSE LOOKS DARKER THAN USUAL – CALVES LOOK BROWNER – LESS RED
"IT DON'T MATTER WHAT THE OTHER KIDS SAY, JUNIOR – MAMA THINKS YOU IS CUTE
COW DULL, LIGHT BROWN ABOVE, DARK BROWN ON MUZZLE & UNDERPARTS – LIGHT LEGS (NOT AS BLEACHED AS INDIVIDUAL ABOVE)
EARS BLUISH-GRAY ON BACK –
JUNE 22, '54 NENANA R.

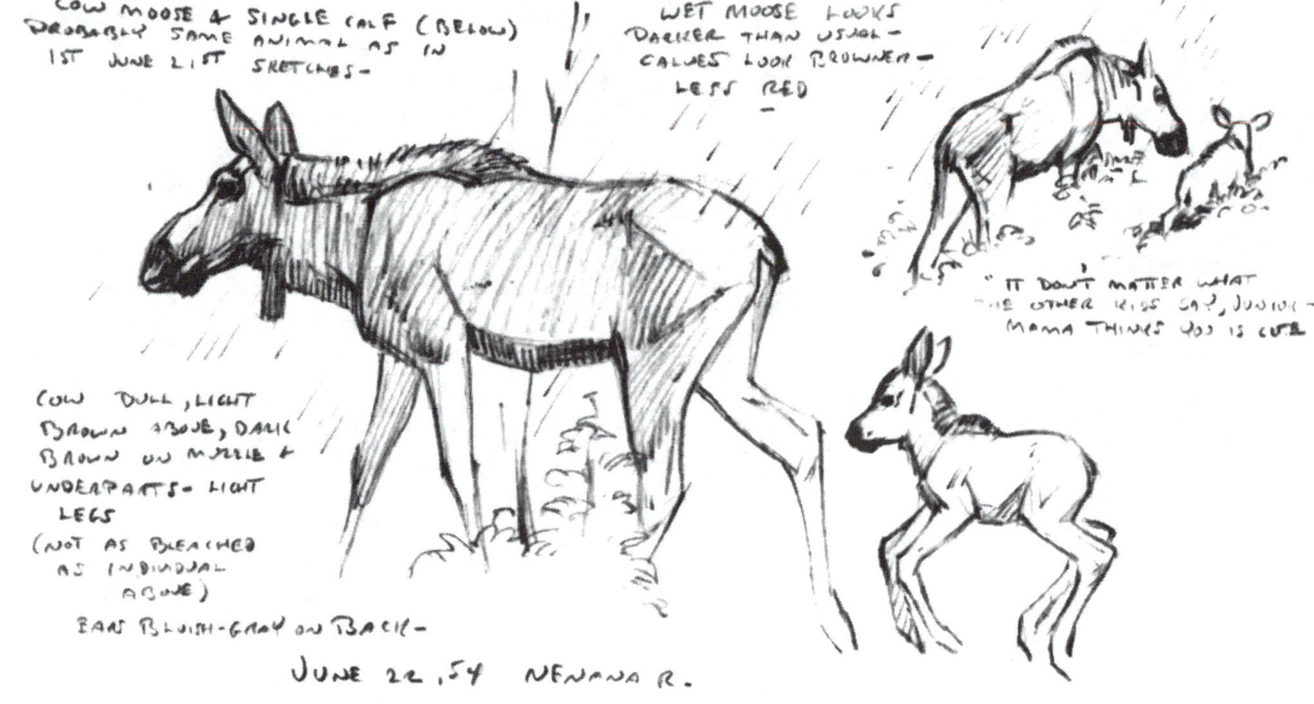

4

CHARLES C. SCHWARTZ

Reproduction, Natality and Growth

Three primary biological factors collectively influence the number of moose living in a particular area: (1) reproduction, (2) mortality and (3) carrying capacity of the area's habitat. Reproduction is the production of new individuals. Mortality represents the death of individuals attributable to all the agents within the environment. Carrying capacity is the number of healthy animals that an area can support over a given period of time. This chapter focuses on the first of these factors, moose reproduction—its anatomy and physiology, the female and male reproductive cycle and, finally, factors that influence moose reproduction, and common techniques to monitor it.

Also, from the moment of conception until death, a moose constantly changes size and shape. Whereas reproduction and mortality affect the dynamics of moose populations, growth, development and weight changes affect the dynamics of individual moose. The rate at which a female calf moose matures influences when she is first able to reproduce and, ultimately, lifetime fecundity (production). Growth and development determine the ability of males to become dominant breeders. Weight loss during winter affects individual fitness. Because moose live where food quality and quantity vary dramatically from summer to winter, they gain and lose weight seasonally. These weight changes affect individual health and survival. This chapter, therefore, in addition discusses growth and development in moose.

Reproduction

The Barren Cow

Several decades ago, knowledge about moose reproduction was derived mainly from field observations of cows with and without calves (Pimlott 1959). The percentage of cow moose seen with calves usually was low. The paucity of mothering cows was a phenomenon often discussed by woodsmen and, as pointed out by Suber (1940), the term *barren cow* became a common expression. In Nova Scotia, for example, Schierbeck (1929) concluded that, based on a sample of 6,175 cows and only 3,406 calves, at least half the cow moose were without calves. In the Nova Scotia situation, the legislature called for a special hunt in which 561 cows were killed, including 273 that were deemed "worthless barren cows" (Schierbeck 1929: 8).

It was believed that barren cows represented bred females that lost calves or cows that, for some physiological reason, simply could not breed. After reviewing the available records, Peterson (1955: 59–60) concluded that "it seems more than a coincidence that the average rate of calfless cows should be so similar across North America." He suggested that more detailed studies of reproduction, particularly in young animals, were needed. Peterson (1955: 60) concluded that, "without any valid evidence to the contrary, it must, therefore, be accepted that the rate of repro-

Many sportsmen and some biologists once believed that moose had very low reproductive rates. This misconception stemmed from frequent observations of calfless cows in the wild. The animals were described as "barren cows." Studies using radiotelemetry have shown that nearly 90 percent of cow moose produce calves, many of which are lost to predation. The lone cow shown on the Kenai Peninsula of Alaska in late April, probably is pregnant and will give birth in late May. *Photo by Charles C. Schwartz; courtesy of the Alaska Department of Fish and Game.*

duction of moose is quite low. All available records indicate that normally less than one-half of the adult cows (two years of age or older) produce calves each year."

Markgren (1969) examined a number of barren cows harvested during the hunting season in Sweden. Nearly 48 percent of the sample was composed of yearling (20 percent) and 2-year-old moose (28 percent) that had not previously bred. Of the remaining moose examined, many were lactating (18 percent) or had pigmented scars in their ovaries (34 percent), indicating previous ovulation and past productivity. This study led Markgren (1969: 208) to conclude that "lasting improductivity must be a rare phenomenon in female moose."

Neonatal predation is a common event. Up to 70 percent of all moose calves born each year are killed by black or brown bears or wolves (see Chapter 7). These losses result in barren cows in autumn. Observations of these cows and those of nonbreeding yearlings and 2-year-old female moose clearly explain why Peterson (1955) concluded that less than 50 percent of all adult female moose produce a calf. It now is known that barren cows (those incapable of producing a calf) are very rare in most populations (see Schwartz 1992c).

Opportunities for management of any wildlife species are improved when that species' reproductive potential is known. Before reviewing moose reproduction, it is important to define some terms. The maximum potential reproductive output of a species usually is considered its *biological potential.* Actual reproductive output is known as *natality,* which changes from area to area and over time.

Natality also can be considered the inherent ability of a population to increase (Odum 1959). It represents the production of new individuals and often is expressed as a rate—the number of individuals born per unit of time (Dasmann 1964). *Maximum natality* is the theoretical maximum production of new individuals under ideal conditions (Odum 1959), or biological potential.

In moose, natality generally represents the number of calves born in a population each year. It also can be expressed as a rate per individual by dividing number of births by the number of females of breeding age in the population. The resulting value represents the number of calves produced per female per year.

Natality refers to populations; fecundity is in context to individuals. The fecundity rate of a female is measured as the number of live births she produces over an interval of time, generally either 1 year or a lifetime. In population biology the fecundity rate of a given female is trivial information; the data of interest are the mean fecundity rates of each female age class (Caughley 1977). For moose, the fecundity rates of primary interest are those of yearlings and mature cows or, in some cases, old cows. Reproductive output by age class provides insight into the reproductive potential of populations. It allows comparison of populations and judgments about population health and its relationship to carrying capacity (see Chapter 6).

The reproductive potential for moose is determined by the age at first reproduction, litter size, length of reproductive cycle and reproductive life. A clear understanding of the biological potential and environmental factors acting on moose reproduction is important for sound moose management.

The temperate climate of North America subjects moose to food shortages and nutritional deficiencies during winter and a period of lush vegetative growth during the growing season. Like other deer species, moose are seasonal breeders, with conception occurring in autumn, allowing for optimal seasonal (summer) conditions for rearing young (Lent 1974). This process repeats itself each year and often is referred to as the reproductive cycle. The reproductive cycle generally begins at puberty, the time of physiological maturity.

Like other northern deer species, moose are seasonal breeders. Reproduction is timed for autumn to provide optimal conditions for raising young. Because moose live in a white-green world, survival is keyed to food abundance during the growing season. Breeding occurs in late September through early October throughout all of moose range. Timing of rut probably is controlled by day length (photoperiodism) by the pineal gland in the brain. This organ senses light and produces a hormone called melatonin that is known to regulate seasonal reproduction in domestic sheep and goats and white-tailed deer. It is likely that moose are similarly influenced. This healthy calf is the product of timed autumn breeding and spring calving, which allowed its mother to obtain abundant and nutritious food to regain condition and produce milk for the calf. The calf also benefited from both the milk and vegetation to grow rapidly before winter. *Photo by Charles C. Schwartz; courtesy of the Alaska Department of Fish and Game.*

Anatomy and Physiology of Female Reproduction

The reproductive system includes two ovaries, the oviducts, uterus, cervix and vagina.

The vulva is the external portion of the female reproductive system. In the cow moose, it is surrounded by a distinctive patch of white hairs. This characteristic makes it easy to differentiate females from males during winter, when bulls have shed their antlers.

The uterus of the cow moose consists of a cervix or neck, a body and bicornuate horns. The uterus is a very elastic organ weighing less than 0.5 pound (227 g) in a nonpregnant female, but expanding during pregnancy to occupy most of the abdominal cavity. Just before parturition, the gravid uterus of the cow moose, with fluids, tissues and fetal mass, can weigh as much as 180 pounds (82 kg) (Schwartz et al. 1987a).

The ovaries are the operational sex organs of the cow. They are small pink organs located at the end of each oviduct. They produce primary sex cells that eventually develop into mature eggs, called *ova*. They also act as an endocrine gland, producing estrogen and progesterone.

The Female Reproductive Cycle

The reproductive cycle of the cow moose begins with puberty (Figure 54). In humans and other mammals that can produce young anytime during the year, puberty classically is defined as the period when the reproductive organs *first* become functional. In moose, puberty first occurs when

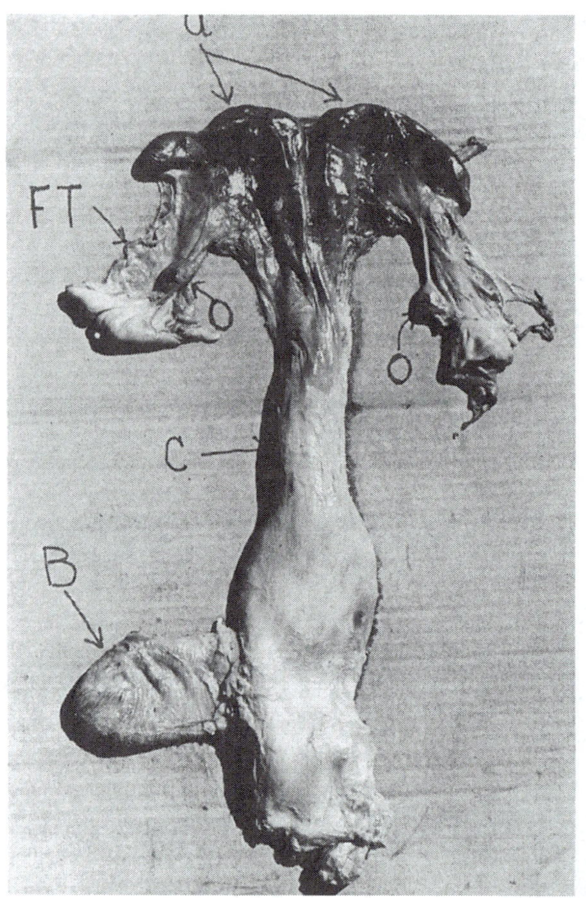

The reproductive tract of a nonpregnant female moose killed just before the breeding season shows paired ovaries (O) and their ducts (FT) connecting to the uterine horns (U), cervix (C), vagina (bottom of photo) and bladder (B). This adult Alaskan cow had almost a 90 percent chance of becoming pregnant based on the average reported for moose. She was capable of having twins, but in poor habitat, her chances of producing only one calf increased. Triplets are rare in moose, even on the best ranges. *Photo by Charles C. Schwartz; courtesy of the Alaska Department of Fish and Game.*

White hairs around the vulva of cow moose enhance biologists' ability to detect the sex of observed animals, and especially during winter surveys when bulls may be antlerless. This sexual characteristic is evident even on young calves. *Photo by Charles C. Schwartz; courtesy of the Alaska Department of Fish and Game.*

the cow is 16 to 28 months of age. Unlike white-tailed deer (Ramsey et al. 1979) and some mule and black-tailed deer (Anderson 1981), moose do not reach puberty as calves. Moose reach sexual maturity at somewhat different ages throughout the species' range in North America, because puberty is influenced by climatic conditions, level of nutrition, heredity, and other extrinsic and intrinsic factors.

After the initial ovulation, hence puberty, almost all cow moose ovulate each year. Sexually mature females begin their estrous cycle in late summer (Figure 55). Stimulated by sex hormones from the anterior lobe of the pituitary, the ovary produces increasing quantities of estrogen and progesterone (Stewart et al. 1985, Monfort et al. 1993). These increased hormones enhance development of the uterus,

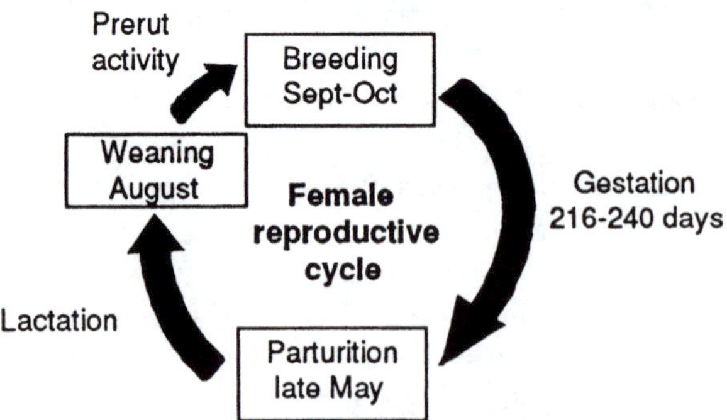

Figure 54. The female moose reproductive cycle. Female moose first become sexually active at 1.4 or 2.4 years of age. Breeding season occurs in autumn. Fetal development begins after fertilization of the egg and continues during a 216- to 240-day gestation period that terminates at calving in late May. Females nurse their calves until late summer or early autumn, when the cycle renews (from Schwartz 1992b).

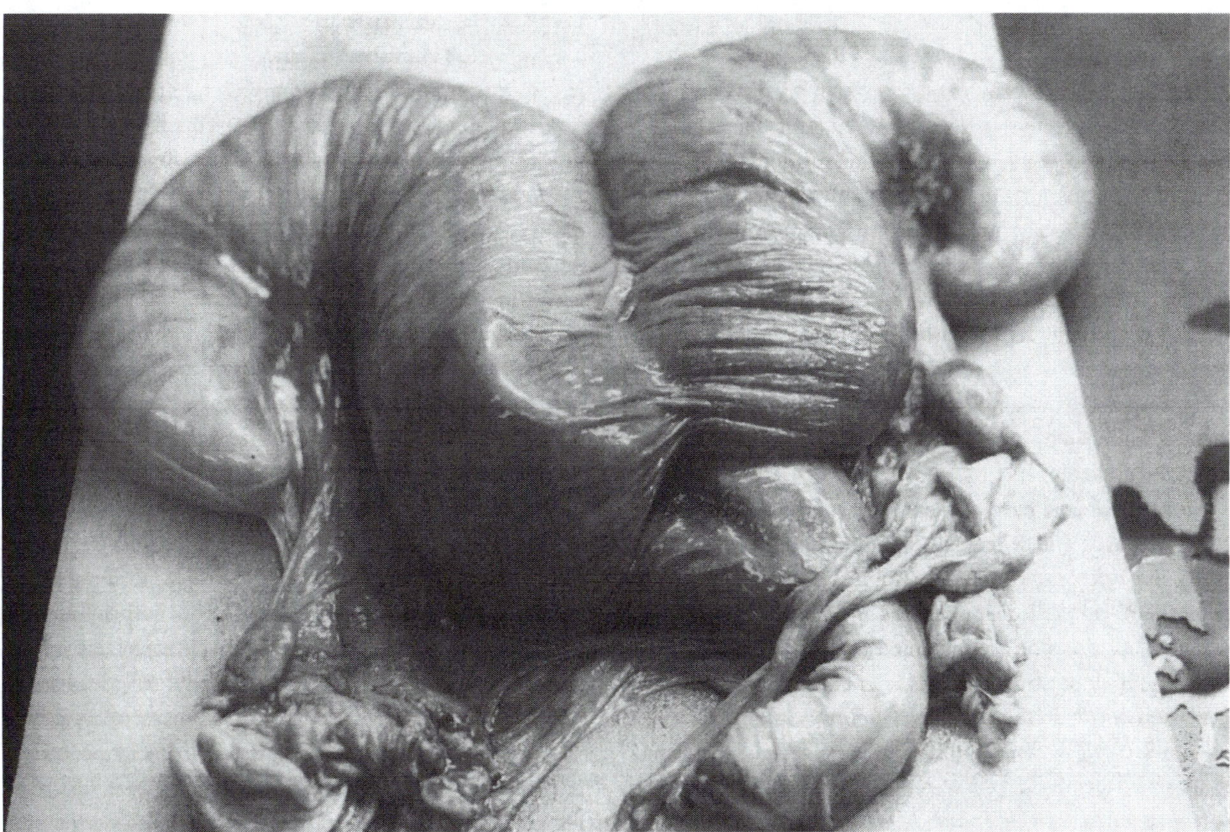

The reproductive tract of a pregnant cow moose collected in midwinter shows the enlarged horns of the uterus. The ovaries appear as egg-shaped objects in the lower left and right center of the photograph. The uterus of a cow moose is very elastic and may weigh only 0.5 pound (227 g) in a nonpregnant female. Its tissue, fluids and fetal mass will expand to weigh as much as 180 pounds (82 kg) just before parturition. *Photo by Sean Berry; courtesy of the Wildlife Branch, British Columbia Ministry of the Environment.*

vagina and oviducts. Primary sex cells in the ovary develop and form the mature ovum within a fluid-filled cavity called the *follicle.* This development occurs during the proestrous period (Figure 55). As the ovarian follicle increases in size, it exerts pressure on the surface of the ovary. Eventually, the ovary wall thins and ruptures. The follicular fluids and ovum are expelled into the oviduct, completing ovulation. In many domestic ruminants, and probably in moose, ovulation is associated closely with "heat" (estrous period), when large amounts of estrogen enter the bloodstream just before ovulation.

Estrus, or heat, is the period of time when a cow will allow a bull to mount. Lent (1974) summarized the literature and concluded that the heat period in moose lasts a day or two. The heat period of moose shows marked variation between individuals, lasting for as little as 1 hour or as long as 36 hours. Most heats last for 15 to 26 hours (Schwartz and Hundertmark 1993).

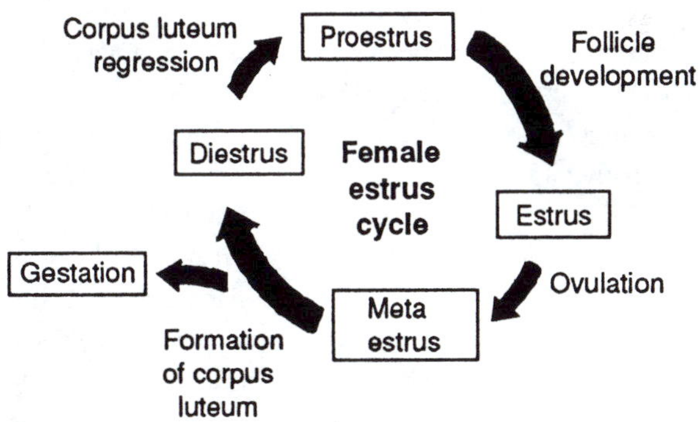

Figure 55. The estrus cycle of the female moose. Follicle stimulating hormone released from the anterior lobe of the pituitary stimulates follicle development in the ovaries of mature female moose. Ultimately these follicles rupture and shed eggs that are fertilized during breeding. The corpus luteum develops within the ruptured follicle and secretes progesterone, which sets the stage for fetal development. If fertilization does not occur, the corpus luteum degenerates and the cycle repeats (from Schwartz 1992b).

The metestrous period is the postovulatory phase during which the corpus luteum is formed, and the vaginal and uterine walls thicken to provide nourishment to the embryo. Progesterone secreted by the corpus luteum blocks additional follicular development, precluding further estrous periods. The corpus luteum is necessary for proper implantation of the fertilized ovum in the uterus, for nourishment of the developing embryo and for development of mammary glands. If pregnancy (ovum fertilization) occurs, the corpus luteum remains intact for all or most of the gestation period. Progesterone levels remain high (Stewart et al. 1985).

If pregnancy does not occur, the corpus luteum regresses, followed by development of a new ovarian follicle. Then the cycle repeats.

Moose probably have a "silent heat" period (Backstrom 1952, Edwards and Ritcey 1958, Markgren 1969, Simkin 1974), like domestic ruminants (Robinson 1959), elk (Morrison 1960) and mule deer (Thomas and Cowan 1975). This first ovulation is not accompanied by an overt estrus or breeding. In domestic ruminants, silent heat happens when ovulation occurs in the absence of a corpus luteum from a previous cycle (Robinson 1959). It is suspected that, in black-tailed deer, elk and red deer, silent heat functions to synchronize estrus and shorten the breeding season (Thomas and Cowan 1975, Bubenik 1982a), although the significance is not fully understood.

Hormones of the Ovarian Cycle

The ovarian cycle in mammals dictates changes in physical structure of the reproductive tract. These changes are brought about through the influence of hormones. The major sites of reproductive hormone production are the pituitary gland and the ovaries (Bronson 1989).

Some hormones affect all body tissues, whereas others act on one organ or gland. In the latter, the organ primarily affected is known as the *target organ,* and the hormone exerting this specialized influence is called a *trophic hormone.* The anterior lobe of the pituitary gland produces hormones that stimulate the reproductive organs. Collectively, these hormones are referred to as *gonadotropins.* Individually they include follicle stimulating hormone (FSH), luteinizing hormone (LH) and prolactin (also referred to as luteotrophic hormone).

At the beginning of the ovarian cycle, increased levels of FSH released from the pituitary cause development of primary follicles in the ovaries. As the follicle matures, it produces estrogen. Estrogen acts on the anterior lobe of the pituitary, stimulating it to produce LH and prolactin. Estrogen also affects female behavior, causing cows to enter heat.

Luteinizing hormone, acting in precise balance with FSH, causes ova maturation and finally ovulation. After ovulation, LH stimulates the ruptured follicle to produce a corpus luteum. In cattle, the corpus luteum is maintained in a functional state by prolactin. This likely is true for moose, but is not documented.

As previously noted, each ovary produces two hormones—estrogen and progesterone. Progesterone is produced by the corpus luteum and later by the placenta. Progesterone is known as the hormone of pregnancy. It prepares the lining of the uterus (endometrium) for implantation of the embryo. Without progesterone, the endometrium would not be sensitized to join the placenta with the embryo. During pregnancy, progesterone prevents additional ovulation by inhibiting LH.

Once pregnancy occurs, there is a shift in the mode of hormonal regulation from the pituitary and corpus luteum to the placenta (Metcalfe et al 1988). The placenta secrets chorionic gonadotropin, which causes the corpus luteum of pregnancy to persist. Chorionic gonadotropin also inhibits the pituitary, thus preventing additional follicle development and ovulation.

If pregnancy does not occur, no chorionic gonadotropin is produced and pituitary hormones are not inhibited. However, both progesterone and estrogen inhibit secretion of FSH. Thus, estrogen helps check its own production. As estrogen declines, LH production also declines and the corpus luteum begins to degenerate. With degeneration of the corpus luteum, concentrations of progesterone and estrogen reach low levels, their inhibition of FSH is lost and the process repeats itself.

Estrus and Ovulation

Most deer species (Sadleir 1982), including moose, are polyestrous (Edwards and Ritcey 1958, Markgren 1969), which means many heats or cycles. If moose do not become pregnant after mating, the estrous cycle begins anew. On the basis of the work of Soviet scientists who backdated embryos to conception date, Lent (1974) concluded that duration of the estrous cycle was only 20 to 22 days. Edwards and Ritcey (1958), using a similar technique, estimated the cycle length at 30 days.

On the basis of studies by Schwartz and Hundertmark (1993), the estrous cycle of moose is widely accepted to be 22 to 28 days. Schwartz and Hundertmark measured the length of the cycle by keeping a group of cow moose with a bull that had been given a vasectomy. Heat was determined by observed breeding. Because the vasectomized bull could not impregnate the cows, they continued to recycle, allowing opportunity to quantify the estrous cycle. The average length was 24.4 days, with most cycles lasting 24 to 25 days (Figure 56). Females breeding for the first time (primiparous females) had a shorter cycle (23.7 days) than did females bred in previous years (pluriparous females) (24.5 days).

A shorter cycle in primiparous females also occurs in domestic cattle (Hansel 1959), but not white-tailed deer (Knox et al. 1988). Length of the estrous cycle in moose appears to be less variable than in black-tailed deer (18–30 days) (Wong and Parker 1988), red deer (13–22 days) (Guiness et al. 1971) or white-tailed deer (21–30 days) (Knox et al. 1988).

If not bred, moose experience as many as six recurrent estrous cycles (seven heats) during a breeding season that potentially extends into late March (Schwartz and Hundertmark 1993). Cessation of the cycle probably is triggered by increasing day length, because it occurs shortly after the vernal equinox. Although none has been identified, the feedback mechanism likely involves the pineal gland and the hormone melatonin (Bittman et al. 1983, Adam 1992).

Studies of black-tailed deer (Wong and Parker 1988), white-tailed deer (Knox et al. 1988), fallow deer (Asher 1985) and red deer (Guinness et al. 1971) all have found that the estrous cycle lengthens with each new cycle. In black-tailed deer, for example, the first cycle may be only 20 to 22 days in length, whereas the fifth cycle may last 27 to 30 days. Conversely, the estrous cycle of moose does not lengthen with each additional cycle and is remarkably consistent among cycles and individual animals (Schwartz and Hundertmark 1993)

Onset of estrus in moose is signaled by increased attentiveness by the bull. Often the bull remains close to an estrous female. Cow moose do not show change in activity just before estrus (Schwartz and Hundertmark 1993), whereas Ozoga and Verme (1975) measured a 28-fold increase in pacing by confined white-tailed deer before estrus, similar to increased behavior exhibited by black-tailed deer (Cowan 1956, Wong and Parker 1988).

Although there are no reliable overt indicators of estrus in female moose, there are marked differences among moose cows in intensity of estrous behavior (see Chapter 5). In some individuals, the heat period is conspicuous; they will stand for mounting five to seven times during their heat. Other females exhibit no overt signs of estrus and mate only once (Schwartz and Hundertmark 1993). The best indication of possible receptivity of a cow is a reduction of her personal space or *intimate zone* (Bubenik 1987). Anestrous females seldom allow bulls to chin (Dodds 1958) their rump or forequarters. They tend to avoid contact with bulls. When courted, anestrous females generally avoid the approach of bulls, whereas females approaching estrus allow chinning and vulva licking by bulls. This activity appears to be incipient to mounting, as suggested by Lent (1974). Reduction in personal space is evident in some females but subtle or lacking in others (Schwartz and Hundertmark 1993).

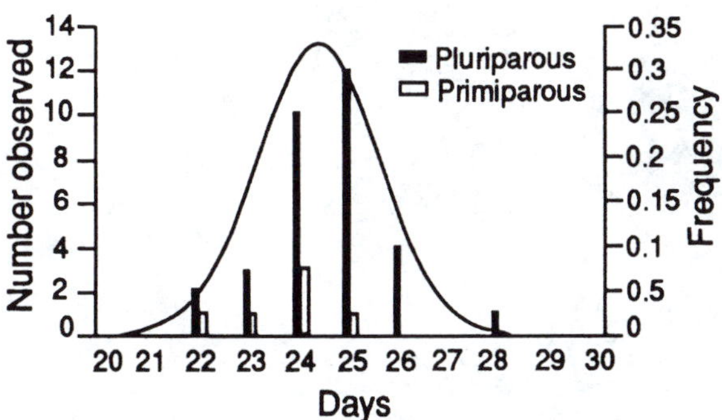

Figure 56. Length of the estrus cycle in moose (sample size, 38; mean, 24.4 days). Primiparous females (those giving birth for the first time) have a significant shorter cycle than do pluriparous females (previous mothers). Original data from Schwartz and Hundertmark (1993).

Gestation

The literature regarding mean length of gestation of moose is confusing. Markgren (1969), in Sweden, calculated a gestation length of 234 days, based on the difference between time from mean dates of breeding and parturition. Peterson (1955: 99) stated that "the gestation period is generally conceded to be approximately eight months, 240–246 days." The gestation of two yearling moose in Saskatchewan was measured as 216 days (Stewart et al. 1987). It was 242 days for three cows in Michigan (Verme 1970). Schwartz and Hundertmark (1993), working with captive moose, measured gestation length in 24 cows over a 5-year period. Actual breeding and parturition dates were observed. Mean gestation for all females was 231 days, with a range of 216 to 240 (Figure 57). No difference was found in the length of

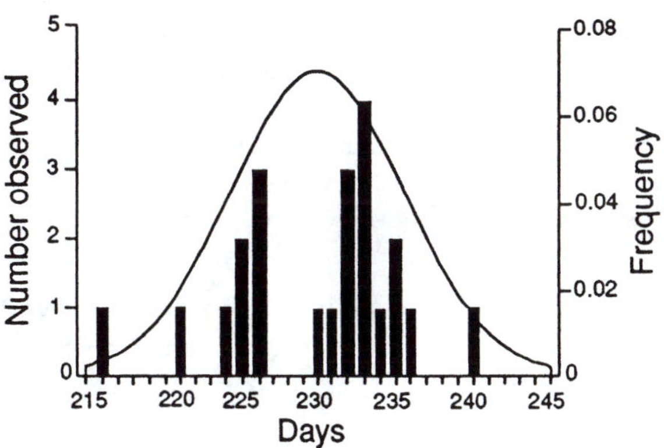

Figure 57. Length of the gestation period in moose (sample size, 22; mean, 231 days). Original data from Schwartz and Hundertmark (1993).

Courtship behavior in moose involves both the male and the female. A bull moose cautiously approaches a cow with her calf from the previous year. The bull has a head high position allowing him to detect chemical signals emitted from the female (top left). If the cow is not in estrus, she will avoid intimate contact and walk away. If the cow is near her peak in estrus, she allows the bull to make contact with her vagina and test her receptiveness. The bull uses "flehming" to detect the presence of the chemicals called volmerones. By raising the head and curling the upper lip, the bull opens the Jacobson's organ and allows entry of the volmerones (bottom left). The Jacobson's organ is the structure that senses the volmerones permitting the bull to detect the females estrus. Detecting that the female is in heat, the bull approaches from the rear and "chins" the female's rump (top right). Such courtship behavior provides sexual stimulation to both male and female. Before mounting, the bull develops an erection in preparation for copulation. Chinning also allows the bull the opportunity to align himself with the female. When the bull is properly aligned with the female mounting occurs (bottom right). The entire sequence of mounting, intromission, and ejaculation takes place in a few seconds. When copulation is complete, the bull slides from the female's back. Mounting may occur several times during the female's heat, or she may only breed once. The bull remains close to a cow in estrus, but will seek other mates soon after the female refuses mounting. A cow is usually in heat for less than a day. *Photos by Gustav Norra.*

gestation between primiparous and pluriparous females, between single and twin litters, or between cows bred during their first or second estrous cycle. Gestation lengths measured in six captive moose in Sweden—226–236 days (Cederlund 1987)—also agree with these data.

The Schwartz and Hundertmark (1993) findings represent the largest data base of measured gestation lengths for moose or any other North American cervid (see Sadlier 1982) except white-tailed deer (Verme 1969). Their observations suggest that the 216-day gestation lengths documented for two yearling moose by Stewart et al. (1987) and the 240 to 246 days reported by Peterson (1955) and Verme (1970) represent extremes. The Schwartz and Hundertmark (1993) results compare more favorably with the estimate of 234 days presented by Markgren (1969) and Cederlund (1987) for Swedish moose.

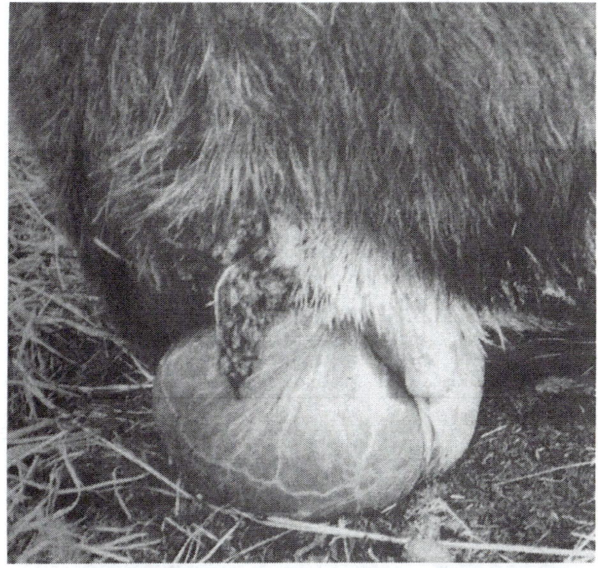

After a gestation period of about 231 days, a cow moose will seek a secluded spot to give birth. Birthing sites are often close to water, on peninsulas or islands. Such secluded locations offer some protection from predators. Cows generally give birth while lying down. When the birth process begins, the placenta-filled sac passes through the vagina (top left). The calf generally is born head first, although breech births do occur. The cow may stand just as the calf is being delivered allowing gravity to aid in the birth process. The calf when first born is covered with the placental membranes (bottom left). After giving birth, the cow moose will eat the placenta and associated tissues. She then cleans the calf by licking it dry (right). Such cleaning ensures that a close bond develops between the cow and her calf. The cow and calf generally remain at the birth site for a few days to about a week. By this time the calf is strong enough to travel. *Top left photo by John Carnes. Bottom left and right photos by Debra Shuey.*

Age at First Breeding

Among cow moose older than calves, considerable variation is found in the age at which first reproduction occurs. Most ovulate for the first time at either 16 or 28 months of age (Table 10), although cows in populations on very poor range may not breed until 40 months (Albright and Keith 1987). Although some moose calves ovulate (Simkin 1965, Addison 1974), unlike white-tailed deer fawns, most are not sexually mature and do not breed (Edwards and Ritcey 1958, Pimlott 1959, Rausch 1959, Peek 1962, Simkin 1965, Houston 1968, Schladweiler and Stevens 1973, Addison 1974, Boer 1987a, Modafferi 1992, Schwartz and Hundertmark 1993).

The age at first ovulation varies considerably between populations. None of 15 yearling cow moose examined in British Columbia were pregnant. When that fact was combined with observational data, Edwards and Ritcey (1958)

Table 10. Reproductive rates for yearling and adult moose throughout North America[a]

Subspecies/ location and/or characteristic	Yearling: Ovulation percentage (sample)	Yearling: Pregnancy percentage (sample)	Adult: Ovulation percentage (sample)	Adult: Pregnancy percentage (sample)	Twinning percentage[b]	Collection: Method[c]	Collection: Dates	Source
A. a. americana								
New Brunswick	50[d] (6)	39 (13)		79 (33)	26/?	CL, FC	Nov–June	Boer (1987a)
Newfoundland	60[e] (42)	46 (78)		81 (239)	14/?	FC	Nov–May	Pimlott (1959)
Newfoundland		0[f] (2)	60 (56)		1/?	DO	May–June	Albright and Keith (1987)
Quebec								
South	14 (57)		87 (54)		55/48	CL	Oct	Crête and Beaumont (1986)
Central	59 (44)		81 (174)		65/53	CL	Oct	
North	33 (33)		62[g] (112)		33/27	CL	Sept–Oct	
Parks	47 (30)		58[g] (57)		45/26	CL	Sept–Oct	
A. a. shirasi								
Montana	48 (51)	32 (22)	92 (304)	86 (73)	16/?	CL, FC	Nov–May	Schladweiler and Stevens (1973)
Montana	0 (2)		71 (14)			CL	Oct–Nov	Peek (1962)
Wyoming	0[h] (18)		4[h] (86)			CL	Sept	Houston (1968)
Wyoming	17–25[i] (12)	6[i] (35)	89 (72)	90 (41)	4ih	CL, FC, PA	Nov–May	Houston (1968)
A. a. andersoni								
Alberta		29 (99)		87 (355)	17/13	FC	Dec–March	Blood (1974)
British Columbia		0 (15)		76 (80)	25/19	FC	Oct–May	Edwards and Ritcey (1958)
Ontario	35 (68)		86 (140)			CL	Oct	Simkin (1965)
Ontario		57 (7)		97 (37)	54	FC	May	Bergerud and Snider (1988)
Ontario	51 (42)					CL	Oct	Addison (1974)
Ontario								
1984				(31)	64/?	DO	May	Addison et al. (1985)
1981–1983				(24)	17/?	DO	May	Addison et al. (1985)
Michigan[j]				78 (97)	37/10	DO	May–June	Aho and Hendrickson (1989)
Manitoba								
1978–1980		0 (1)		86 (37)	28/24	FC	Nov–Dec	Crichton (1988b)
1986		62 (8)		90 (41)	24/22	FC	Nov–Dec	
1986–1991[k]	43 (14)	43 (6)	97 (141)	92 (141)	16/15	CL,FC	Nov–Dec	Crichton (1992c)
A. a. gigas								
Alaska								
Alaskan Peninsula				84 (57)	80	PAP, DO[l]	April–June	Faro and Franzmann (1978)
Central				100 (27)	52	PAP, DO	April–May	Gasaway et al. (1992)
Central		0 (3)		88 (52)	32	PAP, DO	April–May	Gasaway et al. (1983)
Nelchina Basin				81 (140)	38/?	PAP, DO	April–Jun	Ballard et al. (1991)
Southcentral	22 (9)	11 (9)	100 (36)	94 (78)	26/24	CL, FC	Oct–May	Rausch (1959)
Kenai Peninsula								
yearlings	100 (3)	22 (18)				CL, FC	Nov–May	Schwartz and Hundertmark (1993)
prime cows			100 (51)	97 (85)[m]	27	CL, FC	Nov–May	
old cows			100 (7)	14 (7)[m]	0	CL, FC	Nov–May	
1947 burn				(49)	22/?[n]	DO	May–June	Franzmann and Schwartz (1985)
1969 burn				(102)	71/57[n]	DO	May–June	
Yukon				89 (129)	28	PAP, DO	April–June	Larsen et al. (1989b)

[a] Age refers to the time at breeding, therefore a yearling female would have her second birthday during her first parturition.

[b] Percentage of twins produced for those adult females giving birth/percentage of twin births for all adult females.

[c] Ovulation identified by counts of corpora lutea (CL); pregnancy determined by fetal counts (FC), rectal palpation (PAP) or direct observation (DO).

Table 10 continued

[d] Three of six yearlings each had one corporus luteum and one fetus.

[e] Samples were from 42 nonpregnant yearlings; 9 of 25 paired (36 percent) ovaries contained primary corpora lutea; 60 percent of all yearlings were in breeding condition, the 60 percent was derived by adding the percentage pregnant with the percentage ovulating that were not pregnant.

[f] Less than 1 percent of 107 cows observed with twins; no yearlings or 2-year-old cows observed with calves their next birthday; age at first breeding was 3.5 years in this population.

[g] Some ovaries were collected before peak rut, which would bias the ovulation downward.

[h] Twinning was estimated for observations of cow/calf pairs. The value likely is lower than twinning measure in utero.

[i] Three of 12 tracts contained Cls, whereas 5 of 35 (14 percent) tracts of 2.5-year-olds contained corpora albicantia, but Houston (1968) claims 17 percent ovulated. Yearling pregnancy was determined by distended appearance of two uteri from 35 2.5-year-old females.

[j] Translocated herd.

[k] Data calculated from Crichton (1992c: Table 6). Values differ slightly from values in text calculated with different sample size.

[l] Twelve of 15 cows observed during peak calving were accompanied by twins.

[m] Prime age cows were ages 2–15 (n = 85) and old cows were >15 years of age (n = 7).

[n] Sample may contain some yearlings.

concluded that female moose in Wells Gray Park, British Columbia, did not breed until age 2. Yet, 60 percent of the yearlings examined by Pimlott (1959) in Newfoundland had ovulated and 42 percent were pregnant. Studies of other moose populations indicate that some yearling females either were pregnant and/or had ovulated (Table 10).

Although yearling ovulation rate is a good indication of population condition, not all ovulations result in pregnancy. Ovulation rates range from 0 to 100 percent (mean, 49 percent) but pregnancy rates from the same areas have not been shown to exceed 62 percent (Table 10). This suggests a fairly high ova loss in yearling moose. Boer (1992a) listed fecundity rates of 18, 41 and 64 for yearling moose from populations throughout North America above, near and below habitat-carrying capacity, respectively.

Age at first reproduction and sexual maturity in moose generally is related to body size (Saether and Haagenrud 1983). Well-nourished moose grow faster and tend to reach adult body mass sooner than undernourished or malnourished moose (Schwartz et al. 1987a). Yearling cows on a high nutritional plane generally breed and produce a single calf (Saether and Haagenrud 1985, Schwartz and Hundertmark 1993); twins rarely are produced (Pimlott 1959, Serafinski 1969, Blood 1974). Long-term records for cow moose kept in semicaptive conditions suggest that cows breeding as yearlings produce fewer calves (1.5 per year) during their breeding lives (>10 years) than do cows that first breed at age 2.4 (1.9 per year) (Schwartz 1992b). Also, studies in Sweden (Saether and Haagenrud 1985) clearly showed that cow moose that conceived as yearlings weigh less than nonreproducing females of the same age the following year. Moose that first reproduce as yearlings apparently have reduced potential of further growth. Primiparous 2-year-old moose produce significantly more calves (1.6) than do their yearling counterparts (1.0) (Schwartz and Hundertmark 1993).

Ovulation and Pregnancy Rates of Adults

On most ranges, ovulation and pregnancy rates are insensitive indicators of range quality, except in extreme circumstances. Ovulation rates in adult moose are consistently high, ranging from 71 to 100 percent (Table 10). Loss of ova averages about 4.4 percent, but has been reported as high as 9 percent (Schwartz and Hundertmark 1993).

The age at which a cow moose breeds is related to her growth rate and body size. Female moose calves do not breed and, as yearlings, produce offspring as is the case for white-tailed deer females. Yearling moose ovulation and pregnancy rates are influenced by range quality. In areas with excellent forage, virtually all yearlings ovulate, and up to 62 percent may become pregnant. On poor range, yearlings may not breed. Yearlings generally give birth to singletons. *Photo by Charles C. Schwartz; courtesy of the Alaska Department of Fish and Game.*

In his review of moose reproduction in North America, Boer (1992a: 2) concluded that "adult pregnancy rates averaged 84.2 percent and were remarkably consistent." Gasaway et al. (1992) presented data that showed a declining pregnancy rate as populations approached carrying capacity (Table 11). But with the exceptions of Sandy-M-Town, Newfoundland, and the Moose Research Center in Alaska, both of which were above carrying capacity, pregnancy rates exceeded 75 percent and, in most cases, ranged from 82 to 100 percent.

Most cow moose produce either a single or twin calves (Table 11). Twinning varies widely across North America and likely is correlated to habitat quality and the relationship of each moose population to carrying capacity of their habitats (Gasaway et al. 1992). Triplets have been reported, but their occurrence is rare (Hosley and Glaser 1952, Peterson 1955, Pimlott 1959, Moll and Moll 1976, Franzmann and Schwartz 1985).

The literature lacks detailed information on repeated observations of calf production for individual females. Because most adult pregnancy rates are high, most cows likely produce a calf or calves each year. Long-term observations from a group of captive moose substantiate this: average calf production was 1.5 to 1.9 calves per female per year during the life of each cow (Schwartz 1992b). Fertility remains high through the life of most females, but maximum reproductive output probably occurs between ages 4 and 7 years (Sylven 1980, Saether and Haagenrud 1983, Schwartz and Hundertmark 1993). However, observations by Albright and Keith (1987) from very poor range on the south coast barrens of Newfoundland suggest that many females do not produce calves in consecutive years. Only 3 of 17 cows 4 years of age or older were seen with a calf in consecutive years, and only 1 of 7 cows was with a calf in 3 consecutive years.

Yearling pregnancy rates are highly correlated with twinning rates in adult cow moose, suggesting that they may be influenced by similar factors (Boer 1992a). The regression equation (Figure 58) indicates that both variables—yearling pregnancy and adult twinning—change at approximately the same rate, as the slope of the line was close to 1 (1.09). This relationship suggests that yearling pregnancy may serve as an indicator of reproductive performance in populations.

Franzmann and Schwartz (1985) speculated that twinning rates in adults are an indication of nutritional status of

Table 11. Productivity of moose in relation to K carrying capacity (KCC) for populations from North America[a]

Population	Year	Age of cows (months)	Percentage twinning (sample)	Percentage pregnant (sample)	Relationship to KCC	Source
Interior Alaska	1984	>29	52 (27)	100 (28)	?	Gasaway et al. (1992)
Innoko River, Alaska	1988	>29	90 (10)	100 (17)	Below	J.S. Whitman (unpublished data)
Rochester, Alberta	1975–1978	>29	88 (8)		Below	Mytton and Keith (1981)
Alaska Peninsula	1977	>29	80 (15)	84 (57)	Below	Faro and Franzmann (1978)
Kenai Peninsula, Alaska	1982–1983	>19	71 (102)		Below	Franzmann and Schwartz (1985)
Westcentral Alaska	1988–1989	>19	56 (61)		Below	A.J. Loranger and T.O. Osborne (unpublished data)
Pukaskwa Park, Ontario	1975–1979	>29	54 (37)	97 (37)	Below	Bergerud and Snider (1988)
Elk Island, Alberta	1960–1964	>29	50 (28)	82 (34)	Below	Blood (1974)
Southcentral Alaska	1977–1980	>29	41 (64)	88 (59)	Below	Ballard and Taylor (1980), Ballard et al. (1982)
East Newfoundland	1953–1956	>29	41 (29)	87 (38)	Below	Pimlott (1959), W.E. Mercer (unpublished data)
Central Alaska	1975–1978	>29	32 (35)	88 (52)	Below	Gasaway et al. (1983)
Southcentral Alaska	1950s	>29	28 (87)	94 (93)	Below	Rausch (1959)
Southern Yukon	1983–1985	>29	28 (58)	84 (43)	Below	Larsen et al. (1989a)
New Brunswick	1980–1986	>29	23 (52)	79 (33)	Below	Boer (1987a), A.H. Boer (unpublished data)
British Columbia	1952–1956	>29	25 (80)	76 (80)	Near	Edwards and Ritcey (1958)
Kenai Peninsula, Alaska	1977–1978	>19	22 (49)		Near	Franzmann and Schwartz (1985)
Elk Island, Alberta	1959–1973	>29	12 (216)	84 (258)	Near	Blood (1974)
South Newfoundland	1973–1975	>19	2 (88)		Near	W.R. Skinner (unpublished data *in* Albright and Keith 1987)
South Newfoundland	1982–1984	>19	1 (107)		Near	Albright and Keith (1987)
Sandy-M-Town, Newfoundland	1953–1956	>19	3 (87)	74 (116)	Above	Pimlott (1959)
Moose Research Center, Alaska	1973–1975	>29	0 (22)	60 (37)	Above	Franzmann et al. (1976d)

[a] Ranking in relation to carrying capacity based on the original authors' comments in papers or personal communications cited in Gasaway et al. (1992), where original table was created.

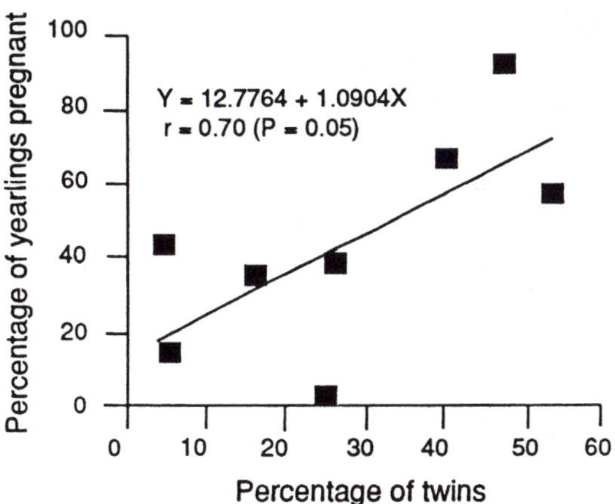

Figure 58. Relationship between twinning rates for adult female moose and pregnancy of yearling females (from Boer 1992a).

the moose population. The proportion of females with twins varied considerably among populations. Gasaway et al. (1992) showed that, in populations below carrying capacity, twin births range from as high as 90 to as low as 25 percent of all births for cows older than 29 months. Twinning rates in populations near carrying capacity range from approximately 5 to 25 percent, whereas twinning occurred less than 5 percent in populations above carrying capacity (Table 11). Twinning rates appear to be independent of pregnancy rates in adults (Boer 1992a).

Measuring Reproductive Conditions

Several methods are used to monitor reproductive performance in moose. The most commonly used techniques are discussed here.

OVARIAN SECTIONING

Cheatum (1949) was one of the first wildlife scientists to recognize the usefulness of the ovaries as indicators of reproduction in white-tailed deer. Since Cheatum's early work, many wildlife biologists have used ovarian sections to determine ovulation rates and previous pregnancy rates in a wide range of mammals, including moose.

Once the mature follicle ruptures and the ovum is discharged, the empty follicle goes through a series of changes. First, it fills with clotted blood and is referred to as a *corpus hemorrhagicum* (red body). The clotted blood is replaced with connective tissue, lutein cells and fibrin, and is referred to as a *corpus luteum* (yellow body). Finally, the corpus luteum degenerates in size leaving a small whitish scar, the *corpus albicans,* on the surface of the ovary.

Ovulation in moose can be detected by examining ovaries from samples collected during and after peak breeding season. Collected ovaries usually are fixed in 10 percent formalin solution to preserve the integrity of the tissue. Formalin also "hardens" the ovaries, making it easy to slice them in thin (1 mm) longitudinal sections with a sharp razor blade or scalpel. Within the sectioned ovary, primary corpora lutea can be identified macroscopically. Depending on the date of collection, primary corpora lutea indicate recent ovulation and possibly pregnancy. Because a corpus luteum forms in each ruptured follicle after ovulation, corpora lutea counts during or shortly after the rut reveal ovulation rates but not necessarily pregnancy. If examination of ovaries occurs several months after breeding, comparison of corpora lutea counts with fetal numbers serves as a good indication of ovulation rates. The difference between corpora lutea counts and the number of fetuses in uteri of the sample is a good indication of ova loss (Robinette et al 1955, Crichton 1992c, Schwartz and Hundertmark 1993).

Corpora albicantia also can be identified and counted. They are indication of pregnancy and/or ovulation in previous years (Addison 1974). Errors occur when all scars are

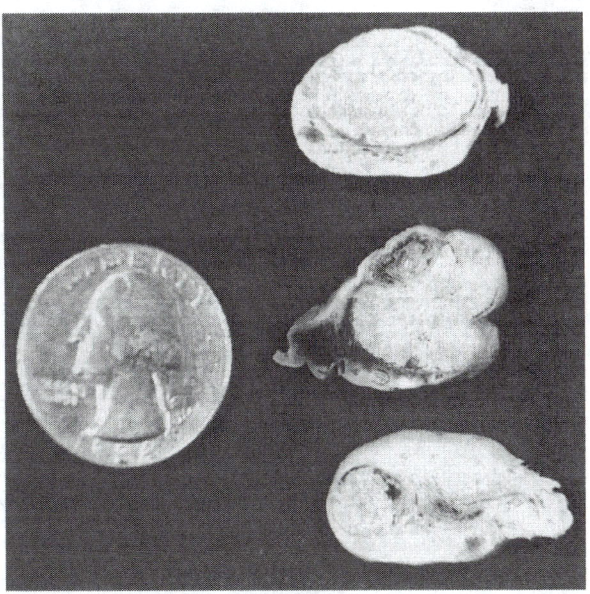

Biologists can determine female moose ovulation and pregnancy by examining sectioned ovaries. Ovaries collected from moose after the breeding season may contain corpora lutea. The corpus luteum forms in the ruptured follicle after ovulation. If the cow becomes pregnant, the corpus luteum produces progesterone, which maintains pregnancy. The corpus luteum is a firm yellow structure that can be identified easily and distinguished from maturing follicles that are clear and fluid filled. This close-up view of a sectioned ovary came from a pregnant cow moose during winter. The large corpus luteum is easily seen, confirming pregnancy. A partially developed follicle also is apparent on the bottom ovary near the left. *Photo by Charles C. Schwartz; courtesy of the Alaska Department of Fish and Game.*

assumed to represent pregnancy, because similar structures form after ovulation even if pregnancy does not occur (Markgren 1969). Microscopic examination is necessary to distinguish between the two.

Simkin (1965), following the criteria described by Gibson (1957), provided a detailed explanation of microscopic ovarian structures commonly found in moose. Primary corpora lutea resulting from ovulated follicles were large and granular. They generally measure at least 0.4 by 0.7 inch (10 × 18 mm) (Rausch 1959). Degenerating corpora lutea of estrus from recently ovulated follicles undergoing regression are unpigmented and no more than half as large as a primary corpus luteum. Mature (Graafian) follicles are more than 2 mm in diameter and appear like a water-filled blister on the surface of the ovary. Corpora albicantia vary depending on their age. Scars of pregnancy 6 months after parturition are neither distorted nor compressed. They are roughly spherical in shape with many thick-walled blood vessels and very little connective tissue. The scar is composed mainly of pigmented lutein cells with a thick connective tissue capsule surrounding it. Because of a gradual reduction in scar size with age, the most recent ones are largest. Scars of pregnancy 18 months after parturition are considerably distorted and compressed. The thick-walled blood vessels are numerous. Many connective tissue fibers and cells are present among the pigmented lutein cells, which are few in number. The degenerating corpora lutea of estrus leaves a scar of irregular outline, consisting mainly of hyaline material. Large blood vessels are present on the periphery and only a few small ones are within the body of the scar.

Neither the macroscopic technique nor the microscopic examination described by Simkin (1965) has been subjected to controlled evaluation by using females with known reproductive histories. Much of what is assumed to be true comes from studies with domestic animals. Probably it is safe to postulate that the physiological processes witnessed in cattle, sheep and goats occur in a similar fashion in wild ruminants. More work is necessary to understand the relationship between corpora lutea counts and fetal numbers in moose, and between counts of corpora albicantia and fetal production in moose. Until that time, counts of corpora lutea are most useful for gross comparisons among populations. Counts of corpora albicantia can serve as indicators of previous reproductive activity, especially for detecting reproduction in yearlings and possibly silent heat (Simkin 1965, Markgren 1969), although they also can serve as a gross estimate of herd productivity (Addison 1974).

RECTAL PALPATION

Rectal palpation has been used to diagnose pregnancy in several moose populations (see Table 10). It requires immobilization and is an invasive procedure. Basically, a trained and experienced veterinarian, wearing a plastic sleeve, slides his or her arm into the rectum of the moose. By feeling for certain anatomical structures, it is possible to determine if the moose is pregnant. It is difficult to determine the number of fetuses in some cases (Gasaway et al. 1983, Ballard et al. 1991, Gasaway et al. 1992), and technician error can result in misdiagnosis (Ballard et al. 1991). Haigh et al. (1982) could detect twin fetuses from singletons.

BLOOD ASSAYS

Because hormone concentrations change during gestation, it is possible to detect pregnancy using blood serum (Haigh et al. 1982, Stewart et al. 1985). Progesterone tends to increase from conception to about 50 days into gestation and then levels out (Figure 59). Concentrations at peak vary among individual moose and can be influenced by the number of fetuses (Haigh et al. 1982, Stewart et al. 1985). Pregnant animals have progesterone concentrations in blood serum that exceed 3.0 nanograms per milliliter (ng/mL), whereas in nonpregnant animals, the level is less than 0.5 ng/mL. Unfortunately, pregnancy determination using progesterone cannot determine the number of fetuses present.

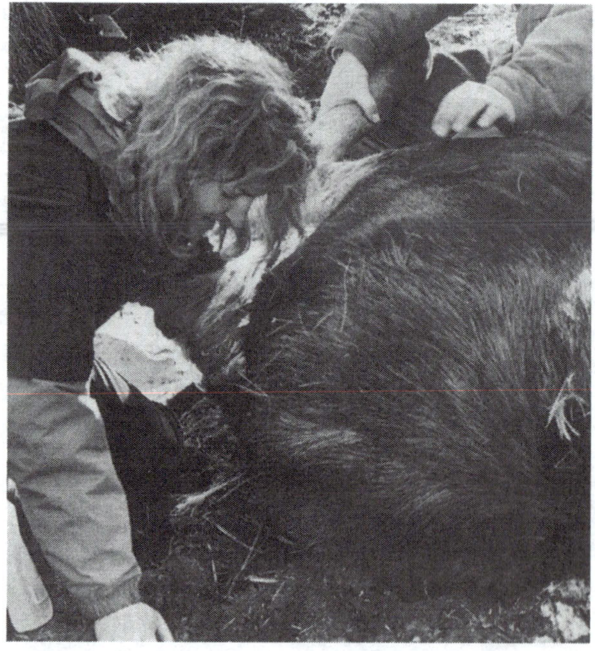

A veterinarian or specially trained biologist can use rectal palpation to detect pregnancy in cow moose. The technique requires cleaning fecal material from the rectum of the animal and inserting the arm into the rectum and large colon. New techniques using radio-immunoassay have eliminated the need for such an invasive procedure, and pregnancy detection now is possible from a fecal or urine sample. *Photo by Charles C. Schwartz; courtesy of the Alaska Department of Fish and Game.*

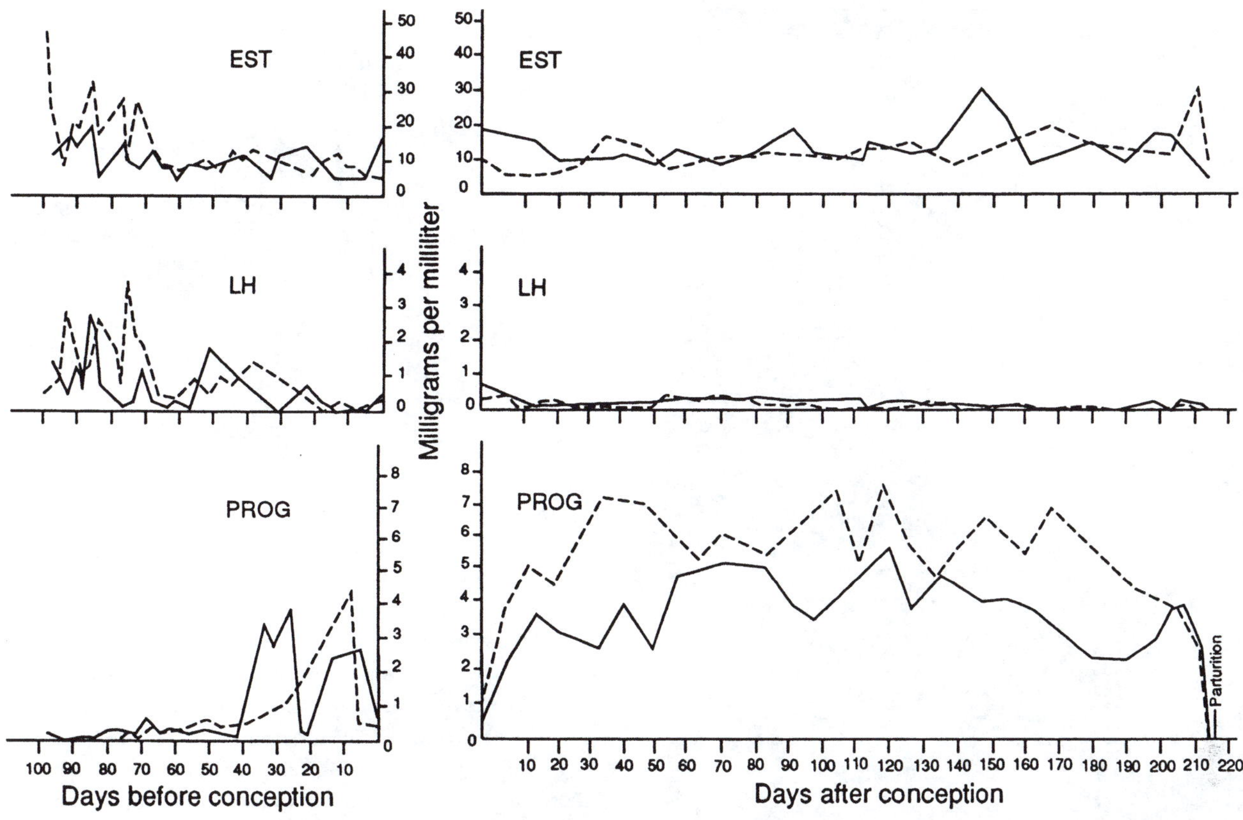

Figure 59. Changes in the concentrations of progesterone (PROG), luteinizing hormone (LH), and estrogen (EST)—the major hormones of reproduction in moose (Stewart et al. 1985). EST and LH are found in high concentrations before estrus, and EST spikes during the day of heat. PROG concentrations build and remain high during the gestation period, whereas LH concentrations are barely detectable.

FECAL AND URINARY STEROID METABOLITES

Urinary and fecal steroids have been used successfully to detect pregnancy in moose (Monfort et al. 1993). Concentrations of estrogen conjugates in urine increase from less than 5 to more than 50 ng/mL of creatine during the final month of gestation, making them useful in detection of pregnancy. Similarly, fecal progesterone concentrations in pregnant animals generally exceed 2,000 ng/mL, whereas those in nonpregnant moose are less than 500 ng/mL. Both methods are noninvasive, and fecal collections are relatively easy. However, similar to blood assays, this technique detects only pregnancy but not the number of fetuses.

REAL-TIME ULTRASOUND

Ultrasonic scanning is used commonly in domestic animal production to determine stage of pregnancy (Bretzlaff et al. 1993). Recently, the application of real-time ultrasound also has been applied to nondomestic species to detect pregnancy (Lason et al. 1993, Lenz et al. 1993). T.R. Stephenson (personal communication: 1994) applied the technique with a 210 portable scanner to cow moose at the Moose Research Center, Alaska. Photographs of the scan showed three diagnostic features: (1) vertebrae of the fetus, (2) cotyledons and (3) amniotic fluid. Technological advances in this type of equipment have made it possible to use real-time ultrasound in the field, and more use of it is likely in the future.

Anatomy and Physiology of Male Reproductive Structures

The male reproductive system of moose consists of two testes, contained in the scrotum, accessory glands and ducts, and the penis (Figure 60). The testes are the main reproductive parts of the male, just as ovaries are in the cow. The testes produce spermatozoa or sperm and testosterone, the male sex hormone (see Chapter 2).

Each testis contains a mass of seminiferous tubules surrounded by fibrous tissue (tunica albuginea). The cells of Leydig, which secrete testosterone, are located in this connective tissue between the seminiferous tubules. Germ cells in the seminiferous tubules produce sperm. These sex cells undergo a series of mitotic (normal cell division) and

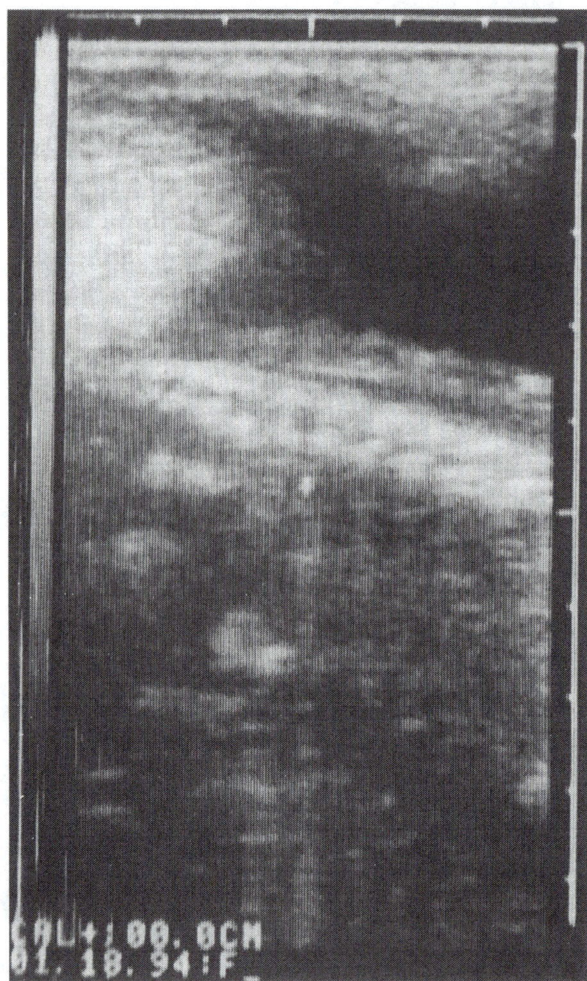

Real-time ultrasound has been used on cow moose to determine pregnancy. This photograph of the scan made on a pregnant cow moose at the Moose Research Center, Alaska, shows diagnostic criteria: (1) the vertebrae of the fetus just above the arrow in the center of the image; (2) a cotyledon in the upper left above the vertebrae; and (3) the dark amniotic fluid above the vertebrae. This technique likely will be used more in the future because the new equipment is portable and reliable. *Photo by T. R. Stephenson; courtesy of the Department of Fish and Wildlife Resources, University of Idaho, Moscow.*

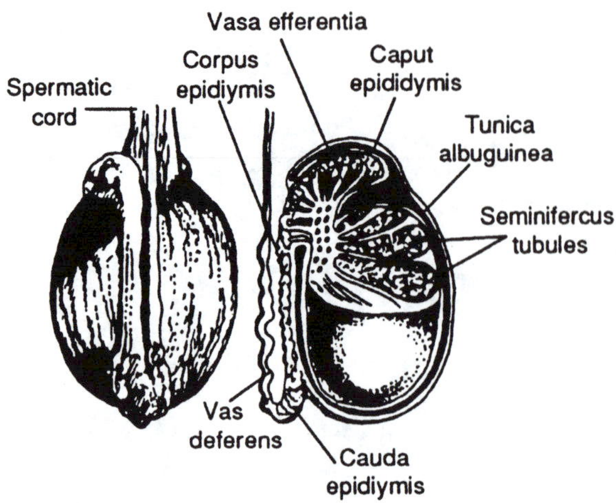

Figure 60. Testis of male moose (in cross section at right), showing a portion of the epididymis (from Bubenik and Timmermann 1982). Each testicle contains a mass of seminiferous tubules that are lined with germinal epithelium containing primary sex cells where sperm are produced. The seminiferous tubules are surrounded by a fibrous capsule called the tunica albuginea. Spermatozoa pass from the seminiferous tubules to the head of the epididymis (caput epididymis) by way of the vasa efferentia. The epididymis is a long, convoluted tube that connects the vasa efferentia to the vas deferens at the tail of the epididymis (cauda epididymis). Sperm are stored in the epididymis until the time of ejaculation when they are propelled by the muscular vas deferens to the ejaculatory duct in the urethra. The vas deferens leaves the tail of the epididymis, and passes through the inguinal canal (opening between the scrotum and body cavity) as part of the spermatic cord (general structure containing blood vessels, nerves and other tissues). The testes of bull moose are rather small for such a large animal.

meiotic (cell division where chromosomes are halved) divisions that are collectively referred to as *spermatogenesis*. After spermatogenesis, mature sperm pass from the seminiferous tubules into the epididymis, a long, convoluted tube where sperm are stored. In mature moose, sperm reserves in the epididymides are thought to be relatively low compared with those of red deer (Bubenik and Timmermann 1982), based on spermatozoa counts.

Testicular weight and size reflect sperm and testosterone production, which varies with age of the bull (Bubenik and Timmermann 1982). Studies in central Ontario found that prime bulls (ages 6 to 9) produced more sperm earlier in the breeding season (late August to early October) than did yearlings (Bubenik and Timmermann 1982), although there was considerable variation among individuals. Variation in testicular weight also was noted by Rausch (1959), but a generalized curve suggested that peak testicular mass for bulls older than yearling occurs around early October and coincides with the rut (Figure 61).

The Male Cycle

The annual reproductive cycle for the bull moose (Figure 62) starts in spring, with initiation of antler growth. It continues through summer, when the antlers develop fully. With increasing production of testosterone, antlers ossify and the velvet is shed (Figure 63; see also Chapter 2). At about the same time, the testes enlarge and spermatogonia (germ cells) begin the production of sperm. With antlers developed fully and sperm cells active by early autumn,

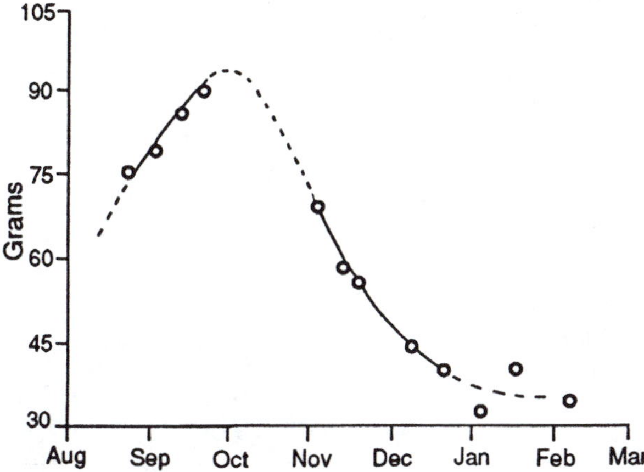

Figure 61. Changes is testicular mass for bull moose, August to February (from Rausch 1959). Testicular mass increases during the breeding season to accommodate sperm production. Testicles decrease in size after the rutting season when levels of testosterone are lowest, antlers are shed and the bulls are sexually inactive.

Bull moose lack white vulva hairs and generally are all black when viewed from the rear. Compared with large bovids, bull moose have relatively small testicles relative to body mass. Current theory suggests that male moose probably have small semen reserves and consequently cannot breed a large number of cows in a relatively short period of time. Although the theory is unproven, most management biologists agree that maintaining an adequate bull/cow ratio helps to ensure timely breeding. *Photo by Charles C. Schwartz; courtesy of the Alaska Department of Fish and Game.*

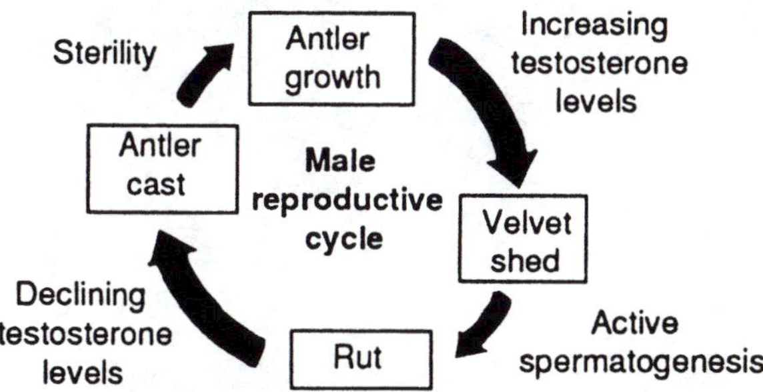

Figure 62. The male moose reproductive cycle (from Schwartz 1992b). An increasing testosterone level stimulates antler growth in early spring. Antler development is complete by late summer, and the velvet is shed just before the breeding season. After the rut, a declining testosterone level is responsible for antler shedding. There is a period during late winter when the male is sterile.

bulls are prepared for the rutting season. After the rut, testosterone declines rapidly, ultimately causing antler casting (see Chapter 2), followed by a period of sexual inactivity until early spring, when the cycle renews (Figure 62).

Puberty in bull moose occurs the autumn after their first birthday. Although yearling bulls have not attained mature body weight (Schwartz et al. 1987a), spermatogenesis (hence puberty) has been reported (Peek 1962, Houston 1968, Bubenik and Timmermann 1982). Yearling bulls usually are excluded from active breeding by the more mature dominant males (Lent 1974, Knowles 1984, Bubenik 1987), but otherwise are capable of impregnating females (Schwartz et al. 1982).

Although testicular weight in calves increases rapidly in autumn, no spermatogenesis occurs (Bubenik and Timmermann 1982). Increased cellular activity is associated with development of the cells of Leydig, also known as *interstitial cells,* which produce testosterone (Frandson 1965). Testosterone is required for the formation of the pedicle in the calf (Bubenik and Timmermann 1982, Gross 1983), and pedicle development is requisite to antler development in subsequent years.

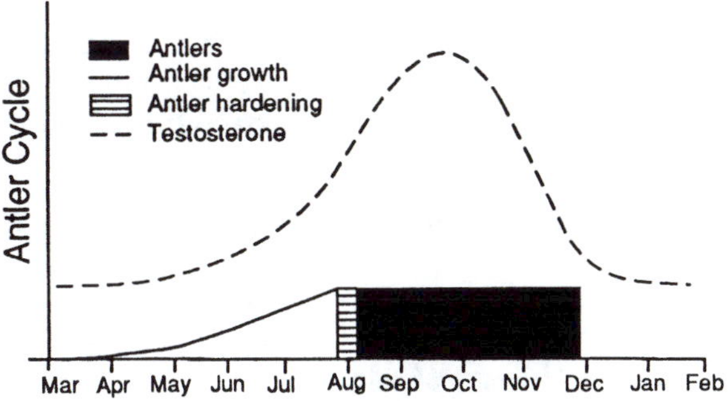

Figure 63. The antler/testosterone cycle in bull moose (from Schwartz 1992b).

Spring and summer are when male moose prepare for the breeding season. They begin to grow antlers in early spring and development continues throughout summer. During this time, bulls consume large quantities of forage. They replenish lost body reserves from the previous winter and store a large quantity of fat. These fat reserves serve as an energy source during the rutting season. With bulls' increasing level of testosterone, antlers harden and velvet is shed. The testicles enlarge during the breeding season as active spermatogenesis begins. By early autumn, bull moose are physically prepared for rigors of the breeding season. During the rut, sexually active males do not eat for up to 3 weeks and possibly longer. This hypophagia results is some bulls losing as much as 20 percent of their body weight. *Photo by Charles C. Schwartz; courtesy of the Alaska Department of Fish and Game.*

Spermatogenesis and sperm storage are greater in mature prime bulls than in younger and senescent males, but the total volume is small for such a large animal (Bubenik and Timmermann 1982, Bubenik 1985). Copulation and ejaculation in the bull during mounting last only a few seconds (Schwartz 1987). Bulls are capable of successive ejaculations within a single day, but the number and frequency are not clearly documented. Schwartz and co-workers (1987) observed one mature bull successfully mount a female five times over a 7-hour 49-minute period. The shortest time between successful copulations was 32 minutes.

Male Sex Hormones

As the male calf matures, changes in physical stature and maturation of the reproductive organs are brought about through the influence of hormones. The major sites of hormone production are the pituitary gland and the testes. LH released by the anterior lobe of the pituitary gland stimulates the Leydig cells of the testes to produce androgens—a class of steroid hormones (Bearden and Fuquay 1984). The main androgen in the mature male is testosterone. The functions of testosterone include development of secondary sex characteristics, maintenance of the male reproductive system, expression of male sexual behavior and spermatogenesis. Testosterone regulates the release of gonadotropic hormones in a fashion similar to progesterone release in the female. High concentrations of testosterone inhibit the release of FSH and LH—a negative feedback control. Conversely, when testosterone levels are low, high levels of FSH and LH are released. This reciprocal action of testosterone with the gonadotropic hormones regulates normal reproductive function of the bull (Bearden and Fuquay 1984).

Testosterone levels measured in two 3.5-year-old bulls were low in July (0.4 ng/mL), at peak in September (1.25–4.0 ng/mL) and in rapid decline after the rut (<0.3 ng/mL) (Hundertmark et al. 1989). Levels were lower than expected when compared with those of roe deer, where concentrations reached 7 ng/mL at peak rut (Sempere and Boissin 1983). Lower values in moose might have been associated with sampling, as neither bull moose was associated with cows during the rut.

From an ethological perspective, the role of the bull moose in courtship and breeding is fairly well documented (Lent 1974, Bubenik 1987; see Chapter 5). The bull's role as a primer of rut synchronism is poorly understood. Chemical compounds play a major role in ungulate reproduction (Muller-Schwartze 1991), and may serve as signaling compounds (Bubenik et al. 1979) and possibly as primers (Schwartz et al. 1990).

A recent study by Schwartz et al. (1990) isolated primary sex pheromones (androstenones) from the saliva of bull moose. These compounds belong to a special group of steroids that have no hormonal activity but are released in the saliva before mating. In red deer and swine, they are responsible for synchronizing and promoting male and female courtship behavior. Although the concentrations in moose are relatively low, Schwartz et al. (1990) speculated that these compounds may synchronize the rut, and possibly are responsible for inducing estrus in the female.

Breeding and Calving Season

Many mammals have evolved seasonal reproductive patterns that ensure adaptation to predictable annual changes in the environment. Moose exhibit marked seasonal changes in reproductive behavior that reflect adaptation to the yearly fluctuations in food availability to ensure favorable conditions for rearing young. Day length may provide the clue to annual timing of the breeding season. The likelihood that seasonal breeding patterns in ruminants are influenced by day length first was discussed by Marshall (1936, 1937). A link has been documented between the pineal gland (a light-sensitive organ) and its secretion of the hormone melatonin and seasonal reproduction in domestic sheep (Bittman and Karsch 1984), white-tailed deer (Plotka et al. 1982) and domestic goats (Hansen 1985). Detailed reviews of the neural and endocrine basis for seasonal reproduction and the link to the pineal melatonin rhythm in animal reproduction are found in Ebling and Hastings (1992), Yellon et al. (1992) and D'Occhio and Suttie (1992).

Length of the breeding season is relatively short for most moose. Because it is difficult to determine the exact date of breeding under natural conditions, few studies provide detailed information. Thomson (1991), Crichton (1992c) and Schwartz and Hundertmark (1993) provide the best data sets. All concluded that moose exhibit a very well-defined breeding season, as judged by conception dates and the spread in observed breeding. The mean date of breeding in the British Columbia study (Thomson 1991: Figure 4) ranged from October 5 to 10, with a standard deviation of about 5 days. The average day of breeding in the Manitoba study (Crichton 1992c) was September 29, and 93 percent of all females were bred by October 12. The average breeding date from Alaska was October 5, with a range from September 28 through October 12. There was very little difference among years in all studies, suggesting that photoperiod, rather than weather, influences rut timing.

By backdating embryos, Edwards and Ritcey (1958) concluded that 89 percent of all females conceived within a 10-day period; 5 percent became pregnant before peak rut; and 6 percent were impregnated after the peak. In a similar study, Schwartz and Hundertmark (1993) estimated that 88, 11 and 1 percent of the fetuses they examined from wild moose were conceived during the first, second and third estrus, respectively. They did not detect any calves conceived before the peak of the rut (see also Edwards and Ritcey 1958).

Out-of-season breeding is relatively rare, with only a few reported cases (Markgren 1969, Coady 1974b, Maehlum 1981, Schwartz and Franzmann 1981, Ballard et al. 1991). Schwartz (1992b) reported one birth in mid-August. The cow was not with a bull until mid-December. Similarly, Coady (1974b) reported on a cow moose, shot on June 27, carrying a fetus that was 7 to 8 weeks from term. On the basis of a gestation of 231 days, conception probably occurred in early December.

Because gestation length is relatively constant in moose, breeding can be estimated by looking at the length of the calving period. Birthing chronology for moose populations across North America is fairly consistent (Figure 64). Some

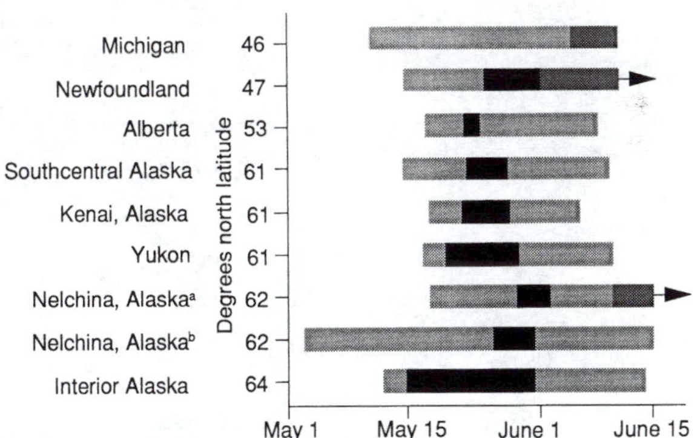

Figure 64. Birthing chronology of moose in North America. The peak birthing period (solid bar) occurs during a relatively short period of time, suggesting that the breeding season also is short, which corresponds to empirical observations. Out-of-season births (indicated by arrow) are rare, but do occur. Information sources are: Newfoundland (Albright and Keith 1987); southcentral Alaska (Rausch 1959, 1965); Michigan (Verme 1970); Alberta (Hauge and Keith 1981); Kenai, Alaska (Schwartz and Hundertmark 1993); Yukon (Larsen et al. 1989b); Nelchina, Alaska[a] (Ballard et al. 1991); Nelchina, Alaska[b] (Van Ballenberghe 1979); and interior Alaska (Gasaway et al. 1983).

researchers have reported a higher incidence of twins early in the birthing period (Bailey and Bangs 1980), suggesting that twins tend to be born sooner than singletons. Thus, it has been inferred that the gestation period for twins is shorter than that of single calves, although the work of Schwartz and Hundertmark (1993) disproves this idea. More plausible explanations are that prime cows (those most likely to produce twins) breed earlier, predation during the calving season reduces the survival of twins (see Chapter 7) or both. Observations by Addison et al. (1985) found no evidence that twins were more frequent earlier in the calving season.

Calf Sex Ratio

A number of investigators have reported differences in the sex ratio of moose calves at birth (Franzmann and Schwartz 1986, Albright and Keith 1987, Boer 1987a, Larsen et al. 1989b, Ballard et al. 1991, Crichton 1992c). Verme (1969, 1983) found, in captive white-tailed deer, that the nutritional status of a doe in the breeding season influenced fetal sex ratios; poor nutrition skewed conception toward males. The same is true for food-stressed wild reindeer (Skogland 1986). In a wild population of Rocky Mountain mule deer, heavier mothers and those with higher kidney–fat indices tended to produce male-biased litters (Kucera 1991). DeGayner and Jordan (1987) concluded that live weight of female white-tailed deer was not a reliable predictor of sex ratio in either single or twin litters, and that nutrition had no proximal effect on the sex ratio of offspring. Older (≥4.5 years) does produced significantly more female offspring. Crichton (1992c) compared the sex ratio of moose fetuses conceived before peak rut with those conceived after it. There were significantly more females (66 percent) con-

Moose are seasonal breeders and most cows become pregnant during a short 2-week period in autumn. Such synchronized breeding results in a similar calving period and may help reduce the overall predation rate on calves. However, because moose are polyestrous, cows will recycle and ovulate again if they did not become pregnant during the previous heat period. Out-of-season births are rare, but have been recorded as late as mid-August. Biologists have developed techniques to estimate the date of conception by measuring the length of a fetus and comparing it to standard curves. These embryos were all collected from a group of cow moose during early December in British Columbia. The variation in size is a result of different conception dates. Conception timing is an important component of population management. *Photo by Kenneth Child; courtesy of British Columbia Hydroelectric.*

ceived early in the rut but more males (55 percent) conceived after peak rut.

Trivers and Willard (1973) hypothesized that, among polygamous mammals (such as moose) where males exhibit greater variance in reproductive success than do females, maternal investment should be greater in sons than daughters. Williams (1979) extended this idea and proposed a model for optimization of sex ratios of offspring in species in which litter size usually is two, and in which males are slightly more energetically expensive to produce than females. According to this model, as the ability to provide maternal investment increases, sex ratio/litter size should go from one female to one male to twin females to mixed-sex twins to male twins. McGinley (1984) and Gosling (1986) further discussed this hypothesis. Data from red deer (Clutton-Brock and Iason 1986), reindeer and caribou (Kojola and Eloranta 1989, Thomas et al. 1989) support the hypothesis of Trivers and Willard (1973). However, after reviewing the literature, Williams (1979) concluded that there was no empirical support for adaptive variation in sex ratios of offspring in outcrossed vertebrates.

Shifts in fetal sex ratio can significantly alter population growth rates for moose (Reuterwall 1981). Changes in fetal sex ratio from 40 to 60 percent males reduced population growth rate by 8.9 percent. Significant shifts from the expected 50:50 sex ratio of moose calves have been reported to favor females in some cases (Boer 1987a, Larsen et al. 1989b) and males in other cases (Franzmann and Schwartz 1986, Albright and Keith 1987, Ballard et al. 1991).

Schwartz and Hundertmark (1993) analyzed reproductive records from a captive population of moose at the Moose Research Center in Alaska. They found no deviation in the sex ratio of calves (1.0 male: 1.05 females) at birth. Excluding primiparous females, heavy-bodied cows did not produce more twin litters than lighter cows did, nor did they produce more male calves. Primiparous females weighing more than 835 pounds (379 kg) produced twins, whereas those weighing less produced a single calf, but gender of the calves was not related to maternal weight. Also, no correlation has been found concerning calf gender and the previous year's breeding success of cow moose, as was found with bison (Rutburg 1986).

Schwartz and Hundertmark (1993) suggested that sex ratios in fetuses of newborn calf moose did not conform to the sex ratio theory of Trivers and Willard (1973). The possibility existed that, for the few cases where sex ratio was skewed in favor of either males or females, the results merely represent the 5 percent of cases where a type I error is committed (wrongly reject the null hypothesis) (Clutton-Brock and Iason 1986). Schwartz and Hundertmark (1993) recommended that managers detecting significant shifts in

The sex ratio of moose calves at birth usually is 50:50, although skewed ratios favoring either males or females have been reported. The scientific literature contains several theories regarding skewed sex ratios at birth. In white-tailed deer and reindeer, for example, females in poor nutritional status tend to produce male offspring. Skewed fetal sex ratios can significantly alter population growth. There currently are no clear reasons why sex ratio in some moose populations deviate from 50:50. *Photo by Charles C. Schwartz; courtesy of the Alaska Department of Fish and Game.*

calf sex ratio should continue to sample to be sure that the evidence is adequate to make such a claim before adjusting population models to reflect deviations from parity. Currently, there is no empirical evidence suggesting that maternal age, maternal mass, previous breeding parity or population density affect variations in litter category (singleton or twin) or sex ratio of offspring.

Reproduction and Genetics

Data presented by Peek (1962), Houston (1968) and Stevens (1970) suggest that Shira's moose may have a lower reproductive rate than do the other subspecies. Twinning rates in adult moose reported by these researchers ranged from 0 to 4.5 percent. Houston (1968: 70–71) postulated that some genetic influence over twinning might occur in the Jackson Hole area of Wyoming.

Why are twinning rates low for the Shira's moose? And is there a possible genetic component limiting their rate of reproduction? The three early studies showed that twinning in adults was low, and Houston (1968) also found only a 6 percent pregnancy rate in yearlings. However, work by Schladweiler and Stevens (1973) in Montana listed twinning rates in adults at 16 percent (Table 10) and yearling pregnancy at 32 percent. Such a wide variation in twinning probably discounts a genetic influence (Houston 1968). More likely, twinning rates are low in Shira's moose because of the relatively stable environment in which they live. Peek (1974a) hypothesized that the essential winter habitats presently occupied by Shira's moose are a more permanent part of the landscape than are the seral stages of boreal forest commonly used by the other subspecies (see Chapter 11). Moose populations at or near carrying capacity generally have low twinning rates (Table 11). These rates are linked with range quality and nutrition.

Logging and fire are important ecological forces throughout much of the moose range in North America. Early seral forests contain many deciduous tree species that represent high quality food to moose. Moose in such habitats tend to have high reproductive rates. There is a premium placed on rapid colonization of quickly created habitats that favor the cow producing the greatest number of offspring (Geist 1974a, Peek 1974a).

In the western United States, much of the relatively stable habitat occupied by Shira's moose is not favored by conditions that effect early seral forest. Consequently, the Shira's moose lives in a region and habitat in which one would not expect to see large swings in reproductive performance. This led Peek (1974a: 137–138) to conclude that "twinning rates among Shira's moose appear lower than for the other subspecies, but the cause is in doubt. The existence of heritable differences in fecundity between the several subspecies would be difficult to demonstrate."

Reproduction, Breeding Strategies and Herd Composition

A number of factors significantly affect moose population composition, hence reproductive behavior and reproduction itself. Where there is a biological surplus of moose in North America, moose hunting seasons have been established (see Chapters 17 and 18). In areas with heavy hunting pressure, it sometimes is necessary to restrict or direct harvest at certain segments of the population (e.g., bulls only). Intense harvest of bull moose can result in a highly skewed sex ratio in the population; ratios as low as 5 to 12 bulls per 100 cows have been reported from areas with heavy hunting pressure in Alaska (Bishop and Rausch 1974). In unhunted populations, adult sex ratios range from close to parity (92 to 100 males:100 females) on Isle Royale (Peterson 1977) to a skewing toward females (29 males:100 females) as in Denali National Park (Miquelle et al. 1992). Also, the rigors of mating often result in higher natural mortality rates among males, again resulting in sex ratios slightly skewed toward females (Peterson 1977).

The breeding strategies of moose are different between the tundra (Alaskan/Yukon) and taiga (Eastern, Northwestern and Shira's) subspecies, and there likely are gradations between the two. Lent (1974) concluded that the size of a rutting group was strongly influenced by the nature of the habitat. Bubenik (1987) concluded that tundra moose tend to be relatively polygamous breeders and form assemblages during the rut. At these assemblages, moose congregate in large rutting groups in semiopen habitats (generally subalpine) where dominant males can monopolize female groups. As a consequence, one dominant bull can service many cows during a short breeding season. Bubenik contrasted aggregations of the tundra moose to the comparatively monogamous strategy of forest-dwelling taiga moose, in which a bull will remain with a single female or a small group of females for one to several days. Although this breeding system is not fully documented (Lent 1974), Bubenik (1987) proposed that one male may breed with only a few females during the rut period.

Because of the breeding differences between tundra and taiga moose, population structures can differ between the subspecies, therefore management strategies vary as well. Common to all areas is maintaining a proper adult sex ratio to ensure timely mating of all breeding females. As discussed earlier, the normal breeding period for moose is relatively short. It has been hypothesized that, when the bull/cow ratio becomes very skewed, with few bulls, not all females will be bred (Rausch et al. 1974) or that some females may not breed or conceive during their first estrus (Rausch 1965). A consequence of late breeding is late-born calves (Schwartz and Hundertmark 1993), which likely do not survive harsh winters.

Thomson (1991) investigated the relationship between traditional measures of productivity (pregnancy rate and cow/calf ratios) and bull/cow ratios. He found no evidence that differences in pregnancy rates or cow/calf ratios were correlated with bull/cow ratios. No evidence was found to suggest that greatly skewed sex ratios favoring cows resulted in changes in pregnancy rates. No relationship was found between bull/cow ratios and cow/calf ratios in a winter inventory. These findings led Thomson (1991: 40) to recommend that "the Wildlife Branch should continue to investigate the relationship between bull age, sex ratio and reproduction to get a clearer picture of their significance to B.C. moose populations." Obviously, more research is needed to document which ratios are necessary to accomplish desired population objectives.

What is an adequate bull/cow ratio to ensure timely breeding? Currently, there is no clear answer to this question. Historically, biologists measured breeding success by evaluating the number of pregnant females in the population or by a measure of calves per 100 females in autumn counts. High rates of calf production have been detected in Alaskan moose populations with bull/cow ratios ranging from 110 to fewer than 5 bulls per 100 females (Bishop and Rausch 1974) (Figure 65). However, calf/cow ratios in autumn are not the most appropriate measure of timely reproduction in moose.

As discussed earlier, moose are a polyestrous species, and females not bred during their first heat will recycle. Most females get pregnant and produce a calf during the breeding season. Hence, pregnancy rate and calf production measured in autumn are not the best statistics when evaluating timely reproduction in a moose population. *It is more important to know when a cow is bred and how many fetuses she is carrying than simply to know that she is pregnant.* Likewise, because neonatal losses attributable to predation tend to be relatively high in moose populations, autumn calf/cow ratios are not adequate measures of the effects of bull/cow ratios on rut synchrony and timing. And finally, because a cow/calf ratio contains two variables (cows and calves), it can change in response to changes in either variable. Using cow/calf ratios to show short-term calf loss assumes no change in cow numbers, which probably is appropriate, but this may not be the case for longer term changes where adult mortality also can vary (see Chapter 17).

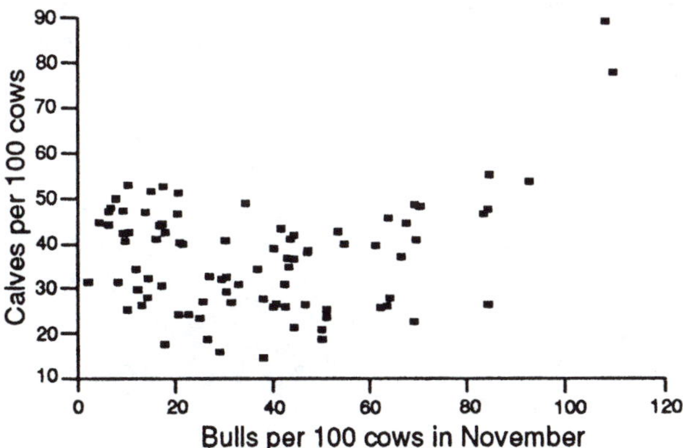

Figure 65. Cow/calf and bull/cow ratios from southcentral Alaska (data from Bishop and Rausch 1974). No relationship between the two variables is evident.

TUNDRA MOOSE

R.M.P. Ward (personal communication: 1994) in the Yukon and Schwartz and Hundertmark (1993) in Alaska compared reproductive success in moose populations with widely divergent bull/cow ratios. Bull/cow ratios in the Yukon varied from 27 to 105 bulls per 100 cows in autumn counts. Ward observed 129 collared cows during calving seasons from 1983 to 1986, where bull/cow ratios varied from 27 to 66. They found no indication that the skewed bull/cow ratio had any effect on pregnancy rates or time of conception. They did not detect a correlation between bull/cow ratios and cow/calf ratios. There was no apparent delay in conception (i.e., second estrous breeding) because 98.5 percent of all cows observed gave birth in a 21-day period.

Similarly, Schwartz et al. (1992) monitored bull/cow ratios in an Alaskan moose population following restrictions in harvest that resulted in the bull/cow ratio increasing from 16:100 to 29:100 over a 5-year period. During this change in population composition, they did not detect a change in calf/cow ratios in autumn counts, nor did they detect a change in pregnancy rates in utero as bull numbers in the population increased. During the study, 83 percent of all females conceived during their first estrus; 16 percent conceived in the second estrus. Because of small sample size, Schwartz et al. did not detect a difference in conception timing among years. They did notice a high incidence of late conception in 1 year after a severe winter, but attributed it to poor body condition of females rather than to a low bull/cow ratio.

For the Alaskan/Yukon moose, there is no clear data set available to suggest that skewed sex ratios favoring females result in reduced reproductive rates within populations. Van Ballenberghe (1983) did demonstrate that the relative rate of increase in a moose population is greater when the population is skewed toward females. By altering the population structure to favor females, net calf production increases. In a population at carrying capacity, the range can only support a finite number of moose. For example, one can compare two ranges both capable of supporting 1,000 adult moose. On range A, the adult moose population is 15 percent male (150 bulls:850 adult cows). Range B has an even sex ratio with 500 bulls and 500 cows. Both areas have a reproductive rate of 1.2 calves per female. Annual calf production in area A is 1,020, but only 600 in area B. Clearly, the output of calves from the population with the skewed sex ratio is greater. Although the example involves a simple mathematical relationship showing that a population with more females produces more calves, there are biological (behavioral) consequences associated with a skewed sex ratio (see Chapter 6).

Moose management objectives throughout most of Yukon and Alaska intentionally skew the ratio of adults toward females. Exact adult ratios vary considerably among areas (see Chapter 17), but most jurisdictions set minimum bull/cow ratios as part of their management plans. Lower limits generally are based on many factors, one of which is adequate breeding. There is no finite limit suitable for all populations. One must consider not only the ratio of males to females, but also moose density and distribution within the managed area. Widely distributed populations with very low densities may require higher ratios to ensure adequate reproduction, whereas high density populations may not. No set ratio is apparent, but ratios commonly reported range between 20 and 30 bulls per 100 cows. This lower limit appears to be above the minimum required to ensure synchronous breeding by all females during their first estrus.

TAIGA MOOSE

As discussed earlier, moose in the boreal forests tend to form pairs during the breeding season. And of this pair-breeding strategy, Bubenik (1987) hypothesized that the female monopolized the bull during her receptive phase, that one bull was capable of fertilizing only a few females during a short, synchronous rut and that a skewed bull/cow ratio might influence timely reproduction in the forest moose.

Child and Aitken (1989) and Aitken and Child (1992) provided the most complete data set on changes in moose reproduction with changes in composition of a moose population in central British Columbia. Hunting regulations were adjusted in 1981 to exert heavy pressure on calves, moderate pressure on bulls and light pressure on adult females. From 1981 through 1985, the number of mature bulls in the population increased. After another regulation change (postrut male season open to all hunters) for the 1986 hunting season, bull abundance declined because of heavy hunting harvest. On the basis of analyses of conception dates and kill dates of bulls, the rut remained synchronous over the entire study period. However, there was a reduction in the amount of variation about the mean date of conception as the number of mature bulls in the herd increased (Child and Aitken 1989). Also, the proportion of cows bred in the second or subsequent estrus declined from 17.5 percent when bull abundance was low, to 7.7 percent as bull abundance increased. Aitken and Child (1992) further demonstrated a positive and significant relationship among the incidence of twinning in females and the population's bull/cow ratio, specifically concerning mature bulls. The researchers concluded that the positive relationship between gross productivity and the mature bull component argues strongly for the nonparental importance of mature bulls during the breeding season. Their data support the long-proposed management strategies of Bubenik (1972) that, not only is a high bull/cow ratio important for adequate and timely reproduction, but that the male component of the population also must contain an adequate number of prime bulls.

Growth and Development

From conception to death, growth in moose is in constant flux (see Franzmann et al. 1978, Schwartz et al. 1987a), but the rate at which an animal grows is not constant throughout its life. According to Brody (1945: 497), "growth may be divided into two principal segments, the first of increasing slope, which may be designated as the *self-accelerating phase of growth,* and the second of decreasing slope, which may be designated as the *self-inhibiting phase of growth.*" Moen (1973) indicated that the first phase of growth starts with the mother and young as a single unit (i.e., pregnant cow) that gradually become separate units (calf independent of cow). A young moose grows rapidly on a milk and forage diet. The transition from phase one to phase two is associated with weaning. The second phase of growth continues for several years in moose but the rate is slow (Franzmann et al. 1978, Schwartz et al. 1987a). There are seasonal increases and decreases that are associated with the breeding cycle, production of young and winter range conditions.

Growth in moose can be described quantitatively with mathematical equations fitted to curves. Any data set representing growth can be fitted rather exactly if the proper function is chosen, but the resulting equation usually is cumbersome and of little comparative value, because no two equations are alike. Consequently, researchers have developed more generalized equations with few independent variables (Ricklefs 1976). Several researchers (e.g., Bertalanffy 1960) have developed mathematical equations that describe growth rates with a single constant. Some of these equations are suitable for moose (see Schwartz et al. 1987a, Schwartz and Hundertmark 1993).

Three basic periods of growth are found in moose: (1) the prenatal phase, beginning at conception and ending at birth; (2) the suckling phase, from birth through weaning; and (3) the maturity phase from postweaning until death. During the first two periods, the calf is either wholly or partially dependent on the cow for sustenance. These two periods are self-accelerating, whereas phase three is self-inhibiting, as previously defined.

Prenatal Growth

It is difficult with the naked eye to locate the developing moose embryo in the uterus before the twentieth day of gestation (Markgren 1969). From that time on, embryonic development is fairly rapid. As the embryo develops so does the uterus. The anatomical structure associated with pregnancy and embryonic development is the placenta. The placenta consists of an arrangement of membranes that carry nutrients from the cow to the calf and waste products in the opposite direction. In moose and other ruminants, the placental attachment to the uterine wall occurs through structures known as *cotyledons* (uterine side) or *caruncles* (placental side) (Frandson 1965). The two structures (cotyledons and caruncles) are collectively referred to as *placentomes.* The usual number of placentomes in moose is 7 to 12 for each embryo, although it can range form 7 to 27 (Markgren 1969). The size of each placentome can range from the size of a golf ball to slightly larger than a tennis ball. The shape usually is oblong rather than round.

Prenatal growth can be divided into two phases of development—embryonic and fetal. Embryonic development is that period when internal organs develop, sometimes referred to as *organogenesis* (Markgren 1969). The embryo tends to reduce its strongly curved position and assumes a C-shape, and finally straightens itself out. During the fetal development period, growth and external differentiation occur.

There are several published works that provide measurements of moose fetuses (Edwards and Ritcey 1958, Rausch 1959, Markgren 1969, Schwartz and Hundertmark 1993), although the early works did not fit curves to the data. By plotting these measurements against collection time, a basic curve of prenatal growth can be constructed. Because the age of embryos or fetuses cannot be determined accurately, it can only be approximated based on indirect evidence. An example of such information was presented by Schwartz and Hundertmark (1993) for prenatal growth in Alaskan moose (Figure 66). Prenatal growth is best described using two basic equations. Body mass and hindfoot length assume self-accelerating shapes, whereas total length and forehead/rump length change at a relatively constant rate and are easily described with a straight line. Because twin calves weigh less at birth than do singletons, a separate model must be fit for twin and single calves (Schwartz and Hundertmark 1993). Hindfoot length, forehead/rump length and total length are notably consistent among singletons and twins, therefore one line is useful in describing these characteristics for both. According to Schwartz and Hundertmark (1993), hindfoot length is the best criterion to describe prenatal growth in moose, whereas change in mass is the poorest (Markgren 1969).

Constructed curves of prenatal growth can be used to predict conception dates. Edwards and Ritcey (1958) confirmed that some moose calves are conceived during periods other than the usual breeding season. Schwartz and

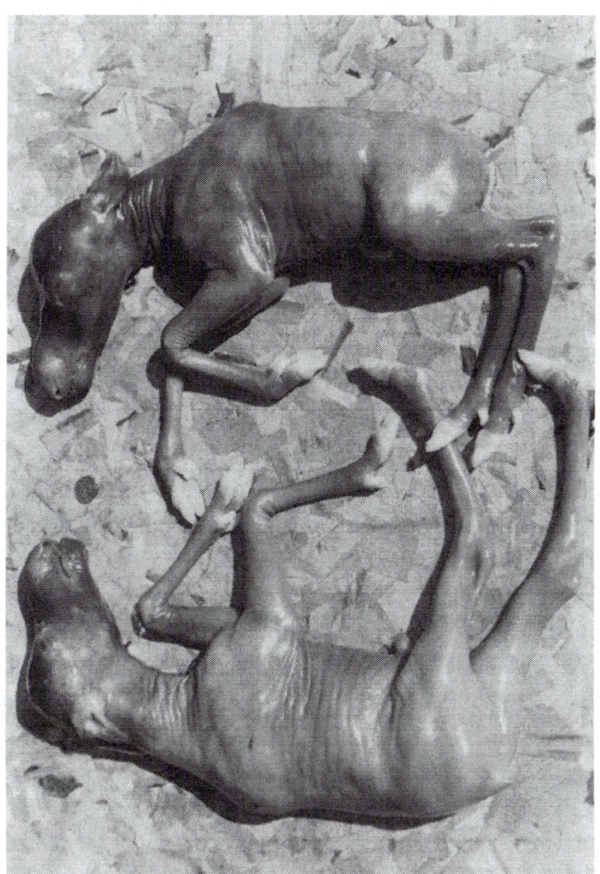

Prenatal growth of moose begins at conception and continues through the gestation period. During early phases, the developing embryo assumes a half-moon shape. The period of embryonic development is generally considered the period of organogenesis; it lasts about 6 weeks. Internal development is accompanied by changes in the external morphology. The period of fetal development is when growth and external differentiation occur. Sex determination is possible by 7 to 8 weeks. These fetuses are about 20 weeks old. They are still naked except for thin hairs on their muzzles and around the eyes. The heads and ears are well developed, as is the triangular spot on the muzzle. By 24 weeks of age, the fetus is covered with hair, and by 27 weeks, the fetus appears fully developed. Parturition generally occurs during week 33. *Photo by Tim Timmermann; courtesy of the Ontario Ministry of Natural Resources.*

Hundertmark (1993) clearly demonstrated this fact; by using a growth model fit to hindfoot measurements, they were able to detect second and third estrous breeding (Figure 67). By offsetting the model by the mean length of the estrous cycle (24.4 days), it became clear which fetuses were conceived during later cycles. Schwartz and Hundertmark were able to predict accurately 95 percent of the known second estrus calves they measured. The only error occurred with one twin fetus, which was exceptionally small, result-

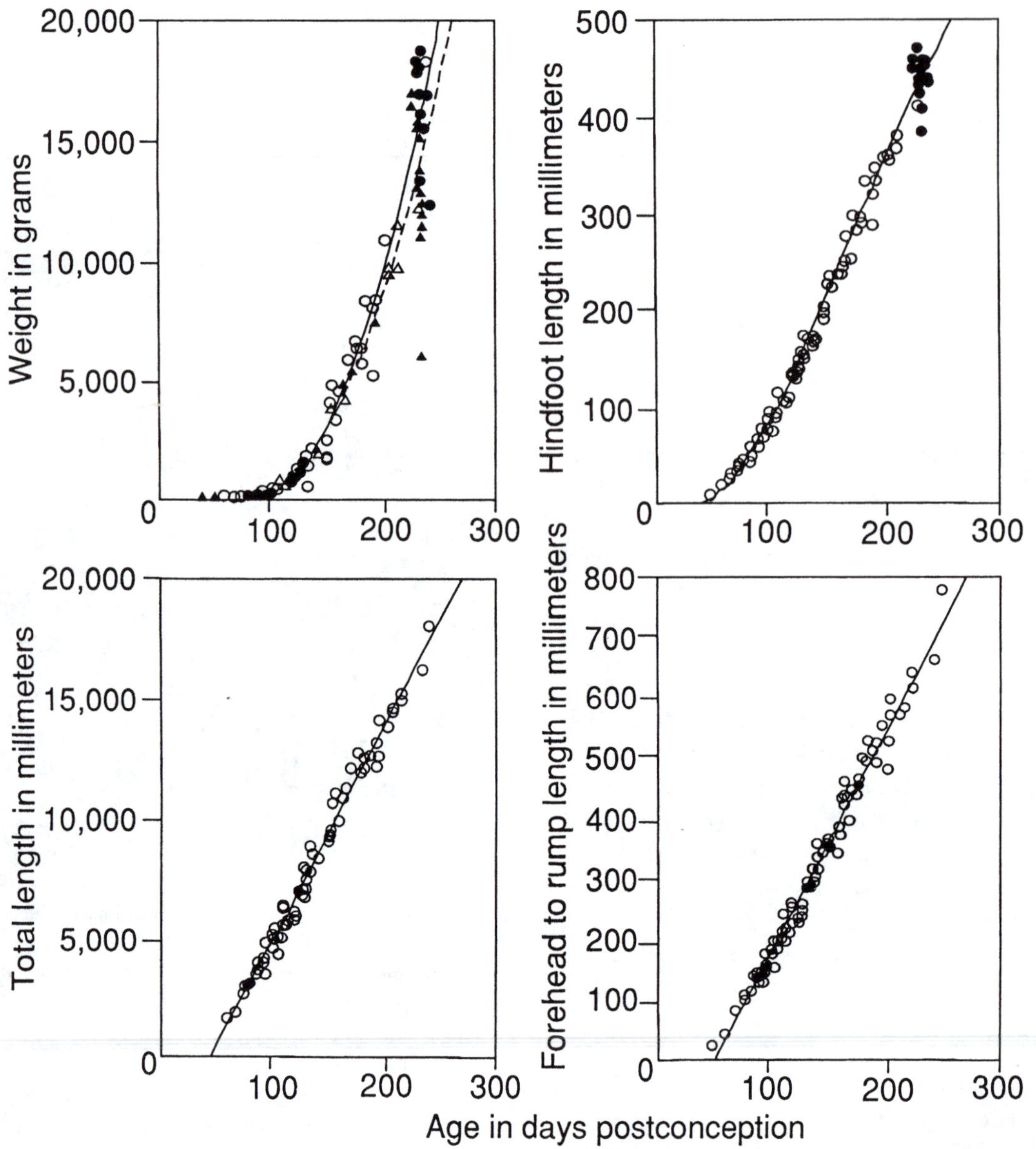

Figure 66. Correlation of age to weight, hindfoot length, total length and forehead to rump length for moose fetuses (data from Schwartz and Hundertmark 1993). Age represents the day postconception, with October 5 representing day 1. Solid lines and circles represent single calves; dashed line and triangles represent twins. Open circles and triangles indicate fetuses from wild cow moose from Alaska's Kenai Peninsula; closed circles and triangles indicate fetuses from tame cows from the Alaska Moose Research Center.

ing from improper in utero development; the other twin was classified correctly.

Birth Mass

Single moose calves weigh more than twin calves at birth, but male and female calves weigh the same (Schwartz and Hundertmark 1993). Mass of calves born to primiparous females is not different from that of calves born to pluriparous females. Sexual dimorphism at birth (i.e., mass), with males being larger, occurs in white-tailed deer (Verme 1989), red deer (Clutton-Brock et al. 1981) and mule deer (Kucera 1991). For both species with multiple births (mule and white-tailed deer), size differences between male and female singletons have not been detected.

Very little is known about how birth mass affects the survival of a moose calf. In elk, calves that weigh more than 35 pounds (15.9 kg) at birth have a 95 percent chance of

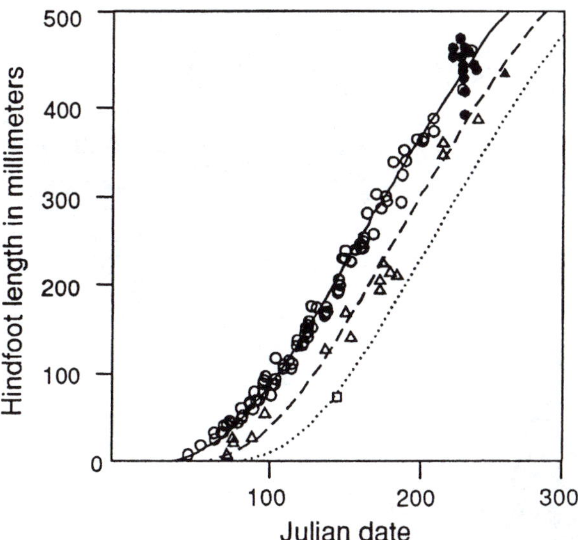

Figure 67. Age/hindfoot length relationship for fetuses collected from wild cow moose from the Kenai Peninsula, Alaska (from Schwartz and Hundertmark 1993). The solid, dashed and dotted lines represent fetuses conceived in the first (circles), second (triangles) and third (squares) estrus, respectively. By plotting hindfoot measurements and collection dates, managers can determine the number of calves conceived during each estrous cycle.

surviving to 1 month of age, whereas those weighing less than 25 pounds (11.3 kg) at birth have a greater than 50 percent chance of dying within a month (Thorne et al. 1976). The only published information about moose is anecdotal (Schwartz and Hundertmark 1993). During an experiment to measure the energy required by moose in winter, Schwartz et al. (1988b) subjected two cow moose to malnourishment during late winter. Both cows lost considerable amounts of weight and gave birth to unhealthy calves almost 2 weeks later than normal. Both calves died shortly after birth. Their body mass (average, 31 lb [14 kg]) was not different from normal birth weight reported in the same area (28–36 lb [12.7–16.4 kg]). It is known that maternal nutrition affects growth and development of the moose fetus, but the significance of cow undernourishment during gestation and for calf survival are not fully understood. The elk study by Thorne et al. (1976) was conducted under controlled conditions, in the absence of predators. Most moose populations in North America are subjected to predation, particularly on neonates (see Chapter 7). If poor maternal nutrition results in weak, nonviable moose calves, some losses attributed to predation actually may be nutrition related. Predation simply may be the consequence rather than the cause.

Lactation

Early growth and development in moose calves are related directly to the quality and quantity of milk consumed. Moose milk contains about 25 to 32 percent total solids, 5 to 12 percent fat, 14 to 19 percent crude protein and 2 to 6 percent lactose (Franzmann et al. 1975a, Renecker 1987b), and about 1.5 kilocalories per gram of energy (Oftedal 1981). Moose milk composition changes markedly during lactation (Renecker 1987b), and serial sampling is necessary to track changes in quality and quantity produced. No such data exist for moose.

Knorre (1959, 1961) studied milk intake in moose calves in Russia by weighing them before and after nursing. Maximum milk intake ranged from 0.4 to 0.5 gallon (1.5–2.0 L) per day during June, and had decreased to approximately 0.13 gallon (0.5 L) per day by August. Total milk consumption from birth to weaning per calf ranges from 26.4 to 52.8 gallons (100–200 L). At about 1.5 to 2 months of age, the diet of calves consists primarily of solid foods. Gasaway and Coady (1974) also noted large amounts of herbaceous material in rumens of calves in Alaska by late June. Captive-reared moose begin consuming solid foods at 2 to 3 weeks of age and eat large quantities by age 2 months (Schwartz 1992b).

Calf and Adult Growth

Growth from birth to weaning in moose is best described as an accelerating period of development. During this phase, body size and mass are changing rapidly (Figures 68 and 69). The self-accelerating phase of growth usually ends about the fifth month, which corresponds with the weaning period, the breeding season and a decline in forage quality associated with autumn (Rausch 1959, Franzmann et al. 1978, Schwartz et al. 1987a).

During the first winter of a calf's life, the animal commonly loses weight and body dimensions (girth and length) (Figures 68 and 69). This is a direct result of negative energy balance in winter (see Chapter 14). In calves fed supplemental diets during winter, these declines are not seen (Schwartz et al. 1987a), suggesting that the growth process is restricted by environmental constraints rather than a physiological mechanism such as seen in adults (Schwartz et al. 1988b).

Growth curves fitted to data collected from several animals over their lifetime provide a realistic pattern of the growth phase in moose calves and adults (Figure 70). The growth phases for male and female calves are similar. The average weight gains range from 1.3 to 1.6 percent per day. This accelerating phase of growth continues for about

Moose milk is a rich source of nutrients. Its percentage composition of fat, protein and ash is higher than in the milk of domestic cows, goats, pigs or sheep, and its energy content (expressed as kilocalories per gram) is considerably higher (Robbins 1983). Among that of wild North American ungulates reported, including Dall's sheep, elk, mountain goat, white-tailed deer and black-tailed deer, moose milk is at least nutritionally comparable. *Photo by William E. Ruth.*

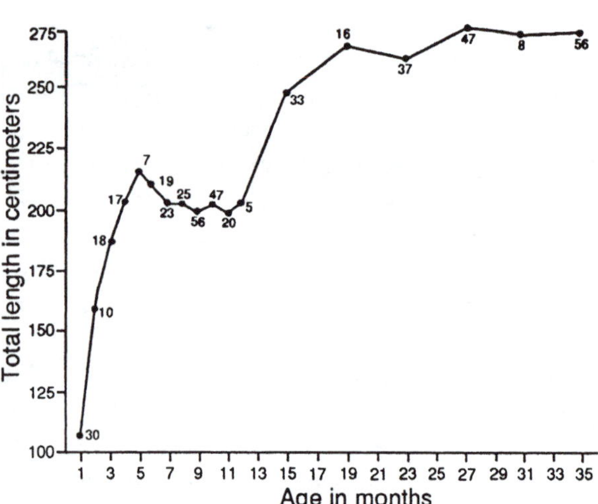

Figure 68. Relationship between total length and age in Alaskan moose (from Franzmann et al. 1978). Graph numbers = sample sizes.

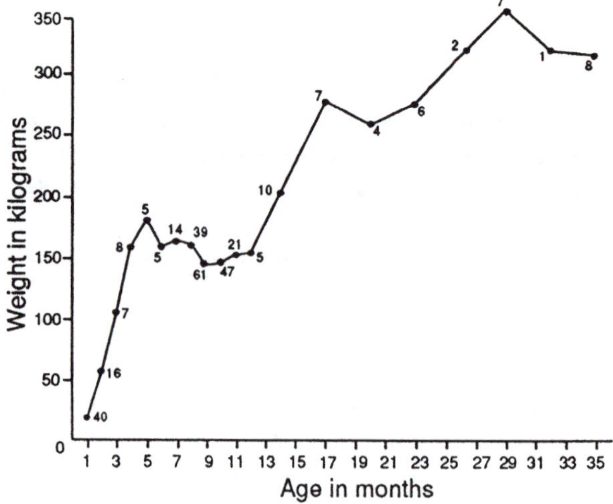

Figure 69. Relationship between body mass and age in Alaskan moose (from Franzmann et al. 1978). Graph numbers = sample sizes.

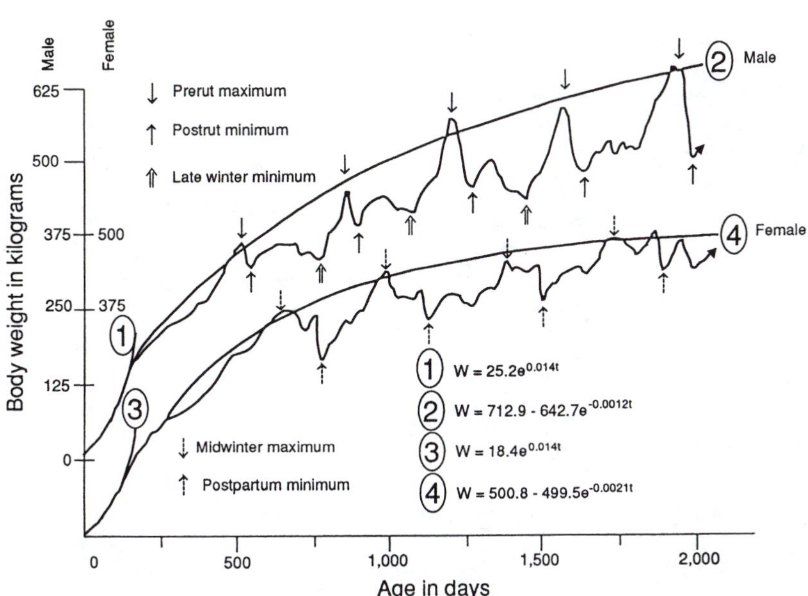

Figure 70. Relationship between body mass and age from birth to maturity in male and female moose (from Schwartz et al. 1987a). The equations represent the self-accelerating (1 and 3) and self-inhibiting phases of growth (2 and 4). Seasonal peaks and troughs in body mass occur at different times for bulls and for cows.

150 to 165 days. Growth rates decline after that point, which coincides with the breeding season and a change in forage quality. The declining phase in growth during the calf's first winter marks a change in the growth patterns for the rest of its life. There is a shift from the accelerating phase to the self-inhibiting phase. Consequently, although the calf is not sexually mature, the remainder of its growth period can be described by a different equation (Figure 70).

The inhibiting phase of growth for males and females is different, with the decline in growth rate for females two times greater than that for males (0.215 percent versus 0.116 percent). A consequence of this difference is that females reach maximum body mass earlier in life and weigh less than males. The maximum weight reached for female Alaskan moose is about 1,100 pounds (500 kg) with an average of 880 to 990 pounds (400–450 kg); males can exceed 1,650 pounds (750 kg) and average 1,435 to 1,545 pounds (650–700 kg). Females reach maximum weight at age 4 years, whereas males continue to grow until 7 to 9 years of age. Although moose continue to gain weight their entire life, very little gain occurs in females after age 4 years, and for males after age 9 years. As moose reach senescence, they lose body mass and become smaller.

Each year, moose gain and lose weight corresponding to the seasonal changes in food abundance and quality. This weight change is predictable and even occurs in animals that are offered food ad libitum. The cycle appears to be linked to an endogenous rhythm associated with changes in seasonal intakes and metabolic rates. This seasonal pattern of gain and loss of weight is different for males and females.

Males reach their peak in body mass and condition just before the breeding season, when they are shedding antler velvet. The shift from weight gain to loss for bulls begins during the rut. During this period, it is not uncommon for males to cease eating for up to 18 days (Kellum 1941, Schwartz et al. 1984, Miquelle 1991). Bulls can lose from 12 to 19 percent of their prerut body weight during the breeding season (Franzmann et al. 1978, Schwartz et al. 1987a). After the rut, bulls regain some of their lost weight, mainly by eating voraciously, but also by replenishing some of their fat reserves. Bulls lose additional weight during winter. These losses are dependent on remaining fat deposits and the length and severity of winter. Overwinter weight loss can range from as low as 7 percent to as high as 23 percent of prerut body mass (Schwartz et al. 1987a). Weight gains over summer range from 33 to 41 percent of postwinter lows.

Cows differ from bulls in that the former's maximum weight is reached in early winter (Franzmann et al. 1978, Schwartz et al. 1987a). Minimum weights occur shortly after parturition. Cows lose an average of 77 pounds (34.9 kg) giving birth to a single calf and as much as 139 pounds (63.1 kg) when giving birth to twins. This loss represents not only the weight of the expelled calf, but also the associated fluids and membranes of pregnancy. Females lose an average of 15 to 19 percent of their body mass from the midwinter peak to their postpartum low. Summer gains range from 25 to 43 percent, with 2-year-old females showing the most gain.

Weight gains in female moose accompanied by calves differ from weight gains of cows without offspring. Gasaway and Coady (1974) indicated that the metabolic demands of lactation are illustrated by lower body weights in lactating cows. The average body weight of nonlactating adult females during late June in interior Alaska was 12 per-

Moose continue to change in size and shape throughout their lives. The best information on moose weights comes from the Alaskan/Yukon subspecies. Single calves weigh more (36 lb [16 kg]) than twins at birth (30 lb [13.6 kg]), but males do not weigh more than females. From birth to maturity, males grow larger than females. Adult bulls can weigh in excess of 1,600 pounds (725 kg), whereas adult cows reach maximum body weight at about 1,100 pounds (500 kg). The body weight of an adult moose is constantly changing in response to food quality and availability, seasons, weather and reproductive demands. Understanding the dynamics of moose weight is useful in predicting population size and condition relative to habitat-carrying capacity. *Left photo by Charles C. Schwartz; courtesy of the Alaska Department of Fish and Game. Right photo—an award-winning scene from 1925—by C. J. Albrecht; courtesy of the Wildlife Management Institute.*

cent greater than that of lactating moose, despite longer total body length indicating larger average body size in the lactating females. By the end of June, lactating females gained approximately 45 pounds (20 kg), whereas nonlactating females gained approximately 155 pounds (70 kg) above the average spring weights of 770 pounds (349 kg). Repeated weights from semicaptive moose indicated that calf rearing "costs" were 8 to 18 percent of a cow's July/August weight (LeResche and Davis 1971, see also Skuncke 1949, Verme 1970).

Jordan et al. (1970) concluded that seasonal body weight fluctuations of moose amounted to only 6.6 percent for females and 10.3 percent for males. However, Rausch (1959) and LeResche and Davis (1971) reported seasonal body weight fluctuations of 20 percent and 15 to 30 percent, respectively, for moose in southcentral Alaska. Verme (1970) found that a "winter-killed" bull moose in Michigan lost 33 percent of its prewinter weight. The same bull lost only 22 percent of its weight the previous year and survived.

Conclusion

Our understanding of moose reproduction has improved considerably during the past two decades. We now know, for example, that the concept of the barren cow is untrue. And we have adequate data to suggest that reproductive rates are keyed to habitat quality. But there still is much to learn.

How habitat quality affects reproductive performance is not understood. For example, how do summer and winter range qualities influence ovulation rates, calf production and survival? Does nutritional intake of a cow during summer influence ovulation in autumn? How does a cow's reproductive history influence future reproduction? Does nutritional intake of a cow during winter influence the survival of her calf during summer? Are high rates of calf predation a sign of a moose population above carrying capacity? When are neonatal calf losses compensatory or additive mortality?

New techniques allow accurate determination of cow moose pregnancy, but in most instances, pregnancy is an uninformative statistic. What really is needed to be known is conception timing and the number of calves conceived. Real-time ultrasound may provide insight, but a simple bioassay or other test would be most helpful and eliminate the need to handle animals.

We just now are beginning to unravel the mysteries and importance of the role of the mature bull in breeding. Many of Tony Bubenik's theories (see Chapter 5) have gone untested and unproved. No one disagrees with management for an adequate bull/cow ratio. But what is "adequate"? Still unclear is how many cows a bull can successfully breed during any given day or the rutting season. And we lack a clear understanding of the role of mature bulls in estrous timing and ovulation in females.

We understand how moose grow and develop from conception to death. But good comparative data sets for the various subspecies across North America are lacking. How do subspecies vary in shape, size and rates of growth? Such information may shed light on moose evolution and subspeciation.

BULL BROWN ABOVE, BLACK BELOW — WHITE OR VERY LIGHT GRAY STOCKINGS — POT-BELLIED LOOK OF SUMMER GONE —

COW & CALF GRAYER ON NECK & SHOULDERS

(MEMORY) BULL, COW & CALF MOOSE RESTING IN WILLOW & ALDER BESIDE LITTLE TUNDRA POND, 75-MILE AREA, McKINLEY PARK, ALASKA — SEPT. 23, '56

AS TRIO STARTED UP BANK, BULL LEADING, COW SLIPPED IN SNOW & WENT DOWN ON HER HINDQUARTERS, BUT QUICKLY REGAINED HER FOOTING —

(MEMORY & REF., MAY 18)

BULL IN BAD SHAPE — L EAR A TATTERED FRAGMENT; L EYE WITH OPAQUE PUPIL (R EYE SEEMED RUNNY, BUT NOT BLIND); L ANTLER STARTING KNOB OR TINE, & SMALLER THAN R — WHOLE AREA APPEARED DAMAGED, INFECTED. BULL WITH UNEVEN COAT — APPEARED TO BE BARE PATCHES ON BELLY ON R SIDE.

6-MI., McKINLEY PARK, ALASKA

MAY 17, '65

(SEE FIELD NOTES)

ENTIRE EYE AREA ON L SEEMED DISTENDED POSSIBLY HAIRLESS SKIN AROUND EYE —

(MEMORY)

DARK COLORATION — LACKS GOLDEN TONES OF OLD BULL —

THOROFARE PASS, W OF EIELSON, McKINLEY PARK, ALASKA — SEPT. 26, '56

YOUNG BULL MOOSE — PROBABLY YEARLING

COMPACT BODY — LACKS SWAYBACK OF COW, BUT HAS LEGGY, JUVENILE LOOK COMPARED TO MASSIVE OLD BULL

(THIS WOULD BE SAME AGE AS "JASPER" — SEE MAR. '53 SKETCHES)

5

ANTHONY B. BUBENIK

Behavior

Prerequisite to understanding moose behavior is an awareness of basic differences between humans and moose in their perceptions of environmental cues, thresholds and planes. It is necessary, therefore, to stress why they transmit, perceive and respond differently.

Perception and Responsiveness in Humans

The large cortex of the human brain accounts for intelligence and abstract thinking. By nature and through education, humans are conceptualists (Carr 1972), typical egoists, and selfish users (Lorenz 1982, Bubenik 1989a, 1992) of ecosystems and their living resources (Bubenik 1989b). Conceptualism is one of the main characteristics that differentiates humans from all other animals. It also is the reason why humans have problems properly communicating with other animals, and why false interpretations about the expressions and responses of those other animals are made so frequently (Von Uexküll 1921). As logical and abstract thinkers, the human conceptualist has difficulty with zoosemiotics, the science of communication (excluding that of humans), which deals with complicated and configurational codes for which species are "programmed" (Sebeok 1962, 1963, 1965). Nonetheless, humans also use innate expressions as codes of emotional stages, and transmit and respond to them subconsciously (Darwin 1872). As mere conceptualists, humans use complicated mental, analytic and synthetic processes to create from perceived records a proper image and corresponding concepts (i.e., conclusions) (Carr 1972).

As stereoscopic creatures, humans perceive three-dimensional space. Distance is an important parameter. Most people discriminate the colors of the visible spectrum. Subconsciously, or by experience or tradition, humans give certain values or meanings to colors. Color variation may have similar meaning to other animals, but this is difficult to prove (Brutt 1979). Like other animals, humans are predisposed to delicate visual discrimination of "gestalts" (integrated group of body figures and movements)—a heritage common with other mammals (Lorenz 1959). Lacking the reflective layer of tissue called the *tapetum lucidum* behind the retina, humans have poor vision in low light intensity.

Human hearing is limited by a relatively high threshold for faint noise and low vibrations. This is because of the small size of the external ear and low sensitivity of the ear's sensory epithelium. Therefore, stereophony for long distances is limited in humans.

Humans produce sound when exhaling, and the resonance from the chest cavity is limited in comparison with that of ungulates when they vocalize by inhaling (Bubenik 1952, 1966a). Nevertheless, people can imitate voices or noises relatively well, and can use them in dialogue with many macrosomatic (scent-oriented) animals as long as the sounds are not accompanied by airborne human scent, which may elicit flight response in an experienced animal.

Human olfaction (scent perception) is poor, as is human

interpretation of the meaning of scents. Nonetheless, many people can respond subconsciously by behavioral and physiological mechanisms to such pristine scents as sex pheromones (Vandenbergh 1983, Leo and Agemian 1986, Voland and Engel 1990). Odors are important indicators of emotional status, sexual readiness, and antagonistic or altruistic mood in individuals, social classes and populations.

Humans tend to evaluate animal expressions without respecting the horizons in which animals communicate. It is no wonder that the dialogues of moose are poorly understood and frequently misinterpreted.

Perception and Responsiveness in Mammals

Most animals, other than and unlike humans, are perceptionists (Carr 1972). They sense and respond to species-specific signals as single or complex codes. Whether the receiver responds to the signal in question depends primarily on its momentary level of responsiveness and on the signal's releasing power (Lorenz and Leyhausen 1973). The speed with which the innate cues operate is not error-free. Animals experience different emotional stages that influence their perception levels and may overrespond to signals. To avoid such erroneous responses and their consequences, species innately respond to super-stimuli or key-releasers (Lorenz 1954). These are cues or signals so dominant that their perceptor suppresses other lower-ranking stimuli (Tinbergen 1990). Key-releasers may indicate kinship, sex, emotional stage or social rank, and are powerful tools for studying behavior. Because odors can be super-releasers, their importance may remain undetected by humans and result in improper interpretation of animal behavior.

Sensoric Range and Thresholds: The Neurohormonal Background of Behavior

Both genders of moose experience seasonal changes of sex hormonal secretion (see Chapter 2). Also, responsiveness to sexual signals depends on the level of circulating sex hormones that trigger or inhibit the activity of those midbrain nuclei responsible for a different kind of behavior (McEwan et al. 1979). Duration and secretion intensity of the sex hormones depend on age, social rank, infrastructure and daily patterns of hormonal secretion. All these conditions determine when, how long and with what intensity each individual and social class of both sexes will release and answer corresponding signals with a full or incomplete pattern. For example, the pattern of sex hormone secretion determines when and why animals will be or will not be responsive or present on rendezvous sites.

Some environmental cues—for which moose do not have innate programs—may be perceived but do not elicit responses. Therefore, the cue either does not pass through the neuronal lock or, because of an extra high threshold, the stimulus is not registered and the animal may not respond. Such behavior (or lack of behavior) has nothing to do with poor sight, hearing or smell, as frequently is assumed by the unacquainted observer.

Umwelt: The Intimate World of Moose

Until ethology, the scientific study of behavior, developed as a special science, it was considered as part of ecology (Eibl-Eibesfeld 1970). This link still exists, but is almost neglected, at least where moose ecology and behavior overlap. For example, food or thermal shelter are not the only parameters that determine where moose can or should live, despite human impressions about the quality of a niche (suitable living place) (Whitaker et al. 1978). People sometimes wonder why an excellent feeding ground is not used by moose, or is frequented by only one sex or even one social class (Bubenik 1984). The answer lies in the "Umwelt concept" (Von Uexküll 1921, 1937, Wickler 1972, Von Uexküll 1982, Bubenik 1984). Von Uexküll (1921, 1937) demonstrated that an animal can respond to and only will search for the environmental signals to which it is "programmed." The Umwelt concept is the basis for the "theory of meaning" (Anderson et al. 1984) (i.e., of species-specific cues to which an animal can respond). Umwelt translates to the "world around," which means the intimate world of the animal. Theoretically, because of different species-specific responses to environmental cues, an environment can have many Umwelts.

Umwelt is not equivalent to environment, habitat, niche or ecotype (Whittaker et al. 1973, Whitaker et al. 1978), because it is determined only by cues to which the species is programmed and receptive. This depends on neurohormonal factors (Perron 1981). Knowing the most important environmental cues of the species, one can search for, predict or even develop necessary Umwelts where they were absent. This should be considered in moose behavior and ecology. The Umwelt of many ungulates is not only species specific, but also gender, social class and season dependent (Bubenik 1984, 1989b). In North America, there are two different habitat ecotones—the boreal taiga and the periglacial forest–tundra—each preferred by different moose subspecies (Gauthier and Larsen 1985; see Chapter 2).

Umwelt of Taiga Moose

Presently, there are no studies on moose distribution and habitat preference from the aspect of the Umwelt concept.

From a habitat standpoint, the boreal taiga moose is genetically adapted to the taiga; both sexes avoid long and distant stays from forest cover. The feeding grounds are relatively open clearcuts or burns with early plant succession. The mean width of the preferred feeding ground strips is only about 333 feet (100 meters) (Hamilton and Drysdale 1975). The importance of forest is evident from the behavior of a frightened moose that escapes into forest instead of running into open landscape. Being selective browsers, moose aggregate on regenerating forest lands. The general and locally varying plant selection of moose is known, but not the thermal and wind conditions that cue them to optimize the Umwelt. Above 60°F (14°C) moose presumably become uncomfortable because of the low ceiling of their upper critical temperature (Renecker et al. 1978, Schwab and Pitt 1991). Aquatics and mineral licks are used (Best et al. 1977, Fraser 1980, Belovsky 1981, Adair et al. 1991), but may not be essential. Powder snow generally is avoided by cows (dams) with calves (because it impairs calf mobility), but not by bulls and barren cows (Peterson and Allen 1974, Mech 1987, Peterson 1989). Breeding areas in the taiga are elevated grounds, usually bordered by water or scattered ponds. Elevated ground or shores are preferred scent-marking, vocalization and listening sites of both sexes in the rut. Preferred calving grounds are small islands not far from mainland.

Umwelt of Tundra Moose

Studies of the Umwelt parameters of tundra moose also are lacking, but these moose are specialized for the periglacial forest–tundra habitat and do not try to inhabit the relatively dense taiga belt. Tundra moose differ from taiga moose in that the former prefer open country, use antlers as long-distance visual signals, seek escape into the open and exhibit specialized breeding strategy and mother/daughter behavior in the rut.

Sociability

The social relationship of moose with encountered conspecifics or other species is an important Umwelt role. Deer species vary from high to low sociability but none is solitary, unsocial, asocial or antisocial (Bubenik 1971, Wickler 1972). Hence, identifying moose as a solitary species is a misnomer (see Houston 1974). "Individualistic," referring to a low degree of sociability, seems more appropriate (Bubenik 1971). Individualistic species can be characterized generally as those comprised of large, private individuals with variable individual space (Portmann 1953, 1961, Hall 1967, Bubenik 1984). They are not antagonistic toward the opposite sex. The perimeter of its space is guarded and indicated by gestalt (integrated acts, gestures, olfactory and eventually acoustic cues) (Hall 1959). This perimeter may be marked visually or olfactorily, if the species is territorial; this is not the case for moose. Trespassing into an individual's space is considered an intentional challenge. Rank of the offender determines whether the provocation will evoke a threat or a retreat. Moose are the largest representative of individualistic deer (Knowles 1984, Lynch and Morgantini 1984, Bubenik 1987). Unfortunately, the range of seasonal, social class, sexual dimorphic and rank-dependent dimensions of the individual space of moose are not exactly known by humans, but they evidently are well understood by moose.

Sight or Vision

A mature, standing moose hears, sees and smells a level above the height of an average-size human. They can search for signals by moving the head from the ground to a height of almost 8 feet (2.4 m), similar to a horse (Figures 71 and 72). Unfortunately, there are no data concerning optics, thresholds for light intensity and color, or color meaning in moose. Thus, to get even an approximate picture of visual perception and acuity in moose, biologists currently must rely on studies of ungulates with similar optics, and for which such information is available.

Although unknown, sight distance capability in moose probably is less important to stereoscopic distance assessment than stereophonic hearing and possibly smell, which are more accurate. The angle of the horizontal field of vision, based on the angle of the eye is shown in Figure 73. The vertical field of vision, compared with humans and elkhound is shown in Figure 74. With decreasing light intensity, the shape of the normally elliptical and horizontally kept pupil changes to circular. Red deer, with a circular pupil, discriminate better vertically than horizontally (Backhaus 1959), which may be important for assessment of conspecifics' antler size. With a narrow overlapping of the field of vision where the image is not too sharp, the elliptical and permanently horizontal field of sight make moose almost a monoscopic animal. They must move the eyeballs for assessment of an object at very close range. To see small objects, such as a wolf attacking parallel to the body, the moose must twist the eyeballs downward (Figure 75). This explains why moose nod in encounters with dogs or wolves, or tilt the head to see in the "blind spot" around the forelegs (Figure 76), and why moose usually do not pay attention to the area high above their heads (e.g., tree blinds).

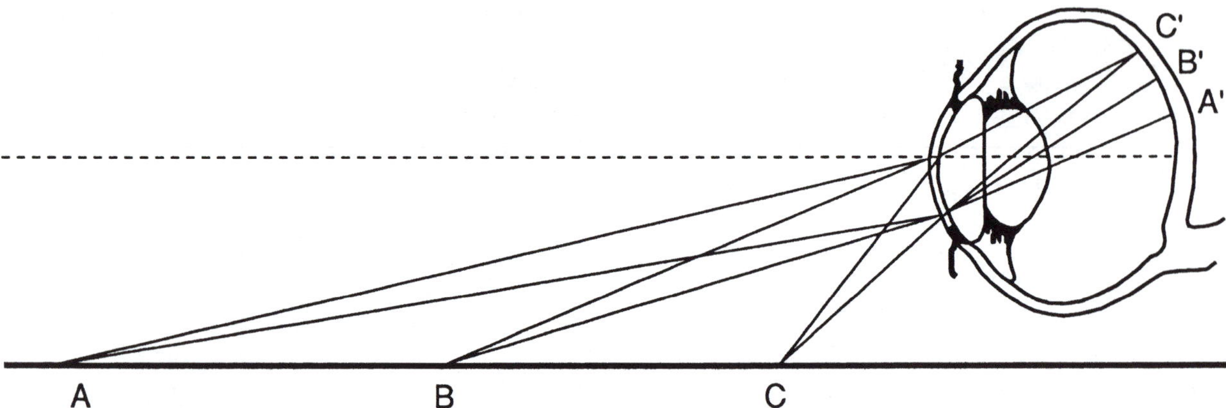

Figure 71. To improve visual clarity of objects at different distances, a moose must move its head or eyeballs up or down to adjust for the appropriate focal length of the retina. Because the lens is rigid, the eye is astigmatic. If a moose wants to focus on a distant object (A), the image must be projected at the retina field (A'). Closer objects, at B or C, must be projected above the horizontal axis to achieve shorter focal lengths of fields B' and C'. The dashed line is the horizonal axis of the eye. *Adapted by A.B. Bubenik from a poster on vision of the horse; courtesy of G. Hager, Veterinary University of Vienna.*

Color and Contrasts

Because all investigated ungulates have cones and rods in the retina (Johnson 1901, Witzel et al. 1978, Weiss 1981), it is likely that moose also see colors. Red deer and white-tailed deer distinguish six colors—a slightly reddish-violet, red, orange, yellow, green and blue (Backhaus 1959, Witzel et al. 1978). They cannot discriminate between yellow and gray at a brightness of 0.1 lux with a black background, and it is extremely difficult for them to discriminate the tones of blue. Although deer are not colorblind, they have more rods than do other ungulates. This may be the reason for low discrimination for blue (Witzel et al. 1978, Alexander and Stevens 1979). The prevalence of rods with the reflec-

Figure 72. Probable angles of lateral fields of visibility (sharp, hazy, blurry and blind) of a moose holding its head rigid and looking straightforward. This assumes that moose vision is similar to that of horses. *Adapted by A.B. Bubenik, from a poster on vision of the horse; courtesy of G. Hager, Veterinary College of Vienna.*

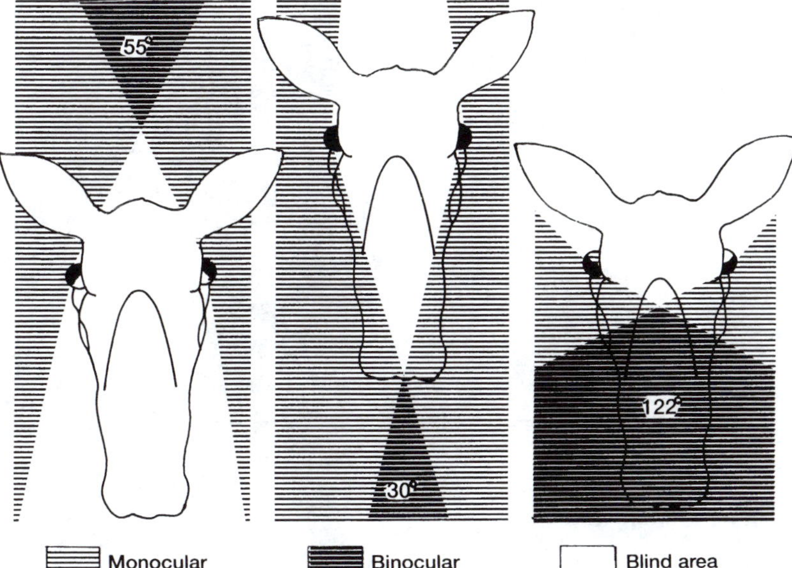

Figure 73. Probable width of horizontal sight horizons of moose with the head held forward and moving the eyeballs back and forth. The drawing is based on the possible angles of horizontal vision with ellipsoid pupils (from Bubenik 1987).

tive *tapetum lucidum* behind them enhance vision in very poor light and is the reason why pupils reflect the light from a spotlight. Whether the different wavelengths of light have the same meaning for moose as for humans is not known. The more numerous rods of the eupecoran (equipped with antlers or horns) ruminant's eye provide better discrimination of contrasts. This may be the reason for accentuation of the head- and rump-pole, maturation stage and sex in deer. (Head-pole and rump-pole refers to the front and back anatomical aspects that cue visual and behavioral responses.) These developed more contrastingly than color patterns (Portmann 1960, 1961, Bubenik et al. 1977).

Gestalt Recognition

There are no studies of friend or foe discrimination in ungulates except for investigations to assess antler rank, as done with antlered dummies of moose (Bubenik 1987), red deer, elk, caribou and fallow deer (Hediger 1946, 1976, Bubenik 1973a, 1975b, 1987, 1989b). In all these species, the

Figure 74. Simulation of lateral visions in a moose, human and elk-hound, to show how much the moose, with a narrow field of vision, must change its head position, in contrast to humans or dogs with round pupils and circular fields of vision (from Bubenik 1987).

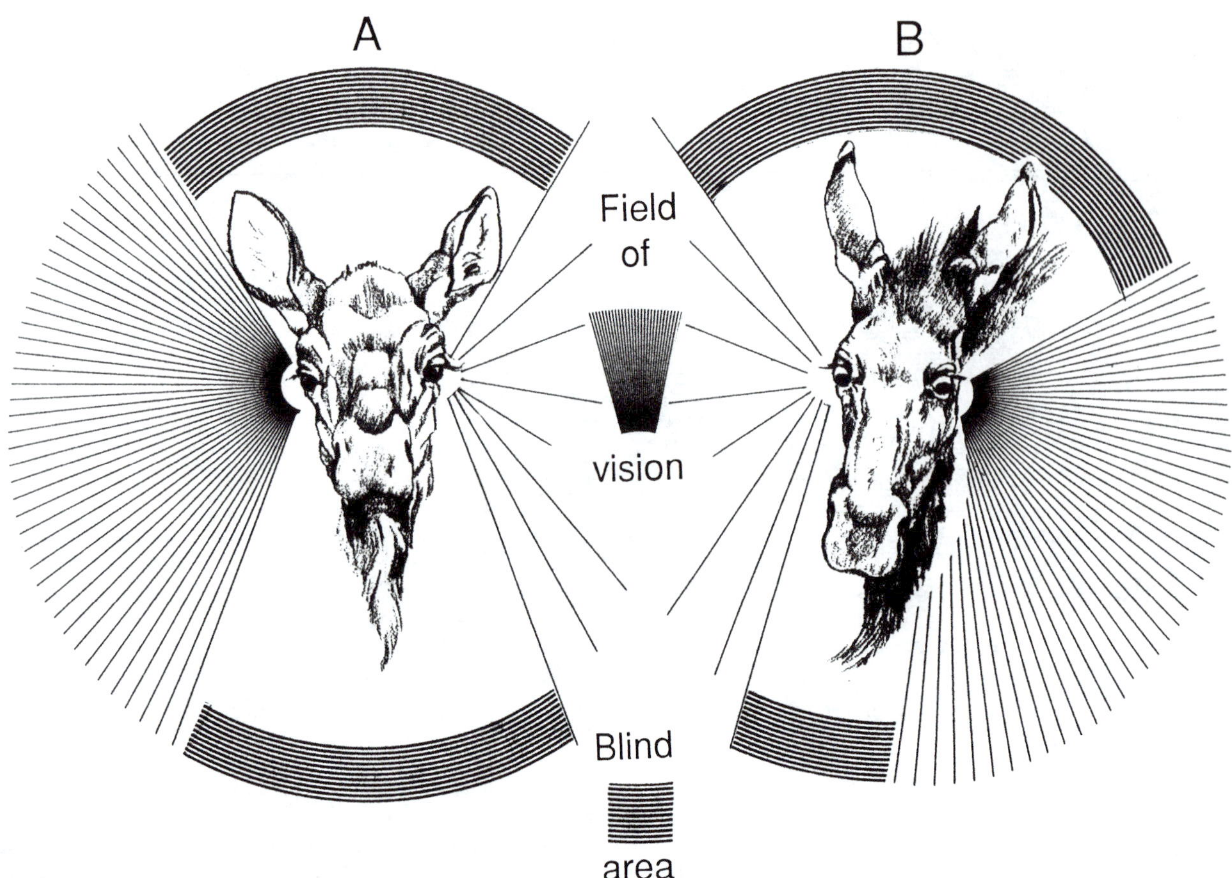

Figure 75. The "blind and blurry" areas of a moose looking straightforward (A), or with tilted head (B) to see an object very close to its front legs or flanks. *Illustration by A.B. Bubenik.*

Figure 76. The moose being pursued by a predator must tilt its head downward and/or move the eyeball to keep the pursuer at ground level in his vision. *Illustration by A.B. Bubenik.*

females must have an innate ability to discriminate visually the rank of antlers (Bubenik 1987). I was surprised that 2-year-old tundra moose bulls could assess relatively small differences in antler size and responded with appropriate behavior. In taiga moose, rank difference between the antlers may need to be quite conspicuous to avoid sparring for rank.

In a short pilot study with a head-pole of a cow on my chest, I elicited investigative behavior from a 3-year-old bull. I do not know if he could recognize the lack of a long torso or four legs. Whether moose have such visual recognition and memory for conspecific images, as was found in red deer or domestic sheep (Hediger 1946, Kendrick 1990), needs to be tested.

Coloration

Moose calves are born without spotted coats, which are common in neonates of other deer species. The brown color of moose calves matches well with bare ground at the birth site. The facial coloration in both taiga and tundra moose in winter coat exhibit sexually dimorphic patterns that may relate to the timing of sex hormone secretion and its acme in blood plasma (Bubenik et al. 1977), similar to analogies to white-tailed deer (Smith 1976, Bubenik and Bubenik 1985) and the hindquarters and scrotum of chamois (Dietrich 1979, Meile and Bubenik 1979). Histochemical studies in tupaia or rodents (Von Holst 1969, Wilson and Spaziani 1976) have shown that testosterone enhances melanocyte activity and helps synthesize from tyrosine the melanine, a pigment responsible for hair darkening (Wilson and Spaziani 1976). This may explain why, at the beginning of the rut, only the highest ranking prime bulls have completely black faces and cows' faces remain light brown. The progress and degree of pigmentation vary according to age (Figures 77 and 78).

Relying on these analogical changes in pigmentation of certain hair areas, it might be hypothesized that there are testosterone receptors around the hairs on the faces of bull and cow moose and the scrotums of bull moose. Similar receptors, but also sensitive to female sex hormones, may be around the anuses and vaginas of cows. Estrogen enhances melanocyte activity and helps synthesize melanin from tyrosine (Wilson and Spaziani 1976). Therefore, as long as their ovaries produce estrogen, cows retain not only a brown face, but also a white vulva patch. Cow moose with testosterone in the blood (inactive ovaries or adrenal tumors), therefore, should have dark faces and no vulva patch. Conversely, a cryptorchid or castrated moose should retain a brown face and also may develop a likeness to a cow's vulva patch.

There is at least concensus view that, in any subspecies of moose, a rump patch is absent in the classic interpretation (Peterson 1955, Mitchell 1970, Crête and Goudreault 1980, Franzmann 1981, Bubenik 1987, see also Guthrie 1971). However, Bowyer et al. (1991) hypothesize that the classic rump patch in the tundra moose is replaced by the light reflexing hair of the upper part of the body.

Leg coloration varies in all moose populations, but whether this has any meaning for moose is unknown.

Bell

The bell is a gender dimorphic feature in moose (Timmermann 1979; see also Chapter 2). It has communicative importance at least in the tundra moose (Geist 1986). During extreme cold, the tail of a bell can freeze off and, consequently, seldom is present in older bulls (Franzmann 1981, Timmermann et al. 1988). However, the bell continues to grow through prime age.

Why the bell is tail-shaped in cows and disk-like in bulls is unknown. Timmermann et al. (1988) speculated that the large surface of the bell in bull moose developed to transfer the urinary pheromones from rutting bulls to the nostrils of the "chinning" cow (Lent 1974, Frund and Bubenik 1982), or to dissipate their volatile components into the air. Miquelle and Van Vallenberghe (1985) expressed the view that a cow does not intentionally search for contact with a bull's bell, but they agree that the bell is a carrier of olfactory cues. In their view, the bell evolved primarily as a visual cue whose size and shape may be a secondary indicator of sex, relative to age and rank, especially during the antlerless period.

Penis and Prepuce

The penis and prepuce of moose deserve attention not only as pheromone dissipators, as in the case of fallow deer (Kennaugh et al. 1977), but also as a possible optical cues of potency. I suggest that the white or gray coloration of prepucial hair may relate to some easy oxidizing urine compounds that, in turn, relate to age and testicular activity.

Communication

Communication among moose involves not only single vocal or visual expressions, but frequently the combination of both, as well as olfactory elements. Some of these signals may be relicts, which are either no longer understood or answered. Others are signals that can be suppressed by trial-and-error conditioning in very experienced individuals. Detailed descriptions of behavioral gestures were provided by Bubenik (1987) and Cederlund (1987).

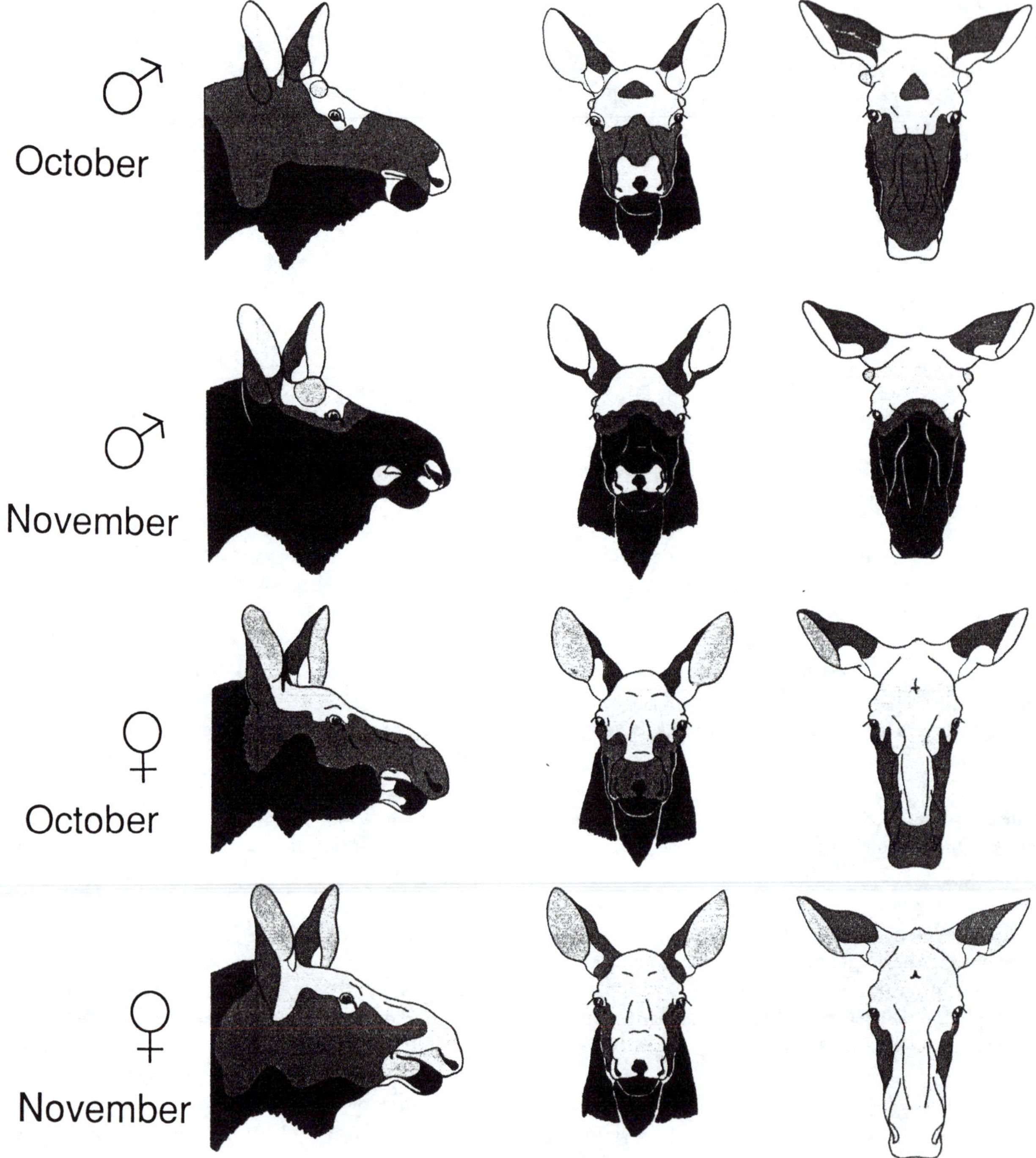

Figure 77. Pigmentation patterns of the head of moose calves at the beginning of winter. *Illustration by A.B. Bubenik; from Bubenik (1987).*

Hair Erection

An agitated moose raises not only the guard hair of the neck and withers, but also patches of hairs along the spine and on the rump and flanks (Figure 79). An agitated cow moose protecting her calf from a predator may assume an aggressive posture along with bristling of hair, particularly of the neck and withers (Figure 80). This bristling of hair patches is common to all cervid species. In the more evolved species, including moose, piloerection seems to be a relict to which conspecifics do not respond.

Gestures

Gestures are optical signals that form the basis for moose communication. Most are configurational cues in which

Figure 78. Sex differential of head pigmentation of prime moose before the rut. Darkness of the hair varies from light brown, through brown to almost black, because of different levels and durations of testosterone or estrogen secretion *Illustration by A.B. Bubenik; from Bubenik (1987).*

scent plays a role. In general, because of their complexity and speed of expression, they seem to be short-distance signals.

ROLLING EYES

Moose eyes can be simultaneously or independently diverted, converted, turned upward or downward, displaying the white part of the bulb. This diverse capability is considered an intimidation pattern, but actually is a normal physical attribute.

EARS

Each ear is independently movable in almost 360 degrees around the ear base and more than 90 degrees perpendicular to the body. The ears' position indicates an emotional stage (Figure 81).

TONGUE

A sexually aroused and aggressive bull frequently flips his tongue with a "smack" sound. This gesture is not identical to the sideward presentation of the tongue in red deer (A.B. Bubenik 1982a). The tongue smack is a typical "appetitive" gesture (Craig 1918, Tinbergen 1990) that indicates a highly emotional level, correlated to consummation of the sex drive or attraction of cows. It also may serve to dissipate salivary pheromones. It is accompanied by a sound called grunting, or croaking (Lent 1974), neither being onomatopoeic (expressing the true nature of the sound) because each resembles a hiccup.

HEAD POSTURE

A standing moose raises its head when searching olfactorily for airborne scents or visually. A frontal approach with the head displayed laterally is an indication of friendly or submissive mood, whereas the head down with the ears backward is an agonistic expression (Geist 1966, Bubenik 1987, Cederlund 1987) (Figure 81).

FLEHMEN

Flehmen or lip-curl is a dramatic gesture performed by bull moose to enhance exposure of the Jacobson's or vomeronasal organ in the palate (see Chapter 2). It is accompanied

The dewlap or "bell" of male moose has communicative importance at least among tundra bulls. Bulls splash urine from rutting pits onto the bell, which then helps transmit scent. In the northern portions of the species' range, a bull that has the tail of the bell (top left) probably is a submature or early prime-aged animal. The bells of mature bulls commonly freeze during severe or prolonged cold and drop off (top right). Nevertheless, a bell will continue to grow in breadth through a bull's prime age. *Top left photo by Joe Van Wormer. Top right photo by J. W. Jackson; both scenes courtesy of the Denver Museum of Natural History. Bottom photo by Victor Van Ballenberghe; courtesy of the USDA Forest Service.*

Figure 79. Patches of bristled hair of an agitated (but not frightened) bull moose. *Illustration by A.B. Bubenik; from Bubenik (1987).*

Figure 80. An agitated moose protecting her calf from a predator may assume an aggressive posture, along with bristling of hair patches, particularly of the neck and withers. *Illustration by A.B. Bubenik.*

An agitated moose raises not only guard hair of the neck and withers, but also patches of hair along the spine, on the rump and flanks. Laid-back ears also are a sign of agitation and imminent aggression. The Kenai Peninsula, Alaska, moose shown is ready to charge. *Photo by Charles C. Schwartz; courtesy of the Alaska Department of Fish and Game.*

by an elevated head with open mouth and rapid breathing (Figure 82).

ANTLERS AS VISUAL SIGNALS

Antlers play a role in long-distance visual signaling only in open landscape, such as tundra, over water and above the tree line. In autumn, because of the low level of the sun and the almost colorless front side of the palms, moose antlers can reflect light for miles. Both sexes are known to register responses by an investigative approach. For short distances, the antler angle toward the observer is an important gesture for assessment of rank, expression of dominance or inferiority, invitation for sparring or challenge for a fight (Bubenik 1987, 1989c). A frontal display with neck and head held perpendicular generally is a proposal for sparring. The neck- and head-down posture is part of an attempt to intimidate and generally is accompanied by thrashing, discussed later. A lateral display of antlers signals lower rank and a friendly mood. An elevated neck with almost perpendicular palms waving with the animals swaying gait is a gesture of highest threat.

TORSO POSTURE

The broadside (lateral, contralateral and frontal) display of the torso in big deer is considered an important signal in all

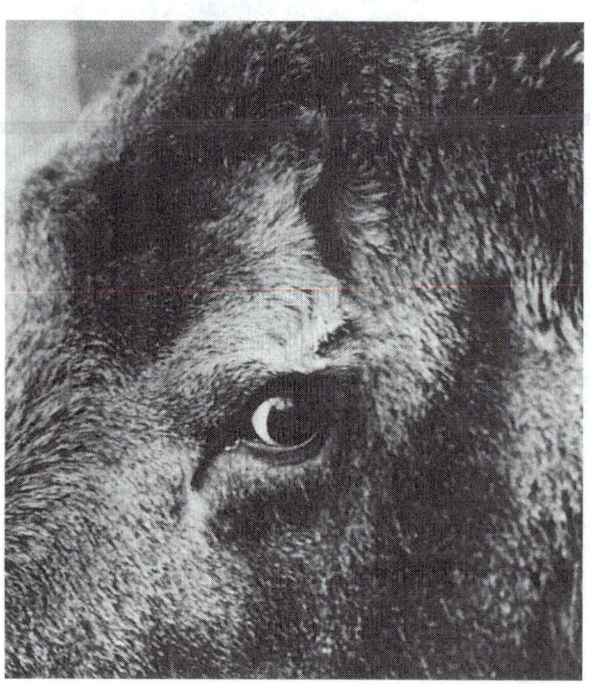

The eyes of a moose can be simultaneously or independently diverted, converted, and turned up or down, displaying the white portion of the eyeball (left). This eye movement has been considered an intimidation pattern, but is rather a physical concomitant symptom. The right scene features the normal eye position. *Photos by Charles C. Schwartz; courtesy of the Alaska Department of Fish and Game.*

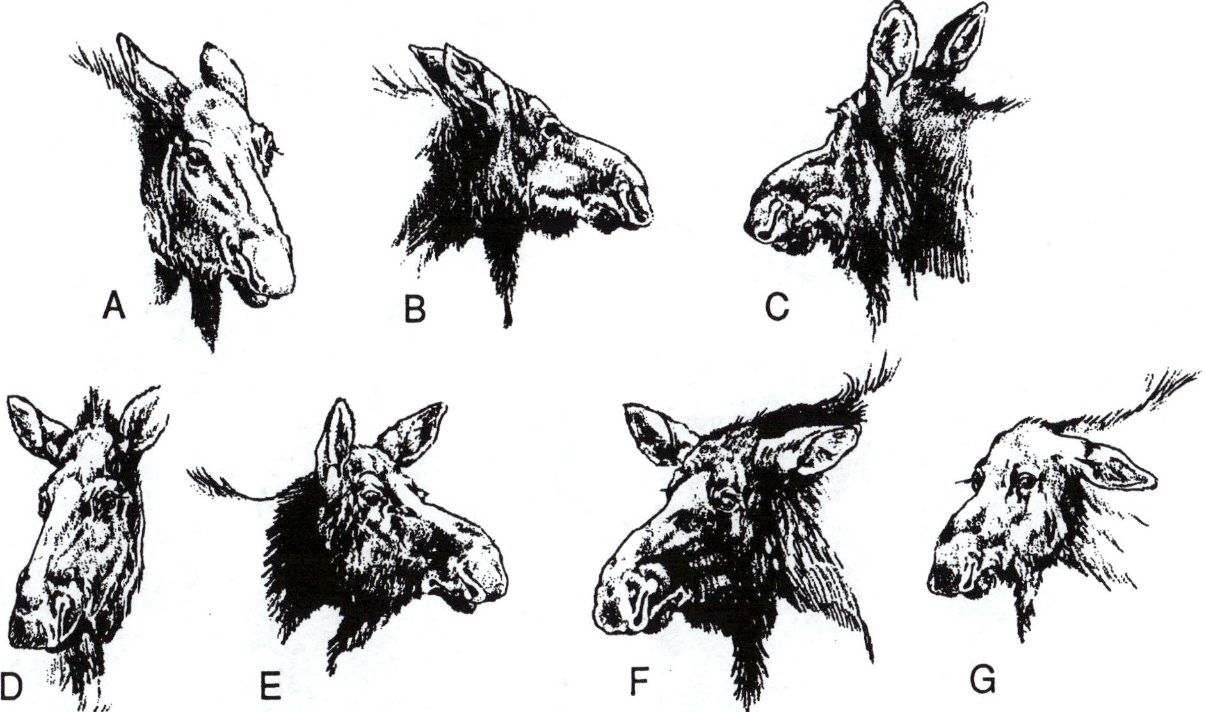

Figure 81. Ear positions of moose either concentrating on behavior of a peer or receiving acoustical signals from different angles or levels above the ground. (A, B and C) Either aversive approach or listening to signals from behind; (D and E) alertness or focusing on frontal cues or signals; (F) expression of concentration, or possibly searching for poorly located signals around the body; (G) display of readiness for attack, or listening to sounds close to the ground. *Illustration by A.B. Bubenik.*

ungulates (Geist 1966, Lent 1974, Knowles 1984, Walther 1984). However, studies with antlered dummies of both taiga and tundra moose and other large deer species (Bubenik, 1972, 1975b, 1987) have shown that the antlered head-pole operates as a super-releaser with inherited responses, regardless of torso orientation or how many legs the dummy has. In other deer with large antlers, the head-pole may equally compete with the rump patch and flight-releasing effect of human scent.

REST AND SLEEP POSTURE

Moose can rest while standing or bedded. When standing, the head and neck are relaxed, but the ears are in steady motion. When bedded, the moose can have all four legs under the body or extended. A favored resting and sleeping position of antlered bulls is on one side of the body, with legs stretched and one antler touching the ground (Geist 1963, Zannier-Tanner 1965). It is unknown whether a moose, with its neck bent backward and nose touching the groin, is receptiveness to tactile cues (Bubenik 1965b, Hediger 1969). When standing up, moose first raise the hindquarters and then the frontal part of the body, which forms a special hoof-print pattern.

BEDDING ACTIVITY

When they feel secure, moose keep seasonal bedding sites. In summer, they choose elevated terrain that enables favorable visibility. When possible, grassy spots exposed to wind that inhibit mosquitoes and flies are preferred. In winter, moose prefer to use powder snow areas in mixed forests, under large conifers. Franzmann et al. (1976a) found a daily winter bedding rate of 5.5 and 5.4 for males and females, respectively.

GAIT

Undisturbed moose walk or pace slowly. Sometimes, when two moose move in tandem, one walks and the other paces (Müller-Using and Schloeth 1967). Otherwise, moose trot or gallop (Taylor 1978). Play-related behavior, such as stotting (stiff-legged jump), gambolling (run and jump), rushing (run and stop) (Cederlund 1987) and "pleasure dancing" (all of the above) are a consequence of hyperactivity or "overflow" activity also reported in a calf around its mother (see Berry 1967). I observed gambolling with stotting in submature moose feeding communally on aquatics.

Gambolling is initiated by a single animal but can elicit

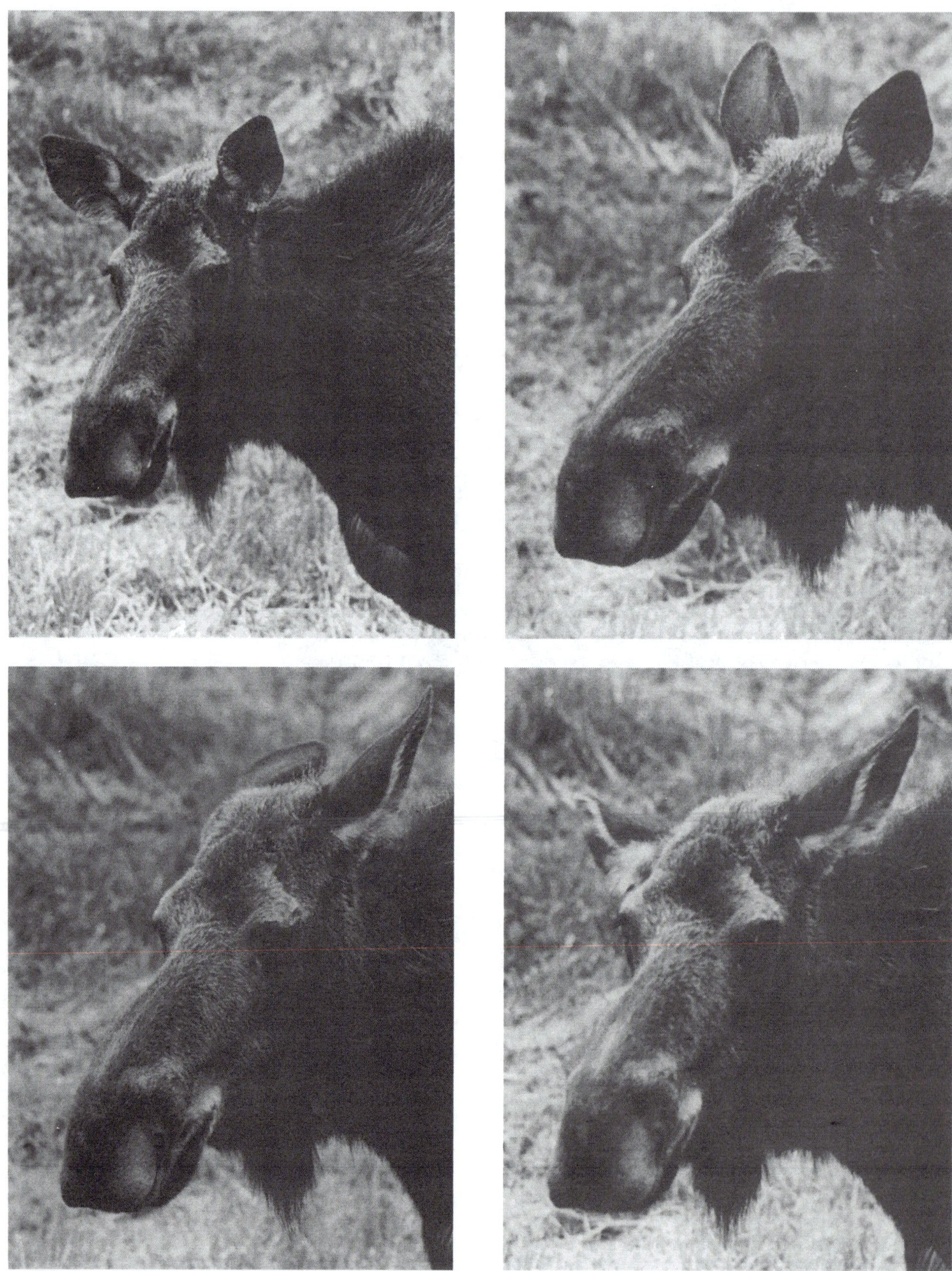

The ears of a moose are independently movable in an almost 360-degree arc around the ear base and more than 90 degrees perpendicular to the body. In the detection of external stimuli, such radar-like listening capability compensates for or supplements relatively poorly developed sight. *Photos by Charles C. Schwartz; courtesy of the Alaska Department of Fish and Game.*

Figure 82. A bull moose raises its head and retracts its tongue during "flehmen." Saliva drips profusely from the open mouth and the upper lip is curled. *Illustration by A.B. Bubenik.*

copying or mimicking behavior by conspecifics in sight (Scott 1967, Mason 1979). When gambolling, a moose makes short runs with stotting, head held down, backbone concave, and kicking and bouncing wildly like a bronco. It is a behavior common in all perissodactyls and artiodactyls.

Dixon (1938) estimated that a moose trots at a speed of 6 miles per hour (9.6 km/hr). C.H.D. Clarke (*in* Peterson 1955) clocked a moose running at a speed of 19 miles per hour (30 km/hr) for 30 seconds. A speed of 35 miles per hour (56 km/hr) was measured by Cottam and Williams (1943).

Lip-curling, or flehmen, helps bull moose expose his vomeronasal or Jacobson's organ to the pheromones produced by an estrous cow. The bull can detect the cow's estrous status by the presence and intensity of the pheromones. *Photo courtesy of the Wyoming Game and Fish Department.*

Moose do not like to jump, and do not jump directly from a gallop. They check the obstacle, then rear up the body on hind feet and thrust themselves over the hurdle.

Peterson (1955) reported on a technique moose use to move over deep snow by using the forelegs as snowshoes and moving in a kneeling position. Stanwell-Fletcher (*in* Peterson 1955: 109) stated that "The forelegs packed the snow before the weight was brought to bear, and only the hind legs sank deeply."

Analyzing slow motion cinematography from Newfoundland, I saw a tactic used by moose to move on crystalline snow in late winter. The metatarsi were raised so high that they were plowing with elevated hooves and spread dewclaws through the snow surface, almost to the forelegs. In this manner, the snow was compressed enough to make jumping forward relatively easy.

Distinct from other moose gaits is the moving with swaying forelegs as a gesture of resolute threat.

SWIMMING

Moose are powerful swimmers and, among North American deer, are exceeded in speed and endurance only by caribou (Peterson 1955). Adult moose generally swim with the whole body submerged, and only the head and part of the neck above water. However, yearlings may have most of the back exposed when swimming. Peterson (1955) referenced Baltzer (1933), Merrill (1920) and Stone (1924), who witnessed moose swimming distances of 8 to 12 miles (12–20 km). The swimming locomotion is a pacing or fast-walk

Moose antlers frequently serve as visual signals. At short distances, antler angle toward the observer is an important gesture for assessment of rank, expression of dominance or submission, invitation to sparring or challenge to fight. A frontal display, with the neck and head held perpendicular, usually is a proposal for sparring. The neck and head down display (above) is a gesture of intimidation, generally associated with thrashing. A lateral display of antlers generally signals lower rank and a nonthreatening attitude. A moose waving its antlers, with palms nearly perpendicular, an elevated neck, combined with the animal's swaying gait, signifies pending attack. *Photo by Victor Van Ballenberghe; courtesy of the USDA Forest Service.*

stroke. The ability of moose to spread their hooves undoubtedly aids the swimming (paddling) ability.

In shallow water, moose run with great speed and are not reluctant to go into deep water to elude predators. Peterson (1955: 108) wrote: "Water is definitely one of the preferred elements in the habitat of moose. When feeding on submerged vegetation they occasionally dive for plants in water over 18 feet [5.5 m] deep. They were frequently seen to submerge so completely that not a ripple remained in the water near where they went down. In the majority of cases the rump would float on the top and break water before the animal raised its head. Occasionally animals were seen to make at least a 180-degree turn while completely submerged, and at other times would seem to roll to one side while attempting to stay under. The average length of submergence was slightly under 30 seconds. The greatest time actually checked was 50 seconds, although some appeared to remain under slightly longer."

Yearling moose were observed jumping into a lake at the Moose Research Center, Alaska. From a standing position at the edge of the lake they jumped into the water and completely submerged. Our interpretation of this and their subsequent swimming action was that they enjoyed doing it (anthropomorphic view). They undoubtedly benefitted from the cooling effect and insect avoidance. Moose regularly swam across a portion of Coyote Lake, Alaska, rather than walk around the large bay (A. Franzmann personal communication: 1994). Despite their apparent acceptance of water in their Umwelt, death by drowning is not unusual, particularly for calves (Edwards 1957, Filinov 1974; see Chapter 8).

Daily Patterns

An important part of moose behavior is the monthly pattern of daily (circadian) activity for both taiga and tundra moose. Daily activity, as demonstrated for the individualistic roe deer (Bubenik 1960) and the gregarious red deer

When moose rest, they usually lie on the chest with legs tucked beneath the body, and the head forward or directed toward the flank and groin (top). Another favored resting position for bull moose is to lie more on the side and with legs stretched. The head may be held upright (center) or with an antler resting on the ground (bottom). The latter typically represents an animal in deep sleep. *Photos by Charles C. Schwartz; courtesy of the Alaska Department of Fish and Game.*

(A.B. Bubenik 1960, 1982b, Bubenik and Bubenikova 1967, Briedermann 1971, Georgii 1981), serves as a timetable for behavioral and physiological observations.

All available studies in moose deal only with activity budgeting within 24-hour periods which is different from the term "circadian," referring to 24-hour interval patterns or rhythms (see Aschoff 1954, 1958). The position of the sun above or below the horizon determines activity patterns of northern ungulates. The patterns also are sex and age dimorphic and depend on the time of intense maternal care or of the post-breeding period (Bubenik 1965b, Bubenik and Casnocha 1965, Hartl 1965). Hence, the concept of activity budgeting is too simplified for behavioral considerations, but it is useful for study of energetics. Activity budgeting is an account of the mean amount of time spent by a distinct number of moose for distinct activities per day as a mean of many days and eventually months. For behavior of moose, whose habitat reaches from 40 degrees to 67 degrees north, it is necessary to know both the basic pattern of circadian activity and how, in the subarctic, the moose cope with up to 24 hours of daylight in summer and 24 hours of darkness in winter. The notion cannot be rejected that, under such extreme conditions, a switch in endogenous rhythm is not possible (Aschoff 1965a, 1965b, 1979).

Without good timetables of circadian periodicity for moose of both sexes, social classes and latitudes involving all activity patterns, such as locomotion, feeding, ruminating, resting, sleeping and vocalizating, knowledge of moose behavior would be limited and incomparable.

Territoriality

Peterson (1955) denied any territoriality in moose. Others may accept it in view of the behavior of dams on calving sites, the scent and acoustic marking of mating areas by cows in the taiga, or the digging of pitholes in breeding areas by bulls on the tundra. The question of territorial behavior in moose is confounded by the ambiguous definition of territoriality, as expressed by Weckerly (1992). For moose, "temporary expropriation of the ground" may be a more appropriate characterization than "territoriality" for behavior during certain reproductive periods (see Chapter 10).

Acoustic Communication

Human interpretation of the vocalizations and other sounds of moose is complicated by the fact that moose often communicate at frequency signal levels lower than audible or distinguishable to the human ear. Also, some of the acoustic signals of moose are prompted, conditioned or expressed in concert with smells or visual manifestations. Dis-

Moose are strong swimmers and have been recorded swimming distances of 8 to 12 miles (12.9–19.3 km). Adults typically swim with most of the body submerged and only the head, part of the neck, and sometimes the top of the shoulder above water (top). Younger animals (bottom photo features two calves with a cow in the lead) ride higher in the water. Moose can run with great speed in shallow water, and are not reluctant to go into deep water to escape predators. *Top photo by Joe Van Wormer; courtesy of the Denver Museum of Natural History. Bottom photo by William W. Dunmire; courtesy of the U.S. National Park Service.*

Moose readily submerge underwater to feed on aquatics and can remain underwater for up to 50 seconds. When they dive for food, they remain submerged by paddling and perhaps by releasing air from their lungs. However, they sometimes have to struggle to stay down. Calves less than 1 week old are capable swimmers. *Photo by Albert W. Franzmann; courtesy of the Alaska Department of Fish and Game, Soldotna.*

crimination is compounded further by variation in signal issue or response among individual moose, populations and subspecies.

Hearing

The excellent hearing of moose results from: (1) the size of the reflective surface of the external ear of about 67 square inches (432 cm²) (Bubenik 1987), compared with about 1.1 square inches (7.1 cm²) in humans, and (2) better stereophony because of the distance between the external ears of about 14.2 inches (36.1 cm), compared with about 7.1 inches (18 cm) in humans. Also, the external ears of moose are independently movable, thus enlarging both the angle and the space of perception (see Figure 81). Therefore, thresholds for programmed calls or conditioned noises could be extremely low. My unpublished records with high-fidelity recorders have shown that moose communicated vocally over a 2.0-mile (3.2-km) distance using seeking or hiccup calls, which are imperceptible to humans. In lake habitats or in the Rocky Mountains, the moan of a cow moose can be heard by human ear at a distance of 1.25 miles (2 km). One of the reasons is that the relatively low frequencies of moose vocal expressions penetrate easier through forested terrain (see Lent 1974) and are carried further than higher tones.

Echolocation

Deer have excellent ability to locate the source of sounds, but this ability is especially well developed in bull moose. The plane of a bull's antler surface may operate as a parabolic antenna (Bubenik 1987). This idea was tested in a primitive experiment in which microphones were placed in the ears of a taxidermic bull head first with, then without antlers (Frund and Bubenik 1988). The noticeable difference in response of a milliampermeter coupled with the recorder showed an impact of antlers on the fidelity of echolocation.

Calls

Most moose vocalizations originate during exhalation, but it is possible that some sounds, like the hiccup, originate from inhalation, as is the red deer call (Bubenik 1966a). Among all deer, only the cow moose has a richer repertoire of vocalizations than the male of the species. And only a few sounds are common to both sexes. The first descriptions of voices were published by Lent (1974) for the tundra moose. Bogomolova et al. (1982) later presented 20 different sonograms for European moose. Their basic pattern does not differ from that in North American moose. However, differences in tuning because of dialects (Lent 1974, cf. Bogomolova et al. 1982) or age of individual, as it is normal in eupecoran (Bubenik 1952, Bowyer and Kitchen 1987, Hall et al. 1988) could not be excluded. Using sonograms, Bogomolova et al. (1982) described the following voices, except the distress call.

SQUEAK

The first and short sound of a newborn calf, emitted during the attempt to get on all four legs and/or searching for the udder. It is a high-pitched *"eek"* sound.

SMACK

The nonvocal sound produced by a calf eager to suckle or having just suckled. It also is the second component of the hiccup of a bull, and resembles the sound a person makes smacking his or her tongue against and away from the upper palate.

SEEKING CALL

The seeking call was described by Lent (1974) as a "whine." It is the call most frequently used to attract other moose, regardless of season. It may be a friendly "greeting," but can change to an overt moan during courting.

I found seeking call used widely for appeasement and, contrary to Lent's view, not limited to 2-year-olds. Its pattern develops gradually during growth of the calf and used life long. In the neonate, it is a short, faint sound, easy to imitate, but difficult to describe, relying on soft and nasal *"i-i-e-h-n."* After a calf is a few weeks old, this vocalization is extended in length and intensity, and may evoke a similar response or a guttural "grunt" by the dam, which sounds like a deep *"ehm."*

DISTRESS CALL

The distress call usually is heard from the young calf "crying for help." A hard, loud and nasal *"i-i-e-h-n,"* it is the most powerful call to which a dam and other cows respond by an immediate approach and various threat expressions.

SNORTING

Lent (1974) identified the snorting as a "bark." It is a short nasal sound, emitted in discomfort or an antagonistic mood, and sometimes is heard when other moose trespass the individual space of the vocalizer. It signals the lowest degree of vocal threat and is used by both sexes. The sound is similar to the snorting of elk, but quite different from the "barking" of red deer, roe deer and muntjac.

GNASHING

I heard the gnashing sound only from cows grinding their teeth. It signaled minor threat or displeasure. Bogomolova et al. (1982) recorded it also during parturition, perhaps as an expression of pain or in intense labor.

MOANING OR WAILING

Moaning or wailing is the call of an estrous cow. It has many variations, probably dependent on emotional stage and age. According to Lent (1974), its mean duration was 1.9 seconds, with a range of 0.6 to 7.5 seconds. Bogomolova et al. (1982) reported a 5- to 6-second duration. It could be described as a nasal, soft *"a-uh-o-o-o-ah"* with the last syllable abruptly ending. Yearling females moan less frequently and intensively than mature females. However, the frequency and intensity of moaning depends mainly on the phase of estrus and the presence or absence of a bull.

Whether a cow's moaning the day after parturition is attributable to a sudden rise of estrogen, as in red deer (Bubenik 1965b), has not been determined, but it is possible.

HICCUP

The hiccup, also known as a "grunt" (Altmann 1959) or "croak" (Lent 1974), is a nonvocal sound produced only by the bull. It results by the sudden opening and closing of the mouth, with a clasping after-sound like a smack, very similar to the sound of a sexually aroused capercaillie cock. Lent (1974) presumed that it is produced by air suction. I support this notion, because air is exhaled after a hiccup. Therefore, I suggest the term "hiccup" as more onomatopoeic (resembling the sound a human makes by a sudden, involuntary contraction of the diaphragm that closes the glottis at the moment of inhalation). The hiccup is sometimes repeated every few seconds with a duration of 0.18 seconds and varying degrees of intensity (Lent 1974).

ROAR

The roar is the maximum vocal threat used by both sexes to intimidate an approaching moose. It is a loud gutteral *"r-r-r-a-a-ah."* Bulls seldom roar during a severe contest or after victory (Bogomolova et al. 1982). More often, I have observed that a courting taiga bull may roar when followed too closely by a "satellite" bull and the courting bull does not want to risk an attack, thereby luring the intruder away from his cow. It has a tremendous intensity, lasts about 3 to 4 seconds and sounds like that of a tiger or a lion. Cows may use the roar when disturbed by humans. The roar was noted from both cows and bulls when people approached their pens at the Moose Research Center, Alaska (A.W. Franzmann personal communication: 1992).

Noises

When feeling secure, moose are very communicative and produce physical noises as well as vocalizations.

FRAYING AND THRASHING

Fraying in bull moose refers to rubbing antlers on trees; thrashing occurs when antlers are swept back and forth on brush and shrubs (A.B. Bubenik 1990). When thrashing, antler sounds are not heard. But during fraying, the resonance of the antlers is heard along with the coarse rubbing of tree bark. Fraying and thrashing may differ in duration, and both noise-producing behaviors can be interrupted by soil scraping. All seem to reflect hyperactivity or removal of antler velvet.

SCRAPING

Scraping the ground frequently is used alternatively with thrashing. It serves for both sexes as an attractant, but mostly for cows by bulls that urinate into the scrapes. Taiga moose scrapes generally are shallow; they are not repeatedly used, because the bulls do not have their own mating area. Moose cows also scrape and sometimes urinate into the depressions, either in the mating area or when two cows share the same island (E. Addison personal communication: 1988). Otherwise, scraping seems to be induced by "overflow" activity (Bubenik et al. 1978b, 1979) or as intentional signals that "advertise" location.

When a bull creates a rutting pit, the digging noises change in tone when the pithole becomes filled with urine. The bull stamps in it to squirt the soupy material on his antlers. This sound attracts cows, who run toward the bull and, by head bobbing and attempting to drink the urine, encourage him to urinate more. Pit digging and subsequent responses appear to be ritualized behavior persisting from the suckling time (Wickler 1961, Lorenz 1966).

ANTLER CLASHING

Sounds produced by antlers when bulls spar and fight tend to attract other bulls. The reason for this attraction is not well understood, but it seems logical that the attraction of noncombatants is the aspect of excitement, attendant to hormonal, social and hierarchical factors of rut. Moose hunters try to imitate antler clash with varied success, because the sound from antlers that rub or bounce together differ from sounds produced by fraying or thrashing (see Chapter 18). The natural sound of antler clashing between bulls is accompanied by thrashing, fraying and ground

Antler "fraying," or rubbing, on trees (left) or shrubs (right) by moose is associated with the shedding of velvet and the onset of rut. Shed antler is white, but from fraying and thrashing, the antlers attain a characteristic dark brown stain. *Left photo by Charles C. Schwartz; courtesy of the Alaska Department of Fish and Game. Right photo by Judd Cooney.*

Fraying and "thrashing" vegetation by bull moose is initiated by shedding of velvet. Thrashing is the sweeping of antlers through brush and shrubs (top), and does not have the sound (like the rubbing of tree bark) and resonance that antlers produce when fraying. Fraying can be almost violent in intensity or subtle scratching (bottom) against a single branch. Antler fraying and thrashing can be comfort or hyperactivity ("overflow") behaviors, or both. *Photos by Charles C. Schwartz; courtesy of the Alaska Department of Fish and Game.*

Shed moose velvet may be found hanging from tree branches during the velvet shedding period. It does not survive long in the forest because scavengers are quick to find and consume it. Such velvet is a source of protein. If shed velvet is found, rutting activity likely is occurring in the vicinity. *Photo by Charles C. Schwartz; courtesy of the Alaska Department of Fish and Game.*

scraping. The clashing sound also has a more-or-less specific chronology. It begins with cautious antler touching, broken by intervals of silence. After a few cautious tests, the antlers clinch and are rubbed together. After another period of silence, a noisy clash and rattling accompanies vigorous back-and-forth thrusting. When sparring or fighting is finished, the losing bull quickly retreats, followed by a snort or rarely a roar by the winner.

INTENTIONAL TWIG BREAKING

Twig breaking is done by bulls to indicate their presence. It is accomplished either by striking branches or pressing a dry twig to the ground (Figure 83). This advertising strategy is used mainly by bulls approaching out-of-sight cows or during serious attack. Twig breaking may be accompanied by thrashing or scraping.

SPLASHING AND URINATION

During the rut, moose are interested in sounds of and coming from water. They respond to the sound of urine striking water or splashing on the ground because of flehmen. They may search for a urine pool and, if it is found, they readily sip the urine. This sipping sound is very attractive to moose of both sexes and easy to imitate (see Chapter 18).

Chemoreception by Nose, Vomeronasal Organ and Tongue

An adult moose releases olfactory signals (odors) from ground level to a height of approximately 7 to 8 feet (2.1–2.4 m) and a moose can perceive such signals at a distance of hundreds of yards, depending on the signal's intensity,

Figure 83. Bull moose occasionally break twigs by squeezing them between the front hooves, signaling either a friendly or an antagonistic approach. *Illustration by A.B. Bubenik.*

and wind and other atmospheric conditions. Olfactory cues originate from interdigital gland secretions, traces of worn hoofs, urine and feces, tarsal glands, saliva, antorbital glands, velvet and urinary pheromones (hence airborne vomerones) (Blissit et al. 1990). Their volatility and location determines the vertical level (height) of olfactory search and corresponding gestures.

The sexually active pheromones and vomerones are very intense indicators of the stage of a moose cow's estrous cycle. Hence, their perception by a rutting bull elicits an almost immediate erection of the penis and eventually a search for the cow. The approximate 7.7 inches (19.6 cm) between the parallel nostrils of mature bull moose may enable "stereo-olfaction," which could explain the precision locating of a urine pool. If the cow is in the "proceptive" phase of estrus, the bull initiates vulva licking and eventually sipping of urine from a urine puddle or directly from the vulva lips. Urinary scent components also are trapped by "squat urination" on hairs of the cow's tarsal glands. This helps the bull to track an estrous cow and stimulates licking and scenting. This is followed by the "receptive" behavioral phase, when breeding occurs. The vomeronal components of the cow apparently help create a potent scent aura that sexually stimulates both partners.

Norwegian elk-hounds are trained to hunt (track) European moose (called "elk"). They locate and follow moose by the scent of moose urine trapped among hairs of the tarsal glands and subsequently left on the ground vegetation. An elk-hound on fresh track and close to a moose (100–200 yards [91–183 m]) instinctively checks the scent horizon or level of the tarsal gland. *Photo by Anthony B. Bubenik.*

Track Scenting

Although secretions (year-round) from interdigital glands are on discarded traces of worn hooves, moose seldom are seen tracking with nose to the ground (Chapman 1985). The apparent low interest for track scenting may be explained by the animal's short neck, which makes this behavior strenuous. This also may explain why moose particularly target scent left by the tarsal glands on higher vegetation. The tarsal gland cresol components evaporate slowly.

Saliva as an Olfactory Cue

Bull moose salivate profusely during the rut. Saliva continually drips over the chin, soaking the bell and dampening the bull's track. Bull moose saliva has a musk odor, indicating the possible presence of androstenoles. However, the potent sex pheromones (also in saliva) are the androstenones, whose influence on the cow moose in estrus cannot be excluded (Schwartz et al. 1990). Because androstenones concentration seems to correlate with testosterone levels and, in turn, with the rank of the bull, there is probability that saliva composition of the bull may contribute to female choice (Bubenik 1987).

Salivary pheromones undoubtedly are present in the pitholes of rutting bulls, from which they splash the antlers. This may be another way to dissipate urinary pheromones and enhance the scent aura between bull and cows.

Antler Perfuming

Moose hunters know that antlers of a rutting bull have a specific and lasting musk odor. Field observations have shown that antler scent seems important for allure of bulls (Bubenik 1987). The shape of the bull moose penis does not permit urination toward the antlers, as bull elk are able to do (Haigh 1982). Therefore, moose use the rutting pit to splash and perfume the antler (Figure 84).

Pain Responsiveness

Compared with white-tailed deer (Bubenik et al. 1988), moose have a high, but stable, year-round level of blood endorphin (Franzmann et al. 1981). Because endorphin suppresses pain (Hill and Chapman 1989), it may make moose less sensitive to superficial injuries than are species with lower endorphin levels.

Social World of the Moose

Prime-age moose (see Chapter 2) have practically no natural enemies that pose a significant threat. This may explain

Adult moose release olfactory (odor) signals from ground level up to approximately 7 to 8 feet (2.1–2.4 m). With favorable atmospheric conditions, these signals can be perceived by other moose several hundred yards away. Olfactory cues originate from interdigital gland secretions, traces of worn hooves, urine and feces, tarsal glands scent, saliva, secretion of antorbital glands, antler velvet and volatile odors of urinary pheromones (vomerones). Recently shed antler velvet has a distinct and even lingering musky odor that can be detected by humans and is of special interest to estrous cows (above). *Photos by Charles C. Schwartz; courtesy of the Alaska Department of Fish and Game.*

why they are reluctant to yield to moving objects, such as wolves, automobiles and trains. In intraspecific communication, they defer to rank as long as it can be distinguished visually or assessed by sparring or fighting.

Polarization of Bodily Accents and Individual Distance

Generally, in horned and antlered animals, but particularly in moose, the evolution of visual signals of behavior enhanced polarization of bodily accents. In essence, the head-pole and rump-pole are the primary sites for visual indication of attitude, portending stereotyped behavior (Figure 85). Antler size enhances the radius of individual space of bull moose. In cow moose, the individual space depends more on age and parity, but it is greater than that of any antlered bull.

Individual Space of Cow and Calf

As previously discussed, moose are not territorial in the classic sense of the term. Although males and females are not agonistic toward one another, they select different Umwelts. In this regard, cows behave more individualistically than bulls. Individualism of moose is expressed more by independence than by isolation. However, a cow will permit her calf to enter her individual zone until the offspring becomes a short yearling. At this time, she drives the yearling away from her and her calving ground. Cows establish

Figure 84. To disperse his pheromonal aura (scent), a bull moose frequently soaks the dorsal palms of his antlers with mud by stomping with his forelegs in a pithole containing his urine. *Illustration by A.B. Bubenik; from Bubenik (1987).*

Figure 85. Moose evolved to focus visually on the head-pole for signals. Other cervids—from left to right, roe deer, fallow deer and elk—have head-pole signals, but also large rump-pole patterns that also function as optical signals. *Illustration by A.B. Bubenik.*

greater distance to other cows with calves than to barren cows.

Conspicuous differences exist in cow/calf relationships between taiga and tundra moose during mating season. Among taiga moose, a calf will stay within the mating area in visible proximity to its dam and her attending bull. In the tundra, many barren and lactating cows build harems around selected bulls, and calves generally are not present. V. Van Ballenberghe (personal communication: 1992) witnessed calves around such a harem, but they are reluctant to join it. Van Ballenberghe suggested that some cows with calves may be bred by satellite bulls at the periphery of the harem. Yearlings bulls are tolerated by the "harem master" (dominant bull).

A calf's sharing of its dam's individual space is basic to the calf's survival against predators and deep snow. An orphan calf presumably will be more tolerated by a bull than by a cow (Jolicouer and Crête 1988), although adoptive behavior has been documented for cow moose (Franzmann and Bailey 1977). According to Jolicoeur and Crête (1988), survival of orphaned calves depends on the absence of wolves and on winter severity. Without wolves, orphans survive almost as well as nonorphans if they can use trails of other moose or can associate with them (Markgren 1975).

Individual Space of Bulls

The individual space at the head-pole area of bulls with hard antlers is greater than during other antler phases or for cows. Antlers apparently become involved in a bull's individual space around the third year of life when their size begins to contribute to the animal's rank.

How frequently antlers are used as weaponry depends on a bull's aggressive nature and his emotional status, which is controlled by the level of testosterone. In all deer, the onset, peak and decline of blood plasma testosterone depends on individual fitness, maturation status and pheromonal stimulation or depression (see Chapter 2). Hence, it seems logical (although not documented) that the highest ranking bulls have the highest testosterone level. Peak testosterone levels in yearling and 2-year-olds are likely too low for overt aggressive behavior.

Compared with taiga bull moose, tundra bulls seem to have a higher threshold for antagonistic behavior, resulting in a more ritualized sparring and courting pattern. This relatively complacent behavior of tundra bulls is corroborated during the rut, when large harems of cows are shared without violence by two bulls of equal rank. Unfortunately, there are no records on how such a compatible—or at least nonaggressive—relationship is established.

Individual Space of Antlerless Bulls

After the rut, testosterone secretion in bull moose drops sharply toward the level necessary for antler casting. Consequently, bulls may exhibit cow-like behavior. Several days before antler casting, bulls fight as cows do, by standing on their hindlegs and striking with their forelegs toward the opponent's head (Geist 1963, 1966; Figure 86).

Because of low or an almost zero level of testosterone, antlerless bulls behave more unselfishly toward peers of their own sex, but still are not frequently found with cows. During this period, frequently in November to December, bull moose may temporarily aggregate (Peek et al. 1974).

Bulls remain more-or-less compatible and nonaggressive until testosterone increases sufficiently to induce velvet antler mineralization. This occurs during the last third of the antler growth period. Then, the bulls behavior again becomes aggressive and antagonistic.

Encounter and Appeasement Ritual

The difference in social rank among prime bulls and 1- or 2-year-olds is so marked that the inferior males are not perceived by prime bulls as threats. However, the closer the individual rank, the more cautious experienced bulls are to one another, until they decide to test rank by contest. Hence, the precontest ritual is exhibited only in prime bulls. In precontest rituals, a frontal approach generally indicates antagonistic attitude, whereas a laterally held head usually signals inferiority or appeasement. In bull moose, as in other deer, an aggressive attitude is reinforced by tongue-flipping, frequent urination and profuse salivation.

If a bull wants to enter the individual space of a higher ranking peer, he will ritualistically stop ahead of the high-ranking bull's individual zone and, standing with head down, patiently offer his antlers for sparring. Depending on the dominant bull's attitude, he will accept a playful sparring either by responding with a similar gesture or by holding his ground, but with antlers turned diagonally toward the slowly oncoming partner, who will sniff the antler. After that, the dominant bull will turn his head and offers the other antler for an olfactory check. Often, an olfactory check is repeated on the first antler. Then, the inferior bull can stay and even browse, head on head, with the dominant one (Figure 87).

Figure 86. After the rut, bull moose inclined to fight tend to do so by striking out (flailing) with forelegs, in the usual combative manner of cow moose. *Illustration by A.B. Bubenik.*

Figure 87. A rarely seen appeasement behavior is performed by an inferior (submature) bull that desires to stay close to a dominant bull. The former will approach cautiously and sniff the antler points of the dominant animal. If the dominant bull tolerates this behavior, the submature bull, having acknowledged the other's rank, usually is permitted to remain nearby. *Illustration by A.B. Bubenik; from Bubenik (1987).*

Antagonistic Behavior Between Bulls

The antagonistic attitude of bull moose differs among individuals and seasons, and can manifest itself as sparring, fighting or scrambling, depending on intent and outcome.

SPARRING

Sparring is the conditioning to opponents' antler shapes and dimensions. Secondarily, it avails opportunity for visual assessment, by observing the outcome of sparring, of peers' antler ranks. In prime bulls, the urge to spar declines with the oncoming of rut. In submature bulls, sparring continues through the whole rut, frequently to quell activity overflow in the absence of available estrous cows.

Initial sparring is directed at bushes and solitary young trees. A bull pushes toward them with the broad side of his antlers to test their spread and his own physical strength and endurance. Using the antlered decoys, and recording encounters between tundra bulls, I found that submature bull moose appropriately deferred to antler rank and displayed their heads laterally (Figure 88). Taiga bulls, however, evaded approach only when the differential of antler rank was considerable. Equal-rank and bulls that spend much time in display and parallel walk may separate without contest (Figure 89). Sparring among bulls with equal antler spreads is a symmetric contest that ends when one or both partners becomes tired or the aggression subsides.

Sparring generally is preceded by thrashing dry branches or even large poles, apparently to gain familiarity of the antler spread and form. Sparring also may serve to confirm a bull's antler rank. Sparring usually is harmless, but among bulls of similar rank, it can escalate to a fight (Peek et al. 1986). Sparring with inanimate objects has many advantages: a "partner" is always available; the immobile but flexible vegetation can be challenged without any time-consuming ritual; it can be "subdued" without the bull sustaining injury; and guaranteed victory boosts confidence.

Once velvet is shed (I have witnessed it subsequently eaten), and an approximate knowledge of antler shape is achieved, intraspecific sparring increases. Bulls learn to recognize rank and not to accept a contest with or challenge a bull if they are not sure about his relative rank. If an on-

Young bull moose mimic the aggressive, rank-testing behavior of rut bulls, including postures, eye contact and mock sparring. These "contests" are learning experiences and do not escalate to fights. These youngsters do not represent threats to their peers and their "play" usually is tolerated by prime bulls in the vicinity. *Photos by Leonard Lee Rue III.*

Figure 88. Studies using cape-mount decoys or dummies worn by a human have shown that dominant bulls tend to react with disinterest toward bulls of obviously inferior rank (mass and antler size), assuming the inferior acts in a submissive manner. Here, the "inferior" (person in decoy) moves lateral to and not facing the dominant—usual expressions of nonaggressive attitude and relatively low social rank. *Illustration (from 8 mm film sequence) by A.B. Bubenik; from Bubenik (1987).*

Opposite page Interactions by closely "ranked" moose bulls during the rut often are quite ritualized. Such an encounter frequently begins with a lateral display of antlers (top), although usually at a closer distance. Note that the right eyeball of the right bull is converted (as may be the left eyeball of the left bull), to maintain vigilance against the other bull's aggression. Head swinging (second photo) signals a challenge. The right bull's eyeball is cast downward, signifying that he is holding or controlling his ground. A clinching of antlers is lined up as the bulls circle one another, maintaining eye contact (third photo). The clinch becomes a shoving contest, reflecting the animals' comparative strength and resolve, but also the leverage of the clinch itself. Here (bottom photo), the left bull has the advantage from an asymmetric clinch, although it is the younger, slightly smaller bull (note the bell tail). The right bull soon disengaged and sauntered off, and the left bull did not pursue. Ritualized encounters enable bulls to test themselves without resorting to full-scale fights that, energetically, are to the benefit of neither. Nevertheless, fights to the death do occur. *Photos by Jim Faro; courtesy of the Alaska Department of Fish and Game.*

Figure 89. The parallel walk of equal or nearly equal bulls is an attempt to resolve conflict by intimidation rather than energy depleting and potentially dangerous sparring or fighting. *Illustration by A.B. Bubenik; courtesy of G. Sedlacek.*

Figure 90. An intense warning approach by a dominant bull is expressed by a stiff walk and side to side casting of the head. The behavior usually is elicited by an equal-rank bull unwilling to depart or show submissiveness. *Illustration by A.B. Bubenik; from Bubenik (1987).*

coming bull is nodding its head, he is gesturing for playful sparring. However, a head held high with a swaying gait signals overt aggression and is accepted only by a challenger of equal or higher rank and corresponding ritual (Figure 90). This scenario has exceptions, such as an overt attack without a ritualized approach (Seton 1927), which may be due to individual character and/or hyperactivity.

SCRAMBLE

Sometimes soon after the rut, and possibly more frequently in tundra moose than in taiga moose, three or more submature bulls start a noisy scramble (sparring, running, vocalizing), which may attract more bulls and arouse the interest of some low-ranking prime bulls. After observing this for a short time, the low-ranking prime bull may enter the group and the submature bulls quickly calm down and leave.

FIGHT

A serious fight may develop between equally ranked bulls after sparring. It may be initiated to assess rank and possibly to gain access to a cow or cows for mating. Fights without ritual are rare. Sudden, unprovoked attacks generally are initiated by highly agitated (emotional) challengers. Usually, however, a fight is preceded by protracted ritual intended to intimidate and to assess antler rank. The bulls roll their eyes to enhance stereoscopy (Figure 91). Then, one or both bulls step into the clinch and commence pushing back and forth until one contestant relents (Altmann 1959, Frund and Bubenik 1988). When being shoved, they roll the eyes backward to detect obstacles from behind that might trip them or otherwise impair countermovements.

The victorious prime bull usually does not attempt to strike or gore a retreating opponent's body, as less-ritualized submature bulls may try. Occasionally, a fight is interrupted by such displacement activity as thrashing, or parallel walking with heads diverted. In these cases, eye contact is maintained, evidently to watch for renewed aggression (Figure 91). A fight loser does not discontinue rutting activity. Instead, he will wander off and seek other challenges, often at the perimeter of the dominant bull's mating area. After a period of recovery, and in the absence of other breeding prospects, a defeated bull actually may rechallenge the dominant, previously victorious bull.

Vicious and prolonged fights between bulls are perhaps not as rare as Peterson (1955: 97) concluded: "One is led to believe that most such meetings end with a little sparring

Figure 91. Before sparring and clinching antlers, bulls convert their eyes to improve stereoscopic vision to assess their opponent's antler architecture. *Illustration by A.B. Bubenik; from Bubenik (1987).*

and much bluffing." It is true that the majority of encounters are of this nature. An initial clash after some display and posturing may be the "fight," and end in a matter of seconds. Nevertheless, there are scattered reports of serious and prolonged fights, with some ending in death. Serious fighting may continue for hours, with interruptions. From an aircraft over the Caribou Hills of the Kenai Peninsula, Alaska, J. Davis (personal communication: 1995) observed an area of regrowth aspen that appeared to have been "bulldozed." On closer examination, he saw two mature bulls at a standoff. In a few minutes, they began clashing, and this lasted about 10 minutes. Neither bull was intimidated by the other, and their combat apparently had been going on for some time, based on the appearance of the vegetation and ground in the area. They still were engaged when Davis had to leave the area. V. Van Ballenberghe (personal communication: 1995) has observed breeding behavior at Denali National Park, Alaska, for many years, and reported that fights generally last several minutes, except for rare occasions when fights continue for hours, with interspersed periods of rest.

Serious fights may result in facial cuts, lacerations on shoulder and torso, lost eyes, broken ribs and shoulder blades, and internal hemorrhaging. Also in the Caribou Hills of Alaska's Kenai Peninsula, postrut captures of moose collared in the early 1970s showed that about 10 percent of the bulls had severe head or facial injuries (A.W. Franzmann personal communication: 1995). Three bulls had severely damaged eyes, including one that was virtually blind.

Scarring of bulls as a result of fighting is common; fatalities are rare but not unusual (Knorre 1959, Kozhukhov 1959). When antlers lock during sparring or fighting, death from exhaustion, starvation or dehydration can result (see Chapter 8).

PRECAST USE OF ANTLERS

After the rut, with declining testosterone level, bulls rub or "mock-spar" their antlers on solid, frequently dead tree stems, apparently to relieve irritation from the swelling of the skin-rim under the coronet attendant to approaching antler cast. This activity is common in all antlered deer.

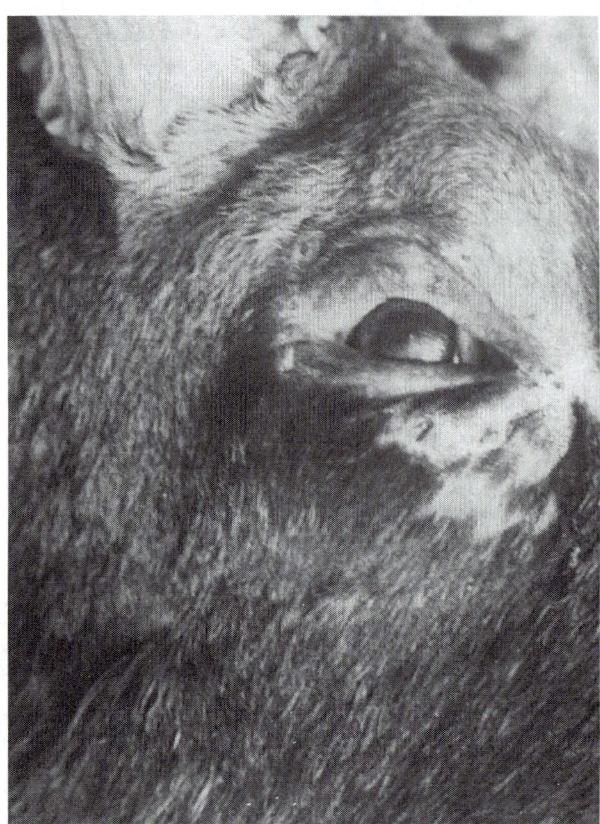

The most weapon-oriented portions of moose antlers are the points of the brow-tines, developed as single points, forks or multiple points protruding from a brow-palm. If the inner distance of the brow-points is greater than distance between the eyes, an asymmetric clinch (left) is possible and an inner point may hit the opponent's face (right). One or both eyes may be pierced. This occurs mostly with submature bulls whose antlers sometimes develop with inner brow points far apart. In prime bulls, most of these points have a narrower distance between them. *Left photo by Judd Cooney. Right photo by Anthony B. Bubenik.*

Cow Versus Bull

Out of the rut, a cow does not like to be approached by a bull of any age, and will attempt to intimidate the intruder by all possible gestures, such as laying her ears backward or downward, gnashing teeth and snorting. If gestures are unsuccessful, a cow may rush toward the bull, with forelegs flailing.

Cow with Calf Versus Yearling Cow

A cow with a newborn calf may be disturbed by its calf of the previous year, now a yearling attempting to maintain its bond with its dam. The dam will not tolerate the interference, and will chase the yearling away until the latter "gets the message." These encounters generally are harmless.

Interspecific Encounters

A mature high-ranking bull or cow generally is not intimidated by any other species, terrestrial vehicles or aircraft. Moose stand their ground rather than flee more often and longer than most other large ungulates will. The readiness of a moose to confront moving objects that trespass its individual space must be considered from the standpoint of the animal's physical strength and ecological "sense of security" (i.e., it fears little, so has limited notion of interspecific threat). Reluctance and unwillingness of moose to flee from dangerous situations are not necessarily evidence of poor sight. Most moose attacks of humans occur as a consequence of a cow defending her calf's space. Also, some attacks on humans by bull and cow moose may be the result of misidentifying people for upright grizzly bears (Figure 92).

Fitness and experience are decisive factors for moose survival when encountered by predators (see Chapter 7).

Heel Rubbing and Squat Urination

Heel rubbing is performed using a squatting posture (Geist 1966): the hind legs are put forward, and the heels are touched and rubbed during urination. In general, the tarsal glands saturated by urine leave an odiferous and easily de-

Figure 92. An upright human and a grizzly bear have similar outlines. A conditioned, hasty, self-protective aggression against a perceived bear by moose may explain some moose attacks on people. *Illustration by A.B. Bubenik; from Bubenik (1987).*

tectable track for moose. This behavior is common to all odocoileid deer, except the pudu (A. B. Bubenik 1990).

Squat urination is performed by moose year-round, but it seems to occur most frequently during the rutting period. At that time, urine must contain some pigmentation components, because the white, bushy hair becomes brown or dark brown in color. The different intensity of coloration indicates that the unknown components may be present in different concentrations. It also may relate to how intensely or often a moose performs heel rubbing.

Squat urination is either intimidation or submission behavior during times of conspecific agitation. However, V. Crichton (personal communication: 1993) witnessed a bull squatting without urinating. In Crichton's view, the squatting bull may have been intimidated by the posturing and displays of a dominant bull in the area.

Sexual Behavior

Sexual behavior in moose is predicated by the social structure of the population in question, and it can be organized or disorganized. I believe that organized populations have a social structure dominated by mature bulls, which ensures timely and successful breeding of cows during early estrous periods. This assures that calves are born in time to attain maximum growth by winter.

Disorganized moose populations, in my view, are characterized by disruptive factors that prevent the dominant

Grooming in moose is limited to licking newborn calves, scratching where old hairs persist and occasionally rubbing flanks or buttocks with antlers. Bleeding wounds, mainly above the heels and caused by tabanid flies, and velvet antlers often are cooled in water. *Photo by Joe Van Wormer; courtesy of the Denver Museum of Natural History.*

bulls from breeding with receptive cows during early estrus. The outcome can be (1) impregnation by bulls that may not be the population's fittest specimens, and (2) delayed calving that can prevent young from achieving adequate size, strength and fat reserves before winter. Yearling bulls produce sperm and can successfully breed female yearlings (Schwartz et al. 1982) and, in populations with low bull/cow ratios, a high proportion of young bulls are active breeders (Thomson 1991; see also Chapter 4).

Reproductive rate is not the only measure, or even the foremost measure, of population well-being. Genomic fitness is an important variable, and there is enough evidence about different fitness levels of individual moose, populations and generations (Wallace 1975, 1977, Gasaway et al. 1983, 1992, Lomnicki 1988, Bubenik and Pond 1992) to prompt concern about the impact of precocious breeders (Pemberton et al. 1988) on long-term population well-being.

(Moose and other ungulates have the following social classes: calves; yearlings—the youngest and most easily assessed age class among submatures; teens, or the older submatures; primes, or matures; and seniors [see Altmann 1959, 1961b, 1963b, Smith 1976, Bubenik et al. 1977, 1979, Bubenik 1984].)

Female Choice

I hypothesized that female choice of mates is an important factor among individualistic deer species, including moose (Bubenik 1987). Female choice emphasizes the importance of this kind of selection (Vandenberghe et al. 1989, Bakker and Milinski 1991). It differs from the strict Darwinian concept of selection by victors (Halliday 1981).

From a genetic viewpoint, features of choice, such as antlers, are not a visual liability or mechanical handicap in high-quality phenotypes, and their size and shape are in harmony with the genome for which the female selects (Anderson 1982, Gibson and Jewell 1982, Bateson 1983, Kodrick-Brown and Brown 1985, Kirkpatrick 1987, Manning 1989, Ryan 1990, Barton and Turelli 1991, Geist 1991). The allure of a male, measured by response of a female, is best demonstrated if a large assortment of prime males is present.

In disorganized moose populations, it is primarily bull body size and breeding experience and, secondarily, antler size that determine which bulls mate. In such populations, mate choice is narrow, and a cow may have to select merely the most aggressive suitor or a submature bull (Clutton-Brock et al. 1982, Bowyer 1986).

As a mating strategy, female choice can conserve the energy resources of the males. Being "chosen" is energetically advantageous, because the bulls deplete strength by fighting.

Rutting Behavior of Bulls

Rutting behavior in bull moose often is referred to as "recurrent puberty" (see Lincoln 1971). This is a misnomer, but does serve to explain the annual male reproductive cycle. Puberty is the age at which reproductive organs become functionally operative, and this is a singular event. After the rut, a bull's testes experience a long period of quiescence until the next rut (see Chapter 2). Testicular activity and subsequent rutting behavior depend on physiological and ecological factors. Physiological factors primarily involve body mass, antler size and endocrine fitness. The most important ecological factor is climate, which correlates to weather and growing season length.

SEXUAL ADVERTISING

The extent of sexual advertising by a bull moose depends on his maturity (age, body mass, antler size and shape, breeding experience, etc.). After velvet shedding, a prime bull broadcasts his copulatory readiness by hiccup-calls, pawing of and urination in the pitholes, profuse salivation, and saturation of antlers with urinary pheromones (see Figure 84).

YEARLINGS AND TEENS

There are no reliable records about the timing of breeding interest in submature bulls (Altmann 1957). In well-organized moose populations, yearling bulls are not potential breeders. And, in my view, all bulls younger than 5 years of age in such populations should be considered teens.

With each breeding season, a bull's testes and epididymides enlarge (Bubenik and Timmermann 1982), possibly suggesting greater semen reserves. In red deer, when submature and mature stags are equal in number, the submature males avoid entering the breeding areas or do not try to test the mature stags for access to females (Bubenik 1984). However, when teen stags outnumber prime stags, the former attempt to breed. From this, I conclude that the teen behavior depends on the strength of the scent aura before the rut, and the ratio of submature to mature males (Bubenik 1984, Pemberton et al. 1988, Verme 1991). Such breeding hierarchy and circumstance also have been shown in bovids and rodents (Vandenbergh 1983, 1987), and are probable in moose.

The ritualistic courting behavior of moose bulls does not develop completely until they are at least 3 years old, which is a general rule in large ungulates. In human terms, moose courting behavior serves to "seduce" a cow into willing copulation. Unless the visualized sequence of bull behavior is well-developed (as subsequently discussed), apparently from experience, cows are reluctant to accept the suitor readily. Attempts to force a cow to copulate—usually by submature males—is disruptive to progress of the rut.

PRIME BULLS

Although every prime-age bull is a potential breeder, not every one will be chosen as a mate. This can lead to the situation whereby a cow remains faithful to a subdominant bull (Bubenik 1987) or to a submature bull of lesser rank. Being *chosen* is different from being *accepted* as a mate (Kodrick and Brown 1985). There may not be opportunity of choice (Clutton-Brock 1982, Bubenik 1987, Hirotani 1989, Cowlishaw et al. 1991). My observations of white-tailed, roe and red deer (unpublished), and of moose (Bubenik 1988) indicate that a female will reject a mate who does not suit her. This phenomenon is common in domestic mammals (Hurnik 1987).

Characteristic behaviors of a prime bull are highly ritualized until the courted cow is receptive to copulation. A courting bull displays a "low-stretch posture" (Figure 93) and is likely to inhale the cow's vaginal discharges into his vomeronasal organ. Knowles (1984) indicated that this is not an expression of subdominance, as it may be in sheep (Geist 1971). The bull does not use the "head-neck stretch" behind the cow until she enters the proceptive phase. To accelerate estrus, a bull uses his pheromones and tactile contact with the vulva. Because of relatively greater semen reserves, a prime bull not only must be ready for but also more capable than submature competitors of frequent ejaculations or breeding more cows. How many successful mountings a bull can perform is unknown (Schwartz and Hundertmark 1993). Also unknown is how much time, after ejaculation, a bull needs to restore a reserve of viable semen (Bubenik and Timmermann 1982). My observations indicate that, after initial ejaculation, the second is not less than one hour later. And subsequent ejaculations, or at least ejaculatory thrusts, are marked by progressively longer intervals.

Because prime bulls are sexually active before submatures, they have the greater chance of mating with cows in their first estrus and producing offspring that will have time to attain adequate fitness before winter. Pemberton et al. (1988) indicated that, in red deer, the most resistant generation originated from prime stags and hinds who mated first.

SENIOR BULLS

Adequate data are lacking on rutting behavior in senior bulls. From antlerogenesis studies, I conclude that the testosterone receptors of the velvet and possibly in the midbrain are not active or sensitive enough to permit senior bulls to become aroused sexually for long or at all. Furthermore, the relatively solitary proclivity of seniors is evidence of a low degree of aggressiveness.

Mating Area

When a taiga cow approaches the attracting phase, she moves into her mating area, or breeding Umwelt. Mating areas, possibly traditional, often are characterized by at least one elevated ridge, possibly with a shoreline or a few small ponds, and good visibility for at least 110 yards (100 m). Determined from my experience (Bubenik 1987) and that of Rykovskiï (1964), the area covers 247 to 494 acres (100–200 ha). Seldom do two cows have their areas so close that one in estrus can stimulate the other to estrus by the former's behavior, suggesting that their estrous periods may be not synchronized. Cows tend to remain faithful to chosen bulls and do not allow other cows to share them (Strickland 1989), but there are exceptions.

If a cow has calves, those offspring will remain with her during the breeding season. These calves keep an appro-

Figure 93. A courting bull moose uses a "low-stretch" posture to smell and taste the vomerones around the vulva of a cow. *Illustration by A.B. Bubenik.*

priate distance from a courting bull, who does not chase them away.

A bull entering a cow's breeding area may be rejected if she is not receptive. The cow will avoid the bull, but not necessarily attempt to drive him away.

Estrus

During each estrus, a cow may pass through three different visual phases of sexual behavior (Beach 1976).

ATTRACTING PHASE

After a cow "expropriates" a mating area, she patrols it day and night. During the attracting phase, she scent marks with urine, preferably in water and along shorelines, scrapes the ground slightly, and emits moaning calls. All these signals advertise approaching estrus, with the intention of luring suitors and a prospective mate.

My trials, using the antlered decoys, (Bubenik 1987) have shown that a cow moose is influenced by antler size and will leave one prospective mate for another if the size and display of antlers of the "newcomer" indicate higher rank. However, if the visual impression is not supplemented with olfactory cues, the cow will lose interest and may return to the original mate. When a bull finds a cow during the attracting phase, he will follow her, exhibit flehmen when she urinates, hiccup occasionally, but not attempt tactile stimulation. During this phase, the cow feeds sporadically.

PROCEPTIVE PHASE

The proceptive phase likely begins shortly before rupture of the follicle. The cow tolerates a bull's approach and permits such physical contact as vulva licking or the bull's chin placed on her rump. Later, the body contact can advance toward the neck and lips. The cow then calls less frequently, but her moans are longer and multisyllabic, and she is increasingly attentive to the bull's proximity.

According to Knorre (1959), the proceptive phase may last up to 3 days, during which time feeding ceases. The bull hiccups become more frequent, and chinning is reversed (i.e., initiated by the cow), who also licks the bull's lips and cheeks. She encourages attention by wallowing in the bull's pitholes and stimulating his urination by head butts. She also may attempt to drink his urine to maximize the pheromonal (scent) aura (Figure 94). And she may hold her tail horizontally, at which time vaginal discharges are frequent and visible (Figure 95). During this phase, the bull attempts mounting, but the cow does not allow intromission.

RECEPTIVE PHASE

The receptive phase is of short duration—about 6 to 18 hours. The cow takes over the initiative in caressing and, between wallowing in pitholes and chinning, she also urinates frequently and offers herself for vulva licking and mounting. She also moans intensely.

Actual breeding may involve several pre-ejaculatory mounts. Intromission with ejaculatory thrust lasts only a few seconds. Knorre (1959) and Schwartz and Hundertmark (1993) agreed that a receptive cow allows many mounts and does not run away from under the bull after the ejaculatory thrust. She stays put, moans softly and then lets the bull slide from her back. Bogomolova et al. (1982) produced sonograms of that special moan emitted by cow during full intromission and ejaculation.

It is not known how many times a cow is serviced. Repeated demands by the cow to be bred are accommodated, after variable intermissions, but it is unknown whether all or some of them resulted in ejaculation.

Probably not longer than 18 hours after the first successful mounting, a cow loses interest in mutual caressing and separates from the bull.

Serial-mating Strategy of Taiga Moose

Merrill (1920) opposed the notion assumed by Knorre (1959) that taiga moose are monogamous. Nonetheless, the term *monogamy* is ambiguous (Wickler and Steibt 1983). In my view, the breeding "strategy" of taiga moose is better described as serial "pair mating" inasmuch as, while one cow exploits the time and sexual energy of a bull until she is bred or sated, other cows are prevented from competing for the bull's attention (Bubenik 1985). Such a strategy prevails unless the population's sex ratio is skewed. A cow with a breeding area may determine when, where and how long a bull will remain with her, as well as which bull she will mate. Her individualism likely explains why her estrus is spontaneous and has an individual timetable.

Behavior of the Rutting Taiga Bull

Taiga bulls search for cows in heat by traveling several miles and, by hiccups and thrashing, signaling his sexual arousal (Figure 96). He also paws shallow pitholes in which he urinates, but they are randomly located and seldom found on the same spot in successive years. Sometimes, a submature bull will accompany a prime bull, but at a distance that prevents provoking the dominant animal. In my experience, the taiga bull antlers are not as well impregnated by urinary pheromones as in the tundra moose, probably because the travel and sexual urgency prohibit expenditure of time at muddy pitholes.

For all cows to be bred during the short, 3-week rut, the serial-mating strategy requires a relatively high number of

Figure 94. A cow moose in a mating area will stimulate a prospective mate to urinate frequently. She also will wallow in the bull's pitholes and rub his mouth and face, push her head under his chin and attempt to drink his urine to enhance her pheromonal scent. *Illustration by A.B. Bubenik; from Bubenik (1987).*

Figure 95. At the outset of courting, a cow moose will maintain a distance from the attentive bull, but may display her sexual inclination by lifting her tail and looking back at the bull. *Illustration by A.B. Bubenik.*

prime bulls. This may not be found in heavily hunted populations or those for which the harvest strategy targets prime bulls in particular.

Communal or Harem-mating Strategy of Tundra Moose

Serial mating of the taiga moose, with the 3-week rut, is unsuitable for climatic conditions of the periglacial tundra. Here, the rut and calving period must be short to ensure that calves are born early enough to capitalize on the short summer vegetative period. Therefore, the tactics of serial mating are replaced by communal or harem mating in tundra moose, but the individualistic behavior of tundra cows remains, because they too choose particular bulls. Whether a particular bull "suits" each of the cows or whether the "mate copying" is involved (Dugatkin 1992) is unknown. Nonetheless, each cow in a harem remains individualistic and antagonistic to the others. The bull has communal domain over the cows.

Mating Areas and Breeding Behavior of the Tundra Bull

Toward the end of August, a prime tundra bull settles in a mating area of about 3.9 square miles (10 km^2). On the basis of my observations at Denali National Park, Alaska, these areas are traditional and likely secured by contest.

In early September, the bull begins to scent-urinate on trails and in pitholes in a low-squat position. Some of the pitholes are almost as large as a bathtub, 1.6 by 4.3 feet (0.5 X 1.3 m) long and 8 to 10 inches (20–25 cm) deep. Early in the rut, they are shallow and overgrown—indication of annually repeated use.

Pit digging often is interrupted when a bull assumes a low-squat urination posture, which can occur several times before actual urinating. In the intervals between urination, the bull resumes digging or pawing (see Miquelle 1991), and sometimes checks the pit odors (Figure 97). (A complete behavioral sequence of scent-urination is shown in Frund and Bubenik [1988].)

About 62 percent of the time after they urinate in pitholes, bulls subsequently cover themselves with urine by splashing and pawing. About 46 percent of the time, they wallow in the pits (Miquelle 1991). The intensity of the urine scent correlates with the age and timing of testicular activity of the bull (Bubenik et al. 1978b). Prime bulls produce and maintain the cresol-scented urine from the beginning to the end of the rut, in contrast to cows, who have cresols in urine only during estrus (Bubenik et al. 1978b).

Submature bulls are more likely than other prime bulls to respond to a pit made by another bull. Only 23 percent of bulls approached a foreign pit, of which 53 percent wallowed and saturated themselves with the foreign urine. Wallowing in another bull's pithole is done almost exclusively by submature bulls. Prime bulls aggressively at-

Figure 96. A taiga moose bull may aggressively follow and court a cow who may not yet be in the receptive phase of estrus, therefore will continue to be evasive. *Illustration by A.B. Bubenik; courtesy of R.B. Addison.*

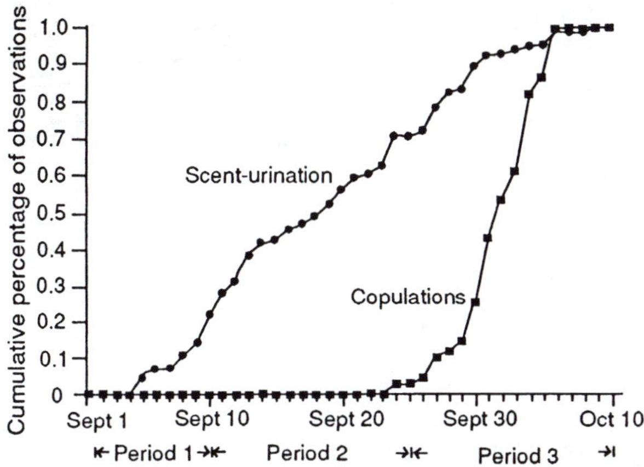

Figure 97. Frequency of observed scent-urination and copulation by bull moose in Denali National Park and Preserve, Alaska, 1981–1983 and 1986 (from Miquelle 1991).

tempt to prevent submatures from using their pits (Miquelle 1991).

A submature bull approaching a harem often is confronted and intimidated by a low-antler threat from the cow most advanced in the estrous cycle. Yearling bulls are tolerated as long as they do not try to contact a cow. Satellite bulls generally remain on the perimeter of a harem, sparring, sometimes with the yearlings. I did not observe submature bulls trying to wallow in a pithole of a prime bull when he was with the harem. Satellite bulls seldom stay long, and soon walk away.

If a prime bull enforces a contest, it is attentively observed by all cows. After such a contest is concluded and if both bulls recuperate in the vicinity of the cows, it is the alpha cow who cautiously approaches both victor and loser and takes scent from their antlers. She then decides with whom she and obviously also the other cow likely will mate.

Two prime bull moose may share a harem when it contains eight or more cows. Harems with 15 cows may have two or three prime bulls of equal rank, each accompanied by three to five cows in synchronous estrus. It is evident that when two or three cows in heat try simultaneously to "seduce" a bull, he cannot spend as much time with one as frequently as taiga bulls do with a taiga cow. Thereby, a tundra bull distributes his semen reserves more economically than is the case with serial mating. During the 8 to 10 days of breeding in a harem, a tundra bull probably can fertilize as many or more cows than the taiga bull does during the 3-week rut.

This led me to speculate that the presence of many cow moose stimulates a bull's testicular activity more than is experienced by a serial-mating bull, as has been observed in sheep (Iglesias et al. 1991, Price et al. 1991) and cattle (Amann et al. 1974). If so, semen production might be higher in tundra bulls than taiga bulls, which may explain why the harem-mating strategy is an adaptation to the Arctic climate.

Breeding Behavior of the Tundra Cow

A cow in a mating area continually attempts to intimidate other cows, by head, ear, eye and lip gestures, teeth gnashing, snorting and foreleg drumming. She stimulates a courting bull to urinate frequently and will wallow eagerly in his pitholes. She will caress his mouth and face, and push her head underneath his chin. An impressive behavior during the rut is the attraction of cows to a bull lying in his pithole (Figure 98).

Figure 98. A group of tundra cow moose in estrus may communally court a bull during the peak of the rut by circling him as he sits in his pithole. The bull's behavior increases his pheromonal aura. *Illustration by A.B. Bubenik; from Bubenik (1987).*

Opposite The creation of pitholes by rutting bull moose is fairly ritualized behavior. The digging of and urinating, stamping and wallowing in these depressions serve to advertise, both orally and olfactorily, a bull's location and mating inclination. The behavior, in essence, is sexual advertising. Although such a pit is being dug by pawing (top left), the bull will periodically pause to urinate in the depression from a low-squat position (left center). Rutting bulls salivate extensively, so salivary pheromones are added to urinary pheromones in the pithole, creating a pungent scent aura that attracts estrous cows (left bottom). By subsequent stamping, pawing and dipping (top right), the bull saturates his antlers, bell and underbelly, increasing his pheromonal "presence." A cow attracted by scent and sounds usually approaches on the run and attempts to drink the bull's urine, causing him to urinate more (right center). Androstenones in the saliva and urine may indicate a male's potency and influence a cow's selection of him as her mate. About half the time, bulls will wallow in their pits (left bottom), although this is more pronounced and frequent with tundra bulls than taiga bulls. *Photos by Joe Van Wormer; courtesy of the Denver Museum of Natural History.*

Indirect Indicators of Short or Long Rut

The rutting period of moose should be as short as possible. Under ideal conditions, with a high proportion of prime bulls in a population, submature bulls will spend more time sparring among themselves than in courting cows, which otherwise would reduce or delay the cow's sexual arousal and keep them away from mating mature bulls. Skewed social class ratio between mature and submature bulls and between cows and bulls may affect the fitness of the offspring as is known in red deer (Pemberton et al. 1988, Albon et al. 1992). Indicators of such instability include heightened sexual activity with worn hooves (Bubenik et al. 1978b) and overt rutting behavior of submature bulls early in the rutting period (Bubenik et al. 1979, Bubenik and Timmermann 1982). Another indicator is a prolonged rut with recurrent estrous periods.

Prolonged rutting periods delay antler cast in prime bulls. In a well-organized population, antler casting of mature bulls begins in late November and is completed during December (see Chapter 2). An aerial census performed during December should detect a synchronous antler cast within 48 hours (A.B. Bubenik 1968a, 1990, Topinski 1975, Malende 1988). The antler cast of submatures, which frequently is asynchronous, occurs later. Younger bulls retain their antlers later, even through March. Cast antlers of bulls experiencing a short rut and, thus, in good fitness, should have a convex antler seal; the seals of bulls close to the end of prime age should have a flat seal, and the seniors a concave seal (Bubenik et al. 1978a; see Chapter 2). These indicators may not be detected until the calving period is prolonged or delayed.

Aberrant Sexual Behavior of Bull Moose

On September 22, 1986, the Canadian *Globe and Mail* newspaper reported that a young bull moose in Newfoundland attempted to mount a dairy cow. Such reports actually are not uncommon, and are similar to those reported from red deer stags, which has led to the speculation that red deer and cattle hybridize (Adamatz 1928). Such behavior may be related to urinary pheromone production. The urinary pheromones and vomerones of a dairy cow may contain similar or identical substances as those in the red deer hind or the cow moose when in heat.

Bull moose that court dairy cows invariably are scared off or dispatched. The novelty of their attention to livestock

Prime bulls in well-organized populations begin to cast their antlers in late November, and casting generally is completed during December. Antler cast in prime bulls usually is synchronous, within 48 hours. The antler cast of submatures, which is frequently asynchronous, occurs later. Younger bulls retain their antlers the longest, even through March. These two submature bulls in Idaho have retained their antlers well into the new year, and the one on the right has an asynchronous cast—only one antler remains. *Photo by Lloyd Oldenberg.*

Bull moose with "newfound harem." Wherever dairy operations occur in or near moose range, there is possibility of a bull moose being attracted to a herd animal or the herd itself. The bull likely will attempt to stay with the cows for some time. The attraction probably is attributable to the similarity of urinary pheromones and vomerones of dairy cows and cow moose. The bull moose above settled in with a herd of Holstein dairy cows on a farm near Morristown, Vermont. He gained national notoriety in 1984 as the "lovesick moose." Initially, this moose drove off an angus bull from another herd. The Vermont Fish and Game Department translocated the bull moose to a 10,000-acre (4,048-ha) wildlife management area. The bull left the area and roamed 5 miles (13 km) to the Morristown farm. The dairyman was chased by the bull and consequently spent a few hours in a tree. He was understandably upset and called the Vermont Department of Fish and Game. Conservation officers were sent to the farm, but unable to haze the moose away, finally had to shoot it. The episode is a classic example of conflict with native wildlife, behaving like wildlife, in areas where humans have settled. *Photo by Jim Hall; courtesy of the Vermont Department of Fish and Game.*

is short-lived, mostly because it is disruptive to the dairy operation and potentially dangerous to the cow and dairy farmer.

Maternal and Calf Behavior

Maternal Behavior

The literature is notably limited on moose calving and maternal care, perhaps because of the reclusive nature of cows with calves. Nevertheless, documentations by Stringham (1974), Altmann (1958, 1963a), Knorre (1949, 1959) and Bogomolova et al. (1992) are suggested for review.

CALVING GROUNDS

There is no known general, important relationship between mating areas and calving sites (grounds) except that they are not the same locations.

Cows move to calving sites several days or sometimes weeks before calving.

Altmann (1958, 1963a) noted three characteristics typical of moose calving sites in Wyoming were secluded shelter, available browse and close proximity to water.

Addison et al. (1990) studied calving grounds in Ontario's Algonquin Provincial Park, a landscape rich with lakes and small islands, and similar to that of Quetico Provincial Parkland, also in Ontario, where I made observations. These parks seem to be representative of most calving grounds on the Canadian shield. Addison et al. found that, in these parks, moose cows select widely diverse calving sites (Table 12) and such selection may depend on the age or social status of cows. Islands frequently are chosen. The islands apparently are relatively safe from predators, because killing of calves on the calving ground is uncommon. Site cover and visibility appear to be unimportant. Because of her

Table 12. Habitat distribution of calving sites of moose in Algonquin Park, Ontario (from Addison et al. 1990)

	Number of beds		
Habitat	Islands	Mainland	Total
Mature hardwood	17	2	19
Mature mixed (conifer/hardwood)	5		5
High-density conifer	5	3	8
Low-density conifer	6	1	7
Mature hardwood/conifer edge	5	2	7
Early successional hardwood, mixed or edge	3	1	4
Open or open conifer edge	2	1	3
Other	1		1
Total	44	10	54

height and sight plane, a cow can monitor effectively a calving ground surroundings for disruption or danger.

Trespassing of the calving space by other moose is prevented by threatening gestures and vocalizations. A.W. Franzmann (personal communication: 1995) reported a 2.5-acre (1-ha) calving site near his home at Soldotna, Alaska. It was used six consecutive years and presumedly by the same cow, who arrived at the calving ground from 2 to 7 days before giving birth. She was accompanied to the area by her young of the previous year, which were prevented by the cow's intimidations from approaching her closer than about 50 yards (46 m) once she was on the site.

Given a moose cow's innate sense of physical security, little attempt if any is made to hide neonates. However, as soon as the cow and calf begin to move, the probability of calves being killed increases, because the offspring is left bedded quite frequently far from the foraging mother.

PARTURITION

I observed two moose parturitions at the Langenberg Zoo in Switzerland. One involved a primiparous (first birth) cow giving birth to a single; the other was of a 4-year-old birthing twins. In both cases, parturition was short, the calves were dropped within 15 to 20 minutes of the onset of labor. The twins were born at 30-minute intervals. Bogomolova et al. (1992) gave an average parturition time of 31 minutes, during which the mean heart rate of the cow increased from 68–85 to 190–230 beats per minute. Similar information was reported by Cederlund (1987).

The primiparous cow was devoted to cleaning and drying the calf. The pluriparous dam cleaned and dried her twins in intervals, between which she swallowed the cotyledons, pulled some amnion membranes out of the vagina, cleaned the ground of amniotic fluids and ate the wetted soil. Bogomolva et al. (1992) reported that, in primiparous cows, not all elements of maternal care are well developed, such as: removal of amniotic fluids; increased reaction to newborns; bleats or calls; increased salivation; licking and grooming the calf; sucking out urine, feces and cleaning the anus; and calf defense. Lack of these characteristics is considered a sign of immaturity and may lead to disinterest in the calf, or even to physical abuse and abandonment.

Within the confines of the Langenberg Zoo, both cows I observed stayed close to the parturition site for about 3 hours.

Moose cows begin cleaning and drying their calves almost immediately after giving birth. The amniotic membranes with cotyledons generally are consumed (left) and the ground is cleaned of amniotic fluids. This may help eliminate scent that could otherwise attract predators or insects. Also, the licking of a calf almost certainly initiates essential bonding and it serves as stimulus to "activate" within minutes the calf's struggle to rise and seek nourishment (right). Getting up and walking may take a calf several hours. *Left photo by Victor Van Ballenberghe; courtesy of the USDA Forest Service. Right photo by Albert W. Franzmann, courtesy of the Alaska Department of Fish and Game, Soldotna.*

From the moment her first calf was born, only the pluripara prevented us from approaching and touching her calves. The primipara retreated and let us weigh her half-dried calf. At the Moose Research Center in Alaska, tame cows with newborn calves allowed human approaches, whereas wild primiparous and pluriparous cows did not (C. Schwartz personnal communication: 1992).

Calf Behavior

Cederlund (1987) detailed the initiation of certain behaviors by captive, hand-reared moose calves (Table 13).

The partially dried newborn moose, with blue pupils (coloration disappears in 3 to 4 days), searches for the udder, and both calf and dam communicate with seeking calls or weak moans. A calf instinctively searches for the udder along the dam's underside as early as about 1.5 hours after birth (Stringham 1974). A calf may have to be pushed in the right direction by the dam's nose.

Cows remain within 55 yards (50 m) of her calf or calves for 5 to 7 days postpartum (Bogomolova et al. 1992). This coincides with my observations and those of Cederlund (1987) and Stringham (1974), which indicate that, in the first days of her calf's life, a moose cow remains within visual or vocal distance of her newborn. A.W. Franzmann (personal communication: 1995) reported that cows with newborns

Table 13. First-observed behaviors of captive and hand-reared moose calves (from Cederlund 1987)

		Calf sex/age in days							
Behavioral categories	Behavioral elements	M/2	F/2	M/3	M/3	F/10	F/5	M/2[a]	F/1[b]
Comfort	Yawning	6	4	7					
	Muzzle licking	2	2	3	3	10	5	2	
	Body care, lying with licking or head rubbing	6		7	7				
	Body care, lying scratching with hind hoof	7		7	19				7
	Body care, standing with licking or head rubbing			5					
	Body care, standing, scratching with hind hoof	13		7	19			60	
	Stretching of body	16	14	11	8	11			
	Shaking of body	8	6	3	3				
Foraging	Nibbling	5	3	3	3			3	
	Lateral jaw movements	5	3	4	4			2	21
	Eating solid fodder	11	8	5	5	10	7	12	
	Soil ingestion	8	6	4	4			2	7
	Cortex biting			8	7	11			
	Ruminating								
Digging movement	Treading								
	Pawing	8		27	4	38	28		
	Freezing			4	4	11			
	Running away, escape			5				4	
	Nondirected striking with front legs[c]							7	
	Directed striking with front legs, to object/individual[c]	6/		12/7	/10	/10	28/28		
	Nondirected striking with hind legsd								
	Directed striking with hind legs, to object/individuald[c]			/11		/10			
	Nondirected butting[c]								
	Directed butting, to object/individual[c]	/67		11/9	/6	/88			
	Chasing[c]					10			
Sexual	Head rubbing against object/individual[c]	16/					28/	/30	
	Head on other's back[c]	32		12					
	Nasogenital sniffing	16		12	11				
	Lip curling (flehmen)	77		11			62		
	Mounting[c]	32		7			57	30	
Locomotive	Rushing[c]	5	14	4	9	38	28	7	
	Gambolling[c]	8	8	4	4			7	
Alertness	Attention posture					11			
	Scenting	6	4	5	5				

[a] From Markgren (1966).

[b] From Stringham (1971, 1974).

[c] Play-related behavior.

on a calving ground near his home in Alaska remained on the site for about 3 weeks postcalving.

When the calf leaves the calving ground, it displays a "lying-out" type behavior (see Walther 1966). However, I believe the calf from birth on lacks the "freezing" behavior typical of precocious young of other wild ungulates when scared or surprised. Therefore, a moose calf on the calving ground is not a "hider." The lying out merely is the resting posture of a young animal yet to gain coordination of its limbs. Altmannn (1958) and Stringham (1974) emphasized that the moose calf is not well-coordinated until at least 4 days old. Lack of coordination, however, does not mean that moose neonates cannot move quickly and capably. Also, Peterson (1955) reported seeing a calf less a week old swimming behind its mother.

Before suckling, a moose calf may execute one or more head butts—a common signal in ungulates that may prepare the udder to release milk. Of nursing bouts by calves less than 2 days old, 93 percent were from a reclining position and the rest were from a squatting position (Stringham 1974). Calves 2 to 8 days of age nursed while standing 72 percent of the time, while squatting 10 percent and while reclining 18 percent. Calves older than 8 days always nursed while standing. Until 8 days old, calves nursed from between the dam's hindlegs 17 percent of the time; thereafter, they suckled only from the flank. In the first 2 weeks of lactation, nursing bouts averaged 5 minutes 18 seconds duration. In the third week, the average was 8 minutes 20 seconds (Bubenik 1965b). Stringham (1974) emphasized that there was not a strict schedule of nursing. The duration and frequency of nursing bouts decreased rapidly with age, from approximately one per hour the first week to eight per day at week 9.

It is important for calf digestion that rumen microbes are transferred into the digestive tract of the calf by contact with the mother. In addition, a moose calf occasionally may perform coprophagy (seek and eat mother's or sibling's feces) (Cederlund 1987), as reported in other deer (Bubenik 1965b).

During the first few days after birthing, a dam may nurse the calf while lying, by raising her leg to expose the udder. Suckling from behind between the legs is less common than from the side, and twin calves may suckle together. There are no reports of suckling behavior in triplets, but Frund and Bubenik (1982) showed (via cinematography) how triplets compete for the nipples.

From time to time, the 3- to 4-week-old moose calf performs a playful mounting of the mother (Altmann 1963a). I have observed moose siblings mount each other (Bubenik 1965b).

Weaning

Weaning usually occurs in mid-September (Dodds 1955, Denniston 1956, Knorre 1961). Suckling after this time may

After a cow and her offspring depart a calving ground, a week or more after the birthing, the cow will venture farther away than was previously the case to feed periodically. During these maternal "respites" from guarding and nourishing the calf (or calves), the young typically rest in a "hider" posture, as seen above. In fact, moose calves are not the classic hiders that certain other young ungulates (such as elk and pronghorn neonates), which presumably "freeze" in a pronate position to reduce the chance of detection by predators. Because frightened or surprised moose calves usually do not react by instantly dropping into lay-out positions, the hider posture in moose is believed to be merely how these calves rest before they gain full coordination of their limbs. *Photo by Joe Van Wormer; courtesy of the Denver Museum of Natural History.*

In the first few days after birthing, a cow moose may nurse the calf while lying down. *Photo by Charles C. Schwartz; courtesy of the Alaska Department of Fish and Game.*

The most traditional position for moose calf nursing is from the side of the cow. Twins often simultaneously nurse from the two teats of the producing mammary glands. *Photo by Charles C. Schwartz; courtesy of the Alaska Department of Fish and Game.*

Nursing from behind and between the rear legs of a cow moose is occasional and apparently at the convenience of the calf. *Photo by Albert W. Franzmann; courtesy of the Alaska Department of Fish and Game, Soldotna.*

There are many records about the differences in aggressive behavior of moose cows toward anything that penetrates their individual space. Usually, however, the individual space of a cow with a calf (or calves) includes that of the offspring as well as herself. Until a calf is too tall (usually by October), a calf will seek protection under its mother's belly. Thereafter it will stay ahead of the dam or at her side. When alarmed, some calves rush to their mothers; others flee. The younger her offspring, the more vigilant a cow must be to circumstances or sensory stimuli that would not seriously threaten or intimidate her in the absence of young. *Photo by Joe Van Wormer; courtesy of the Denver Museum of Natural History.*

occur (see Altmann 1958), but it is discouraged by the dam. The growing calf gets increasingly aggressive (butting and biting) and its nursing appears to become painful to the mother, which may account in part for the weaning (Lent 1973). Cows discourage suckling by agonistic behavior including threats, kicks and moans.

Development and Duration of Cow/Calf Bond

A firm "mother/newborn" bond requires constant contact for the first 7 to 8 days (Bogomolova et al. 1992). During that time, a calf may follow any moving object or person and elicit contact by emitting faint seeking calls. Mother and calf or calves spend the winter in secluded Umwelts to minimize surprise by wolves (Denniston 1956, Altmann 1958, 1959, 1961a, LeResche 1966, Novak 1978). The cow/calf bond remains intact until the following spring when the mother drives her calf (by then a yearling) away in preparation for the former's next birthing. If the cow is pregnant again, the yearling prefers to share at least the mother's home range, but will not return if continually forced away (Bogomolova et al. 1992).

In conclusion, I must stress that there is a great deal yet to be learned about moose behavior, mainly as it concerns the behavioral physiology of both well-organized and disorganized populations. The chemistry that may be behind differences in breeding behavior of the taiga and tundra moose in general, and of moose of different maturation stages and individual statuses in particular, certainly needs further investigation.

(LIFE)
PROPORTIONS-
GENERAL
CONSTRUCTION
OF MOOSE'S
FACE-
RUMP - SHOULDER =
HEIGHT AT SHOULDER
(IN MOSS)
(LIFE -
COMP.
FROM
REF)
RUBBING ANTLERS
IN SPRUCE
(LIFE & MEMORY)
(MEMORY)
ITCH
AUG. 13, '56
IGLOO CREEK,
MCKINLEY PA..
SLEEPING (POSE FROM LIFE)

6

VICTOR VAN BALLENBERGHE AND
WARREN B. BALLARD

Population Dynamics

A population may be defined as a group of potentially interfertile organisms of the same species occupying a particular space at a particular time (Krebs 1978). Time and space boundaries of populations generally are loose and arbitrarily fixed in practice by investigators. Populations of animals consist of individuals of different ages and sexes, each with its own reproductive potential and life span. However, a population has group characteristics that are statistical measures inappropriate for individuals. These include population size, natality (birth rate), mortality (death rate), immigration (movement into the population) and emigration (movement out of the population).

Population analysis deals with numerical attributes of populations (e.g., numbers, sex and age ratios, and rates of increase) together with characteristics of the animals (e.g., age, sex and life span) and properties of the environment (e.g., forage supplies, weather and predator numbers) that determine these values.

Caughley (1977: 1) referred to several levels of detail that characterize population dynamics: "A simple approach is to treat individuals as if they were identical, to express the numbers in the population as an average over several years, and to investigate why the average has this value. One step removed from this elementary approach is the study of fluctuations in total numbers as related to changes in the environment. At a more advanced level the study might be aimed at expressing the rate of change in numbers as a difference between birth rate and death rate, the difference being related to environmental influences. More detailed again is the study which discards the simplifying but unrealistic assumption that the animals in the population are identical. Such an approach recognizes that the environment does not act directly on numbers as such but indirectly through its influence on fecundity at each age and survival over each interval of age. Environmental influences would be related directly to these attributes rather than to numbers or to change in numbers. At the final stage of this progression each individual is recognized as unique. The dynamics of the population are reported as the sum of the demographic reactions of each individual. That approach may seem ideal but with it generality is lost in a porridge of detail. Between the oversimplification of declaring all animals identical, and the self-defeating undersimplification of reducing a population to its fundamental particles, there lies a broad middle ground forming the ecologist's methodological domain."

Understanding the dynamics of a population requires knowledge of how many individuals it contains, how fast it is increasing or decreasing, its rate of production of young, and its rate of loss through mortality. Such properties are called *population parameters*. Population statistics are estimates of parameters. The choice of which parameters to use is difficult. Their usefulness was rated by Caughley (1977) as follows: (1) ease of estimation; (2) extent to which they collectively describe the significant characteristics of a population; (3) ability to extrapolate beyond the data from which they were calculated; (4) directness of their relationship to population processes; and (5) their generality—the extent to which they apply to all kinds of populations, not just to some specific populations. Parameters that are a compromise among these attributes include age-specific survival, age-specific fecundity, age distribution, sex ratio and numbers or density. Statistics including rate of birth, rate of death and rate of increase can be calculated from estimates of those parameters.

Caughley (1977: 2) emphasized that those interested in population dynamics must be ecologists above all else, and understand the components of the environment that influence the animals they study: "I strongly suspect that the deepest insight into a population comes from studying how age-specific survival and fecundity are influenced by the conditions in which the animals live."

Accordingly, in this chapter, we review current literature on population dynamics of moose from an ecological perspective and attempt to minimize lengthy tabulation of data from populations that serve mainly as snapshots in time. Although various aspects of moose population dynamics have been studied for decades, and much additional data on other members of the deer family are also now available, understanding of several elements of moose population dynamics is still incomplete. We concur with McCullough (1979: 84) that studying ungulate population dynamics is a complex endeavor: "The complexity of the population dynamics of a large herbivore in the natural world is overwhelming. There is the initial difficulty of determining population size, given the margin of error inherent in most wildlife census techniques. Even when the change is real, it is difficult to determine which one or set of interacting variables produced the change. There is random variation, density-dependent effects, seasonality, fluctuations in carrying capacity (including those produced by the herbivore), time lags in herbivore population and vegetation response, human hunting, nonhuman predation, parasites and diseases, and catastrophic events, to mention only the more obvious ones. This situation is not conducive to understanding and mainly accounts for the slow progress in the field."

Nevertheless, biologists understand many aspects of ungulate population dynamics and, in recent years, have become increasingly perceptive in unraveling the relationships between population parameters and the ecological variables that drive them.

Abundance

The number of animals in a population—one measure of abundance—is of limited usefulness if geographic population boundaries are vague or nonexistent (Caughley 1977: 12). Many moose populations have poorly identified boundaries. A more useful measure of abundance is the number of animals per unit of area, or absolute density. A third measure of abundance is relative density (i.e., the density of one population relative to another). Which measure of abundance is most useful depends on the ecological questions of interest. Caughley (1977) suggested that problems of sustained yield harvesting and studies relating behavior, reproduction, survival, emigration and immigration to numbers or density demand estimates of population size or absolute density. Problems linked to utilization of habitat, rate of increase, dispersal and reactions of a population to management treatments often can be solved with estimates of relative density. Questions about genetics, zoogeography, behavior and population management may require no estimates of abundance. "The majority of ecological problems can be tackled with the help of indices of density, absolute estimates of density being unnecessary luxuries" (Caughley 1977: 12).

Despite this, much effort in North America has been devoted to determining absolute densities of moose. Various methods, often involving aircraft, have been used to census moose populations. Few studies have compared different aerial techniques for censusing moose. Crête et al. (1986a) compared helicopter and fixed-wing counts of radio-collared moose to estimate sightability during early winter. Observers in fixed-wing aircraft saw 29 percent of the moose that were counted by helicopter. A model consisting of three linear regressions predicted helicopter counts using fixed-wing counts and percentage area covered by moose track networks. This model was cross-validated for seven surveys. Predicted helicopter counts differed at worst by 23 percent from observed mean counts. However, Crête et al. concluded that inadequate sample size and poor stratification produced imprecise density estimates for all surveys.

Several other methods have been used to estimate moose population size. These include pellet group counts, ground observations, hunter harvest reports, and seasonal trend counts. Rolley (1982), Peek (1982) and Wolfe (1982) discussed these and outlined specific methods used in several areas of North America (see Chapter 17).

Density

The density achieved by a specific moose population at a particular time is the result of complex interactions between those animals and their environment, as will be addressed in detail in the section on population regulation. Krebs (1978: 140–141) listed approximate densities of selected organisms in nature, ranging from diatoms to humans, and reported a range of figures spanning more than 12 orders of magnitude. Mammal species display a wide range of densities depending, in part, on body size and trophic level. Here, we review density estimates of moose obtained in various locations to demonstrate the range of values observed and to illustrate the approximate densities that typify this species. Moose densities generally are determined in winter when the animals are most visible.

Moose densities often are difficult to compare, because of problems of population scale and habitat continuity. Local densities—for example, those of moose concentrated in winter habitat—may be much larger than those of moose populations occupying broader areas. Similarly, densities calculated from extensive land areas not suitable for moose may be unusually low. Gasaway et al. (1992) used sites greater than 772 square miles (2,000 km^2) of generally continuous moose habitat to compare moose densities across a broad area of Alaska and the Yukon Territory. They noted that smaller sites exhibited high variability in prey and predator densities and in habitat quality. Posthunt (late autumn or early winter) moose densities were used to enhance comparability further. The mean density of 20 moose populations in Alaska and the Yukon Territory was 0.38 per square mile (0.148/km^2; range, 0.045–0.417/km^2) in areas where predation was thought to be a major limiting factor of moose. Densities of 16 other populations in the same area where predation was not limiting averaged 1.7 per square mile (0.66/km^2; range, 0.17–1.44/km^2). Ballard et al. (1991) provided 29 moose density estimates in Alaska, including some of the same populations studied by Gasaway et al. (1992), and they ranged from 0.13 to 3.2 per square mile (0.05–1.24/km^2).

From 2 years of censuses in northeastern Minnesota, Peek et al. (1976) reported moose densities of 1.6 to 2.0 per square mile (0.62–0.78/km^2; range, 0.18–1.05/km^2). This population had not been legally hunted for about 50 years and habitat did not appear to be limiting.

Elsewhere, Crête (1987) reported moose densities ranged from 0.8 to 5.2 per square mile (0.3–2.0/km^2) in six unhunted moose populations in eastern Canada. Page (1989) indicated the naturally regulated population at Isle Royale, Michigan, fluctuated between about 2.6 and 5.8 per square mile (1.0–2.25/km^2) from 1960 to 1986. Houston (1968) reported a winter density of about 0.13 per square mile (0.05/km^2) for Shira's moose in Wyoming. And Cederlund and Markgren (1987) reported a density of 6.5 per square mile (2.5/km^2) for a Swedish population with negligible predator influence.

This review of selected references suggests that moose populations exist at densities as low as 0.13 per square mile (0.05/km^2) and may reach 6.5 per square mile (2.5/km^2)—a 50-fold range of density. Many North American populations fluctuate between 1.3 and 2.6 moose per square mile (0.5–1.0/km^2). We emphasize that local densities may exceed 13 per square mile (5/km^2) and erupting populations may temporarily exceed 6.5 per square mile (2.5/km^2), but are unlikely to remain at high densities for extended periods (Page 1989).

Fecundity

Fecundity rate in a population is the number of live births produced per year by reproducing individuals. The number of female live births per adult female is a standard parameter of population dynamics that permits numerical analysis of the female portion of the population. The male segment may have different dynamics and require its own fecundity and mortality tables.

Ideally, fecundity rates of each female age class, from birth to the oldest observed age, can provide a fecundity table for a given population at a specific time. Data required include mean litter size and sex ratio at birth. Sample sizes

In some moose populations, more than half of the adult females produce twins, resulting in a high fecundity rate for a species of large body size. High twinning rates are indicative of a population occupying productive habitat year-round. Twinning rates may be difficult to determine in populations where there is a high degree of predation on young calves by brown and/or black bear and wolves. *Photo by Victor Van Ballenberghe; courtesy of the USDA Forest Service.*

for each age class should be sufficient (a minimum of 30) to estimate fecundity accurately. In practice, however, such data seldom are available for moose populations. Age classes of females frequently are combined, usually as calves, yearlings and adults. Observations of wild moose births are rare, therefore birth rates often are estimated by examination of the reproductive tracts of killed females, by examination of live females captured in ecological studies or by neonate surveys conducted within 2 or 3 weeks of mean birth dates. Similarly, sex ratio at birth generally is not determined directly, and must be estimated from other data or assumed to be even.

Production of moose calves is the end product of a complex chain of biological processes including estrous cycles, rutting behavior, mating, fertilization, gestation, prepartum events and parturition (Boer 1992a; see Chapter 4). Fecundity is related to sexual maturation and a broad array of ecological factors (e.g., forage supplies, forage quality and weather) that affect the physiological status of females. These factors influence ovulation, pregnancy rates, litter size and fetal sex ratios. Ultimately, fecundity and subsequent survival of young determine recruitment rates and population trends, which are important factors in moose population dynamics.

Because birth rates generally cannot be estimated directly, various studies throughout the world have examined female moose reproductive tracts to develop indices of fecundity. Ovulation rates (ova shed per female), pregnancy rates (percentage of gravid uteri for various age classes), embryo rates (number of embryos per female) and twinning rates (percentage of twin embryos in total gravid uteri) are examples of indices of fecundity. Verme (1974) cautioned that ovarian structures are difficult to "read" precisely, and hunting seasons (when most female tracts are collected) often coincide with breeding. Therefore, fecundity indices obtained early in the reproductive cycle are likely to underestimate true birth rates.

Reproductive tract studies have shown that female moose 4 to 5 months old do not mate, therefore do not produce calves at 1 year of age. Females 16 to 17 months old may or may not mate depending on body weight (Saether 1987) and produce calves at age 2 years. Most moose calves are produced by females 3 years of age or older. Female moose can continue to produce calves to the end of their life span, sometimes exceeding 18 years.

Litter size in moose ranges from one to three (Peterson 1955), but litters exceeding two are extremely rare (Coady 1982). Twin births are common in some moose populations. Pimlott (1959) reported only one occurrence of twins in 32 uteri of pregnant yearlings in Newfoundland and reviewed data from Sweden suggesting that twins were rare in yearlings there as well. Boer (1992a) reported that twinning percentage in 12 moose populations across North America ranged from 4 to 80 percent. Gasaway et al. (1992) similarly reviewed 21 studies where twins accounted for 0 to 90 percent of observed births or in utero counts. It is well documented that nutritional factors are related to the ability of cow moose to produce twins (Franzmann and Schwartz 1985). Wide variation in nutritional condition accounts for observed differences in twinning rates in space and time (Ballard et al. 1991; see Chapter 4).

Fecundity rates (total live births per female) have been estimated for moose populations by combining data on pregnancy rates and twinning rates. Schladweiler and Stevens (1973) estimated 1.0 calf per female for a Montana population, with pregnancy and twinning rates for yearlings and adults of 32 and 0 percent and 86 and 16 percent, respectively. Mytton and Keith (1981) computed a rate of 1.31 calves per female for a moose population in Alberta. Rolley and Keith (1980) actually observed 1.06 calves per female in winter in Alberta—the highest such value in the moose literature. Albright and Keith (1987) estimated a remarkably low fecundity rate of 0.4 calf per female for a moose population in Newfoundland where females 4 years old or less apparently failed to reproduce and the twinning rate was less than 1 percent. Gasaway et al. (1983) calculated a rate of 1.11 calves per female for a population in central Alaska, assuming a 32 percent twinning rate. Gasaway et al. (1992) provided a rate of 1.38 calves per female for another Alaskan population, in which adult pregnancy and twinning rates were 100 and 52 percent, respectively.

Albright and Keith (1987) indicated that, commonly, 53 to 58 percent of moose born are males. Boer (1992) summarized six studies in Canada and Alaska where fetuses and neonates varied from 39.2 to 62.0 percent male. Overall, 50.8 percent of 1,350 such animals were males. In addition, Reuterwall (1981) reported the weighted average of very large samples of calves shot in winter in Sweden was 57.4 percent male. It is apparent that the sex ratio at birth may be skewed toward males, but factors accounting for this are complex and poorly understood (Boer 1992; see also Chapter 4). A broad, general theory linking progeny sex ratios, maternal condition and reproductive success was presented by Trivers and Willard (1973). They predicted that females should invest heavily in progeny that have higher reproductive success when success varies widely between sexes and is influenced by parental investment. For polygynous large mammals, variation in reproductive success usually is greater among males than females. Trivers and Willard (1973) proposed that females in below average condition should produce disproportionate numbers of female offspring. This appears to be the pattern for red deer (Clutton-

Brock et al. 1984) and American bison (Rutberg 1986), but not for white-tailed deer (Verme 1983) and wild reindeer (Skogland 1986). Caley and Nudds (1987) proposed that local resource competition and male dispersal explained production of male-biased sex ratios by nutritionally stressed females in the genus *Odocoileus*. Competition, therefore, should be distinguished from sex ratio adjustment mediated by variations in fitness between the sexes.

Mortality

Moose populations increase by the addition of calves born each year and decrease by the loss of animals. Death can occur from the moment of birth. A partial list of mortality factors includes predation, starvation, hunting, accident, drowning, vehicle collision, parasitism and disease. Mortality can be divided into two categories—deaths caused by humans or human activity (most often hunting) and those attributable to natural factors. McCullough (1984) referred to three categories of mortality including prerecruitment, chronic and traumatic. Chronic mortality was defined as routine, inevitable death due to the effects of old age, parasites, nonvirulent diseases, malnutrition and debilitating accidents. Understanding the causes and rates of mortality of all kinds, as well as understanding the causes and rates of its complement, survival, in a moose population is an essential component of interpreting that population's dynamics.

Ungulate biologists have used various methods to measure mortality. Hayne (1984) summarized several including: (1) close and direct monitoring of a single cohort (sex and age group) of animals; (2) recovering a series of marked animals; (3) ascertaining the age at death for a reliable sample of each cohort in a population; and (4) determining the age distribution of a population and making assumptions about its history. Each method is fraught with problems and difficulties. Ideally, age-specific death rates for each cohort can provide a mortality table for a given population at a specific time. In practice, however, such data seldom are available for moose populations. As with fecundity, age classes frequently are combined, often as calves and adults, and direct observations of dying moose are rare. Mortality causes must be inferred from field evidence, often sketchy, and rates of mortality for cohorts often must rely on small sample sizes, especially for older animals.

During prolonged winters of deep snow adult male moose suffer higher mortality rates than do females because the former have lower fat reserves as a result of rutting activities. Calves of the year are even more vulnerable, due to their lack of fat reserves because growth is a higher physiological priority. Severe winters in parts of Alaska have resulted in nearly 100 percent calf mortality. *Photo by Victor Van Ballenberghe; courtesy of the USDA Forest Service.*

Adult Mortality

If suitable data are available (e.g., an adequate, relatively unbiased sample of each age class), the mortality pattern of a population may be summarized in the form of a life table showing age-specific mortality rates reducing one large, hypothetical cohort. Life tables may take several forms. Caughley (1977) described one form with six columns: (1) age at 1-year intervals; (2) of the original total born, the number of animals surviving at each year; (3) the proportion of the cohort still surviving at a given age, a l_x series; (4) the probability of death of each age interval, d_x; (5) the mortality rate, q_x, calculated as d_x/l_x; and (6) the survival rate, p_x, or the complement of each respective q_x value. The l_x, d_x, q_x and p_x columns present the same information in four different ways and are algebraically related. Caughley (1977) characterized q_x as least affected by sampling bias, giving the most direct projection of the mortality pattern, and being the best schedule for comparisons within and between species.

Wolfe (1977) and Peterson (1977) constructed life tables for moose by using data from carcass remains (including many wolf-killed moose) from Isle Royale, Michigan. Remains were collected over a 17-year period and totaled 259 males and 216 females. It was assumed that the Isle Royale moose population size was stationary and the age distribution remained stable over the time interval of the collections. Both assumptions are essential to life table analyses. Sampling must adequately represent mortality during the entire year, and accurate age determination is assumed. Small sample sizes for certain age classes are a problem for long-lived animals. Unusually high survival or mortality in specific cohorts may produce nonrepresentative survival rate estimates for the population as a whole (Peterson 1977).

The Isle Royale data from an unhunted population resulted in separate life tables for males and females (Table 14). Female and male moose that survived their first year had mean life expectancies of 7.8 and 7.0 years, respectively. Bulls died by age 15.5; females lived no longer than 19.5 years. Between the ages of 1 and 6, age-specific annual mortality was 11 percent or less, but rose rapidly thereafter. Moose older than 10 had age-specific annual mortality rates exceeding 20 percent. Peterson (1977) combined an independent first-year mortality estimate of 72 percent with age-specific mortality estimates of adults that produced a U-shaped mortality curve of moose typical of most large mammal species (Caughley 1966; Figure 99). Average annual male mortality for all cohorts was 13.3 percent, that for females was 12.0 percent (Peterson 1977).

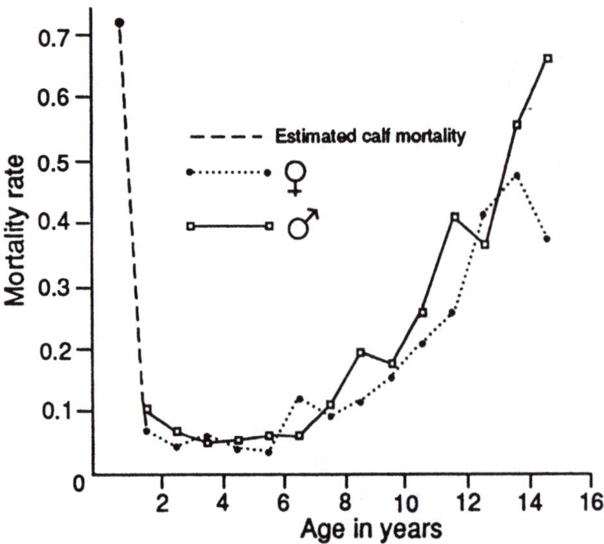

Figure 99. Age-specific mortality curve for moose of Isle Royale, Michigan. This is a typical U-shaped curve for large mammals with high mortality during the first year of life and during old age, and low mortality for young adults (from Peterson 1977).

Boer (1988) also constructed a life table for moose. His was based on hunter-killed moose in New Brunswick. Because certain animals are known to be more vulnerable to hunters, the New Brunswick data were adjusted with calculated vulnerability factors for young males and females.

Table 14. Life table for Isle Royale, Michigan, moose (from Peterson 1977)

	Males					Females				
Age (years)	Number of remains	d_x	l_x	q_x	L_x	Number of remains	d_x	l_x	q_x	L_x
1–2	28	108	1000	0.108	946	16	74	1000	0.074	963
2–3	16	62	892	0.070	861	9	42	926	0.045	905
3–4	12	46	830	0.055	807	12	56	884	0.063	856
4–5	12	46	784	0.059	761	8	37	828	0.045	810
5–6	13	50	738	0.068	713	7	32	791	0.040	775
6–7	12	46	688	0.067	665	21	97	759	0.128	711
7–8	20	77	642	0.120	604	15	69	662	0.104	628
8–9	30	116	565	0.205	507	15	69	593	0.116	559
9–10	22	85	449	0.189	407	19	88	524	0.168	480
10–11	26	100	364	0.275	314	21	97	436	0.222	388
11–12	29	112	264	0.424	208	20	93	339	0.274	293
12–13	15	58	152	0.382	123	23	106	246	0.431	193
13–14	14	54	94	0.574	67	15	69	140	0.493	106
14–15	7	27	40	0.675	27	6	28	71	0.394	57
15–16	3	12	13		7	3	14	43		36
16–17						2	9	29		25
17–18						2	9	20		16
18–19								11		11
19–20						2	9	11		6
Total	259					216				

Boer's life table illustrated the effect of hunting and other mortality factors on lowering adult survival. It reported weighted average annual mortality rates were 0.47 and 0.25 for males and females, respectively.

None of these life tables provided information on moose calf mortality, primarily because calf remains were seldom recovered.

Although life tables are a potentially powerful tool for analyzing age-specific mortality, the data needed to use them may be unavailable for many moose populations, or the restrictive assumptions required may not be met. McCullough (1979: 76–83) discussed several difficulties of applying life table analysis to an enclosed white-tailed deer population where virtually all dead animals were recovered. He worked with 19 years of data, but found that small sample sizes and shifting age distributions reduced the value of life table analyses, even in a small, controlled environment.

Survival rates can be estimated directly from radio-collared moose by a variety of methods (see White 1983, Heisey and Fuller 1985, Pollock et al. 1989). The Kaplan-Meier (1958) procedure, for example, accounts for radio-collared animals lost because of radio failure, collar loss or emigration, and it allows new animals to be added after the study has begun (Pollock et al. 1989); both are important considerations in most moose survival studies. Survival rates and variances can be estimated seasonally or annually. Assumptions include: (1) sex and age classes are randomly sampled; (2) survival is independent for different animals; (3) radio-collaring does not influence survival; (4) censoring (dropping individuals due to collar loss, etc.) mechanisms are unrelated to an animal's fate; and (5) newly tagged animals have the same survival rates per unit of time as do previously tagged animals. The procedure is simple, flexible and allows staggered entry of tagged animals (Pollock et al. 1989). Precision varies with sample size and mortality rate; a high mortality rate allows for lower sample size (Schwartz and Franzmann 1991). The Kaplan-Meier procedure provides unbiased survival estimates in contrast to other methods that do not allow for staggered entry of individuals or assume constant mortality rates and causes over the entire study period.

Ballard et al. (1991) provided data on mean annual survival of adult female moose in Alaska radio-tracked during a 10-year period. From 25 to 80 moose per year were followed; hunting of cows was prohibited. Annual survival averaged 94.8 percent in this population. Data for yearling females spanned 4 years, with 2 to 22 individuals per year wearing radio-collars, and annual survival averaged 95.1 percent. Annual survival of adult and yearling males averaged 75.4 and 90.9 percent, respectively, with hunting the major mortality factor.

Radio-collaring moose allows biologists to determine the fate of individual animals over long periods of time. From such records, mortality rates can be determined by a variety of methods. *Photo by Victor Van Ballenberghe; courtesy of the USDA Forest Service.*

Bangs et al. (1989) followed 51 radio-collared females in Alaska for a 6-year period and reported a 92 percent mean annual survival. Although annual survival among age groups declined from 97 percent (ages 1–5) to 84 percent (ages 16–21) these differences were not significant, perhaps attributable to small sample sizes. Hunting was not a significant mortality factor.

In the southwestern Yukon, Larsen et al. (1989a) radio-tracked 33 to 41 adult females for 3 years and reported 91 percent mean annual survival. Again, hunting was not a significant mortality factor.

Various other studies using radio-collared moose have reported mean annual survival of adults ranging from 75 to 94 percent, depending on the extent of human hunting (Hauge and Keith 1981, Mytton and Keith 1981, Gasaway et al. 1983, 1992). Gasaway et al. (1983) reported annual survival rates during a 3-year period of 100, 93 and 79 percent for radio-tracked adults aged 1 to 5, 6 to 10 and more than 10 years, respectively, for a population in Alaska in which mortality due to wolf predation was experimentally reduced.

Juvenile Mortality

Corresponding with times when moose populations are surveyed, calf moose mortality may be divided into two periods—birth to 5 to 6 months, and from about 6 months to 1 year. These periods also correspond with times when calves are particularly vulnerable to different kinds of mortality including bear and wolf predation in the first period, and hunting and wolf predation in the second period.

Various indirect methods were used before the advent of radio-telemetry to estimate calf mortality during the first 6 months of life. These mainly relied on comparing calf/cow ratios at birth (assuming various natality rates) with such ratios in autumn, when ground or aerial surveys could sample large numbers of animals. Unknown mortality rates of adult females during summer made ratio comparisons difficult.

Gasaway and Dubois (1987) cautioned that mortality rates estimated from calf/cow ratios are subject to large errors, because biased aerial survey techniques often underestimate the occurrence of calves during both summer and winter. By following the fate of calves born to radio-collared females, or by collaring the neonates themselves, telemetry studies allow better estimates of neonatal mortality. Overwinter calf mortality can be estimated by comparing aerial survey data during early and late winter, by following calves of radio-collared cows (Van Ballenberghe 1979, Hauge and Keith 1981, Gasaway et al. 1983) or by following collared calves.

Neonatal mortality of moose populations varies greatly, depending on several factors, most notably the extent of predation. Several studies of radio-collared neonates have documented that predators may account for up to 79 percent of neonate deaths, and that survival during the first 8 weeks of life may be as low as 17 percent (Franzmann et al. 1980a, Ballard et al. 1981a, 1991, Larsen et al. 1989a, Osborne et al. 1991, Gasaway et al. 1992). Further losses during the first year of life may result in annual survival rates as low as 0.1 (Van Ballenberghe 1987).

Annual moose calf survival in unhunted populations occupying predator-free environments has been estimated as high as 0.67 (Mytton and Keith 1981). Rolley and Keith (1980) reported 1.06 calves per female in Alberta during winter in a population where 74 percent of females were accompanied by calves and 41 percent of calf/cow groups contained twins. These remarkable statistics suggest that mean annual survival of calves in this population approached 1.0.

Survival of calves can be monitored by radio-collaring neonates and determining their fates. Several such studies in various areas of North America have demonstrated that early mortality can be high, especially where predation occurs. Other causes of early mortality in calves include primarily accidental deaths and drowning. High twinning rates—up to 70 percent—have been recorded in expanding moose populations. *Photo by Victor Van Ballenberghe; courtesy of the USDA Forest Service.*

Telemetry studies of calves generally have not detected sex differential survival (Larsen et al. 1989a, Ballard et al. 1991, Osborne et al. 1991). However, Ballard et al. (1991) documented 88 percent overwinter survival of female calves versus 55 percent survival of males during a severe winter in Alaska when starvation was the most important mortality factor.

Survival rates of twin moose calves compared with that of singles may be lower where predation is intense (Osborne et al. 1991). However, Franzmann and Schwartz (1986), Larsen et al. (1989a) and Ballard et al. (1991) found no differences in mortality rates of singletons and twin calves.

Relationship of Mortality Factors

One method of ranking the relative importance of various mortality factors is *k*-factor analysis, first developed for insect populations by Varley and Gradwell (1968) and applied to ungulate populations by Sinclair (1977) and Houston (1982). The *k*-value is the "killing power" of a particular mortality factor on a log scale. Examples of *k*-factors for moose calves include reduction in fertility, predation, human hunting and starvation in winter. The sum of the *k*-factors represents the total annual reduction in numbers of young. Each can be calculated as $\log_{10}$ (initial number/final number) at each stage of mortality (Houston 1982: 58). A plot of *k*-values against time in years indicates which factor contributes most to the sum of all *k*-values and to what extent these change over time. Houston's (1982) analysis clearly indicated the importance of neonatal mortality in an elk population. We are unaware of applications of *k*-factor analysis to moose population dynamics, despite its potential for separating the significance of various mortality factors.

Survivorship Schedules and Ecological Conditions

McCullough (1979: 76–83) discussed the significance of survivorship in relation to ecological variables. He noted that survivorship is a variable, not a constant, and specific schedules for a population represent responses to a given set of environmental circumstances. Environments vary over space and time, and a given life table represents only one of an almost infinite number of possible circumstances. Time-specific survivorship may not be broadly applicable to other areas or other times.

By plotting survival curves of cohorts during different time periods (Figure 100), Page (1989) illustrated the application of these concepts to the unhunted Isle Royale moose population. He found five distinct survivorship patterns in Isle Royale moose from 1956 to 1974, during time periods when weather factors, moose population density and predation varied greatly. Survivorship of moose aged 1 to 7 years varied about fivefold during these different periods. Clearly, survivorship in this unhunted population of long-lived animals exhibited marked differences over time despite relatively stable numbers. Survivorship patterns of moose in hunted populations and those inhabiting unstable habitats also are expected to exhibit considerable variation.

Immigration and Emigration

Movements of individuals into and out of moose populations can be difficult to assess, and the effects on population increases and declines often are assumed to be negligible. However, Rolley and Keith (1980) reported a shift from net ingress to net egress as an Alberta moose population increased more than 40-fold during a 13-year period. The observed finite rate of increase was 6 percent greater than that predicted from survival and fecundity alone during the first part of the time interval, and 8 percent lower than predicted in the later years. Because adult survival was high and recruitment did not decline, the drop in observed rate of increase was thought to be caused by a shift to net egress.

Studies of dispersal have shown that moose may be philopatric or may disperse from their natal home ranges as young adults (see Chapter 10). Gasaway et al. (1989) and Cederlund and Sand (1992) reported limited dispersal (high philopatry) and significant overlap of maternal and offspring home ranges among individual moose radio-tracked for as many as 11 seasons after abandonment by their dams. In contrast, Houston (1968) and Ballard et al. (1991) re-

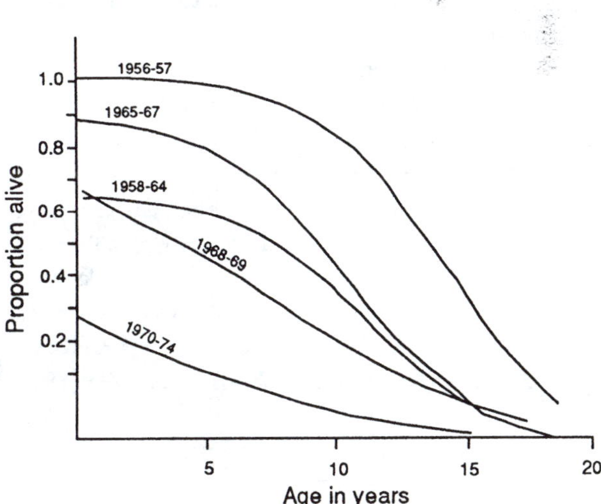

Figure 100. Survival curves for moose of Isle Royale, Michigan, during five different time periods, 1956–1974. This illustrates high variation in survival of moose aged 1 to 7 during periods when weather factors, moose population density and predation varied greatly (from Page 1989).

ported significant, permanent dispersal movements. Ballard et al. (1991) found that no male offspring remained fully within its dam's home range and that 9 of 15 offspring fully or partially dispersed (partial overlap with mother's home range). Several factors, including population density and social stress, are suspected to determine dispersal rates. Dispersers appeared to move to areas of lower density with fewer males in southcentral Alaska (Ballard et al. 1991). Clearly, more research is necessary to clarify the relationship of dispersal to the population dynamics of moose in different environments.

Population Structure

Animal populations subject to varying fecundity and mortality schedules assume different sex ratios and age distributions. Adult sex ratios may be skewed as the result of differential mortality between the sexes or a differential sex ratio at birth. Prevailing age distributions reflect (1) the effects of unusually strong or weak cohorts as they move through the population, and (2) the varying effects of mortality rates that correlate with environmental changes. For example, a series of strong (large) cohorts in time will inflate the older age classes and produce an "inverted pyramid" age distribution, in contrast to the more typical pyramid, wherein increasing mortality with age reduces the size of older cohorts. Similarly, unusually mild winter weather may reduce mortality of older animals, thereby changing the relative abundance of older individuals in the population. This myriad of variation results in an almost infinite number of possible combinations of the parameters that constitute population structure.

Sex ratios estimated at different stages of development include primary (at conception), secondary (at birth) and tertiary (juvenile or adult). Age distributions may include 1-year age classes that span the entire interval from 0 to 1 year to the oldest animals observed. However, these commonly are lumped into calf, yearling and adult groups, and further separated by sex, for a total of six categories. Population ratios usually are expressed as the number of calves/males/females converted to a base of 100 females. A statistic of emphasis is the percentage of calves or percentage of yearlings in the population, and this is used as an index of recruitment (i.e., the number or rate of new animals entering a population at an age when the "newcomers'" mortality rate approximates that of adults in the population).

Gasaway and Dubois (1987) listed four characteristics of a good population parameter estimator: (1) it minimizes variance (i.e., it is precise); (2) it is unbiased (i.e., the mean of a large number of estimates will equal the population mean); (3) it is robust (i.e., insensitive to violation of assumptions); and (4) it calculates normally distributed estimates for common sample sizes. Gasaway et al. (1986) detailed the steps necessary to incorporate these guidelines into field surveys of moose population structure.

Moose population structure may be estimated by field surveys, commonly using fixed-wing airplanes or helicopters (see Chapter 17). Visibility biases similar to those involved with density estimation may complicate population structure estimates, especially in areas of dense forest cover. Moose behavior also confounds estimates of population structure (Van Ballenberghe 1979, Gasaway et al. 1985a). Moose segregate by sex and age (Peek et al. 1974, Miquelle et al. 1992), and large groups are seen more readily than small ones. Therefore, sex and age composition estimates can vary with search intensity and scope. Samuel et al. (1992) reported that groups, rather than individuals, are the minimal sampling unit appropriate for wildlife surveys. Biased ratio estimates and variances result from methods that involve assuming that individual animals or groups are seen with equal probability. Samuel et al. discussed visibility models used to provide corrected estimates of ratios and variability that incorporated errors due to sampling, visibility bias and visibility estimation.

Most population structure data for moose in North America have resulted from sampling unknown portions of populations with aerial surveys of low search intensity. Moose population structure data obtained as a by-product of aerial census are desirable because the procedure involves random sampling to minimize bias and high search intensity to increase sightability (Gasaway and Dubois 1987). In addition, precision estimates can be computed for population structure statistics. Nevertheless, researchers still should heed Caughley's (1966) warning that attainment of truly representative samples of ungulate population structure may be practically impossible.

Age distributions of moose populations can also be estimated from kill data and life tables either by combining fecundity and mortality schedules or by averaging successive l_x numbers as Peterson (1977) did for Isle Royale. This results in a hypothetical age distribution at a point midway in each age group, assuming equal recruitment each year and constant mortality rates. Standing age distributions also have been estimated by capture of large, representative samples of moose populations, generally for the purpose of telemetry or other marking studies (Ballard et al. 1991).

Inferring survivorship from age distribution data is a widely used technique that is subject to serious bias if the population in question was changing when the age structure developed (Eberhardt 1988). Changes in population trajectory are difficult to infer from standing age distributions (Caughley 1977). Shifts in age distributions reflect changes in fecundity, mortality or both, not mortality rates alone. Furthermore, when an age distribution is not stable, additional change does not necessarily imply changes in survivorship or fecundity. The age distribution may be converging toward a stable form in response to previous survival and fecundity parameters. For analyses of age distribution data, Eberhardt (1988) provided equations to replace the assumption of constant population size with one of a constant rate of increase. However, constant survival rates and a stable age distribution still must be assumed.

Page (1989) emphasized the dramatic changes in age distributions that have occurred at Isle Royale since 1958. The feedback of density-dependent predation effects severely impacted age distributions of moose. Under certain conditions, predation and declines in moose fecundity reduced recruitment to very low levels. Prolonged low recruitment robbed the prime-age classes of the replacements necessary to sustain them. Eventually, the age distribution became heavily skewed toward older animals with high mortality and low fecundity, and a population decline was inevitable.

Finally, the proportions of males, females and calves, and the relative numbers of each sex observed in moose populations are highly variable. As with fecundity and mortality, there are no sex and age ratio norms for the species. Tertiary sex ratios may range from even (Isle Royale, Peterson 1977) to highly skewed toward females, at times dropping lower than 10 males per 100 females attributable to hunting, such as in certain populations in Alaska (Bishop and Rausch 1974). In Alaska, even unhunted populations may contain more adult females than males, apparently due to higher mortality rates of males associated with injuries and energy demands during the rut (Miquelle et al. 1992). The percent-

Adult sex ratios in moose populations may range from even to highly skewed toward females as a result of higher male mortality attributable to hunting or various environmental factors. Few moose populations in northern areas have even sex ratios. This scene from a rutting group shows the more constricted, unfilled abdomen of the bull, versus the cow. Bulls during rut may not eat for several weeks and may lose up to 30 percent of their body weight. *Photo by Victor Van Ballenberghe; courtesy of the USDA Forest Service.*

age of calves in early winter populations can range from about 7 percent where predation is very intense (Van Ballenberghe 1987) to 44 percent for populations where hunting is negligible (Rolley and Keith 1980). Such variations in population structure can have profound influence on growth rates of moose populations, and have important implications for management (see Chapter 17).

Rate of Increase

Animal populations increase and decrease over time; long-term stationary populations are rare. Populations increase when fecundity and survival (and perhaps immigration) allow more animals to enter a population than leave; populations decline when the reverse occurs. Reliably estimating rates of change in abundance and determining the factors responsible for change are fundamental problems of population dynamics.

Caughley (1977) discussed methods of calculating rate of increase and defined several different measures. Finite rate of increase (λ) is the ratio of numbers in two successive years, the coefficient of annual population growth, or growth multiplier. Exponential rate of increase (r) is the natural logarithm (ln) of λ. Van Ballenberghe (1983) illustrated equations to compute λ and r, the interrelationships between them, and their use to compute population size over time (Table 15). For example, a moose population of 100 animals increasing at a finite rate of increase (λ) of 1.20 would grow to 172.8 animals at the end of 3 years (100 X 1.2 X 1.2 X 1.2 = 172.8). An exponential rate of increase of 0.18 corresponds to λ = 1.20. However, the methods of computing population size using exponential and finite rates are different (Table 15).

Table 15. Equations used to compute finite (λ) and exponential (*r*) rates of increase and to compute population size if *r*, λ, initial population size and number of years of increase are known

I.	$\lambda = (N_t/N_0)^{1/t}$
II.	$r = (\ln N_t - \ln N_0)/t$
III.	$N_t = N_0\lambda^t$
IV.	$N_t = N_0 e^{rt}$
V.	$r = \ln\lambda$
VI.	$\lambda = e^r$
where:	λ = finite rate of increase
	r = exponential rate of increase
	N_t = number of individuals in year t
	N_0 = number of individuals in initial year, or year zero
	ln = natural logarithm
	e = base of natural logarithms = 2.71828

Caughley (1977) defined *observed rate of increase*—the rate at which a population grows over a specific time interval. No assumptions of constant rates, stable age distributions or resource supply are required. *Survival–fecundity rate of increase* is the rate of growth resulting from a stable age distribution appropriate to present schedules of age-specific survival and fecundity. This usually differs from actual or observed rates because age distributions are rarely stable. Survival–fecundity rate of increase eliminates the effect of age distribution and reveals a population's current capacity to increase. *Intrinsic rate of increase* is the rate of growth of a population with a stable age distribution in the absence of crowding when no resource is in short supply.

Gasaway et al. (1986) stressed the importance of defining underlying models of population growth when calculating rate of increase. *Exponential growth,* the most common model, implies that the number of animals added is a constant percentage of population size. *Linear growth* implies that a constant number (i.e., an ever-decreasing proportion of the population) is added. Animal populations generally grow exponentially. Methods of computing rate of increase differ according to the underlying model of growth. Gasaway et al. (1986) provided equations for both models.

Meaningful rates of change occur only when populations increase or decrease significantly. Population estimates are not exact measures of population size; statistical procedures can calculate the probability that a change in abundance occurred and determine the probable rate of change. Detecting changes in population size is linked to the precision of the estimates. Gasaway et al. (1986) provided statistical tests to determine differences between population estimates.

Gasaway and Dubois (1987) illustrated how sampling error produces variation among population estimates that leads to large differences in estimated rates of increase when no change was actually occurring. They stressed that even with the most precise moose population estimates available, changes of about 20 percent or more are required for detectable changes in rate of increase. Imprecise population estimates require moose numbers to increase by 50 to 100 percent before a change can be detected. When no measure of precision of population estimates is available, statistical confidence in how much or how fast a population changed is lacking, but other data may suggest that real change actually occurred (Gasaway and Dubois 1987).

Keith (1983) reviewed observed rates of increase for several moose populations in North America where food was adequate, large predators were scarce or absent and hunter harvest was negligible. Values of λ for these increasing populations ranged from 1.15 to 1.20, with a mean of 1.23.

Van Ballenberghe (1983) reviewed additional data on various populations and concluded that λ might approach 1.42 for populations with unstable age distributions, no emigration, maximum fecundity and annual survival rates of 0.95 for adults and 0.8 for calves. He suggested that this rate could not be achieved by moose in the northern portions of their range in North America, mainly because calf survival is lower there.

Ecological factors affecting rate of increase of moose include density-dependent effects resulting from food competition, such as changes in age of sexual maturity, litter size, first-year survival, adult survival and sex ratios of young and adults. Modeling of ungulate population dynamics (see Eberhardt et al. 1982, Nelson and Peek 1982) suggests that, within a narrow range of fecundity rates, adult survival has the greatest relative order of magnitude of effect on rate of increase for long-lived species. Adult parameters affect up to 20 cohorts in contrast to subadult parameters that affect only 1 or 2.

Changes in fecundity and survival of young strongly influence rate of increase. Van Ballenberghe and Dart (1982) reported that reducing moose calf survival from 0.50 to 0.25 during the first 6 months of life decreased λ of a model moose population from 1.14 to 1.05—a significant change. Calf survival rates as low as 0.25 in areas where predation is intense are not unusual (Van Ballenberghe 1987). In contrast, high calf survival may counterbalance the effects of low fecundity. Albright and Keith (1987) reported that a Newfoundland moose population remained stationary despite high adult mortality (overall annual survival of 0.75) and low fecundity (0.4 calves/female), apparently attributable to relatively high (0.69) annual first-year survival.

Changes in age distributions also may strongly influence rate of increase. Highest growth rates can be expected when most adults are in cohorts in which fecundity and survival are also high; life table (see Peterson 1977) and fecundity (see Saether and Haagenrud 1983) data indicate that these cohorts span the approximate age interval of 2 to 8 years. For the Isle Royale moose population, Page (1989) provided evidence of dramatic age distribution shifts that had strong effects on that population's rate of increase.

Using simple models, Van Ballenberghe (1983) illustrated the effects of changing adult and calf sex ratios on rate of increase of moose. Finite rate of increase nearly doubled by changing adult sex ratios from 1:1 to 1:7; therefore the population contained high numbers of adult females. Similarly, Cederlund and Sand (1991) reported that changing the adult sex ratio of a model moose population from 1:1 to 1:3 changed λ from 1.39 to 1.55. They concluded that, because of low variation in fecundity among age classes, sex ratios were a more important determinant of λ than were age distributions.

The effect of dispersal on rate of moose population increase was illustrated by Rolley and Keith (1980) for a population in Alberta. As the population increased from 0.08 to 1.94 per square mile (0.03–0.75/km^2) over a 13-year period, dispersal changed from net ingress to net egress accompanied by stabilization of the age distribution. Despite extremely high recruitment (44 percent of the winter population was calves), λ was relatively low (1.12–1.03) during a 3-year period of net egress.

Hatter and Bergerud (1991) provided equations illustrating relationships among rate of increase (λ), adult mortality (M) and recruitment (R) that allow crude determinations of rate of increase when population estimates are unavailable. In the absence of hunting, $\lambda = (1-M)/(1-R)$. When hunting is additive to natural mortality, the equation can be modified to include hunting mortality. If estimates of λ and R are available, $M = 1 - \lambda(1 - R)$. R is estimated by yearlings/yearlings + adults in annual samples of moose seen during aerial surveys in which yearling males are identified by antler size, and sex ratios of yearlings are assumed even. Hatter and Bergerud provided several examples where field data were available to use the equations. Gasaway et al. (1992) used them to estimate parameters for an Alaskan moose population that erupted, declined and remained at low densities for a prolonged period. Hatter and Bergerud (1991) stressed that precise, unbiased estimates of R and M are required to estimate λ accurately, and that estimates of λ from R and M should be cross-checked with those derived from absolute density surveys.

Limiting Factors and Population Regulation

Animal populations do not increase indefinitely or typically decline to extinction. Generally, they fluctuate, often within a relatively narrow range of densities. Populations are regulated with density-dependent effects on mortality and fecundity: as density increases mortality increases and fecundity declines. Eventually recruitment and mortality strike a balance and numbers remain constant. When mortality exceeds recruitment, numbers and density decline, thereby setting the stage for reduced density-dependent effects on mortality and fecundity and the cycle repeats.

In addition to those related to density, various other factors influence animal populations. For example, weather conditions such as snow depth are density-independent factors and may have significant effects on mortality and fecundity. Whether density dependent or density independent, all mortality factors are limiting, but only those demonstrably density dependent are potentially regulatory (Sinclair 1989). The interplay between regulatory factors that cause a population to attain equilibrium and limiting factors that set the position of the equilibrium was illustrated with graphic models by Sinclair (1989).

Identifying the factors that limit and regulate moose numbers has proven to be a complex problem. A population at any given time and place reflects the composite effect of all limiting and regulating influences; rarely is it possible to measure the effect of any single factor or to rank that factor's importance relative to other factors (Connolly 1981). However, identifying limiting factors and modifying the environment to manipulate them are central to moose population management. For opening new horizons for moose managers, Gasaway et al. (1986) emphasized the importance of developing precise, unbiased methods for estimating population parameters. Their statements may be applied equally to the development of new methods of identifying limiting factors and assessing their importance. "With a better grasp of the dynamics of a moose population, management options become more apparent" (Gasaway et al. 1986: 1).

Life History Traits and Carrying Capacity

r selection and *K selection* are terms that describe organisms with radically different biological and ecological attributes (MacArthur and Wilson 1967). Organisms labeled r-selected maximize rate of population growth; they are small and short-lived, produce very large numbers of young and give little care to offspring. K-selected species maximize competitive abilities in crowded environments, and are adapted to relatively stable habitats at densities close to carrying capacity. Evolution of large body size, long life-span, low reproductive rate, and extended parental care characterize the K-selected species. Obviously, these two life history "strategies" represent the ends of a continuum, and many species do not fully meet all the necessary criteria to be extreme r- or K-strategists. However, the concepts are useful in categorizing numerous ecological and biological traits of organisms, and much population theory is based on this conceptualization (see Pianka 1970, Stubbs 1977).

McCullough (1979) classified North American ungulates as K-selected. He stressed that most population theory was based on r-selected organisms, resulting in problems when applying this theory to K-selected species. A major difficulty arises from the assumption that all individuals are qualitatively equal. McCullough (1979: 2) wrote: "Ignoring qualitative differences in K-selected species can result in problems. Typically, these populations exist in relatively low numbers in a rather close relationship to resources. Competitive relations are intense, and density dependence touches all aspects of the life cycle. Qualitative differences among individuals in the population are enormous. They include differences due to age, sex, and health, and these factors influence survivorship and reproduction. These animals typically possess individual identities in highly complex social structures. Social standing, as expressed through dominance hierarchies, systems of territoriality, etc., mediates competitive interactions and further influences survival and reproduction. Extended parental care influences not only size and condition of offspring, but frequently social status as well. Social alliances based upon kinship or mutual interest crop up in unexpected ways. These factors have important population consequences, and they moderate the numerical responses one would expect if all individuals in the population were qualitatively equal. No population parameter can safely be assumed to be a constant."

The concept of carrying capacity is important in understanding population regulation of K-selected species, including moose. Caughley (1976, 1979) characterized the growth pattern typical of an ungulate population as an eruption, a crash and a convergence to a steady-state density resulting from the dynamics of the ungulates interacting with the dynamics of the plants on which they feed. The equilibrium achieved is termed *ecological carrying capacity* (K). Caughley (1979: 7) noted: "We tend to think of carrying capacity largely as a density of animals, but we could equally think of it as a characteristic density and composition of vegetation." Exploitation by predators or humans results in lower standing crops of ungulates and higher standing crops of plants than those at K.

Evidence from long-term studies of moose populations in North America supports the idea that populations erupt, crash and then stabilize at various densities depending on prevailing ecological conditions. Moose at Isle Royale colonized the island about 1910 and erupted in the absence of predators to high densities and crashed by the early 1930s (Mech 1966). A wildfire in 1936 set the stage for a second eruption by increasing forage supplies. By 1948, when wolves colonized the island, a second moose population decline already had begun in response to moose overutilization of terrestrial and aquatic plants. Since 1958, when long-term studies of moose/wolf dynamics were initiated, the moose population has fluctuated between densities of about 2.6 and 6.5 per square mile (1 and 2.5/km^2), in response to predation and competition for high-quality forage (Messier 1991).

On the northern Kenai Peninsula in Alaska, moose were rare before the late 1800s, perhaps because of high predator populations and lack of wildfire (Lutz 1960). Moose increased after several human-caused fires beginning about 1885. Overbrowsing was noted as early as 1913, and sharp declines of moose occurred in the mid-1920s and late 1930s

(Bailey 1978). In 1947, a 473-square mile (1,225 km²) area burned on the northern Kenai Peninsula and created optimum habitat for moose. The moose population increased about 13 percent per year for 12 years after the fire (Spencer and Hakala 1964, Bailey 1978). Between 1949 and 1969, moose increased from about 2,000 to about 9,000 (Bangs and Bailey 1980a), until a sharp decline beginning in the early 1970s. By 1975, moose numbered only 43 percent of those present in 1971 (Franzmann et al. 1980a). As plant succession advanced and habitat quality declined in the area burned in 1947, the moose population density changed from 9.3 per square mile (3.6/km²) in 1970–1971, to 3.4 per square mile (1.3/km²) in 1981–1982 to 0.8 per square mile (0.3/km²) in 1986–1987 (Schwartz and Franzmann 1989). Loranger et al. (1991) reported a linear decline of about 9 percent per year in winter moose density in the 1947 burn between 1964 and 1990 (17 to 43 years postfire) (Figure 101).

Peek et al. (1976) reviewed historical trends of moose numbers in northeastern Minnesota between 1885 and 1972. Moose were scarce in the mid-1800s but became common about the time caribou disappeared in 1890. Crude estimates were made of moose numbers from 1923 to 1970. The population generally increased from 1923 to 1933, with a decline thereafter until about 1950. Numbers then increased from about 500 to more than 4,000 by 1968, apparently in response to a hunting ban and favorable habitat created by logging.

These case histories and numerous others across the circumpolar range of the species indicate that moose populations exhibit a basic response to changes in habitat quality. Geist (1971, 1974a) characterized moose habitat as permanent or transient—the former being climax plant communities largely unaffected by fire and the latter representing unstable, short-lived tree and shrub communities that emerge from burns. Over the species' evolutionary history, moose typically have occupied limited areas of permanent habitat, but have rapidly colonized large areas of transient habitat created periodically by fire. Many of the life history traits of moose that influence population dynamics, including dispersal patterns and fecundity, were viewed by Geist (1974a) as evolving under selection pressures influenced by this regime.

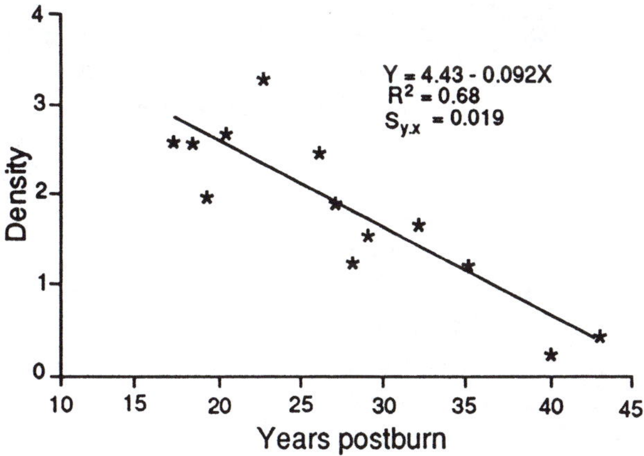

Figure 101. Relationship of forest age and moose density (moose/km²) during winter in the 1947 burn on the Kenai Peninsula, Alaska. A linear decline of about 9 percent per year is evident from 17 to 43 years postburn (from Loranger et al. 1991).

Although the concept of ecological carrying capacity (K) is fundamental to understanding regulation of moose populations, few studies have attempted to estimate the density of moose at which K occurs. Crête (1989) suggested two approaches: (1) estimate density in an undisturbed area where moose and vegetation have coexisted long enough to stabilize; and (2) estimate forage production and daily food requirements of moose, thereby computing an estimate of the maximum moose numbers the vegetation could support. Crête (1989) reported that one unhunted area in Quebec supported 4.7 to 5.2 moose per square mile (1.8–2.0/km²)—a record high density for that area of Canada. K was postulated to exceed this density, as recruitment remained relatively strong and emigration apparently reduced density. Estimates of both browse production and consumption suggested that deciduous browse could support 4.9 moose per square mile (1.9 moose/km²); including coniferous browse suggested 91 moose per square mile (35/km²)—an unrealistically high density. Crête (1989) recognized that K for moose likely was not constant, because of variations in snow depth that affect forage availability, interannual variations in forage quality partly because of variation in summer rainfall, and forest succession that influences long-term trends in forage production.

Density-dependent Effects

Studies throughout the world on ungulates as diverse as African buffalo (Sinclair 1977), red deer (Clutton-Brock et al. 1987), elk (Houston 1982), wild reindeer (Skogland 1985) and white-tailed deer (McCullough 1979) have documented the effects of population density on fecundity, mortality, recruitment and population dynamics of those species. However, disagreement still occurs on the importance of density dependence in nature, on unambiguous demonstrations of it as measured by population growth rates and on methods of detecting it (McCullough 1990). Despite this, ungulate populations clearly display a wide range of responses to changes in density. These responses influence population dynamics and have important implications for population regulation.

Klein (1981) summarized the population consequences of high density in North American deer populations.

1. The food habits of deer change as they are forced to feed on less desirable forage. Overall diet quality declines and the digestible component is reduced. The deer experience reduced growth rates, smaller body size, poorer body condition and reduced fat reserves before winter. Their vulnerability to disease, parasites, starvation and predation increases.
2. Where food quality is limiting, delayed sexual maturity and lowered conception rates result. And under such conditions, fertility appears to be influenced more by reduced conception than by in utero mortality after conception. Decreased postpartum fawn survival has been associated with poor nutrition of the mothers, and heavy losses of fawns during their first winter are common on ranges where competition among deer for available food is strong.
3. When food becomes limiting because of high deer densities, ensuing competition, range deterioration or other reasons, the age of sexual maturity is delayed, fertility is lowered, fawn survival decreases and the population becomes numerically skewed toward older animals. Such populations, although usually stable or declining, can respond rapidly through population increase if habitat conditions are markedly improved. Under optimal range conditions deer populations are skewed toward young animals. And although such populations usually are increasing, they are vulnerable to heavy mortality and population decline if habitat conditions deteriorate through overgrazing, plant successional changes or other factors.
4. Under poor range conditions, males suffer heavier mortality than do the females, and the population becomes skewed toward females. This factor, although often associated with the deterioration of range and population decline, nevertheless increases the reproductive potential of the population. This potential, however, rarely is realized.

Relationships between density and recruitment of white-tailed deer were documented by McCullough (1979) on the George Reserve in Michigan, a fenced 1,146-acre (464-ha) site where deer were introduced in 1928. Deer density was controlled by hunting. Net recruitment rate declined linearly as density increased and became curvilinear as the population approached K (Figure 102). The productivity curve (number of recruits as a function of population size) or net recruitment reached a maximum at about 0.56 K and declined sharply to zero at K. Although similar data on free-

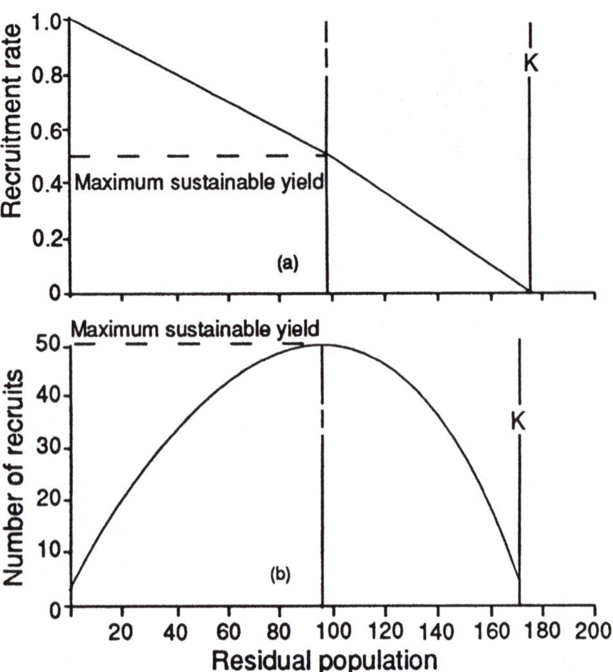

Figure 102. Recruitment (a) and productivity (b) curves of the George Reserve white-tailed deer herd in Michigan. Recruitment rate declined linearly as density increased to the point of maximum sustained yield (I), and curvilinearly thereafter. The parabolic productivity curve shows that the number of recruits peaks at maximum sustained yield density, and falls to zero at carrying capacity (K) (from McCullough 1984). Although similar studies with enclosed moose herds have not been done, the shapes of these curves apply to all members of the deer family.

ranging populations of cervids are lacking, the general pattern of declining recruitment as density increases appears typical. For example, Houston (1982: 45) reported that calf recruitment of Yellowstone elk declined from about 40 percent of cows with calves when total elk numbers were less than 7,000 to about 26 percent when numbers exceeded 7,000. There was about a threefold difference in percentage of cows with calves between the highest (low density) and lowest (high density) recruitment estimates.

McCullough (1979: 87–88) emphasized the usefulness of focusing on recruitment in analyses of deer population dynamics. He noted that a population integrates a broad complex of environmental variables and population responses, with recruitment as the net result. Furthermore, the same outcome in recruitment can result from a large number of combinations of variables for the animals and their environment. Sex ratios, survivorship, age distributions and all other parameters vary greatly from year to year. Treating each separately results in high variance. Recruitment—the functional outcome of the integration of all variables—

shows scarcely more variance at low density than does any other single parameter. However, McCullough also emphasized that, at high density, the productivity curve (Figure 102) has wide confidence limits because most of the response depends on survivorship generated by competition for resources. These studies of deer and elk responses to density indicate a general pattern for members of the deer family, including moose.

Recruitment data for the Isle Royale moose population suggest complex interactions between moose density and environmental variables, and illustrate the difficulty of determining the relative importance of any one variable in space and time. Peterson (1977) reported that twinning rates generally were correlated with population density. They ranged from 6 percent in 1929 and 1930 when moose numbers were very high, to 48 percent in 1964 when numbers were much lower and known to be increasing. Page (1989) indicated that recruitment declined linearly from about 1960 to 1971 as moose numbers increased from about 750 to about 1,300 (Figure 103). During this interval, recruitment declined from about 18 percent to zero. Similarly, Messier (1991) indicated that summer calf/cow ratios—an index of recruitment—were correlated negatively with moose density during the period 1970 to 1983. However, Page (1989) also reported that recruitment after 1972 was inversely density dependent (Figure 103) as a result of predation. Messier (1991) reported no measurable effect of snow accumulation on either calf/cow ratios or the percentage of twins. He suggested that moose density and wolf predation primarily determined recruitment. No direct estimates of fecundity were available for this moose population; recruitment was estimated from observed calf/cow ratios or from reconstructed population parameters (Page 1989), thereby compounding the difficulty of determining density effects on recruitment. Furthermore, strong effects likely are undetectable until moose populations reach very high densities, an infrequent event for many populations.

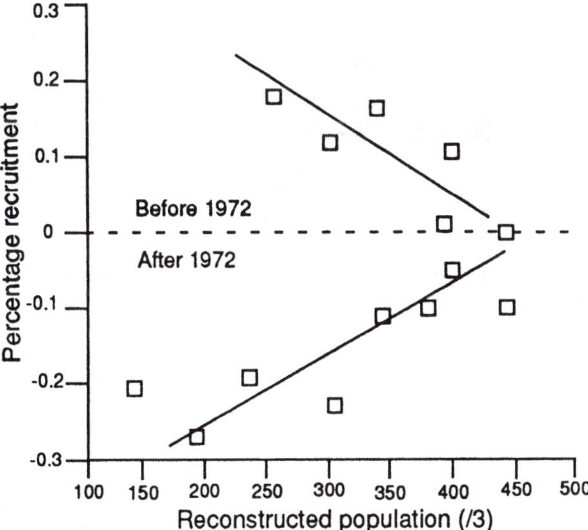

Figure 103. Recruitment as a function of moose population size on Isle Royale, Michigan. Before 1972 density was increasing and recruitment declined linearly because of density-dependent effects. After 1972, recruitment was inversely density dependent as a result of predation (from Page 1989). The moose population was "reconstructed" based on the ages of recovered skeletal remains resulting from wolf predation and natural mortality; the numbers on the x axis are one-third of actual.

Studying moose in Alberta, Rolley and Keith (1980) reported no significant relationships between moose density and various indices of recruitment, including the percentage of females with calves, twinning rates and the percentage of calves in the population during a 13-year interval when density increased more than 40-fold. However, moose density at the end of this interval was 1.9 per square mile (0.75/km^2)— a relatively low density compared with many other populations including those of Isle Royale, which have reached densities as high as 6.5 moose per square mile (2.5/km^2).

Limiting Factors

Limiting factors for moose populations include various density-dependent and density-independent influences. Houston (1968) graphically summarized a myriad of factors affecting a Wyoming moose population (Figure 104), including a variety of limiting factors that subsequent studies also have identified as important. Habitat quality and food resources (Caughley 1976, Sinclair 1977, Houston 1982) and predation (Bergerud et al. 1983, Gasaway et al. 1983, Van Ballenberghe 1987) often are considered major limiting factors of other ungulates as well, but several other environmental factors may influence populations. Gasaway et al. (1992) identified the factors of nutrition, weather, hunting, disease and predation in limiting low-density moose populations in Alaska and Yukon, and evaluated their roles. Other moose population studies have considered some or all of those as limiting factors (Houston 1968, Peek et al. 1976, Peterson 1977, Fuller and Keith 1980, Ballard et al. 1991).

Messier and Crête (1984) and Gasaway et al. (1992) summarized the population consequences of nutritional stress sufficient to limit moose population growth. Low fecundity, reduced recruitment and low survivorship of adults accompanied by retarded body growth, poor physical condition and starvation before old age all represent signs of acute nutritional stress that may or may not be associated with high density. Gasaway et al. (1992) emphasized that reduced recruitment and poor adult survival are not necessarily exclu-

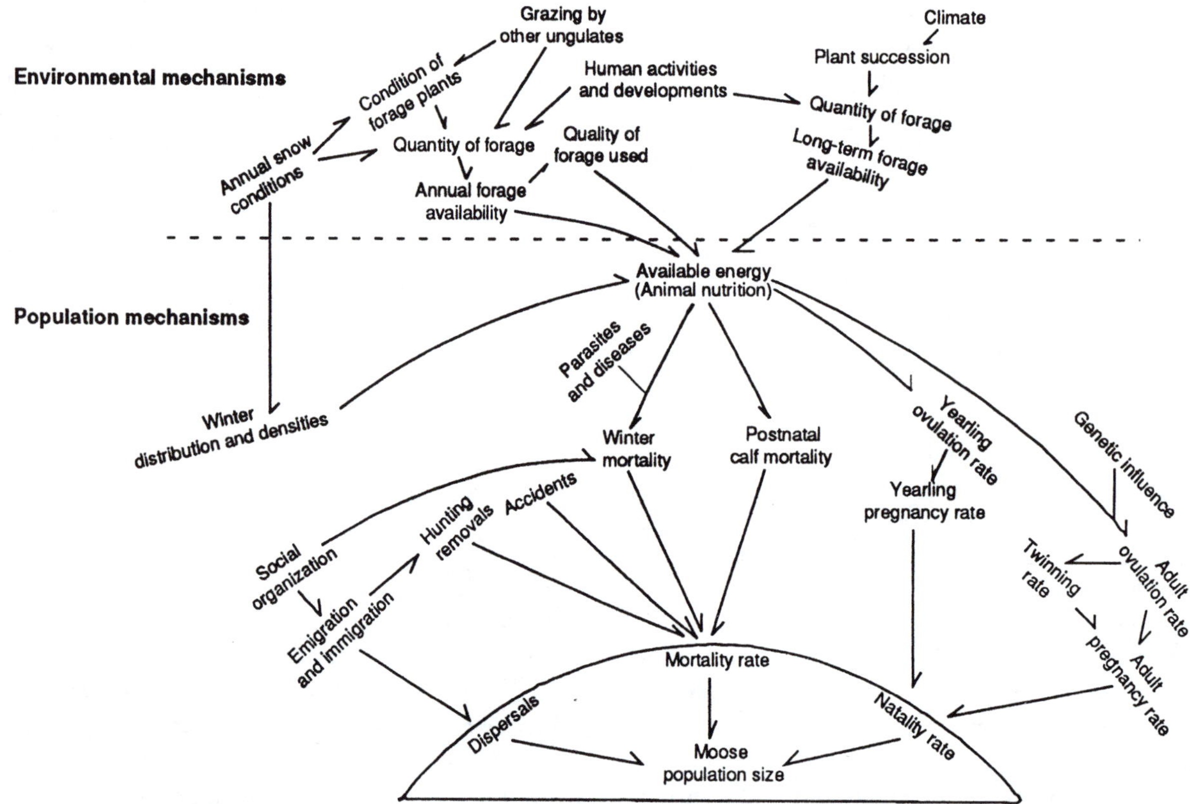

Figure 104. Model of factors influencing Jackson, Wyoming, moose population dynamics. Environmental mechanisms influence forage quantity and quality, which in turn affect animal nutrition. Nutrition influences natality, mortality and dispersal through a variety of complex pathways (from Houston 1968).

sive to nutritional stress. Deep snow and predation also may be involved. Population effects induced by nutritional stress also tend to be accompanied by high browsing rates on preferred species and perhaps significant plant mortality and altered rates of plant succession. Absence of these influences may implicate major limiting factors other than nutrition as being significant.

Franzmann and Schwartz (1985) reviewed data suggesting strong correlations between body condition of female moose (as influenced by habitat quality) and twinning rates in several diverse moose populations. In an area known to contain abundant and high-quality food resources on the Kenai Peninsula, Alaska, up to 70 percent of cows with calves had twins shortly after parturition. This contrasts to other populations in which twinning rates as low as 5 percent were reported (Pimlott 1959, Houston 1968, Markgren 1982), but some of the estimates may have considered postnatal mortality. The data suggest that twinning rates exceeding 40 to 50 percent are not characteristic of moose populations strongly limited by nutrition.

Messier (1991) emphasized that, as a limiting factor, forage quality may significantly influence moose population growth. However, poor-quality forage, if in unlimited abundance, actually may slowly permit continuous moose population growth and consequent high density. Also, competition for high-quality food among moose at high densities could depress population growth, and thus be a regulating factor. Messier's (1991) analysis of moose population dynamics at Isle Royale indicated that forage competition, rather than wolf predation, regulated moose numbers at high density.

Weather conditions, primarily snow accumulation and summer rainfall, may have a variety of direct and indirect effects on moose and their habitat, and thereby influence population dynamics. Plant growth and nutritional quality during summer affect moose body growth and fat storage. Summer may be a critical season for moose because the size of fat and protein stores determines how long animals survive in negative energy balance during winter (Schwartz et al. 1988b). Summer weather conditions, particularly rainfall, are known to influence moose habitat quality (Bo and Hjellord 1991), and growing season length has been shown to affect recruitment (Stewart et al. 1977).

Peterson (1977) and Mech et al. (1987) stressed the role of snow accumulation as a limiting factor for Isle Royale moose. Coady (1974a) indicated that 35 inches (90 cm) of

Snow buries forage of moose and restricts moose movements when deep. Mortality may be extensive in moose populations during winters of deep snow. *Photo by Victor Van Ballenberghe; courtesy of the USDA Forest Service.*

snow represent a critical depth for adults, in that movements are restricted such that accessibility to adequate food may be limited or prevented. Bishop and Rausch (1974) documented substantial winter mortality, mainly of calves, for several Alaskan populations at high density, when snow exceeded a critical depth. Peterson (1977) documented poor calf survival at Isle Royale during winters of deep snow, primarily associated with increased wolf predation. He stressed residual effects on calves born after severe winters; early growth and development apparently were retarded, and survival in subsequent years was reduced. Coady (1982) reviewed differential mortality resulting from deep snow and malnutrition and concluded that calves, older adults (8 years and older) and adult males probably could tolerate less nutritional stress than other classes of moose.

Hunting is a major limiting factor of many moose populations throughout the world. That hunting pressure can reduce moose population density was documented by Crête et al. (1981a). Crête (1987) reported harvest rates as high as 25 percent in Quebec, where natural mortality apparently was low. He also reported moose harvest rates ranging from 2 to 17 percent for various other parts of North America. In concert with other factors, high harvest rates have contributed to moose population declines in Alaska (Gasaway et al. 1983). In addition, hunting can significantly reduce the number of bulls, perhaps below the level at which first-estrous conception is reduced (Bishop and Rausch 1974). When fewer than 10 bulls per 100 adult cows occur because of heavy hunting pressure, some cows simply may not encounter a bull early in the mating season. In some European environments, where severe winters, predation and nutritional stress are absent, moose harvests as high as 50 percent of the winter population are sustainable (Cederlund and Sand 1991). Most North American moose populations harvested at this rate would decline sharply.

Moose are known hosts for a variety of diseases and parasites (Lankester 1987; see also Chapter 15), but these have seldom been implicated as major limiting factors affecting population growth. The moose winter tick and meningeal worms causing neurological disease may both result in mortality of individual moose in North America. Both are absent in northern Canada and Alaska. Although neurological disease allegedly has caused declines of moose in eastern North America, Nudds (1990) suggested that documentation was insufficient. Gasaway et al. (1992) evaluated the role of disease as a limiting factor of moose in an Alaskan population, and found no evidence of exposure to 10 bacterial and viral diseases, as indicated by antibody tests of sera samples.

Determining the limiting effects of predation on moose numbers is an elusive problem (see Chapter 7). Necessary data are difficult to obtain, controlled experiments are complicated by problems of scale and human influence on moose, predators and habitat often confounds interpretations. Despite this, several North American studies have indicated that predation by gray wolves and bears may have significant limiting effects on moose populations under certain ecological conditions (Fuller and Keith 1980, Bergerud et al. 1983, Gasaway et al. 1983, Page 1989, Messier 1991, Gasaway et al. 1992, Van Ballenberghe and Ballard 1994). However, studies also have suggested that other limiting factors, mainly weather and forage quality, are equally important limiting factors, or that published evidence is inadequate to conclude that predation is a major limiting factor (Peterson 1977, Mech et al. 1987, Thompson and Peterson 1988, Boutin 1992).

Predation may exert strong limiting effects on moose by reducing juvenile or adult survivorship. Ballard and Larsen (1987) and Van Ballenberghe (1987) reviewed nine case histories in North America where predation by wolves and bears caused high rates of juvenile or adult mortality. Annual survival of calves was as low as 0.1 in certain populations; predators caused up to 79 percent of neonatal mortality. Predator reduction experiments improved calf survival about threefold and increased finite rates of increase from about 1.0 to 1.23.

Various other approaches to determine the influence of predation on moose populations include estimating annual kill rates by predators and comparing these with other moose population losses and gains (Fuller and Keith 1980, Messier and Crête 1985, Ballard et al. 1987). Yet another approach has been to compare moose densities in areas where predators are limited by humans with moose densities in areas where predators are not limited (Gasaway et al. 1992). Gasaway et al. (1992) compared losses to predation (31 percent of postcalving numbers) with hunting (1.5 percent) and other losses (6 percent) to illustrate the strong limiting effects of predation on a moose population in Alaska (Figure 105).

Eruptions of moose in the absence or scarcity of large predators and long-term failure of moose to increase to high densities in the presence of naturally regulated predator populations also have been interpreted as evidence of strong predator limitation (Peek 1980, Van Ballenberghe 1987, Gasaway et al. 1992). As with nutritional stress, the

Remains of an adult female moose killed and consumed by a brown bear in Denali National Park, Alaska. Brown bears frequently cover carcasses with ground debris to discourage scavengers. Over time, these bones will be scattered about the area as scavengers visit and carry off their find. *Photo by Victor Van Ballenberghe; courtesy of the USDA Forest Service.*

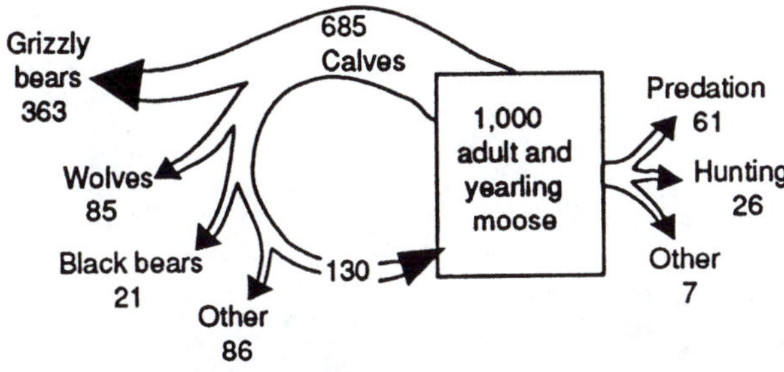

Figure 105. Losses to an Alaskan moose population from various mortality factors. Recruitment (13 percent) is low, primarily because of high rates of bear predation. Predation takes 31 percent of the postcalving population, whereas hunting accounts for only a very small fraction (1.5 percent) of the population loss annually (from Gasaway et al. 1992).

limiting effects of predation may vary with moose density (Messier and Crête 1985, Messier 1991). Messier and Crête (1985) reported that, in Quebec, wolf packs were unable to persist at moose densities less than 0.5 per square mile (0.2/km^2) but predation was strongly limiting at moose densities of 0.5 to 1.0 per square mile (0.2–0.4/km^2).

Theories of Population Regulation

Caughley and Krebs (1983) described two basic kinds of mammalian population regulation. Intrinsic regulation, or self-regulation, occurs when animals lower their rate of increase behaviorally or physiologically in response to rising density. Extrinsic regulation implies that inadequate food resources, severe weather, hunting, predators or disease intervene to reduce density. Caughley and Krebs (1983) concluded that intrinsic regulation was characteristic of many small mammal (<66 pounds [30 kg]) populations, whereas extrinsic regulation was a feature of large mammal population dynamics. Peek (1980: 220) indicated that "general agreement prevails that ungulate populations are externally controlled by changing environmental conditions rather than through intrinsic regulating mechanisms." Intensive studies of population regulation by Sinclair (1977), Peterson (1977), McCullough (1979), Houston (1982) and others have suggested no evidence of intrinsic regulating mechanisms in several diverse ungulate species.

A broad, general view of population regulation in the context of community ecology was presented by Hairston et al. (1960). They suggested that whether or not organisms are predator- or resource-limited depends on their position in the food chain. In mature, terrestrial communities, predators presumedly keep herbivores in check, thereby allowing green plant biomass to accumulate—the "world is green" proposition. This view has been challenged and debate continues in the ecological literature (Matson and Hunter 1992) in terms of top-down (predation) versus bottom-up (resources) forces in population and community ecology. For ungulates, colimitation by both predators and resources has been postulated (see Sinclair and Norton-Griffiths 1979, Peek 1980, McLaren and Peterson 1994). Hunter and Price (1992) suggested that ecologists ask not whether resources or predators regulate a particular population, but what factors modulate resource limitation and predation in the system. Ecologists must determine when and where predators or resources will dominate in regulating populations.

Several perspectives on natural regulation of ungulates (i.e., regulation of numbers in environments where human influence is minimal) occur in the literature (Peek 1980, Caughley 1981), but are somewhat complicated by semantic and conceptual difficulties (see Boyce 1991). However, Van Ballenberghe (1987) and Van Ballenberghe and Ballard (1994) emphasized the importance of understanding moose population dynamics in undisturbed ecosystems as an aid to interpreting events in human-influenced environments where moose populations fluctuated in complex and difficult-to-predict patterns. They stressed that naturally regulated environments are rare. In the strictest sense, no contemporary ecosystems are pristine, but some exist where habitat alteration and harvesting of moose and predators have been minimal for several generations of moose.

Page (1989) described ecosystem dynamics of moose, wolves and balsam fir (an important moose browse species) on Isle Royale, which is essentially a naturally regulated island ecosystem in Lake Superior. Moose were said to be regulated both by predation from the trophic level above (wolves) and by food availability in the level below. Changes in moose population parameters occurred over prolonged time intervals, as indicated by low recruitment in the 1970s, resulting in an age distribution skewed toward older animals. Age distributions inferred from a reconstructed population, based on remains of recovered moose, confirmed shifts in mean ages over time. During the 1950s and 1960s, the Isle Royale moose population density increased. As wolf numbers increased, predation of moose became more intense, promoting moose population instability suggestive of long-term cycles. Similarly, moose browsing on balsam fir resulted in old-age tree stands and very low recruitment to the trees' own population. Page (1989) referred to this system as an inverted pyramid, with decreased survivorship and increased reproductive rates in higher trophic levels—the opposite of what was expected and this was partly responsible for the postulated cyclic behavior.

Isle Royale is ecologically unusual in several respects, including its island nature and consequent restrictions on dis-

An adult female moose foraging during early summer. Food supply plays an important role in regulating moose populations through a complex set of interactions with nutrition and its influence on reproduction. Note the incisor teeth on this cow; these occur only in the lower jaw. Her incisors show some wear, which is influenced by age and by presence of abrasives on the vegetation. Moose from the Matanuska Valley area of Alaska show considerable tooth wear because of the presence of glacial silt blown onto their vegetation. Cementum annuli on permanent incisors are also used to age moose. *Photo by Victor Van Ballenberghe; courtesy of the USDA Forest Service.*

persal of both moose and wolves. Also, its well-studied moose population is not subject to predation by either black or brown bears, a unique occurrence in North America.

Van Ballenberghe (1987) concluded that, where alternate prey was scarce, predation by bears and wolves on moose produced low-density equilibria in naturally regulated environments. This implies that predation is density dependent and regulatory at moose densities far below K. This contrasts with cycles or recurrent fluctuations in which moose populations periodically escape strong limitation by predators and increase to densities where forage competition triggers a decline that may be extended by inversely density-dependent predation. Even more complex responses, resulting in two or more stable states, may occur with predation regulating moose numbers at low densities and forage competition regulating at high densities.

When humans strongly influence moose, predators or habitats, moose populations can decrease to low densities and mimic multiple stable states, cycles or recurrent fluctuations, or show no consistent patterns (Van Ballenberghe and Ballard 1994). A case history from Alaska (Gasaway et al. 1983) provided a good example of a situation in which predators were reduced, moose consequently erupted, the wolves then recovered, and hunting and deep snow triggered a moose population decline that also was extended by predation. Predator control was deemed necessary to allow the moose relief from the limiting effects of predation. In the absence of hunting, predator control or both, this moose population would have displayed considerably different dynamics.

This chapter's brief review of theoretical concepts of population regulation, related to empirical evidence from moose populations, suggests that understanding of important factors limiting and regulating moose populations still is incomplete. Long-term case histories are few, and especially so concerning naturally regulated ecosystems. And there are few experimental studies. A combination of case history studies, long-term population monitoring and large-scale experiments are required to enhance understanding of limiting and regulating factors. Such studies should be replicated in different geographic areas to assess the influences of varying ecological factors.

(MEMORY)
YOUNG BULL – VELVET –
DULL BROWN ON HEAD &
BODY – VERY DEEP BROWN
ALONG MANE, SHOULDER, &
SPINE – ESP. ON SHOULDER
& RUMP – DULL GREYED
BROWN OF NECK BECOMES
MORE GOLDEN ON SIDES
OF BODY – FORELEGS,
FROM KNEE DOWN, &
HIND LEGS FROM
BACK OF RUMP DOWN
LIGHT GRAY –
FEEDING ON LONG, STRINGY
WATER PLANTS IN POND
E OF WONDER LK AREA –
McKINLEY PARK, ALASKA
AUG. 20, '54 –
2 HUGE BULL MOOSE BROWSING ON
WILLOWS – MOVED AWAY OVER BANK,
WHERE A THIRD BULL JOINED THEM
AS THEY CROSSED IGLOO CREEK &
CLIMBED THE RIDGE – BULL MOOSE
MOVE PONDEROUSLY &
IN APPARENTLY
IN SLOW MOTION.
IGLOO CREEK
McKINLEY PARK,
ALASKA –
AUG. 22, '54
(POSES FROM LIFE –
COMP. FROM MEMORY)
RICH GOLDEN OR REDDISH BROWN
SHADING ALMOST TO BLACK ON
UNDERPARTS – RICH BROWN VELVET –
HIGH SHEEN ON LEGS & FRONT SHOULDER –

7

WARREN B. BALLARD AND
VICTOR VAN BALLENBERGHE

Predator/Prey Relationships

The cow moose is nearly 18 years old and pregnant with her twenty-third calf. The last deep-crusted snow of a harsh winter has recently melted. The cow is in poor physical condition, but with spring green-up and more and better forage available, she will improve quickly. She is carrying one calf, and the fetus is smaller than normal because of the cow's nutritional stress during the long winter.

Shortly after birth, the slow-moving calf begins nursing. This peaceful scene is interrupted suddenly by the cow's scenting of an age-old enemy, a grizzly bear. She tries to move the calf to a more defensible position, but the bear charges and grabs the calf. The cow attempts to ward off the bear by kicking it with her front legs. The bear drops the dead calf and briefly chases the cow. The cow eludes and maneuvers around the bear, but is momentarily distracted by the appearance of two bear cubs. The brief distraction allows the bear to charge again. Grabbed by the neck, the cow attempts to break the bear's hold but is pulled to the ground and, after two swats to the head and shoulders, is dead. The bear then begins to feed on the cow, and her cubs start to consume the calf. This is a common scenario, which has been repeated with minor variations for eons.

Wolves, grizzly bears and black bears are the principal predators of moose. In the aforementioned scenario, the cow was old, weak and nearing the end of her lifespan before she was killed. Investigators would analyze whether she might have died from some other cause if she had not been killed by the bear. Also, her calf was weak. Biologists would look for evidence explaining the extent to which the death was "compensatory" (i.e., the animal was predisposed to death from other causes) or "additive" (i.e., the animal would have survived if it had not been killed by that source of mortality). However, the paramount question is whether a particular type of mortality limits or regulates a moose population.

The study of predator/prey relationships is one of the most controversial aspects of moose population dynamics. Connolly (1978) indicated that a selective review of the literature can be interpreted so as to reinforce any perspective on predation. Although that may have been partially true for moose/predator relationships 20 and more years ago, population dynamics studies since then have helped to clarify the role of predation.

In this chapter, we review select case studies that have greatly increased knowledge of moose/predator relationships. Not coincidentally, these studies also serve as basis for arguments about predators limiting or regulating moose populations. We define *limiting factors* as those that alter the rate of population increase. *Regulating factors* are a subset of limiting factors that are solely density dependent and keep a population within its "normal density ranges" (Messier 1991).

Case Histories

Figure 106 locates the primary study/case history sites discussed in this chapter.

Figure 106. Location of moose/predator study areas discussed in the chapter.

Denali National Park, Alaska

One of the few areas of Alaska that has not been subjected to moderate or greater moose hunting pressure or habitat alteration is Denali National Park. Within the park, ungulate and their predator population sizes continue to fluctuate under nearly natural conditions. Pioneering work by Murie (1944, 1981) provided most of the baseline data, but intensive studies of predator/prey relationships were not undertaken until the past two decades.

Data on moose population trends in the park are incomplete (Van Ballenberghe 1987). Haber (1977) asserted that the moose population increased between 1971 and 1973 after the severe winter of 1970–1971, primarily because of ingress from neighboring areas. On the basis of recruitment indices and ground observations conducted during 1979 to 1983, Van Ballenberghe (1987) believed that the moose population declined during that time period. Haber (1977) reported moose/wolf ratios of 17–26:1 within a 580-square mile (1,500-km^2) area from 1970 through 1973. During the late 1970s and 1980s, Denali's wolf numbers declined, but moose calf survival remained low nevertheless (Van Ballenberghe 1987).

Walters et al. (1981) proposed a model of predator/prey relationships based largely on Haber's (1977) observations of two wolf packs. This model suggested that moose populations can stabilize at different equilibria in response to natural and/or human-caused changes. Haber (1977) asserted that wolf predation had no impact on moose under natural conditions and became important only after excessive hunting or natural catastrophes.

Van Ballenberghe (1980) criticized Haber's (1977) assertions and Walters et al.'s (1981) proposition of multiple equilibria. The latter was deemed inappropriate for moose and shown to contain several qualitative flaws. Although no moose calves were radio-collared in either the Haber (1977) or Walters et al. (1981) studies, and in contrast to Murie's (1981) 1955 to 1970 observations that bears seldom killed moose calves, Van Ballenberghe (1987: 449) reported that observations of grizzly bears preying on newborn moose calves have become relatively common "in recent years." From 1974 to 1982, the large numbers of sightings of bears preying on newborn moose calves in park areas of low wolf numbers and low annual moose recruitment provide strong circumstantial evidence that grizzly bears appreciably limited moose calf recruitment.

Along the Sanctuary River in Denali National Park in Alaska, photographer Michio Hoshino observed a moose calf race from shoreline brush into the river, closely pursued by a young grizzly bear. In midstream, both animals struggled against the current, but eventually regained land. The calf ran back into the river, again pursued by the bear. This time, the calf's mother rushed to interpose herself between her offspring and the predator (top). As the calf again fought the river's current, the cow and young bear had a stare-down (center). The bear broke off the attack (bottom) and both cow and calf safely reached the opposite shore. A larger grizzly in deeper water likely would have prevailed. *Photos by Michio Hoshino; courtesy of Minden Pictures.*

Van Ballenberghe (1987) compared bear densities with observations of moose predation by bears reported by Murie (1981) and Dean (1976), and concluded that bear density was not linked to changes in bear predation on moose calves. Bear densities ranged from 9.8 to 10.6 per 100 square miles (3.8–4.1 / 100 km^2) from 1959 to 1973 (Dean 1976). Observations of bears preying on moose calves were rare before 1959, but common in 1973. Accordingly, Van Ballenberghe (1987) concluded that predation by brown/grizzly bears is a learned behavior.

In 1986, Mech et al. (1995) began a radio-telemetry study of wolves and found that moose were their predominant prey. The calves and adult moose killed by wolves during winter had low marrow fat values, which indicated they were in poor general condition. There also was a high incidence of debilitation among the moose prey, which indicated a predisposition to wolf predation. After winters of deep snowfall, the wolves reportedly preyed on weakened caribou. Increased vulnerability of prey allowed the wolf population to increase.

Radio-collaring wolves has made it possible for biologists to obtain the data now used for assessment of predator/prey relationships. Radio-collared wolves provided information on pack movements, dispersal, kill rates, territories, population size, mortality and reproduction. *Photo by Russ Dixon; courtesy of the Alaska Department of Fish and Game.*

Eastcentral Alaska

Most Alaskan moose populations increased in the 1950s and early 1960s in response to mild winters, frequent wildfires and predator control. After 1960, the eastcentral moose population declined (Gasaway et al. 1992). Unlike other Alaskan moose population (Kenai Peninsula and Matanuska Valley) decreases that were caused by severe winters and hunting, that of the eastcentral population was associated with wolf population increase after the termination of predator control programs. The moose population decline ended in 1976, at which time wolf numbers declined and moose survival increased.

From 1981 to 1983, Gasaway et al. (1992) reduced wolf populations within a 5,985-square mile (15,500-km^2) study area to test the hypothesis that wolf predation was limiting moose population growth. Wolf populations were reduced annually by 28 to 58 percent, but the reduction program had no statistically significant ($P > 0.05$) impact on moose calf survival. Gasaway et al. concluded that (1) wolf control was not severe enough or long enough to cause a significant increase in moose calf survival, and (2) significant increases in moose calf survival would have occurred had both wolves and grizzly bear populations been reduced at the same time.

Predation by grizzly bears was the largest cause of early moose calf mortality and an important source of adult moose mortality in 1985 and 1986 (Boertje et al. 1987, 1988). Predation by both bears and wolves accounted for 31 percent mortality of the moose population annually (see Figure 104). Hunting had an insignificant impact on the moose population. Reductions in wolf densities were not severe. Brown bear densities were considered high at 4.1 per 100 square miles (1.6 / 100 km^2), and bear predation probably accounted for the lack of moose population increase during the period of wolf control. Gasaway et al. (1992) concluded that, over large portions of interior Alaska and the Yukon, predation by bears and wolves limited moose populations to low densities that were well below habitat-carrying capacity. Grizzly bear predation on moose was not related to moose density. Consequently, bear predation could have greater impacts on low-density moose populations than on high-density populations (Boertje et al. 1988, Ballard et al. 1990, Ballard 1992a).

Eastcentral Saskatchewan

Moose populations in eastcentral Saskatchewan exhibited trends during the 1970s similar to those of other North American moose populations that were limited by predation (Stewart et al. 1985b). Early winter calf/cow ratios declined from 60 to 80:100 in the mid-1970s and to 30 to 40:100 by the late 1970s and early 1980s. Black bear predation on newborn moose was thought to be the primary cause of reductions in calf survival (Kowal and Runge 1981).

Stewart et al. (1985b) attempted to remove all black bears from 35- and 50-square mile (91- and 130-km^2) areas in 1983 and 1984, respectively. Bear densities were estimated at 34 and 47 bears per 100 square miles (13.1 and 18.1/100 km^2) in the two study areas, respectively. No other predator species occurred there. After bear removal, moose calf/cow ratios—40 to 80/100 cows in Area 1 and 39 to 87/100 in Area 2— in each area doubled those of the previous year. After control ceased the following year, calf/cow ratios declined and were not different from those of unmanipulated areas. Stewart et al. concluded that control did not provide lasting relief from predation.

If Stewart et al. (1985b) had continued with control, or at least continued to monitor the status of the moose population, a different conclusion may have emerged, because a one-time increase in calf survival can result in a large increase in calf/cow ratios for 1 to 2 years. The increase then is followed by a decline in calf/cow ratios in subsequent years because nonreproductive yearlings are counted as cows. Calf/cow ratios are higher than precontrol levels, but lower than those during and immediately after control efforts (Figure 107).

Stewart et al. (1985b) estimated that black bears killed 40 to 48 percent of the moose calves. They indicated that, when high-density bear populations were heavily hunted, moose calf/cow ratios ranged from 60 to 80:100. They also believed that black bear predation could be managed through manipulation of bear hunting seasons and bag limits.

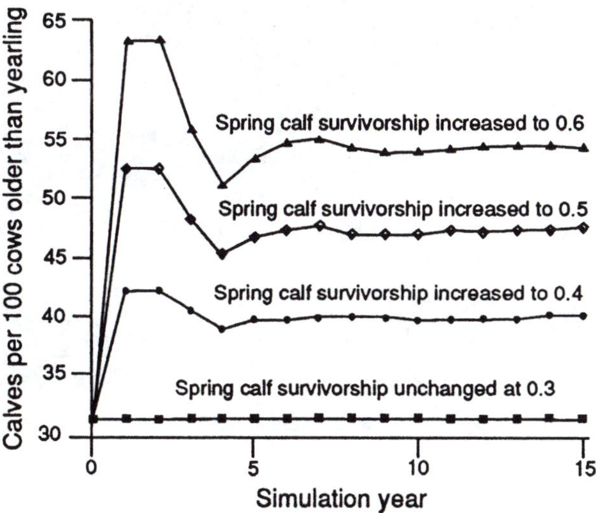

Figure 107. Modeled impacts on moose calf/100 adult (≥2 years old) cow ratios of increasing moose calf survival from 0.3 to 0.4, 0.5 and 0.6 during 1 year (from Miller and Ballard 1992).

Isle Royale, Michigan

Isle Royale has been studied longer than any other large predator/prey system in North America. Moose colonized the 210-square mile (544-km^2) island about 1910. The population quickly increased in the absence of predators, and it crashed by the early 1930s (Van Ballenberghe 1987). Moose increased again after a forest fire and, by 1948, the population was again declining when wolves colonized Isle Royale. In 1958, Mech (1966) initiated a long-term study of the island's wolves and moose, and concluded that wolf and moose populations reached a state of dynamic equilibrium in the late 1950s and early 1960s. Wolves preyed on heavily parasitized, diseased, old or otherwise inferior moose. Peterson (1977) re-examined the data and concluded that moose numbers increased during the 1960s and wolves did not limit the population. Wolf numbers remained relatively stable throughout this period. In the 1970s, calves and young adult moose became vulnerable to wolf predation because of severe winters. Moose numbers declined from about 1,200 to 600, and wolf numbers increased from 20–22 to 41 (Peterson 1979). Wolf numbers continued to increase and peaked in 1980, although the moose population already had declined (Peterson and Page 1983, Peterson 1992). Wolf numbers declined after 1980 to a record low of 12 animals by 1988, and the moose population increased to somewhat more than 1,600 (Figure 108).

Peterson (1977), Mech et al. (1987), Peterson and Page (1988), and McRoberts et al. (1995) all maintained that moose/weather/forage interactions were the main determinants of moose density on the island, and that wolf density and predation were of secondary importance. Peterson and Page (1983: 252) asserted that the relationship between wolves and moose was best described as "unstable cyclic interactions." Mech et al. (1987) suggested that deep snows during consecutive winters created a cumulative impact on the physical condition of moose and predisposed them to predation. Bergerud et al. (1983) disagreed with the conclusions reached by Peterson (1977) and Peterson and Page (1988). Bergerud et al. found no correlation between snow depths and the percentage of moose calf kills in winter, or with snow conditions faced by cows during gestation and the subsequent percentage of moose calves in the wolves' winter kill. They concluded that wolf predation significantly limited moose calf survival on Isle Royale.

Messier (1991) modeled the significance of food competition, wolf predation and snow accumulation as limiting factors of Isle Royale moose. He concluded, contrary to Mech et al. (1987), that snow accumulation had no measurable effect on moose numbers. Messier (1991: 377) wrote that "wolf predation and food competition explained 80 percent

Figure 108. Wolf and moose fluctuations on Isle Royale National Park, Michigan, 1959–1993 (from Peterson 1992). Moose population estimates from 1959 to 1981 were revised based on population reconstruction from recoveries of dead moose. Moose estimates from 1982 to 1993 are based on aerial surveys.

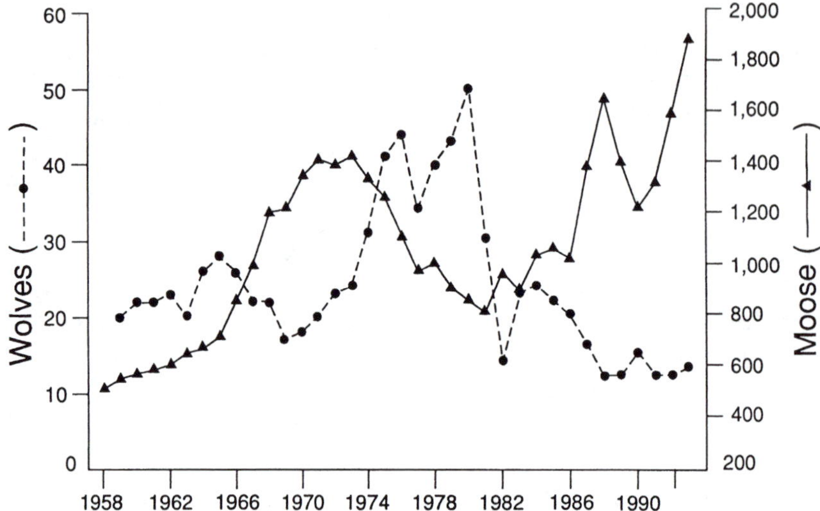

of the inter-annual variation of moose abundance. . . . Competition for food, but not wolf predation and snow, had a regulatory impact on moose." McRoberts et al. (1995) argued that Messier's analyses were flawed because his results were an artifact of his misuse of the Mech et al. (1987) snow depth data, and because of his smoothing of the ungulate population data, which removed 22 to 51 percent of the variability from the original data. McRoberts et al. (1995) concluded that snow depths affected adult ungulates cumulatively from winter to winter and resulted in impact on offspring after the third winter of deep snow. F. Messier (personal communication: 1993) reiterated no evidence for such effects and, because McRoberts et al. (1995) did not find any 1-year effect for snow, and the 3-year effect was not consistent among demographic parameters, there was cause for doubt in the McRoberts et al. conclusions.

The Isle Royale studies have provided a wealth of useful information about predator/prey relationships, but these findings must be viewed with caution for several reasons. First, the fact that Isle Royale is a relatively small island may limit application of the findings to mainland situations. Second, Isle Royale is the only area that is limited to a single predator species and one ungulate species. All mainland ecosystems have alternate ungulate species and contain one or both species of bears (brown/grizzly and black bears), which are also effective predators of moose (Ballard et al. 1981).

Deep and persistent snow accumulation, especially for consecutive winters, may create a cumulative impact on the physical condition of moose and increase their susceptibility to predation. Deep snow that forms a crust may support a pack of wolves, but not a moose (above), making defense or escape virtually impossible. *Photo by Ron Modaferri; courtesy of the Alaska Department of Fish and Game.*

Kenai Peninsula, Alaska

The circumstances involving historical moose population fluctuations on the Kenai Peninsula are similar to those in southcentral and interior Alaska, with the following exceptions: wolves were absent from the Peninsula from 1915 to 1960 (Peterson and Woolington 1982); the moose popula-

Predators have a low success rate in relation to attempts to kill adult moose. Mech (1970) reported that if a moose stands its ground upon attack (top), it usually is able to fend off wolves. Bubenik (1987) observed that fully mature and healthy moose stand their ground when confronted by wolves, whereas younger, inexperienced moose generally run and are killed. It is important for a young moose to use its mother's defensive behavior. In many cases, moose run to rivers or lakes to escape pursuing predators. The bottom photo shows typical escape behavior after a wolf encounter, with a cow moving behind her calf. *Top photo by Durward Allen. Bottom photo by Rolf Peterson; courtesy of Michigan Technological University.*

tion grew until 1970–1971 because of forest fires (Spencer and Hakala 1964, Bishop and Rausch 1974); and the Peninsula's climate is moderated by maritime conditions that often result in relatively mild winters (Van Ballenberghe 1987).

During the winters of 1971–1972 and 1972–1973, entire calf crops were lost because of deep snows and reduced forage availability (Bishop and Rausch 1974, Oldemeyer et al. 1977). High winter calf mortality continued into the early 1970s, and wolf populations increased to densities of 2.8 to 5.2 per 100 square miles (1.1–2.0/100 km^2) (Peterson et al. 1984). Moose population increases after the 1947 and 1969 burns peaked in 1971 and 1984, respectively, and then began declining (Schwartz and Franzmann 1991). Hunting of cow moose accelerated the decline in some areas. Wolves preyed disproportionately on calves and old adults (Peterson et al. 1984). Also, many of the moose calves killed by wolves in winter were malnourished (48 percent had 10 percent or less bone marrow fat) and adults exhibited a relatively high degree of debilitating factors (e.g., periodontitis, arthritis, broken legs) that could have predisposed them to death (Peterson 1982, Peterson et al. 1984).

Franzmann et al. (1980a) were the first to determine that black bears could be a significant cause of moose calf mortality. Of 47 newborn moose calves captured and radio-collared during 1977 and 1978, black bears killed 16 (34 percent), brown bears killed 3 (6 percent), wolves killed 3 (6 percent) and unknown predators killed 1 (2 percent) (Figure 109). Franzmann et al. questioned whether the Kenai Peninsula moose population could remain stable or increase in the face of predation on newborn calves by black bears and brown bears, year-round predation on calves and adults by wolves and other forms of mortality. But despite those pressures, the moose population increased in the early 1980s (Van Ballenberghe 1987).

Franzmann and Schwartz (1986) and Schwartz and Franzmann (1989, 1991) compared causes of calf mortality in the unproductive 1947 burn and in the productive high-quality habitat within the 1969 burn. Black bears killed 34 and 35 percent of the calves, respectively, whereas wolves and brown bears killed 5 to 13 percent, respectively. Total calf mortality before winter from all causes ranged from 51 to 55 percent. Moose densities were four times greater in the 1969 burn area (959 per 100 square miles [370/100 km^2]), where the population was increasing, versus the 1947 burn (259 per 100 square miles [100/100 km^2]) where the moose population was declining. The investigators concluded that the moose population in high-quality habitats were impacted less significantly by predation than were populations in poorer quality habitats because of higher reproductive performance (i.e., higher twinning rates and a high proportion of cows with calves).

Schwartz and Franzmann (1989) inferred from their studies and those of others (e.g., Gasaway et al. 1983, Ballard and Larsen 1987) that predation could influence both the rate of change and absolute density of a moose population. For example, after moose habitats are improved by wildfire or mechanical methods, predation may retard the subsequent rate of moose population increase, and peak moose population densities could be lower where predator populations are lightly harvested or naturally regulated versus those that are moderately harvested or managed. Where predator populations are near carrying capacity, a peak moose population would decline at a greater rate and lower density. The opposite effect was postulated for a predator-free environment.

Nelchina Basin, Alaska

Similar to other populations within Alaska, moose and caribou populations in the Nelchina Basin of southcentral Alaska increased in the 1950s and peaked in the early 1960s (Ballard et al. 1991, Ballard 1992a). Increases were attributed to favorable range conditions created by frequent wildfires, low hunter harvests and low numbers of predators (Bishop and Rausch 1974, Van Ballenberghe 1987, Ballard et al. 1991).

Between 1948 and 1953, federal programs reduced wolf populations to low levels—12 to 25 wolves in a 23,784-

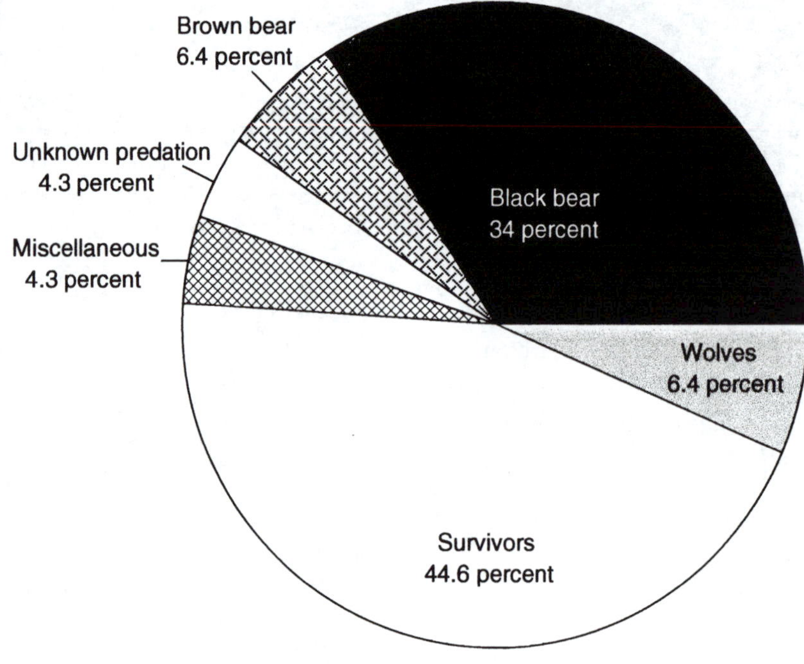

Figure 109. Causes of moose calf mortality in the 1947 burn on the Kenai Peninsula, Alaska, during 1977 and 1978 (from Franzmann et al. 1980).

square mile (61,600 km^2) area (Rausch 1967, 1969, Van Ballenberghe 1981, Bergerud and Ballard 1989). Use of poison during wolf control activities probably also reduced grizzly and black bear populations to low levels. The wolf control efforts ended in 1954 and wolf populations increased to relatively high levels.

Rausch (1967, 1969), Van Ballenberghe (1981), Ballard et al. (1987) and Bergerud and Ballard (1989) reviewed the history of wolves in the Nelchina Basin and, although actual numbers were not known, Rausch (1967, 1969) estimated the population at 350 to 450 wolves in 1965 (2.0 to 2.6 per 100 square miles [0.8–1.0/100 km^2]). Wolf densities peaked at 2.7 per 100 square miles (1.0/100 km^2) by 1975; it declined through 1982, as a result of state wolf reduction programs and intense public hunting and trapping (Ballard et al. 1987). Spring wolf densities gradually increased after 1983 through 1986, then declined slightly in 1987 through 1989. However, by 1990, the wolf population reached the highest density ever recorded (2.6–6.0 per 100 square miles [1.0–2.3/100 km^2]) (Tobey 1989, Becker and Gardner 1991, Ballard 1992b).

As wolf and bear populations increased, the Nelchina Basin moose populations declined to record low numbers by 1975 (Ballard et al. 1991). The decline was triggered by several severe winters, and exacerbated by excessive human harvests and predation. Wolf population dynamics, food habits and predation rates were studied with the aid of radio-telemetry during 1975 to 1986 (Ballard et al. 1987). Moose composed 70 percent of all ungulate prey. Wolves preyed on calves approximately in proportion to calf occurrence in the moose population during May through October. Proportionately more calves were killed during winter, but adult moose were the most common year-round prey. According to bone marrow fat analyses, adult cows killed by wolves were not near starvation, although they were older than the average adult in the population (9.5 versus 7.7 years, respectively). Wolf packs killed about one adult moose equivalent per 7 to 16 days in summer and one equivalent per 5 to 11 days during winter (Ballard et al. 1987).

From 1976 to 1978, wolf populations within a 2,804-square mile (7,262 km^2) experimental area were reduced by 42 to 58 percent annually by the Alaska wildlife agency. When wolf control ended, wolf populations returned within 3 years to precontrol levels, with annual finite rates of increase ranging from 1.04 to 2.40 (Ballard et al. 1987). Despite significant correlations between spring wolf densities and autumn calf/cow moose ratios, improvements in moose calf survival were not as great as those of interior Alaskan moose populations (Gasaway et al. 1983, Ballard et al. 1987).

Modeling of the moose population suggested moose increased at annual finite rates of 1.03 to 1.06, whereas aerial census data suggested finite rates ranging from 1.02 to 1.23 (Ballard 1992a, 1992b). Modeling also suggested that the

Federal wolf reduction efforts before 1954 were undertaken to allow for increases of both moose and caribou in Alaska's Nelchina Basin. Aerial shooting and poisoning effectively limited the wolf population and enabled rise in the Basin's moose population. When the reduction efforts were terminated, both population trends reversed. *Photo courtesy of the Wildlife Management Institute.*

moose population would not have increased if wolf populations had not been reduced from 1975 levels. Wolf predation was not limiting moose population growth from 1976 through 1984, but probably was limiting before 1976. Because calf survival increased less than originally projected, biologists sought to determine the causes of mortality directly by collaring neonates (Ballard et al. 1979).

During 1977 through 1979 and 1984, 198 newborn calves were collared and monitored (Ballard et al. 1981a, 1991). Grizzly bears killed 87 (44 percent) of the calves, and wolves killed only 6 (3 percent) (Figure 110). Ninety-six percent of the natural mortality occurred before the second week of July each year. These studies by Ballard et al. were the first to document that predation by grizzly bears can be a significant factor limiting moose calf survival.

Subsequent telemetry studies indicated that grizzly bears killed moose calves at the rate of one per 7 to 12 days during late spring and early summer, and were a significant year-round predator of adult moose (Boertje et al. 1988, Ballard et al. 1990, Ballard 1992a). Temporary translocation of 60 percent of the grizzly bears from a 1,326-square mile (3,436 km^2) study area in 1979 resulted in a significant improvement in moose calf survival—from 34 to 58 calves:100 adult cows—for 1 year (Figure 111) (Ballard and Miller 1990).

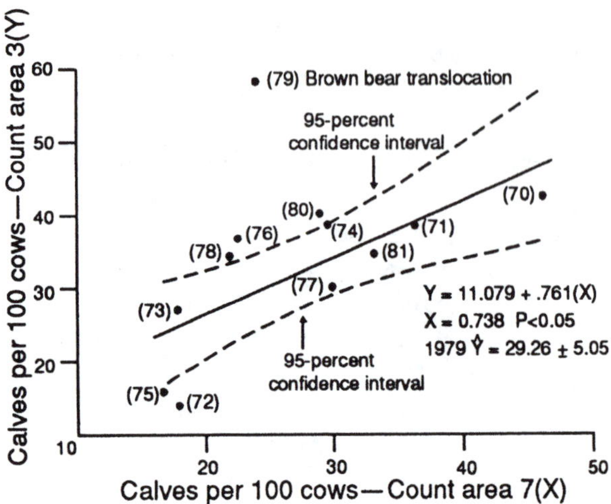

Figure 111. Relationship between autumn moose calf/cow (>2 years old) ratios in Count Area 3, where grizzly bear densities were lowered by 60 percent in 1979, in relation to adjacent Count Area 7, where bear densities were not manipulated (modified from Ballard and Miller 1990). Regression line and 95 percent confidence interval are for 1970 through 1981 and exclude 1979.

Grizzly bear hunting seasons were liberalized after 1979 in an effort to improve moose calf survival (Miller and Ballard 1992). Bear densities were reduced by 36 percent over a 7-year period (1980 through 1987), with no significant ($P >$ 0.05) increase in moose calf survival (Miller and Ballard 1992). From 1978 through 1986, wolf densities increased within the area where bear reductions were most pronounced (Ballard 1992b). Population modeling suggested that, at the observed rates of wolf predation, numerical increases in wolf numbers were sufficient to mask improvements in moose calf survival that might have resulted from bear reduction (Ballard 1992b). Four other alternate but

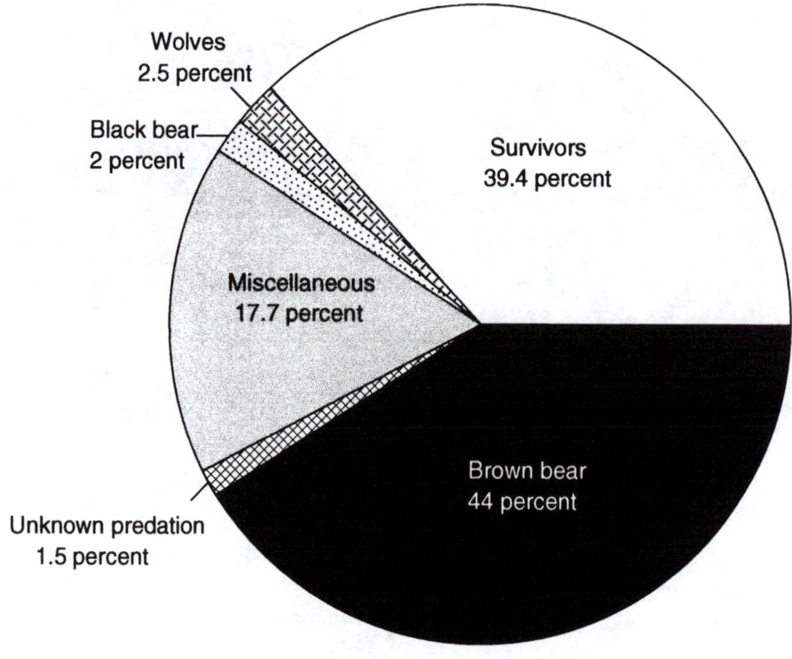

Figure 110. Causes of moose calf mortality in the Nelchina Basin of south-central Alaska during 1977 through 1979 and for 1984 (from Ballard et al. 1991).

untested explanations included (1) bear predation may have been more compensatory at higher moose densities than at lower moose densities; (2) remaining bears may have increased their kill rates on calves; (3) hunter-induced reductions in bear density may have had different influences on predator/prey relationships than did removal from all sex and age classes; and (4) small to moderate increases in moose calf survival could be difficult to detect owing to sampling variation in moose counts.

Ballard et al. (1991) concluded that the moose population increased after 1975 because of the combined effects of mild winters, reduced hunter harvest and reduced predation. Reductions in bear density after 1980 to allow the moose population to increase appeared to have been unnecessary, because the population had been increasing for 5 years before the liberalized bear harvest program (Ballard 1992b, Miller and Ballard 1992).

Alberta

Moose ecology was studied in two areas of Alberta during the 1970s, and the results provided a good indication of contrasting predator/prey relationships.

In the first study, moose/wolf relationships were studied in northeastern Alberta from 1975 through 1978 in a 9,562-square mile (25,000 km^2) area near Fort McMurray (Fuller and Keith 1980, Hauge and Keith 1981). Moose and wolf densities averaged 47 and 1.6 per 100 square miles (18 and 0.6/100 km^2), respectively, resulting in a moose/wolf ratio of 30:1. Annual moose calf mortality was 73 percent, of which 29 percent was attributable to wolf predation (Hauge and Keith 1981). Black bears were suspected of killing many neonates, but no quantifiable estimates were available. Calf mortality was highest (39 percent) the first month after birth. Calf/cow ratios declined from an average of 72:100 at birth to about 40:100 by late February and March.

Fuller and Keith (1980: 600) compared estimated annual wolf kill (15 percent) with estimated annual recruitment (19 percent), and concluded that "wolves were clearly the major limitation on this apparently stationary or slowly declining moose population." Van Ballenberghe (1987) and Boutin (1992) pointed out problems with their assessment. For example, wolf kill estimates may have been inflated, and accurate estimates of moose population size were difficult to obtain. Also hunting mortality, which likely was underestimated, accounted for about half of the annual increment. Therefore, hunting and wolf predation appeared to be the primary limiting factors, whereas the role of black bear predation was unknown.

In the second Alberta study area (111 square miles [287 km^2]), where wolves and brown bears were absent and black bears scarce, Rolley and Keith (1980) and Mytton and Keith (1981) reported one of the highest reproduction and calf survival rates observed in North America. Initial production was estimated at 167 calves per 100 cows. Over a 13-year period (1965 through 1978) there was a winter average of 106 calves per 100 cows and a mean twinning rate of 41 percent (Rolley and Keith 1980). Annual calf and adult mortality rates averaged 16 and 33 percent, respectively. The investigators concluded that high reproductive rates resulted from abundant high-quality browse, and that high calf and adult survival rates were attributable to an absence of predators.

Pukaskwa National Park, Ontario

Bergerud et al. (1983) censused moose and wolf populations in Pukaskwa National Park from 1975 through 1979. To ascertain why the moose population had not increased based largely on circumstantial data, four hypotheses were tested—low reproduction, starvation, egress and predation. The moose population appeared to have relatively high reproductive rates, few moose starved and few moose emigrated from the area. Observed wolf/moose ratios and estimated rates of predation based on a review of the literature led Bergerud et al. (1983) to conclude that the park's moose population was limited by wolf predation.

Thompson and Peterson (1988) argued that the data were not precise and alternate hypotheses not tested. Bergerud and Snider (1988) countered that the conclusions were consistent with other studies where wolf predation limited moose population growth.

Quebec

Moose/wolf relationships were studied by Crête et al. (1981a, 1981b), Crête and Jordan (1982a, 1982b), and Messier and Crête (1984, 1985) within the 5,257-square mile (13,615 km^2) La Verendrye Reserve in southwest Quebec. They concluded that winter forage did not limit moose population growth, and that food shortages could not explain why the moose population was at an apparent equilibrium density of about 104 per 100 square miles (40/100 km^2). However, they concluded that the wolf predation rate was density dependent within a moose density range of 44 to 104 per 100 square miles (17–40/100 km^2). Wolf and black bear densities in this area were estimated at 3.6 per 100 square miles (1.4/100 km^2) and 65 per 100 square miles (25/100 km^2), respectively. Bears were not considered to be important predators of moose in this area.

To test the hypothesis that wolf predation was limiting the moose population, wolf populations were reduced by

40 percent annually in a 270-square mile (700-km^2) area from the winter of 1981 through winter 1984 (Crête and Messier 1984). Early winter calf/cow ratios increased from an average of 40:100 for 2 years before control to 52:100 for 2 years after control. Ratios in a nearby control (i.e., unmanipulated) area did not increase. Crête et al. (1981a), Crête and Messier (1984) and Messier and Crête (1985) maintained that the results of their experiments were consistent with the hypotheses that wolf predation limited moose population growth and that wolf predation caused moose populations to stabilize at equilibrium densities of 104 per 100 square miles (40/100 km^2).

Crête and Jolicoeur (1987) also tested the hypothesis that wolves and black bears were limiting moose population growth. Wolves and black bears were reduced in separate areas—270 and 139 square miles (700 and 360 km^2), respectively—and compared with a third area (348 square miles [900 km^2]), where there was no concerted predator reduction. Wolf densities were reduced from 5.4 to 2.0–2.8 per 100 square miles (2.1 to 0.8–1.1/100 km^2). Black bear densities were reduced from about 60 to 41 per 100 square miles (23 to 16/100 km^2), or about a 20 to 30 percent reduction. Before the predator reductions, moose population density was possibly as high as 78 to 104 per 100 square miles (30–40/100 km^2) (Crête and Jolicoeur 1987). After the reductions, calf/cow ratios (110:100) were 38 percent higher in the bear removal area than in either the wolf reduction or control (unmanipulated) areas (80 calves:100 cows each), but the differences were not significant. Crête and Jolicoeur (1987) attributed the lack of statistical differences to insufficient reductions of predator populations, annual and regional variations among study areas, sampling error, small sample sizes and small study areas.

Southwest Yukon Territory

Moose populations within many areas of the Yukon exhibited trends similar to those described by Gasaway et al. (1983, 1992) for interior Alaska. Using methods described by Ballard et al. (1979) and Ballard and Larsen (1987), Larsen et al. (1989a) studied the causes of moose calf mortality. As in most of the Alaskan studies, the Yukon wolf densities were lowered as part of a reduction program. Grizzly and black bear densities were both estimated at 4.1 per 100 square miles (1.6/100 km^2), and wolf densities in late winter ranged from 0.6 to 1.5 per 100 square miles (0.22–5.9/100 km^2) (Larsen et al. 1989a, 1989b). At 86 per 100 square miles (33/100 km^2) moose were the most abundant ungulate prey (Ballard and Larsen 1987).

Ninety-five (81 percent) of 117 radio-collared moose calves died in a 1-year period (Larsen et al. 1989a). Predators killed 63 (66 percent), including 40 (42 percent) by grizzly bears. Between calf births and June 20, bears and wolves accounted for 60 and 10 percent of the mortalities, respectively. After June 20 and through the remainder of the year, wolves accounted for 54 percent of the mortalities, and grizzlies caused 27 percent. Grizzly bears, wolves and hunting accounted for 50, 26 and 9 percent, respectively, of the annual mortalities (n = 534) of the total moose population. Larsen et al. (1989a) and Ballard (1992a) concluded that predation by grizzly bears was the dominant cause of moose mortality and the major factor limiting moose population growth in the Yukon.

Tanana Flats, Alaska

Moose populations in the Tanana Flats exhibited trends similar to those of the Nelchina Basin population from the 1950s through the early 1970s. Wolves were controlled by the federal government from 1954 to 1960, and peak numbers of moose occurred in the late 1960s and early 1970s (Gasaway et al. 1983). Wolf populations peaked at the same time that moose were declining in the late 1960s and early 1970s.

Wolf reduction efforts—at a rate of 38 to 61 percent annually—were initiated in 1976 and continued through 1982 (Gasaway et al. 1983). Bear predation was not identified as a significant mortality factor on Tanana Flats, but a large number of calves was lost to unknown causes between birth and autumn moose counts (Boutin 1992). After wolf reduction, calf and yearling moose survival increased two- to fourfold, and adult mortality was reduced from 20 to 6 percent annually. Moose numbers increased from about 2,800 in 1975 to more than 5,000 by 1982 and to 8,000 to 10,000 by 1988 (Gasaway et al. 1983, McNay 1989a). After termination of wolf reduction activities in 1982, wolf numbers increased to 180 to 220 by 1983 (McNay 1989b). Wolf numbers stabilized at about 195 during 1985 through 1987, and slowly increased thereafter.

Gasaway et al. (1983: 34) proposed three general categories of moose/wolf interactions to help guide wildlife managers in areas where moose are the primary prey of wolves: "First, at 20 moose/wolf, predation is usually sufficient to cause a decline in moose abundance and low survival of both calves and adults (Peterson and Page 1983).... Secondly, at 20–30 moose/wolf, predation can be the primary factor controlling numbers of moose; whether the moose population remains stable or declines is largely dependent on the combined effect of other factors influencing the dynamics of the moose population, including hunting, food supply, alternate prey, and winter severity (Peterson 1976, Fuller and Keith 1980, Peterson and Page 1983). . . .

Carcass remains of a freshly (<12 hours) predator-killed moose usually indicate the type of predator that made the kill, because of certain typical or stylized feeding patterns of the predators. Black bears, for example, typically leave the hide inverted (top), and all parts but the hide, lower legs and head are eaten. Grizzly bears sometimes only consume the brains and viscera. In these circumstances, the ears, eyes and tongue also are taken. When bears totally consume a carcass, usually all that remains are small bone fragments, hoof sheaths and bear scat containing hair, bone fragments and flesh. Wolves (below), on the other hand, generally do not consume the brains, eyes, tongue or viscera. Characteristically, they leave ribs and long bones chewed on the ends. In some cases of carcasses found more than a week or two after abandonment by the predator, identification of the killer may be masked to some extent by scavenging. *Top photo by Charles C. Schwartz; courtesy of the Alaska Department of Fish and Game. Bottom photo by Luray Parker; courtesy of the Wyoming Game and Fish Department.*

Table 16. Causes of mortality and survival rates of radio-collared moose calves to November in relation to observed predation rates and predator densities in North America (modified from Ballard (1992)

	Southcentral Alaska				Kenai Peninsula, Alaska						
	Areas 1–3	Area 1	Area 4	Pooled areas	1947 Burn	1969 Burn	Southwestern Yukon[a]	Eastcentral Alaska	Saskatchewan	New Brunswick	Newfoundland
Years	1977, 1978	1979	1984	1977–1984	1977, 1978	1981, 1982	1983, 1985	1984	1982	1983, 1985	1983, 1988
Numbers of calves	124	28	46	198	47	74	117	33	12	11	88
Percentage mortality caused by											
Grizzly bear	41.9	42.9	52.2	44.0		6.4		2.7	41.9	51.5	
Black bear			8.7	2.0	34.0	35.1	3.4	3.0	50.0	9.1	30.0
Grizzly and black bears						2.7					
Wolf	1.6		6.5	2.5	6.4	1.4	17.9	15.2			
Unknown predation	2.4			1.5	4.3	2.7	2.6				
Miscellaneous	4.8	14.3	13.1	8.1		5.5	6.0	12.1		9.1	
Unknown	3.2		2.2	2.5	1.3	1.4	9.4				
Percentage survival	46.0	42.9	17.4	39.4	44.6	48.5	18.8	18.2	50.0	81.8	70.0
Density[b]											
Moose	168.4	181.3	231	168.4	259	958.3	57	45.3	116.6	?	777
	(65)	(70)	(89.2)	(65)	(100)	(370)	(22)	(17.5)	(45)		(300)
Grizzly bear	6.2	2.6	7.3	6.2–7.3	3.1–7.3	3.1–7.3	4.1	4.1	0	0	0
	(2.4)	(1.0)	(2.8)	(2.4–2.8)	(1.2–2.8)	(1.2–2.8)	(1.6)	(1.6)			
Black bear	0	0	23.3	0–23.3	53.1	66.8	4.1	2.1–2.8	51.8–103.6	mod.?	147.6
			(9.0)	(0–9.0)	(20.5)	(25.8)	(1.6)[c]	(0.8–1.1)	(20–40)		(57)
Wolf	0.47–0.93	0.59	0.72	0.47–0.93	2.85	2.85	1.1	1.0	Low?	0?	0
	(0.18–0.36)	(0.23)	(0.28)	(0.18–0.36)	(1.1)	(1.1)	(0.41)[d]	(0.40)			
Calf moose kill rate by[e]											
Grizzly bear				0.097			0.085[f]	0.143			
Black bear			0.025		0.019[f]	0.103[f]					
Adult moose kill rate by[e]											
Grizzly bear				0.023			0.022				
Sources[g]	1	2	3	3, 4	10, 12	9, 12	11	7, 8, 15	5,13	6	14

[a] Causes of mortality and survival rates are annual estimates.

[b] Number per 100 square miles (100 km^2).

[c] Black bear densities not estimated but thought to be similar to grizzly bear densities (D. G. Larsen personal communication).

[d] Average late winter density for 1983 and 1985.

[e] Kill per bear per day.

[f] Assumed all mortalities occurred between birth and mid-July, i.e., 60-day period. Derived by dividing estimated number of calves killed by 60 days.

[g] 1 = Ballard et al. (1982), 2 = Ballard and Miller (1990), 3 = Ballard et al. (1990), 4 = Ballard et al. (1991), 5 = Beaulieu (1984), 6 = Boer (1988), 7 = Boertje et al. (1987), 8 = Boertje et al. (1988), 9 = Franzmann and Schwartz (1986), 10 = Franzmann et al. (1980a), 11 = Larsen et al. (1989a), 12 = Schwartz and Franzmann (1991), 13 = Stewart et al. (1985), 14 = W.E. Mercer (personal communication), and 15 = R.D. Boertje (personal communication).

Thirdly, at >30 moose/wolf, predation can be significant but not necessarily limit growth. Moose populations are likely to remain stable or increase if they are below ecological carrying capacity and if other sources of mortality are not exceptionally great (Mech 1966, 1970; Peterson 1977; Allen 1979; W. Ballard, unpubl. data; R. Peterson, unpubl. data). . . . However, regardless of how much above 30 moose/wolf a ratio is, high additive mortality (from hunting, predation, severe winters, or other factors) can cause a decline."

Effects of Predation

The 11 case histories reviewed in this chapter constitute the principal data used to evaluate the effects of predation on moose. Predator/moose relationships are of interest to many biologists because moose are circumpolar, and because distribution and habitats for the most part have not been impacted to the degree experienced by other native ungulates.

Interpretation of predation impact is hampered by misuse of various terms and definitions, lack of testable hypotheses and tedious examination of alternative explanations. Wildlife biologists use the terms *limiting, regulating* and *control* interchangeably throughout scientific literature, causing considerable confusion and misunderstanding (see Sinclair 1991, Boutin 1992, Van Ballenberghe and Ballard 1994). The three terms have distinct meanings and important implications for moose population dynamics.

Sinclair (1991) and Messier (1991) stressed that *limiting* factors are composed of both density-dependent and density-independent factors, as discussed in Chapter 6. *Regulating* factors, on the other hand, are composed solely of density-dependent factors and are a subset of limiting factors (Messier 1991). Consequently, all mortality factors whether density dependent or density independent can be limiting factors, but only those that are density dependent can be regulatory (Van Ballenberghe and Ballard 1994). Sinclair (1991) stressed that factors that set the position of equilibrium are limiting factors. He also indicated that, by definition, predation of any sort is limiting. However, this also depends on whether such predation is totally compensatory or additive. If predation is largely compensatory, replacing another form of mortality, it may not be limiting. If such mortality is totally additive, then predation may be limiting to some extent.

The term *control* often has been used synonymously with regulation (see Thompson 1929, Nicholson 1933, Lack 1954, Milne 1957). However, Keith (1974) defined control as the maintenance of a population through a combination of density-dependent and density-independent factors. Nevertheless, some investigators have used Keith's definition interchangeably with regulation (e.g., Gasaway et al. 1983, Ballard and Larsen 1987, Ballard et al. 1991), and this has created confusion. Therefore, we recommend that, in reference to population dynamics, only the terms limitation and regulation be used.

Predation as a Limiting Factor

Some biologists question whether predation actually limits moose population growth. For example, Boutin (1992: 125) stated: "Contrary to reviews by Ballard and Larsen (1987) and Van Ballenberghe (1987) this critique suggests that evidence for predation acting as a major limiting factor in most moose populations is less than convincing."

We suggest that the evidence for predation acting as a major limiting factor in many moose populations is strong but the evidence that predation regulates moose populations is debatable.

PREDATION ON CALVES

Predation is a major cause of mortality of calves in many moose populations (Table 16). Mortality studies using radio-telemetry have demonstrated that predators kill large numbers of elk calves (Schlegel 1976), caribou calves (Adams et al. 1989, Whitten et al. 1992) and moose calves (Franzmann et al. 1980a, Ballard et al. 1981a, Larsen et al. 1989a, 1989b, Ballard et al. 1991, Osborne et al. 1991). The largest proportion of annual mortality in many moose populations occurs among neonates within the first 6 weeks of life (Figures 104 and 112). Black bears are known to kill 2 to 50 percent of newborn moose calves, grizzly bears have been shown to kill 3 to 52 percent and documentation indicates that wolves take 2 to 18 percent (Table 16). Gasaway et al. (1983) demonstrated through predator removal studies that, in some cases, wolves apparently kill more than 18 percent of moose calves in a given population.

Depending on the magnitude of other mortality factors, wolf predation is a significant source of mortality even at levels considerably lower than 18 percent. In the Nelchina Basin of southcentral Alaska, radio-telemetry studies and analyses of wolf scats suggested that wolves annually killed 3 to 9 percent of the moose calves born during late spring and early summer (Ballard 1992a, 1992b), yet spring wolf densities were inversely correlated with autumn calf/cow ratios (Ballard et al. 1987).

Despite heavy bear predation, reductions in wolf densities, coupled with mild winters and bull-only hunting, resulted in a moose population increase of 3 to 6 percent annually (Ballard et al. 1991, Ballard 1992a, 1992b). Evidence that both grizzly bears and black bears can be significant

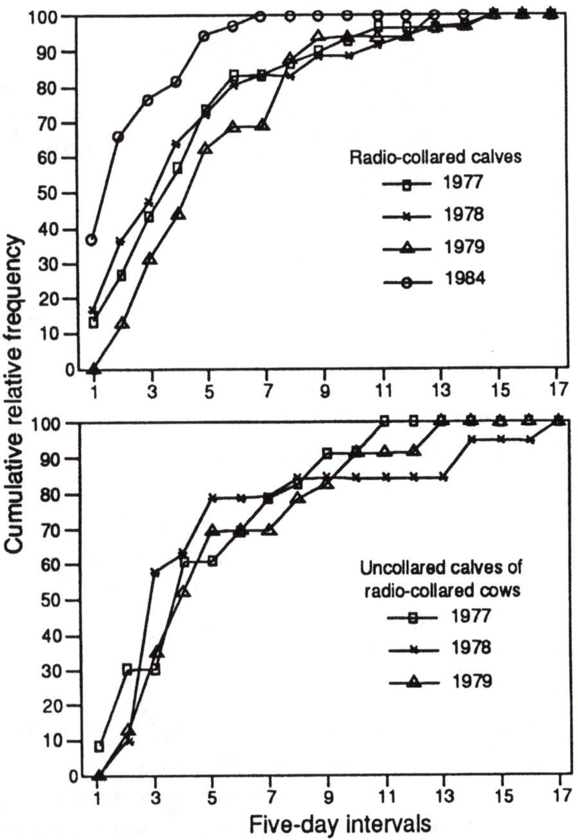

Figure 112. Dates of mortalities of radio-collared and uncollared moose calves within Game Management Unit 13 of southcentral Alaska, 1977 to 1984 (from Ballard et al. 1991). First 5-day interval begins on May 26 and the last 5-day interval ends on August 3 (interval 14). Intervals 15 to 17 correspond to August through September.

causes of moose calf mortality (Franzmann et al. 1980a, Ballard et al. 1981a, 1991) confounds attempts to understand and manage the predator/prey relationships.

On the basis of review of several North American predation studies involving bears, Ballard (1992a) suggested that grizzly bears are a significant source of moose calf mortality when bear densities exceed 4.1 per 100 square miles (1.6/100 km^2) (Table 16). The grizzly bears' kill rates range from 0.143 to 0.85 calf per bear per day.

Where they outnumber grizzly bears and wolves by factors of 10 and 30, respectively, and their densities exceed 52 per 100 square miles (20/100 km^2), black bears are a significant cause of moose calf mortality (Ballard 1992a). They prey at rates of 0.02 to 0.09 moose calf per bear per day (Table 16).

Mortality attributable to predation can be additive or compensatory depending on the relationship between a moose population and the carrying capacity of its habitat (Ballard 1992a). The Nelchina Basin, Alaska, in the 1970s and early 1980s, and eastcentral Alaska and southwestern Yukon from 1970 to 1992 represented circumstances of moose populations well below habitat-carrying capacity and where predation was largely or totally additive.

Perhaps the best example of bear predation being largely compensatory was on the Kenai Peninsula of Alaska (Schwartz and Franzmann 1991, see also Ballard 1991, 1992a). Black bears killed 35 percent of the moose calves in some areas. But in other areas where survival of calves to autumn was high, they were not a significant cause of calf mortality. However, in the latter areas, many calves starved even during mild winters. In such cases, reduction of black bear predation may do little to improve calf survival. However, if bear predation and winter losses are largely additive sources of mortality, assuming that birth processes are density independent, then a case could be made that predator reduction might increase moose population growth.

Ballard and Larsen (1987), Boertje et al. (1988), Larsen et al. (1989a), Ballard and Miller (1990), Ballard et al. (1990) and Schwartz and Franzmann (1991) concluded that bear kill rates occur independent of moose densities (see also Ballard 1992a). On the basis of these studies, Boutin (1992: 124) erroneously concluded that bear predation is density dependent. Grizzly bear kill rates did not increase with moose density, but black bear kill rates did (Ballard 1992a). Evidence that wolf predation on newborn moose calves is density dependent is weak at best and warrants further study. Virtually no information is available on the condition of neonatal moose calves killed by bears, although weights and blood values (Franzmann et al. 1980b) and results of removal studies (Ballard and Miller 1990) indicated that most calves were healthy, because other calves survived when relieved of predation pressure.

If bear kill rates on moose calves are density independent, as most investigators believe, they can have significant impact on both low- and high-density moose populations. Gasaway et al. (1992) indicated that density-independent processes, such as bear predation, can play a major role in determining the range of densities within which moose populations fluctuate. If other factors, such as wolf predation or food competition, regulate moose population growth, then bear predation could be extremely important in determining the density that moose populations attain.

Predation by wolves usually has its greatest impact on calf moose during winter. Although wolves may kill only 3 to 9 percent of the calves during summer, they can exert a strong limiting effect on moose populations beginning in September and lasting through winter, and particularly during January through May, when deep snow makes calves vulnerable to predation. Ballard et al. (1987) indicated wolves killed calves disproportionately to their occurrence

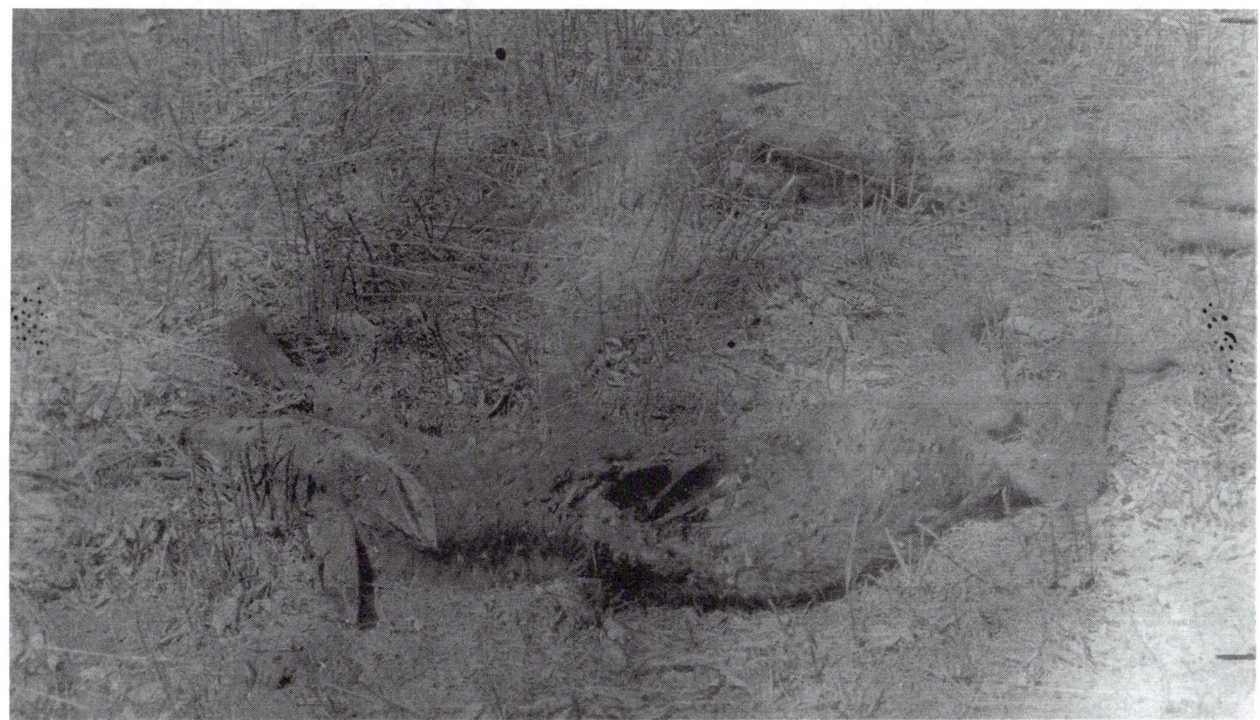

Black bears can kill up to 50 percent of the calves in a moose population (top photo features the carcasses of twin moose calves killed by a black bear), but only rarely take adult moose. Usually, when a black bear is seen feeding on the carcass of an adult moose (bottom), it is scavenging. Hunters and others who find recently killed moose partially covered ("buried") with brush, are advised to retreat, because either a black or brown bear may be resting nearby and can be quite protective of food caches. *Top photo by Olaus Murie; courtesy of the U.S. National Archives. Bottom photo by John Oldemeyer.*

in the moose population during November through April; calves made up 40 percent of the observed kills. Marrow fat values of wolf-killed moose calves during winter were higher than those of calves dying of starvation. Gasaway et al. (1992) demonstrated that wolves accounted for about 15 percent of first-year mortality in eastcentral Alaska (Figure 104).

PREDATION ON ADULT MOOSE

Although bears prey primarily on calf moose, recent studies indicate that grizzly bears also can be significant predators of adult moose. Grizzly bears can prey on adult moose at a rate of about 0.02 moose per adult bear per day (Ballard 1992a). Boertje et al. (1988) found no significant differences ($P > 0.05$) in killing rates on adult moose between spring, summer and autumn, and that adult grizzlies killed adult moose at rates of 0.6 to 3.9 per bear per year. Boertje et al. (1988) reported that adult bruins killed more adult moose than did adult sows. Ballard et al. (1990) found no significant differences ($P > 0.05$) in bear killing rates among years, sexes, ages or sows with young. Although black bears may occasionally kill adult moose, they are not considered significant predators in western North America (see Ballard 1992a).

Wolf predation may reduce annual survival of adult moose and may be compensatory or additive depending on the individual moose populations' relationship to habitat-carrying capacity. Wolves generally kill relatively old adults that may (Peterson et al. 1984) or may not (Ballard et al. 1987) be in poor physical condition or have some other form of debilitation. Deep snow accumulations can heighten the vulnerability of moose to predation (Coady 1974a). And by

Where moose and grizzly or brown bear ranges overlap, grizzlies can be a significant predator of moose. They may kill from 3 to 52 percent of the moose calves in a population. They can also be effective predators of adult moose and may kill from 0.6 to 3.9 adult moose per year. Bear predation can be additive or compensatory mortality depending on a moose population's relationship to habitat-carrying capacity. At right, a grizzly feeds on a 3-week-old moose calf in Denali National Park, Alaska, in early June 1996, about an hour after the kill. Before then, the calf and its mother were pursued by the bear for about 0.5 mile (0.8 km) before being overtaken as they were entering a stream. The cow was unable to fend off the bear's attack on the calf. *Left photo, also in Denali, by William E. Ruth. Right photo by Victor Van Ballenberghe.*

stressing cows, hence unborn calves, such conditions may affect moose vulnerability to predation by wolves in other seasons (Mech et al. 1987).

In areas where moose are the principal prey, wolf packs kill moose at rates of one adult per 7 to 16 days in summer and 5 to 11 days in winter. Peterson et al. (1984) suggested that wolf kill rates were lower in summer than winter, but James (1983) and Ballard et al. (1987) concluded that there were no differences.

Van Ballenberghe (1987) noted that the data on wolf predation impact on adult moose survival were equivocal (see Chapter 6). Peterson (1977) reported annual moose survival rates of 0.87 for adults and yearlings where wolves were the principal cause of mortality. Bangs et al. (1989) indicated an adult survival rate of 0.92 where wolves and hunting were the primary mortality factors. For an unhunted predator-free population in Alberta, Mytton and Keith (1981) reported an annual survival rate of 0.84. And in another moose population subjected to intensive predation and hunting in Alberta, adult survival rate averaged 0.75 (Hauge and Keith 1981). Gasaway et al. (1983) reported that the adult moose survival rate improved from 0.8 to 0.94 after wolf densities had been lowered in the Tanana Flats, Alaska. Ballard et al. (1991) reported an average adult cow survival rate of 0.95 in a moose population for which wolf densities were reduced, winters were relatively mild and hunting pressure was on bull moose.

We conclude that wolf predation can lower adult moose survival rates, but the magnitude of this effect varies by population.

Predation as a Regulating Factor

Four models currently are under consideration to describe the interactions between moose and their predators (Van Ballenberghe 1987, Boutin 1992, Van Ballenberghe and

Wolves are a major predator of moose. They are most effective as an organized pack. In habitats where moose are their principal prey, a pack kills at rates ranging from one adult moose per 5 to 16 days. This rate varies with the size of the pack, which may contain as few as 2 and as many as 22. *Photo by Rolf Peterson; courtesy of Michigan Technological University.*

Ballard 1993). Three of the models assume that moose densities are below habitat-carrying capacity, but the fourth does not. Descriptions of each hypothesis and how the hypotheses potentially affect moose populations are discussed below.

RECURRENT FLUCTUATION HYPOTHESIS

Moose populations under this model are characterized by fluctuating densities that are not at equilibrium (Figure 113). Changes in moose density can occur from many factors, including weather, hunting, food quality and accessibility, but predation is the primary factor that limits density most of the time. Predation on moose is inversely density dependent at high moose densities, and is not regulatory. Moose densities occasionally can avoid the effects of predation and reach high levels such that competition for food initiates a moose population decline. The decline could be accelerated or extended by inversely density-dependent predation. Such systems usually are characterized by one principal predator species and one prey species, but multiple predator/prey systems also can be included in this hypothesis. After a moose population is altered, it does not return to a predictable density.

LOW-DENSITY EQUILIBRIA OR SINGLE STABLE-STATE EQUILIBRIA HYPOTHESIS

Moose populations are regulated by density-dependent predation at low densities for extended time periods (Figure 114). They do not escape the constraints of predation for many decades. Food competition is not important, because the population never reaches high densities. Moose populations regulated by predation reach a low-density equilibrium and do not increase unless predation is reduced. When reduced predator populations recover to precontrol densities, moose populations also return to precontrol densities. Moose populations under such systems usually are subject to predation from multiple species, such as wolves, black bears and brown bears, and may occur sympatrically with such other prey as caribou, deer and beaver.

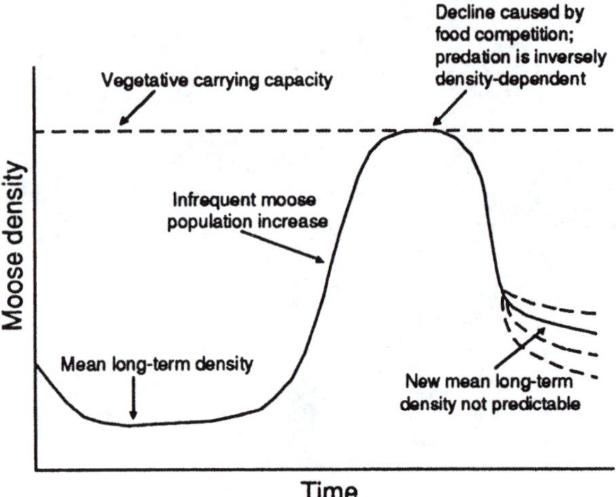

Figure 113. Conceptual model of moose density regulation under the recurrent fluctuations hypothesis.

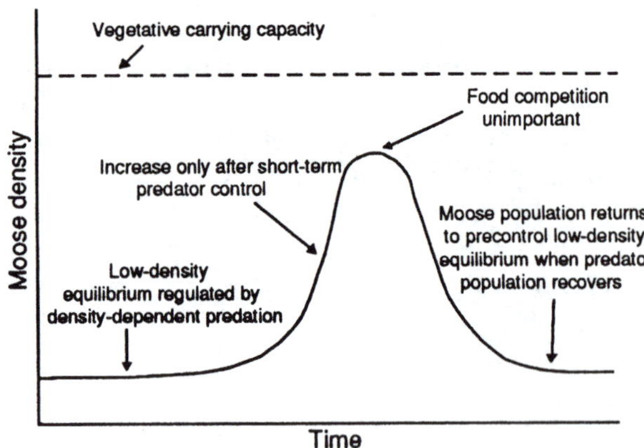

Figure 114. Conceptual model of regulation of moose densities under the low-density equilibrium hypothesis.

PREDATOR-PIT, MULTIPLE EQUILIBRIA, OR TWO-STATE MODEL HYPOTHESIS

Moose populations fitting this model are regulated by density-dependent predation at low moose densities and by food competition at high moose densities (Figure 115). Either through natural phenomena or predator control programs, moose sometimes are relieved of a low-density equilibrium caused by predation. They then increase to a higher equilibrium, where food competition limits growth in a density-dependent fashion. Predation at the low-

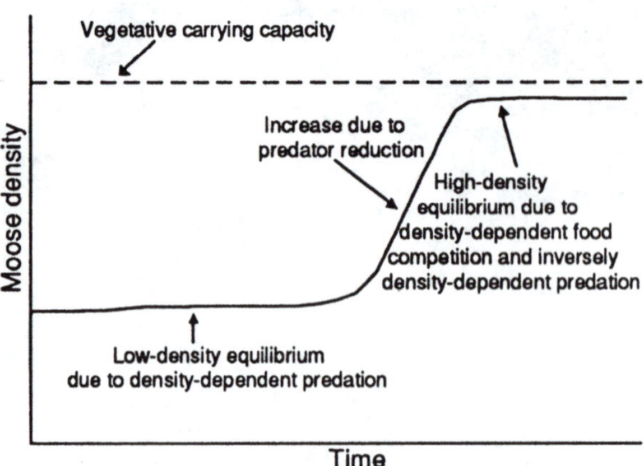

Figure 115. Conceptual model of regulation of moose densities under the multiple equilibrium hypothesis.

density equilibrium is density dependent (regulatory), but is inversely density dependent (nonregulatory) at higher moose densities. When predator populations are reduced and the moose population increases to a higher equilibrium, the latter remains at the higher level even if the former returns to precontrol levels (unless a decline is caused by some natural catastrophe). Such systems are characterized by multiple species of predators and prey. The term *predator-pit* refers to the range of moose densities above the lower equilibrium densities in which the moose population cannot increase because of intense predation and negative population growth (Figure 115).

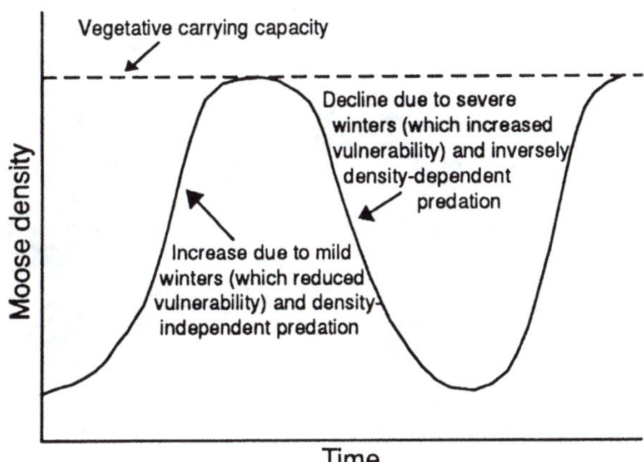

Figure 116. Conceptual model of regulation of moose densities under the stable limit cycle hypothesis.

STABLE LIMIT CYCLE HYPOTHESIS

Moose populations fitting this model exhibit regular oscillations with 30- to 40-year cycles (Figure 116). Weather influences viability and vulnerability to predation of cohorts from birth to adulthood. Predation is density independent (nonregulatory) during moose population increases and inversely density dependent (also nonregulatory) during moose population declines. Moose/weather/forage interactions regulate moose densities. No predator-pits occur with this hypothesis. Such circumstances, or systems, are characterized as having only one species of predator and one prey.

The above four models are the center of debate among biologists about which most accurately describes predator/prey relationships in North America. For example, moose/wolf relationships on Isle Royale are described as exhibiting characteristics of all four hypotheses (Table 17). Several controversies exist. We categorized each of the previously reviewed studies based on literature and our assessment of the evidence supporting each hypothesis (Table 17).

Gasaway et al. (1992) recently compared moose densities from interior Alaska and Yukon, involving populations whose predators (both bears and wolves) were lightly exploited (naturally regulated) and those whose predators

Table 17. Classification by regulation hypothesis of the 11 major moose/predator studies conducted in North America since 1965[a]

Regulation hypothesis	Study area	Source
Recurrent fluctuation model	Kenai Peninsula, Alaska	Our evaluation
	Nelchina Basin, Alaska	Our evaluation
	Tanana Flats, Alaska	Gasaway et al. 1983
	Northeastern Alberta	Fuller and Keith 1980, our evaluation
	Isle Royale, Michigan	Keith 1983, Van Ballenberghe 1987, Van Ballenberghe and Ballard 1994
	Eastcentral Saskatchewan	Our evaluation
One-state equilibrium, low-density equilibria or predator-limited model	Denali National Park, Alaska	Van Ballenberghe 1987
	Interior Alaska	Gasaway et al. 1992
	Isle Royale, Michigan	Bergerud et al. 1983
	Pukaskwa, Ontario	Bergerud et al. 1983
	Quebec	Messier and Crête 1985
	Yukon Territory	Gasaway et al. 1992
Predator-pit, multiple equilibria or two-state model	Denali National Park, Alaska	Haber 1977, Walters et al. 1981
	Isle Royale, Michigan	Messier 1991
	Quebec	Messier and Crête 1985, Messier 1991
Stable-limit cycle model	Isle Royale, Michigan	Peterson and Page 1983, Peterson et al. 1984, Mech et al. 1987, Messier 1991
	Pukaskwa, Ontario	Thompson and Peterson 1988

[a] Classifications were based on inferences made by each study investigator and by our own evaluations based on findings from each study.

Moose populations can be held at low-density levels when predators such as wolves are not managed and have alternative prey. This situation is referred to as a "low-density equilibrium," and was described for a large area of interior Alaska. *Photos by Rolf Peterson; courtesy of Michigan Technological University.*

Besides preying on moose, both bears and wolves ravage moose carcasses and hunter-killed moose remains. Wolves can derive a significant portion of their seasonal diet from starved moose during deep-snow winters. *Photo by Michael H. Francis.*

were held below carrying capacity. The latter had average moose densities of 171 per 100 square miles (66/100 km^2) (n = 16; range, 44 to 373 per 100 square miles [17–144/100 km^2]), whereas the former averaged 38 moose per 100 square miles (14.7/100 km^2) (n = 20; range, 11.7 to 108 per 100 square miles [4.5–41.7/100 km^2]). Bergerud (1992) reported similar density relationships in Canada. Both Bergerud (1992) and Gasaway et al. (1992) concluded that predation held moose at low-density equilibria.

In review of 27 studies in which moose were the primary prey, Messier (1994) concluded that wolf predation was density dependent at the low range of moose density. He suggested moose populations would stabilize in the absence of predators at 518 per 100 square miles (200/100 km^2), at 337 per 100 square miles (130/100 km^2) in simple wolf/moose systems and at 52 to 104 per 100 square miles (20–40/100 km^2) in multiple predator/prey systems. Consequently, areas such as Isle Royale are regulated by recurrent fluctuation processes, whereas areas in interior Alaska and the Yukon fit the low-density equilibria described by Gasaway et al. (1992).

We believe case history evidence supports the hypothesis that predation can and does limit many, but not all moose populations. Recurrent fluctuations in simple wolf/moose systems and single stable-state equilibria in multiple predator/prey systems appear to be the most promising hypotheses for explaining moose/predator relationships. However, these conclusions must be qualified.

As mentioned earlier Gasaway et al. (1983), Keith (1983) and Fuller (1989) proposed using wolf/ungulate ratios for predicting potential impacts of predation. Although these ratios are useful for initial, cursory analyses (best used for simple wolf/moose systems), they can be misleading. They do not account for changing prey availability and vulnerability, surplus killing (Kruuk 1972), changing predator and prey age structures, changing predation rates that may vary severalfold, bear predation and changing vegetation/ungulate relationships (Ballard 1991). Messier (1994) further stated that wolf/ungulate ratios do not adequately reflect changing predation rates because they do not consider the functional response (change in diet) of predators. For these reasons, biologists should interpret predator/prey ratios cautiously.

Ballard and Larsen (1987) presented a conceptual model of the relationship between habitat-carrying capacity, predation and human harvest levels. Predation influences the level and the rates of population decline and causes moose populations to reach lower levels than expected from diminution of habitat quality alone. No recent studies document sufficient increases in productivity and survival as a result of increased forage to offset the effects of predation after a moose population decline. The same apparently holds true for mule deer, because no compensatory increases in productivity or survival occur in relation to forage, but there may be a density-dependent relationship with mortality (Bartmann et al. 1992).

Schwartz and Franzmann (1989) indicated some ways that predation can influence rates of change and the absolute densities of moose (Figure 117). After habitat improvements, the rate of increase of a moose population is retarded, and peak densities are lower when predator populations are at or near carrying capacity. After a moose population peak, rates of decline are greater and densities lower in an undisturbed predator population. Higher moose densities exist at all stages of plant succession in predator-free environments.

McCullough's (1979, 1984) model for deer suggests that, when ungulate populations are at or near habitat-carrying capacity, predation mortality is largely compensatory. In contrast, when moose populations are well below carrying capacity, predation mortality is largely additive to other natural causes of mortality (Gasaway et al. 1983).

Spatial relationships between moose and predators may be an important aspect affecting the predator/prey relationship. Van Ballenberghe (1987: 456) summarized the concept as follows: "Bergerud et al. (1983) reasoned that mean densities of moose resulting from predation by wolves may vary in different areas due to differences in escape habitats for moose. Certain areas with abundant shorelines and islands may support higher densities of moose with identical natality and survival rates to lower density populations with widely spaced individuals in lieu of escape features in the environment. Lower prey densities result in increased predator search times and less efficient predation."

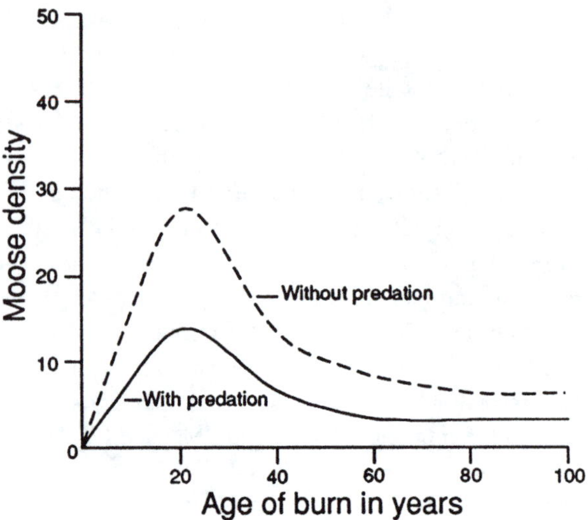

Figure 117. Proposed conceptual relationships between forest succession, predation and moose densities on the Kenai Peninsula, Alaska (from Schwartz and Franzmann 1989).

Van Ballenberghe (1987) and Van Ballenberghe and Ballard (1994) stressed the importance of verifying or testing hypotheses in undisturbed ecosystems. With the exceptions of Isle Royale and Denali National Park, all of the ecosystems studied to date are affected by various degrees of human harvest and/or habitat alteration. Even Isle Royale and Denali are not truly pristine ecosystems.

Few predator populations in North America have not been subjected to some type of human influence. Wolf packs, including those of national park systems, have been subjected to various levels of harvest or their population dynamics have been influenced by changes in rates of immigration from and emigration to adjacent areas. Similarly, nearly all grizzly and black bear populations experience such direct human influences as legal hunting, poaching and habitat alteration. The historic ranges of most unprotected wolf and brown bear populations have been reduced by approximately 50 percent (Ballard and Larsen 1987). Black bears still occupy most of their historic range, but human harvests, habitat alterations and reductions of competing predators and prey may have influenced black bear population dynamics.

Substantial changes in the distribution of wildlife populations in North America have forced biologists to study smaller, more variable species (e.g., insects, small mammals, coyote/jack rabbit relationships) to understand predator/prey dynamics. However, such relationships and dynamics may have little application to large ungulate/large carnivore systems. Before predator/prey populations in altered ecosystems can be understood, much less managed, understanding of such systems in natural or near-natural settings is imperative. The northern ecosystems of Canada and Alaska offer the best opportunity for understanding these relationships, because most of the habitats there remain relatively undisturbed.

Biologists use a number of different methods to study predator/prey relationships. The most common methods are: following radio-collared predators and observing the types of prey and rates of kill; manipulation of predator densities and then monitoring prey response; monitoring mortality of radio-collared prey; and monitoring concurrent predator/prey population numbers, densities and distributions. Often, several indices of recruitment and mortality are emphasized. As noted in Chapter 6, many measurements and indices have statistical assumptions that rarely are met in biological studies. Many of the estimates may be inaccurate—they either contain no measure of accuracy or are measured with relatively large variances (±20 percent). Differences in study methods, lack of accurate and precise measurements of population characteristics and studies of variously modified ecosystems could lead to inappropriate conclusions. Consequently, comparisons between different studies may be inappropriate.

Related to the above problems is the question of study longevity and adequate funding. Aside from the Isle Royale project, all other moose studies have been of relatively short duration (<10 years). Many were designed and implemented to address an immediate management problem, such as a declining moose herd. Once a problem appears to be solved or the immediate cause identified, the study usually is terminated by administrators forced to divert funds to other pressing problems. Unfortunately, moose studies often end before the nature of moose population increases and declines is understood fully. Also, many of the studies implemented by management agencies are not designed to test more than elementary hypotheses. For example, most studies have tested the hypothesis that predators limit moose population growth; few have tested hypotheses of population regulation.

Van Ballenberghe (1987), Sinclair (1991), Skogland (1991), Boutin (1992), Messier (1994) and Van Ballenberghe and Ballard (1994) described tests needed to verify or refute the various regulation hypotheses (Table 18). These investigators stressed that management agencies (in contrast to universities) often are in a unique position to conduct the large-scale studies necessary to understand predator/prey relationships better. However, as Van Ballenberghe and Ballard (1994) pointed out, although agencies may have the ability to conduct different types of studies and test hypotheses, sociopolitical considerations often dictate what is done—or not done. For example, Sinclair (1991) and Boutin (1992) advocated that, as a test of the low-density equilibria hypothesis, managers purposefully reduce moose

Radio-collaring moose calves, to secure a variety of biological data, also serve to locate animals killed and identify the type of predator. The radio-collared calf in the left scene was found to have been killed by a grizzly. Radio-collars placed on the newborn moose calf's neck must be expandable to accommodate growth and be designed to fall off in approximately 6 months. To obtain the information on causes of moose calf mortality, the calves must be monitored by signals sent from the collar transmitters to receivers in aircraft. *Left photo by Warren Ballard; courtesy of the Alaska Department of Fish and Game. Right photo by Albert W. Franzmann; courtesy of the Alaska Department of Fish and Game, Soldotna.*

As human settlement and activities continue to encroach on moose habitats, particularly winter range, harassment of moose by domestic and feral dogs is likely to increase. Dogs can prey on moose, and when severe winter conditions persist, their harassment can cause undue stress on individual moose, particularly pregnant cows. *Left photo courtesy of the National Archives of Canada. Right photo by Francis Dickie; courtesy of the Wildlife Management Institute.*

Table 18. Predictions from hypotheses to explain moose population dynamics (from Boutin 1992)

	Hypotheses			
	Predator limitation	Predator regulation	Predator-pit	Stable-limit cycle
Experiment 1: Under variable ecological conditions				
A. At low to intermediate moose densities the proportion killed by predators will be	inversely density dependent	density dependent	density dependent	dependent on condition, not density
B. At high densities the proportion killed will be	inversely density dependent	inversely density dependent	inversely density dependent	inversely density dependent
C. Densities will show	fluctuations	a low-density equilibrium	two equilibria (high and low)	cycles
Experiment 2: Following experimental reduction of moose				
A. Proportion killed by predators will	increase	decrease	decrease	depend on cohort removed[a]
B. Moose density will	decline	increase	decline	follow initial trajectory[b]
Experiment 3				
During a sequence of predator reductions followed by predator recovery, moose will	increase and then decline	increase and then decline	increase to an upper equilibrium	populations will drift out of phase[c]

[a] Removal of strong cohorts (born under good conditions) would increase the proportion lost to predators as more individuals of the remaining weak cohorts could be preyed on; removal of weak cohorts would produce the opposite effect.

[b] If moose densities were reduced while the population was undergoing an increase, the population would continue to increase, and vice versa if the reduction was done while the population was declining.

[c] A long-term predator removal should change the period of the oscillation and thereby cause the experimental moose population to drift out of phase with the control population.

populations without reducing predators. Public opposition to such an experiment, which would have to be conducted in a relatively large area, would severely hamper such a project (see Harbo and Dean 1983, Miller and Ballard 1992, and Stephenson et al. 1995 concerning public debate on wolf and bear management). Van Ballenberghe and Ballard (1994) pointed out that Gasaway et al. (1983) were able to monitor the type of situation (Tanana Flats) advocated by Boutin (1992) and Sinclair (1991), but that the research was not geared as such.

Van Ballenberghe and Ballard (1994) reiterated that many of the aforementioned problems and actual differences in populations subjected to various human influences could result in moose populations exhibiting characteristics of any or all four of the major models. Moose populations below habitat-carrying capacity occurring in areas of low predator density probably are not limited or regulated by predation. In such cases, weather and hunting primarily determine moose population density. Even if hunting is a significant mortality factor, neither predation nor food resources may be limiting. Very likely, no single simple model will characterize predator/moose relationships under all ecological conditions (Van Ballenberghe and Ballard 1994).

Biologists must continue to focus on conditions that lead to regulation or limitation of ungulate populations. How and when predation may limit (but not regulate) a moose population are significant determinations for managers attempting to satisfy demands for consumptive use. To understand more fully and manage predator/prey interactions properly, hypotheses of population regulation and limitation need to be tested or verified and new hypotheses formulated. We conclude that more well-designed observational and experimental long-term studies are necessary.

LINE OF VERY DARK HAIR ALONG NECK & HUMP— THEN LIGHTER ON SHOULDERS & UPPERPARTS, SHADING INTO VERY DARK BROWN—
ABOUT A BUCKETFUL OF WATER POURED OVER HIS HEAD EACH TIME HE RAISED IT
FORAGING ON THE BOTTOM OF A SMALL POND— EARS USUALLY OUT— OCC. UNDER
VENUS RISING FROM THE WAVES—
HALF-SWIMMING— RIDING HIGH IN WATER
WHAT'S THAT?
BULL MOOSE IN VELVET— FED FOR 45 MINUTES OR SO OCCASIONALLY WATCHING US BUT USUALLY IGNORING US— FINALLY TROTTED AWAY OVER THE RIDGE, SILHOUETTED AGAINST THE SUNSET
(COMPLETED FROM MEMORY)
MOOSING MODELLING FOR A CALENDAR
JULY 5, '54
POND NEAR WONDER LAKE
McKINLEY PARK, ALASKA—
SLEEPING BEAUTY (MEMORY)
"ROCKY" CHEWING HIS CUD— (LIFE)
MT. LEMORAY LODGE JUNE 2, '54
MOOSE'S NECK IS TOO SHORT TO REACH LOW PLANTS — HE EITHER BENDS AWKWARDLY (RIGHT), OR KNEELS
ANTLERS JUST STARTING TO GROW— AND THEY ITCH— (LIFE & MEMORY)
HEAD RAISED— SNIFFING
CURIOUS ABOUT SOMETHING

8

KENNETH N. CHILD

Incidental Mortality

Accidents kill moose. Leopold (1933) considered accidents to be decimators of wildlife that, in concert with hunting, predation, diseases, parasites and starvation, could reduce population numbers directly or affect productivity through alteration of sex and age ratios.

By drowning, entrapment by vegetation, abandonment or as the result of some mishap at the birth site, moose calves suffer heavier mortality than do adults. Adult moose, on the other hand, are more prone to die in collisions, fall into deep depressions or tumble into natural sinkholes, salt licks and gulleys, break through thin ice and drown, succumb to exhaustion after prolonged struggles in deep mud, quicksand or snow, and get trapped in reservoir debris or sometimes in deadfalls in logging slash. Moose die in rapidly advancing forest fires, and many are killed in unusual circumstances, including by the "good intentions" of people. Peterson (1955) suggested that "accident" should be given its rightful place in the demography of moose populations because of the substantial numbers of moose killed annually by various mishaps.

Incidental mortality was previously regarded as of little consequence to moose populations or to management programs and sustainability of annual harvests. However, in some years, the large number of accidental moose deaths necessitated review of management programs and harvest objectives. More important, with increasing public demands for consumptive and nonconsumptive wildlife-related recreational opportunities, managers now recognize that all mortality factors are important and need to be considered in management programs. Accidents are not the least of the "decimators," as previously thought; it is a source of mortality that can have serious consequences on the welfare of local moose populations and recreational opportunities.

Various accidents and mishaps are described in this chapter that claim the lives of thousands of moose each year in North America. Many of the incidences described are taken from anecdotal reports and observations of wildlife biologists and researchers, newspaper articles and a variety of literature sources. The magnitude of losses can be significant, although quantifiable data are lacking in most cases. For some populations, we can infer that the impact of accidents is potentially of large magnitude and possibly long-lasting in effect.

Collision Mortality

Transportation Corridors

Moose are adaptable to artificially disturbed habitats, and therefore, are often found in close proximity to roads, major highways and railways. But the association is far from compatible.

Human transportation corridors alter important habitat components, such as forage, water and cover, and often intersect migration and daily travel routes. In winter and spring, roads and railways provide alternate travel routes for moose to traditional ranges and their sides provide a

Table 19. Relationship of usual highway and railway characteristics to ideal characteristics in moose habitat of the subboreal spruce ecotype

Characteristic	Ideal	Railway	Highway
Opening width	<109 yards (100 m)	33 yards (30 m)	33–55 yards (30–50 m)
Vegetation	Early	Early	Early
Landscape	Seral	Seral	Seral
Water	Natural	Ponds	Ponds/pools
Salt licks	Natural	?	Salt pools
Distance to cover	55 yards (50 m)	16 yards (15 m)	16–27 yards (15–25 m)
Snow-free habitat	Forest	Railbed	Roadbed
Snow intercept	Forest canopy	Plowing	Plowing

good mix of preferred browse species. Both offerings invite problems. Because transportation corridors provide maximum "edge," and many of the features found in habitats (Table 19) associated with glaciated landforms and riparian habitats (Eastman and Ritcey 1987), railways and roads are like an "ecotonal trap," and present an environment of high collision risk to moose (Damas and Smith Co. 1983, Child et al. 1991).

MOOSE/VEHICLE COLLISIONS

Moose/vehicle collisions are a serious problem in North America (Grenier 1973, Child and Stuart 1987, Child et al 1991, Del Frate and Spraker 1991, McDonald 1991, Oosenbrug et al 1991, Schwartz and Bartley 1991). Collision mortalities of moose nearly doubled in Newfoundland from 1983 to 1989 because of increased traffic volume, greater traffic speeds and increased truck transport after the provincial railway ceased operation in 1987 (Oosenbrug et al 1991). G. Redmond (personal communication: 1992) reported that, of 259 nonhunting mortalities of moose in New Brunswick, 163 (63 percent) were due to collision with trains and vehicles. In Maine, K. Morris (personal communication: 1992) similarly reported that, in 1990, 572 moose/vehicle collisions were reported, but fatalities were suspected to be higher because drivers failed to report all such accidents or moose kills.

In a survey of seven U.S. states and nine provinces and one territory in Canada (Table 20), nearly 3,000 moose/vehicle accidents occur annually in North America. During winters of heavy snowfall, the number of collisions reported in Alaska and British Columbia may triple the number of fatalities occurring in an average snowfall season and be even higher if train kills are included. These estimates are conservative, however, because not all collisions result in a fatality; some moose are never found at the scene of the accident, and not all collisions are reported (Child and Stuart 1987).

Moose/vehicle collisions occur most frequently (1) at dawn and dusk, (2) under conditions of poor visibility, (3) on straight and relatively flat stretches of highway, (4) at vehicular speeds in excess of 50 miles per hour (80 km/hr), and (5) because of driver inattention (Stuart 1984, Del Frate and Spraker 1991).

Solutions are fairly obvious—discourage moose from the rights-of-way and give drivers ample warning of the po-

Table 20. Moose collision fatalities in portions of North America and correlation to annual allowable harvest

State or province	Estimated population	Annual allowable harvest	Number of known fatalities by[a] Vehicles	Trains	Percentage of annual allowable harvest
Alaska	Unknown	Unknown	Unknown	9–725	Unknown
Alberta	118,000	21,000	231	Unknown	3.7+
British Columbia	178,000	11,610	713	494	10.4+
Idaho	3,800	260	20	7	9.6
Manitoba	Unknown	4,700	12	12	0.5
Maine	22,000	1,000	600	Unknown	60.0
Minnesota	7,500	1,000	70	18	8.8
Montana	Unknown	500	30	0	6.0
New Brunswick	15,000–25,000	1,850	150	35	10.0
Newfoundland	100,000	20,000	460	Unknown	2.3
New Hampshire	4,000	100	196	0	196.0
Nova Scotia	3,000	Unknown	5	1	Unknown
Ontario, northcentral	28,000	3,400	108	11	3.5
Quebec	75,000	10,500	265	Unknown	2.5+
Vermont	1,500	Unknown	78	2	Unknown
Yukon Territory	50,000	2,000	15	NA	0.7

[a] These are minimum "point-in-time" estimates.

Moose/vehicle accidents occur in part because moose are attracted to roadsides where cutting and mowing facilitate new-growth browse (top left), and to the roads themselves that provide relatively easy travel lanes for moose, especially in winter. Salting highways in winter to prevent icing also results in artificial "licks" that attract moose (top right). Compounding the problem is the fact that moose are particularly active at dawn and dusk when visibility is poorest for motorists. Also, by virtue of their size and appetites, moose are less deterred from daily movements by snowfall and deep snow than are other wild ungulates. So motorists must be alert in severe winter weather to the sudden appearance of moose on roads. *Top left photo by Charles C. Schwartz; courtesy of the Alaska Department of Fish and Game. Top right photo by Donald M. Jones. Bottom photo by Kenneth Child; courtesy of the British Columbia Ministry of the Environment.*

tential danger. Roadway signs identifying moose crossing sites and concentration areas are strategically located to warn motorists of collision risks and their need to drive cautiously. In some states and provinces, the local media daily urges motorists to be alert when moose frequent the roads. Vegetation sometimes is cleared along roads to improve roadside visibility. The resultant improvement in visual horizons provides motorists ample opportunity to anticipate potential problems, by increasing their reaction time when a moose is observed (Del Frate and Spraker 1991), thereby minimizing the element of surprise for moose and motorist alike (Poll 1989). But, more permanent solutions

Roadway signs often are placed at moose concentration areas and crossing sites to warn motorists and prepare them for sudden stopping. "Give moose a brake" signs in Alaska (top right) and "Brake for moose" signs in New Hampshire help educate motorists about the serious hazards of moose on roadways and the need to reduce speed in concentration/crossing areas. *Top left photo by Linda Steiner. Top right photo by Gino Del Frate; courtesy of the Alaska Department of Fish and Game. Bottom photo by Mary Ford.*

to roadkills have been sought through planning and design of transportation corridors, through vegetation management and forest clearing, by erection of physical barriers and by traffic control measures.

McDonald (1991) reported that moose-proof fencing, one-way gates, an underpass and lighting of a section of the Glenn Highway in the Matanuska Valley of Alaska reduced moose/vehicle accidents by 70 percent overall and 95 percent within the fenced portions of the highway. Lighting reduced collision kills by 65 percent. Underpasses permit passage of moose beneath the highway to feeding areas and do not alter seasonal distribution patterns relative to the highway corridor.

Exclusion fencing, one-way gates and underpasses are strategically located on the Coquihalla Highway in the Okanagan region of British Columbia and successfully direct moose migrations to seasonal ranges without collision losses (Clayton Resources Ltd. 1989). Fencing is effective, but because of high installation and maintenance costs, its use is generally discouraged except possibly in known moose concentration areas or at chronic collision sites adjacent to highway corridors (Schwartz and Bartley 1991).

Liberal hunting seasons are advertised in some areas or recommended in others to reduce moose numbers and lower the probability of moose/vehicle collisions and consequent human injury and property damages (Damas and Smith Co. 1983, McDonald 1991, Schwartz and Bartley 1991).

MOOSE/TRAIN COLLISIONS

Moose/train collisions kill thousands of moose each year, and have been a problem ever since rail lines were laid in North America's moose range. These collisions result in significant property damage, resource losses, train derailments and costly mechanical problems. Inconvenience to rail corporations, delays in "road time" and damage to railcars and locomotives can be substantial (T. Brooks personal communication: 1992).

At least 200 moose die on the railways in British Columbia each year as they migrate to seasonal ranges (Child et al. 1991). The number of collisions increases during autumn and early winter months, peaks in January and February and then rapidly decreases. Fifty-four percent of all collisions occur in January and February, about one month after the peak in moose/vehicle accidents (Figure 118). In win-

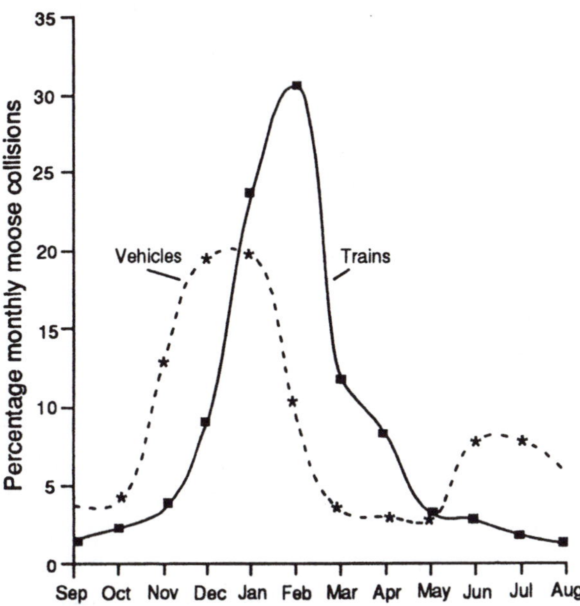

Figure 118. Average monthly moose collisions on roads and railways in central British Columbia, 1983–1990 (after Child et al. 1991). Vehicle collisions peak in mid-December, about 1 month before the peak in train fatalities. Intensive forestry practices, highway salting, traffic volume and speed, and deep snow conditions may increase collision risks to moose, trains, autos and passengers alike.

Fencing, underpasses and one-way gates on the Glenn Highway, Alaska (above), and on the Coquihalla Highway, British Columbia, reduce moose/vehicle collisions and accommodate daily movements of moose. Construction costs—$80,465 (Canada) per mile ($50,000/km)—for such protections along the Coquihalla Highway have discouraged their more widespread use in the province and elsewhere in North America's moose range. Maintenance cost for such installations normally are covered by contractual agreements for general roadway upkeep. *Photo by Charles C. Schwartz; courtesy of the Alaska Department of Fish and Game.*

ters of deep snow, reported moose kills on railways in the central interior of British Columbia have numbered more than 1,000 animals (Child 1983a, Child et al. 1991).

From 1963 to 1990, 3,054 moose were killed on the Alaska Railway with annual losses ranging from 7 to 725 animals (Modafferi 1991). In the winter of 1989–1990, because of deep snow, known moose fatalities (731) on roads and rails exceeded previously recorded levels by almost 100 percent. In the Willow-Talkeetna area during the same winter, the number of rail kills reported represented a 70 percent loss of moose from the resident population (Schwartz and Bartley 1991). Muzzi and Bisset (1990) reported that about 40 to 50 moose are struck and killed each year by trains along a 140-mile (225-km) section of track between Sioux Lookout and Armstrong, Ontario. In the Gogama District, Ontario, 152 incidental moose mortalities were reported from 1980 to 1991, of which 29 (19.0 percent) were attributed to train collisions (D. Dasti personal communication: 1992).

The frequency of moose/train collisions varies seasonally. Most kills occur in springtime, shortly after the calving season, and in midwinter (January and February), when snowfall and snowpack conditions "fence" moose adjacent to railway corridors and encourage them to travel the plowed railway berm (Rausch 1956, Child 1983a, Hatler 1983, Child and Stuart 1987, Muzzi and Bissett 1990, Becker and Grauvogel 1991, Child et al. 1991, Modafferi 1991). In Alaska and central British Columbia, moose annual fatalities on railway corridors can exceed 800 and 1,000 moose, respectively, as depth of snowpack and frequency of snow clearing increase (Damas and Smith Co. 1983, Hatler 1983, Child et al. 1991, Modafferi 1991). Similar relationships have been established for moose/train collisions in Norway (Jaren et al. 1991), where 95 percent of the collisions occur in winter, including 80 percent from December to February, when migratory moose encounter railroads and road networks in valley bottoms. In all areas reporting moose/train conflicts, heavy snowfall and deep snow reportedly augment the collision problem (Andersen et al. 1991).

Railways and roadways usurp moose habitat, intersect migration corridors and possibly interfere with daily activity routines of moose. Collision mortality of moose in some areas of North America may rank with hunting and predation as a major mortality factor. Moose kills on the railways of Alaska and British Columbia can number in the thousands in winters of deep snow. *Photo by Brock Gable; British Columbia.*

Roadway shoulders plowed high and vertical with compacted snow can act as berms that trap young or weak moose on the roads where, without browse, the animals can starve if not first struck by vehicles. "Channels" plowed or left unplowed at intervals through such physical barriers may give only temporary relief from these frequently fatal situations. *Photo by Brock Gable; British Columbia.*

Moose find easy travel in winter on railway berms and will move to feeding areas in traditional ranges adjacent to the railway corridor. Much in the same fashion when confronting wolves (Bubenik 1987; see also Chapter 7), healthy, adult moose are inclined to stand their ground in the face of oncoming trains and vehicles, whereas immature moose are more likely to flee (Stringham 1974, Child 1983a, Hatler 1983, Surrendi 1983). Moose that react defensively likely are killed in collisions more frequently than those moose that move to escape such predicaments. At night, some moose apparently are "hypnotized" in the glare of the headlights of a locomotive, negating the normal flight response (Rausch 1956). Or if already running, some moose evidently are reluctant to leave the lighted path to enter darkness (Geist 1963).

If flight is unsuccessful, moose stand their ground and fight. With head held in a low threatening posture (Geist 1963), moose will charge cars and trains. Sometimes, a moose will "box" a passing train with its forelegs (Child 1983a). But, most are killed on impact. If not killed outright, most moose suffer serious injuries, amputations or excessive blood loss and die of shock some distance from the railway corridor. Many injured moose are dispatched by rail officials.

In Alaska, carcasses are salvaged and distributed to charitable organizations. In most areas, however, because of inconvenience, remoteness of the collision site and high

Seeking a travel route of least resistance in areas of deep snow, moose are drawn to plowed railways. When confronted by a train, a moose may stand its ground or flee. Those that confront the threat (left) and even become aggressive invariably are killed. All too often, those that attempt to flee are reluctant to escape to the deep snow, and choose instead to run along the tracks (right). Almost always, the result is collision and death for the moose. *Photos by Michelle A. Barnes.*

recovery costs, carcasses are left in the field for the scavengers (Child and Stuart 1987).

MOOSE / AIRCRAFT COLLISIONS

Although moose / aircraft collisions are rare, they do occur. On January 7, 1988, Alaska's *Peninsula Clarion* read: "Sleepy moose killed when airplane comes in." The pilot reportedly did not see the resting or sleeping bull moose on the runway of the Soldotna airport "until the last minute," when he made attempts to avoid hitting it. The moose was decapitated by the plane's left propeller. The plane sustained damage to the left landing wheel door and left wheel gear. The runway has since been fenced for "public security and safety reasons." Many close calls on runways of several other airports in Alaska have necessitated extensive and expensive fencing.

In December 1990, the *Peninsula Clarion* again reported that a MarkAir commuter aircraft clipped a moose on the runway at the Kenai Municipal Airport. The DeHavilland Dash-6 sustained minimal damage in the encounter. The adult cow was injured on the top of her shoulder hump where the propeller struck. From the appearance of the wound, and drying condition of the blood, the moose did not seem "worse for the wear." When pursuers approached, she ran off. The cow survived the collision but it is not known whether she died of injuries inflicted by the propeller blades.

A less serious aircraft / moose encounter occurred on December 22, 1992, near Chapleau, Ontario (V. Crichton personal communication: 1993). An aircraft chartered by the Ontario Ministry of Natural Resources crashed into woods near the shoreline of a frozen lake. Except for some bruises and a broken arm, all occupants of the plane survived the crash. The aircraft, however, sustained extensive damage, with both wings torn from the fuselage. When the plane finally came to rest and the occupants gained their composure, they heard loud "bumping and kicking" sounds against the fuselage. Peering outside, they witnessed a moose scramble from beneath one of the severed wings and run into the forest. Although some blood was spotted on the moose, it did not appear to be seriously injured. Reports suggested that the crashed aircraft had unknowingly come to rest on or near the moose.

Implications of Collision Losses

The recurrent nature of collisions may result in (1) locally declining moose populations; (2) loss of recreational opportunities; (3) restrictive hunting regulations; (4) rising costs to rail corporations and the motoring public; (5) increased risks to human life and property; (6) escalation of insurance premiums; and (7) public dissatisfaction with moose presence on roads and with the seeming inaction of management agencies to seek solutions to the collision problem.

Throughout many portions of the North American moose range, air transport of people, goods and services is commonplace if not essential. Consequently, there are numerous airfields, particularly for small aircraft, in the midst of moose habitat. Airplane / moose collisions are rare, but those that occur tend to be serious for all parties. Full fencing is impractical for the vast majority of airports. Moose that cross or browse or loaf on grass runways or the mown strips between runways must be chased away. *Photo by Mason Marsh; courtesy of the* Peninsula Clarion *(Kenai, Alaska).*

Rail corporations attempt to minimize moose / train collisions by improving snow maintenance, creating pathway escape routes in the snow, erecting physical barriers, manipulating light and horn signals, reducing train speed in known moose concentration areas, planking bridges and trestles, positioning light reflectors at strategic locations and managing vegetation along the railway to reduce its attractiveness to moose (Rausch 1956, Child 1983a, Modafferi 1991, Schwartz and Bartley 1991). In Alaska, pilot cars are driven ahead of trains to chase or scare moose from the tracks.

Some success has been reported for some of these measures. Reducing train speed is most effective when practiced in conjunction with vegetation management along rail berms (Becker and Grauvogel 1991, Schwartz and Bartley 1991). In Norway, when vegetation was removed 66 to 98 feet (20–30 m) on each side of railbeds, train kills were reduced by 56 percent (Jaren et al. 1991). And when trains were operated at slow speeds in known moose concentration areas, collisions were reduced further. Unfortunately, limited success was reported for most of these same measures in British Columbia (Gilchrist 1986), where collision mortalities were of increasing management concern (Child 1983a).

Solution has been sought through planning and design of transportation corridors, vegetation management, construction of physical barriers and traffic control measures. Management of vegetation gives only temporary relief and likely augments collision risk by maintaining early seral vegetation along roads and railways. Salting of roads to ensure driver safety compounds the problem (Fraser and Thomas 1982), and placing salt blocks adjacent to roadways to lure moose from the roads is ineffective (Jaren et al. 1991, Schwartz and Bartley 1991). In areas where herbicides are used for intensive silviculture, alteration of vegetation landscapes might displace moose into more favorable habitats near transportation corridors. Light reflectors (mirrors) have been positioned at strategic locations along highways and more recently along railways. Fencing, although effective, is impractical for extensive use because of high costs (Schwartz and Bartley 1991). Most methods promise only short-term solutions (Child 1983a, Gilchrist 1986).

For long-term solution to this problem, study of moose behavior is needed specifically to determine (1) sensory stimuli that elicit flight and (2) whether this avoidance behavior can be directed toward trains and vehicles when the same stimuli are simulated electronically (Child 1983a, Damas and Smith Co. 1983, Child et al. 1991). Almkvist et al. (1980) investigated a variety of repellents to keep moose off roads and railways. Light, sound and different scents were tested, but their effectiveness was doubtful because moose habituated to most of the sensory stimuli perceived.

Nevertheless, the use of sound in combination with other sensory stimuli to frighten moose from roads and railway berms might reduce collision. Moose can be frightened by horn blasts from locomotives and vehicles. Damas and Smith Co. (1983) suggested that an acoustic key stimulus probably exists for moose because similar stimuli have been described for most of the deer species. They further suggested that, if this key acoustic stimulus were electronically simulated, moose / train and moose / vehicle collisions could be reduced substantially as moose would avoid approaching locomotives and vehicular traffic. The use of sonic devices on locomotives in Ontario (Muzzi and Bissett 1990) and investigations with similar devices on automobiles in British Columbia (Child and Foubister 1986) supported this suggestion—moose displayed avoidance and collisions were seemingly reduced. However promising this might be, the interference of background noises and habituation of moose to the stimulus might weaken response thresholds (Lavsund and Sandegren 1991).

Behavioral stimuli elicit adverse reactions in ungulates. For example, white-tailed deer avoid trails "marked" with plywood dummies painted to simulate flagging behavior (Bayford 1975). The simulated cue (flagging) is a warning signal that is species specific to white-tailed deer and a key stimulus to which whitetails innately respond. A.B. Bubenik (personal communication: 1982) suggested that placing moose dummies at strategic locations along transportation corridors might elicit similar avoidance by moose and reduce collision. More research is needed to discover the key stimuli for moose.

Interestingly, manufacturers of Volvos in Sweden and BMWs in Germany are experimenting with ultraviolet (UV) headlamps. UV is not visible to the human eye, but when UV radiation falls on some substrates, they fluoresce and emit visible light. Road signs coated with fluorescent material are known to "leap out" of the darkness when caught in the invisible glare of UV headlights. UV technology has promise because it would provide drivers increased safety on the roads because moose coats fluoresce naturally in UV light. Collisions might be more easily avoided because driver-alert distances and time to react would be greatly increased.

Currently, there is no easy or practical long-term solution to elimination of moose collision mortality, except possibly by a combination of vegetation management and traffic controls (Schwartz and Bartley 1991). Increasing public concerns for life and property demands action. Transportation officials, enforcement agencies, insurance companies and rail corporations are improving their cooperative efforts to improve documentation of collision events, address mitigation needs and support research. Obviously, collision is gaining recognition as an important accidental mortality factor for moose.

Fighting and Sparring

When moose fight, generally the attack is a short, fast rush, ending in antler clashes or with one moose striking its opponent with front legs and hooves (see Chapter 5). In the rush attack, moose often charge their opponent only to halt abruptly and strike the ground with their hooves. Both sexes share some of the same agonistic behaviors; some are permanently expressed in females, but are displayed by bulls only during the annual rutting season. Bulls may fight by rising and striking with their front legs, as cows often do, or they may clash antlers, as is common during the rut and in early winter before antler casting. Cows with their yearling calves are especially aggressive in the spring, before calving (Geist 1963).

Threat usually is expressed by high or low head positions with gestures of a charge or by ritualized antler presentations (see Chapters 4 and 5). These gestures maintain individual distance and communicate rank to reduce social conflict, optimize distribution of resources and ensure maximum reproductive success (Bubenik 1987). Sometimes, these antagonistic interactions can result in injury or death of one or both combatants.

Mature bull moose compete for mating rights to receptive cows. Interactions of two mature bull moose during sparring battles can inflict fatal injury. Body wounds vary from slight cuts and scrapes to deep open gashes (Geist 1963, Bubenik 1982b), but the full extent of wounding is mostly indiscernible. Large gashes can be easily detected but small puncture wounds are hidden. Wounds are seen mainly on the face, neck, haunches and occasionally, along the rib cage. Some bulls leave trails of blood in their tracks or carry fresh or frozen blood stains on their antler tines.

Sparring and fighting are highly ritualized, presumably to conserve energy and prevent unnecessary conflict and injurious consequence (Bubenik 1982b). Usually sparring is a shoving match, but some contests "go the distance." Such confrontations often result in injury to the forehead and eye regions. At times, antlers are "locked" together, especially in contests between bulls of similar social rank and with asymmetrical antlers. Bulls are known to lose an eye in these contests. The occurrence of bulls with one "opaque blue eye" in the hunter harvest or in the wild is more commonplace than one would expect (Bubenik 1987).

Fighting fatalities of some adult bulls result because of (1) traumas inflicted during the rut or during prerut sparring matches, (2) inability of bulls to free themselves from locked antlers or (3) mortal wounds received during antler wrestling bouts preceding antler drop. In Nova Scotia, T. Nette (personal communication: 1992) suggested that fighting injuries likely are responsible for many of the adult bull deaths reported during the rut. Nette described an instance in which he inspected the carcass of a mature bull that, once skinned, evidenced multiple, festering puncture wounds and a portion of the small intestine protruding from the lower abdomen and being "pinched off" by the gradual healing of the abdominal wall at the site of the wound. Nette supposed that an animal in this condition quickly would become "wolf bait" or a bonanza for scavengers. M. Wilton (personal communication: 1992) suggested that rut wounding is a serious mortality factor on bull moose in Algonquin Park, Ontario. He indicated that "delayed" mortalities, although unspectacular and hard to diagnose, might be more prevalent in the wilds than one would expect.

Bull moose engage in sparring and fighting during the rut. Much of this behavior is dominance testing and otherwise ritualistic. Nevertheless, serious injuries to the eyes and face can occur, and sometimes are eventually fatal. One form of nonlethal injury is a hypertrophic condition (above)—a bony protuberance from the skull—resulting from repeated forehead butting. *Photo © by Michelle Barnes.*

Exhaustion, dehydration, severe hemorrhaging and in-

Intense sparring or fighting during the rut by prime bulls of similar size can result in locked antlers and consequent death from broken necks, suffocation or starvation. This unusual scene, photographed in 1939 near Telegraph Creek, British Columbia, features full skeletons of bulls that died with their antlers locked. Typically, moose killed in this manner eventually are scavenged and most of their bones scattered. And it is expected that at least one of the animals ultimately would have fallen on its side. The above scene, although a possible discovery, likely was reconstructed to some extent. *Photo by George Hossick; courtesy of the U.S. National Archives.*

fection may be the actual cause of many fighting fatalities. Skeletal remains of distant battles or full carcasses of recent conflicts sometimes are discovered in lakes, ponds, mineral licks, streams, creek bottoms or in trampled snow "fighting arenas." Competing bulls with locked antlers also may succumb to drowning, suffocation or broken necks. C. Todesco (personal communication: 1992) found two mature bulls that starved to death, and two others that had "fought hard in shallow water, but slipped into deeper water and drowned."

No one knows how many bull moose die in sparring fights from injury or because of antlers locking together. In an attempt to evaluate this form of mortality among bulls, Peterson (1955) determined from a questionnaire of 139 moose hunters that 41 percent of the hunters had observed instances of bull moose dying in fighting contests. Cowan (1946) believed such deaths to be "fairly frequent," but gave no further quantification. Although such losses probably are numerically incidental for most moose populations, they usually represent elimination of prime bulls of superior characteristics and, therefore, a potentially significant elimination of a population's more formidable genetic stock.

Abandonment

Cows hide their calves in seclusion to conceal detection by predators, but if necessary, will resort to aggressive behavior and threat gestures to ward off predators (see Chapter 5). Moose give birth lying down and first nurse their young from the recumbent position until the calves can stand and suckle easily from the flank or rear position. When cow/calf pairs nurse lying down, the calf usually stops suckling, but if nursing occurs from a standing position, the cow terminates the nursing bout (Stringham 1974). A cow and calf will "lie apart," when the cow leaves its calf to feed. Some of these separations may last up to 4 hours (LeResche 1966). It is during these intervals that accidents can happen to calves, including kidnapping by people, killing by predators and abandonment by their dams.

Neonatal mortality has been attributed to predation, stillbirth, exposure, disease, accidents and abandonment.

On December 20, 1939, a wildlife agent with the Alaska Game Commission flew over Farewell Lake on Rainy Pass and spotted two bull moose nearby with locked antlers. By the time the agent and some assistants landed at the scene, one bull was down (top). The men were able to get the other animal down (bottom) and free the two by sawing off their antlers. The first downed moose died from wounds. The stronger bull, once freed and back on its feet, tried to attack its rescuers before being driven off. *Photos by Jack Benson; courtesy of the Wildlife Management Institute.*

M. Wilton (personal communication: 1992) reported that mortality of calves in Algonquin Park, Ontario is significant up to 6 months after calving. He indicated that predators (black bears and wolves) take a large number of the newborns, but some calves die at the birth site as well. Trampling of a neonate by its dam also may occur, resulting in injury and crippling of the calf. Wilton believed that some calf carcasses found at birth sites were a consequence of inadvertent smothering by their dams. Wilton suggested that, "when one considers the high loss of calves, any calf lost, regardless of the cause, is significant." He believed that, in Algonquin Park, moose calves probably experience a higher accidental mortality rate than do adults.

In Nova Scotia, T. Nette (personal communication: 1992) observed a "very excitable cow/calf pair running through a logjam along the shores of a large river. The calf emerged from the other side of the logs with a fence stake-sized pole protruding horizontally from its abdomen." The calf likely became the victim of a predator.

Accidental mortality of moose and calves in particular is difficult to quantify because (1) accidents often occur in unusual, remote places, (2) injured animals are prime targets of predators and (3) carcasses can be difficult to find even when not quickly devoured or scattered by scavengers.

In a 3-year study of calf mortality on the Nowitna and Koyukuk National Wildlife Refuges in Alaska, Osborne et al. (1991) determined that, of 151 documented fatalities, 8 (5 percent) of the calves died from causes other than predation. Some calves were found in lakes, and Osborne suggested that these calves may have died of injuries sustained in predator attacks or simply drowned.

Drownings accounted for 18 percent of calf losses in southcentral Alaska (Ballard et al. 1991), and Franzmann and Schwartz (1986) reported that, of 74 radio-collared calves monitored in predation studies on the Kenai Peninsula, 4 (5 percent) of the calves died after natural abandonment or from drownings or unknown causes.

Cow/calf pairs gradually increase their temporal and spatial distances as postcalving progresses and the calves become more active (LeResche 1966, Stringham 1974). Calves may become lost or abandoned at this time. Others are assumed to be orphans and "kidnapped" by well-intentioned and caring people. Gamefarms and zoos are sought as "surrogate mothers" for the orphans. In such cases, calves often are returned to the site of capture with hopeful expectations that the dam might return and claim her offspring. Some reunions are successful. But, those calves not successfully reunited with their dams usually are "sacrificed." No statistics are available to quantify the number of calves "rescued" by people. But, in light of the regularity of occurrence and the tendency of people to adopt wildlife orphans, despite admonitions against the practice, these calf losses are mostly unnecessary.

Drowning

Hydroelectric Projects and Reservoirs

Flooding of watersheds for hydroelectric power can have significant implications for moose, including (1) permanent loss of habitat; (2) displacement and disruption of seasonal movements; (3) increased vulnerability to natural predators (by concentrating moose); (4) blockage of access to traditional calving and winter ranges when mass starvation may increase; (5) ice shelving, open water, thin ice and floating debris may cause direct mortality of swimming moose; (6) collisions on service roads; (7) snow drifting impedes movements or increases risk of collision; (8) increases in legal hunting and poaching as animals concentrate about reservoirs; and (9) loss of habitat due to gravel excavations and reservoir operation (Ballard et al. 1988).

Describing the impacts of the Ootsa Reservoir on resident and migratory moose in the central interior of British Columbia, Edwards (1957: 2) wrote: "Migrations involving hundreds of moose between summer and winter ranges were disrupted. Submerged trees and floating debris presented perilous conditions to migrating moose or to moose attempting to escape rising water levels. During flooding, moose became trapped on hills and ridges. Considerable numbers of swimming moose drowned and many more fell through the reservoir ice. Moose trapped by floating debris and by rising water levels were a common sight; swimming and escape were virtually impossible. Perhaps the most important influence of reservoirs on moose is that of the barrier effect of floating debris intercepting migration routes. Formerly, hundreds of moose crossed lakes and followed the river courses on the ice in early winter to reach traditional winter ranges characterized by shallow snowpack, abundant browse and good aspect. However, the return migration in the spring was not on ice but forced to the water. In the reservoir, moose were observed to concentrate on the new shoreline near their traditional crossings. Migration was confused. Local overbrowsing resulted, and many moose were forced to winter on inferior ranges and were unable to return to their summer ranges. Moose populations decline under such conditions."

Thousands of moose have died in reservoir developments in British Columbia and comparable numbers have probably died in similar hydroelectric projects across the species' range in North America. The two most detrimental impacts of hydroelectric impoundments are permanent loss of highly productive habitat and displacement of moose to less productive sites.

Sumanik and Harrison (1968) suggested that possibly 5,000 moose died by drowning, from displacement, starvation, predation or hunting because of the flooding of Peace, Finlay and Parsnip river lowlands by construction of the Bennett Dam on the Peace River in the late 1960s. In a detailed evaluation of the impacts of the Bennett Dam on moose, Bonar (1975) argued that, because of lost winter ranges, drowning and other less direct effects of the hydro development, the moose population may have declined 70 percent, from about 12,500 animals at preflood to fewer than 4,000 when completely flooded.

Thin ice on reservoirs takes its toll on moose numbers. R. Bonar (personal communication: 1992) observed "holes in the ice" on the Revelstoke Reservoir during flooding by the Mica Dam in British Columbia. He found several locations where moose had fallen through thin ice, and felt that moose started to fall through the ice on the new reservoir as soon as it was being filled. He counted at least 30 holes through the ice in the first winter of reservoir development and twice that many in the subsequent year of flooding. Moose fatalities from drowning and falling through thin ice likely are commonplace on large reservoir developments, except in Manitoba, where no fatalities have been reported in hydro reservoirs because efforts were taken to drive moose by aircraft and boat from the impoundment areas before full flooding was achieved (V. Crichton personal communication: 1992).

In studies of the effects of the Mica Dam on fish and wildlife species, Peterson and Withler (1965) suggested cow and calf moose dependence on the wetland areas during the early summer months, and most of the area's moose population depended on bottomland for winter range. The 310 miles (500 km) of riparian habitat and 4,760 acres (1927 ha) of sloughs and wetland edge in the mainstream reservoir area probably supported about 400 moose or 50 percent of the population estimated to reside in the area occupied by the reservoir. Fifty percent of the moose population was impacted by many of the same factors hypothesized by Ballard et al. (1988) to impact moose negatively. More explicitly, Peterson and Withler (1965: 38) wrote: " . . . some are displaced; many die; others are just shot."

The impacts of hydroelectric developments on moose are long term. Unfortunately, however, the impacts on moose during their construction and operation never have been fully quantified, because either postimpoundment studies are deemed uncomparable to data collected before flooding or no preimpoundment studies have been conducted to enable comparison. Consequently, estimates of losses are conservative and mostly speculative. However, the effects likely are cumulative, and moose populations associated with the reservoir area will decline through a combination of direct and indirect mortality factors, many of which are described elsewhere in this chapter. Fortunately there now is more willingness to seek compensation and mitigation to restore and protect moose populations occupying habitats around such reservoirs (see Davidson and Dawson 1990).

Lakes, Ponds, Rivers and Ice

Drowning plays no favorites; it claims the lives of adults and young alike. Peterson (1955) indicated that moose calves are especially susceptible to drowning, and that drowning is a significant cause of adult deaths during early or late winter as a result of falling through thin or "rotting" ice on lakes, streams, ponds and reservoirs.

R. Bonar (personal communication: 1992) observed a cow moose in northeastern British Columbia defend the carcass of its calf, which probably was "not much more than a week old." The calf apparently drowned when it followed its dam across flooded marshlands, became exhausted and hopelessly entangled in matted vegetation.

In the Yukon, accidents claim the lives of approximately 8 percent of all reported calf deaths, primarily from drowning or exposure. Some drownings occur when calves run to water to escape pursuing predators and, for some reason, cannot regain shore. Other calf drownings have been attributed to the currents and debris in flooded creeks and streams after heavy rainfalls or surface runoff from spring thaws (Larsen et al. 1989a).

In lakes and rivers of Algonquin Park, Ontario, carcasses of moose that apparently have fallen through the ice at spring break-up are found periodically.

Moose that fall through ice sometimes can pull themselves out by their forelegs, gripping the leading ice edge. More often than not, however, they are not so fortunate. M. Wilton (personal communication: 1992) reported witnessing a bull moose that fell through thin ice on a beaver pond. "After several repeated attempts, the bull could not successfully free himself. He became exhausted and eventually drowned." K. Morris (personal communication: 1992) reported that, in Maine, a moose fell through thin ice on a swamp pond and, with its head and three legs submerged, could not gain sufficient purchase to escape its predicament, and drowned.

At freeze-up, moose fall in rivers, lakes or reservoirs, and weakened by fatigue and cold waters succumb to hypothermia and drown. Frozen carcasses and partially scavenged skeletons are discovered in lake or river ice, testa-

Moose calves are susceptible to a wide variety of incidental mortalities, including drowning, entrapment in floating vegetation, abandonment and mishaps at the birth site. Drownings are especially frequent and usually result when calves are swept away in freshette floods when crossing swollen creeks or streams after heavy rainfall and surface runoff from spring thaws or when chased by predators to dangerous water. *Photo by Kenneth Child; courtesy of the British Columbia Ministry of Environment.*

ment to a futile and exhausting struggle to escape the cold waters and fragile ice as they try to "plough" their way through (Peterson 1955: 193). The observations of R. Bonar (personal communication: 1922) of a cow moose and coyotes on a hydro reservoir in the southern interior of British Columbia are worthy of mention: "Moose started falling through the ice on the new reservoir almost as soon as it was filling. One cow went through and struggled for more than a day to free itself. We hauled her out on the ice, but she was too exhausted and died overnight. Another moose went through the ice way out in the middle [of the reservoir]. I watched her for most of the day. Two coyotes were bedded down within 10 feet of her. The next day, tracks showed she had escaped, wandered about 50 yards, and fell through the ice again. She was still there. I watched her for several hours in the morning, and she got out and made it to shore. The coyotes waited for nothing. Moose seemingly have no concept of thin ice conditions. However, they seem to be pretty good at getting out—they hook their bent forelegs on the ice in a bedding position and haul themselves out."

Bonar reported other cases of moose falling through thin ice on the shores of Goldstream River and Downie Creek in the Revelstoke area. Some died because of injuries, others drowned or starved, and others successfully escaped. Bonar was of the opinion that most moose (95 percent) caught in such precarious circumstances actually escape.

Peterson (1955: 194) related an account of a bull moose in Ontario that had struggled for some time and because of the weight of ice that had accumulated on its antlers from splashing, "he was no longer able to keep his head-up and finally drowned in deep water."

The frequency of moose falling through newly formed or thawing ice in Manitoba is unknown. Incidences of moose caught in the ice at spring break-up are on record but most deaths are attributed to hypothermia or drowning (V. Crichton personal communication: 1992). Moose sometimes are unable to get to shore because of shelf ice and either swim downstream to a more favorable landing site or drown.

In central British Columbia, bear hunters patrol the rivers in early spring looking for floating moose carcasses and scavenging black and grizzly bears. Many of the moose possibly fell through thin ice, got caught in an ice jam or were trapped by shelf ice and, unable to extricate themselves, drowned (B. Meisner personal communication: 1989).

Hosley (1949) reported that heavy mortality was known to occur in Alaska when moose ventured out on thin ice on lake outlets, rivers and creeks. C. Schwartz (personal communication: 1992) recalled the springtime discovery of 18 moose carcasses around the perimeter of Hidden Lake on the Kenai Peninsula. He indicated that the lake had patches of thin ice that were produced by upwelling of warmer waters from springs on the lake bottom. As in other cases, moose fall through the ice, and those unable to find or clamber onto safe ice become exhausted and drown.

Although it is difficult to quantify the numbers of moose lost to drowning, the history of these mishaps suggests that such mortality is common. In Ontario, P. Bewick (personal communication: 1992) reported that, in 83 fatalities of moose caused by various accidents, 22 (27 percent) drowned. Undoubtedly, many moose successfully escape from such predicament, but it is logical to assume that many others—those in poor physical condition or if wounded or suffering from disease or the merely young—are quite vulnerable to drowning mortality in waters that otherwise are a mainstay of their habitat. A similar differential mortality would likely

Attempting to cross rivers and lakes, moose sometimes encounter steep and/or unstable barriers—such as logs, undercut banks and ice ridges (above)—that prevent coming ashore. The animals are forced to return to the shore of embarkation or swim farther downstream. If water currents are severe, the barriers lengthy and the distance great, even adult moose, which are exceptional swimmers can succumb to hypothermic shock and drown. *Photo by Kenneth Child; courtesy of the British Columbia Ministry of Environment.*

apply to moose mired in deep bogs, muds, mineral licks and snow (see Murie 1934, Munro 1947, Hatter 1948, Chamberlain 1977).

Slips, Trips and Falls

Moose fall from precipices and sustain serious injury, die on impact or by broken necks. Every year in late winter in Newfoundland, for example, moose lose their footing on ice and fall from cliffs above the coastline (B. Collins personal communication: 1992). Some survive unscathed, but are trapped on ledges between the cliff face and ocean. Others sustain fractures, and some die of starvation unless rescued by fishermen. In Maine, K. Morris (personal communication: 1992) documented a bull moose falling 700 feet (213 m) from a cliff into a lake and two others that had either "jumped, fell or were pushed off the cliffs" to their deaths. In Scotia and Browning Township, Ontario, two moose were found at the base of a cliff, one dying from various injuries sustained in the fall and the other likely sustained a broken neck (D. Dasti personal communication: 1992). Of the Kenai Peninsula, Alaska, G. Del Frate (personal communication: 1992) noted that "much of the Peninsula is surrounded by bluffs that vary in height from 18 to 200 feet [5.4–61 m], and during winter, moose have been observed falling from the bluffs after feeding too close to the edge."

In the Chapleau area of Ontario, a moose frightened by a vehicle, jumped from a rock cliff and died of fatal internal injuries in a creek bed (C. Todesco personal communication: 1992). W. May (personal communication: 1992) observed a cow moose tumbling from a rock-cut on a major highway in Wawa, Ontario: " . . . she apparently could not see until it was too late (a skid track showed an attempt to stop), the cow lost her footing and fell to her death." In a similar instance, May reported that a bull moose ran into the face of a rock-cut on a highway. "Apparently, the moose was confused and blinded by vehicle headlights, ran head-first into the cut face of the rock bluff in its escape. Although, it survived the impact; it suffered a severely broken jaw and could not get up. It died later."

Moose have been rescued from abandoned or newly dug wells, natural bogs, mineral licks, beaver ponds and, in some instances, even from deep snow drifts from which they could not extricate themselves. Most moose caught in these situations succumb to exhaustion, hypothermia, starvation, suffocation or drown. Predators often salvage the carcasses, making determination of the exact cause of death

On December 5, 1994, residents of Soldotna, Alaska, spent more than 3 hours trying to save a cow moose that fell through ice on the Kenai River. The ice was too thin to support the animal's weight, but too thick to enable it to plow its way to shore. With the help of a utility company's equipment, a citizen eventually was able to approach close enough to lasso the cow. The exhausted animal was hauled to shore with the aid of a truck. A rescuer cleared the animal's airway and she wobbly retreated to safety. Researchers suspect that there are countless incidences each year of moose falling through thin ice, and that the vast majority of these moose either break their way to shore or are able to hunker back onto solid ice. However, ice conditions as in the Kenai River case will doom young, infirm or otherwise weak moose. *Photo by Mason Marsh; courtesy of the* Peninsula Clarion *(Kenai, Alaska).*

difficult or impossible. C. Schwartz (personal communication: 1992) related an instance in which he rescued a calf moose from an oil well access pipe on the Kenai Peninsula. The pipe was about 3 feet (0.9 m) in diameter and 10 feet (3.0 m) deep. Schwartz entered the pipe, placed ropes around the calf and lifted the animal out with a fork lift (only to find out later that the pipe contained dangerous levels of toxic gases). The calf rejoined its mother after its release and probably survived the ordeal. All access pipes were since covered to prevent similar occurrences and ensure public safety.

In 1979, in the Sioux Lookout District, Ontario, a hunter heard a moose "crying" at a vacated logging campsite (A. Mathews personal communication: 1992). He discovered a calf moose in an uncovered well that was approximately 4 feet (1.2 m) square at the surface and about 6 feet (1.8 m) deep. The calf was rescued.

In Alberta, one or two moose that have fallen victim to a sinkhole are found each year near Grand Prairie. The sinkhole is almost quicksand-like. Moose get mired in the clay-like soils and, after much struggling, die of exhaustion or starvation (D. Moyes personal communication: 1992).

V. Crichton (personal communication: 1992) related an account of a radio-collared cow moose vanishing from visual contact on Hecla Island, Manitoba. Radio signals indicated that the animal was alive but she could not be found. Several days later, a ground crew discovered the cow confined in a limestone depression. It apparently had fallen into the cave-like depression and was hidden from view by freshly fallen snow that had canopied its position.

In 1950, a trapper discovered several dead moose around a clay-bottomed pond near Thackery Creek, about 13 miles (21 km) south of Lake Abitibi in Ontario (Chamberlain 1977). The pond erupts geyser-like, spewing mud and water 13 to 16 feet (4–5 m) into the air. In 1971, a conservation officer visited the site and found three dead cow moose and a set of twin calves still alive and active nearby. Both calves died within the week. Skulls and skeletal remains of at least 10 other moose were found within 100 feet (30 m) of the pond. The moose evidently became mired in the bottom sediments that, like quicksand, quickly anchored them in place. Death was by drowning or exposure. The pond has been fenced.

Large depressions from strip mining and gravel borrow pits claim the lives of some moose each year in the Kirkland Lake District of Ontario (R. Kervin personal communication: 1992). Moose fall into these excavations, break their necks or die of exposure or starvation. In the Chapleau District, moose have been found mired in mud holes where they probably died of starvation, drowning or suffocation (C. Todesco personal communication: 1992).

In Newfoundland, moose have been rescued from abandoned or newly dug wells, peat bogs and, in at least one instance, from a cavern in a snow drift (S. Oosenbrug personal communication: 1992). In another instance, a radio-collared moose that had died of starvation or by suffocation was found in the moist bottom muds of a bog (B. Collins personal communication: 1992).

Mining exploration "trenches" are deathtraps for moose. Cattle have fallen into the open trenches of the Trans Canada Pipeline and it is likely that some moose may have experienced the same fate. Most abandoned mines and well sites now are fenced in Ontario as a condition of work permits to address human safety and liability (W. May personal communication: 1992).

On occasion, hunters find moose "trapped" or "wedged" in the decking of wooden bridges on logging roads in southeastern British Columbia (A. Wolterson personal communication: 1992). Some are dragged free but are so severely injured, they are shot and the carcasses salvaged. Unplanked railway trestles are known to "trap" moose occasionally in Alaska (R. Modafferi personal communication: 1991) and British Columbia (D. McDonald personal communication: 1991).

Damage Prevention and Control

Moose venture into cities, towns and villages, trespass private property and, in so doing, sometimes raise concerns for human safety and property. They eat exotic plants and shrubbery, break fences, chase dogs and threaten people. Some moose must be immobilized and returned to the wilds. Others, if harassed by dogs or curious onlookers, are injured and usually euthanized. Human/moose conflicts are most common in winter when snow conditions drive the animals into residential areas.

Hand feeding of moose is of concern, especially for public safety and, of course, for the survival of the moose. G. Del Frate (personal communication: 1992) observed that many reported and unreported instances occur where moose injure people trying to feed, photograph or pet them. In one instance, a 5-year-old boy was injured when he tried to ride a moose. Human/moose conflicts all too often result in the death of the moose or serious injury to the onlookers.

It is not known how many moose are destroyed to protect human life and property, but it occurs often enough to suggest that such "damage control" could represent substantial numbers of adult moose in some years. G. Del Frate (personal communication: 1992) opined that these instances could "become a serious problem in the future." He suggested that, "as areas are developed and habitat destroyed, moose management in and around towns and cities will become more challenging." The media should advise curious people, however well-intentioned, to keep a safe distance from wild moose under any circumstance. Human proximity and behavior can prompt normal defensive responses from moose that can cause very serious damage to humans, property and, ultimately, the moose.

Fire

Fire is a periodic and generally favorable influence on moose habitat. It rejuvenates forest succession and is a key factor to high productivity in moose populations.

The extent of incidental mortality of moose to natural or

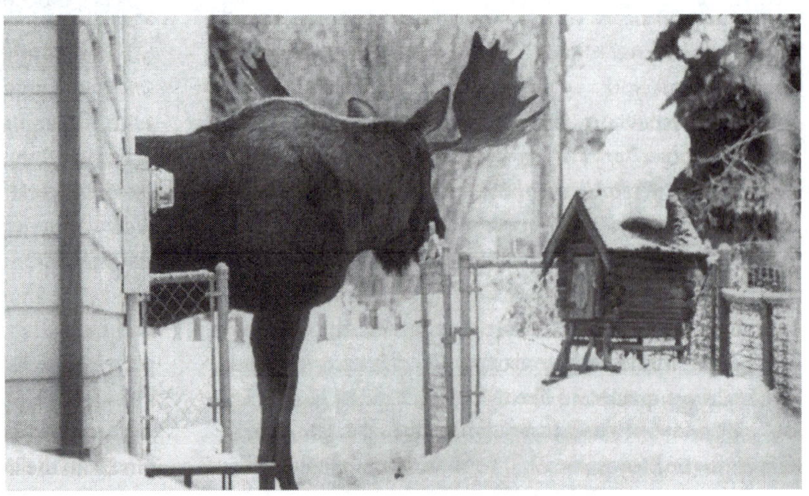

The appearance of moose in cities, towns and villages is a harbinger of human/moose conflict. Property damage, harassment, inconvenience and injury to people and moose are likely to occur. Because of the potential for damage or injury, people are justifiably intolerant of moose in their yards and streets. But such situations are one price of residing in moose habitat. Another is the cost of removing the animals by hazing, translocation or killing. *Photo by Bob Hallinen; courtesy of the* Anchorage *(Alaska)* Daily News.

Of great concern to wildlife managers are cases of habituation of moose to human proximity and activity. Equally disconcerting is the anthropomorphizing or cavalier attitude that people can have about wildlife with which they come in contact. Fascination with moose or disregard for their wildness can lead to serious, even fatal consequences. During a 15-month period from 1993 to 1995, two people in Anchorage, Alaska, were killed by moose drawn into the city by deep snows. Hand feeding of moose was outlawed in 1993 by the Alaska Board of Game. Moose habituate readily to handouts and often become aggressive when food is not proffered. Furthermore, well-intentioned feeders can become too incautious about maintaining a respectable distance from moose. *Photo by Bob Hallinen; courtesy of the* Anchorage *(Alaska)* Daily News.

prescribed fires is unknown, but generally is believed to be insignificant. Moose are quite capable of moving away from most fires. But occasionally a fast-moving hot fire will trap or disorient moose, causing death by asphyxiation or burning (V. Crichton personal communication: 1992).

In a wildfire in the Chapleau district of Ontario in 1948, moose were so badly burned that firefighters had to kill them (Peterson 1955). V. Crichton (personal communication: 1992) indicated that a hot spring fire in 1976, advancing at a rate of 3 miles per hour (4.8 km/hr), killed moose in Manitoba. Some were found dead; others were so badly burned that they were shot. W. May (personal communication: 1992) related a similar occurrence in Red Lake, Ontario, where nine moose had succumbed to a rapidly advancing

Record snowfalls in Anchorage, Alaska, during the winter of 1994–1995 brought an unusually large number of moose into the city, setting the stage for injuries and damages that escalate public intolerance of the animals and necessitate costly removals of moose. The high snow levels, coupled with people shoveling snow from their roofs to reduce the weight load enabled the occasional moose to find itself a rooftop vantage point. This did nothing to endear the animals to homeowners. *Photo by Bruce Bartley; courtesy of the* Anchorage Daily News.

fire. Eight of the moose were found "huddled together" on a hilltop, suffocated from the intense heat. The ninth, a single bull, collapsed and died in the middle of a forest road, apparently from smoke inhalation.

The number of "fire" fatalities to moose likely is related to season, moose population density, habitat type, fuel load, availability of escape terrain and prevailing winds. The fact that some animals are killed by fire or forced to move invariably is of lesser importance than the long-term improvements in moose range that fire normally creates in North America's temperate and boreal regions (Leopold and Darling 1953, Peterson 1955, Pimlott 1961).

Deep Snow and Avalanches

Moose have starved to death in deep snowbanks after eating all available food. Cows with calves break through crusted snow in windfall areas, get trapped or mired in the fallen debris and, unable to extricate themselves, starve to death, according to R. Bonar (personal communication: 1992). Bonar also reported the case of a bull moose "walking on top of a strong crust in over [79 inches] 200 cm of snow in late February, and slipped down under a large hemlock log in a recent clearcut. The bull became jammed under the log in an upside down position and could not get out. There was little evidence of a struggle, and rigor mortis had not fully set in. The bull probably died of rumen aspiration."

Fatalities also have occurred in the central interior of British Columbia, especially in winters of heavy snowfall, when moose become trapped in snow trenches created during their foraging forays. Gradually weakened through malnutrition, the moose are unable to negotiate the trench walls and starve as the food supply is gradually depleted.

Moose carcasses have been found in an avalanche chute near Revelstoke, British Columbia. R. Bonar (personal communication: 1992) suggested that moose may be frequent victims of avalanches, and that these losses might be more frequent than historically reported. No data are available to determine the magnitude of these "snow deaths" or their significance as a mortality factor on moose populations in the Rocky Mountains or elsewhere for that matter.

Entanglement

Rarely stymied from bulldozing through thick cover, moose likely do not visually distinguish between natural

Snow depth, crusting and duration are major factors affecting moose well-being on winter range. Under severe conditions, moose have been "trapped" in foraging trenches. Weakened and suffering from malnutrition, they often are unable to extricate themselves from their snow "prisons" and starve. *Left photo courtesy of the Alaska Department of Fish and Game. Right photo by Kenneth Child; courtesy of the British Columbia Ministry of the Environment.*

vegetation and most artificial cords, cables, wires and ropes. At least some moose encounters with the latter elements cause entanglement and death.

In the Dryden District of northwestern Ontario, a bull moose was found tangled in discarded cable from an abandoned power line. The bull was anchored securely to a tree by the steel cable, which was firmly attached to its antlers. The bull tried to free itself, but the struggle merely strengthened the attachment. It was cut free, but was so exhausted and stressed that it died 24 hours later about 0.5 mile (0.8 km) from the point of release (W. May personal communication: 1992). After this fatality was discovered, Ontario Hydro removed 20 miles (32 km) of the discarded cable to reduce the potential of "snaring" moose.

In Manitoba, a 3-year-old bull moose was caught by its antlers in discarded telephone cable in an old Abitibi logging camp (V. Crichton personal communication: 1992). The wire was tightly wrapped around the muzzle, neck and eyes, indicating not only had the bull put up a considerable struggle to free itself, but had broken one antler in two places in the process. The bull died because of exhaustion and injuries sustained in the struggle.

Similar "snaring" events have been reported in the Yukon where bull moose get tangled in discarded telegraph wires left from World War II (D. Larsen personal communication: 1992). Although the exact number of moose that have died this way is undetermined, the losses are thought to have been insignificant.

In Newfoundland, a bull moose was tangled in a length of rope hanging from a communications tower (B. Collins personal communication: 1992). Another bull was caught in telephone cable, and a third bull was found snared in wire fencing. All three animals were successfully released.

K. Morris (personal communication: 1992) reported that moose occasionally are found tangled in abandoned electrical wiring in Maine and, once chemically immobilized, are released. Most survive, but a few die of stress-related complications.

A cow moose in the Kenora District, of northwestern Ontario, had a discarded telephone cable tightly wrapped around her neck (V. Critchon personal communication: 1992). She suffocated in the struggle to get free. Moose skeletons and antlers sometimes are found wrapped in wire, suggesting the possibility of foiled poaching attempts (A.G. Mathews and G. Eliuk personal communications: 1992).

In Chapeau, Ontario (V. Critchon personal communication: 1992) and near Prince George and Quesnel, British Columbia (W. Sharpe personal communication: 1992), moose have been found caught by their muzzles or by a leg in wire snares set for wolves. Caught by their muzzles, the moose likely pushed their muzzles through the loop of the snare while investigating the sets and, when lifting their heads,

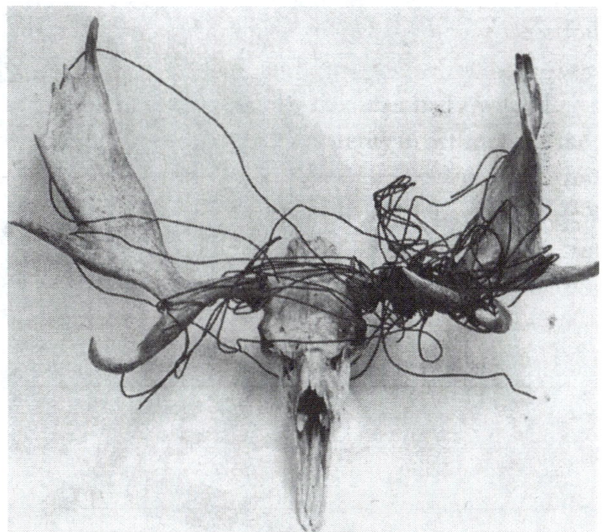

Discarded communication cables, guy wires, ropes and snares entrap moose. Moose sometimes are found with a "bouquet" of wire tightly wrapped around the antlers. (Left) A moose carcass found in 1952 on Glover's Island in Grand Lake, Newfoundland, was tangled in copper wire used as telephone lines by pulp cutters on island camps supplied by boats. The lines were suspended from tree to tree. When bull moose encountered the lines, they shook their heads to disengage the obstruction from their antlers, which is typical behavior. However, the copper wire did not break and simply "wrapped up" the animal. The losses of lines and moose were serious problems. (Right) The skull of a moose that perished after entangling in steel wire anchored to a tree. *Left photo by Robert A. McCabe; courtesy of the University of Wisconsin-Madison Department of Wildlife Ecology. Right photo by Mike Buss; courtesy of the Ontario Ministry of Natural Resources.*

caught the snare. When moving or when attempting to free themselves, the noose would tighten, causing suffocation.

Farm fences also ensnare moose. When jumping fences, moose sometimes trip on the top wire strand and get tangled or "hog-tied" in the fencing and sustain severe lacerations, or die by broken neck, starvation or shock. Because of the generally remote location of many fences in moose range, many associated injuries and fatalities are believed to go undetected.

In Ontario, fences are considered a daily hazard to moose (M. Vukelich personal communication: 1992). In the Timmins area, a moose hung itself in a five-strand fence when attempting to cross. Each year in Alaska, 15 to 20 moose are known to be killed by "entanglement and injuries suffered because of fences, stray rope or wires" (M.G. McDonald personal communication: 1992). Comparable numbers likely occur elsewhere.

To reduce entanglement and injury, top-railing and wooden fences are recommended on fence lines located in known moose concentration areas, especially those that intercept migration routes or parallel transportation corridors.

Crippling and Wounding Losses

An unknown number of moose are shot, wounded and die unrecovered by hunters. However, if hunter shooting skill tests are any indication, crippling losses of moose may be considerable (Timmerman 1977, 1987). Results obtained from voluntary testing of 5,500 Ontario moose hunters between 1976 and 1988 indicated as many as 30 percent of shots taken at life-size targets 50 yards (46 m) distant could have resulted in crippling wounds. Newfoundland reported a 40 percent failure rate among 3,000 hunters who were unable to place two of three shots inside a marked target representing the vital organs (heart and lung) on a life-sized target. In New Brunswick, where first-time moose hunters are required to hit a 2- by 2-foot (61 X 61 cm) target with two of three shots at 50 yards (46 m), the failure rate was

Occasionally moose are caught in snares set for other animals. Trappers near Chapleau, Ontario, had strategically placed 10 or more wolf snares around a small balsam thicket and about 2 feet (0.6 m) off the ground. Apparently, two moose ambled down a nearby road and into the thicket. They disturbed the sets and one moose escaped, but the other caught its muzzle in one of the snares. In its struggle to escape, the moose merely tightened the snare. And when the animal backed around the balsam and fell, pinning its right foreleg between the snare wire and tree trunk, only to securely hogtie itself. The snare was firmly locked about the nose, but did not seem to restrict air passage. The moose probably died as a result of stress and from choking on rumen contents that it was unable to clear because of its position. *Photo by Tim Moody; courtesy of the Ontario Ministry of Natural Resources.*

Under most circumstances moose easily leap most fences. In deep snow, leaping is relatively difficult and there is possibility of slipping on the takeoff. A short leap of a multistrand fence can entangle a hind leg. If a moose so caught can regain its front footing, it usually can break free. But if caught by too many wire strands firmly attached to fenceposts, or "hung" in deep snow, the scenario likely ends in a fatality. Even moose that free themselves may have sustained serious injury to one or more of its limbs. Wood railings on the top of fences or mesh fencing can reduce such accidents. *Photo by Kenneth Child; courtesy of the British Columbia Ministry of the Environment.*

about 14 percent. In all cases, wounding of moose is highly likely with the subsequent retrieval of some moose doubtful, because not all wounds are fatal (Timmerman 1977, Buss et al. 1989).

Wounding or crippling losses are not well documented, and considered of superficial importance in calculating allowable harvest (Buss et al. 1989). Nevertheless, Gasaway et al. (1983) arbitrarily set a 15 percent adjustment for unreported kills and wounding losses in Alaska. Fryxell et al. (1988) applied a 20 percent correction to the legal harvest of moose in Newfoundland to account for wounding losses. And Boer and Keppie (1988) made a 10 percent allowance for wounding and crippling losses in harvest levels when simulating a managed moose population in New Brunswick. More study is needed to determine a reliable correction factor for these losses, and encouragement should be given to improve the shooting ability of people licensed to hunt.

Miscellaneous Mishaps

An unusual cause of death of two moose occurred on separate occasions in Maine and Newfoundland. Of the Newfoundland instance, Bill Collins (personal communication: 1992) stated: "I saw one moose 'impaled' on a wrought-iron fence. The fence was about 6 feet [1.8 m] high and the palings were shaped like the top of a lance. The fence surrounded an abandoned graveyard. It appears that for some reason the moose attempted to jump the fence even though there were other access routes from the graveyard. The fence paling pierced the animal's rib cage and entered the heart–lung region." K. Morris (personal communication: 1992) also wrote of a moose impaled on an ornamental wrought iron fence.

At least two instances of moose being "stuck in a snowball" have been reported. In Saskatchewan, M. Killaby (personal communication: 1992) found a cow moose with a bulge of snow attached to her left front leg. "It appears," wrote Killaby, "that the cow had suffered a compound fracture and snow began to accumulate [freeze] around the wound. As time passed, the snowball increased in size possibly because of wet sticky snow conditions or accumulation of snow at site of a wound or injury or by combination of both factors. The animal, in an emaciated condition, was dispatched." Killaby was certain that the moose would have died "sooner than later" of starvation or by predation. C. Schwartz (personal communication: 1992) reported similar occurrences on Alaska's Kenai Peninsula, where two moose were found with large "snowballs" anchored to their forelegs—one of the moose had slay-foot and the other had a wound.

For all their strength and dexterity in dense cover, moose are not immune to being tripped up by vegetation, sometimes even fatally. Near Thunder Bay, Ontario, a moose was found dead with its front right foreleg caught in the crotch of a birch tree (H.R. Timmerman personal communication: 1992). Biologists speculate that the moose may have lost its balance while browsing and, falling backward onto its right side, caught its front foreleg in the crotch of the tree, making it impossible to right itself. The moose was believed to have died of starvation. Two other bulls met similar fates on Isle Royale (R. Peterson personal communication: 1992). One starved to death after catching a foreleg between two birch trees. The other apparently caught its foreleg in a sapling, tripped and tumbled to its death over a cliff.

A most unusual event resulted in the deaths of two moose in southeastern Alaska in July 1992. According to the *Anchorage Daily News,* two Gustavus fishermen saw what they described as dark spots in the water, heading from Chichagof Island to the mainland. The men rowed their skiffs closer and realized the spots were moose, paddling to

Wet sticky snow can accumulate on moose at the sites of wounds or injuries, particularly on the legs. A huge "snowball" formed at a fracture on the left foreleg of this cow. The snowball enlarged as the cow moved and rested. When euthanized, the cow was in extremely poor condition and suffering from malnutrition. *Photo by Marlon Killaby; courtesy of Saskatchewan Parks and Renewable Resources.*

Because of their strength and long legs, moose can negotiate in areas and conditions that preclude movements by other North American deer. This ability also exposes moose to many opportunities for missteps, trips, tumbles and slips that can be injurious or fatal. This is especially true in winter, when deep snows cover aspects of dangerous footing, thereby changing the height of the travel plane. The bull moose above caught its foot in the crotch of a birch tree and evidently fell on its side against the anchored foreleg. Unable to rise from its fallen position, the animal subsequently starved. *Photo by P. Nunan and E. Swift; courtesy of the Ontario Ministry of Natural Resources.*

shore. Then came "a lot of splashing," said one fisherman. "When killer whales are after food, they are so amazingly fast," said the other. The fishermen watched the "frothing display" and, after the first moose was eaten, pondered the fate of the second. It was spotted about 1 mile [1.6 km] away and "swam right in the middle of the kelp bed. It took about 10 minutes to work itself into the middle of the kelp and it was safe from the pod of three or four killer whales," explained one fisherman. The orcas spent about an hour trying to find a way into the kelp, "but every time they'd get in there, they'd get tangled up in the kelp." The whales finally left. When the fishermen returned to check on the fate of the moose, it had drowned, apparently too exhausted or too enmeshed to escape the grip of the kelp.

In June 1993, the *Peninsula Clarion* newspaper reported "It's a whopper: Garden hose lands moose." Near Kenai, Alaska, a cow moose got tangled in a garden hose that was looped over a pole on the bow of a boat parked in a yard. The hose tightened around the moose's neck and, before anyone could assist, the animal choked to death. In the process, she dragged the boat into two parked vehicles.

These few examples of unusual fatalities illustrate that moose can and do die in bizarre ways. Although these "freak accidents" may be extremely rare in the wild, they do constitute additional losses that, in concert with other mortality factors, may influence local moose population levels.

Management Considerations

Accidents injure, maim and kill substantial numbers of moose and can result in significant property damage, human injury and death. Property damages caused by moose amount to millions of dollars annually in North America (Damas and Smith Co. 1983, Child and Stuart 1987, Del Frate and Spraker 1991, Oosenbrug et al. 1991). Moose/train and moose/vehicle collisions kill and cripple thousands of moose each year across North America (Child and Stuart 1987, Child et al. 1991, Del Frate and Spraker 1991, Oosenbrug et al. 1991). Humans are seriously injured, and fatalities are on record in Alaska (4), British Columbia (10), New Brunswick (1), Newfoundland (6), New Hampshire (11) and Quebec (5). If traffic volume, vehicle speed and moose populations increase locally, human injury and deaths and moose fatalities will increase (Damas and Smith Co. 1983, Child and Stuart 1987). Public education and driver awareness and caution probably are the best measures presently available to reduce collision risks and costs.

Railway and roadway kills of moose are related primarily to habitat condition, innate behavior and snowfall. Various mitigation measures, such as vegetation management, widening of transportation corridors, adjustments of travel speed, improved signage and construction of physical barriers offer only temporary or partial relief. Further research on moose behavior, particularly human/moose interaction situations may offer better solutions, but greater commitment to such investigation is needed from government agencies, transportation officials, insurance companies and conservation groups.

In North America, collision fatalities represent approximately 6 percent of the annual allowable harvest of moose, ranging from 0.5 percent in Manitoba to more than 10 percent in British Columbia, Maine and New Hampshire (Table 20). In New Hampshire, incidental mortalities are almost twice the conservative harvest level "politically" set to support introduction of recreational hunting (K. Bontaites personal communication: 1992).

Leopold (1933) and Peterson (1955) contended that incidental fatalities may decimate local moose populations and can impact productivity by altering age and sex ratios. R. Wheeler (personal communication: 1992) reported that, of 88 incidental mortalities recorded from 1987 to 1991 in Kapuskasing, Ontario, 32 (36 percent) were adult males,

All too often, moose/vehicle collisions result in death. Moose frequently are killed outright or are injured sufficiently that they will die shortly afterward of injuries. Conservatively, 3,500 moose are killed annually on North American roadways. Humans also die in such accidents, and that cost is incalculable. And even when the collisions are not lethal, property damage usually is significant. In this scene, the young bull was killed and the vehicle was "totaled," but the driver miraculously suffered only minor injuries. *Photo courtesy of the Alaska Department of Public Safety.*

35 (40 percent) adult females, 5 (6 percent) were adults of unknown gender and the remaining 16 (18 percent) included 5 male calves, 6 female calves and 5 calves of unknown sex. The ages of the adults killed were unknown, negating any attempt to assess the impact of these losses on population structure or productivity, but the high mortality of adults of both sexes suggests that at least some deleterious impact was likely. And, if a high proportion of adults is broadly typical of incidental mortalities, then these losses conceivably could represent significant impacts to resource well-being, population growth and productivity, and local recreational opportunity.

In relation to the percentage of the annual harvest represented by collision losses and the wide spectrum of other fatal mishaps, accidents are, as Leopold (1933) suggested, decimators of moose. In fact, incidental mortality could claim the lives of more moose than that attributed annually to predation, parasites, diseases or a combination of all three factors. For example, in a 10-year study of nonhunting mortality of moose in the Sudbury area of northeastern Ontario, K. Morrison (personal communication: 1992) reported that, of 1,673 moose fatalities, 986 (58.9 percent) died in collisions, 83 (4.9 percent) died in various accidents and mishaps, 179 (10.7 percent) were poached, 41 (2.5 percent) were killed by wolves and possibly dogs, 286 (17.1 percent) were taken by subsistence hunters and 98 (5.9 percent) were killed by unknown causes (Table 21). Losses of moose to accidents and various mishaps were twice the number of moose killed by predators. And if collision losses are included, total incidental mortalities (1,069) were almost double the combined losses (604) to predation, subsistence, poachers and unknown causes. Nevertheless, an objective assessment of the impacts of accidents on the biology, demography and sociobiology of moose is impossible until more and better data are collected (Child and Stuart 1987, Child et al. 1991, Modafferi 1991, Schwartz and Bartley 1991).

Too often, incidental mortality has been ignored in management programs or in setting harvest levels (Child and Stuart 1987). This has occurred because its impact on population structure usually has been "trivialized" or "written off" as demographically insignificant by discounting annual fatalities against population estimates or, in the case of collisions, against linear distance of right-of-way or a measure of track or road length (Child et al. 1991). These descriptions tend to mask or minimize the magnitude of moose losses and population impact and, as such, argue against need for corrective action or adjustment in management strategy. Peterson (1955) contended that, because incidental mortality kills substantial numbers of moose each year, accidents and various mishaps should be given importance approximately equivalent to that as losses to predation and diseases combined. However, given the numbers of moose killed throughout the North American range each year by nonhunting factors, the better equivalent comparison of accidental mortality may be the legal harvest, at least for some moose populations.

Managers recognize that accidents can impact moose through population reductions and management programs through loss of recreation opportunities. In British Columbia, annual harvests, license issues and seasons are adjusted to account for reported collision losses (Child et al. 1991). Hunting closures or more restrictive regulations may be necessary in areas where moose losses exceed the legal harvest (Oosenbrug et al. 1986). Hunting seasons are closed in Alaska in areas where incidental mortality (collisions) has contributed to local population declines (Schwartz and Bartley 1991). And in Ontario, because of the numbers of roadway and railroad kills (about 9 percent of annual adjusted harvest), conservative hunting seasons and harvests

Table 21. Summary of nonhunting mortality of moose in the northeastern region of Ontario from 1983 to 1991

Year	Road	Rail	Poaching	Predation	Subsistence	Other	Unknown	Total
1983	70	10	17	4	13	14	21	149
1984	81	39	18	6	11	12	14	181
1985	91	16	18	1	22	7	11	166
1986	95	28	47	1	42	9	7	229
1987	89	25	18	11	24	7	8	182
1988	112	18	19	4	43	11	16	223
1989	103	17	13	3	34	13	5	188
1990	92	16	10	7	56	4	13	198
1991	70	14	19	4	41	6	3	157
Total	803	183	179	41	286	83	98	1,673
Percentage	47.9	10.9	10.7	2.5	17.1	4.9	5.9	

are advertised (A. Bisset personal communication: 1992). But the influence of accidents as a mortality factor still are assessed subjectively and adjustments to regulations, seasons or recreation are assigned conservatively in most areas (Child and Stuart 1987).

As a mortality factor on moose, accidents may (1) impact population structure, (2) alter sex and age ratios and (3) reduce productivity. Importantly, in light of the socioeconomic values of moose (see Bisset 1987, Wolfe 1987), these fatalities should not be "written off" as compensatory mortality (Eastman and Hatter 1983) or reasoned to be of little consequence to management programs or recreation. Finally, such mortality will continue to claim the lives of moose as human activity encroaches on moose habitat.

(BELOW) BULL SHEDDING
VELVET – RAGGED BROWN STRIPS
ON L ANTLER – CLEAN AREAS
BONE WHITE –
75-MILE AREA –
TWO BULLS & COW BROWSING IN DRAW
JUST ABOVE THOROFARE BAR, MCKINLEY PARK, ALASKA –
AS WE APPROACHED, THEY RETREATED DOWN &
AROUND SLOPE, THE COW IN THE LEAD –
BULL WITH LARGEST ANTLERS (L) HAD FLAP-SHAPED
DEWLAP, NO BELL – SMALLER BULL HAD BELL AND
FLAP – SEPT. 4, '56
(MEMORY)

NOTE MASSIVENESS
OF UPPER FORELEG
WIDE, FAN-LIKE FORWARD
PALMS – MANY TINES
AUG. 25, '56
(COMP. FROM MEMORY)
BULL MOOSE IN VELVET –
BROWSING & FEEDING IN
WILLOW & ALDER THICKETS
70-MILE AREA, MCKINLEY
PARK, ALASKA –
OPEN TUNDRA –
ROLLING, BROKEN
COUNTRY SEVERAL
MILES FROM
ANY TIMBER –
PONDS,
STREAMS –
AUTUMN
COLORS WELL
STARTED
(MEMORY) SEPT. 13, '56 CAMP DENALI,
KANTISHNA AREA,
ALASKA –
ANTLERS DARK ON
UNDERSIDE, GOLD OR
PINKISH-GOLD ON
UPPER SURFACE –
DARK-APPEARING MOOSE
(CLOUDY – JUST AFTER HEAVY RAINS)
GRAYED, PURPLISH- OR REDDISH-BROWN
ABOVE –
BLACK OR
DEEP BROWN
BELOW –
SLIGHT
HUMP
WALKING THROUGH BROWNISH-REDS & PURPLES
OF BLUEBERRY & BIRCH, GOLD WILLOWS –
MOUTH
OPEN –
SEPT. 15, '56
KNOLLS SW
OF MOOSE CR.
BRIDGE –
LIGHT-COLORED
MOOSE – YELLOW-
BROWN UPPER-
PARTS, ESP. LIGHT ON HUMP –
DARK PURPLE-BROWN BAND
ON SIDES, FLANKS, ETC. – LIGHT
GRAY-WHITE LEGS – DARK
STRIPE DOWN NECK &
SHOULDER –

9

KRIS J. HUNDERTMARK

Home Range, Dispersal and Migration

The ways in which moose use their environment both spatially and temporally are of great interest to resource managers. The dynamics of animal movements and distribution in space and time are integral to behavioral, ecological, genetic and population processes. Thus, the attributes of the space occupied by individual animals both annually and seasonally *(home range),* patterns of movement within home ranges, establishment of new home ranges by young moose and colonization of new habitats *(dispersal)* and movements between seasonal ranges *(migration)* must be considered in comprehensive management programs. Although much effort has been expended in describing these behaviors in moose, little effort has gone into explaining their ecological significance. This chapter provides a basic understanding of these concepts and summarizes the research on these topics relevant to moose.

Knowledge of moose home range and movements has developed only recently. Early reports of home range size of moose (e.g., Seton 1927, Murie 1934) were highly speculative and based on localized observations of recognizable individuals. Peterson's (1955) classic monograph on the moose reveals little on this subject, and what can be discerned is, with the benefit of hindsight, inaccurate. Indeed, Peterson speculated that moose migrations were limited to elevational shifts by populations inhabiting the mountainous regions of western North America, and that the home range of a moose in good year-round habitat could have a radius of 2 to 10 miles (3.2–16.1 km). As interest in the spatial and temporal dynamics of moose movements and their relevance to management has increased, so has our ability to study them.

Collection of reliable information on home range and movements of moose was enhanced dramatically with the advent of chemical immobilization (see Chapter 16) and radiotelemetry (see Cochran 1980, Mech 1983, Kenward 1987). The application of chemical immobilization to wildlife research allows biologists to mark animals, thus assuring positive identification during relocation. The usefulness of this technique, however, is limited by lack of visibility of the mark because of dense vegetation, or due to observer bias; the probability of seeing a particular marked animal often is a matter of pure chance (Bailey et al. 1983). Therefore, studies using this technique usually are characterized by limited numbers of relocations of study animals. If enough relocations are obtained solely by visual contact, the estimate of home range should not differ from that produced by radiotelemetry (Phillips et al. 1973). Unless each relocation attempt is successful, this estimate might be questionable because of the potential of missing data to change radically the shape and size of the home range estimate. Development of compact and lightweight narrowband transmitters mounted on neck collars has improved greatly the ability to locate animals. Under good conditions, radio-equipped animals can be relocated with a great degree of accuracy, regardless of whether they can be located visually.

Recently, global positioning system (GPS) technology has been applied to the study of animal movements and

Numbered neck collars allow moose to be located and positively identified from a distance. However, in the absence of radiotelemetry devices, the probability of observing a particular moose often is governed by pure chance and the density of foliage between the observer and the moose. *Photo by Albert W. Franzmann; courtesy of the Alaska Department of Fish and Game, Soldotna.*

habitat use. Specially designed GPS receivers attached to animals by collars determine their location by analyzing signals received from satellites in Earth orbit. Locations, and the times at which they were obtained, are stored in a receiver's memory until they are downloaded directly to a data receiver by an FM link or are relayed to a ground station through a separate satellite system such as ARGOS. As with any animal tracking system, GPS is prone to locational error and also is relatively expensive. Nonetheless, this technology offers many advantages to biologists studying animal movements and habitat use, including the ability to locate animals at night and during weather conditions that normally would preclude obtaining locations using aircraft. Rempel et al. (1995) assessed the utility of GPS for tracking moose in boreal forest habitats and recommended that biologists choose a tracking system based on an evaluation of the relative costs and inherent locational error of different systems in relation to study objectives.

Home Range

The concept of home range can be traced to Seton (1909), but most biologists recognize Burt (1940, 1943) as having provided the first working definition of the term. Burt (1943: 351) wrote that home range is "that area traversed by the individual in its normal activities of food gathering, mating, and caring for young." Although this definition has been modified to some extent by many biologists, it still serves as the basis for the generally accepted concept of home range as the area used by an animal during its routine activities (Jewell 1966). Nonetheless, it remains subjective, inasmuch as the investigator must determine what constitutes "routine" activities. Implicit in the definition is an assumption that this area provides an animal with the resources necessary to sustain life (e.g., food, water and shelter). However, in some cases, this assumption may not hold.

The home range represents for its occupant a familiar area in which to feed, rest, escape from predators and meet its other life requirements. Moose tend to exhibit a strong fidelity, or philopatry, to their home range; they tend to return to the same seasonal range and/or remain in the same home range for years. The strength of this bond was demonstrated by Bailey and Franzmann (1983). They documented the mortality of moose in the years after four 1.0-square mile (2.6-km^2) enclosures were constructed on the Kenai Peninsula of Alaska. Moose were captured and intro-

Before radiotelemetry, many wildlife agencies used neck collars to identify individual moose. There were attempts to involve the public in reporting sightings. Public response was good, but there was poor distribution of sightings—moose that frequented an area near human traffic were reported hundreds of times, but too many moose were not reported at all. The color poster above was 15 by 20 inches (38.4 X 51.2 cm). It became a collector's item, and copies disappeared rapidly. It was replaced by a newsprint poster that was not as attractive, but served the same purpose and longer, since copies were not "collected" as much or as quickly. *Photo by Albert W. Franzmann; courtesy of the Alaska Department of Fish and Game, Soldotna.*

Lightweight and weatherproof transmitters attached to collars allow biologists to relocate moose even in dense conifer forests where the animals cannot be easily observed. When tracked from an aircraft, the signal from a transmitter can be detected over a distance of many miles, and the battery life is in excess of 2 years. The band width of the transmitted pulse is narrow enough that many collars with distinct frequencies can be tracked with a receiver containing only one crystal. For example, a receiver containing a crystal capable of detecting frequencies between 150.000 and 151.999 MHz could conceivably distinguish between 200 transmitters with signals differing by 0.01-MHz increments. In practice, however, a greater separation of frequencies is desired. *Photo by Charles C. Schwartz; courtesy of the Alaska Department of Fish and Game.*

duced into the enclosures to supplement the resident population initially enclosed by the fence. Introduced moose died at a faster rate than did residents. Calves born to resident moose survived at a greater rate than did those born to introduced moose. Mean age at death did not differ between introduced and resident moose, but all introduced moose died within 2 years. Also, introduced moose paced fence lines near their points of introduction. Bailey and Franzmann speculated that increased expenditure of energy associated with pacing, as well as time taken away from feeding and resting, contributed to the increased mortality in this group. No overt aggression by resident moose toward introduced moose was observed, which reinforced the observation of Peterson (1955) that moose are not territorial. Strong affinities developed for their original home ranges and the apparent desire to return to those ranges likely reduced survival of the introduced animals to below that of resident moose.

Much effort is dedicated to measuring the size of home ranges, yet surprisingly little effort is expended discussing the ecological ramifications of this parameter. Animals establish a home range of a size appropriate to ensure survival through critical biological periods (Lindstedt et al. 1986). Therefore, the ultimate factor determining home range size seems to be energetics (McNab 1963). Essentially, a species with greater metabolic demands will have a larger home range, because it must range farther to consume enough food to meet its needs. As metabolic rate scales to body size (see Chapter 14), home range size and body size should be related, when compared among species in similar habitats. In general, moose would be expected to have large home ranges relative to those of smaller herbivores. Comparative home range estimates include meadow vole (0.05 acres [0.02 ha]), snowshoe hare (135 acres [55 ha]), pronghorn (4.5 square miles [11.4 km^2]) (Swihart et al. 1988) and 11 square miles (28.5 km^2) for moose in Maine (Leptich and Gilbert 1989).

Using the theoretical relationship between home range size and energetics, McNab (1963) predicted that animals living in colder climates have larger home ranges than do those living in warmer climates because of the former environment's greater metabolic demands of survival. It follows that animals living in habitats characterized by relatively poor productivity (lower carrying capacity) have larger home ranges than do conspecifics living in more productive habitats (see Eisenberg 1981). McNab (1963) also postulated that animals have larger home ranges in winter than in summer because of the increased metabolic demands of cold weather. Schwartz et al. (1988b), however, demonstrated that moose have lower metabolic rates in winter because of behavioral and physiological changes.

Mace et al. (1983) noted that, although numerous studies reported correlations between body size and home range size, none confirmed McNab's (1963) hypothesis that energetics is the ultimate factor that determines home range size. However, there is evidence for moose that the distance moved by an animal over a given period of time—which can be related to home range size—is related directly to forage biomass (Lynch and Morgantini 1984, Miquelle et al. 1992).

Methods of Estimation

Many different methods have been proposed to estimate home range size; each has its limitations (Table 22). The most commonly used technique is the minimum convex polygon (Mohr 1947). This technique involves connecting the outermost locations for a given animal to define the limits of its home range (Figure 119). The term *convex* is an important modifier, because it imposes the condition that all angles of the circumference of the polygon must point away from the interior. Only one convex polygon can be constructed from any set of locations, whereas more than one polygon can be constructed if at least one of the angles is concave (Figure 119). Thus, of all the polygons that can be constructed from a set of locations, only the convex polygon is a repeatable measure.

Table 22. Techniques and computer programs for analysis of home range data (adapted from White and Garrott 1990 and Anderson 1982)

Technique	Programs[a]	Comments	Reference
Minimum convex polygon	McPAAL, DC80, HOMERANGE, HOMER, CALHOME	Most commonly used method; assumes home range shape is convex	Mohr (1947)
Adjusted minimum convex polygon		Provides method for eliminating unused areas from estimate	Krausman et al. (1989)
Harmonic mean	McPAAL, DC80, HOMERANGE, TELEM/PC, HARMONIC, CALHOME	Nonparametric; based on distance of relocations from user-defined grid system; identifies core areas; results can be influenced by size and placement of grid	Dixon and Chapman (1982)
J-T ellipse	McPAAL, DC80, HOMERANGE, TELEM/PC, HOMER, CALHOME	Assumes bivariate normal distribution	Jennrich and Turner (1969), see also Koeppl et al. (1975)
Weighted ellipse	HOMERANGE	Modification of J-T ellipse; weights each relocation by its distance from mean of all relocations; less influenced by outlier observations	Samuel and Garton (1985)
Dunn ellipse	HOMER	Assumes bivariate normal distribution; not constrained by assumption of independence of observations	Dunn and Gipson (1977)
Kernel	KERNELHR, CALHOME	Nonparametric; good statistical properties; identifies core areas	Worton (1989)
Fourier smoothing	McPAAL	Nonparametric; most accurate when estimating the area encompassing 50 percent of use; sample size bias with higher percentages; more of an index than an estimator	Anderson (1982)

[a] Sources for computer programs are: McPAAL, M. Stuwe, Smithsonian Institution, Washington, D.C.; DC80, J. Carey, University of Wisconisn, Madison; TELEM/PC, S. Sheriff, Missouri Department of Conservation, Columbia; HOMERANGE, E.O. Garton, Universiy of Idaho, Moscow; HOMER, G.C. White, Colorado State University, Fort Collins; and HARMONIC and KERNELHR, D.E. Seaman and R.A. Powell, North Carolina State University, Raleigh; CALHOME, J. G. Kie, USDA Forest Service, Pacific Northwest Research Station, LaGrande, Oregon.

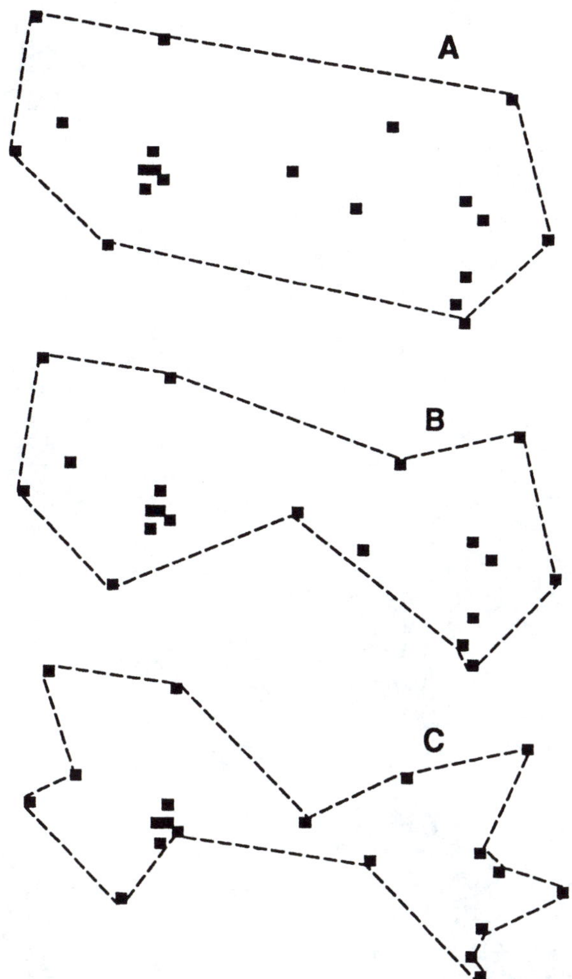

Figure 119. The minimum polygon estimator of home range (Mohr 1947) applied to simulated relocations of a single animal. (A) The standard use of this technique, in which all angles are convex; (B and C) examples of polygons with at least one concave angle that can be constructed from the same data.

A problem arises when the boundaries of a home range are delineated, in that some of the area within may not fit the definition of home range (i.e., spaces not used by the animal during its normal activities but surrounded by spaces that are). Thus, the estimated size of the home range is inflated. Some investigators have attempted to adjust their estimates by creating "concave" polygons (see Gasaway et al. 1980, Grauvogel 1984) that exclude unused areas through inclusion of internal angles in the home range circumference. This technique has merit biologically, but as stated previously, it is not repeatable. Krausman et al. (1989) proposed a technique that eliminates *null* areas within the home range boundary by identifying those habitat patches within the perimeter that are used less than 3 percent of the time, and within which animal use always is less than habitat availability. These null areas are removed from the estimate of home range, and the revised estimate is termed *adjusted home range*. These techniques assume that the excluded areas have no value to the home range occupant. Before using these methods, biologists should consider carefully whether this assumption is valid.

Recognizing the limitations of the convex polygon technique, biologists have developed home range estimators that are based on a probabilistic approach to analysis of movements (see Worton 1987). The computational requirements for many of these estimators are rigorous, and numerous computer programs exist to assist biologists in these calculations (Table 22) (see Larkin and Halkin 1994). Some of these estimators are based on the assumption that the relocation data are normally distributed and therefore, permit calculation of confidence levels. As home ranges are at least two dimensional, the data must be distributed normally in both dimensions—a term known as *bivariate normality*. This type of home range configuration can be viewed as a center of activity in the geometric center of the home range surrounded by concentric isopleths (contours within which there is a specified probability of finding the animal at any given time) of decreasing probability of animal occurrence as the isopleths progress outward from the center (Harrison 1958). These techniques, sometimes referred to as *parametric* (see White and Garrott 1990), produce circular or elliptical representations of home range. Animals seldom use their home range in a manner approximating a normal distribution; thus, parametric estimators are limited in their application. Tests have been developed to assess the bivariate normality assumption of animal relocations (Smith 1983, Samuel and Garton 1985), and when data for all animals in a study meet this assumption, application of these techniques should be considered because of their statistical properties. In the only published test of the underlying distribution of moose home ranges, Garton et al. (1985) found that data (for two bulls) did not conform to bivariate normality.

Other techniques have the inherent advantage of not being limited by a priori assumptions concerning the shape of the home range. Thus, concave areas of the perimeter can be delineated in a nonsubjective manner, and multiple or noncentral core areas can be defined. This is a distinct advantage in an estimator because animals move throughout their environment in relation to a host of stimuli that do not always produce predictable patterns. These techniques are termed *nonparametric*. A disadvantage of nonparametric estimators is that they offer no estimate of variance by which the reliability of the estimate can be judged. Two nonparametric techniques that are being used with increasing frequency are the *harmonic mean* (Dixon and Chapman 1980) and *kernel* (Worton 1989) estimators of home range.

Radio tracking of moose can be accomplished on foot (top left), by vehicle (top right) and by aircraft (bottom). The antennae are directional—the signal from the transmitter will be detected most readily when the front of the antenna is pointed directly at the transmitter. When tracking by aircraft, an antenna is placed on each wing strut and these antennae can be monitored simultaneously or separately. Differential signal strength from the opposing antennae, when combined with aircraft position, allow the pilot to home in quickly on a telemetered animal. Although less expensive, ground tracking can be less accurate than aerial tracking if triangulation of two or more observations is required to determine the animal's location. *Top left photo by Charles C. Schwartz; courtesy of the Alaska Department of Fish and Game. Top right photo by Ron Batten; courtesy of Lotek Engineering, Inc., Newmarket, Ontario. Bottom photo by Kris J. Hundertmark; courtesy of the Alaska Department of Fish and Game.*

The harmonic mean technique involves superimposing a grid, defined by the user, over a map of relocations and measuring the distance from each relocation to each grid node (the intersection of the horizontal and parallel lines of the grid). The harmonic mean of these distances is calculated for each node. The harmonic mean differs from the arithmetic mean by allowing those observations closer to the node to have more influence over the mean than do the observations farther away. These means are then used to construct contours that describe the home range. This technique is compromised by the fact that the estimate of home range size is unstable with reference to grid size and placement (Worton 1987), although Samuel et al. (1983) proposed a method to correct for this bias. Worton (1987) considered the harmonic mean technique as an inferior form of the kernel estimator.

The kernel estimator is complex in its derivation and application, but a simplistic description will serve to illustrate the underlying concept. A bivariate probability density function, or kernel, is assigned to each observation, which has the effect of converting the sample of relocations being analyzed to an estimate of the utilization distribution for the "population" of all the locations the animal occupied. For instance, if a normal distribution is chosen as the kernel, the probability function for each location would be shaped like a bell, with the tallest part of the bell—representing the greatest probability of occurrence—centered on the observation. The probability functions of observations that are close to each other will overlap and the heights of the bells (probability densities) at these overlapping points are added together. By adding all of the probability densities for an animal's relocations and smoothing them to remove irregularities, a utilization distribution for the entire home range is produced.

The form of the distribution chosen as the kernel has little influence on the estimate of home range, but the spread of individual kernels ("band width" or "smoothing parameter") has a dramatic effect, much the same as the width of bars in a histogram can either indicate gross trends in data (a few wide bars) or display fine-scale variation (many narrow bars) (J.A. Baldwin personal communication: 1993). Therefore, the choice of band width is an important consideration. In addition, two approaches to assigning band width can be used. The band width either can remain constant for all relocations (fixed kernel) or vary by assigning wider band widths to more isolated relocations (adaptive kernel). Worton (1989) reported that the adaptive kernel was more accurate than the fixed kernel, but Seaman (1993) observed that a fixed kernel with band width determined by least squares cross-validation performed better than the adaptive kernel. Least squares cross-validation also can be used with the adaptive kernel (Worton 1989).

Caution must be observed when comparing estimates of home range size from studies using different estimators (Figure 120). Boulanger and White (1990) compared the precision (amount of variation) and bias (accuracy) of five different home range estimators applied to four different home range shapes using computer simulation. They determined that none of the estimators accurately predicted the size of the simulated home ranges with the exception of one parametric technique (Jennrich and Turner 1969) applied to a simulated elliptical home range. Furthermore, for any given shape of home range, the five estimators yielded significantly different estimates of size. Boulanger and White (1990) concluded that, overall, the harmonic mean estimator was the least biased, yet one of the least precise estimators. This study illustrates the danger in making comparisons of home range estimates derived by different methods. In direct comparison using simulated home ranges, the kernel estimator was less biased than the harmonic mean estimator and exhibited equivalent precision (Seaman 1993).

Autocorrelated Data

The primary problem with the home range estimators is that there is a relationship between the size of the estimated home range and the number of relocations. Estimated home range size tends to increase with sample size when using certain techniques, including minimum convex polygons, but other techniques show the opposite trend (Boulanger and White 1990). Ballard et al. (1991), using minimum convex polygons, determined that estimated size of home range of Alaskan moose began to level off after 60 to 90 relocations had been obtained for an individual animal (Figure 121). They estimated that 75 percent or more of the home range size was described if 40 or more relocations were obtained. Nonetheless, obtaining an adequate number of relocations can be problematic, because an assumption inherent in all home range estimators (with the exception of the Dunn estimator [Dunn and Gipson 1977]) specifies that all relocations must be independent. In other words, there must be enough elapsed time (Δt) between relocations so that the location of an animal at time $t + \Delta t$ must not be predictable from its location at time t. This time constraint limits the number of useful observations that can be collected for an animal because use of observations that are not independent, or autocorrelated, can result in underestimation of home range size (Swihart and Slade 1985a, Solow 1989).

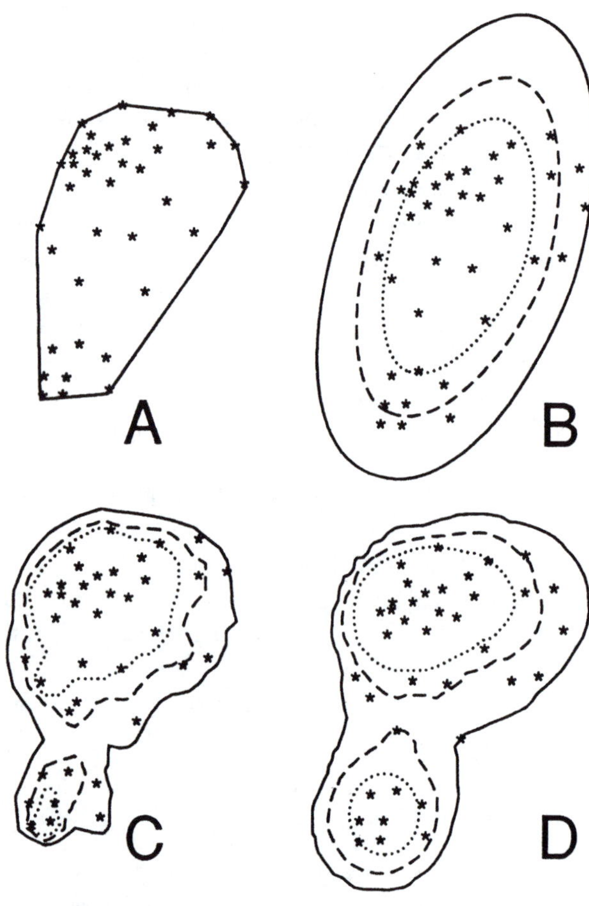

Figure 120. Estimated shape and size of a home range depends on the technique used to define the boundary. A simulated home range was analyzed with four different techniques using program CALHOME (Kie et al. 1996). Home range A represents a minimum convex polygon (Mohr 1947), and has an area of 12.3 square miles (31.8 km²). Home ranges B, C and D represent 95 percent bivariate normal ellipse (Jennrich and Turner 1969), harmonic mean (Dixon and Chapman 1980) and adaptive kernel (Worton 1989) estimates of the same home range, respectively. Estimates of home range size are: B = 27.5 square miles (71.3 km²); C = 14.4 square miles (37.2 km²); and D = 19.2 square miles (49.6 km²). Inner contours within B, C and D represent 60 percent (dot) and 80 percent (dash) use.

A rule to use in establishing independence of observations is to determine whether enough time has elapsed between relocations to have allowed the animal to have traversed its home range completely (White and Garrott 1990). This determination, however, requires more information than just the potential rate of movement of an animal and the size of its home range; the ecological aspects of movement must also be considered (Swihart et al. 1988). Trophic level, sex, age, habitat productivity, social interactions, weather and species mass all influence the rate at which animals use their home ranges (Lindstedt et al. 1986).

Swihart and Slade (1985b, see also Schoener 1981) developed a test to determine whether observations are independent; this test was refined by Solow (1989) for use with small sample sizes or when the underlying distribution of locations is unknown. These tests evaluate the null hypothesis that distances moved between sequential locations do not differ from distances between nonsequential locations. Data that cause this hypothesis to be rejected should be removed from the analysis when using home range estimators based on probabilistic approaches; otherwise, the assumption of independent data is violated and estimates of home range shape and size will be biased, particularly when using small sample sizes. Polygon techniques are rel-

Figure 121. Relationship between the estimate of moose home range size and number of relocations, using the minimum convex polygon technique (Mohr 1947), from Ballard et al. (1991). This relationship is one of the fundamental disadvantages of this technique because it makes comparisons difficult.

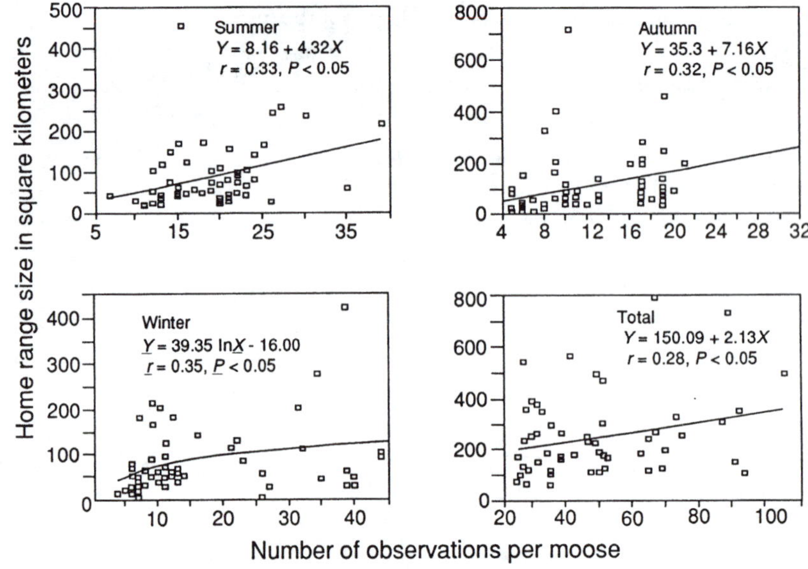

atively immune to this problem because only the locations on the perimeter of the home range are used in the analysis. The robustness of other home range estimators with respect to violation of the independence of data assumption needs to be evaluated.

Any presence of autocorrelation in a data set, however, must be interpreted from an ecological perspective. Analysis of annual home range might yield a high degree of autocorrelation if seasonal trends in range use are present. Defining seasonal ranges based on movement patterns would allow more locations to be used (Nicholson et al. 1997).

Alternative Indices

If the objectives of a project require analysis of movements but not necessarily determination of home range size, the investigator should consider using another index of movements. Various alternatives to home range exist and do not suffer from the many problems associated with home range estimation. Moose movements have been characterized by mean distance between successive locations (Roussel et al. 1975, Addison et al. 1980, Hauge and Keith 1981), maximum separation of all relocations or home range length (Van Ballenberghe and Peek 1971, Mould 1979, Doerr 1983, Gasaway et al. 1983) and distance of autumn or winter relocations from summer home range centers (Houston 1968, Albright and Keith 1987). Garner and Porter (1990) expressed movements as mean distance traveled per day. These indices can be effective in analyzing ecological aspects of range use.

A powerful technique recently applied to movements data (Hölzenbein and Marchington 1992) is the multiresponse permutation procedure (Biondini 1988) that detects differences in distribution between sets of data without assuming any underlying distribution and without delineating home range boundaries. Other multiresponse procedures (Slauson et al.

Moose habitat often is patchy, from both vegetative and physiographic perspectives. The ways in which moose respond to habitat, weather and social stimuli can be analyzed through the study of movements. The type of terrain pictured here conceivably could be inhabited by a migratory population occupying the higher elevations in summer and sharing the lowland habitat with a resident (nonmigratory) population in winter. These populations may have different rates of reproduction and mortality and may occupy different areas during the hunting season. *Photo by Albert W. Franzmann; courtesy of the Alaska Department of Fish and Game, Soldotna.*

1991) have been used to test for independence of observations and home range fidelity (Nicholson et al. 1997).

Characterization of Home Range

Concept of Movements

"Movements," as used here, refers to changes in the distribution of animals through space and time within a season. By this definition, home range can be considered as a measure of movements. Two specific types of movement not included in this definition are dispersal and migration, which are addressed later in this chapter.

Movements by moose differ greatly depending on the season. Daily distances traveled seem to be greatest in summer and least in winter (Phillips et al. 1973, Best et al. 1978, Joyal and Scherrer 1978, Garner and Porter 1990). One of the reasons for increased movements in spring and summer is the search for aquatic vegetation (Krefting 1974a, Joyal and Scherrer 1978) or mineral licks (Best et al. 1977, Tankersley and Gasaway 1983). Movements by bulls during the rut increase dramatically, whereas those of cows decrease (Houston 1968, Phillips et al. 1973, Garner and Porter 1990). Immediately after rut, this trend is reversed (Phillips et al. 1973). During winter and as a means of conserving energy, large males tend to move less than small males and females (Miquelle et al. 1992).

Daily movement patterns differ between the sexes. Before the rut, bulls in northwestern Minnesota traveled nearly equal distances during day and night—0.3 miles (0.5 km) per 24 hours (Phillips et al. 1973). Early in the rut, bulls moved more during the day (0.5 miles [0.8 km]) than during the night (0.3 miles [0.5 km]), but late in the rut, move more at night (0.7 miles [1.1 km]) than during the day (0.2 miles [0.3 km]). In early summer, cows move more at night (0.4 miles [0.6 km]) than during the day (0.2 miles [0.3 km]).

Some moose use submerged aquatic vegetation (top) and mineral licks (bottom) in spring as sources of macroelements, particularly sodium and calcium. These nutrients are important components in physiological processes and, if scarce, can have an effect on the health of moose. The distribution of these areas across the landscape can often be more important than energetics in determining moose movements and home range sizes. *Top photo by Albert W. Franzmann, courtesy of the Alaska Department of Fish and Game, Soldotna. Bottom photo by L. Godwin; courtesy of the Ontario Ministry of Natural Resources.*

Seasonal movements of moose are determined in part by the animals' sex and age, because of the different ways in which moose of various age and sex classes interact with their environment and with one another. Mating strategies, nutrition and predator avoidance are also factors that influence movements. *Photo by Len Rue, Jr.*

They move equally day and night (0.2 miles [0.3 km]) in late summer, and more at night (0.25 miles [0.4 km]) during the rut than during the day (0.1 miles [0.16 km]).

Activity Centers and Core Areas

Activity centers and *core areas* are terms used to describe some form of central tendency in movement patterns, but are distinct concepts. Hayne (1949) defined activity center as the mean of all relocations when plotted on a two-dimensional grid. Thus, it is a point estimate of the center of the home range, as determined by the distribution of movements. This measure is useful when determining distance between annual or seasonal home ranges, as well as when determining dispersal and migration distances. Core areas (Kaufmann 1962) are one or more areas within the home range in which occurred a disproportionately high percentage of observations, indicating that the habitat in these areas is important in satisfying an animal's needs. As such, characterizing these areas and the movements from these areas to other areas is important if the ecology of the animal in question is to be understood (Mohr and Stumpf 1966). The harmonic mean and kernel estimators are two techniques that can be used to identify core areas by defining, for example, the 50 percent isopleths of home range use.

Home Range Development

Moose calves accompany their dams for the first year of life and usually are driven off before the next calving season (Altmann 1958, Stringham 1974; see Chapters 4 and 5). At this time, many yearlings become independent of their mothers, but some may remain in close proximity well into their second year of life (Denniston 1956, Altmann 1958). In southcentral Alaska, Ballard et al. (1991) determined that separation occurred between 10 and 28 months of age, with most separations occurring between 12 and 16 months.

In Sweden, Cederlund et al. (1987) reported that the distance between cows and their offspring progressively increased for the first month after separation; thereafter, it became stable. Studies from North America, however, generally suggest that young moose spend much time wandering, and do not establish home ranges until they are at least 2 years old (Houston 1968, Addison et al. 1980). The eventual size of the home range is not established until the second year of independence from the dam (Gasaway et al. 1985b) and is related directly to the size of the dam's home range (Ballard et al. 1991; Figure 122).

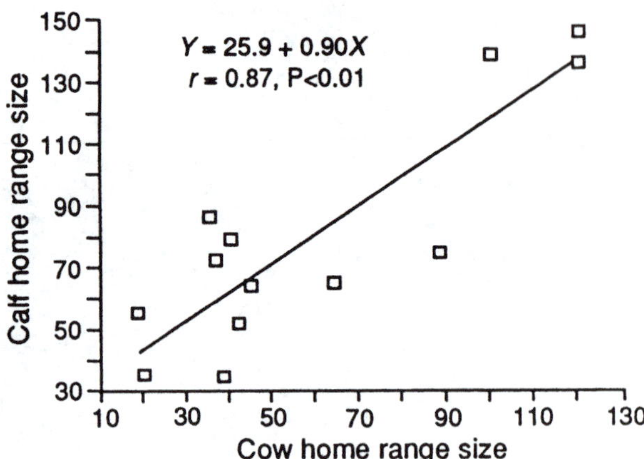

Figure 122. Home range size (in square kilometers) of moose cows is directly related to the ultimate home range size of their offspring. The offspring of cows with large home ranges tend to have large home ranges themselves (from Ballard et al. 1991).

Philopatry and Tradition

The degree of philopatry exhibited by a population conveys information about its stability. Garner and Porter (1990) observed little home range philopatry among bulls (no cows were monitored) in a pioneering moose population in New York. After moving long distances to rutting areas, bulls established winter ranges in suitable habitat nearby rather than moving to more distant areas used previously. Nonphilopatric behavior in moose might be expected in populations with low densities occupying areas of homogeneous habitat, and in populations pioneering previously unoccupied habitat. In established moose populations, philopatry is more prevalent (Houston 1968, Ballard et al. 1991), but may vary. Taylor and Ballard (1979), for example, observed greater fidelity to summer range than to winter range in an established population in Alaska.

Home Range Size

In a review of home range size in moose, LeResche (1974) stated that seasonal home ranges generally do not exceed 2 to 4 square miles (5–10 km^2). At the time of that review, however, few published studies that used telemetry to track moose were available. Recent studies indicate that seasonal home ranges often can exceed these sizes (Table 23).

Two studies from Alaska generated the largest estimates of home range size, although one of these (Grauvogel 1984) included migratory locations in the estimates of seasonal ranges, which can increase home range size significantly. Moose in southcentral (Ballard et al. 1991) and northwestern Alaska (Grauvogel 1984) had average seasonal home ranges no less than 36 square miles (92 km^2). With the exception of home ranges of nonmigratory adults in the latter study, total home range sizes exceeded 100 square miles (259 km^2). In contrast, estimates of annual home range size for moose in northwestern Minnesota were no greater than 1.4 square miles (3.6 km^2) (Phillips et al. 1973).

Certainly such great variation could not be attributed to random chance; rather, some external factors must influence size of home ranges. The size of a moose home range varies with the sex and age of the animal, season, habitat quality and weather.

SEASONAL VARIATION

Seasonal ranges, when they exist, represent partitioning of the environment based on behavioral and energetic constraints. Migratory moose (those that use separate winter and summer ranges) use distinct seasonal ranges because they attempt to optimize their nutrient intake on summer range, but winter conditions on these ranges may preclude occupation during some or all winters. Moose that remain on the same range during winter and summer are termed *resident* or *nonmigratory*, and do so because environmental conditions permit their residence. A third seasonal range, associated with mating, occurs in autumn, but many investigators define this as part of the summer range. Also, some investigators have identified distinct spring ranges (e.g., Lynch and Morgantini 1984).

Because some moose have been observed to limit their movements in winter—the so-called yarding behavior (Peterson 1955)—it may seem logical to assume that moose winter ranges are smaller than summer ranges. An examination of the literature indicates that seasonal range sizes vary. Hauge and Keith (1981) and Ballard et al. (1991) determined that summer and winter home ranges were not different in size. Studies finding that winter ranges were smaller than summer ranges include those of Addison et al. (1980) and Garner and Porter (1990). Doerr (1983) and Lynch and Morgantini (1984) observed winter ranges larger than summer ranges. Apparently other factors are involved that make invalid generalized assumptions concerning the effect of one factor, such as season, on home range size. This should come as no surprise to biologists who routinely sort through the many environmental and social factors that have an effect on moose behavior. In this instance, habitat quality, terrain, population density, and the sex and age composition of the population all may have had an influence on home range size. It seems, then, that McNab's (1963) hypothesis concerning relative sizes of winter and

As parturition nears, pregnant moose cows drive away their offspring of the previous year. In so doing, a cow must be aggressive, which begins with antagonistic body gestures, including ear signals (top) and threatening posture. The yearling, unaccustomed to its dam's displays of disaffection and reluctant to depart, invariably must be chased away by the cow (center and bottom). A banished yearling will remain in close proximity to its dam and the newborn or rejoin the cow if the calf is lost. Permanent separation usually occurs sometime during the offspring's second year. *Photos by Charles C. Schwartz; courtesy of the Alaska Department of Fish and Game.*

Table 23. Moose home range estimates from studies using radiotelemetry

Study area	Migratory status[a]	Sex	Age	*n*[b]	Mean home range size[c] Total	Winter	Summer	Other	Method[d]	Reference
Alaska										
Kenai Peninsula	M	M	Adult	4	53.1 (21.6–71.4)				MCP	Bangs et al. (1984)
					137.5 (56.0–185.0)					
	N	M	Adult	3	19.9 (14.3–24.7)					
					51.7 (34.0–64.0)					
		F	Adult	30	49.3 (9.8–169.9)	24.3 (4.9–71.1)	13.7 (0.9–58.7)		MCP	Bangs and Bailey (1980b)
					127.7 (25.4–440.0)	*62.9 (12.7–184.1)*	*35.7 (2.3–152.0)*			
Seward Peninsula[e]		M	Adult	17	289 (35–746)	103 (9–538)	119 (17–511)		MCP	Grauvogel (1984)
					749 (91–1,932)	*267 (23–1,393)*	*308 (44–1,323)*			
		F	Adult	20	234 (56–601)	77 (8–201)	81 (16–264)			
					606 (145–1,557)	*199 (21–521)*	*210 (41–684)*			
	M		Adult	22	362 (91–746)	120 (14–538)	125 (16–511)			
					938 (236–1,932)	*311 (36–1,393)*	*324 (41–1,323)*			
	N		Adult	7	84 (35–135)	39 (14–86)	36 (17–58)			
					218 (91–350)	*98 (36–223)*	*93 (44–150)*			
	I		Adult	8	131 (78–229)	47 (8–129)	81 (23–215)			
					339 (205–593)	*122 (21–334)*	*210 (60–559)*			
Southcentral		M		11			11.2 (0.2–48.3)	3.7 (1.3–7.3)[f]	MCP	Modafferi (1982)
							29.0 (0.4–125)	*9.6 (3.3–18.9)*		
		F		28–29			4.3 (0.4–25.6)	1.7 (0.3–4.1)[f]		
							11.2 (1.0–66.2)	*4.5 (0.7–10.6)*		
	N	F		19–43	112 (42.9–303.9)	43.6 (3.9–166)	39.8 (8.9–176.1)	60.6 (16.6–178.4)[g]	MCP	Ballard et al. (1991)
					290 (111–787)	*113 (10–430)*	*103 (23–456)*	*157 (43–462)*		
	M	F		2–4	195 (101.5–285.7)	56.8 (5.8–144.8)	101.5 (23.2–240.2)	124.3 (34.4–368.7)[g]		
					427 (274–580)	*147 (15–375)*	*263 (60–622)*	*322 (89–955)*		
Southeastern	N	F	Adult	5	11.0 (3.5–19.9)	4.4 (1.2–11.7)	5.5 (0.9–11.4)		MCP	Doerr et al. (1983)
					28.4 (9.0–51.4)	*11.4 (3.2–30.3)*	*14.2 (2.2–29.6)*			
Alberta										
Central				16–45		5.8 (0.8–20.8)	6.2 (0.4–13.1)		MCP	Mytton and Keith (1981)
						15.0 (2.0–54.0)	*16.0 (1.0–34.0)*			
Northcentral		M		16–33		19.9	8.5	12.8/10.1[h]	MCP	Lynch and Morgantini (1984)
						51.6	*22.1*	*33.2/26.1*		
		F		29–66		18.1	8.8	9.9/5.9[h]		
						46.8	*22.7*	*25.6/15.4*		
		M	Yearling	5–12		22.6	11.5	7.4/6.7[h]		
						58.6	*29.7*	*19.4/17.4*		
		F	Yearling	2–4		23.0	1.9	4.1/10.2[h]		
						59.6	4.9	*10.7/26.3*		
						59.6	4.9	*10.7/26.3*		
		M	2	5–9		24.2	10.0	20.3/18.5[h]		
						62.6	*25.9*	*52.7/47.9*		
		F	2	4–10		16.2	2.9	21.5/4.2[h]		
						41.9	*7.5*	*55.8/10.9*		

Table 23 continued

Study area	Migratory status[a]	Sex	Age	n[b]	Mean home range size[c]: Total	Winter	Summer	Other	Method[d]	Reference
		M	Adult	3–12		12.0 *31.2*	4.5 *11.7*	16.6/7.1[h] *43.1/18.3*		
		F	Adult	23–52		18.1 *47.0*	10.4 *27.0*	8.3/6.1[h] *21.6/15.9*		
Northeastern	M			7–12		11.6 (1.2–42.9) *30.0 (3.0–111.0)*	14.3 (6.9–37.5) *37.0 (18.0–97.0)*		MCP	Hauge and Keith (1981)
	N			10	37.4 (23.2–70.6) *97.0 (28.0–71.0)*					
Maine	N	F	Adult	5		10.0 (4.4–16.7) *25.8 (11.4–43.2)*			MCP	Crossley and Gilbert (1983)
		M		3			10.8 (10.2–17.6)[i] *28.0 (26.4–45.5)*		MCP	Thompson (1987)
		F		9			12.5 (5.8–48.8)[i] *32.3 (15.0–126.3)*			
		6M/7F	h	13	10.8 (0.8–23.2) *27.9 (2.0–60.0)*				MCP	Leptich and Gilbert (1989)
Minnesota, northwestern		M	Adult	10–13	1.2 (0.4–2.1) *3.1 (1.0–5.4)*	4.2 *10.9*			HRF	Phillips et al. (1973)
		F	Adult	9–13	1.4 (0.3–2.9) *3.6 (0.8–7.5)*	4.9 *12.7*				
		M	Adult	13		5.7 (1.0–8.6) *14.8 (2.6–22.3)*			MCP	
		F	Adult	9		6.7 (2.1–15.6) *17.4 (5.4–40.4)*				
New York, northern	M	M	Adult	4		2.9 *7.5*	14.0 *36.3*		MCP	Garner and Porter (1990)
						6.7 *17.3*	21.8 *56.5*		HM	
Northwest Territories, Mackenzie Valley	N	F	Adult	29	67.2 (15.4–363.8) *174 (40–942)*	22 *57*	26.3 *68*	51[g] *132*	MCP	Stenhouse et al. (1994)
Ontario, northwestern	M	1M/2F	Adult	3		2.2 (0.8–4.6) *5.7 (2–12)*	16.5 (2.3–34.7) *42.7 (6–90)*		MCP	Addison et al. (1980)
	N	F	Adult	1	5.4 *14.0*					

[a] M = migratory; N = nonmigratory; I = intermediate.

[b] Number of moose used in study; two numbers indicate minimum–maximum numbers of moose used to generate different seasonal estimates.

[c] Top (roman) entries are expressed as square miles; bottom (italic) entries are square kilometer equivalents; entries in parentheses represent the respective numerical range, when known.

[d] MCP = minimum convex polygon; HM = 95 percent harmonic mean; HRF = home range fill (Rongstad and Tester [1969]).

[e] Estimates include migratory locations.

[f] Breeding home range.

[g] Autumn home range.

[h] Spring/autumn home ranges.

[i] Median value.

summer ranges is not easily tested in moose because of the many confounding factors involved.

GEOGRAPHIC VARIATION

The broad distribution of moose in North America allows for geographic comparisons of home range size. As stated previously, some of the largest home ranges reported for moose, whether seasonal or total, were reported by Ballard et al. (1991) for a mountainous region of southcentral Alaska (Nelchina Basin). In this region, 31 percent of the habitat was considered unsuitable for moose, but it was interspersed within the home range boundaries defined by minimum convex polygons. Use of the adjusted home range estimator (Krausman et al. 1989) reduced estimates of size accordingly, but the estimates remained among the largest ever reported. Grauvogel (1984) and Stenhouse et al. (1994) also reported very large home ranges in other northern moose populations (Table 23).

Although moose and white-tailed deer occupy the same range in northeastern North America, moose tend to winter in different habitats. And although yarding is more common in whitetails, some moose populations concentrate during winter in areas of relatively homogeneous habitat with low elevational gradients to take advantage of lower snow depths and/or abundant browse. Balsam fir is a common forage in such yards and often shows signs of excessive browsing, and a "hedge line" may be apparent. *Photo by S. E. Aldous; courtesy of the U.S. National Archives.*

Comparing estimates of annual home range sizes for nonmigratory moose in Alaska indicates a pattern of geographic variation. Grauvogel (1984) estimated size of annual home ranges of moose on the Seward Peninsula in northwestern Alaska at 84 square miles (218 km^2). These estimates were similar to those of Ballard et al. (1991) of 112 square miles (290 km^2) in the Nelchina Basin in southcentral Alaska. Bangs et al. (1984) described annual home ranges of 19.9 square miles (51.7 km^2) for moose on the Kenai Peninsula, and Doerr et al. (1983) reported an estimate of 11.0 square miles (28.4 km^2) for moose inhabiting the Stikine River drainage in southeastern Alaska. Estimates of home range size generally increase from south to north.

Elsewhere, moose in northeastern Alberta had mean annual, summer and winter home range sizes of 37, 15 and 12 square miles (97, 38 and 30 km^2), respectively (Hauge and Keith 1981). In northcentral Alberta, Lynch and Morgantini (1984) reported summer and winter range sizes of 13 and 19 square miles (33 and 48 km^2), respectively. However, comparison of the results of these two reports is difficult because Lynch and Morgantini (1984) defined four seasonal ranges based on calendar dates, whereas Hauge and Keith (1981) defined two seasonal ranges based on patterns of animal movements. In Wyoming, Houston (1968) noted that most adult moose had winter and summer ranges of less than 1.5 square miles (3.9 km^2), but the mean number of relocations per animal per season was small—less than five in winter and six in summer. In nearby Idaho, Ritchie (1978) reported mean summer home range sizes of 7.4 square miles (19.3 km^2) for three cows and 16 square miles (42 km^2) for two bulls. In Ontario, Addison et al. (1980) calculated a mean "nonwinter" range size of 2.6 square miles (6.8 km^2) for nonmigratory moose, and winter and nonwinter home range sizes of 5.7 and 42.7 square miles (14.8 and 111 km^2), respectively, for migratory moose. In New York, Garner and Porter (1990) indicated mean summer ranges of 14.2 square miles (36.3 km^2) and mean winter ranges of 2.9 square miles (7.5 km^2).

Plotting winter and summer home range sizes against the approximate latitude of the study area indicates that sizes of summer ranges remain relatively stable below 60 degrees north latitude, and sizes of winter ranges may increase somewhat with latitude over this range (Figure 123). At these latitudes, seasonal home range sizes did not exceed 20 square miles (51.2 km^2). Above 60 degrees north latitude, however, sizes of seasonal ranges increase dramatically. The variance of seasonal range size also seems to increase with latitude, indicating that other factors are also influencing

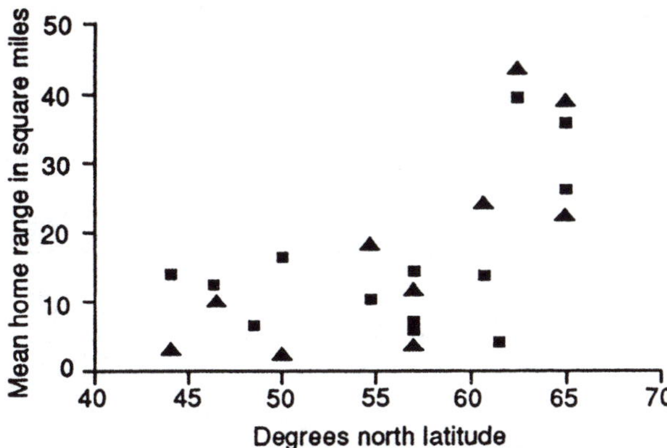

Figure 123. Between 40 and 60 degrees north latitude, mean sizes of winter (triangles) and summer (squares) home ranges of moose remain relatively stable, although winter ranges may increase slightly with latitude. Above 60 degrees, home range sizes can increase dramatically. This trend may reflect differences in dispersion and productivity of seasonal habitats.

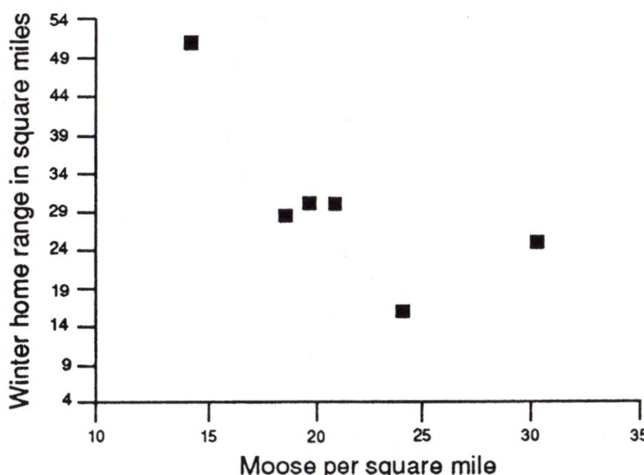

Figure 124. In Sweden moose winter range size decreased with increasing population density (Sweanor and Sandegren 1989). Although this relationship was not statistically significant ($P = 0.08$) the trend is worth noting. This type of a relationship may not be observed in most North American moose populations because the Swedish population was characterized by extremely high density.

home range size. Dispersion and productivity of habitat might be two factors involved.

DEMOGRAPHIC VARIATION

Increasing population density results in either a greater degree of home range overlap or a compression of home range sizes. The degree of home range sharing in herbivores is related to body size (Damuth 1981). Therefore, moose would be expected to, and do, exhibit a considerable amount of home range overlap. As home ranges overlap, the occupants compete directly for food resources. Because an animal will occupy a home range only large enough to meet its nutritional requirements, increasing the number of animals sharing these resources may act to increase individual home range sizes. Alternatively, because moose are more solitary than gregarious, mutual repulsion might be expected, thus causing smaller home ranges.

Sweanor and Sandegren (1989) examined changes in home range size and population density in a Swedish moose population over a 6-year span and noted a trend toward smaller home ranges at higher densities (Figure 124). Although they concluded that the relationship was nonsignificant ($P = 0.08$), the trend suggests further study is warranted. Mytton and Keith (1981) reported moderate densities in central Alberta that were three to five times higher than those observed by Hauge and Keith (1981) in northern Alberta. Seasonal ranges studied in central Alberta were smaller than those studied farther north, indicating a trend similar to that observed by Sweanor and Sandegren (1989), although differences in habitat may have been a factor as well.

There is no conclusive evidence supporting the existence of density-dependent home range size in moose. Nonetheless, increasing densities eventually result in home ranges unable to provide the nutrient requirements for optimal physical condition of the occupants. Hence, productivity and survival would be expected to decline.

Houston (1968) proposed that the extent of movements by females during the moose mating season was related to the density of males. This view was supported by LeResche (1974), who observed large-scale movements by females during this period in a population with a low bull/cow ratio compared with limited movements by cows in the population with many bulls studied by Houston (1968). Sizes of summer and breeding home ranges for females in studies by Modafferi (1982) and Lynch and Morgantini (1984), however, were similar despite bull/cow ratios of the former being less than half that reported in the latter. (It should be noted that Modafferi (1982) reported a ratio of 23 bulls:100 cows, which was greater than that observed by LeResche (1974) and represented a density of breeding bulls thought to be high enough to ensure successful breeding.)

WEATHER AND HABITAT VARIATION

Few data are available on the influence of weather on moose home range size, but reports that mention weather-related responses cite snow depth as the factor involved. Van Ballenberghe and Peek (1971) observed a cow occupying an area of approximately 6 acres (2.4 ha) during a month-long period of deep snow. Ritchie (1978) observed bulls limiting their movements to areas no larger than 10 acres (4 ha) for weeks at a time during winter. In northern Maine, Thompson (1987) observed a median size of winter home ranges of 2.7 square miles (7.1 km^2) during a year in which snow depths were low. But during the subsequent winter in which snow depths were often greater than 28 inches (70 cm), home ranges were 0.6 square miles (1.6 km^2). Miquelle et al. (1992) reported that distances traveled per foraging bout and rate of travel were less for moose in a winter of deep snow than during a winter of shallow snow. Moose in southeastern Alaska reacted to deep snow on their winter ranges by confining their activities to those areas within their ranges that had significantly lower snow depths (Hundertmark et al. 1990).

In Sweden, Sweanor and Sandegren (1989) observed a significant ($P < 0.03$) relationship between size of winter range and the number of days during which snow depths exceeded 28 inches (70 cm) (Figure 125), which is the approximate upper limit of tolerance for moose (Kelsall and Prescott 1971, see also Coady 1974a). Sweanor and Sandegren (1989) reported no correlation of home range size and duration of snow depths greater than 16 or 10 inches (40 or 25 cm). Of these depths, 16 inches (40 cm) triggered migration in that population (Sandegren et al. 1985), and 10 inches (25 cm) buried dwarf shrubs, an important winter food item. Thus, snow depth can influence home range size. Moose will restrict their movements or select areas with relatively shallow snow rather than expend the energy necessary to travel through deep snow.

McNab's (1963) theory that home range size is determined ultimately by energetics suggests that moose living in more productive habitats would use smaller home ranges, or at least have smaller core areas. The analysis of home range characteristics in relation to productivity of the

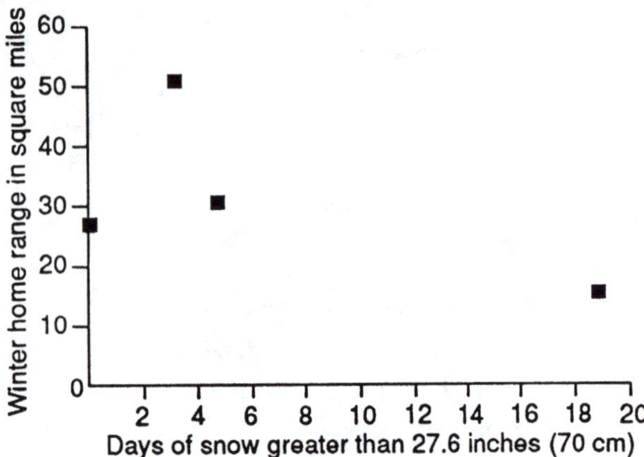

Figure 125. Influence of snow depth on the winter home range size of moose in Sweden (Sweanor and Sandegren 1989).

habitat on anything more than a qualitative basis is difficult because of the nature of the variables involved as well as other potentially confounding factors (Greenwood and Swingland 1983). For example, because moose habitat often is a matrix of different habitat types, home range size would be determined in part by juxtaposition of those types and the interspersion of unsuitable habitat (e.g., Ballard et al. 1991).

Leptich and Gilbert (1989) reported that home range sizes of moose in Maine varied from 0.8 to 23 square miles (2–60 km^2), with much of the variation being explained by distribution of aquatic feeding sites. Those home ranges that encompassed such a feeding site were small, whereas the larger home ranges were long and narrow with activity centers at both ends, one of which was the aquatic feeding site. Ritchie (1978) speculated that the relatively large home ranges he observed in Idaho, when compared with other studies conducted in nearby locations, were attributable to relatively unproductive lodgepole pine habitat. Taylor and Ballard (1979) believed that terrain influenced home range size, because they observed small home ranges for moose living in mountainous habitat and larger home ranges for moose living in nearby lowlands.

SEX AND AGE CLASS VARIATION

LeResche (1974) summarized age- and sex-specific trends in home range size of moose. He noted that cows with newborn calves restricted their movements for the first few weeks, after which they gradually expanded their home range size to approximately the same size as those of other adults. Other studies suggest that home ranges of cows with calves are similar to those of other adults (Lynch and Morgantini 1984, Leptich and Gilbert 1989), but do not indicate whether movements were examined intensively over the period in question. Ballard et al. (1980) reported that cow/calf pairs had smaller summer home ranges than did other moose, and that calf movements increased exponentially with age during the first 6 weeks of life (Figure 126). Peek (1962) observed a 2-mile (3.2-km) movement by a cow and calf within 3 days after birth. However, he captured and tagged the calf before the movement, which may have been responsible for the movement.

Ballard et al. (1980) conducted a unique study to determine the effect of brown bear densities on movements and home ranges of cow/calf pairs during the 6 weeks after parturition. Brown bears were translocated from part of the area during the latter part of the study. Home ranges of cow/calf pairs were significantly smaller in the experimental area after removal of bears, and also were smaller than those in a control area nearby, suggesting that movements

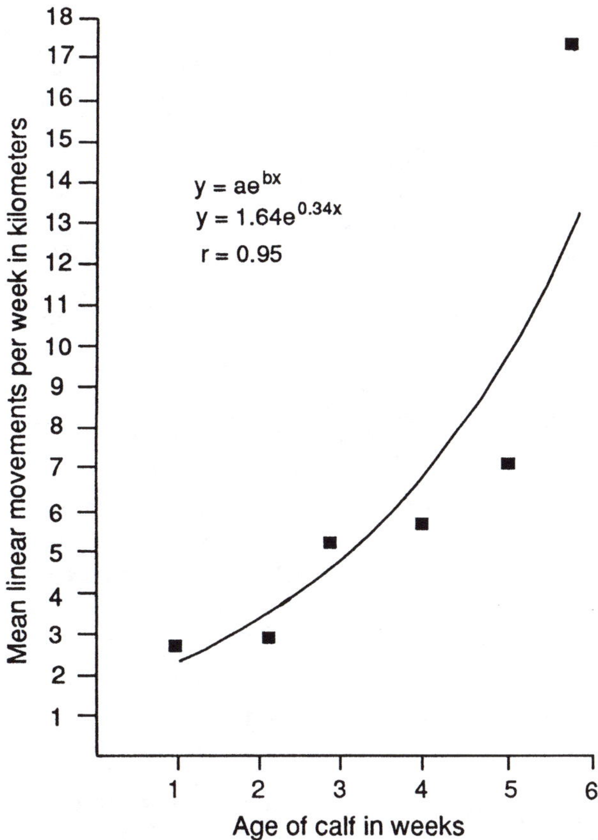

Figure 126. Weekly movements of moose calves and their dams. Immediately after calves are born, these movements are small. As calves grow, their movements increase and the animals' home ranges reach normal size by midsummer (from Ballard et al. 1980).

and home range size were influenced, to some extent, by predator avoidance strategies.

Home range sizes of yearlings can vary greatly from those of other age classes. Houston (1968) observed yearling home ranges that were much larger than those of adults, but also indicated that moose in his study may not have established home ranges until they were 2 years old or older. Addison et al. (1980) reported that a yearling exhibited a wandering lifestyle and showed no signs of establishing a home range in its second winter. Lynch and Morgantini (1984) presented many sex/age comparisons of home range size and determined that yearling and 2-year-old males had larger home ranges than did females of the same ages, and that the largest seasonal home ranges of any sex/age group were exhibited by 2-year-old males in autumn. These large home ranges exhibited by young animals, particularly males, are attributable largely to dispersal and the establishment of a new home range (to be discussed later).

Before describing a home range for an animal, one must be certain that the animal indeed has a home range and is not merely wandering randomly in search of an area in which to settle (Burt 1943, Garton et al. 1985). Thus, discussion of home ranges for animals prone to dispersing is problematic.

When differences in home range size are attributed to sex, male moose always are found to occupy larger areas. Ballard et al. (1991) reported that males had significantly larger ($P < 0.05$) home ranges than did females. Ritchie (1978) speculated that cows had smaller home ranges than bulls, but his sample included only three cows and two bulls. Lynch and Morgantini (1984) reported no differences between the sexes, but did note a tendency of bulls to occupy larger winter and spring home ranges. Phillips et al. (1973), Taylor and Ballard (1979) and Hauge and Keith (1981) noted no such differences between the sexes. Likewise, Sweanor and Sandegren (1989) reported no such differences between sexes or among age classes in Sweden. On the other hand, Cederlund and Sand (1994) observed that annual home range sizes for bulls in Sweden increased significantly ($P < 0.05$) with age, whereas those of cows did not. If this relationship exists in moose populations elsewhere, the age structure of telemetered bulls in the sample may influence results of sex-specific differences in home range size.

A notable difference between the sexes occurs during rut, when males tend to move more than females. Houston (1968) noted that most relocations of bulls during the rut were greater than 2.5 miles (4 km) from the center of summer ranges, whereas most relocations of cows were within 1 mile (1.6 km) of their home range centers. In Alberta, Hauge and Keith (1981) expressed differences between the sexes as the proportion of movements exceeding 3.1 miles (5 km) between biweekly relocations. On an annual basis, 26 percent of bull movements and 19 percent of cow movements significantly exceeded this distance ($P < 0.01$). The primary reason for this disparity was the increase in movements by bulls during the rut (Figure 127). Bulls also exhibited distinct increases in movements during autumn and spring migrations, whereas cows migrated at a seemingly slower pace. The direct relationship between age and annual home range size for male moose in Sweden (Cederlund and Sand 1994) mentioned previously was the result of significant relationships between these parameters in autumn ($P < 0.01$) and winter ($P < 0.01$). Cederlund and Sand attributed these relationships to greater energetics demands of bulls and greater activity of bulls during the rut.

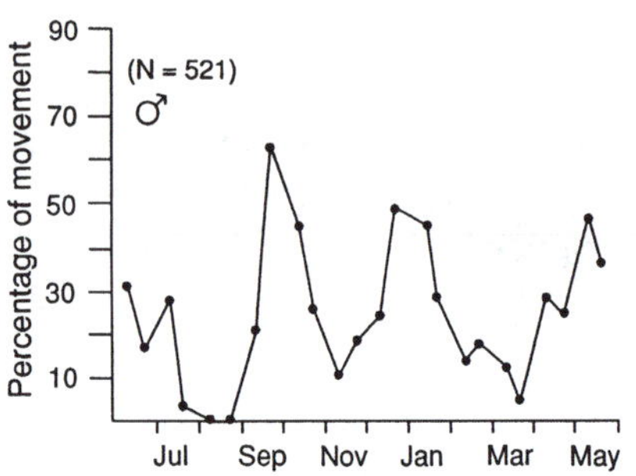

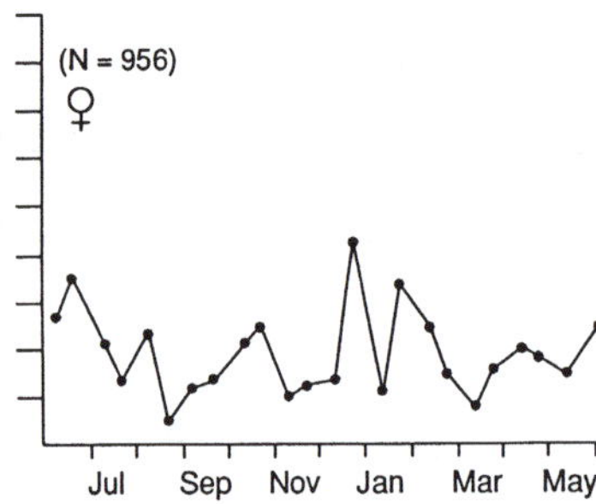

Figure 127. Percentage of movements over 2-week intervals that exceeded 3.1 miles (5 km) for moose in Alberta (from Hauge and Keith 1981). Movements by bulls peaked during the rut, and during winter and spring migrations. Differences in cow movements were less pronounced, with the exception of a peak during winter migration.

The different approaches used to study home ranges make comparisons among moose studies difficult. Results often cannot be compared because of differences in defining seasonal ranges (among other reasons), thus suggesting the need for a unified conceptual framework for discussing home range. Attributes of home ranges must be examined by describing how moose respond to their environment, and comparisons between studies or areas must be expressed in terms of the ways in which the environments in question differ. Sanderson (1966: 231) recognized this when he wrote that "biologists should lose some of their preoccupation with the shape and size of home range. . . . Movement patterns are established and regulated by the density of the species, food supply, reproductive activity, the quality and physiographic arrangement of the habitat, and no doubt many other factors. Thus . . . shape and size of movement pattern are probably of little or no consequence." This approach would preclude, for example, defining seasonal ranges by subjective calendar dates, because moose do not respond to a calendar per se, but instead respond to environmental stimuli that generally are unpredictable and only roughly correlated with time. Only

Young moose, particularly males, often wander extensively during the first year after separation from their dams. This behavior is important in establishing individual home ranges and in population dispersal. Research biologists must exercise caution in attempting to define and analyze movements of young moose in relation to population distribution and density, as the animals may not yet have a prescribed home range per se. *Photo by Charles C. Schwartz; courtesy of the Alaska Department of Fish and Game.*

Bull moose tend to move significantly greater distances to mating areas than do females. The difference undoubtedly is related to the polygamous nature of bulls searching for multiple mates. Increased movements, however, can make rutting bulls more vulnerable to harvest. These moose, in Mount McKinley National Park of Alaska, are of the Yukon/Alaskan subspecies, which commonly form breeding assemblages. *Photo by Victor Van Ballenberghe; courtesy of the Institute of Northern Forestry, Anchorage, Alaska.*

The increase in daily movements by bulls early in the rut, their decreased wariness at this time of the year and their willingness to engage in combat pose risks for these bulls as well as for humans who encounter them. This bull wandered into a residential area in New Hampshire during the rut. He was killed shortly after this photograph was taken. *Photo courtesy of UPI/Bettmann Photograph, New York.*

when all biologists adhere to such a principle will the information generated by disparate studies be useful to moose biologists everywhere.

Territoriality

Although moose are not territorial by nature, adults will defend an area from other moose at certain times of the year (see Chapter 5). Altmann (1958) described a "sliding territory" in cow/calf groups during the summer that diminishes at the onset of mating season. This term describes the intolerance of cows with calves to close approaches of other moose, but the area is more of an "intimate space" (Bubenik 1987) that constantly surrounds the cow/calf group, rather than being an area with established and marked boundaries. Therefore, it does not conform to the traditional definition of "territory" (see Weckerly 1992). Denniston (1956) noted that dominant bulls defend their rutting areas from other bulls during the mating season, but a more likely explanation involves bulls defending cows, not a specific area.

Group Association and Sexual Segregation

The ways in which moose, either individually or in groups, partition their habitats and associate with other moose can be informative in determining the needs of the various segments of the population. Houston (1968) referred to moose as "quasisolitary," and noted that large groups are uncommon. The solitary nature of moose was demonstrated by Berg and Phillips (1972), who studied interactions of moose with overlapping winter ranges in Minnesota. In most instances, whenever two moose were observed close to each other, they had moved away from each other by the next day, usually in opposite directions and often to the extreme margins of the habitat patch. However, moose in Alaska exhibit a greater degree of tolerance for conspecifics than those living elsewhere. The adaptive significance of this tendency likely involves an antipredator strategy in open habitats (Molvar and Bowyer 1994).

Cows with calves are consistently the most solitary members of the population, probably because of predator avoidance, and almost always occur alone (Figure 128) (Peek et al. 1974, Hauge and Keith 1981, Miquelle et al. 1992). Phillips et al. (1973) noted that cows with calves moved mostly at night during the rut; the investigators hypothesized that this was a strategy to avoid bulls. Pregnancy rates, however, indicate that virtually all cows breed annually irrespective of the presence of a calf at heel (Schwartz and Hundertmark 1993). Therefore, it seems that a more plausible explanation would be that cows with calves were avoiding groups of moose, and not bulls per se. Cows with calves in Denali National Park, Alaska, occur on the fringes of mating groups and leave the groups immediately if disturbed (R.T. Bowyer personal communication: 1993). Cows without calves are more solitary than bulls during the calving season, but steadily become more gregarious through summer (Miquelle et al. 1992). Males are consistently gregarious in summer. Associations between large males and cows are more common during the prerut, rut and postrut than at any other time (Peek et al. 1974, Miquelle et al. 1992).

Peek et al. (1974) cited evidence from moose populations in three states that indicated that group size increased with increasing population density. On the other hand, Rolley and Keith (1980) reported no evidence of this in a rapidly growing population in Alberta. This presumed density dependence of group size may have been caused instead by differences in distribution of cover in the three study areas, because the densest population also occurred in the most open habitat. Indeed, Peek et al. (1974) documented larger groups above treeline than below in the Alaskan population. LeResche (1972) and McDonald (1991) noted that large bulls in Alaska segregated from other moose along an elevational gradient during late summer and early winter, with large bulls inhabiting areas at higher elevations (Figure 129). This segregation likely is due to the increased energy demands of large bulls, which cause them to segregate from females to exploit areas with high forage biomass (Miquelle et al. 1992).

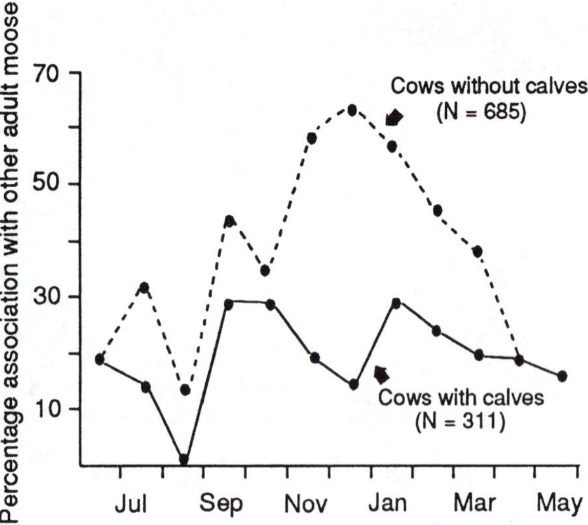

Figure 128. Moose cows with calves tend not to associate with other moose. Cows without calves, due either to pre- or postnatal loss, segregate early in the summer but are more social as the year progresses (from Hauge and Keith 1981).

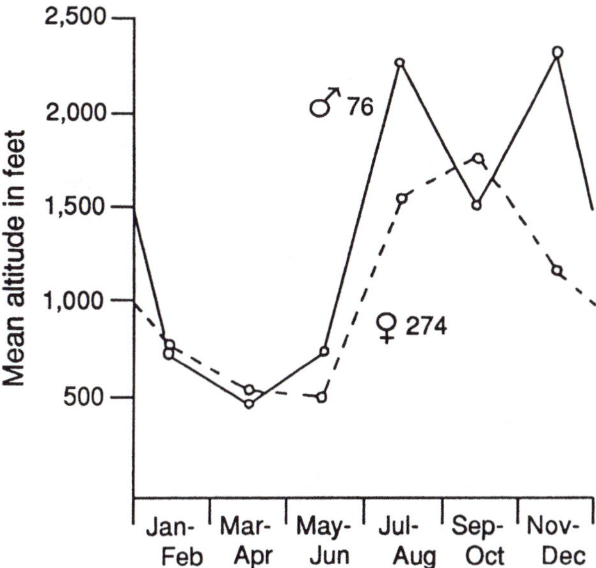

Figure 129. Moose of different sex and age classes tend to segregate at various times of the year. On the Kenai Peninsula, Alaska, cows and bulls segregate by elevation in summer and autumn (from LeResche 1972).

Dispersal

Concept of Dispersal

Dispersal can have a dramatic effect on such population dynamics as changes in population size, spatial distribution, colonization of new habitats and gene flow (see Baker 1978, Greenwood 1980, 1983, Horn 1983, Shields 1983, 1987). Different researchers use this term in different ways and to describe various processes. Howard (1960) defined *dispersal* as the movement of an animal from its natal range to where it will potentially breed. Greenwood (1980) more appropriately terms this transition as *natal dispersal,* and defined three additional types as (1) *breeding dispersal*—movement of an animal between successive breeding sites; (2) *gross dispersal*—permanent movement of animals away from their natal range, whether or not they breed; and (3) *effective dispersal*—which pertains to those animals included in the gross dispersal that breed successfully (cf. Shields 1983, 1987). The last two definitions are used primarily when examining aspects of population genetics and will not be considered further here. Determining the threshold of spatial

Habitat choice by moose differs among males and females during much of the year. The higher energetic demands of males, owing to their larger body size, requires them to select habitats of maximum forage quality and quantity, whereas cows—and especially those with offspring—may compromise choice of areas of maximum foraging in favor of those that also afford shelter or escape cover. *Photo by Charles C. Schwartz; courtesy of the Alaska Department of Fish and Game.*

separation between dispersal and philopatry is subjective, and it certainly differs in the literature, making direct comparisons among studies difficult.

The degree of dispersal exhibited by members of a population can differ among species, among populations of the same species, and within the same population over time. Taylor and Taylor (1977) described the ultimate factor controlling dispersal as the balance between two opposing strategies for maximizing fitness. The first is a strategy that causes an animal to maximize the resources available to it by separating itself from competitors; the second is a strategy causing the animal to make maximum use of available resources, thus enhancing congregation with conspecifics. The prevalence of one of these opposing strategies depends on population density in relation to habitat productivity. Impetus for dispersal is low in a low-density population (relative to carrying capacity) because adequate resources are available within the natal range. Once population density reaches a certain point, it is advantageous to leave the natal range in search of new habitat. Therefore, the increased probability of mortality associated with leaving a familiar area would be offset by the increased probability of finding conditions more conducive to successful reproduction. This does not mean that there is no dispersal in low-density populations; indeed, dispersal occurs at any density depending on the circumstances, but among different segments of the population.

Lidicker (1975) distinguished between dispersal from populations that are below carrying capacity and growing (presaturation dispersal) and those that are at or near carrying capacity (saturation dispersal). Dispersal in moose populations has been described in this context (e.g., Gasaway et al. 1985b, Cederlund and Sand 1992). A more refined approach, however, was proposed by Stenseth (1983), wherein dispersal was classified as *adaptive* and *nonadaptive*. Adaptive dispersal includes presaturation dispersal as well as "ambient" dispersal, which is low-level dispersal that occurs independent of population density. Nonadaptive dispersal involves animals that are forced from natal or established home ranges by social factors inherent in high-density populations. Stenseth (1983) argued that adaptive dispersers would be young, healthy animals as well as reproductively active adults, whereas nonadaptive dispersers would consist of nonreproductive adults and juveniles that would face a high probability of mortality after dispersing.

But which sex group should disperse, or should both? Greenwood (1980) argued that competition for resources and for mates determines which sex is predisposed to disperse. The polygynous mating system of moose dictates that, within a normal population structure, young males will not gain an opportunity to breed, and that the burden of rearing young lies exclusively with females. Thus, unless the nutritional quality of the habitat is poor, females are likely to exhibit philopatry to assure a resource base for rearing young. Conversely, young males are likely to disperse. The proximate factors causing dispersal of juvenile male moose is not known, but in white-tailed deer the presence of the male's mother was an important factor in dispersal tendency (Hölzenbein and Marchington 1992), presumably through agonistic encounters. Dispersal of young males prevents excessive inbreeding and reduces competition for resources among family members, which enhance fitness particularly in long-lived species such as moose (Waser and Jones 1983).

Characteristics of Dispersal

Although there is not an overwhelming abundance of literature concerning dispersal in moose, existing data give credence to characterizations of this concept by Stenseth (1983) and Greenwood (1980). Dispersers predominantly are juvenile males, but other age and sex groups are represented as well.

In two separate studies from Alaska, Ballard et al. (1991) observed that 5 of 15 offspring established home ranges that did not overlap with those of their dams in a high-density population (1.6–2.1 moose per square mile [0.6–0.8/km^2]), whereas Gasaway et al. (1985b) noted that only 1 of 36 offspring dispersed to such a degree in a moderately dense (0.5–2.1 moose per square mile [0.2–0.6/km^2]) but growing population. Ballard et al. (1991) noted less home range overlap between males and their dams than with females and their dams, and that dispersers tended to establish new home ranges in low-density areas. Conversely, Cederlund et al. (1987) reported no difference in rates of dispersal between male and female yearlings in Sweden, but noted that high hunting pressure on males within the study area may have influenced the rates. Lynch (1976) observed higher dispersal of subadults than of adults in a population in Alberta.

Saunders and Williamson (1972) ear-tagged moose in Ontario and recorded the distances between the tagging sites and subsequent locations of recovery by hunters. For subadult and adult moose of either sex, they detected no differences in these distances, and concluded that dispersal was not related to age or sex. However, examination of their data indicates that subadults, particularly males, had a much greater propensity for movement than did adults. Furthermore, it is unclear whether Saunders and Williamson's methodology would have accounted for recovery of animals that dispersed as yearlings.

Rolley and Keith (1980) reported the effects of dispersal in a rapidly growing population in Alberta. In a 14-year span, their study population increased by a factor of 40. They compared the observed finite rate of increase, based on population growth, and the hypothetical finite rate of increase based on the reproductive potential observed in the females. During the early years, the population grew faster than could be explained by reproduction. In the later years, the population grew slower. Temporal differences in the two rates (Figure 130) were attributed to an initial period of net immigration, followed by a period of net emigration. Thus, the investigators concluded that, in the early phase of population growth, individuals from surrounding populations were dispersing into their study area. Rolley and Keith also concluded that, in the later phase of population growth, when density presumably was approaching or had exceeded carrying capacity, more moose were dispersing from the study area than were dispersing into it.

The evidence on which Rolley and Keith (1980) based these arguments was circumstantial (they had no direct evidence of dispersal), but the scarcity of predators and the documented lack of a decline in reproduction with increasing population density added support to their claims. The mechanism behind this process is easy to discern. An increase in density of moose would, at some point, cause an increase in home range overlap (Sweanor and Sandegren 1989). In turn, this would cause an increase in social interactions. Because moose usually are solitary, this increased interaction likely would cause intraspecific aggression that would initiate dispersal.

One consistent characteristic of dispersal in moose populations is that dispersers generally move short distances from their natal range, often establishing a home range adjacent to or slightly overlapping the natal range. Gasaway et al. (1985b) reported a mean dispersal of 1.9 miles (3 km) after the first year of independence for 18 moose, and only 1 of these moose had a home range that did not overlap that of its dam. Home ranges of dispersers tended to stabilize after the first year of independence (Figure 131) (Gasaway et al. 1985b, Cederlund and Sand 1992).

The significance of limited dispersal in moose was demonstrated in a study of moose colonization of a recent burn in central Alaska (Gasaway and DuBois 1985, Gasaway et al. 1989). In 1980, approximately 193 square miles (500 km^2) of black spruce/aspen lowland habitat burned. Seventy-five percent of the area was moderately to severely burned and 10 percent was unburned. The burn occurred during a long-term study of moose movements allowing comparison of pre- and postburn use of the area by resident moose. Gasaway and DuBois (1985) demonstrated that moose living in or near the burn area were not displaced by the fire. In a study of moose movements during the 5 years after the fire, Gasaway et al. (1989) stated that, of those moose that had prefire contact with the burned area, six increased their use of the area after the fire and two decreased

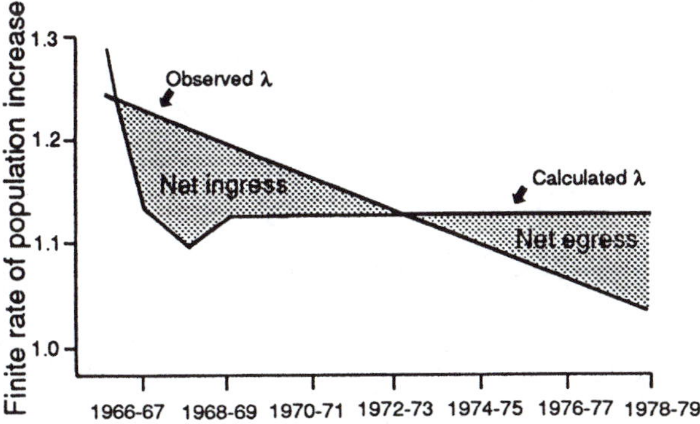

Figure 130. Immigration and emigration can have a significant effect on moose population growth. In a rapidly growing population in Alberta, Rolley and Keith (1980) attributed differences between observed and calculated rates of increase to periods of ingress and egress.

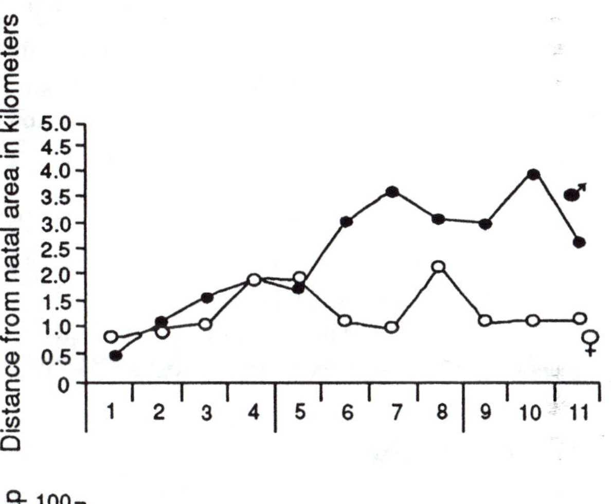

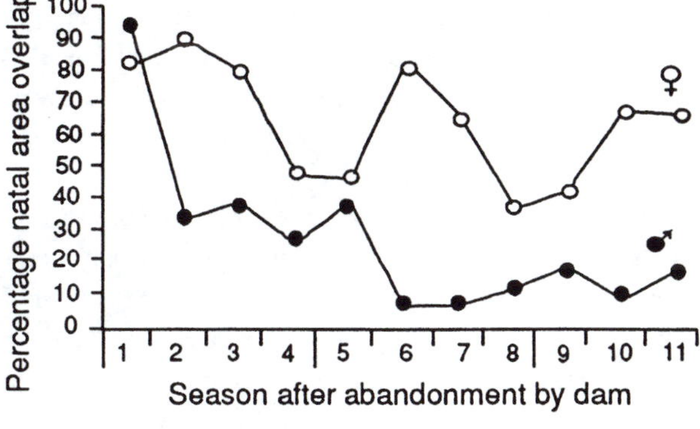

Figure 131. Male moose tend to disperse farther and more often than do female moose. Cederlund and Sand (1992) expressed this phenomenon as distance (in kilometers) dispersed from the natal area (top), and percentage home range overlap with their dams (bottom) in 11 seasons after abandonment by the dams.

their use. Eight moose had home ranges adjacent to the burned area, and none of these made use of the area after the fire, despite being located at times within 0.1 to 3 miles (0.2–4.8 km) from the edge of the burn. Overall, radio-collared moose in this study increased their use of the burn significantly during the first year after the fire and continued this pattern of use for the 5 years of monitoring after the fire. There was no evidence, however, of increased use of the burn by dispersing yearlings, which is in contrast to the high immigration into a burned area observed by Peek (1974b) in northeastern Minnesota. Differences observed in these two studies may be accounted for by the unequal pre-fire densities of the two populations, with Gasaway et al. (1989) reporting 0.26 moose per square mile (0.1/km^2) and Peek (1974b) reporting 2.3 moose per square mile (0.9/km^2) in an area close to the burned area he studied. Also, Gasaway et al. (1989) believed that adequate forage existed outside the Alaska burned area, therefore the moose in that area's population may have had little incentive for dispersing.

Despite the preponderance of information on short-range movements, longer moose dispersals have been documented. Mytton and Keith (1981) observed four young moose (one male and three females) in Alberta disperse distances of at least 31 miles (50 km), and another young bull dispersed 155 miles (250 km) from the study area over a 2-year period. Ballard et al. (1991) reported a dispersal distance of 110 miles (177 km) by an adult cow. Such long-distance dispersals can facilitate the exploitation of vacant habitats, but they can also carry a high probability of mortality by exposing individuals to unfamiliar or unsuitable habitat (Miller et al. 1972, Garner and Porter 1990).

Colonization of New Areas

Dispersal is the mechanism by which moose colonize new areas. Knowledge of this phenomenon is derived from studies associated with introductions of moose into new areas, which also provide insight into adaptive or presaturation dispersal from these events.

For example, moose recently colonized the Yakutat forelands of Alaska, an area of excellent moose habitat completely isolated by the St. Elias mountains and the Gulf of Alaska. In the late 1920s or early 1930s, retreat of the Alsek glacier opened a corridor between the forelands and the interior (British Columbia and Yukon Territory). This corridor was located at the eastern end of the forelands. On the basis of anecdotal information, moose reached western-most portions of the forelands sometime in the 1940s, some 50 miles (80 km) distant, and from this point colonized two isolated areas across Yakutat Bay within the next 10 years (W.B. Dinneford personal communication: 1993). This information yields a potential rate of range expansion of approximately 5 miles (8 km) per year.

That rate is similar to that given by Mercer and Kitchen (1968) for overall range expansion in Labrador—6 miles (9.7 km) per year—but less than the maximum rate of 8 to 15 miles (13–24 km) per year during the peak of expansion. The Labrador population had a reported density of 0.05 moose per square mile (0.02/km^2), indicating that population density may not be a factor in population expansion.

Nowlin (1985) monitored an introduced population in Colorado at the end of the first summer after their introduction in Colorado. The population occupied 51 square miles (132 km^2) at a density of 0.05 moose per square mile (0.02/km^2). The farthest relocation from the point of introduction was 7 miles (11 km). During the second summer, the area occupied by the moose was 249 square miles (645 km^2) and the farthest relocation was 21 miles (34 km) from the point of introduction. The population density during the second summer was 0.16 moose per square mile (0.06/km^2). A summary of range expansion rates from across North America is contained in Table 24.

Migration

Migration is the movement of an animal between seasonal (i.e., mating and nonmating) ranges. The traditional view of the adaptive value of migration is that it has evolved as a strategy to minimize the deleterious effects on food resources that are limited seasonally (Baker 1978). Sinclair (1983) argued that this explanation can be refined by viewing migration as a mechanism for placing an animal in an

Table 24. Range expansion rates for some North American moose populations (adapted from Gasaway et al. 1985b)

Area	Period	Rate of expansion, in miles (km) per year	Reference
Alaska			
Upper Kobuk and Kivalina rivers	1880–1960	3 (5)	Coady (1980)
Yakutat forelands	1920–1950	5 (8)	W. B. Dinneford (personal communication)
Alberta	1965–1972	1.3 (2)	Rolley and Keith (1980)
Colorado	1978–1980	11 (18)	Nowlin (1984)
Labrador	1953–1961	6 (10)	Mercer and Kitchen (1968)
	1880–1950	8 (13)	
	1949–1961	15 (24)	
Newfoundland	1904–1934	7 (11)	Pimlott (1953)
Ontario	1895–1955	4 (6)	Peterson (1955)
	1875–1895	15 (24)	

Natural or translocation (above) colonizing of new range by moose is facilitated by dispersal. Some dispersal occurs regardless of moose population density, therefore even low-density populations will expand their range if suitable habitat exists. Accordingly, dispersal rates must be measured directly, because they cannot be inferred from changes in population density alone. Rates of range expansion can vary from approximately 1 to 15 miles per year (1.6–24.1 km/year). This young moose is being sedated before crating and after capture in this trap at the Moose Research Center, Alaska. The moose was subsequently translocated. *Photo by Donna Franzmann.*

optimal *mating* environment. Sinclair asserted that the impetus behind migratory behavior is to place the animal in an area with an abundance of high-quality food before mating. In large part, such "positioning" can enhance the reproductive output of that individual and consequently its fitness.

Unlike dispersal, migration is directional in nature, and thus predictable. Whereas dispersers may wander at random in search of new home range, migrating moose move with purpose because they tend to be philopatric to seasonal ranges and migration routes are traditional. Although moose do not exhibit the mass migrations seen in caribou and some African ungulates, an example of simultaneous spring migration was reported in Newfoundland by Albright and Keith (1987: 385), where "groups of two to eight moose were moving northward over the largely snow-free barrens. . . . It was impossible to estimate the total number involved, but for 15 to 20 kilometers [9.3–12.4 miles] moose were seldom out of sight."

Migratory Versus Nonmigratory Populations

Most moose populations contain migratory and nonmigratory segments. The proportion of migrators versus nonmigrators in a stable population will reach an equilibrium at which the relative benefits of leaving are equal to the benefits of staying (see Sinclair 1983). Mytton and Keith (1981) speculated that migratory cows had higher fecundity than did nonmigratory cows in an Alberta population. If this were true, the proportion of migrators in this population would be expected to increase to some point and then stabilize. Stability would be achieved because the summer range of migrators would become more crowded to the point that a density-dependent decline in productivity would occur. The relative numbers of migrators would increase through their own increased productivity, not through the process of nonmigrators becoming migrators, because migration in moose seemingly is a learned and traditional behavior.

Some populations are totally nonmigratory, likely because they live in isolated or homogeneous habitats (Mould 1979). Moose must achieve a benefit from migration and, if snow depths and availability of forage are uniform over an area, a nonmigratory population, as opposed to a migratory population, would be expected to reside there.

Patterns of Migration

The typical pattern of migration observed in moose populations consists of movements between a common winter range and separate summer ranges. In mountainous areas, these winter ranges usually are at lower elevations than areas used at other times of the year. However, migration by moose is not limited to populations living in mountainous regions, although these are characterized by longer migration distances than are populations residing in relatively even terrain.

MIGRATION DISTANCES

The straight-line distance between winter and summer ranges can vary within and among populations, and is mainly a function of habitat dispersion or terrain (Addison et al. 1980, Sandegren and Sweanor 1988). Movements by migrating individuals can be expressed as elevational or horizontal, but LeResche (1974) emphasized the "ecological distance" of migration. LeResche was viewing migration as a movement along an environmental gradient, and that the

Moose do not exhibit mass migrations as classically seen in some African ungulates and North American elk, pronghorn and caribou. However, these moose were observed during a simultaneous migration in spring on the south coast barrens of Newfoundland. Favorable weather conditions likely resulted in a sudden onset of cues that triggered this migration. Usually, the timing and speed of migration varies enough within a population that moose do not travel in large groups. Cows travel either alone or with their offspring, and bulls sometimes can be seen in groups of two or three. Although normally solitary, moose often form groups after the rut. Author Hundertmark once observed more than 60 moose congregated in a meadow approximately 300 acres (121.5 ha) in size before the onset of winter migration, but these moose did not migrate together. *Photo by Lloyd B. Keith; courtesy of the University of Wisconsin-Madison Department of Wildlife Ecology and the Ottowa Field Naturalist's Club.*

physical distances involved in this movement were a function of the gradient.

Ballard et al. (1991) measured migration distances of 9.9 to 57.8 miles (16–93 km) in a mountainous region of Alaska. In a separate study in the same general area, Van Ballenberghe (1977) reported distances between 5 to 58.4 miles (8–94 km). Other studies in western North America have reported migration distances of 5 to 20 miles (8–32 km) in Wyoming (Houston 1968), a mean distance of 8 miles (13 km) in Idaho (Ritchie 1978) and a maximum distance of 40 miles (64 km) in British Columbia (Edwards and Ritcey 1956). In the relatively even terrain of Alberta, Mytton and Keith (1981) observed short migrations, with bulls moving an average of 8.3 (±2.2 SE) miles (13.4 ± 3.6 km) and cows moving an average of 4.2 (±0.4) miles (6.8 ± 0.7 km). In other areas with nonmountainous terrain, Phillips et al. (1973), Addison et al. (1980) and Roussel et al. (1975) measured migration distances of 8.7 to 21.1 miles (14–34 km) (northwestern Minnesota), 1.2 to 8.1 miles (2–13 km) (northwestern Ontario) and 4.4 miles (7 km) (Quebec), respectively.

TIMING AND SPEED

Moose tend to leave summer ranges when snow depth increases or when travel to rutting areas is necessary. As would be expected, timing of migration can vary considerably between years. Indeed, instances have been reported in which traditionally migratory populations have not migrated or delayed movement to winter range in a year of little or no snowfall (Houston 1968, Van Ballenberghe 1977, Ballard et al. 1991). Movement to winter range generally is triggered by snow accumulation (see Coady 1974a), therefore timing can vary considerably depending on weather.

The speed of migration to winter range also varies. Van Ballenberghe (1978) noted that different moose took from 10 days to 6 weeks to cover the same distance. This tendency for winter migration to be a gradual process seems to be common (Edwards and Ritcey 1956, Stevens 1970, Phillips et al. 1973).

TRADITION

Migratory routes and the tendency to migrate are traditional in moose (Berg 1971, LeResche 1972). Migratory moose tend to use the same routes year after year. A study in Norway described archaeological evidence showing that a migration route had been used by moose for at least 5,000 years despite recent habitat deterioration that made the winter range unsuitable to the point that the migratory segment of the population showed a decrease in productivity compared with that of the nonmigrants (Andersen 1991b).

Migratory behavior seems to be learned as calves accompany their dams on annual migration routes. The migratory pattern of cows is exhibited by their offspring, and the distance migrated by adults is directly related to that of their mothers (Sweanor and Sandegren 1988). Studies of calves separated either experimentally or naturally from their dams before migration reported that the calves established winter ranges separate from their mothers (Markgren 1975, Mytton and Keith 1981, Sweanor and Sandegren 1989).

UNUSUAL PATTERNS OF MIGRATION

Gasaway et al. (1983), Mytton and Keith (1981) and Andersen (1991b) described unusual migration patterns in which moose summer in lowland habitat and winter at higher elevations in Alaska, Alberta and Norway, respectively. In central Alaska, Gasaway et al. (1983) observed that a number of moose populations move from high elevation winter ranges to a central lowland summer range in February and April. The lowlands also were home to a nonmigratory population. Migratory moose leave the lowlands in August through October and return to winter ranges, some of which contain their own resident (nonmigratory) populations.

That migration pattern runs counter to those generally observed for moose that make an elevational shift. The reasons behind this phenomenon have not been documented, but a potential explanation was described by M.E. McNay (personal communication: 1993) as follows. Essentially, migrating moose may select lowland spring and summer areas because they provide better forage and calving habitat than uplands. The lowlands are characterized by poorly drained, black spruce forests interspersed with bogs and lakes. These conditions provide abundant aquatic vegetation, which is sought by moose in spring (see Chapter 11), as well as being relatively secure habitat for calving. The amount of herbaceous vegetation available in these areas sustains both resident and migratory moose throughout the summer, but as the vegetation matures, uplands provide a relatively greater source of winter browse. Snow does not seem to be a factor in migration in these areas, because normal snow depths usually are well below the critical level of 27.6 inches (70 cm) (Kelsall and Prescott 1971), but when deep snow does occur, it usually is deeper on the lowlands. Common characteristics of upland and lowland habitats for both this central Alaska area and the Alberta study area (Mytton and Keith 1981) indicate that a similar mechanism may be involved. Phillips et al. (1973) also observed this phenomenon in Minnesota, in which approximately 20 percent of the moose population on the Agassiz National Wildlife Refuge left the area in winter and returned in spring.

These examples seem to run counter to the general pattern of migration, because not all animals leave the summer range with the onset of winter weather. However, they are consistent with Sinclair's (1983) view that migration serves to place an animal in an optimal mating environment.

Management Implications

Importance of Movements

Although moose movements can be evaluated in several ways, identification of home ranges is necessary when determining the areas occupied by a population. This type of information is useful when mitigating for the effects of habitat alterations stemming from human development and resource extraction activities, such as hydroelectric projects (Taylor and Ballard 1979, Modafferi 1982) and timber harvesting (Doerr 1983, Hundertmark et al. 1983). Designation of critical habitat areas and biological reserves is also facilitated by these analyses.

Movements also must be considered in the construction of physical barriers that may block traditional migration routes, which can cause disruption of population dynamics and increase incidental mortality. Moose attempting to cross under elevated sections of the Trans-Alaska oil pipeline occasionally were detoured temporarily from their route, and sections of pipe elevated less than 4 feet (1.2 m) above the ground and ditches deeper than 10 feet (3 m) effectively blocked passage (Van Ballenberghe 1978, see also Eide et al. 1986, Sopuck and Vernam 1986). Placement of road and rail corridors that bisect traditional corridors of movement can result in high incidental mortality (McDonald 1991, Modafferi 1991; see Chapter 8), and moose are unlikely to alter their movements to avoid such dangers (Andersen 1991a, 1991b).

Seasonal movements of moose often cause them to come into conflict with humans. In New England, moose disrupt maple sap collection by moving through stands of sugar maple and tearing down the tubing used to transport the sap to collection points. In many areas of Alaska and elsewhere, moose habitat disturbed because of residential and commercial developments, particularly along river valleys, creates an abundance of vegetation in early seral stages that attracts moose in winter. These wintering moose pose a public safety problem owing to their proximity to highways, and also create problems for homeowners because of

Studies designed to predict the effects of habitat alteration on moose populations must consider the animals' movements. Delineation of seasonal ranges, migration routes, rutting and calving areas, as well as the shape, size and location of home ranges and core areas all are important variables indicating how moose use their environment. This area, on the Kenai Peninsula of Alaska, supported nonmigratory moose and some wintering migratory moose. After mechanical rehabilitation of the vegetation, the number of moose observed in autumn surveys in this area doubled within 15 years. *Photo by Albert W. Franzmann; courtesy of the Alaska Department of Fish and Game, Soldotna.*

Moose use traditional migration corridors. Construction of potential barriers to migration such as fences, highways or pipelines can disrupt population dynamics, and therefore must be carefully planned to minimize disruptive impacts. In designing the Trans-Alaska pipeline, some areas of the line were buried (left) to permit movement and others were raised (right) to allow movement of animals underneath. *Photos by Victor Van Ballenberghe; courtesy of the Institute of Northern Forestry, Anchorage, Alaska.*

the propensity of moose to browse on ornamental trees and shrubs. When their populations exceed carrying capacity and a winter with deep snow occurs, moose can exhibit yarding behavior in these areas, and communities must cope with many starving moose in their midst.

The solitary nature of cows with calves can lead to underrepresentation of this group in aerial surveys. Management surveys to assess sex and age composition of a population normally are conducted as soon as an area is completely covered with snow to assure good contrast between moose and the landscape. Because cows with calves are more likely to be solitary at this time of year, biologists cannot rely solely on assessing composition of groups of moose that are readily observable from the air. Systematic searching of areas without large groups of moose must be accomplished with the same intensity as in areas with large groups, to be certain that cows with calves are observed. Nonetheless, all but the keenest observers are likely to undercount solitary animals.

Harvest

Movements by various segments of a moose population during the hunting season can effect their vulnerability to harvest. Yearlings and 2-year-olds are up to twice as vulnerable to harvest as are older age classes (Boer 1988), probably because of the increased movements they exhibit during dispersal and establishment of home ranges. Mature bulls are more vulnerable to harvest during the rut, because their heightened activity, including movement, is likely to bring them into contact with hunters at the same time when their innate wariness is diminished. As many moose hunting seasons correspond with the rut, there exists a potential for overharvest of this age group (Wilton 1992).

Local variation in population density can be caused by differential harvest and low recolonization rates by dispersers. Wildlife managers must consider this when framing hunting seasons. Hunter access as well as methods of harvest must be regulated so that hunting mortality is proportional to the surplus available in local populations, not regional estimates (Cederlund and Okarma 1988).

Population Identification

In wildlife science, "population" is used arbitrarily to identify a group of animals being studied. Sometimes, however, population is used to represent a group of animals composed of migratory and nonmigratory segments on a common seasonal range or a group of animals inhabiting an arbitrarily defined geographic area. If a population is recognized as a group of interbreeding individuals (Mayr 1970: 82), then the migratory and nonmigratory segments likely are not part of the same population because they have separate seasonal ranges for part of the year, and likely distinct breeding ranges. LeResche (1972) recognized this distinction when he identified moose populations on the Kenai Peninsula in Alaska. Discussing migratory and nonmigratory moose, Andersen (1991b) identified the different segments of his study population as *demes*— groups of animals within which breeding occurs. In this context, population refers to a group of animals being studied and that can be defined by some geographic criteria, and is the terminology that will be used here. This distinction is more than a semantic exercise. To comprehend the dynamics of the groups of animals being studied or managed, biologists must understand how demes within a population segregate during the mating season.

Migration places different demes in different breeding habitats, and can result in the demes having different levels of productivity (Mytton and Keith 1981, Andersen 1991b, Sweanor and Sandegren 1991). Furthermore, the locations of migratory and nonmigratory demes during the hunting season subject them to different levels of hunting pressure and harvest (Ritchie 1978, Bangs et al. 1984). Both of these factors, if not understood, can confound attempts to manage populations. For example, predator control in an area of central Alaska not only increased calf recruitment in the resident (nonmigratory) moose deme, but also increased recruitment in demes that summered in the area but wintered elsewhere (Gasaway et al. 1983). Aerial surveys of winter range detected this increased recruitment, but without knowledge of the migratory and nonmigratory components of the population, the reasons for the increase could have been misinterpreted.

LeResche (1972) noted that different demes inhabiting the northern Kenai Peninsula of Alaska had different sex and age structures. The migratory deme, inhabiting uplands in the summer, was characterized by mature bulls and cows with low productivity. A resident deme, inhabiting a highly productive lowland burn, was characterized by young bulls and productive cows. LeResche attributed these differences to the better habitat and greater hunting pressure in the lowland burn. If these demes were observed only when they were on common winter range, the forces driving population trends would be difficult to discern.

Effects of Dispersal on Adjacent Low-density Populations

The limited dispersal tendencies exhibited by moose must be considered when attempting to promote recovery of low-density populations. Although some dispersal occurs in all populations, the amount reported is inadequate to affect

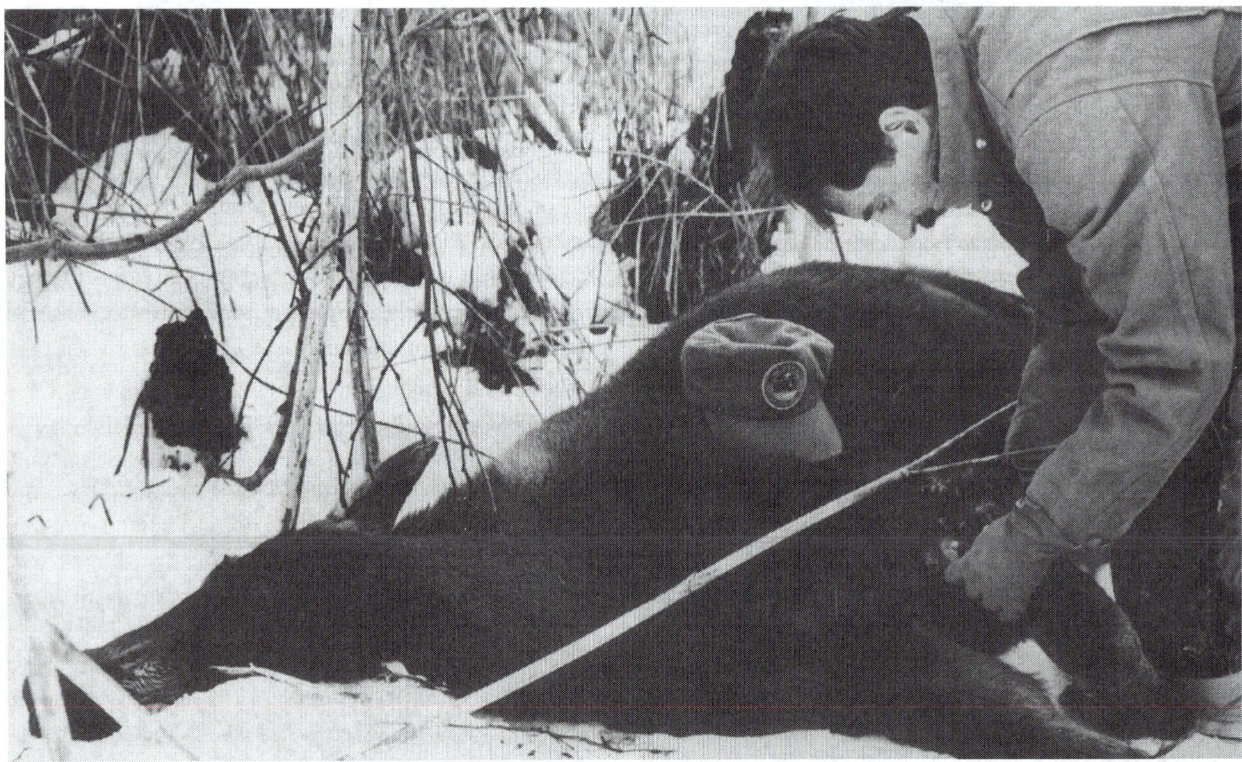

Moose winter range in North America is increasingly "invaded" by human development, which leads to conflicts that rarely are resolved in favor of the moose. In regions that experience extensive snow accumulations, moose tend to move or migrate to lower elevations and thermal pockets relatively protected from severe weather. Not coincidentally, those are the areas to which humans are attracted for housing, businesses and other facilities and services. For humans, such location is a matter of convenience and economics. For moose, those areas otherwise represent more accessible food. When weather and human habitation combine to limit scarce foods in winter, moose may resort to unusual substitutes. In the bottom photo, a biologist examines the same moose shown in the top scene. That moose died of starvation. During necropsy, its rumen was found to be impacted with human food, plastics, tinsel and other trash, and its bone marrow indicated that fat reserves had been depleted. *Photos courtesy of the* Homer *(Alaska)* News.

significantly the density of adjacent populations (Goddard 1970, Gasaway et al. 1989). This factor must be considered when evaluating the effect of differential harvest rates between areas that are easily accessible by hunters and those that are relatively inaccessible (Boer 1990). Recovery of heavily hunted populations must be a function of reproduction rather than immigration from surrounding populations. This principle also applies to habitat management aimed at increasing a moose population. Managers cannot assume that significant numbers of moose will immigrate

Where moose move in winter habitats occupied by people, a variety of efforts are used to alleviate probable dangerous conflict. In Homer, Alaska, for example, human residents have planted willow food plots outside of the city. Even so, when moose enter public or residential areas, they must be hazed or lured away or euthanized. *Photo courtesy of the* Homer *(Alaska)* News.

Cow moose that calve near urban areas and human transportation corridors are faced with the added stress of human curiosity and approach to view and photograph. Above, the cow's body language is a prelude to aggressiveness. It is a warning that many people do not recognize or appreciate. Moose may acclimate somewhat to the presence of people, but many biologists believe that what is construed as acclimation actually may be escalated association stemming from an increase of human sprawl into core areas and along migration routes of traditional moose range. *Photo courtesy of the* Homer *(Alaska)* News.

to these managed areas and take up residence unless the areas are placed strategically along corridors of moose migration (Gasaway et al. 1989).

Concluding Thoughts

Analysis of movements should be a primary consideration in any attempt to understand the ecology of a moose population and, accordingly, options of management. The knowledge gained in the pursuit of the "when" and "where" of movements will provide biologists with a solid foundation on which other studies can be based. Sanderson (1966) lamented the paucity of effective techniques for studying mammal movements and the lack of a unified approach to this subject. In the past three decades, a proliferation of useful, new techniques have allowed a more refined approach for such study. I believe, however, that a more concerted effort toward answering the "why" of moose movements is needed before a unified model of home range and movements can be constructed.

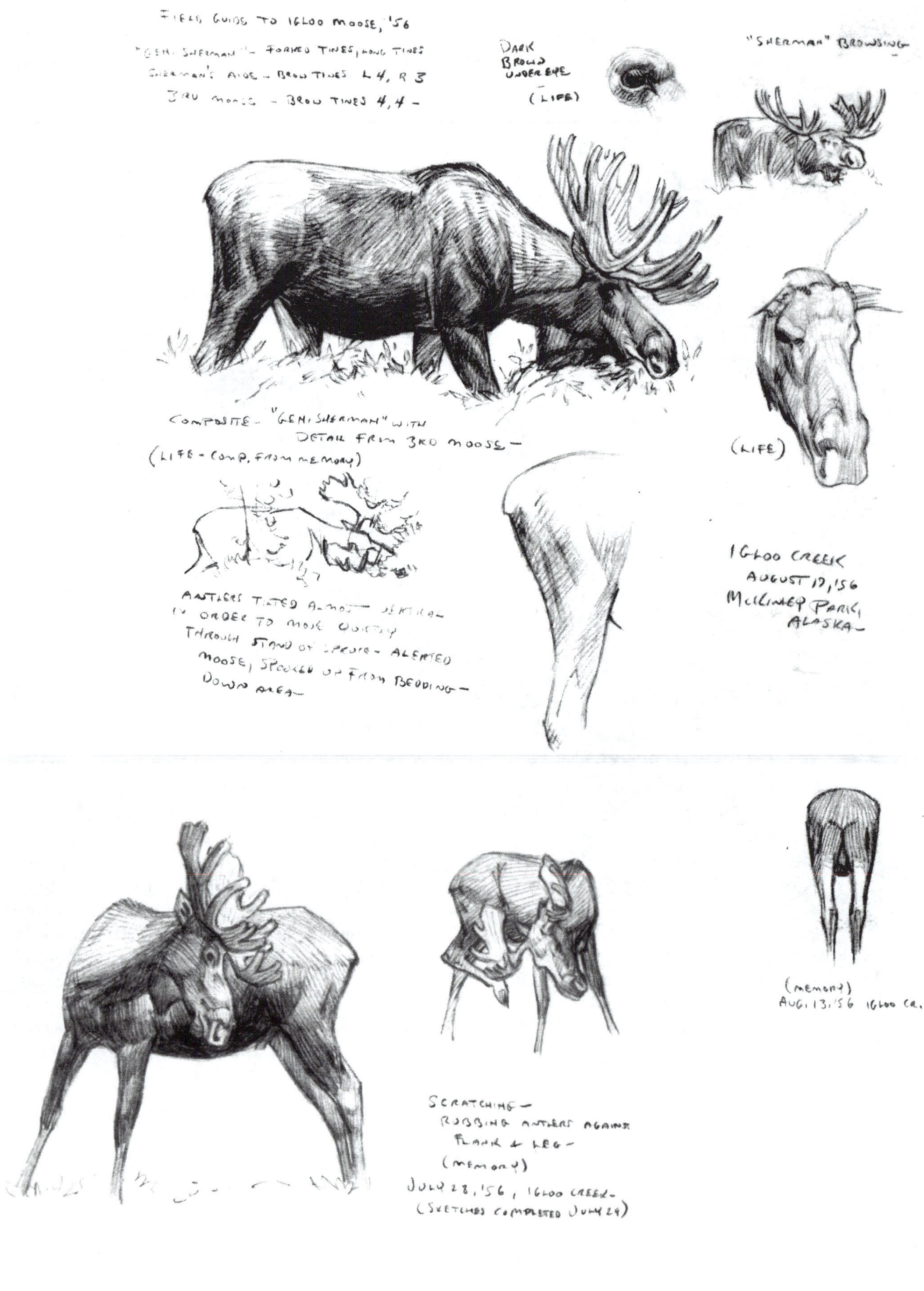

FIELD GUIDE TO IGLOO MOOSE, '56

"GEN. SHERMAN" – FORKED TINES, LONG TINES
SHERMAN'S AIDE – BROW TINES L 4, R 3
3RD MOOSE – BROW TINES 4, 4 –

DARK BROWN UNDER EYE –
(LIFE)

"SHERMAN" BROWSING –

COMPOSITE – "GEN. SHERMAN" WITH DETAIL FROM 3RD MOOSE –
(LIFE – COMP. FROM MEMORY)

(LIFE)

ANTLERS TILTED ALMOST VERTICAL IN ORDER TO MOVE QUIETLY THROUGH STAND OF SPRUCE – ALERTED MOOSE, SPOOKED UP FROM BEDDING-DOWN AREA –

IGLOO CREEK
AUGUST 17, '56
McKINLEY PARK, ALASKA –

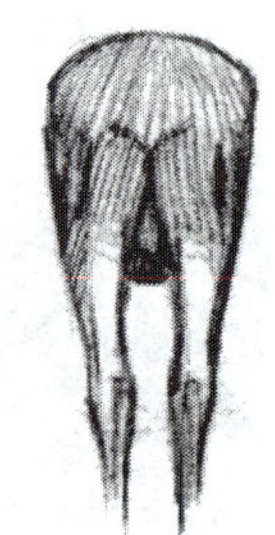

(MEMORY)
AUG. 13, '56 IGLOO CR.

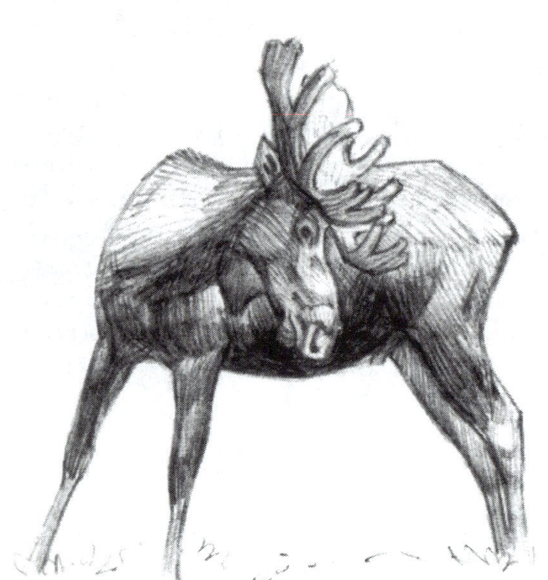

SCRATCHING –
RUBBING ANTLERS AGAINST FLANK & LEG –
(MEMORY)
JULY 28, '56, IGLOO CREEK –
(SKETCHES COMPLETED JULY 29)

10

ARNOLD H. BOER

Interspecific Relationships

Throughout the boreal forest biome of North America, moose live in a wide variety of habitats, feed on a wide array of foods and face unique combinations of winter severity and topography. Although usually associated with boreal forest, moose are found in many other forest types (Telfer 1984). They share habitat spatially and temporally with a variety of herbivores, including caribou, mule deer, white-tailed deer, bison, elk, musk-oxen, mountain sheep, snowshoe hare and beaver. Interaction with those species is inevitable as they form part of the environmental backdrop and generate selective pressures on moose. The manner in which and extent to which moose use local habitats reflect, in some measure, the effects of interspecific interactions.

Interspecific interactions between moose and other species take one or more of the following general forms:

1. *Competition.* This is a negative form of interaction and is related to common or joint exploitation of essential resources such as food.
2. *Parasite-mediated competition.* This involves the presence of a shared parasite that affects differentially the competitors and influences the outcome of interspecific interaction.
3. *Predation.* This form of interaction involves predator and multiprey systems in which moose numbers influence or are influenced by the number and distribution of other prey species. This interaction can influence moose both directly or indirectly (see Chapter 7).
4. *Commensalism.* One species benefits whereas the other is relatively unaffected in this form of interaction. An example of such an interaction is the creation by beavers of ponds that become important sources of aquatic and succulent emergent vegetation for moose. The attraction of moose to such habitats has little or no influence on the beaver population.

Because of the diversity of habitats, species combinations and abundance of sympatric species, a variety of competitive mechanisms operates throughout moose range. Some generalities may be drawn, however, and these may help explain ecological processes and moose population dynamics in a particular area.

Of the interspecific interactions possible, competition is the most obvious and profound influence on moose distribution and habitat use patterns. Competition occurs in two forms: (1) resource competition in which both competitors use a common but limited resource such as food; and (2) interference competition whereby one species harms the other when seeking a resource not in short supply (Birch 1957). Interspecific competition occurs over food, water and space resources. Outcomes of competitive interactions are not always predictable because conditions influencing the form and intensity of competition vary locally.

According to Hardin's (1960) "competitive exclusion principle," complete competitors cannot coexist. Hence, animals live in dynamic tension with competitors, and the form of the relationship is affected by shifts in environmental conditions that may favor one over the other. Organisms increase competitive ability if they evolve interference mechanisms or become more efficient exploiters of a common resource. Competition is a conceptual explanation of the interactive processes among species and, although theories such as the Lotka and Volterra equations (see Krebs 1978) attempt to predict and quantify competitive outcomes, they do not account adequately for all species and competition circumstances. Because wildlife species evolve mechanisms to avoid or reduce competition, competition is difficult to investigate in the field (Pianka 1976). It is likely intermittent and not always evident (Weins 1977).

Within any ecosystem, animals occupy niches that define their role in the biotic community (Whittaker et al. 1973). Temporal and spatial variations in behavior, habitat use, food preferences and feeding strata help to segregate competitors and minimize competition. *Niches* are expressions of collective environmental pressures, including competitive processes and other forms of interspecific interactions acting on a specific population.

Interspecific Relationships with Large Ungulates

On the Serengeti Plain, African ungulates share geographic range through a well-developed system of resource sharing (Jarman and Sinclair 1979). In North America, a similar model of coexistence and resource sharing among sympatric ungulates can be found in Elk Island National Park near Edmonton, Alberta. This 61-square mile (192-km^2) park provides an excellent example of complementary habitat use among the four principal ungulates that occupy the park—moose, white-tailed deer, elk and bison (Blood 1974).

Resource partitioning mechanisms facilitate coexistence of sympatric species of large mammals; they may take the form of spatial or temporal segregation, species-specific preferences for forage plants and plant parts and different feeding heights (Stelfox and Taber 1969, Hudson 1976). Ecotone (transitional area between two different plant communities) and habitat interspersion are important to herbivore distribution and, consequently, habitat selection by ungulates. In northern North America, snow is another important factor in determining distribution. The morphological characteristics and evolutionary adaptations of the various species produce different abilities and strategies to cope with snow.

Despite relatively dense populations and the unique species combination, the general distribution of ungulates in Elk Island was shown to correspond with habitat usage elsewhere (Cairns and Telfer 1980). All four species demonstrated selective habitat use, and although sympatric, temporal and spatial segregation minimized direct competition.

A Shira's moose calf and a cow elk view one another with evident curiosity but without apparent alarm. *Photo by Judd Cooney.*

Moose are primarily browsers and associated with habitats containing a high proportion of the shrubs preferred for browsing. Shrubland and shrub meadow were the habitat types in the park most used by moose throughout the year (Holsworth 1960, Cairns and Telfer 1980). Among the other habitats available, sedge meadows were used extensively in spring and summer, as were aspen forests in winter.

White-tailed deer are primarily browsers but also feed on available grasses and succulent forbs (Cairns and Telfer 1980). Whitetails, the smallest resident ungulate of the park, were also the most restricted by deep snow. Aspen forests are selected by the deer when snow depths began to restrict movement and food resources. In spring and summer, their use of forests diminished in favor of upland grass sites.

Elk are primarily grazers, but readily use available browse (Houston 1982). The principal elk habitats in the park were upland sites. Use of shrubland and shrub meadows increased in winter, and sedge and shrub meadows were used in spring and summer in conjunction with upland grass sites.

Bison are grazers and, opting for grasslands, they exhibited the greatest habitat specificity of the four ungulates (Cairns and Telfer 1980). Among the four species, moose and bison have the most divergent food preferences. Consequently, differences in habitat selection by moose and bison are the most pronounced. Moose usually are associated with forested and shrubland areas, whereas bison used upland grass areas and spent little time in aspen forests or shrubland. Bison distribution was limited, but bison did occur in a few areas inhabited by moose (Figure 132).

Elk and deer habitat selections overlapped that of both moose and bison. However, although some overlap occurred, the four species generally are segregated both spatially and temporally. Differential food preferences and snow-level tolerances minimized direct competition among these species.

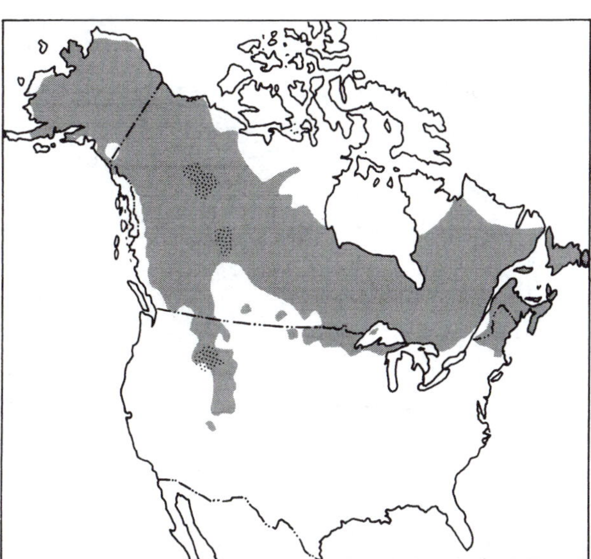

Figure 132. Separate and conterminous ranges of overlap of moose (solid screen) and bison (dot screen) in North America (after Banfield 1974).

Elk Island National Park also provided an example of commensalism. Browsing by moose, and to a lesser extent by white-tailed deer and elk, may have helped to stabilize forest edge and to maintain upland grassy openings on which bison depend (Cairns and Telfer 1980).

Bison (left) are grazers, whereas moose (right) are browsers, therefore their habitat preference, predicated mostly on food resources, precludes much interaction and competition in the few areas of North America where their ranges overlap. *Photos by Albert W. Franzmann; courtesy of the Alaska Department of Fish and Game, Soldotna.*

White-tailed Deer

The relationship between moose and white-tailed deer is one of the most complex interactions among large herbivores. The northern limit of the white-tailed deer range overlaps much of the southern portion of the moose range (Figure 133). Wherever the two species coexist, there is potential for competition for common resources, and changes in habitat and seasons tend to give advantage to one over the other at a particular time. Both deer and moose are browsers and compete for similar food items at certain times of the year. However, differences in forage preference, seasonal spatial segregation and abilities to cope with snow and cold temperatures are mechanisms that help partition resources between these species.

The northern and altitudinal limits of deer ranges are set primarily by winter conditions, and the distributions of local populations reflect such conditions. Moose distribution is determined primarily by browse availability and distribution (Telfer 1978a, Prescott 1974). Therefore, two mutually exclusive factors determine the extent of interaction and competition between these species.

In northwestern North America, particularly in mountainous areas, moose and deer are segregated more than in other portions of sympatric range. Spatial segregation in these shared areas reflects altitudinal influence on vegetative communities and snow depths. Topographic variations in mountainous areas can abruptly change plant species and snow characteristics within short vertical distances.

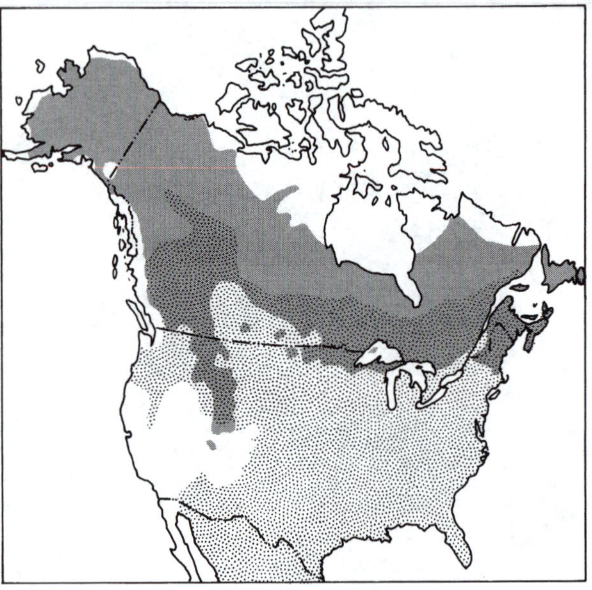

Figure 133. Separate and conterminous ranges of overlap of moose (solid screen) and white-tailed deer (dot screen) in North America (after Banfield 1974).

In New Brunswick, Telfer (1970) found differential temporal uses of habitat by deer and moose, particularly in winter. Because of their smaller size, whitetails are more readily impeded by deep snow and low temperatures than are moose (Telfer and Kelsall 1979). Hence, deer select softwood cover during winter (Verme 1965).

Snow depths are lower, the snow is denser and average ambient temperature is higher in the mature softwood stands that comprise a deer-wintering area than in adjacent or more open forest stands. Deer-wintering areas often are associated with cedar swamps, spruce and balsam fir forests and stands of large hemlock, each of which provides thermal cover (Boer 1978, Rounds 1981, Crawford 1982). Moose tend to favor areas with abundant balsam fir and white birch, which are important winter foods for moose, and are less often associated with the softwood stands selected by wintering deer (Kearney and Gilbert 1976, Rounds 1981). Hence, moose and deer selection patterns for overstory cover and understory browse suggest a mechanism of resource partitioning on snow-covered ranges.

Differences in winter cover requirements contribute to variation in moose and white-tailed deer diets in boreal forests. Moose generally exist at much lower densities and are more evenly distributed than deer, therefore the opportunity for food-related competition is low.

Where the two species occur in the same areas during winter, moose and deer *may* compete for browse, but direct, proximal competition seems to occur infrequently and usually only in late winter when moose shift to softwood cover (Telfer 1970, Peek et al. 1976), perhaps in response to increasing snow depth or to escape rising ambient temperature. Irwin (1975) concluded that, although moose and deer used similar forage plants in a Minnesota burn area, competition for food did not seriously affect their interactions, presumably because forage was superabundant. Ludewig and Bowyer (1985) also reported that moose and white-tailed deer winter diets were comparable in Maine. Moose used 11 plant species and deer used 13 species; 7 species were common to both. Schoener's (1968) index indicated a 41.2 percent overlap in deer and moose diets.

Although competition for food between white-tailed deer and moose is not commonplace, the species do have an ecologically intriguing and potential serious relationship in the northeastern portion of their sympatric range. This relationship centers on the effect of deer on moose numbers through parasite-mediated competition involving the meningeal worm (see Chapter 15).

Both deer and moose are infected by accidentally ingesting infected terrestrial gastropods—the intermediate hosts—while feeding (Anderson and Prestwood 1981). Whitetails rarely react adversely to this parasite, *Pare-*

Wherever the ranges of white-tailed deer and moose overlap, there is little if any competition for food resources between these two species. However, in the northeast, moose on summer habitat occupied by whitetails face increased risk of ingesting the parasite *Parelaphostrongylus tenuis*—resulting in a condition often fatal to moose. *Photo by Michael H. Francis.*

laphostrongylus tenuis (Anderson 1963, 1965a; Chapter 15). In the moose, however, penetration by the meningeal worm into nervous tissue is the norm rather than the exception (Gilbert 1974). The life cycle of meningeal worm is rarely completed in moose, which are thought to be of little importance to transmission of this parasite (Bogaczyk 1990).

Before the twentieth century there was little overlap of white-tailed deer and moose ranges in eastern North America (Peterson 1955, Taylor 1956). Forest cutting and land clearing during the early 1900s allowed white-tailed deer to expand their range northward into moose range (Anderson 1972). As the northern deer population increased, the numbers of moose declined, and many moose were observed with symptoms of neurological disease (Anderson and Lankester 1974). There is general agreement that the meningeal worm caused these moose declines. Furthermore, this parasite continues to be an important mortality factor of moose along the southeastern fringe of moose range (Karns 1967, Anderson 1970, Gilbert 1974, 1992, Boer 1987b). Moose continue to persist in many areas despite meningeal worm-induced mortality (Bogaczyk 1990, Garner and Porter 1991, Beach 1992). Beach (1992) indicated that moose contact with gastropod intermediate hosts is less than deer–gastropod contact, which suggests that meningeal worm transmission to moose is less likely than to deer. Also, as cleared and cutover landscapes reforest and become less attractive to whitetails, the changing habitat may play a role in reducing meningeal worm-related deaths of moose (Cole 1981, Nudds 1990, 1992). Improved forest cutting practices should reduce transmissions to both deer and moose (see Kelsall and Telfer 1974, Davis and Franzmann 1979).

Mule Deer

Although the overlap of ranges is not extensive, mule deer and moose share habitats along the southern fringe of boreal forest in southwestern Canada and in the Rocky Mountains of the western United States and Canada (Figure 134). Moose are associated primarily with the boreal forest, whereas mule deer are associated typically with open and shrubby areas and adjacent forest ecotone. There is little published work on the interactions between these two species (see Mackie 1981); competition between mule deer and moose is not thought or expected to be great.

Winter is the most energetically stressful time of year for both species. When preferred and highly nutritious foods become scarce because of deep snow and cold temperatures, there is potential for competition, but it rarely materializes. As with white-tailed deer and moose, the different abilities of mule deer and moose to cope with snow spatially segregate these potential competitors. Mule deer are seasonally

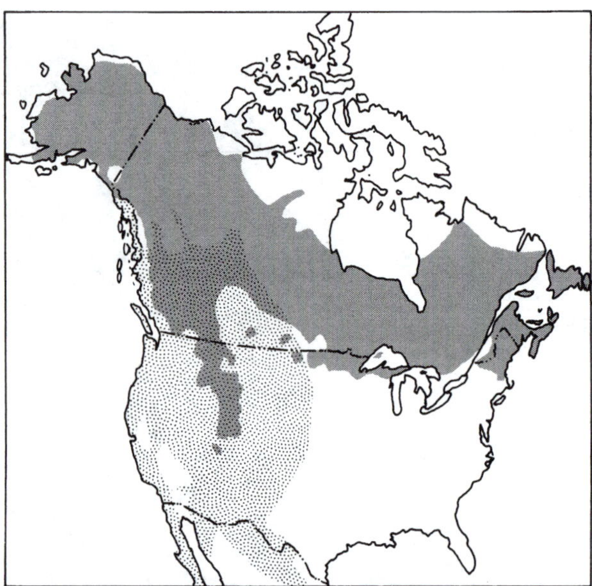
Figure 134. Separate and conterminous ranges of overlap of moose and mule deer in North America (after Banfield 1974).

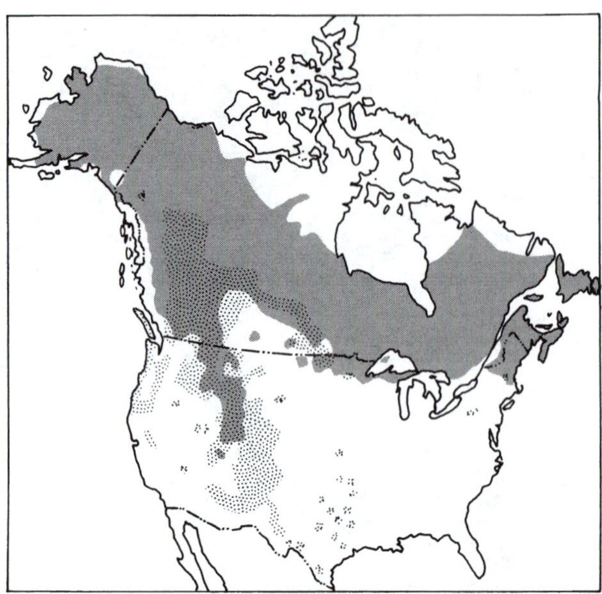
Figure 135. Separate and conterminous ranges of overlap of moose and elk in North America (after Banfield 1974).

migratory and congregate in valley bottoms during winter months. Moose remain dispersed as singles or in small groups throughout the year and can exploit browse plants in range vacated by mule deer in winter. Consequently, although both species rely on similar plant species, they seldom compete for browse in the same area during winter.

Selection of different food plants minimizes competition during summer. Mule deer use a wider variety of foods than moose do, including succulent forbs and grasses (61 percent of summer diet in forested habitats) as well as browse (Deschamp et al. 1979). The summer diet of moose in areas of range overlap with mule deer consists mainly of browse and aquatics (Flook 1964, see Eastman and Ritcey 1987).

Elk

Elk share a broad portion of moose range in the western United States, southwestern Canada and in localized areas in southcentral Canada (Figure 135). According to Flook (1964), they have the potential to outcompete moose for food. Elk are grazers primarily, but their food preferences shift to reflect food availability and they appear to be successful browsers as well (Houston 1982). Flexible foraging habits, ability to use a wide variety of terrain types and high fecundity allow elk to be an effective competitor.

Moose, mule deer and mountain sheep comprised the ancestral ungulate community of Banff and Jasper national parks before 1917 (Cowan 1950). In 1917, and again in 1920, elk were introduced into the parks. Their food preferences overlapped those of the other species already occupying the range, and as elk numbers increased, competition for preferred food resources increased. Elk dominated competition for food among the ungulate community of the parks, largely because of their diverse diet selection (Cowan 1950). Although the causal linkage between increasing elk numbers, competition for food and declining moose and mule deer numbers has not been studied experimentally, elk apparently can outcompete moose (Cowan 1950, McMillan 1953a). Mountain sheep numbers have remained relatively stable throughout the period except in marginal habitat. Presumably, elk and mountain sheep food preferences have been divergent enough to minimize competition between these two species.

Moose were introduced into Colorado in 1978, 1979, 1987, 1991, 1992 and 1993. From these introductions moose have spread throughout western Colorado and now compete with resident elk populations. Competition for willow forage is seasonal, either summer or winter, depending on the particular site (R. Kufeld personal communication: 1995).

In most other areas where elk and moose ranges overlap, wintering elk appear to prefer open grass or brush/grass vegetation with thermal cover (heavy brush or trees) nearby (Knight 1970). Snow depths on the open grass areas used by elk usually are low because of southern exposure and wind action. Moose have a greater tolerance for snow than elk do, therefore snow depth may be a significant factor influencing proximal distribution of the two species in winter (Stevens 1974). Moose also tend to select riparian habitats with abundant deciduous browse, particularly willow species (Houston 1968). Elk seem to avoid the deep snows of these habitats. Moose wintering at low densities have a competitive edge over herding species such as elk,

Elk extensively share moose range in western Canada and the United States. Competition for food generally is not significant, because elk, as grazers primarily, and moose, as browsers, tend to occupy different habitats, particularly during periods of resource scarcity, such as winter. Competition between the species is possible where elk numbers are allowed to increase dramatically. When the preferred foods of elk are limited, they readily shift to other foods, some of which are those utilized by moose. *Photo by Michael H. Francis.*

because moose can use scattered winter forage more efficiently than large concentrations of elk can (Houston 1968).

However, competition between moose and elk does not necessarily require that both are present in the same area at the same time. High elk numbers using riparian habitat in summer may reduce the amount of browse available to moose during the subsequent winter when elk have migrated elsewhere (Martinka 1969, Telfer and Cairns 1986).

Considering the ecology of the two species under normal conditions, elk and moose occupy relatively discrete niches. Some interspecific competition may be necessary to maintain their respective niches; each species is successful to varying degrees in exclusion competition on portions of their common range.

Caribou

Caribou inhabit much of the boreal coniferous biome in North America, and moose are sympatric with both barren-ground and woodland caribou over much of this range (Figure 136). Despite this broad range overlap, very little information is available on moose–caribou interactions.

Direct competition between moose and caribou apparently is limited and of insignificant effect on other species (Davis and Franzmann 1979). According to Seip (1992), an increase in moose in southeastern British Columbia provided wolves with alternative prey, resulting in an increase in wolves and increased predation on woodland caribou. Before this event, wolf densities were low because prey, primarily woodland caribou, migrated to higher elevations in winter and were largely inaccessible to wolves. Before moose numbers increased in the lower elevations, high densities of wolves could not be sustained by a low prey base.

Food preferences of moose and caribou coincide to some degree, although the diet of the caribou appears to be more specialized. Caribou consume forbs and deciduous vegetation in summer and lichens in winter (Darby and Pruitt 1984, Servheen and Lyon 1989). Moose primarily consume browse, but also use forbs and deciduous vegetation during summer (Dodds 1960, Eastman and Ritcey 1987).

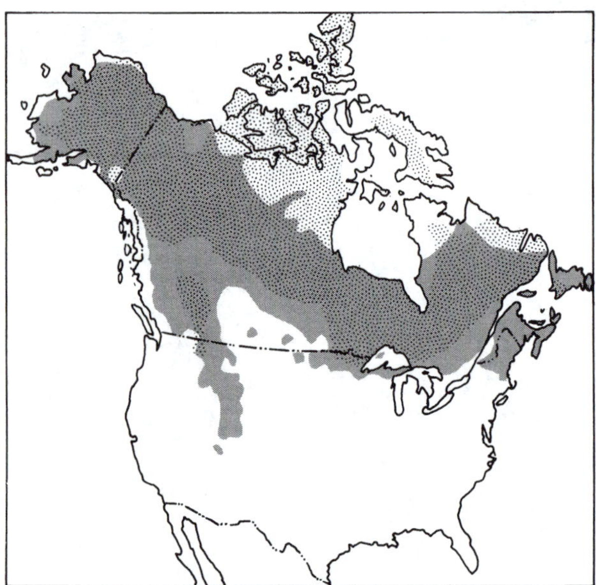

Figure 136. Separate and conterminous ranges of overlap of moose (solid screen) and caribou (dot screen) in North America (after Banfield 1974).

Introduction of moose into Newfoundland apparently had little detrimental effect on native caribou herds (Peterson 1955). There, caribou feed largely on lichens and fungi, with seasonal preferences for sedges, forbs and some shrubs (Bergerud 1972). Balsam fir—the most important browse species for moose in Newfoundland (Dodds 1960)—was a minor component (4 percent) of caribou diets.

Habitat partitioning and discrete altitudinal preferences may serve to keep moose and caribou populations segregated and their interactions nominal and inconsequential. Woodland caribou prefer mature coniferous forest for most of the year and open areas during the rut and in late winter (Bergerud 1974, Fuller and Keith 1981, Darby and Pruitt 1984, Servheen and Lyon 1989). On the other hand, moose prefer a more open habitat, with a dense shrub layer in winter (Cairns and Telfer 1980), and usually are associated with riparian habitats in summer (LeResche et al. 1974). In mountainous regions of northern British Columbia, caribou select subalpine and alpine areas, whereas moose generally avoid alpine areas (Boonstra and Sinclair 1984).

In multiprey systems, moose and caribou populations may influence each other indirectly. Increasing moose numbers in western and central portions of Alaska's Denali National Park may have resulted from increased availability of caribou as alternate prey for wolves (Singer and Dalle-Molle 1985). In the eastern section of that park, the migrating caribou are available as prey for only brief periods of time, and therefore they are not a particularly important factor of the area's prey base. And, in that area, moose populations have declined. Moose are the primary prey of wolves in other areas of Alaska as well (Gasaway et al. 1983), although Coady

Moose (left) and caribou—both woodland (right) and barren-ground subspecies—overlap ranges that cover much of North America's boreal coniferous biome. Despite some common foods—mainly forbs and certain deciduous vegetation—there is little evidence of direct competition between moose and caribou. This is explained by the vastness of the biome, the species' mobility and separate migratory/movement behaviors, separate seasonal habitat acclimations or preferences, segregations by cover type and altitude, and other factors. On the other hand, the presence or absence of the two species can serve to concentrate predation on one or the other. *Photos by Charles C. Schwartz; courtesy of the Alaska Department of Fish and Game.*

(1980) attributed an increase in moose numbers in northern Alaska to a preference by wolves for caribou. Alternate prey abundance influences predation rates on moose, but the system is complicated by seasonal movements of migratory caribou. Following an increase in caribou numbers in three study areas in Alaska, wolf predation on moose increased in one, decreased in another and remained stable in the third (Boertje et al. 1992, in press).

Although mechanisms that determine which of the two species is most influenced by the presence of the other as alternate prey in a given area is not always clear, the species have competitive association through predation.

Mountain Sheep, Mountain Goat and Musk-oxen

Mountain sheep and mountain goats occupy the altitudinal fringe of moose range, and musk-oxen occur on the latitudinal fringe (Figures 137, 138 and 139). Competition between moose and mountain sheep and mountain goats probably is negligible because of discrete diets and spatial separation of preferred habitat (Cowan 1950, Wolfe 1974). Similarly, moose and musk-oxen are unlikely competitors because of their widely disparate habitats and food preferences; musk-oxen live in the arctic tundra, above treeline.

Mountain sheep and musk-oxen may be interactively linked to moose through their role in multiprey systems, in which the availability of alternate prey species can help to maintain predator numbers at higher densities than could a single prey species.

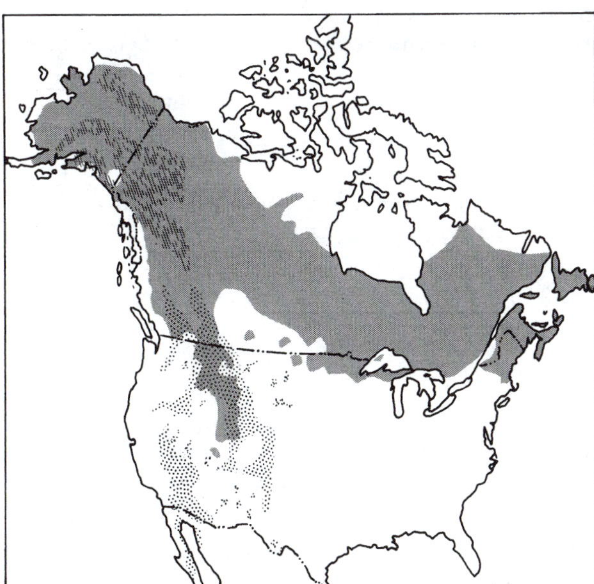

Figure 137. Separate and conterminous ranges of overlap of moose (solid screen) and Dall's (stipple screen) and bighorn (dot screen) mountain sheep in North American (after Banfield 1974).

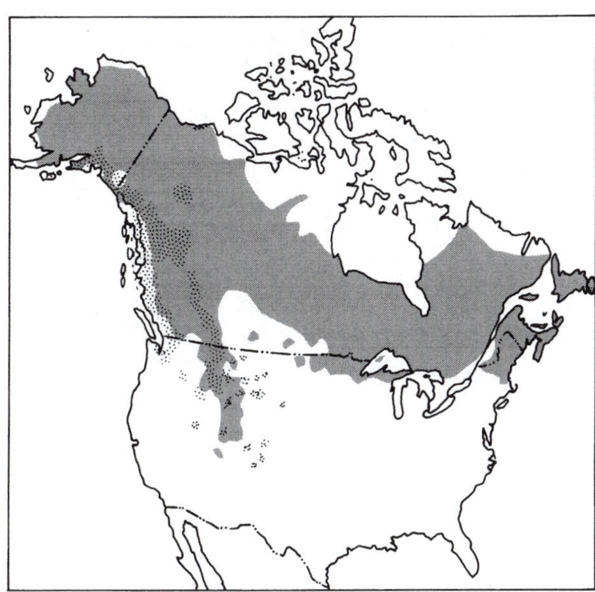

Figure 138. Separate and conterminous ranges of overlap of moose (solid screen) and mountain goat (dot screen) in North America (after Banfield 1974).

Interspecific Relationships with Small Herbivores

Snowshoe Hare

Moose and snowshoe hare are sympatric throughout much of their North American ranges (Figure 140). Their dietary preferences appear to overlap considerably, providing a

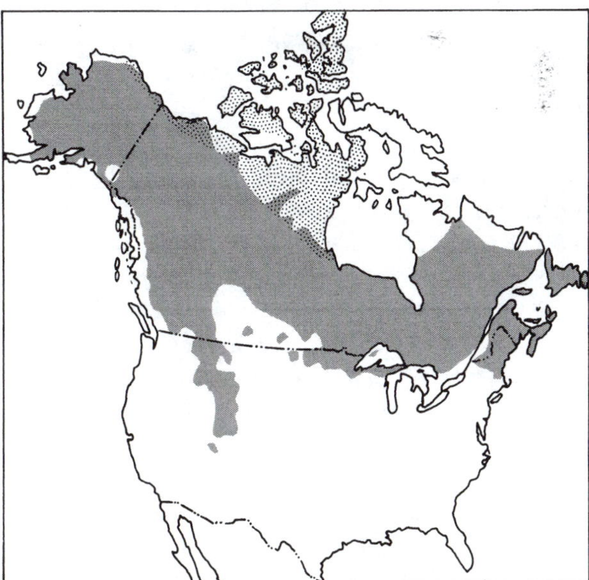

Figure 139. Separate and conterminous ranges of overlap of moose (solid screen) and musk-oxen (dot screen) in North America (after Banfield 1974).

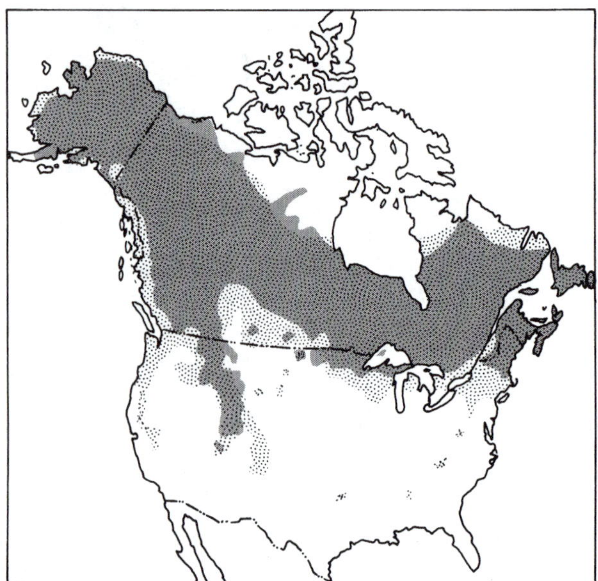

Figure 140. Separate and conterminous ranges of moose (solid screen) and snowshoe hare (dot screen) in North America (after Banfield 1974).

basis for potential competition. Moose in Newfoundland browsed on 30 woody plant species, 27 of which also were used by hares (Dodds 1960). The most important of the hardwood species to moose, white birch, was the most important of all species for hares (Wood 1974, Oldemeyer 1975). Telfer (1972) reported a high degree of winter browse overlap among moose, white-tailed deer and snowshoe hare in New Brunswick. Conifers accounted for nearly half of the browse taken by the hares and both ungulate species, with hares showing a preference for spruce, and ungulates preferring balsam fir. Red maple and wild raisin formed a substantial portion of browse taken by all three species.

Despite the extensive overlap of moose and snowshoe hares, there are few reports of moose populations being adversely affected by competition with hares. Wolff (1980), for example, suggested that a moose population decline in the interior of Alaska during the early 1970s was caused, in part, by high hare densities. The greatest decline in moose numbers occurred in 1970–1971, the prepeak year in the hare cycle. At that time, moose mortality approached 50 percent, and production fell to 6 yearlings per 100 cows. Conversely, on the Kenai Peninsula of Alaska, competition between moose and snowshoe hares was not deemed a serious influence on moose numbers and well-being (Oldemeyer 1975).

Bergerud and Manuel (1968) believed that in central Newfoundland reductions of available birch browse and conifer cover attributable to moose browsing probably reduced snowshoe hare densities.

Although moose and snowshoe hares display generally similar dietary preference, their shared food resources may be partitioned. In light of the two species' dramatically dissimilar body sizes, the most obvious partition is that of foraging height (Telfer 1972, Oldemeyer 1975). However, Belovsky (1984) found that moose and hare consumed vegetation of similar height 91 percent of the time during winter and summer. He concluded that height may be inadequate to explain food partitioning because (1) twig utilization is moderated by snowfall, inasmuch as hares can walk on top of snow, and (2) both hares and moose prefer

Snowshoe hares and moose are sympatric through much of their individual ranges. Both are browsers and their diets overlap considerably. However, there are few reports of competition for food causing adverse effects on either species. In the relatively warm seasons, the animals occupy areas of extensive browse resources and their utilization of common plants is partitioned by height. Moose fed at heights greatly above the reach of hares. Only in winter periods of considerable snowfall is there significant opportunity for competition, because the hares, traveling atop the snow cover, sometimes reach browse at levels that moose feed at. Also, a moose exclosure at the Moose Research Center in Alaska shows all browse up to 3 feet (0.9 m) removed from a stand of young birch. This was accomplished by snowshoe hares that were not excluded from the stand. As snow depths increased the previous winters, hares progressively consumed the higher browse. *Photo by Albert W. Franzmann; courtesy of the Alaska Department of Fish and Game, Soldotna.*

the most nutritious plants, which often are younger and consequently closer to the ground. During a period of high density of snowshoe hares on the Kenai Peninsula, Alaska, in the early 1970s, a browse line caused by hares was evident. As snow depth progressed so did the browse line, up to 3 feet (0.9 m) (A. Franzmann personal communication: 1993).

Partitioning at the stand level may serve to limit competition between the two species, with hares preferring dense cover and moose selecting more open habitats. However, in late winter, moose tend to move into denser cover to avoid deep snow, which can result in greater competitive interaction. Dodds (1960) speculated that heavy moose browsing on balsam fir in cutovers reduced winter cover for hares, thereby preventing hares from inhabiting affected cutovers during the early years after cutting. Reciprocally, there also is potential for competition in years of peak hare density when they invade the more open moose habitat (Wolff 1980).

Beaver

Geographic distribution of moose closely approximates the northern range of beaver (Figure 141). Although the morphologies and life histories of these two herbivores are vastly different, the interspecific relationship between the species is an intricate combination of competition and commensalism. Also, beaver and moose often are linked in predator/multiprey systems.

Beaver activity creates habitat conditions beneficial to moose. Damming watercourses by beavers creates aquatic communities, and the associated vegetation is heavily browsed by moose in summer (Wolfe 1974). Submergent and emergent aquatics may be an important source of sodium and other minerals required by moose (Jordan 1987). Similarly, the cutting of mature aspen, a preferred food of beaver, promotes suckers and sprout growth, increasing browse availability to moose.

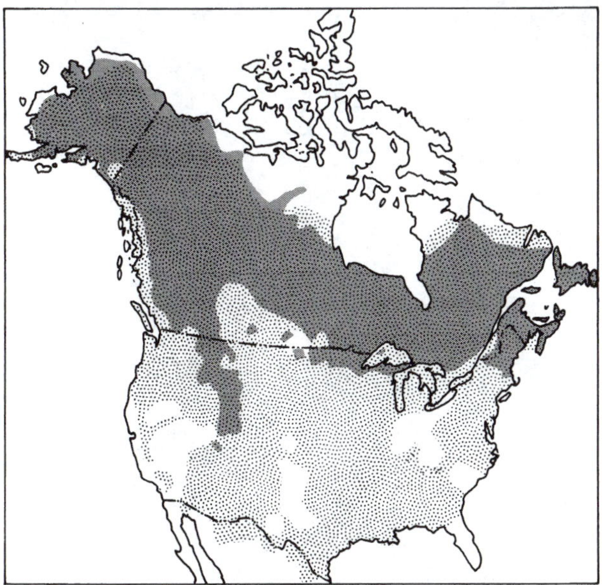

Figure 141. Separate and conterminous ranges of moose (solid screen) and beaver (dot screen) in North America (after Banfield 1974).

Moose use of aspen and other riparian and aquatic plants can represent competition with beavers. Moose browsing on regenerating aspen sprouts or birch may inhibit the growth of large trees required by beaver (Wolfe 1974, Shelton and Peterson 1983).

Correlation of beaver population size to moose numbers was evident in Newfoundland (Bergerud and Manuel 1968). In areas of moose density exceeding 12 per square mile (4.63/km^2), beaver colony density was as low as 0.1 per square mile (0.04/km^2). And, in areas where moose density was less than 6 per square mile (0.23/km^2), beaver colony density was as high as 1.2 per square mile (0.46/km^2).

Where moose, beaver and wolf ranges overlap, beaver may subsidize wolf populations during summer months (Mech and Karns 1977). As prey, beaver may reduce predation pressure on moose calves and other ungulates during ice-free periods. However, when beaver are not available in winter, wolves may focus on moose. Consequently, high beaver numbers in an area may increase losses of moose to predation in winter (Shelton and Peterson 1983).

Other Small Herbivores

Although locally important in a few areas, interactions between moose and livestock are of little consequence at a landscape level. Moose and cattle compete for willow forage in summer when cattle are moved into riparian areas at higher elevations in Colorado (R. Kufeld personal communication: 1995). Similar competition between cattle and moose during summer grazing periods, both directly for forage and indirectly for agricultural land, occurs in British Columbia (D. Blower personal communication: 1995) and Manitoba (V. Crichton personal communication: 1995).

Moose are occasionally a nuisance by damaging fencing for cattle and horses in Vermont and New Hampshire (C. Alexander and K. Bontaites personal communications: 1995). Accounts of amorous bull moose courting dairy cows have been reported in newspapers, but these incidents have occurred on the fringe of moose range and probably involve yearling bulls dispersing into areas without cow moose.

Summary

A survey of biologists working with moose in the jurisdictions encompassing its range is summarized in Table 25.

Table 25. Interspecific relationships of moose and certain other North American wildlife

State or province	Elk			White-tailed deer			Mule deer			Bison			Caribou			Musk-oxen			Beaver			Snowshoe hare			Other (cattle)			Source[b]
	O	R	P	O	R	P	O	R	P	O	R	P	O	R	P	O	R	P	O	R	P	O	R	P	O	R	P	
Alaska													Y	Pr	Uc							Y/S	Cp/Pr	Uc/S				Chuck Schwartz
																												Rod Boertje
																												Kris Hunbertmark
Alberta	Y	Pr	S	Y	Pr	S	Y	Pr	S				Y	Pr	S				Y	Cm	S	S	Cp	S				Gerry Lynch
British Columbia	Y	Cp	+				S	Cp	S				S	Cp/Pr	S										S	Cp	S	Dan Blower
Colorado	S	Cp	+	Y	Pm																	Y	Uk	Uc	S	Cp	++	Roland Kufeld
Manitoba	Y	Pr	S	Y	Pm	S							Y	Uk	Uc							Y	Uk	S	S	Uk	S	Vince Crichton
Michigan				S	Cp/Pm	S													S[c]	Cm	S	S[c]	Cp	S				Rolf Peterson
Minnesota				Y	Pm/Pr	S													S/Y[d]	Cm	Uc							Bill Peterson,
																												Mark Lenarz
Montana	S	Uk	S	S/Y	Uk	S	S	Uk	S													S/Y	Uk	S	S	Cp	S	Gary Olson
Newfoundland													S	Cp	S							Y	Cp	S				Shane Mahoney
New Hampshire				Y	Cp/Pm	+																			S	N	Uc	Kristine Bontaites
New York[e]				Y	Uk	+																Y	Uk	+				Alan Hicks
Northwest Territories										Y	Pr	–	Y	Pr	Uc	Y	Pr	Uc				Y	Uk	Uc				Bas Oosenbrug
Nova Scotia				Y	Pm	+																S	Cp	Uc				Tony Nette
Ontario				Y	Cp/Pm/Pr	+							S	Pm/Pr	S –				Y	Cm	+	Y	Cp	S				John McNicol
Quebec				S	Pm	S													S	Cm	S							Réhaume Courtois
Saskatchewan	Y	Cp	+	Y	Cp/Pm	+							Y	Pr	–							Y	Cp	+	S	Cp	S	Rhys Beaulieu
Vermont				Y	Pm	+													S	Cm	+				S[f]	N	+	Cedric Alexander
Yukon																						Y	Cp	S				Richard Ward

Interspecific relationship with moose[a]

[a] O = range overlap: Blank = none; S = seasonal; Y = year-round.

R = relationship: Cp = competitive; Cm =commensal; Pm = parasite-mediated; Pr = predation influence; N = nuisance; Uk = unknown.

P = future potential for or prospect of interaction: S = same; – = decrease; + = increase; ++ = serious increase; Uc = uncertain.

[b] Personal communication: 1995.

[c] Isle Royale National Park.

[d] Beaver interaction is both seasonal (creation of aquatic feeding sites) and year-round (regeneration of browse supply in areas with very high beaver populations is sometimes significant).

[e] Moose densities are extremely low.

[f] Competition is for space not food; potential moose range occupied by livestock brings species into conflict when moose tear down fences and horses and cattle escape.

White-tailed deer and snowshoe hare appear to have the greatest interaction geographically, followed by elk and caribou. Ecological relationships between and among sympatric species are complex; the nature and consequences of the interactions are dependent on characteristics of the shared habitat, relative population sizes and dynamics of the species involved. There is a danger in oversimplifying relationships, but a general pattern does emerge.

The parasite *P. tenuis* seems to dominate the relations between moose and white-tailed deer. Competition for browse between snowshoe hare and moose may occur at the high densities associated with the peak of the hare cycle. Elk also may compete directly with moose especially for willow forage, a staple winter food of moose where the two species overlap. As prey for wolves and bears, elk may affect moose indirectly, by maintaining relatively high predator populations. The same appears to predominate the interrelationship between caribou and moose.

There are surprisingly few recent studies of interspecific relationships between moose and other large and small herbivores. For many of the species (i.e., musk-oxen, mountain sheep, bison and livestock), this may mean that the relationship with moose is considered of relatively little consequence. However, the potential effects of white-tailed deer, snowshoe hare, elk, caribou and beaver sympatric with moose are more intriguing ecologically. Across landscapes being shaped by human use and by climatic change as well as temporal and spatial fluctuations of the interacting species, the interspecific relationships themselves can be expected to vary in relative ecological significance. Understanding processes in such a dynamic system will require carefully structured and long-term studies.

GOLDEN WASH ALONG HUMP – OVERLAY OF BLACK HAIRS (OR DARK BROWN)
RICH REDDISH-BROWN ON FACE, SHOULDER – SHADING TO DEEP PURPLE-BROWN · LIGHT BROWN ON LEGS BELOW – SMALLER BULL LIGHTER – MORE · YELLOWISH OR TAN

(COMP. FROM MEMORY)
ANTLERS OF LARGEST BULL REACHED BACK ABOUT EVEN WITH SHOULDER WHEN HEAD WAS IN LINE WITH BODY (FORWARD).
SLEEK, GLOSSY SUMMER COATS (ESP. DARK AREAS)

BROWSING IN WILLOWS (GREEN) – IGLOO CREEK, McKINLEY PARK, ALASKA
AUG. 12, '55

(MEMORY)

YOUNG COW MOOSE – TWO-TONE COLORATION (LIGHT BROWN + DEEP PURPLE-BROWN – BLACK APPEARING) IN WILLOWS ALONG STREAM BED – ABOVE TIMBERLINE ON SABLE PASS –

SUSPICIOUS MOOSE – LEFT OFF BROWSING TO STRIDE THROUGH WILLOWS, HEAD UP – (SMALLER BULL)
AUG. 12, '55 IGLOO CREEK

11

JAMES M. PEEK

Habitat Relationships

Moose successfully occupy a large variety of holarctic habitats. They exhibit patterns of habitat use that, during some periods, make them "generalists," i.e., able to use a variety of habitats equally well, and "specialists," i.e., adept at using certain habitats and not others, during other periods (Rosenzweig 1981). Categorizing moose as generalist or specialist ignores the ways that the species adapts to survive in many habitats, and can be misleading for habitat management. Moose might be considered "selective generalists"—capable of using forage and other habitat components in higher proportions than they occur in the environment, and adept at selecting seasonally advantageous habitats.

The following terms associated with ecological relationships of wildlife generally can be applied to moose in particular. *Habitat* simply is the place a moose lives—the sum of the inanimate and the living environment that comprises the space occupied. It represents especially an animal's food, water, shelter and space. *Habitat availability* relates to the accessibility and viability of a certain area for use by a moose. *Habitat use* generally relates to occupation by moose of a particular habitat relative to other available habitats. *Habitat preference* refers to the differential occupation or utilization of a certain habitat type and that type's occurrence within the home range of an individual moose or a moose population. A habitat may be "preferred" in some areas, although it may not occur within the range of other populations. For example, moose invariably select available aquatic habitats, but they also occupy areas where these habitats do not occur (Telfer 1967a).

Habitat selection implies that, of an array of habitats available, certain ones will be occupied and/or used more than others. Such selection by moose can be evaluated statistically. Measurements of use may be compared quantitatively with expected use according to that habitat type's relative availability within the environment or subject's range. Habitats that receive extensive use but, in terms of size and accessibility, are not used disproportionately to their occurrence may be very important to moose nevertheless.

Habitat requirement refers to a certain component or seasonal aspect that must be present in a habitat or the area cannot sustain moose. For example, moose living along the river delta systems draining to the Arctic Ocean are dependent on riparian communities; they likely would not colonize these high latitude habitats in the absence of a shrub complex on which they feed.

Habitat-use patterns are related primarily to the presence, absence and dynamics of forage and cover suitability and availability, but also are correlated to the presence of predators, pathogens, interspecific competition and conspecific population density. Physical factors, including snowfall accumulation, fire history (as it relates to forage and cover), surface water and temperature all affect moose habitat use. Human activity also influences habitat selection and use by moose.

There is sufficient knowledge of moose habitat-use patterns to outline the general circumstances for most regions. However, there always will be a need to examine how these patterns are elaborated in local situations, how they change as habitat components change, and whether such changes ultimately are beneficial, detrimental or inconsequential to moose. Substantial knowledge also exists to explain why moose use certain habitats and not others, although more understanding of cause/effect relationships affecting habitat use is desirable.

Characterization of Moose Habitats

Geist (1971) recognized both stable and transitory habitats as important in the evolution of moose. Permanent habitats are those that persist through time without alteration in their kind or condition, such as riparian willow/poplar communities and high-elevation shrub/scrub communities that do not succeed to different kinds of vegetation. Alluvial habitats and stream valley shrub habitats are dynamic in the sense that flooding and consequent stream bed alteration produce a constantly changing system. Although these shrub communities may change with disturbance, they are relatively permanent. However, some are successional to deciduous or coniferous forest if left undisturbed (Houston 1968, Viereck 1970).

Telfer (1984) categorized the full range of moose habitats to consist of boreal forest, mixed forest, large delta floodplains, tundra and subalpine shrub, and stream valleys (Figure 142). Boreal forest habitats are considered fire-controlled and likely represent the major environments in which moose evolved (Peterson 1955, Kelsall and Telfer 1974). In the mixed coniferous/deciduous forests of the Maritime Provinces of Canada and the northeastern United States, insect outbreaks and plant diseases influence moose habitat by creating forest openings.

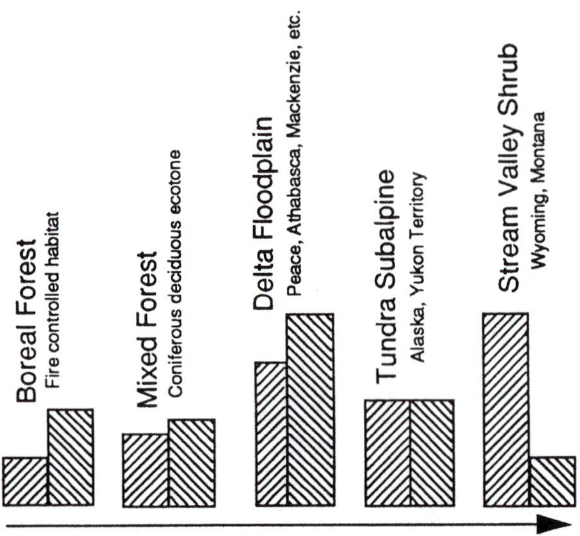

Figure 142. Hypothesized relative moose population density and stability in the species' major North American habitat types (after Telfer 1984).

The Koyukuk River drainage in Alaska, like other floodplain alluvial habitats, supports an abundant moose population because it features an abundance of nutritious food, primarily in the form of feltleaf willow stands of three ages. The different age stands were produced by regeneration after changes in the river channel. Willow, foreground, immediately back from the riverbank deposition of sand and silt, is 1 to 3 years old. Both of two successive stands are 1 to 3 years older. The upland trees are white spruce and paper birch. Although the nature of this riparian habitat changes, the river also changes levels and channels, eventually creating new riparian space and habitat. *Photo by Charles C. Schwartz; courtesy of the Alaska Department of Fish and Game.*

Alluvial habitats along floodplains and deltas provide a high biomass of forage. Here, deciduous shrubs proliferate because of the annual influx of nutrients from waterways and sufficient soil moisture. Alluvial river systems in North American moose range include the Saskatchewan, Peace-Athabasca, Slave, Yukon, Copper, Mackenzie and Kuskokwim. The Jackson Hole region in Wyoming and the Big Hole and Red Rock lakes regions in Montana also may be included in this large-scale floodplain classification, although they essentially are stream valley systems characteristic of mountainous regions. These habitats are more permanent than the important forage-producing habitats in boreal or mixed forests.

Stream valley shrubs characteristic of the Rocky Mountains from Colorado and Utah through British Columbia and Alberta to the Yukon Territory and Alaska represent another identifiable moose habitat. These habitats contain permanent vegetation and are a smaller scale representation of the floodplain / delta habitats of larger rivers.

The aspen parkland region of northwestern Minnesota, Manitoba and Saskatchewan is dominated by aspen, willow and balsam poplar, and is transitional to hardwood and conifer communities. However, it is comparable to the alluvial habitat category because of its high forage biomass and permanence (Berg and Phillips 1974). The northernmost aspen parklands are relatively permanent moose habitat because of a short fire cycle, but they still represent an ecotonal condition between prairie and coniferous forest (Rowe 1982).

The climax tundra and subalpine shrub communities at higher latitudes and elevations in the Northwest Territories, Yukon and Alaska are more stable in time and space than the alluvial and riverine habitats (Viereck and Dyrness 1980). Prehistorically, when moose colonized recently unglaciated areas (Geist 1971, Kelsall and Telfer 1974), they likely encountered plant communities similar to the subalpine shrub habitat they now occupy.

Transitory habitats of moose include boreal forests where fire creates seral shrub communities that provide extensive forage. Geist (1971) hypothesized that "islands" of permanent habitat found along watercourses and deltas, and the high elevation dwarf shrub communities, serve as refugia where moose populations persist and from which they expand into transient habitats created by fire. The common occurrence of fire in boreal forests is considered sufficient to promote adaptations favoring dispersal of yearling moose to newly created habitats. Cronin (1992) suggested that the high mobility of moose leads to the genetic homogeneity he observed. However, moose likely occupied refugia farther south as well (Peterson 1955), and other reasons may exist for the tendency of moose to disperse.

Alluvial habitat in the Jenny Creek floodplain of Denali National Park and Preserve in central Alaska is important winter habitat for moose because of its high biomass, including abundant feltleaf willow. These habitats, although smaller in size than the floodplains of major rivers, support extensive moose populations. *Photo by R.A. Riggs.*

Important winter range for Shira's moose is typified by riparian willow stands, such as shown on the upper Ruby River in Montana. *Photo by James M. Peek; courtesy of the University of Idaho, Moscow.*

Shrub/scrub communities at high altitudes, such as featured here in Denali National Park and Preserve, Alaska, are examples of stable, relatively permanent moose habitats uninfluenced by fire. These low shrub communities are especially preferred by moose during the rut. *Photo by James M. Peek; courtesy of the University of Idaho, Moscow.*

Seral shrub communities created by logging or burning boreal forest constitutes important moose habitat. These bulls are using a cutover stand in northeastern Minnesota, opened in much the same fashion as would have occurred if the area had been subject to a wildfire. Use of these communities by moose after the rut and before periods of deep snow in midwinter is very important. *Photo by James M. Peek; courtesy of the University of Minnesota, St. Paul.*

Peek's (1974a) classification of winter habitats of the Yellowstone subspecies included willow bottom/stream conifer complexes along high-gradient streams (Table 26). Floodplain riparian communities such as in the Jackson Hole, Wyoming, region (Houston 1968) and the Big Hole in Montana can support a year-round resident moose population as well as a wintering population. In portions of central Idaho and Montana, moose may occupy primarily mature coniferous forest (Stevens 1970, Pierce and Peek 1984, Matchett 1985). These areas occur where riparian habitat is less extensive, and are distinguished from high elevation ranges in spruce/fir communities by their greater diversity of forage species. Unique habitats occupied by the Shira's moose include the aspen communities in southeastern Idaho and adjacent Wyoming and Utah, and in the arid sandhill region where desert shrubs, such as bitterbrush, and dryland riparian shrubs, such as chokecherry, are important in the diet (Ritchie 1978).

Peek (1974a) postulated that stream valley habitats within the range of the Yellowstone moose are sufficiently permanent to have influenced the behavior, physiology and morphology of moose that evolved with them. However, based on an examination of mitochondrial DNA, Cronin (1992) suggested that moose were remarkably similar in genetic make-up across North America. This could indicate that any observed differences are likely phenotypic responses to local habitat and climatic regimes. Differences in breeding systems (see Lent 1974) and aggregation patterns (see Peek et al. 1974) may be related to moose adaptation to permanent versus transient habitat as well as open versus forested terrain. However, in light of Cronin's (1992) findings, such adaptations likely reflect phenotypic rather than genetically related responses.

Fire-influenced Habitats

Since the earliest efforts to understand the ecological relationships of moose in North America, the species has been

High-elevation coniferous forests provide winter moose habitat in central Idaho and southwestern Montana. Moose are able to subsist on the conifers, bounding from one tree well to the next. Subalpine fir, Douglas-fir, lodgepole pine and Pacific yew are the conifer species used in these severe wintering situations. These habitats are particularly critical for moose in areas where willow bottoms are scarce or absent. *Photo by James M. Peek; courtesy of the University of Idaho, Moscow.*

associated with postfire habitats. Aldous and Krefting (1946) recognized the importance of fire in maintaining shrub habitats that were responsible for an increase of moose populations on Isle Royale, Michigan, in the 1940s and 1950s. Spencer and Chatelain (1953) observed an increase in moose populations on the Kenai Peninsula in Alaska after an extensive wildfire in 1947. Peterson (1953) emphasized that favorable moose habitats in Ontario were seral forests with mixed coniferous/deciduous stands. The British Columbia studies by Hatter (1949) and Cowan et al. (1950) reported that the variety of palatable deciduous species typical of the seral brush fields produced most of the forage for moose.

Geist (1971) argued that the high reproductive potential of moose compared with that of mountain sheep and the proclivity for subadult moose to disperse were related to the transient nature of boreal forest habitat. Fires create high-quality forage sources that are best used by populations capable of rapidly colonizing postfire habitats. The association of moose with early successional stages of postfire succession in boreal forest habitat is commonly accepted. However, whether or not moose occupying transient boreal forest habitats have better developed mechanisms for dispersal than do moose occupying more permanent habitats is untested. Comparative studies of the dispersal of yearling bulls and cows are needed (see Chapter 10).

Bendell (1974) examined the relationship between moose and postfire environments, using the Kenai Peninsula of Alaska as an example. The region had a long history of fire, and moose populations increased as postfire shrub communities developed (Lutz 1960, Spencer and Hakala 1964). However, Bendell (1974) pointed out that there were lags in response, and other factors interacted along with the postfire vegetative condition to affect moose response. First, an irruption of moose in 1920 was not correlated with production of shrubs resulting from fires that preceded the irruption. Also, 14 years after a large burn, when shrub production should have been high in the region, the moose population was down to pre-1920 levels. Finally, the population response to the 308,750-acre (124,960-ha) wildfire of 1947 lagged behind the presumed stage of highest shrub production.

Subsequently, Oldemeyer and Regelin (1987) discovered that vegetation on the Kenai Peninsula differed greatly

Table 26. Percentage winter habitat use by Shira's moose

		Percentage use							
Location	Habitat category	Willow	Aspen	Conifers	Open	Other	Total moose	Monitoring technique	Source
Gravelly Range, Montana	Riparian	79	11	10	0	0	85	Aerial observation	Stevens (1967)
Rock Creek, Montana	Riparian	34		27	18	21		Track location	Stone (1971)
Centennial Valley, Montana	Floodplain riparian	93	3	3			413	Ground observation	Dorn (1970)
Big Hole Valley, Montana	Floodplain riparian	90	1	0			223	Aerial observation	Stevens (1967)
Jackson Hole, Wyoming	Floodplain riparian	68	6	13	9	4	3,343	Ground observation, marked animals	Houston (1968)
Gallatin Mountains, Montana	Conifer	10	42	39	9	0	976	Aerial observation	Stevens (1970)
Yaak River, Montana	Conifer	0	0	100	0	0	56	Radiomarking	Matchett (1985)
Central Idaho	Conifer	0	0	100	0	0	155	Radiomarking	Pierce and Peek (1984)

among soil types (Table 27). Density and frequency of willow and paper birch—staples in the moose diet—were higher on loamy soils than on gravelly soils. Aspen and mountain cranberry—palatable to moose but not as prevalent as willow or paper birch—were more abundant on gravelly soils. Thus, shrub populations that have undergone similar burning regimes may respond differently depending on soil type, and moose response may alter their foraging and other habitat use accordingly. Predation and hunting also may have interacted to prevent direct moose response to presumed increases in forage.

Radio-collared moose used remnant forest more than burned forest on a 86,450-acre (34,985-ha) 11-year-old burn (1969) and a 308,750-acre (124,960-ha) 33-year-old burn (1947) on the Kenai (Bangs et al. 1985). Moose using the 11-year-old burn were located most frequently within 100 yards (91 m) of the forest edge, which Neu et al. (1974) also reported for a Minnesota burn. Use of portions of shrub fields near conifer cover is expected, especially when snow is deep and restricts moose movement. Schwartz and Franzmann (1989) reported that, after a 1969 burn, moose increased to high levels approximately 15 years later, when browse plants reached maximum production, although population peaks were lower when wolves and black bears were present (Figure 143). The increase was ascribed to high production and low mortality, with some initial shifting of home ranges from adjacent high-density populations. Where fire was absent for at least 25 years, moose densities on the Kenai Peninsula were sufficiently high to cause the forage base to shift from a multispecies complex to a much less diverse community dominated primarily by white birch (Oldemeyer et al. 1977). Kelsall et al. (1977) concluded that the optimum successional stages for moose were 11 to 30 years after burning in boreal forests.

The relationship of moose population characteristics to postfire vegetative responses has received little attention.

A fire of about 483 square miles (1,250 km^2) on Alaska's Kenai Peninsula in 1947 created excellent moose habitat. Because the wildfire burned during a cool summer, an extraordinary amount of edge cover resulted. The interspersion of aquatic area, burned forest and unburned mature tree stands produced a moose population of 9.3 per square mile (3.6/km^2). The peninsula has subsequently burned several times, but not as extensively as in 1947. *Photo by Charles C. Schwartz; courtesy of the Alaska Department of Fish and Game.*

Conceptually, populations that exhibit low productivity and exist in marginal habitats are unlikely to respond by increasing as much as would populations in productive high-quality habitats. Two evaluations have produced different results, which may be attributable to small sample sizes and

Table 27. Density of shrubs and saplings by soil phase on the Kenai National Wildlife Refuge in Alaska (after Oldemeyer and Regelin 1987)[a]

		Density in stems per acre (ha)				
Soil phase	Number of stands	Black spruce	White spruce	Aspen	Willow	White birch
SO1—loamy on nearly level to rolling topography	45	405 (164)	781 (316)	914 (370)	6,952 (2,813)	2,673 (1,082)
SO2—loamy on hilly to steep topography	62	2,492 (1,008)	165 (67)	213 (86)	1,315 (532)	520 (210)
SO5—gravelly on hilly to steep topography near outwashes	36	2,378 (962)	18 (7)	1,666 (674)	231 (93)	292 (118)
SO5—gravelly on hilly to steep topography near moraines	32	2,557 (1,035)	309 (125)	2,238 (906)	827 (335)	1,065 (431)

[a] These stands include burn sites and those that were mechanically crushed to rejuvenate forage.

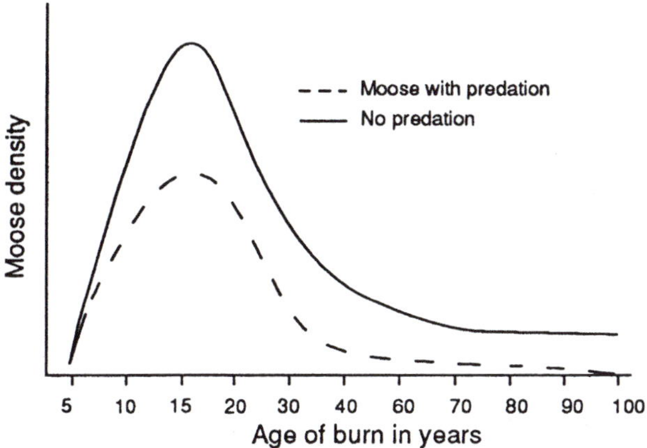

Figure 143. Relationship of moose density and forest succession in the presence and absence of predators after a burn on the Kenai Peninsula, Alaska (after Schwartz and Franzmann 1989). The extent to which predation limits moose population growth affects moose response, ranging from no response to some level represented within the range indicated by the dotted line.

different analytic methods, but also may be real (Peek 1974b, Gasaway and Dubois 1985, Gasaway et al. 1989). Nevertheless, conclusions about moose population response beyond these case studies are questionable.

Assuming that Peek's (1974b) evaluation represented a population response, then a higher proportion of yearling bulls existed on the Little Sioux Burn in northern Minnesota than adjacent to the burn. Statistical treatment showed significant differences in habitat selection, with more use of the burn periphery than in adjacent unburned portions (Neu et al. 1974). Samples were small, and supporting information would have been desirable (Figure 144).

Studies of moose inhabiting portions of the Tanana River basin in Alaska, involved 7 and 14 radio-collared moose

Vegetative response to fire in the boreal forest is fairly predictable but varies according to fire intensity and duration, which generally are conditioned by atmospheric conditions, fuel type and accumulation and soil characteristics. Peak forage production usually occurs after about 15 years postfire. Thereafter, stems of the maturing aspen, birch and willow regrowth exceed the reach of browsing moose, and the moose population productivity and density consequently decline. *Photo by Charles C. Schwartz; courtesy of the Alaska Department of Fish and Game.*

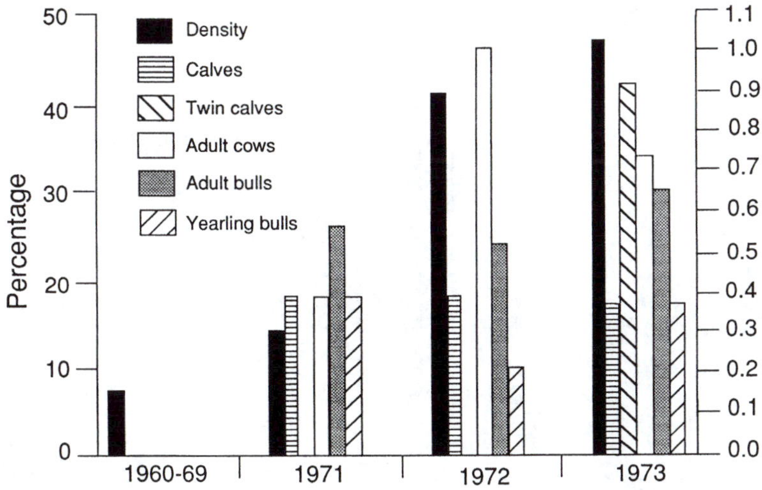

Figure 144. Midwinter sex and age composition and density of a moose population before and after the autumn 1971 Little Sioux burn in Minnesota (Peek 1974b). Twin calves are expressed as a percentage of all calves.

(Gasaway et al. 1989). No movement onto the burns or other significant population responses were documented in the Alaska study areas, and the habitat-use pattern was essentially the same before the burn as it was for the 5 years after the fire (Figure 145). Subadult moose were not exploiting the Tanana River burn at a higher rate than adults did. Gasaway et al. concluded that, contrary to the results of the Minnesota investigation, subadult moose did not exploit the Tanana River burn at a higher rate than adults. These studies suggest that, for burns to be colonized promptly, they must occur where moose are abundant. Burns along major migration corridors where density of nonmigratory moose populations are low but migratory populations are high also may be rapidly colonized.

Both the Alaska and Minnesota cases were fortuitous because fires occurred when and where investigators could quantify responses. The investigations did not provide information that can be reliably correlated to other situations, even in the same area at another time. However, if conclusions of both studies accurately reflected the separate moose population responses, then the Minnesota population produced dispersers that were colonizing the burn, whereas the Alaskan population did not colonize the burn there. The relatively high moose densities adjacent to the Minnesota burn may have accounted for the rapid colonization of that burn. Vegetative response in Minnesota was remarkable, with willow shoots growing 10 feet (3 m) high within 2 years after the burn on some sites.

Predictions of moose population response to postfire environments must consider population characteristics as well as biotic regeneration. Schwartz and Franzmann (1989) suggested that predation may influence population responses in such areas. Although habitat quality may increase after fire, low-density moose populations with high predation mortality may not be able to respond to improved habitat conditions. Furthermore, dispersion may be minimized if forage is adequate.

One hypothesis for yearling moose dispersal is that the animals may be attracted to a combination of food and cover in new areas that resemble that of the natal home range from which they strayed or were driven by social pressure or predation. Also, if the original home range lacks suitable habitat resources, then dispersal to an area of improved resources may simply be adaptive. Another set of hypotheses indicates that predation and conspecific intolerance may encourage yearling dispersal regardless of the natal home range habitat quality. Likely, one or more factors influence moose selection of home range (see Chapters 6 and 9).

Franzmann and Schwartz (1985) suggested that increased calf production was responsible for increases in moose density observed 13 years after the 1969 Kenai Peninsula burn, at a time when browse production was peaking on that burn (Oldemeyer and Regelin 1987). Subsequently, Schwartz and Franzmann (1989) concluded that (1) habitat within the 1969 burn was deteriorating, (2) an area burned in 1947 supported only a remnant moose population and (3) habitat alteration again was needed on both areas to assure a high density moose population and the complementary component of predators.

On Isle Royale, where moose had persisted for an 80-year period at relatively high density, Pastor et al. (1988) reported that forest composition and soils were affected by browsing. In exclosures that protected against browsing for at least 40 years, mountain ash, mountain maple, trembling aspen and white birch were more abundant than outside the exclosures. The intensive browsing created relatively open stands dominated by white birch or white spruce (Figure

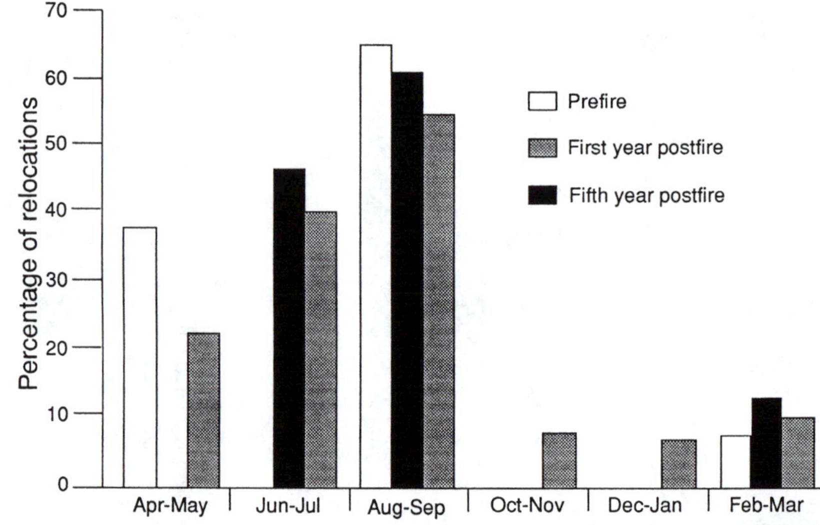

Figure 145. Locations of radio-collared moose before and after the Tanana Flats, Alaska, fire of 1980 (Gasaway et al. 1989). The postfire data were taken from May 1980 to March 1985. The data do not suggest that moose were responding to the fire at any season at 1- and 5-year intervals after the fire, when compared with the prefire distribution. The rate and degree of plant growth after burning influences moose population response to use of the burn.

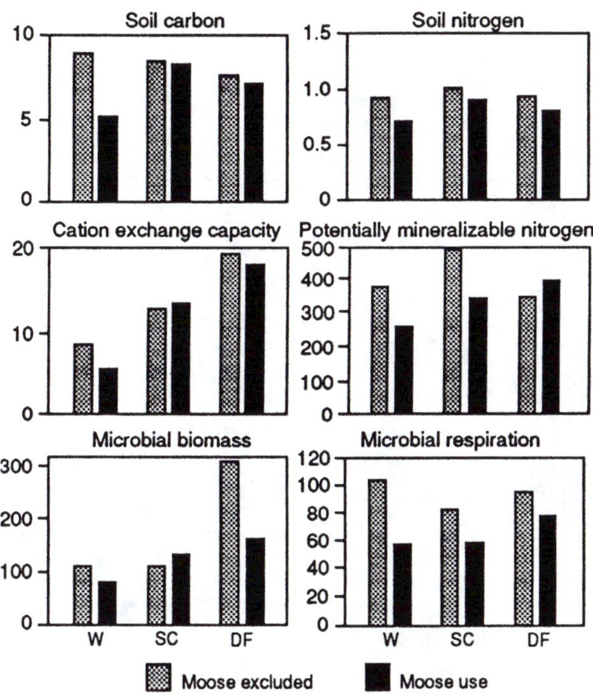

Figure 146. Comparison of soil properties on sites used by moose and areas from which moose were excluded for 40 years on Isle Royale, Michigan (Pastor et al. 1988). The three sites are Windigo (W), Siskiwit Camp (SC) and Daisy Farm (DF). Measurements of soil carbon (percentage), soil nitrogen (percentage) and microbial respiration (micrograms of carbon dioxide–carbon per gram of total carbon) were consistently higher on exclosed sites. Cation exchange capacity (milliequivalents per 100 grams of soil), potentially mineralizable nitrogen (milligrams of nitrogen per gram of total nitrogen) and microbial biomass (micrograms of carbon per 100 grams of soil) were higher in two of three excluded sites. These measures of soil fertility and microbial activity suggest that persistent heavy browsing pressure can reduce soil substrate quality and, ultimately, forest structure and productivity.

Early seral stands of vegetation caused by fire or logging produce a wide diversity and large amount of high-quality food for moose. The intensity of disturbance primarily determines the rate and productivity of the regrowth. In early summer, moose prefer stands of seral vegetation, along with aquatic areas and upland stands of mature forest. Use of these open areas declines during summer, but increases again after the rut. Moose continue to use these areas of high forage biomass during winter, until deep snows hinder movements and access to the plants. *Photo by Charles C. Schwartz; courtesy of the Alaska Department of Fish and Game.*

146). Soil carbon, nitrogen and mineralizable nitrogen were reduced in heavily browsed areas, wherein less mineralizable nitrogen was available and microbial biomass was reduced. The general decline in nitrogen availability could eventually reduce soil productivity and accelerate succession toward spruce—possibilities that warrant more investigation. In communities dominated by spruce/birch/fir, moose browsing was reducing the rate of succession on Isle Royale, according to Risenhoover and Maass (1987), who also reported greater opening and diversity of browsed forest compared with unbrowsed forest. Only American yew and red-osier dogwood appeared to have declined as a result of browsing. The Isle Royale findings were consistent with reports from Quebec (Bedard et al. 1978) and Newfoundland (Bergerud and Manuel 1968), where growths of paper birch and balsam fir were retarded by browsing, resulting in a more open forest.

Risenhoover and Maass (1987) reported major temporal shifts in moose concentrations on Isle Royale. An extreme burn in 1936 provided important winter range for moose from 1940 through the mid-1950s, but became less important as succulent vegetation grew out of reach (Krefting 1974a). By 1970, there was very little use of this burn area during winter (Peterson 1977); the moose occupied areas where forage supplies were greater and more readily available. Still, Risenhoover and Maass (1987) reported that, whereas reduced vertical growth of vegetation was a result of browsing, nothing indicated that stem densities of preferred species were adversely affected by moose browsing (Table 28).

Table 28. Changes in average height in inches (cm) of selected plant species in exclosure (E) and control (C) plots at four sites in Isle Royale National Park, Michigan, 1949–1982 (after Risenhoover and Maass 1989, Krefting 1974a)[a]

		Daisy Farm		Windigo		Siskiwit Camp[b]		Siskiwit Lake	
Species	Year	C	E	C	E	C	E	C	E
Balsam fir	1949	6.1 (15.5)	7.3 (18.5)	15.8 (40.1)	7.3 (18.5)				
	1966	65.9 (167.4)	59.8 (151.9)	119.4 (303.3)	24.4 (62.0)	51.2 (130.0)	21.9 (55.6)		
	1982		227.2 (577.1)	65.7 (166.9)	343.4 (872.2)	21.9 (55.6)	84.2 (213.9)	16.4 (41.6)	
Mountain maple	1949		7.3 (18.5)	14.6 (37.1)	14.6 (37.1)			7.3 (18.5)	10.9 (27.7)
	1966				68.3 (173.5)	36.6 (93.0)		45.1 (114.5)	15.8 (40.1)
	1982				47.0 (119.4)	20.3 (51.6)		44.7 (113.5)	12.0 (30.5)
Serviceberry	1949		12.2 (31.0)		12.2 (31.0)	20.7 (52.6)			
	1966		60.9 (154.7)	60.9 (154.7)	47.6 (120.9)		12.2 (31.0)		
	1982			100.2 (254.5)	38.1 (96.8)		14.9 (37.8)		
White birch	1949	7.3 (18.5)	18.3 (46.5)	18.3 (46.5)	17.1 (43.4)	41.4 (105.1)	34.1 (86.6)	45.1 (114.5)	45.1 (114.5)
	1966	95.1 (241.5)	50.0 (127.0)	97.5 (247.6)	48.8 (123.9)	232.9 (591.6)	120.7 (306.6)	229.2 (582.2)	192.6 (489.2)
	1982		405.3 (1,029.5)	260.0 (660.4)	55.2 (140.2)	560.0 (1,422.4)	199.3 (506.2)	577.3 (1,466.3)	568.9 (1,445.0)
Red-osier dogwood	1949	12.2 (31.0)	4.9 (12.4)	10.9 (27.7)	10.9 (27.7)	9.8 (24.9)	20.7 (52.6)	12.1 (30.7)	
	1966	71.9 (182.6)	26.8 (68.1)	53.6 (136.1)	19.5 (49.5)	51.2 (130.0)	46.3 (117.6)		
	1982	38.9 (98.8)	30.0 (76.2)			37.3 (94.7)	27.8 (70.6)	41.6 (105.7)	
White spruce	1949	3.6 (9.1)		29.2 (79.2)	63.4 (161.0)		12.2 (31.0)		
	1966	34.1 (86.6)		192.6 (489.2)	190.2 (483.1)		147.5 (374.6)		
	1982	71.5 (181.6)	350.0 (889.0)	566.8 (1,439.7)	620.0 (1,574.8)		124.6 (316.5)		
Aspen	1949	34.1 (86.6)	13.4 (34.0)	23.2 (58.9)	17.1 (43.4)	34.1 (86.6)	58.5 (148.6)	36.6 (93.0)	
	1966	23.6 (59.9)	70.7 (179.6)	110.9 (281.7)	48.8 (123.9)	341.4 (867.1)	206.0 (523.2)		
	1982	1,200.0 (3,048.0)	15.6 (39.6)			303.1 (769.9)	184.8 (469.4)		21.6 (54.9)
Mountain ash	1949		4.9 (12.4)	20.7 (52.6)	15.8 (40.1)			52.4 (133.1)	
	1966		30.5 (77.5)	101.2 (257.0)	41.4 (105.1)			353.5 (897.9)	63.4 (161.0)
	1982		29.8 (75.7)	40.1 (101.8)	38.5 (97.8)	19.8 (50.3)	9.6 (24.4)		50.8 (129.0)

[a] Significant differences between exclosure and control plots at $P < 0.05$.

[b] Data for Siskiwit Camp collected in 1950.

Whether Isle Royale represents a case history of a natural grazing system or not seems debatable. The so-called "Krebs effect" (Krebs et al. 1969)—confining animals into a small space causes them to overuse resources—may apply, although Isle Royale, at about 210 square miles (544 km^2), seems a fairly sizable space. A major concern is that exclusion of fire for more than 50 years has produced an atypical pattern of moose habitat use, with resultant environmental effects that would not occur under more natural conditions (Heinselmann 1973). The fact that Isle Royale has supported high densities of moose for more than half a century may reflect their inability to disperse from the island to other suitable habitat. In the natural situation, lack of forage abundance or quality, typically attributable to severe winters with deep snow in boreal forest, forces moose to disperse to other habitats. This has been observed to some extent on Isle Royale, but may be more frequent in larger blocks of habitat.

Many moose populations in eastern Canada winter in "yards" (Proulx 1983, des Meules 1964, Telfer 1967a, 1967b, 1970). Proulx (1983) reported that 42 yards in the boreal forest of southern Quebec were characterized by a mix of relatively closed canopied, multilayered conifer forests (41–80 percent) with heights of 30 to 60 feet (9.1–18.3 m), interspersed with young stands rich in browse. White birch, balsam fir and white spruce were the dominant overstory species. Yards tended to be situated on gentle slopes, which helped to reduce the animals' energy expenditures in deep snow (Table 29). Fire, insect outbreaks and diseases were natural forces that created diversity within these forests, which since has been diminished by logging.

Overstories in study areas of southwestern Quebec investigated by Crête and Jordan (1982a, 1982b) were extensively altered by spruce budworm epizootics and contained high quantities of palatable shrubs. The investigators examined 18 moose yards, with densities ranging from 0.15 to 1.00 moose per square mile (0.4–2.6/km^2). In January, moose preferred open stands with abundant browse. In March, moose preferred closed stands. Stand selection and

Table 29. Percentage vegetative composition of four typical winter yards of moose in southern Quebec (adapted from Proulx 1983)

	Winter yards			
General aspect	1	2	3	4
Mean area	0.02 (0.05)[a]	0.04 (0.1)[a]	0.01 (0.03)[a]	0.01 (0.03)[a]
Coniferous forest	32	27	24	9
Deciduous forest	22	13	10	14
Mixed/coniferous[b]	9	16	21	24
Mixed/deciduous[c]	28	27	32	39
Nonwoody areas	2	2	4	3
Regeneration	7	15	9	11

[a] In square miles (km^2).

[b] Mixed, dominated by conifers.

[c] Mixed, dominated by deciduous trees.

browsing intensity were not related to moose density. Winter forage was not limiting to moose in these areas. Messier and Crête (1984, 1985) subsequently concluded that wolf and black bear predation was depressing these populations to levels below that which would be indicated if forage was a limiting factor.

The Cobequid Hills portion of the Atlantic Uplands of Nova Scotia are subject to damage from wind, fire (Rowe 1982), tree diseases and insects. These forms of disturbances create stands of palatable browse species such as white birch, red maple, aspen and red oak. Telfer (1967a, 1967b) reported that moose winter yards in this region were at higher elevations, in deeper snows and more discontinuously coniferous than yards in the region occupied by white-tailed deer. Telfer (1967a) noted that the spatial separation of moose from white-tailed deer in winter may help reduce the incidence of moose mortality attributable to a brain worm parasite that is indigenous to the deer population but is pathogenic to moose (see Chapter 15). Telfer (1967a) reported low communal habitat use by moose and deer during late winter and early spring when brain worm infection is most virulent. Also, although Proulx (1983) reported that winter yards in Quebec typically are on level ground with little relationship to exposure, Telfer (1967b) indicated that south facing slopes were primary moose winter range in the Cobequid Hills region.

Telfer's (1967b, 1970) investigations in central New Brunswick indicated habitat separation by deer and moose similar to that in Nova Scotia's Cobequid Hills (Figure 147). The forests of central New Brunswick consist of red spruce, black spruce, balsam fir and larch, and have been burned and logged extensively. Kelsall and Prescott (1971) reported that moose were wintering in the high elevation areas of Fundy National Park in snow depths that precluded use by white-tailed deer. In this area, the greater the snow depths, the greater the separation of species. This phenomenon was contrary to the findings of Jenkins and Wright (1987) in northwestern Montana, where increased dietary overlap in similar habitats was found between moose and white-tailed deer during more severe winters.

Moose habitat-use patterns in boreal forest were investigated in northeastern Minnesota by Peek et al. (1976). Historically, fire was a significant ecological force, but at the time of investigation in the 1970s, logging was more important in influencing habitats. Nevertheless, habitat-use patterns exhibited by moose were similar to those in a mosaic created by fire. Sufficient diversity of the forest pattern allowed for a wide latitude in selection of habitats. In early summer, aquatic areas, aspen/birch and upland stands of

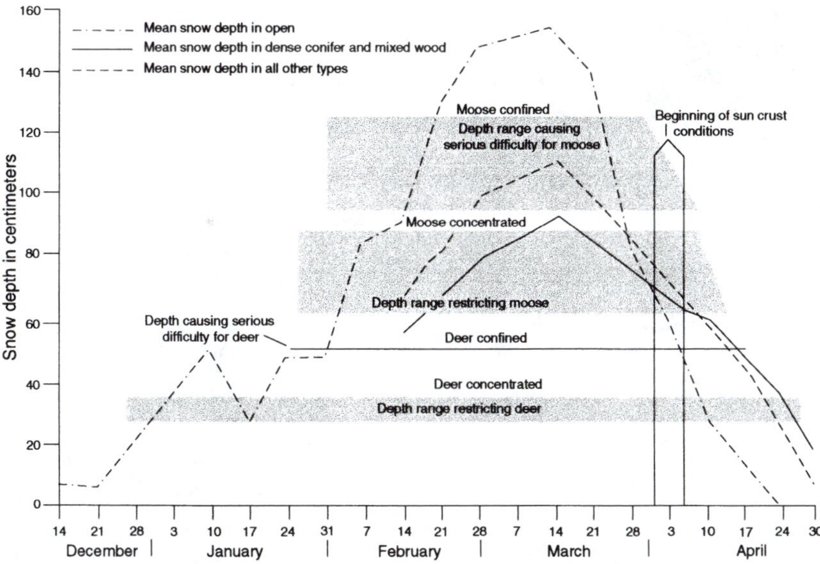

Figure 147. Winter 1966–1967 snow depth information for the Acadia Forest Experiment Station in New Brunswick (Telfer 1970), featuring a typical snow pattern for moose range in the maritime provinces and the relative hindrance effect of snow depth on deer and moose.

mature balsam fir received extensive use. The more open stands with shorter and perhaps younger trees were preferred habitat. In mid- and late summer, moderately stocked upland stands of aspen and white birch and lowland stands of black spruce, aspen and balsam fir on lowlands were intensively used. By late summer, the more mature stands appeared to be preferred. During prerutting activities (September 7–21), sparsely stocked stands were preferred, probably because the relatively open forest enhanced visibility.

Moist, lowland habitats were important during the actual rut. After the rut, moose preferred open habitats containing the highest forage biomass. In early winter, the open communities still were preferred, but as snow accumulated during midwinter, moose gradually shifted to closed canopy stands dominated by balsam fir and black spruce. The use pattern (Figure 148) was one of high forage-producing, low canopy habitats in spring and early summer and again in late autumn and early winter, shifting to denser cover in late summer and in midwinter. Habitat use appeared to be governed primarily by the availability of palatable forage except when winter severity forced occupation of closed canopy forest stands. In this subboreal forest region where vegetation is influenced by the atmospheric moderations of Lake Superior, moose use of spruce/fir communities typical of those farther north was common, whereas pine-dominated stands more typical of mesic conditions in the region were used much less frequently.

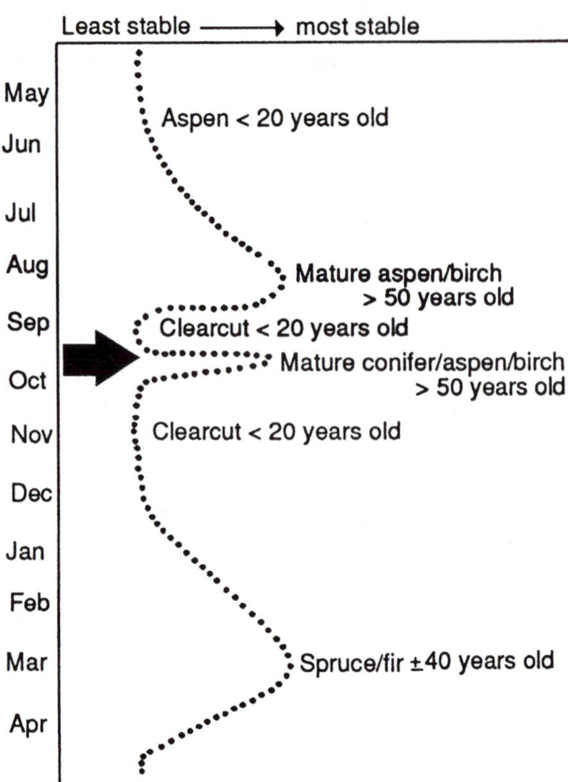

Figure 148. Selection of various moose habitats according to their relative stability in northeastern Minnesota (Peek 1974a). The arrow identifies habitat use during the rut, when older forests are preferred. Although older forests are considered more stable than those in younger, more rapidly changing successional stages, they also are subject to periodic fire; therefore, the entire range of habitat must be considered dynamic.

Upland stands of mature aspen and white birch are used extensively by moose in late summer and autumn. These communities afford a reasonable amount of forage, particularly along their edges, and some degree of cover. The overstories retard plant development, and thus provide palatable forage when more open areas are desiccated. *Photo courtesy of the Alaska Department of Fish and Game.*

That moose shift into closed canopy habitat as winter progresses is supported by a number of investigations from across North America, including from Alberta (Rolley and Keith 1980), British Columbia (Eastman 1974), Wyoming (Houston 1968), Montana (Stevens 1970), eastern Minnesota (Peek et al. 1976), western Minnesota (Phillips et al. 1973), Ontario (Thompson and Vukelich 1981), Quebec (Crête and Jordan 1982a) and Nova Scotia (Telfer 1970).

Western Canada features a variety of moose habitats, ranging from the classic boreal lowland forest typical of interior Alaska and Canada to the moist, mountain forests influenced by the Pacific Ocean (Eastman and Ritcey 1987). Lowland boreal forest habitats extend eastward across Canada. Boreal and related subalpine forests occur in mountains of the cordillera of northwestern Canada. Boreal forests are characterized by long, cold winters with soft, relatively shallow snow, and fire. The dry interior forests of southcentral British Columbia support low-density migratory moose populations. Habitat-use patterns in that region are similar to those farther south, described by Stevens (1970). The wet interior region, which includes Wells Gray Provincial Park, is a spruce/fir forest at low to mid-elevations and cedar/hemlock vegetation in valley bottoms. Moose are primarily migratory in this mountainous region of deep snows. Fire is characteristic of spruce/fir habitats but not of cedar/hemlock. In mountainous terrain, regardless of the region, moose tend to occupy the higher elevations in late summer and early autumn, migrating into shrub fields of old burns or logging areas or into floodplain riparian vegetation in late autumn, and remaining there through the winter. Stevens (1970) reported that willow, aspen and conifer cover were used by moose as winter range in southwestern Montana. In early winter, moose may be found at high elevations in subalpine fir stands, but as snow depth increases, most move into floodplains. This pattern appears to be similar to that documented by Edwards and Ritcey (1956) for Wells Gray Provincial Park. In north-central British Columbia, moose preferred the high biomass 10- to 25-year-old shrub fields and partially logged coniferous forests (Eastman 1974). The productive life of burns in terms of forage production for moose is up to 35 years in this region.

The argument that habitat selection by moose is primarily associated with forage availability is emphasized further by an investigation of habitat use relative to an active open-pit mine in northcentral British Columbia (Westworth et al. 1989). Highest densities of preferred forage were found in a 10-year-old clearcut, which also was where the highest density of moose pellet groups were found. Telfer (1978a) reported that moose utilization of three Alberta study areas was related positively to biomass of winter browse. In mountainous southwestern Alberta, where high winds minimize snow accumulation, aspen groves and shrublands dominated by willows and water birch were important habitats for moose (Telfer 1978a).

Fire is an important source of moose habitat in boreal forests. Fires burn at different intensities, exhibit irregular patterns and produce a variety of plant communities. Investigations have suggested that each kind of community—including closed canopy conifer forest, mature deciduous stands of aspens and birches, aquatic areas and high forage-producing shrub fields—contributes to the quality of moose habitat. Logging, if accomplished in time, place and with sensitivity to conform with moose habitat needs, can create the plant community diversity that historically was afforded by fire. Of course, there are important local ecological differences that need to be recognized before logging is initiated. Also, creation or maintenance of diverse habitats for moose at all times of the year ensures viable habitat for a wide variety of other migratory and resident wildlife species as well.

The major gap in most previous studies is scale. Large-bodied mammals, such as moose, have large home ranges (McNab 1983), presumably required to meet all their life requisites (see Chapter 10). A moose's home range typically will have a mix of habitats with some communities being more important than others. Peek et al. (1976) suggested that a township-sized block (36 square miles [93.2 km^2]) of suitable moose habitat should contain 40 to 50 percent shrubfields, 35 to 55 percent mature aspen/birch forest and 5 to 15 percent spruce/fir stands.

Nonfire and Limited Fire Habitats

Although moose usually are associated with habitats that are frequented by fire, they also commonly occur in areas where fire is infrequent.

Investigations in central Idaho (Pierce and Peek 1984), the Copper River Delta of Alaska (MacCracken 1992), southeastern Alaska (Doerr 1983) and habitats at or above timberline in central Alaska (Miquelle and Van Ballenberghe 1989, Risenhoover 1989) illustrate that high-density moose populations can exist in areas where fire either is not a major influence or is detrimental to habitat supply. River delta systems, shrub/scrub communities at the northern fringes of moose habitat and some of the most mesic maritime forests of eastern Canada (Telfer 1984) are not as subject to wildfire as is the major portion of moose range.

In central Idaho, moose habitat is densely forested, steep terrain (Peek et al. 1987), where narrow high-gradient streams limit riparian vegetation. The double canopy forests typically contain Pacific yew as a subcanopy beneath taller conifer forests that are 150 years old or older and may be dominated by mixes of grand fir, Douglas-fir, western larch, lodgepole pine and Engelmann spruce (Crawford and Johnson 1985). Yews provide both food and cover for moose, especially in winter (Pierce 1984, Pierce and Peek 1984). The double canopies intercept significant quantities of snowfall and thus provide moose with the most suitable winter habitat in this region.

Where yew is not abundant in other areas of central Idaho, spruce/fir communities with more than 55 percent overstory closure are major wintering areas (Peek et al. 1987). Very often, overstories are mixtures of seral and climax species including subalpine fir, Engelmann spruce, western larch, lodgepole pine and Douglas-fir. Compared with areas occupied farther north and east, shrubfields

Retention of stands of mature conifers is critical to survival of moose where riparian communities are scarce in deep snow habitats of the Rocky Mountains. *Photo by W.E. Steuerwald; courtesy of the USDA Forest Service.*

created by logging or fire in this deep snow region are not used as extensively by moose.

In western British Columbia and portions of southeastern Alaska, moose are supported in coastal forests dominated by Sitka spruce, western hemlock and western red cedar (Eastman and Ritcey 1987). In southern portions of Alaska, moose occupy Douglas-fir and subalpine fir communities. These wet forests typically have deep, heavy snows, therefore moose use dense conifers and winter along riparian zones in floodplains. Winter cover is a primary limiting factor for moose distribution in this type of habitat.

In southeastern Alaska, the Thomas Bay moose population occupied a mixture of clearcuts, riparian habitats and hemlock/spruce forest (Doerr 1983). Radio-collared animals used glacial riverwash soils and tended to avoid muskeg. Hemlock/spruce forest adjacent to clearcuts and riparian zones were used extensively. Clearcuts were used more than unlogged forest. Unlogged river terrace forests were preferred over river terrace clearcuts, and clearcuts less than 30 years old were used more than high-volume old-growth forests. Moose along the Stikine River were found primarily in floodplain riparian habitats, including spruce river terrace forest, cottonwood forest, riparian sedge muskeg or mixtures.

In southcentral Alaska, the Copper River Delta is another moose habitat where fire is infrequent. Thelenius (1990) reported that the 1964 earthquake changed much of the delta vegetation from a wetland comprised of emergent plants and shrubs to a tall shrub community, as the uplifted portions of the delta drained. MacCracken (1992) reported that tall alder/willow, low sweetgale, Sitka spruce and wet forb communities are important to moose in the area. The taller shrub communities and woodland spruce serve as winter cover.

In the interior, mountainous portions of Alaska, shrub-dominated communities above timberline also are important moose habitat. Among the upland shrub communities recognized by Viereck and Dyrness (1980), those dominated by diamondleaf willow and Richardson willow appear to be most important to moose (Risenhoover 1989). Also important to moose were alluvial willow stands dominated by feltleaf willow, littletree willow, Richardson willow and some balsam poplar (Miquelle et al. 1992). Risenhoover (1989) concluded that, based on the high quality and digestibility of the dominant willows, these riparian communities can support high densities of moose in winter.

Riparian communities in the northwestern United States represent another moose habitat situation. These communities are subject to periodic wildfire, but the shrubs resprout rapidly, and the major effect of fire is to increase productivity temporarily. Thus, these riparian communities are essentially stable from the standpoint of moose habitat. At most, periodic fires have caused temporary displacement of moose, until vegetation has regenerated and forage supplies increased within a few years.

In the Jackson Hole, Wyoming, area, 60 percent of

Streamside shrub / scrub vegetation generally is not influenced by fire, and is usually dominated by alder and willow that provide moose with both cover and food. Taller shrubs and conifers serve as winter cover. Major river delta systems are important moose habitat. *Photo by Albert W. Franzmann; courtesy of the Alaska Department of Fish and Game, Soldotna.*

moose observations were in riparian willow communities (Houston 1968). The blueberry / Geyer's willow community received the highest use and were common in the year-long moose diet. Those species also characterized the more stable portions of riparian willow communities, whereas communities dominated by interior willow were relatively unstable and less abundant.

Aquatic Habitats

In North America, moose commonly are associated with lakes, streams and ponds throughout the late spring through autumn seasons, but their use of the aquatic habitats varies geographically. Peterson (1955), Murie (1934) and DeVos (1958) believed that the presence of palatable forage rather than insect avoidance was the major reason for moose attraction to aquatic areas. However, Ritcey and Verbeek (1969) reported that horseflies and deerflies may cause moose to use deeper water. Peterson (1955) concluded that declines in aquatic feeding by moose were correlated with reductions in forage quality or palatability, because no depletion of biomass of palatable species was found (see also Peek et al. 1976). In Ontario, substantial use of aquatic vegetation occurred in late June, July and early August, with June 3rd as the earliest recorded date of aquatic feeding by moose (DeVos 1958). A marked decline in aquatic feeding in late August occurred in Ontario's Algonquin Provincial Park (Peterson 1955). The earliest recorded feeding by moose on aquatics on St. Ignace Island was June 4th, and feeding in Lake Superior was begun a month later.

Peek et al. (1976) reported that use of aquatic habitats by moose in northeastern Minnesota was initiated earlier, reached its highest intensity earlier, but may last longer than farther north in Ontario. These observations support Peterson's (1955) phenology / palatability hypothesis. Also, use of aquatics was observed before the appearance of large flies, which become common in late summer in northeastern Minnesota. However, mosquitoes and blackflies occurred in highest densities in June and may have contributed to the amount of time moose spent in water. Peek (1971a) reported that moose observed in water either were feeding or standing with their bodies exposed to insect attack, rather than being submerged as might be expected if water was being used to elude insects.

Moose frequently use aquatic areas of the Bowron Lake Park, British Columbia, in mid-August (Ritcey and Verbeek 1969). In reviewing Russian work, Heptner and Nasimowitsch (1967) reported that aquatic areas are occupied during the entire warm season, from the moment of aquatic plant emergence at the end of May until the end of Sep-

Moose seek aquatic habitats for drinking water, insect relief, aquatic foods and thermoregulation. It is not unusual for a moose to submerge completely in search of palatable pond weeds and other aquatic plants. During June and July, aquatic macrophytes are an important component of the moose diet. *Photo by Ernest Oberholter; courtesy of the Wildlife Management Institute.*

As evidenced by aerial view of patches and trails through weed beds, moose readily consume nutrient-rich emergent aquatic vegetation in early summer and at least until the plants' palatability and forage quality diminish in late summer. *Photo by L. Godwin; courtesy of the Ontario Ministry of Natural Resources.*

tember. MacCracken (1992) reported use of aquatic plants by moose year-round on the Copper River Delta, Alaska, with peak use occurring from May through August. Joyal and Scherrer (1978) reported use of aquatic habitat from June 28th to September 15th in western Quebec, with a peak in early August. Variations in timing of use and peak use are attributable to ice-free conditions and perhaps differences in forage palatability between areas.

Kelsall and Telfer (1974) concluded that moose are limited in their southern distribution by summer temperatures that make thermoregulation difficult. Such heat sensitivity suggests that moose also use water to cool themselves, and may explain the longer period of use of aquatic habitat in Minnesota than in Ontario. However, it does not necessarily explain the long period of aquatic habitat use in western Quebec.

Through feeding activity, moose can alter density and composition of aquatic vegetation, as Murie (1934) and Krefting (1951) observed on Isle Royale. Studies of exclosed aquatic vegetation on Isle Royale (Aho and Jordan 1979) and Sibley Provincial Park, Ontario (Fraser and Hristienko 1983), similarly demonstrated that density and composition of aquatic plants can be altered dramatically by moose foraging. These observations suggest that moose find aquatic forage to be highly palatable, and lends further support to argument that forage is the primary reason for moose gravitation to and use of streams and ponds.

Jordan (1987) reviewed the idea that sodium is the primary nutrient sought by moose when foraging on aquatics. First, sodium concentrations (1,000 to 9,000 parts per million dry weight) in submerged and floating aquatics are known to be higher than in terrestrial vegetation (10 to 60 ppm dry weight) on Isle Royale (Botkin et al. 1973). Fraser et al. (1984) reported aquatic sodium concentrations of 3,000 to 20,000 ppm in Sibley Park. Jordan (1987) concluded that the high concentration of sodium is adequate to explain the attraction of moose to aquatic plants, and explains how the species may live in habitats where an essential nutrient is otherwise critically scarce. However, Fraser et al. (1982) concluded that the lengthier feeding period observed on Isle Royale than on the mainland may reflect a greater depletion of aquatics compared with that of most mainland aquatic habitats. In addition, Risenhoover and Peterson (1986) reported that mineral licks on Isle Royale provided a more concentrated source of sodium than did its aquatic plants, based on observed ingestion rates at licks.

MacCracken (1992) hypothesized that moose foraging in aquatic habitats was a more efficient means of obtaining nutrients than was foraging on land. On the other hand, Belovsky (1978) argued that moose foraging in water was less efficient because of the increased energy cost of locomotion and large volume of water ingested with aquatic plants. For his study area in the Copper River Delta, MacCracken (1992) concluded that aquatic plants contained higher amounts of crude protein and gross energy, when adjusted for digestibility, than did terrestrial plants, and forage biomass was also greater in the aquatic habitats. Estimates of foraging bout length were lower in aquatic habitats than in terrestrial habitats, suggesting that aquatic foraging by moose was more efficient. Also, maximum utilization of aquatic plants occurred before peak production of aquatic plants, as similarly occurred in Ontario and Minnesota studies (Peterson 1955, Peek et al. 1976, Fraser et al. 1982). The circumstantial evidence suggested that increased foraging efficiency in aquatic habitats could be another explanation of their use by moose.

Where accessible, aquatic habitats obviously are important to moose. Their forage base with high biomass, high digestibility, high sodium concentrations, opportunity for relief from biting insects and hot weather all may be reasons for their preferred use by moose. Most likely, reasons vary in combination both temporally and spatially.

Habitat Use and Snow

The influence of snow on moose habitat-use patterns has received considerable attention. Snow characteristics of ecological significance include temperature, density, hardness and depth (Peek 1986). Snow temperatures fluctuate less than and never get as low as ambient air temperatures, thus providing insulation against temperature extremes.

Considered chionophyls, or "snow lovers" (see Pruitt 1959), moose are better adapted to snow environments than any other large ungulate. Their size and strength enable them to negotiate snows deeper and dense and to reach vegetation higher than other native herbivores. Even so, snow depths of 28 inches (70 cm) or more impedes movements sufficiently to cause moose to seek better cover, yard microhabitat or winter range at lower elevation. Snow density also affects movements. Light, powdery snow permits greater ease and range of movement than does wet, packed snow. Crusted snow can make movement extremely difficult and cause injuries to lower legs and hooves. A crust capable of supporting 14.2 pounds per square inch (1,000 g/cm^2) will support most moose (Kelsall and Prescott 1971). *Photo courtesy of the Alaska Department of Fish and Game.*

Density—the relative compaction of individual snow crystals—typically increases during snowpack maturation through the winter period. Hardness is the degree to which individual snow crystals fuse to one another as snowpack matures. Hardness and density jointly influence the ability of snowpack to support an animal that travels through or on the snow. Snow depth is the most commonly measured parameter, and rules of thumb concerning depths that significantly limit travel for moose include 39 inches (1 m) (Des Meules 1964) and 28 to 38 inches (71–97 cm) (Kelsall 1969). In Mount Robson Park, Bowron Lakes and Wells Gray Park in British Columbia, fluctuations in moose populations that were juxtaposed by deep snow winters (Edwards 1956a) demonstrated the importance of snow depth as an influence on moose populations.

Fundy National Park, New Brunswick, has provided an example of how moose occupy habitat in a deep snow belt (Kelsall and Prescott 1971). Snow depths there are nearly always greater in open areas until late winter thaws melt the snow exposed directly to the sun more rapidly than snow protected by tree canopies. When depths have approached 38 inches (97 cm), moose have been confined to areas where forest canopies are high. Kelsall and Prescott (1971) and Peek (1971b) in northeastern Minnesota found no relationship between snow density and canopy closure, but both investigations indicated that hardness was higher in the more open areas (Figure 149). Peek (1971b) found that during one mild winter in northeastern Minnesota, moose shifted into denser canopies when only 12 inches (30.5 cm) of snow was on the ground. This suggested that the greater hardness of snow in open habitat may cause moose to shift to closed canopy forests where travel is less energy demanding, even if depths are not excessive. Midwinter activity reduction, reduced forage intake and greater use of heavier cover that ameliorates snow and weather influences are moose responses to existence in harsh winter environment.

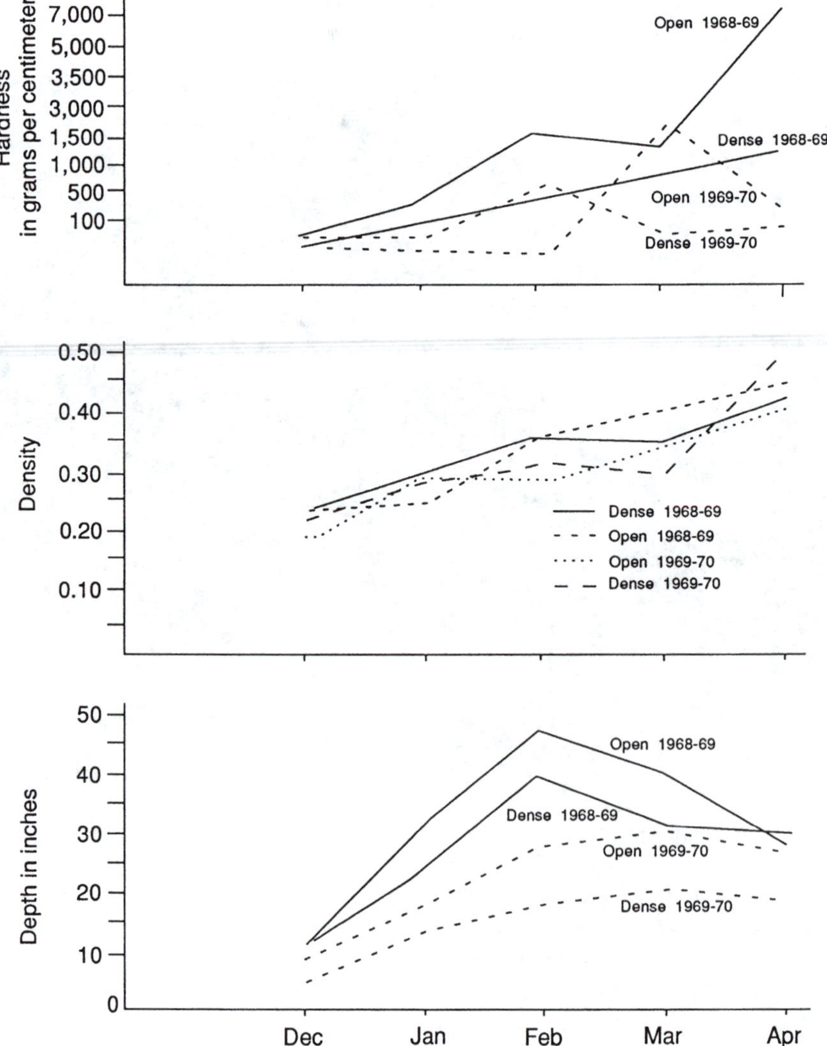

Figure 149. Snow characteristics during two winters in northeastern Minnesota (Peek 1971b). The deeper snowfall of the 1968–1969 winter showed more dramatic changes in hardness (crusting) and depth, but not density, relative to the lesser snowfall of the 1969–1970 winter. In both winters, snow hardness and depth were lower under dense forest than in openings.

The distribution of snow within a forest influences moose habitat-use patterns. Snow falling on tree branches, especially the branches of fine-needled firs and spruces, tends to be retained and is called "qali" (Pruitt 1959). Qali produces a lower snow depth immediately beneath the tree canopy called a tree well or "qamaniq" (Pruitt 1959). Also, if qali drops from the trees, it produces a dense and hard snowpack on the ground, which provides more support for moose traveling beneath the canopies. In the Yellowstone and central Idaho moose ranges, the qamaniq offers moose opportunity to seek forage protruding through the lowered snow depth. In addition, the tree itself, especially subalpine fir or Douglas-fir, may be browsed. Moose that winter at high elevations in subalpine fir forests seek out microsites within closed canopies where (1) snow depths are lower, (2) the dropped qali offers somewhat improved ground surface support and (3) access to palatable shrubs and trees is greatest. This behavior has also been observed in north-central British Columbia (D. Eastman personal communication: 1992).

In deep snow habitats where conifers are absent, such as in shrub/scrub tundra or riparian communities, moose still use the best microsites offered by combinations of shrub canopies and topographical situations that reduce snow depths. However, the principal adaptation simply is to reduce energy expenditure. In whatever form available, overstory cover aids this adaptation.

Other Influences on Habitat Use

This chapter advances the view that moose select habitat primarily for the most abundant and highest quality forage. Because these resources are unequally distributed in space and time, moose habitat may be considered as a series of patches of different kinds and sizes, with the value of each type of patch varying through the year. However, the total year-round value of a diverse habitat should be emphasized, even if each part is only critical at one season or another. The caveat to this general proposition is that sufficient size of habitat, and possibly of each patch of any given habitat, must be accessible to make an area most suitable for occupation by moose. As a corollary, if a certain critically important community, such as shrub fields, is unavailable in sufficient quantity, then the overall ability of a habitat to support moose may be reduced even if it contains a highly diverse set of plant communities except for that critical one.

The typical pattern of moose habitat selection includes open upland and aquatic areas that provide best forage in early summer, then more closed canopy areas that provide best forage as summer progresses and plant quality changes. In autumn after the rut and into winter, intensive use again is made of open areas where the highest biomass of dormant shrubs—the remaining major source of palatable forage—occurs. Closed canopy areas are used in late winter when forage is naturally at its lowest value and quantity for the year. The nature of the cover used at this time will provide the best protection available, and may range from tall shrub communities to tall closed canopy conifer stands. Metabolic activity in moose generally corresponds to this pattern—highest in summer and lowest in midwinter.

Within this general pattern, habitat use may be modified by rutting activity, calving, predators, presence of white-tailed deer carrying the brain worm parasite, interspecific competition, snow depth and hardness, population density, sex and age of individuals, and aggregation patterns. Selection of isolated habitat, such as islands in lakes or rivers for calving and early postparturition cover by cows, appears primarily a means to defend against predators—a conclusion that seems obvious although not exhaustively demonstrated. However, other factors, such as competing ungulates and the presence of parasitized deer, are difficult to demonstrate without expensive experimentation. Even if experimental evidence is forthcoming, application to the wide variety of habitats used by moose will still need confirmation in the field.

Examples exist of situations where forage supplies are not driving habitat selection. One such example is the decrease in feeding activity by bulls in autumn when they virtually cease to eat for several weeks during the height of the rut (Schwartz et al. 1984, Miquelle 1990). However, habitats selected by cows for feeding still may influence habitat selected by rutting bulls at the time rutting areas are initially established or at intervals thereafter. Cows may select a rutting area for its forage production or for other reasons including simply because it is within the home range. Also involved may be nonforage habitat factors, such as soft ground from which to paw rutting pits and open terrain in which other moose can be more readily seen.

Adaptations and Habitat Use

The annual metabolic cycle of moose coincides with cycles for forage quality, food intake and body weight (Regelin et al. 1985). Those cycles also coincide with habitat selection. In the annual cycle, forage quality is at the lowest ebb in midwinter (Ammann et al. 1973) and moose food intake is concurrently low. Midwinter habitats that favor conservation of energy are the ones used. These habitats typically represent the best cover available within the range, including closed canopy conifers in the boreal forest (Peek et al. 1976), and the tallest and densest shrub cover in riparian systems (Berg and Phillips 1974). The cover assists moose in

Moose choose different habitats throughout the year. A typical pattern includes the use of open uplands (top left) and aquatic areas (center left) in early summer, closed canopy forests (bottom left) where forage quality is high during mid- to late summer, and open habitats (top right) during the rut. During winter, moose select open areas (bottom right) with abundant forage until deep snow forces them into forest. Conifer stands provide natural cover during all seasons. *Top left photo courtesy of the U.S. Bureau of Land Management. Center left photo by W. May; courtesy of the Ontario Ministry of Natural Resources. Bottom left photo by J. M. Greany; courtesy of the U.S. Fish and Wildlife Service. Top right photo courtesy of the Alaska Department of Fish and Game. Bottom right photo by Albert W. Franzmann; courtesy of the Alaska Department of Fish and Game, Soldotna.*

Habitat selection by moose can be influenced by the needs for food, shelter and/or protection from predators. During the calving season, cow moose select isolated and secure habitats at the expense of abundant forage. Extremely protective cow moose will vigorously defend their calves against any perceived threat, even the unwary hiker who may accidentally venture too close. *Photo by Charles C. Schwartz; courtesy of the Alaska Department of Fish and Game.*

conserving energy by reducing convective and radiant energy losses (see Chapter 14). At this time, moose also minimize activity. This is especially evident with mature bulls that enter the winter period in relatively poor condition after the rut, occur in small groups and attempt to isolate themselves from other sex/age groups to minimize social interaction (Miquelle et al. 1992). When winters are especially severe, the tendency in moose to conserve energy by reducing activity and using high-quality cover is more pronounced than when moderate winters with less snow allow more activity at less energy expense.

Midwinter energy conservation by moose gradually is succeeded in the spring by an increase in activity and a shift to habitats that provide an abundance of high-quality forage. Such sites may be shrub fields or streams and ponds. The period of rapid antler and body growth, calving and most intensive lactation coincides with use of habitats that produce the largest amounts of most nutritious forage. As summer progresses and changes in plant phenology occur, with the attendant changes in forage quality, moose shift to plant communities that produce the most nutritious forage. These typically are communities in which plant development is retarded because of varying amounts of overstory closure, such as mature deciduous forests of birch or aspen. Bo and Hjeljord (1991) documented lower levels of leaf tannins in palatable forage during a cloudy, wet summer, compared with such levels in a relatively sunny summer. This finding suggests that one possible reason that moose may use those more closed canopy forests in late summer is lower concentrations of bacteriostatic-inhibiting compounds in forage found there. After the rut, bulls shift back into shrub fields, probably because of the presence of high quantities of forage as they switch from rutting to foraging as the predominant activity.

Sexual Segregation and Habitat Use

Differential use of habitat by sex and age classes of moose formerly was of concern because it could affect the accuracy of sex and age ratio information obtained by direct observation (Peterson 1955, Pimlott 1959). Pimlott (1959) reported that observational biases significantly influenced cow/calf ratio data collection. He attributed this in part to the "secretive behavior" of a cow when accompanied by a calf. Females classified as "barren" may have simply hidden their young calves. However, Peek (1962) noted that bulls and cows without calves made greater use of open cover than did cows with calves during summer in southwestern Montana. Peek considered this sufficient to overestimate bulls and cows without calves in the population. Therefore, not only might hidden calves be unobserved, but cows with calves might be less observable than other sex and age groupings. In northcentral Alberta, Mytton and Keith (1981) reported that bulls used lowlands more often than cows do in summer and late autumn, and mature aspen forests were used more often in winter than by cows. MacCracken (1992) also reported different habitat-use patterns by sex; cows used closed canopy tall shrub habitat in summer more than bulls did, and bulls used sweet gale communities more in winter than cows. Roussel et al. (1975) reported greater movements by adult males than adult females in Laurentides Provincial Park, Quebec, implying a greater diversity of habitats would be used by males than females. Leptich and Gilbert (1989) reported that, in summer, females use lowland habitats more frequently than males, and males use uplands more frequently than females.

Peek et al. (1974) reported a greater tendency for cows without calves than for cows with calves in Alaska, Minnesota and Montana populations to aggregate with other moose. These observations suggested that cows with calves

Shrub fields with a high density of forage biomass typically are the best cover available in riparian systems of moose range. During winter, moose reduce their intake of food, have a lower metabolic rate and conserve energy. Snow depth permitting, moose will remain in these areas during much of the winter. They may form small loose groups and develop an extensive trail system as winter progresses and snow accumulates. Wintering group size and habitat use are related to the sex and age classes of the moose. Cows with calves tend to select secure forest habitats and disassociate themselves from other moose. Large bulls form small bachelor groups after the rut or occasionally associate with females. Cows without calves, and young bulls tend to form the largest groups often in open habitats (above). Moose in these areas are easy to see and count. Biologists are aware of this varying usage of habitats by moose and take this fact into consideration when conducting composition surveys. Counting only large groups in open habitats tends to result in low bull/cow and calf/cow ratios that are atypical of the population at large. *Photo by Charles C. Schwartz; courtesy of the Alaska Department of Fish and Game.*

are less observable because detection is correlated to group size (Cook and Martin 1974). Moose occupying the most open habitats on the northern portions of their range occurred in larger groups than did moose occupying the denser boreal and mountainous forests farther south. Miquelle et al. (1992) showed that in central Alaska, small bulls typically occurred in larger groups than did large males, except during rut and postrut periods. In northeastern Ontario, cows with calves used sites with extensive cover and substantial amounts of forage, but did not exhibit much use of open areas (Thompson and Vukelich 1981).

The evidence suggests that cows with calves select denser cover and are more secretive than other age groups. Mature bulls may occur in smaller groups and less frequently with females than do younger bulls. Moose inhabiting forested terrain generally occur in smaller groups than do moose occupying relatively open habitat. Moose population density also may be expected to have an influence on group size and habitat selection.

Differences in selection of habitats by the sexes have survival value in several ways. First, in forested habitats, the probability of being observed by predators may be reduced if group size is reduced, but the limited, dispersed and patchy distribution of forage may override or interact with predator avoidance strategy to affect group size. In relatively open terrain individuals in larger groups may be less vulnerable to predators than are lone individuals. Second, the secretive behavior and selection of denser cover by females with calves also may relate to predation avoidance (Thompson and Vukelich 1981). Third, reduced postrut activity and avoidance of large groups by prime bulls may be consequences of high energy expenditures during the rut, dictating reduced activity and forage competition through the winter, which, in turn, may reduce exposure to predation (Miquelle et al. 1992).

Two other results of differential habitat selection by sex and age classes throughout the annual cycle are (1) more complete occupation of available habitat and (2) broader distributed population than might otherwise occur. In terms of population welfare, this implies that a smaller portion of a population may be affected by mortality factors related to local weather, habitat and predation. Perhaps, management efforts to maintain relatively high density and longevity may promote the general welfare of populations in highly fluctuating environments and, if so, should be considered a fundamental conservation strategy.

Interpreting Habitat-use Information

Interpretation of habitat-use information often is difficult. If an assessment of habitat preference is to be made, in favor of a simple description of the presence and frequency of animal occurrence in different habitats, then what is available to a population must be defined (Porter and Church 1987). This definition sometimes is difficult to apply, because environmental parameters vary as conditions change.

Westoby (1980) concluded that if forage supplies were a primary driving force for habitat selection, then the relative availability of forage would have to be very low before consumption would be affected. Accordingly, the assumption that habitat use is related to forage supply becomes moot except at those low levels. However, because forage quality and palatability affect availability in terms of digestible nutrients, precise definitions of availability again are needed.

Also, the implied assumption that habitat utilization is correlated with carrying capacity (Fretwell and Lucas 1970) has been questioned, because examples show otherwise (Van Horne 1983). If the correlation holds for moose, then alternative assumptions involving habitat selection, such as predation and/or parasitism, may be more valid in predicting habitat use than are forage quality and quantity.

Some evidence suggests that, at certain levels of selection, habitat utilization is affected by predation or parasitism. Johnson (1980) distinguished four levels of selection: (1) overall range; (2) home range; (3) habitat components within the home range; and (4) forage supplies. The ecological separation of moose and white-tailed deer in winter (see Telfer 1967a, Karns 1967) is a characteristic example of selection at the levels of home range and habitat components. Telfer (1967a) found that, although patches of suitable winter habitat for moose could be found many places, populations were maintained on uplands primarily because deer left for nearly half the year. The latter's departure reduced the amount of feces harboring meningeal worms. Moose that dispersed into surrounding lowlands, where the deer had moved, experienced a high rate of meningeal worm infection and resultant mortality. Similar findings occurred where moose/whitetail habitat overlap was investigated in New Brunswick (Telfer 1970); and adequacy of forage supplies apparently minimized the potential for interspecific competition.

Karns (1967) reported similar relationships in northern Minnesota where areas with low moose densities supported high deer densities; the reverse was true of other areas. One assumption from these findings is that in the absence of whitetails, moose would use winter habitats at lower elevation where snow depth is relatively low and forage supplies relatively abundant. In fact, if moose are inclined to select for such lower elevation habitats in winter, the actuality of whitetail presence and moose exposure to meningeal worm infection probably would eventually reduce or eliminate the portion of the moose population that selected for the winter habitat used by deer. The balance of the moose population would be conditioned to select for the type of winter habitats where overlap with deer is low, thereby reducing the probability of infection (see Chapter 15).

The differences in habitat-use patterns between moose and deer in the eastern portions of moose range also may be related to other conditions. After a wildfire in northeastern Minnesota, moose density increased more rapidly than did whitetail density (Irwin 1975). The fire occurred in a deep snow area and removed much cover, suggesting that winter habitat of deer was considerably diminished, whereas the shrub field habitats important to moose were enhanced. Apparently, sufficient coniferous cover was left to provide suitable late winter habitat for moose but not for deer. Also, diminished cover may have increased deer vulnerability to wolves enough to prevent recolonization of the burn by deer (Mech and Karns 1977), consistent with similar dynamics on Vancouver Island in areas where coniferous cover used by wintering black-tailed deer had been fragmented and diminished by logging (Hatter 1988).

But the question persists as to why moose select deep snow belts when areas of shallow snow occur nearby, if competition with deer for food is not the cause, because the larger and relatively mobile (in snow) moose can outcompete white-tailed deer for available forage. A series of reasons in concert or separately can be hypothesized: competition is intermittent; deer diets are more diverse at other times of the year; habitat use in midwinter is not influenced by particularly strong selection forces; meningeal worm infections have gradually isolated moose populations in those habitats where risk of infection is minimal; and wolf predation tends to favor moose survival over that of deer in deep snow country (see Chapter 9).

Nudds (1990) considered this information to be weak evidence that parasitic infection was influencing moose distributions. However, deer and moose share range most frequently at times of the year when forage is most abundant and snow depths are not confining. The potential time for infection is late winter and early spring when habitat overlap is minimal. Finally, where moose and whitetail ranges overlap, forage supplies appeared not to limit moose, strongly implying that the associations between deer, other moose, parasites, snow conditions and possibly wolves are the factors of moose distribution in winter. Whether wolf predation affects moose winter distribution, confining moose to deep snow areas where deer are not as likely to be present, has not been sufficiently evaluated.

Stevens (1974) reported similar ecological separation between moose and elk in Montana and Wyoming, with moose invariably occurring in areas of deeper snow. Meningeal worms are not a factor in this region. And, food habits of elk are more diverse during periods when snows do not restrict availability. Forage availability may or may not limit moose, but the likely reason for the separation is that in times of forage scarcity, elk are more adaptable to a wider range of forages, hence competitive, than moose. Thus, competing ungulates may affect habitat selection in terms of the food base, although this may never be demonstrated conclusively (Wiens 1977; see also Chapter 9).

Predation also may depress moose populations and influence plant succession and moose habitat-use patterns. In Quebec, Messier and Crête (1985) concluded that wolf predation on unhunted moose stabilized populations at the relatively low density of approximately 1.0 moose per square mile (0.4/km^2). No evidence from the Quebec study areas suggested that forage supplies were heavily browsed (Crête and Jordan 1982a, 1982b), and comparisons of body condition between moose populations with densities ranging from 0.4 to 1.00 per square mile (0.15–0.4/km^2) made from hunter- killed moose in autumn (Messier and Crête 1984) were done during the time of the year when differences between populations would likely be minimal.

High-density moose populations can negatively impact preferred browse plants. Moose tend to rebrowse the same plants each year. Continued overuse results in hedged plants, resulting in numerous, small annual growth stems and poor vigor. Stands of willow—a preferred browse species throughout the range of moose—often show heavy use. *Photo by Charles C. Schwartz; courtesy of the Alaska Department of Fish and Game.*

The Quebec densities were considerably lower than densities of three to five moose per square mile (1.2–1.9/km^2) obtained in Minnesota at a time when browsing intensity was not at levels that would suggest that forage was limiting moose populations or adversely affecting the forage sources (Peek et al. 1976). On Alaska's Kenai National Moose Range, forage supplies on the 1947 burn apparently were being altered when moose densities approached nine per square mile (3.5/km^2) in 1970 (Schwartz and Franzmann 1989). On Isle Royale in Lake Superior, moose densities roughly estimated at five to six per square mile (1.9–2.3/km^2) (Murie 1934) were altering forage supplies. Subsequently, as shrub fields resulting from the 1936 burn matured, moose densities of approximately five per square mile (1.9/km^2) were capable of further diminishing the shrub diversity (Risenhoover and Maass 1987). Where moose populations are below five per square mile (1.9/km^2) in habitat of high forage production, the populations probably are not regulated by forage. However, should forage supplies deteriorate through succession or overuse, moose densities may decline or significantly accelerate diminished forage abundance and composition.

Is moose habitat selection primarily density related or do some habitats receive preference regardless of moose density? Hobbs and Hanley (1990) pointed out that erroneous conclusions may be drawn from habitat-use information used to infer carrying capacity under the assumption that preferred habitats necessarily are those where highest density of animals occurs. Hobbs and Hanley concluded that frequency distributions of animals reflect habitat-carrying capacities only when animals are spatially distributed in a random manner, environmental conditions permit long-term equilibrium between animal populations and limiting resources, and use/availability data are obtained after equilibrium is achieved. In the fluctuating environments of moose range such conditions are seldom met, if ever.

Generally habitat-use patterns for moose are obtained by comparing frequency distributions of tracks, pellet groups or radiolocations within each habitat parameter of interest with the availability of that parameter to the population. Van Horne (1983) reported that in the application of habitat-use information, there is an assumption of positive correlation of species abundance and habitat quality. Therefore, high frequencies of moose observations in high forage-producing shrub fields presumably would indicate that the shrub fields are high-quality habitat. Also, moose commonly are observed to shift to dense cover in midwinter, suggesting that those habitats, which vary considerably across the range of moose, are high-quality habitat.

Perhaps the critical question is whether or not patterns of habitat use change as moose population density changes.

Would a moose population at one density select shrub fields as frequently as would a moose population at a different density? If the shrub fields provide important forage at a critical time, the answer would be yes. Van Horne (1983) pointed out that seasonally preferred habitats may be critical for individual survival and reproduction.

Generally, moose habitat is variable because of the temporal and spatial unpredictability of weather conditions. Van Horne (1983) also reported that unpredictability in the environment may allow for population density increases in low-quality habitat. This may occur with moose if circumstances reduce use of high-quality habitats, such as when sudden storms produce deep-snow conditions that preclude moose movement into preferred cover where such cover is fragmented.

The third condition Van Horne (1983) identified in which population density and habitat quality might not be directly related is the circumstance of habitats being so patchy that moose are forced to move between preferred patches (microhabitats) to such an extent that they alternate their population density between low- and high-quality habitats. An example is the highly patchy conditions in shrub/scrub tundra (Van Ballenberghe 1992), where the taller, preferred willow forage occurs along drainages and is interspersed by extensive stands of bog birch and low willow. During the rut, large concentrations of moose may occupy the usually less preferred latter habitat where visibility and opportunity to interact are less impeded than would be the case in the denser vegetation of the preferred, high forage-producing areas. The lower shrub/scrub habitat could be used to facilitate rutting activity.

Undoubtedly, a variety of habitats and the flexibility to use any part of them accommodate moose best. The kind and amount of cover available in winter affect the energy loss associated with winter conditions and, subsequently, the amount of time in spring and early summer required to build energy reserves, skeletal growth, adequate lactation, antler growth and, consequently, the timing and condition of animals entering the autumn rut. The variety of forage available in spring and summer affects condition of animals entering the rut. The nature of late autumn and early winter habitat influences the condition of animals entering the severe times of winter. All of these factors are mediated through population density, and when densities become too high, then population conditions (e.g., habitat, animal health, productivity) generally decline.

To be of most value, habitat selection information must be obtained over a period of years, so that the pattern of use may be related to weather variables that are the major influence of change in habitat use, such as snow conditions, which are highly variable among years. Also, nutrient content of plants changes over time and from stand to stand (Cowan et al. 1950, Peek et al. 1976, Jelinski and Fisher 1991). These factors interact with others to influence vulnerability of moose to predation, parasitism and competition with other cervids. Certain communities, including high forage- producing shrub fields, closed canopy forests and aquatic foraging areas are key components of moose habitat, but a variety of others enhance the value of a range by providing forage and cover during critical periods. However, correlations between any given habitat and its carrying capacity for moose must account for the inherent environmental variability and the adaptability of the species to that variability.

Hobbs and Hanley (1990) concluded that forecasting consequences of habitat change depends on a knowledge of underlying processes that control animal response to change. This assumes understanding of a number of the processes that cause moose to select one habitat over another, and that science is capable of predicting habitat use in general terms. However, given inherent environmental variability and the species adaptability, using habitat preferences to infer carrying capacity probably is less worthwhile than detecting patterns of use, illustrating variation between seasons, years and areas, and determining reasons for variation. To ensure provision for moose habitat into the future, challenges are to define the range of adaptability of moose to the naturally changing environment, and to understand how human activities affect this adaptability.

Conclusion

Understanding of moose habitat availability, preference, requirements, selection and use are essential to managing for the species in an increasingly anthropogenic world. A significant shortcoming to date has been an inability to determine accurately the limits of adaptability of moose (and most other species) to habitat alteration. Although moose are highly adaptable, limits to the species' ability to withstand change in a specific habitat are poorly understood. Such changes often have an imperceptible effect at first, but become critical as they accumulate. There are time lags of varying degrees in population response and, at any given time, factors other than habitat change may limit and override the ultimate consequences of habitat deterioration. What are the thresholds? How are they to be determined when research resources are scarce, experimental evaluations difficult to conduct, and interacting factors hard to eliminate? Moose are adaptable, but there are limits to that adaptability.

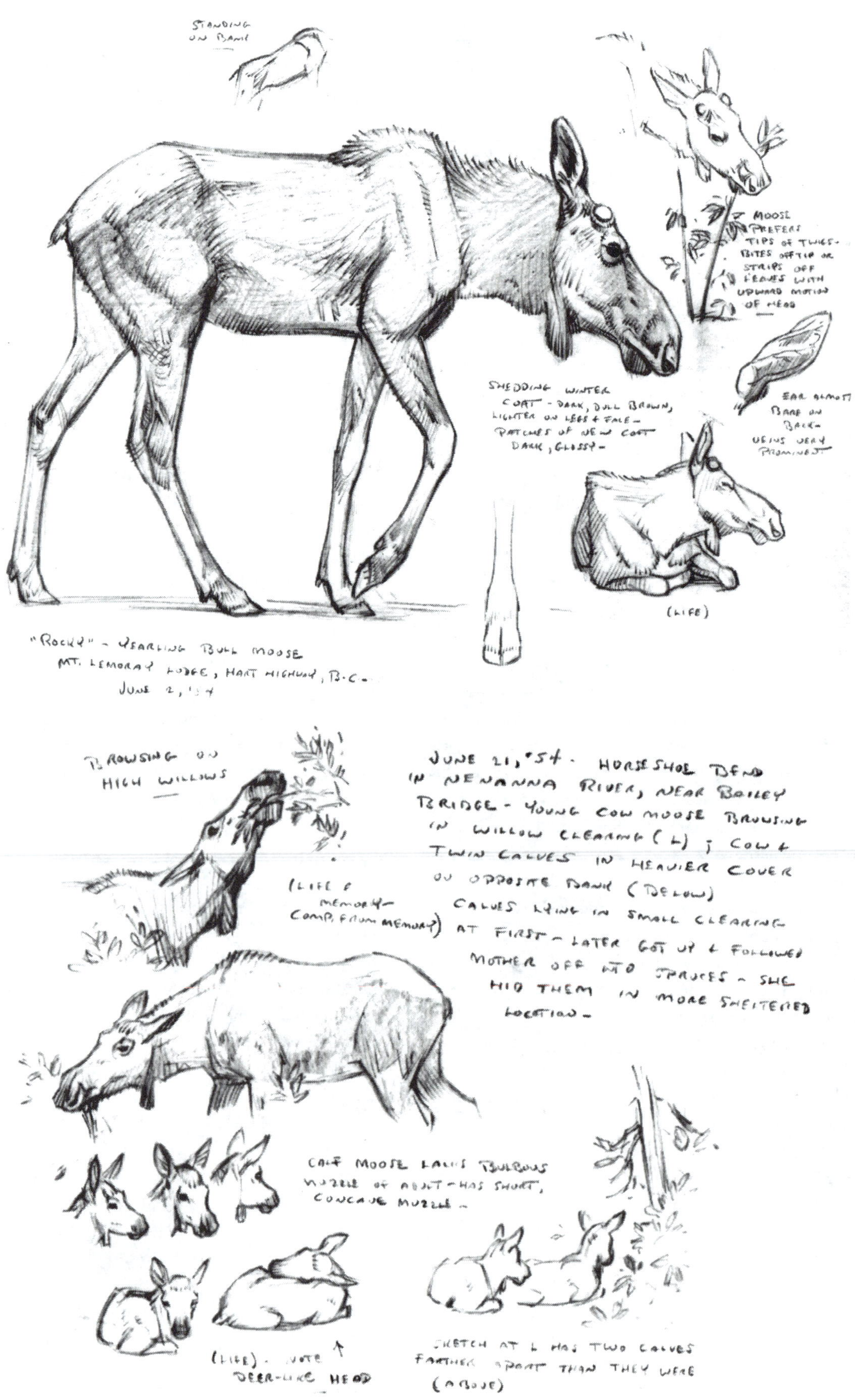
STANDING ON BANK
MOOSE PREFERS TIPS OF TWIGS - BITES OFF TIP OR STRIPS OFF LEAVES WITH UPWARD MOTION OF HEAD
SHEDDING WINTER COAT - DARK, DULL BROWN, LIGHTER ON LEGS & FACE - PATCHES OF NEW COAT DARK, GLOSSY -
EAR ALMOST BARE ON BACK - VEINS VERY PROMINENT
(LIFE)
"ROCKY" - YEARLING BULL MOOSE
MT. LEMORAY LODGE, HART HIGHWAY, B.C. -
JUNE 2, '54
BROWSING ON HIGH WILLOWS
(LIFE & MEMORY - COMP. FROM MEMORY)
JUNE 21, '54 - HORSESHOE BEND IN NENANNA RIVER, NEAR BAILEY BRIDGE - YOUNG COW MOOSE BROWSING IN WILLOW CLEARING (L); COW & TWIN CALVES IN HEAVIER COVER ON OPPOSITE BANK (BELOW)
CALVES LYING IN SMALL CLEARING AT FIRST - LATER GOT UP & FOLLOWED MOTHER OFF INTO SPRUCES - SHE HID THEM IN MORE SHELTERED LOCATION -
CALF MOOSE LACKS BULBOUS MUZZLE OF ADULT - HAS SHORT, CONCAVE MUZZLE -
(LIFE) - NOTE DEER-LIKE HEAD
SKETCH AT L HAS TWO CALVES FARTHER APART THAN THEY WERE (ABOVE)

12

IAN D. THOMPSON AND ROBERT W. STEWART

Management of Moose Habitat

Many North American wildlife agencies manage moose as a featured species because of recreation, aesthetic and economic considerations. Moose prefer a combination of young and old forest, but chiefly the former. Any process that converts some stands of timber to young age classes may improve habitat for moose. An integral part of a management program for moose is the use of timber-harvesting procedures to create or maintain components of the species' habitat. The goals are to enhance the value of logged forests as moose habitat and to extend the time period during which moose will use a given forest area, while concurrently minimizing the loss of commercial wood fiber (Ontario Ministry of Natural Resources 1988c). Where industrial logging is infrequent, natural disturbances, such as fire, are necessary to replenish moose habitat. Habitat management for moose requires knowledge of the species' biology, including habitat preferences, as well as forest ecology and silvicultural treatments.

Moose Habitat Requirements

Habitat for moose can be viewed as a series of annual components that may differ depending on the animal's sex or age, life history event and the season. Moose have basic requirements for food, provided by deciduous shrubs, some conifers, forbs and aquatic plants, and for shelter from deep snow and extreme temperatures, provided by mature forest stands. Lack or low abundance of these various habitat components lowers the overall capacity of a given area to support moose. Habitat choice may be influenced by predator avoidance behavior. Predators may force moose to select areas where detection by predators is less probable than in more preferred habitats (Edwards 1983, Stephens and Peterson 1984).

Individual components of moose habitat include: areas of abundant high-quality winter browse; shelter areas that allow access to food; isolated sites for calving; aquatic feeding areas; young forest stands with deciduous shrubs and forbs for summer feeding; mature forest that provides shelter from snow or heat; and mineral licks (Peterson 1955, Allen et al. 1987, Eastman and Ritcey 1987, Joyal 1987, Thompson and Euler 1987). Habitat management often is used to enhance, improve or create these individual components.

In summer, and for much of the winter, moose select young forest stands with the highest density, biomass and quality of food (Crête 1977, 1989, Telfer 1978a, 1989, Joyal 1987, Saether and Andersen 1990). The relationships between biomass, quality and diameter of twigs browsed are complex, but studies suggest that moose can modify their foraging behavior to adjust to habitat conditions (Hanley 1982, Saether and Andersen 1990). Regardless of the ability to alter foraging behavior, the higher the quality of forage, the greater is the fitness of the animal through improved condition before the winter (Regelin et al. 1987a) and at the end of the winter (Saether and Andersen 1990). Therefore, highly productive forest sites provide the best moose habitat (Oldemeyer et al. 1977, Franzmann and Schwartz 1985).

Moose tend to select forest stands that offer the highest density, biomass and nutritious foods. When one or more of these aspects are lacking from habitat within an animal's home range, the moose may alter its foraging behavior by consuming less nutritious food and/or expanding its home range. *Photo by Len Clifford*

Most work suggests that the juxtaposition of food and shelter in late winter can be an important and possibly limiting feature of moose habitat (Coady 1974a, Crête 1977, Telfer 1978b, Welsh et al. 1980, Thompson and Vukelich 1981, Eastman and Ritcey 1987, Peek et al. 1987). Energy expended by moose to move in deep snow can be greater than energy derived from food intake (Renecker and Hudson 1985a); plowing through snow to get to forage may result in a negative energy balance. Because of deep snow, moose become restricted in their ability to move, therefore food under or near shelter provided by a conifer canopy is important (Coady 1974a, Peterson and Allen 1974, Poliquin et al. 1977, Welsh et al. 1980).

In some parts of moose range, winter yards are used when snow is deep. Such yards are selected primarily based on cover characteristics and topographical features (Proulx 1983, Eastman and Ritcey 1987). In montane areas, moose move down to overwinter in river bottoms vegetated with preferred foods (primarily willows) (Peek 1974c, Eastman and Ritcey 1987). On mountain sites where Pacific yew occurs as an understory species, moose use the double canopy stands as preferred winter range (Peek et al. 1987). Cows with calves often avoid high-density yards, possibly as a predator avoidance mechanism; nonetheless, they require even more stringent conditions for shelter and food because of the small size of calves (Thompson and Vukelich 1981).

A long-term examination of moose habitat after logging was initiated in Alberta in 1956 (Stelfox et al. 1976, Stelfox

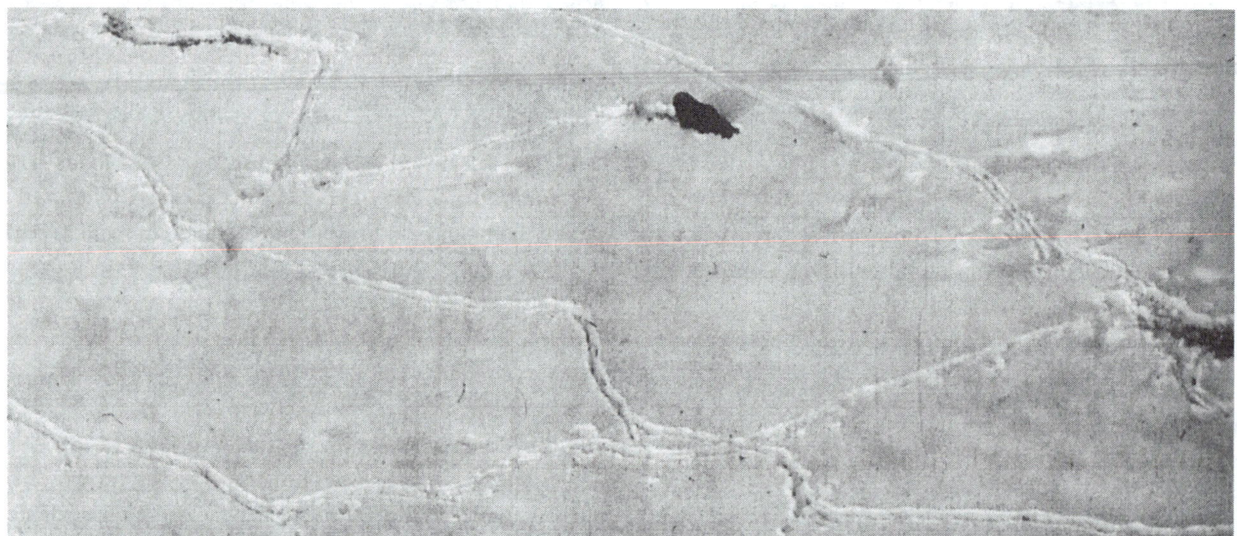

The energy expended by moose traveling through deep snow may result in a negative energy balance or inability to replace energy lost by forage intake (see Chapter 14). In areas noted for heavy snowfall, the availability of food near or under overstory forest shelter can be an important aspect of winter survival. Very large clearcuts of forest do not benefit moose as much as smaller cuts where forest edge is near. In some areas during winter available food may be covered with snow, forcing moose to dig or plow through the snow to locate it; this also adds to the energy expended. In winter, the nutritional quality of forage diminishes, but moose compensate by lowered metabolism and using stored energy from fat (see Chapter 14). Their daily consumption of food actually drops in winter. *Photo by Ron Modaferri; courtesy of the Alaska Department of Fish and Game.*

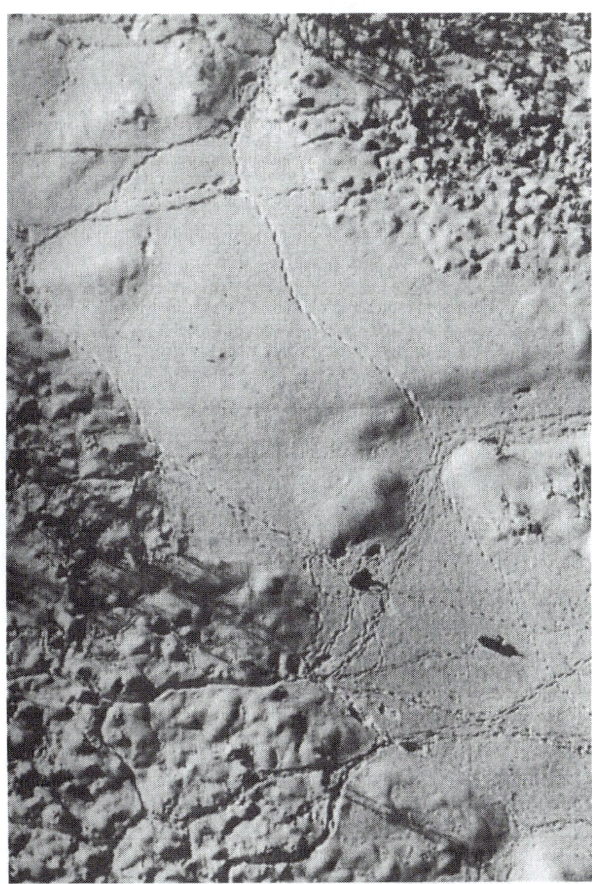

In mountainous areas in winter, moose move into river bottoms vegetated with preferred foods (primarily willows). When weather permits, moose will rest close to their food supply, exposed to the sun and in contours that give relative protection from wind. Moose depend primarily on their acute hearing to detect predators, and apparently do not select resting sites on the basis of visual advantage. *Photo by Ron Modaferri; courtesy of the Alaska Department of Fish and Game.*

1981). The study was conducted on 5.8 square miles (15 km^2) using strip cuts in white spruce-dominated forests, with the uncut portions removed 12 years after the initial cutting. Results suggested that regrowth of adequate winter shelter began only 20 to 25 years after logging, and that moose did not use the area during winter before that time (Stelfox 1981). Welsh et al. (1980) showed that moose in boreal Ontario migrated from young forest stands created by clearcutting to mature forest as much as 50 miles (80 km) away. Late winter cover clearly is an important component of moose habitat.

Several researchers stressed the importance of calving habitats. Sites are chosen by females to avoid detection of calves during and briefly after parturition (Peterson 1955, Edwards 1983, Stephens and Peterson 1984, Cederlund et al. 1987). Unfortunately, the paucity of research on the attributes of calving habitats makes characterization imprecise. Jackson et al. (1991) reviewed the literature and concluded that isolated stands of timber in bogs or clearcuts or on islands or peninsulas commonly are used calving sites. Certainly, there is some evidence to suggest that sexual segregation by habitat choice exists in moose, not only during calving, but also at other times of the year (LeResche et al. 1974a, Leptich and Gilbert 1989, Miller and Litvaitis 1992).

Throughout eastern and central parts of the moose range, summer habitat choice is influenced by extensive availability of aquatic feeding areas (Peterson 1955, Peek et al. 1976, Fraser et al. 1980, Jordan 1987). Aquatic plants provide a high protein content as well as a source of concentrated sodium needed by moose to compensate for a sodium deficit incurred in the winter months and resulting from increased potassium uptake from green vegetation (Jordan et al. 1973, Jordan 1987). Wherever they occur, mineral licks are another source of sodium used by moose (Best et al. 1977, Fraser and Reardon 1980, Tankersley and Gasaway 1983). Other essential elements may be obtained at licks and from aquatic plants, but no assessment of the relative importance of such resources has been made (Franzmann et al. 1975d).

The importance to moose of new green vegetation, with its high protein content, cannot be overemphasized. The protein and minerals are required by moose to replenish body weight lost during winter (Schwartz et al. 1987c). Spring movements throughout the moose range are influenced by the need to feed on new vegetation of high quality. If aquatic forage is unavailable, then habitats chosen contain areas with lush growth, such as subalpine meadows (Eastman and Ritcey 1987), shrub and riparian willow communities (Peek et al. 1976, Doerr 1983) and, in Newfoundland, timber landings, fens and alder communities (I. Thompson personal files). Moose seek habitats where their high protein requirements can be filled in spring and summer as rapidly as possible (Belovsky 1978).

Role of Habitat Management

Habitat management is one aspect of a moose management program. Other aspects include monitoring population size and distribution fluctuations, establishing hunting regulations, assessing hunter-harvest mortality and possibly regulating predators. In the development of a moose habitat management initiative, managers must be concerned with cost effectiveness. For example, they must judge how much to invest in aerial monitoring of a population's age and sex ratio versus on-the-ground habitat manipulation versus conservation law enforcement versus research on predation and other mortality influences, etc. The compro-

Cow moose choose calving sites that offer maximum cover for their newborn calves to prevent detection by predators. These sites generally are isolated stands of timber in bogs or cut-overs, or on islands or peninsulas. The area generally provides good visibility of the surrounding area and adequate cover for the calf to lie and rest as the mother feeds nearby. In the management of forest habitats providing moose populations with these types of sites is an important consideration. *Photo by Albert W. Franzmann; courtesy of the Alaska Department of Fish and Game, Soldotna.*

Throughout much of the range of moose in North America aquatic feeding areas are an important source of concentrated nutrients in spring and summer. They also are used to escape insects and to cool off (thermoregulate). These areas must be identified and considered in short- and long-term habitat planning for moose. *Photo by Albert W. Franzmann; courtesy of the Alaska Department of Fish and Game, Soldotna.*

Moose obtain sodium and other mineral nutrients at natural mineral licks (top left) as well as from aquatic areas. The location and patterns of use of these licks are important information for managing habitat. Use of mineral licks by moose is commonplace in spring and summer, but less so in winter (top right). Moose also use artificial mineral licks, most of which are created inadvertently by the salting of roadways in winter. The result is that soils along the roads have a high concentration of sodium, which attracts moose. And this circumstance can lead to serious accidents (bottom). The planning of road construction and maintenance in moose range must consider such consequences. *Top left photo by Peter Duff; courtesy of the U.S. National Park Service. Top right photo by Rolf Peterson; courtesy of Michigan Technical University. Bottom photo by Charles C. Schwartz; courtesy of the Alaska Department of Fish and Game.*

mise usually involves a commitment to low-cost habitat manipulations through arrangements made with forest products companies, by convincing those enterprises of the benefits of increased moose numbers. The manipulations generally include provision of buffer strips, creation of travel corridors and retention of forest stands or blocks within clearcuts (Figure 150). In some cases, such as in Ontario and Quebec, guidelines are used in a general manner. Elsewhere, in British Columbia and Alberta, for example, individual projects have been undertaken to deal with discrete areas.

The implicit assumption of habitat management is that moose population levels can be increased through the creation, maintenance or other manipulation of high-quality habitat. In this context, moose populations are recognized at two levels: (1) local populations or herds that can be affected by individual cuts or habitat alterations; and (2) landscape-level populations that are the sum of all local herds with a broad management unit of perhaps thousands of square miles.

The most convincing evidence that habitat management could have a positive influence on moose populations is the fact that historically moose populations have increased when habitat has been created or rejuvenated. For example,

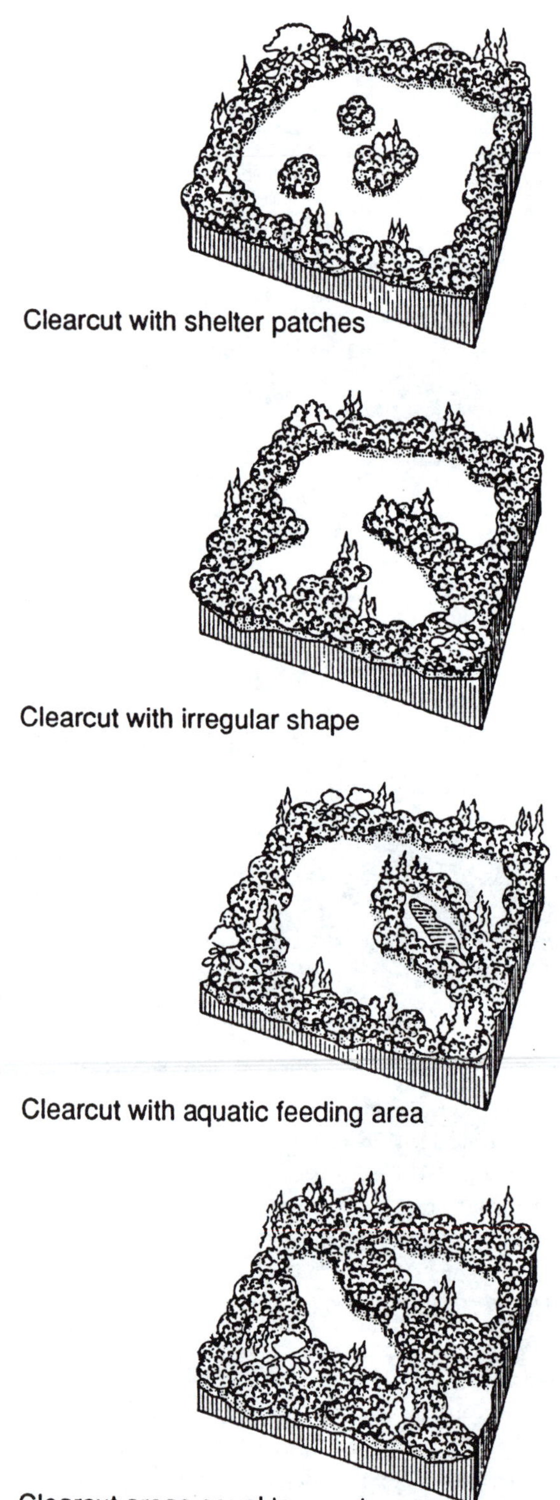

Figure 150. Representative schematics for local area management of moose habitat, particularly in boreal forest where commercial clearcut logging occurs. The designs emphasize protection of aquatic feeding areas from siltation, provision of escape cover, minimal distances between uncut edges and irregular lines-of-sight, and maximum browse close to uncut edges (from Hogg 1989).

until 1890, moose were reportedly scarce or absent in much of northern Ontario (Peterson 1955). Thereafter, and despite heavy hunting pressure, moose from more southerly habitats expanded into large areas of northern Ontario as a result of logging operations and fires associated with construction of the trans-Canada railway. Expansion of the range of moose in interior British Columbia followed the same pattern—new habitat created in the wake of human disturbances.

There is evidence from portions of Alaska that moose populations increased dramatically as a result new habitat created by wildfires in 1947 and 1969 (Spencer and Hakala 1964, Schwartz and Franzmann 1989). A similarly large increase in the number of moose occurred on Isle Royale, Michigan, after a 1936 fire that burned more than 39 square miles (100 km) of the island (Hansen et al. 1973). Moose were introduced into a predator-free environment in Newfoundland in 1878 and 1904, and in habitat created by pulpwood logging (Pimlott 1963), their numbers grew rapidly to a level where the animals damaged postlogging regeneration in some areas of the province and altered forest successional patterns (Bergerud and Manuel 1968, Thompson et al. 1992).

Evidence is less persuasive that management directed at various individual components of moose habitat has had an effect at either population level—local or landscape. Further, distinguishing habitat effects from the confounding variables of predation and hunting mortality generally has not been possible. Data often suggest that moose use habitats created specifically for them, but the necessary control experimentation to deal with confounding variables has been lacking. However, some studies have shown differences in the use of managed habitat by moose. Areas of spruce/aspen/birch forest in Alaska "crushed" to improve browse quantity for moose have been used extensively (C. Schwartz personal communication: 1992). Brusnyk and Gilbert (1983) reported that moose used uncut shoreline reserves of standing timber adjacent to clearcuts more so than areas without such reserves. In a similar study, Mastenbrook and Cumming (1989) showed that moose preferred edge provided by cover corridors—109- to 219-yard-wide (100–200-m) strips of uncut timber to locations without edges within the clearcuts. The need for shoreline reservations around aquatic feeding areas has been contentious because the reserves tie up large timber volumes that otherwise would be harvested. Timmermann and Racey (1989) indicated that moose may need shoreline reserves only at the actual feeding site to act as escape cover. There remains a need for management-level cause-and-effect experimentation to test the effectiveness of intentional moose habitat alteration. Although some management programs have been

As soon as green-up commences in spring, after winter's low ebb of protein availability in browse, moose maximize nutrient intake by foraging and consuming more daily. Moose seek out the sites with earliest new plant growth, and managers need to ensure that such sites and moose access to them are protected. *Left photo by Albert W. Franzmann; courtesy of the Alaska Department of Fish and Game, Soldotna. Right photo by H.R. Timmermann; courtesy of the Ontario Ministry of Natural Resources.*

successful in this regard at local levels, there remains a general lack of evidence to support the hypothesis that at the unit level, moose populations have been increased through directed management of habitat over the long term.

Scale of Habitat Management

The scale or extent to which moose habitat is managed is an important consideration. Scale in management can be temporal and/or spatial, corresponding to local- or landscape-level population circumstances. Among the questions of scale is whether a manager should manipulate moose habitat everywhere and all the time. Also, should managers be concerned only or primarily with the immediate supply of suitable habitat relative to some population objective? Similarly, should moose habitat be actively managed only in specific areas and for local populations or should area management be predicated mainly on objectives for an entire unit?

Moose require young successional forest for most of the year. Consequently, the value of a given group of forest stands to moose is transient, lasting perhaps less than 30 years in boreal forest. In most areas where moose exist, fiber production is the dominant land use, therefore moose are a "by-product." Accordingly, moose should be managed at a spatial scale large enough to ensure that population objectives can be achieved at a time scale consistent with that of optimal forest rotation. This does not preclude local management for individual habitat components. The particular habitats that might be managed at the local scale include mineral licks, traditional aquatic feeding sites and river bottoms.

Establishing Objectives for Moose Habitat Management

Wildlife management programs at whatever scale and time frame require a set of clear objectives that address both short-term and long-term population goals. The intended result of habitat management by objective is habitat that brings about desired moose population characteristics (e.g., abundance, distribution, condition) in a specified period of time.

Objectives set for the short term, such as 10 to 20 years, should deal with habitat attributes at the stand level (stand level refers to characteristics, usually as derived by aerial

Moose prefer seral stage vegetation that emerges after such disturbances as wildfire and logging. Extensive logging in Canadian woodlands and the forests of the northeastern and upper Great Lakes states during the early 1900s opened vast tracts of landscape (top left), some of which enabled expansion of moose range. Timber and pulp demands continue and are accommodated by better technology (top right) and, in most instances, heightened sensitivity to the forest wildlife values. The selectivity, size, configuration and distribution of logged areas individually and collectively have significant effect on the rate and timeliness of moose pioneering or reoccupation of the sites (bottom). *Top left photo courtesy of the Wildlife Management Institute. Top right photo by Merle Prosofsky. Bottom photo by S.E. Aldous; courtesy of the U.S. Fish and Wildlife Service.*

photography, of a forest stand of 25 to 250 acres (10–100 ha). Examples of short-term objectives include: (1) protection of moose colonizing a recently logged or burned area from excessive hunting pressure for a period of 5 to 8 years until the regrowth reaches a height sufficient to provide escape cover and to allow roads to deteriorate enough to reduce ease of hunter access; (2) provision of shelter reserves adjacent to aquatic feeding areas or mineral licks; (3) maintenance of particular calving sites (e.g., leave a peninsula uncut); (4) improvement in the quality of a specified number of hectares of moose winter habitat through mechanical means or prescribed burning; and (5) diversion of moose from dangerous areas such as roads and railways.

Longer term objectives would be applied at the unit

Block clearcuts have potential for creating or improving moose habitat by opening maturing forests and allowing regrowth of low-growing high-quality browse plants. However, the viability of cuts as moose habitat depends on such factors as proximal relationship to associate cover, water, mineral licks, aquatic vegetation and human disturbances. Merely cutting forest to stimulate regrowth does not constitute habitat management for moose. *Photo by Bob Stewart.*

Five- to 20-year-old clearcuts for timber harvest can provide good early winter habitat for moose. Regenerating deciduous woody vegetation, along with coniferous trees, particularly young residual conifers, provides both abundant browse and some shelter. *Photo by J. Nistico; courtesy of the Ontario Ministry of Natural Resources.*

level, and for a period as much as 80 to 150 years. Examples of long-term objectives include: (1) provision of a specified area (a portion of the unit) in suitable late winter habitat, thermal cover and/or early winter habitat for a period of one full rotation of the forest; (2) maintenance of hunter access (expressed as average miles of road per square mile); and (3) creation of a specified habitat mosaic that would optimize or maximize the production of moose to a predetermined density. Careful planning and coordination among managers of local populations or herds are necessary to ensure that a unit's moose population reaches and stabilizes at a desired level.

In prior millennia, wildlife was the most influential mechanism of creating or revitalizing moose habitat throughout the species' range in North America. Within this historic time, timber harvest and other human disturbances have been at least as important in affecting moose habitat, except in the far north—such as Alaska's Kenai Peninsula (above)—where fire suppression either was not practiced or effective. Prescribed burning is a common habitat enhancement or regeneration technique, but is difficult to accomplish on a landscape-level population scale. *Photo by Charles C. Schwartz; courtesy of the Alaska Department of Fish and Game.*

Evaluation of Habitat Potential

To make a decision when and where to apply habitat manipulation for moose requires knowledge of the capability of the area to support moose. To manage habitat piecemeal or by trial and error across a landscape, year after year, in attempt to create moose habitat is unduly costly and ineffective. Such efforts waste time, resources and public credibility, as well as jeopardizing the moose and negating any broadly based, colateral biodiversity objectives. Managers should carefully select areas that will respond to habitat manipulation by producing the greatest (or otherwise optimal) number of moose per unit of effort or expense. To assess moose habitat potential, all annual components required must be considered. Van Horne (1983) pointed out the necessity to understand fully the habitat requirements of an animal before making predictions based on occurrence alone. For example, if late winter range is missing or restricted, the overall capacity of an area to support moose is diminished, regardless of the quality of summer habitat.

The capacity and capability of land to support wildlife species are not uniform (climate, topography, mortality influences and stages of vegetation development being equal). There is ample evidence for white-tailed deer that productivity, physical size and population are dependent to a large extent on quantity and nutrient quality of the available food, which, in turn, is directly related to soil type and nutrients (Klein 1964, Verme and Ozoga 1980, Hobbs and Swift 1985). Similarly for moose, Schwartz et al. (1985) demonstrated a positive linear relationship between daily digestible energy intake and weight gain. Moose may select areas with the highest available density of preferred foods (Crête 1977, Telfer 1978a), but they also can distinguish between high- and low-quality forage (Peek 1974c, Thompson et al. 1989). Peek (1974c) suggested that assessment of habitat capability in boreal forest could be based on moose productivity. Franzmann and Schwartz (1985) provided evidence to support that contention by demonstrating that good habitat is reflected by moose twinning rates and/or the number of barren cows. Moose in highly productive habitats produce more calves than do those in poor habitats. Other researchers (e.g., Peek 1974c, Telfer 1974, 1978a, Crête and Jordan 1982b) have applied methods, such as key browse species or twig biomass, to assess moose range, but these techniques are labor intensive and limited to relatively small areas. Thus, landscape capability to support moose is not uniform over large areas.

Early attempts to assess the capability of land to support moose were made by Stewart (1970) and Houser (1972) in Ontario. Houser (1972) suggested that higher capability lands supported more moose. Unfortunately, no other variables were controlled (hunting pressure, vegetative age structure, amount of uncut forest or topography). Nevertheless, the data indicated that some form of land classification based on soil fertility could be used to determine capability of broad areas to support moose. Racey et al. (1989) used the Northwestern Ontario Forest Ecosystems Classification (see Sims et al. 1989) to interpret moose habitat capability. They suggested that prime summer and early winter habitats would occur on medium to rich sites, except the wettest or driest sites. Racey et al. predicted that late winter shelter could be supplied on medium to poor sites as long as the stand was 70 to 80 percent stocked. However, none of their predictions has been tested.

To improve browse quality and quantity, Fleco rollers (top) pulled by tracked vehicles and LeTourneau tree crushers (bottom) were used on the Kenai Peninsula, Alaska, to disturb vegetation that had grown beyond a productive state for moose. Formerly productive sites for browse species favored by moose provided the best results. Conversely, the sites that formerly were unproductive provided the poorest results. Site selection for mechanical rehabilitation of browse must consider the potential productivity of browse species. *Photo by Albert W. Franzmann; courtesy of the Alaska Department of Fish and Game, Soldotna.*

River bottom habitats tend to be especially beneficial to moose. Vegetative growth is regularly interrupted by flooding or shifting channels. Consequently, seral stage vegetation normally is continuous, and the moose populations that occupy such habitats are more stable than those that reside in less frequently disturbed areas. *Photo by Ron Modaferri; courtesy of the Alaska Department of Fish and Game.*

Critical habitats—such as aquatic feeding areas, mineral licks, calving sites and stands of seral vegetation—are most used by moose where there is forest shelter adjacent or nearby. Managers need to allow for shelter reserves in programs of habitat disturbance to improve the food base for moose. *Photo by Albert W. Franzmann; courtesy of the Alaska Department of Fish and Game, Soldotna.*

Moose productivity—such as twinning rates—is one measure of a population's habitat quality; there are additional physiological indices. Efforts to assess habitat quality directly, particularly that of food resources by such measurements as annual growth (left) and twig biomass (right), are labor intensive and spatially limited. *Photos courtesy of the Alaska Department of Fish and Game.*

Concentrating efforts on managing all aspects of habitat for moose should occur on the most productive sites rather than on poorer areas, with the possible exception of providing for some late winter sites. Land or ecosystem classification methods provide a framework for comparative assessment of the potential productivity of different areas for moose.

Current Management

Most jurisdictions with large moose populations engage in some form of vegetation management program to maintain or increase moose populations. However, habitat management generally is not a major component of moose management for a number of reasons:

1. Lack of definitive spatial and temporal links between habitat supply and expected population responses.
2. Professional judgments that moose populations are limited by such factors as climate, weather, predation, hunting and parasites more so than by habitat.
3. Successes of moose population manipulation programs (hunting strategies) and predator-control programs in achieving desired population effects.
4. High costs associated with vegetation management programs without predictability of desirable returns.
5. Relative temporal and spatial stability of moose populations across most of their range in the absence of designed habitat interventions.
6. Commercial timber harvesting has increased the supply of early successional habitats preferred by moose.
7. Reliance on wildfires to accomplish goals of successional setback in remote areas.

Through general guidelines applied to timber management planning or by means of small habitat management projects, state, federal and provincial wildlife agencies actively invest in moose management (Table 30). However, few agencies have clearly articulated objectives for moose habitat management in terms of area of moose range needed or created, or for expectations at the local population level.

British Columbia

To increase productivity and size of certain moose herds in British Columbia, general habitat management goals included: (1) creating/maintaining habitat mosaics; and (2) enhancing edge, setting back older succession and improving the nutritional value of winter forage (K. Child personal communication: 1995). Prescribed burning and

Table 30. Moose habitat management programs by state and provincial wildlife agencies, 1992

Jurisdiction	Habitat management	Objectives	Procedural guidelines	Effects monitoring
British Columbia	Yes	No	No	Yes
Alberta	Yes	No	No	Yes
Saskatchewan	No	No	No	No
Manitoba	Yes	No	Yes	Yes
Ontario	Yes	No	Yes	No
Quebec	Yes	No	Yes	No
Nova Scotia	No	No	No	No
New Brunswick	No	No	No	No
Newfoundland	No	No	No	No
Alaska	Yes	Yes	No	Yes
Wyoming	Yes	Yes	No	No
Idaho	Yes	Yes	Yes	No
Minnesota	No	Yes	Yes	No
Wisconsin	No	No	No	No
Maine	No	No	No	No

mechanical overstory removal are methods used on a site-specific basis to improve the yield of preferred browse species on moose winter range. Most habitat manipulations are conducted on a project-specific basis, funded by the British Columbia Habitat Conservation Fund Program or through compensation agreements with the British Columbia Hydro and Power Authority. Although moose management objectives are set project by project, British Columbia does not manage moose by objective on a provincial scale.

Alberta

Alberta wildlife managers have conducted several experimental habitat manipulations, but not on a large scale. Manipulations involve mechanical shearing of vegetation, and burning to promote regeneration of deciduous browse species. General wildlife-sensitive timber harvest-planning guidelines exist, but none is specific to enhancement of moose habitat.

Manitoba

Manitoba has conducted several small-scale habitat manipulations through prescribed burning and mechanical means (V. Crichton personal communication: 1995). These were monitored and proved successful in increasing local moose numbers. Also, the province established a set of forest management guidelines for wildlife. The major component of these guidelines relative to moose is a provision for commercial timber harvests to retain at least 50 percent of the cover in known wintering areas.

Ontario

Ontario developed forest management guidelines for moose habitat. They are applied locally when timber management plans are developed, but there are no habitat objectives, and habitat created through guideline compliance generally is not monitored.

Quebec

Quebec uses guidelines in timber management planning to leave uncut coniferous or mixed forest patches of 7.4 acres (3 ha) or more within clearcuts. The area of uncut patches must be at least 4 percent of any clearcut, and the cuts should not exceed 618 acres (250 ha). The patches are left to supplement the amount of available cover, and to reduce line-of-sight for hunters in open areas.

New Brunswick

New Brunswick leaves buffer stands of trees around bogs for calving habitat, but does not practice active habitat manipulation. No forest harvest guidelines or monitoring programs are in place (G. Redmond personal communication: 1995).

Alaska

A variety of habitat manipulation techniques to promote moose habitat has been used in Alaska, including mechanical disturbance, prescribed burns and application of herbicides and fertilizers. However, the responses of moose forage to these investments in most areas have not been carefully evaluated (Oldemeyer and Regelin 1980).

Minnesota

The moose is 1 of 34 indicator species used by the state in managing its ecosystems (W. Russ personal communication: 1995). Specific timber harvest guidelines were developed for these indicator species, such as relaxation of the maximum allowable clearcut size from 40 to 200 acres (16–81 ha). This guideline was established to simulate change at a natural scale, and thus promote ecosystem regeneration.

Integrated Resource Management

There are two basic approaches to integrated resource management. One entails operational guidelines to regulate cut-block sizes and configurations to achieve desired landscape patterns. The second emphasizes specific habitat objectives by adjustment of operational plans. Guidelines are a set of suggested modifications to timber harvests that may be generally applied. Habitat objectives refer to a specified amount of area that will be modified through timber harvest for moose habitat improvement. The two approaches (guidelines and objectives) have definitive but significantly different implications for future forest landscape designs. In particular, the use of guidelines does not guarantee the long-term occurrence of sufficient habitat to maintain moose. The application of either approach should result in benefits to moose by producing more suitable forest landscape vegetation patterns in the short term.

Forest Wildlife Habitat Guidelines

Written guidelines are intended to constrain timber harvest management by defining and delineating the spatial and temporal distribution and the shape, size and interspersion of cut and uncut stands to create a forest that contains the essential habitat elements necessary to sustain moose and other designated wildlife. In Ontario and Manitoba, for example, timber companies are expected to incorporate individual wildlife species management guidelines in their timber harvest designs. However, guidelines are not laws, and application of such guidelines requires close cooperation of government biologists and industry foresters.

Forest Wildlife Habitat Objectives

Wildlife population objectives may be expressed as a specific amount of habitat necessary to sustain the population in question at a level that permits a certain degree of recreational hunting or viewing opportunities. New Brunswick (Patch et al. 1986), Saskatchewan (Saskatchewan Forest Habitat Program 1992) and Minnesota (W. Russ personal communication: 1995) have examples of specific projects or management programs based on the general theme of management by habitat objectives.

Habitat Management by Guidelines

Guidelines generally apply to six forest management activities—road construction, harvest, regeneration practices, tending, insect and disease control, and zoning for reserves. For example, identification and avoidance of important wildlife habitats are major objectives of road corridor selection, and emphasis often is placed on avoiding moose winter range. Closure of logging access roads also has been used to reduce hunting pressure on moose immediately after an area has been logged. In Saskatchewan, road retirements and closures, coupled with a road corridor/game preserve program that restricts hunting within 437 yards (400 m) of selected roads, resulted in a 60 percent increase in moose populations within 2 years (Greif 1992).

Timber harvest guidelines generally focus on cut-block sizes and shapes, although some consideration is given to scheduling harvest adjacent to initial cuts. Cut-block sizes recommended by various jurisdictions range from 50 to 346 acres (20–140 ha), depending on forest type (Table 31).

The primary strengths of guidelines include that they are:

1. relatively straightforward and easy for management agencies to implement, particularly when the number of featured species being managed is low;

Table 31. Cut-block size recommendations in moose habitat guidelines by state and provincial wildlife agencies, 1992

	Cut-block size in acres (ha)		
Jurisdiction	Conifer	Hardwood	Comments
Alberta	59–79 (24–32)	247 (100)	
Saskatchewan	99 (40)	297–346 (120–140)	
Manitoba	None	None	Distance to thermal cover less than 219 yards (200 m), and line of sight less than 437 yards (400 m)
Ontario	198–321 (80–130)		Distance to cover less than 219 yards (200 m)
Quebec	618 (250)		4 percent of the cut must remain in islands greater than 7.4 acres (3 ha)
New Brunswick	309 (125)		
Nova Scotia	124 (50)		
Minnesota	200 (81)		

2. readily understood and accepted by professional resource managers and the public; and
3. easy to identify and enforce compliance.

On the other hand, a number of weaknesses are inherent in the guideline approach to habitat management, including that:

1. guidelines rarely are connected to explicit forecasts of effects, therefore it is impossible to evaluate the outcomes relative to either stated or measurable objectives;
2. guidelines generally require approvals at the departmental level and, therefore, become institutionalized and extremely difficult to change; and
3. application of habitat guidelines for multiple species on the same land base (which usually is the case) may require evaluation of additional criteria to determine management priorities.

General Guidelines to Create Moose Habitat

Several investigators and agencies have formulated prescriptions to create moose habitat in North America. Among the first prescriptions was an attempt by Thomasson (1973) to provide areas of at least 10 square miles (25 km^2) in Ontario. That prescription called for 15 percent of an area to be capable of producing suitable aquatic vegetation, 25 to 40 percent of the area should be imperfectly drained to produce favored shrub species, the remaining 45 to 60 percent should support mixed wood forest, and tree height in shrub habitats should be maintained at less than 10 feet (3 m). A similar description was drafted by Peek et al. (1976) for northern Minnesota, featuring townships (36 square miles [93 km^2]) as the management units. Peek et al. proposed that 40 to 50 percent of the area (township) be maintained in clearcut less than 20 years old, with 5 to 15 percent in mature spruce/balsam fir, and 35 to 55 percent in water and mature trembling aspen/white birch stands. Allen et al. (1987) subdivided the latter category to include 5 to 10 percent in suitable wetlands. In Alaska, Oldemeyer and Regelin (1987) based moose management prescription on areas of 4,940 acres (2,000 ha) or more. They suggested that 40 percent should be in cover stands and widths of shrub-producing areas not exceed 656 yards (600 m), with 219 yards (200 m) as an optimum width. Current habitat management guidelines in Quebec call for residual stands of 7.5 acres (3 ha) each or more, to cover 3 to 4 percent of cut-blocks (Crête 1977), or 20 mixed wood stands of 5.0 to 7.5 acres (2–3 ha) each per 4 square miles (10.4 km^2) (Girard and Joyal 1984).

Montane habitats represent special cases because management is aimed primarily at maintaining winter shelter. Eastman and Ritcey (1987) categorized several wintering areas in British Columbia and generally advocated the use of cut-block size restrictions, usually under 247 acres (100 ha), with maximum distances between cut edges of generally less than 328 yards (300 m). They proposed logging by a series of three cuts, with 10 years or more between subsequent cuts. Peek et al. (1987) recommended that for areas dominated by Pacific yew in the understory, logging should remove less than 50 percent of the canopy—40 to 70 trees per acre (99–173/ha)—and maintain 50 to 60 percent of the yew subcanopy, using a rotation of about 210 years, with no less than 45 percent of the total area in trees younger than 90 years. They further suggested that no more than 14 percent of the entire winter range should be logged during any 30-year period.

Possibly the most extensive set of guidelines for managing moose habitat are those of the Ontario Ministry of Natural Resources (1988c) (Figure 151). The guidelines are to be applied generally to all commercial timber harvesting if the planned individual cuts exceed the limits specified by the guidelines, or if cutting is planned in winter concentration areas of moose. Different guidelines apply to Ontario's two major forest regions—Great Lakes/St. Lawrence and boreal. For the former, 15 percent of the total area managed should be maintained in mature conifer patches of 7.4 to 12.4 acres (3–5 ha); cuts in deciduous areas should be large enough to stimulate growth of deciduous shrubs. For boreal forests, the guidelines suggest cut-blocks of 198 to 321 acres (80–130 ha), with buffer zones between cuts and residual stands within cuts. The shelter patches of 7.4 to 12.4 acres (3–5 ha) of mixed wood, with at least 33 percent in conifer and a basal area of 48 square feet per acre (11 m^2/ha), 70 percent of which should be immature conifers, should be left to reduce the linear distance between cut edges to 328 yards (300 m). If the shelter patches are meant primarily for late winter use, then the stocking should be 70 percent or more of conifers.

In addition to these general guidelines, specific prescriptions are given for winter concentration areas of moose—e.g., cuts never greater than 437 yards (400 m) in width—and reserves are dictated around calving sites and aquatic feeding areas. Prescribed burning is recommended as the preferred site-preparation technique.

A major criticism of the Ontario guidelines is that they do not contain a temporal scale; there is no direction for managers that enables a perspective beyond the prevailing local population circumstance to future moose populations for a given management unit. Lacking are insights into such matters as the long-term sufficiency of late winter habitat; instead, the only guided strategy is to preserve existing

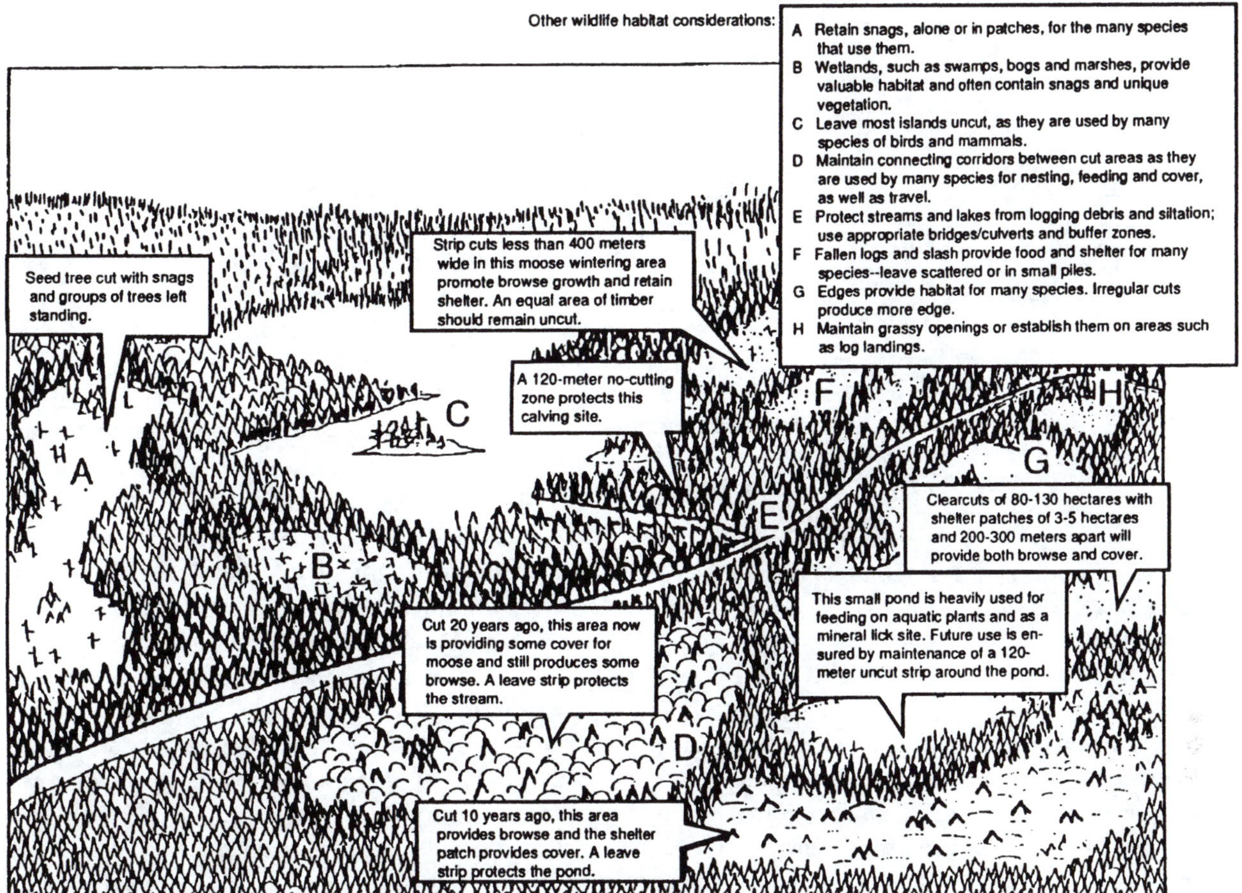

Figure 151. Expected effects of the application of moose habitat guidelines over an area of several hundred square kilometers of boreal forest in Ontario. The result is a landscape that is clearcut, yet the contiguous nature of normal logging procedures is broken by interspersion of uncut patches and strips (from Ontario Ministry of Natural Resources 1988).

areas. Eventually, stands preserved as winter habitat will die and break up, particularly in the boreal region. Planning is needed at temporal and spatial scales appropriate to the individual type of forest. A second and more generic criticism of the guideline approach to moose management is its lack of provision for research into the effects of guideline application to broad areas, not with respect to moose populations, but also in terms of overall biological diversity and sustainability.

Habitat Management by Objectives

Habitat management by objective requires that managers first identify priorities and assess the technical and economic feasibility of possible planning alternatives. For example, is it reasonable to manage for a substantial moose population increase on jack pine sites, on a defined land base, giving the existing knowledge of relationships between logging disturbance and subsequent browse regeneration on such sites?

A flow chart generated at the University of Arizona (see Covington et al. 1988) for the Terrestrial Ecosystem Analysis and Modeling System (TEAMS) provides a good conceptual overview of a natural resource decision support model that can be used in analysis of objectives (Figure 152). The model provides a framework for comparison of outputs of different operational management strategies, including economic evaluation. The analyses theoretically could be applied to any set of operational forest management plans.

The use of objective-setting procedures affords the manager substantial flexibility in the design phase of a management program, inasmuch as the number of alternative designs that could satisfy the objectives are potentially large. Timber harvest planning is a process that requires identification of specific stands eligible for harvest to meet wood supply requirements. The potential number of combinations of stands that could produce a desired spatial and temporal distribution to maintain a given moose population is a function of all of the stands eligible for harvest.

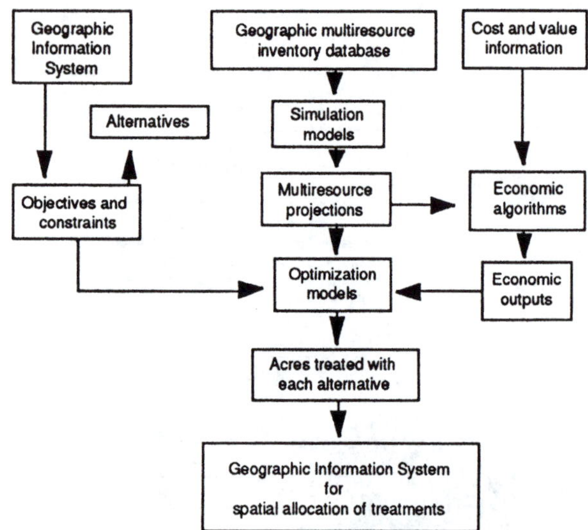

Figure 152. Flow chart for a forest management decision support system (Covington et al. 1988). Such charts demonstrate and organize the conceptual components of broad land management models.

The complexity of the challenge for forest managers increases with the number of features to be managed. Thus, if the goals are to manage moose, timber, plant diversity and neotropical songbirds, the potential number of alternative harvest strategies increases dramatically over managing for any single resource. The complexity of management by objectives can be viewed as an impediment to effective decision making. The magnitude of the number of choices may make identification of a preferred solution problematic. However, proponents argue that the selection need only be limited to a reasonable number of production possibilities, and that the advantages afforded by flexibility of the system far outweigh the risk of failing to identify the optimum solution.

Evaluation of the efficacy of an objective-based forest management design requires quantitative data. Habitat values must be measured or estimated in some manner and used to compare alternative strategies. The linkage of moose habitat suitability to wood supply models is the essence of the approach. A manager must assign indices of value to moose habitat for each forest stand, and then develop models to permit habitat assessment at appropriate temporal and spatial scales. A guidelines approach does not require any commitment to comparative statistics, and the manager only has to implement the guideline-based design on the land base. However, the existence of quantitative moose habitat evaluation procedures is needed regardless of which approach to management is taken. The literature on moose contains numerous investigations describing the features of moose habitat and their use, but there is a deficiency of meaningful correlations among various habitat configurations and expected population levels.

Wildlife Habitat Models

Wildlife habitat models attempt to define relationships between a species and its environment through mathematical functions (Schamberger and O'Neil 1986). Such models provide the manager with a tool to predict some attributes of a wildlife population by using a set of habitat components or measures that link habitat values to either quantitative or qualitative values for the population. Impetus for the development of applied wildlife habitat models began with the adoption of legislative acts on National Environmental Policy, Endangered Species, Forest and Rangelands Renewable Resources Planning, National Forest Management, and Federal Land Policy and Management in the United States (Berry 1986). These laws require land managers to predict the effects of resource use on species of wildlife and their habitats. The complexity of models is obviously related to the degree to which the natural system is to be perturbed, but as Salwasser (1986: 419) noted, "models are useful, but they must not be viewed as permanent statements of truth."

Because models represent simplifications of the natural system, tradeoffs in both accuracy and confidence are made when models are designed and applied. The more complex the model, the higher the uncertainty of accurate prediction. Given that interrelationships in forest ecosystems are complex, accounting for all ecosystem processes that define wildlife populations is neither possible nor reasonable.

The most widely accepted wildlife habitat modeling system is habitat evaluation procedures (HEP), developed by the U.S. Fish and Wildlife Service in the late 1970s and early 1980s, in response to the legislative requirements. The HEP approach relies on habitat suitability index (HSI) models that recognize that basic requirements of food, cover and physical habitat serve to define the potential of a land base to support wildlife populations. HSI models are not models of carrying capacity, and they explicitly exclude a host of factors that may interact to limit population size (Figure 153).

Habitat quality is expressed through values known as habitat suitability indices. Managers use data and expert opinion to develop equations that predict the suitability of habitats for wildlife based on a small number of variables. Habitat suitability for each variable being measured (tree canopy closure, successional stage of stand, amount of edge, etc.) is determined by an assigned value ranging from 0 (unsuitable) to 1 (optimal). An HSI model is constructed using some mathematical treatments of these habitat suitability indices.

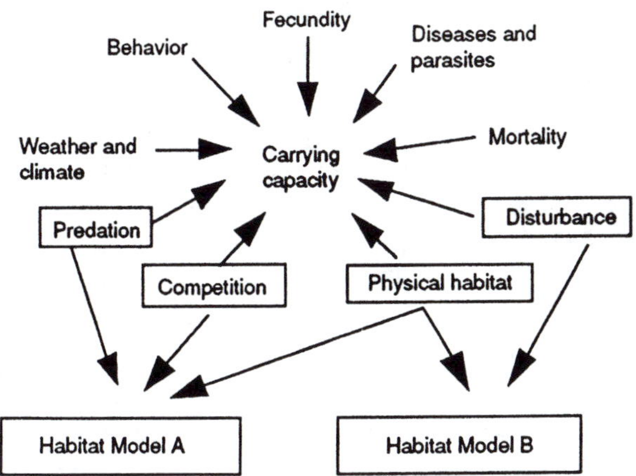

Figure 153. Factors that affect "carrying capacity"—the concept that any given area can sustain a certain number of animals of the same species for a particular length of time. "Habitat" (i.e., food, water, shelter and space) represents only one of many factors that influence carrying capacity, and the weight of individual factors differs from area to area (Schamberger and O'Neil 1986).

Another type of model is the habitat capability model. It differs from HSI in that habitat variables are used to predict animal density. For example, the pattern recognition model (PATREC) uses hypothetical conditional probabilities to estimate present and predict future animal populations (Berry 1986).

Numerous other modeling efforts are evolving, some of which include linkages between wildlife habitat models and models of vegetation dynamics and succession. Others capitalize on the spatial technology features of geographical information systems (GIS) to permit the display of data base attributes through mapped outputs. GIS enables the manager substantially increased latitude in defining and manipulating habitat variables, and dramatically expands the scope and accuracy of the products. Progressive technological advancements of both hardware and software are evolving so rapidly that it is virtually impossible at this time to track and evaluate all the potential options.

Habitat Suitability Models for Moose

An approach to the problem of assessing habitat for moose has been taken by Allen et al. (1987) through their development of a moose habitat model. The strategy has enabled an assessment of the predictions from models for individual habitat components (Adair et al. 1991, Allen et al. 1991). A modeling approach to habitat management provides several desirable features. First, as with any model, it forces the biologist to think in a reductive sense about how moose respond to individual components of their habitat. Second, the model permits predictions of how moose use their habitat and allows testing of those predictions, with subsequent adaptive reworking of model components. On the other hand, habitat suitability models are static, providing information for a point in time; they cannot be applied at a broad scale, through time to landscape-level populations.

Three HSI models for moose in the Lake Superior region were developed by Allen et al. (1987, 1991) (Table 32). Model I evaluates abundance and quality of growing season and dormant season food and cover, and is believed to be most applicable to small project areas. Model II defines an assumed relationship of cover type composition to moose habitat quality, and is recommended for rapid evaluation of large areas. Both models I and II are based on the assumption that habitats with food of sufficient quality, inter-

Table 32. Model life requisites, data requirements and outputs for habitat suitability index (HSI) models I and II, for moose in the Lake Superior region (from Allen et al. 1987)

HSI model	Life requisite	Data requirements	Output
I	Growing season browse	Annual browse production, diversity and quality	Potential number of moose/km^2 that can be supported by growing season browse
	Aquatic forage	Area of nonforested, nonacidic wetlands	Potential number of moose/km^2 that can be supported by aquatic forage
	Growing season cover	Area and species composition of forest cover	Potential number of moose/km^2 that can be supported by growing season cover
	Dormant season browse	Annual browse production, diversity, quality, and distance to dormant season cover	Potential number of moose/km^2 that can be supported by dormant season browse
	Dormant season cover	Height, density and species composition of forest cover	Potential number of moose/km^2 that can be supported by dormant season cover
II	Cover type composition in relation to overall habitat quality	Percentage of area in the following cover types: shrub and forested $<$20 years old; spruce/fir forest $\geq$20 years old; deciduous or mixed forest $\geq$20 years old; and nonforested wetlands	Index of habitat quality ranging from 0.0 to 1.0 where 0.0 = unsuitable and 1.0 = optimum

spersed with a suitable amount and quality of cover, have the potential to support moose populations that will increase at faster rates and stabilize at higher densities than will habitats without these features. Relatively sophisticated, model III uses GIS to account for habitat interspersion, and was designed using concepts of the dormant season component from model I. Model III uses a matrix of late winter cover and forage suitability indices as source of algorithms defining optimum late winter interspersion and habitat, using the area within 109 yards (100 m) of high-quality forage and cover (Table 33). Cover is defined as having a suitability index greater than or equal to 0.5 (Table 34). The model was tested with good success on 189 square miles (490 km^2) of the Superior National Forest in northeastern Minnesota (Allen et al. 1991).

Approaches taken to integrate moose and timber management plans for pilot projects in Saskatchewan and New Brunswick have paralleled the Superior National Forest study. Forest inventories generally provide the baseline data for forest productivity and moose habitat assessment. Most commercial forest inventories contain stand-level information to indicate tree species composition and associations, canopy closure, age or maturity classes, and usually soil texture and drainage. Other information may be available for forest successional patterns, growth and yield curves for commercial tree species and, in some cases, site classification systems used to relate vegetation community patterns to site features (soil fertility and moisture). This information then can be interpreted in a model for moose habitat.

Forest inventory is evolving toward the use of common data bases by forest and wildlife managers. The intent of inventory and modeling efforts is to establish data tables that

Table 33. Model III dormant season forage suitability indices for various forest types and size/density classes (after Allen et al. 1991)

	Stocking percentage by size/density class									
	Nonstocked	Seedlings			Pole timber			Saw timber		
Forest type	<16	16–39	40–69	>70	16–39	40–69	>70	16–39	40–69	>70
Jack pine	0.5	1.0	1.0	1.0	0.5	0.2	0.1	0.1	0.1	0
Red pine	0.5	1.0	1.0	1.0	0.5	0.2	0.1	0.1	0.2	0
White pine	0	0.2	0.2	0.2	0.1	0.1	0.1	0	0	0
Balsam fir/trembling aspen/paper birch	0.1	1.0	1.0	1.0	0.5	0.5	0.5	0.1	0.1	0.1
Black spruce	0	0	0	0	0	0	0	0	0	0
Red spruce/fir	0	0.2	0.2	0.2	0.1	0.1	0.1	0	0	0
Northern white cedar	0	0.2	0.2	0.2	0.1	0.1	0.1	0.1	0.1	0.1
Larch	0	0	0	0	0	0	0	0	0	0
Lowland brush	1.0	1.0	1.0							
Upland brush	1.0	1.0	1.0							

Table 34. Model III dormant season cover suitability indices for various forest types and size/density classes (after Allen et al. 1991)

	Stocking percentage by size/density class									
	Nonstocked	Seedlings			Pole timber			Saw timber		
Forest type	<16	16–39	40–69	>70	16–39	40–69	>70	16–39	40–69	>70
Jack pine	0	0	0	0	0	0	0.1	0.1	0.2	0.3
Red pine	0	0	0	0	0.5	0.2	0.1	0.1	0.2	0.3
White pine	0	0	0	0	0.1	0.1	0.2	0.2	0.3	0.3
Balsam fir/trembling aspen/paper birch	0	0	0	0.1	0.1	0.5	0.5	0.8	0.8	0.8
Black spruce	0	0	0	0	0.1	0.1	0.2	0.2	0.3	0.3
Red spruce/fir	0	0	0	0	0.5	0.8	0.8	1.0	1.0	1.0
Northern white cedar	0	0	0.1	0.1	0.2	0.5	0.5	0.8	1.0	1.0
Larch	0	0	0	0	0	0.1	0.2	0.2	0.2	0.2
Lowland brush	0	0	0							
Upland brush	0	0	0							

assign appropriate habitat suitability values to various types of forest stands. The degree of sophistication is a function of the quality and quantity of stand information (forest type, yield and diversity of understory forage species). The resolution and quality of model output obviously improve as input data quality is enhanced.

The Superior National Forest data on forage habitat suitability was generated from browse preference information. The forest inventory was assumed to describe the availability of dormant season forage within each cover type, based on relationships similar to those described in Figure 154. Portions of the Allen et al. (1991) stand data tables that were generated by application of forage and suitability curves are shown in Tables 33 and 34. In general, the forage suitability indices were assumed to decline with increasing stand age and density. Allen et al. (1987) also considered that interspersion of cover and food was important in optimum late winter habitat. Optimum interspersion was defined as the area bounded within 109-yard (100-m) interfaces between high-quality forage or high-quality cover. GIS data bases are needed to calculate interspersion indices.

The Superior National Forest HSI provides the opportunity to estimate habitat suitability for moose within a defined land base at a point in time. However, complete habitat management requires models capable of temporal prediction. The latter can only be enabled by linking the habitat model to an appropriate model of vegetative succession. Disturbance by fire or logging initiates forest regeneration. The path of succession is influenced by history, climate, physiography, disturbance type, site characteristics and silvicultural treatments. Attention to forest dynamics is particularly important where the joint production of timber and moose are objectives, because the value for each changes relative to the other as the forest matures.

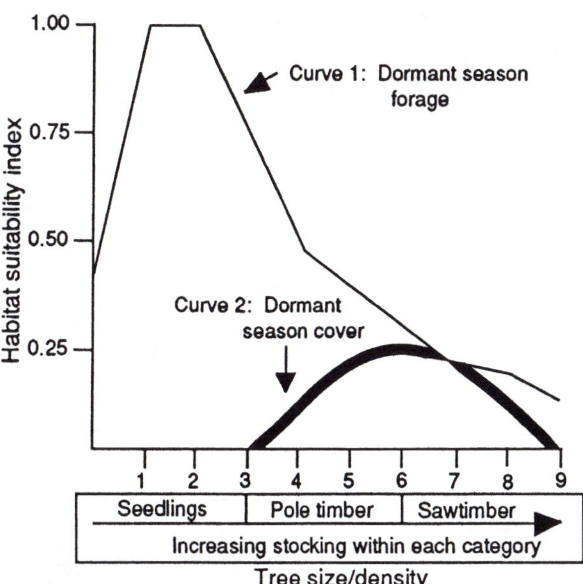

Figure 154. Relationship of dormant season forage and cover to habitat suitability index for stocking and stem size in aspen-dominated stands (adapted from Allen et al. 1991).

A simple model that simulates the natural effects of forest succession on the availability and distribution of forage and cover is illustrated in Figure 155. Logging operations may promote habitats that increase the availability of forage at the expense of cover, but successional processes tend to favor the expansion of cover at the expense of forage. The optimum solution for maximizing the production of moose may be one that provides the most attractive spatial and temporal distribution of food and cover, assuming all other life requisites are met. Objectives to maintain a population of moose on a defined land base in perpetuity would demand that the future supply of habitat could be predicted.

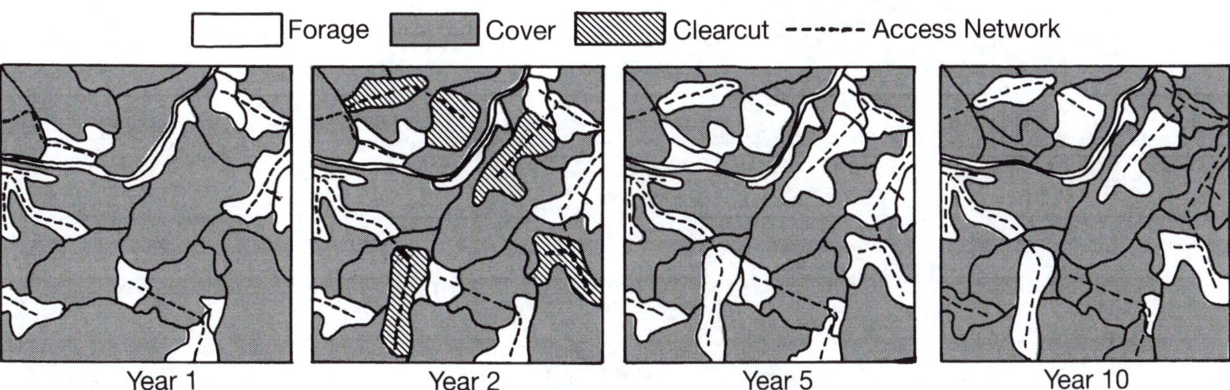

Figure 155. Habitat information systems can (and should) be used to plan and monitor spatial and temporal aspects of disturbances to and regeneration of moose habitat, such as created by commercial logging. The size of the habitat plan and other characteristics, including topographic relief, are aspects of input by habitat information systems.

The adoption of a habitat-based approach is not without risks and uncertainty. O'Neil and Carey (1984) provided the following cautions with respect to the use of variables to describe habitat:

1. Habitat descriptors may not account adequately for a species' tolerance of physical conditions, life history requirements and affinities for particular types of environments or for the relationship of the species to other members of an ecological community.
2. Particular variables from one area cannot predict migratory species' population responses from another area because of the influence of either dissimilar or geographically separated habitats.
3. The influence of weather on a population or the complex interactions of weather, prey, predators, competitors and parasites all will alter the predictive capability of habitat variables.
4. Consistently low and highly variable populations of a particular species are difficult to predict.
5. Habitat descriptors cannot account for the factors controlling populations that may vary among and within types of ecoregions and from year to year.
6. Population responses that are subject to stochastic events that may be genetic, demographic or environmental in nature and, hence, hard to predict using habitat variables.

Adaptive Management Strategies

The complexity of forest ecosystems, combined with the large area under forest management regimes, introduces substantial uncertainty and risk into forest planning. Variability in the precision and accuracy of information, based on statistical sampling, generates error when the quantity and quality of the data are limited. Constraints on the data base are related to both practical and resource limitations, and are compounded by the time frame over which forest management activities occur. However, to abandon progressive action in the face of uncertainty would obviously stall management. Therefore, an adaptive management strategy is proposed to address uncertainty in planning. The concept was developed by Holling (1978), and its application to integrated forest management was expanded by Baskerville (1985).

Adaptive management is management with negative feedback control that Baskerville (1985) characterized by nine steps:

1. A quantifiable goal is chosen for management of a natural system.
2. Links among the goal, the cause–effect system dynamics, and the possible management actions are explicitly stated, usually in a model, for a system forecasting tool.
3. With the model forecasting tool, a set of actions is designed to regulate the system toward the goal. The way each member of the action set causes an effect and the way all actions collectively cause the system goal to unfold are stated quantitatively.
4. The action set is implemented by the natural system.
5. At specified times, the progress of the system toward the goal is measured in the terms specified in step 1. The actions and their results in the cause–effect sense also are measured in the manner in step 3.
6. The measured progress toward the goal is compared with the explicit forecast from step 1, and the deviations are noted. The results of action in terms of cause–effect are compared with the explicit forecast in step 3, and the causes of deviations are determined.
7. On the basis of comparisons of the forecasts with actual system performance, adaptations are designed. The goal is reevaluated, based on the measure of progress toward the goal, to ensure that it is attainable and, if necessary, it is changed. Wrong forecasts of system performance indicate an incorrect structure of the dynamic cause–effect relationships in the forecast. In this case, change must be made in the model (hypothesis) of structure of the system dynamics, and new forecasts are prepared. In turn, this leads to design of a new action set necessary to reach the goal under the new statement of cause–effect relationships.
8. The new action set is implemented in the natural system.
9. The process is returned to step 5.

The significance of an adaptive management as a strategy is the attention to learning by doing, and treating each action as an experiment with a clearly identified hypothesis. Uncertainty and risk are progressively reduced as more is learned about forest responses to harvesting practices. For example, imposing shelter patches has cost consequences for a timber company. If research were to show that patches of a certain size were of no value to moose, then the management strategy would be changed.

Some Examples of Moose Habitat Projects

In Alaska, moose habitat is created primarily by wildfire, such as the large burns in 1947 and 1969 on the Kenai Peninsula (Spencer and Hakala 1964; see also Chapter 11). Because of natural forest succession, the habitat within those two areas became decreasingly productive for moose (Schwartz and Franzmann 1989). An experiment was undertaken to improve moose range for moose by crushing mature mixed woods and aspen stands using "LeTourneau tree crushers" to produce browse (Oldemeyer and Regelin 1987). Between 1975 and 1978, 10.5 square miles (27 km^2) of forest were crushed in four areas. Although the total experimental area was not large enough to show an effect at the population level, there were some meaningful results. On the sites where birch and aspen were dominant, the program was successful in increasing food available for moose. In some cases, the moose response was so extensive that successional patterns were retarded (C. Schwartz personal communication: 1992). Poor results in terms of browse production were achieved on some sites, but this could have been anticipated given the original stand types and autecology of the dominant deciduous species. The crushing study clearly showed the need for managers to understand forest ecology fully before habitat manipulation. An important conclusion from the tree-crushing study was that the production of browse plants could be accelerated by as much as 8 years over the rate expected after fire (Regelin et al. 1987a, C. Schwartz personal communication: 1992). A secondary effect of browse rehabilitation on the Kenai Peninsula was a reduction in predation on moose calves by black bears (Schwartz and Franzmann (1983). Tree crushing apparently reduced escape habitat for the bears (large trees for climbing), and lower bear density resulted in lower calf mortality.

In Ontario, G. Eason (1989, personal communication: 1992) attempted to protect a local moose population through a modified timber harvesting program that resulted in substantial escape cover from hunters. The program was designed to provide greater cover for moose compared with that resulting after normal clearcut logging. The prescription was for dispersal of 0.4-square-mile (1-km^2) shelter patches and cuts over the 77-square-mile (200-km^2) study area. Owing to an already excellent mixture of cover and food, moose density on the study area was high before logging. The moose population declined subsequent to harvesting, as would be expected given hunting and lack of food in newly logged areas. However, the decline was much less than normally associated with newly clearcut areas, and aerial censuses suggested that the local moose population was about 75 percent of prelogging levels. Adjacent clearcut areas had moose densities two-thirds lower than on the experimental site. The population remained at the same postcut density up to 5 years later. An important result of the study was the observation that moose were not equally distributed over the study area. Moose were strongly associated with the largest residual patches of 494 acres (200 ha) or greater. If there is to be a population response to the creation of abundant edge, cover and browse on the study area, it should occur over the next decade as browse production peaks. Unfortunately, the long-term results of this study are confounded by substantial site preparation accomplished by forest managers to reduce and eliminate deciduous species in favor of conifers.

A winter aerial view of a portion of Alaska's Kenai Peninsula featuring an area mechanically altered with LeTourneau crushers. The treatment included leaving "islands" to serve as cover. Initially, because the crushed area was too large and flat, moose were reluctant to move away from the area's edge. But as regrowth gained height, moose used the area entirely. *Photo by Albert W. Franzmann; courtesy of the Alaska Department of Fish and Game, Soldotna.*

An ideal example of moose habitat enhanced by patch cutting southwest of Disraeli Lake in the Thunder Bay District of Ontario features a high degree of edge created by the admixture of deciduous and coniferous regeneration adjacent to logging roads. Repopulation by moose of a treatment site depends on many variables, including but not limited to size and shapes of the treatment, ecotype, snow conditions, climate, vegetative type and stages, treatment method, proximity moose population size, and hunting pressure. In mixed boreal forest, as featured, an immediate response by moose to the availability of down limbs and forb regrowth would be expected, assuming the presence of a moose population. This use would taper off after the first growing season and decrease to about year 5, when regrowth of deciduous browse plants would start becoming productively useful to moose. This would continue to about year 20. Any increase in food availability that is additive will potentially increase the health and productivity of the moose population. If the moose population is not regulated by harvest or otherwise, or if browse production does not increase, the result would be a decline in moose population health and productivity. To maintain a high level of browse production and reasonable stability of the moose population, retreatment of the area would have to be undertaken every 20 years. This also is the treatment rotation for coastal lowland boreal forest land of Alaska's Kenai Peninsula. Other rotation patterns apply to other areas. *Photo by H.R. Timmermann; courtesy of the Ontario Ministry of Natural Resources.*

Under a habitat redevelopment program in Alberta, several areas of public lands are being managed to increase local ungulate populations. An example from the westcentral portion of the province in 1987–1990 was carried out to enhance existing habitat by increasing browse for moose on 188 acres (76 ha) at a cost of $107 (Canadian) per acre ($265/ha) (Thomson 1990, M. Brumen personal communication: 1992). Pretreatment surveys indicated that food probably was a limiting factor for moose. The dominant browse species in the area were willows and aspen, and the plants either were too tall or browsed so heavily that only large diameter twigs remained. Heavy equipment was used either to bulldoze or cut vegetation in open areas among residual stands of conifers. Although the earliest treatments occurred 3 years before assessment, the data showed that browse production by willows had already exceeded that available on untreated sites. These browse production projects were small and affect only the local population, but they illustrate the kind of habitat manipulation that is possible for a particular forest type to improve habitat for moose.

Research Needs

Importantly, the appropriate scale for moose management is an entire management unit of, for example, 386 square miles or more (1,000 km^2). Therefore, the most important research question is whether habitat management pro-

An uncut knoll amid a 1,600-acre (648-ha) jackpine clearcut in the Dog River area of Ontario's Thunder Bay District. Clearcutting followed by scarification to expose the soil is a common silvicultural practice to promote regeneration. Large clearcuts, however, usually do not provide a good combination and variety of edge favored by moose. A modified cut in patches or strips over a number of years would have produced the desired variety, edge and retention of cover. *Photo by H.R. Timmermann; courtesy of the Ontario Ministry of Natural Resources.*

grams have effect on landscape-level moose population. Large-scale experimentation, as advocated by Walters and Holling (1990), should be a priority for research into the effectiveness of moose habitat management. Such programs must be designed to allow control of the many factors affecting moose populations. They must involve carefully formulated hypotheses and be adequately replicated. Minimum requirements to measure effects are preexperiment population, predator and habitat data; postexperiment time-trend data on populations; monitoring of habitat variables including food and shelter; monitoring of mortality sources; and measurement of weather. Large-scale research management programs must be followed consistently for at least 30 years. No published experiments to date have met many or all of these criteria.

Many factors limit and regulate moose populations (Messier 1991; see also Chapter 5). To determine whether habitat management can be conducted so as to optimize and sustain moose populations, further intensive research is needed into the ways that limiting factors such as disease, parasitism, predation and weather interact with the spatial arrangement of food, shelter and other critical components of moose habitat. In particular, the dispersion of moose and wolves on a landscape can be altered by the placement of travel corridors and retention of forage patches and riparian buffers, and therefore, alter the dynamics of predation. Research into the consequences of management actions on moose populations is needed. Careful testing of hypotheses is required to examine interactive effects of habitat change on factors that may limit moose population growth.

(LIFE + MEMORY)

JULY 11 – WONDER LAKE –
BULL FEEDING IN SHALLOWS NEAR N END,
WITH McKINLEY OUT (BUT HAZY) IN DISTANCE –

TOO MANY PHOTOGRAPHERS –

SUMMER MOOSE
HAS WEIGHT IN BELLY
RATHER THAN NECK &
SHOULDERS

(LIFE &
MEMORY)

BROWN WITH TAN GRIZZLING
ON NECK & SHOULDERS – RICH BROWN ON
BODY – DEEP UNDERPARTS –

BULL MOOSE
FEEDING IN SHALLOWS, WONDER LAKE
McKINLEY PARK, ALASKA – JULY 11, '56

(MEMORY)

ONE OF 3 BIG BULLS FEEDING
ON SLOPES ABOVE TIMBER – IGLOO CREEK,
McKINLEY PARK – JULY 27, 56

RESTING & CHEWING CUD

13

LYLE A. RENECKER AND CHARLES C. SCHWARTZ

Food Habits and Feeding Behavior

Daily patterns of moose use of time and space explain how this animal satisfies hunger, remains fit, avoids thermal stress, maintains security from predation, and reproduces. Because many of the moose's life cycle prequisites interact daily, trade-offs often occur, because most requirements are more critical at certain times than at others. The day-to-day needs of moose for food and thermoregulation are most often pre-empted in favor of other activities that accommodate fitness and mating. However, the survival instinct is satisfied most on a daily basis by optimizing food consumption at minimal risk and effort. In this regard, a basic constraint for most moose is an abundant food supply of low quality.

Patterns of Diet Selection

Foraging Strategy and Ruminant Physiology

Characteristic of the ruminant is a four-chambered stomach, consisting of a rumen, reticulum, omasum and abomasum (Figure 156). The rumen is the first compartment that food enters when swallowed. Its walls are thin and lined with finger-like projections called "papillae" (Robbins 1983). The papillae increase the surface area of the rumen for absorption of fermentation by-products. As food particles are reduced in size, through rumination, then regurgitated for further mastication and reswallowed, they pass into the honeycombed reticulum. Further reduced in size in the reticulum, particles of ingesta move from the ruminoreticular vat into the weir-like omasum, which separates out the highly acidic digesta for entry to the abomasum (true gastric stomach for enzymatic digestion) (Robbins 1983). Large particles are retained in the ruminoreticulum for further breakdown.

For moose, as with other mammals, milk digestion is the initial nutritional investment of the neonate. A muscular esophageal groove, which extends from the esophagus to the abomasum, allows the suckling offspring to function monogastrically (as with one stomach). Milk flows from the esophagus through the groove to bypass the fermentation chamber (where energy loss would occur because of anaerobic fermentation) to enter the gastric abomasum. As a pre-ruminant (before forage is consumed), the nursing neonate lacks substrates (solid forage) and bacteria in the rumen or the lower intestine to generate energy for survival. Therefore, milk is required as the primary energy source. As the animal grows, its esophageal groove becomes well-developed and muscular. Hofmann (1985) speculated that it provides an avenue for soluble sugars from forage to bypass the rumen to be absorbed directly in the small intestine.

In ruminants browse, forbs and grass are held in the large-chambered rumen of the stomach until adequate nutrients are extracted from the fibrous material and the plant particles are small enough to pass through the ruminoreticular to the omasum. On the bases of feeding habit specialization and design of the digestive anatomy, ruminants are classified into three main groups (Hofmann 1973): *browsers* (herbivores that eat primarily shrubs and trees); *grazers* (herbivores with a diet that comprises mainly grasses); and

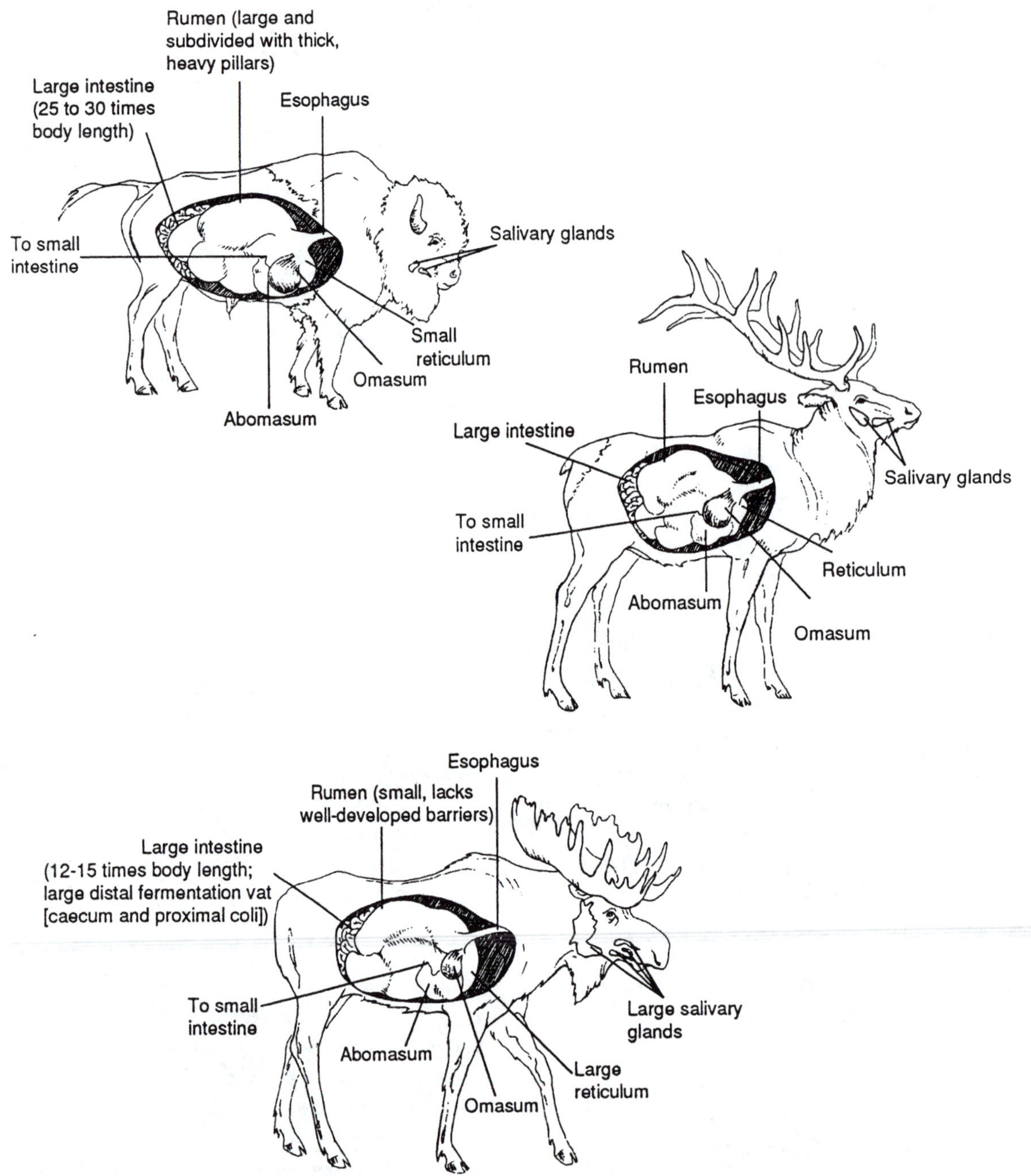

Figure 156. Anatomy of the digestive system of various ruminant species is influenced by the type of foods they eat (from Renecker and Hudson 1993). Roughage eaters, such as the American Plains bison, consume a diet of grass, which is high in fiber. Contrasting with the roughage eaters are the concentrate selectors, such as moose, which eat foods that are easily digested and low in fiber but high in lignin. Intermediate feeders, such as elk, eat a mixture of grass, forbs and browse in relation to their availability and quality. The general anatomy and morphology of the intermediate feeder are better adapted for greater digestive capacities and more fibrous diets than are those of concentrate feeders (Robbins 1993). Intermediate feeders also can change the mucosal absorptive surface of the rumen seasonally in response to the amount of fiber in the diet. Roughage eaters have small salivary glands and liver, but large rumens and long intestinal tracts (25–30 times their body length) to accommodate their poor quality, high fiber diets. Passage of digesta through the stomach compartments is slow to accommodate a lengthy fermentation rate. Concentrate selectors tend to have the opposite physical characteristics, with a small intestine (12–15 times body length) and rapid passage of digesta through the stomach with a high fermentation rate. Concentrate selectors have developed relatively large salivary glands in relation to body weight, and these maintain high flow rates to increase the buffering capacity of the rumen during rapid fermentation and perhaps accelerate the passage of sugars through the rumen–reticulum for absorption in the small intestine. Moose stomach redrawn from Hofmann and Nygren (1992) and Hofmann (1985).

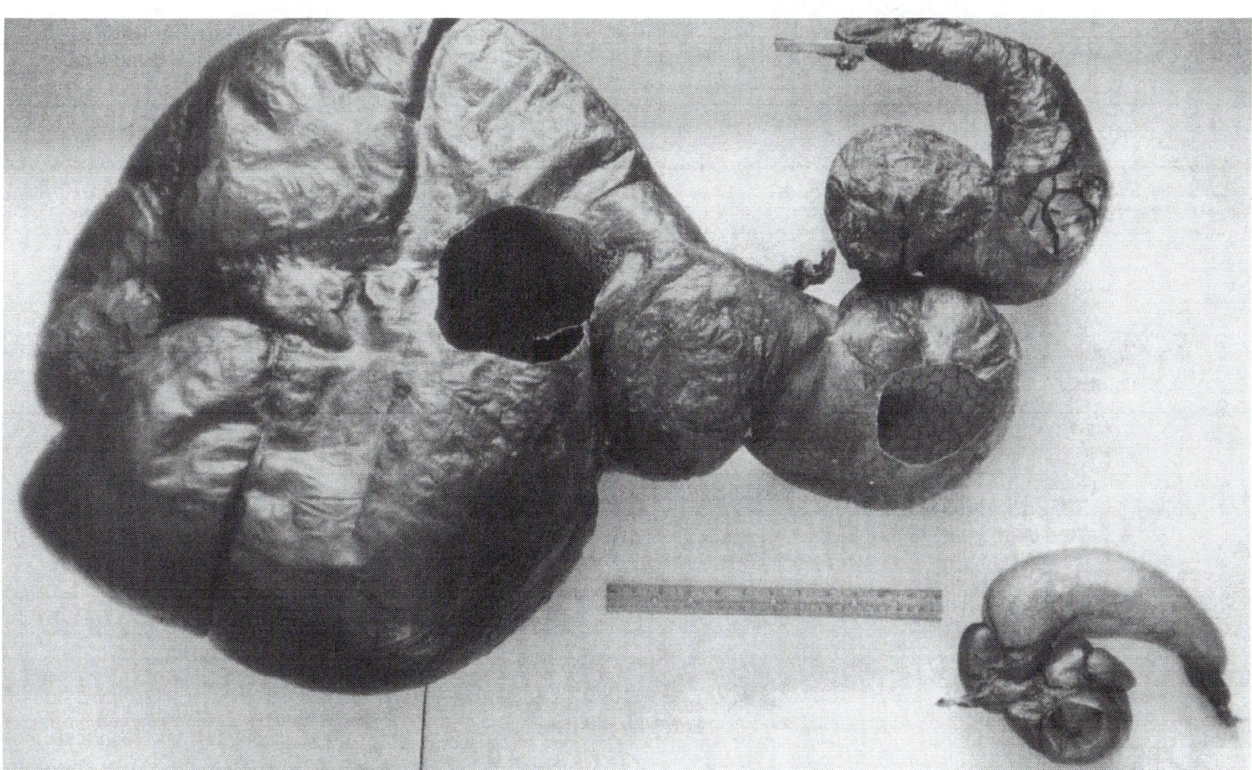

A member of the suborder Ruminantia, the moose possesses a four-chambered stomach. In the adult moose, the rumen (large chamber on the left) serves as a fermentation vat where bacteria break down food particles and volatile fatty acids are absorbed through the wall. The reticulum is the small chamber attached to the rumen. It sometimes is called the "honeycomb" because of the appearance of its lining (visible through the cut-out hole). Digested food from the rumen–reticulum complex (rumenoreticulum) passes into the omasum (directly above the reticulum) where excess water and soluble nutrients are removed. The horn-shaped abomasum (upper right), or "true stomach," is the site of acid digestion similar to the monogastric stomach. From the abomasum (point of clothespin), the digested food passes into the small intestine, which is the primary site for enzymatic digestion and absorption. The digesta then enters the cecum and large intestine where some plant fiber fermentation, absorption of water and soluble nutrients and bacterial vitamin synthesis occur. The approximate percentage of total volume each chamber occupies is 88 for the rumen, 8 for the reticulum, 2 for the omasum, and 2 for the abomasum. Calves are born with a well-developed abomasum (lower right) that efficiently digests their milk diet. The other three stomach chambers are undeveloped at birth and slowly enlarge as the calves begin eating vegetation. In the stomach of a 1-week-old calf (lower right), the percentage of total volume is 29 for the rumen, 6 for the reticulum, less that 1 for the omassum and 64 for the abomasum. *Photo by Charles C. Schwartz; courtesy of the Alaska Department of Fish and Game.*

mixed or *intermediate feeders* (those ruminants that survive on a mixture of grasses, browse and forbs depending on availability and quality) (Figure 157). The moose is a browser.

Besides moose, the browser group of ruminants includes black-tailed deer, white-tailed deer, roe deer, dik-dik, duiker, kudu and giraffe. These herbivores consume mainly tree and shrub leaves, but browsers in northern latitudes eat woody twigs during winter. Nevertheless, these species also have a relatively small ruminoreticular chamber and must search for high-quality foods that will pass rapidly through the digestive system. A comparison of dietary habits of some North American ruminants is presented in Table 35.

Browsers consume plant species (browse) and parts (twigs and foliage) high in cell-soluble sugars that ferment readily in the rumen. They generally avoid plants that are fibrous and require extensive breakdown in size before passage from the rumen. Browsers have narrow muzzles, prehensile lips and tongues that allow them to select high-quality but small plant parts. The rumens of these selective feeders generally are small and lack barriers that would delay passage of digesta, but they have abundant papillae to help absorb volatile fatty acids (energy source) produced by rapid fermentation. The rumenoreticular contents of browsers range from 6.5 to 14.8 percent of live weight, compared with 11 to 16 percent (average 13.3 percent) for grazers (e.g., American bison, buffalo and wildebeest) and an 11.5 percent average for intermediate feeders (e.g., elk, red deer, reindeer) (Kay et al. 1980). To help maintain an optimal

Figure 157. Classification of various Cervidae on the basis of morphophysiological feeding type (Hofmann 1985). Most cervids are intermediate feeders and show some flexibility in adaptation to forage type. Bison would be equal to domestic cattle. The moose is the largest cervid, therefore, also the largest concentrate selector. Feeding frequency illustrates the relative number of feeding periods or bouts that occur daily for concentrate selectors, intermediate feeders and roughage feeders.

Neonate moose calves obtain energy for growth from their dams' concentrated milk. By 3 weeks of age, calves begin sampling forage. Ultimately, their survival depends on the introduction of specialized bacteria populations into the rumen to digest hard forage. As the calf grows, solid food becomes a greater portion of their daily energy needs. By 3 months of age, moose calves generally are weaned. *Photo by James Keith Rue.*

Table 35. Average food habits of certain wild ruminants (Van Dyne et al. 1980)

	Percentage of diet		
Species	Shrubs/trees	Forbs	Grass
Bighorn sheep	21	14	65
Bison	4	5	91
Caribou	46	22	32
Desert mule deer	59	28	13
Elk	17	14	69
Moose	90	8	2
Mountain goat	10	29	61
Musk-oxen	11	2	87
Pronghorn	43	42	15
Reindeer	16	49	35
White-tailed deer	60	30	10

environment for digestion in the rumen, browsers have relatively larger salivary glands—about 0.18 percent of body weight, compared with 0.06 percent for grazers and 0.12 percent for intermediate feeders (Kay et al. 1980)—and higher flow rates of saliva than the other ruminant groups. This higher production of saliva may aid in the flow of soluble sugars and particles to bypass the ruminoreticulum to the small intestine through an esophageal groove. In this way, browsers avoid the loss of energy to by-products of anaerobic fermentation that would occur in the rumen environment.

"Generalist" browsers—such as deer, including moose—ingest moderate amounts of a variety of secondary plant compounds in comparison to "specialist" browsers which consume high levels from only a few plant species (McArthur et al. 1991). Studies have shown that generalist browsers have evolved tannin-binding proteins in their saliva to combine with and precipitate the tannins (Austin et al. 1989). Hagerman and Robbins (1992) found that moose produce specific salivary proteins that only bind linear-condensed tannins common in their key foods, such as willow, aspen and birch. It appears that this is an adaptive response of moose to a feeding niche with presence of secondary plant metabolites in browse from shrubs and trees.

Body Size and the Bell-Jarman Principle

Attempts have been made to classify feeding styles of ruminants on the basis of body size and how food is collected (Bell 1970, 1971, Jarman 1974, Jarman and Jarman 1974). Large body size is a common characteristic of ruminants that occupy open environments (i.e., grasslands) where food is more abundant but generally lower in quality. This larger size probably evolved in response to a need to reduce the risk of predation, ensure mating success through dominance and increase the animal's ability to travel rapidly because of the lack of abundant cover. Thus, a relationship exists between body size, habitat type and abundance of food. Compared with smaller animals, ruminants with a large body size have lower energy requirements relative to body weight.

Consider, for example, the comparative energy requirements of the 11-pound (5-kg) dik-dik and the 1,323-pound (600-kg) bison. The dik-dik needs about 11.6 kilocalories of energy per pound of body weight$^{0.75}$ per day (88 metric kilojoules of energy/kg of body weight$^{0.75}$/day). Therefore, total daily body needs of the dik-dik are 128 kilocalories of energy or 11 pounds times 11.6 kilocalories of energy per pound of body weight$^{0.75}$ (440 metric kilojoules of energy; 5 kg X 88 metric kilojoules of energy/kg body weight$^{0.75}$). The bison, on the other hand, requires about 0.2 kilocalorie of energy per pound of body weight$^{0.75}$, for a total daily body requirement of 265 kilocalories of energy or 1,323 pounds times 0.2 kilocalorie of energy per pound of body weight$^{0.75}$ (2 metric kilojoules of energy/kg body weight$^{0.75}$, for a total daily body requirement of 1,200 metric kilojoules of energy). Thus, the dik-dik has a higher relative requirement than the bison (11.6 versus 0.2), but a lower absolute need for the entire body (128 versus 265). This was calculated from the interspecies equation for energy cost for basal metabolism (BMR) of $70W^{0.75}$ ($293W^{0.75}$).

In contrast, larger ruminants, such as moose, require greater absolute quantities of lower quality but more abundant food and therefore, their limitation becomes one of logistics and the fact that there is only so much time in a day that can be spent consuming food (Figure 158). It would be impossible for a large-bodied ruminant to consume enough food if its strategy was to forage on sparse but high-quality items, simply because too much time would be spent in search rather than harvest effort. Conversely, small selective feeders require high-energy foods (i.e., fruits and leaves) because of the high maintenance cost of their body, but they have a relatively limited rumen capacity relative to body size. The relationship can be thought of in terms of one ruminant having a motor with a high idling speed and requiring a small but relatively constant injection of high octane fuel, whereas the larger animal idles at a lower speed and uses lower octane fuel, but requires a greater quantity on a daily basis.

This pattern is the basis of the "Bell-Jarman principle," which implies that ruminant size will determine differences in food selection among various herbivore species (i.e., browser versus grazer). The differences depend on forage quantity and quality. As a result, limitations of each ruminant species and its feeding strategy will be determined by tradeoffs between the time available to carry out daily activities, the energy cost of those activities, and the limitations on eating food, such as ruminoreticulum capacity. Small ruminants are expected to spend more time searching for high-quality foods to meet their relatively high metabolic requirements. Nevertheless, the costs—energy in this case—must not exceed benefits. In comparison, large grazers do not have the same need for high-quality forage, but do require greater amounts of food. The intermediate feeder is more flexible and adjusts the seasonal mixture of dietary items relative to availability and quality.

Moose appear to violate the Bell-Jarman principle. As previously indicated, a large body size is a common characteristic of ruminants that inhabit open habitats, such as grasslands, that support abundant but lower quality food resources. Therefore, it is unusual, at least according to principle, that moose, largest of the cervids, live as browsers

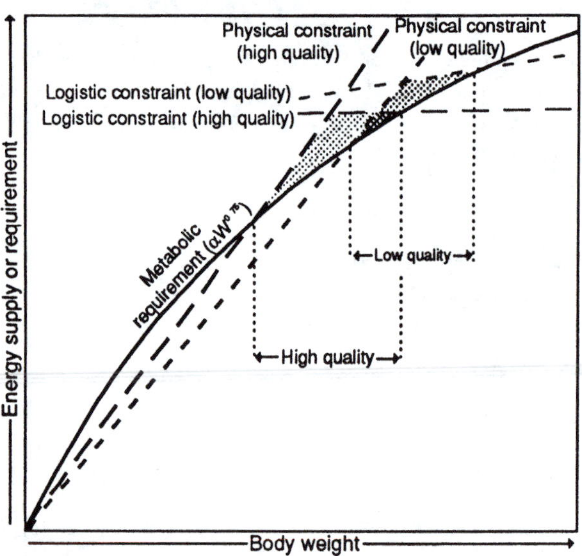

Figure 158. Constraints on adaptive body size of herbivores exploiting both high- and low-quality food resources (from Hudson 1985). Metabolic or energy requirements scale to body weight to 0.75 power. Hence, as an animal gets larger (more body weight), its energy requirements are less per unit of body weight and the metabolic requirement line assumes a curvilinear shape. Physical constraints scale to the power 1.0 and are linear. The logistical constraint refers to the animals' ability to harvest food and is independent of body size for high-quality forages (hence, the flat line). However, the ability to harvest food improves for large animals eating poor quality diets, perhaps because incisors grow with increases in body size. The relationship between metabolic requirements, digestive capacity and foraging efficiency places limits on the range of adaptive body size expected for ruminants that attempt to exploit any particular food resource.

In aspen-dominated boreal forests of western Canada, three ruminant species have habituated to lifestyles that minimize competition for food resources. The relationship features a "mixed species grazing guild," involving bison (top left), grazers of grasses and sedges, elk (top right), intermediate or mixed feeders that alternate use of browse, forbs and grasses in relation to food quality and availability, and moose (bottom), primarily browsers. Competition only occurs between elk and moose during severe winters, when the more dietary flexible elk will outcompete. *Top left photo by Wainwright Studio; courtesy of the Wildlife Management Institute. Top right photo by Herman B. Wilsey; courtesy of the Wildlife Management Institute. Bottom photo courtesy of the Wyoming Game and Fish Department.*

in northern latitudes where food is limited and widely spaced (but clumped) and the seasonal period of green growth is brief. In these boreal regions, large body size is a clear advantage for moose, in terms of coping with predators and periods of extreme cold and deep snow. It also imposes limitations on activities through heat stress during warm periods and restricts the southern expansion of their range. However, their size may create a tenuous energy budget. For example, because of their large size, moose must consume large quantities of food. This food is high in indigestible lignin as well as soluble sugars (high source of energy). Because moose require large quantities, they must find, consume and digest food at a rapid rate. For the moose to be a successful forager, thermal stress must be reduced and not interfere with the time required to locate and eat nutritious food.

Although the Bell-Jarman principle was proposed to explain how body size affects social behavior and diet in some ruminant species in East Africa (Bell 1970, Jarman 1974), it is a generalization and does not explain all exceptions there and elsewhere, such as moose. Geist (1974b) argued that the principle is applicable to moose. To survive and be successful, moose must obtain enough (large quantities) highly digestible, low fiber (nongrass) and high lignin food by choosing feeding patches that permit high rates of intake from a variety of dietary items. The ability of the moose to balance daily tradeoffs (i.e., behavioral decisions, day-to-day living, thermoregulation, travel, feeding) probably has enabled it to survive and thrive in boreal regions (Figure 159). Simulation analyses have shown that energetic constraints attempt to hold moose within parameters of the Bell-Jarman principle, but the species' large size is too important in their environment. Thus, there are also other factors that control size and feeding habits (Renecker and Hudson 1992a). It is most appropriate to think of the operating niche of a moose in terms of forage and nutrient distribution (Figure 160).

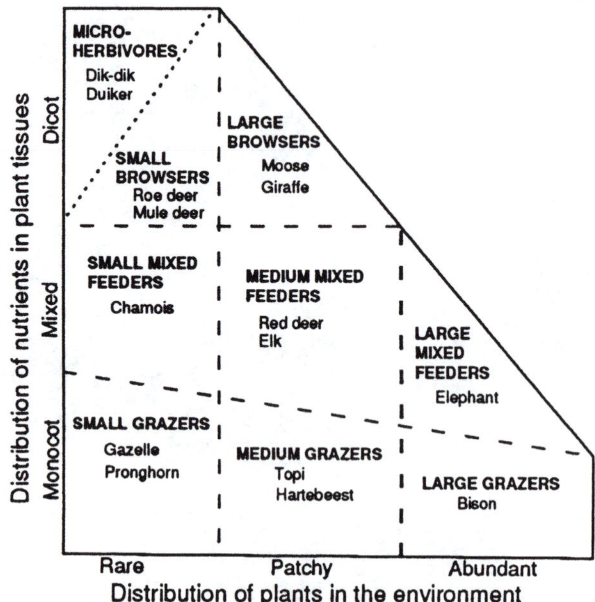

Figure 160. Feeding niche of ungulates, as defined by the distribution of nutrients in plant tissues and the spatial distribution of plants in the habitat (from Hudson 1985).

Forage Characteristics in Northern Ecosystems

The moose is the largest member of the deer family. It possesses a unique set of ecological and physiological adaptations for exploiting forage resources in the harsh environmental conditions of northern circumpolar latitudes (Telfer 1984). In North America, moose occur in a variety of habitats, including montane forest and riparian communities in the Northern Rocky Mountains, mixed deciduous hardwood forests of northcentral and northeastern portions of the United States, and aspen-dominated boreal forests of Canada and boreal forests of Canada and Alaska (see Chapters 3 and 11). Within the past century, moose have expanded their distribution northward, and now occur in arctic riparian areas beyond treeline. Moose also presently use riparian and willow stands along their southern limit of distribution.

The feature most common to all of these ecoregions and most characteristic of moose habitat is abundant woody vegetation. As browsers primarily, moose make considerable year-round use of the foliage and twigs of both deciduous and evergreen shrubs. The availability of woody plants is especially important to wintering moose. Snow depths and conditions in northern forested regions negatively affect animal mobility and limit options for obtaining food.

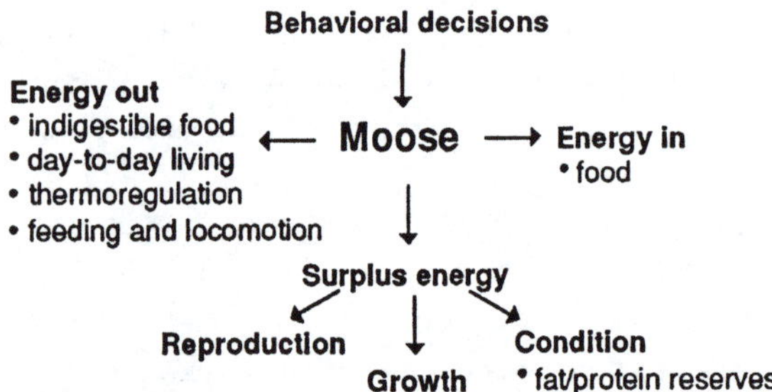

Figure 159. The occupational (day-to-day) energy budget of a moose, illustrating tradeoffs that occur daily.

Forage Availability and Sociability

Although large body size increases an animal's mobility, reduces susceptibility to certain predators and benefits thermoregulation in cold climates, it creates bioenergetic limits and imposes constraints on patterns of resource use. Because of their large body size, moose require large absolute amounts of forage to sustain themselves. Dispersion and density of moose, therefore, are related to forage abundance and quality, and conspecific competition for food. Limited availability of suitable forage forces moose into a "individualistic" or fairly solitary lifestyle. This helps to reduce competition.

Miquelle et al. (1992) observed sexual segregation of moose in Alaska's Denali National Park, where high forage biomass and only a moderate browsing pressure eliminated food availability as the primary cause of the social grouping. The Miquelle et al. study indicated that segregation was greatest in winter and with large adult bulls. The investigators hypothesized that, because of sexual dimorphism and the larger body size of breeding bulls and their postrut energy deficit, these animals were more prone than other cohorts to winter starvation. Shank (1985) reported that male chamois become solitary in winter to avoid excessive energy allocation to social interaction. This social pattern appears to be true for mature bull moose, whose optimal strategy is to maintain a more solitary lifestyle, reduce activity levels, decrease travel distances, increase bite size and inhabit an area of high food availability.

During summer, moose cows with calves remain solitary and in the forest (where biomass may be lower) to reduce predation risks (Miquelle et al. 1992). On the other hand, bulls in Denali are more social and tend to aggregate during summer in open areas of high forage biomass.

Food Habits

"Preferred" foods are those that are consumed in greater proportion to their presence in the habitat. "Principal" foods (staples) are those that, in the absence of much choice, are consumed most. For example, moose tend to select willow whenever and wherever it is available, therefore it is generally considered a preferred food. In many areas of the boreal forest, paper birch makes up the majority of the winter diet of moose because it is the most of what is available and palatable. Paper birch, therefore, represents a principal food. In areas where willow and birch occur in equal amounts, willow is both the principal and preferred food. "Palatability" defines plant characteristics that influence selectability.

To understand food habits, methods for their estimation must first be considered. The method of choice depends on the animal species studied, habitat characteristics and questions asked (Schwartz and Hobbs 1985). One of the most common techniques for estimation of principal foods is examination of rumen contents from sacrificed animals, hunter kills or roadkills. One of the biases with this technique is the differential rate of disappearance of foods in the rumen (for example, forbs may disappear and digest more completely than woody stems, thus may not be reflected correctly in relation to their actual proportion in the diet).

Alternatively, animals fitted with esophageal fistulas can be used to sample boluses being swallowed. However, this technique requires sampling from tame hand-reared animals, and necessitates time to maintain fistulas.

Many of the most recent studies have relied on examination of plant cuticular material in the subject animal's feces. The technique eliminates the need for tame animals. It is noninvasive and relatively inexpensive, but it does require laboratory experience because microhistological features of the fragments are difficult to identify.

Feeding habits also can be determined through direct observation of free-ranging animals under field conditions. This is perhaps the most insightful technique, but there sometimes is difficulty seeing all plants eaten especially

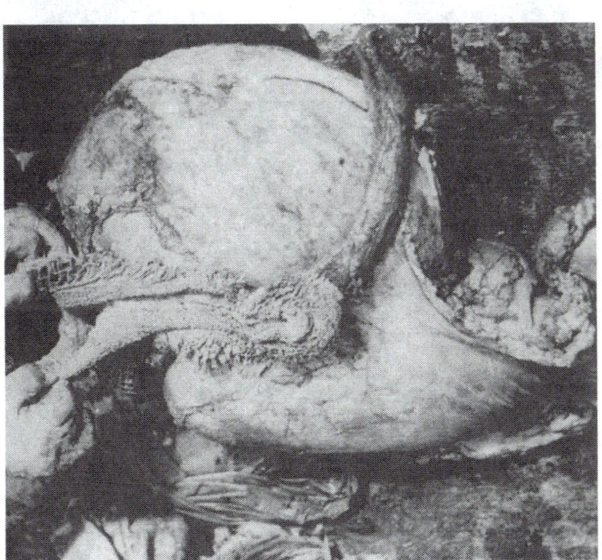

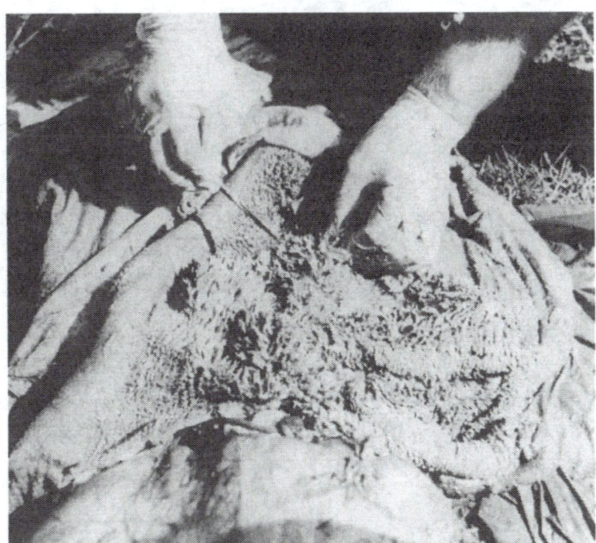

In all young ruminants, including moose, there is an esophageal groove (left) from the posterior end of the esophagus (where it enters the rumen) along the rumen wall to the abomasum (true stomach). This allows milk consumed by neonates to bypass the rumen to the abomasum, preventing bacterial digestion of milk in the rumen, which would result in an energy loss. It once was believed that the esophageal groove disappeared when the young ruminant began to consume and digest forage. However, research has shown that it remains intact in adults. The probable purpose of this groove is to help flush high-quality cell solubles (available energy) past the rumen into the abomasum and small intestine for absorption. The wall of the rumen also is covered with papillae (right) that serve as extensions of the wall and increase its surface area, aiding in absorption of nutrients. The papillae enlarge during summer when there are more solubles in green forage. *Photos by Lyle A. Renecker; courtesy of the University of Alaska, Fairbanks.*

when the subject takes small selective bites or consumes numerous species with one bite. However, information gained by direct observation can be verified or supplemented by subsequent examination of feeding sites.

As opposed to determination of principal foods, preferred foods usually are ascertained by "pen trials." In these experiments, the researcher can control the type availability of foods, but is limited by the artificial circumstances. Often, cafeteria-type trials are performed, from which voluntary intake can be determined for perhaps a single forage. The results do not account for the associative effects of mixed consumption of forage species. Herbivores eat many species of plants and in different proportions and mixes.

Moose consume plant parts from a variety of species and forage types. However, the majority of moose foods are classified as browse (woody plants). Basically, this means a diet of primarily twigs during winter and striped leaves from deciduous species during summer. Moose preference for different species varies among seasons, probably in response to plant quantity and quality. Of the North American food studies we reviewed, moose sampled a total of 221 different plant species and/or genera (Table 36). However, they usually ate high quantities of only a few species (Table 37), such as aspen, birch and willow, encountered throughout their range. Moose are inclined to sample new plants they locate. They also can be extremely selective, such as consuming fruits of western snowberry, as documented by Renecker and Hudson (1986a), but this apparently is related to the palatability of the available options.

Fallen leaves probably are among the largest food resources available to moose during autumn, winter and early spring in aspen-dominated boreal forest and parkland regions of western North America. Tables 36 and 38 show the plant mix in the moose diet and the range of plant species consumed in the various regions of North America, respectively. Fallen leaves are an important food in forest regions where aspen canopy is a major component. In autumn, fresh fallen leaves have a higher level of crude protein (CP) and digestibility (about 8 percent and 49 percent, respectively) than do woody twigs (about 7 percent and 35 percent, respectively) (Renecker and Hudson 1988). By eating fallen leaves in early winter, moose continue to have higher levels of dry matter digestibility than could be realized from woody twigs alone (Renecker and Hudson 1985a). In addition, this food resource can be exploited with a minimal amount of travel because of the stand's high forage bio-

Scientists have learned a great deal by observing the diet habits, activity, foraging rates and habitat selection of semitame moose. Combined with other data, such as heart rate and environmental parameters, these observations can be used to estimate living costs of the subject animal. *Photo by Lyle A. Renecker; courtesy of the University of Alaska, Fairbanks.*

Understanding food preferences is useful in understanding the relationships between moose and their environment. However, forage preferences of moose are difficult to measure under natural conditions because not all foods are equally abundant. Simply because a plant species makes up the majority of the diet does not necessarily mean it is preferred. To differentiate principal foods from preferred foods, cafeteria-type feeding trials are used with captive animals. These experiments provide the subject animals with an equal amount of an array of foods. By randomizing the location of forage within the feed bunk and conducting repeated trials, it is possible to determine the preferred plant species among those available and at that time of year. *Photo by Charles C. Schwartz; courtesy of the Alaska Department of Fish and Game.*

mass. Because most food habit studies are conducted by indirect means, such as fecal fragmentation analysis, the contribution of fallen leaves probably is underestimated in the autumn, winter and early spring diets of moose throughout aspen-dominated range. Of course, feeding on fallen leaves declines when deep snow or crusts make this food resource inaccessible and when more nutritious options become available in spring.

In subalpine and subarctic habitats of Alaska (Risenhoover 1989, Machida 1979) and montane regions of North America (Peek 1974c), willow is an important food source for moose. In Alaska, these riparian willow habitats are relatively productive year-round (LeResche et al. 1974a, Risenhoover 1987) as a source of high-quality food that can support a high density of moose. Similarly, productive willow stands are present in the drainages of Wyoming's Grand Teton National Park and throughout the Rocky Mountains.

Bark stripping has been observed for a number of ungulate species, including red deer (McIntyre 1972), Rocky Mountain elk (Krebill 1972) and moose (Murie 1934, Renecker and Hudson 1985a, Jezierski 1987, Miquelle and Van Ballenberghe 1989). This feeding behavior can impact commercial forestry by inhibiting maturation of harvestable timber. Bark normally does not comprise a large portion of the moose diet and usually is considered "starvation" (last resort) food. In Alberta, moose were observed stripping bark mainly from aspen during early spring (Renecker and Hudson 1985a). This pattern of food selection corresponded with warmer temperatures and the apparent movement of soluble sugars up the stem of the plant. Extensive bark stripping by moose also was observed during winter in the Susitna River drainage (E. Packee and J. Andrews personal communications: 1992) and Denali National Park of Alaska (Risenhoover 1987), and corresponded with relatively high snowfall, high winter density of moose and a limited winter food resource.

In Denali National Park, bark stripping also has been associated with a low availability of twigs during calving,

Table 36. Forages consumed by moose throughout North America

Common name	Scientific name	Location[a]	Reference[b]
Trees and shrubs			
	Aceraceae		
Maple	*Acer* spp.	MN, O	64, 67
Douglas maple	*A. glabrum*	WA	53
Box elder	*A. negundo*	MB, MA	42, 58
Mountain maple	*A. spicatum*	MN, MB, NS, MA	8, 10, 18, 25, 38, 43, 51
		PQ, IR, ON, MT	58, 61, 62, 63, 65, 68, 64
			69, 59
Sugar maple	*A. saccharinum*	NS, IR	38, 43, 61
Red maple	*A. rubrum*	NS, PQ, IR	9, 43, 61
Striped maple	*A. pensylvanicum*	MA, PQ	10, 62
	Aacardiaceae		
Poison ivy	*Rhus radicans*	MA	42
	Aquifoliaceae		
Mountainholly	*Nemopanthus mucronata*	NF	26
	Caprifoliaceae		
Honeysuckle	*Lonicera* sp.	IR, AB	30, 66
American fly honeysuckle	*L. canadensis*	MT, WY, NS, IR	43, 34, 50, 54, 61
Utah honeysuckle	*L. utahensis*	WY	34
Bearberry, honeysuckle	*L. involucrata*	AB	39
Glaucous honeysuckle	*L. dioica*	AB	39
Elderberry	*Sambucus* sp.	BC, NF	1, 59
Elderberry	*S. pubens*	IR	43
Red elderberry	*S. racemosa*	AK	46, 47
Western snowberry	*Symphoricarpos occidentalis*	AB, MB	25, 39
Moosewood	*Viburnum edule*	AK, AB, BC, ON	39, 41, 47
		MN, IR	43, 46
Hobblebush	*V. alnifolium*	PQ	9, 42
European cranberry	*V. trilobum*	AB, MB, AK, MN	25, 37, 58
Highbush cranberry	*V. pauciflorum*	BC	56
Witherod	*V. cassinoides*	NF	26
Nannyberry	*V. lentago*	MB	25
	Celastraceae		
Myrtle pachistema	*Pachistima myrsinities*	WA, BC	1, 53, 57
Mountainholly	*Nemopanthus mucronata*	NF	26
	Cornaceae		
Pagoda dogwood	*Cornus alternifolia*	IR	43
Red-osier dogwood	*C. stolonifera*	ON, PQ, MN, WY	1, 19, 25, 33, 38, 39, 43
		MT, BC, MB, AB	51, 54, 34, 48, 30, 58
		IR	63, 64, 65, 67
	Corylaceae		
Alder	*Alnus* spp.	AK, WA, NF, IR, MN	26, 45, 27, 66, 58, 59
Thin-leaf alder	*A. tenuifolia*	MT, AB	50
Speckled alder	*A. rugosa*	NS, MB, NF, IR	25, 43, 61
Sitka alder	*A. crispa*	BC, PQ, NF, AK	1, 11, 24, 40
Birch	*Betula* sp.	MN, NWT, SK, MB, AK	24, 25
		MB, AK	44, 67
Kenai birch	*B. kenaica*	AK	46, 47
Bog birch	*B. glandulosa*	AB, PQ, AK, WY, BC	1, 10, 32, 43, 45, 35, 33
		MN, NWT, MT, ON	48, 64, 58
Dwarf arctic birch	*B. nana*	AK	35
Water birch	*B. occidentalis*	UT, AB	52
Paper birch	*B. papyrifera*	IR, ON, MT, WY	1, 12, 20
		BC, NF, MN, AB, MA	25, 38, 39, 41, 43, 45
		MB, PQ, IR, AK	34, 50, 54, 48, 59, 62
			63, 64, 65, 58, 69

Table 36 continued

Common name	Scientific name	Location[a]	Reference[b]
Yellow birch	*B. allegheniensis*	NS, IR, PQ	16, 38, 43, 61
Hazelnut	*Corylus californica*	BC	1, 56
Beaked hazelnut	*C. cornuta*	AB, ON, PQ, MB	8, 18, 19, 20, 25, 38, 39
		NS, IR, MN	41, 43, 61, 65, 66, 58
	Elaeagnaceae		
Silverberry	*Elaeagnus commutata*	MT	50
Russett buffaloberry	*Shepherdia canadensis*	AB, MB	25, 39, 41
	Ericaceae		
Labrador-tea	*Ledum* sp.	AK, NWT, SK, MB	37, 44
Menziesia	*Menziesia ferruginea*	WY	34
Rhododendron	*Rhododendron* sp.	NF	59
Cascades rhododendron	*R. albiflorum*	BC	1
Cranberry	*Vaccinium* sp.	MB	25
Cowberry	*V. vitis-idaea*	AK	17, 27, 45
Grouse whortleberry	*V. scoparium*	MT, WY	34, 50, 54
	Fagaceae		
American beech	*Fagus grandifolia*	MA	42
	Myricaceae		
Sweetgale	*Myrica gale*	NF	26
	Oliaceae		
Black ash	*Fraxinus nigra*	IR, MN	43, 58
	Pinaceae		
Subalpine fir	*Abies lasiocarpa*	WY, MT, BC, AB	1, 34, 32, 51, 50, 33
			54, 56
Balsam fir	*A. balsamea*	IR, ON, MB, NF, NS	1, 8, 22, 19, 25, 26
		MA, PQ, MN, BC	42, 43, 58, 59, 61, 62
			63, 64, 65, 69
Tamarack	*Larix laricina*	MN	58
White spruce	*Picea glauca*	MB, MA, IR, NF	25, 42
Black spruce	*P. mariana*	MN, IR	65, 58, 43, 59
Pine	*Pinus* sp.	AB, SK, MB, NWT	44
Lodgepole pine	*P. contorta*	WA, MT, BC, AB	1, 53, 54, 30
White pine	*P. strobus*	IR	43
Douglas-fir	*Pseudotsuga menziesii*	MT, WY, BC, AB	1, 32, 56
White cedar	*Thuja occidentalis*	MN, MA, IR, ON	42, 43, 64, 58
Western red cedar	*T. plicata*	BC	1
Hemlock	*Tsuga* sp.	MN	67
Canada hemlock	*T. canadensis*	MA	42
	Ranunculaceae		
Clematis	*Clematis* spp.	BC, AB	30
	Rhamnaceae		
Ceanothus	*Ceanothus* sp.	MT, WY	34, 50
			54
Snowbrush	*C. velutinus*	WY, WA	33, 33
Buckthorn	*Rhamnus* sp.	MT, WY	34, 50, 54
	Rosaceae		
Serviceberry	*Amelanchier* sp.	MN, PQ, ON, IR	12, 19, 23, 43, 48, 67
		AK, BC, NF	59
Saskatoon serviceberry	*A. alnifolia*	AK, WY, MT, BC	1, 25, 30, 39, 41, 35
		AB, MB	33, 51
Hawthorn	*Crategeous* sp.	MA	42
Common ninebark	*Physocarpus opulifolius*	IR	43
Cherry	*Prunus* spp.	BC, ON, AB	19, 30
Common chokecherry	*P. virginiana*	AB, ID, MN, IR, MB	25, 38, 39, 43, 49, 51
		MT	

Continued on next page

Table 36 continued

Common name	Scientific name	Location[a]	Reference[b]
Pin cherry	*P. pensylvanica*	AB, NF, ON, MA PQ, IR, MN, MB	11, 20, 25, 38, 43, 59 62, 63, 64, 66, 67, 69
Bitterbrush	*Purshia tridentata*	WY, ID, MT	33, 53
Rose	*Rosa* sp.	AB, MB, AK, MT	25, 39, 46, 47, 54, 30
Prickly rose	*R. acicularis*	AK	17
Raspberry	*Rubus* spp.	NF, MN	59, 58
Allegheny blackberry	*R. allegheniensis*	NS	61
Red raspberry	*R. idaeus*	IR, MN, AB, AK MB, NF	25, 39 46, 47, 58, 67
Western thimbleberry	*R. parviflorus*	WY	34
Wild raspberry	*R. strigosus*	AB, MB	25
American burnet	*Sanguisorba canadensis*	NF	26
American mountain-ash	*Sorbus americana*	PQ, ON, NF, IR	10, 13, 18, 26, 38, 43 63, 65, 64, 69
Greene mountain-ash	*S. scopulina*	IR, PQ, MA, MN, AK WY, MT, BC	33, 35, 33, 48, 62
Spirea	*Spiraea* spp.	MB	25
Broadleaf meadowsweet spirea	*S. latifolia*	NS	61
	Saliaceae		
Quaking aspen	*Populus tremuloides*	MA, NF, MB, AK, AB WY, MT, BC, IR, MN NWT	1, 13, 21, 25, 27, 29 39, 42, 43, 45, 50, 54 34, 48, 30, 58, 59, 62 63, 65, 66, 67
Balsam poplar	*P. balsamifera*	AK, MB, MN, AB, NWT	25, 29, 39, 41, 43, 46 47, 58, 67
Black cottonwood	*P. trichocarpa*	BC	1
Bigtooth aspen	*P. grandidentata*	MA	42
Willow	*Salix* sp.	IR, PQ, ON, NF, MT WA, BC, MB, SK, NWT AK, AB, WY, MN	1, 13, 18, 19, 25, 29 39, 41, 43, 44, 45, 59 46, 47, 33, 50, 51, 54 34, 48, 63, 64, 65, 67
Bebb willow	*S. bebbiana*	MB, MT, BC, MN, PQ NWT	10, 25, 29, 32, 35, 56 68
Hoary willow	*S. candida*	NWT	29
Grayleaf willow	*S. glauca*	NWT, AK	24, 28, 29
Richardson willow	*S. lanata*	AK	24
Halherds willow	*S. hastata*	AK	
Beaked willow	*S. depressa*	AK	35
Scouler willow	*S. scouleriana*	AK, BC, MT	35, 36, 51, 56
Littletree willow	*S. arbusculoides*	AK	24, 35, 36, 35
Barclay willow	*S. barclayi*	AK	35, 35
Sandbar willow	*S. interior*	AK, WY	34, 36
Feltleaf willow	*S. alaxensis*	AK	24, 28, 36
Planeleaf willow	*S. pulchra*	AK	36
Barren-ground willow	*S. niphoclada*	AK	36
Blueberry willow	*S. pseudocordata*	WY	34
Blueberry willow	*S. myrtillifolia*	MT	32, 51
Planeleaf willow	*S. planifolia*	MT, PQ, AK	10, 24, 32
Willow	*S. lemmoni*	MT	31
Undergreen willow	*S. commutata*	MT	31
Drummond willow	*S. drummondiana*	UT, MT	52, 51
Geyer's willow	*S. geyeriana*	MT, UT	32, 52, 55
Wolf willow	*S. wolfii*	MT, WY	32, 34, 55
Pussy willow	*S. discolor*	MT, MN, MB, NWT	25, 29, 31, 68
Willow	*S. maccalliana*	NWT	29
Park willow	*S. padophylla*	NWT	29
Tall blueberry willow	*S. novae-angliae*	AK	24

Table 36 continued

Common name	Scientific name	Location[a]	Reference[b]
Basket willow	*S. purpurea*	MB	25
Slender willow	*S. gracilis*	NWT	29
Shining willow	*S. lucida*	WY	34
	Saxifragaceae		
Currant	*Ribes* sp.	AK, MT, MB, AB	25, 39, 50, 51, 34
American red currant	*R. triste*	AK	48
American black currant	*R. americanum*	AK	48
Northern gooseberry	*R. oxyacanthoides*	AB	39
	Taxaceae		
Canada yew	*Taxus canadensis*	NF, MN, IR	43, 60, 65
Western Pacific yew	*T. brevifolia*	BC	1
	Tiliaceae		
Basswood	*Tilia* sp.	MN	67
Fallen leaves;		AB	39
Bark			
	Corylaceae		
Birch	*Betula* sp.	NWT, SK, MB	44
	Aceraceae		
Mountain maple	*Acer spicatum*	ON	71
Red maple	*Acer rubrum*	PQ	63
	Rosaceae		
Pin cherry	*Prunus pensylvanica*	PQ	63
American mountain-ash	*Sorbus americana*	PQ	63
	Saliaceae		
Quaking aspen	*Populus tremuloides*	ON, AB, AK, MB BC, PQ	1, 39, 63
Balsam poplar	*P. balsamifera*	AB, AK, MB	39
Willow	*Salix* spp.	AB, AK	39
Forbs			
	Alismataceae		
Arrowhead	*Sagittaria* sp.	MN	3
	Araceae		
Wild calla	*Calla palustris*	MN	3, 58
Skunk cabbage	*Lysichiton kamtschatkense*	BC	1
Skunk cabbage	*Symplocarpus foetidus*	MN	3
	Balsaminaceae		
Snapweed	*Impatiens* sp.	IR	66
	Ceratophyllaceae		
Hornwort	*Ceratophyllum* sp.	WY	34
	Compositae		
Common yarrow	*Achillea millefolium*	ON, AB	39
Bigleaf aster	*Aster macrophyllus*	IR	66
Elk thistle	*Cirsium foliosum*	MT	50
Bull thistle	*C. vulgare*	MT	31
Canada thistle	*C. arvense*	AB	39
Fleabane	*Erigeron* spp.	AB	39
Western coneflower	*Rudbeckia occidentalis*	MT	50
	Cornaceae		
Bunchberry dogwood	*Cornus canadensis*	NF, BC	1, 26
Bunchberry	*C. suecica*	NF	26
	Droseraceae		
Roundleaf sundew	*Drosera rotundifolia*	AK	27

Continued on next page

Table 36 continued

Common name	Scientific name	Location[a]	Reference[b]
	Geraniaceae		
Sticky geranium	*Geranium viscosissimum*	MT	50
	Haloragaceae		
Watermilfoil	*Myriophyllum* spp.	WY	55
	M. verticillatum	ON	5
	Juncaceae		
Rush	*Juncus* spp.	AK	48
Seaside arrowgrass	*Triglochin maritima*	NWT	29
	Leguminosae		
Peavine	*Lathyrus* sp.	AB	39
Lupine	*Lupinus* sp.	MT	31
Nootka lupine	*L. nootkatensis*	AK	23
Clover	*Trifolium* spp.	AB	39
Alsike clover	*T. hybridum*	AB	39
	Lentibulariaceae		
Bladderwort	*Utricularia* sp.	WY	55
Common bladderwort	*U. vulgaris*	ON	5
	Liliaceae		
Canada beadruby	*Maianthemum canadense*	NF	6
Claspleaf twistedstalk	*Streptopus amplexifolius*	BC	1
	Nymphaeaceae		
Watershield	*Brasenia* sp.	ON	71
Waterlily	*Nuphar* sp.	AB	72
Large yellow waterlily	*Nymphaea advena*	IR, ON	66, 70
Yellow waterlily	*N. varigatum or mexicana*	MN, ON, PQ	15, 6, 58, 71
Fragrant waterlily	*N. odorata*	IR, ON	66, 71
	Onagraceae		
Willow-herb	*Epilobium* sp.	WY	34
Fireweed	*E. angustifolium*	AK, BC	1, 17, 24
Red willow-herb	*E. latifolium*	AK	24
	Pontederiaceae		
Mudplantain	*Heteranthera* sp.	WY	55
	Polypodiaceae		
Ferns		NF	26
Woodfern	*Dryopteris* sp.	IR	66
	Potamogetonaceae		
Pondweed	*Potamogenton* spp.	AK, BC, IR, MN ON, WY	2, 5, 66, 58, 70, 55
Northern pondweed	*P. alpinus*	ON, MN	6, 7
Ribbonleaf pondweed	*P. epihydrus*	ON	6
Leafy pondweed	*P. foliosus*	WY, ON	5, 34
Richardson pondweed	*P. richardsonii*	BC	2
Robinson pondweed	*P. robbinsii*	BC	2
Grassleaf pondweed	*P. gramineus*	BC	2
Floating-leaf pondweed	*P. natans*	BC	2
Large-leaf pondweed	*P. amplifolius*	BC	2
Fineleaf pondweed	*P. filiformis*	ON	6
Fennel-leaf pondweed	*P. pectinatus*	AB	72
	Ranunculaceae		
Common marsh marigold	*Caltha palustris*	IR	66
Watercrowfoot buttercup	*Ranunculus aquatilis*	WY	34

Table 36 continued

Common name	Scientific name	Location[a]	Reference[b]
	Rosaceae		
Cloudberry	*Rubus chamaemorus*	AK	27
	Scrophulariaceae		
Paintbrush	*Castilleja* spp.	AB	72
	Sparganiaceae		
Burreed	*Sparganium* spp.	BC, MN, AK	2, 58, 55
	Typhaceae		
Cattail	*Typha* spp.	PQ, AB, MN	14
Common cattail	*T. latifolia*	AB	39
	Urticaceae		
Slim nettle	*Urtica gracilis*	AB	39
	Zosteraceae		
Common eelgrass	*Zostera marina*	ON	70
Lichens		AK	45
Foliose	*Peltigera* sp.	AK, MB	25, 27
Fruticose	*Cladonia* sp.	AK, NWT, SK, MB	37, 44
Foliose and fruticose	*Cetraria* sp.	AK	37
Moss		MB, AK	25, 45
Horsetail			
	Equisetaceae		
Horsetail	*Equisetum* spp.	BC, MN, MB, SK NWT, IR, NF, ON, AK	2, 25, 44, 66, 70, 55
Water horsetail	*E. fluviatile*	BC, MN, ON	4, 2
Mushrooms		IR	66
	Boletus sp.	AK	27
Grasses and sedges			
	Gramineae		
Grass	*Gramineae* spp.	NF, IR, BC, AK, MB	1, 26, 37, 39, 25, 45 27, 34, 51, 54, 66
Wheatgrass	*Agropyron* spp.	WY, MT	55
Brome	*Bromus* spp.	WY	34
Bluegrass	*Poa* spp.	WY	34, 55
Wildrice	*Zizania aquatica*	MN	58
	Cyperaceae		
Sedge	*Carex* spp.	AB, NF, IR, BC, AK MB, NWT, SK, MB	1, 25, 37, 39, 41, 44 27, 66
Spikerush	*Eleocharis* sp.	MN	4
Bulrush	*Scirpus* spp.	ON	70
Water bulrush	*S. subtermalis*	ON	5

[a] MN = Minnesota; ON = Ontario; WA = Washington; MB = Manitoba; MA = Maine; NS = Nova Scotia; PQ = Quebec; IR = Isle Royale; MT = Montana; NF = Newfoundland; AB = Alberta; WY = Wyoming; BC = British Columbia; AK = Alaska; NWT = Northwest Territories; SK = Saskatchewan; UT = Utah; and ID = Idaho.

[b] 1 = Eastman and Ritcey 1987; 2 = Ritcey and Verbeek 1969; 3 = Jordan 1987; 4 = Belovsky and Jordan 1978; 5 = Fraser et al. 1982; 6 = Fraser and Hristienko 1983; 7 = Aho and Jordan 1979; 8 = Crête and Bedard 1975; 9 = Poliquin 1978; 10 = Joyal 1987; 11 = Audet and Grenier 1976; 12 = Crête and Jordan 1982a; 13 = Bourque 1982; 14 = Crête and Jordan 1981; 15 = Boudreau and Bisson 1983; 16 = Proulx and Joyal 1981; 17 = Oldemeyer and Regelin 1987; 18 = Brusnyk and Gilbert 1983; 19 = Thompson and Vukelich 1981; 20 = McNichol et al. 1980; 21 = Hamilton et al. 1980; 22 = Thompson 1988; 23 = Thompson and Euler 1987; 24 = Risenhoover 1989; 25 = Trottier et al. 1983; 26 = Butler 1986; 27 = LeResche and Davis 1973; 28 = Risenhoover 1985; 29 = Penner 1978; 30 = Barrett 1972; 31 = Stone 1971; 32 = Dorn 1970; 33 = Harry 1957; 34 = Houston 1968; 35 = Spencer and Hakala 1964; 36 = Milke 1969; 37 = Svendsen 1990; 38 = Miquelle and Jordan 1979; 39 = Renecker 1987a; 40 = Miquelle et al. 1992; 41 = Cairns 1976; 42 = Ludewig and Bowyer 1985; 43 = Risenhoover 1987; 44 = Thomas 1990; 45 = Steigers and Becker 1986; 46 = Spencer and Chatelain 1953; 47 = Aldous 1944; 48 = Hosley 1949; 49 = Chadwick 1960; 50 = Knowlton 1960; 51 = Stevens 1970; 52 = Wilson 1971; 53 = Poelker 1972; 54 = Smith 1962; 55 = McMillan 1953b; 56 = Cowan et al. 1950; 57 = Ritcey 1965; 58 = Peek 1974c; 59 = Dodds 1960; 60 = Pimlott 1963; 61 = Telfer 1967a; 62 = Dyer 1948; 63 = DesMeules 1965; 64 = Peterson 1953; 65 = Krefting 1951; 66 = Murie 1934; 67 = Manweiler 1941; 68 = Peek 1971a; 69 = Pimlott 1953; 70 = de Vos 1958; 71 = Peterson 1955; and 72 = Renecker and Hudson 1993.

Table 37. Diet breadth of moose in North America

	Number of different foods consumed at various locations						
Frequency of food habit reported	Trees/ shrubs	Bark	Fallen leaves	Forbs	Sedges/ grasses	Moss/ lichens/ mushrooms	Horsetail
1	66	4	1	50	6	4	0
2	26	2	0	7	1	2	0
3–5	26	2	0	3	1	1	1
6–7	6	1	0	1	0	0	0
8–10	6	0	0	0	1	0	1
>10	3	0	0	0	0	0	0

Table 38. Seasonal foraging patterns of moose throughout North America

	Percentage composition of diet					
Location/season	Graminoids	Forbs	Shrubs/ trees	Lichens/ moss	Leaf litter	Reference
Alberta						
Ministik Wildlife Research Station						
Autumn	6	10	55	0	29	Renecker 1987a
Winter	0	1	72	0	27	
Early spring	1	0	58	0	41	
Late spring	1	0	99	0	0	
Summer	1	25	74	0	0	
Elk Island National Park						
Late autumn	3	17	80	0	0	Renecker 1989
Early winter	0	5	95	0	0	
Midwinter	1	13	96	0	0	
Late winter	0	0	100	0	0	
Early spring	0	2	98	0	0	
Cypress Hills Provincial Park						
Autumn	0	3	93	0	0	Treichel 1979
Autumn	0	0	100	0	0	Barrett 1972
Northeastern Alberta						
Autumn	0	0	100	0	0	Nowlin 1978
Winter	0	0	100	0	0	
Rock Lake						
Winter	0	0	100		0	Millar 1953
Utah						
Uinta Mountains						
Winter	0	0	100	0	0	Wilson 1971
Montana						
Gravelly Range						
Summer	1	71	29	0	0	Knowlton 1960
Minnesota						
Red Lake						
Winter	0	0	100	0	0	Peek 1974c
Alaska						
Kenai Peninsula						
Early spring	0	0	100	0	0	Regelin et al. 1987a

Table 38 continued

Location/season	Percentage composition of diet					Reference
	Graminoids	Forbs	Shrubs/ trees	Lichens/ moss	Leaf litter	
Late spring	Trace	5	95	0	0	
Early summer	0	15	85	0	0	
Midsummer	4	28	68	0	0	LeResche and Davis 1973
Late summer	0	5	95	0	0	Regelin et al. 1987a
Early autumn	Trace	8	91	0	0	
Midautumn	0	0	100	0	0	
Early winter	0	0	100	0	0	
Midwinter	0	0	100	0	0	
Winter	0	Trace	100	0	0	LeResche and Davis 1973
Seward Peninsula						
Summer	Trace	0	100	Trace	0	Svendsen 1990
Susitna River Basin						
Winter	4	2	74	18	0	Steigers and Becker 1986
Denali National Park						
Winter	0	0	100	0	0	Risenhoover 1989
Northwest Territories, Saskatchewan, Manitoba						
Taiga forest						
Winter	0	0	100	0	0	Thomas 1990
Manitoba						
Riding Mountain National Park						
Summer	4	1	95	Trace	0	Trottier et al. 1983
Autumn	13	1	86	Trace	0	
Winter	1	0	99	0	0	
Spring	2	Trace	98	0	0	

when cows were located in aspen/spruce forests with an apparent low forage biomass (Miquelle and Van Ballenberghe 1989). Bulls avoided these areas because of the low food availability. Although bark stripping is associated in many instances with a low availability of forage, bark appears to be reasonably nutritious (Renecker and Hudson 1985a) at a time in early spring when other foods are extremely low in quality. However, the ultimate question of tradeoff is whether the energy required to acquire a bite of bark (Miquelle and Van Ballenberghe 1989) is compensated and more by the acquisition of digestible nutrients.

As previously noted, the nutritional characteristics of woody plants vary among species and seasons. Renecker and Hudson (1989b) and Schwartz (1992a) found that spring and summer diets of moose were 150 to 300 percent more nutritious than winter diets (see also Chapter 14).

Foraging Ecology

Biomass

Forage biomass, from a human perspective, may be quite different than herbivore perception of accessible food in a given habitat (Table 39). More often than not, the latter rather than the former explains an area's moose density. Accordingly, the expressions "available biomass" and "usable biomass" more adequately reflect quantifiable amounts of forage for moose in time and place.

Moose are unlikely to remain in areas where woody vegetation is sparse or woody plants widely dispersed (both being circumstances of low available biomass). Dense and expansive stands of woody vegetation, such as in riparian habitats, with high usable biomass tend to attract concen-

Table 39. Estimates of forage biomass of moose habitat in parts of North America

		Biomass in pounds per acre (kg/ha)				
Location	Habitat type	Winter	Spring	Summer	Autumn	Source
Quebec	Tolerant hardwoods	43.7–79.4 (49–89)		193.6–286.4 (217–321)		Crête and Jordan 1982a
	Open forest	17.0–215.9 (19–242)		65.1–257.8 (73–289)		
	Tolerant hardwoods					
	Closed forest	21.4–88.3 (24–99)		65.1–257.8 (73–289)		
	Intolerant hardwoods					
	Open forest	51.7–64.2 (58–72)		211.4–257.8 (237–289)		
	Closed forest	25.0–135.6 (28–152)		69.6–269.4 (78–302)		
Isle Royale, Michigan	Deciduous maple	14.3 (16)				Risenhoover 1987
	Deciduous aspen	83.9 (94)				
	1986 burn	2.7 (3)				
	Mixed forest	64.2 (72)				
	Coniferous forest	237.3 (266)				
Denali National Park, Alaska	Spruce	20.5 (23)				
	Spruce riparian	150.8 (169)				
	Open spruce/willow	116.0 (130)				
	Willow riparian/lowland	207.0 (232)				
	Upland willow	48.2 (54)				
	Upland birch	71.38 (80)				
	Alluvial willow	134.7 (151)	991.2 (1,111)			Miquelle et al. 1992
	Tall upland willow	162.4 (182)	610.3 (684)	962 (1,079)		
	Low upland willow	160.6 (180)	1,050.1 (1,177)	805.6 (903)		
	Birch/willow	103.5 (116)	526.4 (590)	379.2 (425)		
	Alder/willow	83.9 (94)				
	Alluvial spruce	63.3 (71)	330.1 (370)	527.3 (591)		
	Spruce/willow	114.2 (128)	210.6 (236)	426.5 (478)		
	Aspen/spruce	47.3 (53)	66.9 (75)			
Alberta[a]	Upland aspen	91.9 (103)		827.1 (927)	2,364.3 (2,650)	Renecker and Hudson 1986a
	Forest edge	1,486.4 (1,666)		2,080.6 (2,332)		
	Poplar lowland	447.9 (502)				
	Willow	306.9 (344)		2,718.5 (3,047)	436.3 (489)	
	Old field succession	54.0 (63)				
	Steep incline/aspen forest			835.1 (936)		
	Willow/sedge			2,761.4 (3,095)	5,242.5 (5,871)	
	Cattail/sedge			3,144.1 (3,524)		
	Scrub poplar/grassland				3,452.8 (3,870)	
	Sedge meadow				4,279.9 (4,797)	
	Grassland snowberry				1,609.5 (1,804)	

[a]Minitik Wildlife Research Station.

trations of moose, at least until browsing pressure depletes the forage supply.

Productivity and species composition of woody plants can be significantly impacted by high and/or repeated moose browsing, as experienced in Alaska's Susitna River Valley (E. Packee and J. Andrews personal communications: 1992) and on the Kenai Peninsula (Oldemeyer 1981). Although initially attractive, forage in these areas became rapidly depleted, forcing moose to disperse to other habitats of lower usable forage biomass.

Forage biomass varies with seral age of forests. Parker and Morton (1978) demonstrated that in Newfoundland available woody biomass (primarily balsam fir) increased from about 178 pounds per acre (200 kg/ha) in 2-year-old clearcuts to more than 1,784 pounds per acre (2,000 kg/ha) by 8 years, at which time it peaked and subsequently declined gradually. On the Kenai Peninsula, Alaska, important browse species (mainly paper birch, but some aspen and willow) peaked about 15 years after fire (Spencer and Hakala 1964). Oldemeyer and Regelin (1987) estimated the

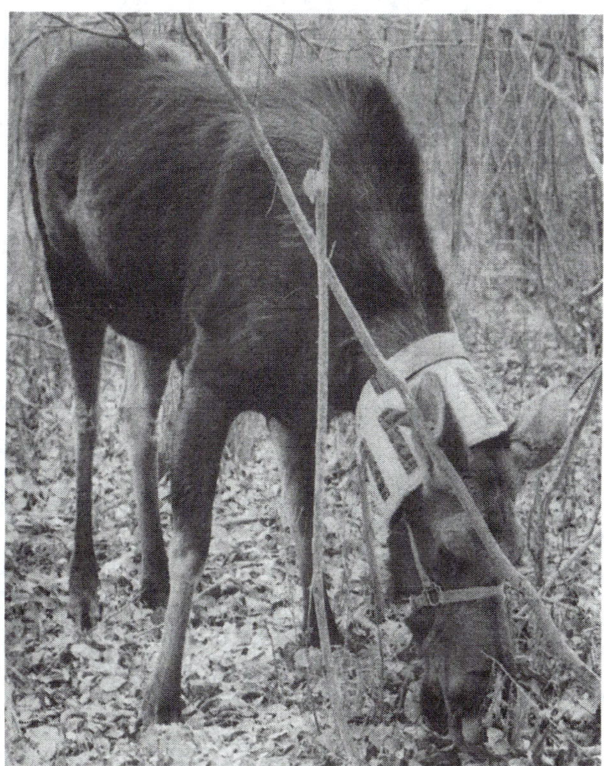

Throughout their circumpolar range, moose meet their nutritional requirements by consuming a wide variety of plants and plant parts. Some are considerably more important in some seasons than others, in part as a result of different seasonal food values of the plants and in part because of relative availability. Fallen leaves, for example, can have a crude protein value of more than 12 percent in autumn. Although some value is lost through autumn and winter from leaching, these leaves remain as nutritious as many twigs and can be extremely abundant. However, deep snow and/or ice crusting during winter can make fallen leaves unavailable. *Photo by Lyle A. Renecker; courtesy of the University of Alaska, Fairbanks.*

Stripping bark from trees by moose is a feeding behavior normally associated with food shortage in late winter when preferred foods are unavailable. However, during the warming months of spring, trees (mainly aspen) mobilize food stores from the tree roots to leaf buds, resulting in "sap" flows that contain high concentrations of soluble sugar. The soluble sugars are easily absorbed by the digestive system of the moose. This is available and highly palatable to moose at times when food is of short supply and poor quality. *Photo by Lyle A. Renecker; courtesy of the University of Alaska, Fairbanks.*

biomass of important browse species in seral stands of forest; browse production measured at 3, 10, 30 and 90 years postburn was 33, 1,246, 354 and 4 pounds per acre, respectively (37, 1,399, 397 and 4 kg/ha, respectively). Similarly, in interior Alaska, aspen forage was most abundant 1 to 5 years postburn—176 pounds per acre (198 kg/ha) in a 1-year-old stand. Birch (149 pounds per acre [167 kg/ha] in 11-year-old stands) and willow (59 pounds per acre [66 kg/ha] in 16-year-old stands) provided the most browse (highest biomass) 10 to 16 years postburn (Wolff and Zasada 1979) in winter. Stands more than 25 years of age generally had less than 22 pounds of forage per acre (25 kg/ha). Moose density often is associated with food abundance (Eastman and Ritcey 1987, Joyal 1987, Oldemeyer and Regelin 1987, Thompson and Euler 1987, Schwartz and Franzmann 1989), unless other forms of mortality (i.e., hunting, predation, disease and parasites) prove limiting.

Populations of moose on the Kenai Peninsula declined at a rate of about 9 percent per year after peaks in forage abundance at about 15 years postburn (Figure 161). If population growth was traced from time 0 to 15 years postfire, the line would have a slope of about 0.167—from 1.3 to 7.8 moose per square mile (0.5–3.0 km^2) in 15 years. This suggests that vegetative quality improves rapidly and moose population growth occurs at about 1.8 times the rate of decline. Obviously, the rate of decline is not linear at the older stages of succession or the population would go extinct. The curve probably is nonlinear with an asymptote at about 0.3 to 0.7 moose per square mile (0.1–0.3/km^2), which is the density seen in mature forest (Loranger et al. 1991). This scenario illustrates the links between forest age, food abundance and moose density.

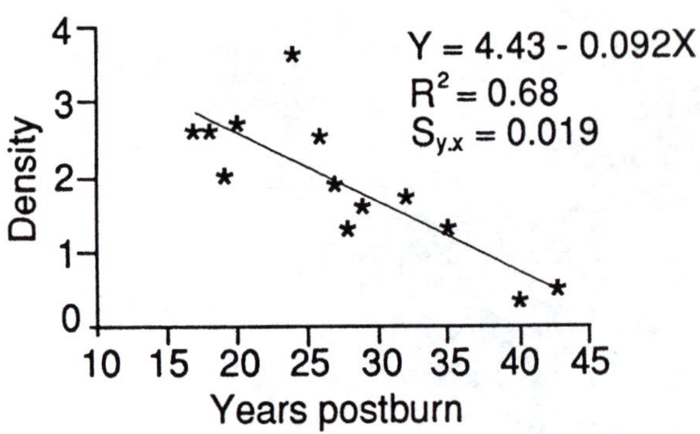

Figure 161. Moose density (moose/km²) is linked closely to food abundance (from Loranger et al. 1991). In boreal forests on the Kenai Peninsula of Alaska, forage quantity peaked about 15 years after forest fire and then declined as the forest matured. Moose numbers declined at about 9 percent per year from 1964 to 1990 after a burn in 1947.

Effects of Snow

Unfavorable snow characteristics tend to make locomotion difficult and restrict activity of many wild ungulates. However, the relative energetics cost of movement is less for moose than for other smaller ungulates (Taylor et al. 1970).

Generally, northern ungulates do not tolerate snow depths greater than chest height and are impeded when snow is knee deep (Kelsall and Prescott 1971). With a chest height 64 percent greater than that of white-tailed deer (Kelsall 1969), moose are morphologically well adapted to northern environments where snow may be present for 4 to 8 months of the year. Moose have the advantages of larger mass, height and longer legs than other boreal browsers such as white-tailed deer. This permits them to remain active and exploit widely dispersed food resources (Telfer and Kelsall 1984). The ability of moose to travel in snow is governed by the snow's depth, density and hardness (compaction). Moose and deer have similar foot loadings—about 8.5 to 11.4 pounds per square inch (600–800 g/cm²)—and therefore, similar sinking depths (Kelsall 1969). The loadings compare with 4.3 to 8.5 pounds per square inch

The long legs of moose serve them well when traveling in deep snow. Moose have small hooves in combination with legs nearly twice as long as those of white-tailed deer, yet comparable "loading" (i.e., distribution of weight per step). Consequently, moose are much more capable of enduring deep snow to find and reach available browse. Under soft snow conditions, snow accumulation of 2.0 feet (0.6 m) or less does not inhibit moose from the standpoints of mobility or energetics. However, because moose can persist during winter where other ungulates cannot, they are more often constrained by compacted or crusted snow conditions than are other species (see also Chapters 9 and 11). Compaction or crusting of snow at almost any depth places an extreme energetic limitation on moose, beside dramatically reducing their foraging range. *Photo by Luray Parker; courtesy of the Wyoming Department of Game and Fish.*

(300–600 g/cm^2) for roe deer, 2.8 pounds per square inch (200 g/cm^2) for chamois and 2.0 pounds per square inch (140 g/cm^2) for reindeer (Nasimovich 1955). Thus, in addition to strength, the large-bodied moose is advantaged by long legs and small hooves for travel in deep soft snow to access clumps of browse (Schwartz 1992a).

In northern forests, harsh conditions and winter storms periodically produce times (of variable duration) of food shortage (inaccessibility) for moose. The energetic costs of locomotion in snow increases exponentially in accumulations greater than 23.6 inches (60 cm). During exceptionally deep snow conditions (31.5–39.4 inches [80–100 cm]) on Isle Royale, moose move into bottomland areas and reduce daily movements. Risenhoover (1987) attributed energy savings to the occupation of these areas, because of the animals' reduced activity there. However, these savings are not explained merely by the increased cost of locomotion, but rather, by reduced overall mobility. Although white cedar trees in such bottomlands intercept some snow, snow depths in these areas apparently do not offer moose better conditions for travel. Average snow depth in bottomlands is similar to other forested areas, but is more variable. Snow tends to drift to greater depths between trees than beneath them. Occasionally, snow is shallower beneath large white

Deep snow during winter can concentrate moose on winter range. The consequent intense browsing can affect hardwood regeneration, because nearly all trees are browsed; few, if any, "escape" to produce a mature forest. In the Susitna Valley of Alaska, successive years of intense winter browsing on previously cleared homestead lands resulted in "brooming" of the regenerating trees, creating a browse line, which represents negative impacts to the timber industry and possibly to the tourism industry. When browsing is intense over many years, it can have negative effects on the ability of the stand to produce regrowth. *Top photo by Luray Parker; courtesy of the Wyoming Department of Game and Fish. Bottom photo by Jonathan Andrews; courtesy of the Department of Forest Sciences, University of Alaska, Fairbanks.*

During early winter, moose will paw (crater) through snow (left) to reach vegetation or fallen leaves on the forest floor. This cratering activity continues until the snow pack becomes too deep or crusted. On the Kenai Peninsula of Alaska, lowbush cranberry is an important moose food that is available in mid to late successional forests where deciduous browse is sparse (right). Moose forage on this plant as long as it is accessible. And, although this cranberry is less nutritious than many deciduous browse species, it comprises up to 50 percent of the moose diet on depleted ranges, and up to 21 percent on good range. *Photos by Albert W. Franzmann; courtesy of the Alaska Department of Fish and Game, Soldotna.*

cedars, but these trees often have several crust layers, making footing difficult (Risenhoover 1987). Increased use of these areas by moose has been attributed to the usable biomass (although lower quality) of woody forage, thereby reducing the need for extensive feeding movements.

In addition to increasing the costs of locomotion, deep snow reduces the availability of food growing near the ground. On Isle Royale, moose spend considerable time (hence, energy) cratering to access and forage on small seedlings and saplings. The effect of snow on moose foraging is easily apparent the following spring by the noticeable browse line on balsam fir seedlings at approximately 23.6 inches (60 cm). Moose cratering in deep snow produces lower average feeding rates because of the amount of time spent uncovering forage. However, in uncrusted snow less than 3.5 inches (9 cm) deep, moose in aspen-dominated boreal forest can access an abundance of fallen leaves, which requires a minimum of movement and energy expenditure.

Selectivity

Moose are quite selective about the plant and plant parts they consume (Risenhoover 1987), especially when forage quantity and quality are high (Vivas and Saether 1987, Saether and Andersen 1990). Moose foraging in Denali and Isle Royale national parks consume only a small portion of the available biomass. Moose consumption of individual woody plants, plant parts and species varies among habitats and appears to be related to forage palatability and relative abundance (Risenhoover 1987).

In Norway, studies showed that moose moved more rapidly between trees in stands of birch where stem density was high (Vivas and Saether 1987). In such situations, the animals selected fewer bites per stem and smaller bites per twig. In terms of energetics, this strategy implies a high digestibility because smaller twig portions were eaten (Hjeljord et al. 1982, Vivas and Saether 1987) (Figure 162).

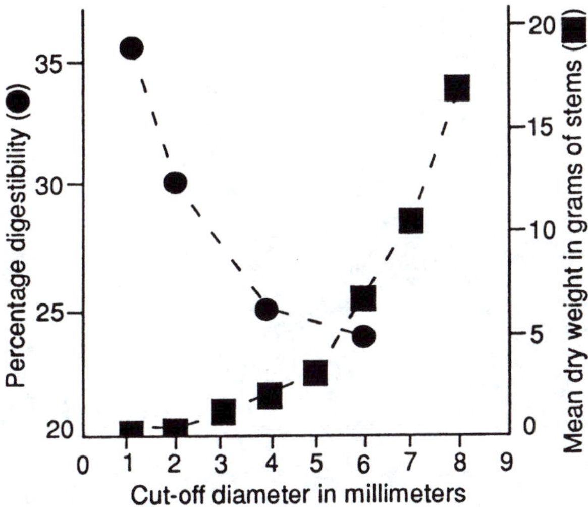

Figure 162. As moose eat larger diameter stems, they consume more dry matter per bite (from Vivas and Saether 1987). However, there is a tradeoff. As bite size increases, the nutrient quality of each bite decreases, and so does dry matter digestibility. These relationships between bite size, in vitro dry matter digestibility of the diet, and yield of dry matter influence how a moose uses available forage.

In lower quality and less dense stands of birch, moose used a smaller area for foraging, were less selective of stem size (eat to a larger stem diameter) and consumed more of the available twigs (Saether and Andersen 1990). However, when Norwegian moose were offered a variety of winter food choices of a higher quality than birch, they mainly ate the other species and/or only the thin shoots of birch. This suggested that nutrition plays an important role in moose food selection (see Chapter 14).

Research efforts in Alaska's Susitna River Valley addressed the impact of moose browsing on aspen, birch and balsam poplar hardwoods (E. Packee and J. Andrews personal communications: 1992). On agricultural clearcuts of 10 to 160 acres (4–65 ha), moose intensively used all woody browse species to such a degree during winter that few browsable stems were left. This impact appears to have been a regular occurrence for trees as old as 25 years. In the stands of regrowth, no comparable unbrowsed trees could be found, and even those trees considered escapees showed evidence of previous breakage by moose. Packee and Andrews indicated that breakage of aspen in these stands was significant enough to indicate that the photosynthetic vigor of the stand was impacted and the trees were dying. In these stands, moose broke stems up to 2.4 inches (6 cm) in diameter and 13 feet (4 m) high during winter by pulling or walking them down. These stems also were stripped of bark. Moose numbers in the area appeared to be in excess of the winter food supply, as demonstrated by the repeated browsing that occurs in low snow years. In this case, moose showed no selectivity for different species, individual plants or plant parts. There was no evidence of the presence of plant chemical defenses. Only lack of access defended the stems in these previous clearings from moose predation.

Similar types and degrees of impact were observed on the Kenai Peninsula (Oldemeyer 1983), and to a lesser degree in the Tanana River Valley (E. Packee and J. Andrews personal communications: 1992).

Although some studies have shown that moose tend to select highest quality food from usable biomass in the best available habitats (Saether and Andersen 1990), this tendency has exceptions, especially when animals do not capitalize on the most nutritious plant species. Selection of winter forage sometimes does not correspond to the presence of other more nutritious foods (Danell and Ericson 1986, Lundberg et al. 1990). Instead, selection may be more a function of forage availability overall or of forage availability at different tree heights. In Norway, moose removed a greater proportion of available biomass from Scots pine below 10 feet (3 m) in height versus more removal from European mountain ash above 10 feet (3 m) in height (Saether 1990).

Foraging Movements

Moose are highly mobile ungulates and often traverse a considerable distance/area while searching for forage during a 24-hour period. Risenhoover (1987) reported that moose foraging movements during winter provided a sensitive indicator of the amount of acceptable foods along a foraging path. Distance traveled by moose during foraging increases exponentially as the duration of the bout increases, suggesting that low feeding rates and/or low amount of usable biomass causes animals to spend more time moving, searching for forage and less time eating. Also, according to Risenhoover (1987), the average foraging movement rates of moose can be strongly predicted by the amount of usable biomass along its chosen foraging path. Risenhoover's investigation indicated that the combination of food selectivity and movements while foraging may provide our best predictors of the balance made by moose after all tradeoffs shown in Figure 159 have been assessed. However, Vivas and Saether (1987) in Norway indicated that their moose moved randomly between plots of different sapling densities. There was no indication that plots of low usable biomass were visited more frequently than plots with a high usable biomass. But the moose moved longer distances within the plots of high usable biomass, and appeared to be more selective in what they browsed.

Intensive winter browsing by moose yarded or otherwise limited in movement because of prolonged deep, dense or crusted snow can cause serious damage to the food plants' stem or kill the plants outright. Trunk breakage and brooming of aspen by moose, such as on Alaska's Nenana Ridge, complicates the land managers' ability to use hardwoods for forest stand regeneration. At right, a 4-inch (10.2-cm) balsam fir near Grand Marais, Minnesota, was browsed by moose to a height of 10 feet (3 m) during winter when the snowpack was so deep that the animals could not crater to understory, reach overstory of birch and aspen or move for an extended period to a better foraging area. *Left photo by Dan Golden; courtesy of the Alaska Division of Forestry. Right photo by S.E. Aldous; courtesy of the U.S. National Archives.*

Landscape

Behavioral responses of moose to changes in forage biomass, quality and distribution need to be addressed (Hjeljord et al. 1990). During the course of a feeding period, moose may respond to the abundance of forage at the individual plant level, at the feeding site level (group of plants) or to the distribution of patches of habitat. Likewise, over the course of several days of foraging movements, moose may disperse over long distances and forage within several habitats while searching for resources, interacting with conspecifics, and/or avoiding predators. Collectively, these daily movements, when combined, produce seasonal patterns of a home range.

Functional Response

One of the strongest determinants of feeding rate—the product of bite rate (bites per minute) and bite size (pounds or grams of dry matter per bite)—is usable forage biomass, measured pounds per acre (kg/ha). For grazing ruminants, usable biomass provides a reasonable predictor of animal feeding performance (sources). In a grass sward, the dispersion of forage is relatively continuous, and individual bites vary little in their physical and chemical qualities reducing the importance of selectivity to feeding efficiency (i.e., nutrient ingestion rate). The feeding rate (g/min) of an animal is determined largely by bite size, which is a function of length of the basal leaves (flower or tiller) of the grass stem (i.e., the longer the leaves, the larger the bite size). Thus, feeding rate is relatively constant during foraging bouts for the nonselective grazer compared with that of a selective browser such as the moose.

Available woody plant biomass is less useful as a predictor of the feeding rate of browsers (Renecker and Hudson 1986a, Risenhoover 1987, Shively 1989, Shipley and Spalinger 1992, Spalinger and Hobbs 1992), or for intermediate feeders consuming a mixed diet including browse (Wickstrom et al. 1984, Hudson and Watkins 1986). However, the relationship between usable biomass and moose intake in forested habitats is stronger when the diet is considered in proportion to the relative time required to consume individual plants and plant parts of different species (Renecker and Hudson 1986a; Figure 163). Individual leaves and twigs occur in patchy distribution in three-dimensional space (on individual or clumps of shrubs). They tend to vary more in their nutritional qualities and therefore, subject animals to more options, which results in variability of bite size. These characteristics would favor selective feeding and allow animals to vary bite size to balance tradeoffs between diet quality and feeding rate. Differences in the relative value of individual plants are likely to produce changes in moose feeding behavior as the supply of preferred plant parts becomes depleted. Thus, the acceptability of forage—measured as the density of individual bites or biomass per square foot or meter—is likely to change over time as the supply of nutritious twigs or leaves is consumed, and the feeder's willingness to eat alternative plant parts increases.

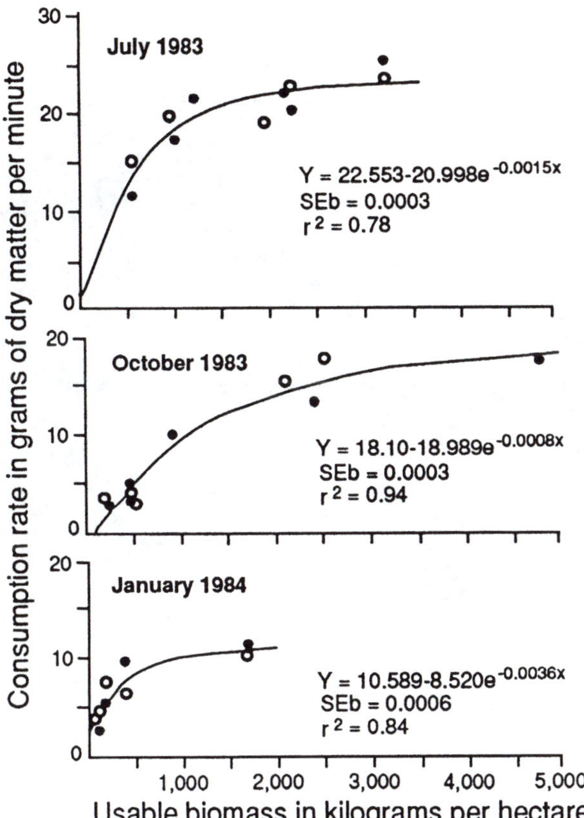

Figure 163. The amount of forage a moose can consume is related to food abundance (from Renecker and Hudson 1986a). As usable forage biomass increases, so does intake until forage quantity no longer limits eating rate. Maximum intake rates (asymptote of the curve) tend to be higher in summer, when forage quality is higher, than in winter. SEb = standard error of b coefficient.

Apparently, these characteristics (palatability, nutritiousness, abundance, structure) of woody plants are not clearly reflected simply by measures of available biomass.

However, different relationships that hold for grass and shrub habitats and for structure and quality of forages can impose constraints on feeding rates. In addition, different ungulates perceive and respond to resources in different but unique manners (Figure 164).

The response of an animal to changing food availability often is expressed as the biomass at which intake is reduced to 50 percent of maximum asymptote. Therefore, this adjusts food supplies to constraints of supply, structure and quality. For moose, the critical usable biomass that demands a 50 percent drop in consumption rate are 446 pounds per acre (500 kg/ha) in July, 847 pounds per acre (950 kg/ha) in autumn and 134 pounds per acre (150 kg/ha) in January (Renecker and Hudson 1986a). This compares with 484 pounds per acre (542 kg/ha) (Wickstrom et al. 1984) and 1,070 pounds per acre (1,200 kg/ha) (Hudson and Nietfeld 1985) for elk on grassland, and less than 89 pounds per acre (100 kg/ha) for mule deer on sparse pastures.

Bite rate for browse species is strongly influenced by features of the plant, such as plant shape and leaf area. For many herbivore species, bite size is related to width of the incisor bar. Because moose strip leaves from shrubs and trees during summer, they probably have a higher intake rate than predicted of a grazer with an equal mouth width.

Handling time (from bite to swallow) appears to vary with the food that is selected (Roese et al. 1991). Unlike species that consume foods of similar quality, browsers, such as moose, must select from patches of highest available food quality and/or from those of highest available forage biomass. In this way, moose can sample from a greater number

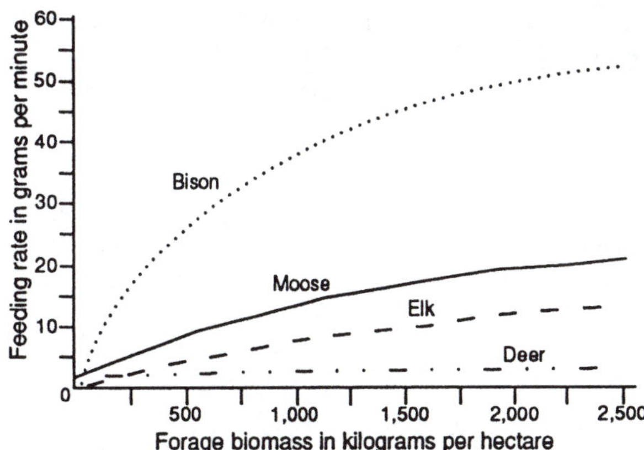

Figure 164. Comparative intake rates, averaged from summer and autumn seasons, of four ungulate species of different body size (Hudson and Watkins 1986, Renecker and Hudson 1986a, Hudson and Frank 1987, Wickstron et al. 1984).

As selective feeders, moose tend to avoid consuming nonpalatable, nonnutritious plant foods. They also can vary their bite size and are able to forage within a high vertical span, to take advantage of the best (most nutritious) plant parts or strata. In terms of foraging efficiency, such feeding behavior or strategy usually compensates more than adequately for a low foraging rate, which is emphasized by nonselective feeding within a narrow vertical plane, such as by grazers. *Top left photo by Tim Lewis Rue. Top right photo by Albert W. Franzmann; courtesy of the Alaska Department of Fish and Game, Soldotna. Bottom left photo courtesy of the Denver Museum of Natural History. Bottom right photo by H. R. Timmermann.*

of trees (Lundberg and Danell 1990). Through simulation analysis, Spalinger and Hobbs (1992) showed that the method used by a wild ruminant to handle food for swallowing has a functional importance on maximum intake rate, which is closely related to stem density and size.

Bite Rate and Size

The limits on bite rate of herbivores are the logistical time to grasp, secure, moisten, masticate and swallow food items. Except in communities where food is scarce, bite rates are inversely related to bite size and perhaps fiber content. Bite rate also can be related to the usable biomass and spatial arrangement of items. Also, obstacles to accessing food, such as snow, snow crusts and ice, can depress bite rate.

How changes in bite size influence bite rate and to what extent herbivores discriminate in the selection of food vary among animal species and individuals. Bite rate and size also vary seasonally, by available plant species and among study areas (according to the spatial arrangement of food) (Renecker and Hudson 1986a, Risenhoover 1987). Renecker and Hudson (1986a) found a significant difference in both bite rate and size among seasons. Their studies in the aspen-dominated boreal forest of Alberta showed a compensatory relationship between bite rate and size of moose during summer but not in other seasons. In the other seasons, this relationship may be obscured by changing food quality and the greater emphasis on locating suitable forage.

During winter, bite size may be the major factor determining intake rates of individual browse species (Risenhoover 1989). As bite size of woody twigs increases, animals require more handling time to process the food, and bite rate declines. However, the proportion of time required to process the food (per ounce [g]) declines as bite size increases. Therefore, moose can maximize intake rate by maximizing the bite size of woody stems during winter (Figure 165).

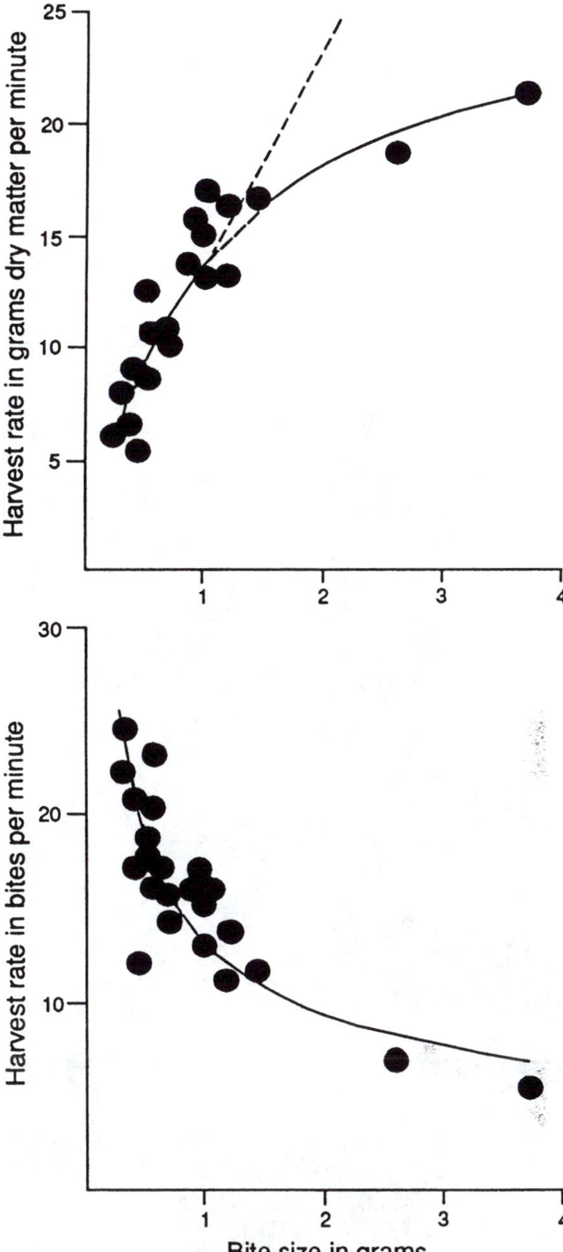

Figure 165. When moose take large bites, their intake rate (unit weight per minute) increases (from Risenhoover 1987). However, as bite size increases, the bite rate (bites per minute) decreases because of the increased time required to chew the additional material. Nevertheless, the amount of handling time required per gram of food declines as bite size increases, thus intake rate can be maximized by taking large bites of forage.

Browsing Diameter

Mean plant diameter at point of browsing can vary among individual animals (Crête and Audy 1974), plant species and habitats (Peek et al. 1971, Joyal 1976, Hubbert 1987, Risenhoover 1987, Renecker and Hudson 1992b). All the factors that affect mean plant diameter at the point of browsing can influence diet quality. Preferred species are browsed to a greater diameter (Peek et al. 1971, Joyal 1976), and moose demonstrate a clear food preference for the largest available twigs (Risenhoover 1987). It appears that, for most browse species, twigs from 0.09 to 0.2 inches (0.23–0.51 cm) are browsed more frequently than would be expected based on twig availability (Risenhoover 1987).

Morphology and twig location appear to be important criteria of selection by moose. Moose appear to prefer twigs at the top of plants, as opposed to lateral stems (Vivas and Saether 1987). In Sweden, stems of birch at the top of 8.2- to 9.8-foot (2.5–3.0 m) tall plants were longer, thicker and had a higher fresh weight than did lower twigs (Danell 1983).

In Sweden, moose tended to decrease bite diameter of Scots pine stems as the number of visits to the stand increased, suggesting that rebrowsing of shoots was rare (Edenius 1991). Other studies indicated that diameter of browsing is strongly related to twig diameter of current annual growth (Bergstrom and Danell 1986), which declines in size from top to bottom in Scots pine (Kozlowski 1971).

Edenius (1991) suggested that moose first eat upper larger twigs and then shift to smaller lateral twigs in subsequent visits to a stand.

Variation in point of browsing is common for the same plant species in different areas, as suggested for balsam fir—0.13 to 0.44 inch (0.32–1.11 cm)—in a Quebec study site (des Meules 1962) versus 0.08 to 0.3 inch (0.2–0.76 cm) in Nova Scotia (Telfer 1969). Point of browsing probably is more strongly influenced by twig morphology than by feeding intensity of moose on a plant species. In Norway, moose browse twigs in low-density stands to a larger diameter than in high-density stands of birch (Vivas and Saether 1987). In Alaska, studies have shown that point of browsing of paper birch does not vary as moose stocking rates increase from 4.6 to 12.3 moose per square mile (1.8–4.7/km^2) and level of birch utilization increased from 23 to 66 percent

Moose will browse twigs to various diameters. The point where a moose chooses to bite off a stem (diameter of browsing) is related to forage quality, concentration of secondary plant compounds, handling capacity of the mouth, palatability, availability of other foods and time of year. During winter, moose tend to rebrowse the same trees each year, removing the previous summer's stem production (current annual growth), thus retarding the trees' vertical growth. In food patches of high-quality forage, moose will browse a large number of available stems and then move to another food patch. They often return to the same food patches during winter and browse the previously uneaten stems. They rarely rebrowse stems where the current annual growth has been previously removed unless conditions severely limited the food supply. *Top left photo by Lyle A. Renecker; courtesy of the University of Alaska, Fairbanks. Top right photo by Charles C. Schwartz; courtesy of the Alaska Department of Fish and Game. Bottom photo courtesy of the Utah Division of Wildlife Resources.*

of current annual growth (Hubbert 1987). Moose adjust to their own increased population density by browsing plants more intensively (i.e., remove more current annual growth stems), but do not revert to eating poorer quality food (Hubbert 1987). Exceptions occur at very high densities or when moose are confined to winter range for long periods by deep snow, browse diameter (Renecker and Hudson 1992b) and utilization increases, bark stripping becomes more common (Schwartz 1992a), and diet quality declines.

Activity Patterns

Like other wild ruminants, moose spend most of their time feeding and sleeping or resting/ruminating. Somewhat less amounts of time are spent in grooming, movement/search and social interaction (Renecker and Hudson 1993). Activity patterns reflect the animal's behavioral adjustment to daily and seasonal variations in the environment and energy demands or constraints (Bunnell and Gillingham 1985). Although many factors determine activity patterns, daily feeding and resting are dictated largely by the cycles of rumen filling and emptying. Much can be learned about regional and seasonal patterns of activity in moose by studying factors influencing the feeding and ruminating bouts.

Free-ranging moose in the aspen-dominated boreal forest of western Canada showed seasonal patterns of daily rumination, but not of feeding (about 10 hours per day) (Renecker and Hudson 1989a). Their feeding occurred during 5 to 8 bouts per day (more in late spring and summer) (Renecker and Hudson 1992a). In contrast, moose in the interior of Alaska appeared to forage from 6.2 hours daily in early May to 9.5 hours daily in July (Van Ballenberghe and Miquelle 1990). And, during winter, reduced daily feeding time to 4.9 hours (Figure 166).

Because feeding by moose is preempted by other activities, it is limited to about a maximum of 13 hours daily (Renecker and Hudson 1992a). Compared with larger conspecifics, smaller moose require lesser absolute amounts of food daily and therefore, less time for daily rumination. In this context, theoretically the smaller animal can more easily maintain a positive energy balance in summer and less negative balance in winter than a larger moose because of the lesser likelihood of preemption of the former's daily foraging time (Renecker and Hudson 1992a). However, it is probably adaptation to predators, deep snow, cold, etc., that forces moose to be larger as this species exploits the arctic limits of its range, consistent with Bergmann's rule of body size, which holds that the largest forms of animals are found in temperate and arctic latitudes. The overall success of moose in balancing seasonal energy budgets probably is related to how they mitigate thermal discomfort in relation to their foraging efficiency.

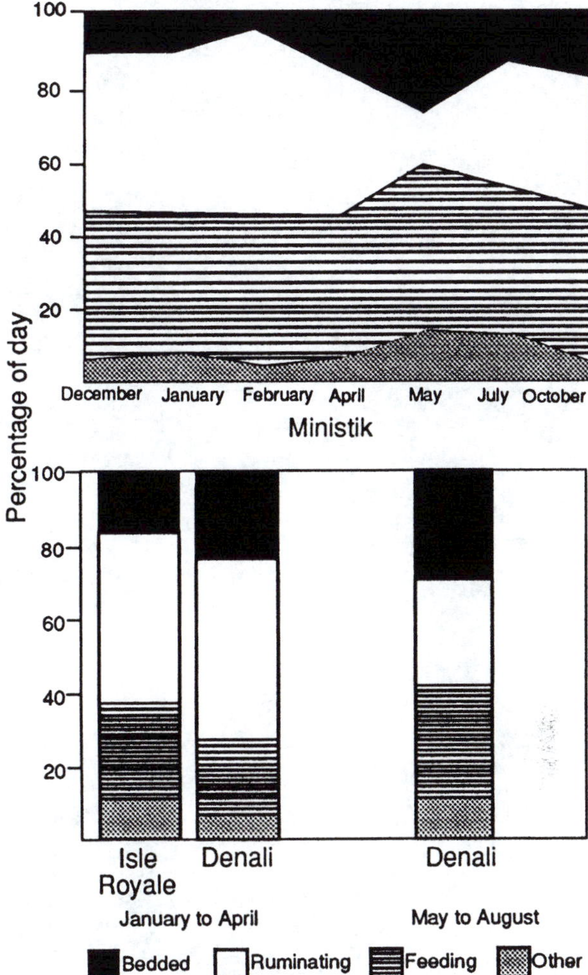

Figure 166. Moose spend most of each day feeding, resting and ruminating. Their seasonal activity budgets vary in response to changing forage quality and abundance. This activity budget was constructed by observing free-ranging moose at the Ministik Wildlife Research Station in Alberta (Renecker and Hudson 1989a), Isle Royale National Park, Michigan (Risenhoover 1987) and Denali National Park and Preserve, Alaska (Risenhoover 1987, Van Ballenberghe and Miquelle 1990).

Thermoregulatory Needs and Feeding

A fundamental problem for moose is consumption of enough nutrition to meet the demands of changing thermal conditions of their environment. Because of a large body size and good insulation, moose are extremely tolerant of cold temperatures. This larger size conserves heat and reduces energy needs during winter (also in agreement with Bergmann's rule of body size) when quality and quan-

As with most ruminant species, the principal activities of moose are feeding and resting. During summer, moose consume large quantities of highly digestible leaves (top left) and aquatic plant parts. Because this material is easily digested, their resting/feeding cycle is short (top right). During winter, when moose consume a diet of coarse, relatively poor-quality woody vegetation (bottom left), they must spend many hours ruminating (bottom right). Rumination involves regurgitating and rechewing (masticating) partially digested foods. This mastication further breaks down food particles, increases their surface area, and enhances bacterial digestion, but it requires additional processing time. As a consequence, the resting/feeding cycle is increased during winter. *Top left photo by Len Clifford. Bottom left and both right photos by Luray Parker; courtesy of the Wyoming Department of Game and Fish.*

tity of available forage is insufficient to meet the costs of day-to-day living. Furthermore, bedding with their legs tucked under the body or in powder snow can further reduce heat loss during cold periods.

Ambient air temperatures above 23°F (–5°C) in winter and above 57°F (14°C) in summer can be stressful and discomforting to moose (Renecker and Hudson 1986b, 1989b, 1990b; see Chapter 14). The high energy cost of heat stress leads to a reduction in overall activity. This is especially true during moderating weather conditions in winter, when moose have maximum pelage density and fat accumulation, males do not have highly vascular antler velvet to aid cooling and winter energy expenditures return to a value near the interspecies mean of basal metabolism. Reduced activity in summer because of warm weather influences feeding time and therefore, consumption. Although moose have less insulation during summer, they still have a low surface area-to-volume ratio because of their large size. This insulative characteristic serves to conserve heat, not expel it rapidly. Heat stress begins to occur when moose cannot lose sufficient heat by convection, conduction or radiation. Moose increase breathing rates and begin to pant to expel excess heat. If heat stress becomes more severe, open-mouthed panting will occur and breathing rate will increase to more than 70 breaths per minute in adult moose. To escape high temperature and gain "comfort" while foraging in montane regions, moose may seek cool mountain valleys. In all habitats, they also can remain bedded during daylight hours and forage at night or feed in open areas or wetlands where the effects of wind and water reduce heat discomfort. Yet to be answered is, whether during periods of unusual heat, moose visit aquatic habitats principally to feed or whether there is a significant thermoregulatory aspect to the site selection.

There are many days each year when moose are exposed to heat sufficient to depress foraging activity and weight gain (Renecker and Hudson 1986b). In Alberta, moose were observed to bed in an open sedge meadow habitat on a hot late afternoon in July. The water there was 8 inches (20 cm) deep. Energy expenditure and respiration rate were reduced by 9 and 29 percent, respectively, below the daily average of a bedded/resting moose not lying in water (Renecker 1987a).

During warm, sunny periods in late autumn and winter, as on warm summer days, moose sometimes rest or sleep in an exaggerated sprawl that helps with cooling at a time when their insulative characteristics are at peak. Other wild ungulates rarely have the luxury of such relaxation because of the threat of predation—a minimal prospect for adult moose. Such thermoregulation by moose not only has comfort benefits, but aids in maintaining energy balance in the resting/feeding cycle. *Photo by Victor Van Ballenberghe; courtesy of the USDA Forest Service.*

Shade provided by aspen or poplar cover types probably is attractive to moose during periods of high ambient temperature. During daytime, when ambient temperatures approached 79°F (26°C), moose in Alberta remained bedded in the shade, with their legs extended (Renecker and Hudson 1989b, 1990b). The animals forfeited a feeding bout in favor of remaining cool through inactivity. Through model simulation, Renecker and Hudson (1992a) showed that the lost feeding bout on such a hot day could not be compensated for in the days to follow. The probable result was a loss in weight. The negative influence of heat stress on foraging activity of moose is greater in adults than in subadult moose because of the former's larger mass.

Foraging Theory and Resource Selection

Herbivores invariably are faced with the tradeoff between optimal feeding rate and optimal diet digestibility, because the most nutritious forage (such as forbs and willow and aspen leaves) tend to be those most uncommon. Research has focused on attempts to understand optimal diets in a plant community, optimal choice of food patches, and an optimal time allocation in different patches. Two main optimal foraging strategies have been identified (Pyke et al. 1977). First is time minimization. It describes patterns of foraging that allow animals to meet their basic requirements in the least possible time and is expected to operate when animals are persistently exposed to risk of predation or cold stress while foraging. The second strategy, energy maximization, describes behavior that maximizes nutrient capture (see Westoby 1974, Belovsky 1978). In harsh environments, food items tend to be of equal value, therefore the strategy of energy maximization is reduced to the basic logistics of feeding in terms of locating a food patch, searching for food items and consuming the plant parts. As the disparity between forage availability and animal requirements widens, diet expansion is expected as the animal becomes less selective (Ellis et al. 1976).

Theory about optimal diet choice (McArthur and Pianka 1966) assumes the forage type is encountered according to its relative abundance, but only a proportion will be sought. Foods are ranked according to their net value relative to the cost of locating and handling each forage. Search cost is expected to decline as diet selection increases, but it also may increase if the animal selects uncommon or inaccessible forage items. The optimal forager would be expected to choose forages that can be located and consumed at a cost less than some average expense to search for, locate and consume plant parts.

Models of optimal patch choice (McArthur and Pianka 1966) are explained on the basis of an optimal number of feeding sites rather than foods. In this static model, costs to search for new patches and travel between feeding sites are included. Travel cost would decline for the forager as more feeding sites are included. The optimal solution would be to expand foraging range to include more habitat types until the cost of increased search time is greater than the savings from reduced travel.

Also, in regard to the optimal foraging, the timing of departure or abandonment of a particular patch is explained by the "marginal value theorem" (Charnov 1976). Within a patch, the herbivore eventually will reduce the food supply and therefore, depress intake. As intake is reduced because of the depleted supply, an asymptote is reached. The marginal value theorem indicates that the herbivore then should leave the patch when intake rate drops to the average rate of acquisition for the environment (all available patches).

Consistent with the marginal value theorem, moose select patch and food totally on the basis of forage availability (Haukioja and Lehtila 1992). First, a patch is selected to begin the foraging bout (Lundberg and Astrom 1990). Moose probably do not have the ability to do more than generally appraise a patch on the basis of its forage density rather than its particular plant species. Food quality may be appraised after browsing commences and, if deemed suitable, moose then would continue to forage in that patch until intake dropped below asymptote for the area (Astrom et al. 1990). However, this theory relaxes some of the logistical constraints of optimal foraging models, such as search and handling times. From patch analysis, Haukioja and Lehtila (1992) suggested five hierarchical food selection criteria used by moose: (1) general plant quality; (2) forage density; (3) sufficient quantity for a bite (during the foraging period, moose "capture" bites rather than species); (4) size of tree and branch density (generally less browsing occurs on small trees with few branches [see Bergstrom and Danell 1986]); and (5) some food will not be consumed (browsing does not remove all food). The last two criteria will not hold true under high concentrations of moose, as described earlier for the Susitna River Valley of Alaska.

The problems of applying these foraging theories to large, generalist herbivores are summarized by Westoby (1974). In reality, search times are small and similar for all food items and the probability of a bite taken is essentially one, or 100 percent. Large herbivores, such as moose, also show partial preferences that are not predicted by these models. Westoby contended that the optimal strategy of the large generalist herbivore is to obtain the best mix of nutrients. In addition, herbivores also must continually sample novel foods in small amounts to adapt their diet in relation to forage quality and palatability in a changing environment.

Moose tend to occupy a "home niche" where food resources are not continuous or particularly dense. Instead, they inhabit "patchy" areas, such as aspen-dominated boreal forest (left) and mixed-age deciduous forest in proximity to conifer stands (right), both ideally adjacent to water. To exploit forage resources, moose must move from patch to patch, and their long legs make such travel relatively easy, especially when and where snow covers the ground for 6 months or more. *Left photo by Robert J. Hudson; courtesy of the Department of Animal Sciences, University of Alberta, Edmonton. Right photo by L. Godwin; courtesy of the Ontario Ministry of Natural Resources.*

Each plant contains a unique spectrum of nutrients as well as secondary (chemical) defenses, as previously noted. This presents a dilemma for the herbivore. A herbivore, such as the moose, must maximize intake, yet balance the nutrient content and avoid toxicities (Schoener 1971, Pearson, 1974, Westoby 1974). Belovsky (1978) applied linear programming to this ecological problem to predict the diets of moose as either energy maximization or time minimization. He concluded that under the conditions of his study moose clearly behaved as energy maximizers.

In the modeling of ecological systems, attempt should be made to formulate systems on the basis of biological constraints of the real world. Hobbs (1990) suggested that Belovsky's (1978) attempt to explain moose behavior through linear programming presented a model that works but had no biological basis. Hobbs argued that a foraging ruminant is faced with opportunities that change and require adjustments in behavior to optimize intake and "lifestyle," which ultimately relate to productive functions of the animal. Furthermore, he contended that as model parameters deviate from those used in simulation, reliability declines. In response to Hobbs, Belovsky (1990) maintained that the digestive constraint on dry matter intake in his linear programming model was necessary and the unexplainable success of the model in the prediction of moose behavior suggested further testing. Although the approach broke new ground, caution must be taken to identify the correct constraints that drive the model. Simulation models can soon become too large and unmanageable as the complexity increases.

Renecker and Hudson (1992a) examined the influence of body size on seasonal energy balance and time spent foraging. On the basis of metabolic needs and digestive capacity of moose, the investigators' simulation of foraging by moose concluded that the animal's large body size imposes longer foraging times and energy budgets that must be balanced more carefully. This probably is achieved by efficient foraging and thermoregulatory tradeoffs (i.e., choice of whether to feed and, if so, where). Optimization models are valuable in the explanation of feeding behavior of moose and perhaps how body size is related to the success of their lifestyle (see Figure 159).

Feeding Strategies

Food Acquisition

If northern boreal habitats are viewed in light of the Bell-Jarman principle, there should be a clear relationship between size of an animal and the food it eats. Because food often is in short supply and often widely distributed in patches, much travel is required to obtain sufficient quantity of highly digestible food. In this environment, a browsing ruminant would be expected to have a smaller body requiring smaller quantities of nutrients than is the case for moose. Accordingly, the Bell-Jarman principle should be viewed in the spectrum of animals from which it was derived. A more acceptable interpretation of the relationship between feeding strategy, forage quality, food availability and animal body size was presented by Hudson (1985), who described the occupational pattern and lifestyle of moose in terms of food quality, patchiness of the habitat and food scarcity.

The overall feeding strategy of any animal should be that of maximizing nutrient intake, but it may be more important for browsing ruminants. In fact, studies have shown that moose spend time feeding in habitats in proportion to the nutrient return (Figure 167) (Renecker and Hudson 1993). For example, the returns can be crude protein or digestible dry matter, and moose will occupy habitats where the availability of plants permits high foraging rates. However, other factors such as heat stress, insect harassment, snow melt/freeze cycles and seasonal changes in food availability influence where, when and how long moose feed.

In summer, only relatively small differences in the quality of food are found between the patches in boreal forest/shrubland. Therefore, everything else being equal, there is less time spent traveling to find food, so the overall energetic cost is less. As a result, moose are relatively nonselective in the foods they choose, compared with their dietary habits in autumn. Nevertheless, nutrient quality is higher in the shade than the sun, and moose always choose the better "menu" available to them. During autumn and likely also during spring, they can maximize intake of high-quality food by choosing habitats that provide the best forage choices. This is possible because of the vast differences in food quality during those seasons, when some plants begin to turn green sooner than others (i.e., on south-facing slopes and in sunny

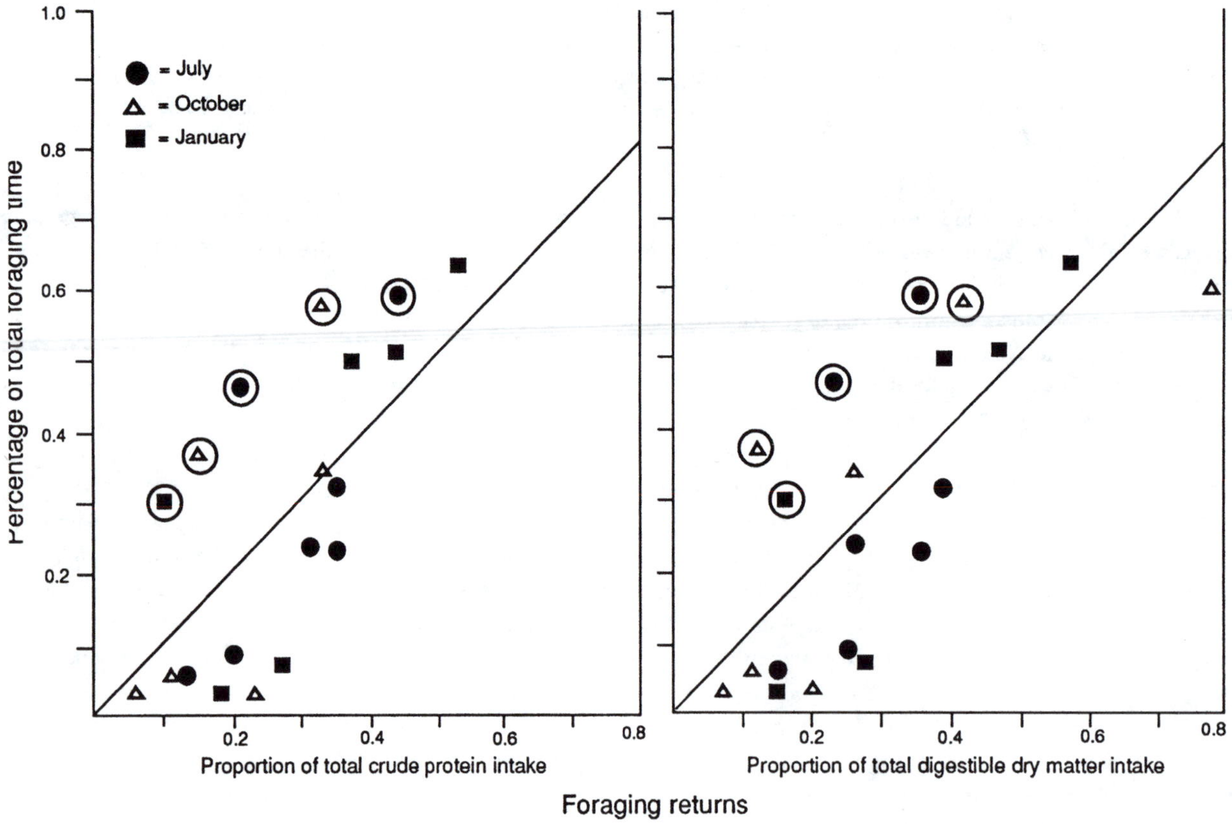

Figure 167. Relationship between the percentage of foraging time spent by moose in major habitats and foraging returns expressed as grams of crude protein and grams of digestible dry matter intake consumed per minute, divided by the total for all habitats at the Ministik Wildlife Research Station in Alberta. Large circles indicate habitat use during periods of high ambient temperature, insect harassment, selective foraging or crusted snow. The solid line represents the isometric line (see Renecker and Hudson 1992b). Symbols above this line represent habitats with less efficient foraging than those below it.

areas) in spring, and some plant parts (e.g., aspen leaves) remain better longer in autumn. There is efficiency in changing foraging areas (patches) when the cost of moving is offset by the location of more and/or better browse. This happens when forage quality or quantity differs markedly between vegetation types (i.e., spring and autumn). Such movement among patches can reduce the overall rate of food intake, but the improved availability of green plant parts and fruits increases the nutrient return. Simply, the moose compensates for a reduction in total amount of food consumed by selecting only high-quality forages.

As food quality drops in winter, moose will continue to spend time eating according to the quality and availability of the food. Because all food at that time is of relatively poor quality, selectivity of forage is no longer an important way to offset the limitations of supply. Therefore, in winter moose travel as little as possible and consume readily available food. This behavior is in response to the reduction in forage biomass and the impediments of snow conditions. Moose use readily available foods such as fallen leaves as long as snow crusting or deep snow do not make this resource inaccessible.

As discussed earlier, how moose spend their time is strongly influenced by the thermal environment. Moose remain cool during periods of high temperatures by selecting open, aquatic and/or wetland feeding sites. Although the energy returns from the plants eaten in these areas may be low, compared with plants in other communities, the cooling effect of water allows moose to tolerate higher temperatures and reduces loss of body weight. Moose also suppress feeding activity during winter when temperatures become warm.

Figure 168 shows why moose are able to sustain and generally thrive in northern habitats despite persistent tradeoffs. High ambient temperature and resulting heat stress add complexity to the site and food selections. Although a smaller size would be more appropriate for the foraging behavior of a browser in the northern latitudes, Renecker and Hudson (1992a) showed through computer simulations that moose are driven to become as large as possible probably to withstand such usual forces as cold, snow and predators.

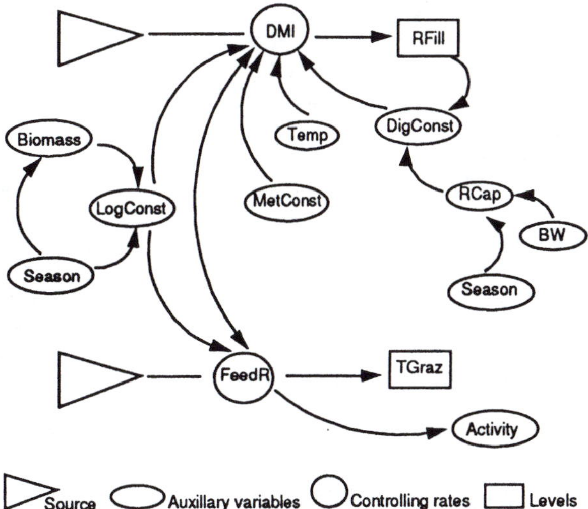

Figure 168. Daily forage consumption—or dry matter intake (DMI)—of a moose is determined by sets of factors that influence total browsing time (TGraz), feeding rate (FeedR). Browsing time appears to change most adaptively to food quality, whereas feeding rate is controlled by factors that influence the logistics of finding food (LogConst). In turn, the time required to find another bite depends on food supply (Biomass). Moose seem to increase DMI by feeding longer rather than faster. Because feeding and resting (along with ruminating) are the major activities, they must be considered in the energy budget. With the exception of high-quality forages, DMI is constrained by the physical capacity of the rumen (RCap), which varies with body weight (BW) and season (Season). In turn, this influences digestion (DigConst), or the time required for the rumen microbes to break down the food sufficiently and reduce rumen fill (RFill) so another meal can be eaten. Basically, the potential DMI of a moose is the difference between rumen capacity and rumen fill, which is regulated by digestive rate. Other variables that influence DMI include ambient temperature (Temp) and the metabolic constraints (MetConst) or energy needs of the animal (Renecker and Hudson 1992a). For moose, high temperatures can cause moose to forego eating in favor of resting to remain cool. Under such conditions, DMI is reduced. The size, feeding behavior and visceral structure of the animal all influence intake, as the moose must select a diet that will provide the energy for its survival and productivity at a given metabolic body weight.

LIGHT GOLDEN-BROWN ON SHOULDER & NECK, ALSO WASH OVER TOP OF HEAD—
DARK PURPLISH-BROWN ALONG RIDGE OF BACK & ON UNDERPARTS—
REDDISH-BROWN SIDES
LIGHT INSIDE EARS (LIKE NECK)
SPLASHING
IN THE GRASS—
(L) YOUNG MOOSE WITH CONSPICUOUSLY PALE AREA ON SHOULDER— NENANA R. NEAR SPURGIN'S CABIN, APRIL 11, '62 (MEMORY)
MOOSE PLUNGING AWAY FROM ROAD THROUGH DEEP SNOW
BRIGHT WHITE
APP. COW MOOSE— MORE SLENDER THAN ANIMAL TO L—
(MEMORY) FEB. 19 '55
HEAVY-SET, THICK-NECKED MOOSE— APP. BULL—
MOOSE ARE DEEP, RICH RED-BROWN
PAXSON LAKE, RICHARDSON HIGHWAY— ALASKA—
OPEN SPRUCE FOREST BORDERING LAKE—

14

CHARLES C. SCHWARTZ AND LYLE A. RENECKER

Nutrition and Energetics

Moose populations throughout North America are limited by a variety of critical factors. At any given time, predation, hunting, disease and/or weather influence the dynamics of the population. However, food and the animal's ability to process it ultimately regulate all populations. Consequently, understanding how moose interact with their environment when selecting, harvesting and processing foods is vital to their management. This is especially true when the moose population in question is at or near the carrying capacity of their habitat.

Nutritional needs of moose vary throughout the year and by sex and age of the animal. The availability and quality of the foods fluctuate from summer to winter and are influenced by distribution, topography, weather, and successional stage and influences. Consequently, to comprehend the dynamics of moose habitat and make appropriate and timely decisions regarding management of these habitats, at least a basic understanding of moose nutrition is necessary. A fundamental question in evaluating the nutritional quality of moose range is how well (or poorly) the extant habitats provide for the needs of individuals living there. Schwartz and Hobbs (1985) suggested that bioenergetics (the energy transformation in living systems applied to mechanisms at either the cell, organism or population level) provides a "currency" for quantifying the resource supply in units that correspond with animal demand. This allows stepwise movement from rudimentary estimates of food habits to more refined predictions of the nutritional status of moose, and finally to habitat-carrying capacity (Figure 169). Energetics is important in the biochemistry, physiology and ecology of species (Braefield and Llewellyn 1982).

Evaluating forage resources for moose from a nutritional point of view is difficult because not only must the existing foods be identified, but their amount, availability, current nutritional value and even safety also must be assessed (Schwartz and Hobbs 1985). For moose, the nutritive value of plant foods is related to the essential nutrients they contain, including carbohydrates, fats, water, protein, vitamins and minerals.

Energy is the capacity for activity or function. It is necessary for biochemical transformations, muscle contractions, nerve impulse transmissions, excretion processes and all other bodily functions (Robbins 1993). Energy commonly is measured as the heat released on combustion, expressed in calories or joules (1 cal = 4.184 J). Moose obtain virtually all of their energy by consuming plants. The energy content of various plants differ because their types and concentrations of specific chemical compounds vary. Most carbohydrates and proteins contain 3.8 to 5.7 kilocalories per gram, whereas fats, waxes and oils contain more than twice that amount of energy.

Water comprises about 65 to 75 percent of the ingesta-free body mass in adult moose, more so in neonates and calves (Robbins 1983, Schwartz et al. 1988a). Water is an integral part of the structure of soft tissues and cell contents, and it has the unique capability to absorb and dissipate heat. Water acts as a transport medium for nutrients and waste products within and from the body. Without water, proper

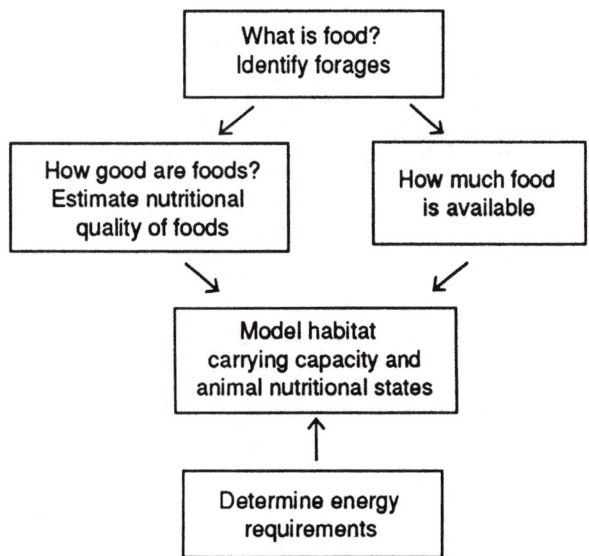

Figure 169. Evaluation of forage quality and quantity is important when predicting moose habitat value on a nutritional basis. To evaluate stocking rates (carrying capacity) of a particular habitat, biologists must determine which plants represent moose foods, the nutritional value of those foods and their availability by season. Stocking rate then can be estimated from the amount and quality of foods divided by the nutrient requirements of moose (from Schwartz and Hobbs 1985).

bodily functions cease. Yet, although water is essential, it rarely is considered a nutrient by most moose biologists, perhaps because it is associated more with a physical rather than chemical role, or because water rarely is a limiting factor for moose.

Most ungulate nutritionists agree that digestible protein and digestible energy are the most important nutrients (see Moen 1973, Wallmo et al. 1977). Proteins are composed of amino acids linked by peptide bonds (Crampton and Lloyd 1959), and make up the majority of soft body tissues. All animals require amino acids for proper bodily functions. For ruminants, including moose, their rumen microbes synthesize amino acids from organic and inorganic nitrogen. Consequently, amino acid assays are not critical to understanding the contribution of plant foods to the protein requirements of moose; crude protein determinations generally are sufficient.

Minerals are a diverse group of nutrients with many essential functions. Minerals required in large amounts (mg/g) are called "macroelements," and include calcium, phosphorus, sodium, potassium, magnesium, chlorine and sulfur. Minerals required in small amounts are referred to as "trace elements," and they include iron, zinc, manganese, copper, molybdenum, iodine, selenium, cobalt, fluoride and chromium (Robbins 1993). With the exception of a few limited studies—on sodium, copper and chromium—very little is known about the mineral requirements of moose.

Until about 1900, carbohydrates, fat, protein and certain minerals were considered to be the only dietary requirements (besides water) for normal bodily functions (Lloyd et al. 1978). It was known that other substances in natural foods were important for maintaining good health. These substances were originally termed "accessory food factors" and later called "vitamins" (Crampton and Lloyd 1959). Vitamins are organic compounds that occur in food in minute amounts and are distinct from carbohydrates, fat and protein. They are divided into two groups based on their solubility in water. The fat-soluble vitamins include A, D, E and K, whereas the water-soluble forms are B-complex and C vitamins. Moose obtain their fat-soluble vitamins either by manufacturing or ingesting them in the foods they eat. For moose and other ruminants, their rumen bacteria synthesize the B-complex vitamins. Moose do not require vitamin C.

Forage analysis procedures focus on describing protein content, digestible energy, minerals and fiber characteristic of plants. Analyses are also performed to identify secondary chemical compounds that inhibit digestion or are otherwise harmful to the animal. Several systems of forage analysis have been developed for foods—primarily grasses and cultured legumes—eaten by domestic herbivores (Van Soest 1982). In many instances, the chemical and physical properties of foods eaten by moose differ greatly from the forages fed to domestic livestock. Consequently, application of some of the techniques developed for domestic animals can lead to erroneous conclusions when applied to forage consumed by wild herbivores, and special care must be applied to the interpretation of such conclusions.

There are two major groups of forage analysis procedures: (1) those that measure characteristics of forage within the animal (in vivo) and (2) those conducted in artificial, laboratory settings (in vitro). Laboratory analysis is further divided into systems of chemical and in vitro simulation of digestive processes (Schwartz and Hobbs 1985).

Forage Quality

Plant foods are unequal in their capacity to provide essential nutrients that meet the demands of maintenance (an animal that is neither gaining nor losing weight is considered "at maintenance"), growth and reproduction in moose. To understand the nutritive complexity within a plant community, a review of the structural dynamics of a plant, starting at the cellular level, is the initial step.

Plant cells are surrounded by a relatively inflexible wall,

The nutritional quality of moose food plays a major role in population dynamics. Animals on high-quality forage have high reproductive and survival rates and positive population growth. Animals forced onto overstocked range are generally in poor physical condition and may be forced to eat foods of low quality. *Photo by Luray Parker; courtesy of the Wyoming Game and Fish Department.*

which is largely composed of the carbohydrate cellulose, along with hemicellulose and lignin. This cell wall forms an envelope that surrounds and contains the cell constituents, or cytoplasm. It also provides structural support to the plant. In young rapidly growing plants, the cell wall is thin and flexible. This condition is characteristic of the leaves and fruits of grasses and other flowering plants. In older, more mature tissue, particularly those bearing weight, the cell wall becomes thick and hard and the volume of cytoplasm decreases. This is the situation in the stems of tall grass and woody plants. For the feeding moose, division of food into cell contents and cell walls is important because it affects the availability of protein and energy. Moose cannot manufacture cellulase, the digestive enzyme necessary to break down the cell wall. However, moose can digest cellulose because of their symbiotic relationship with rumen microorganisms that produce cellulolytic enzymes (Van Soest 1982). The cell wall, therefore, constitutes a physical obstacle to the extraction of the cell contents.

Chemically the cell contents are separated from the cell wall using detergent analysis on sequential samples (Van Soest 1963a, 1963b, 1965, 1967; Goering and Van Soest 1970). The cell wall residue contains cellulose, hemicellulose, lignin and cutin-suberin (Figure 170).

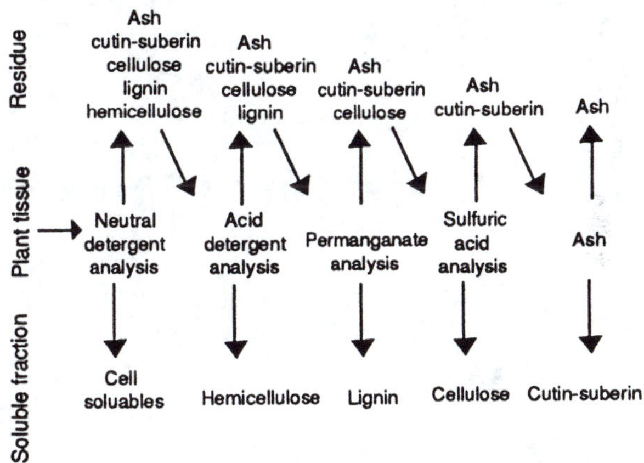

Figure 170. The Van Soest system of plant fractionation. Plant tissues are subjected to a sequence of chemical analyses, whereby certain components are dissolved during a reflux process (soluble fraction). These fractions and their residues (undigested portion) represent chemical components found in plants that are potentially digestible to the moose. For example, cell solubles are a major source of energy and protein and nearly 100 percent digestible, whereas lignin is undigestible and represents a barrier to digestion. By subjecting moose foods to such chemical analyses, biologists can determine the relative nutritive value of a plant to a moose without actually feeding it to an animal.

Chemically, fiber composition varies considerably between grass, forbs and shrubs/trees. In general, the cell wall constituents of grasses are high in cellulose and low in lignin (Van Soest 1973). Grasses also have much more hemicellulose than legumes do (Gillard 1965). The cell wall content of most moose browse is high in lignin (Oldemeyer et al. 1977). The ratio of lignin to cell wall constituent is much lower in grass (0.05–0.07) than in browse (0.29–0.47) (Schwartz et al. 1980). Thus, the proportion of lignin making up the fiber fraction of moose forage is relatively high. Expressed as a percentage, the structural carbohydrates of grasses are greater than 90 percent cellulose and hemicellulose, whereas lignin comprises up to 50 percent of the structural carbohydrate in browse.

These differences in composition of the structural carbohydrates explain why grasses are very different from browse. Cellulose provides strength yet flexibility to grass leaves and stems. Lignin provides rigidity to woody vegetation. It makes stems firm and erect, and although woody vegetation is flexible, it is much more rigid than grass. Forbs, depending on their structure, fall somewhere in between.

Once the cell wall is digested or broken down, cell contents also are available for digestion, providing additional energy. Therefore, a moose must extract both protein and energy from a food supply containing various amounts of lignin.

Diet

In the range of moose, plant species respond to seasonality by growing during short summers and entering a state of dormancy during long winters. As plants change seasonally, so does their nutrient quality. Plants begin their growth phase in early spring, long before actual green-up occurs.

In deciduous trees and shrubs, stored nutrients are transported from roots to buds and twigs. This translocation represents the first noticeable change in the nutrient quality of moose foods. As spring advances, new cells are formed in the buds. Buds swell and new leaves emerge. This active growth represents a time of rapid increase in the plant's nutritive value. The actively growing tissues of a plant contain many cell soluables but few structural carbohydrates. Cell walls are thin and expanding. As the season progresses, plant tissues mature and more structural carbohydrate is manufactured. Plants become less nutritious to the moose. Finally, in preparation for dormancy, soluble nutrients are transported back to the root system, and the leaves die and

Bark stripping by moose occurs on poor winter range or when the animals are trapped in areas because of deep snow. Moose strip bark from trees by taking the lower incisor teeth and racking them perpendicular to the tree trunk. This is different than barking by elk, which normally bark trees by racking the lower incisors vertically along the trunk. *Photos by Charles C. Schwartz; courtesy of the Alaska Department of Fish and Game.*

fall off. In conifers, the leaves remain on the tree, but generally are well defended with secondary plant compounds (Bryant et al. 1991) that deter foraging moose. After leaf drop in autumn, deciduous browse remains relatively constant in its nutritive quality, although winter rains can leach some soluble components.

In general, spring and summer foods of moose are 1.5 to 3 times more nutritious than winter foods, depending on which constituent is examined (Schwartz 1992a) (Table 40; Figure 171). Summer diets contain excess digestible energy and protein, whereas winter diets generally are insufficient to meet maintenance requirements (Schwartz et al. 1987b, 1988b, Renecker and Hudson 1989b).

Plant nutrient quality can be affected by growing condition. For example, plants growing in shade tend to have a higher protein and water content than do the same species growing in full sunlight (Regelin 1971, Hjeljord et al. 1990). Shading delays phenology and affects plant physiology. Energy input from the sun has a strong influence on secondary compounds, which are greater in plants growing in the open than in those growing in shade (Hjeljord et al. 1990). There also is a reduction in palatability of plants with high secondary compounds. And the nutrient content of plants tends to remain higher over a longer growing season during cool wet summers compared with those of hot and dry summers (Bo and Hjeljord 1991).

Diet quality also varies with the intensity of range use and plant age. Crude protein content of moose browse on winter range receiving no use versus heavy use declined from 6.1 to 5.8 percent in willows and from 8.8 to 7.4 per-

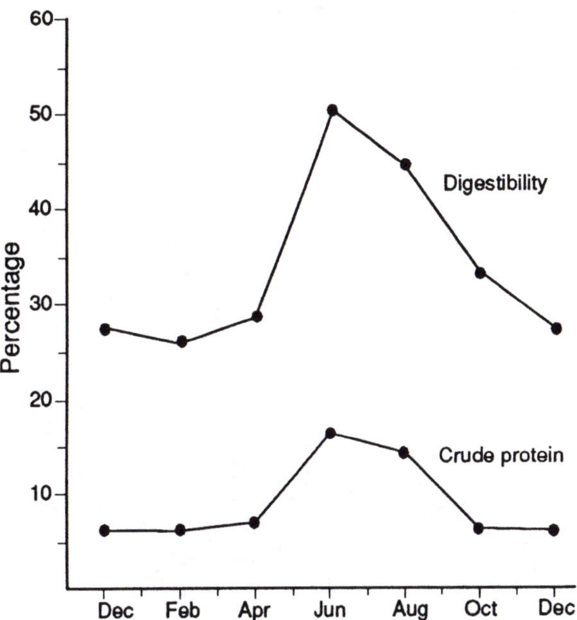

Figure 171. The quality of moose food changes with the seasons. Plants are most nutritious during the early part of the growing season when cells are rapidly expanding as the plant is growing. As the plant nears maturity, forage quality declines. Dormant plants in winter have the lowest nutrient content. This seasonal pattern of crude protein content and dry matter digestibility of moose foods illustrates the typical pattern (from Regelin et al. 1987a).

Table 40. Percentage nutrient composition of foods consumed seasonally by moose (data from Schwartz 1992a)

	Winter			Spring			Summer			
Location	Protein	DMD[a]	CWC[b]	Protein	DMD[a]	CWC[b]	Protein	DMD[a]	CWC[b]	Source
Alaska	6–7	32–41	55–57							Risenhoover (1989)
	6–7	26–29	65–67				15–18	47–67	29–46	Regelin et al. (1987a)[c]
	7–8	20–50	56–77							Schwartz et al. (1988c)[d]
Alberta	5–7	44–52	49–52	19–20	65–69	31–34	13–14	64–65	35–43	Renecker and Hudson (1985a)[e]
Michigan	5–7	31–38	56–68							Risenhoover (1987)
Norway	5–10	28–49								Hjeljord et al. (1982)[d]
Quebec							10–14			Crete and Jordan (1982a)[f]
Wyoming	5–8									Houston (1968)[g]

[a] DMD = dry matter digestion.

[b] CWC = cell wall constituents.

[c] Data for winter were from December, February and April. Data for summer were from June.

[d] Winter samples of common browse species. Values presented do not represent diets.

[e] Estimates of dry matter digestion were determined by nylon bag technique, and the data were presented in a bar graph. Therefore, values were estimated from this chart. Winter values were the range for December, January and February. Spring values were from May, whereas summer values were from July.

[f] Samples were collected at the end of the growing season. Values presented are for leaves of major browse species and do not represent diets.

[g] Crude protein content was listed for major browse species and does not represent diet protein.

Forage quality and quantity change seasonally within moose habitat. Actively growing plant tissues represent the highest quality foods available to moose. *Left photos by Charles C. Schwartz; courtesy of the Alaska Department of Fish and Game. Top right photo by Joe Van Wormer; courtesy of the Denver Museum of Natural History. Bottom right photo by Luray Parker; courtesy of the Wyoming Game and Fish Department.*

cent in paper birch (Spencer and Chatelain 1953). Similarly, nutrient content of browse declines with age of twigs. Cowan et al. (1950) reported a marked decline in crude protein content from 8.0 to 5.1 and 4.1 in hazel browse that was 1-, 2- and 3-year-old growth, respectively.

Winter diets of moose primarily consist of woody twigs from deciduous trees and shrubs and some conifers. Most of the nutrients used by moose are contained in the buds and bark, and in the case of conifers, also in the leaves. Except for conifer leaves, virtually all of this material—the available nutrients—is on the surface, and the undigestible or hard-to-digest carbohydrates (cellulose, hemicellulose and lignin) represent the center. As twig diameter increases, the ratio of surface nutrients to the core declines, and so does the nutritive value to moose.

Hubbert (1987) analyzed the nutrient content of paper birch at various twig diameters and showed that, as twig diameter increased, the available cell contents declined. Also,

Moose food quality varies with the intensity of range use and forest age. Heavily browsed plants (top left) contain little usable food and are low in crude protein. In early seral stands, there usually is an abundance of high-quality food for moose. As forests mature, many trees grow out of the reach of moose. To reach palatable shoots during winter moose commonly pull down trees to feed on the top shoots (right). Such foraging often results is stem breakage (bottom left). An increase in stem breakage on moose winter range is a signal that habitat quality is declining. *Left photos by Charles C. Schwartz; courtesy of the Alaska Department of Fish and Game. Right photo by Joe Van Wormer; courtesy of the Denver Museum of Natural History.*

the structural carbohydrates cellulose and hemicellulose increased with increased twig diameter, whereas lignin and ash remained relatively constant. Digestion declined by more than 30 percent as twig diameter increased (Figure 172).

Hjeljord et al. (1982) found that nutrient content declined with diameter at the point of browsing. Digestibility of twigs cut between 0.08 and 0.4 inch (2–10 mm) ranged from 50 to 29 percent for great willow, from 46 to 28 percent in European mountain ash and from 22 to 9 percent for silver birch.

Diet Quantity

Long-term stability in energy balance of moose is thought to be predicated on the amount of fat reserves and the rates of utilization and replenishment (Mrosovsky and Posley

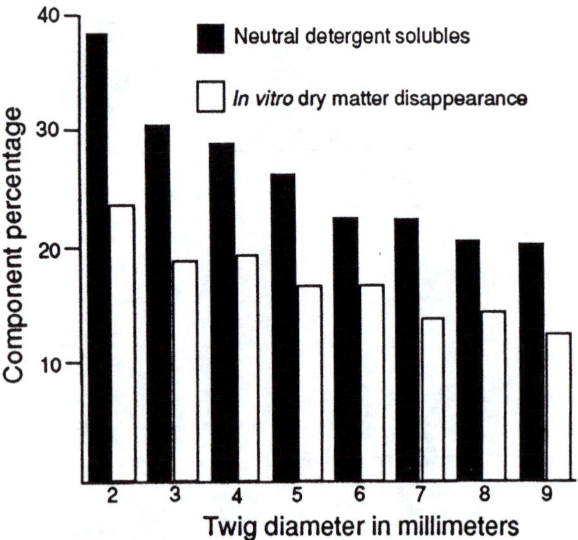

Figure 172. Not all moose food is equal in quality. As moose eat larger diameter plant stems, nutrient quality and digestibility decline. Although moose consume more food when they eat larger diameter browse, there is a change in cell wall constituents and in vitro digestibility, as illustrated here with paper birch stems (from Hubbert 1987).

Moose and elk have affinity for aspen bark. But unlike elk, moose do not commonly bark standing trees. However, when a mature aspen falls, moose readily consume the bark (above). Foraging moose will return to these trees each day until no more bark remains. This behavior can cause problems for moose population sampling conducted by pellet group censuses, because of the unusually high concentration of fecal groups at or near such fortuitous food sources. *Photo by Charles C. Schwartz; courtesy of the Alaska Department of Fish and Game.*

1977, Arnold 1985, Renecker and Samuel 1990, Watkins et al. 1991). But just as nutritive quality of moose foods changes with the seasons, so does the daily amount of food that a moose eats. The quantity of food that a moose can consume during any foraging bout is a function of both rumen volume and fill (see Chapter 13). Consumption of food is regulated and limited by the animal's physiological and metabolic requirements. The animal response involves the consequences of the quantity of indigestible residues within the digestive tract and the absorption of nutrients that enter the metabolic system. Consequently, food intake is regulated by the physical capacity and function of the digestive system and physiological mechanisms of the animal (Church 1971, Van Soest 1982, Robbins 1993).

Depending on the season and the animal's condition, regulation of intake changes from a matter of the size (capacity) of the gastrointestinal tract to a matter of physiological demand for particular foods and nutrients or elements. Moose do not meet minimum energy requirements when forced to consume foods of low nutritive values. The volume of the gastrointestinal tract and the rate at which poor-quality food moves through it limit the quantity of food that can be eaten and processed at any given time. As forage quality improves, moose can eat sufficient quantities of food to meet or exceed their daily energy requirements. Simply, the greater the nutritive quality of food, the less moose have to eat (Schwartz et al. 1988b).

Regulation of food intake ultimately comes from the brain, but the nature of the control system is not well understood. Short-term hunger involves nerve impulses from stretch receptors in the lining of the rumen, changes in body temperature, intestinal or circulatory hormones, fatigue and changes in plasma concentrations of metabolites during feeding.

Physiological Scaling

Organisms normally adapt to their physical and biological environments through genetic changes brought about by natural selection. Accordingly, individuals differ within and between species in such features as body size and shape, age at first reproduction, and manner of growth and reproduction. Many bioenergetic, productive, life history and ecological traits of wild herbivores scale allometrically (relative size and shape) (Hudson 1985). In the most general sense, the allometric equation conforms to the power function $Y = aX^b$, where X and Y are size-related measures and a and b are constants (Reiss 1989). In most instances, X represents body mass or weight (BW). For example, antler growth in cervids, including moose, scales hyperallometrically ($0.04BW^{1.33}$) (Robbins 1993). Capacities of organs that function volumetrically (e.g., lungs, heart, digestive tract) scale approximately to $BW^{1.0}$ and are said to be isometric (equal). Isometry is a special case of allometry (Reiss 1989). Volumetric rates (e.g., respiratory gas exchange, metabolic rate, cardiac output) scale to $BW^{0.75}$. Finally, biological frequencies (e.g., heart, fermentation and respiration rates) scale to $BW^{-0.25}$ (Hudson 1985).

Allometry allows comparison of animals of vastly different sizes and shapes. By taking the log of each side of the power function ($\log Y = \log a + b \log X$), plotted data form a straight line, conforming to the linear equation $Y = mX + c$. Interspecific comparisons, when scaled with the log function, allow determination of close similarities among different species with increasing size. Deviations from the expected norm represent adaptations to specific environments and provide insight into the species adaptive strategies for occupying a particular niche or environment.

Intake of dry matter or energy and metabolic rates commonly is expressed as a function of body weight to the 0.75 power allowing for interspecific comparisons.

Close observation of tame moose provides a useful way to study feeding behavior and food habits. Intake rates can be monitored by estimating the amount of forage eaten based on the quantity consumed per bite for each species of plant and then the number of bites. *Photo by Charles C. Schwartz; courtesy of the Alaska Department of Fish and Game.*

Seasonal Intake Rates

Moose (Schwartz et al. 1984), white-tailed deer (Ozoga and Verme 1970) and caribou (McEwan and Whitehead 1970) regulate seasonal intake of dry matter and digestible energy even when offered unlimited amounts of high-quality food year-round. Vast amounts of high-quality foods are available in summer; during winter, food is of poor quality and limited abundance or availability. Moose evolved with this seasonal change in food quality and abundance, and have developed an inherent rhythm of food consumption that juxtaposes the cyclic rhythms of their natural environment. To the casual observer, the animals may seem to adjust food intake with food availability, but the relationship is not that simple. Long-term stability of food intake is rarely observed in moose. Intake is influenced not only by the quality and availability of foods, but also by physiological demands of reproduction, age, disease and parasites, general body condition and, likely, photoperiodism.

The quantity of food an animal consumes generally is expressed as the amount eaten in a given time period. Measuring intake rates for a wild herbivore such as moose is not an easy task, and various techniques have been used to do so. Probably the best and most frequently used technique is observation of tame moose in natural environments. Both hand-reared tame animals (Regelin 1987a, Renecker 1987a) and tractable wild moose (Risenhoover 1987) are used. Close observation of animals indicates what the subject animal eats, and it allows for measurement of the amount of plant material taken with each bite. By counting the number of bites and estimating the mass of each bite, total intake by plant species and plant part can be calculated.

Researchers also have estimated the amount of forage consumed by moose during winter by following their tracks in snow and examining the amount of biomass removed from browsed trees (Vivas and Saether 1987). The technique relies on the premise that there is a reasonable relationship between the diameter at point of browsing and biomass of vegetation removed (Risenhoover 1987, Hubbert 1987). Measurement of the diameter of browse twigs then can be converted to biomass consumed. Radiotelemetry often is used with this technique, because it is important to relate intake to time. Animal locations at known intervals are determined, and foods consumed between those points can be estimated. The technique suffers because it cannot be used on winter ranges with many animals or heavy browsing, because differentiation with previous browsing by other animals is difficult. Also, where many moose are feeding, tracking an individual animal is difficult or impossible.

Intake also can be estimated by feeding an animal an inert marker. If the digestive system is in a steady state, then input of a marker should equal output. Marker output is controlled by rumen volume or the amount of food in the gut and rate of passage of that food. Once the marker in the digestive system reaches equilibrium with the flow of food (or an animal is given a constant amount of the indigestible marker several times a day for several days), administration of the marker can cease and feces can be collected. Dry matter intake is calculated from the egestion curve (Krysl et al. 1984). Rare earth elements, such as dysprosium, are valuable markers because they are nonradioactive, easily collected and analyzed through neutron activation analysis (Kennelly et al. 1980).

Renecker and Hudson (1985c) administered a single pulse-dose of dysprosium to free-ranging moose. Fecal samples collected over a 4-day period were analyzed for marker concentration. By regressing marker concentration against time, the volume of indigestible dry matter in the gut was calculated at time of administration of the marker from the peak concentration of dysprosium (in milligrams per gram of indigestible dry matter) by least squares analysis. Once the original dilution of dysprosium was calculated, fecal output was estimated. Because the original amount of marker given and the dilution after mixing with food in the gut were known, it was possible to correct the indigestible diet (fecal output) for digestibility and estimate dry matter intake. Estimates of dry matter intake for moose using this pulse-dose technique of dysprosium were within 7 to 10 percent of estimates of intake made by the bite count technique and a ratio technique (total fecal collections combined with estimates of digestibility) (Renecker and Hudson 1985c). This technique has promise, but attention must always be given to the influence of plant phenology on retention time and passage of ingesta.

Food intake studies using confined animals under controlled conditions not only provide estimates of the quantity of certain foods consumed, but usually can be designed to estimate forage digestibility (Hjeljord et al. 1982, Schwartz et al. 1984, 1988a). The intake by animals given a free choice of clipped food approximates normal intake rates. However, this technique eliminates the animal's normal forage selection process. Plants clipped and offered to moose may not represent what the moose would choose if confronted with the same choices in the wild. However, such problems can be overcome. For example, Danell et al. (1991) transplanted entire plants in known arrangements and density in study plots. Such studies not only allow researchers to measure how much forage is removed, but they also can evaluate foraging behavior as it relates to stem density and composition of available foods.

Intake of dry matter is influenced by the digestible energy content of the diet. Captive moose eating foods of varying energy content consumed different amounts of dry matter, but similar amounts of digestible energy. These moose consumed enough food to meet a seasonal caloric requirement rather than maximizing dry matter fill (Schwartz et al. 1988b).

Food intake generally is expressed as grams of food or energy consumed per unit of body weight raised to the 0.75 power ($BW^{0.75}$). On the assumption that intake is a function of metabolic requirements, measurements of food intake are related to metabolic body mass. Also available are data that demonstrate that intake of herbivores is linearly related to body size (rumen volume scales as $BW^{1.0}$), so that the amount of food an animal consumes also can be expressed as a percentage of total body weight (Van Soest 1982).

Food intake in moose peaks in summer and decreases to a nadir in late winter (Schwartz et al. 1984, Renecker and Hudson 1985a, Schwartz et al. 1987c; Figure 173). Moose eat approximately 2.6 to 3.5 percent of their body weight per day in dry matter during summer, but only 0.5 to 1.3 percent in winter (Schwartz et al. 1984, Renecker and Hudson 1985a). These equate to approximately 116 to 142 grams

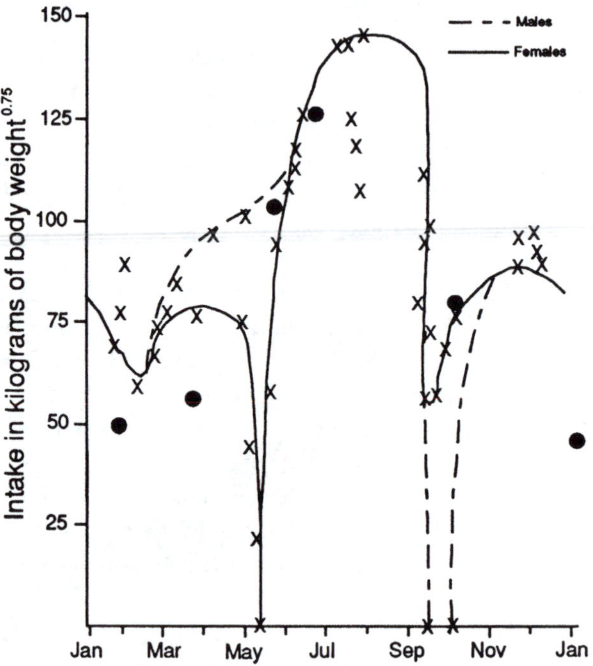

Figure 173. Moose eat different amounts seasonally. Peak rates of food consumption occur in spring and early summer when forage quality and digestibility are high. Bulls do not eat during the autumn rut and cows reduce their intake during estrus. Winter intake rates are about half those during summer (from Schwartz et al. 1987c). Baseline graph (crosses) from Schwartz et al. (1984); intake data (solid circles) from Renecker and Hudson (1985a).

per kilogram (g/kg) of $BW^{0.75}$ per day in summer and 30 to 51 per day in winter.

Such estimates of dry matter intake are useful in many respects, particularly for assessing and managing habitat. If, for example, a 1,000-pound (454-kg) cow moose on summer range in Alaska needs to consume 3 percent of her body weight in dry matter daily then she must eat 30 pounds (13.6 kg) of dry matter. However, all of the leaves she eats contain water. The water content of actively growing forage ranges up to 30 percent, so to obtain 30 pounds (13.6 kg) of dry matter, this cow would have to eat almost 43 pounds (19.5 kg) (30/0.7 = 42.9) of food each day.

Digestibility and Nutrient Absorption

The processes of ingestion, mastication and digestion are, in part, functions of the anatomical structures of the digestive system (see Chapter 2). The ability of moose to obtain essential nutrients from their foods not only is a function of how efficiently they digest and assimilate nutrients, but also of how rapidly they process forage and what the associated energy costs are.

The principal functions of the gastrointestinal tract are to digest and absorb nutrients and eliminate waste. Although the functions may be similar in very diverse species, the nature of their foods differ markedly and so do their digestive systems. Moose belong to the suborder Pecora—the group of animals commonly referred to as ruminants. Ruminant animals have digestive systems that are specialized to process herbaceous foods high in structural carbohydrate. The ruminant stomach provides extensive pregastric microbial fermentation. The moose contributes both nutrients and a suitable anaerobic environment where the microbes flourish. In exchange, the microbes digest the complex carbohydrates cellulose and hemicellulose, which the moose could not otherwise digest. In addition, the moose obtains energy from the waste products of bacterial fermentation and digests the bacteria themselves. These bacterial waste products are commonly referred to as "volatile fatty acids." Most volatile fatty acids are absorbed directly from the rumen through finger-like structures on the surface of the rumen known as "papillae." The concentrations of volatile fatty acid produced by moose are similar to those of other ruminants consuming high fiber diets (Gasaway and Coady 1974).

The stomach of the ruminant animal is divided into four compartments—the rumen, reticulum, omasum and abomasum (see Chapter 13). The rumen and reticulum form a common vat, and digesta flows freely between these compartments. It is in this rumenoreticular complex where most of the microbial fermentation of foods occurs. The omasum functions mainly to remove excessive water from rumen contents; it also acts as a filter, preventing large undigested materials from entering the true stomach, or abomasum. The abomasum operates similar to the simple stomach of a monogastric (one-stomach) animal, where acid and enzyme digestion occurs.

Mastication and its Influence on Digestion

Feeding times of moose increase when requirements for energy and nutrients are high (i.e., spring and summer) or when food is in low supply. But, when food quality is low, the ruminant is left with less time to feed because it requires more time to break particles down in size. This is especially important for the moose, which is a selective browser. For the moose, the capability to eat the next meal depends strongly on diet digestibility and passage (Renecker and Hudson 1992a).

The ruminant relies on chewing as its primary method for physical breakdown of plant material. By breaking down food into smaller pieces, the surface area increases for microbial digestion, and the food's size is reduced to accommodate swallowing and egestion. Concentrate feeders would be expected to have a low capacity for digestion of foods rich in fiber (i.e., grass) (Van Soest 1982), because they have adapted to faster passage rates (Huston 1978). Large particles of grass high in cell wall content are more resistant to passage from the rumen than are more lignified foods that break down easily when masticated. Longer rumination times required by grass foods would imply restrictions on the flow of digesta in a browser, such as the moose, with a tube-like gut.

Rumination time, or the time required to rechew food eaten earlier, called a "bolus" or "cud," is controlled largely by total cell wall content in the diet. Renecker (1987a) noted that, with changes in diet quality and intake, moose adjust their chewing rate of a regurgitated bolus from 24 to 107 chews per minute. As forage matures in summer and moose consume lower quality food, the chewing required to break down particles within a bolus increases (Figure 174). When food is succulent during the early growth period, Renecker (1987a) found that moose chewed food at about 62 chews per bolus, compared with a high of 133 in winter.

Food particles must be reduced in size before they can pass out of the rumenoreticular vat. During winter, moose consume largely diets of woody twigs that remain in the rumen longer. These diets also require proportionately more time for chewing and particle size breakdown per feeding bout than do meals eaten during summer (Renecker and Hudson 1989a). With the higher quality, more digestible foods of spring and summer, moose require less time chewing/ruminating food because larger fragments are propelled

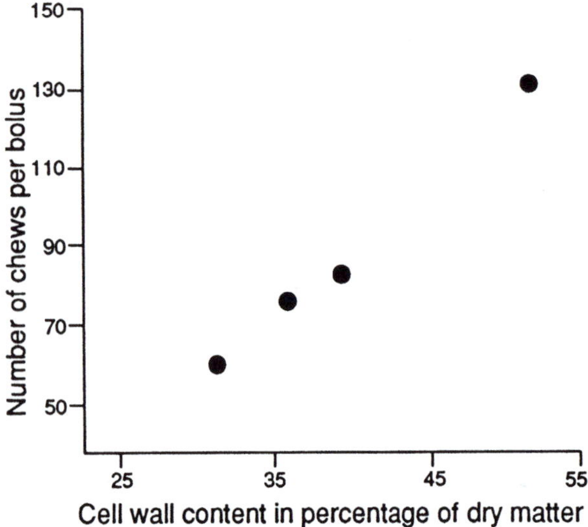

Figure 174. Moose masticate their food to aid digestion. As the fiber component (cell wall constituents) increases, the animal must chew each bolus more to ensure proper digestion (from Renecker 1987a). Consequently, chewing rates serve as a rough index to forage quality.

and flushed through the gastrointestinal tract (Renecker and Hudson 1990a). The shorter rumination times of moose during late spring are correlated to higher forage quality and digestibility, greater concentration of cell solubles, and lower cell wall contents in these diets (Renecker and Hudson 1985a, 1989a).

Rumen Turnover Time and Rates of Passage

Rate of passage, in a strict mathematical sense, refers to the flow of undigested matter within and through the entire gastrointestinal tract per unit of time (Robbins 1993). On the other hand, rumen turnover (the rate at which food passes out of the rumen) is a function of both digestion and passage (Van Soest 1982).

To extract maximum energy from ingested foods, ruminants must optimize the tradeoff between retention and passage (Foose 1982). Retention of food in the rumen allows for more complete digestion, whereas rapid passage allows for more food to be processed.

A major characteristic of moose and some other ruminants is a prolonged retention of food in the rumen complex (Hume and Warner 1980). Rate of passage is controlled by morphological structures (rumen pillars, reticulorituminal folds, reticulo-omasal orifice, etc.) and by rate of particle breakdown in the reticulorumen (Van Soest 1982). For other ruminants, especially concentrate selectors and intermediate feeders, a short retention time is common (Hofmann 1985). Rate of passage becomes important when foods contain high quantities of undigestible materials.

Schwartz et al. (1980, 1988d) reviewed components of voluntary intake and passage rate as they relate to moose. Several factors control passage rate, with the most important being excretion and absorption. Mertens (1973) concluded that the lower tract of ruminants did not limit passage. Movement of foods out of the rumen is controlled by the reticulo-omasal orifice, which limits passage of large particles. As dietary fiber increases, voluntary intake and rate of passage decline (Van Soest 1982). Likewise, cell wall content and lignin are inhibitors to physical breakdown of forage in the rumen and, thus, of rate of passage (Figure 175). However, most foods consumed by moose are high in the fiber component lignin (Schwartz et al. 1980), and although lignin content is related directly to food particle size, it is related inversely to fecal particle size (Van Soest 1966). The largest, most lignified food particles are transformed into the smallest, most lignified fecal particles. Furthermore, although foods high in lignin require greater rumination time, particles produced are more optimally shaped for passage. Because moose consume a diet that is composed of fiber mainly in the form of lignin, particle breakdown occurs at a faster rate than would forage containing a similar concentration of fiber in the form of cellulose.

Schwartz et al. (1980) speculated that moose, as selective browsers, are able to deal with poor quality forage by passing digesta rapidly and by digesting mainly the soluble components of browse. Rapid passage is possible because highly lignified browse, when masticated, shatters into large

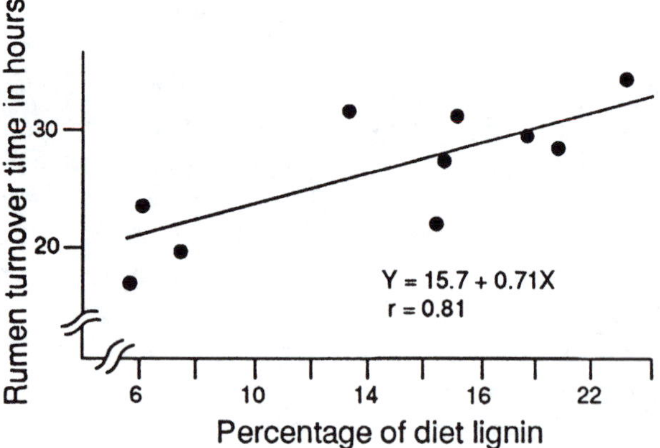

Figure 175. Lignin, which provides structure to woody material, impedes digestion. As lignin content of forage increases, the time it must remain in the moose rumen for digestion also increases. Thus, there is a strong relationship between rumen turnover time of liquids (in hours) and diet lignin (from Schwartz et al. 1987c). Data from Hjeljord et al. (1982) and Schwartz et al. (1988c).

To maximize the nutritional value of food consumed the moose must chew plant material extensively. After feeding bouts some food is regurgitated and masticated, which reduces the plant material size, enhancing the digestive process, thereby making it more efficient. *Photo by Joe Van Wormer; courtesy of the Denver Museum of Natural History.*

cuboidal particles suitable for passage (Mertens 1973, Milchunas et al. 1978). In a comparative study of cattle (a grazer), elk (a mixed or intermediate feeder) and moose (a concentrate selector), Renecker and Hudson (1990a) addressed the Schwartz et al. (1980) speculation. Renecker and Hudson fed aspen browse, alfalfa hay and grass hay to these three ruminant species, and concluded that the ability of browsing moose to extract adequate energy from forage of low digestibility depended on rapid passage rate. Relative to cattle or elk, moose were most sensitive to diet. The moose passed browse diets more rapidly than cattle, but retained grass hay and alfalfa longer than either elk or cattle did. Rapid passage of browse was achieved by moving large particles out of the rumen. On the basis of critical particle size theory, the threshold range of particles was highest for moose, with the probability of passage increasing for high lignin ratios.

In a similar study and using the nylon bag technique, Renecker et al. (1982) compared the digestive kinetics of grass/alfalfa, alfalfa and alfalfa/browse mixtures in cattle, elk and moose. They demonstrated only slight differences in digestion of dry matter, which suggested that the fermentation environment in the three ruminant species was basically the same. The volume and rate of digestion of the three forages by the three ruminant species were nearly identical (Figure 176). Digestive efficiency in moose appears to be regulated more by forage selection, rumination, gut morphology and mechanisms controlling rate of passage than by microbial digestion in the rumen.

In moose, there is a close association between the flow of liquid and solid materials from the rumen (Hubbert 1987,

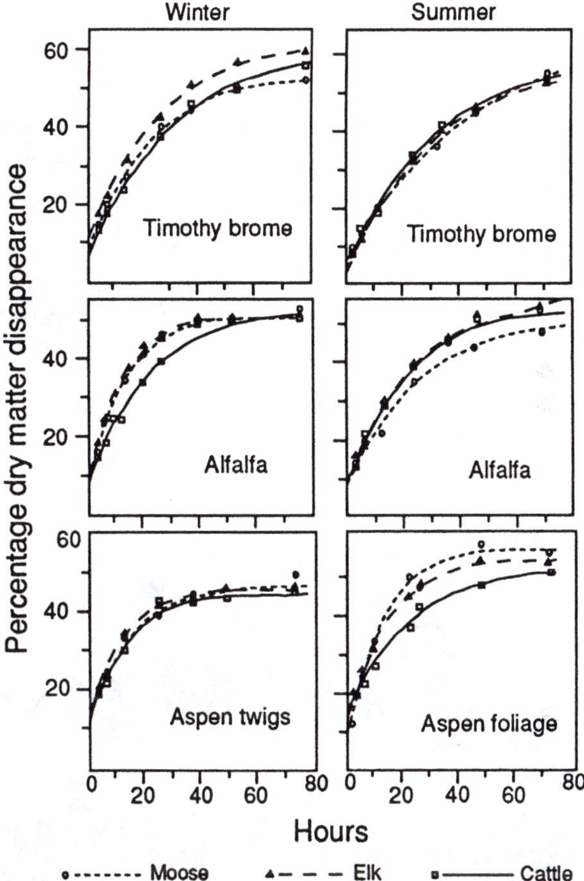

Figure 176. Cattle, elk and moose digest forages at different rates because the morphologies of their rumen complexes differ. Cattle are classified as grazers and "roughage eaters" and efficiently process such forages as timothy and brome. Moose are browsers, and do not digest grass efficiently. Elk are intermediate or "mixed" feeders and can digest both grass and browse seasonally. Consequently, when cattle are fed browse (lower right) or moose are fed grass (upper left) neither processes the materials as efficiently as the other. Elk are more capable of handling either forage. The chemical nature of browse is different from grass. Dry matter digestion of browse reaches maximum (asymptote) much earlier than does grass. This is because most of the nutrients in browse are cell soluble materials, whereas grass contains large quantities of cellulose, which require a long period to digest (from Renecker 1987a).

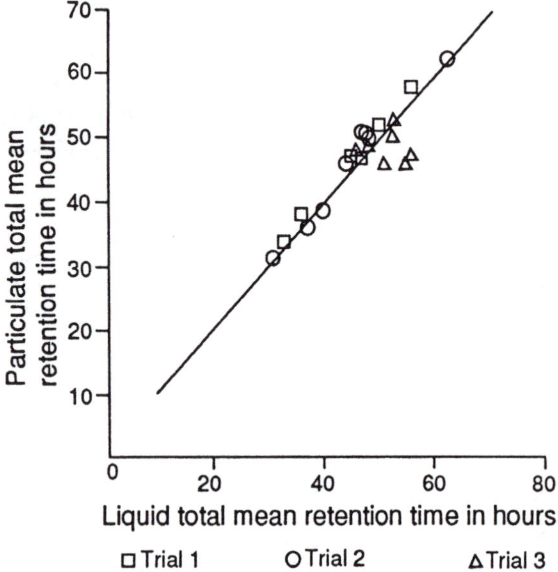

Figure 177. The morphological structure of the moose stomach allows both solid and liquid materials to pass at approximately the same rate (from Hubbert 1987). This contrasts to the stomachs of cattle, in which the liquid phase passes more rapidly than the solids, thus preventing a loss of important cellulosic bacteria essential for fiber digestion. Because moose mainly digest the cell soluble components of browse, liquid retention is less important.

Renecker and Hudson 1990a; Figure 177). Hubbert (1987) discussed the rumen turnover times of liquids and solids in cattle, domestic sheep and moose. His liquid-to-solid rumen turnover time ratios were 0.964 for moose, 0.788 for sheep and 0.395 for cattle. Therefore, the liquid material in cattle (a grazer) flowed faster than the solids; in moose (a browser) the differences were small; and in sheep (a mixed feeder) the turnover time was between that of cattle and moose. Grazers retain solids longer because cellulose and hemicellulose require prolonged digestion. In browsers, retention of highly lignified fiber would be inefficient and costly.

Rumen turnover times for moose fed single species of mixed browse diets range from 21 to 34 hours, and from 9 to 28 hours for a 50:40:10 ratio mix of common timothy, meadow fescue and red clover (Hjeljord et al. 1982). Rumen turnover times for a pelleted ration, a mixture of a pelleted ration and aspen browse and a mixed browse diet were 17.6, 18.7 and 31.2 hours, respectively (Schwartz et al. 1988d). These values represent the wide range of times measured, and serve as baseline for further research.

Forage Digestion

Digestion in moose is dynamic and involves the inflow of food into the rumen and the outflow of liquids, bacteria

During spring and early summer foods eaten by moose are high in both nutrients and water. Moose consume large quantities of these foods and often produce soft, poorly formed feces that resemble those of domestic cattle (left). During winter, when moose eat woody browse low in moisture, pellets are produced (right). Pellets are popularly called "moose nuggets," and have been used to make a number of novelty items for the tourist trade. *Photos by Charles C. Schwartz; courtesy of the Alaska Department of Fish and Game.*

and undigested food residues. Rate of digestion, the quantity of food digested per unit time, is related to food quality, deficiencies, excesses and availability of nutrients (Van Soest 1982). Generally, the soluble sugars and other cell contents are fermented rapidly by the rumen bacteria. The less soluble cell walls, their fiber and protein tend to ferment much slower. Dry matter digestion is inhibited by lignin (Figure 178). Richards (1976) speculated that lignin may inhibit digestion by either preventing attachment of microbes to substrate or by preventing the enzyme from attaching to the cell wall linkages (Morris 1984). Renecker and Hudson (1990a) confirmed this and showed that digestibility of the potentially digestible neutral detergent fraction decreases with increasing cell wall lignin (Figure 179).

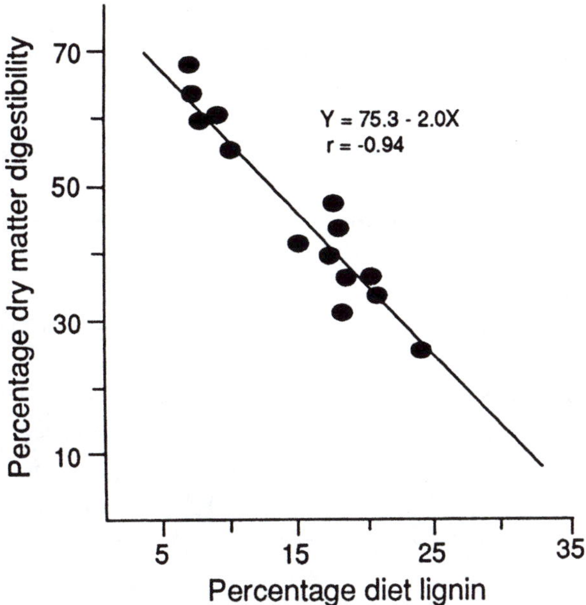

Figure 178. Relationship between diet lignin and digestion of dry matter (from Schwartz et al. 1987c). Note, as the lignin concentration increases, digestibility declines rapidly.

To optimize energy intake, moose have an adaptive feeding process. Foods are more digestible in spring and summer than in winter (Table 40). Summer diets are composed mainly of highly digestible leaves of deciduous browse, forbs and aquatic plants. Intake rates are high, and moose produce (anabolize) large quantities of fat. Winter diets are composed almost exclusively of woody browse (see Chapter 13) and low in digestibility. Moose must burn (catabolize) fat to meet an energy deficit. Although summer and winter foods have different rates of digestion (Table 40), both contain indigestible lignin. Woody browse has a highly lignified cortex, and most of the digestible components are contained in the outer bark and buds. Moose digest this outer material efficiently and pass the indigestible cortex rapidly. However if forage is limited, as in winter, slowing the rate of passage may be beneficial by allowing for some digestion of the woody browse cortex (Hubbert 1987).

IN VIVO TECHNIQUES

The first step in measuring the relative value of foods to moose is to estimate the amount retained in the body. Estimates of forage digestion usually are obtained by conducting conventional digestion and balance trials. The technique involves acclimating a group of animals to the test forage for several days to weeks before trials are begun.

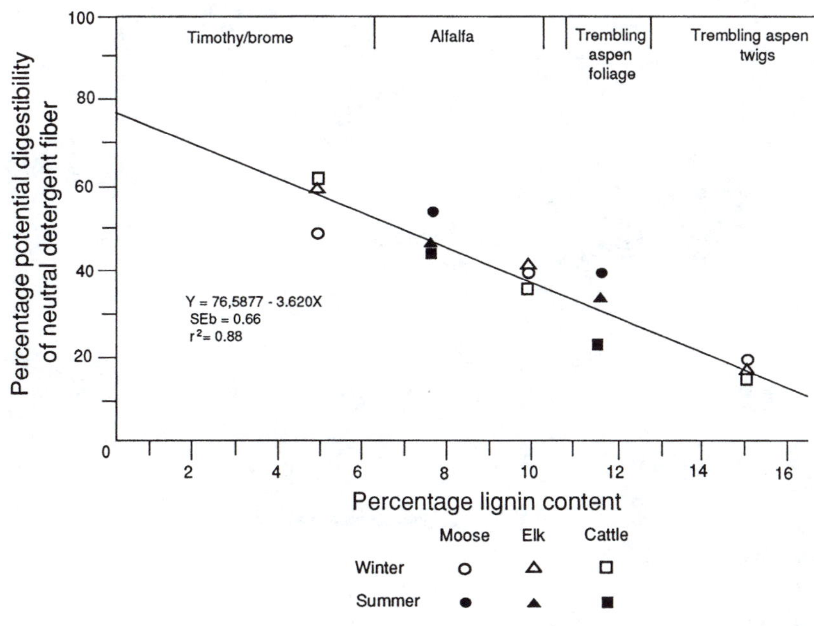

Figure 179. For moose, lignin not only inhibits the digestion of cell soluble materials, it also binds with other forms of fiber in plant tissues (from Renecker 1987a). Therefore, as diet lignin increases, the digestibility of neutral detergent fiber decreases.

During this pretrial period, daily forage consumption is measured. After the acclimation period, animals are placed in stalls designed for the collection of feces and urine. Animals generally are fed at approximately 90 percent of their pretrial intake for several days to adjust their fecal and urine output to a constant intake. After this adjustment period, recordings are taken of daily dry matter consumed and total fecal and urine outputs. The fraction of forage energy, dry matter or other nutrients (i.e., protein, cell wall) digested is calculated as:

Apparent digestion coefficient = 100 X Intake – Feces / Intake.

True digestibility, by contrast, is estimated as:

True digestion coefficient = Intake – (Feces – Endogenous) / Intake.

The important difference between the two is that true digestibility removes the influences of materials of animal origin (endogenous) added to the feces during digestion. Discrepancies between true and apparent digestibilities can be substantial, depending on the magnitude of elimination of endogenous products contained in digestive enzymes, mucosal sloughing and bacterial bodies. The amount depends on the physical and chemical characteristics of the forage items (Hungate 1966, Robbins et al. 1975, Mould and Robbins 1981, 1982).

The primary advantage of in vivo balance trials for evaluating forage is the strong and direct inference they provide to nutrient assimilation within the animal (Holecheck et al. 1982). However, digestion trials are time consuming and frequently require collection of large quantities of test forage. It is extremely difficult to collect even common forage items needed to feed moose. Confinement necessary to collect feces and urine also can influence the subject animals' digestive processes (Mautz 1971), and clipped forage may not represent what a free-ranging animal chooses to eat.

IN VITRO TECHNIQUES

In vitro digestion systems reduce the cost and time necessary for conducting complete digestion and balance trials. Estimates of both dry matter and energy digestion of moose foods can closely agree with in vivo values (Hjeljord et al. 1982, Schwartz et al. 1987c, 1988c). Two in vitro systems commonly are used to evaluate forages—the Tilley and Terry (1963) two-stage digestion system and the Goering and Van Soest (1970) system. The Tilley and Terry two-stage digestion approximates apparent dry matter digestion, whereas the Goering and Van Soest system more closely approximates true digestibility. Frequently, dry matter digestion of browse predicted by the Tilley and Terry

To measure how efficiently moose process various foods, researchers must rely on feeding test rations to confined animals. By measuring the amounts of food consumed and feces produced, the digestibility of forage can be calculated. Such calculations can serve as reasonable approximations of forage quality. *Photo by Charles C. Schwartz; courtesy of the Alaska Department of Fish and Game.*

technique is lower than the Goering and Van Soest system (Milchunas et al. 1978, Hobbs et al. 1981, 1983), which assumes that nearly 100 percent of the cell wall contents are digestible (Robbins et al. 1975). Also, in vitro estimates can be influenced by the inoculum donor and its diet (see Schwartz and Hobbs 1985), but highly soluble forages such as browse are relatively insensitive to variation in inoculum (Milchunas and Baker 1982).

Partitioning of Forage Energy

Understanding how the energy contained in plants is used by a moose requires an understanding of how energy flows within the animal. Traditionally, energy partitioning in ruminants divides gross energy ingested into digestible, metabolizable and net energy fractions (Figure 180). Gross energy (or combustion energy) in plant samples is routinely between 4.2 to 4.5 kilocalories per gram of dry matter (Golley 1961, Milchunas et al. 1978). It may be higher when plants contain high levels of fats, oils or waxes. However, variation in gross energy content of plants is of little significance to a moose. Instead, the quality of individual foods is determined by how efficiently this gross energy is used.

Energy partitioning is evaluated with in vivo balance trials. Methods for balance trials involve feeding known amounts of forage of known composition and collecting the waste products from the animal, including urine, feces and gases. Because the principal energy loss as gas is methane, which is difficult to measure, its contribution often is estimated indirectly (Schwartz and Hobbs 1985).

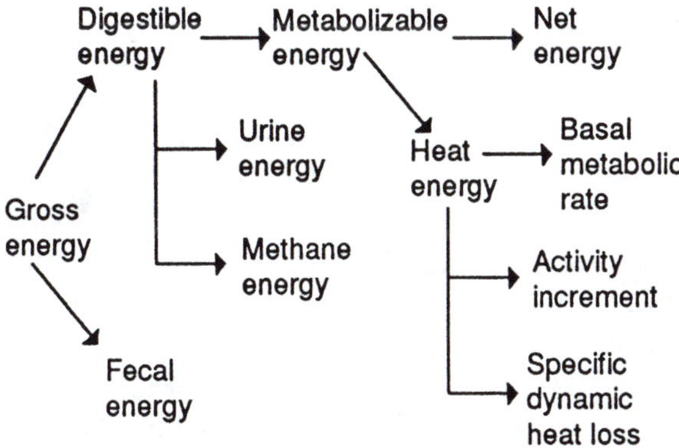

Figure 180. Moose cannot use all the energy contained within the forage they consume. Energy is lost as undigested material in the feces. Energy consumed less (minus) fecal energy loss is termed "apparent digestible energy." Digestible energy less the energy lost in urine and methane from bacterial fermentation represents the potential energy of metabolism. Increments of metabolizable energy are used to maintain normal body functions (basal metabolism), enable work (activity increment) and digest additional forage (specific dynamic heat loss). The remaining or net energy represents fuel for reproduction and growth.

Digestible Energy and Dry Matter

FECAL ENERGY LOSSES

Apparent digestible energy or dry matter represents the difference between what the animal consumes and what is lost in the feces. The indigestible portion of the gross energy consumed is the most variable component of food quality, and it depends on forage chemistry. Digestible energy (gross energy minus fecal energy) also depends on forage chemistry, which is highly dependent on the plant growth stage, species and the site (Regelin et al. 1987a). There are only a few estimates of in vivo dry matter digestion and even fewer determinations of energy partitioning with moose. Digestible energy ranges from 25 to 70 percent in moose forage (Table 41).

The percentage of digestible dry matter contained in moose forage can be converted easily and accurately to the percentage digestible energy using the equation $Y = -2.42 + 1.04X$ (Schwartz et al. 1987c), which is similar to equations that apply to elk (Mould and Robbins 1982), mule deer (Milchunas et al. 1978), white-tailed deer (Robbins et al. 1975) and roe deer (Drozdz and Osiecki 1973). A similar equation was derived by Schwartz and Hobbs (1985) from data presented by Hjeljord et al. (1982).

Metabolizable Energy

Metabolizable energy from ingested foods is the energy retained in the animal after losses associated with urine and methane are deducted from apparent digestible energy intake. It represents the usable energy of forage that can be converted into body tissue, maintain basic body function (basal metabolism), and fuel activity, digestion, reproduction and other life processes (Moen 1973). It also represents about 78 percent of the apparent digestible energy of a mixed browse diet and 84 to 90 percent when moose are fed a pelleted ration (Table 41; Schwartz et al. 1987b). Variation in metabolizable energy is a function of dietary components and other factors.

URINARY ENERGY LOSS

Loss of dietary energy in urine is largely dependent on diet; it increases with increased levels of terpenoids, phenols and other secondary plant compounds in forage (Robbins 1993). Urinary energy loss (Table 41), expressed as a percentage of gross energy intake, usually ranges from 1.7 to 6.1 percent in moose, with browse diets at the high end. Urinary energy loss ranges from 2.6 to 16.2 percent of apparent di-

Table 41. Partitioning of gross energy (GE) of moose fed various diets[a]

	Diet								
	Percentage protein				Pellets				
Trait	8	11	13	16	Nov.	March	Sept.	Pellets/browse	Browse
Daily intake GE/kg body weight$^{0.75}$	241.6	240.8	244.0	242.1	361.7	190.6	337.2	241.4	186.4
Fecal energy (percentage GE)	47.2	46.1	46.0	44.1	35.6	30.5	43.1	43.3	62.3
Urine energy									
Percentage GE	1.8	3.2	4.0	4.1	1.7	3.0	1.7	1	6.1
Percentage ADE[b]	3.4	5.9	7.4	7.3	2.6	4.3	3.0		16.2
Methane (percentage GE)	3.5	4.7	4.7	3.1	4.8	4.8	4.8		2.1
Methane[c]	7.4	7.5	7.5	7.6	8.1	8.4	7.6		6.5
Heat (percentage GE)	11.4	7.0	9.4	5.0					9.1
Metabolizable energy	47.6	45.9	45.2	48.5	57.9	61.8	50.3		29.5
Kilocalories/gram	2.01	1.93	1.93	2.05	2.52	2.63	2.10		1.50
Net energy									
Percentage GE	36.5	39.4	37.9	41.8					20.4
Kilocalories/gram	1.54	1.66	1.61	1.77					1.04

[a] Original data from Schwartz et al. (1981, 1983) and Regelin et al. (1981b); original table from Schwartz et al. (1987c).

[b] ADE (apparent digestible energy) = GE minus fecal energy.

[c] From Baxter (1967).

gestible energy (Schwartz et al. 1987c), similar to values reported for other ruminants (Robbins 1993).

Increased energy loss in urine often is associated with increased levels of protein in the diet. In studies aimed at estimating the dietary protein requirements of moose, Schwartz et al. (1987b) demonstrated that, for each gram of urinary nitrogen, 8.3 kilocalories of energy are lost in the urine. The value is higher than reported for white-tailed deer fawns (6.7 kilocalories) (Smith et al. 1975), deer aged 3 to 17 months (7.7 kilocalories) (Holter and Hayes 1977) and yearlings (6.1 kilocalories) (Holter et al. 1979), suggesting some variation among species.

Metabolizable energy coefficients also are influenced by the level of energy intake. Animals consuming foods that do not fulfill their maintenance requirement must burn fats and muscle to make up the energy deficit. When a moose burns muscle, not all of the energy in this protein is usable. The unusable energy is excreted in the form of urea. This urea represents urinary energy not associated with ingested food. Consequently, it is possible to correct a metabolizable energy coefficient by adding back the energy associated with the urea lost during the catabolism of muscle tissue. Nitrogen balance corrections can significantly increase the metabolizable energy coefficients when an animal's intake of energy is far below that required for maintenance. For example, nitrogen-corrected metabolizable energy for white-tailed deer increased from 10 to 92 percent above unadjusted estimates because these animals lost significant energy not associated with the diet (Robbins 1993).

METHANE ENERGY PRODUCTION AND LOSS

Estimating the metabolizable energy contribution from a food for moose generally is limited by measuring methane production. Methane is a product of bacterial fermentation in the gastrointestinal tract. Methane, propionate and, to a lesser extent, unsaturated fatty acids are hydrogen "sinks" (stores). Their production or hydrogenation allows for continued fermentation without significant production of hydrogen. However, methane cannot be oxidized and therefore, constitutes a loss of energy (Robbins 1993). Methane is lost either by belching (eructation) or passed through the lower gut. Methane production in moose consuming browse represents about 2.1 percent of their gross energy intake, which is about one-third the energy lost in urine (Schwartz et al. 1987c). Methane production is often predicted for domestic stock from standard regression equations using digestible energy as the independent variable (Blaxter 1965). However, this standard equation overestimates actual levels of methane production in moose by 3 to 4 percent (Schwartz et al. 1987c).

Net Energy

Net energy represents the efficiency with which metabolizable energy can be used for maintenance and production (Robbins 1993). Of all the efficiency estimates, the net energy coefficient is the most difficult to estimate and understand, because it varies with diet and production state (e.g.,

Moose store large quantities of fat during the summer. Those that consume food in amount and/or quality below the animals' maintenance requirements eventually use both fat and muscle to make up for the energy deficit. Consequently, body condition can serve as an indicator of range quality. Malnourished moose are particularly susceptible to disease and predation, as well as starvation. *Photo by Charles H. Willey; courtesy of the Vermont Department of Fish and Wildlife.*

pregnant, lactating, growing) of the animal (Reid et al. 1980). Net energy is calculated by subtracting energy losses associated with specific dynamic heat loss, basal metabolism and activity. Net energy coefficients for maintenance and production usually are determined by feeding animals housed in a thermoneutral environments at different levels of intake. Energy balance (heat production) is plotted against metabolizable energy intake (Figure 181). When the regression involves energy balance as the dependent variable, the slope of the regression is the net energy coefficient. When heat production is the dependent variable, 1 minus the slope is the efficiency estimate (since heat production represents a loss of energy) (Robbins 1993).

HEAT INCREMENT

Energy loss as heat represents the heat expended in fermentation, digestion, rumination and radiation. Heat of fermentation can represent energy lost or it can serve as a source of heat to maintain the animal's homeothermic state. Heat is theoretically partitioned into basal heat production, as in a fasting animal, and that due to dietary intake, which can be considered a feeding cost (Van Soest 1982). The only available estimates of heat increment of digestion in moose (Table 41) range from 5.0 to 11.4 percent of gross energy intake (Schwartz et al. 1987c). Heat increment is influenced by diet, metabolizable energy intake and animal size (Van Soest 1982). It generally increases as metabolizable energy intake increases (Robbins 1993).

According to Robbins (1993), the energy coefficient measured below maintenance is the efficiency with which metabolizable energy substitutes for tissue energy in meeting maintenance requirements. Because heat increment is a curvilinear function (Figure 181), an entire range of net energy coefficients is possible. Similarly, the net energy for gain is curvilinear and decreases with increased intake. Linear functions often are used below and above maintenance to estimate both maintenance and gain. Net energy coefficients for fat and milk production and for hair and fetal

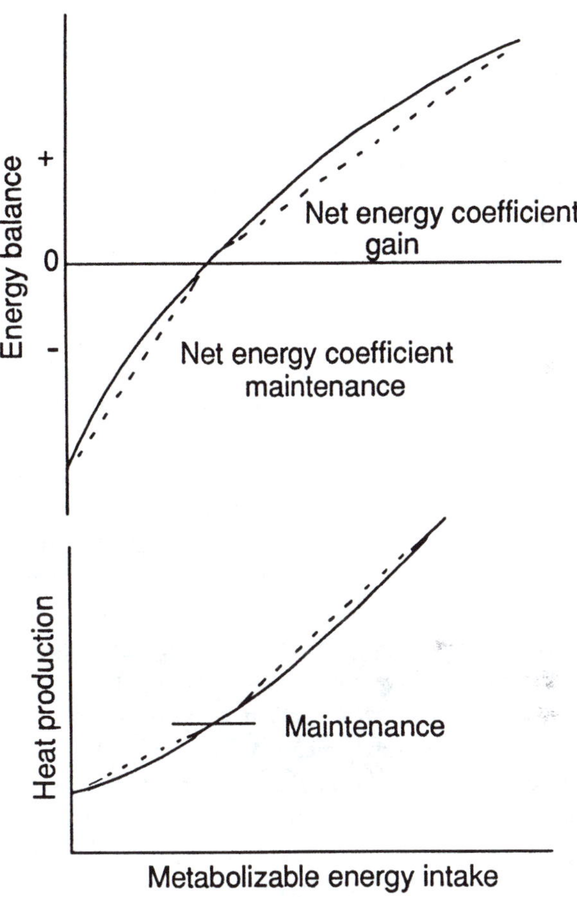

Figure 181. Experimental relationship between metabolizable energy intake, heat production and energy balance, and the linear approximations used to estimate net energy coefficients (from Robbins 1983). When energy intake equals zero (upper graph), the animal meets its maintenance energy requirements. As energy intake exceeds maintenance, any additional energy can be used for growth and reproduction (net energy coefficient of gain). Animals below maintenance must burn fat and muscle protein to make up their energy deficit. Two separate lines (linear approximations) are used to describe the curve above and below maintenance because efficiency of maintenance is greater than that for gain. The lower graph curves in the opposite direction because heat production represents a loss of energy.

growth are simply the efficiencies to which the metabolizable energy of feed and tissue can be converted into the gross energy of these tissues.

There are few net energy estimates for foods eaten by moose. Net energy coefficients expressed as percentages of metabolizable energy ranged from 76.6 to 86.3 percent for pelleted diets varying in protein content and was 69.3 percent for moose fed a mixed browse diet (Schwartz et al. 1987c). Net energy coefficients for wild ruminants tend to be higher than similar estimates for domestic ruminants (Robbins 1993).

Nutrient Quantity

The efficiency of forage utilization by moose is only part of the information needed to understand the species' nutritional relationship with its environment. Knowledge of the amounts of various nutrients required not only for basic body function, but also for growth and successful reproduction also is necessary.

Basal Metabolism

There is some confusion in the literature regarding the terminology used to describe metabolic rates of wild ruminants. This has led to misinterpretations about the methods employed to measure metabolic rate and valid comparisons of different data sets. To understand metabolic rates in moose, one first must understand how those rates are measured and what fraction of total metabolism each measurement represents.

Several terms are used to represent various metabolic measurements in wild ruminants. *Basal metabolic rate* (BMR) is the minimum energy cost of an animal that is lying at rest in a thermoneutral environment and in a postabsorptive state (Brody 1945). BMR theoretically translates to the minimal energy cost to maintain life. BMR represents a baseline expenditure of energy to which additional costs of energy expenditure are added.

BMR frequently is called standard fasting metabolism (SFM). There is a relationship between SFM and body weight that scales to the 0.75 power. For SFM in most mammals, the empirical measure of BMR is 70 kilocalories per kilogram $BW^{0.75}$ per day (Kleiber 1975). The so-called "mouse-to-elephant curve" represents a broad generalization for most species, but many mammals have significantly lower or higher metabolic rates than predicted by this equation (Robbins 1993).

Kleiber (1975) suggested that measurement of SFM should take place after a prolonged period of feeding at maintenance levels. He recommended this because metabolic rate is largely determined by the metabolically active tissues, including the liver and digestive organs (Arnold 1985). These tissues increase in size with increasing levels of food intake (Smith and Baldwin 1974, Koong et al. 1982). Hudson and Christopherson (1985) suggested that the term *standard metabolic rate* (SMR) be used for large herbivores conditioned at maintenance prior to fasting. However, wild ruminants are in a constant flux, gaining and loosing weight seasonally, and there are very few times during the year when they are at maintenance. Consequently, estimates of metabolic rate often are made with no maintenance adjustments, and are termed *fasting metabolic rate* (FMR) (Hudson and Christopherson 1985).

Measuring metabolism in wild ruminants is difficult because it often is impossible to meet the empirically defined conditions. The necessary time to fast a ruminant to obtain a postabsorptive state depends on the size of the animal, its level of forage intake, and passage rate of food (Marston 1948), which may exceed 48 hours (Marston 1948, Blaxter 1962, Wesley et al. 1973, Kleiber 1975). The only data available for moose suggest that a postabsorptive state is reached after about 2 days of fasting (Renecker and Hudson 1986b) (Figure 182), although this estimate is a function of diet quality, rumen fill and rate of passage.

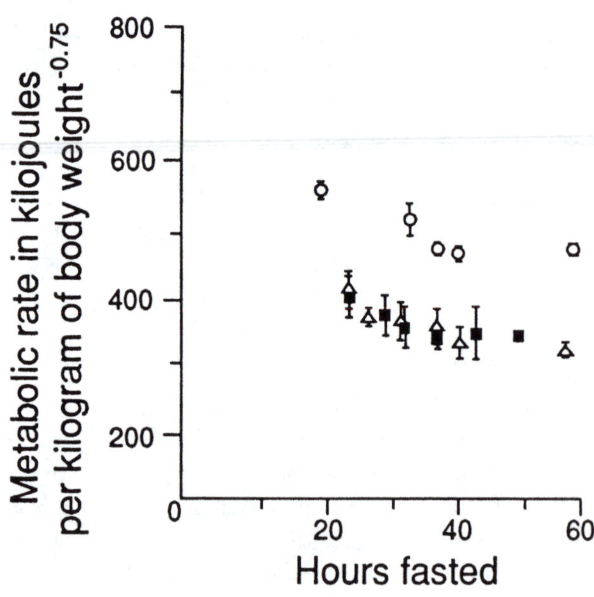

Figure 182. Effect of fasting on mean metabolic rates of two cow moose in March (solid squares), November (open triangles) and April (open circles) (data from Renecker and Hudson 1986b). Metabolic rate changes very little after approximately 48 hours of fasting because heat production due to rumen fermentation has declined.

Fasting young moose, particularly easily stressed neonates, makes realistic measurements impractical. Conditioning wild moose to accept the necessary confinement for measurements is difficult to accomplish and requires hand-raising animals and extensively training them (Regelin et al. 1981a, 1982). Even when accustomed to the measurement procedure, wild ruminants, by their very nature, seldom remain in a relaxed state for extended periods of time. Similarly maintaining moose at a maintenance level of food intake may not be possible because of their inherent seasonal rhythms of forage intake.

Because of difficulties measuring SMR and FMR accurately in wild ruminants, energy expenditure of recumbent, nonfasted animals in a defined thermal environment is commonly estimated (Silver et al. 1969, Pauls et al. 1981, Regelin et al. 1985, Schwartz et al. 1991). Such measurements of energy expenditure are referred to as *resting metabolic rate* (RMR) (Hudson and Christopherson 1985). RMR is greater than FMR because it includes heat of digestion. And RMR is greater than SMR because it is influenced by changes in BMR associated with changes in organ size.

There are no available estimates of BMR in moose where the experimental protocol meets the defined criteria. There are, however, estimates of theoretical basal metabolism that closely approximate BMR. Theoretical BMR is calculated by regressing metabolizable energy intake (X) against resting metabolism (Y). The point where metabolizable energy intake equals zero (intercept) represents theoretical BMR, because heat increment (heat of digestion) also equals zero (Reid and Robb 1971) at this point. Hubbert (1987) calculated theoretical BMR for a group of moose fed varying qualities of food during winter (Figure 183). His estimate of 68.8 kilocalories per kilogram $BW^{0.75}$ per day was almost identical to the interspecific mean of 70, discussed earlier (Kleiber 1975). In a similar study, Schwartz et al. (1988b) calculated theoretical BMR for a group of moose fed varying quantities of food during winter (Figure 184). Their estimate of 72.7 closely agreed with that presented by Hubbert (1987) and also was close to the interspecific mean of 70.

Regelin et al. (1985) estimated heat production for recumbent fasted moose in an open circuit respiration chamber with temperatures maintained between 32 and 39°F (0–4°C). These values should closely approximate FMR, although the chamber temperature was slightly above the thermoneutral zone of 19 to 23°F (–2 to –5°C) of moose in winter but within the thermoneutral zone for moose during summer (Renecker and Hudson 1985b). The lowest estimate of FMR, obtained in late March, was 85 kilocalories per kilogram $BW^{0.75}$ per day, whereas the highest value, obtained in July was 133.1 per day (Figure 185).

The cyclic nature of FMR measured by Regelin et al. (1985) clearly mimicked seasonal intake of dry matter (Figure 185). Differences between their FHP and RHP estimates averaged about 12 percent, varying from –5 to 23 percent. The magnitude of change in RMR for summer to winter far exceeded any contribution that heat increment could make to BMR, even with increasing levels of food intake during summer.

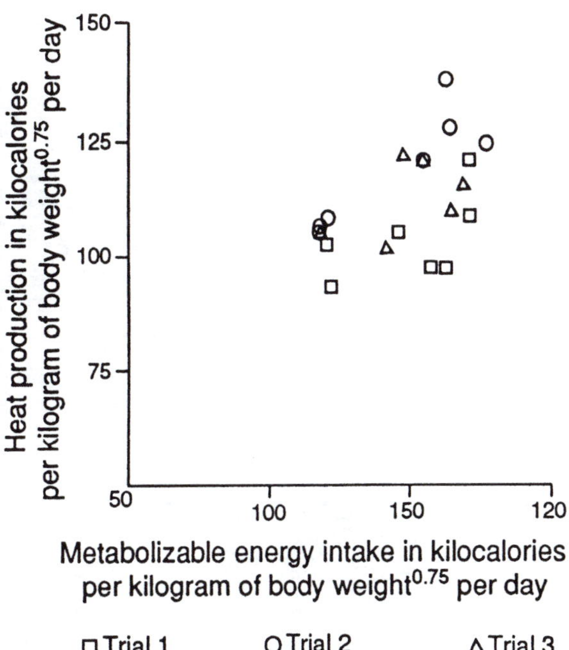

Figure 183. Metabolic rate is linked to size of organs (i.e., liver, digestive tissues) and the previous level of forage consumption. Consequently, there is a positive relationship between the previous 28 days' metabolizable energy intake and heat production in moose (data from Hubbert 1987).

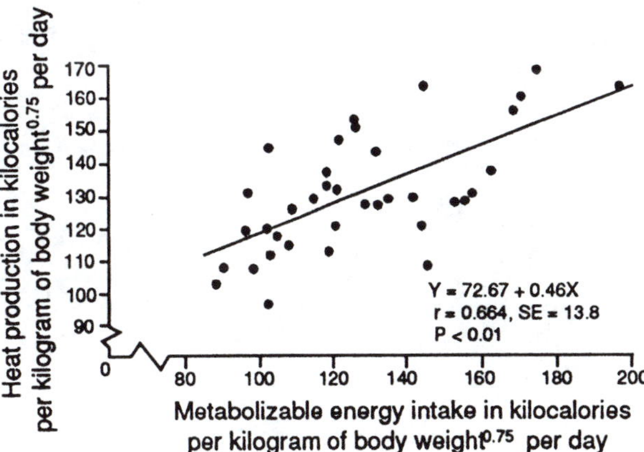

Figure 184. Relationship between metabolizable energy intake and heat production for moose (from Schwartz et al. 1988b). The intercept of 72.7 represents a close approximation of theoretical basal metabolic rate.

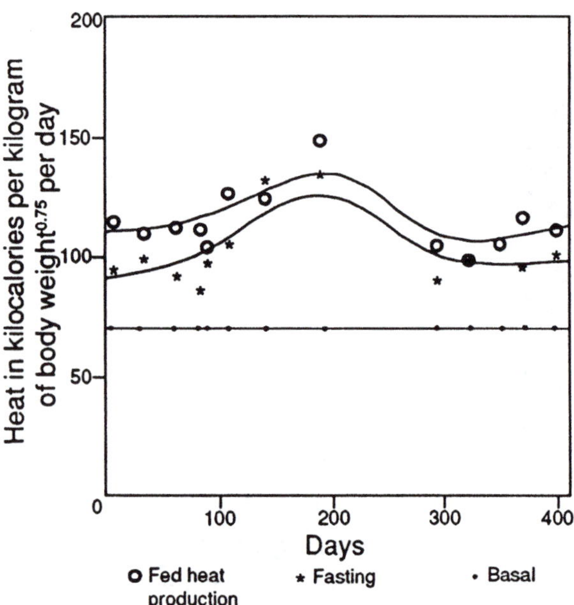

Figure 185. Change in metabolic rates in moose by season (data from Regelin et al. 1985). Basal metabolism probably remains constant at around 70 kilocalories per kilogram of body weight$^{0.75}$. Fasting and fed heat production estimates vary seasonally, with peak in midsummer and low in late winter.

Hubbert (1987) tested the relationship between previous level of metabolizable energy intake and its effect on resting metabolism. He found a significant relationship between resting metabolism and the intake of metabolizable energy in the previous 28 days. Part of this positive relationship undoubtedly was associated with an increased heat increment as level of food intake increased. However, based on the data presented by Regelin et al. (1985), the heat increment alone apparently was not responsible for the marked difference noted. Hubbert (1987) felt his results suggested that metabolic adjustments to the level of intake are a long-term process. His findings supported the hypothesis that cellular metabolism adapts to substrate supply slowly, and therefore, takes considerable time to return to basal or reference level of heat production. These data also supported the contention that the increase in FMR and RMR from winter represents an increase in metabolically active tissues associated with increased levels of forage intake. There appears to be about a 28-day delay between the metabolic response and food consumption rate. FMR and RMR cycle with forage intake, but with about a month delay.

Estimates of heat production range from 1.2 to 2.1 kilocalories per kilogram BW$^{0.75}$ per hour for adult moose and

To understand the energy metabolism of moose, biologists measure oxygen consumption and carbon dioxide production. This is accomplished by means of a mask system with one-way valves that permit inhaled air to enter and exhaled air to be forced through a tube in collection bags (at left above). Later, this "collected" air is pumped through an oxygen analyzer to measure the amount of oxygen consumed. *Photo by Chris Hanslik; courtesy of the Department of Animal Science, University of Alberta, Edmonton.*

1.9 per hour for calves fed a pelleted ration (Renecker et al. 1978, Renecker and Hudson 1986b). Regelin et al. (1985) detected a narrower difference of 0.95 to 1.19 kilocalories per kilogram $BW^{0.75}$ per hour between fed and fasted Alaskan moose on the same diet. Differences probably were related to feeding level.

Effects of Season

Seasonal or circannual cycles of energy metabolism are very noticeable in northern ruminants. Silver et al. (1969, 1971) measured heat production in white-tailed deer, which was 50 percent higher in summer than winter. Seasonal differences are somewhat less in caribou (McEwan and Whitehead 1970). In bighorn sheep, metabolic rates decline from autumn to midwinter, then rise more than 40 percent to a peak in early summer (Chappel and Hudson 1978). In adult moose, RMR increases about 118 percent between winter and summer (Renecker and Hudson 1986b). As mentioned elsewhere in this chapter and in Chapter 13, these rhythms seem linked to fluctuations in forage quality, growth rate, hormone activity and photoperiod. Beyond the question of what regulates what, is the one of adaptive importance. From the work of Renecker and Hudson (1986b), the winter RMR for moose is close to the interspecific mean and is not depressed. But the summer metabolic rate is high. This type of cycle relates to the need for the animal to grow rapidly and fatten during the brief period of summer, when food quality is high.

Energy Costs of Activity

To survive, moose must move about in their environment, seeking food, shelter and mates and avoiding predators. All of these activities demand energy and have specific costs associated with them. Understanding the relative costs of various activities is useful for estimating animal requirements and making relative determinations of nutritional carrying capacity (Moen 1973). Nutritional carrying capacity is the ratio of range nutrient supply divided by the nutrient demands of individual animals (Mautz 1978).

COST OF STANDING

As in many other wild ruminants, energy cost of standing in moose is high compared with the accepted value of 10 percent for domestic livestock (Blaxter 1962). Renecker and Hudson (1986b) suggested an increment of 25 percent for adult moose and 35 percent for calves (Renecker et al. 1978). From adult moose, Regelin et al. (1986) measured increments that ranged from 12 to 148 percent, depending on the level of alertness and movement. The estimates for calm animals averaged 22 percent above resting metabolism.

Most researchers use the convention developed for domestic livestock and express cost of standing as a percentage of RMR. Because RMR in moose varies seasonally, simple percentages may not be appropriate for this species (Regelin et al. 1986). Therefore, values for incremental costs of activities in wild ungulates are best expressed as rates. The cost of standing in adult moose averages 0.23 ± 0.01 kilocalories per kilogram BW per hour or 1.01 ± 0.07 kilocalories per kilogram $BW^{0.75}$ per hour. There is no apparent difference in these costs associated with season or sex of the animal (Regelin et al. 1986). Renecker and Hudson (1983, 1989a) estimated the cost of standing at 0.19 kilocalories per kilogram BW per hour using calibrated heart rate index, and 1.0 kilocalories per kilogram $BW^{0.75}$ per hour using indirect calorimetry (Renecker and Hudson 1986b).

A comparison of the cost of standing for different species reveals a wide range of values (Table 42). Domestic livestock have the lowest energetic costs associated with standing, perhaps attributable to minimal anxiety or alertness compared with wild ruminants. Large ruminants tend to have a lower standing cost than do smaller ones, but there are exceptions (e.g., bighorn sheep). Like metabolic rate measurements, there likely are differences in experimental protocol among studies. This likely accounts for the high values listed for white-tailed and mule deer. Although some differences may have an anatomical basis, the role of alertness must be considered. Measurements of wild ruminants almost always are made opportunistically, and experimental animals are more likely to lie down when they are relaxed. Hence, standing measurements often include effects of arousal as well as posture.

COSTS OF OTHER ACTIVITIES

Like other ruminants, moose spend most of the day searching for food, eating and ruminating (Risenhoover 1987, Renecker and Hudson 1989a, Van Ballenberghe and Miquelle 1990). These activities vary (Renecker and Hudson 1989a) according to seasonal forage abundance, quality, digestibility and passage rate. Feeding bouts are longer during summer than winter when forage availability and digestibility are high (Risenhoover 1987, Renecker and Hudson 1989a, Van Ballenberghe and Miquelle 1990). Moose in Alaska's Denali National Park are active 12.8 hours per day in summer but for only 5.8 hours per day in winter. This represents a 2.2-fold increase in the daily activity period from winter to summer.

Renecker and Hudson (1989b) provided the only estimates for various energy expenditures in moose (Table 43). Above that for standing, the energy costs associated with foraging are 0.02 to 0.56 kilocalories per kilogram BW per

Table 42. Energetic cost of standing in several ruminant species (data from Regelin et al. 1986)

Species	Percentage	Kilocalories per hour per kilogram	Kilocalories per hour per kilogram$^{0.75}$	Source
Roe deer	22	0.52[a]	1.1	Weiner (1977)
Pronghorn	21	0.80	1.9	Wesley et al. (1973)
Bighorn sheep	18	0.18	0.5	Chappel and Hudson (1979)
White-tailed deer[b]	63	0.92	2.4	Mautz and Fair (1980)
Mule deer	72			Kautz et al. (1982)
Elk	30	0.45[c]	1.5	Gates and Hudson (1979)
Elk calves	25	0.44[d]	1.2	Parker et al. (1984)
Moose	22	0.23	1.0	Regelin et al. (1987b)
Moose	29	0.19[e]	0.8	Renecker and Hudson (1983)
Moose calves	35	0.65	1.9[f]	Renecker et al. (1978)
Reindeer	10			White and Yousef (1978)
Domestic cattle	19	0.14	0.57[g]	Vercoe (1973)
Domestic sheep	13	0.12		Webster and Valks (1966)

[a] Estimated weight of 44 pounds (20 kg).

[b] Standing included movement up to 50 percent of time.

[c] Estimated weight of 276 pounds (125 kg).

[d] Estimated weight of 110 pounds (50 kg).

[e] Estimated weight of 716 pounds (325 kg).

[f] Estimated weight of 187 pounds (85 kg).

[g] Estimated weight of 602 pounds (273 kg).

hour in adult moose and 0.52 per hour in calves (Renecker et al. 1978, Renecker and Hudson 1989b). The values are comparable for those of other ungulates, such as elk (0.27 to 0.88 kilocalories per kilogram BW$^{0.75}$ per hour) (Wickstrom et al. 1984), bighorn sheep (0.43–0.46) (Chappel and Hudson 1978) and domestic sheep (0.45–0.54) (Graham et al. 1974, Osuji 1974). The energy increment for foraging in free-ranging moose cows ranges from 5.5 to 31 percent. Incremental expenditure of browsing is highly correlated with resting metabolism and remains at 47 to 55 percent—a relatively constant proportion above resting (Renecker and Hudson 1989b).

The cost of lying/ruminating over resting is 14 ± 1 percent or 0.5 ± 0.1 kilocalories per kilogram BW$^{0.75}$ per hour, varying seasonally (Renecker and Hudson 1989a, 1989b). In winter, when chewing rates are low (72 ± 4 chews per minute), increments are 13 percent or 0.36 kilocalories per kilogram BW$^{0.75}$ per hour. In summer, costs increase with rising ambient temperature and chewing rates (81 ± 4 chews per minute) to 26 percent or 1.96 kilocalories per kilogram BW$^{0.75}$ per hour. Incremental costs of rumination are correlated highly with resting metabolism. Energy expenditures of resting/ruminating remain about 13 to 16 percent above resting metabolism, and are remarkably constant. Expenditures while lying and ruminating correspond to those determined for elk (Pauls et al. 1981). If rumination is considered to be an expenditure above lying with head up, then a 7 ± 1 percent increase for cow moose is consistent with reports for domestic sheep (Graham et al. 1974).

Thermoregulation

Homoiothermic (warm-blooded) animals must maintain body temperature within fairly narrow limits despite changes in environmental temperatures (Parker et al. 1984). For moose, normal body temperature is between 101 and 102°F (38.4–38.9°C) (Franzmann et al. 1984a). Over a wide range of thermal conditions, animals can control heat exchange within the environment largely by physiological and behavioral adjustments (Figure 186). Near the upper critical temperature, additional heat is released through such mechanisms as panting. Below the lower critical temperature, metabolic rate increases to produce heat to offset that lost to the environment.

Maintenance of a fairly constant body temperature results from an equilibrium between the rate at which heat is exchanged with the external environment. This exchange between the animal and the environment (Figure 187) must eventually sum to zero, and can be represented as: Heat loss – Heat gain = 0 or QR + QHC (Ta – Ts) + QEX + QESW

Table 43. Energy expenditures for various activities of two nonpregnant/lactating, free-ranging moose cows, from December 1982 to January 1984 at the Ministik Wildlife Research Station, Alberta, Canada (data from Renecker and Hudson 1989b)

	Metabolic rate (kcal/kg body weight$^{0.75}$)													
Activity	December		February		April		May		July		October		January	
Bedded														
Dozing	47.6	56.5	46.0	49.0	46.0	43.1	84.9	78.7	154.4	131.0	61.9	69.0	49.4	45.6
Ruminating	51.9	60.7	50.6	51.5	52.3	43.9	97.5	85.8	123.2	145.6	62.8	71.5	51.0	49.8
Alert	53.6	60.2	50.6	48.5	46.4	43.9	92.5	79.9	159.4	146.4	66.1	72.4	52.3	51.9
Grooming								80.8			73.6	77.4	54.8	
Other		59.8	49.8		49.8	45.2	140.6	74.1	166.5	163.6	70.3	76.3	54.0	57.3
Feeding														
Bark					63.2	51.5				185.4				74.9
Stripping														
Cratering	61.5	74.1	62.8	60.3	58.6	61.1					89.5	136.4	62.8	65.3
Grazing								155.6	166.1	203.8		93.7		
Browsing														
Low	62.3	73.6	60.3	63.2	60.3	47.7	118.4	129.7	223.4	199.6	90.8	100.8	69.9	72.0
Middle	64.0	74.5	59.8	62.3	62.8	57.7	109.6	126.4	219.2	209.2	74.9	148.5	67.8	72.5
High	64.0	78.7	56.9	62.8	62.8	50.2	111.7	145.6	205.0	204.2	79.4	141.4	76.6	85.8
Miscellaneous														
Standing														
Alert	64.0	69.9	69.4	59.8	63.2	51.9	145.2	168.2	243.5	302.1	88.3	130.1	78.6	76.1
Ruminating			59.0	64.0			141.4		172.0	172.0			63.6	69.9
Grooming							297.0		183.3		137.2		78.2	
Drinking/eating snow					66.5	61.9	163.2	154.0	249.8	249.8	87.4	190.8	79.4	90.4
Walking	79.1	86.6	80.8	69.0	76.1	77.0	168.2	185.8	355.6	480.7	139.8	307.5	171.5	215.9
Running							380.7	363.6	549.3	1,468.6			195.0	271.5
Other													68.6	70.7

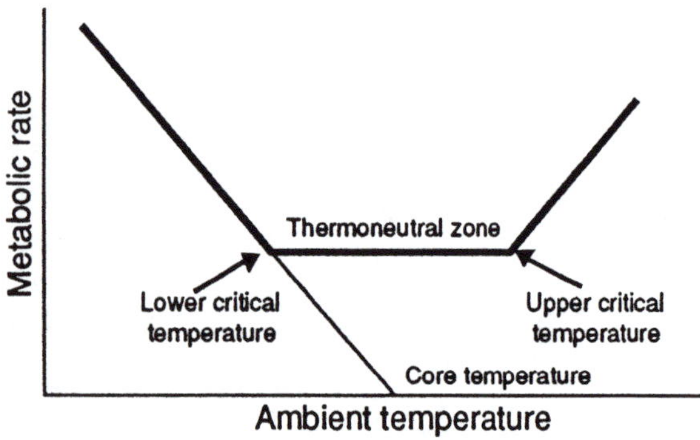

Figure 186. Energetic response of moose to a changing thermal environment (from Renecker and Hudson 1992b). Below the lower critical temperature, moose must increase metabolic rate to maintain body temperature. Above the upper critical temperature, moose increase panting to dissipate excess heat. There are no additional metabolic requirements to maintain body temperature within the thermoneutral zone.

± QC = M ± SK, where QR = radiation; QHC (Ta – Ts) = convection; QEX = evaporation through respiration; QESW = evaporation through sweating; QC = conduction; M = energy produced in animal metabolism; and SK = heat storage.

Heat flow from the animal usually is represented by radiation, convection, evaporation and conduction, whereas heat input is from metabolism and radiation.

Heat exchange usually is measured in units of energy per unit of area. Thus, it is important to know with a fair degree of accuracy the surface area of animals. But measurement of the surface area of an irregularly shaped animate object such as a moose is difficult. This measurement can be obtained by dividing the animal into geometric shapes, such as cones, spheres and cylinders, or by photometric techniques.

Simple calculation of total surface area may not be appropriate for understanding thermal exchange and animal thermoregulation. At any one time, only a portion of an animal is exposed to direct sunlight or other types of radiation. As a result, the calculations of these partial body values are difficult.

Radiation can represent either a net gain or net loss to or from the animal, respectively. Heat is lost in the absence of radiation, such as at night, or when environmental circumstances (e.g., cold temperatures in winter) counteract radiation. Differences also can result from upward and downward radiation exchanges in different forest cover types (i.e., an aspen forest in autumn after leaf fall versus a coniferous stand of spruce). During the day, solar radiation provides a net gain of heat from the environment, which can help offset the effects of cold in smaller ruminants or be a heat load that must be dissipated. Direct solar radiation and radiation scattered by dust particles in the atmosphere or on the ground comprise most shortwave radiation. Longwave radiation comes from gases in the atmosphere or the ground, which absorb shortwave radiation and then reflect it.

The heat load or stress imposed on the animal depends mainly on the surface area exposed, size of the animal and

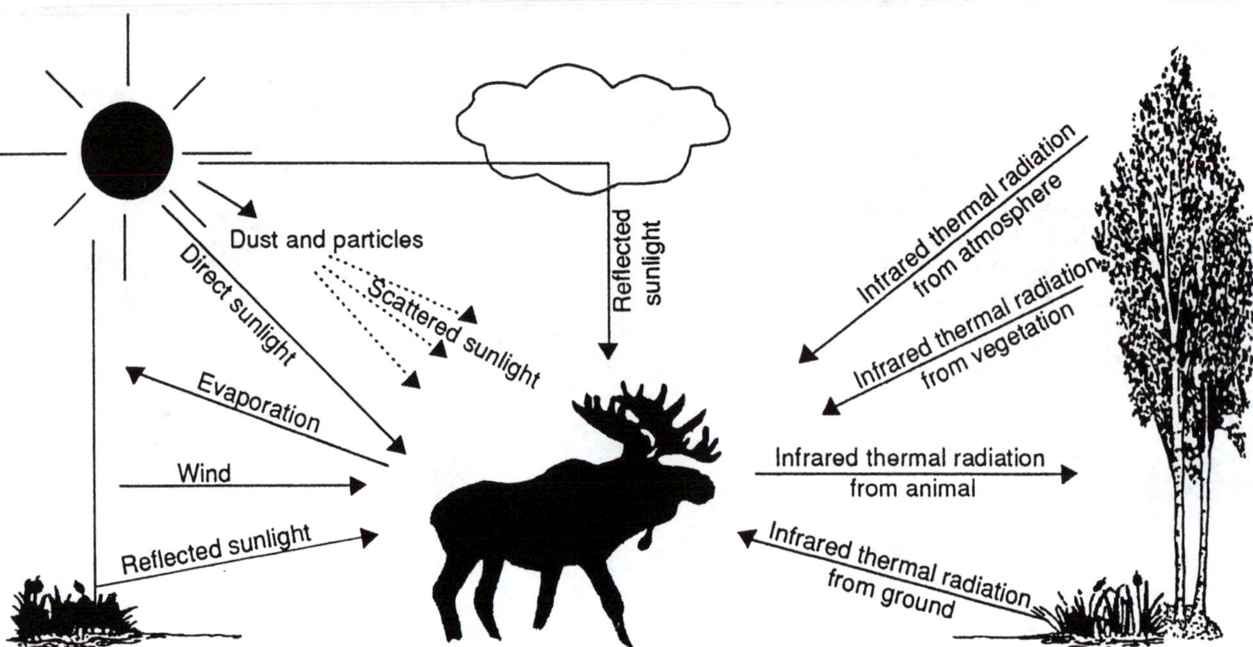

Figure 187. Streams of energy between a moose and its environment.

Moose are homeothermic animals that must maintain a constant body temperature (about 101°F [38.3°C]). Heat exchange occurs by radiation, convection, conduction and evaporation. Heat inputs are metabolism and solar radiation. To monitor the effects of environmental changes on moose metabolism, scientists equip subject animals with heat-rate transmitters that record the number of heart beats per minute. Increases in heart rate are related to increases in metabolism and energy consumption by the animal. *Photo by Lyle A. Renecker; courtesy of the University of Alaska, Fairbanks.*

insulative characteristics of the fur. How much shortwave radiation (measured with a solar meter and animal's profile) is absorbed depends on coat color. Longwave radiation (determined from air temperature) is almost totally absorbed. Renecker and Hudson (1990b) showed that radiant heat load is correlated positively with bedded heart rate of free-ranging moose cows (Figure 188).

Heat exchange by convection occurs via still air and moving or turbulent air. Convection is difficult to measure because it is a function of air velocity and direction and the shape of the animal.

Evaporative loss occurs by sweating or panting. The heat or energy loss can be calculated from estimates of water loss and heat of vaporization of water. Respiratory water loss in the reindeer (Hammel et al. 1962) is about one-third that of domestic cattle (Webster 1971). However, reindeer increase energy loss through evaporation by 10 times during exercise as metabolic rate increases (Hammel et al. 1962).

Conduction of heat is important in several ways. First, heat can be transferred through the hair coat. Then, heat can also be transferred from the animal to some surface, such as the ground (e.g., a moose lying on a cool, wet surface would cause more heat transfer than in a moose lying on a warm, dry surface). A foremost purpose of hair is to maintain a blanket of air within the pelage. Air is a poor conductor and therefore, a good insulator. However, as wind speed increases the hair fibers are moved and their density changes. As the hair is flattened by wind, the insulating properties decrease and thermal conduction increases (cooling the animal). At colder temperatures, the insulation of hair can be increased through piloerection, which is when the hairs stand straight up to increase overall depth of the coat.

Insulation by the animal's tissues is determined by factors such as body weight, conformation, skin thickness and fat deposition. Also, such physiological factors as rate of blood flow to or away from the skin, extremities and tissues near the skin surface are important in the animal's insulative ability. At subfreezing temperatures, blood flow to the skin and extremities increases to prevent tissues from freez-

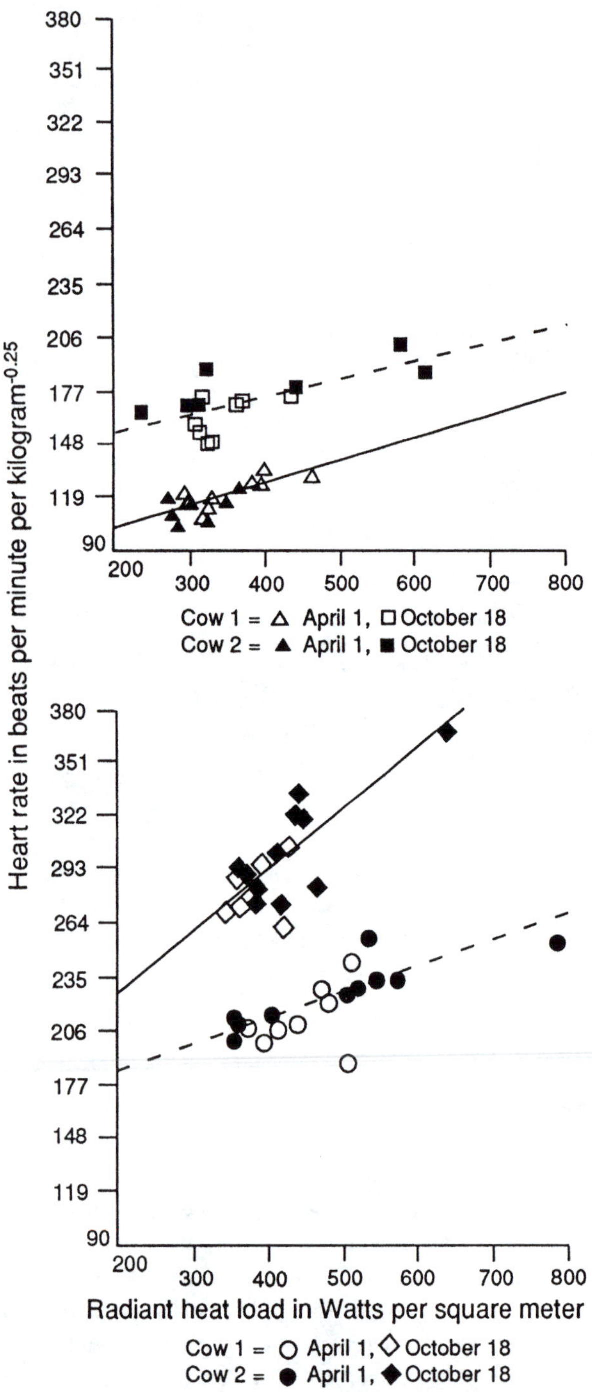

Figure 188. Relationship between heart rate and radiant heat load in two cow moose (data from Renecker and Hudson 1990b). The upper graph represents data from April (triangles) and October (squares); the lower graph represents data from May (circles) and July (diamonds). As the heat load increases, heart rate goes up as the animal dissipates the excess heat through increased respiration.

ing. Therefore, the properties of external insulation are important to keep skin temperature above freezing and prevent tissue damage and possible frostbite.

Animals also may store heat to adjust to rapid changes in ambient temperature. This is especially true for the moose. The variables that define the heat storage capability of an animal are the extent of the temperature change, body size and specific heat of tissues (i.e., mean body temperature).

Weather is the result of the atmospheric interaction of temperature, wind, moisture and radiation. Adaptations to winter weather usually are evaluated by determining the lower critical temperature below which an animal must increase metabolic rate to maintain its normal body temperature.

Northern wild ungulates, such as moose, are quite resistant to cold, perhaps because of selective pressures and adaptations during the colder Pleistocene periods. Lower critical temperatures approximate –4°F (–20°C) for larger species (Table 44). For some of the largest northern ungulates, such as moose and bison, lower critical temperatures and insulation have not been precisely determined, because energy expenditures continue to drop with temperature as low as is normally observed. This is especially true for moose, which appear comfortable in winter weather. However, smaller members of the deer family (and, to a certain extent, moose calves) tend to remain bedded in extreme cold, to reduce surface heat loss. Moose calves assumed a bedded posture as temperature approached their lower critical temperature of about –22°F (–30°C) (Renecker et al. 1978).

Perhaps because humans are especially sensitive to cold, more attention has been placed on understanding cold stress than heat stress in wild ungulates. For large ruminants such as the moose, however, there are more days of exposure to heat stress great enough to interfere with feeding and affect weight gains than of cold stress days of simi-

Table 44. Critical temperatures of wild ruminants during winter

Species	Critical temperature (°C)		Source
	Lower	Upper	
White-tailed deer	5	25	Parker and Robbins (1985)
Mule deer	–20	5	
Caribou	<–40	20	
Elk	–20	15	
Pronghorn	–16		
Bighorn sheep	–10		
Moose	<–30	–5	Renecker and Hudson (1986b)
Mountain goat	<–30		Krog and Monson (1954)
Bison	<–30		Christopherson et al. (1978)

Maintaining a constant body temperature is critical to the survival of moose in their cold, northern environment. Low temperature and wind can make it necessary for a moose to produce extra heat to maintain core temperature and stay warm. Because of their lower metabolism and greater capacity for fat storage, most adult moose can cope with extreme cold better and longer than calves can. Moose calves must produce additional heat when temperatures drop below about –22°F (–30°C), the known, low critical value for moose calves, measured under controlled conditions using a tame moose calf in a wind tunnel at Ministik Wildlife Research Station, Alberta. *Photo by Lyle A. Renecker; courtesy of the University of Alaska, Fairbanks.*

lar effects. Moose have a great difficulty dissipating surplus heat during warm temperatures. During summer, thermal stress for moose begins at about 57°F (14°C) and open mouth panting at 68°F (20°C) (Renecker and Hudson 1986b, 1990b) (Figure 189). A particularly difficult time for

With large body size, long legs, pelage of long and hollow guard hairs, and dense shorter hairs that help them conserve heat, adult moose are well adapted to cold winter weather. Snow often accumulates on moose and does not melt because of the good insulative qualities of the pelage. Adults are known to withstand temperatures of –40°F (–40°C), but scientists have been unable to measure the low critical ambient temperature for moose during winter. *Photo by Lyle A. Renecker; courtesy of the University of Alaska, Fairbanks.*

heat stress is spring when moose still are in winter coats. In winter, moose can experience heat stress when temperatures rise above 23°F (–5°C).

Energy costs decline as animals get larger. The large moose minimizes heat loss by a low surface area-to-volume ratio. In contrast, smaller ungulates, such as deer, have a greater propensity to lose heat, and must develop other energy-conserving mechanisms (Eisenberg 1983) to offset the high energy requirements. Effects of heat loading are great for the moose, as would be expected because of their large size. However, the moose has an advantage that is provided by their long legs, which increase the surface area for thermoregulation in warm temperatures. As shown in Chapter 13, moose throughout their boreal range utilize aquatic sites, the emersion in which helps them keep cool (Renecker and Hudson 1989b, 1990b). Also, whereas body temperature relaxes the constraints of cold temperatures, the stress imposed by heat actually limits the southern distribution of moose (Kelsall and Telfer 1974).

Maintenance Energy Requirements

Maintenance energy requirements are defined as the total calories required by a moose neither to gain nor lose body weight. For captive animals, where food ingestion rates can easily be measured, weight balance can be used to estimate maintenance requirements. When a range of intake (X) is plotted against weight change (Y), the X intercept at zero gain or loss represents the maintenance requirement (Figure 190).

Intake of digestible energy to meet maintenance requirements is about 149 kilocalories per kilogram $BW^{0.75}$ per day for captive moose (Schwartz et al. 1988b) and 179 per day for free-ranging moose (Renecker and Hudson 1985a). Intake of metabolizable energy to meet maintenance require-

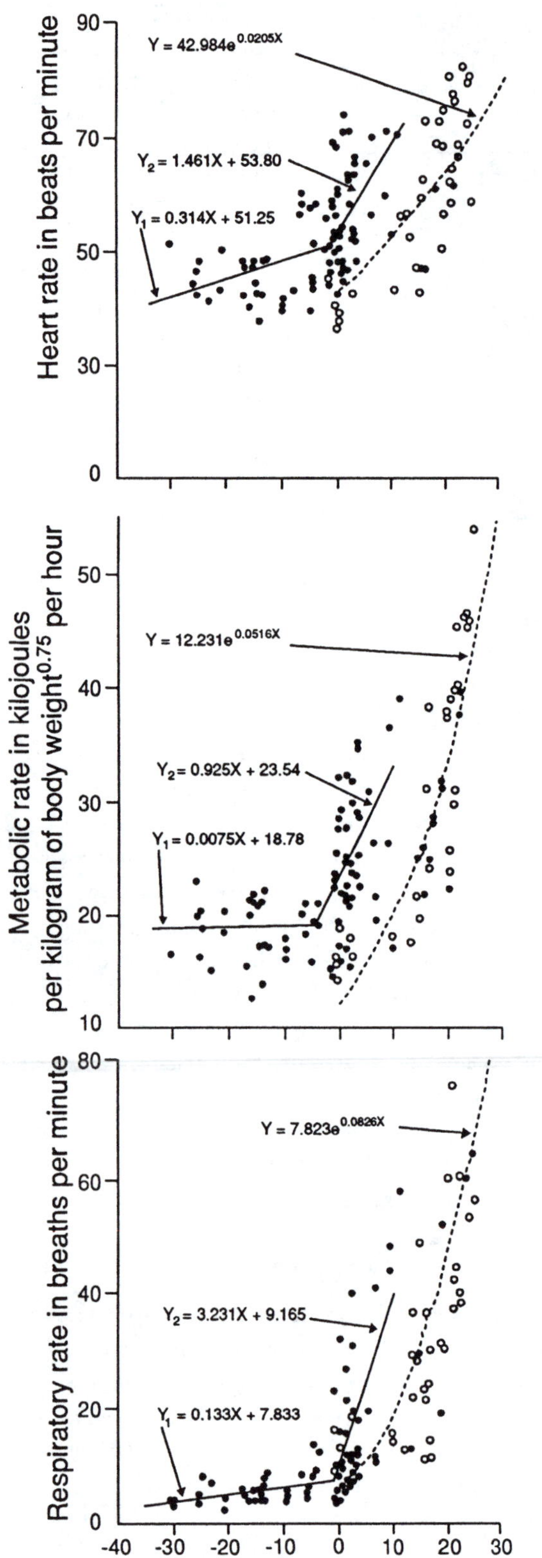

Figure 189. Effect of ambient temperature on respiratory rate, energy expenditure and heart rate of two moose cows during winter (solid circles) and summer (open circles) (from Renecker and Hudson 1986b). As temperatures exceed the thermoneutral zone, energy expenditure increases as the animal attempts to dissipate excess heat.

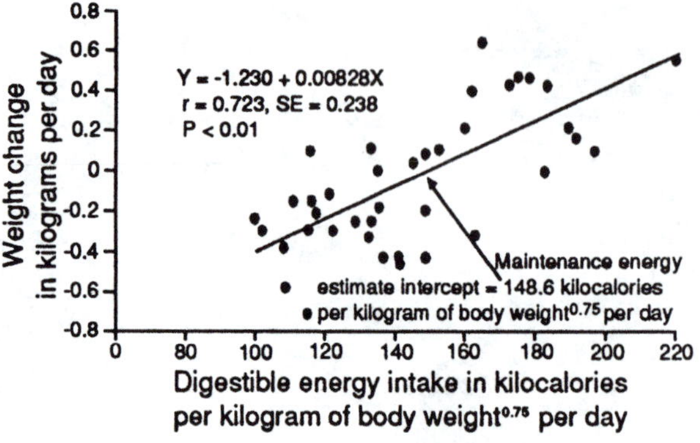

Figure 190. Relationship between daily digestible energy intake and weight change for moose (data from Schwartz et al. 1988b). The intercept of 148.6 represents the amount of digestible energy necessary for maintenance.

ments is about 135 kilocalories per kilogram $BW^{0.75}$ per day for captive moose (Schwartz et al. 1988b) and 140 per day for free-ranging moose (Renecker and Hudson 1985a). The higher estimates for free-ranging moose represent the additional energy costs for foraging.

Maintenance energy requirements for moose are similar to those of white-tailed deer (158–160 kilocalories per kilogram $BW^{0.75}$ per day of digestible energy and 131 per day of metabolizable energy) (Ullrey et al. 1969, 1970). These requirements can be used as a reasonable baseline when evaluating diets and intake rates. For example, assume poor quality winter browse contains about 4.5 kilocalories per gram of gross energy and metabolizable energy is 28 percent of gross energy, or 1.26 kilocalories per gram (4.5 X 0.28). A 772-pound (350-kg) cow moose ($BW^{0.75}$ = 81) with a metabolizable energy requirement of 131 kilocalories per kilogram $BW^{0.75}$ per day would require 10,611 kilocalories per day (131 X 81 = 10,611) to be at maintenance. Therefore, she would need to eat 17.8 pounds (8.5 kg) of browse dry matter to meet her daily energy requirements (10,611 / 1.26 = 8,421 g, or about 8.5 kg). In winter, moose consume about 1 percent of their body weight per day in dry matter, so this cow would be expected to eat about 7 to 8 pounds (3.2–3.6 kg) of browse containing 4,375 kilocalories (3,500 X 1.25) of metabolizable energy. She would be in negative energy balance and would have to make up this energy deficit by burning fat. On good quality winter range, browse may contain 2.2 kilocalories per gram of metabolizable energy. The same cow eating 7 to 8 pounds (3.2–3.6 kg) of browse would get 7,700 kilocalories of energy from the food she eats. She would also be in negative energy balance, but her fat supplies would last almost 43 percent longer than that of the cow eating the poor diet. Accordingly, maintenance energy estimates can be used to calculate energy deficit and evaluate forage quality.

While making use of the sun's radiant energy during late winter and early spring, moose conserve energy by bedding in snow in the lee of coniferous windbreaks. A biologist, at right, measures the depth of this moose bed in the Thunder Bay District of Ontario. *Left photo by Joe Van Wormer; courtesy of the Denver Museum of Natural History. Right photo by John McNicol; courtesy of the Ontario Ministry of Natural Resources.*

Protein Requirements

Ingested sugars, fats and complex carbohydrates (fiber) are metabolized in the body to release energy. Metabolizable calories are merely the gross calories corrected for incomplete digestion. The body has no mechanism for altering the digestibility of forage or its own basal metabolic rate. Metabolizable energy represents the body's fuel supply. With protein, a different situation exists. Ingested proteins can serve as a source of energy, or they can be broken into individual amino acids and used as building blocks for tissues and chemicals.

Protein requirements for moose are influenced by age, sex, reproductive status, time of year, energy intake and other factors. Because of their rapid growth rate, young animals require more protein than adults do. There are no estimates of protein requirements for calf moose, but the requirements probably are similar to the 13 to 20 percent value estimated for white-tailed deer fawns (French et al. 1956, Ullrey et al. 1967). The minimum daily protein requirement for winter maintenance in adult moose is about 5 to 7 percent. The exact amount is dependent on the quantity of forage consumed. Other protein requirements, such as for hair or antler growth, gestation and lactation, can approach twice maintenance. As with energy, not all protein eaten by a foraging moose is used. Also, rumen microbes are capable of using nonprotein nitrogen to manufacture amino acids. As a consequence, consideration of nitrogen requirements of moose are more important than identification of proteins. Furthermore, the Kjeldahl procedure (Association of Official Agricultural Chemists 1965) used to estimate protein content of forage actually measures the amount of nitrogen.

Apparent Nitrogen Digestion

Forage nitrogen is lost from the body through feces and urine. Like energy partitioning, apparent digestible nitro-

Because of their rapid growth rates young animals require more protein in their diet than do adults. Calf moose require between 13 and 20 percent crude protein in their diet, which they obtain from milk and forage. *Photo by Charles C. Schwartz; courtesy of the Alaska Department of Fish and Game.*

gen represents the difference between ingested nitrogen and that lost in the feces. Fecal nitrogen originates from both undigested forage nitrogen plus nitrogen from other sources. The nitrogen from these other sources is termed *metabolic fecal nitrogen* (MFN). MFN represents nonabsorbed digestive enzymes, intestinal cellular debris, undigested rumen microbes associated with fermentation, mucus and any other nitrogenous product of immediate animal origin egested in the feces (Robbins 1993).

Apparent digestion of crude protein increases as more protein is eaten. Orskov (1982) reported that, over a range of diets, nondietary fecal nitrogen varied from 72 to 92 percent of the total nitrogen excreted in the feces. Consequently, apparent digestible protein is a poor way to evaluate the dietary contribution of protein to moose. MFN must be separated from total fecal nitrogen to provide a reliable estimate of true protein digestion. MFN can be calculated by regressing digestible nitrogen (Y) against dietary crude protein (X) (Robbins 1993). For every 100 g of dry matter consumed by moose, 0.458 g of nitrogen is lost as MFN (Schwartz et al. 1987b). With this correction factor, diets with apparent protein digestion ranging from 50 to 68 percent had a true protein digestion of 85 to 89 percent. Estimates of MFN from the literature for domestic and wild ruminants range from 0.35 to 0.78 g nitrogen per 100 g of dry matter consumed (Robbins 1993).

Maintenance Protein Requirements

The minimum intake required to meet the winter maintenance estimates for nitrogen balance in moose is a function of diet quality and quantity. As previously noted, for every 100 g of food consumed, 0.458 g of nitrogen will be lost as MFN regardless of the nitrogen content of the food. Also, true protein digestion is about 87 percent. Consequently, the minimum protein requirement necessary just to offset MFN losses can be calculated as $(0.458 \times 6.25)/0.87 = 3.29$ percent. Therefore, a moose consuming a diet containing 3.29 percent crude protein just meets the MFN losses. Hence, for any food, regardless of its protein content, 3.29 g of 100 g of dry matter merely offsets MFN losses. Any increase in dietary crude protein above 3.29 percent can go toward maintenance.

Schwartz et al. (1987b) estimated the winter maintenance requirement for adult moose of apparent digestible nitrogen at 0.254 g per kilogram of $BW^{0.75}$. This converts to an apparent digestible crude protein content of 1.59 g (0.254×6.25). For example, a diet containing 5 percent crude protein contains 1.5 g of apparent digestible crude protein (5 − 3.29 = 1.71; 1.71 × 0.87 = 1.49) that can be used for maintenance. A moose consuming winter browse with 5 percent crude protein meets its maintenance by consuming dry matter at a rate of 107 g of food for every kilogram $BW^{0.75}$ (1.59/0.01486). Similarly, dry matter intake can be calculated for maintenance for other dietary crude protein contents. The curve shown in Figure 191 demonstrates that, as protein content of forage decreases toward the 3.29 percent, the amount of dry matter required to maintain nitrogen balance increases at an accelerating rate. This rapid increase of intake required as crude protein declines in the diet poses a problem for moose during winter, because intake of poor quality forage is limited by digestibility and rate of passage. Diets containing less than 3.29 percent crude protein cannot meet maintenance requirements, because no matter how much food is eaten, the level of protein intake is less than losses associated with MFN, and no nitrogen is left over to meet losses in urine.

Protein in browse consumed by moose on winter range usually is low. On Alaska's Kenai Peninsula, for example, diets contained between 5.4 to 7.1 percent crude protein (Regelin et al. 1986). Estimates of dry matter digestion for these diets range from 23 to 38 percent. Estimates of dry matter intake for moose on natural forage (Renecker and Hudson 1985a) in winter range from 38 to 60 g/kg $BW^{0.75}$. Therefore, a diet with 5.1 percent crude protein would require a dry matter intake rate of 101 g/kg $BW^{0.75}$. That exceeds what would be predicted for normal winter intakes of dry matter. These data suggest that, in winter, adult moose probably eat forage that contains levels of crude protein near the critical value shown in Figure 191, where dry matter intake necessary to meet maintenance nitrogen requirements is impossible to attain.

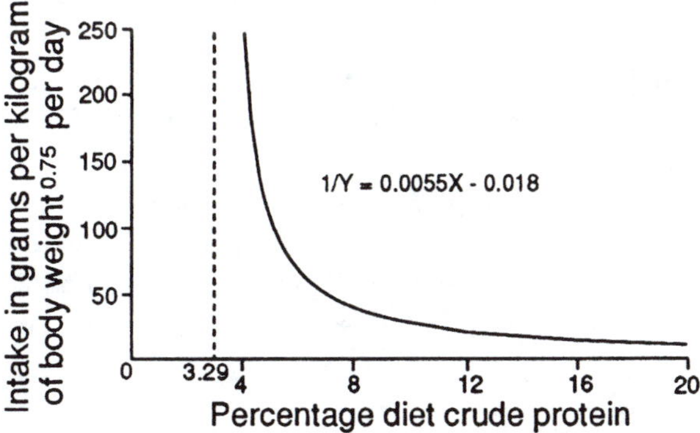

Figure 191. Relationship between crude protein in the diet and intake of dry matter required to meet maintenance protein requirements for moose (from Schwartz 1987b). Crude protein lost as metabolic fecal nitrogen represents the first 3.29 percent of consumed protein.

No estimates are available for maximum intakes of forage by moose in winter. Intake rates of tame moose fed a pelleted diet were near 70 g/kg $BW^{0.75}$ (Schwartz et al. 1987b). If intake rates for moose on natural forage can be assumed not to exceed 70 g, then forage must contain a crude protein content of 5.9 percent to meet their minimum requirements.

Energy Costs of Reproduction and Lactation

Production of offspring is energetically expensive. Growing a fetus and suckling neonates exert substantial energy demands on the mother (Oftedahl 1985). Studies have been conducted to estimate the energy requirements of gestation and lactation in domestic animals, but no data are available for moose.

GESTATION

Energy requirements for maintenance in domestic ruminants range from 100 to 130 kilocalories per kilogram $BW^{0.75}$ per day at maintenance to 140 to 170 kilocalories per kilogram $BW^{0.75}$ per day in late pregnancy, with a further rise to 239 to 270 kilocalories per kilogram $BW^{0.75}$ per day in early lactation in nondairy animals (Oftedal 1985). Using data presented by Reid (1968) for domestic livestock and growth rate information for fetal moose presented by Rausch (1959), Gasaway and Coady (1974) estimated that metabolizable energy costs of gestation for Alaskan moose increased from 875 kilocalories per day in March to 5,250 kilocalories per day at term in June. Expressed per unit of metabolic body weight, the metabolizable energy requirement was 10.6 to 63.5 kilocalories per kilogram $BW^{0.75}$ per day. Gasaway and Coady's (1974) estimates are nearly identical to those of Oftedahl (1985), if the incremental cost of maintenance is added (131 kilocalories per kilogram $BW^{0.75}$ per day).

LACTATION

Parturition ends the energetic demands of pregnancy, but initiates an even more energy-demanding phase—lactation—which is two to three times more costly than gestation (Robbins 1993). In well-nourished ruminants, milk production rises to a peak and then declines over a long period as suckling young are gradually weaned to solid foods. Calculation of the energy requirements of lactation requires data on the energy content of milk and milk yield at various stages (Oftedal 1985). Unfortunately, for moose, few data are available on milk composition, and the only data on milk yield are from Russia, where domestic moose cows have been milked.

Moose milk contains about 1.5 kilocalories per gram

(Franzmann et al. 1975a, Oftedahl 1981, Renecker 1987b)—a value that is lower than those for black-tailed deer (1.8) and reindeer (1.7), but higher than red deer (1.4) and elk (1.1) (see Oftedal 1985). Milk composition changes quite markedly during the lactation cycle, and serial sampling is necessary to track changes in quality and quantity produced.

Wild moose cows produce an estimated 220 to 330 pounds (100–150 kg) of milk per calf per lactation, whereas domesticated moose can produce up to 948 pounds (430 kg) (Knorre 1959, 1961, Yazan and Knorre 1964). The yield for wild moose is close to that of red deer (231–359 pounds [05–163 kg]) and reindeer (260 pounds [118 kg]), but well below that reported for elk (904 pounds [410 kg]) (see Oftedal 1985).

Maximum daily milk consumption of 3.3 to 4.4 pounds (1.5–2.0 kg) occurs in June and decreases through August to 1.1 pounds (0.5 kg) per day (Knorre 1959, 1961, Yazan and Knorre 1964). Weaning time is highly variable (Dodds 1955, Altmann 1958, Stringham 1974), but milk output drops substantially by August.

If moose milk contains an average energy content of 1.5 kilocalories per gram, at peak lactation (0.53 gallon = 2.0 L = 2,000 g) the cow is losing 3,000 kilocalories per day. Milk production and secretion also have energetic costs. Consequently, the efficiency of milk secretion often is expressed as a proportion of metabolizable energy intake above maintenance. Efficiency is related to dietary quality, but in general, ranges from 60 to 70 percent (Oftedal 1985). Using the midrange (0.65) of this efficiency estimate, a cow moose producing 4.4 pounds (2 kg) of milk per day must consume about 4,615 kilocalories (3,000/0.65) of metabolizable energy to meet her lactation requirements. Of course, these costs would be less in earlier and later stages of lactation.

Expressed in metabolic terms for a 770-pound (350-kg) cow at peak lactation, her milk yields 37 kilocalories per kilogram of $BW^{0.75}$ per day (3,000 kcal/$350^{0.75}$ = 3,000/80.9). On a comparative basis, this value is much lower than what has been reported for other ruminants, including elk (72), reindeer (79), red deer (76) and black-tailed deer (118). We believe the estimate of milk energy content is correct and, therefore, question the reported milk yield at peak lactation. One possible answer is that the Russian subspecies of moose is smaller than North American forms. However, even if the cow's body mass is reduced to 550 pounds (250 kg), her milk yield, on a metabolic basis, still is only 48 kilocalories per kilogram $BW^{0.75}$ per day.

Such an exercise in metabolic scaling demonstrates the usefulness of allometry. By making such comparisons, detection can be made from existing data of potential outliers that may require additional testing. The milk production estimates used in the above calculations obviously need further examination.

Cow moose produce a highly nutritious milk, which allows for the necessary rapid growth of calves if they are to survive the cold weather and deep snows of winter. Lactation is even more energy demanding for cow moose than is pregnancy. *Photo by Charles C. Schwartz; courtesy of the Alaska Department of Fish and Game.*

E. Reese and C.T. Robbins (personal communications: 1992) estimated milk output for two cow moose nursing single calves. Peak milk output occurred 21 to 31 days postparturition at 12.1 pounds (5.5 kg) per day. Initial output was around 106 ounces (3,000 g) per day and declined to around 53 ounces (1,500 g) per day by 4 months. Estimates of milk output for a cow nursing twins were about 67 percent greater than for those of a cow with a single calf. Using the above value, at peak milk production, the energy loss associated with lactation was 8,250 kilocalories. Scaling this value to metabolic weight, a 770-pound (350-kg) cow moose has a daily energy yield in her milk of 102 kilocalories per kilogram $BW^{0.75}$ (8,250/80.9), which closely corresponds to the estimate for black-tailed deer. We believe this value approximates the actual value at peak lactation more closely than do the aforementioned Russian data.

Vitamin and Mineral Requirements

Little is known about the vitamin and mineral requirements of moose. Much of what is known is from a review of mineral element studies on North American moose by Flynn and Franzmann (1987). Also, seasonal variation in moose hair element values were reported by Franzmann et al. (1975c). Whether these variations are driven entirely by dietary intake or something else (i.e., hormones), has yet to be determined. Variability in moose hair element levels is detectable between geographic locations (Franzmann et al. 1977), but no factors are identified to explain why these differences exits. A specific question that must be investigated deals with the mobilization of large amounts of calcium in bull moose during antler development. The potential applications of mineral data for moose management, law enforcement and habitat management were projected by Franzmann et al. (1975d), but more research and base line data are needed before such applications are practicable.

SODIUM

Considerable attention has been given to the natural craving moose seem to exhibit toward sodium. Sodium has been identified as the major mineral in licks as well as aquatic plants (Jordan 1987).

COPPER

Flynn et al. (1977) were the first to recognize a potential copper deficiency in moose on Alaska's Kenai Peninsula by surveying copper in hair. The low copper levels detected later were correlated with faulty hoof keratinization and a decline in reproductive rates. About 1 to 3 percent of the moose population was identified with the deficiency. Low levels of blood copper also supported the contention that this population was suffering a copper deficiency.

Moose suffering from low or chronic copper deficiencies have hair with decreased tensile strength and faulty hoof keratinization. Associated hoof abnormalities greatly affect moose mobility, particularly during winter, when wet snow can ball on the deformed hoof (see Chapter 8). In wild populations, limited by predation, the extent and potential of such a problem are not seen because these affected individuals represent easy prey.

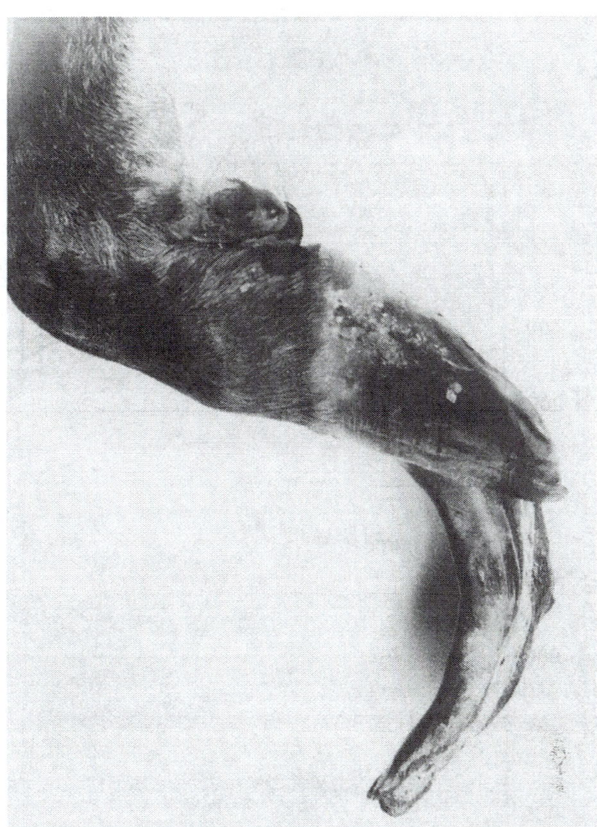

Copper is an essential trace element required by moose. Where copper is deficient, such as the Kenai Peninsula, Alaska, faulty hoof keratinization can occur. Although moose can survive with the resulting abnormality, they are predisposed to predation or starvation because of reduced mobility. *Photo by Albert W. Franzmann; courtesy of the Alaska Department of Fish and Game, Soldotna.*

Energy Budgets

Summing Energetic Costs

By comparing energetic costs with intake rates and metabolic efficiencies for moose, determination can be made of energy flows within the animal and its environment. Such an exercise provides an estimate of the species' ecological metabolism (i.e., total daily energy expenditure).

Renecker and Hudson (1989b) used such a technique to predict the annual energy expenditures for two cow moose in the aspen-dominated boreal forest of central Alberta. They established an index in captive moose, whereby metabolic rate could be predicted from radiotelemetered heart rate ($Y = 4.655e^{0.0071X}$, where Y = predicted energy expenditure; X = heart rate normalized to body weight$^{-0.25}$) in free-ranging animals. Daily energy budgets then were computed from the rate of activity-specific energy expenditures and the time a moose spent in each activity during a 24-hour day at different times of the year (Figure 192).

It also is important to relate the daily energy costs in different seasons to energy gained from food. By determining the seasonal cycle of nutrient balance—namely, daily energy costs by season versus energy consumption—essential understanding is gained concerning the tradeoffs and actions by the moose that permit it to survive (Figure 193). Gross energy consumed was calculated from seasonal estimates (Renecker and Hudson 1985a) and the calorie content of browse (4.8 kcal/g dry matter) (Hjeljord et al. 1982). Gross energy consumed was converted to digestible energy

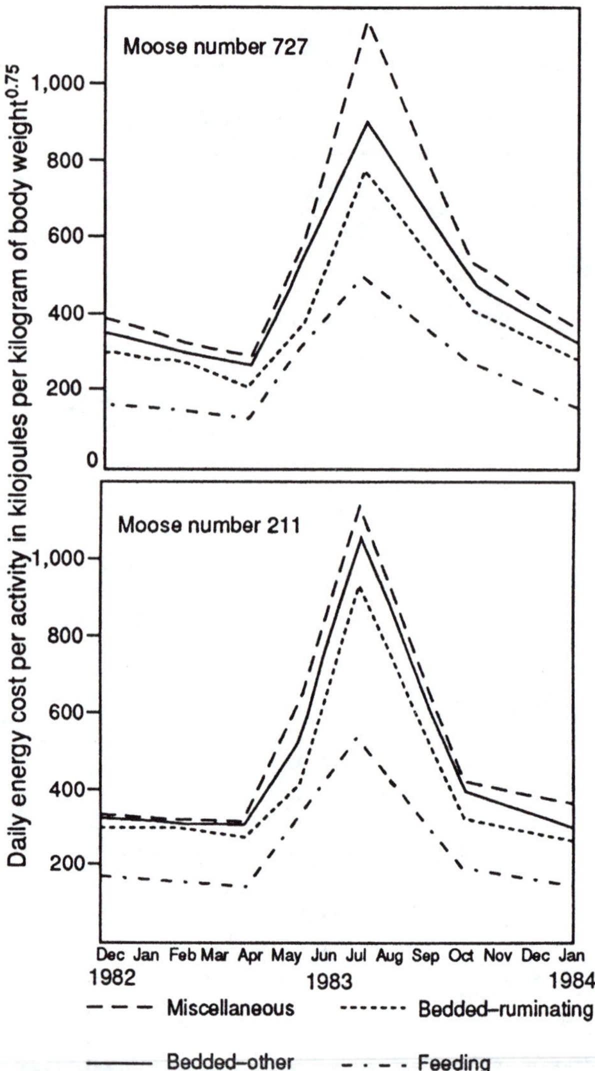

Figure 192. Energy costs of various seasonal activities of moose (data from Renecker and Hudson 1989b).

using the nylon bag technique (Renecker and Hudson 1985a). The metabolizability of digestible energy intake was determined by multiplying the computed value by 0.82 (Robbins 1993). Finally, the metabolizable energy available for maintenance or gain (that actually is used by body tissues) was ascertained by applying an efficiency coefficient of 71 percent (Hubbert 1987). The estimates from Renecker and Hudson (1989b) were for nonpregnant, nonlactating female moose, so energy costs for pregnancy and lactation (in this volume) must be added to these values to estimate the costs for a cow moose producing a calf.

As noted previously, quality and availability of food decline during winter, the amount of energy available to free-ranging moose is insufficient to meet daily energy needs; this results in weight loss (Renecker and Hudson 1985a). With the growth of new forage in spring, both dry matter intake and food quality increase. This creates opportunity for a surplus of energy for gain in protein and fat stores. The positive energy balance for lactation and improved body condition appears to continue to late autumn. During summer, heat stress from high ambient temperatures and/or insect activity can result in higher costs (Figure 192), which can possibly result in no weight gain or even weight loss.

Body Composition

Changes in body composition (i.e., the amounts of fat, protein, water and mineral in the body) and fat metabolism in moose are dynamic processes. Large gains and depletions are associated with the summer flush of forage and the winter decline of food availability and quality, respectively. Seasonal weight dynamics of moose have been associated with reduced diet quality and forage availability (Renecker and Hudson 1986a, Schwartz et al. 1984). Estimates of body composition of moose suggest that moose enter winter with large amounts of body fat (20 to 26 percent), with significant differences between males and females (Schwartz et al 1988a). Males actively involved in breeding lose large amounts of body fat during the rut, because they quit eating (Schwartz et al. 1984; see Chapter 4). Females continue to improve body condition until early winter (Schwartz et al. 1984, 1988a, Renecker and Hudson 1986a), and then lose body mass and fat.

Stored body fat and protein represent the reserves that allow moose to survive a winter energy deficit. Fat and protein are metabolized during winter and used to supplement the energy deficit associated with reduced energy intake. Energy balance, therefore influences survival. Summer ranges govern growth rate of calves and fattening in adults, whereas winter weather and ranges dictate depletions. Excesses in energy during summer influence reproduction; moose on high-quality range breed earlier and tend to produce more twins (Peterson 1955, Franzmann 1978, Franzmann and Schwartz 1985).

Management Application

Bioenergetics information helps wildlife biologists estimate the relative quality of moose range and predict the outcome of management actions that influence range quality. Managers often are asked to justify in monetary terms (costs versus benefits) the value of habitat enhancement programs or to mitigate for habitat losses. Such justification can be accomplished using bioenergetics. Estimating nutritional carrying capacity—i.e., the ratio of range nutrient supply divided by the nutrient demands of individual animals (Mautz 1978)—allows accurate prediction of the con-

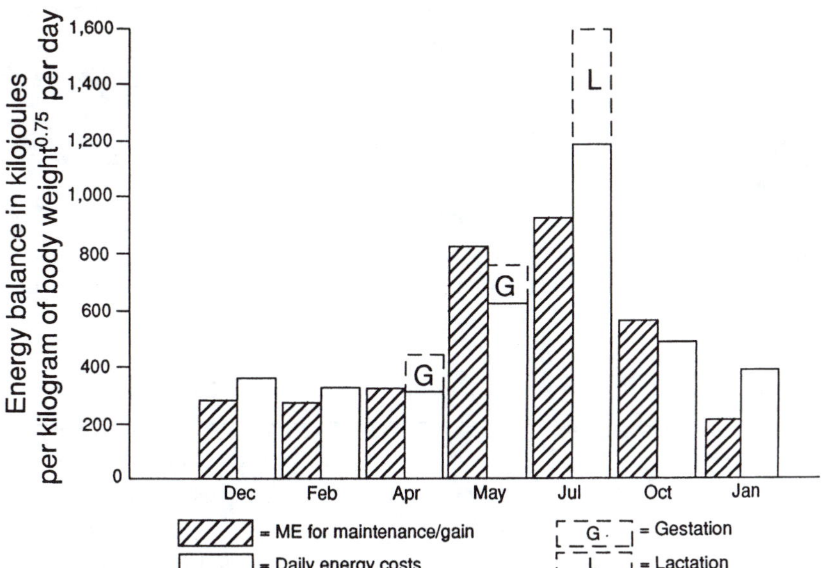

Figure 193. Relationship between metabolizable energy for maintenance or gain and daily energy costs for free-ranging moose (data from Renecker and Hudson 1989b).

sequences of management actions in terms which can be related to animal productivity and eventually to population dynamics. Evaluating ranges from a nutritional carrying-capacity perspective was first advocated for wildlife by Moen (1973), Robbins (1973) and Wallmo et al. (1977).

By relating animal requirements to spatial characteristics of habitat, forage quality and quantity can be integrated into an estimate of habitat value. The approach is useful because, by expressing range supply in units of energy or nutrients, individual bioenergetics requirements can be considered in carrying-capacity estimates (Schwartz and Hobbs 1985). Estimates of carrying capacity in North America have been derived for elk (Hobbs et al. 1982), white-tailed deer (Potvin and Huot 1983), mule deer (Wallmo et al. 1977) and moose (Regelin et al. 1987b, Hubbert 1987, Becker 1988, Miquelle et al. 1992).

Studies by Becker and Steigers (1987) and Becker (1988) showed the usefulness of this technique. In the late 1970s, the State of Alaska proposed to construct two large hydroelectric dams on the Susitna River in southcentral Alaska. The impoundments were going to be built in the heart of a large winter range of moose. To assess the potential impacts of this project on moose, the Alaska Power Authority contracted the Alaska Department of Fish and Game to provide field data, analysis and recommendations. The Department of Fish and Game had several options. First, short-term losses could be determined simply by estimating the number of moose using the impoundment area before flooding and later determining the fate of those animals as a result of dam construction. The option had several flaws. Moose populations are not always at carrying capacity. Predation, hunting and severe winter weather can reduce moose populations to well below carrying capacity (Gasaway et al. 1992). Counting numbers of moose under these conditions invariably results in an underestimation of the potential carrying capacity of the area in question. Consequently, the Department chose instead to estimate the nutritional carrying capacity of the impoundment area, which more closely reflected the number of moose the range could carry if the animals were not limited by other factors. This option provided a more realistic estimate of what would be lost in terms of moose habitat.

Detailed estimates of food habits, diet quality and forage abundance (food model) were integrated with a moose energetics model (animal model) (Hubbert 1987). Together, the two simulation models provided estimates of the potential carrying capacity of the impoundment areas. Additional studies were necessary to determine the availability of forage for moose during winter, because areas were located in mountainous terrain where elevations changed from 900 to 3,400 feet (274–1,036 m). Carrying-capacity estimates were calculated for February through April—the time when moose were forced out of the mountains into the Susitna River Valley. Integrating both available food and other moose needs, the stocking rate of the basin under consideration for flooding (25 square miles [65 km^2]) would be reduced by an estimated 404 moose in a mild winter. In a severe winter, when virtually all moose were forced to lower wintering areas, the stocking rate would be reduced by an estimated 674 moose. Although the dams were never constructed, the simulation modeling and analysis demonstrated the utility of predicting carrying capacity using nutritional energetics. Becker (1988: 11) concluded that "this process is very flexible, and it allows one to generate moose carrying-capacity estimates under different biological assumptions. In addition, it is very easy to develop new carrying-capacity estimates when the project design has been changed."

The process also is useful when evaluating habitat management programs. For example, by mechanically treating vegetation, root suckering can be stimulated to increase forage production. By knowing the per acre cost of treatment, the increase in forage quantity and quality and the expected life of a project (from a moose browse production standpoint), the cost to support one additional moose per unit of time (e.g., during the 4 months of winter) can be easily calculated. Such costs estimates simplify evaluation of cost-to-benefit ratios of moose habitat improvement projects.

Strategies for Survival

Schwartz (1992a) summarized the selective nutritional processes or strategies that enable moose to survive in the periodically harsh northern climate to which they have adapted. Diets consist mainly of leaves and stems of woody browse—foods that predominate in the boreal forest. Moose are selective feeders, spending most of their time eating and ruminating in habitats rich in available forage. The digestive system of the moose is adapted to process woody browse efficiently. Large bites and lengthy chewing breaks down the food material into particles capable of passing quickly out of the rumen. Diet fiber is high in lignin and not retained in the rumen.

Moose have a higher metabolic rate and eat more in summer when food quality and availability are high. The opposite is true in winter. The energy and protein requirements of moose are similar to those of other ruminants.

Fat reserves of moose are cyclic in response to seasonal rhythms of intake and metabolism. These rhythms appear to be controlled by feedback mechanisms linking body condition and fat reserves. Arnold (1985: 82) reviewed the mechanics of intake control and noted that "long term stability of energy balance is thought to be controlled by the size of the fat reserves . . . [and] many species in temperate and arctic areas appear not to have stability in energy balance even in a constant nutritional environment." Furthermore, "regulation of forage intake by free-ranging wild herbivores is through both internal controls concerned with digestion, rate of passage of digesta through the digestive system, and set points probably including energy balance and body condition" (Arnold 1985: 97–98).

Based on their observations of captive moose, Schwartz et al. (1988b) proposed the following. When high-quality food is available during summer, moose consume large quantities. The body condition of bulls peaks in early autumn before the rut. Peak body condition for cows occurs later, and is associated with natural declines in forage quality and availability. Peaks in body condition activate mechanisms (set point) that depress intake. Depressed intake occurs in captivity even when animals are offered high-quality food ad libitum. Reductions of intake occur in the wild concomitantly with natural declines in food quality and availability. Decreased intake results in lower metabolic rate (Hubbert 1987) and a shift to a negative energy balance. Body stores are depleted to a low in body condition (set point) in spring. The process then reverses. Set points, which result in metabolic shifts, correspond with environmental changes in food quality and availability (Regelin et al. 1987a, Renecker 1987a, Renecker and Hudson 1988). They vary among individuals and levels of intake. Body condition likely influences intake during winter. If the lower set point is reached before change in food availability (i.e., green-up), animals in the wild starve. However, under confined conditions where food is available, moose that reach the lower set point before spring increase intake, whereas moose in good condition (above the set point) do not (Schwartz et al. 1988b). These mechanisms are not rigidly fixed, and they exhibit plasticity when animals are stressed.

The annual cycle of food intake, fat metabolism, metabolic rate and body mass dynamics is not driven simply by food quality and availability. It represents a complex interaction between internal physiological regulators and the external environment. Physical condition of moose is keyed to environmental quality, but deeply rooted in physiological and morphological adaptations of the species. Nutritional energetics play an important role in the survival of individual moose and ultimately influence population dynamics and survival of the species.

Previous page: Michael H. Francis *Derrick Hamrick* ▲ *H. R. Timmermann* ▼ *Donald M. Jon*

▼ *Victor Van Ballenberghe* ▲

William E. Ruth ▼

Victor Van Ballenberghe ▲ Michael H. Francis ▼ Donald M. Jones ▼ Tom Walker ►

Bill Marchel Charles H. Willey ▼ William E. Ruth ▲ Michael H. Francis ▼

▼ *Charles H. Willey* ▲

iam E. Ruth ▲ Ted Williams ▼ Next page: Michael H. Francis

Previous page: Victor Van Ballenberghe

Donald M. Jones ▲

Tom Walker ▼

Bill Marchel ▲

Albert W. Franzmann ▲

Michael H. Francis ▼

Tom Evans ▲ Michael H. Francis ▼ Bill Silliker Jr. ▲

r Van Ballenberghe ▲ Michael H. Francis ▼ Next page: Donald M. Jones

15

MURRAY W. LANKESTER AND
WILLIAM M. SAMUEL

Pests, Parasites and Diseases

Moose are one of the largest and most formidable inhabitants of the northern forests. Nevertheless, even some of the largest and healthiest moose are weakened by disease. The number of moose affected and the severity of a disease depend in part on the agent involved. Viruses and bacteria are more likely to cause epidemic-like events, whereas worms and other parasites cause fewer deaths. Parasites and other agents of disease need not always kill to have an impact. "Subclinical effects" are those not readily observed. Their impact can be significant but difficult to detect. For example, if a diseased moose's behavior interferes with its interest or ability to breed, it does not contribute to the next generation. If such an animal is abnormally unwary of predators or vehicles, its otherwise nonlethal infection may nonetheless lead to demise.

As disconcerting as it may be for a hunter, naturalist or even a disease expert to find a moose dead or dying from disease, the circumstance of one or several such animals does not necessarily indicate or forecast population calamity. Because most animals die "before their time," they tend to compensate by reproducing in excess. Moose have coevolved with disease organisms and other "natural causes" of death under traditional circumstances. Resulting deaths are thereby accommodated from a population standpoint. Given modern hunting methods, human alteration of habitat and encroachment of domestic livestock, diseases of wildlife occasionally assume greater importance than would be expected otherwise.

A major difficulty in assessing disease impact is recognizing when a large proportion of an animal population is affected and at what point the species' ability to persist over the long term is at risk. To overcome this potentially significant problem, the population size and number of animals affected must be known. Getting an accurate count of live animals, even those as large as moose, is a difficult and expensive proposition, but estimating the number of animals that might have died of a particular disease is even more so.

Even if the number of diseased animals dying can be determined, there is the question of whether those deaths should simply be added to all other mortality causes, including predators and hunters, or if they represent some degree of compensatory mortality. In other words, are these animals weakened by disease the ones that will die, thus sparing healthy individuals? This seems logical, but it is not all that simple. For example, predators did not evolve relying exclusively on dead and dying prey, as scavengers do. Some predators can take healthy moose, and the enhanced availability of sick ones may stimulate increase in the number of predators and increase or prolong the impact of disease. Clearly, the many causes of mortality are inextricably linked, and the benefits of being able to quantify their individual effects are undoubtedly diminished by the difficulty and cost of doing so.

Hunters are those most likely to see parasites on or in game animals. Alarm at such discovery is the natural response. Such hunters invariably share the concern of the biologist about what the presence of parasites means to the health of the animal population that provides them with enjoyable recreation and a prized harvest. Many hunters also will have reservation about the edibility of meat from an animal with identified parasites. The prospect for wastage is considerable. Hunters need to know that virtually all wild animals have parasites (including viruses and bacteria) in their bodies, but few are diseased. Parasitization and a diseased condition are not synonymous. *Disease* is defined here as any abnormal condition that reduces life span or reproductive capability. By far, the majority of parasites occur in their host without causing recognizable disease.

Only a few of the known agents of disease in moose are suspected of having the potential to impact moose populations significantly. These include the meningeal worm *(Parelaphostrongylus tenuis)* in eastern North America, winter ticks *(Dermacentor albipictus)*, possibly brucellosis and, in restricted western areas, the arterial worm *(Elaeophora schneideri)*. Parasitic organisms that are transmissible to humans, or that may make moose meat inedible, are even fewer in kind and number. Brucellosis (*Brucella* spp.) and *Toxoplasma gondii*, although rare, should be included in this category. Also included is the tapeworm *Echinococcus granulosus*, although human health risk involves association with the final host, wolves and other canids, rather than moose, the intermediate host. Larval stages in the lungs of moose can infect dogs. The resulting eggs passed in dog feces are infectious to humans.

The relatively good health enjoyed by most moose populations is explained in part by their history and biology. Moose are relatively recent arrivals from Asia (70,000 to 10,000 years ago; see Chapter 1). Many of the species presently parasitizing moose in North America have not been reported from Old World moose and, thus, must have been acquired from indigenous wild ungulates or domestic animals (Anderson and Lankester 1974). Unlike these hosts, moose usually occur at relatively low densities—from 0.3 to 1.6 per square mile (0.1–0.6/km^2)—and, for much of the year, adults tend to avoid each other. Hence, the probability of transmission of parasites and disease agents from one individual to another by direct contact is minimized. Also, severe winter conditions in their northern environment kill some of the infectious stages of parasites that might otherwise build up on moose range. These would include such forms as the short-lived, direct life cycle, abomasal nematodes. Even if excessive mortality occurs, the ability of well-nourished cow moose to produce twins each year, gives the population considerable capacity to recover. Moreover, moose have a remarkable ability to overcome bacterial infections after injury. Equally remarkable is their ability to recover from broken bones.

Infectious Diseases

Viruses, bacteria and, to some extent, protozoa comprise the so-called infectious organisms that affect moose, although such organisms are not necessarily infectious to humans. Some are, but most of those described here are not. Many agents within this group of organisms are contagious—meaning that they are easily spread from one host to another by direct contact or in food or waste products. The host may be another moose, related wildlife or domestic animals. Also, unlike some others, infectious organisms may multiply profusely in a previously unexposed host, and the course of infection may progress rapidly. Generally, infectious organisms have the greatest potential to cause disease and death in susceptible hosts. However, hosts that survive may develop complete or partial immunity against reinfection.

Domestic animals are reservoir hosts for many of the agents infectious to moose. But the extent of their impact on the health of moose populations is largely unknown, in part because diagnosis of specific viruses and bacteria requires specialized expertise that is not available in most wildlife agencies. Ideally, dead moose are found and reported by local officials and diagnosed by trained pathologists. This may occur where moose make frequent contact with domestic animals. However, many moose deaths occur in unpopulated or fairly inaccessible habitat and the remains vanish before detection.

The usual procedure for diagnosis of viral and bacterial infections is testing blood samples for the presence of antibodies produced in response to past exposure to such agents. Many tests are used, with each being fairly specific to a particular virus or bacterium. A positive test may mean that the animal still is infected and a carrier of the organism. More likely, it reveals that the moose has survived and completely overcome the infection. Affording a degree of protection against future reexposure, antibodies remain in the blood serum for some time. The proportion of seropositive animals in a population reveals nothing about how many animals died or how debilitated the survivors may have been after exposure. However, knowing the number of surviving reactors can be useful in assessing the potential impact of an infectious disease on a host population and the possible importance of the host in transmitting the infection to other wild or domestic animals. A thorough understanding of these factors often can be obtained only by conducting controlled experiments with captive animals (see Chapter 16). Otherwise, the temptation is to infer outcome from experience with disease in domestic animals. The variable responses of wild species to disease make this corelationship assumption very risky.

The availability of trained, captive moose is vital to address certain questions about moose health, in addition to other aspects of their biology. Moose can be trained to step onto a weigh scale and have blood taken from the jugular vein (above) or from the saphenous vein in the hind leg. *Photo by H.R. Timmermann; courtesy of the Ontario Ministry of Natural Resources.*

Viruses

Arboviruses

Wild moose are seropositive to several viruses (see Lankester 1987), but there is no direct evidence that any moose has died as a result of exposure.

The viruses causing epizootic hemorrhagic disease (EHDV) and bluetongue (BTV) are transmitted by sand flies (biting gnats)—both cause almost identical diseases. These viruses injure the inner lining of blood vessels, causing clogging of smaller vessels throughout the body (Hoff and Trainer 1981, Wobeser 1989). Lethargy, fever, swelling of the head and neck and blueness of membranes in the mouth and around the eyes are characteristic symptoms. Death may follow in 8 to 36 hours. EHDV is particularly virulent in white-tailed deer; an outbreak may kill up to 90 percent of the animals in a local population. Die-offs usually occur in late summer and early autumn, when gnats are abundant. It is generally thought that the distribution of EHDV in moose parallels that of white-tailed deer. But moose and other native cervids show only mild (inconsequential and short lived) signs of the disease.

No clinical disease was seen in moose experimentally exposed to EHDV (Hoff and Trainer 1978). The only wild moose found seropositive to EHDV (6 percent) were from Alaska (Zarnke et al. 1983); none was found in a subsequent study of Alaskan moose (Kocan et al. 1986b).

BTV has been detected in livestock throughout the continental United States, but has been rare or absent in Canada and Alaska in recent times (Thorne et al. 1982, Wobeser 1989). It has caused major losses in white-tailed deer, pronghorn, and wild and domestic sheep. Cattle are believed to be the most important reservoir host of BTV because they seldom exhibit signs of the disease, although the virus persists in circulating blood where it can be picked up by biting gnats (Gibbs 1982, Thorne et al. 1982). The only moose ever found seropositive to BTV were from the Chapleau District of Ontario (Trainer and Jochim 1969).

Other arboviruses detected serologically in moose include snowshoe hare virus (SSHV), St. Louis encephalitis virus (SLEV), western equine encephalitis virus (WEEV), Norway virus (NORV) and Jamestown Canyon strain of California encephalitis virus (CALV). All are transmitted by mosquitos from other wild or domestic animals, but no known disease results in moose.

Bovine Respiratory Viruses

Three widely distributed viruses are known to cause respiratory disease in domestic cattle and in some wild cervids. These include infectious bovine rhinotracheitis (IBRV), parainfluenza type 3 virus (PI-3) and bovine viral diarrhea (BVDV). Moose from Alaska to Minnesota have been reported to be seropositive to all three (see Lankester 1987), but no disease is recognized. Kocan et al. (1986b) found up to 33 percent of moose from the Alaska Peninsula to have antibodies to BVDV; 17 percent were positive for IBRV.

Miscellaneous Viruses and Spirochaetes

Wild or domestic sheep and goats are the main source of two other viruses to which moose are exposed. Malignant catarrhal fever (MCF)—caused by a herpes-type virus—is a fatal disease of bovids and cervids throughout the world. Symptoms include diarrhea, discharge from the eyes and nose and erosion of the lining of the alimentary tract. The form in North America is presumed to be acquired from domestic sheep, hence the name "sheep-associated MCF." The disease was diagnosed in a captive moose in Wyoming (Thorne et al. 1982, Williams et al. 1984). Two additional cases in captive moose calves were listed by Schwartz (1992b); ibex and domestic sheep held in the facility were the suspected source.

Contagious ecthyma is a pustular skin condition caused by viruses of the parapox group (Wobeser 1989). Lesions developing into proliferative growths with scabs are most common on the lips and muzzle but also occur on the udder and legs. Transmission is by direct contact. Humans can become infected. With treatment for secondary bacterial infection, lesions usually resolve. Moose were shown to be susceptible to experimental infection, but resulting lesions were considered mild (Zarnke et al. 1983).

Rabies appears to be rare in moose; Irvin (1970) listed them as an incidental host. Negri bodies were found in brain tissue of an animal in Utah that was unusually lethargic and placid (Anonymous 1979). A rancher who tried to feed the moose received postexposure treatment for rabies, as did eight wildlife personnel and veterinarians who conducted the necropsy.

Leptospirosis is an infectious disease caused by the spirochaete *Leptospira interrogans*. Almost 200 different serovarieties are recognized, and they infect a variety of rodents, carnivores and ungulates. Humans can become infected. Clinical signs, including high fever, develop about 2 weeks after infection. Initially, spirochaetes are present in the circulating blood, but later may be detectable only in the kidneys. Leptospirosis is self-limiting in most mammals with spirochaetes being shed in the urine for only a limited time (Thorne et al. 1982). Abortion may be the most serious result. Antibodies to seven different serovars have been reported widely from moose (Kocan et al. 1986b).

Fibromas and Other Tumors

The most common tumor of moose is a benign, wart-like growth protruding from the skin. Larger growths may become pedunculated and hang by a slender stalk. They may be smooth or cauliflower-like, bald or haired, single or multiple, and range in size from 0.4 to 4 inches (1–10 cm) or more in diameter. These neoplasms are correctly termed "infectious cutaneous fibromas" and, in deer are known to be caused by a popova virus (Cosgrove and Fay 1981). Infection is by direct contact or possibly by biting flies serving as vectors. The fibromas may become lacerated and secondarily infected by bacteria, but otherwise cause the animal no harm. Eventually, they regress and disappear. They are removed on skinning a harvested moose, and do not affect edibility of the meat.

Other firm, massive tumors called "lymphosarcomas," originating from cells of the lymphoid system, have been reported within the abdominal cavity of moose (Garner and Schwartz 1969). A large cauliflower-like growth termed a "myxoma" was seen in a moose calf in Ontario (Lankester and Bellhouse 1982).

Bacterial Diseases

Brucellosis

Brucellosis is an important disease of domestic animals and is caused by the bacterium *Brucella*. Three species occur in

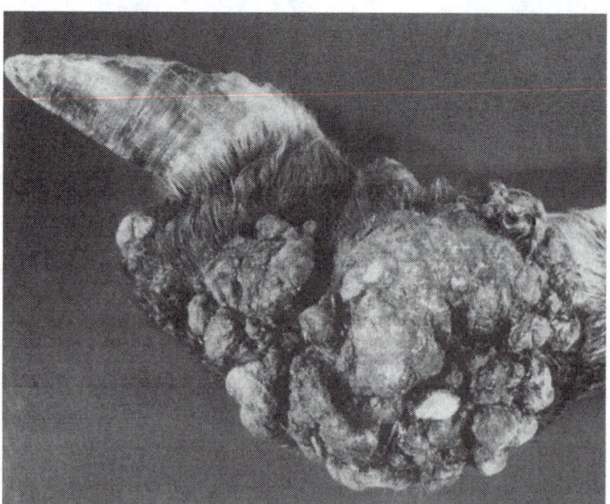

Fibromas on head, neck (left) and forelimbs (right) of moose may be smooth or wart-like, bald or haired, single or multiple, and range in size from 0.5 to 4 inches (1.3–10.2 cm) or more in diameter. Fibromas are benign tumors of the skin that may become lacerated and infected by bacteria, but otherwise seldom cause problems for moose. *Left photo by F.J. Johnson; courtesy of the Ontario Ministry of Natural Resources. Right photo by D.K. Onderka, courtesy of Alberta Agriculture.*

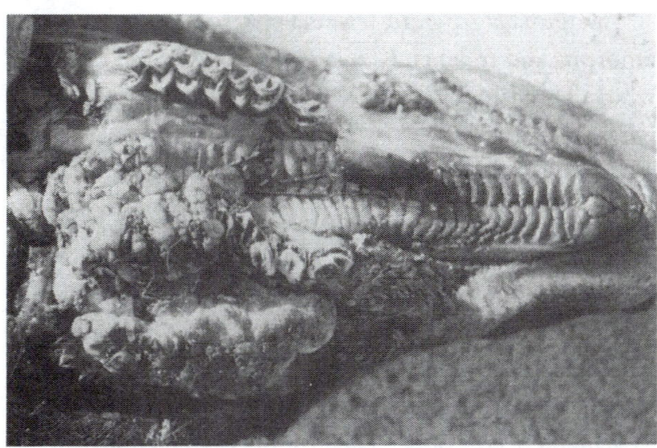

The upper jaw of a calf moose, with a large, whitish, cauliflower-like neoplasm (myxoma) that has displaced the animal's cheek teeth and would severely impair feeding. *Photo by M.W. Lankester.*

North America, each associated principally with cattle *(B. abortus)*, swine *(B. suis)* and sheep *(B. ovis)*. Humans can become infected. Knowledge of infection in wild ungulates is largely from serological evidence of exposure, but *B. abortus* is established in certain bison and elk herds, and *B. suis* biovar 4 occurs widely in barren ground caribou and reindeer (Tessaro and Forbes 1986, Wobeser 1989). *Brucella suis* biovar 4 was cultured from four of six cattle held for 30 days with naturally infected reindeer (Forbes and Tessaro 1993). No clinical or pathological signs of disease were evident in the cattle killed 2 months after separation. Infection with *B. suis* is very rare in white-tailed and mule deer (Rinehart and Fay 1981). As yet, the importance of brucellosis in moose is undetermined.

Upon infection, the *Brucella* bacterium initially involves the male and female reproductive tracts, later spreading to the lymph nodes and joints. Enlargement of the testes is commonly seen. Abortion, stillbirth and sterility are the most significant aspects of the disease in cattle. Swollen or stiff joints also may result. Brucellosis becomes a chronic disease in adult animals; mortality is rare. The organism is very contagious. Most infections probably occur by the oral route through contact with aborted fetuses, afterbirth, vaginal discharge or skin abscesses. It is suspected that the bacterium also may be transmitted by blood-sucking flies or venereally in semen (Wobeser 1989).

B. abortus has been isolated from dead or dying moose in Minnesota, Montana and Alberta (see McCorquodale and DiGiacomo 1985). Yet, the absence of seropositive animals in other areas where the disease occurs in cattle indicates that infection of moose may be infrequent. Some investigators have suggested that moose may be highly susceptible and die quickly after infection, leaving few if any seropositive survivors detectable in the population (Jellison et al. 1953, Corner and Connell 1958).

Research has attempted to resolved this controversy. Two moose were experimentally exposed to *B. suis* biovar 4 originating from reindeer in Alaska, but the results were contradictory. Rausch and Huntley (1978) saw no clinical disease in a calf whose antibody titer peaked after 2 months and then declined. The antibody titer in a calf infected by Dieterich et al. (1991) peaked after 56 days and remained high. The animal showed signs of fever, elevated white blood cells, lack of appetite and reluctance to rise from a lying position. Bacterium was recovered from blood and internal organs when the animal was killed. Dieterich et al. (1991) believed that the calf would have died of the infection had it been in the wild.

Forbes et al. (1996) experimentally infected four moose with *B. abortus* biovar 1 from cattle. All became infected. None showed clinical signs, with the exception of one that died suddenly at 85 days after inoculation. The other three were killed at 77 (two) and 166 days. Internal lesions were indicative of endotoxemia, and the bacterium was isolated from all four animals. Forbes et al. concluded that moose are highly susceptible to *B. abortus* infection, but under field conditions, death probably comes rapidly, making moose a poor reservoir of infection for other wild or domestic animals. They cautioned that substantial risk to moose hunters may exist in areas where moose can become infected by cohabiting with reservoir hosts such as elk and bison.

Miscellaneous Bacteria and Pus Pockets

Other bacteria have been reported from moose. Each is potentially harmful and may cause death directly or indirectly. However, the small number of cases of each described in the literature suggests that infection is rare and probably of little significance to the overall health of moose populations.

Incidental bacterial infections reported in moose include listeriosis caused by *Listeria monocytogenes*. This bacterium occurs widely in other wild and domestic animals, and is spread by direct contact and by biting arthropods. Infection may lead to encephalitis, sometimes resulting in circling behavior (Archibald 1960).

Johne's disease is a chronic, wasting condition caused by infection of the lower digestive tract with *Mycobacterium paratuberculosis*. It has been diagnosed twice in moose (Soltys et al. 1967, R. Lewis personal communication: 1992).

Anthrax is an acute, fatal, systemic disease caused by *Bacillus anthracis*. Most hooved animals are susceptible, and mortality rates are high in epizootics. Animals become infected by ingesting resistant bacterial spores that survive for long periods in soil. Infection in moose has been seen only at the time of an outbreak of the disease in bison in Wood Buffalo National Park (Choquette 1970).

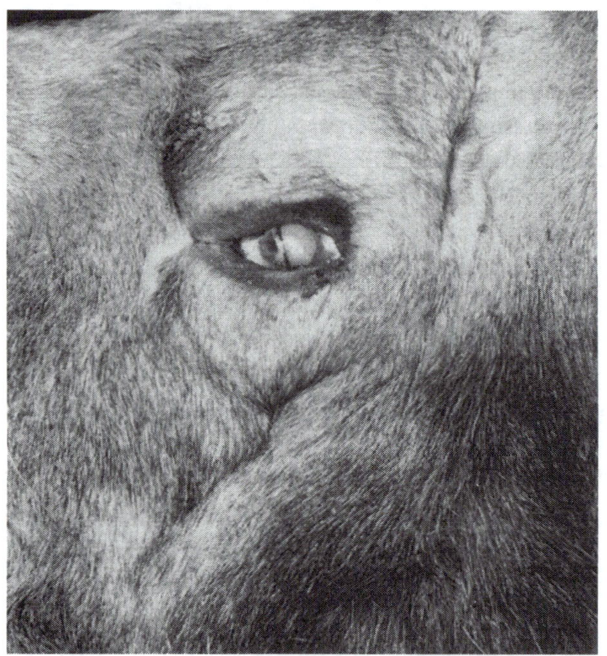

Opaqueness of the cornea over the eye or just of the lens (cataract) is not uncommon in moose. The original cause usually is impossible to ascertain, though injury and infection, such as *Moraxella* sp., are known to be involved. *Moraxella* is a bacterial infect, usually acquired from contact with other animals. It can lead to corneal opacity, but no particular agent could be associated with the case in this photo. *Photo by M.W. Lankester.*

Tularemia is an acute, feverish infection primarily of lagamorphs and rodents. It is caused by *Francisella tularensis,* which is transmissible to humans. A small percentage (1–7 percent) of moose in eastern Canada have been found seropositive (Borque and Higgins 1984).

Erysipelas is a cutaneous and systemic infection caused by *Erysipelothrix rhusiopathiae*. It was identified in moose that died in Algonquin Park, Ontario, in late winter, with hair loss caused by ticks (Campbell et al. 1994). With increased evidence of the importance of winter ticks as a significant mortality factor in moose, any possibility that this bacterium also might play a role deserves careful examination.

Pinkeye is a bacterial infection of the cornea and eyelid common in cattle. It is caused by *Moraxella* sp., and has been isolated from a moose in Wyoming, where infection also occurred in deer and pronghorn (Thorne et al. 1982). Infection may eventually cause rupture or opacity of the cornea. Involvement of both eyes is most common, but where vision is affected more seriously in one eye, the animal may circle in the direction of the better eye.

Actinomycosis, caused by *Actinomyces* sp., is manifested by abscess formation in the lower jaw resulting in varying degrees of bone erosion and deformation of the lower jaw. Infection in moose probably is associated with worn teeth and food impaction in older animals (Ritcey and Edwards 1958).

Foot rot, caused by *Fusobacterium necrophorum,* leads to necrotizing lesions of the feet and abscess formation in the oral cavity and internal organs. It may be more common in moose than the few reported cases suggest (Lankester 1987). Chronic or healed cases may exhibit swollen lower joints.

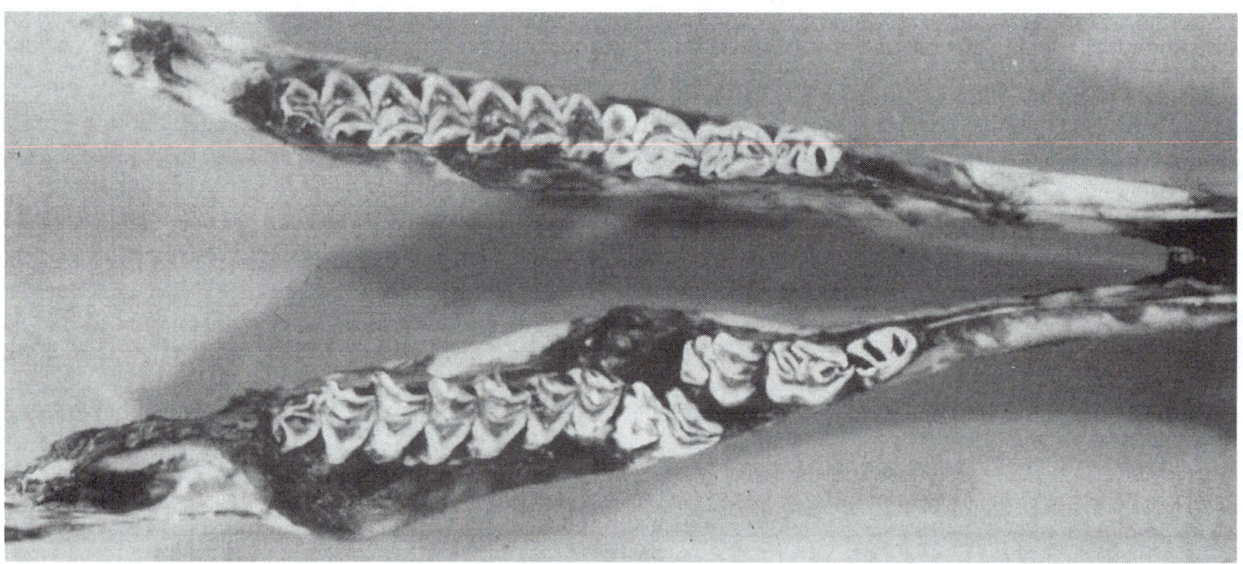

An actinomycosis infection of the lower jaw of moose causes distortion of the mandibular bone, misalignment of teeth and uneven tooth wear. The pain and malocclusion experienced by the animal shown no doubt hampered its feeding. Affected animals may be underweight. *Photo courtesy of Alberta Agriculture.*

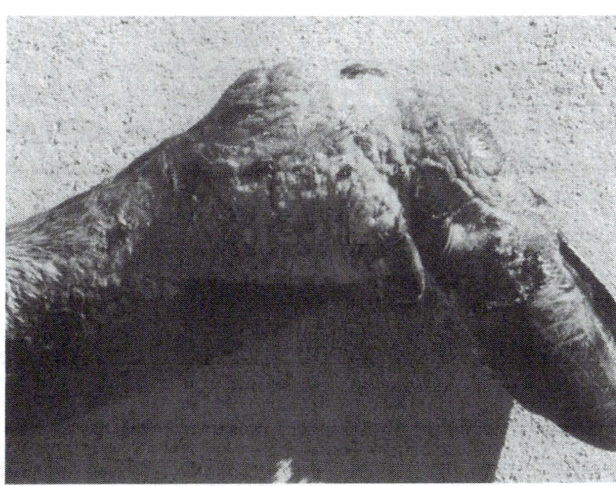

Footrot in a moose causes necrotizing lesions of the soft tissue of the foot, often resulting in subsequent infection, swelling of the joints and pronounced lameness. *Photo by H.R. Timmermann; courtesy of the Ontario Ministry of Natural Resources.*

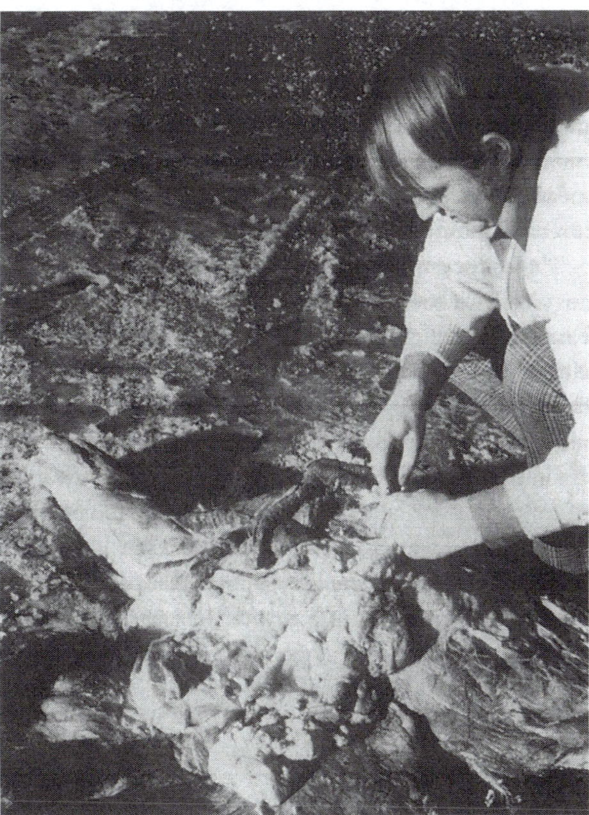

Large abscesses or pus pockets (above in the front shoulder of a bull moose) are localized bacterial infections that usually result from fighting or other traumatic injury. Hunters often are concerned about the effect of these infections on the edibility of meat from the general area. If the surrounding muscle appears normal in texture and color, and the carcass looks otherwise normal, careful removal of the abscess(es) with liberal trimming, will salvage the meat. *Photo by H.R. Timmermann; courtesy of the Ontario Ministry of Natural Resources.*

Hunters commonly report the presence of pus-filled abscesses in the skeletal muscle of moose. Thorne et al. (1982) provided the only published description of so-called pus pockets from which the bacteria *Staphylococcus aureus*, *Acinetobacter* sp., and *Actinomyces* (= *Cornybacterium pyogenes*) were isolated. Often large and located in the flank region, muscle abscesses were said to be common in moose of Wyoming. Only male moose in one sample had abscesses suggesting that the lesions resulted from puncture wounds inflicted during the rut. Such abscesses usually are localized and reasonably well isolated from surrounding muscle by a capsule of fibrous tissue.

Protozoans

Toxoplasmosis

Toxoplasma gondii is a single-celled organism that is spread by domestic, feral and wild cats (Miller et al. 1972). It is shed in cat feces as minute oocysts. These can infect almost any domestic or wild animal, including moose. Such animals serve as intermediate hosts, in which microscopic cysts containing bradyzoites (slow multiplying forms) occur in muscle and other tissue. In this form, the parasite can be passed from one animal to another by carnivorism. Herbivores are exposed (but less frequently), probably by eating oocysts scattered on vegetation. Humans become infected by eating meat containing bradyzoites. Infections in large hosts will be at a dead end. Only smaller prey species convey the infection to cats, the essential host, in whose intestine the parasite reaches sexual maturity.

The distribution of *T. gondii* is worldwide in animals and humans (Dreesen 1990). Detection is based on a specific blood test. Most infections in humans are asymptomatic or, at most, promote mild, flu-like complaints. An immunity to reinfection results. Many people who test positive cannot remember a particular illness that might have followed the occasion of their being infected. A more serious disease (congenital damage to the brain and eyes) is possible, although rare, in the human fetus if a previously unexposed mother becomes infected during pregnancy. Uncooked or undercooked pork is thought to be the most important domestic source of infection (Dubey 1986). Although up to 40 percent of human adults in North America have antibodies to *T. gondii*, evidence of clinical illness is rare (Dubey 1986, R.H. Nottenboom personal communication: 1996).

T. gondii was first diagnosed in a moose in Montana (Dubey 1981). Subsequently, it was detected serologically in 23 percent of moose in Alaska (Kocan et al. 1986a). It also has been detected in 15 percent of 125 moose in Nova Scotia (Siepierski et al. 1990). The extent of this parasite's distribution in moose is unknown, but it may be fairly common and, for this reason, moose meat should not be eaten rare. As well, hands should be washed after handling raw meat. Complete freezing at household freezer temperatures and thorough cooking kill this parasite (Dubey 1986).

The zoitocysts of at least three other species of coccidians also have been found in the muscle tissues of moose. None is thought to be harmful to humans or moose. The elongated, whitish cysts of some forms are large enough to be seen with the naked eye. Their presence in the meat is referred to as "sarcosporidiosis." Sporulated oocysts that are infectious to moose are spread in the feces of a variety of carnivores. An undetermined species of *Sarcocytis* has been reported widely in moose (see Lankester 1987) and a species named *S. alceslatrans*, by Dubey (1980) was shown to reach sexual maturity in a coyote. *Hammondia heydorni*, known from cattle and dogs, also has been reported in moose (Dubey and Williams 1980).

Other Protozoa

Two other protozoan parasites have been reported in moose. A flagellate *(Trypanosoma cervi)* living in the blood is widespread in cervids, and may be transmitted by horse flies (Kingston 1981b). An amoeba, resembling *Entamoeba bovis*, was found in the intestinal cecum of moose (Kingston 1981a). Neither appears to cause disease.

Trematodes

Trematodes are parasitic flatworms that inhabit the digestive tract and associated organs of vertebrates. Most use aquatic snails as intermediate hosts. Final hosts, such as moose, usually acquire the parasite by ingesting the infective form along with their food. An alternate name used for this group of organisms is "flukes."

Fascioloidiasis

A condition in moose, fascioloidiasis—aptly described as "liver rot"—is caused by the large American liver fluke, *Fascioloides magna*. Chronically infected livers with numerous large capsules containing black, pasty material are a memorable sight, but whether the parasite is detrimental to moose has not been clearly demonstrated. This fluke is one of the largest trematodes known; mature ones can be up to 3.1 inches long and 1.4 inches wide (7.8 by 3.5 cm). They are reddish on top and a mottled grayish-yellow on the bottom. Most specimens found in moose liver are immature and usually smaller, dark red and without eggs (Lankester 1974). Because the parasite seldom reaches maturity in moose, it cannot perpetuate itself in this host (Lankester 1974). Therefore, fascioloidiasis in moose is expected only in areas where moose share range with a variety of other susceptible cervids in which *F. magna* can successfully complete its life cycle.

Normal hosts in which flukes mature and pass eggs include white-tailed deer and possibly mule and black-tailed

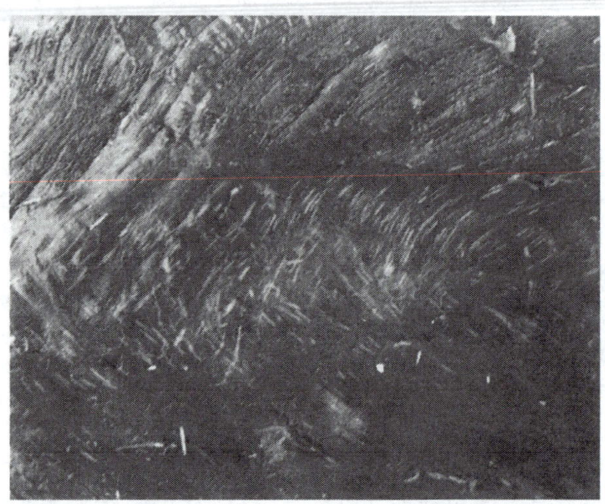

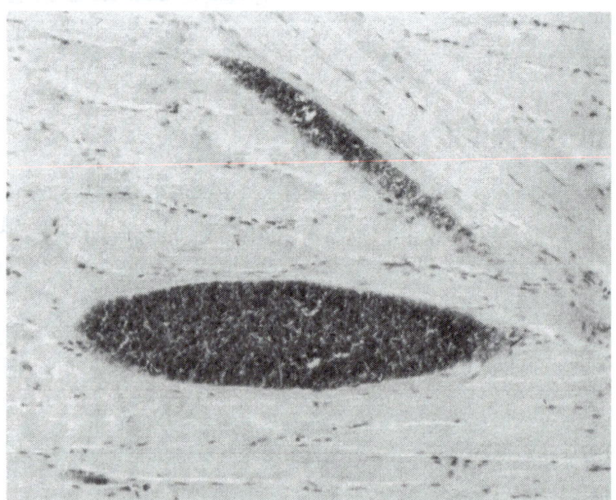

Sarcosporidiosis is represented by tissue cysts (whitish, elongated bodies, as much as 1 mm in length) that can infect all muscle tissue (cardiac, striated and smooth) of moose (in esophageal muscle at left). Although thorough cooking kills this parasite, the flesh of heavily infected animals is aesthetically displeasing to most hunters, and heavily infected meat is not recommended for consumption, due to unknown health effects. Domestic animals with heavy infections similar to this are condemned at slaughter. Microscopically, sarcosporidiosis tissue cysts (right) contain hundreds of bradyzoites that are infective to carnivore predators or scavengers. They are not thought to affect the health of moose. *Left photo by W.M. Samuel. Right photo by J.L. Mahrt.*

deer, elk and caribou (Wobeser et al. 1985, Lankester and Luttich 1988). In these cervids, flukes migrate for a time in the liver, apparently in search of another individual with which to pair. Eventually, two flukes become enclosed in each capsule. The capsules are continuous with bile ducts that allow eggs and metabolic wastes to drain out. Although flukes are hermaphroditic (male and female organs in the same worm), cross-fertilization apparently is preferred. Fluke eggs are flushed with the flow of bile into the intestine and are discharged with the feces (Foreyt 1981).

Details of the parasite's subsequent development have been known for many years (see Swales 1935). Eggs hatch in water, releasing a ciliated larva (miracidium) that seeks out and penetrates an aquatic snail (species of the family Lymnaeidae). Multiplication occurs in the snail, and several hundred individuals of another larval form (cercariae) are released back into the water. Cercariae attach to aquatic plants and form a protective cyst around themselves. Cervids are infected when the aquatic vegetation is eaten.

In moose, bison, cattle, sheep and goats, maturing worms wander extensively in the liver, creating bloody tunnels or tracts. The resultant tissue damage is fatal to sheep and goats. In moose and bovids, considerable fibrous scar tissue and thick-walled capsules develop. The capsules may become calcified and are closed, preventing any eggs from leaving through bile ducts. A black-brown liquid-to-pasty material fills the capsules. Although eggs may be seen in this exudate, they do not appear in feces (Lankester 1974). Previous investigators reporting *F. magna* eggs in moose feces may have mistaken eggs of the rumen fluke *Paramphistomum* spp. that are similarly large. Infected moose livers may be twice the normal size and appear lumpy. Roughened, whitish, fibrous tissue and varying amounts of jet black pigment usually are visible on the surface. When the

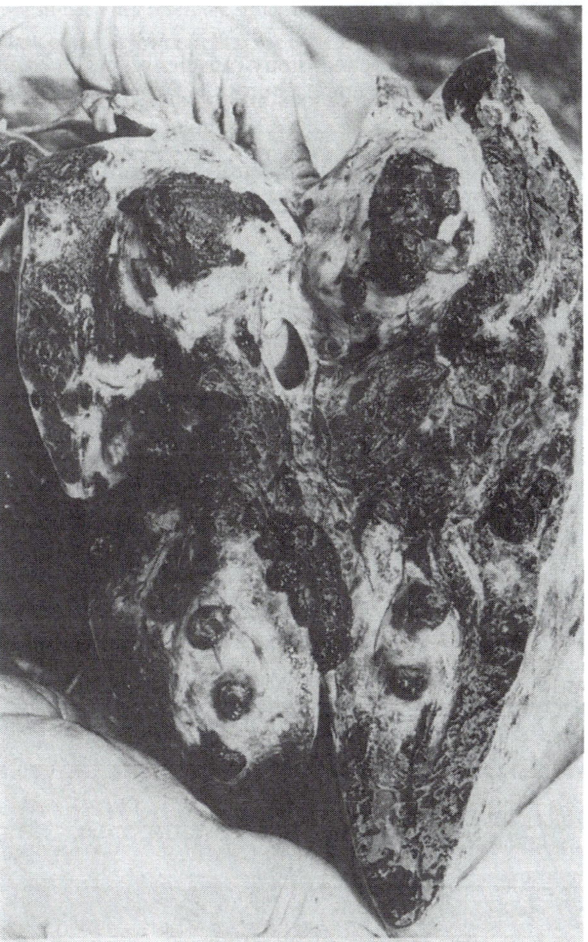

Large American liver flukes found in moose usually are immature, dark red in color and seldom larger than the smallest specimen shown above (left) in a caribou. Large American liver fluke infection in moose (right), commonly called "liver rot," features numerous blood-filled migration tunnels and thick-walled, fibrous cysts (often calcified) containing blackish, pasty material in the liver. Live flukes are found in moose only infrequently, despite the extensive tissue damage. Washing thick slices of damaged livers over a screen will reveal any flukes present. *Photos by M.W. Lankester.*

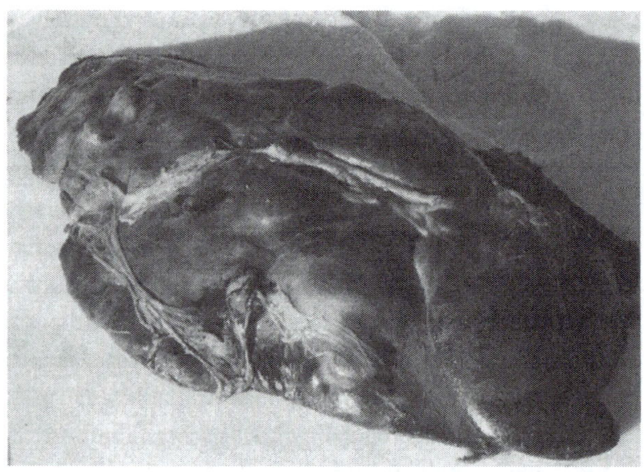

Whole moose liver with lumpiness and patches of black pigment indicate *Fascioloides magna* infection. Despite the infected liver's appearance, no clinical disease is clearly recognizable in most wild cervids. Some investigators have suggested that heavy infections accompanied by malnutrition may contribute to mortality. *Photo by M.W. Lankester.*

liver is cut, the thickened, stony capsules resist the knife. In recent infections, copious amounts of foul-smelling, blackish-brown material may flow out. Despite the appearance of the infected organ, no clinical disease attributable to *F. magna* is clearly recognizable in most wild cervids (Foreyt 1981, Wobeser et al. 1985, Lankester and Luttich 1988). However, several investigators have suggested that heavy infections accompanied by malnutrition, especially in deer and moose, may contribute to the death of animals (Fenstermacher and Olsen 1942, Cowan 1946a, Cheatum 1951).

The large American liver fluke is native to North America. Interestingly, it was first discovered and named in Italy by Brassi (1875), who found it in elk imported from North America to a zoological park near Turin (Erhardova-Kotrla 1971). Infected moose have been reported in Ontario, Manitoba, Alberta, British Columbia, Minnesota and Michigan (Lankester 1987).

Another fluke, *F. hepatica,* an inhabitant of liver bile ducts of domestic ruminants, has not been recovered from moose. However, serological tests by Brindle et al. (1979) suggest that some moose in Quebec have had contact with this parasite, probably by way of infected cattle. Whether this parasite is of any significance to the health of moose is unknown.

Paramphistomiasis

Two species of rumen flukes occur in North American moose. These include *Paramphistomum cervi* and *P. liorochis.* The former also is found in European cervids (see Kennedy et al. 1985). The latter species is indigenous to North America and probably evolved with white-tailed deer. Both are similar in appearance and can be separated only by detailed microscopic examination. Their individual distributions are not known, nor is the extent to which they may concurrently infect the same moose. Virtually all adult moose in an area can be found infected. On average, moose in northwestern Ontario each had about 3,500, with a few moose having more than 28,000 (Snider and Lankester 1986). Intensity of infection increases with age to about 4.5 years, and then begins to decline. No clinical disease has been attributed to the presence of rumen flukes, with the possible exception of unusually large numbers in calves (Seyfarth 1938).

Rumen flukes are pinkish in color. They aggregate in a pouch of the rumen adjacent to the outlet into the reticulum (Figure 194). Flukes remove the terminal paddle-like part of the papillae that line the rumen and attach by a posterior sucker to the remaining stub. When contracted, they

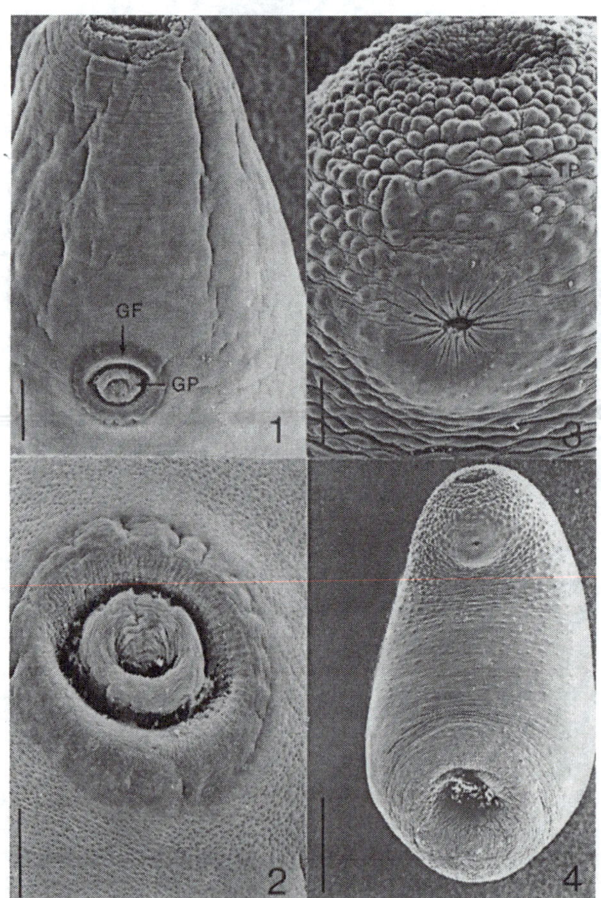

Scanning electron micrographs of two different species of rumen flukes occurring in moose distinguishes *Paramphistomum cervi* (1 and 2) from *P. liorchis* (3 and 4) by the appearance of the genital opening and the tegumental papillae seen in the latter. In 1, GF = genital fold; GP = genital pore. *Photo by M.J. Kennedy; courtesy of Alberta Agriculture, and reprinted from Kennedy et al. 1985.*

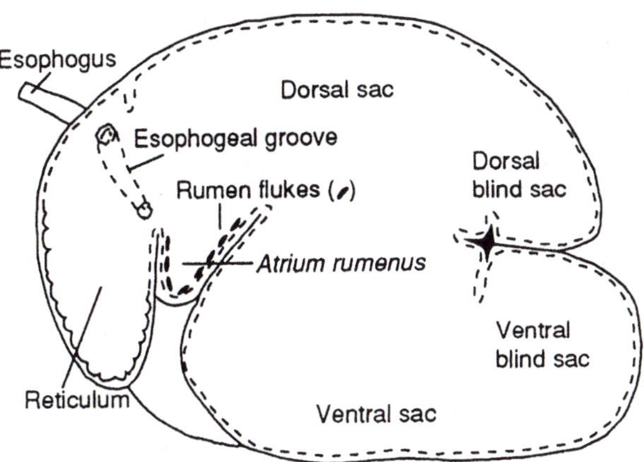

Figure 194. (Left) Lateral view of moose rumen showing location of rumen flukes, *Paramphistomum* spp., in the *atrium rumenus;* also showing the reticulum, esophagus, esophageal groove, and dorsal sac, dorsal blind sac, ventral sac and ventral blind sac of the rumen. *Illustration by J.B. Snider.*

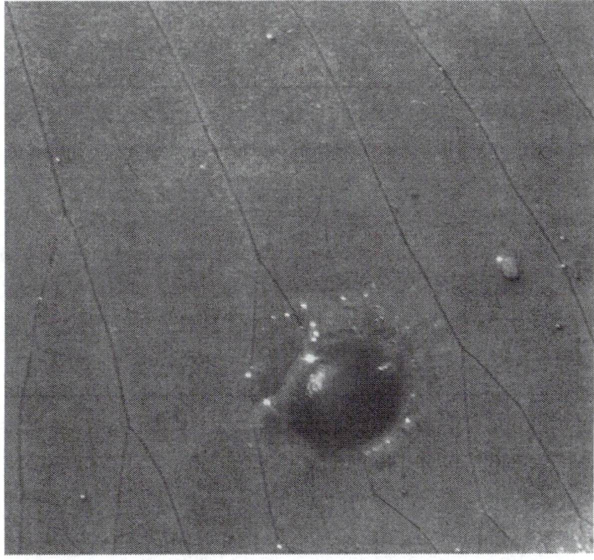

Metacercarial cysts of the moose rumen fluke, *Paramphistomum* spp. commonly occur on the underside of lilypads that moose consume. They resemble small, black, sand grains. (The large, dark object at lower right is the egg case of a leech.) *Photo by M.W. Lankester.*

are pear-shaped and about 0.3 inch long by 0.1 inch wide (7.3 by 2.6 mm). In warm and fresh rumen contents, active worms can stretch to twice that length.

The developmental cycle of *Paramphistomum* spp. is similar to that of the liver fluke. Eggs passed in feces hatch in water, and the miracidium penetrates a snail. Only the ramshorn snail (*Helisoma* spp.) is known to be a suitable intermediate host (Hoeve 1982, Snider 1985). Darkly pigmented cercariae emerge from infected snails during daylight. They swim toward light and encyst on the underside of floating vegetation, where they are subsequently eaten by moose.

This parasite has adapted efficiently to living in a north-

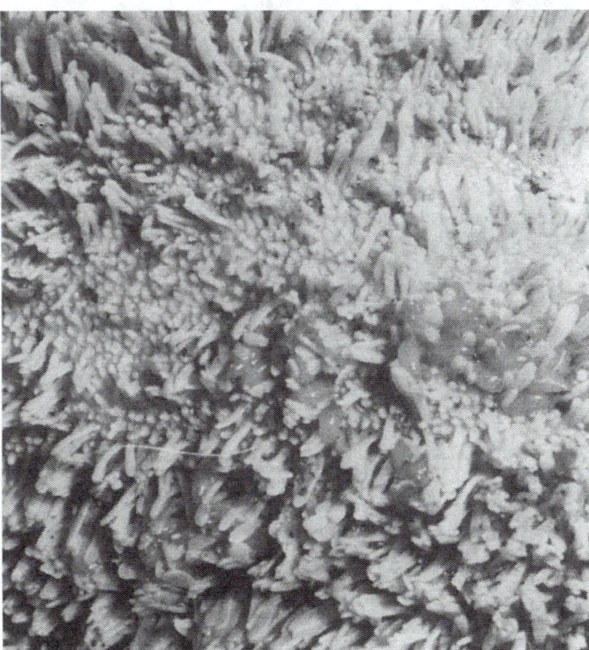

Pear-shaped rumen flukes, loose among rumen contents (left), attach to papillae lining the rumen (right) of moose. The paddle-like portion of the papillae is lost, and flukes attach to the base. *Left photo by M.W. Lankester. Right photo by H. R. Timmermann.*

ern host. No eggs are produced by worms during winter months, which represents a valuable energy-conserving adaptation because their eggs cannot survive freezing. Individual worms live at least 3 to 4 years. During their life, they repeatedly become gravid in spring and stop producing eggs in early autumn. Before experimentation was terminated, this cycle was repeated for 3 years in captive moose (Snider 1985). It is not known whether rumen flukes detect the onset of autumn and spring by seasonal changes in the moose's diet or as a result of cues sensed through changes in the host's hormones that are regulated by photoperiod.

The size of the moose rumen—containing up to 7.9 gallons (30 L) of ingested food—makes regular detection of flukes difficult. Undoubtedly, this has discouraged study of the parasite and its potential effects on wild moose. In domestic animals, the most severe pathology occurs shortly after infection, when ingested larvae (metacercariae) excyst in the duodenum and begin migrating to the forestomachs

Examining the contents of a moose rumen for paramphistome flukes, parasitologists face a daunting task, particularly if the subject is a large adult moose. Above, the rumen is from a small subadult. *Photo by M.W. Lankester.*

(Jubb et al. 1985). Establishment of adult flukes in the rumen and the removal of patches of papillae generally are thought to be of little consequence. However, Seyfarth (1938) suspected that the severe denudation of rumen papillae was responsible for the death of moose calves in Germany that were heavily infected with *P. cervi*.

Other Flukes

Zygocotyle lunata is a small (0.28 by 0.12 inch [7 by 3 mm]) fluke that commonly occurs in the intestine of ducks and, occasionally, in moose. As in paramphistomes and liver flukes, larvae of *Z. lunata* emerge from snails (*Helisoma* spp.) and encyst on aquatic vegetation. When ingested by moose, they pass along the digestive tract until reaching the cecum, where they mature and reside. *Zygocotyle* has been reported only from moose in Alberta and Ontario (Samuel et al. 1976, Stock and Barrett 1983, Hoeve et al. 1988). More diligent searching may show this parasite to be as common in moose as it is in the waterfowl with which they share aquatic areas. The numbers of worms found in moose usually is low to moderate (<120), and no known disease results. Worms, although smaller in moose than in waterfowl, reach maturity in moose and pass eggs.

Cestodes

Cestodes, or "tapeworms," are dorsoventrally flattened, whitish, cosmopolitan parasites that live as adults in the alimentary tract or associated ducts of many vertebrates. Moose of North America harbor six species of tapeworms, three of which live as nonpathogenic adults in the small intestine and bile ducts, and three that live as nonpathogenic larvae in various organs and muscle. This latter group—*Echinococcus granulosus, Taenia hydatigena* and *Taenia ovis krabbei*—is well known to hunters because these parasites are readily observed in the lung, liver and muscle, respectively (Samuel 1972a). Each autumn, wildlife agencies in moose country deal with two commonly asked questions regarding these parasites: "Is the meat fit to eat?" and "Can I become infected?" The answers are "yes" and "no," respectively. The adult stage of these tapeworms is found primarily in wolves and usually less so in coyotes, dogs and other carnivores. In general, wherever wolves and moose share range, these tapeworms are prevalent in moose (Table 45). With the possible exception of *E. granulosus* hydatid cysts in moose lungs, adult and larval tapeworms are not thought to harm moose.

Hydatid Cysts

The hydatid cyst is the larval stage of the tapeworm *E. granulosus*. Wolves are the final host of the parasite. The three-

Table 45. Prevalence of infected larval tapeworms in moose in parts of North America with or in the absence of wolves

Location	Percentage of moose infected			Source
	Echinococcus granulosus	*Taenia hydatigena*	*Taenia ovis krabbei*	
With wolves				
Alaska	20			Rausch 1959
Alberta[a]	52	63	47	Samuel et al. 1976
	73	61		Pybus 1990
British Columbia	68	84		Ritcey and Edwards 1958
Minnesota	40	40		Wallace 1934
Ontario[b]	67	74	74	Addison et al. 1979
Quebec	44			McNeill and Rau 1987
Saskatchewan	30			Harper et al. 1955
Without wolves				
Alberta[c]	0	13	0	Samuel et al. 1976
	0	5	0	Stock and Barrett 1983
Alberta[d]	15	37	0	Samuel et al. 1976
Newfoundland	0	9		Threlfall 1967, 1969
Ontario[e]	0	0	0	Hoeve et al. 1988

[a] Boreal Forest (northern Alberta).

[b] Boreal Forest (northern Ontario).

[c] Cypress Hills Provincial Park (southeastern Alberta).

[d] Elk Island National Park (central Alberta).

[e] Larose Forest (eastern Ontario).

segmented adult tapeworms are 0.12 to 0.24 inch (3–6 mm) long and occur in the intestines. In North America, the wild ungulates that most commonly serve as intermediate hosts for the hydatids are moose, elk and barren ground caribou. They become infected by eating eggs associated with canid feces or contaminated vegetation (Figure 195). Hydatid cysts, found primarily in the lungs but occasionally in the liver, vary in size from the shape and size of a pea to that of a golf ball, and have an average weight of 1 ounce (3 g) (Messier et al. 1989). They contain thousands of larval forms (protoscoleces) resembling sand grains.

Hydatids are common in lungs of moose from areas with wolves; they generally are absent from areas without wolves (see Anderson and Lankester 1974, Lankester 1987; Table 45). McNeill and Rau (1987) and Messier et al. (1989) reported that prevalence, numbers of hydatid cysts and mean total weight of cysts increased with moose age in Quebec, suggesting that moose acquire infection throughout their lives. These indicators of parasite population size also increase with moose density (Messier et al. 1989)—a response concomitant with increasing wolf densities and predation rates on moose.

Hydatid cysts in the lung apparently render the host more susceptible to predation by reducing stamina of the host during pursuit. McNeill and Rau (1987: 420) stated that "a pulmonary infection, which may be of little consequence in a resting or browsing moose, may compromise the vigor of a stressed animal." Messier et al. (1989: 218) added: "We anticipate that individuals harboring heavy pulmonary infections are recognized as vulnerable by wolves."

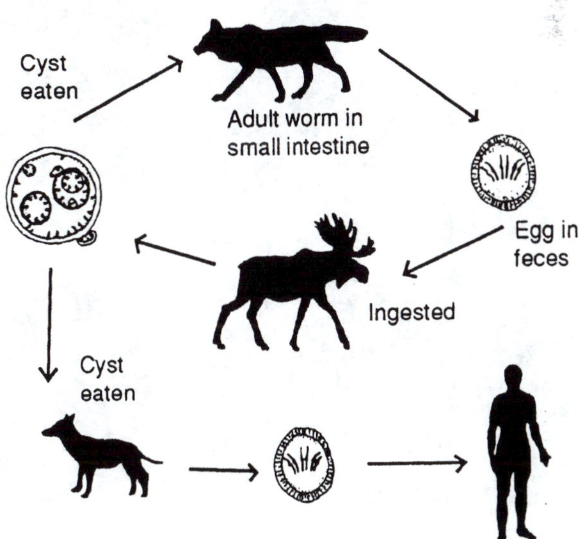

Figure 195. Life cycle of the tapeworm *Echinococcus granulosus*. Only the eggs of the tapeworm in feces of wolves, coyote or domestic dogs are infectious to humans. *Illustration by J.L. Samuel.*

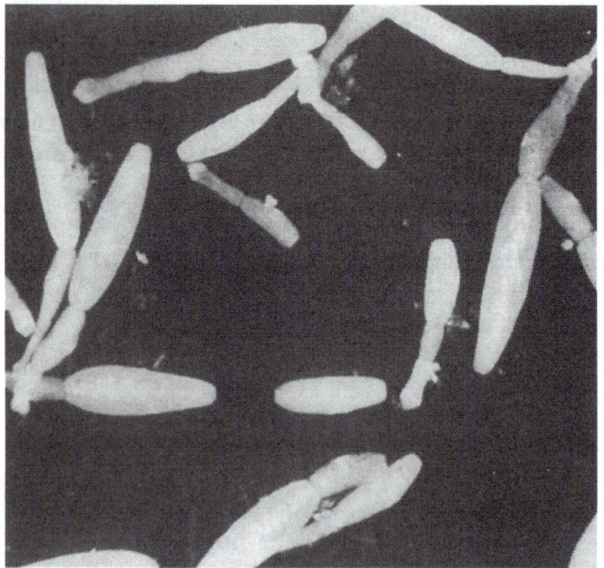

Adult *Echinococcus granulosus* tapeworms from the intestine of a wolf. Eggs passed in wolf feces and scattered by rain on vegetation, develop into hydatid cysts after consumption by moose. *Photo courtesy of the Ontario Ministry of Natural Resources.*

E. granulosus is the only tapeworm of moose that can infect humans, but humans do *not* become infected through direct handling of moose carcasses. Only the eggs of *E. granulosus* passed in canid feces are infectious to humans. Dogs generally are considered to be the source of human infection. Thus, fresh lungs and liver of moose should not be fed to dogs (complete freezing or cooking kills tapeworm larvae). In addition, precautions should be taken (e.g., wearing gloves and mask) when handling wolves or coyotes and especially their feces. A detailed review of the biology of this parasite was completed by Rausch (1993).

Cysticerci

Two additional tapeworms, *Taenia hydatigena* and *T. ovis krabbei,* mature in the intestine of wild canids and use moose as an intermediate host. Their life cycles are similar to that of *E. granulosus,* except that their larvae, called "cysticerci," are found in the liver and muscles, respectively. Cysticerci of *T. hydatigena* are marble-sized and found beneath the membrane covering the surface of the liver. The cysticerci of *T. o. krabbei* are smaller (pea sized), and located in muscle of the heart (Addison et al. 1979) and often throughout skeletal muscle (Samuel 1972b). As with *E. granu-*

Golfball-sized hydatid cysts may be seen on the surface of moose lung (left) and throughout the parenchyma of cut moose lung (right). *Echinococcus granulosus* is the only tapeworm of moose that can infect humans, but humans are not infected by handling a moose carcass. Only the eggs passed in canid feces are infectious to humans, so fresh lung and liver from moose should not be fed to dogs. *Left photo by H.R. Timmermann; courtesy of the Ontario Ministry of Natural Resources. Right photo by M.W. Lankester.*

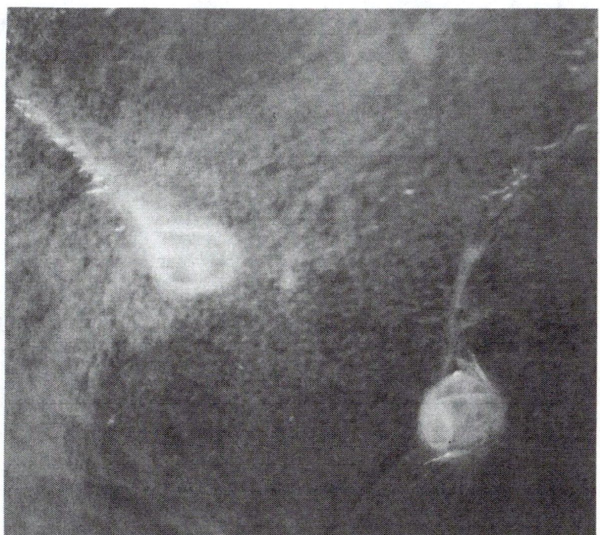
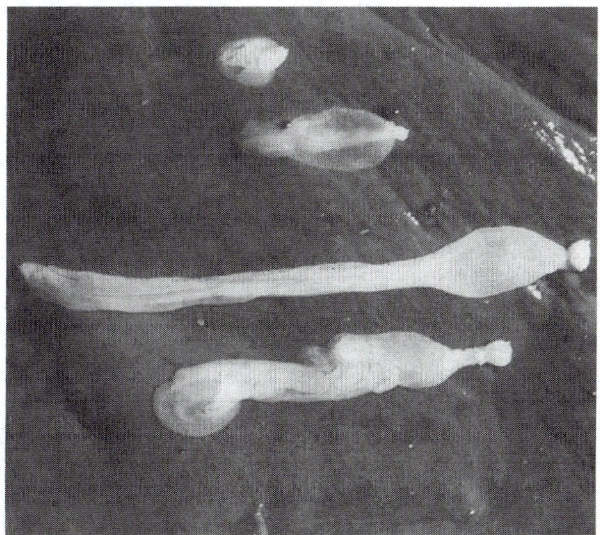

Cysticerci of *Taenia hydatigena* (thick-necked bladder worms) may be visible beneath the serosal membrane covering the liver of moose (left; note the migratory tract made by the lower right parasite). Wolves and coyotes are final hosts of this tapeworm. The cysts in moose may be aesthetically displeasing to hunters (right; removed cysts on a moose liver surface), but they are not infective to humans. *Left photo by W.M. Samuel. Right photo by E.M. Addison; courtesy of the Ontario Ministry of Natural Resources.*

losus, the prevalence and intensity of these parasites are correlated positively with host age (Addison et al. 1979).

Tapeworm cysticerci probably are not significant pathogens for moose. Addison et al. (1979) found no relationship between numbers of parasites and physical condition of moose. However, Messier et al.'s (1989) aforementioned "anticipation" for *E. granulosus* also may apply for moose with very high numbers of *Taenia o. krabbei* cysticerci. The presence of cysts in muscle or liver may be aesthetically displeasing to hunters, but the fact that these parasites *cannot* infect humans must be emphasized.

Moniezia spp. and the Fringed Tapeworm

Adult *Moniezia* spp. and the fringed tapeworm *Thysanosoma actinioides* are common parasites of many wild and domestic ruminants. Seldom are fringed tapeworms longer than about 1.0 foot (0.3 m), but *Monezia* species may be 3 to 10 feet

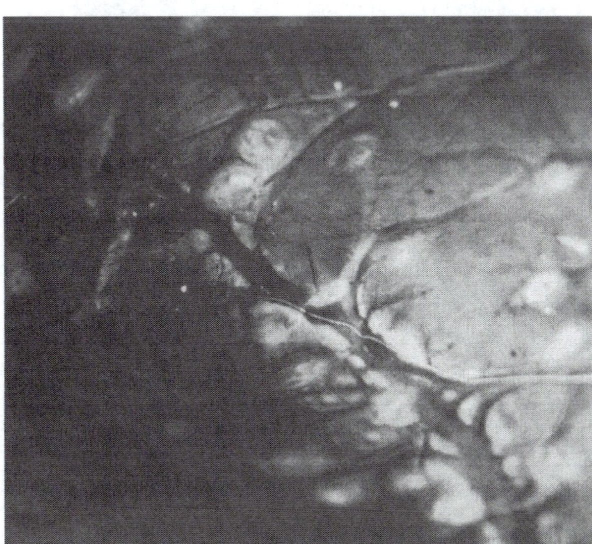
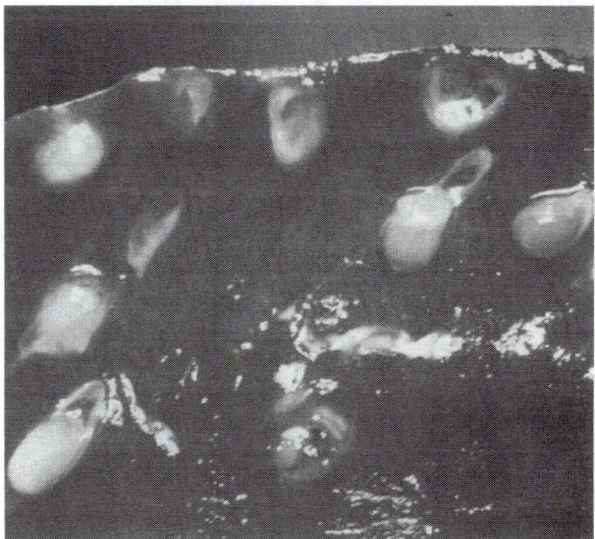

Cysticerci of *Taenia ovis krabbei* may be visible on a whole moose heart (left) and throughout cardiac muscle of a cross-sectioned ventricle (right). The presence and intensity of these parasites are correlated positively with host age. This parasite generally is not considered serious for moose because it does not invade muscle cells (note the lack of muscle pathology around the parasites). *Left photo by H.R. Timmermann; courtesy of the Ontario Ministry of Natural Resources. Right photo by E.M. Addison; courtesy of the Ontario Ministry of Natural Resources.*

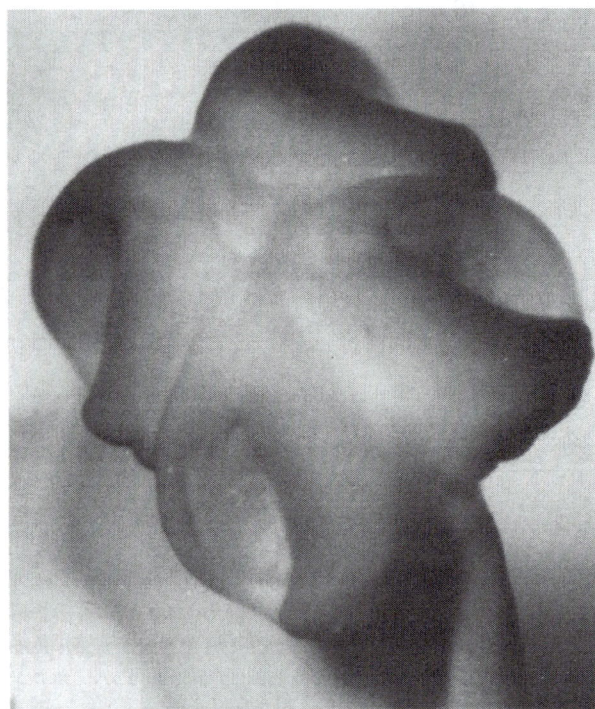

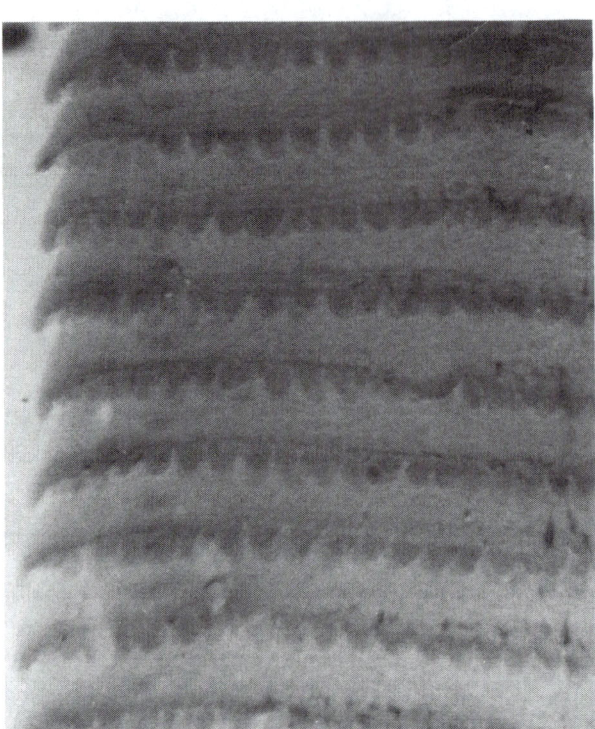

A fringed tapeworm, *Thysanosoma actinioides*, taken from a moose bile duct features four prominent suckers of the head or scolex (left) and "fringes" along the posterior border of each segment (proglottid) (right). No grossly visible pathology has been reported in the bile duct or small intestine of infected moose. *Photos by R. Mandryk; courtesy of the University of Alberta.*

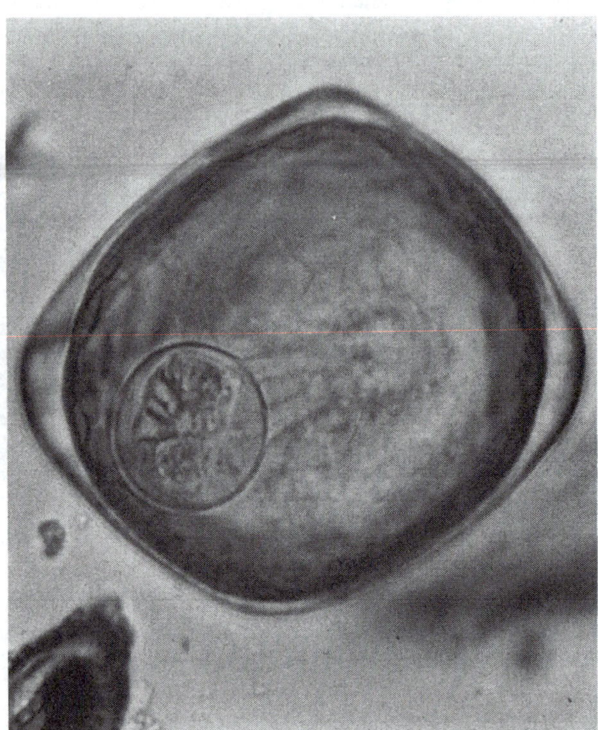

Within the eggs of the tapeworm *Moniezia* spp.—the only large tapeworm in moose—is the embryo of *Moniezia* with six hooklets. Numbers of adult worms in moose are few, hence, infection is considered benign. *Photo by W.M. Samuel.*

(0.9–3.0 m) in length (Thorne et al. 1982). Eggs escaping from tapeworm segments (proglottids) passed in host feces can be detected using a variety of flotation techniques. The larva contained within the egg must be eaten by a small invertebrate. Free-living mites are the required intermediate host of *Moniezia*, whereas psocopterous insects (barklice) probably transmit the fringed tapeworm. Both North American species of *Moniezia*—*M. benedeni* and *M. expansa*—have been reported from the intestine of moose, but neither is considered harmful. *Thysanosoma actinioides*, which lives in the bile ducts and small intestine, is the least common. Stock and Barrett (1983) found up to 48 fringed tapeworms in a moose, but suggested that infection is benign.

Nematodes

Nematodes, or roundworms, are elongate, thread-like worms varying in length from about 0.25 to 4.0 inches (0.6–10.2 cm). The development of all species is basically the same. They hatch from an egg and develop through four larval stages, each concluded by shedding or molting the external cuticle. The fifth stage matures into an adult male or female worm. Variations on this theme are seen in different nematodes parasitizing moose. The elaphostrongyline, or tissue nematodes of cervids, undergo the first three larval stages in terrestrial snails or slugs. Moose are infected when they ingest these gastropods along with vegetation. The lungworm *Dictyocaulus* and the gastrointestinal nematodes develop to the third stage in soil and litter. They too are ingested with vegetation.

In the case of filarioid nematodes, first-stage larvae (microfilariae) are drawn from an infected host in blood taken by various biting flies (female flies require blood protein to develop their eggs). In the fly, microfilariae develop through two molts to the third, infective stage that is injected into another moose when the fly takes its next blood meal. Adult filarioids mature in a variety of locations within their mammal hosts; females usually are situated close to blood or lymphatic vessels where they can release their microfilariae.

The nematodes of greatest importance to moose health are the brain worm, or meningeal worm *Parelaphostrongylus tenuis,* and to a more limited extent, the European tissue worm *Elaphopstrongylus* spp.

Parelaphostrongylosis

A neurologic disease known as "moose sickness" (parelaphostrongylosis) is caused by the meningeal or brain worm *P. tenuis*. The original common name "meningeal worm" refers to the location of the adult worms on or in the meninges, the membranes that cover the brain and spinal cord of mammal hosts.

Sick moose may show any or all of the following symptoms: weakness in the hindquarters; turning of the head and neck to one side (torticollis); fearlessness; lethargy; rapid eye movement; apparent blindness; circling; or inability to stand. White-tailed deer are the usual host of the parasite; 50 to 90 percent may be infected, each with an average of three worms, but signs of disease are rare.

Parelaphostrongylosis probably contributed to some of the moose declines reported in eastern North America in the first half of this century, but whether it actually was a major cause appears impossible to prove (Whitlaw and Lankester 1994a). Its significance today is nearly equally difficult to determine. In some areas managed largely by hunting, deer and moose numbers can remain fairly stable over a decade or so (Whitlaw and Lankester 1994b). In others, as deer increase in number, moose decline, and when deer decrease, moose increase. The role played by *P. tenuis* in this inverse relationship still is the subject of ongoing research, and because the parasite can kill moose and several other cervids native to North America, it is dealt with in some detail here.

The parasite occurs in white-tailed deer of the eastern deciduous forest biome and deciduous/coniferous ecotone of eastern and central North America. It is absent in the West. The precise limits of its most westerly distribution are poorly known, but it has been found in white-tailed deer in western Manitoba and in the United States east of a line projected south from western Minnesota, through central Oklahoma and into the extreme eastern portion of Texas. The central grasslands may have been a barrier preventing *P. tenuis* from moving westward with whitetails. For detailed reviews, the reader is referred to Anderson (1971, 1972), Anderson and Lankester (1974), Anderson and Prestwood (1981), Lankester (1987), Samuel (1991b), and Anderson (1992).

In white-tailed deer, infective third-stage larvae of *P. tenuis* are released from the foot tissue of an infected gastropod that has been incidentally ingested with food taken low to the ground. They penetrate the wall of the true stomach (abomasum) and enter the abdominal cavity (Anderson and Strelive 1967) (Figure 196). Migrating along nerves in the body wall toward the back, the larvae take about 10 days to reach the vertebral canal. They enter nerve

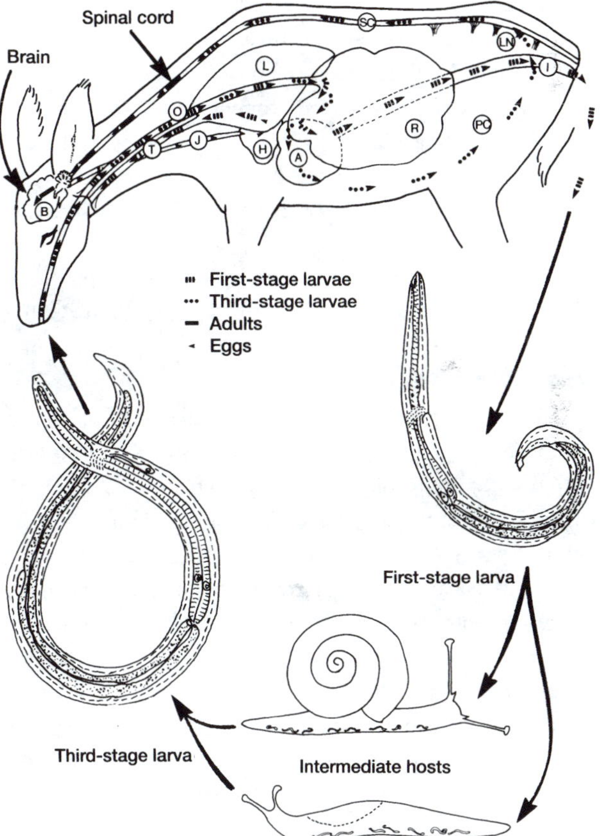

Figure 196. Life cycle of the brain worm *Parelaphostrongylus tenuis*. B = brain; T = trachea; O = esophagus; J = jugular; L = lung; H = heart; A = abomasum; SC = spinal cord; R = rumen; PC = peritoneal cavity; LN = lumbar nerve; and I = intestine. *Illustration by U.R. Strelive, from Anderson (1992); reprinted with the permission of Commonwealth Agricultural Bureau, Cambridge.*

tissue of the spinal cord and begin to develop in the dorsal horns of the gray matter. The third-stage molts to the fourth, and the fourth molts again to the early fifth or subadult stage.

Larvae grow from a length of about 0.06 inch (0.16 cm) to about 3 inches (7.6 cm) as adults. By 40 days after infection, most of the maturing males and females have left the spinal cord and are found in the fluid-filled space between the cord and the covering dura membrane (Anderson 1963, 1965b). They migrate anteriorly into the cranium over the brain and become associated with large veins and venous blood sinuses in the dura. Worms commonly are located within the cavernous and intercavernous sinuses in the floor of the cranium surrounding the pituitary gland.

Female worms release eggs from the vulva near the posterior end of the body. The eggs are swept away in the blood and travel by way of the jugular vein to the right side of the heart and then through the pulmonary artery to the lungs. In the lungs, eggs in the one- or two-cell stage lodge in fine blood capillaries and are surrounded by fibrous tissue, forming tiny nodules. A first-stage larva develops within the entrapped egg and eventually breaks out into the air spaces of the lung. They are carried up out of the lungs in mucous moved by cilia, the so-called bronchial escalator that lines the airways. Reaching the oral cavity, they are swallowed and can be found in the feces 90 to 137 days after a deer gets infected (Rickard et al. 1994). First-stage larvae in feces are about 0.01 inch (0.3 mm) long and have a crooked, pointed tail with a subterminal dorsal spine. Young, recently infected deer pass more larvae than do older animals, but most larvae are passed in spring by deer of all ages (Peterson and Lankester 1991).

Larvae occur only on the outside of the deer fecal pellet, in the covering layer of mucous (Lankester and Anderson 1968). They are readily removed by rain and melting snow, and disperse in the soil. Being resistant to deep freezing and limited periods of drying (Shostak and Samuel 1984), some probably can survive for several months. These larvae must penetrate into the foot tissue of a terrestrial snail or slug (gastropods) to develop the infective, third-stage larva.

A number of different terrestrial snails and slugs are capable of serving as intermediate hosts, but a few species in particular probably are most important as sources of infection in the wild. These include the small, dark brown to black, native slug *Deroceras laeve; D. reticulatum,* a slug introduced from Europe; and the small woodland snails, *Discus cronkhitei, Zonitoides* spp., *Succinea* spp. and *Cochlicopa* spp. (see Lankester and Anderson 1968, Platt 1989, Lankester and Peterson 1996). All are fairly common and widely distributed throughout most of deer range in eastern North America. Transmission to deer may occur mostly in autumn when larvae acquired by gastropods during spring and summer have reached the infective stage and when young, susceptible deer feed low to the ground. *Deroceras laeve,* for example, remains active in autumn until temperatures are close to freezing (Lankester and Peterson 1996). Third-stage larvae, however, can survive winter in gastropods (Lankester and Anderson 1968, Platt 1989). Therefore, any species of gastropod that lives for 2 or 3 years can be a source of infection throughout the snow-free seasons whenever conditions are suitable for gastropod movement.

The prevalence (frequency) of *P. tenuis* infection in gastropods is dependent on a number of factors, including the density and residence time of deer disseminating larvae. For example, on Navy Island in southern Ontario, where deer existed year-round at a high density of 120 per square mile (46/km^2), almost 5 percent of snails and slugs were infected (Lankester and Anderson 1968). In northeastern Minnesota, where the density of whitetails on summer range was around 10 per square mile (4/km^2), only 0.04 percent of gastropods were infected (Lankester and Peterson 1996). Infection rates were four times higher (0.16 percent) in a wintering yard where deer aggregated at 130 per square mile (50/km^2) for almost 5 months each year.

The lower rate of gastropod infection near the northern limits of deer range may be related to lower seasonal temperatures, which could affect survivorship of first-stage larvae and shorten periods of gastropod activity. Nonetheless, most young deer in the northern Minnesota population still became infected by the end of their first or second summer (Slomke et al. 1995). This probably is explained simply by the large volume of vegetation eaten by deer, particularly during autumn when much of it is taken close to the ground while certain species of gastropods still are active (Lankester and Peterson 1996).

Infected gastropods usually harbor only a small number of third-stage larvae—each has about three on average. However, *Deroceras laeve* has been found on a few occasions with many more (up to 97). This particular slug may be more likely than other gastropods to be attracted to fresh deer feces, rather than encountering *P. tenuis* larvae on dried feces or dispersed in the soil.

Almost 90 percent of deer more than 1 year old in northeastern Minnesota are infected. Yearlings have an average of three adult worms in the cranium, and 7- to 15-year-olds have an average of four (Slomke et al. 1995). Obviously, infection does not build up over time. In fact, because nematodes closely related to *P. tenuis* are known to be long lived, it is hypothesized that worms acquired in the early life of a deer are the same ones found there years later. Therefore, meningeal worms in a deer are approximately the same age as the deer itself. Supporting evidence for this belief is pro-

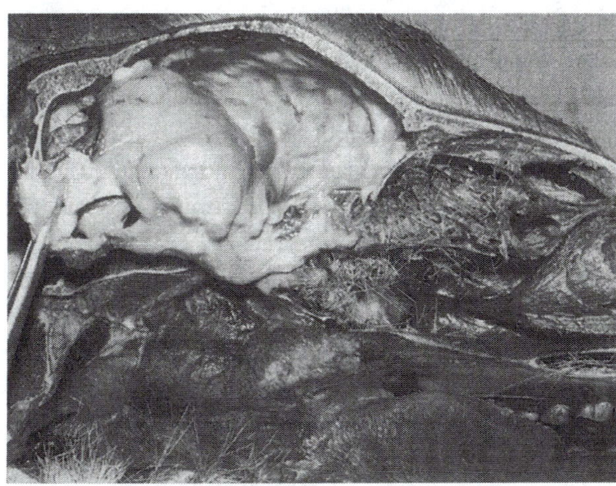

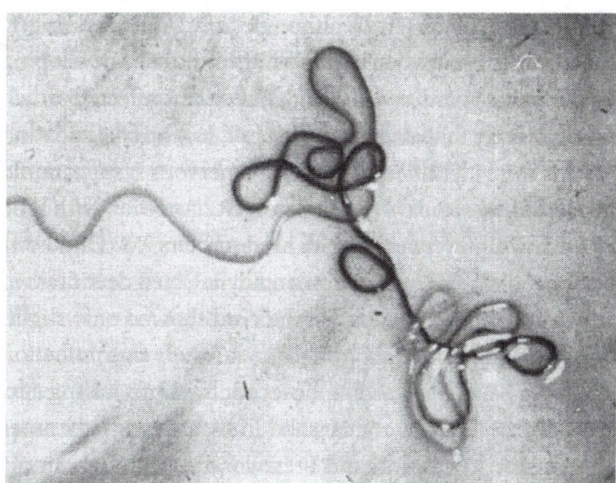

Above left is the head of a white-tailed deer cut sagittally and the brain resected to reveal a meningeal worm *Parelaphostrongylus tenuis* (arrow), on the roof of the cranial vault within the subdural space. Above right is a close up of the same nematode, which is entangled by fibrinous strands loosely attaching the parasite to the underside of the dura membrane. *Photos by William Rees.*

vided in the form of another curious observation. As many as one-third of infected deer have brain worms of only one sex and, of course, no larvae are passed in feces. If repeated infection were possible, both male and female worms eventually would be expected in all deer. But, the proportion of these sterile, unisexual infections is the same in both young and old animals. This suggests that an initial exposure to the parasite promotes a fairly effective immunity to further infection. Because the average number of worms in the head of a deer is similar to the average number of infective larvae per snail, protection may begin to develop shortly after the ingestion of a single infected gastropod (Slomke et al. 1995).

Deer infected in their first summer probably are refractory to further infection by the following spring. For this reason, winter deer yards with higher densities of infected gastropods may be relatively unimportant in the transmission of the parasite to deer. The ground usually is snow covered when deer aggregate in autumn and, by the time snails and slugs are available in spring, many deer may be immune (Lankester and Peterson 1996). Increased knowledge of this parasite's biology in its normal host, the white-tailed deer, has greatly improved assessments of its impact on other species.

P. tenuis causes neurologic disease in moose as well as in several other cervids and bovids. Why the disease is more pronounced in these animals, compared with deer, is not entirely clear. In moose, for example, worms develop in the gray matter of the spinal cord at about the same rate and reach the same maximum size as they do in deer (Anderson 1964, 1965b). However, the amount of tissue damage and host cellular reaction are somewhat greater in moose, and worms remain longer than 40 days in the nerve tissue while continuing to grow. Also, deer have a more effective natural or acquired immune protection that kills invading larvae. In experimental infections using very large doses, the success of infective larvae reaching the spinal cord was less in deer than in moose (Anderson 1963, 1964, 1965b). The ratio of the number of larvae given to that reaching the cord was on the order of 20:1 in deer and 5:1 in moose.

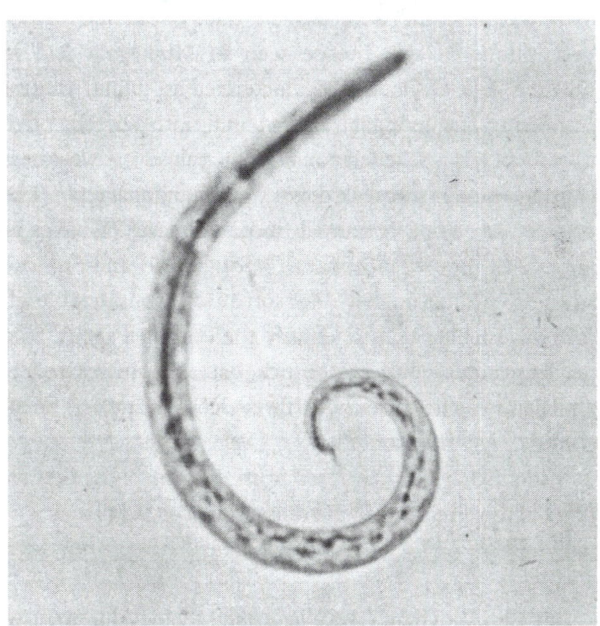

A first-stage larva of *Parelaphostrongylus tenuis* passed in the feces of an infected deer clearly shows the bend or crook near the end of the tail, plus a small projection or spine that gives the larva its descriptive name, "dorsal-spined larva," is barely visible at this magnification. It is located at the point where the terminal tail crook starts. *Photo by R.C. Anderson; courtesy of the University of Guelph.*

Further experiments, although using large dosages, showed that even when the numbers of *P. tenuis* developing in the spinal cord are similar, moose are affected much more severely than deer. A moose calf, in which 41 subadult worms were identified in the central nervous system, could not stand after 60 days (Anderson 1964). Another, with nine worms, was very weak in the hindquarters (M. Lankester personal files). Yet, an experimentally infected deer fawn in which 26 larvae developed in the cord showed only slight, brief signs of lameness in the fourth week after infection (Anderson 1965b). Also in moose, adult worms having successfully reached the cranial subdural space may later enter brain tissue (Anderson and Prestwood 1981). Such an occurrence may explain some cases of moose sickness seen late in winter—4 to 6 months after gastropods were last available.

Anderson (1970) suggested that the greater tolerance of white-tailed deer to infection is indicative of a long coevolution between this host and the parasite. Susceptible species, such as moose, caribou, elk and fallow deer, are relatively recent arrivals from Eurasia. Mule deer and black-tailed deer, although native to North America, also are susceptible. But these western species of deer presumably always have been isolated from this eastern parasite.

An unexplained neurologic disease in moose was identified as early as 1912 (Fenstermacher and Jellison 1933). Thomas and Cahn (1932: 230) provided one of the earliest descriptions of sick moose seen in Minnesota and in Quetico Park of Ontario, characterized as "blind, aimless wandering, staggering, bumping blindly into perfectly obvious obstacles, weakened condition, paralysis . . . weakness causing animals to break down in the hindquarters." The disease was variously named "moose disease" (Wallace et al. 1933), "moose encephalitis" (King 1939) and "moose sickness" (Cameron 1949, Benson 1958). Considerable effort was spent trying to identify the causative agent. Suspects included moose winter tick, bacteria, viruses, toxins and dietary deficiencies, but little evidence of any was forthcoming (see Anderson 1964).

More than 50 years elapsed from the time of the first reports of the disease until the cause of moose sickness was finally revealed by R.C. Anderson. The discovery unfolded stepwise.

Initially, Anderson (1963) described the development of the then little-known meningeal worm, *P. tenuis,* occurring in the cranium of white-tailed deer. In this host, he found that the worm required an initial period of development within the spinal cord. Therefore, it had the potential to cause nerve damage in deer or related hosts. Analyzing published reports of the mysterious neurologic disease of moose, Anderson recognized that it only occurred in areas where moose shared range with deer. He also noted that earlier investigators had reported the presence of unidentified nematodes in the brain of sick moose, but had not attributed any particular significance to those findings.

These pieces of information led Anderson (1964) to hypothesize that *P. tenuis* might be the causative agent. The hypothesis was quickly validated by an experiment in which two moose calves were given infective larvae of *P. tenuis*. Both calves developed the classic signs of moose sickness. These experimental findings were verified shortly thereafter by recovering adult *P. tenuis* from the brains of sick moose collected from several areas occupied by deer (Smith et al. 1964, Anderson 1965a, Loken et al. 1965).

Earlier investigators had associated the occurrence of moose sickness with declines in moose numbers, particularly in Maine and Nova Scotia (Lamson 1941, Cameron 1949, Benson 1958). But finding that the agent of the disease was a parasite spread by white-tailed deer for the first time pointed to a possible relationship between moose sickness and high deer densities.

Following Anderson's discovery, Telfer (1967b) suggested that parelaphostrongylosis, coincident with high deer populations, may have caused the moose decline seen in Nova

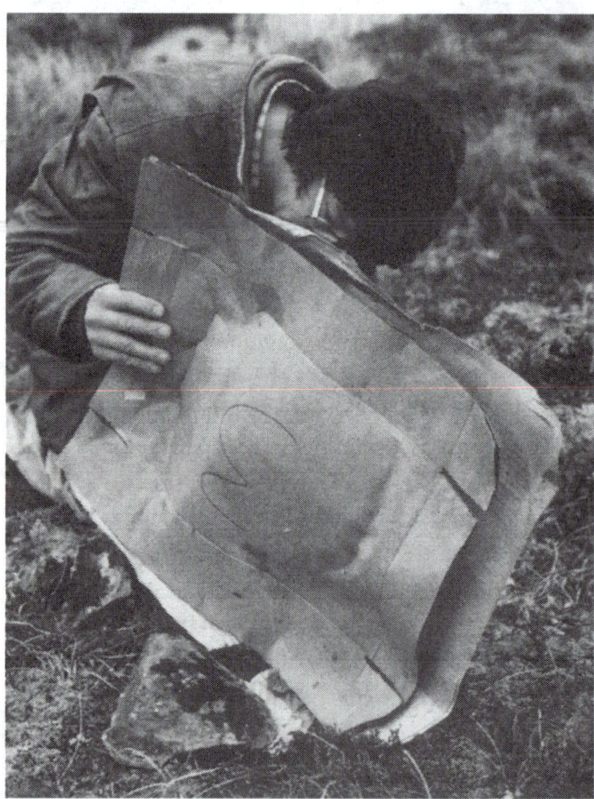

Checking beneath moistened cardboard sheets is a simple but effective method of quantitatively sampling gastropod species on ground vegetation. *Photo by M.W. Lankester.*

Scotia during the late 1940s and early 1950s. Karns (1967) suggested that high densities of infected deer were responsible for a major moose population decline in Minnesota in the mid-1920s and early 1930s. Subsequently, Prescott (1974) concluded that *P. tenuis* had had a greater impact on moose populations in eastern North America than any other factor. This explanation for historic moose declines generally has been accepted for a number of years (see Lankester 1987), but increasingly, there are good reasons to examine the idea more closely (Nudds 1990, Whitlaw and Lankester 1994a, 1994b).

During the 1980s, moose and deer increased simultaneously in some eastern provinces and states (Clarke and Bowyer 1987, Upshall et al. 1987, Thomas and Dodds 1988, Bogaczyk 1990, Boer 1992b). Noting the increase in moose in Nova Scotia and finding *P. tenuis*-like larvae in moose feces, some investigators were led to suggest that moose may have evolved a greater tolerance, or that the parasite had become less pathogenic than in former times (Clarke and Bowyer 1987). However, estimates of moose and deer numbers in these areas were mostly anecdotal, and the extent to which the two cervids occupied precisely the same areas was not known. Also, dorsal-spined larvae had been reported previously in moose feces (Anderson 1965a), and the surprisingly high number of moose found passing larvae (Clarke and Bowyer 1987) could not be confirmed by others (see McCollough and Pollard 1993). Nonetheless, an increase in moose numbers in areas with deer was inconsistent with prevailing ideas about the parasite's pathogenicity in moose.

Other investigators were similarly motivated to explain how moose sometimes are able to coexist in the presence of what was considered to be a mortal disease. Various hypotheses were put forth to suggest how moose might be separated temporally or spatially from habitat used by infected deer. These include areas where the two species were proposed to be separated by using habitat at different elevations for part of the year (Telfer 1967b); the existence of "refugia" where moose might experience a lower rate of infection (Telfer 1967b, Gilbert 1974); and habitat heterogeneity that reduced overlapping use of the habitat (Kearney and Gilbert 1976). Some of the evidence was compelling. Nonetheless, any expectation these researchers may have had that some form of isolation was required to explain the persistence of sympatric moose probably stemmed from the belief that the parasite is extremely pathogenic to moose. If infection is always lethal, then it is reasonable to expect all resident moose in an area eventually would disappear, unless they were isolated from infection to some degree.

Both Gilbert (1974) in Maine and Karns (1967) in Min-

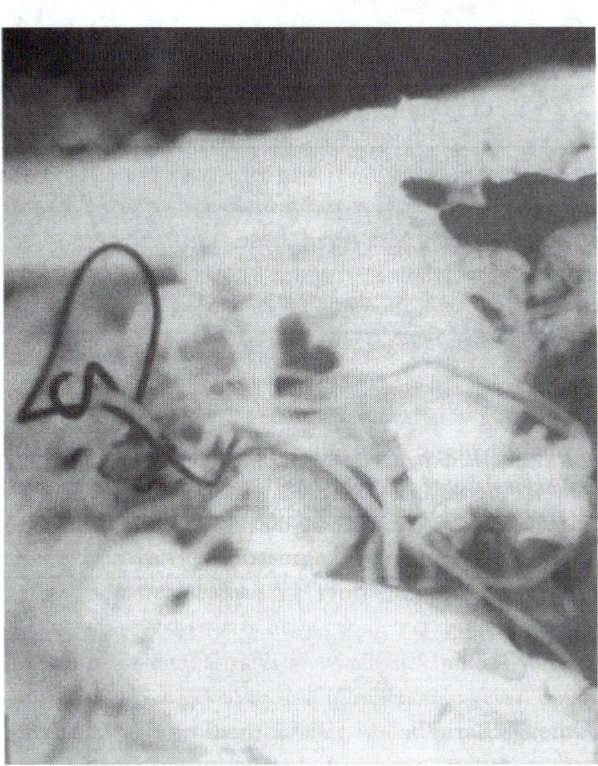

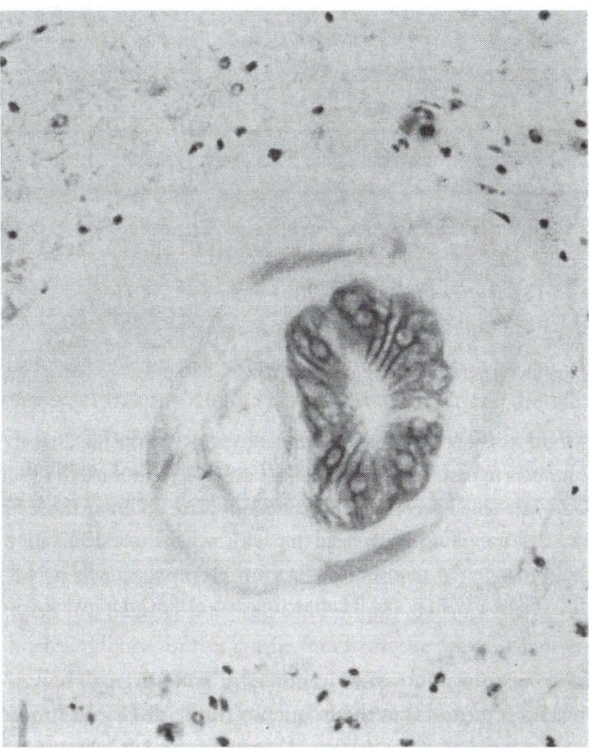

A knot of four or five *Parelaphostrongylus tenuis* (left) on the floor of the cranium (intercavernous sinus) of a moose that exhibited the signs of neurologic disease. At right is a cross section of a meningeal worm in white matter of the spinal cord of a moose showing physical damage to surrounding tissue. *Left photo by M.W. Lankester. Right photo by R.C. Anderson.*

Moose sickness is a neurologic disease caused by the meningeal (brain) worm *Parelaphostrongylus tenuis*. Historic declines in moose numbers in eastern North America have been attributed to this disease. However, reevaluation of the circumstances and recent findings on white-tailed deer/moose interactions place doubt on those findings but do not disprove them. Among the clinical symptoms of moose sickness are tilting of the head (top left), weakness or debilitation of hindquarters (top right), and circling movement, as indicated in the bottom scene by moose tracks in snow observed aerially. *Top left photo by W.J. Peterson. Top right photo by M.W. Lankester. Bottom photo by R. Williams, W. May and P. Lattner; courtesy of the Ontario Ministry of Natural Resources.*

nesota reported that the frequency of the disease in moose was related to deer density. However, the long history of moose sickness in Minnesota may have lessened the likelihood that Karns (1967) would contemplate the existence of isolating mechanisms to explain the persistence of moose. Although Karns believed that *P. tenuis* was responsible for moose declines seen in Minnesota when deer numbers were high, he felt that moose could be expected to remain at acceptable levels, from a manager's point of view, as long as deer did not exceed 12 per square mile (4.6/km^2).

Roy C. Anderson and assistant Murray Lankester, in 1963, with one of the two experimentally infected moose calves used to prove that *P. tenuis* from white-tailed deer was the cause of moose sickness. *Photo by Don Robinson.*

In 1985, Michigan reintroduced moose. Despite *P. tenuis* initially causing 38 percent of the observed mortality, the small moose population has continued to increase, despite a deer population density of about 11 deer per square mile (4.3/km²). At least 65 percent of these deer were shown to pass dorsal-spined larvae in their droppings (Schmitt and Aho 1988, Aho and Hendrickson 1989). The introduced moose experienced high twinning rates, and no wolf or bear predation was suspected. This valuable experiment demonstrated that a protected moose population can grow, despite some mortality attributable to *P. tenuis,* at least when other factors are favorable.

Whitlaw and Lankester (1994a, 1994b) reexamined historic and more recent data on changes in cervid numbers coincident with reports of sick moose. A total of 372 cases of moose sickness had been reported during the previous 90 years, with most (329) from Minnesota, New Brunswick, Nova Scotia and Maine. Sick animals were reported during 13 discrete time intervals—almost always when deer numbers were higher than normal for the area.

Moose declined when deer densities were greater than 13 per square mile (5/km²), but some declines occurred when no sick moose were reported. Sick animals were seen in periods when moose numbers were decreasing, but they also were seen when moose numbers were stable, as well as when they were increasing. Furthermore, Whitlaw and Lankester (1994a) were unable to detect any relationship between the annual rates at which sick moose were reported and changes in moose numbers. However, opportunistic sightings of sick animals seldom provide accurate estimates of the impact of a disease on a population. But in the Whitlaw and Lankester retrospective study, reporting rates were the only available historic data that could be used to test for a causal link between moose declines and the disease. The extent to which *P. tenuis* was a key factor in historic moose declines probably never will be known.

Moose cohabiting with infected deer in a variety of circumstances are not extirpated. Whitlaw and Lankester (1994b) reported that moose coexist with infected deer in 45 wildlife management units in Ontario, and have done so at least since 1980. Also, moose still persist in all of those jurisdictions where moose sickness was first observed as long as 90 years ago. Accordingly, isolating mechanisms probably are not responsible in all instances for that persistence. Where the two species are sympatric in Ontario, moose were most numerous when deer were less than 10 per square mile (4/km²), and an inverse relationship was found between moose densities and the numbers of *P. tenuis* larvae being passed by resident deer. However, case studies of recent moose declines served as reminders that other factors—including habitat quality, hunting, wolves and winter ticks—also may be causative factors during periods of decline and cannot be ignored. Clearly, the effects of this parasite on moose populations are more subtle than previously believed, and further study is needed to determine its importance relative to other known moose mortality factors (Whitlaw and Lankester 1994a).

Also possible is that biologists do not yet have an accurate understanding of how the meningeal worm affects individual moose under natural conditions. The parasite always has been considered to be highly pathogenic in this host. Evidence for this comes first from the previously mentioned experimental infection of two young moose calves (Anderson 1964). Both were given many more *P. tenuis* larvae (164 and 200) than were later found to be common in nonexperimental animals (Lankester and Anderson 1968). Both calves developed debilitating neurological disease and had to be euthanized after 40 and 60 days, respectively (Anderson 1964). Second, wild moose exhibiting neurological symptoms seldom have more than three worms in the cranium. In many, only a single *P. tenuis* can be found (Anderson 1965b, Gilbert 1974, M. Lankester personal files). Unless more worms in sick moose lie undetected in the spinal cord (which is seldom examined), the disease symptoms must be attributed to the observable nerve tissue damage and the few worms that can be recovered. Paradoxically, most deer in any endemic area acquire at least two to three worms before they reach one and a half years of age (Peterson and Lankester 1991). Because moose persist in such areas, they may avoid infection at the rates experienced by deer, perhaps by feeding in a different way or in a different place. Or, moose may not be as susceptible to this parasite as has been believed.

It was reported years ago that some moose examined with adult worms in their cranium did not exhibit recognizable signs of neurological disease (Archibald 1967, Gilbert 1974). Also, some survive long enough for adult worms to mature and pass larvae in the host's feces (Anderson 1965a), which could be 4 or 5 months after infection. Clinically normal moose may not be as common, however, as some literature indicates (see Clarke and Bowyer 1987, Thomas and Dodds 1988). Unless the animals are observed to pass them, field-collected feces containing *P. tenuis* larvae cannot be assumed to be from normal moose.

A study of *P. tenuis* in elk calves indicated that the severity of the disease in this species is dose dependent (Samuel et al. 1992). All elk given 125 or more larvae died. However, of eight animals given 25 or 75 larvae, seven shed larvae in their feces, six showed neurological signs and only two died. Interestingly, no disease was seen in five elk each given 15 larvae, although two or three adult *P. tenuis* were found in their craniums.

How moose respond to low-dose infections of *P. tenuis*, similar to those encountered in nature, has been studied only recently (M.W. Lankester personal files). Two moose were hand-raised for 5 months before being given three larvae each. Some lameness in the front legs and hindquarter weakness were observed within 6 weeks of infection. By 3 months, the abnormal signs had almost disappeared in both animals, yet single adult worms were found on the brain and spinal cord. Four other moose were not infected until 9.5 months of age. Two given 10 and 15 larvae exhibited slight but persistent front lameness and hindquarter weakness, and one and three worms, respectively, were recovered at necropsy after 95 days. Two moose given five larvae each showed no lasting neurological signs and no worms could be recovered. One of the moose given five larvae resisted a challenge infection of 15.

These results indicate that *P. tenuis* infection is not necessarily lethal to moose. Apparently, the outcome is dose dependent as well as age dependent. Initial nerve tissue damage caused by one or two developing larvae may not be permanent. Some worms may be killed, and the moose may be refractory to a later infection. However, it remains to be explained why wild moose with only one or two worms still may exhibit debilitating neurological signs. Possibly, the classic signs of moose sickness in these animals result from a chronic inflammatory response to a few persistent worms or the parasite's eggs on the meninges of the brain, rather than from the more acute effects of worms developing in the spinal cord.

The meningeal worm probably has a greater range of effects on moose than was previously suspected. Some may be extremely subtle and difficult to detect. Clearly, the impact of *P. tenuis* on populations will only be known completely when survival and reproductive rates of individual moose can be measured in relation to their experience with the parasite. A blood test indicating whether a moose is infected or has previously been exposed to *P. tenuis* is essential to such an effort.

Elaphostrongylosis

The so-called tissue worms, *Elaphostrongylus* spp., also can cause neurological disease in moose. Infection of moose is common in parts of Norway and Sweden where animals with symptoms of neurological disease are frequently seen (Stuve 1986, Steen and Roepstorff 1990).

Three different species of this parasite are recognized (Steen et al. 1989), but they are not strictly host specific. The precise effect of each on moose is yet to be determined. Both *Elaphostrongylus cervi*, found commonly in red deer, and *E. rangiferi*, found in reindeer, can develop in moose (Stuve and Skorping 1987). The third species, *E. alces*, occurs most commonly in moose of Sweden. It also can develop in reindeer, although fewer worms reach maturity in that species (Steen 1991).

Only *E. rangiferi* occurs in North America. So far, it is limited to Newfoundland. Probably it was introduced there with infected reindeer imported from Norway (Lankester and Fong 1989). In Newfoundland, young caribou have shown signs of severe neurological disease (Lankester and Northcott 1979), but the disease has not been observed naturally in moose. Moose on caribou range acquire the worm, but first-stage larvae are not present in feces, at least from older animals (M. Lankester and D. Fong personal files).

The potential for *Elaphostrongylus* spp. to cause neurological disease exists because of the route taken by developing worms in the host. Like *P. tenuis*, Elaphostrongylus migrates initially into the spinal canal, but it does not appear to enter nerve tissue. Later, worms appear on the surface of muscles of the chest region and the limbs. They may move out of the spinal cord by following nerves that lead to muscles in those parts of the body, although this has not yet been confirmed.

A moose calf given *E. rangiferi* from caribou of Newfoundland developed severe neurological disease within 3 months (Lankester 1977). One adult worm was found within a ventricle of the brain; others were found on the brain and on muscles of the chest and forelegs. This experiment did not determine whether *E. rangiferi* can produce larvae in moose (Lankester et al. 1990), but a similar study completed by Steen (1991) using *E. rangiferi* from reindeer in Sweden indicated that a few larvae are passed by infected calves.

Related nematodes, the muscle worms *P. odocoilei* and *P. andersoni*, occur primarily in deer and caribou. The former, which occurs in mule deer and black-tailed deer, was developed experimentally in moose (Platt and Samuel 1978, Pybus and Samuel 1980). The latter, *P. andersoni*, occurs widely in white-tailed deer and caribou (Lankester and Hauta 1989), but has not been tested in moose, and natural infections have not been reported. There is no evidence that either is likely to cause disease in moose. But the possibility of cross-reactivity between these and the brain worm may complicate future efforts to develop a badly needed serological test for *P. tenuis* infection in moose.

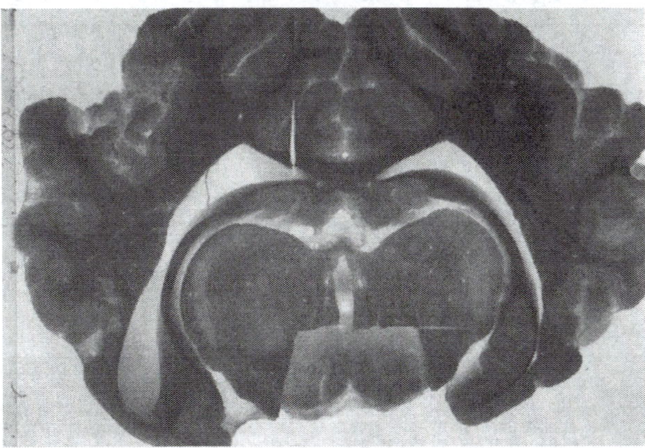

The tissue worm *Elaphostrongylus rangiferi* was introduced into Newfoundland with infected reindeer. Under experimental conditions, this neurotropic nematode can cause severe neurologic disease in moose. This disease has yet to be reported in wild moose in Newfoundland. Worms are found in the spinal canal and the brain as they migrate to their final location in the skeletal musculature of the shoulders and limbs. A maturing *E. rangiferi* can be seen (above) within the right, lateral ventricle (arrow) of the brain of an experimentally infected moose. *Photo by M.W. Lankester.*

Lung Nematodes

The lung nematode *Dictyocaulus viviparus* is widely distributed in moose, but usually does not cause recognizable disease in this host. The long, slender white worms are found in the smaller bronchi where they may be bathed in a mucinous, frothy exudate. Their biology is known largely from studies of related species and from strains in sheep and cattle. They have a direct life cycle. Larvae pass in droppings, quickly develop to the infective stage in herbage and are ingested with ground vegetation. They migrate via the lymphatic system from the intestine to the lungs, where they reach adult size in about a month. In sheep and cattle, most worms are expelled within 3 to 6 months (Jubb et al. 1985). The severity of the bronchitis induced by adult worms and lung lesions caused by eggs and larvae depends on the susceptibility of the host and the number of infective larvae ingested. Disease, when seen in domestics, occurs in young animals exposed for the first time on heavily contaminated pastures. Lung worm infections in moose are also seen most frequently in calves and yearlings (Pybus 1990).

Clearly there are strain differences between *D. viviparus* found in wild cervids and domestic cattle. Cross-infections can occur, but strains in domestics apparently infect wild ungulates more readily than is the reverse case (Presidente

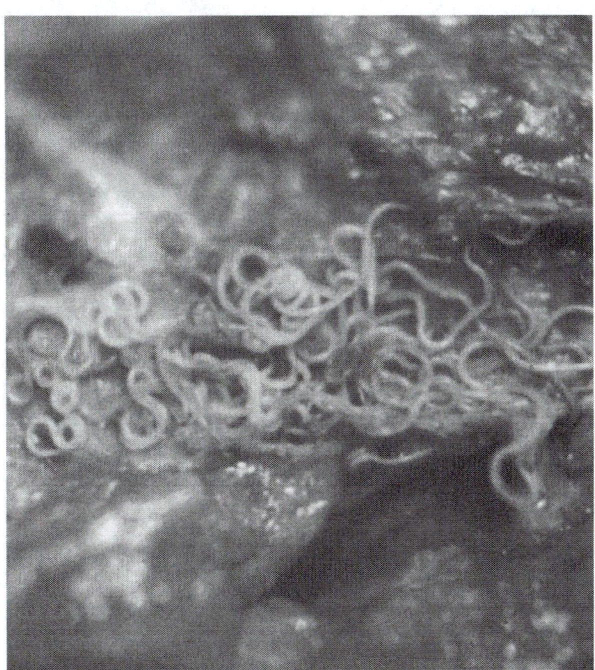

Lungworms, *Dictyocaulus viviparus,* are shown above in the smaller airways of the lung of elk. The same nematode is seen in the lungs of moose, but the percentage of animals infected and the number of worms found usually are much lower than in elk. *Photo by W.M. Samuel.*

et al. 1972). This is true of moose strain *D. viviparus* (Gupta and Gibbs 1971).

Orthostrongylus macrotis has been reported from moose only in Alberta (Samuel et al. 1976, Stock and Barrett 1983). It is found in the bronchioles of the lungs of mule deer, and produces first-stage larvae detectable in feces. Unlike *Dictyocaulus,* it requires development in a terrestrial snail or slug before reaching the infective stage. It is not known to cause disease in moose.

Gastrointestinal Nematodes

A small group of nematodes inhabits the abomasum and the duodenum. Most are members of the Trichostrongyloidea, which have a direct life cycle. All are 0.2 to 1.2 inches (0.5–3.0 cm) in length and about the thickness of a hair. They cause enteritis, with scouring (watery diarrhea) in young domestic animals when heavy infections are acquired on contaminated pastures. No disease is known in wild moose.

Found in moose throughout the circumpolar region, *Nematodirella alcidis,* the thread-necked strongyles, are the most ubiquitous of the gastrointestinal nematodes. They inhabit the first 3 to 4 feet (0.9–1.2 m) of the small intestine. Although 50 to 100 percent of moose may be infected, the numbers of worms present in individual animals usually is relatively low, averaging less than 100, with maxima of about 1,000 (Stock and Barrett 1983, Fruetel and Lankester 1988, Hoeve et al. 1988).

N. alcidis is fairly specific to moose, and its biology appears to differ from that of related nematode species in domestic animals. Typically, the worms occur in low numbers. Also, they develop in moose of all ages (Freutel and Lankester 1988). Up to 80 percent of *N. alcidis* worms present are arrested in a subadult stage, possibly only developing to maturity at the death of established adult worms. This strategy ensures a small but steady production of parasite eggs throughout the year. It is not surprising, therefore, to find that the eggs of *N. alcidis* are resistant to freezing and desiccation. Hence, those produced during winter and dry periods are not wasted, but resume development to the infective stage when conditions are favorable.

Nematodes inhabiting the abomasum of moose include species of *Trichostrongylus* and the medium stomach worms *Ostertagia* (see Stock and Barrett 1983). These, as well as the whipworm *Trichuris* spp., in the cecum and colon, usually occur in small numbers and are not commonly reported. They are of no known consequence to the health of wild moose. An exception may be *T. discolor.* It normally is a parasite of domestic cattle, but was recovered by Hoeve et al. (1988) from moose in an area in Quebec almost surrounded by agricultural land. This species has been implicated as the cause of mortality in dairy cattle in the area (Frechette et al. 1973). Worms may erode the lining of the cecum and colon, causing bloody diarrhea. Heaviest infections occur in young animals. The eggs are long-lived and resistant to a range of weather conditions. They are more likely to accumulate in a farming or captive situation. *Trichuris* may be numerous

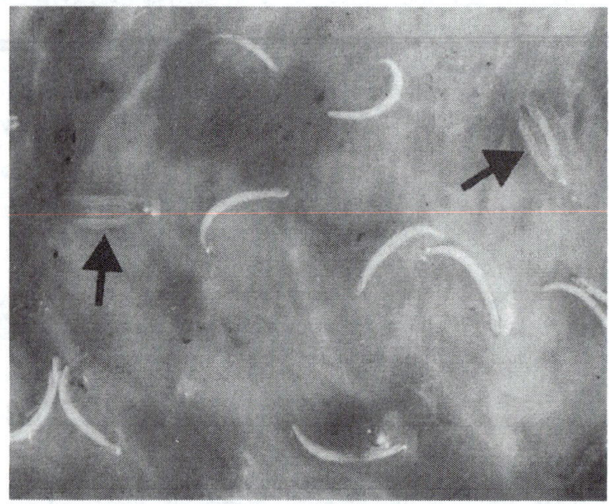

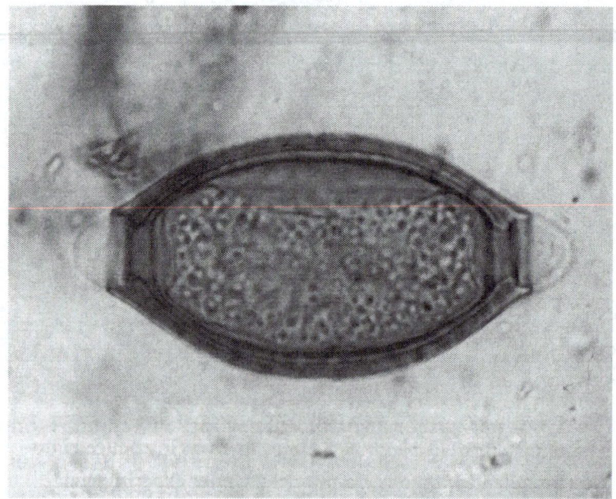

Whipworms, *Trichuris* spp., are so named because the stout posterior half of the worm resembles the handle of a bull whip (left). The slender anterior "whip" is intimately associated with the mucosa lining the cecum, but is not visible here. The trematode *Zygocotyle lunata* (left; arrows) is a parasite of ducks that also may be reasonably common in moose. The orange-brown eggs of *Trichuris* spp. (right) are about 78 micrometers long and have a distinct bulbous opercular plug. They readily float in sugar solution used for fecal analysis. Although whipworms can be numerous in captive moose, numbers are few in wild moose, and are considered of little consequence. *Photos by W.M. Samuel.*

in captive moose (Williams et al. 1984) and have been associated with deaths (W. Samuel unpublished data, R.J. Hudson personal communication: 1993). *Trichuris* is of no concern in wild moose.

Filarioid Nematodes

The filarioids are tissue-inhabiting nematodes that have evolved a specialized means of transmission (Anderson 1992). Their first-stage larvae (microfilariae) are released by female worms into tissue spaces of the skin or into the bloodstream and are transmitted to other hosts by biting insects. Of four filarioids known from moose, only *Elaeophora schneideri* is known to cause disease in this host.

Arterial Worms

Mule deer are believed to be the normal host of the arterial worm, or "blood worm," *Elaeophora schneideri*. In elk, it causes a significant disease called "clear-eyed blindness." Transmission is by horse flies (representatives of *Hybomytra* and *Tabanus*). Moose can become infected (Worley 1975), but the parasite's distribution—in black-tailed deer on the southern tip of Vancouver Island, in blacktails and mule deer in western states, and in white-tailed deer in southern and southeastern states—precludes extensive contact with moose. Serious threat to moose exists only in the northwestern states where moose share range with infected mule deer during the horse fly season.

Fewer than a dozen cases of moose infection by blood worms are known. All were in Montana, Wyoming, Utah and Colorado, in fairly isolated localities with growing populations of Shira's moose and mule deer on common range. Clinical disease with blindness and circling has been documented only twice in moose (Worley et al. 1972, Madden et al. 1991). Although some female worms become gravid in a few moose, it remains to be demonstrated whether moose serve as a significant reservoir of infection for other cervids.

Adult female blood worms are 4.7 by 0.03 inch (12 by 0.08 cm), and locate in the smaller branches of the common carotid and internal maxillary arteries (Hibler and Adcock 1971). Microfilariae released by females are swept away by the superficial temporal arteries and become concentrated in capillaries of the skin in the region of the forehead and face (Hibler and Prestwood 1981). There they are available to horse flies seeking a blood meal. Adult worms live for about 4 years. Moose and elk seem to have little resistance to infection, and greater numbers (up to 40) of developing worms are found in these hosts than in mule deer (1–3) (Worley 1975, Madden et al. 1991).

Legworm or Footworm

These white worms live beneath the skin around the joints of the front and back legs and feet. Hunters become concerned about edibility of moose meat when many adult legworms can be seen under the skin. This is not surprising, given the large number of worms often present and the considerable brown, fibrous tissue encasing them. However, edibility is not affected. The numbers of *Onchocerca cervipedis* increase with the age of moose and, in heavily infected animals, worms may be seen in the brisket and belly areas. Female worms are up to 8 inches (20 cm) long, and males (which are rare) are about one-fourth that length. Apparently infection is of little or no consequence to moose.

Legworms have been reported from moose only in western Canada and Alaska, but their distribution probably is more extensive. They are known to occur, for example, as far east as Manitoba (V. Crichton personal communication: 1992). The same species commonly is encountered in mule and black-tailed deer of the western states and has scattered distribution in whitetails across the United States (Hibler and Prestwood 1981).

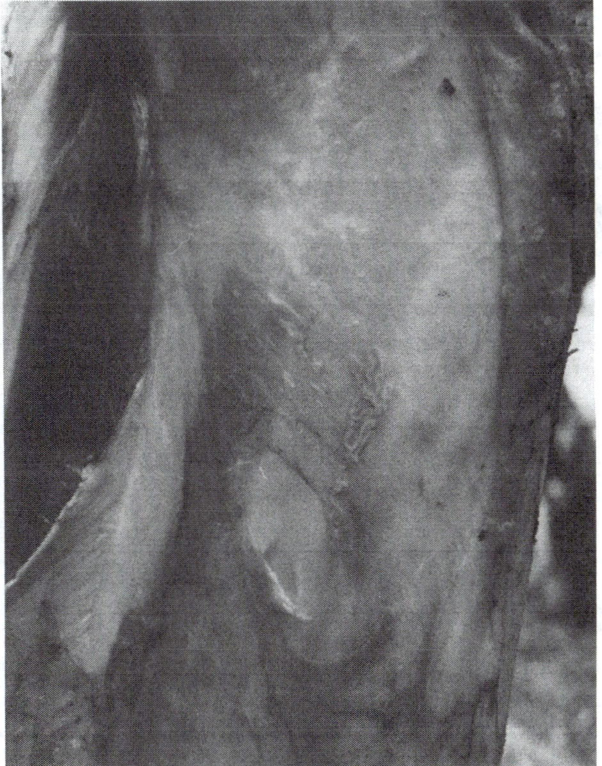

Legworms, *Onchocerca cervipedis*, sometimes are seen when a moose carcass is skinned. Aggregations of worms occur in subcutaneous tissue, often around the knee joint of the front legs. But when numerous they can be found almost anywhere beneath the skin or among muscles. *Photo by Vince Crichton.*

Most of our knowledge of the biology of this filarioid nematode in moose comes from the work of Samuel et al. (1976) and Pledger et al. (1980) in Alberta, where up to 64 percent of moose were infected. Eighty percent of the worms (98 percent female) found were beneath the skin of the lower forelegs, with most of the remainder in the lower hind legs. Black flies are the known intermediate host and vector of *O. cervipedis* in Columbian black-tailed deer (Weinmann et al. 1973) and very likely transmit the infection among moose (Pledger et al. 1980). Feeding activity by black flies on wild moose in Alberta was concentrated on the inner and outer aspects of the legs, particularly the hind legs. Microfilariae could be detected in skin of the legs only during June and July.

Abdominal Worm

Setaria yehi is a slender, whitish worm, up to 4 inches (10 cm) long, occurring free in the body cavity of moose and other cervids throughout North America (Anderson and Lankester 1974). A mild, chronic peritonitis can sometimes be seen on organs in the vicinity of worms, but these worms probably are harmless otherwise. Microfilariae circulate in the blood. In white-tailed deer the prevalence of infection declines with age (Prestwood and Pursglove 1977), but in moose, it is similar for all age groups (Samuel et al. 1976).

The transmission of *S. yehi* has not been studied, but several other related species are known to use mosquitos (Anderson 1992). However, stable flies, *Haematobium stimulans*, should not be discounted. They have been shown to transmit *Setaria cervi* among cervids in Europe and Asia (Osipov 1966). Although abdominal worms from North American cervids generally are thought to be *S. yehi*, a specific diagnosis is difficult. Samuel et al. (1976) suspected that some specimens recovered from moose in Alberta's Elk Island National Park may have been *S. labiatopapillosa*, which were common in bison living in the area.

The presence of the abdominal worms *Setaria* in the abdominal cavity of moose may cause a mild chronic peritonitis on organs in the vicinity of the worms. *Photo by W.M. Samuel.*

Rumen Filarioids

Rumenfilaria andersoni is a fourth filarioid nematode found in moose. Worms are coiled loosely in vessels in subserosal connective tissue between folds of the rumen. No pathology is associated with infection. The worm has so far been found only in moose of northwestern Ontario, and neither the location of microfilariae nor the required vector is known (Lankester and Snider 1982).

Ticks

Ticks are external parasites that feed on the blood of most vertebrates, except fish. They have three parasitic blood-feeding life stages—larva, nymph and adult. Ticks cause blood loss, itching, inflammation and skin ulcerations. They also transmit microbial agents of disease.

Only one species, the "winter," "moose" or "elk" tick *Dermacentor albipictus* is known to be important relative to moose in much of moose range south of about 60 degrees north latitude. Winter ticks spend the entire winter on wild ungulates, mainly moose, elk, mule deer and white-tailed deer. Where moose and winter ticks overlap in distribution, moose are the most frequently and severely infested host (Welch et al. 1991). These ticks do not occur in Alaska, but results of an experiment showed that they can survive and presumably could become established in that climate if introduced (Zarnke et al. 1990). The winter tick also does not occur on Newfoundland.

Epizootics of winter ticks, concurrent with moose mortality, have been reported in many localities (Blyth and Hudson 1987, Lankester 1987, see also Anderson and Lankester 1974). Humans represent a poor host for winter ticks, and rare is the hunter who complains of acquiring ticks from moose.

In the winter tick, three parasitic stages occur on a single host. Adult female ticks drop off moose at the end of March and in early April (Figure 197), and begin to lay eggs. Eggs do not hatch until later in summer. Moose become infested

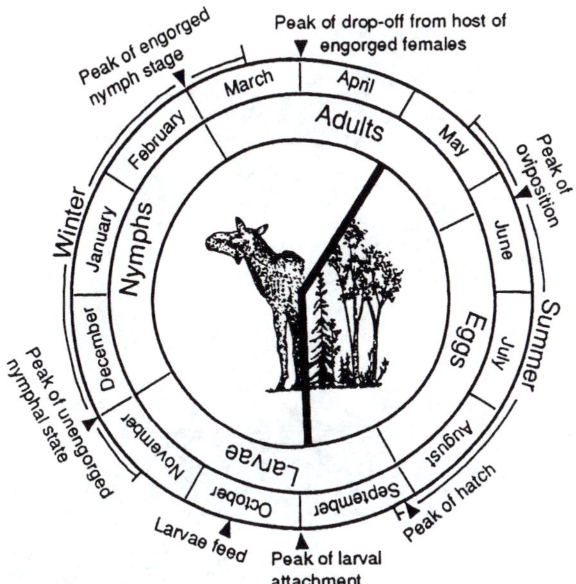

Figure 197. Life cycle of the winter tick *Dermacentor albipictus*. *Illustration by University of Alberta Technical Services.*

in late summer and autumn when they contact clumps of tick larvae ("seed ticks") on the tips of vegetation. They are less than 0.04 inch (1 mm) long and have only three pairs of legs, rather than four pairs as seen in the nymphs and adults. Peak numbers of aggregating larvae coincide with a very active period for moose—the breeding season.

Once on moose, the larvae feed on blood and, by November, molt to nymphs about 0.07 inch (1.7 mm) long. In January and when about 0.2 inch (5 mm) in length, nymphs begin blood feeding. They peak in mid-February and molt to become adult females (dark, reddish-brown, no pattern on back, 0.26 inch [6.5 mm] long) and adult males (white cross-hatch pattern on dark brown back, 0.28 inch [7 mm] long). The grayish "grape-sized" ticks, approximately 0.06 by 0.35 inch (1.5–8.9 mm) in size, weighing 0.02 ounce (5 g), seen on moose in late winter and spring are engorged females. Their numbers peak in late March through April, and all drop off by mid-May. Moose are tick-free during the summer.

Most moose (south of 60 degrees north latitude) probably become infested each year of their life. Samuel and Welch (1991) recovered ticks from all 212 moose hides examined from northern British Columbia, Alberta and Manitoba. The estimated average number of ticks per moose was 32,500, or 9.23 ticks per square inch ($1.43/cm^2$); 39 moose had more than 50,000 ticks. In comparison, average numbers of ticks for hosts in Elk Island National Park were 37,000 on moose, 3,400 on elk, 1,450 on white-tailed deer, and 175 on Plains and wood bison.

One consequence of winter ticks taking blood from moose is the resulting excessive grooming that damages the winter hair coat in a characteristic pattern. Grooming and hair loss usually begins in late February, apparently in response to blood feeding by nymphal ticks. Damage to the hair coat in response to blood feeding by adult females begins in March on the neck and shoulders and progresses rapidly to the withers and tail region (Samuel et al. 1986, Samuel 1991a).

Samuel and Welch (1991) reported that 89 percent of 724 moose observed in Alberta, Manitoba, Wyoming, Utah and Maine (1978 through 1990) emerged from winter with tick-induced hair damage. In one moose population near Rochester, Alberta, an average of more than 40 percent of hair along the sides of the body was damaged or lost prematurely by the end of 8 of 13 winters. As well as damaging the hair coat, excessively grooming moose are restless, develop chronic anemia and low serum protein, and have reduced visceral fat (McLaughlin and Addison 1986, Glines and Samuel 1989, Samuel 1991a).

The significance of the costs of grooming and attendant energy loss to pelage damage as a result of harboring winter ticks is not known. One could logically assume that extensive hair damage occurs at a time when moose experience thermal stress and increased metabolic demands during periods of severe cold. Glines and Samuel (1989) reported hypothermia in an experimentally infected captive moose. Welch et al. (1991) found no detectable effects of winter ticks on fasting metabolic rates or weight changes of captive moose during a relatively mild winter, but things might be different in nature during a cold winter. Welch et al. suggested that, because most grooming and hair damage occur near the end of winter, tick-induced hair damage may impose only nominal thermoregulatory costs on moose. Expenditure of energy might be a more important cost of grooming, particularly for pregnant cows or undernourished moose (see Garner and Wilton 1993). Mooring and Hart (1995) have documented that one cost of grooming by tick-infested impalas is distraction from vigilance for predators.

A positive consequence of grooming is that many ticks are removed before completing their development or before female ticks can take a complete blood meal. Barker et al. (1990) estimated that grooming by cattle reduces the potential yield of eggs from female *D. albipictus* by as much as 98 percent. Ticks from moose (and presumably cattle) that have fed only partially lay fewer eggs (Addison and Smith 1981).

The exact relationship between winter ticks and moose populations is not clear. Although direct evidence of winter ticks causing major mortalities of moose is lacking, data from experimental (McLaughlin and Addison 1986, Glines

Winter ticks *Dermacentor albipictus* typically cluster on moose in late winter and early spring on the belly and inguinal region (left) and around the anus (top right). In western Canada, the number of ticks on moose commonly exceeds 30,000. The largest ticks are blood-fed adult females (bottom right). *Left and top right photos by M. W. Lankester. Bottom right photo by W.M. Samuel.*

and Samuel 1989), observational (Cowan 1951, Berg 1975, Garner and Wilton 1993) and other field studies (Samuel and Barker 1979, Samuel and Welch 1991) strongly support the idea that winter ticks kill many moose. In Elk Island National Park, Alberta, there have been five epizootics of ticks associated with major losses of moose during at least 10 winters since the early 1930s (Blyth and Hudson 1987). Each die-off occurred when moose were at peak numbers and winter weather usually was more severe than long-term averages. Dead animals typically were malnourished, with much tick-related damage to the hair and many thousands of winter ticks. This suggests that a complex interaction of winter ticks, host density and nutrition, and weather is involved in die-offs.

March and April are months of nutritional and energetic stress for moose (Renecker and Hudson 1989b) and the times when nymphal and adult tick life stages of *D. albipictus* feed on moose blood. On the basis of domestic animal literature, one would expect that moose of lower than normal nutritional status would have a weakened immune response, thus groom less in response to feeding ticks and, as a result, have more ticks. Thus, ticks and host nutrition appear to play major roles in tick-associated die-offs (DelGiudice et al. in press).

One might expect that when moose densities are high, so too are tick numbers, especially because winter ticks cycle from moose to vegetation to moose. Such could be the course of events in a moose population growing in number, but when the population is at or near carrying capacity, factors other than moose density probably are more important in determining tick abundance from year to year. Survivorship of the egg, larval and adult stages while off the host is largely determined by weather. For example, when tick larvae are passed from vegetation to moose by direct

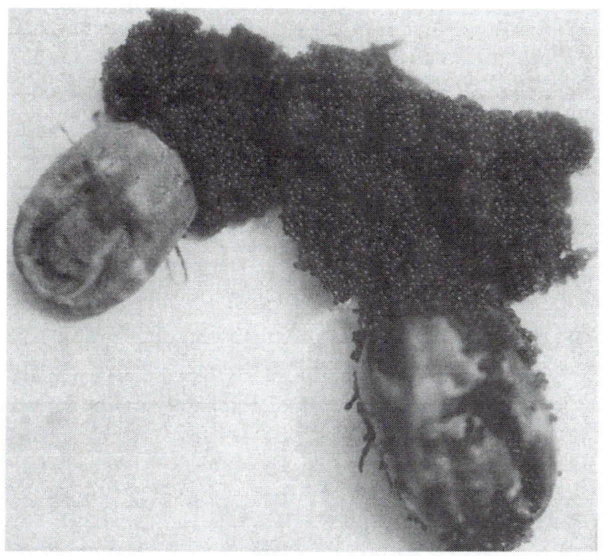

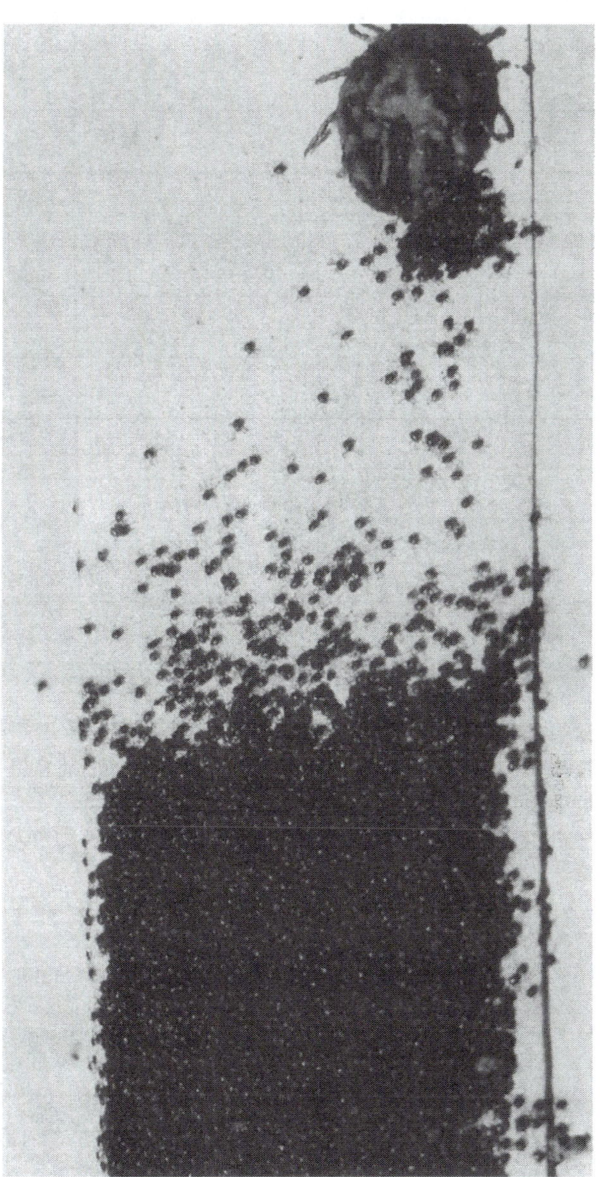

Two *Dermacentor albipictus* ticks and egg masses (top left), produced in a laboratory after being removed from moose in April. Larvae of *D. albipictus* hatching (right) in a laboratory from the eggs near the spent female at top. In nature, eggs are laid in May–June and hatch in August–September. Large clusters of winter tick larvae (seed ticks) can be seen near the tips of grass blades (bottom left) and woody shrubs in autumn. Moose become reinfested with winter ticks each year by direct contact with the larvae, usually between early September and mid-November. *Top photos by M.W. Lankester. Bottom left photo by E.M. Addison; courtesy of the Ontario Ministry of Natural*

Parasitic life stages of *Dermacentor albipictus* (left) removed from a moose showing relative sizes (left to right) of a blood-fed larva, an unfed nymph, three engorging nymphs, an unfed adult male, an unfed adult female and two blood-fed adult females. At right adult winter ticks (left to right): male; partially fed female; and engorged female. *Left photo by W.M. Samuel. Right photo by M.W. Lankester.*

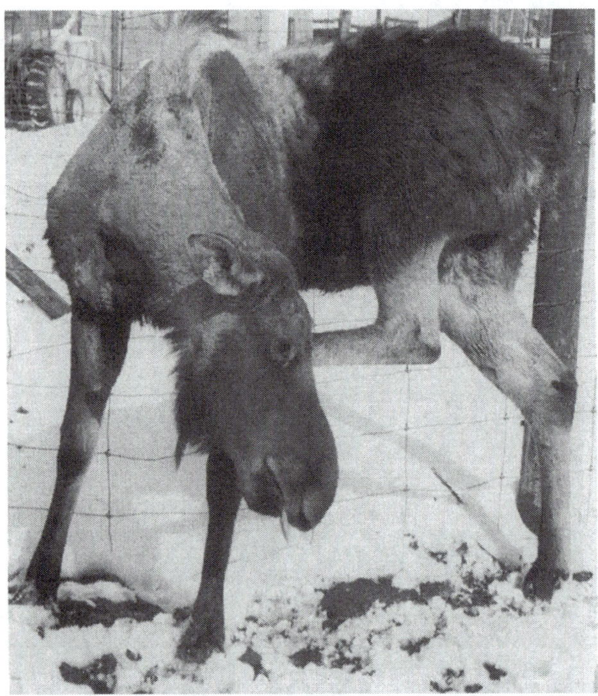

Moose with winter ticks spend much time grooming in February, March and April. Licking and scratching with the hind foot are two of several tick grooming behaviors. Note the extensive damage to the above cow moose's winter coat. *Photo by W.M. Samuel.*

contact in autumn, early snows and cold temperatures would shorten the transmission period by burying and killing larvae, respectively (Samuel and Welch 1991). Likewise, presence of snow on the ground in late winter/early spring resulted in decreased survival of engorged adult female ticks dropping from moose (Drew and Samuel 1989). Wilton and Garner (1993) suggested a link between mean April temperature and tick-induced hair loss and mortality of moose the following late winter/early spring. When April temperatures are warmer than normal, hair loss and moose mortality increase the following year, presumably because more adult female ticks dropping from moose survived to lay eggs.

Flies

The true flies are members of the order Diptera (Figure 198). Most flies are attracted to moose for a blood meal. With the exception of the muscid flies, only females are involved. Carbon dioxide and heat given off by moose are the important attractants. Various flies have daily and seasonal periods in which they are most active. For some, the season is relatively brief, mainly in spring and early summer. Others, such as the moose fly, may be found with their warm-blooded hosts for much of the snow-free period.

Approximately 90 percent of moose observed during late winter in various portions of their range show varying degrees of hair damage or loss related to infestation by winter ticks. Hair damage and loss begins on the neck and shoulder and extends anywhere on the body that the moose can scratch, lick, or rub. In water, moose may gain some relief from the ticks. The Isle Royale bull shown above has a whitish appearance from the broken hair shafts. Such animals often are referred to as "ghost moose." *Photo by R.O. Peterson.*

A young moose, its coat severely damaged, was found dead in May 1938 near Winton, Minnesota. The carcass was covered with winter ticks, but the cause of mortality was undetermined. Biologists suspect that heavy tick infestations may cause or contribute to moose mortality as a result of exposure (hair loss), energy expenditure (excessive grooming), reduced vigilance for predators and/or susceptibility to other pathogens, but definitive evidence is insufficient. *Photo by C.M. Aldous; courtesy of the U.S. National Archives.*

Flies can affect the health and well-being of mammals in at least three ways. First, the larvae (maggots) of some flies require a period of parasitic development inside the host. Second, biting species may be vectors of bacterial infections, blood protozoans and nematode parasites. And third, flies may harass and annoy animals to the point where they lose weight trying to avoid them, cause them to retire from preferred habitat and distract them from serious hazards.

The commonly held belief that moose move to roadside areas in spring to avoid flies may be incorrect. At least in Quebec and Ontario, moose are attracted to roadsides that have a high concentration of highway salt, presumably contributing to the frequency of vehicle/moose collisions (Grenier 1973, Fraser and Thomas 1982; see Chapter 8). There also are differing views on whether moose enter water in the summer to avoid flies or cool off (Wolfe 1974). Wallowing and lying in mud may protect the legs and belly from flies (Wormer 1972, Sein 1985). Although no quantitative studies exist, experience with captive animals suggests that moose appear to be bothered less by biting flies than are caribou (Sein 1985).

Flies whose larvae require development inside the body of a mammal include such forms as screwworm flies, warble flies and throat bots. Throat bots are the only members of this group known to have been reported from moose of North America.

Throat and Nasal Bots

Bots are large bee-like flies that produce live young. First instar larvae are forced individually from the posterior ovijector of females while in flight. Species that strike mule deer in California gain a position in front of the nose where they cannot be seen (Anderson 1975). It has long been assumed that female flies squirt larvae directly into the nostrils, but an experimental study by Cogley and Anderson (1981) indicated that larvae deposited on the muzzle, instead crawl into the mouth and along the palate or tongue to reach the lateral pharyngeal pouches at the back of the throat. Five species of *Cephenemyia* are known in cervids of North America (Bennett and Sabrosky 1962). They seem to be restricted more by geography than by host specificity. Within the distributional range of a fly several species of cervids may be infected.

There are only a few incidental reports of throat bots in moose. Bennett and Sabrosky (1962) recorded *Cephenemyia jellisoni* from moose in Montana and Ontario, and *C. phobifera* from moose in Ontario. *C. jellisoni* is primarily a western species found in deer and elk, whereas *C. phobifera* is found in the East and in white-tailed deer.

The life cycle of *C. phobifera* was studied by Bennett (1962). Larvae placed in the nasal sinuses eventually migrate to lateral pouches at the back of the throat. There,

Figure 198. "The Biting Fly Derby" shows Dipteran pests (jockeys) of North American moose (racers), and some other insects (racegoers) of moose range. Insects were not drawn to scale. From the key: *Jockeys* 1 = horse fly; 2 = blackfly; 3 = mosquito; 4 = biting midge; 5 = moose fly; 6 = deer fly; 7 = snipe fly; 8 = stable fly; *Racegoers* 9 = mite; 10 = darkling beetle; 11 = firefly; 12 = field cricket; 13 = ripple bug; 14 = carrion beetle; 15 = scarab beetle; 16 = yellow-jacket wasp; 17 = longhorn beetle; 18 = moose tick; 19 = tiger beetle; 20 = carpenter ant; 21 = leafhopper; *Racers* 22–29 = moose. *Illustration by and courtesy of Barry Flahey, Manotick, Ontario.*

Moose sometimes lie in mudholes in midsummer, perhaps to get relief from flies biting the lower legs. Note the moose flies *Haematobosca alcis* on the upper hind leg of the moose above. *Photo by M.W. Lankester.*

Several bot species—*Cephenemyia jellisoni* and *C. phobifera*—occur rarely in moose. As in caribou, bot flies deposit larvae on the muzzle, and the larvae crawl into the nose or mouth to attach, in the latter case, at the back of the throat, as shown above. *Photo by M.W. Lankester.*

they grow to 1 inch (2.5 cm) in length before being sneezed or coughed out to pupate in the soil. In Algonquin Park, Ontario, there are two generations per year, with adult flies on the wing in early and late summer.

Little evidence exists to suggest that throat bots are significant pathogens, despite the large numbers of mature maggots (more than 100) often seen in the pharynx of hosts such as white-tailed deer and caribou. In black-tailed deer infected with *C. apicata* and *C. jellisoni*, the retropharyngeal recesses were greatly enlarged and the epithelial lining eroded, but these nonsupportive lesions healed rapidly after the larvae left (Cogley 1987). A similar result might be expected in moose infected with large numbers of fly larvae. Gaab (1948) associated bots with swelling beneath the jaw of a calf moose that later died.

Moose Fly

Moose flies, *Haematobosca alcis,* are present wherever moose occur in North America (Anderson and Lankester 1974), and apparently do not bother with other cervids. They are not found in Eurasia (Pont 1986). Unlike most biting dipterans, adult moose flies are closely associated with their host at all times. The flies can be seen throughout the summer in dense swarms over the rumps of moose or resting on hairs of the hind quarters. Also, unlike most other biting insects, both male and female moose flies feed. *H. alcis* resemble ordinary house flies (Muscidae), although are marginally smaller. Other close relatives include the stable fly *(Stomoxys calcitrans)* and horn fly *(Haematobia irritans)*, which is a serious pest of cattle.

Five hundred or more *H. alcis* may accompany a single moose. Possibly stimulated by gases released by the moose when defecating, gravid females descend and deposit eggs into crevices of the fecal matter (Burger and Anderson 1974). Adult flies become active in mid-May in the Thunder Bay area of Ontario, and persist into September, when their numbers decline sharply with the onset of freezing temperatures (Lankester and Sein 1986).

There is no evidence that the moose fly is a serious pest, as was believed by Peterson (1955). Moose appear to pay them no heed. But raw, wet skin lesions on the backs of the legs of moose, just above the hock, have long been attributed to the feeding activity of moose flies (Murie 1934, Peterson 1955). Peek et al. (1976) suggested that horse flies were responsible for these sores. Also, in Ontario's Algonquin Park, (M. Runtz personal communication: 1992) leeches have been seen attached to the region of the hock of moose where lesions appear. On the other hand, moose kept in pens on Alaska's Kenai Peninsula get similar sores on the hocks (C. Schwartz personal communication: 1993), but horse flies are rare there and leeches do not have access. Although it has never been proven that the moose fly initiates the lesions above the hocks, their mouth parts probably are capable of doing so. Very clearly, however, the continual

Male and female moose flies *Haematobosca alcis* stay on moose throughout summer. If a moose is startled and moves off quickly, the flies will rise like a cloud (left), but quickly catch up to the moose and realight (right); note the open sores on the sides of the hind legs just above the hock joint. Female moose flies leave the rump area briefly to lay eggs in freshly deposited moose droppings. *Photos by M.W. Lankester.*

feeding activity of this fly prevents the healing of lesions. For example, open sores first appeared on a captive female moose in mid-June (Lankester and Sein 1986). Through July and August, up to 12 spherical wounds 0.6 inch (1.5 cm) in diameter were present on each back leg. Large numbers of *H. alcis*, arranged radially like the petals on a daisy, aggregated at the periphery of sores, feeding on oozing fluids. When fly numbers declined in early September, the lesions quickly healed.

Other Biting Flies

Few detailed studies have been conducted on the biting flies of moose, with most information coming from captive animals (Table 46). Among the flies receiving little attention is the stable fly, *Stomoxys calcitrans*. Both males and females seek blood. Stable flies are not strictly pests of barnyards, as their common name might imply. Large numbers occur as a result of domestic animal and human activities, but stable flies also are found in remote areas, far from places habitually occupied by livestock or humans.

In a captive situation, stable flies were common on a moose calf kept on straw bedding, while few such flies were attracted to a nearby cow (Sein 1985). The moose calf exhibited less reaction to bites to the lower legs than did caribou calves, which flinched and stomped continually when stable flies were active.

Smith et al. (1970) identified and documented the seasonal succession of 24 species of horse flies and deer flies attracted to moose in Algonquin Park, Ontario (Table 46). Tabanids commonly feed on the back legs above the hock, the forelegs, beneath the belly and on the face (Sein 1985).

Table 46. Biting flies (*Diptera*) collected from moose

Species	Location	Source
Muscidae		
Haematobosca alcis	Minnesota	Snow 1891
	Michigan	Murie 1934
	Ontario	Peterson 1955, Stone et al. 1965
	Newfoundland	Stone et al. 1965
	Michigan	
	Minnesota	
	Alaska	
	Wyoming	Burger and Anderson 1974
	Alberta	Pledger et al. 1980
	Ontario	Lankester and Sein 1986
Stomoxys calcitrans	Ontario	Sein 1985
Tabanidae		
Chrysops (deer flies)	Ontario	Smith et al. 1970
C. celvus		
C. cincticornis		
C. cuclux		
C. excitans		
C. frigidus		
C. lateralis		
C. mitis		
C. montanus		
C. niger		
Hybomitra (horse flies)		
H. affinis	Ontario	Smith et al. 1970
H. arpadi		
H. criddlei		
H. epistates		
H. illota		
H. lasiophthalma		
H. lurida		
H. microcephala		
H. nuda		
H. trepida		
H. trispila sodalis		
H. typhus		
H. zonalis		
Tabanus (horse flies)		
T. marginalis	Ontario	Smith et al. 1970
T. nigripes		
Simuliidae		
Simulium (black flies)		
S. venustum	Minnesota	Olsen and Festermacher 1942
S. pictipes	Minnesota	Nicholson and Mickel 1950
S. venustum	Alberta	Pledger et al. 1980
S. decorum		
S. arcticum		
S. vittatum		
S. aureus		
S. meridionale		
S. furculatum		
S. pugetense		
S. latipes		
S. croxtoni		
S. jenningsi		
S. euryadminiculum		
Prosimulium (black flies)	Alberta	Pledger et al. 1980
P. formosum		
P. decemarticulatum		
P. exigens		
Culicidae (mosquitos)		
None identified		

"When the sun is very hot in summer these insects [horseflies] are most active," wrote Lugger (1896: 168), "and thousands of them fly about, and frequently torment the larger animals to such an extent as to make them perfectly wild and frantic. The moose and deer attacked by such tormentors, lose all fear of man and plunge into rivers and lakes to escape their attacks: they soon become very poor, as they have no rest to feed excepting at night." Lugger apparently discounted the possible annoyance factor of nocturnal mosquitos. *Photo by Michael H. Francis.*

Pledger et al. (1980) recorded 15 species of black fly attracted to moose in Alberta (Table 46), but only 6 took blood (*Simulium decorum, S. venustum, S. vittattum, S. arcticum, S. aureum* and *Prosimulium formosum*). Most black flies fed on the lower legs, belly, brisket and around the anus, where hair is short and thin.

There appear to be no published reports identifying the species of mosquitos that feed on moose.

Skeletal Pathology

Senior moose experience skeletal changes that begin to interfere with mobility, sometimes disastrously so. Limping moose were rarely seen during a long-term study on Isle Royale, Michigan (Peterson et al. 1982), yet the high frequency of joint disease in wolf-killed animals suggested a very effective culling process. Joint problems in moose may occur after an injury, a persistent bacterial infection in the synovial fluids (infectious arthritis) or, more commonly, as a result of wear and deterioration of the cartilaginous articulating surfaces. The latter is termed "primary degenerative joint disease" (Sokoloff 1969), and may be largely a natural consequence of aging. In the advanced stages of joint disease, identifying which of these might have initiated the condition usually is impossible.

In moose, degenerative joint disease affects males more so than females, and the frequency increases after 7 years of age (Peterson 1977). Older animals on Isle Royale became increasingly vulnerable to wolves and the high proportion of arthropathy in predated animals suggested that joint disease was an important predisposing factor. The hip (coxofemoral) joint is most commonly affected, but excessive articular wear and osteophyte proliferation also are seen on the moving surfaces between the vertebrae in the lower back (Wobeser and Runge 1975, Peterson 1977). Although less severe, articular wear is evident in joints of the foot and lower leg, especially in the rear legs (Timmermann and Lankester 1978). Striking differences may be seen between populations. In a comparative study by Peterson et al. (1982), most degenerative joint disease seen in moose on the Kenai Peninsula of Alaska involved the lumbrosacral joint in the back, whereas a majority of cases seen on Isle Royale involved the hip joint.

The frequency of joint disease in a moose population may change over time and presumably influence predator/prey dynamics. Peterson (1988) examined the remains of 752 moose killed by wolves over a 30-year period. All were at least 7 years at the time of death. In the interval between 1955 and 1987, there was a steady increase in the prevalence of joint disease in both males and females. Poor nutrition of

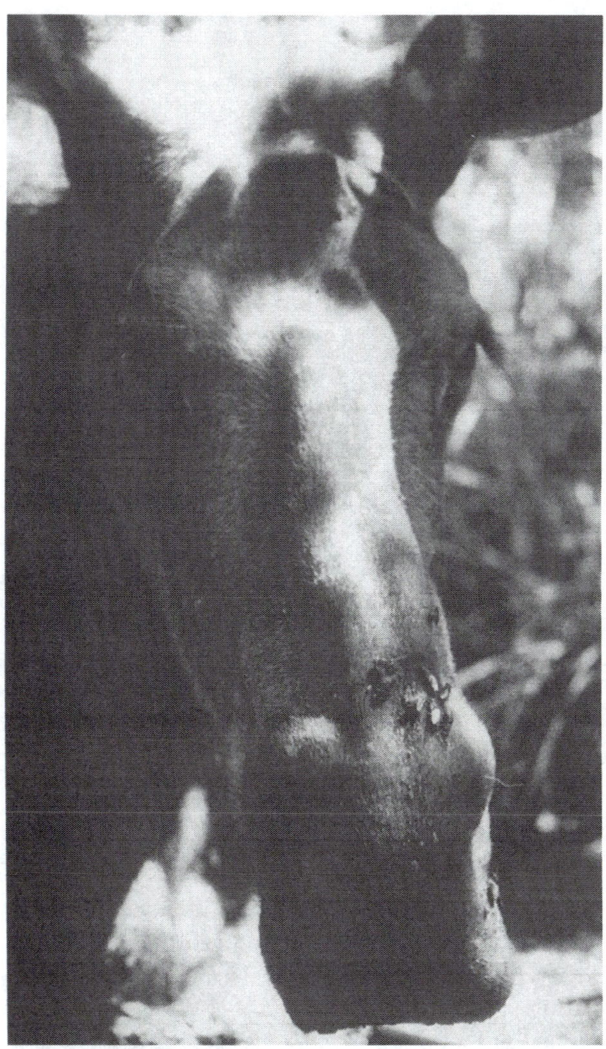

Horse flies (above) and other biting flies bite and obtain a blood meal from moose, particularly where the hair is thin and short. Moose appear to be amazingly tolerant of moderate numbers of these flies. *Photo by M.W. Lankester.*

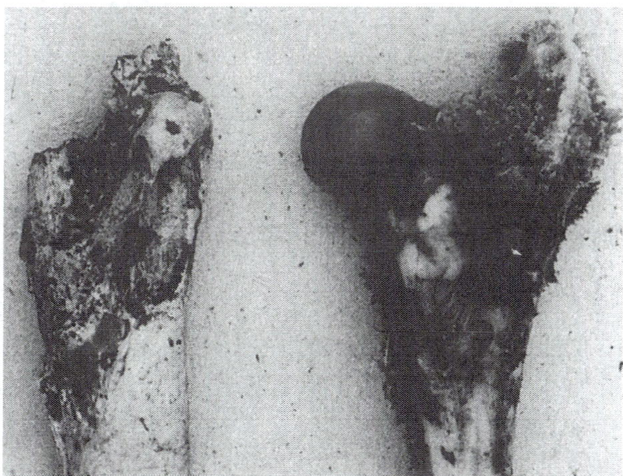

An arthritic femur (left) versus a normal femur (right) from moose on Isle Royale National Park, Michigan. Such degenerative joint disease, or "arthropathy," likely is a consequence of natural aging and has been postulated to be a predisposing factor to predation. *Photo by R.O. Peterson.*

pregnant females and growing calves was hypothesized to lead to abnormalities in joint cartilage that increased the frequency of degenerative joint disease later in life.

The loss of bone density, or osteoporosis, has been detected in moose using computerized tomography scans (Hindelang et al. 1992). An initial study suggested a link between measurable osteoporosis in long bones and more pervasive skeletal pathology.

Summary

Parasites are a part of the life of every wild and domestic animal. For moose, the list of parasites is long and varied, but not unusually so. Most of these specially adapted organisms are reasonably well-behaved companions. Yet, in a manner more subtle than that of large predators, parasites similarly serve in an evolutionary sense to strengthen moose as a species by making survival just a little tougher for the weaker, less well-adapted and supernumerary individuals. Because of persistent, selective pressure from parasitic organisms throughout evolutionary history, healthy moose have an amazing ability to recover from a wide range of injury and infection.

Of all the parasites described here, only a few are presently thought to have the potential to regulate or limit moose numbers. In particular, the winter tick *D. albipictus* can cause appreciable late winter die-offs, as a result of a complex interaction of the tick, moose densities and nutrition, and weather. The meningeal worm *P. tenuis* kills some moose and may affect others in as-yet unrecognized ways. Unlike that of the winter tick, the impact of the meningeal worm is not a function of moose densities, but of the density of cohabiting white-tailed deer. It cannot, therefore, regulate moose populations in the strictest sense of this word, but may be an important limiting factor to moose population growth.

Mortality factors seldom act alone to influence animal populations and, except for hunter-killed animals, few individuals die of a single cause. The challenge remains to measure and comprehend fully the significance of the role played by parasites and diseases in these interrelated living systems.

COLOR & SIZE RELATIONSHIP OF YOUNG CALF SEEN NEAR PARK HDQ., MAY 29, '62. COW WAS TRAVELLING AT A WALK & CALF KEPT UP EASILY AT A LOOSE-JOINTED TROT, ONCE RUNNING AHEAD BRIEFLY ~ AWKWARD AND "LEGGY" IN ITS GAIT, BUT STEADY ON ITS FEET. (COMPLETED FROM NOTES & REF., MAY 30). ROCK CRK., MCKINLEY PARK, ALASKA.

THE FIRST CALF OF THE YEAR SEEN AT DENEKI LAKES — JUNE 4, '62 — NIBBLED AT WILLOWS IN PASSING — ITS MOTHER APPEARED TO BE THE SAME COW SKETCHED ON MAY 10, WITH A PEAK OF HAIR BACK OF HUMP & THE CALF HAD A SMALLER VERSION OF THIS SAME "COWLICK" — (FROM MEMORY & NOTES)

(NOTE — ON CALVES ABOVE, MUZZLE SHOULD BE SHORTER, "BOXIER"; AND EARS BIGGER — JUNE 10, '62)

16

ALBERT W. FRANZMANN

Restraint, Translocation and Husbandry

Restraint, health and husbandry of moose are so closely interrelated that concern for and attention to one cannot be to exclusion of the others. Moose are restrained to collect biological material, check their health, deploy transmitters and translocate them. Successful wildlife restraint has many facets requiring ecological, biological and behavioral familiarity with the species in question. It varies with each type of restraint but becomes especially complex with drug use. The use of dart guns for chemical restraint requires an understanding of ballistics and gun and human safety. Persons involved should be qualified in first aid and cardiac/pulmonary resuscitation (CPR). Most drugs used require veterinary supervision; consequently, most wildlife resource agencies have a veterinarian on staff or on call to supervise drug procurement, storage and use.

The discussion of restraint in this chapter is an introduction to the subject. It neither substitutes for detailed training in wildlife restraint nor prepares one to perform the procedures. Persons not fully trained in wildlife restraint have attempted some of these procedures with disastrous results. Even highly trained and experienced persons cannot foresee all the problems that may surface during a procedure, but they certainly are better prepared to handle those that arise. Many state, provincial and federal resource agencies require staff training in wildlife restraint before field involvement.

Restraint may be accomplished by physical or chemical means, or a combination of both. Physical restraint of adult moose has limited application because it involves the use of heavy-duty corral and fence-line traps, squeeze gates and chutes. Calves to 1 month of age may be caught and restrained without physical structures, depending on reaction of their mothers.

Chemical restraint may be used with physical restraint, but most often is used alone. Within the past 30 years, refinements of immobilization drugs and the systems to administer them to large mammals have made chemical restraint a routine procedure (Franzmann 1982). Nevertheless, persons involved must be qualified to monitor the health of the animal and apply sound husbandry practices in the process.

Monitoring health during restraint involves applying traditional veterinary techniques to evaluate effects that restraint has on various physiological systems of the subject. However, consideration for the health of moose goes beyond the application to restraint. Animals respond to changes in their environment that affect their health. Measures of animal health serve as indicators to environmental change, often called the "indicator animal concept" (Franzmann 1985). When properly applied, this concept is useful to assess population health and thereby measure environmental change.

The term *husbandry,* referring to the production and care of animals, is applied most often to the care of domestic livestock. There are occasions when moose raised in confinement may require certain husbandry practices unique to this species for successful "domestication." Historically, moose served as beasts of burden and for production of milk and meat. The most extensive attempt at moose hus-

"Domestication" of moose, including as "beasts of burden," appears occasionally in historic records from across the species' range in North America (see also Chapter 1). Photographic documentation is not reliable indication of the success of such efforts, which was minimal. Nearly all older photographs feature young animals. It was and is very difficult to maintain tame moose much past 2 years of age because of digestive problems associated with feeds other than natural browse. Today, moose can be maintained on nonbrowse feeds, using formulated rations that have a high lignin content. The special dietary needs of moose and the amount of food needed makes moose farming or ranching an expensive proposition and generally not commercially feasible. Moose are tamed and maintained today primarily for exhibition and research purposes. *Left photo by E. Brown; courtesy of the Provincial Archives of Alberta. Right (circa 1908) photo by P.S. Hunt; courtesy of the Anchorage Museum of History and Art.*

bandry occurred in Russia on three experimental farms that began operating in the late 1940s. Since then, approximately 450 moose have been domesticated there. In recent years, one justification for these operations was production of milk that supposedly has unique medical properties in treating stomach ulcers and radiation lesions (Timmermann 1991). These experimental farms were downgraded or closed during the peristroika period for lack of success and limited financial support. In North America, moose are domesticated primarily for wildlife research and zoological exhibition. Moose ranching or farming has not been successful because of the types and amount of feed needed and expense of facilities. No matter what the application, good husbandry practices are necessary.

Physical Restraint

Aboriginal people used various traps and pitfalls to capture moose and kill them for consumptive use. Before development of satisfactory drugs and their delivery systems, physical restraint was the only method available to capture moose for translocation (Pimlott and Carberry 1958) or marking for identification (Ritcey and Edwards 1956). Moose were trapped for translocation almost exclusively using corral type traps baited with browse and/or hay (Figure 199). Trapped moose were crated for shipping directly from the trap or from a holding corral built near the traps. From 1934 to 1953, 230 moose were captured in corral traps in Michigan, Wyoming, Newfoundland and Alberta. Of those, 133 were translocated, 35 escaped and 62 died (Pim-

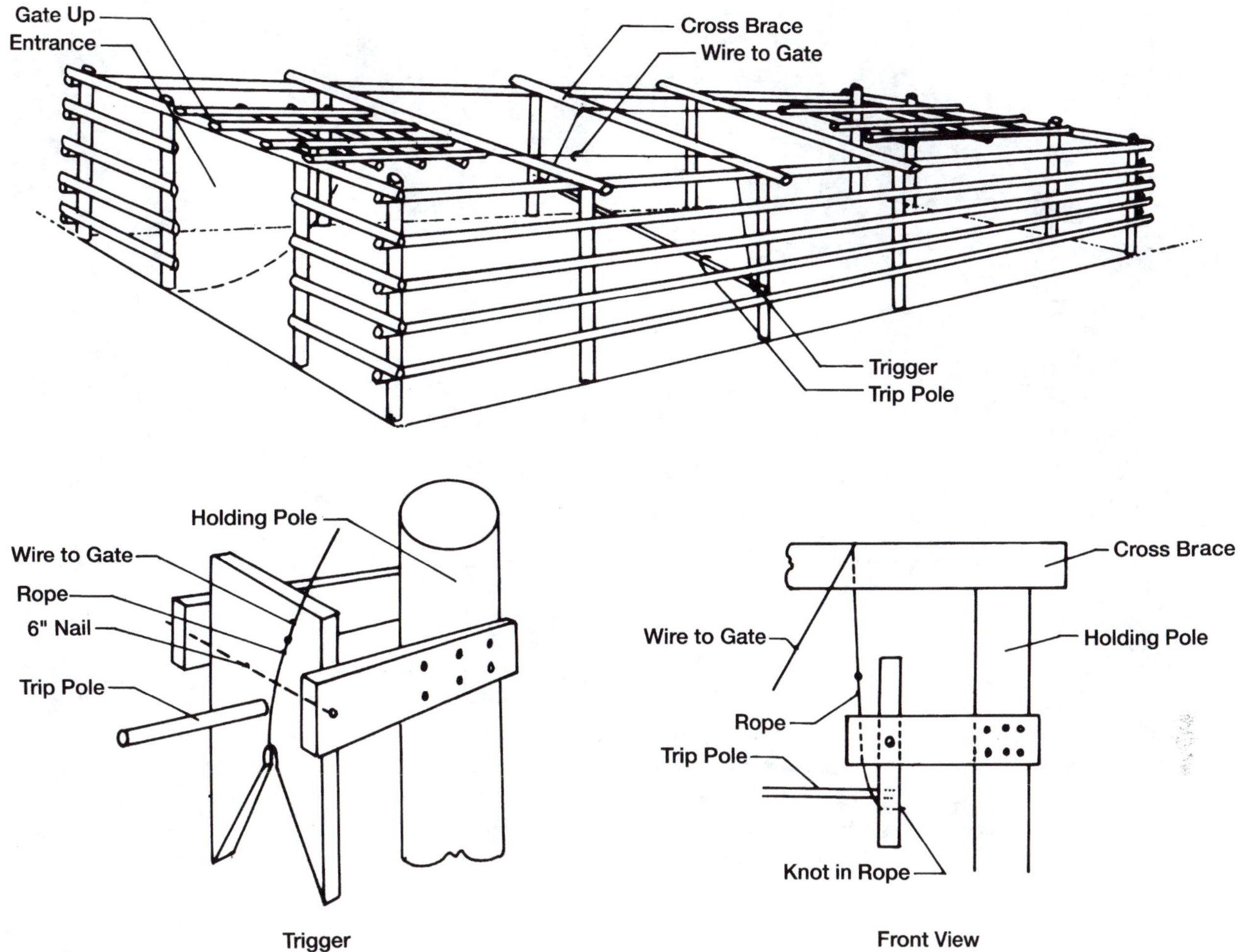

Figure 199. Diagram of a corral-type trap (top) with double gates for moose, and of the tripping device (bottom left and right) for the gates. This type of trap, with various modifications for each situation, requires baiting with hay. Some traps of this basic design have a squeeze chute at the end of a runway where the animal can be confined and physically restrained for brief procedures. In the absence of a chute, the moose usually are immobilized for handling and processing. *Illustration from Pimlott and Carberry (1958).*

lott and Carberry 1958). With the limited technology available at the time, such a high mortality rate was unavoidable. Even so, the successful translocations resulted in establishment of new populations that subsequently expanded into new habitat. Today, a 25+ percent mortality rate would be unacceptable, because of the availability of chemical restraint and improved methods of monitoring and medical care for captive wild animals.

In the past, captured moose were ear tagged in corral traps and then driven into holding chutes (Ritcey and Edwards 1956). Amphibious helicopters now are used to force moose into deep water, allowing a capture crew to tag and/or collar them (Simkin 1963). Motor boats can be used for the same task (Roussel and Pichette 1973). Snowmobiles are useful in deep snow to approach calf moose for tagging (Roussel and Pichette 1973).

The combination of physical and chemical capture using fence-line traps and drugs began in the late 1960s at the Moose Research Center in Alaska. The Center is a research facility operated by the Alaska Department of Fish and Game located on the Kenai Peninsula. The facility has four 1-square mile (2.6 km^2) enclosures constructed with 8-foot (2.4-m) high woven-wire fences. Exclosures are in each enclosure, and support facilities are around its perimeter (Figure 200). Two smaller pens hold experimental hand-reared moose.

Around the facility, fence-line traps (Figure 201) are used to capture and process moose for various research pur-

The health of a moose population ultimately is reflected by its productivity. Generally, single calves (left) are produced, but twinning (right) occurs at greater proportion in populations that are healthy and have an abundance of food resources. Triplets have been recorded, but are rare. Also, how an animal responds to its environment is an indicator of the quality of its habitat. Indicators are gained from physiological and morphological measurements, as well as from number of calves produced. Biologists using these indices are applying the "indicator animal concept." *Left photo by Albert W. Franzmann; courtesy of the Alaska Department of Fish and Game, Soldotna. Right photo by R.O. Peterson; courtesy of Michigan Tech University.*

poses. Moose traveling along the fence line trip a string and gates on each end drop, trapping the animal for subsequent immobilization. This method has proved useful in restraining hundreds of moose since 1970. As more effective drugs became available, the procedure improved, but the basic trap design has not changed significantly (Franzmann et al. 1982).

"Mortality" transmitters are used to study causes of moose calf death (Franzmann et al. 1980a, Ballard et al. 1981a). These transmitters emit a routine signal when placed on the calf. When the calf is alive and active, the transmitter beeps approximately every second. If the calf remains motionless for more than an hour or so the signal pulse will double or triple, depending on the setting chosen. A fast signal indicates that the calf is likely dead. By locating the transmitter, biologists can determine the cause of death (Ballard et al. 1979).

Chemical immobilization of moose calves is unnecessary and could lead to abandonment by the mother. The best procedure is to force a cow away from her calf. A capture team quickly captures and collars the calf, while a helicopter keeps the cow away from the capture crew. The cow usually returns quickly to the calf, but if not, can be encouraged by herding with the helicopter (Ballard et al. 1979).

Physical restraint as an adjunct to chemical restraint may consist simply of moving or holding a body part in place for a certain procedure. For example, a rope halter can hold the head when blood is collected from the jugular vein. The legs of a moose may require restraint with hobbles to protect personnel. Rope slings often are necessary for lifting moose to obtain a milk sample or weight data. Blindfolding immobilized moose may help prolong sedation and ear plugs may be used to minimize noise disturbance. Hand-reared semidomesticated moose used for research and/or exhibition can be physically restrained using techniques developed for domestic animals. Net-guns that shoot a canister-held net now are used on many species (Morkel and La Grange 1993), including moose (Carpenter and Innes 1995), except those of the Alaskan/Yukon subspecies. Fowler (1978) outlined the many types of restraint tools, ropes and equipment and their potential application for both wild and domestic animals.

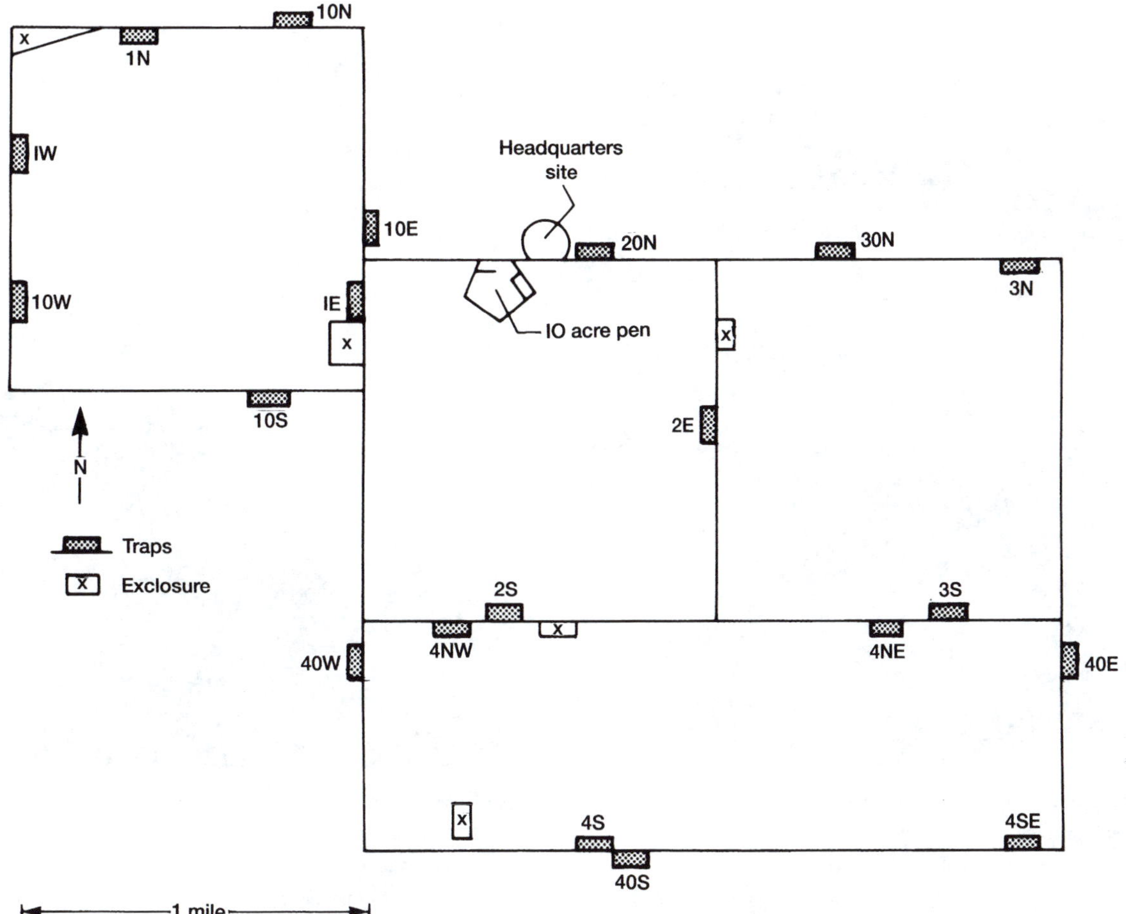

Figure 200. Schematic of the fencing and facilities at the Alaska Fish and Game Department's Moose Research Center near Soldotna (from Franzmann et al. 1976d). Fence-line traps are used to capture moose within the enclosures and around the outside perimeter of the facility. The traps were designed so moose can be moved from one adjacent enclosure to another or from the enclosures to outside or vice versa. Each of the four pens is 1 square mile (2.64 km^2).

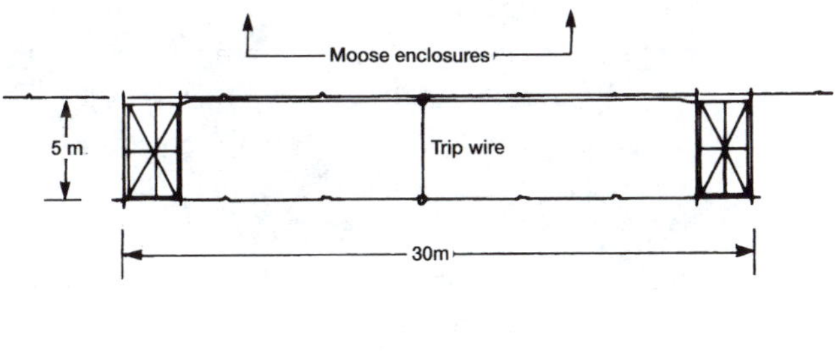

Figure 201. Diagram of a fence-line (top) trap for moose similar to those used at the Moose Research Center, Alaska. Moose follow the fence line through the open gates (at both ends) and, when midway through the trap, a trip-string (bottom) is pulled that allows both gates to drop. *Illustration from LeResche and Lynch (1973).*

Fence-line traps along the fence perimeters and between enclosures at the Moose Research Center in Alaska (left) were designed to capture, process and turn out moose from the surrounding area (see Figure 200 for location of the traps). They also are used to recapture resident moose and to obtain long-term serial data on physiology, morphology and reproduction. Moose are caught in one of the Center's fence-line traps when they walk into it from either end. A string across the trap is tripped, dropping doors on both ends. At right, a person is preparing to use a dart gun to immobilize a captured moose so that it can be processed (weighed, blood sampled, measured, tagged, radio-collared, etc.). The density of moose maintained at the Center has depended on the objectives of studies underway. *Left photo courtesy of the Alaska Department of Fish and Game. Right photo by Albert W. Franzmann; courtesy of the Alaska Department of Fish and Game, Soldotna.*

Chemical Restraint

Chemical restraint of animals is not new technology, but the drugs and methods used to administer them are. Indigenous people in many parts of the world used various plant extracts, such as curare, on spears and arrowpoints to restrain and kill animals for food.

Drugging an animal with a hand-held syringe, after some form of physical restraint, is done with some species. However, this has never been widely applied to moose because of the problems associated with physical restraint, the first generation of drugs required large dosages and effective drugs to immobilize large animals were not available until recently. Modern chemical restraint of wildlife began with the development of projectile darts in the 1950s (Hall et al. 1953, Crockford et al. 1958). Refinement and marketing of the dart and development of a powder-charged rifle made available a remote drug delivery system for wildlife capture. Since then advancements in both drugs and equipment have made better restraint possible. The increased ability to capture moose safely has improved understanding of the species' ecology and, thereby, improved management.

The person who administers a drug to capture or restrain an animal has responsibility for the life of that animal. To assure that the animal receives the best care, the following items must be considered before capture (International Wildlife Veterinary Service, Inc. 1991):

- Is the purpose justified?
- Type of animal
- Type of terrain or escape cover
- Condition of the animal
- Emotional state of the animal
- Safety equipment available
- Ambient temperature
- Time of day

- Which drug(s) would be most effective?
- Will physical restraint alone suffice?
- Is there adequate assistance and professional support?

Considerations with use of chemical restraint are not limited to animal well-being, but also include regulatory, social, medical, safety, economic and political factors.

Drug Delivery Systems

The types of remote drug delivery systems vary, and not all are covered here. For detailed descriptions refer to Fowler (1978), International Wildlife Veterinary Service, Inc. 1991, Bush (1992) and Rohr and McKenzie (1993). The equipment varies from the simple hand-held syringe to the sophisticated powder-charged and CO_2-powered guns (short-range or long-range). The guns propel darts with the selected drug(s) into the animal. Other types of projectile equipment include jab sticks (a standard syringe on a long rod or pole), blow guns, long bows, compound bows and crossbows. Selection of equipment is determined by the physiology of the animal in question and circumstances of the procedure.

For moose, hand-held syringes, jab sticks and blow guns are for short range. Their practical applications are for moose restrained in a trap or for captive animals that can be approached closely. Even then, the hand-held syringe may not be practical, the jab stick and blow gun have limited application, and the blow gun has some application for the experienced person. Short-range CO_2 guns or longer-range powder-fired guns invariably are the choices for darting moose.

For darting, biologists approach most free-ranging moose with a helicopter or occasionally, by snowmobile (Ritchie and Barney 1972), all-terrain vehicle, boat or on foot. The Cap-Chur[R] projectile system (Palmer Chemical Co., Douglasville, GA) is used most often (Figure 202). This system remained in use long after the development of lightweight darts and other projectile systems because it could accommodate the volume of drug needed (up to 12 milliliters [mL]). Before the availability of the drug carfentanil hydrochloride (Wildnil[R], Wildlife Pharmaceuticals, Fort Collins, CO), lightweight darts could not be loaded with the volume of drug needed. This was a problem specific to the United States because the only effective drug licensed by the Food and Drug Administration was etorphine hydrochloride (M-99[R], Lemmon Co., Sellersville, PA). M-99 was a very dilute solution requiring a large volume that could be handled only by Cap-Chur[R] equipment. Canadian veterinarians were able to obtain etorphine hydrochloride in the powdered form or a concentrated product (Immobilon[R], Rickett and Coleman, Hull, England). They were able to use smaller darts and a variety of new projectile guns. When effective concentrated drugs became available for moose, many researchers and managers chose to use the lightweight darts with capacity of no more than 3 mL.

Injecting drugs into wild animals for purposes other than immobilization and biologicals (vaccines, bacterins) may be done with the same dart gun equipment. A system (BallistaVet, Inc., White Bear Lake, MN) is available by which a drug or biological is in a bullet fired into the animal. This bio-bullet penetrates the skin and lodges in muscle. Within a few minutes, the bullet dissolves and the product enters the bloodstream.

The limitations and applications of a projectile system are essential information. There are many choices, and users should spend time practicing to become completely comfortable and capable with the system chosen. Accuracy is important; expertise is necessary to place the dart in heavy muscle tissue. The preferred sites on moose are the heavy musculature of the upper rear leg or the shoulder. The top of the rump is a desirable target for shots from a helicopter. However, during late summer and autumn, a

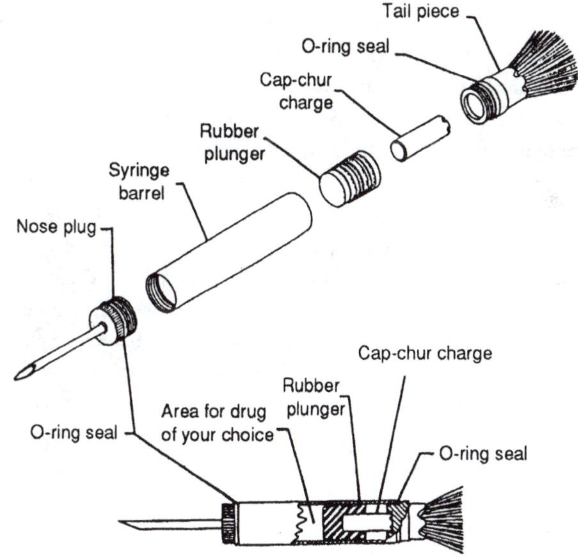

Figure 202. Diagram of the Cap-Chur[R] dart, which was the first successful commercial dart used for immobilizing wildlife, including moose. The dart still is extensively used, but newer, lightweight designs have replaced it when small volumes of drug can be used. *Illustration courtesy of the Palmer Chemical and Equipment Company of Douglasville, Georgia.*

Expandable collars fitted with radio transmitters are used to monitor moose calves. A change in the frequency pulse indicates immobility and, in time, mortality. Biologists then follow the radio signal to the collar (and calf) to examine the site and carcass. The "mortality" collar technique begins with daily monitoring of the calving areas with a helicopter to locate subject neonates. Once a calf (or calves) is located, a crew of two is dropped off nearby. The mother's attention is diverted by the helicopter while the crew quickly slips a collar on the calf and departs. A cow's reaction varies from very aggressive defense of her calf or calves to no defense at all. After retrieving the ground crew and before proceeding to the next calf, the helicopter pilot makes sure that the cow and calf are back together. *Photo by Albert W. Franzmann; courtesy of the Alaska Department of Fish and Game, Soldotna.*

thick layer of subcutaneous fat covers the area. Fat has very little blood supply and should be avoided. Subcutaneous injections, especially in fatty areas, result in delayed absorption. No matter how good the drug, if it is not injected into an area where it can be properly absorbed, it will not perform as expected. There is both science and an art to darting animals; science can be taught, but art requires experience and patience.

Drugs for Chemical Restraint

The modes of action of various classes of drugs help determine the drug needed for a particular objective. Tranquilizers produce a calming effect, but do not immobilize animals. These drugs, in combination with other immobilizing drugs, produce additive and/or synergistic effects. They calm wild animals during transport or during prolonged captivity. Sedative/hypnotic drugs cause various degrees of depression of most central nervous system functions, depending on dosage. This class of drug includes alcohol, barbiturates and ether. A subclass of sedative/hypnotics includes drugs with alpha-adrenergic receptor site activity. Reversal of these drugs requires specific antagonist drugs. Narcotics are powerful pain killers (analgesics) and, with increased dosage, cause central nervous system depression and anesthesia. The effects of this class of drugs also can be reversed with specific antagonists. Most narcotics originate from morphine. Dissociative drugs separate the conscious mind from the sensory and motor control mechanisms of the brain producing rapid analgesia and a trance-like state (International Wildlife Veterinary Service, Inc. 1991). Use of these drugs may result in pronounced muscle rigidity, increased body temperature and convulsions. Therefore, they are combined with tranquilizers or alpha-adrenergic sedatives to modify this effect. Paralytic drugs block transmission between motor nerves and muscles. They have a very narrow safety margin. Animals paralyzed with these drugs have full sensory awareness and can feel pain.

Chemical restraint or immobilization of moose follows a pattern of development similar to that for other ungulates (Franzmann 1982). Nicotine salicylate was the first drug used, but was not well received because of its unpredictable effects. Succinylcholine chloride followed, and was the drug of choice for several years. Both are paralytics that did not anesthetize the animal; both have a very narrow range of effectiveness. Synthetic narcotics replaced them. First was etorphine (Houston 1970), which is 10,000 times more potent than morphine, followed by carfentanil (Franzmann et al. 1984c), which is 20,000 times more potent than morphine. Fentanyl citrate, another morphine-like drug was used to a limited degree in moose because it is less potent than either etorphine or carfentanil. These narcotic drugs require licensing of users by the Drug Enforcement Administration in the United States and by the Food and Drug Directorate in Canada, depending on the country of use (Hebert and McFetridge 1979). Guidelines established for licensing in the United States include drug distribution limited to veterinarians, strict guidelines for secure storage, and accurate keeping of records that are subject to audit.

Despite improvements in immobilizing drugs for moose,

The use of net-guns fired from helicopters to capture large mammals originated in New Zealand during a red deer reduction program. This technique has since been applied to most North American large mammals. The technique for moose was slower to develop because of the animal's size, but it has recently (since 1992) been applied with success for all subspecies except the Alaskan/Yukon moose. Capture by netting requires no drugs and is preferred when drug use is prohibited because capture coincides with hunting seasons or where drugs are not allowed because subsistence hunting is in effect. Narcotic drugs are the most effective for moose immobilization, but not all agencies have personnel licensed and capable of using them, so opt instead for netting. Referring to netting, Carpenter (1996: 3) wrote that "The ability to capture large numbers of moose quickly, safely, and humanely has greatly increased the potential for designing studies for this species." Disadvantages of netting are that it is generally more expensive, not as useful for specific individuals or small groups of moose (problem animals), perhaps not applicable for Alaskan/Yukon moose or adults of other subspecies at peak condition due to size, not ideal for procedures requiring long down time or lengthy immobilization, some sampling procedures for moose in nets may be difficult or compromised, and injury to handlers is more likely. When good replacements for narcotic drugs for immobilizing moose become available in North America, the advantages for netting will decrease, but it will remain a valuable option for capture under certain circumstances. *Photos by Richard P. Smith.*

Powder-charged dart guns (top two guns, inset), carbon dioxide-powered dart guns (middle two guns), and carbon dioxide-powered dart pistols (bottom two) are used to fire projectiles that deliver immobilizing drug to free-ranging moose. The type of gun used in any immobilizing event depends on such factors as species, age and/or sex characteristics, range, drug volume, environmental conditions (principally, weather and topography) and shooting platform. *Photo courtesy of Pneu Dart, Inc., of Williamsport, Pennsylvania.*

none of the drugs currently available meets the 10 criteria that an ideal drug should possess (Franzmann 1982):

1. Rapid absorption and action
2. Concentrated form—small quantity for injection
3. Wide range of tolerance for the animal
4. Safe for handler
5. Reversible with an antagonist drug
6. No side effects
7. Effective anesthesia level
8. Not subjected to Dangerous Drug licensing
9. Cleared for use on animals for food
10. Low cost

By these criteria, etorphine and carfentanil are similar, except that carfentanil is more concentrated. Their negative aspects are: danger to handler; subject to controlled substances licensing; not cleared for use on animals used for food; and high cost. Nevertheless, carfentanil has become the drug of choice because it is available in a highly concentrated form. Dosage for carfentanil in moose ranges from 3 to 6 milligrams (mg) per adult (0.010 mg/kg body weight). The wide dosage range reflects the variability in moose response because of several factors. The first is body weight. Seldom is the true weight of an animal known, and its determination must depend on estimations. Other factors that influence dosage are age, sex, time of day, seasons,

The most common method of immobilizing free-ranging moose is "shooting" from a helicopter, but approach by boat, snowmachine, all-terrain vehicle and on foot also is done. The time required for the moose to become immobilized once injected with the best drugs is around 4 minutes. *Photo by Albert W. Franzmann; courtesy of the Alaska Department of Fish and Game, Soldotna.*

body condition, temperament, food consumed, drug tolerance, drug idiosyncrasy and pathological conditions.

Xylazine hydrochloride (Rompun[R], Haver Laboratories, Shawnee, KS) is a nonnarcotic, alpha-adrenergic sedative, analgesic and muscle relaxant that may be combined with narcotic immobilizing drugs to balance their qualities. It may be used at high dosages for long-time sedation, but prolonged recovery occurs without a reversing drug. This limits its application in moose because the available reversing drug for xylazine (yohimbine hydrochloride, Antagonil[R], Wildlife Pharmaceuticals, Inc., Fort Collins, CO) is not always effective for moose (although it is effective in most other species). Some biologists prefer xylazine mixed with etorphine (Gasaway et al. 1978a), or with carfentanil (Seal et al. 1985) because of xylazine's muscle-relaxing effect. The combined drugs also require less narcotic. The amount of xylazine in the mixture should not exceed 300 mg for an adult moose, or recovery will be delayed. Idazoxan (Ricketts and Coleman, Hull, England) is a new xylazine antagonist that may be effective in moose (Lance 1991).

Drugs used to immobilize moose cannot be discussed without considering what may become available in the future. Lance (1991) identified new pharmaceutical tools for the 1990s, including some that are applicable to moose. The most promising is medetomidine (Farmos Pharmaceutical, Turku, Finland). It proved to be an effective and safe nonnarcotic alpha-adrenergic sedative when combined with ketamine hydrochloride for moose and a wide variety of nondomestic mammals in zoos in Helsinki, Finland and Kolmarden, Sweden (Jalanka and Roeken 1990). Medetomidine is a potent, highly specific, nonnarcotic and safe drug, and can be reversed with an antagonist (antipamezole). A newer short-acting fentanyl derivative (A3080) successfully immobilized elk (Stanley et al. 1988) and a variety of African species (Janssen et al. 1991). These drugs currently are available for experimental and testing purposes only. Most new drugs testing is on captive moose at research facilities, such as the Moose Research Center in Alaska.

Most drugs used to immobilize moose are classified as "dangerous drugs" by federal regulators. Extreme care must be used in handling because a single drop of these drugs can be fatal to humans if injected, splashed in the eye, or makes contact with broken skin. Only experienced and professional people should administer the drugging of moose (or other wildlife). Training courses are available and required by many jurisdictions. *Photo by Heather K. Bickle; courtesy of the Ontario Ministry of Natural Resources.*

Modifying Agents

The two primary drugs used to immobilize moose (etorphine and carfentanil) are narcotics that can be reversed (antagonized) by other drugs to arouse the animal from its anesthetic state. The first product produced as an antagonist to etorphine was diprenorphine hydrochloride (M50-50[R], Lemmon Co., Sellersville, PA), which is given intravenously. The dosage rate is equivalent to twice the milligram dosage of etorphine necessary to immobilize the animal. Overdosages of diprenorphine severely depress the central nervous system, may complicate recovery or even cause respiratory collapse and death.

Naloxone hydrochloride (Narcan[R], DuPont Pharmaceuticals, Wilmington, DE) is a pure narcotic antagonist that has no depressant activity. It has a very short half-life (time for the body to metabolize the drug) of 30 minutes when given intravenously. If part of the narcotic is slowly absorbed, it may enter the bloodstream after metabolization of naloxone, and the animal may become narcotized again—a process known as "renarcotization." Giving high doses of naloxone—and part of it intramuscularly or subcutaneously—helps prevent renarcotization. The dosage of naloxone is equivalent to 100 to 150 times the milligram dosage of carfentanil used to immobilize the animal.

Naltrexone hydrochloride and nalmefene hydrochloride are two new pure narcotic antagonists that show great promise because their half-life is 12 hours versus 30 minutes for naloxone.

Other drugs may be selected to modify the effects of immobilizing agents; when combined they are known as "cocktails." The use of complex cocktails during immobilization requires a very experienced professional. Hyaluronidase (Wydase[R], Wyeth Laboratories, Philadelphia PA.) speeds up absorption of the primary drug (Franzmann and

Arneson 1974, Haigh 1979). A mixture of carfentanil, fentanyl, xylazine and hyaluronidase successfully immobilized 123 moose in the Yukon Territories, and a combination of diprenorphine and naloxone reversed the mixture (Larsen and Gauthier 1989). Other combinations should be critically analyzed in each situation and tested under controlled conditions when possible. Carfentanil, the current narcotic of choice for moose, has been successfully and widely used without any adjunct drugs. However, most biologists prefer using it in combination with xylazine.

A promising new class of drugs used for long-term tranquilizing effect is the long-acting neuroleptics (LANS) used in human medicine for behavioral modification. These drugs used in impala resulted in sustained tranquilization with no effect on feeding or water intake (Gandini et al. 1989). These drugs have significant promise for postcapture use in all species when there is any amount of holding or processing time. LANS have not been tested on moose, but undoubtedly will be when available.

I recommend immobilizing free-ranging moose in late winter and early spring because of cooler weather, increased daylight, presence of snow and easier access to populations concentrated on winter range. At this time, female moose are pregnant, and the effect of drugging on unborn calves has been a concern. Roussel and Patenaude (1975) reported no effects on nine pregnant cows using etorphine. Larsen and Gauthier (1989) using the mixture of carfentanil, fentanyl, xylazine and hyaluronidase on 123 cow moose also reported no ill effects. However, they report that this drug use on cows in the last 2 to 3 months of pregnancy may have lowered postnatal survivorship. Over a period of 15 years, I found no negative effects from immobilizing hundreds of pregnant moose at the Moose Research Center. This included using most of the drugs listed herein and some unlisted experimental ones. The moose at the Moose Research Center were under long-term observation for their reproductive success and production of viable calves. Ballard and Tobey (1981) reported lower productivity of pregnant moose cows immobilized with succinylcholine chloride. They attributed the differences to stress on the moose from the paralytic drug. Van Ballenberghe (1989) disputed the findings of Ballard and Tobey (1981), but acknowledged the need for more data. Other studies on caribou immobilized with etorphine (Valkenburg et al. 1983) and on white-tailed deer immobilized with xylazine (DelGuidice et al. 1986) showed no side effect on pregnant females.

Other drugs may be used on an immobilized animal in emergencies (discussed later). No matter what drug is used on moose, users must be aware that the Food and Drug Administration has not cleared any drugs for use on wild mammals that will be consumed within 45 days of administration. Field immobilization of moose must be done 45 days before a hunting season or after it closes.

Preventive Measures Drugs

Preventive medical treatment for moose may be necessary in some circumstances, and that should be decided with and by veterinary supervision. The types of treatment to consider are: long-acting antibiotics to control infection; vitamin E and selenium to combat capture myopathy; antiparasite drugs (external and/or internal); intravenous fluids for dehydration and shock; general nutritional supplements; and vaccination against certain diseases. The situation in the field will dictate the drug(s) to use. During prolonged immobilization or preparation for moose translocation, special supportive and\or preventive medication may be needed. Also, certain drugs may be needed for moose with special medical problems not necessarily related to capture.

Translocation

Throughout recorded history, humans have moved free-ranging wild animals from one location to another, mostly for human benefit. Burris and McKnight (1973) listed six translocation objectives: (1) increased recreational hunting; (2) additional food supply; (3) economic gain; (4) population reestablishment; (5) preservation of endangered species; and (6) opportunity to view a species in a new setting. Until the mid-1900s, most translocation attempts in North America failed and, in some cases, the introduced species negatively impacted the relocation area. Presently, a few translocations may be politically motivated and a few others undertaken without proper regard to the potential impact, but most are based on sound wildlife management. Riney (1982: 461) summarized the conclusions from an International Union for Conservation of Nature meeting regarding concerns for animal introductions as:

- "Accidental introductions should be minimized and if they do occur, should be promptly investigated to see if practical remedial measures are necessary.
- "Species should be introduced into national parks and equivalent reserves only if they formerly existed in the area.
- "No species should be deliberately introduced into a new habitat unless the presence of the species conforms with, and can be expected to contribute to, the type of land-use assigned to the area and unless the

operation has been carefully and comprehensively investigated and planned."

Riney (1982: 461–462) also listed the important considerations in such planning:

- "No species should be considered for introduction to a new habitat until the factors that limit its distribution and abundance are thoroughly studied by competent biologists and an estimate of the timing of the initial oscillation and an appraisal of its dispersion pattern made.
- "No species should be considered for introduction into a new area, if stock of any comparable native species is present in the area and can be built up, or is no longer available but can be rehabilitated by reintroduction, or extension from a nearby range.
- "The species to be introduced should be reasonably expected to enhance the economic and/or aesthetic value of the area and provide a minimal foreseeable risk of conflict with land-use policies in the area of introduction or in adjacent area to which the species might spread.
- "The individuals to be introduced should be, to the satisfaction of competent authorities, free of unwanted communicable diseases, viruses and parasites.
- "The species, if introduced, should not threaten the continued presence or stability of any native species or distinctive population, other than those that are pests, whether as a predator, competitor for food, cover or breeding sites, or in any other way arising from its characteristics and ecological requirements.
- "The knowledge of how a species could be eliminated, or at least kept under control in its new range, should be available; and means of doing so by a socially and economically acceptable method.
- "Whenever controllable experimental areas are available that contain habitats typical of those into which it is proposed to introduce the species, these should be used to test locally the suitability of the introduction before it is made generally.
- "Animals should never be introduced into areas where one or more elements of their required habitat is downgrading.
- "Both the techniques of capture and of transport should be carefully planned with a view to minimizing mortality and wounding; minimizing disturbance and with full awareness of the costs involved.
- "Immediately following an introduction, continuing studies should be undertaken of the introduced species in its new habitat and especially of its rate and pattern of spread and of its impact on the habitat. The importance of initial close monitoring cannot be overemphasized."

There were fewer translocations of moose in North America than of other large mammals, such as elk. From Yellowstone National Park alone between 1892 and 1967, 15,745 elk were shipped to various locations (Robbins et al. 1982). There were only 545 moose known to have been translocated to establish new populations in all of North America from 1784 to 1993 (Table 47). Others undoubtedly were not recorded.

A common measure of the success of a translocation is the establishment of a reproducing population. Of the 22 recorded translocations of moose, 12 resulted in establishing reproducing populations. Introduced moose now are legally hunted in several states and provinces (Newfoundland/Labrador, Nova Scotia, Alaska, Colorado, Michigan). Only two of the translocations (Colorado 1987, 1991–1993, and Michigan 1985, 1987) used chemical immobilization, the remainder depended on physical restraint or hand-reared calves.

The choice of using physical or chemical restraint for best results is difficult to make based on evaluating past translocations. Many early translocations using physical restraint experienced high mortality rates (Pimlott and Carberry 1958). Recent translocations (e.g., Colorado and Michigan) using both chemical and physical restraint had few or no mortalities. The 59 Ontario moose translocated to Michigan in 1985 and 1987, captured using chemical restraint, increased to about 200 animals by 1992 and have continued to expand in population size and range since then to the point where translocations from this population to other areas is being considered. This occurred despite significant mortalities from brain worm infestations (S.M. Schmitt personal communication: 1994). The Colorado operations used both corral traps and etorphine immobilization. Officials concluded that corral trapping was easier, more efficient, and less stressful and taxing on the animals than was chemical immobilization (Duvall and Schoonveld 1988).

Most mortalities in the early translocations were from "shock-type disturbances" (Pimlott and Carberry 1958). Such disturbances include stress, true shock, capture myopathy and/or hyperthermia. Better management during capture and transport has minimized these syndromes.

The transport of captive animals is an operation for which there is no substitute for experience. Transport of moose in individual crates, in trailer trucks, and as calves in pickup trucks has been successful. Each situation will dictate the preferred method, but the choice should be the

Table 47. Documented moose translocations in North America, 1784–1993

Year	Translocation From	Translocation To	Number	Status[a]	Reference
1784	New Brunswick	Grand Manan Island, New Brunswick	2	0	Dodds 1974
1878	Nova Scotia	Gander Bay, Newfoundland	2	+	Pimlott 1953
1898	Manitoba	Berkshire County, Massachusetts	6	–	Vecellio et al. 1993
1904	Nova Scotia	Howley, Newfoundland	4	+	Pimlott 1953
1894–1895	Canada[b]	Nehasane, New York	11	0	Dodds 1974
1902–1903	Canada[b]	Raquette Lake, New York	12	0	Dodds 1974
1895–1905	Quebec	Anacosti Island, Quebec	38[c]	0	Pimlott and Carberry 1958
1900	Western Canada[bd]	Hokitika, New Zealand	4	0	Donne 1924
1905	Western Canada[b]	Dusky Sound, New Zealand	10[c]	–	Donne 1924, Tinsley 1983
1935–1937	Isle Royale, Michigan	Upper Peninsula, Michigan	77	–	Hickie n.d.
1947–1948	Elk Island, Alberta	Cape Breton, Nova Scotia	18	+	Pimlott and Carberry 1958
1948–1950	Moran, Wyoming	Bighorn National Forest, Wyoming	16	+	Grasse 1950
1949–1958	Southcentral Alaska[d]	Copper River, Alaska	24	+	Burris and McKnight 1973
1953	Newfoundland	St. Lewis River, Labrador	12	+	Mercer and Kitchen 1968
1957–1959	Southcentral Alaska[d]	Kalgin Island, Alaska	6	+	Burris and McKnight 1973
1958–1960	Southcentral Alaska[d]	Berners Bay, Alaska	22	+	Burris and McKnight 1973
1963–1964	Southcentral Alaska[d]	Chickamin River, Alaska	9	–	Burris and McKnight 1973
1966–1967	Southcentral Alaska	Kodiak Island, Alaska	6	0	Burris and McKnight 1973
1978	Uinta Mountains, Utah	North Park, Colorado	12	+	Nowlin 1985
1979	Moran Junction, Wyoming	North Park, Colorado	12	+	Duvall and Schoonveld 1988
1987	Jackson Hole, Wyoming	Laramie River, Colorado	12	+	Duvall and Schoonveld 1988
1985, 1987	Algonquin Park, Ontario	Marquette, Michigan	59	+	Aho and Hendrickson 1989
1991–1993	Colorado, Utah, Wyoming	Southwestern Colorado	106	+	Olterman et al. 1994

[a] Present status: 0 = no moose remain ; – = a few moose remain; + = population established.

[b] Exact location unknown.

[c] Number is an estimate.

[d] Only calves used in translocation.

method that stresses the animal the least. Good planning, logistics and communications are necessary. Some general considerations are (International Wildlife Veterinary Service, Inc. 1991) security, protection from inclement weather, protection from trauma, protection from diseases, protection from stress, access to animal during transport and safety.

Animal Health during Restraint

Attention to the health of an animal during restraint begins with the planning process. First, determination must be made if the objectives can be obtained without restraint and, if restraint is necessary, whether physical restraint alone will accomplish the objectives. The objectives will dictate how long the animal must be restrained, which affects most other aspects of the operation. Consideration then must be given to the species of animal. With moose, the subject's size and susceptibility to overheating are factors, as are the type of terrain where the project will occur and the physical hazards of the area (e.g., steep hills, cliffs, lakes, large rivers). An immobilized moose will travel at least 4 minutes after injection. Planning must cover access-

→

Sequence of translocating a moose from Algonquin Provincial Park, Ontario, to the Upper Peninsula of Michigan. (Opposite top left) Taking biologics and other data, blindfolding and ear plugging an immobilized moose. (Opposite center left) Adjusting the moose in a sling for helicopter transport. (Opposite bottom left) Stabilizing the moose in its sling before lift-off. (Opposite top right) Aerial transport. (Opposite center right) Examination and sling removal of moose at touch-down and transfer site, before crating for transporting by truck to a holding facility on the Upper Peninsula. (Opposite bottom right) Crated moose nearly underway to the Upper Peninsula. This was one of a number of moose translocated to Michigan in 1985. A total of 59 moose were part of that translocation and a subsequent one in 1987. By 1992, the population reportedly had increased to about 200. *Top left and center photos courtesy of Rose Lake (Michigan) Wildlife Research Center. All other photos by Heather K. Bickle; courtesy of the Ontario Ministry of Natural Resources.*

C-GAHZ

C-GAHZ

ing the animal and its travel route options. Moose will be in poorest condition in late winter and early spring and in peak condition in late summer and early autumn. Seasonal condition of moose influences drug dosage. For example, poorly conditioned animals may be easily overdosed with drugs, whereas prime moose may present absorption problems for drugs injected into fat. Season also dictates the presence or absence of snow and frozen lakes that affect access to and recovery of moose.

The success of wildlife immobilization efforts also depends on preplanning for expedient handling and provision for immediate response to medical emergency. Both entail experienced personnel and proper equipment to take and monitor the animal's vital signs, position and comfort the animal, perform a physical examination, tag or mark the animal, protect personnel, and prevent or treat medical emergency. Much of the following, related specifically to moose, is from the "Wildlife Restraint Series" of the International Wildlife Veterinary Service, Inc. (1991). The guidelines provide an insight into the problems associated with monitoring immobilized animals and are not meant to be a course in how to do it. Actual application requires proper technical training and experience with each species.

Vital Signs

Vital signs reflect what the state and/or condition of such major physiological systems as respiratory, circulatory and metabolic. Basic monitoring can be done with a few simple instruments, such as a stethoscope and a thermometer. Every person involved in the capture should have the ability to take vital signs and know when an emergency arises. Obviously, veterinary training and experience benefit the handling of many emergencies, but in the essential interest of efficiency, knowledge by all participants of how to take and monitor vital signs and their acceptable ranges allows the veterinary expertise of the crew to focus on more complex aspects of the operation.

RESPIRATION

Respiration rate of moose can be checked by observing the animal's chest movement or feeling air passage from its nostrils. Rate of respiration is not as important as the depth and regularity of breathing. The normal respiratory rate for moose is 13 to 40 per minute (Franzmann et al. 1984a), but will be slowed by the immobilizing drug and should not fall below six per minute. Shallow and rapid (every few sec-

With any moose drug immobilization effort, there is possibility of mishap, afforded by shot placement, terrain, carelessness and bad luck. In the 4 to 10 minutes after being drugged and in an effort to escape the disturbance, moose sometimes find awkward or dangerous circumstances. Their tendency to escape to water is one such possibility. Fortunately, the moose above went down in shallow water, was subsequently processed, revived and chased back onto land. *Photo by Vernon Lofsted; courtesy of the Kenai (Alaska) Air Service.*

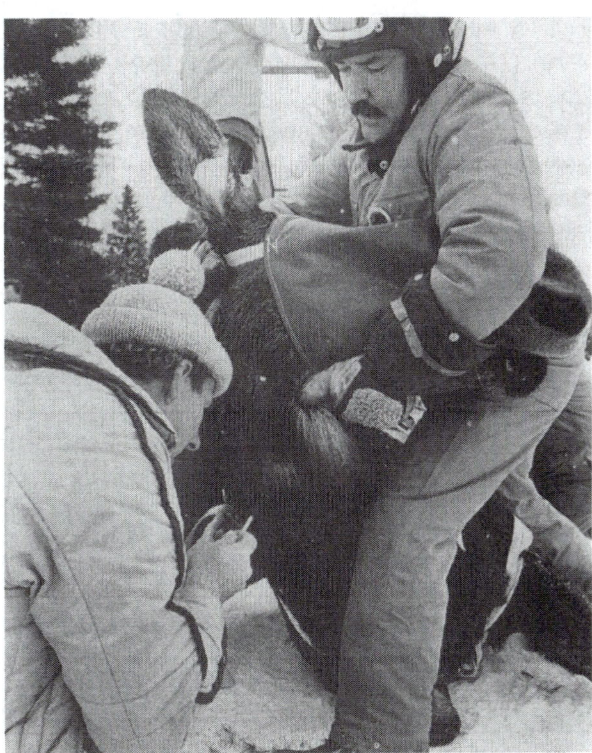

After a moose is located by and darted from a helicopter, and until the animal is down, the aircraft can be used to keep the animal from going into areas that may be hazardous for it and the capture crew. There is a fine line for accomplishing this without causing undue stress to the animal. An experienced helicopter pilot and a crew chief are essential in such operations. Use of a blindfold and ear plugs on an immobilized moose will minimize sensory disturbances and stress to the animal during handling. *Left photo courtesy of the Alaska Department of Fish and Game. Right photo by Heather K. Bickle; courtesy of the Ontario Ministry of Natural Resources.*

onds) respiration signals an emergency. A stethoscope placed on the chest provides an indication if the lungs are clear. Raspiness and gurgling indicate that fluid has entered the lung, which may predispose foreign body pneumonia.

If respirations cease and the heart continues to beat, an open breathing passage must be assured and artificial respiration initiated. For moose, this means compressing the chest at a rate of about 10 times per minute. The moose also may be ventilated by holding the mouth and one nostril closed and breathing into the other nostril (mouth-to-nose resuscitation). If breathing does not commence in a few minutes, emergency drugs should be used. The immobilizing drug should be reversed at the appropriate time dictated by circumstances such as personnel safety.

CIRCULATION

Heart beat should be monitored using the stethoscope, but also the pulse can be felt and even seen at the jugular vein of the moose. Wetting the hair over the jugular will help make the pulse more visible. The normal pulse rate for moose ranges from 70 to 91 per minute (Franzmann et al. 1984a). Stress or excitement will speed the pulse, but it should drop as the animal calms or the drug takes effect.

Mucous membranes are normally pink in color. The mucous membranes of lips, eyes and gums of moose generally are not pigmented and provide a convenient location to check color. If the color turns blue, the animal's blood oxygen has been lowered. If the color is gray to white, it indicates lowered blood pressure. Consideration should be given to immediate reversal of the immobilizing drug. In either case, emergency procedures such as artificial respiration and/or respiratory and circulatory stimulants are needed.

TEMPERATURE

Body temperature is ascertained with a rectal thermometer. The normal body temperature for moose is 101.1 to 102.0°F (38.4–38.9°C) (Franzmann et al. 1984a). Temperature may escalate very rapidly, and hyperthermia results. In the field, when their temperatures approach 105°F (40.6°C), attempts to cool them with cold water should begin. On the other hand, hypothermia, or lowered body temperature, may occur when a moose is immobilized and its sensory stimuli impaired. Because of their large body size, moose are not likely to lose body heat as rapidly as would smaller mammals, so the condition is rare. Nevertheless, impairment of thermoregulation occurs at body temperature below 93.2°F (34°C). At temperatures below 89.6°F (32°C), the animal is likely to go into a coma and not respond to stimuli (Fowler 1978). As the body temperature reaches 95°F (35°C), attempts to warm the animal with blankets or heat source should begin. The use of drugs, including immobilization-reversal drugs, for the control of hypothermia and hyperthermia will be dictated by circumstances.

REFLEXES

Depth of sedation is monitored by reflex stimuli. Response of the animal to capture personnel movement usually is the first involuntary reflex to disappear. This occurs when the animal's eyes do not follow a moving object in front of them. Response to a loud noise may arouse an animal that still has an auditory reflex. Loss of pain with good sedation can be monitored in moose by pinching its nose. Nonresponse to touch of the eyelid (the blink reflex) indicates deep anesthesia. One must be certain that the presence or absence of reflexes is related to the drug and not from a medical problem.

Position and Comfort

An immobilized moose can get into some awkward and dangerous positions or situations. The immobilizing drug should be reversed immediately if the moose cannot be comfortably positioned or is in a life-threatening situation.

I once had a partially immobilized moose wander onto a frozen lake, despite efforts to force it away by helicopter. The ice broke and the moose sank. The water was shallow and the moose kept its head and shoulders above the surface. As the animal became more sedated, it appeared it would soon drop its head and drown. My only solution was to wade out and administer the antidote and hope that it worked quickly. Fortunately, it did, and the moose wandered out of the lake and began feeding as if nothing had happened. (The next emergency was for me to get thawed out.)

Moose should be positioned in sternal recumbency, with the legs naturally folded underneath whenever possible. If this is not possible, the right side down position is best, because the rumen is on the left side. A sedated moose should never be rolled over to change sides, because the action can cause a torsion of the abomasum.

After resolving any life-threatening situation, the comfort of the animal should be a primary concern. The eyes need to be protected from exposure to sun, dirt and other foreign objects. This is accomplished most effectively by covering the eyes with a clean cloth eye wrap after an antibiotic eye ointment has been administered to control infection and keep the eye moist. Because the protruding eyeballs of moose may be abraded, eye wraps should not be left on for more than 10 minutes. Moose should be protected from direct sunlight, by shading the eyes, if the procedure occurs in bright sunlight. Immobilization should not be done during extreme cold—below –20°F (–28.9°C) if the moose will have a down time of more than 15 minutes. Handling, jostling, noise and other disturbances should be kept to a minimum because, even during some stages of sedation, moose can be unduly stressed by these stimuli.

Physical Examination

After stabilization, moose are examined to evaluate their condition. As much information as possible should be gathered without stressing the animal beyond that imposed by the immobilization and prevailing environmental conditions.

The moose first should be "overviewed" from a distance of several yards. It is possible to overlook "the forest for the trees" if attention begins with the animal's various parts. For example, a fractured or dislocated limb may not be noticed right away from close range. On overview assessment of the moose's general physical attitude, condition, conformation and temperament may be as revealing and important as the subsequent close examination.

As ruminants, moose are subject to bloating, which can be aggravated by its position. Again, if possible, moose should be kept lying on their sternum. For a deeply sedated moose, it may be necessary to assist the animal onto its sternum and maintain the animal in sternal recumbency. An animal is less likely to bloat if properly positioned. Normal expulsion of gas from the rumen then can occur and aspiration of fluids into the lung is less likely. Generally, in field operations with a helicopter, two-man ground crews are used. It may be necessary to call in additional help to position an immobilized moose properly. When this has not been possible, I often have administered the reversal drug and not processed the animal.

After the overview, followed by the taking of vital signs, each body system or area should be examined, including skin and integument, lymph nodes, head, neck and back, abdomen, thorax, extremities, respiratory system, circulatory system, reproductive system, nervous system and digestive system. One person, preferably a veterinarian or someone trained by one, should do the monitoring and physical examination. Others perform tasks concerning the preplanned objectives of the immobilization.

Malkmus (1901) first outlined the clinical examination of domestic animals and, as technology improved, many other such texts, relative to domestic and wild animals, have been prepared. For detailed coverage concerning wild ani-

Checking and recording vital signs of an immobilized moose are essential tasks. Any significant variation from normal ranges for heart rate, respiration and body temperature must be assessed to determine the cause and appropriate medical/chemical response. A physical examination is done to determine any injuries, disease or abnormalities, and samples are collected for laboratory analyses that may include histopathology, microbiology, toxicology, parasitology, reproduction, physiology, morphology and aging. The tissues sampled and their care will be determined by the information needed. *Photo by Charles C. Schwartz; courtesy of the Alaska Department of Fish and Game.*

mals, refer to Fowler (1978, 1986) and International Wildlife Veterinary Service, Inc. (1991). The following is a list of the major elements of examination (for immobilized and/or nonmobilized moose).

- Skin—color, moisture, elasticity, swelling, lacerations, ulcers, external parasites (e.g., ticks, lice, fleas, mites). Collect parasites.
- Hair—brittleness, luster, length, alopecia (hair loss), color. Collect hair.
- Lymph glands—swelling.
- Eyes—discharge, color of conjunctiva, clarity of cornea.
- Pulse—rate.
- Heart and circulation—sound, rate, swelling of extremities, ascites (abdominal filling with fluid). Collect blood.
- Respiration—rate, rhythm, intensity, sounds (e.g., wheezing, rattling, blowing, coughing), breath odor.
- Nasal discharge—color, consistency, odor, presence of blood or air.
- Oral cavity—odor, teeth, gum color, abrasions, foreign material. Collect tooth (for aging).
- Abdomen—bloat, rumination, impaction.
- Evacuations—diarrhea, blood, mucous, frequency, form, color, composition. Collect feces (to assess food habits and quality, parasites).
- Food and drink—appetite, deprived, depressed, mastication (chewing), thirst, vomiting.
- Urination—frequency, difficulty, painful, quantity, color. Collect urine.
- Reproductive system—discharge, swelling, check for pregnancy.
- Feet, hooves and antlers—puncture wounds, swelling, hoof and antler growth and abnormalities.
- Mammary gland—lactating, swelling, wounds. Collect milk.
- Skeletal system—fractures, enlargements, measurements.
- Excitability evaluation—based on the animal activity before and during handling on 1 to 5 scale (Franzmann et al. 1975b): 1 = not excited; 2 = slightly excited; 3 = moderately excited; 4 = excited; 5 = highly excited.
- Condition evaluation—for moose, condition-

evaluation criteria established (Franzmann 1977) are based on a 10 to 0 scale:

10. A prime fat moose with thick, firm rump fat by sight. Well fleshed over back and loin. Shoulders round and full.
9. A choice, fat moose with evidence by feel of rump fat. Fleshed over back and loin. Shoulders round and full.
8. A good fat moose with slight evidence by feel of rump fat. Bony structures of back and loin not prominent. Shoulders well fleshed.
7. An "average" moose with no evidence of rump fat, but well fleshed. Bony structures of back and loin evident by feel. Shoulders with some angularity.
6. A moderately fleshed moose beginning to demonstrate one of the following conditions—(a) definitions of neck and shoulders, (b) upper foreleg musculature distinct from chest or (c) prominent rib cage.
5. When two of the characteristics in class 6 are evident.
4. When all three of the characteristics in class 6 are evident.
3. When the hide fits loosely about the neck and shoulders, head is carried at a lower profile. Walking and running postures appear normal.
2. Malnutrition obvious. Outline of the scapula is evident. Head and neck low and extended. Walks normally, but trots and paces with difficulty and cannot canter.
1. Point of no return. Generalized appearance of weakness. Walks with difficulty and cannot trot, pace or canter.
0. Dead from malnutrition/starvation.

To gain experience and evaluation consistency, these condition criteria should be applied to each moose captured no matter what the objectives of the capture. These criteria are used regularly in Alaska when applying the indicator animal concept to evaluation of the health and condition of a moose population.

Capture and Translocation Mortality

The most troubling aspect of any capture procedure is the potential for mortality. Capture-related fatalities have decreased over the years with increased knowledge of the causes and deployment of skilled capture crews. Nevertheless, whenever a wild moose is handled, the possibility of injury and death exists, and the foremost responsibility of persons involved in such procedures is to do everything practicable to minimize that possibility.

The three general causes of injury and death are physical, environmental and psychological/physiological (Spraker 1982). Many physical factors, such as abrasions, contusions, lacerations and fractures, can be minimized by avoidance of terrain (e.g., rocks, cliffs, steep hills, dense vegetation, streams and lakes) that predisposes these problems, and by careful handling and monitoring of captured animals. Environmental causes related to ambient temperature, sunlight, wind and precipitation can be minimized by judicious and flexible scheduling. Psychological/physiological problems, closely related to stress, are more difficult to control, but can be minimized by the aforementioned considerations for physical and environmental factors and by reducing noises and disturbances, time of chase and in traps and transport, and by close monitoring of captured animals. The recent introduction of long-acting neuroleptics will significantly decrease mortality associated with capture and translocation for moose as they have for African species (Ebedes 1991).

Gaining popularity in Africa is a program in which animals are gradually introduced to corral trapping (Harthoorn 1982). The program has application to the capture of moose. The key is to allow time for the subject animals to become familiar with captive feeding and the surroundings before confinement and/or translocation.

STRESS

Stress often is the cause of unexplained problems or death associated with capture and translocation. It is a simple term and explanation for a process that is complex. Selye (1950) described stress as the "general adaptive syndrome" (GAS) and compared it with inflammation. Inflammation is localized reaction to trauma; stress is the general or bodily reaction to trauma. Selye defined the three stages of the GAS as the (1) alarm reaction, (2) resistance and (3) exhaustion.

The alarm reaction is the fear/flight reaction, and every animal pursued or trapped during capture experiences this. Every person is familiar with the effect that adrenalin has on the body. Animals respond similarly, experiencing increased heart rate, constriction of arteries, heart dilatation, respiratory rate increase, hair erection and sugar mobilized in the blood. Every animal can handle this response up to a point, but when experiencing extreme fear, the adrenal cortex becomes exhausted and fainting and/or death may occur.

In the resistance stage, the animal adapts to the stressor (external challenge) and no longer produces adrenalin. The adrenal cortex shrinks but continues to produce hormones that mobilize glucose. This results in the animal appearing normal, and it will return to normal if there is no additional challenge. If the challenge continues, the animal will enter the exhaustion stage.

The exhaustion stage occurs when the adrenal cortex can no longer produce life-sustaining hormones. The animal may die suddenly or enter a vegetative state for several days before death.

Anyone handling wild animals must be familiar with GAS, particularly the stage of resistance. This is when handlers could assume all is well and resume activity that would further stress the animal and put it into the final stage.

Ability to measure stress in a live animal from a free-ranging population generally is poor. However, certain measures are helpful, including the subjective excitability classification (Franzmann et al. 1975b), based on the physical examination. From blood sampling, stress can be measured by corticosteroid and beta-endorphin analyses. During necropsy, adrenal glands can be examined for enlargement that indicates the animal experienced long-term stress (Franzmann and Schwartz 1983a), which can be very important in identifying the epidemiology associated with the animal's death.

CAPTURE MYOPATHY

Capture myopathy occurs in moose (Haigh et al. 1977), and has been observed by most biologists capturing moose. It is a stress-related condition affecting animals following capture, with consequences varying from sudden death to prolonged distress. Capture myopathy is a facet of GAS, so the stress may be of physical, environmental and/or psychological/physiological origin. The disease has four clinical syndromes (Spraker 1982): peracute death; acute death; ataxic myoglobinuric; and ruptured muscle. Acidosis, cardiac fibrillation, circulatory collapse and sudden death characterize the peracute form. The acute form shows acidosis with muscle stiffness, lethargy, rapid pulse, increased respiratory rate and death within 12 hours from pulmonary edema. The ataxic form shows damage to skeletal muscles and internal organs resulting in muscular paralysis; animals may survive for several weeks but do not respond to treatment, and death generally occurs from renal failure. The ruptured muscle syndrome involves no apparent effects until 24 to 48 hours postcapture, but then is characterized by crippling the rear quarters due to bilateral rupture of the gastrocnemius muscle. Affected animals may survive for 3 to 4 weeks, but the process is irreversible and does not respond to treatment.

HYPERTHERMIA

Hyperthermia (increased body temperature) commonly occurs during capture, but is particularly problematic in moose because of their large body size and intolerance of high ambient temperature. The combination of high ambient temperature, muscular activity and moose inability to lose body heat creates a scenario that can be irreversible. Moose may start to show some forms of heat stress when ambient temperatures exceed 57 to 68°F (14–20°C) (Renecker and Hudson 1986b). Moose body temperature generally rises during capture, so must be monitored and kept under control. The process of cell damage begins to occur at 108°F (42.2°C), at which point the body temperature must be lowered immediately. Moose with a temperature of 110°F (43.3°C) do not recover, based on my experience. They cannot be easily cooled down by water, as can some smaller mammals. To prevent hyperthermia, capture and/or translocation of moose should not be undertaken when the ambient temperature is above 50°F (10°C), and prolonged chase should be avoided at any temperature (Franzmann 1982). In my opinion, the ideal ambient temperature (for both moose and crew personnel) for the capture of moose is 20 to 40°F (–6–4.4°C). Overcast skies and no wind would complete the ideal circumstances.

SHOCK

When cardiac output can no longer provide an adequate blood flow to tissues and organs, the shock state occurs. It can occur during capture and result from a severe physical or psychological challenge. The clinical signs are decreased blood pressure, pale mucous membranes, lethargy, cool body surface, coma, rapid breathing, weak and rapid pulse, dilated pupils and decreased body temperature (Fowler 1978) (Table 48). Treatment consists of eliminating the challenge, providing supplemental oxygen and restoring blood volume.

Table 48. Shock classifications important in wildlife restraint (Fowler 1978)

Cardiogenic: failure of the heart as a pump
Ventricular fibrillation—catecholamine response
Cardiac standstill—cholinergic response
Cardiac tamponade—pressure on the heart
Hypovolemic (actual): decrease in blood volume
Hemorrhage—whole blood loss
Plasma loss—contusions, burns
Dehydration—exercise, hyperthermia
Hypovolemic (relative): change in vascular bed, increase capacity
Neurogenic response—pain, fear, anger
Endotoxins—infections
Toxins—drugs

Safety during Restraint

Safety in capture and translocation operations pertains to the human operators and the animals. Concerns for animal safety are met by application of the previously discussed principles of proper restraint and by not exposing the animal to unnecessary physical and psychological stress. In planning an operation, human safety should be the primary concern because there are many opportunities for serious injury of the untrained, ill-prepared or careless person.

When a wild animal responds aggressively to a challenge, it tends to become even more aggressive after being stressed into an alarm reaction (Selye 1950). Moose can cause serious damage with both antlers and hooves. Even immobilized moose should not be considered harmless. They can act reflexively or awaken sooner than expected. Crew members should wear boots and protective clothing that covers the arms and legs. Animals should never be handled by only one person, and someone away from the operation should be aware of where the work is being done, who is involved and what the schedule of operations is.

When drugs are used, safety measures are extremely important to avoid serious problems. The currently preferred immobilizing drug, carfentanil, is so potent that one drop can kill an adult person. The concentration of this drug that one may receive accidentally using carfentanil is greater than any type of narcotic overdose they may have experienced. When using carfentanil or other potent narcotics, the antidote naloxone or naltrexone must be on hand and all personnel know how to administer it in an emergency. Some syringes and darts containing the drug can accidentally inject a person during handling. Care must be taken when loading darts because the drug can be absorbed through the mucous membranes of the eye or mouth and through cuts and abrasions. And a person accidentally injected with carfentanil is at extreme risk and expert medical attention is essential. A good policy is notification of the local hospital or emergency care center that this drug is in use in the field, so that medical staff can be prepared to handle a related emergency. The following outline is for Alaska Department of Fish and Game personnel using narcotic immobilizing drugs (W. Taylor personal files):

I. General safety procedures
   A. The team(s) should review safety and emergency procedures
   B. Work in pairs
   C. Use latex or rubber gloves when handling drugs
   D. Store loaded darts and used darts in non-breakable, leak-proof containers
   E. Personnel should be qualified to administer CPR and first aid

II. Emergency
   A. Narcotic spilled on skin should be thoroughly flushed with water, then the person observed for symptoms of poisoning; usually, no additional treatment is necessary
   B. Accidental injection of narcotic usually results in one or more symptoms of narcotic poisoning, which can be life threatening and requires immediate treatment
   C. Symptoms of narcotic poisoning are
      1. Severe respiratory depression
      2. Severe drop in blood pressure
      3. Dizziness and disorientation
      4. Loss of consciousness
      5. Muscle rigidity
      6. Vomiting
      7. Pin-point pupils

III. Treatment
   A. Maintain open airway and administer CPR if necessary
   B. Administer the antagonist (e.g., Naloxone)
      1. Give 25 mg of naloxone (intramuscularly or intravenously slowly) for every milligram of the narcotic accidentally injected
      2. If no response in 3 to 5 minutes, give an additional dose
      3. Give additional antagonist if clinical signs of narcotic poisoning begin again as the antagonist is metabolized
      4. Diprenorphine can be used as an antidote only if naloxone is not available
   C. Transport the patient to a medical facility immediately, and inform the attending physician of the name and the dose of the narcotic the individual received and the amount antagonist already given

This discussion of the safety problems associated with the use of narcotics to immobilize wild animals is a sobering one, and clearly emphasizes the need for an effective nonnarcotic drug.

Human safety in capture operations has another dimension when helicopters are used, either as a shooting platform or to drive animals into traps or nets. The safety measures are common sense and, if the helicopter operator does his job correctly, everyone involved will be given safety instructions (a Federal Aviation Administration regulation) before the flight (see International Wildlife Veterinary Service, Inc. 1991).

Human safety is always a concern where dart guns are used. These are lethal weapons, and even without considering the consequences of the drug in the dart, the physical damage of darting could be life threatening. Firearms safety and training courses are available through the National Rifle Association and elsewhere, and should be required for all personnel using dart guns in animal capture.

Other Restraint Concerns

Concerns most closely related to the field operations of moose capture and translocation programs are mainly chemical, medical and logistical. But before field operations can begin, other considerations—regulatory, social, medical, safety, economic, political and record keeping—must be understood and addressed.

Regulations

Regulatory concerns begin with drug licensing by the Directorate of Food and Drug in Canada, and the Drug Enforcement Administration (DEA) in the United States. The DEA registers drugs into classes based on their potential for abuse and designates them as controlled substances. Schedule I drugs are illegal drugs in the U.S. (heroin, crack-cocaine, etc.) and have no accepted medical use. Schedule II drugs are potent narcotics, and include those used for moose (carfentanil, etorphine, diprenorphine). There is a special category within Schedule II for these drugs (Schedule IIn) that requires Schedule I security. Schedule IIn drugs can be purchased only by persons holding a valid DEA registration and complying with the following guidelines (International Wildlife Veterinary Service, Inc. 1991):

- Distribution of Schedule II drugs is for veterinarians in zoo and exotic animal practice, persons in wildlife management programs, and researchers. Others working in these programs must obtain the drugs through a licensed and approved veterinarian.
- Storage of these drugs must comply with U.S. Government Class 5 security. Bank safe-deposit boxes comply, but access must be limited to the DEA-approved veterinarian.
- Records must be maintained to ensure full accountability for the drugs received.
- Applicants for Schedule IIn status must be approved by DEA before any Schedule IIn drugs can be obtained.

Xylazine and naloxone are not controlled substances and do not require DEA accountability, but veterinary prescription is necessary. As new drugs become available, their regulatory status must be identified and proper procedures followed for purchase, storage, use and disposal, as outlined by the Animal and Plant Health Inspection Service (APHIS) of the U.S. Department of Agriculture.

Social Acceptability

Social concerns primarily entail public perception of wild animal capture and translocation. Most people who are familiar with the work condone it, but disapprove of poorly designed operations lacking professionalism. However, a vocal minority, mostly affiliated with animal rightist groups, opposes any type of animal utilization or manipulation.

The best approach to dealing with all publics is to make sure that each capture and/or translocation operation conforms to the highest professional standards and uses the most humane methods. A news release issued before any operation, which outlines the project and its objectives, is much better than postoperation explanations. News releases must focus on the objectives.

Another social circumstance that greatly influenced wildlife capture was when "angel dust" became a popular hallucinogenic street drug. This drug is phencyclidine hydrochloride, which was widely used to immobilize carnivores. When abuse of this drug became commonplace in the 1970s, the U.S. Food and Drug Administration classified it as a Schedule II controlled substance. Manufacturers did not want to produce the drug under those restrictions for a limited market, therefore they ceased production. In essence, a combination of social and economic factors prompted testing to find a substitute product, and several years passed before an improved product was marketed. In the interim, many capture programs were compromised because of the lack of a good substitute drug.

Economics

Economic concerns are important but not always obvious. The market for drugs used in wildlife capture is very small. Many products tested on wildlife as experimental drugs never reach the market unless there is a market for domestic or human application. The most famous example of this type is the 1:1 ratio of tiletamine hydrochloride and zolazepam hydrochloride tested in the 1970s. It was an improved replacement for phencyclidine hydrochloride (angel dust). The drug did not reach a market until 1987 when it was marketed as Telazol[R] (A.H. Robbins Co., Richmond, VA) for domestic canine medicine.

Another economic concern for wildlife capture is reflected in the monetary value of the animals involved. For example, a trade of four Indian rhinoceroses captured in Nepal in 1985 was made for 16 trained, working elephants from India. The market value for trained working elephants was approximately $40,000 each. That placed the value of each rhino at $160,000. Also, at a 1990 wildlife auction at Hluhluwe National Park, South Africa, five black rhinoceroses sold for approximately $1 million. White rhinoceros sold for $20,000 each and giraffes went for $4,500 each. And a dentist from the United States paid $7,000 in 1991 for the privilege of shooting a white rhinoceros in Africa with an immobilizing dart, which subsequently was translocated to another area.

The value of a live moose is difficult to establish because there is no legal marketplace in the United States and only a limited trade of gamefarm moose in Canada. Unfortunately, the basis for most determinations of value for North American large mammals is correlation with domestic meat prices, which greatly undervalues the economic and many other benefits of wildlife.

Politics

Political decisions sometimes dictate that a translocation must be attempted. Such interventions tend to occur in response to misinformation and public pressure. These may not be biologically sound but may have to be done in spite of the best advice. This is a difficult situation for the biologist and wildlife veterinarian, because the requirement for such efforts may contradict their scientific training and experience. Fortunately, many states, provinces and most federal agencies now have established policies regarding translocation guidelines and protocols that assure that they are biologically motivated and justifiable (Franzmann 1988).

The best defenses against unwarranted political interference are maintenance of the highest degree of professionalism in all capture and translocation projects, full compliance with government standards and forthright, proactive communication with public interests.

Record Keeping

Records seem to become a concern only when there are none. Every aspect of a capture and/or translocation operation—e.g., date, location, personnel, logistical resources, time, weather conditions, methodology used, biological data, expense, miscellaneous observations, summary—should be recorded. It is imperative to maintain complete records of drugs used (required for controlled drugs) and immobilization procedures on a data sheet (see International Wildlife Veterinary Service, Inc. 1991). During a translocation procedure, a data record for each animal should be maintained by the project's principal investigator and a copy filed with the responsible wildlife agency or agencies. Photographic and/or video documentation of the procedures also are very useful. The records are necessary for evaluating the project and improving on future ones. Reporting of findings also is important as reference for other professionals.

Population Health/Condition

Monitoring the health of an entire wildlife population is an extension of individual animal monitoring. When capturing a wild moose, there is an opportunity to extract information from that animal that relates to the local population's well-being. Also, a captured group or subpopulation of moose can represent a sample from a larger population, and from which meaningful information about the population status can be extracted. Such representations, known as the "indicator animal concept," have received increased research emphasis in recent years. The approach complements but does not replace traditional methods for evaluating wild animal populations. Traditional methods include direct measurements of the habitat and such vital population statistics as sex and age composition, natality, mortality, recruitment and survival. With increased demand for more refinement in managing wildlife populations, as much information as possible needs to be incorporated into the data base for each population. The indicator animal approach is a refinement and potentially improves our ability to predict events in a population.

It is extremely important that the prevalence of and potential for health and disease in wildlife be related to populations rather than individuals. This necessitates discarding the anthropocentric tendency to consider disease first and foremost the collision of a pathogenic agent (bacteria, virus, etc.) and a susceptible individual (Dixon 1978). Also, wildlife disease research in the past concentrated primarily on parasitology and infectious diseases—diseases with specific and identifiable etiologies (causes). This led to focus on individuals, not populations, and on cure and not prevention or preventive medicine. Individual animal orientation can be beneficial to that animal, but not necessarily to the health of the population. Also, identification of an organism potentially challenging the health of an animal should not necessarily be assumed to be the causative factor in the disease process. It may be an additive factor because, when identified as causative, it may result in overlooking another basic problem, such as habitat loss or forced confinement of a population. The distinction between dying with something and from something must be understood.

The balance of this section is a basic primer to health monitoring of moose, with emphasis on the population and not the individual. Information from a population is obtained directly by studying its morphology (form), physiology (function), and behavior (activity). Here, the focus is on form and function. Population health monitoring uses both living and dead animals from which indices of physical condition can be obtained. Physical condition is the most sensitive measurement of an animal population's response to its environment (Riney 1982).

Examination of Live Moose

Handling and examining live animals from a free-ranging population can provide unique insight into that population. Those opportunities arise during studies requiring animal capture, marking and radio-collaring.

The following outlines the approach needed to maximize the population data base.

HISTORY

Too often the most neglected part of the examination is the history, and it may be the most important part of the evaluation. The history of a wild animal will include events that may have caused an alteration in its basic ecological/habitat requirements (i.e., food, water, shelter/cover and space). Such events could include competition for these basics, actual loss or an imbalance of components of the basics. Therefore, someone knowledgeable of this information should be on hand during the examination.

RECORDING DATA

A field form for each subject animal should be designed to include: date (including year), location, species, sex, age, color and markings, method of capture and collectors. Drugs should be identified and quantified. The animal identification number (ear tag, collar, tattoo) must be recorded, as should the frequency of the radio transmitter if used. A portion of the form should be dedicated to recording vital signs and any emergency treatment. Each project and operator can select or design an appropriate form.

PHYSICAL EXAMINATION

This was previously addressed in the "Animal Health During Restraint" section.

MEASUREMENTS AND WEIGHTS

Live body weight of moose is difficult to obtain except at research facilities that have captive hand-reared moose trained to walk onto platform scales. Mechanical devices are not always practical for weighing animals in a free-ranging population, by virtue of cost or population accessibility. Consequently, indices of body weight using measurement/weight relationships provide reasonable estimates (Franzmann et al. 1978). Total body length and chest girth are the measurements that most highly correlated with total body weight, based on measurements of more than 1,300 Alaskan moose. The formulas are: body weight (kg) = –239.7 + 2.07 x total length (cm); and body weight (kg) = –245.3 + 3.14 x chest girth (cm). Karns (1976) tested for correlation of moose chest girth and field-dressed weight, and concluded that such correlation was only a fair estimator. Hundertmark et al. (1992a) reported that a combination of the condition evaluation criteria discussed earlier and total body length measurements improved estimates of total body weight. They found that heart girth measurements were not useful in predicting total body weight. Combining a linear measurement with a condition evaluation is perhaps the best approach because body length changes with age, not with season. Condition status and season are closely related (see Chapter 4). Other body measurements used as predictors of weight in moose include head length (Haigh et al. 1980), hind foot length and shoulder height (Franzmann et al. 1978).

Body weights and indices to predict them have value, but are limited as indicators of animal/population condition because of seasonal variation and daily differences in food and water contents of the digestive tract (Franzmann et al. 1978, Schwartz et al. 1987a). The condition evaluation scale (Franzmann 1977) previously outlined probably is just as useful and obviously is much easier to obtain. When available, body weight information provides supplemental support for other indicators. Persons immobilizing moose should have had experience weighing moose so they may become better at estimating weights when preparing drug dosages.

BODY FAT

Measurements of fat reserves (stored energy) are used to assess condition. Indices of body fat from live moose include the condition assessment scale, measurement of rump fat and ultrasound. Rump fat measurements are obtained by needle probe. This method is useful only for comparisons of animals in good condition, because rump fat is the first to be used up during nutritional stress in caribou (Dauphine 1976).

Early ultrasound equipment was tested on moose to measure fat thickness over the rump, but results lacked precision (Franzmann and Schwartz 1978). Subsequent real-time ultrasound on captive yearling bulls at the Moose Research Center proved accurate in measuring rump fat thickness (Stephenson et al. 1994). The investigators compared fat thickness before the rut (0.2 to 0.9 inch [0.3–2.4 cm]) with thickness after the rut (0.08 to 0.5 inch [0.2–1.2 cm]). They suggested that this in vivo technique exhibited potential for monitoring body condition.

Indices such as the tritiated water technique (Schwartz et al. 1988a) and urea dilution (Hundertmark et al. 1989) lack practical field application and are highly variable.

Research at the Moose Research Center has focused on a technique called "bioelectrical impedance analysis" used in

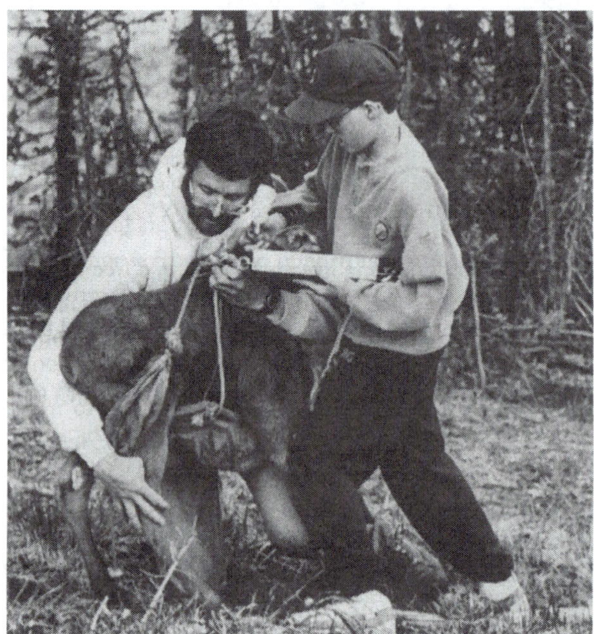

Live weights of moose at specific ages and seasons and over time are important measurements. In captivity, moose calves trained to step onto a scale, by enticement with a nursing bottle (top left), usually are not reluctant to get on a platform scale as adults. Untrained calves can be weighed when harnessed to a hand-held spring scale (top right); the mother of this calf was a hand-reared tame cow that allowed her calf to be approached and handled. To weigh wild, free-ranging adult moose, immobilization is necessary. Techniques include use of a winch/tripod device (Franzmann and Arneson 1975) (bottom left) and, attachment of the animal in a sling to a read-out scale built into a helicopter's lift (bottom right) or by placing a scale between the connection of the lift hook and the sling harness. *Left photos by Albert W. Franzmann; courtesy of the Alaska Department of Fish and Game, Soldotna. Top right photo by Charles C. Schwartz; courtesy of the Alaska Department of Fish and Game. Bottom right photo by Heather K. Bickle; courtesy of the Ontario Ministry of Natural Resources.*

Certain linear measurements from moose are used to estimate weight and as indices of relative population health. Measurement/weight formulas have been developed based on information obtained from captured moose. A combination of a condition-evaluation criteria and total body length improved estimates of total body weight, compared with single linear measurements. Aside from its biological importance, the weight of a moose seems to be the matter of foremost curiosity to general observers. *Photo by Spencer Linderman; courtesy of the Alaska Department of Fish and Game.*

human sports medicine (Hundertmark et al. 1992a). This technique measures the impedance (resistance to alternating current) of hydrated (containing body fluid) body tissues to current of known frequency (Lukaski 1987). Because of an indirect relationship between body water and body fat, measured electrical resistance (influenced by water content in the body part) provides an estimate of body fat. This technique must be further perfected before it has practical field application for moose.

BLOOD

A common characteristic of blood constituents is that they are maintained at functional levels by physiological forces. This is homeostasis. Practitioners of human and animal medicine have used this characteristic to identify pathological conditions in an individual by associating alterations in selected blood parameters. Blood studies in moose attempt this on a population, not an individual basis (Franzmann et al. 1987). To do so, baseline or normal values for various physical and chemical properties have to be determined. This requires samples from hundreds of moose and classification of samples according to sex, age, month sampled, season age class, reproductive status, excitability, condition, drug used, climate, disease, body fat, habitat and location. How these variables influence blood parameters in various herbivores in both controlled and free-ranging studies was reviewed by Franzmann (1985).

Blood is collected from moose by vein puncture using sterile needles (18 to 21 gauge) and sterile syringes, or by using sterile evacuated glass vials (Vacutainer^R, Becton-Dickinson and Co., Rutherford, NJ). I prefer collection from the jugular vein in the neck for venipuncture in moose, but the radial vein in the foreleg works for young calves. Generally four Vacutainer^R vials (15 mL each) are filled with blood, one of which contains an anticoagulant (heparin or EDTA). Vials are identified and protected from freezing and extreme heat until they are taken to a laboratory equipped with electricity, preferably within 24 hours. When possible

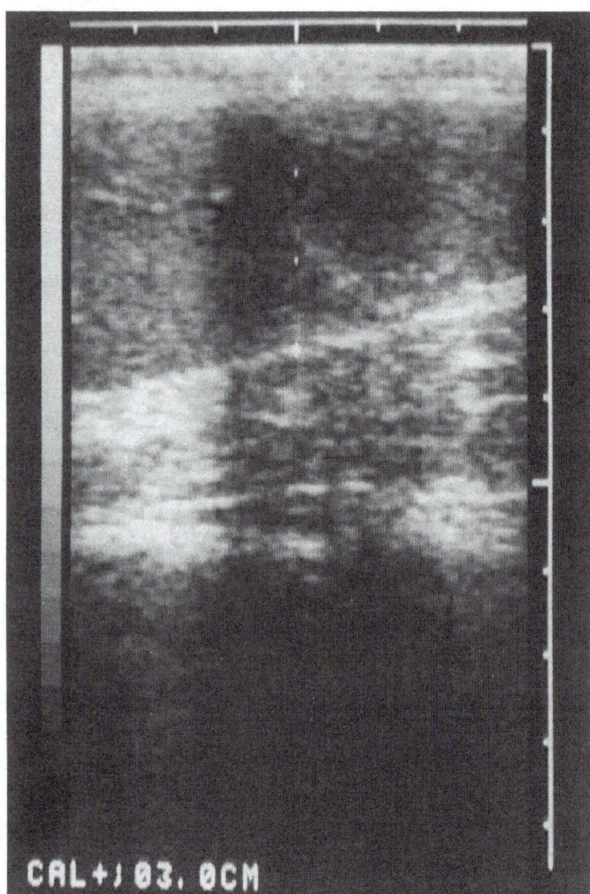

Portable real-time ultrasound imaging exhibits potential to monitor moose body condition. Shown above is a longitudinal cross-section of a portion of the subcutaneous fat layer in the rump region of a live moose. The fat layer is between the two + marks and is not of uniform depth (thicker on the left side). The skin of the moose is the narrow uniform band at the top of the screen, and the lower portion of the screen below the fat is muscle. *Photo by Thomas R. Stephenson; courtesy of the University of Idaho Department of Fish and Wildlife Resources.*

after collection, a blood smear should be taken as described by Coles (1967). The coagulated blood is centrifuged and the serum (liquid part of blood) placed in three separate plastic vials and freeze. The whole or uncoagulated blood is used to find hemoglobin (Hb) and packed cell volume (PCV) values. Hb is the oxygen-carrying red pigment of red blood corpuscles. PCV is the mass of red blood cells obtained by centrifuging a column of whole blood. The Hb is determined using a Hemoglobinometer[R] (American Optical Co., Buffalo, NY), and the PCV is ascertained with a microhematocrit centrifuge (Triac[R], Clay-Adams Co., Parsippany, NJ). PCV and Hb are useful parameters for evaluating condition in moose (Franzmann and LeResche 1978) and can be obtained in a field laboratory with minimum equipment.

The frozen serum has several population health monitoring applications. A part may be used for obtaining a potentially great variety of blood chemistries from a commercial laboratory. Those selected will be determined by the objectives of the study. The blood chemistries best suited for condition evaluation in moose are calcium (Ca), phosphorus (P), and total protein (TP) (Franzmann 1985).

Corticosteroid blood levels help evaluate stress (Franzmann et al. 1975b). Brain peptide (beta endorphin) levels from moose were compared to other species for physiological and behavioral relationships (Franzmann et al. 1981). There were differences in beta endorphin levels among species, but their significance could not be determined. Blood urea nitrogen (BUN) may be selected to assess the level of protein intake and catabolism (body protein tissue breakdown) during starvation (LeResche et al. 1974b).

Other applications of blood chemistry analysis depend on objectives and in research programs these are nearly unlimited (Kaneko 1980, Franzmann 1985). To refocus on condition parameters used for moose, the following have the greatest value: PCV, Hb, calcium and phosphorus. If a single

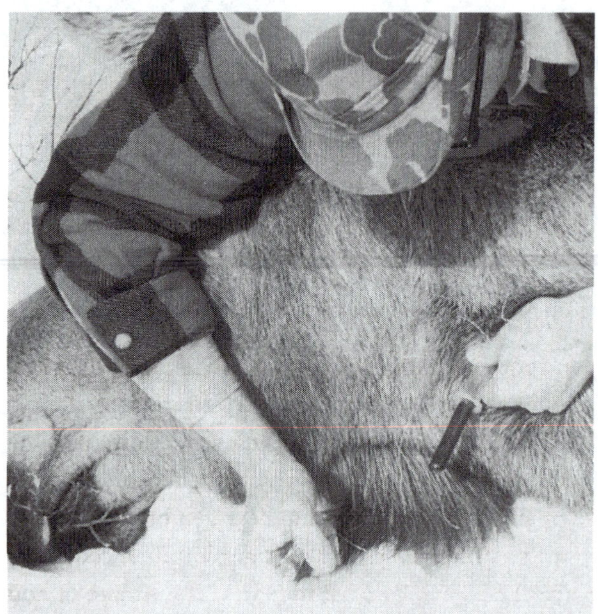

Blood from a moose can be analyzed for a variety of qualities that reflect on its own health and the health of its population, as reflected by packed cell volume (PCV) and hemoglobin (Hb) from whole blood and by calcium, phosphorus and total protein from blood serum. Antibodies in blood provide a record of past or present exposure to diseases. Additional analyses are for stress evaluation, protein intake, catabolism (body protein tissue breakdown) and other physiological measurements that may apply to a research program. *Photo by Albert W. Franzmann; courtesy of the Alaska Department of Fish and Game, Soldotna.*

parameter is used, it should be PCV. This is the test that is applied to evaluate condition in Alaskan moose populations.

Another part of the serum sample may be used to test serologically for various diseases. These tests measure the presence of antibodies produced by the moose in response to the presence of invasive organisms. A positive test means that antibodies are present and the animal has been exposed to the organism causing the disease, but it does not necessarily mean that the animal has an active infection.

Serological tests of wildlife can serve as a check on wild population exposure to various diseases, particularly when and where range is shared with domestic animals. In Alaska, serological tests on moose help to monitor changes in disease status (Kocan et al. 1986b, Zarnke 1991). Such data have particular value over time. If information on a particular population shows no historic evidence of exposure, yet antibodies are subsequently detected, the events that may have led to exposure may be readily identified.

Part of the serum sample should be placed in a serum bank for future use. Some tests that are not possible now may be performed in the future. A serum bank for all species should be maintained by every resource agency.

URINE

Studying white-tailed deer, DelGuidice et al. (1989) concluded that, to assess nutritional status, urinalysis offers advantages over blood analyses. There is no dependable and/or practical procedure for collecting urine from the live moose. However, DelGiudice et al. (1989) used urine collected in snow from white-tailed deer for nutritional assessment. Urea nitrogen, sodium and potassium levels in snow were analyzed and corrected for dilution with snow by using ratios to creatinine. Ratios with creatinine were chosen because the amount excreted in 24 hours is constant. Urea nitrogen-to-creatinine, sodium-to-creatinine and potassium-to-creatinine ratios declined with poor nutrition and elevated with improved nutrition. Using this technique to compare two moose populations on Isle Royale, DelGuidice et al. (1991) found that the creatinine ratios provided a quantitative measure of their physiological status. Urea nitrogen-to-creatinine ratios approaching four and higher suggested a severe decline in nutritional status. Using this technique at the Moose Research Center in Alaska, Hundertmark et al. (1992a) found that potassium-to-creatinine, calcium-to-creatinine and sodium-to-creatinine ratios allowed differentiation between healthy and undernourished moose. The sodium-to-creatinine ratio differed between animals on unrestricted and restricted diets. Hundertmark et al. concluded that urine deposited in snow can be a useful indicator of population condition in moose, but a clear understanding of the physiological processes, based on the creatinine ratios and their inherent limitations, is essential to proper implementation. This approach deserves additional testing and possible refinement because of its "field friendly" approach.

MILK

There are no reports of any relationship between moose milk constituents and condition, health or nutrition, although the significant value of whole milk to animal health is well-known (Cook et al. 1970, Franzmann et al. 1975a, 1976b, Renecker 1987b). The primary applications made from analyses have been the formulation of milk diets for hand-rearing moose calves (Regelin et al. 1982) and assessment of the cost of lactation in energetic studies.

SALIVA

Sex pheromones in male wild boar saliva stimulate estrus and copulation of that species. Similarly, saliva has been collected from male moose to test for their presence. Bull moose produced these pheromones, but at lower levels than did the boars. The role of pheromones in rut synchro-

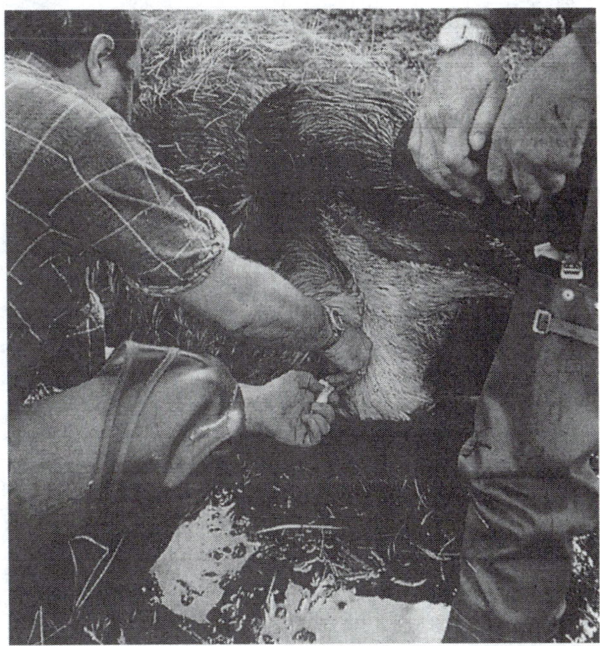

To provide a formula for preparing milk diets used in raising orphaned moose calves, milk is collected from immobilized wild moose cows immediately after parturition and at intervals thereafter because the constituents of milk change over time. The first milk, called "colostrum," is rich and contains many of the antibodies produced by the mother that protect the calf until it is capable of developing its own. Moose calves orphaned before receiving colostrum milk are at a disadvantage. *Photo by Albert W. Franzmann; courtesy of the Alaska Department of Fish and Game, Soldotna.*

nization and inducing estrus requires additional study (see Schwartz et al. 1990). Saliva is not presently used to test for health/condition relationships in moose, but the potential may exist.

HAIR

Hair is an ideal tissue to sample from wild populations because it is easy to obtain, transport and store. Hair should be plucked, not cut, from the peak of the shoulder of moose, where the hair growth rate is most rapid (Flynn et al. 1975). That hair records mineral uptake makes it extremely valuable for delineating mineral status in moose (Flynn et al. 1974, 1975, Franzmann et al. 1975c, 1977, Stewart and Flynn 1978) and detecting copper deficiency (Flynn et al. 1977).

As nutrients, minerals are classified as macro- or microelements (trace elements). The classification corresponds to concentration levels in tissues and requirement (Robbins 1983). The essential (i.e., needed to sustain life) macroelements are calcium, chlorine, potassium, magnesium, sodium, phosphorus and sulphur. The essential microelements are iodine, iron, copper, zinc, manganese, cobalt, molybdenum, selenium, chromium, tin, vanadium, fluorine, silicon and nickel (Underwood 1971). Schwarz (1974) suggested 20 additional elements should be considered and investigated as possibly essential.

Hair is analyzed for every known element by atomic-absorption analysis, including such toxic elements as lead, mercury and cadmium. Animals are indicators of toxic element accumulation in the environment (Cumbie 1975, Huckabee et al. 1973, Raymond and Forbes 1975). An added benefit of hair analyses is that stored hair is stable and the amount of each element does not change over time.

Hair plucked from the top of the shoulder of a moose, where hair growth is the most rapid, is analyzed for mineral studies and toxic metal accumulation. Hair samples may be stored indefinitely and continue to reflect the element content at the time of plucking. The accumulation of toxic elements in a particular environment may be determined best by sampling from the animals that live there, and hair is an ideal and easily collected material (tissue) to accomplish this. *Photo by Albert W. Franzmann; courtesy of the Alaska Department of Fish and Game, Soldotna.*

FECES

Fecal samples may be collected for parasite and food-habit studies. Fecal nitrogen analysis was proposed for evaluating dietary quality in free-ranging wildlife (Arman et al. 1975). Robbins (1983) suggested that the technique should be used as a very general indicator of protein intake, but not as an absolute predictor. Hobbs (1987) concluded that the fecal nitrogen technique does not offer reliable, quantitative predictions of diet quality in herbivores.

TEETH

Incisor teeth of moose provide a technique for age determination; cementum layers correspond to annual growth (Sergeant and Pimlott 1959; see Chapter 4). Annual cementum deposition consists of translucent (light) and opaque (dark) layers when viewed with transmitted light. Opaque layers are deposited during mid- to late winter, and the number of opaque layers corresponds to age of the moose (Gasaway et al. 1978). Extraction of the incisor requires extensive separation of the tooth from the gum using a tool called an "elevator." Loosening the tooth with a large animal extractor assists the process. Moose examined after incisor extraction show no adverse effects. Aging of moose populations has its primary application to population dynamics (see Chapter 6).

Examination of Dead Moose

Procedures for handling and examining dead animals in the field include more than a necropsy. The examination technique will vary with the objectives of the study, time of death, population history, logistics and personnel involved. The procedures to be considered are: (1) circumstances about the area; (2) circumstances about the animal; (3) necropsy; (4) physical condition; and (5) tissue collection.

CIRCUMSTANCES ABOUT THE AREA

When the vicinity of a deceased moose is entered, the scene must not be disturbed, although this may not be as impor-

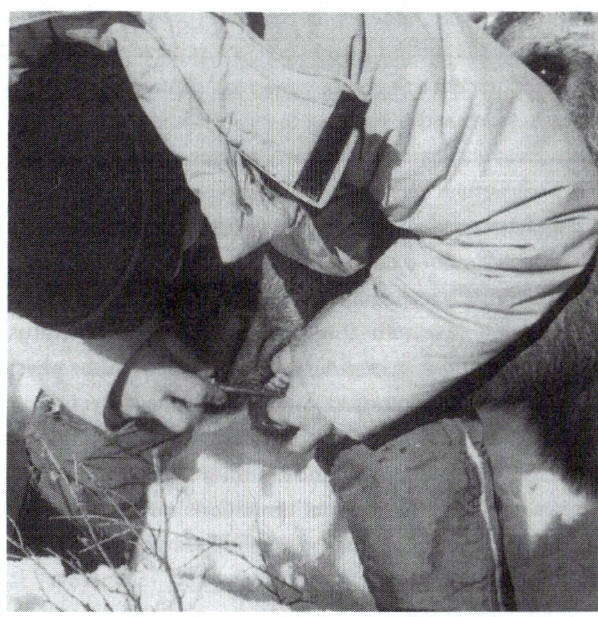

An incisor tooth extracted from an immobilized moose is used to determine the age of the animal by the number of cementum layers in a cross section of the tooth. Hundreds of such removals by biologists and veterinarians studying moose in Alaska have caused no apparent problems for the animals subsequently. No treatment has been shown to be necessary to prevent infection. *Photo by Albert W. Franzmann; courtesy of the Alaska Department of Fish and Game, Soldotna.*

tant if the animal is sacrificed for population sampling. Otherwise, investigators must assume that signs in the area may provide clues to the cause or circumstances of death.

Advisably, the position of the carcass is recorded and the area photographed before further examination proceeds. The area should be investigated for tracks or spoor, beds, fecal droppings, hair, and blood. The terrain and habitat type should be noted. Evidence of physical phenomena, such as lightning strike or high-water marks, should be noted. Once all characteristics and evidence in the proximity are detailed, animals can be examined.

CIRCUMSTANCES ABOUT THE MOOSE

Some steps previously outlined for examination of a live animal are just as appropriate for a dead one. The only difference is that more information may be obtained and more tissues may be available for collection. Before a moose carcass is handled, the investigators must assume that the animal may be a reservoir for human disease, regardless of the apparent cause of death. Anyone working on the animal should wear acceptable protective gear, such as rubber apron, gloves and boots.

Zoonotic diseases (transmissible from animals to humans) can range from localized infections from broken skin to such highly infectious bacterial diseases as anthrax. None of the latter type has been reported from moose.

After examination of the moose carcass is completed, an assessment may be made as to the desirability and feasibility of bringing a pathologist to the scene, taking the carcass to a pathologist or diagnostic laboratory or proceeding with a gross necropsy.

GROSS NECROPSY

Wobeser and Spraker (1980) outlined procedures and equipment for necropsy of large mammals. Several considerations must be emphasized:

- Seek professional assistance when possible.
- Develop and follow a standard necropsy technique.
- Consider that each animal has a disease transmittable to humans and secure protection accordingly.
- Fresh specimens are most desirable.
- Necropsy procedure should not be limited to infectious and parasitic disease identification—having an indicator animal (carcass) from a free-ranging population provides opportunity to assess reproductive and/or nutritional status, food habits, condition, and stress.

Tissue Collection: General Considerations Types of tissue examinations and tests are nearly unlimited, and not all are warranted. Collections should be made based on study objectives and on site assessments, preferably with professional assistance. Some general rules on tissue collection are:

- If possible, collect and deliver to the laboratory the entire carcass (seldom applicable with moose).
- Collect extra specimens when unsure of the quality or quantity from initial collection.
- Refrigerate samples for short-term preservation.
- Place specimens in leak-proof containers, and clearly identify each (e.g., collector, date, location, specimen, sex, age).
- Whenever possible, perishable specimens should be delivered directly to a laboratory (or shipped in a manner to preserve the specimen).
- Avoid contamination of and physical damage to specimen.
- Refer to techniques for preservation of biological material, as outlined by Wobeser et al. (1980).
- When in doubt as what to collect or how to handle specimens, contact and work with the laboratory involved.

Tissue Collection: Histopathology Histopathology is the study of tissue at the cellular level. Some basics for collecting samples for histopathological examination are:

- Collect a portion of the gross lesion (affected area) and adjacent normal tissue.
- For supportive diagnosis, also collect sections from the major organs (i.e., liver, kidney, spleen, lung, heart, stomach, intestine and brain).
- The ideal tissue sample size for fixation (preservation) is 1.0 inch square (2.54 cm) and 0.5 inch (1.27 cm) thick.
- Fix tissue in 10 percent neutral buffered formalin.

Tissue Collection: Microbiology Examination for bacteria, viruses, fungi and yeasts generally is dependent on isolating living organisms. Therefore, prompt delivery of the tissue to the laboratory is essential. Some general considerations are:

- Avoid contamination—sterile collection is imperative.
- Refrigeration is desirable.
- Specimens for bacteriological and fungal examinations may be frozen, but except at –40°F (–40°C), freezing should be avoided for viral specimens.
- For rabies suspects, the entire head of the animal should be refrigerated and delivered to the laboratory.
- Sterile swabs may be used to collect and transport samples for bacterial culture.

Tissue Collection: Toxicology A thorough history should accompany samples in which poisons or toxins may be involved. Basic considerations include:

- Collect blood, liver, kidney, brain, fat and digestive tract contents.
- Tissue may be refrigerated or frozen.
- Pluck hair from the shoulder of the moose (for determination of heavy metal accumulation).
- Specimens collected for pesticide or any organic compound analysis should be contained in aluminum foil, not plastic.

Tissue Collection: Parasitology Some general rules for collection for parasitological examination include:

- Ectoparasites (parasites on skin) are preferred live for identification.
- Helminths (intestinal worms) should be relaxed in cold water or saline, nematodes (round worms) should be fixed in hot 20 percent ethyl alcohol or 5 percent formalin, and cestodes (tapeworms) and trematodes (flukes) are fixed in 10 percent formalin.
- Fecal sample collected for egg identification or quantification (eggs per gram of feces).
- Protozoa may be identified from tissue, blood, and feces, therefore refrigeration is recommended, and blood smears obtained at collection site.
- Various tissues may be collected for specific parasites (e.g., the brain and carotid artery for the arterial worm, brain tissue for the brain worm and muscles for muscle cysts [sarcosporidiosis]).

Tissue Collection: Reproduction The reproductive success of a population is best measured by natality (birth rate), which often is difficult to determine from a free-ranging moose population. The reproductive tract of individual cows may give insight about the population's reproductive status. Kirkpatrick (1980) outlined a detailed account of the physiological indices of reproduction (see Chapter 4).

Fat Indices Obtaining fat indices from the live animal was previously discussed and its limitations noted. From the dead animal, there are many more indices of body fat available, including:

- Whole body fat. Harris (1945) described the order of fat catabolism on declining nutritional intake as follows: (1) rump fat; (2) subcutaneous fat; (3) visceral fat; and (4) marrow fat. There are limitations to measurement of these fats, depending research on objectives and (Robbins 1983, Price and White 1985). However, if an objective is to evaluate early fat loss, measurement of rump fat is useful. Conversely, to assess starvation, measurement of bone marrow fat is useful. Also, whole body fat extraction has limited field application for moose due to the animal's size, and indices of fat are preferred. Fat extraction of body parts as an index of total body fat may be useful because it mirrors whole body fat.
- Femur (upper rear leg bone) marrow fat. Kirkpatrick (1980) reviewed various techniques for measuring femur marrow fat. Marrow fat content is difficult to relate to total body fat except at extreme levels. It does not measure degrees of fat loss, but an animal approaching starvation will have low marrow fat reserves. Femur marrow fat values below 10 percent identify moose that have starved (Franzmann and Arneson 1976).
- Mandible (jaw bone) marrow fat. This index is gaining favor in large mammals over that of femur marrow fat because the mandibles are easier to obtain. Jaw bones are more likely to be left at wolf kill sites, and leg bones frequently are carried off. Snider (1980) compared marrow fat in both femurs and mandibles of moose and concluded that the percentage marrow fat was significantly correlated. Ballard et al. (1981b) made similar correlations between mandible and long bones (femur, metatarsal and metacarpal bones) in Alaskan moose.

- Carcass grading system. Kistner et al. (1980) developed a grading system for mule deer based on the amount of fat visible on the carcass and various body parts. This system has not been used for moose, but could and may be useful at an abattoir where many carcasses may be graded.
- Kidney fat index. This index, based on the relationship between weight of the kidney and adhering fat deposit (Riney 1955), was widely applied to ungulate species, including moose (McGillis 1972). However, the kidney fat index has a very poor relationship to total body fat, and is considered meaningless for predicting body composition (Finger et al. 1981).

Measurements and Weights Carcass weight to whole body weight relationships were reported for moose by Blood et al. (1967), who found that dressed carcass yield was 50 percent of whole body weight. Timmermann (1972) reported whole body weight loss to dressed weight varied from 15 to 29 percent and concluded that there was great variability in percentage weight loss due to field dressing. Because of this variability, dressed carcass weight to whole body weight indices generally are not useful for moose.

Weighing body parts from dead animals may indicate the live condition. Body parts include selected muscle groups (Ringberg et al. 1981) and organs and bones (Price and White 1985). Thymus gland weights indicate live condition in white-tailed deer (Ozoga and Verme 1978). Adrenal gland weights correlate with adrenal gland secretion of corticosteroids (Adams and Hane 1972), therefore can be used as a measure of long-term stress. Short-term exposure to stressors generally does not cause increase in adrenal weights (Kirkpatrick 1980). None of these organ weight indices has been applied to moose to date, but the potential exists. Bubenik and Bellhouse (1980) suggested that brain volume of moose may be valid as a predictor of generation performance and population status.

Blood, Hair and Teeth These are the same as for live moose (discussed earlier).

General

In addition to aforementioned assessment methods used for moose to apply the indicator animal concept to assess population status, reviews by Kirkpatrick (1980), Franzmann (1985), Price and White (1985) and Huot (1988) should be referenced for more details and for other methods that may apply to other mammals. Selection of assessments will be dictated by many factors, including funding, personnel and training level, support personnel and laboratory capability, location, political and social constraints, population statistics, technology available, cultural and religious constraints, and data-processing capabilities.

Each project leader must decide what parameters provide the best information, given the project limitations. Some parameters reflect short-term disturbances and others reflect long-term perturbations. Huot (1988) listed the features that methods for health/condition evaluation should possess:

- sufficiently sensitive to detect minor changes in nutritional state;
- specific enough to suggest deficiency in energy, minerals, proteins and other basics;
- ease of measurement or collection of samples from live or dead animals by staff with little training;
- practicable at different times of year for all sex and age classes;
- uninfluenced by stress of capture, measurement or sampling; and
- objective and repeatable.

Involving other institutions and/or persons with unique expertise, through cooperative agreements, is a practical and efficient way to accomplish some objectives. No single group or institution has all the capabilities, experience and expertise that potentially can be brought to bear. Interpretation of findings may require additional help from a wide diversity of specialists. Their assistance in a team approach should be sought.

The lack of controlled studies in wildlife populations and/or a lack of baseline or normal levels of the parameter being measured confounds data interpretation. Considerable time and many samples are necessary to establish baseline values. This is an important reason for collecting as much information as possible from a population by way of indicator animals when given the opportunity. Establishing or contributing to baseline values should be an objective for every collection. Controlled studies of wild animals present a very different and difficult problem, particularly for an animal as large as moose and that has a poor history of survival in captivity.

Moose Husbandry

The need for tractable captive moose for research purposes is apparent. Records show that moose were and are raised for exhibition and education, but in North America, moose presently are not raised for meat or milk production or as beasts of burden. Until the development of a ration for moose to sustain them over time (Schwartz et al. 1980,

1985), captive moose older than 2 years were rare unless they were fed fresh-cut browse, or had access to browse daily. This also accounts in part for the scarcity of moose in zoos until recently. Maintaining them on browse was simply too costly, and they did poorly on traditional hay-based domestic animal feeds (Thorne 1979). Schwartz (1992b) provided results of a review of moose husbandry in North America based on a survey and experiences at the Moose Research Center in Alaska. From his inquiries, Schwartz found that 29 facilities kept moose, including all four North American subspecies; 26 responded to his questionnaire (14 zoos, 9 research facilities and 3 game farms). Much of this section is derived from that review.

Raising and Feeding Moose Calves

Raising moose calves has been done with varied success, and everyone who has done so has their own idea how best to care for and feed the animals. What is presented here primarily reflects the program followed at the Moose Research Center and elucidated in other published reports.

Because the moose raised at the Moose Research Center are used for long-term studies, it was necessary to develop a ration that would sustain moose in good condition throughout their life. And such a ration was developed at the Center (Schwartz et al. 1980, 1985).

Schwartz et al. (1980) hypothesized that moose on diets high in fibrous material from grasses and forbs (hay based) become bulk-limited because of reduced rates of passage and do not extract enough nutrients to survive. Schwartz et al. believed that a formulated ration should resemble more closely the natural diet of moose in all constituents. The diet subsequently formulated was nutritionally balanced, but it featured a fiber source of sawdust from quaking aspen instead of hay residues (Table 49). The diet was tested over a 5-year period on 11 moose (Schwartz et al. 1985). Five of six females produced calves at 2 years age, indicating a high level of nutrition. Males on this feed successfully bred as yearlings (Schwartz et al. 1982).

The Moose Research Center-developed ration resulted in greatly increased capability to maintain moose in captivity. A similar ration formula now is on the market (Mazuri Moose Maintenance[R], PMI Feeds, St. Louis, MO). Consumption of food by moose at the Moose Research Center varied with season from 0.5 to 1.0 percent of body weight of dry feed daily in winter to 2.5 to 3.0 percent in summer (Schwartz et al. 1984). Not many game ranches have the area or browse available to sustain a viable moose population without supplemental feeding. The costs of feeding moose and building and maintaining structures to contain and handle them are substantial. Reneker et al. (1987) concluded that, although moose are behaviorally predisposed to intensive management, they are too sensitive to disease and too expensive to fence and feed for viable commercial production.

Table 49. Composition of Alaska's Moose Research Center moose ration (Schwartz et al. 1985)

Ingredient	Percentage
Corn, ground yellow	30.0
Sawdust	25.0
Oats, ground	15.0
Barley, ground	12.5
Cane molasses, dry	7.5
Soybean meal (7.4 percent nitrogen)	6.3
Pelaid[R] (Rhodera Inc., Ashland, Ohio)	1.3
Dicalcium phosphate	1.1
Sodium chloride	0.5
Vitamins A, D_3 and E[a]	0.3
Mycoban[R] (Van Waters and Rogers, Anchorage, Alaska)	0.025
Trace minerals and flavor[b]	tr

[a] Each kilogram of feed contained 5,000 international units (IU) of vitamin A, 13,000 IU of vitamin D3 and 44 IU of vitamin E.

[b] Added trace minerals and flavor (mg/kg) were: zinc 65.2; iron 30.5; magnesium 22.0; manganese 16.8; copper 7.6; iodine 1.3; and anise 125.0.

Regelin et al. (1982) reported on raising, training and maintaining moose calves at the Moose Research Center. Soon after their arrival, captured calves were fed a commercial milk replacer (Suckle[R], The Carnation Co., Los Angeles, CA). They were maintained on the milk supplement for 100 days and fed five times per day until 30 days of age, and then daily feedings were reduced until 100 days (Table 50). Daily injections of vitamin B complex were given to compensate for any loss of water-soluble vitamins from the digestive tract associated with loose stools or diarrhea.

Table 50. Milk feeding schedule for moose calves from birth to weaning at the Moose Research Center, Alaska (Regelin et al. 1982)

Days of age	Intake in gallons (L) per day	Feedings per day
1–3	With cow	
4	0.63 (2.4)	5
5	0.79 (3.0)	5
7	0.85 (3.2)	5
11	0.90 (3.4)	5
13	0.92 (3.5)	5
22	1.00 (3.8)	5
25	1.11 (4.2)	5
37	0.87 (3.3)	4
45	0.77 (2.9)	3
67	0.48 (1.8)	2
88	0.24 (0.9)	1
100	Weaned	

Moose to be "domesticated" and trained for research purposes must be acquired as calves. Without calf bonding or acclimation to humans, older moose are intractable and nearly impossible to train. Calves abandoned in spring (late May and early June) represent a primary source. Calves born to tame, matured moose cows is another. Like any orphaned animal, moose require a great amount of care and attention. For the first month, they must be fed at least five times per day, with peak milk consumption of about 8.9 pints (4.2 L) per day at 1 month of age. *Photo by H.R. Timmermann.*

The calves were initially housed in 16 square feet (5.8 m^2) plywood pens with a plastic or plywood roof. Each pen was bedded with sawdust and cleaned regularly. Keeping pens and calves dry was very important. At 6 weeks age, the calves were moved to a 500-square foot (46.5 m^2) holding pen, with access to a 17-acre (7-ha) fenced exercise area.

Water was always available, and calves began drinking at approximately 5 days age. The Moose Research Center pelleted ration also was available ad libitum and the calves began eating it after a few days. They were given fresh-cut vegetation daily, which they began consuming after about 5 days of age. At 10 days, rumination began. By 40 days of age, calves were consuming large quantities of browse. This became an overwhelming logistical problem, so the animals were released into an adjacent 640-acre (259-ha) enclosure where they could browse at will. The calves returned to the feeding facility daily. Calf groups raised later survived solely on the Moose Research Center ration, and provided browse was not a critical part of their diet.

Welch et al. (1985) were successful in weaning 44 of 54 calves raised on a formula of even parts whole bovine milk and evaporated milk with colostrum added. The success was attributed to the milk formula, cleanliness of the facility, detailed observations and data recording, and prompt medical attention. In addition, Welch et al. made the following recommendations:

- Start feeding 0.53 gallon (2.0 L) per day every 4 hours.
- Never feed more than 0.66 gallon (2.5 L) per feeding.
- Provide pelleted ration ad libitum throughout weaning.
- Provide large amounts of natural browse.
- Avoid overcrowding.
- Have as few people as possible involved in direct care and feeding.
- Have personnel on duty at all times.

Feeding moose calves is more art more than science. Practices differ as evidenced by the aforementioned (see also Denniston 1956, Dodds 1959, Yazan and Knorre 1964, Markgren 1966, Addison et al. 1982), and each has experi-

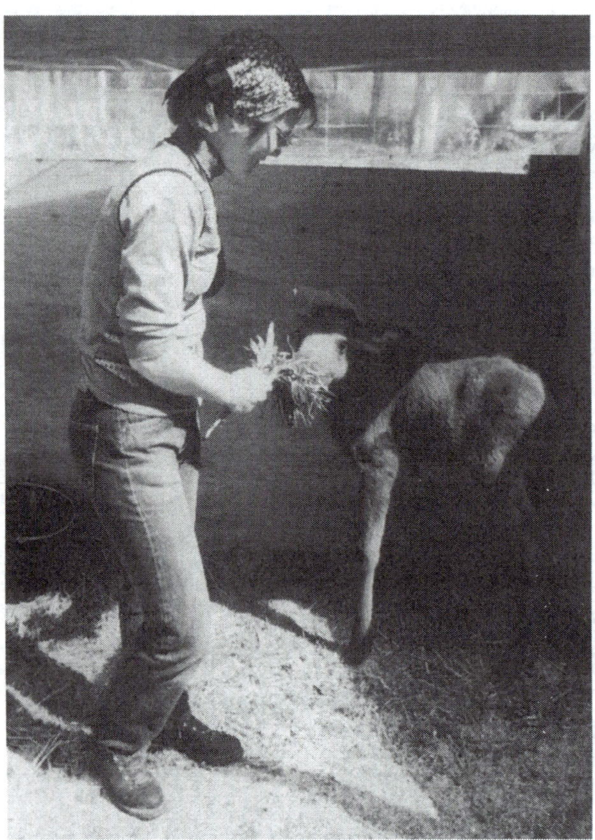

At about 5 days of age, hand-raised moose calves are offered and begin to consume forbs and browse. Rumination can be detected when these calves are about 10 days old. *Photo by Albert W. Franzmann; courtesy of the Alaska Department of Fish and Game, Soldotna.*

enced both success and failure. From the Schwartz (1992b) survey, nine facilities fed a whole cow milk/evaporated milk mixture at ratios ranging from 5:1 to 1:1. Six facilities used commercial milk replacers alone or mixed with whole goat or cow milk. Three used an evaporated milk/water mixture, and one used a commercial doe milk replacer. Each facility was successfully raising calves.

Schwartz (1992b) suggested that the primary disease problem with calf raising was diarrhea. Next was laminitis, or "foundering." Both diseases are related to husbandry or feeding practices. The best treatment, based on close observation of animals and prompt response, is adjustment of feeding practices. Restricted milk intake and/or replacement with an electrolyte solution when first signs of liquid stool appear is helpful. Laminitis results from carbohydrate overload that occurs with extreme changes in amount and quality of intake. Accordingly, controlled and measured intake is the key to problem avoidance. Other disease problems associated with raising moose calves are varied, and typically are handled clinically similar to treatment of domestic young.

Training Moose Calves

Raising moose calves also requires their training to facilitate handling as the animals mature. Procedures for training moose vary with the purposes for maintaining the animals, and recommended methodologies are at least as numerous as the different feeding regimens. Training may include getting the moose to walk onto a scale, lead with a halter, respond to certain signals, and/or walk into a metabolism chamber or onto a digestion crate. In all cases, the sooner training starts, the more likely it will succeed.

Lautenschlager and Crawford (1983) described their experiences training three moose to lead with a halter for loading into a horse trailer. The time required for training to lead and load depended on how well a calf imprinted on its trainer and at what age training began. All three moose eventually were used in vegetation selection studies and were regularly led and transported to feeding fields. However, one of the moose—a male—became impossible to control during the rut; it was necessary to castrate the bull and deantler him each autumn.

In at least several regards, the Lautenschlager and Crawford experiences parallel perspectives gained at the Moose Research Center. In particular, training should begin the day that the calves are confined. Also, the best trainer is likely to be the person who raises and feeds the calf. And for best application of training, that same person should be involved.

In addition, bull moose are not manageable during the rut and can be extremely dangerous. All males retained in the confined areas at the Center are deantlered in autumn to lessen the possibility of injury to handlers.

Regelin et al. (1982) noted that imprinting is very important, and the person assigned to raising a calf should spend long hours devoted to petting, brushing and just sitting with the calf. A feeding bottle can be used to train a calf to go onto a scale or into a building or crate. Some moose at the Moose Research Center were trained to enter a small box in preparation for entry of a respiratory chamber (Regelin et al. 1981a, 1982). Moose used for milk production in Russia required a different training regime (Timmermann 1991).

Each research objective for which a group of calves may be used will require a special kind of training. There are no firm guidelines and, even under the best of circumstances, results are not always those anticipated. A clear lesson from experiences with "domesticated" moose is that they are individuals, with unique personalities and demeanor. When attempts are made to place or train these animals into statistically meaningful groups, the ability to account for this individuality can be lost.

Moose calves in captivity tend to imprint on one person. Rearing calves for nutrition, energetics, reproduction or other study involves much time, patience and what is best characterized as tender loving care. The greatest success is with calves handled by persons willing to devote full time to the animals. Breaking calves to halter is an important part of the associative, bonding and training process. As training progresses and calves become familiar with various procedures, the need for the initial trainer or imprint person diminishes, and other trainers or workers can be used. There does not appear to be any variation in response to training on the basis of moose gender. As calves grow and mature, problems of handler safety increase. By its very size, an adult moose led by halter can cause injury to handlers. Two or more persons should be present when all moose but calves are being handled. Sexually mature bulls should not be handled during the rut. *Photo by Albert W. Franzmann; courtesy of the Alaska Department of Fish and Game, Soldotna.*

Maintaining Adult Tame Moose

Of 26 facilities surveyed by Schwartz (1992b), only 2 kept moose in areas large enough to sustain them on natural browse. The other 24 facilities fed moose a formulated ration, and 21 of these supplemented the ration with natural browse (Table 49). Of 25 facilities that provided the composition of their rations, 18 used sawdust-based diets, 4 used alfalfa-based diets and 3 fed wood fiber and grain. The most common diet was the Moose Research Center ration.

Captive moose do not require closed shelter, but windbreaks are necessary. In contrast, they cannot tolerate high heat and humidity, as discussed earlier. Shelters, ponds or streams, and trees that provide relief from those conditions are required. Zoos south of normal moose range use sprinklers or misters.

To maintain adult moose, most facilities (65 percent) used 7- to 8-foot high (2.1–2.4 m) woven-wire fences; the others used chain link fence (Schwartz 1992b). Some problems have occurred with wild moose or deer jumping into facilities, primarily with drifted snow or during rut.

Predators have been a problem at the Moose Research Center, where both wolves and black bear have killed confined moose. These predators most frequently gained access by digging under the fence, but black bears also climbed fencing. Brown bears rarely enter the facility and have not killed any confined moose.

High voltage (7,900 volt) electric fence (Gallagher-Snell Power Fence, San Antonio, TX) was tested at the Moose Research Center (Schwartz 1992b). This type of fencing kept female moose confined, but required considerable maintenance. Before their removal in autumn, antlers insulated most bull moose from the shock, and this enabled one male to tear down a section of fence. Long winter guard hairs also have some insulative qualities, so moose were only intermittently shocked. And snow acted as an insulator, so alternating positive and negative wires on the fence were required.

A breakthrough food ration for year-round maintenance of captive moose was developed at Alaska Department of Fish and Game's Moose Research Center. Research on tame animals in a controlled environment, such as at the Moose Research Center, enables data not available in any other manner. However, except for research purposes, the costs of food, health care and facilities (plus laws governing the captivity of native wildlife) preclude profitable moose ranching. *Photo by Albert W. Franzmann; courtesy of the Alaska Department of Fish and Game, Soldotna.*

Noninfectious diseases and accidents were responsible for 30 of 45 deaths (67 percent) reported for adults. The most prevalent of these (14 deaths; 31 percent) was chronic diarrhea, primarily in moose fed hay-based rations. This condition may account for a life span of captive moose that reportedly has not exceeded 13 years (Schwartz 1992b). Drug-related deaths accounted for seven deaths (16 percent), primarily from aspiration pneumonia and capture myopathy. Other noninfectious mortality was attributable to laminitis from foundering and arthritis.

Infectious diseases caused 14 (32 percent) of reported mortalities: 9 were of bacterial and viral origin, and 5 were parasitic. The former included pneumonia, systemic infections, gastroenteritis (inflammation of digestive tract) and malignant catarrhal fever (virus carried by sheep, goats and wildebeest). Parasitic diseases included meningeal worm, whip worm, thread worm and hydatid cysts.

Nonfatal health problems for moose in confinement include winter tick, biting insects, overgrown hooves and dewclaws, and injuries to velvet antlers. Insects and ticks can be controlled with pesticides when necessary. Moose flies and black flies may occur in large numbers on moose and often cause open wounds on the hock area with their persistent biting (Lankester and Sein 1986).

These aforementioned diseases are not necessarily representative in type or frequency of those that occur in free-ranging moose. Although each may occur in the wild, con-

The health of a moose population may be reflected in the growth and antler development of yearling bulls. At this age, most wild bulls have only spiked or forked antlers. In captivity and on the ration developed at Alaska's Moose Research Center, yearlings, such as the 15-month-old bull above, have potential to have exceptional body and antler development. *Photo by Albert W. Franzmann; courtesy of the Alaska Department of Fish and Game, Soldotna.*

finement tends to result in biased estimates of incidence in the wild. Effects of confinement on disease and on hierarchical and/or socialization relationships of moose were not evident for animals in the 1-square mile (259-ha) enclosures. Seven generations of moose have been reproduced in those circumstances, and their reproductive rates are similar to those of moose in the wild. Close confinement and/or high stocking rates likely would cause disease, hierarchical and social problems that would affect reproductions. However, because neither situation exists, there are no confirming data. See Chapter 16 for a complete outline of disease syndromes of wild moose.

The exchange of information generated by the Schwartz (1992b) survey provided a baseline. It was particularly useful because nearly everyone keeping moose responded. This type of survey needs to be repeated regularly, and hopefully will stimulate better moose husbandry through the exchange of data and ideas.

Increasing numbers of moose are in captivity. And although six generations of moose have been maintained in captivity, the mortality rate remains higher than it should be. However, most adult mortalities are potentially preventable with good feeding practices, and there is capability now to decrease drug-associated deaths to zero. Calf losses also can be significantly reduced by application of improved husbandry gained from recent research and experience. Husbandry of moose and research derived from captive moose will continue to benefit not only from more knowledge, improved methodologies, better logistical resources and refined standards, but also from broader application of what already is known.

COW (W/CALF)
LAKES, OCT. 1, '61
ALARMED —
OUT OF LAKE —
MOVING ALONG AT RAPID WALK (OR
VERY SLOW TROT), OCCASIONALLY LOWERING
HEAD AS IF SEEKING TRAIL —
UPPER SIDE
OF PALM GOLD-
BROWNISH ON
BROW SECTION
BROWN
UNDERSIDE
PROBABLY LESS
MUSCLE VISIBLE
THAN INDICATED.
FAT — EVEN
HEAD FILLED
OUT —
MOOSE IS
HEAVIEST
SHOULDER —
— FATTER - MORE
WET ANIMAL (RAIN) — HAIR
BUNCHY, ESP. ON RUMP, WHERE
MUCH WHITE EXPOSED
(COMPARE DRAWING ABOVE WITH
AUG. 17, '56 SKETCH —
FALL (RUT) VS. SUMMER
BULL)
FAT, BIG BULL — POSSIBLY OLD, AS
ANTLERS NOT PARTICULAR LARGE
UNUSUALLY LONG BELL — BOTH
DEWLOP & BELL —
GOLD ANTLERS AGAINST
-GREEN FOREST —
DARK LEGS &
AGAINST
GRASS —
REFLECTIONS —
ORANGE WILLOWS
OCT. 2, LAKES, DENALI
HIGHWAY, ALASKA —
MORE ACCURATE SKETCH
OF WALKING POSE ABOVE (SLOWER
WALK) — MORE BEEFY, HEAVY-
OCT. 3, '61

17

H.R. (TIM) TIMMERMANN AND
M.E. (MIKE) BUSS

Population and Harvest Management

Proper management of moose requires knowledge, good judgment and consideration for both the animals and the people who value them. Moose management can help maintain a population of healthy animals for the sake of those animals and for the social and economic benefits associated with them. Reasons for managing moose include:

- They are magnificent animals, deserving of human attention and responsible treatment.
- They add beauty, diversity and interest to the environment.
- They are part of North America's wildlife heritage, and an integral part of a complex ecosystem.
- They have substantial recreational and economic values.
- They are an important source of food, particularly for subsistence users and Native people.
- In the absence of management, they can cause serious forest damage and otherwise conflict with humans.

People will agree or disagree with these reasons according to their individual association with moose. Many urban residents may never see a wild moose. Yet, to know that moose exist and thrive in nature can be sources of comfort and/or intrigue. People who have direct contact with moose have varying opinions about them. A driver who hits a moose may have a decidedly negative feeling about the species' value. A canoeist on a remote waterway may long and fondly remember the moose photographed only yards away. Most hunters regard moose as a grand quarry that, if taken, represents a vast store of highly palatable and nutritious food.

There also are people who wish to see moose and other wildlife "managed" per se by leaving them to their natural destiny despite or in spite of human activity that encroaches on the animals and their habitats. But such encroachment, when combined with other factors, including predation and disease, invariably causes moose populations to decline, sometimes to threateningly low levels.

Most areas inhabited by moose in North America have changed dramatically since the first European explorers arrived. Human population and sprawl have altered much of the natural environment. Such enterprises as agriculture, urbanization, forestry, mining, suburbanization, transportation systems, exurbanization and hydroelectric development all have impacted the land and wildlife. In particular, forest fire control has had a significant impact on moose in the boreal forest. Without natural fires to rejuvenate aging forest, moose are reliant on human-caused disturbance, particularly on timber harvesting, to maintain or create suitable habitat.

Conserving moose seems to be a good idea, but there is confusion regarding the meaning of "conservation." To some, it implies a hands-off approach to natural resources, especially wildlife or forest resources. To others, the definition embodies the notion of "wise use," in itself a vague and dynamic term usually embracing anthropocentric considerations to the exclusion of all others.

For successful conservation of any renewable natural resource, including moose, that resource must be managed, husbanded, treated and considered beyond the scope of the prevailing human utilitarian value system. To a great extent, this perspective was acknowledged in 1980, when the World Conservation Strategy laid out its main principles for conservation—maintaining essential ecological processes, preserving genetic diversity and ensuring the sustainable use of species and ecosystems (International Union for the Conservation of Nature and Natural Resources 1980). These principles do not suppose that conservation precludes "use" of natural resources, nor does use imply that harvest or harness of a resource is predication of its value. On the other hand, conservation does not rule out harvesting and other forms of utilization that supplement and optimize the resource's inherent value, provided, again, the application of sound judgment, and caution. To a large degree, managers have responsibility to accommodate and moderate society's activities (uses) to promote and maintain healthy ecosystems, including those that feature moose populations.

In practical terms, the economic utility of a resource provides the incentive and proscribes the acceptability of management. Moose, as with any other natural resource, must have tangible socioeconomic values before the public, through government, can be expected to allocate money for its management.

Moose hunting provides a readily recognizable value of moose. For example, hunters annually contribute more than $50 million to the Ontario economy (Ontario Ministry of Natural Resources 1980a). Such expenditure not only gives moose hunters a voice, but also compels governments to recognize the importance of moose. Moreover, many more people simply like to view and photograph moose. However, relatively few make special trips solely for that purpose, so their contributions toward identification of an economic value of moose is not as compelling. Nevertheless, their interest in the species must be acknowledged and factored.

Values of northern ecosystems to the vast majority of people are measured in large part by the economic contributions provided by a variety of natural resources. The contribution of moose usually is much less than those of timber harvesting or mining, and no government wants to implement policies that could endanger large numbers of jobs. The forest industry is a major employer in moose range. One in 10 Canadian jobs, for example, depends on forestry, so to believe that forest management activity will be undertaken solely to enhance moose populations is unrealistic. However, a well-managed forest can provide moose and other wildlife values without jeopardizing timber production.

Moose respond positively to sound (scientifically based) habitat and population management. But the objectives of such management should not be considered or oriented toward short-term benefits—for moose or humans. Management aimed at producing an ever-increasing moose population might ultimately lead to reduced aesthetic, environmental and economic values. The moose has an intrinsic value, and managers must be mindful of this. Human activity should be geared to encourage healthy populations of moose because the species deserves to exist and function compatibly in a complex and fragile ecosystem.

Moose, like virtually all members of the deer family, are subject to external limitations, such as starvation, disease and predation, that help control populations. Hunting is a controlled form of predation. In addition to its recreational and economic values, hunting can be a tool of management *if* there is accuracy in its predicted effects (Boer 1991). Modern recreational hunting allows for utilization of the sustainable surplus of a moose population and maintenance of populations commensurate in size and distribution with available habitat and human tolerances. Unhunted and underhunted moose populations often fluctuate askew of the dynamic balance with the available habitat or without a benefit to humans who are members of the same ecosystem, for better or worse (Boer 1991). More important, a selective harvest from predetermined cohorts can help maintain a reasonably natural social structure of a moose population (Bubenik and Pond 1992). A population structure that is socially balanced usually is characterized by optimum recruitment and physical maturation and reduced social and environmental stresses. These conditions are reflected in the well-being of individual moose, the population and the environment (Bubenik 1991). Accordingly, moose population management goes far beyond enhancing forage resources and reducing nonhunting predation so as to maximize the production of furred targets. It undoubtedly will be the only defensible position for wildlife management in the future.

Moose managers are in a unique and highly responsible position. Although moose are not by any means the only game animal in northern forests, their large size and low population density make them significant within the ecosystem. Simply in terms of mass, for example, a single adult moose is the equivalent of at least 600 ruffed grouse or 300 snowshoe hares. Snowshoe hares and ruffed grouse have high reproductive capability, and their populations normally exhibit wide fluctuations. If some environmental factor such as predation or disease removes large numbers of these animals, their populations usually recover quickly when circumstances improve. Moose, conversely, have a much lower reproductive rate, and the loss of a single animal has relatively much greater consequences to its population. Accordingly, management for moose, including its harvest, must be relatively sensitive and refined.

Management Policy, Goals and Objectives

In the early 1990s, two territories and nine provinces in Canada and at least 11 states in the United States were actively managing moose harvests (Timmermann and Buss 1995; Figure 203). In addition, New York, Vermont and Michigan had self-sustaining populations. Eight jurisdictions have specific goals and objectives incorporated into moose management plans. Most objectives were related to maintaining or increasing the moose population and the recreational, social, and economic benefits associated with a harvest, and gaining new knowledge about moose ecology. Goals and objectives generally are formulated for the entire jurisdiction, but may be tailored to specific areas and targeted to specific populations.

G. Eason (personal communication: 1992) categorized a variety of objectives necessary for managing a moose population:

- Population characteristics. Goals and targets for population size, age and sex structure, and distribution are to regulate (increase, stabilize or decrease) populations and focus management effort.
- Biological products. Harvest of meat as a utilitarian and legitimate use of the renewable resource (i.e., moose meat is healthy, low in cholesterol and high in protein).
- Ecological functions. Integration of moose coincident with objectives for other components (i.e., predators, scavengers and vegetative types) of the ecosystem.
- Consumptive recreation. Identification of hunting benefits as opportunities, and specification of harvest success and hunter density to help focus management.
- Nonconsumptive recreation. Identification of goals and targets for viewing opportunities, and specification of viewing success to help focus management effort on these benefits.

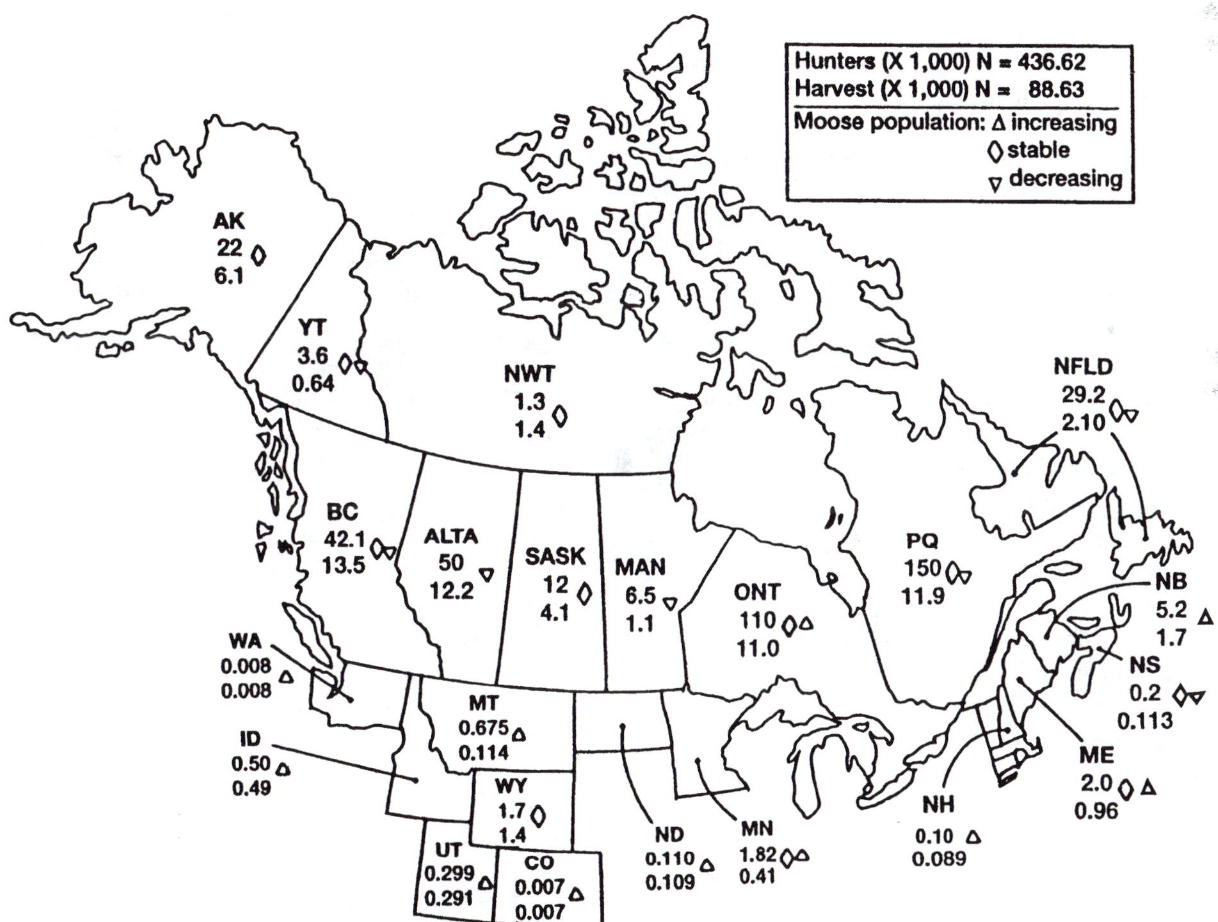

Figure 203. Estimated 1990–1991 moose hunter numbers, harvests and population status for 22 North American jurisdictions (after Timmermann and Buss 1995). Most states and provinces reported stable or increasing moose populations.

- Economic benefits. Meat value in remote and economically depressed areas should be considered when setting harvest objectives. Allowing some commercial outfitting can increase local and regional economic benefits and broaden management support. Regulating fees charged and number of participants can help maintain quality standards.
- Knowledge. Provide up-to-date general public and hunter information. Better understanding of moose and the benefits they provide will help gain and maintain management support. Objectives for monitoring or researching information are useful in directing management effort.
- Conflicts and costs. Specify damage objectives to help direct efforts at reducing highway accidents, agricultural and forestry damage, and moose interactions with deer and caribou.
- Other existence values. Identify ethical and heritage arguments to maintain and reestablish moose throughout their range, even if no other benefits are provided. Consider population restoration and maintenance objectives.

Examples of Management Policies

Most jurisdictions have developed management policies with specific goals and objectives.

SOUTHEAST ALASKA

Goals established (Alaska Department of Fish and Game 1990) include:

- To maintain, protect and enhance moose habitat and other components of the ecosystem.
- To maintain viable populations of moose in their historic range throughout the region.
- To manage moose on a sustained-yield basis.
- To manage moose in a manner consistent with the interests and desires of the public.
- To manage primarily for meat hunting and not trophy hunting of moose.
- To manage for the greatest hunter participation possible, consistent with maintaining viable moose populations, sustained yield, subsistence priority, and interests and desires of the public.
- To provide opportunities to view and photograph moose for the benefit of both hunters and nonhunters (nonconsumptive users).
- To develop and maintain a database useful for making informed management decisions.

Moose harvest objectives for the southeastern Alaska region were established for 1994 based on the 1989 moose population size and harvest data (Table 51).

MINNESOTA

Long-range planning for moose in Minnesota included both a goal and three objective statements as follows:

> Manage the moose resource at a level that yields maximum recreational benefits within acceptable public and environmental limits.
>
> Objective 1—achieve the following winter moose population by 2000: northeastern area—7,000 moose (40 percent increase from 1986); northwestern area—500 moose by 1992 (25 percent increase); and northwestern prairie area—3,800 by 1992 (15 percent increase).
>
> Objective 2—harvest 900 to 1,200 moose every 2 years.
>
> Objective 3—provide additional opportunities for use and appreciation of moose by 1992 (Minnesota Department of Natural Resources 1990, M.S. Lenarz personal communication: 1996).

MAINE

Maine revised its moose management plan in 1991 to cover the period 1991 through 1996 (Maine Department of Inland Fisheries and Wildlife 1991). The goals were to maintain moose numbers at 1985 levels, increase harvest and maintain viewing opportunities. Three objectives were:

- *Abundance*—maintain moose populations at 1985 levels in all Wildlife Management Units through 1996.
- *Harvest*—increase harvest to 1,000 to 1,400 moose per year or to a level needed to maintain populations throughout the state by 1996.
- *Use*—maintain opportunity to view moose, and decrease unsuccessful viewing trips by 50 percent by 1996.

Table 51. Moose harvest objectives for 1994 in Alaska's southeastern region, based on 1989 data

Population and harvest parameters	1989	1994
Posthunt moose population	2,530	2,675
Annual sport harvest	177	231
Number of hunters	1,041	1,215
Hunter days of effort	4,149	5,275
Hunter success (percentage)	17	19

UTAH

Utah initiated a moose hunting season in 1958 (Utah Division of Wildlife Resources 1988). Its goal has been to manage existing moose populations and promote expansion of moose populations into suitable habitats. Objectives include managing for trophy class animals, investigating the effects of antlerless harvests, law enforcement, identification of suitable but unpopulated habitat, and monitoring success of translocated moose.

MANITOBA

Manitoba places emphasis on the social, cultural, subsistence and economic significance of moose. Twelve management principles and objective are central to its management program (Crichton 1992b). Emphasis is on cooperation and communications with all users, as well as recognition of the dependence of subsistence use and the cultural importance of moose to native people. In addition, management strategies focus on the controlled use of technological advances, such as off-road vehicles, aircraft and motor vehicles in moose hunting to prevent overharvest. A status report of all wildlife including moose must be submitted to the Manitoba legislature every 5 years. The 1987 report predicted a shortfall in moose supply and demand would continue to exceed supply in the 1990s.

BRITISH COLUMBIA

Five major goals were identified in a draft provincial moose statement (Hatter et al. 1990). These goals (1) promote the maintenance and enhancement of populations and habitat, (2) provide opportunities to appreciate, study and view moose, (3) provide for opportunities to hunt moose, (4) allow commercial uses through guiding, tourism, taxidermy, arts and crafts, and (5) protect people and property from intolerable levels of damage.

QUEBEC

Quebec identified five major moose management problems in the early 1990s (Anonymous 1992), including (1) declining hunting quality, (2) inadequate harvest control, (3) low densities, (4) regulations that focus on control of hunter activity, rather than herd control and (5) continued harvest of the herd's most productive female breeders (4–12 years of age). A thorough review of moose population characteristics in Quebec's 23 hunting zones by Courtois (1989) resulted in a new management plan proposed for 1993 through 1997. Three general objectives are central to this plan: (1) to preserve and increase moose populations; (2) to preserve opportunities for recreation; and (3) to improve the quality of hunting. The new policy proposed a radical change in the method of controlled hunting. The main components included selective hunting of females in all hunting zones, a general license valid for calves and adult males available to all hunters, and a limited number of licenses to hunt adult females. The overall objective was to stabilize or increase declining moose populations. A management plan focusing on the protection of cows was introduced in 1994 (Courtois and Lamontague 1997).

ONTARIO

In 1979, Ontario's moose management program also experienced declining moose populations and related recreational and economic benefits (Ontario Ministry of Natural Resources 1980a). A series of 72 public meetings (involving more than 7,000 hunters) were held to discuss management options. In 1980, a provincial moose management policy (Ontario Ministry of Natural Resources 1980b) was approved. It included a broad program objective, four program targets, 14 policy guidelines and 15 management goals.

Specific program targets included: (1) to increase the herd from 80,000 to 160,000 by the year 2000; (2) to harvest 18,000 moose per year by 1995 and 25,000 moose per year by 2000; (3) to provide for 875,000 hunter days annually by 2000; and (4) to create sites where more than 1 million people annually have the opportunity to observe moose by the year 2000.

The new policy prompted many changes in moose management. It established parameters for selective harvesting of moose. Other important aspects of the program were designed to increase moose numbers by habitat manipulation, increased law enforcement, selective predator control, and improved methods of monitoring moose populations. Most important, the tradition of unrestricted hunting access to moose of any age or sex ended, and managers were given broader control of the moose harvest.

A survey of 22 state and provincial agencies that manage moose populations revealed that 8 are guided by a specific moose management plan or a written policy and 14 by an unwritten or generalized wildlife policy. The policy objectives generally stress: (1) population protection, stability or enhancement; (2) sustained yield management; (3) recreational viewing and hunting opportunities; (4) recreational and subsistence harvests; and (5) economic benefits through tourism. Other objectives attempt, through controlled and selective harvest, to promote a balanced social organization within the moose herd and thereby optimize productivity and health. This concept of controlling harvest by sex and age was stimulated by Bubenik (1972) and has been used in British Columbia, Saskatchewan, Ontario, and portions of Alaska and Newfoundland (Child 1983b, Euler 1983a, Stewart 1983, Schwartz et al. 1992).

Population Assessment

Effective management of moose populations requires ongoing assessment of a number of parameters that indicate or enumerate their size, composition, growth rate and productivity. The value of these parameters, or indicators, is enhanced when the methods of collection and analyses are comparative and estimates precise enough to afford a reliably accurate indication of conditions in the wild. Most often, interpretation of population assessment information is possible only when trend-through-time data are available. Single surveys rarely provide data on which an assessment of a moose population can be made. Further, the effort and cost of gathering assessment data need to be consistent with the intensity and objectives of management. The more critical or subtle the management intervention contemplated, the more acute the need for precision. Much time, effort and expense have been directed at assessing the status of wildlife populations, only to find that the information provided was far too imprecise to evaluate conclusively if a particular management action had a desired effect. Knowledge of the techniques and evaluation of their results will provide the manager the best opportunity for selecting the necessary parameters and measures to meet specific management objectives.

Moose population size is assessed in three basic ways—total area counts, sample estimates and indices. Total counts or censuses are made by attempting to count all the animals in a particular area. Sample estimates are made by counting a portion of the population and, using statistical procedures, extrapolating to the total count. Indices are relative measures made by counting some object (e.g., fecal pellet groups) or sign (e.g., tracks) related to the animal (Davis and Winstead 1980). Survey objectives may include evaluations of (1) the impacts of harvesting moose by licensed recreational and subsistence hunters, (2) the effects of predation and (3) habitat use.

Both population and sex/age composition estimates are biased if they fail to account for variable sightability and uneven seasonal distribution of moose aggregations (Gasaway and Dubois 1987). Therefore, observed sex and age classes can vary with sampling design, i.e., where, when and how intensive the search. Moose are the least sociable of all North American deer, and often are considered solitary or seasonally individualistic (see Chapter 5). For example, Houston (1974) indicated that, in a year-long study of moose in Wyoming, single animals represented 58 percent of the reported sightings (n = 3,351), two moose represented 26 percent and groups of three or more moose accounted for only 16 percent. In an Alaskan study (Ballard et al. 1991: 26), "average moose group size was about two moose from January through July, increasing to 3.0 in August, 4.9 in September and 7.6 in October. After October, group size decreased to 3.2." Late autumn and early winter aggregations of bulls and cows generally are largest, but they often appear loosely knit and transitory, according to Coady (1982).

A number of jurisdictions—British Columbia, Saskatchewan, Manitoba, Yukon and Alaska—encountered management difficulties in the 1970s. Moose numbers dropped before the rate and magnitude of decline were recognized and before timely and effective solutions were implemented. In interior Alaska, for example, moose declined significantly during the early 1970s and reached a low density before managers were certain the decline was not self-reversing (Gasaway et al. 1983). Had good data on population size, rate of change and mortality been available and action taken, recovery likely could have been initiated sooner. Accurate population data also are a requirement for complex management programs designed to alter social structure, including age/sex composition and to increase populations (Stewart 1978, Bubenik 1981, Child 1983b, Euler 1983a). Other objectives may include ecological research to assess the impact of human interventions on wildlife populations (Gasaway and Dubois 1987).

Predictable population and harvest management requires reliable estimates of major population parameters. Population size, annual rate of change, recruitment, sex and age composition and mortality are the principal parameters necessary. Aerial surveys of moose on representative areas of habitat or range are most commonly used to obtain estimates of these parameters. The size of the moose population, especially in areas readily accessible to hunters, is reflected in part by the number of the animals killed and by the observed frequency of moose or their sign. To determine population parameters, ground and aerial surveys are conducted.

Ground Surveys

PELLET GROUP COUNTS

The index technique of counting droppings or pellet groups over a given area and time period was developed in the late 1930s to estimate population levels of deer (Bennett et al. 1940). Pellet group counts are estimates of the actual or relative number of big game animals or the animal-days of use in a given area (Neff 1968). Such counts can provide an objective measure of substantial population fluctuations and help determine preferred habitat types and seasonal use patterns (Riney 1957). The primary use of pellet group counts for moose has been to obtain information on relative moose densities between areas or years. These indices

Moose population and sex/age composition estimates need to factor the animals' differing sightability and nonuniform distribution, in large measure because moose are the least sociable of all North American deer. Aggregation size varies seasonally, and late autumn and early winter groups of bulls and cows are generally the largest. When group size increases during the autumn rut, associations appear loosely knit and transitory. Increase in snow depth on winter range eventually will force moose to lower elevations and disperse to conifer-dominated stands that provide cover from deep snow. *Top photo courtesy of the Wyoming Game and Fish Department. Bottom photo by W.B. Ballard.*

are used, along with aerial survey and harvest statistics, to estimate population trends.

In northern latitudes, pellet group surveys usually are conducted in early spring (April and May) immediately after snow melt and before full emergence of leaves. At this time, the previous winter's pellet groups are lying on top of the autumnal leaf drop and relatively easy to see. Jordan and Wolf (1980) designated September 20 as the usual onset of deposition and May 25 as the end of the pellet deposition season on Isle Royale, Michigan. Their calculations normally were based on 247 days, less the appropriate number when the count was made before May 25.

Fecal material is classified as "winter" (pellets) or "summer" (nonpelletized). Subjective judgment is made by survey crews regarding "new" groups deposited the preceding winter or "old" groups deposited before then (LeResche 1970).

Most survey plots are circles or rectangles and distributed in some form of random or systematic design. Krefting and

Shine (1960) developed a multiple-random-start systematic sampling technique for conducting deer pellet group counts in northern Minnesota. This system for moose pellet group studies was subsequently used on Isle Royale (Krefting 1974b). In preferred winter habitat, moose pellet groups usually are deposited in a clumped pattern rather than randomly (DesMeules 1965). Sampling in such areas frequently has been carried out along with browse or habitat surveys on winter concentration areas or on permanently marked plots that are periodically cleared (Bergerud and Manuel 1968). Stratified random sampling, as described by Eberhardt (1960) for white-tailed deer in Michigan, is considered desirable whenever possible.

Neff (1968) provided a review of the technique for big game trend, census and distribution information, and discussed its problems. Problems associated with the pellet group technique include observer bias and fatigue, the small area actually sampled, rapid loss of pellets through insect attack and rain, decreased visibility of groups as vegetation advances, size of plot and type of surface terrain. Another problem is the determination of daily deposition rates. Human error, or observer bias, is considered the thorniest problem (Neff 1968). Careful adherence to standards is necessary to minimize bias.

Most estimates of winter defecation rates of moose lie between 13 and 21 pellet groups per day, based on reports from British Columbia, Quebec and Alaska. Edwards (1956b) first suggested using 14.9 groups per moose per 24 hours for animals in Wells Gray Park, British Columbia. Vozeh and Cumming (1960) in Ontario revised the rate to 13.0 and used it widely in the 1960s. DesMeules (1968) reported an average rate of 21.5 depositions per day for two moose (11.9 by a cow and 9.6 by a calf) for a period of 11.3 days in Laurentides Provincial Park Quebec, and suggested the use of a 10.7 depositions per moose per day based on these findings. Jordan and Wolfe (1980) used an average of 20.9 groups per day on Isle Royale, but suggested the rate needed to be determined experimentally.

Pellet group counts were used to estimate moose numbers and determine habitat preferences at the Moose Research Center in Alaska (Franzmann et al. 1976a, 1976c). At the Center, May counts of permanent plots in a 1.0-square-mile (1.6-km^2) enclosure with known numbers of moose resulted in unusually high defecation rates (20.2 groups per moose in 1970–1971 to 28.7 groups in 1973–1974). Eight individual moose were followed during winter 1975, when snow cover allowed accurate counts of daily defecations. Findings suggested an average of 14.6 (range, 10–22) for adult cows and 19.6 (range, 14–25) for adult bulls. Rates were significantly different ($P < 0.01$) between sexes. Oldemeyer and Franzmann (1981) continued work at the Center and reported seven moose deposited a daily average of 16.2 pellet groups per moose in the winter of 1976–1977 and six moose averaged 17.2 the following winter. Oldemeyer and Franzmann suggested the need to clear pellet group plots in both autumn and winter to estimate winter use more accurately. The discrepancy in daily rates may be attributable in part to sampling error due to incorrect pellet aging, differential defecation rates between males and females, and quality and amount of food eaten. A.W. Franzmann (personal communication: 1993) suggested that, given the accuracy of pellet group counts, the use of different correction factors would have less affect on accuracy than would the pellet sampling itself.

Beginning in the late 1950s, Ontario used pellet group counts as a standard technique for investigating relative densities of moose in winter concentration areas. Application of these surveys declined because of problems inherent in the methods, which made interpretation difficult.

Krefting (1974b) used pellet group counts to measure population trends on Isle Royale between 1948 and 1970. Counts were made in spring, beginning in 1948, on 0.01-acre (0.004-ha) plots at 10-chain (200-m) intervals on transects in representative cover types. Similar counts were repeated each spring on a total of 844 plots. Mean numbers of pellet groups per acre (0.4 ha) decreased ($P < 0.01$) reflecting a known herd die-off between 1948 and 1950. Later counts suggested a steady population increase from 1950 to 1970. These data, however, contrasted to the findings of Jordan et al. (1972), who reported population stability from 1959 to 1969. Lautenschlager and Jordan (1993) suggested a combined winter track/pellet group count method for censusing moose in small, relatively easily accessible areas. They believed midwinter defecation rates to be more uniform than those in autumn or late winter, and that such counts could lower sample variation.

In summary, fecal deposition rates clearly are variable among areas, sexes and ages of moose, and both forage quality and availability are important in determining the rate of forage passage and digestibility. Intake rates are scaled to body size so that a larger animal normally will defecate more than a smaller one. Therefore, pellet group counts are best used as an indirect index to moose abundance and not to estimate numbers. The intensity of sampling necessary to gain census precision would negate its use as a practical management tool in field application, according to Franzmann and Schwartz (1982).

LIVE SIGHTINGS AND INCIDENTAL MORTALITIES

Experienced observers have recorded ground sightings of moose in various parts of North America. Pimlott (1959) reported that field staff in Newfoundland gathered 9,881 observations of moose (8,610 of which were identified by sex) in a 6-year period between 1950 and 1955. Several other provinces, including Ontario (Peterson 1955), Quebec (Moisan 1952), New Brunswick (Wright 1956) and Nova Scotia (Dodds 1963), also have used record cards. A major problem, according to Pimlott (1959), had to do with the ability of observers to distinguish among calves, bulls and cows. The Minnesota procedure featured a postcard that was filled out and mailed in by field cooperators in the early 1960s (Timmermann 1974). The data obtained were considered worthwhile, and recommendations were made to continue collections. Fire watchers in towers tabulated the sex of animals sighted in Maine during the summers of 1965 to 1969 (Dunn 1966, 1969). In Ontario, from 1949 to 1955, numbers of moose were estimated from trappers reports compiled annually for each provincial administrative area (Reynolds 1953).

In the early 1990s, at least 9 of 24 agencies used ground counts to help estimate moose population trends. Random sightings by agency staff were routinely recorded in Washington, Wyoming, Montana, Colorado, Massachusetts and Vermont. Maine recorded hunter sightings. Both British Columbia and New Hampshire used volunteer observers. Reports of hunter sighting in Maine "suggest that hunting is altering the sex ratio in some zones, but this is not reflected in the harvest data" (Morris and Elowe 1993: 97). Hunter effort and observations for monitoring population trends were reviewed by Crichton (1993), who cautioned that consistency in collecting such data must be ensured to make valid comparisons between years.

A major weakness of ground counts is the difficulty in detecting moose in dense vegetation. At the Moose Research Center in Alaska, LeResche (1970) reported that eight men spent 312 man-hours counting browsed twigs over a 2-week period in a moose enclosure, but observed a total of only 21 moose, mostly along cleared fence lines. Known densities in each pen averaged 12.0 per square mile (4.6/km^2). Direct ground observations of moose, in general, do not provide reliable population estimates, but are useful in verifying herd composition and population trends gained from other sources.

The numbers and age of known incidental moose mortalities can provide data on population growth, distribution and reproductive output. Hicks (1993) reported that use of roadkills in an index to moose population change may provide insights to the status of unhunted populations, in the absence of other data. Hicks cautioned that roadkill information not be used by itself. Vermont, for example, reported a steady increase in moose mortalities, primarily motor vehicle-related, through the 1980s and a sharp increase in 1991 (Figure 204). The moose population size needed to sustain such an increasing kill rate was estimated to be at least 1,000 moose in December 1991 and suggested that the herd was increasing by approximately 15 percent each year (Alexander et al. 1992). In addition, the high representation of younger age classes (≤4.5 years) suggested a productive moose population. By 1993, the population was estimated at more than 1,500, and the state introduced a limited hunt, with 30 either-sex permits in one management unit (Alexander 1993).

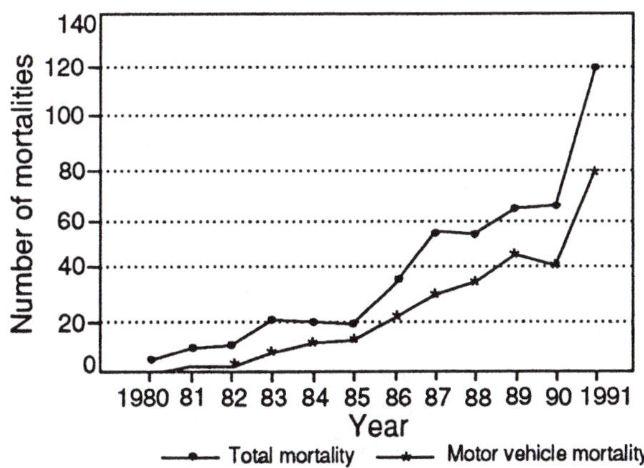

Figure 204. Comparison of Vermont moose motor vehicle mortalities with total mortalities (n = 469), 1980 to 1991 (from Alexander et al. 1993). Reflected in the mortality data was a dramatic moose population increase in the state in the 1980s. The moose became abundant enough that the Vermont state legislature authorized a hunting season beginning in 1993—the first such season since 1896.

HUNTER HARVEST STATISTICS

Information on relative population levels of moose can be obtained by using an index or relative measure of population trend, such as hunter harvest data. Hunter check stations represent an excellent source of information on some management areas. In 1990–1991, about half (12 of 22) of the wildlife agencies of moose states and provinces used check stations to help monitor moose harvests. Posthunt mail survey sampling of licensed hunters was used by 13, whereas 5 used telephone surveys.

Harvest quotas for Montana are determined in part by evaluations of range conditions, age structure of the harvest and past hunting success (Stevens 1971). In Newfoundland, Mercer and Manuel (1974: 662) reported that hunter success and the number of moose seen per day provided "perhaps the best data to indicate population changes." In three management areas, daily success declined 52 percent and moose seen per day decreased 53 percent between 1960 and 1972. Population decline also was indicated by aerial surveys in the same management areas. In Quebec, Crête et al. (1981a) and Crête and Dussault (1987) reported an inverse relationship between hunting effort and moose density, suggesting that declining harvest success was a consequence of declining moose populations. Effort was expressed as days per hunter per moose, which, in turn, was converted to density in some management units.

Lykke and Cowan (1968) felt that total harvest was only broadly indicative of population levels, but gave no indication of the fluctuations in effort that went into the success achieved. Hunting season length, license fee adjustments, transportation costs and weather all affect harvest statistics. In addition, hunters often are inclined to select for an antlered bull over a cow when given a choice. Peterson (1955) and Pimlott (1961) felt that harvest data obtained from vast areas, with only limited access and uneven hunting pressure, were little better than rough trend indicators.

In recent times, hunting statistics have been used in Quebec and Ontario to identify discrete moose populations exhibiting similar characteristics and to develop regression models that predict population parameters (Courtois and Crête 1993, Timmermann et al. 1993). Both hunter success and the percentage of calves harvested appear to correlate with density estimates obtained from aerial surveys.

A more sensitive indicator of population size is the proportion of bulls in the harvest. Lykke and Cowan (1968: 5) reported that in Norway a high percentage of bulls in an either-sex harvest "indicates a good moose population in relation to hunting pressure; the reverse occurs during a population decline." The Norwegian bull harvest averaged 55.5 percent of the total kill, with fluctuations from 50 to 60 percent in normal years. However, Sweden maintained an average bull harvest of 54.5 percent despite a rapid increase in total kill. Fraser (1976) suggested an estimate of the proportion of the annual moose harvest can be made on the basis of age-related changes of animals shot. In the future, there may be potential for indirect population estimation of moose based on survival and reproduction, or by the change-in-ratio method developed for Michigan deer by Eberhardt (1960) (Figure 205).

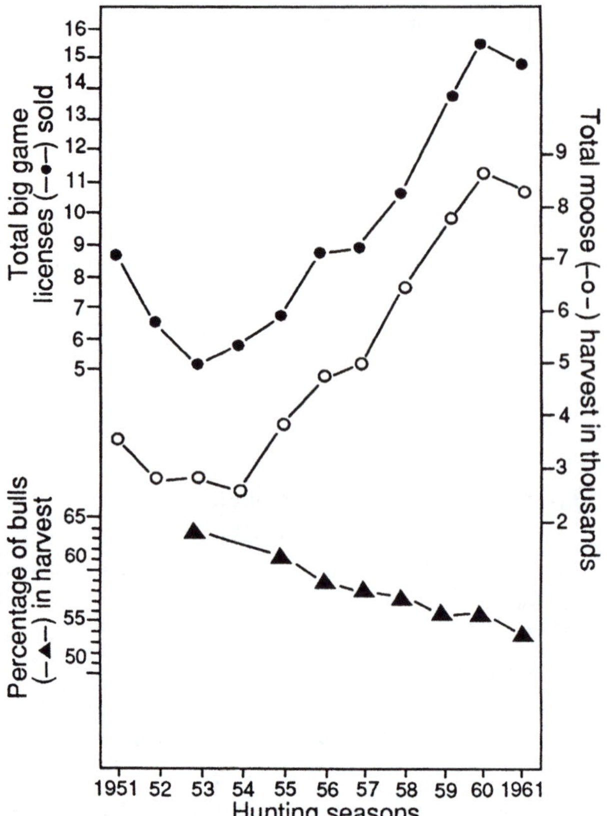

Figure 205. Comparison of moose licences sold and total moose harvest in Newfoundland, 1951–1961. As total moose harvest increased, the percentage of bulls in the kill decreased, reflecting their greater vulnerability to harvest than experienced by cows (from Bergerud 1962).

PHYSICAL AND PHYSIOLOGICAL INDICES

Moose populations can be monitored in part by measures of specific physical and physiological indices, including general conditions of both live and dead moose, morphological measurements and characteristics of blood, body and marrow fat, urine, feces, hair and antlers (see Chapter 16). All methods have limitations, and none should be relied on solely; rather, each should be considered as one of many tools available (Franzmann and Ballard 1993). Managers should recognize that physical and physiological parameters can exhibit considerable variation by sex, age, season, habitat, region, etc., and that, in many cases, a large data base may be required to serve as baseline values for detecting abnormalities.

Aerial Surveys

Counting moose on winter range from aircraft is the most practical method for estimating moose numbers over large areas of North America (Timmermann 1974, 1993, Gasaway et al. 1986, Gasaway and DuBois 1987). However, many variables affect survey accuracy and precision, and most are difficult to quantify.

HISTORIC PERSPECTIVE

In 1945, Aldous and Krefting (1946) made the first recorded estimates of moose numbers based on winter aerial surveys on Isle Royale, Michigan. They flew transect lines or strips covering 30 percent of the island, and estimated 510 moose. Likewise, Bowman (1955) attempted summer aerial counts

Recent clearcuts provide desirable early winter habitat for moose in loose aggregations, sometimes as large as 10 to 20 animals. The attraction of moose to abundant browse in such clearcuts and the openness of these habitats provide ideal conditions for aerial surveying of moose. In late winter, deep snow and crusting conditions force moose into dense conifer cover, reducing the precision and reliability of surveys conducted at that time. *Photo by J. Nistico.*

in Ontario during July 1946 and, although moose were seen, mainly along water courses, the method did not prove to be practical. Much work was necessary before an acceptable field technique was developed and widely applied. Several methods, including transect censusing and intensively searching of sample units or plots, were developed in the 1950s and 1960s (Timmermann 1974). Two approaches generally are used—searching an entire area or sampling a portion of it. Sampling is more commonly used, because estimates can be made for large management areas. These areas usually are stratified for sampling based on different expected densities of moose. Sampling units randomly selected from each strata can be either transects or plots.

TRANSECTS

The line transect or strip method of censusing moose usually involves flights along preselected parallel routes at a constant altitude. Observations are made at a prescribed distance from the flight line, usually after the ground is well covered with snow. Banfield et al. (1955) summarized the transect or strip census method most generally used. In Ontario, moose transect surveys originally were conducted in DeHavilland piston-powered Beaver aircraft. The crew of four consisted of a pilot, navigator and two observers (Trotter 1958). Flight lines were laid out on a map running north and south. The desired degree of coverage dictated the distance between transects. Flying was done at an altitude of 800 feet (244 m) and an air speed of 90 miles per hour (145 km/hr). The two observers spotted moose, while the navigator kept the aircraft on the flight line and recorded

Aerial surveys to estimate moose population density and sex/age composition are conducted by most agencies from December through February when snow cover enhances detection of the animals. Sightability is high in open deciduous-dominated forests (above), but commonly averages no better than 60 to 70 percent in mixed or coniferous stands. Moose management in many jurisdictions continues to be hampered by quality (precise and accurate) population estimates. *Photo by Jack Soeby.*

the location of moose observed. A similar technique was employed in British Columbia, Alaska, Maine, Saskatchewan and Alberta (Spencer and Chatelain 1953, Edwards 1954, Dunn 1966, Lynch 1971, Hope 1972). All described technical and logistic problems associated with transect surveys, including: confinement to relatively flat terrain; general low sightability of moose, resulting in a large proportion of moose missed and a large sampling variance; special high-resolution maps or photos required; and the bias and precision of incorporating a sightability correction factor for moose surveys not yet evaluated (Gasaway and Dubois 1987). Nevertheless, compared with plot sampling, transect sampling has several advantages: lower cost per unit area; less observer fatigue; ability to stratify areas by relative densities and then assign blocks by optimal allocation; ease of location in areas with few landmarks; ability to provide precise estimates of observed moose; and the option of estimating a sightability correction factor for missed moose.

Moose usually are counted in a strip 0.125 to 0.31 miles (0.2–0.5 km) wide on either side of the flight line; the angle of sight is used to determine transect width (Saugstad 1942, Edwards 1952, Rausch and Bratlie 1965, Dunn 1966). Observation is limited to a definite field, the outer margin of which can be marked by guides, one on the aircraft window and the other on the wing strut (Figure 206; Banfield et al. 1955). Transect width is calculated from the tangent of the fixed angle and a fixed altitude. Transect width and length are used to calculate total sample area. Decisions to include or exclude animals from the transect, especially near its boundaries, are critical, because of their effect on the variance of the estimate.

The transect method was discontinued in Manitoba in 1971 (Timmermann 1974), because results lacked the necessary accuracy needed to provide information useful for management. Because of variable results obtained in relatively closed forest cover, Ontario terminated transect use in the late 1950s, in favor of the intensive-search method (Trotter 1958). A few jurisdictions (Idaho, Saskatchewan and Nova Scotia) continue to use transects to monitor long-term population trends, recognizing that estimated densities can be extremely variable among years.

Aerial transects also have been used to obtain moose density and distribution estimates for different areas. This procedure was widely used to stratify areas prior to random quadrat sampling in northern Minnesota, northwestern Ontario, central Alberta and Alaska (Timmermann 1974). Maine used line transects since 1965 to define relative population levels of moose, as a basis for reinstituting a legal harvest season in 1980 (Dunn 1971).

Thompson (1979) and Dalton (1990) incorporated visibility bias into the line transect survey by calculating the distribution of distances from the center line of the survey to animals observed. Dalton (1990) indicated that this "modified" line transect method should be more accurate

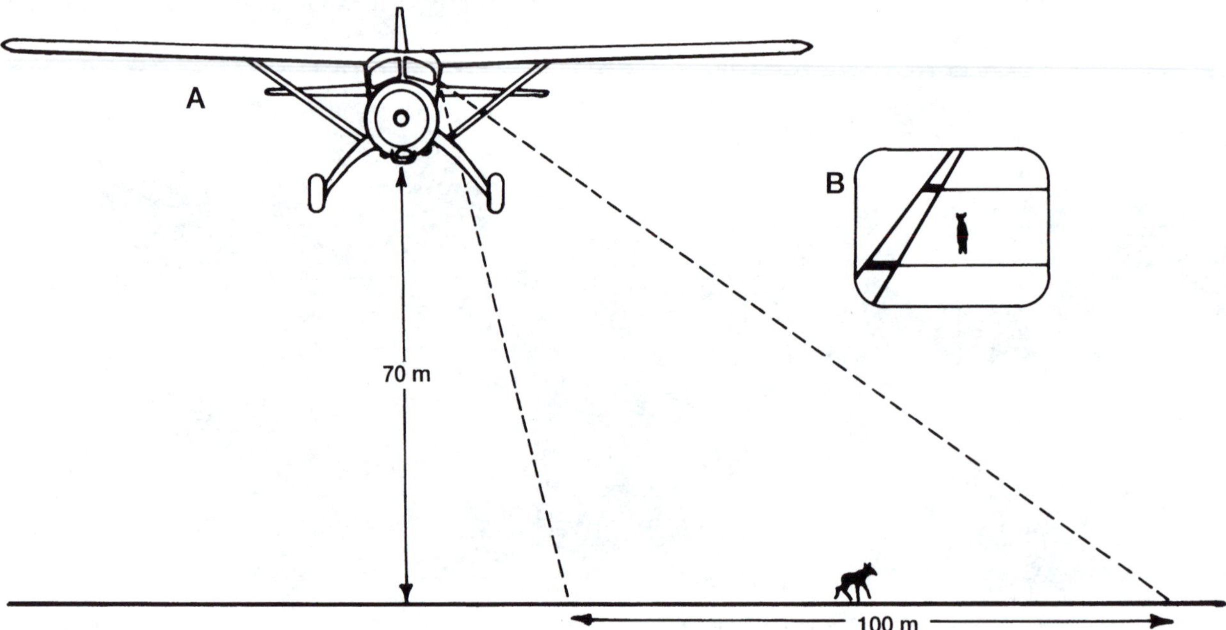

Figure 206. Moose are counted from the air (A) in a transect 0.13- to 0.3-mile (0.2- to 0.48-km) wide on either side of the aircraft. Visual delineation of transect boundaries are aided by strut markers, and transect width is calculated from the tangent of the fixed angle and the fixed altitude (Banfield et al. 1955).

than plot search methods because more information is used, including observed tracks, to find moose. However, line transect census relies on four basic assumptions that must be met to minimize bias and maximize accuracy: (1) moose groups on the center line are not missed; (2) groups are fixed at the initial sighting; (3) distances to observed moose are measured exactly; and (4) observations are independent events (moose are not tallied on more than one transect). The ability of different observers to satisfy these assumptions varies widely. Transect surveys appear best suited to estimate moose populations over large areas with low moose densities.

SELECTED PLOTS OR QUADRATS

Plots are sample units that are uniform squares or rectangles or are irregular in shape and area (Gasaway and Dubois 1987). Aerial surveys of randomly selected plots were first introduced by Ontario in 1958. Population estimates were based on the number of moose actually observed plus those estimated to have been present but not seen on each 25-square-mile (65 km^2) plot (Lumsden 1959). This technique of counting, commonly referred to as the "intensive search" or "orbiting method," was described by Trotter (1958). Pimlott (1961: 248) suggested that, even with intensive search procedures," only about 75 percent of the moose can actually be seen." Such quadrat or block sampling tends to give higher estimates than obtained by flying transects or linear strips. Lynch (1971) reported that, in Alberta, only 67 percent of the moose observed on block surveys were spotted during straight-line counts. Likewise, Evans et al. (1966) in Alaska estimated that only one-fourth as many moose were seen in a given area surveyed using linear strips, as compared with intensively searching plots.

The method consists of dividing the area to be censused into strata on the basis of relative population densities previously determined by a variety of methods, including vegetation, soil type and track counts. Each stratum is divided into sampling units of a predetermined size. A random selection of sample plots in each stratum is made, then each plot is intensively searched. The perimeter of each plot first is flown to identify precise boundaries. Searching involves a circling pattern over fresh tracks or moose sighted, directed along parallel predrawn flight lines (Figures 207 and 208). Recently, both Global Positioning System (GPS) and Geographical Information System (GIS) technologies were used in Alberta to aid in aerial surveys of moose populations (Lynch and Shumaker 1995). GPS receivers were used to: (1) stratify an area by flying lines of latitude; (2) locate and fly survey lines within survey units; (3) record positions of individual moose during stratification; and (4) locate moose groups within survey units. GIS technology was used to plot moose sightings electronically and print physiographic output. Overlap of adjacent flight paths increases with the difficulty of observing the ground.

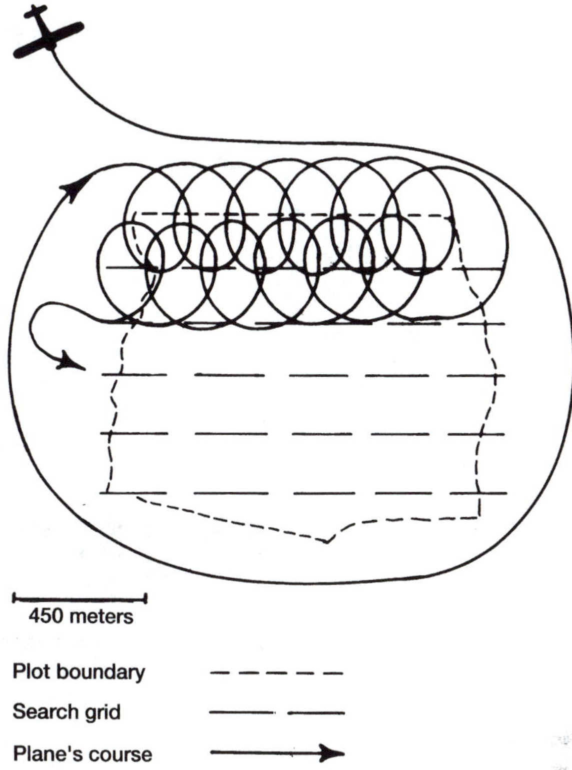

Figure 207. Overhead view of flight pattern for intensive censusing of moose on plots or sample units (from Jordan and Wolfe 1980). Plots are systematically searched by a flight pattern of spirals or circles along parallel courses.

PROBLEMS

The major problem in counting moose in forested habitat is accounting for those not seen. Many factors affect accuracy and precision. The type of aircraft used is a major contributor to precision. The characteristics of aircraft best suited for aerial survey work include maneuverability and slow flight capability. An air speed range of 65 to 90 miles per hour (105–145 km/hr) at an altitude of 200 to 800 feet (61–244 m) is considered ideal. In North America, the Piper Supercub aircraft seems to fit these specifications best (Evans et al. 1966). Larger aircraft including the Cessna 180 and both the piston- and turbine-powered DeHavilland Beaver, although often used, are regarded as too fast and less maneuverable than smaller aircraft for intensive searching (Bergerud and Manuel 1969).

Helicopters have the advantages of low speed, low cruising height, excellent visibility, ease of maneuverability and superior downward viewing. However, they often have a

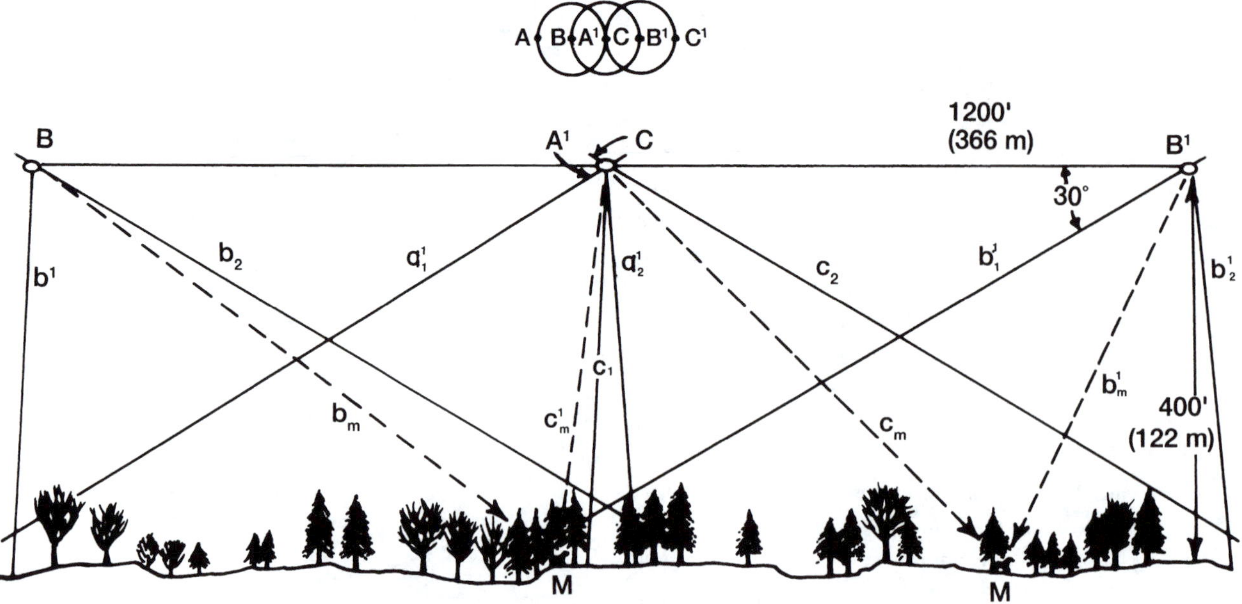

Figure 208. Horizontal profile of aerial search for moose showing the intersection points (capital letters) of a circling aircraft with a vertical plane (from Jordan and Wolfe 1980). The pattern is flown with 50 percent overlap of spirals or circles between rows that are perpendicular to the vertical plane. The diagram boxed at top right, with corresponding symbols (A–C), is a vertical view in smaller scale of circles made on three flight rows where they intersect the vertical plane. The main diagram encompasses the complete circle of row B plus half of rows A and C. Lower case letters with numerical subscripts indicate the limits of viewing imposed outward by the aircraft's wing (shown as a tilted bar), and straight downward by the aircraft's ski. These moose, of course, would be visible from many other angles outside this vertical plane; the observer does not search with a fixed stare, but rather scans back and forth over each section of open ground.

limited range and are much more expensive to operate than fixed-wing aircraft.

Pilot capability is as important as the type of aircraft employed and less suitable aircraft can be effectively used with capable and experienced pilots.

Le Resche and Rausch (1974) quantified the accuracy and precision of aerial censusing using controlled experiments and known moose numbers. Under ideal conditions in Alaska, 49 observers flew 74 replicate counts over four fenced 1-square-mile (2.6 km^2) moose enclosures at the Moose Research Center. Flying 15 minutes over each enclosure, experienced and inexperienced observers saw 68 and 43 percent of the moose present, respectively. Other factors that affected observability were past and recent observer experience, number of observers, weather conditions, habitat and terrain, time of day, and relative moose densities. High moose densities, particularly in mixed or heavy cover, reduced survey accuracy, because tracks were not helpful to locate animals. Le Resche and Rausch (1974) concluded that census conditions must be ideal and as nearly constant as possible from year to year to obtain counts that provide valid estimates of the relative size of the herd and population trends among surveys.

REQUIRED PRECISION

A confidence interval on an estimate of moose counted depends on the variability among plots and the number surveyed (Addison 1970). Plot size can affect total counts. Both Bergerud and Manuel (1969) and Evans et al. (1966) favored smaller plots of 1 square mile (2.6 km^2), because this minimized observer fatigue and reduced precision in boundary determination. Larger plots, however, may lower the error caused by edge effect by reducing the proportional amount of edge sampled. Variance about the mean number of moose per plot tends to decrease with larger plot size. Conversely, smaller plot size tends to increase the variance. Evans et al. (1966) felt that stratification of range into high- to low-density strata would help account for variance found in a heterogeneous distribution pattern. This would reduce the variance and increase survey precision. Bergerud and Manuel (1969) pointed out that plots should not include so many animals that accuracy was lost, or too few to produce many zero plots. Broad confidence intervals obtained from a large number of zero plots usually are due to the fact that moose distribution generally is patchy rather than homogeneous. Addison (1970) submitted that samples seldom are sufficiently large enough to reduce the confidence interval to less than approximately 30 percent. This large spread in population estimate, which may account for only a fraction of the actual number of moose, is caused by the inability to see all the animals present and their patchy distribution (Figure 209).

In Newfoundland, Bergerud and Manuel (1969) developed a method to improve precision. They estimated the number of moose presumed present but unobserved on each survey plot by multiplying the number of track aggregations present on the plot by the mean observed aggregate number of moose. Their goal was to combine track and animal counts into an overall population estimate. Calculations by this method for data from the Shabotik study area in the White River district of Ontario showed a 20 percent increase over the mean calculated without inclusion of tracks (Timmermann 1974).

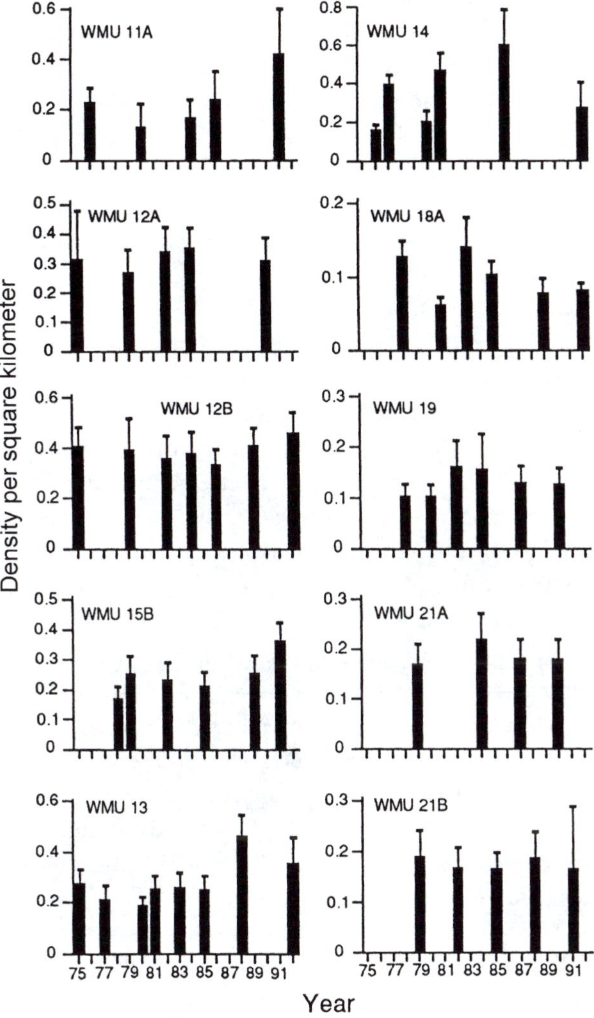

Figure 209. Moose density estimates for 10 wildlife management units in northcentral Ontario, 1975–1992 (from Timmermann and Whitlaw 1992). Capped bars represent 95 percent confidence intervals for the estimated population mean. These surveys collectively represent a cost exceeding $1 million (Canadian) over an 18-year period, and were used to set annual harvest and license quotas in a selective harvest system initiated in 1983.

OPTIMUM CRITERIA

After much trial and error, the following criteria were identified as important to obtaining reasonably accurate population estimate: (1) counts should be made within a short period (2 to 5 days) after a fresh snowfall (Table 52); (2) clear or lightly overcast days preferred; (3) wind speeds less than 10 miles (16 km) per hour; (4) counting restricted to short periods of 2 to 3 hours, ideally to coincide with the period of greatest moose activity, just after sunrise; (5) only experienced observers and pilots used; (6) sufficient time allotted to search each plot fully (a minimum of 4.5 minutes per square mile [1.7 min/km^2]); (7) accuracy increased by using more than one observer and including animals tallied by the pilot; and (8) maximum counts in most forested habitats with adequate snow cover (≥12 inches [30.5 cm]) usually can be obtained in December and January before moose shift into heavier cover (Vozeh and Cumming 1960, DesMeules 1964, Bergerud and Manuel 1969). However, even with attention to these criteria, variability in successive counts of the same area over a single winter period has been documented in several studies (Table 53) (Vozeh and Cumming 1960, Addison 1970, LeResche and Rausch 1974).

Snow depths and condition are important influences on the ability to count moose. Snow measurements should be recorded during each survey so that moose density estimates between surveys can be better evaluated.

Table 52. The influence of habitat selection and search intensity on the sightability of moose during aerial surveys with good, moderate and poor snow conditions (after Gasaway et al. 1986)[a]

	Percentage of radio-collared moose seen during a standard aerial search[b]		
Habitat	Good snow	Moderate snow	Poor snow
Nonspruce[c]	87	83	68
Spruce[d]	60	30	0

[a] Data for early and late winter are combined.

[b] Search intensity of 4 to 6 minutes per square mile (1.5–2.3/km^2).

[c] Includes herbaceous, low shrub, tall shrub, deciduous forest and larch forest.

[d] Includes spruce forest and sparse spruce forest.

Table 53. Seasonal variations in the moose count on the Englehart Management Unit in the Swastika District of Ontario (after Timmermann 1974)

		Number of moose		
Year (winter)	Month	North sector	South sector	Total
1964–1965	December	32	34	66
1964–1965	January	14	14	28
1964–1965	February	14	12	26
1964–1965	March	81	0	81
1967–1968	February	60	50	110
1968–1969	February			59

Weather is a critical factor during aerial moose surveys. Ideal conditions for visibility are fresh snow cover, light overcast and wind speed of less than 10 miles (16 km) per hour. Counts should be made within a short period (2 to 5 days) in December or January before moose shift to heavier cover, and accomplished by experienced observers (more than one whenever possible) and pilots, with adequate time allowed to search each plot. *Photo by Albert W. Franzmann; courtesy of the Alaska Department of Fish and Game, Soldotna.*

SIGHTABILITY CORRECTION FACTOR

The observed number of moose may be highly variable, but is of limited index value unless it can be adjusted for unseen moose in the surveyed area. "Sightability is defined as the percentage of moose seen in the area searched (Caughley 1974), and a sightability correction factor (SCF) must be estimated to produce an estimate of population size with minimum sightability bias. The SCF is the product of an observed SCF . . . and a corrected factor constant . . . for moose missed while estimating SCF" (Gasaway et al. 1986: 31). The variance of total moose observed plus the variance of the SCF are calculated to determine survey precision. "Variance can usually be reduced by (1) stratification of the population by density classes, (2) sampling when the animals are most homogeneously distributed, and (3) increasing the size, number and linearity of the sampling units" (Bergerud 1968: 24). Whenever possible, use of statistical procedures is important to evaluate data precision, given that changes of about 20 percent or more between surveys are required before real population changes can be detected (Gasaway and Dubois 1987).

Three methods commonly used to estimate the SCF are: (1) resurveying a portion of a sample unit either at a higher search intensity (twice the original time), and assuming that this intensity locates all of the moose present (Gasaway et al. 1986), or by incorporating a more efficient aircraft (helicopter rather than fixed-wing); (2) counting the number of fresh tracks where moose are not sighted and multiplying by the mean aggregate number of moose observed (Bergerud and Manuel 1969); or (3) using marked moose in a Peterson Index calculation (Oosenbrug and Ferguson 1992). In the latter method, moose can be radio-collared and/or marked with paint or color-collared. Ratios of radio-collared or marked moose present and observed to total moose observed in the sampling area are calculated to determine the SCF (Table 54). Higher SCFs usually are associated with denser cover and higher moose densities (Peterson and Page 1993).

Areas of future research needed include an analysis of the influence of habitat factors on moose observability, particularly where forest canopy closure is greatest.

CURRENT STANDARD

Much effort has been devoted to determining absolute densities of moose in various parts of their range. The current standard for conducting aerial moose censuses was developed in Alaska in a step-by-step manual (Gasaway et al. 1986). Five essential criteria for estimating population size are considered: (1) eliminate bias and avoid overestimation or underestimation of actual population size; (2) provide an

Table 54. Sightability correction factor (SCF) estimates obtained from North American moose surveys (after Timmermann 1993)

Agency	Mean density per square mile (km^2)	Effort in minutes per square mile (km^2)	SCF estimate	Source
Ontario		5.2–10.1 (2.0–3.9)	1.04–1.06[a]	Novak and Gardner (1975)
	0.47 (0.18)		1.75–2.60[b]	Thompson (1979)
	<1.01 (0.39)	3.1 (1.2)	1.27[a]	Bisset and Rempel (1991)
Alberta			1.14	Mytton and Keith (1981)
Michigan	3.4–6.5 (1.3–2.5)		1.19	Jordan and Wolfe (1980)
	5.2–6.7 (2.0–2.6)	36 (14)	1.35	Peterson and Page (1993)
Maine	1.71 (0.66)	19.7 (7.6)	1.23[a]	Maine Department of Inland Fisheries and Wildlife (1990)
Quebec	0.13–0.47 (0.05–0.18)		1.37	Crête et al. (1986a)
Montana			1.47–2.07[c]	Montana Department of Fish, Wildlife and Parks (1992)
Alaska	7.0–49.5 (2.7–19.1)	15.0 (5.8)	1.48[d]–2.33[e]	LeResche and Rausch (1974)
Newfoundland	1.3–7.3 (0.5–2.8)	6.7 (2.6)	1.59–2.49[c]	Oosenbrug and Ferguson (1992)
	1.6 (0.6)		1.67[f]	Bergerud and Manual (1969)
Colorado			1.74[c]	Kufield (1992)
Yukon	0.21	4.0	1.03–1.11	Yukon Department of Renewable Resources (unpublished files)

[a] Higher intensity search, i.e., ≥2.4 minutes per square mile (1.5/km^2).

[b] Transect survey.

[c] Mark-recapture census.

[d] Experienced observers.

[e] Inexperienced observers.

[f] Mean track aggregate method in intermediate tree cover.

Moose track observed during aerial surveys can be an index to moose density or lead to the location of the animals for actual count. Intensive circling over tracks is required to detect a moose, usually because heavy conifer cover reduces sightability. As a result, observed numbers tend to be highly variable and of limited estimation value unless corrected for moose not seen. A sightability correction factor can be calculated based on an intensive search of a portion of the sample unit or by using marked moose in a Peterson Index calculation. Ratios of marked moose observed on later survey flights to total moose observed can be calculated to determine the sightability correction factor, which usually ranges from 1.5 to 2.0 in heavy cover. *Photo by H.R. Timmermann.*

adequate level of precision (as indicated by the width of confidence intervals) based on realistic measures of sampling errors; (3) suitability for flat, hilly or mountainous terrain; (4) special maps or photos for sampling not required; and (5) affordability.

Methods of estimating population size are based on a stratified random sampling design modified from Siniff and Skoog (1964) and Evans et al.(1966). Bias is minimized by using flight patterns that enable high sightability (percentage of animals seen) and by estimating SCFs for moose not seen. Estimates of precision combine sampling variance of estimated observable moose and sampling variance of the estimated SCF to determine a more accurate overall density estimate. Use of irregularly shaped sample units and varying search patterns adapt the technique to various terrain features. Use of natural terrain features on topographic maps helps to preclude the need for special maps or photos. Rate of population change (increase or decrease) over time is determined from comparison of population estimates separated by one or more years (Figure 210).

Five steps in the census procedure include: (1) define the population of interest and select a survey area; (2) delineate all possible sample units on a topographic map; (3) stratify the area by conducting a preliminary aerial survey to determine relative (high, medium and low) moose density within the area; (4) select a random sample of units and recount some using more-intensive search effort to estimate percentage of moose missed; and (5) calculate an estimate of population size and confidence limits. This method of censusing moose presently is used widely throughout North America (Bailey 1978, Gasaway et al. 1983, Crête 1987, Ballard et al. 1991, Bisset and Rempel 1991, Gasaway et al. 1992, Timmermann and Whitlaw 1992). However, the technique is costly, especially for large areas, because it relies on aircraft. Also, the technique may yield only crude estimates of abundance in areas of dense forest.

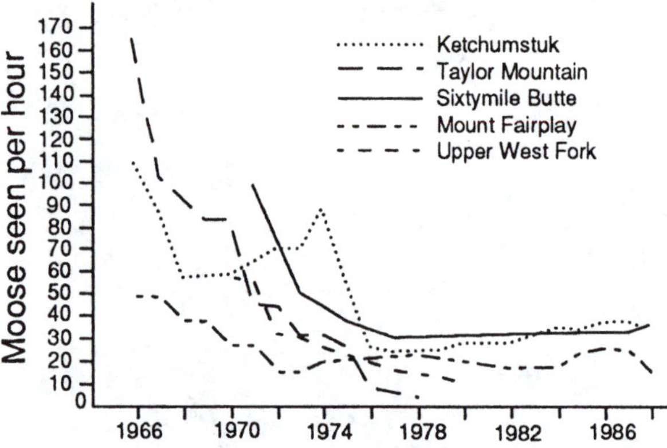

Figure 210. Trends in number of moose seen per hour in five aerial survey areas of eastcentral Alaska, 1966–1988 (from Gasaway et al. 1992). Moose seen per hour is a useful population trend index in some areas if sources for error can be minimized by assuring the same pilot and observer over time, similar seasonal influences, consistent snow conditions, consistent time of day and same areas flown each year.

SURVEY TIMING AND AIRCRAFT TYPE

Effective management of moose is enhanced when populations are assessed on a regular basis. Today, aerial surveys are done annually in 10 state/provincial jurisdictions to monitor population trends and sex and age composition. The balance of moose jurisdictions (12) conduct periodic surveys, e.g., Alberta every 1 to 3 years, Washington every 5 years and Maine as needed. Moose range in many jurisdictions covers large land masses. Surveys in Quebec, for example, are carried out on approximately 231,660 square miles (600,000 km^2) divided into 20 management zones. Fifteen zones are surveyed every 5 years.

Most agencies conduct surveys in December or January. Alaska, Yukon and Northwest Territories prefer November. Utah, North Dakota, Ontario, Quebec, New Brunswick and Nova Scotia surveys may continue into February or March.

Half of the jurisdictions exclusively use helicopters, four employ only fixed-wing aircraft, and six use both. Helicopters are best suited to help determine the age and sex composition of a population by allowing close observation. Helicopters' slower speed, ease of maneuverability, superior downward viewing, and ability to land nearly anywhere to check sign or refuel give them an advantage over most fixed-wing aircraft (Timmermann 1993). Helicopter use yields counts more accurate than those obtained with fixed-wing aircraft, but their greater operating costs often preclude their use (Jordan and Wolfe 1980, Rivest et al. 1990).

SEX/AGE COMPOSITION

The composition of a population or its social structure is important in properly assessing a moose population (Bubenik et al. 1975). Hence, managers should not rely solely on numerical censuses, but consider periodic composition surveys. Composition estimates are influenced by moose behavior, because the animals often segregate by sex and age, making some classes more difficult to observe than others (Peek et al. 1974). The numbers of moose in specific sex/age classes in a population can be estimated from observations made during population surveys. Composition counts generally classify moose into numbers of bulls, cows and calves, so that composition among years and areas can be compared and mortality rate assessed. Ratios are expressed as bulls per 100 cows, calves per 100 cows and calves (single or twin) as a percentage of total classified. Gasaway et al. (1986) cautioned that data from population estimation surveys in Alaska produced higher and more representative calf/cow ratios than did surveys conducted to gather composition data alone.

Survey intervals to detect changes in composition may be as frequent as every 2 years or once in 5 years, depending on the dynamics of the specific population.

Sex and Age Identification Criteria

All moose can be placed into one of the following sex and age classes: antlered bulls; unantlered bulls; adult cows; unknown adults; and calves (either singles or twins). Antlered bulls can be further grouped into several social classes based on antler size and shape (Timmermann and Gollat 1994; Figure 211).

Sex and age characteristics of a moose can be divided into either primary or secondary indicators (Oswald 1982, Crichton 1987b, Timmermann 1993). Observers should strive to combine two or more primary criteria—e.g., antler or pedicel scars, vulva patch size, behavior, bell size and shape—and/or secondary criteria—e.g., aggregation make-up, facial coloration, body conformation, pelage coloration, head position while moving, position of legs when rising from beds. (An "aggregate" is one or more moose that occur within reasonable proximity to each other and among whom there is a behavioral or social bond [Bergerud and Manuel 1969, Peek et al. 1974, Timmermann and Gollat 1994]).

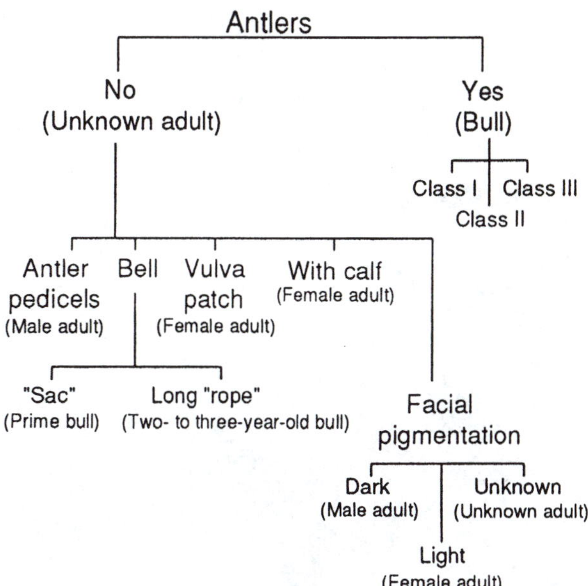

Figure 211. Sex and relative age identification criteria used to classify adult bull and cow moose (after Timmermann 1993).

Primary Criteria

ANTLER AND PEDICLE SCARS

The presence of antlers nearly always indicates a male or bull more than 1 year old (see Chapter 2). Antlered bulls can be placed into one of three classes (Figure 212):

Class 1. Teenager—1.5–2.5 years: spike, fork or multiple fork

Class 2. Subprime—3.5–4.5 years: main palm (not including points) generally does not extend beyond the ear tips; usually single brow tine rather than forked; width of individual palm usually considerably narrower than width of head at widest point; points relatively short and few in number; and antler spread generally less than 40 inches (102 cm)

Class 3. Prime—5.5+ years: main palm (not including points) extends beyond ear tips; usually forked or palmated brow tine; width of individual palm about as wide as width of head at widest point (bulls 11.5+ years may not show this feature);

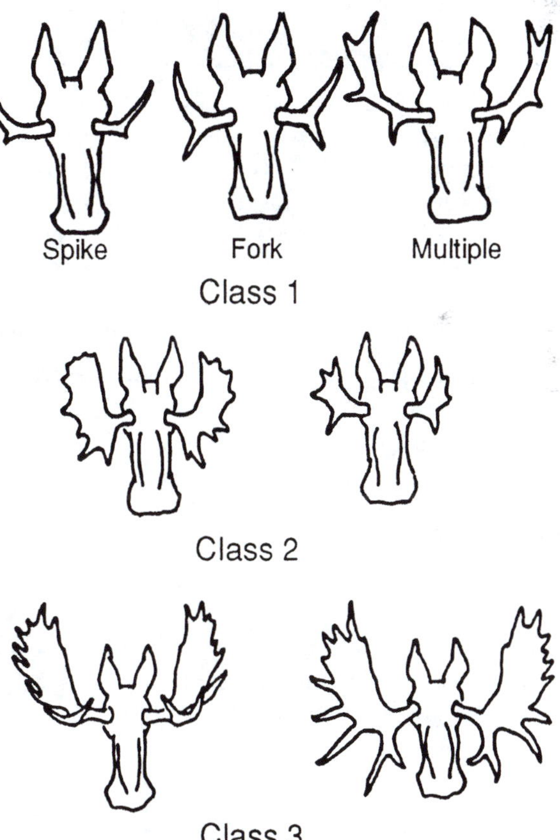

Figure 212. Bull moose antler classes: Class 1 (teenager) 1.5 to 2.5 years old; Class 2 (subprime) 3.5 to 4.5 years old; and Class 3 (prime) 5.5+ years old. *Illustrations by R. Gollat, after Bubenik et al. (1977).*

The correlation of antler size and conformation to moose age is an important visual clue to observers during aerial surveys to assess age structure of the bull segment of the population. Much of the baseline information on these correlations has been from moose immobilized for other purposes or harvested. *Photo by Albert W. Franzmann; courtesy of the Alaska Department of Fish and Game, Soldotna.*

points generally long and numerous, however, on occasion they may appear rounded off at top of palms with few points; and antler spread generally greater than 40 to 45 inches (102–114 cm)

UNANTLERED BULLS

When antlers are dropped (usually beginning late November) an open wound 1.5 to 3 inches (3.8–7.6 cm) in diameter is left at the point of attachment to the skull and flush with the skin (Figure 213). This scar is difficult to see and should not be a criterion unless closely observed (i.e., from a helicopter).

VULVA PATCH

The vulva patch (Mitchell 1970) is an area of light colored hair—approximately 6 by 8 inches (15 X 20 cm) in adults—located around the genital area of nearly all female moose of all ages. This light area is in sharp contrast to the adjacent dark hair and may vary from white to a light tan color

Figure 213. Bull moose generally shed their antlers in late November, leaving open wounds until pedicel scars are formed. Under ideal conditions such wounds or scars may be detected for gender determination. *Illustration by Jim Carson, after Crichton (1987b).*

(Figure 214). Observers should always attempt to determine the presence or absence of a vulva patch on adult antlerless moose. Other than the absence or presence of antler or detachment scars, this patch is the best indicator of sex.

SIZE AND BEHAVIOR

Size variation of individual moose helps identify cow/calf groups or aggregations. Calves also exhibit relatively small ears compared to head size and have a shorter stouter "face" than that of older moose, which display a longer and more bulbous nose (Figure 215). Body size is the most useful criterion for distinguishing calves from adults, although large calves may be mistaken for yearlings, so close attention should be given to head features.

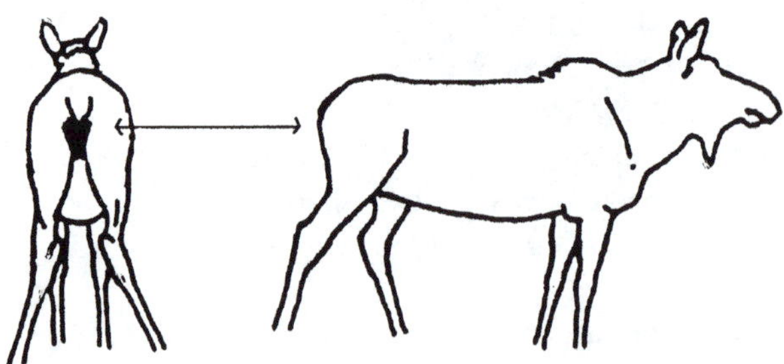

Figure 214. Light-colored hair around the vulva of most female moose can be a useful gender identification for surveys undertaken after bulls have shed their antlers. *Illustration by Jim Carson, after Crichton (1987b).*

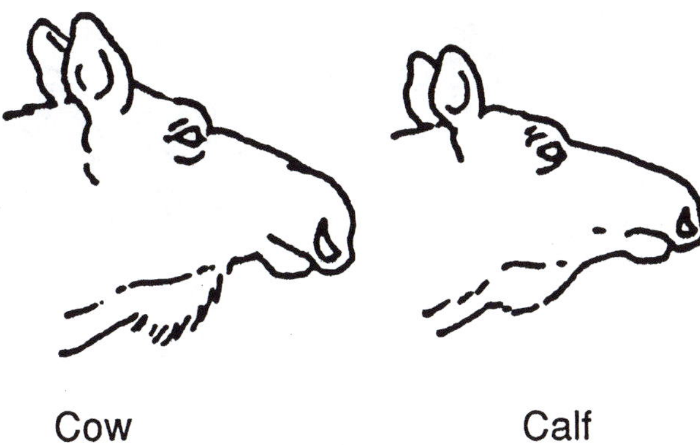

Figure 215. Moose calves may be differentiated from older females by overall size and relatively small ears and short, pointed noses. *Illustration by Jim Carson, after Crichton (1987b).*

Although cow/calf groups almost always are identified by the animals' body sizes, several behavioral traits are helpful. Cows and calves usually are found close together, often bedded side by side and seldom separated by more than 30 to 40 yards (27–37 m). Also, they frequently are nearer to heavier cover than are most other moose aggregates. And a disturbed calf generally will follow directly behind its mother and close at heel.

BELL SIZE AND SHAPE

The shape and size of the bell, which consists of hair-covered skin hanging under the throat, helps to identify the sex of an adult moose (see Chapter 5). Mature bulls tend to display a larger and longer bell than do mature cows. The longest bells, narrow and rope-like in appearance usually are found on juvenile bulls 2.5 to 3.5 years of age (Timmermann et al. 1985). Older prime bulls often display a large, prominent, wide, sack-shaped bell with a narrow "rope" section of varying length (Figure 216).

Secondary Criteria

GROUP COMPOSITION

After the rut, bulls frequently form bachelor groups of 2 to 12 or more animals. Adult cows tend to form groups of two to six, and choose relatively open habitat with good quantities of browse in early winter. Many solitary bulls or cows are prime-age individuals. Cow/yearling groups sometimes reform after the rut, when cows usually have their current calves close by as well.

FACIAL COLORATION

After antler drop, coloration of the nose bridge or face can be a reliable secondary identifier among adults. A light-brown nose bridge, with little or no color contrast between the nose and forehead, is indicative of a cow more than

Long rope
- younger bulls 1.5 to 2.5 years

Narrow rope
- also characteristic of some yearling (to 1.5 years) bulls

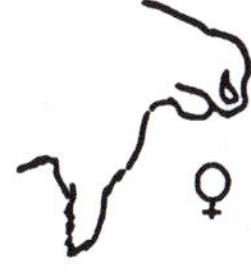

Round with posterior rope
- may be angular also
- also characteristic of medium-aged (2.5 to 3.5 years) bulls and infrequently on older bulls

Round
- generally large but size varies
- found on older (prime) bulls

Angular
- wider than round bell
- less common than round bell
- generally large but size varies
- found on older (prime) bulls

Figure 216. Hair-covered skin under the throat of moose can be used to help identify sex and age of adult moose. Adult bulls have larger and longer bells than do cow moose of equal age. Long, rope-shaped bells generally are found on juvenile bulls (2.5–3.5 years). Prime bulls often have a large, prominent, wide, sack-shaped bell with or without a narrow rope-like section attached. *Illustration by Jim Carson, after Crichton (1987b).*

1 year old. Conversely, the nose bridge of mature bulls generally is dark brown or black, contrasting with the lighter colored forehead area (Figure 217). The overall facial coloration of cows tends to become lighter with age, whereas that of bulls darkens (see Chapter 5). The face of calves show little color difference between sexes.

BODY CONFORMATION AND PELAGE COLORATION

Generally, these indicators are useful only in the case of mature or prime bulls, and should not be used alone. Older animals tend to have a stocky "wedge" outline when observed broadside, exhibit heavy shoulder development, and often are very dark brown or black over the entire torso. Cows, in particular, and younger bulls are slimmer and show more brown body coloration.

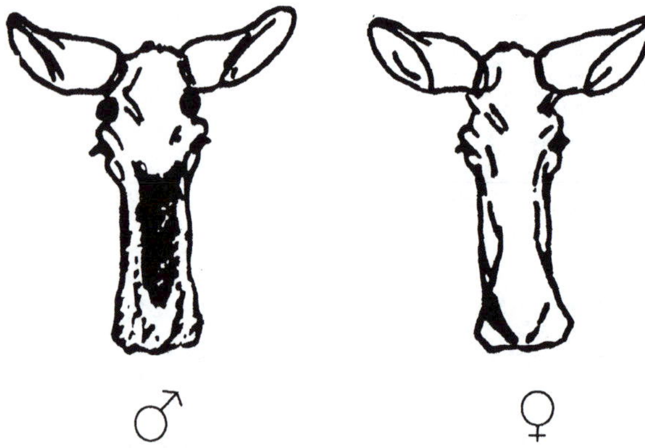

Figure 217. Coloration of the face can be used as a secondary sex identifier of adult moose after antler drop. Bulls tend to have dark brown to black faces, whereas cows have light brown faces. With age, cows tend to lighten and bulls darken in facial coloration. *Illustration by Jim Carson, after Crichton (1987b).*

Figure 218. Position of the head of a moving moose may help detect the animal's gender. Cows tend to move with their head more parallel to the ground and bulls move with their heads more perpendicular. *Illustration by Jim Carson, after Crichton (1987b).*

HEAD POSITION WHILE MOVING

Bulls frequently move with their foreheads somewhat perpendicular to the ground, whereas cows move with their heads more parallel to the ground (Figure 218).

POSITION OF LEGS WHEN RISING

When aroused from beds by searching aircraft, adult bulls frequently move their hind legs forward in a urination-like posture. Only rarely do rising cow moose exhibit this type of posture.

Criteria Comparison

AERIAL SURVEYS VERSUS PELLET GROUP COUNTS

Any survey technique gains credibility when independent indices agree. Pellet group counts, aerial census and hunter observations were compared in central Newfoundland (Bergerud et al. 1968), and all three indices measured a population decline following a purposeful harvest reduction (Table 55).

Pellet group counts were used on a limited basis in the Superior National Forest in northern Minnesota during the mid-1960s on plots flown the previous winter (Timmermann 1974). A deposition rate of 13 per day generally agreed with aerial observations of moose. Jordan et al. (1972) used pellet group counts in conjunction with aerial surveys to study the herbivore/vegetation and predator/prey relationships on Isle Royale. Heavy clumping of pellet groups around blown-down trees that moose stripped of bark resulted in estimates in excess of 10 moose per square mile (3.9/km²), figuring a deposition rate of 20.9 groups per day. Total island densities were estimated as high as 8 moose per square mile (3.09/km²) between 1964 and 1979 by Jordan and Wolfe (1980). They reported that, between 1964 and 1970, aerial and pellet count estimates were not consistently related. Nevertheless, Jordan and Wolfe believed the best aerial count, coupled with pellet count trends calibrated or adjusted by reliable aerial counts, provided a reasonable approach for estimating the Isle Royale moose population and for establishing long-term monitoring procedures. More recently, Jordan et al. (1993) concluded that the pellet method provided reliability similar to that from aerial counts for tracking an increasing trend of Isle Royale moose densities, 1980 to 1993. A need for improvement in sampling design and analytical procedures for both methods was recommended.

Even under ideal conditions, aerial estimates from areas other than Isle Royale tend to be conservative compared with pellet group counts. Using pellet group counts, Simkin (1959) estimated for one study area in northwestern Ontario a density of moose nearly twice that calculated by aerial surveys. He concluded that pellet group counts collected in a consistent manner in different areas provided a useful index to moose numbers, rather than an absolute population estimate. Gibson (1971) compared both pellet

Table 55. Decline of the Noel Paul moose population in Newfoundland after expanded harvests in 1960 and 1961 (after Bergerud et al. 1968)

Year	Winter pellet group per acre (ha)	Aerial census of moose per square mile (km²)	Reported hunter observations of moose per day
1960	190 (470)	11.9 (4.6)	1.6
1961	197 (487)		0.9
1962	123 (304)	6.9 (2.7)	1.1
1963	90 (222)		1.0
1964	13 (32)	4.5 (1.7)	0.9

group counts and aerial counts on Shakespeare Island, Lake Nipigon, Ontario, before and 5 years after hunting had been permitted. Pellet group counts indicated a 96.4 percent decline in moose, whereas aerial counts indicated a 95.8 percent decline.

A comparison between population estimates derived from moose pellet group surveys and aerial surveys in the Kenora Forest District, Ontario, in 1970 suggested that 50 to 59 percent more moose were estimated by the pellet group survey (Gustin 1973). In a more densely forested area of the district, 300 to 400 percent more moose were estimated by pellet group counts than by aerial census surveys.

Vozeh and Cumming (1960) recommended that, if a true comparison to the pellet group count was desired, a series of aerial counts be taken and averaged. They suggested that large differences result when a single aerial survey is compared with the calculated winter moose population estimated from pellet group counts. Vozeh and Cumming illustrated their point by showing that a population estimate from 12 winter aerial surveys was well within 95 percent confidence limits obtained from pellet counts on the same area.

Gustin (1973) used the pellet group technique to estimate population densities of moose on 6,525 square miles (16,900 km^2) of suitable moose habitat in the Kenora Forest District of northwestern Ontario. He estimated that the total moose population was 11,440 to 13,520, or 1.75 to 2.07 per square mile (0.68–0.80/km^2). Aerial survey estimates for the entire district of 12,173 square miles (31,528 km^2) indicated a range of 0 to 25,685 moose with a mean of 7,182 (0–21 per square mile [0–8.1/km^2]). Table 56 lists the variability in estimates that can occur.

Table 56. Estimated Isle Royale, Michigan, winter moose population and density estimates based on pellet group counts and aerial surveys, 1963–1970 (after Jordan and Wolfe 1980)

	Pellet group[a]		Aerial census	
Winter	Island total	Moose per square mile (km^2)	Island total (± 95 percent)	Moose per square mile (km^2)
1963–1964	722	3.39 (1.31)	704±	3.34 (1.29)
1964–1965	814	3.83 (1.48)	848 ± 35.5	4.01 (1.55)
1965–1966	1,048	5.10 (1.97)	712 ± 31.5	3.37 (1.30)
1966–1967	1,409	6.63 (2.56)	530 ± 34.7	2.51 (0.97)
1967–1968	1,062	5.00 (1.93)	1,015 ± 22.7	4.82 (1.86)
1968–1969	1,368	6.45 (2.49)	1,150 ± 23.1	5.46 (2.11)
1969–1970	1,208	5.70 (2.20)	945 ± 25.6	4.48 (1.73)

[a] Calculated using a deposition rate of 20.9 pellet groups per day.

Indirect Estimates

TRACK COUNTS

Counting moose tracks over snow-covered transects was used in the Soviet Union to arrive at an index of moose abundance and distribution (Semyonov 1956, Priklonskiy 1965). Moose tracks can be readily seen from low-flying aircraft. Fresh tracks usually are clearcut with sharply defined shadows, whereas old tracks have indistinct outlines and soft shadows (Oswald 1982).

Track counts also have been used to estimate the number of moose missed by direct aerial observation. For example, in Alaska, Bentley (1961) estimated from fresh tracks that, for every moose seen, at least three were missed. Areas of daily activity of groups or single animals usually are distinct, and only occasionally do groups of tracks overlap. How apparent such overlap is depends largely on how soon after a major snowfall the survey is conducted. In Ontario, Gawley and Dawson (1965) found little correlation between track counts and actual numbers of moose observed. They concluded that, although correlation might by valid under optimal weather conditions, with 4 to 6 inches (10–15 cm) of new snow and absolutely clear weather, these conditions are encountered infrequently and the continued use of the track index technique to determine relative moose densities was not recommended. A similar survey in Alaska in 1962 was abandoned because of difficulty in estimating the area of daily activity and interpreting the data (Evans et al. 1966). Track counts are best used to help determine winter distribution and relative density levels for stratifying range in preparation for aerial surveys (Timmermann 1974, 1993). "Stratification" is the process of delineating the survey area into approximate or anticipated densities of moose based on a presurvey observation of track distribution. There are generally three strata delineated—high, medium and low. This stratification becomes the basis of survey sampling intensity, with the greatest number of plots or survey units selected in the high strata area.

AERIAL PHOTOGRAPHY

In Ontario, Passmore (1963) suggested that winter aerial photography held promise as a moose census tool. He felt that photographs should be of a scale of 400 feet to the inch (48 m/cm) and taken 36 hours after snowfall on bright days in March between 10:00 A.M. and 2:00 P.M.. These conditions often are difficult to meet. Moose become more difficult to observe as they frequent coniferous cover in late winter (Banfield et al. 1955, Vozeh and Cumming 1960, DesMeules 1964, Bergerud and Manuel 1969). Passmore's (1963) ideas generally are not accepted as a census technique because of constraints such as difficult determination of scale and strip width, unfavorable light and atmospheric conditions, lack of knowledge on daily activity patterns of moose in various habitats, complex photo interpretation, and difficulty of replicating survey flight paths. The advantages of aerial photography include a permanent record of animals censused, moderate cost of film for modern cameras, and a large area of range can be covered in a relatively short time period.

THERMAL INFRARED IMAGERY

Infrared-scanning equipment that detects and records heat rather than visible light holds great promise as a big game censusing technique. Croon et al. (1968) first demonstrated actual success in censusing white-tailed deer in Michigan on a snow-covered 2-square-mile (5.2-km^2) area of open deciduous habitat. In Pennsylvania, Graves et al. (1972) found that for censusing deer, limitations of the equipment made large-scale aerial surveys using infrared detection equipment impractical unless the census area was flat and relatively free of obstructing vegetation.

McCullough et al. (1969) summarized the progress made in large animal census by thermal mapping, and discussed problems and limitations of the technique. The salient points included: foliage presents a barrier to infrared; infrared presently is more applicable to open range lands, tundra, low brushy areas or defoliated deciduous forests; variations in weather conditions, apparent temperature of animals and the variability in temperature of inanimate objects in the background affect interpretation; required are a pilot, an aircraft the size of a Cessna 180 or larger, an equipment engineer and a skilled interpreter; and equipment is expensive. The basic problem involves detection of "hot spots," followed by recognition of these hot spots as the desired target. To be detected, an animal must emit enough energy in the appropriate wavelengths and be sufficiently exposed to be distinguished from the background.

In Ontario, infrared thermal imagery was tested as a fire detection device during the late 1960s. It was hoped that high equipment costs could be partially offset by use of the scanner as a big game census tool during winter months. Addison (1970) used an infrared line scanner in an attempt to census moose in northwestern Ontario. Cold temperature in the cabin of the aircraft caused malfunctions in the electrical equipment. Moose were included in some of the imagery, but at a point just before scanner failure as the critical temperature of 14°F (–10°C) was reached. LeResche tried an infrared scanning device and found it entirely unsuccessful for counting moose in Alaska (Timmermann 1974).

Many problems are inherent in the use of infrared thermal imagery for moose censusing. A particular problem is the fact that infrared radiation does not readily penetrate green foliage, therefore evergreens present an effective barrier (Croon et al. 1968, McCullough et al. 1969, Addison 1972, Graves et al. 1972). This can be partially overcome by censusing in early winter, when deciduous foliage is absent and moose are occupying relatively open coniferous and deciduous forest habitats.

Addison (1972) maintained that increased knowledge of moose behavior and seasonal habitat preferences are necessary prerequisites to effective use of infrared imagery in interpreting photography. Surveys should be conducted during periods when animal activity is high, by selecting the optimal times of year and day. Weather conditions, including wind velocity, humidity, solar radiation and ambient temperature, influence the differential amounts of heat emitted by the animals and their background. Addison (1972) suggested that a manual gain control system, which would develop the target's thermal signature, would allow enhancement of animal images in open, treeless habitat. When objects of similar temperatures to moose are indistinguishable from the animals, interpretations of imagery can be difficult (Addison 1971, Gustafson 1993). However, remote sensing has the advantage of consistency, so can be efficiently used in strip censuses. Also, it permits coverage of larger areas than is possible with observational methods during the same time period.

Digital processing and enhancement of the infrared images has provided better sensitivity and resolution (Wyatt et al. 1980, 1984). The use of unmanned aerial vehicles carrying infrared equipment may help reduce problems with detection and recognition by allowing night flights, when solar-loading effects on targets and background are minimized (Gustafson 1993).

Continued improvement in infrared technology is certain, and the results to date offer cautious optimism about airborne thermal infrared systems as a censusing technique for moose and other big game.

Population Estimation Summary

The goals of population estimation should be to derive the best estimate possible, commensurate with the study objectives, and to examine the precision of the estimate—how well assumptions are met and the effect of sampling error (Overton and Davis 1969). The diversity of moose habitats across North America suggests that methods of census and harvest evaluation should be designed to fit the environmental situation. Crichton (1992a) provided 12 parameters identified as useful in assessing moose populations and measuring the impact of various management programs: (1) mean age of harvested adults; (2) percentage of yearlings in the harvest; (3) harvest magnitude by area; (4) moose observed per hour of survey time; (5) percentage of adult males and females in the population; (6) ratio of calves per 100 cows; (7) ratio of bulls per 100 cows; (8) estimates of mortality from poaching, predation and subsistence users; (9) sex of calves in the population; (10) productivity of cows 2.5 years of age and older; (11) percentage of yearling cows breeding; and (12) numbers of moose seen and harvested per day by hunters.

No single source of information should be used to assess population status. All sources available should be carefully weighed to obtain a comprehensive picture of population levels. Some areas will show decreasing population trends and others will show static or increasing trends, and management policies and programs then can be tailored to meet various objectives for each situation. Managers need to select a census method that provides a level of accuracy consistent with their needs and abilities to meet management objectives for the moose population in question. The challenge lies in properly applying the methods of population inventory, recognizing their limitations and using such information as basis for management programs to ensure a healthy and optimally used moose population.

Population Modeling

Population modeling can be a useful tool to predict moose population changes under different management strategies, to evaluate data quality and to identify data gaps. Models also are used to evaluate management systems. They are most instructive when objectives are stated clearly, input data are minimized, assumptions listed and limitations identified. In the early 1990s, most moose range wildlife agencies (15 of 22) used population models to help set harvest quotas, develop indices to population condition or assess population trends. Descriptive models commonly are built by the reconstruction and testing of a variety of parameters that affect populations. Ratio estimates, life tables, cohort analysis, and harvest and population data analysis provide inputs to complex models (Fraser 1976, Boer 1988, Fryxell et al. 1988, Timmermann and Whitlaw 1992, Courtois and Jolicoeur 1993, Ferguson 1993). Estimates of population size, hunting mortality and herd productivity are used to model a population until it approximates known values. Commonly, some model input values are known, whereas others are estimated or default values inferred from published findings outside the immediate study area.

Age at first breeding and the sex ratio of the breeding population are critical parameters for modeling populations of long-lived species such as moose (Moen and Ausenda 1987). In addition, sex ratio at birth, sex-related mortality of neonates, and natural and hunting mortalities by sex and age groups are important data for setting up simulation models.

The model ONEPOP and its many variants (POP 50 and POP II) are widely used to model ungulate populations (Gross et al. 1973, Bartholow 1985). These relatively complex models perform many and varied calculations and are useful for comparing the effects of age and sex selective harvesting within large mammal populations. This type of model was used by Ontario to compare and communicate the benefits of several different hunting options: calf-only hunts, bull/calf hunts, any-moose hunts and selective harvest hunts (Figure 219; Euler 1983b). Model results suggested that a selective harvest strategy based on harvesting a limited number of bulls and cows, while allowing unregulated hunting of calves, provided maximum benefit to hunters and ensured long-term moose population growth. Moose population response to the strategy was evaluated 9 years after initiation (Heydon et al. 1992). Results of modeling harvest and population data showed significant differ-

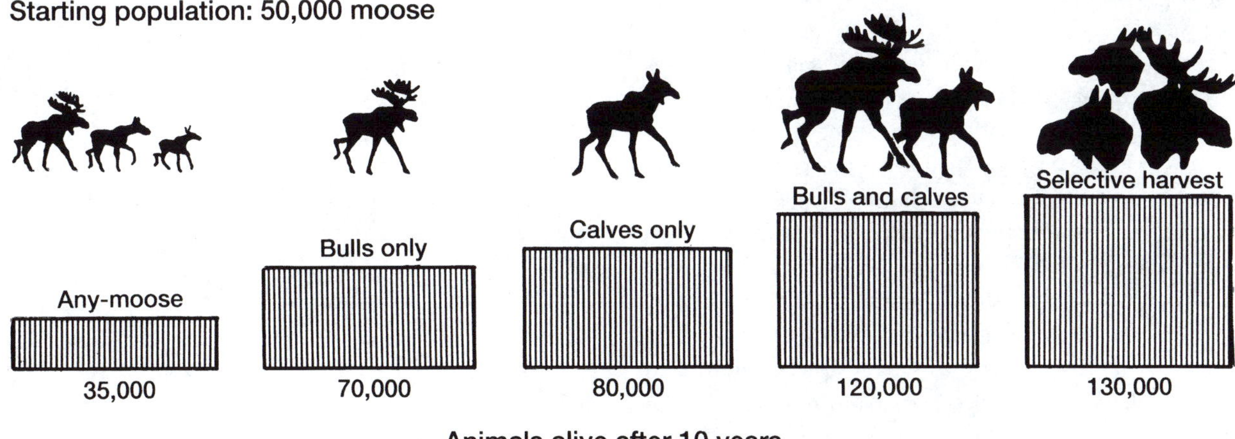

Figure 219. Computer modeling often is used by managers to examine the potential results of various moose harvest options. Illustrated above are computer-simulated projections of moose harvests probable with five different harvest systems (from Euler 1983b).

ences between modeled and surveyed population trends, suggesting assumptions based on provincially averaged population parameters were inappropriate, and that locally generated inputs should be used to reflect actual population trends more accurately.

Several agencies, including Alberta (Lynch 1986), Alaska (Reed 1989), Manitoba (Johnson 1990) and Minnesota (M.S. Lenarz personal communication: 1993), developed "in house" models designed specifically for their respective circumstances. The Alberta Big Game Model, for example, is used as a tool to help predict population trends 1 or more years into the future by applying sex- and age-related mortality and natality rates against a starting population. A spread-sheet version of the Alberta model provides more utility, according to G.M. Lynch (personal communication: 1993). Known population parameters are entered and tested against a variety of harvest strategies to determine population response over time (1 to 20 years). After completing several simulation runs, a harvest rate that produces the desired outcome is selected. Then, with the expected harvest success rate factored in, the expected harvest is converted into a number of permits or hunting opportunities.

The program NEMOOSE is a simulation model designed to evaluate the moose population in northeastern Minnesota (M.S. Lenarz personal communication: 1993). Simulation begins with population numbers estimated in the 1984–1985 aerial survey, mimics changes identified in subsequent aerial surveys and provides opportunity to project future population numbers. In addition, the model includes a subroutine for projecting the harvest based on the number of permits or licenses issued. This combination allows managers to assess potential impacts of proposed harvests against prevailing population objectives.

Crête et al. (1981a) compared hunter harvest rates, predation rates and differences in moose population density between areas in Quebec, to calculate optimum harvest strategies. Moose densities in freely accessed hunting areas were depressed by moderate harvest levels. The Crête et al. model suggested that harvest to the point of maximum sustained yield was the most appropriate strategy. Maintaining a stable sex ratio of about 67 adult bulls to 100 adult cows was believed to be important. Courtois and Jolicoeur (1993) used Schaefer's and Fox's surplus-yield models to estimate optimal moose harvest and hunting effort in Quebec. Analysis suggested maintaining harvest around 1.17 per 10 square miles (0.45 / 10 km^2) in central Quebec and 1.3 to 2.33 per 10 square miles (0.5–0.9 / 10 km^2) in eastern Quebec where predation rates were lower. In northern parts of the province where predation was a factor, Courtois and Jolicoeur recommended that the harvest be less than 0.78 moose per 10 square miles (0.3 / 10 km^2). "Models suggest maintaining effort between 3 and 19 hunting-days / 10 km^2" (Courtois and Jolicoeur 1993: 149).

Moose populations were modeled by Walters et al. (1981), who using data from Alaska's Denali National Park, where mountain sheep and caribou served as alternate prey for wolves. Findings suggested that fluctuations in moose and caribou numbers were linked to occasional severe winters. Also, in some situations, wolves shifted prey selection from sheep to moose, severely reducing moose populations. When hunting mortality was added to the model, an irreversible decline followed, because of increased predation on the remaining moose, forcing them into a classic predator pit, or low stable equilibrium. Stocker (1981) modeled the wolf / ungulate prey relationship and suggested that a reduction of wolves was necessary if sustainable moose harvests were desired (see Chapter 7).

Ballard et al. (1986) developed a moose population model that used parameters routinely measured by Alaskan managers. These parameters included moose distribution, abundance, sex / age composition and impact of brown bear, black bear and wolf predation. Components of this model were divided into a series of events that occur in the annual cycle of a moose population (Figure 220). Natality and mortality components can be altered to meet various assumptions. The model was used to help quantify potential impacts of habitat alteration and better understand moose herd dynamics. Although several limitations affecting its

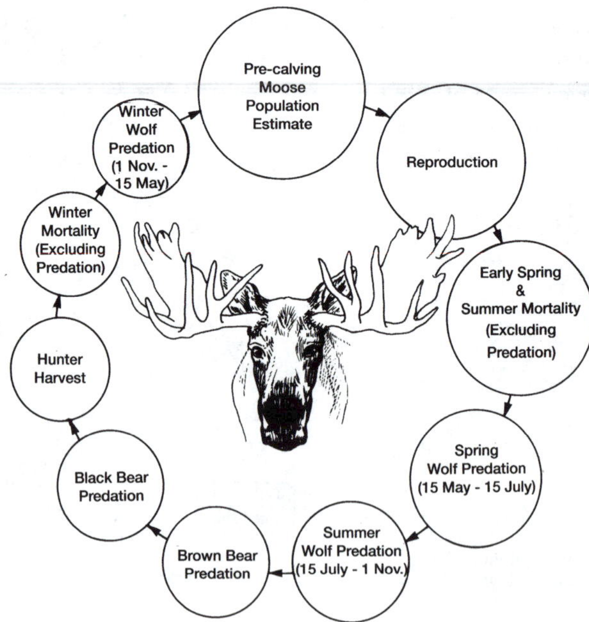

Figure 220. Natality and mortality timing and sequence used in moose population models for Game Management Unit 13 and the Susitna River study areas in southcentral Alaska, 1975–1982 (from Ballard et al. 1986).

usefulness as a predictive tool were identified, the model was instructive for understanding historical population fluctuations and for generation of additional hypotheses.

Boer and Keppie (1988) used a deterministic simulation model to mimic a real moose population and explore the role of hunting in a 925-square-mile (2,396-km^2) area of southeastern New Brunswick. Modeling results suggested that the population would begin to decline with harvest rates above 6 percent if no antlered males were shot, and above 9.5 percent if only antlered males were taken. Maximum sustained yields were estimated in the model to occur at 1.0 moose per square mile (0.4/km^2); below that density, hunting was considered an additive mortality factor. Hunting after the rutting period allowed a harvest rate 10 percent higher than before or during the rut, and there was little benefit in distorting the harvest sex ratio above 60 percent antlered males. This model suggested that, although hunting mortality was an important factor, moose populations in southeastern New Brunswick were chiefly driven by natural mortality factors and poaching of the adult cohort.

Principles of sociobiological control of moose population well-being are described by Bubenik and Pond (1992). Modeling is based on knowledge that living populations are composed of overlapping generations with various fitness levels and a recognition of the inadequacy of using average mortality rates derived from life tables (Albon et al. 1992). Three fitness classes (poor, average and optimal), each with a different set of mortality and recruitment parameters, were modeled. Results showed that some social classes are surplus, because of low natural mortality or overrepresentation. Albon et al. (1992) suggested that sociobiologically only moose that are surplus to maintaining a balanced population and ecosystem should be removed by hunting. Recognizing this, managers can model populations, determine surplus social classes and implement effective harvest strategies. For example, if young moose (1.5–4.5 years) are overabundant, then setting seasons after the rut could be effective in reducing their numbers, as well as reducing the take of prime breeders (Child and Aitken 1989).

Ferguson (1992) modeled the Newfoundland moose population to help forecast future density and predict harvest strategy impact (Figure 221). Because wolves were absent, human hunting rather than natural predation largely determined numbers of moose. Accordingly, hunters were used to control moose population dynamics. Mortality and productivity estimates were pooled for each of 47 moose management units. Modeling suggested that the island-wide winter moose population peaked at 177,000 in 1989 and declined to about 150,000 by 1993. Optimal winter density at two-thirds of vegetative carrying capacity was estimated at 75,000, or 5 moose per square mile (2/km^2), with an annual potential legal harvest of 20,000. Ferguson (1992) concluded that the relationships identified through modeling needed to be confirmed with experimental manipulations of real populations (adaptive management), and better estimates of productivity and recruitment were needed. The model was helpful in identifying inadequacies of Newfoundland's moose demographic data. Subsequently, Ferguson (1993: 99) used a fisheries computer model (CAGEAN) to perform cohort analysis of Newfoundland's age-specific moose kill data, and found that "Cohort abundance estimates generally compared favourably with aerial survey results and indirect indices determined from hunter questionnaires . . . but important limitations exist." Future model use will explore selective harvest strategies that can maintain a healthy, productive and socially balanced population.

The software package Lotus 1-2-3™ was used in Alaska to help predict anticipated declines in the moose harvest and changes in the bull/cow ratio after implementation of a selective harvest system on the Kenai Peninsula (Schwartz et al. 1992, Schwartz 1993). Model inputs included survival coefficients (birth to autumn, hunting season and overwinter) by sex and age class, sex ratio of calves at birth and twinning rate. Initial inputs simulated existing populations before the introduction of selective harvest. Anticipated selective harvest changes were simulated by altering survival coefficients for males, reducing hunting mortality for age classes 2 to 5 and maintaining reproductive output. In addition, the effects of a severe winter on harvest and herd composition were simulated by reducing winter survival rates for calves from 0.36 to 0.10, and reducing adult survival

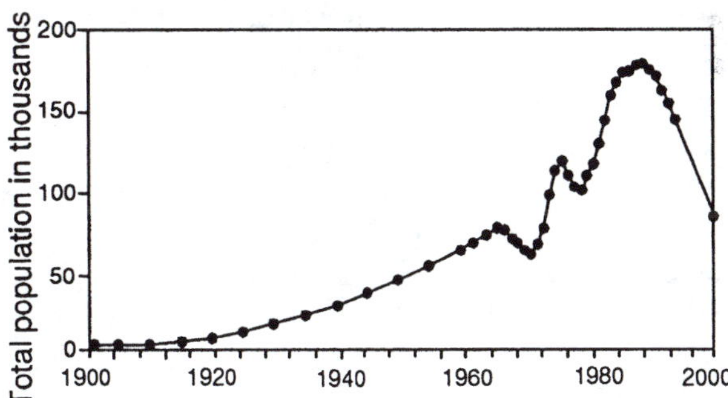

Figure 221. Estimated and predicted winter moose population in Newfoundland, 1905–2000 (from Ferguson 1992): 1900–1991 based on logarithmic regression of 10 moose in 1910 to 72,000 in 1966; 1966–1990 based on extrapolation of a 3-year average of moose seen per day, based on the 1991 estimate of 169,000 moose; projected 1992–1994 population estimates calculated from model; and 1995–2000 based on a management objective of maintaining moose populations at optimum density—5.18 moose per square mile (2 moose/km^2) of forested habitat = 75,000 moose.

during winter by 5 percent for each sex/age class. A severe winter was simulated during year 3 of selective harvest to test changes that occurred during the severe winter of 1989–1990. Modeling results were useful in predicting bull/cow ratios and other population parameters as a result of the change to selective harvest. For example, in Game Management Unit 15A, the model predicted a 43 percent bull harvest decline the first year of selective harvest; the actual decline was 48 percent. The model predicted that effects of the 1989–1990 severe winter would reduce harvest 63 percent from the preselective harvest kill; the actual recorded harvest declined 66 percent. The information generated also was useful in alerting the hunting public to expect changes in hunting success due to changing bull/cow ratios and the impact of a severe winter on future populations.

Page (1987) believed models had little impact on moose management decisions before the mid-1980s. He suggested robustness and completeness are two important criteria in assessing model quality, and simulation models are most valuable if kept simple. An accompanying word manual can assist in model understanding and allow managers to alter harvest strategies and play "what if" games to determine potential results. Page concluded that future modeling should stress simplicity, generality, reality and relevance, and that models should be accessible to managers and provide learning through probing of "what if" scenarios.

Harvest Management

Historic Perspective

Moose were numerous in northern New England and in the western Great Lakes during colonial times and furnished an important source of food and hides for trade (Crête 1987, Bontaites and Gustafson 1993, Vecellio et al. 1993). Moose were plentiful throughout Vermont, for example, and "in many places the early settlers depended upon their flesh for no inconsiderable part of the subsistence of their families" (Thompson 1853: 50). Because they were so conspicuous and easy to kill, they "did not survive" according to Cahalane (1942: 102). With introduction of more effective firearms, moose numbers decreased progressively between the seventeenth and eighteenth centuries as new settlements were established. Unregulated shooting and commercial exploitation before the twentieth century were major factors in local extirpations of moose, especially in the Maritime Provinces and New England (Wolfe 1987). Decimation attributable to gunning was delayed in western North America and is believed to have been less severe, "because this part of the continent was colonized later" (Crête 1987: 554).

By 1850, many jurisdictions began to regulate the taking of moose by generally prohibiting winter hunting. This regulation proved ineffective and eventually, hunting during any season was closed in many provinces and states. In Vermont, moose were protected from harvest from 1876 until a hunting season was authorized in 1993 (Alexander 1993). Seasons in New Brunswick were closed from 1888 through 1890 (Peterson 1955). Bull-only and short hunting seasons were common strategies in the early to mid-1900s. Populations were believed to have fluctuated in some areas and declined markedly in others. In 1914, for example, 2,447 moose reportedly were legally killed in Manitoba; by 1924, the take dropped to 257 (Jackson 1926). From 1932 to 1941, an average of 167 moose were legally shot and reported, according to Manitoba provincial records. Similarly, Saskatchewan reported 1,220 moose legally killed in 1920. Harvests averaged 546 annually from 1921 to 1928, increased to 1,163 in 1930 and steadily decreased thereafter (Peterson 1955). The decline in moose populations occurred throughout much of continental moose range, and many jurisdictions responded by closing seasons—Minnesota in 1922, Alberta in 1932, Maine in 1935, New Brunswick in 1937, Nova Scotia in 1938, Manitoba in 1945, Saskatchewan in 1946 and Ontario in 1949 (Pimlott 1961).

Moose populations consequently recovered, so that more liberal regulations—including short either-sex seasons—were reintroduced in the 1950s and 1960s. For example, Pimlott (1961) reported the first open season for moose of either sex or age was held in 1953 in Newfoundland, Ontario, Manitoba and Saskatchewan, and in 1957 in Alberta. Pimlott cautioned that information on continental moose populations before about 1960 was spotty, and concluded that any clear impressions of past trends from available data were impossible to obtain. Estimates were based on subjective information, such as trappers' reports or roughly calculated annual kills based on game officer impressions. Pimlott (1961: 262) added that "fears that moose would be extirpated have been replaced with the knowledge that, when in good habitat they can withstand heavy hunting pressure."

In summarizing moose management to 1960, Pimlott (1961) believed that inadequate harvest was the most pressing issue. And associated with inadequate harvest, several factors were identified: lack of information on the status and ecology of moose, resulting in fears that populations would be decimated by more intensive harvests; inaccessibility of large areas because of lack of roads or the refusal of logging companies to permit hunter access to their licensed lands; public pressure against liberalizing hunting seasons and against the killing of females; lack of inclination and ability of most hunters to penetrate into a hunting area more than 1 mile (1.6 km) from an established road.

During the 1970s, many moose populations declined from the cumulative and in most cases, coincidental effects of intense hunting pressure, severe winters, habitat maturation and increased protection of wolves in some areas. Managers reacted by closing or shortening seasons, implementing limited sex/age-specific seasons, and regulating access to moderate harvests and enable moose populations to recover and stabilize. *Photo by H.R. Timmermann.*

Modern Era

Cumming (1974a: 675) outlined the evolution of moose management in Ontario as: "(1) protection from suspected overexploitation; (2) development of some use; (3) increased use of an undeveloped resource; (4) sustained yield and optimum use of a resource product; and (5) recreational and economic benefits to the people of Ontario." The evolution is mirrored by similar events in most regions of North America and reflects the demands placed on the moose resource into the 1970s.

Moose populations across broad areas of North America increased into the mid-1960s, (Timmermann 1987). As a result, many agencies liberalized their recreational hunting seasons, and harvests increased when longer either-sex seasons replaced shorter bull-only seasons (Ritcey 1974). During this period of expanded exploitation, recreational harvests peaked in Ontario (14,500) in 1968, Alaska (8,883) in 1971, Saskatchewan (9,000) in 1971, Manitoba (3,926) in 1971 and Newfoundland (11,837) in 1974. (Timmermann 1987). Until the early 1970s, the main focus of most agencies was to stabilize population densities and hunting opportunities.

Increased hunting pressure in newly opened logging areas often results in serious depletions of moose. Signs of overharvest include decreasing annual yields, a younger mean age of harvested animals and consecutive low proportions of bulls in the harvest (Cumming 1974b). Hunting, coupled with severe winters and predation, apparently had a major impact on moose across wide geographic areas during the early 1970s. Significant population reductions were recorded in Alaska, and across much of Canada (MacLennan 1974, Mercer and Manuel 1974, Grenier 1979, Timmermann and Gollat 1982, Gasaway et al. 1983, Timmermann 1987).

In reaction to declining populations, many agencies applied restrictive harvest measures to reduce and control the impact of hunting. Strategies focused on limiting the harvest in specific hunting areas and controlling the sex and age composition of harvest to maintain or increase populations without unduly restricting hunter participation.

Population Density

Moose densities in North America generally range from 0.3 to 1.3 per square mile (0.1–0.5/km^2) in areas where wolves and or bears are unexploited (Bergerud 1992). These densities are considered low, when compared with those of Newfoundland or Scandinavia where wolves are absent and moose densities are on the order of four to five times higher. Where bear and wolves were lightly exploited in Alaska and the Yukon, mean densities of 19 moose populations were 0.4 per square mile (0.15/km^2) versus 1.6 per square mile (0.62/km^2) from 15 populations where bears and wolves were heavily harvested (Gasaway et al. 1992).

Bergerud and Snider (1988: 561) reported that where "moose were the primary prey" of wolves, moose densities averaged 0.6 per square mile (0.25/km^2) for 11 hunted populations in Alaska, northeastern Alberta, British Columbia, northern Ontario and southwestern Quebec. Crête (1987) reported nearly one moose per square mile (0.4/km^2) for four unhunted populations in Quebec, northern Ontario and Alaska. In multiungulate systems where deer and elk are more susceptible to natural predation, moose densities commonly reach about 1.6 to 3.3 per square mile (0.62–1.3/km^2) (Carbyn 1983, Crête 1987, Bergerud and Snider 1988). In contrast, Cederlund and Sand (1991) reported densities of hunted moose in southcentral Sweden, with no other predators, at approximately 4 per square mile (1.5/km^2). This is close to the 5.2 per square mile (2.0/km^2) density considered by Crête (1987) to be at North American food carrying capacity (K) in a predator-free environment (see Chapter 6).

Recently disturbed, high-quality habitats have the potential to produce higher moose densities when predation pressure (among other mortality factors) is low. In most jurisdictions, substantial numbers of moose are removed annually in such areas by hunting, thereby depressing densities (Eason 1985). According to Gasaway et al. (1980), repopulation may be a slow process and is initiated primarily by offspring of an area's survivors. No dispersal was reported by Gasaway et al. at densities of 0.5 moose per square mile (0.2/km^2), whereas Ballard et al. (1991) documented a dispersal rate of 60 percent (9 of 15) in an adjacent study area with a density of 1.55 moose per square mile (0.6/km^2). Management strategies in these areas should focus on retaining a minimum density of 0.5 to 0.7 moose per square mile (0.2–0.3/km^2) to help sustain or increase future populations and align harvest levels with existing densities. Consideration could be given to managing low- and high-density populations separately (Gasaway et al. 1980, Cederlund et al. 1987). Depending on overall objectives, this may require a refinement of management areas.

Harvest Strategies

LEGISLATIVE DIRECTION

The responsibility for managing moose populations and harvests is vested in individual provinces and states. Governments adjudicate moose hunting through the sale of permits or licenses, and associated regulations that limit harvest and access, establish seasons and restrict use of various weapons or equipment. Some hunting regulations may be established by direct state legislative action in the United States. However, most are formulated through proposals that are reviewed and approved or rejected by boards or commissions and then implemented by the appropriate wildlife resource agency. Regulations in Canada commonly are established by action of the Cabinet or Cabinet Minister who heads the provincial resource management agency. In some jurisdictions, a large number of agencies participate by virtue of land tenure, and the means of establishing regulations varies, often resulting in a complex pattern of hunting seasons and regulations.

HARVEST CONTROL OBJECTIVES

Two territories and nine provinces in Canada and 11 states in the United States presently administer moose hunts (Figure 203). Collectively, more than 436,600 moose hunters harvested an estimated 88,630 moose in 1991. A decade earlier, 408,330 hunters cropped 71,050 (Table 57). Harvest demands on moose populations have increased severalfold in the past two decades, resulting in the need for more restrictive hunting regulations. Control of hunting is required to affect the desired allocation of moose harvest among recreational moose hunters, ensure the sustainability of moose populations and achieve other specified management objectives or goals for a particular area, as previously discussed.

ALLOCATION OF HUNTING OPPORTUNITIES

The moose resource in most areas of North America is essentially publicly owned and held in trust by government. In allocating harvest, at least 10 jurisdictions give prime consideration to subsistence use by Native people, as provided under treaty or other legal agreements (Franzmann and Schwartz 1983b). Residents, nonresidents and nonresident aliens are considered in decreasing priority by most jurisdictions. During the 1970s, when jurisdictions sought to reduce the moose harvest, added controls often were placed on nonresident hunters, giving residents priority in the allocation of hunting opportunities. By 1982, increased license fees, resident-only seasons, guide and registration requirements, and limited permits resulted in the reduction of nonresident hunter numbers to one-third of 1972 levels in 11 jurisdictions (Table 57). In 1991, nonresidents still were eligible to hunt in all but seven jurisdictions. A resident guide was required of a nonresident hunter by 8 of 15 agencies, and several agencies required registration with a licensed outfitter to help stimulate additional local economic benefits.

POLICIES AND OBJECTIVES

Moose harvest strategies should be developed from a clearly defined set of policies and objectives. They can be quite varied, depending on population goals and public desires. For example, if the objective is to manage for nonconsumptive use, a trophy hunt or a "calling rut" hunt, regula-

Table 57. Moose hunters, harvest and population estimates for 22 North American jurisdictions, 1972, 1982 and 1990–1991 (after Timmermann and Buss 1995)

	Total hunters (×100)			Nonresident harvest (×100)			Estimated harvest (×100)			Estimated moose population (×100)	
Jurisdiction	1972	1982	1991	1972	1982	1991	1972	1982	1991	1982	1990–1991
Alaska	218	212	220	23.7	11.2	24.1	57	59	61	1,200[a]	1,550
Washington[b]			0.08						0.08		2
Idaho[b]	1	2	5				0.9	1.5	4.9	36	55
Utah	1	1	2.99	0.1	0.08	2.9	0.7	0.9	2.9	14	27
Wyoming	16	16	17		3.1	2.2	13	13	14	79	126
Montana	7	5	6.75	0.34	0.45	0.19	4	3.6	1.14		40
North Dakota[b]		0.3	1.1					0.2	1.09	3	5.5
Colorado			0.07						0.07		4.25
Minnesota	16	35	14.6				3.7	7.6	4.1	90	67
Maine		20	20		2	2		8.8	9.6	200	230
New Hampshire			1.0			0.2			0.89		40
Yukon[c]		44	36		3.0	4.6		10	6.4	35[a]	500
Northwest Territories		10	13		0.25	0.6		1.3	1.4	33[a]	90
British Columbia	550	458	394	56	19.4	18.6	143	128	135	2,400	1,750
Alberta	447	650	500	41.1	12.3	11.5	94	124	122	1,180	1,007
Saskatchewan	130	127	120	10	3	11.7	41	26	41	450	500
Manitoba	102	102	65	19.9	2.3	1	21	17	11	280	270
Ontario	731	887	1,100	125	31	30	138	107	110	800	1,200
Quebec	620	1,260	1,500	30	10	25	68	118	119	750	675
New Brunswick[b]	24	50	52				10	13	17	120	200
Nova Scotia[bd]	10	4	2				4.0	1.6	1.13	40	30
Newfoundland	191	200	292	43.1	9.8	14	110	70	210	700	1,400
Total	3,064	4,083.3	4,366.2	349.24	106.71	145.8	708.3	710.5	886.3	8,410	9,768.15

[a] From Kelsall (1987).

[b] No nonresident season.

[c] Exclusive of Native hunters and harvest.

[d] 1981 data.

tions must ensure low harvest rates of cows and light harvests of bulls 2 years of age or older. This strategy would help enable a longer life expectancy for bulls to reach trophy age and for most cows to attain maximum productivity. On the other hand, managing for quantity (maximum sustained yield) requires maximum productivity, and higher harvest rates of both sexes, including calves. This lowers the mean age of the population.

Selective hunting strategies can be effective in managing moose to realize their reproductive potential and allowing maximum meat and trophy production (Timmermann 1992b). Harvests can be adjusted by modifying season length and timing, restricting use of firearm type or archery equipment, limiting the number of licenses issued, and requiring hunters to be selective by establishing specific sex and/or age class targets. If the objective is to maximize recreational hunting opportunities, then managers can consider liberalizing pre- and postrut seasons, reducing hunter efficiency, requiring several hunters ("party") to share one moose, and targeting for a lower density population, thereby decreasing success rates and allowing more hunter participation per moose shot.

CONTROL CONCEPTS

A variety of strategies are used to regulate harvests, distribute hunting pressure and control success rates. These can be classified as either "passive" or "active" controls. Passive controls usually apply equally to all hunters, do not directly impact individuals, and allow unlimited numbers of hunters. Active controls directly affect individuals by limiting hunter numbers and hunting opportunities. Passive strategies may include manipulation of season length and timing, road access, firearm/archery requirements and license qualification requirements. However, they stop short of limiting the number of moose to be harvested in a given area or the number of hunters. Active strategies usually include limiting license sales or specifying the sex, age or number of the animals that can be taken in specific areas.

The objective of both active and passive strategies is to affect or control the number of moose harvested, and thereby the population through selective harvest of predetermined numbers of animals.

All jurisdictions rely on strategies that combine components of both passive and active controls. During 1991, nine agencies did not limit hunter numbers or total harvests over the majority of their areas. However, all but one used a limited hunter-participation strategy in some management areas (Figure 222). Five agencies combined unlimited nonselective strategy with an unlimited selective antlered strategy, whereas five others used five or more selective or nonselective strategies. Quebec, with 150,000 hunters in 1991, for example, allowed both sexes to be harvested and "encouraged" the shooting of calves, with no limit on hunter numbers during the 1970s and 1980s. In 1994, Quebec introduced limited antlerless quotas to reduce the number of breeding cows harvested. Saskatchewan, Ontario and British Columbia continue to use unlimited entry and sex- or age-specific licenses to maximize recreational hunting opportunities and limit hunter harvest.

Passive Control

LICENSE QUALIFICATIONS

Proof of a previous license or completion of a hunter safety/education course is a basic requirement in nearly all jurisdictions. The two exceptions are New Brunswick and Newfoundland, which require hunters to pass a shooting and written test before a big game hunting license is issued. This requirement encourages shooting practice and improved marksmanship. It also promotes a commitment to learning about moose and their management.

Fee increases for general hunting and species-specific licenses can be used to deter hunter numbers. Resident fees generally are modest and have had little long-term effect in either discouraging or redistributing hunting pressure. In 1991, individual moose hunting license fees for residents

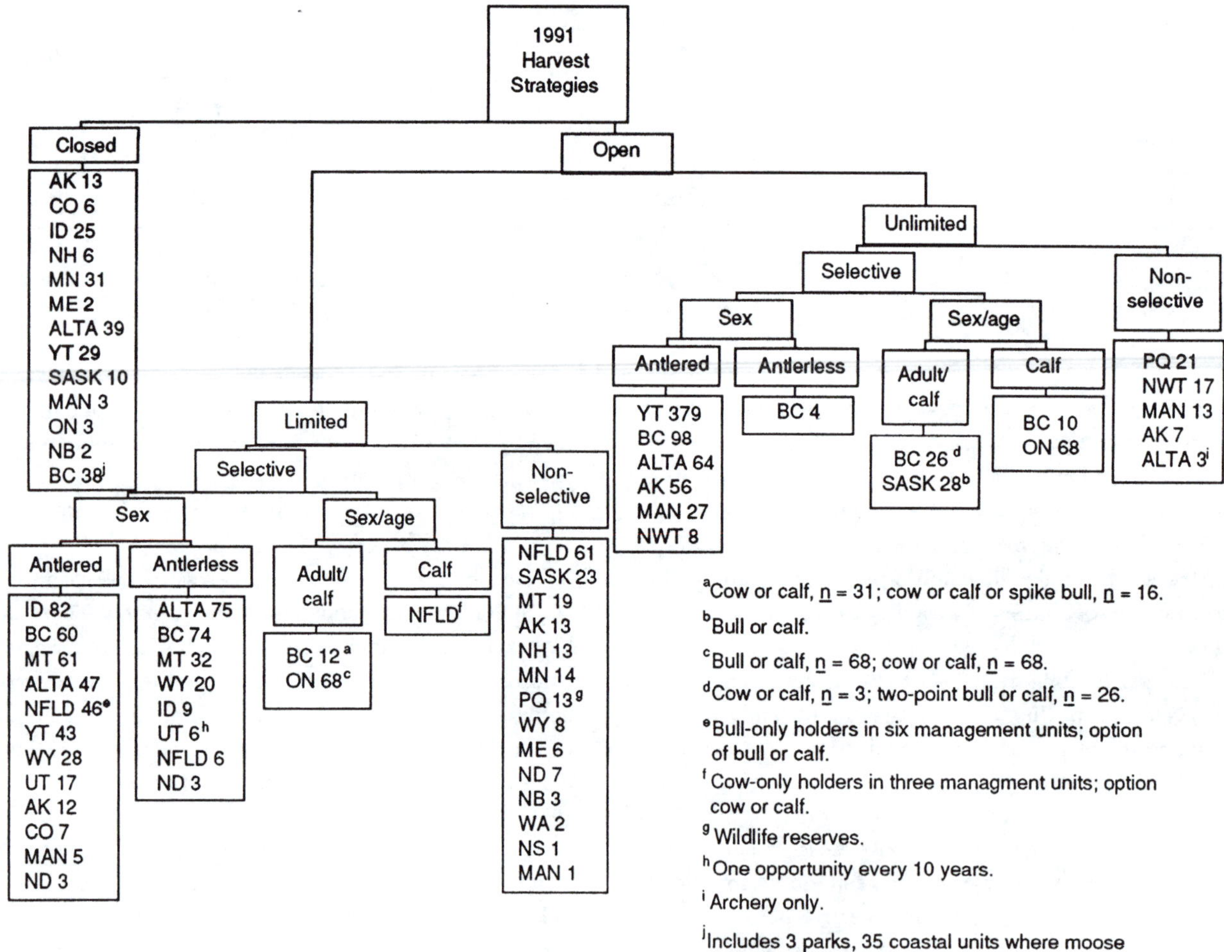

Figure 222. Moose harvest strategies used by 22 North American jurisdictions, 1990–1991 (after Timmermann and Buss 1995). Numerals designate number of management areas or subdivisions.

ranged from $5 in the Northwest Territories to $203 in Colorado, with 12 agencies in the $12 to $35 range and four charging between $50 and $79. Nonresident fees varied from $10 in the Northwest Territories to $1,003 in Colorado, with most others in the $150 to $500 range in 1991. Some jurisdictions use different fees for nonresident citizens and nonresident aliens. In addition to the license fee, trophy fees or export permits may be required from successful hunters.

SEASONS

Opportunities to hunt may be controlled by the length and schedule of hunting seasons. Length of season controls the amount of time available for hunting, whereas scheduling may control success because of seasonal vulnerability of moose to hunting mortality. Hunting season opening dates can influence the magnitude and age structure of the harvest (Timmermann and Gollat 1982). Older prime bulls come into rut earlier than young bulls do, and because rutting bulls are more vulnerable to harvest, seasons held at the peak of the rut will increase the harvest of prime bulls. So as to leave more prime bulls in the population and ensure timely and successful breeding of cows, hunting seasons often are scheduled after the rut peak, when these bulls are extremely wary and much less vulnerable. Early winter aggregations of bulls and cows also can lead to excessive harvests from hunters using snowmobiles and all-terrain vehicles. Managers need to develop harvest strategies that control the harvest of prime breeding bulls and cows during such vulnerable periods.

A secondary effect of hunting season timing may be its limitation on access to moose range by seasonally dependent modes, such as float planes, all-terrain vehicles and snowmobiles. Seasons set during the rut or snow period generally result in higher harvests, because of sex-specific vulnerability (Timmermann and Gollat 1982, Crête 1987). The most liberal 1991 season (243 days in Alaska) extended from August to March. The majority of jurisdictions opted for open seasons between September and December (Table 58).

In a 1992 survey of 19 moose management jurisdictions, Wilton (1992) found that 74 percent of 136 open seasons were coincident with the normal rutting period (September 16 through October 15). Eight jurisdictions used split seasons to help distribute hunting pressure, and allowed early ice-free, open-water and later snow-tracking hunts. Most (17) agencies use a fixed opening date or day of the week, whereas the remaining five agencies provided a weekend opening, which is popular with many hunters. Sunday hunting is prohibited by at least six jurisdictions (Table 58).

Where low density, heavily stressed or highly vulnerable moose populations occur, closed seasons may be used. At least 12 jurisdictions legislated closed seasons in one or more portions of their moose ranges in 1991 (Figure 222). For example, seasons in all 10 of Nova Scotia's former moose areas were closed in 1982 because of a general moose population decline (Timmermann 1987). Manitoba and Newfoundland closed several game areas to recreational hunting because unregulated harvests by treaty Indians and poachers, or both, reduced moose populations.

Hunting success may be limited or reduced by promoting the use of archery equipment, muzzleloaders and shotguns, all of which have a shorter effective range and lower firing efficiency than modern rifles. At least eight agencies have increased hunting opportunities by establishing special archery seasons that overlap the rut, thus facilitating the calling of moose into effective range. In addition, Ontario set aside one wildlife management area exclusively for muzzleloader rifle and archery hunting, and Alaska had one muzzleloader-only area. All jurisdictions regulated the transport of firearms in motorized vehicles and use within specified daylight hours. These regulations pertain mainly to safety and ethical considerations, but they also help to reduce harvest from opportunistic shooting of animals encountered near travel roads and highways and from shooting when low light conditions might result in wounding.

ACCESS CONTROLS

Harvest pressure is directly related to ease of hunter access, and many agencies are concerned about the growing use of mechanized equipment. Road development into moose range escalated substantially from 1950 through 1990, largely as a result of timber and mineral extraction activities. In Ontario, Timmermann and Gollat (1982) found that areas newly accessed by road for resource extraction had disproportionately high moose harvests when compared with roadless areas or those with stable road networks (Figure 223). Local moose population declines associated with increased road access were reported in Alberta (Lynch 1973), Quebec (Bider and Pimlott 1973), Ontario (Eason et al. 1981) and Alaska (Bangs et al. 1984). This situation has prompted managers to apply a variety of regulatory techniques in an attempt to moderate harvest success and prevent local overharvest of moose populations. Some regulations serve the dual purpose of safeguarding persons and property as well as preventing hunters from exercising a widely perceived unethical advantage over moose by using vehicles to locate and/or pursue animals.

Most of the 22 jurisdictions regulate against the discharge of any firearm, including archery equipment, from or across the traveled portion of maintained public roads or highways. Some prohibitions involve no-hunt zones extending a considerable distance from roadways. Also, most

Table 58. Characteristics of moose hunting seasons in North America, 1990–1991 (after Timmermann and Buss 1995)

	Management areas				Season length and timing		
Jurisdiction	Number with moose	Area in square miles (km²) Smallest	Largest	Number with moose hunting season	Maximum number of days	Earliest	Latest
Alaska	94	1.1 (2.9)	20.5 (53)	83[a]	243[b]	August 1	March 21
Washington	4	2.6 (6.8)	2.7 (7)	4	16	November 10	November 25
Idaho	123	0.5 (1.3)	15.1 (39)	98[c]	86	August 30	November 23
Utah	13	3.1 (8)	7.7 (20)	10	16	September 7	December 1
Wyoming	40	1.7 (4.5)	15.8 (41)	37[a]	82	September 10	November 30
Montana	71	1 (2.5)	9.7 (25)	71	90	September 1	November 29
North Dakota	7	1.3 (3.3)	1.6 (4.1)	7[c]	30[b]	September 11	December 25
Colorado	10	0.5 (1.3)	5.9 (15.4)	4[c]	16	November 10	November 25
Minnesota	44	0.5 (1.3)	6.95 (18)	14[cd]	16	September 28	November 30
Maine	8	15.4 (40)	53.3 (138)	6[e]	6	October 1	October 13[g]
New Hampshire	19	1.9 (4.9)	7 (18.1)	11	10	October 15	October 24
Yukon	451	0.2 (0.6)	10.8 (28)	427	90	August 1	October 31
Northwest Territories	18	29 (75)	7,722 (20,000)	18	153	September 1	January 31
British Columbia	189	2 (5.2)	75.7 (196)	154	118[b]	August 15	December 10
Alberta	122	0.8 (2.1)	116 (300)	83[ae]	83[b]	September 9	November 30
Saskatchewan	40	7.7 (20)	463 (1,200)	30[ae]	64[b]	August 26	November 30
Manitoba	38	0.8 (2)	506 (1,310)	35[ae]	112[b]	September 1	December 21
Ontario	71	6.6 (17)	331 (858)	68[ae]	88	September 19	December 15
Quebec	22	8.3 (21.5)	870 (2,252)	21[ae]	92	September 1	December 1
New Brunswick	25	3.2 (8.3)	24.7 (64)	23[ce]	3	September 22	September 24
Nova Scotia	11	2.1 (5.4)	28.6 (74)	1[cef]	5	October 15	October 19
Newfoundland	67	0.45 (1.16)	17.4 (45)	67[af]	79[h]	September 14	December 28

[a] Special archery seasons in some areas.

[b] Split seasons(most are 20–30 days duration).

[c] Resident only.

[d] Northeast region closed—44 areas normally open.

[e] Closed Sundays in some or all areas.

[f] 1981 data(most are 20–30 days duration).

[g] Six-day season; opens first Monday in October.

[h] Special winter season in one area—February 1–15.

agencies prohibit the discharge of firearms within a specified distance of occupied buildings, schools, mining camps, generating stations and dwellings. Others, such as Ontario, provide for the temporary posting against hunting in active work areas, such as logging and road building operations. And most jurisdictions prohibit the carrying of loaded firearms in or on vehicles. The most common exception to this regulation is a stipulation that allows hunters to possess unencased and loaded firearms in a boat or canoe if the craft is not motor powered or while being propelled by a motor. These regulations are similar to those that govern firearm use in watercraft while waterfowl hunting in most areas of North America. The Northwest Territories is the only jurisdiction that currently allows the discharge of a firearm from a snowmobile or three-wheeled motorcycle. Although the wording of the above regulations may differ among jurisdictions, the intent is similar—to ensure public safety.

Regulations promulgated to enhance hunter safety and that of the nonhunting public have the secondary effect of offsetting the distinct hunter advantage availed by improved vehicular access in otherwise remote moose habitats. A third and perhaps more subtle effect is that of enforcing some level of ethics in the hunt by preventing road hunting where the practice is unwarranted or publicly objectionable.

Roadways into moose range represent only one aspect of concern for managers about hunter access. Most jurisdictions either prohibit or limit the use of aircraft, both fixed-wing and helicopters, as a primary means of access. Aircraft access may be allowed by the most direct route to and from the hunting site only, or to specified lakes or airfields. In some jurisdictions, aircraft may be used to transport hunters

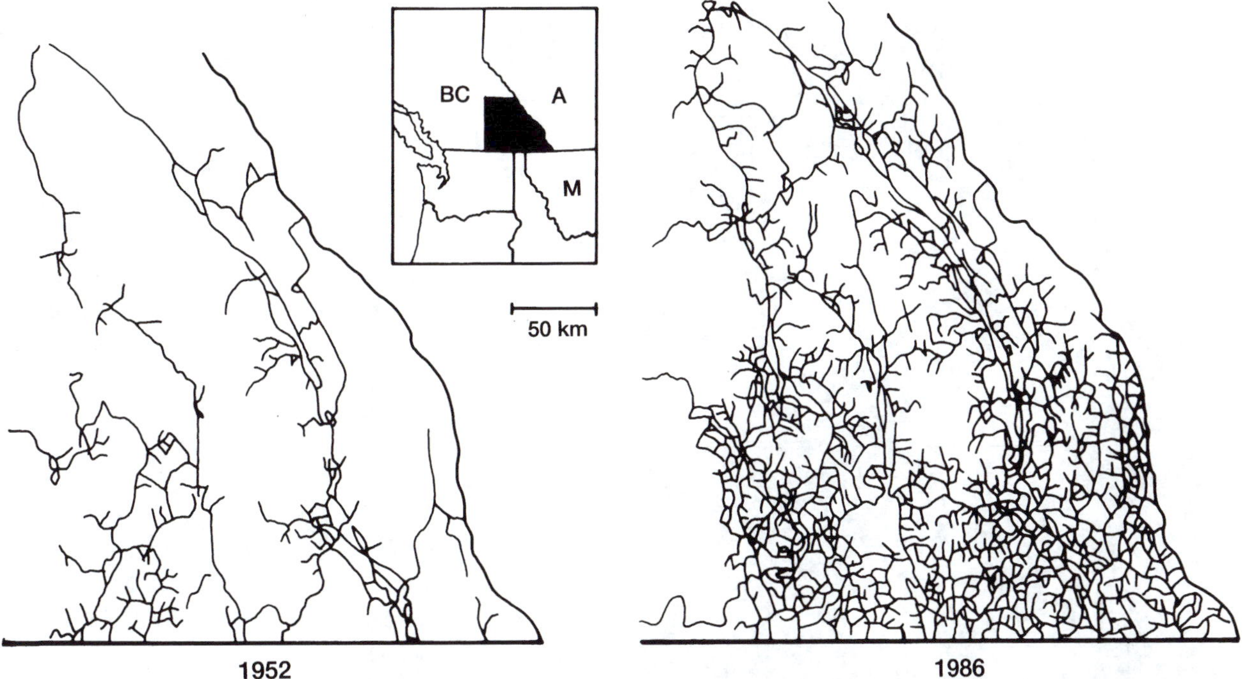

Figure 223. Example of increasing road access in a formerly remote area of moose range in British Columbia (data compiled by R.A. Demarchi, British Columbia Fish and Wildlife Branch, Cranbrook). Road access by conventional and off-road vehicles is a growing concern to managers and a variety of regulations have resulted in an attempt to restrict their use during hunting seasons.

Access to formerly remote portions of moose range by way of secondary roads built for logging operations can quickly and dramatically increase hunter harvest success. Management agencies must be able to assess the impact of such access on the moose population and be in a position to regulate access if harvests begin to exceed management expectations or goals. *Photo by H.R. Timmermann.*

Moose hunting pressure and harvest may be strategically reduced by regulating for archery and/or muzzleloaders. The shorter range single-shot limitations of archery equipment and muzzleloaders, and the consequent additional time and energy required of hunters who use them, discourages participation by casual hunters or those whose primary motivation is trophy or meat. However, many agencies have increased moose hunting opportunities by offering archery and muzzleloader seasons that overlap the rut when patient and skillful hunters may be able to call bulls into close range for well-placed broadside shots. Some of these "special seasons" are opened before or after the regular firearms season to supplement regular season harvests of moose, to accommodate additional recreational opportunity within the sustained yield capacity of the moose population and/or to avail "quality" (reduced competition) hunts. *Photo by Jerry Young.*

to and from the hunting site only (disallowing search for or hazing of game). Others permit hunting only after a specified time after arrival by aircraft at a hunting site (from 6 hours in British Columbia up to 72 hours in North Dakota). Some regulations restrict aircraft use to commercial carriers traveling between public airports on scheduled flights.

In the late 1950s, Ontario designated several large remote areas where aerial spotting of moose was legal and even encouraged. The objective was to disperse hunting pressure to high-density moose areas where lack of road access precluded harvests of the unexploited moose populations. No such regulations currently are in effect in North America.

The widespread use of off-road vehicles or all-terrain vehicles beginning in the 1960s has effectively prolonged the use of temporary access roads resulting from logging and mining activity. Roads, which would have become overgrown in time, preventing conventional vehicle traffic and buffering against undesired intense hunting activity, have been kept open by off-road vehicles. This, too, has prompted a variety of regulations that curtail the use of these vehicles in specified areas during the hunting season. For example, most agencies require that firearms, including archery equipment, be unloaded while in or on a vehicle. Some require firearms to be unloaded and encased. Others, such as in Alberta, allow shooting only when the hunter is beyond a specified distance (e.g., 50 yards [46 m]) from an off-road vehicle. Manitoba and Saskatchewan restrict off-road vehicle use to designated routes or trails in specified areas and only for the retrieval of downed animals. Wyoming and Newfoundland prohibit their use completely for hunting purposes in wilderness areas.

A previously noted refinement in the management of access is the development of road corridor sanctuaries where the discharge of firearms is prohibited or the use of vehicular traffic is restricted over a specified distance or time period. This restriction may create a linear sanctuary, the dimensions of which vary from 5 miles (8 km) either side of the Dalton Highway in Alaska (although archery hunting is legal within the corridor) to 400 yards (365 m) either side of a road in portions of Alberta. Six of 11 Canadian jurisdictions allowing moose hunts provide for road closures and/or sanctuaries with the intent of reducing moose hunting pressure in certain areas. These regulations are moose management oriented, compared with the safety/ethical regulations previously described.

Except the U.S. Forest Service, Bureau of Land Management and Park Service, only Montana among moose range jurisdictions regulates designated road access specifically for the management of hunting activity. In the case of Montana's "Cooperative Travel Management Areas," vehicular access for hunting purposes is restricted to roads in specified areas where private or public landowners have partnership associations with the state or provincial wildlife agency.

Regulations restricting access have met with mixed results. Saskatchewan, for example, attributed moose population increase in some management areas to reduced harvest, resulting from road closures and road corridor game preserves (Beaulieu 1992). In Alaska, hunters adapted to restrictions implemented in 1985 on the use of all-terrain vehicles in the first half of moose season on the lower Kenai Peninsula (T. Spraker personal communication: 1992). These restrictions have been modified to legalize all-terrain vehicle use from August 20 to September 10 and September

In many areas, a high percentage of moose are harvested adjacent to access roads constructed primarily for timber and mineral extraction (left). Compared with roadless or stable road network areas, newly accessed areas produce a disproportionately high moose harvest. To regulate or control moose harvest, access to and from hunting areas can be limited to foot (nonmotorized) traffic or snowmachines (right). Reciprocally, where additional harvest is desired, access restrictions can be loosened, at least temporarily. *Photos by H.R. Timmermann.*

15 and 16, to allow hunters to haul out meat and move their camps (T. Spraker personal communication: 1996). After 1 year of reduced harvest with relatively stable hunter numbers, the harvest again increased as hunters shifted their activity to the first half of the season, allowing all-terrain vehicle use and increased their use of horses as a method of access.

Many of the jurisdictions also provide special consideration for physically disabled persons. These considerations, granted by local managers or in designated areas and at

Modern all-terrain vehicles (ATVs) allow hunters virtually unlimited access to much of North America's moose range. Regulations must be flexible, so as to manage hunter access and encourage ethical restraint in terms of increasingly elaborate and sophisticated mechanizations. *Photo by H.R. Timmermann.*

Extensive use of aircraft by the tourism and outfitting industries in the North has resulted in profitable business development associated with moose hunting. It also has helped to disperse moose hunting effort over large geographic areas. All jurisdictions have regulations concerning the extent and/or location of aircraft transport for hunting purposes, and prohibit aircraft use as a tool of the actual hunt. *Photo by H.R. Timmermann.*

specified times, allow handicapped persons to possess loaded firearms and discharge them from vehicles and roadways. Such accommodation is made to expand recreational opportunity to a broader segment of the public and is not an aspect of moose population management.

Other indirect measures have been used to affect hunting access. Ontario, for example, limits the establishment of licensed fly-in outpost camps in areas designated for remote, roadless hunting and fishing recreation. Once a predetermined density of camps is reached, no further disposition of Crown Land for hunting camps is allowed. Nonresident hunters are required to conduct moose hunting activities through licensed outfitters. The outfitter provides an adult moose permit allocated as a portion of the quota in the management unit in which the outfitter is licensed. Finally, Ontario attempts to protect established outfitters by restricting vehicular access to remote fly-in areas that may become road accessible as a result of logging road development. These temporary road closures generally end in late autumn after float plane operations cease.

In deference to the increased use of mechanized vehicles and access improvements, many hunters still prefer the experience of a wilderness hunt. In moose country, the canoe provides access to many areas and in mountain country, horses often are used.

FIREARM CONTROL

Part of the responsibility of an agency managing a hunting season should be to provide hunters with direction on the humane and effective taking of game animals. This is accomplished directly through education programs and law enforcement and indirectly through regulated restrictions on firearms, archery equipment and ammunition (projectiles).

Moose are the largest deer worldwide, and most jurisdictions have specific requirements, commensurate with the animal's size that are designed to help ensure hunters have "enough gun for the game." In reality, shot placement may be more significant to affecting a humane kill than foot-pounds of energy at impact. This surely is the case with archery equipment, because even the most powerful of bow and arrow combinations produce fewer foot-pounds of energy than a .22 caliber rimfire cartridge.

Nearly all jurisdictions regulate the caliber of conventional rifles, handguns, muzzleloaders, bullet construction, gauge and shot size in the case of shotguns, and draw weight and broadhead construction in the case of archery equipment (see Chapter 18). Few agencies regulate marksmanship by requiring proficiency testing—a situation that is changing, as will be discussed.

For many moose hunters, the greatest satisfactions come from traditional modes and methodologies of the hunt, minus artificialities. Not all hunters have the time, wherewithal, interest, experience or opportunity for such perspective. Many others simply prefer to use and gain considerable satisfaction from the convenience and effectiveness of modern technology and equipment, within legal and ethical prescriptions. Moose hunting affords a great diversity of recreational hunting experiences, and wildlife management agencies must be able to accommodate each reasonably, but in concert with dynamic moose population and habitat conditions and long-term management objectives. *Photo by H.R. Timmermann.*

BAG LIMITS

In all but 9 of the 22 jurisdictions, hunters are allowed to shoot and claim only one moose per hunter per season. Those nine reduce the bag limit to less than one moose per hunter in selected management areas by requiring or encouraging sharing to maximize recreational opportunities. For example, "forced" sharing of a bag limit of one moose per two licensed hunters has been required in 18 Quebec hunting zones since 1979 (Crête 1982). Ontario used a similar system from 1980 through 1982, and introduced a voluntary party-license system for two to eight hunters in 1992. Minnesota has licensed four hunters per permit since 1972, whereas Maine, Alaska, Idaho, Colorado, Manitoba, New Hampshire and Newfoundland issue a limited number of single moose permits for parties of two hunters.

MOOSE HUNTER AND HUNTING EDUCATION

All harvest management strategies require hunter compliance, and most agencies have attempted to educate their hunters to understand, appreciate and comply with these new management strategies (Timmermann 1987, 1992a).

Hunter and other citizens' cooperation was essential to the success of instituting or reinstituting moose hunting in a number of jurisdictions, including Idaho in 1946 (Compton and Oldenburg 1994), Minnesota in 1971 (Karns 1972), North Dakota in 1977, Washington in 1980, Maine in 1980 (Morris and Elowe 1993), Colorado in 1985 (Kufeld 1994), New Hampshire in 1988 (Bontaites and Gustafson 1993) and Vermont in 1993 (Alexander 1993). Establishing moose hunts in jurisdictions with no recent history of moose hunting presented a formidable educational challenge. For example, 3 years before opening its first season, New Hampshire began providing public information on the biology, population status and management needs of moose (Sherrod 1990).

In response to expanding moose populations, a number of states have established moose seasons for the first time or reopened moose hunting seasons after long closures. Included are Utah (1950), Minnesota (1971), North Dakota (1977), Washington and Maine (1980), Colorado (1985), New Hampshire (1988) and Vermont (1993). Above, a North Dakota resident poses with a "once in a lifetime" bull taken during the first modern day season opening in that state. *Photo courtesy of the North Dakota Game and Fish Department.*

Nearly every state and province presently offers hunter safety/education programs for first-time hunters (National Rifle Association 1982). Most of these courses are mandatory and prerequisite to obtaining any type of hunting license. In most cases, a written test is part of the qualification (Table 59). Alberta and British Columbia offer a voluntary education program, but require successful completion of a qualification test.

Most courses emphasize safe firearm handling and wildlife species identification. Additional material includes sections on hunter ethics, responsibility, wildlife management, field care and dressing of game, survival, and first aid (Alberta Energy and Natural Resources 1980, Keiser 1987a, 1987b). However, existing hunter education/safety courses provide little if any specific training for moose hunters.

BASIC HUNTER EDUCATION REQUIREMENTS

As previously noted, a previous hunting license, completion of a hunter education/safety course or test, or attendance of a hunter education course and passing a mandatory test were prerequisites to hunting in 19 of 22 jurisdictions as of 1990–1991 (Table 59). Fees for courses ranged from free to $120. Of the jurisdictions managing moose hunts, only Alaska, Yukon and the Northwest Territories lacked mandatory hunter education requirements for rifle hunters, but each provided voluntary training courses.

Novice big game/moose hunters present a particular educational challenge. Some agencies have taken up the challenge and designed programs to deal specifically with the needs of moose/big game hunters. For example, Alaska and Quebec require all hunters who intend to hunt moose with archery equipment to participate in a bow hunter education course, exam and field skills test. In Alaska, muzzleloader hunters must attend a 1-day orientation course. In many jurisdictions, significant numbers of inexperienced hunters take up moose hunting each year because moose hunting opportunities are limited to once in a lifetime experiences or only after a substantial waiting period. Hence, novice moose

Table 59. Mandatory (M) and voluntary (V) hunter education requirements, as of 1990, in 22 North American jurisdictions that manage a moose hunt (after Timmermann and Buss 1995)

	Course type		Exam required			
Jurisdiction	General[a]	Specific[b]	General[a]	Specific[b]	Shooting test	Fee
Alaska	V	M[c]		M[c]	M[c]	
Colorado	M		M			$5–10
Idaho	M		M			$2
Maine	M					$0–2
Minnesota	M	M	M	M[d]		$5
Montana	M		M			
New Hampshire	M	M	M	M[d]		
North Dakota	M		M			
Utah	M		M			$4
Washington	M		M			
Wyoming	M		M			$0–5
Alberta	V		M			
British Columbia	V		M			$14–120
Manitoba	M		M			$1–10
New Brunswick	M	M[c]	M	M[c]	M[c]	
Newfoundland	V	M[e]		M[e]	M[e]	
Northwest Territories	V					
Nova Scotia	M		M			$10
Ontario	M	V[f]	M	V	V	$20–40
Saskatchewan	M		M			$0–50
Quebec	M	M[g]	M	M	M[g]	$22–25
Yukon	V					

[a] General hunter safety/education course.

[b] Specific big game or moose hunter education.

[c] Big game archery/muzzleloader course.

[d] Prehunt orientation session.

[e] Once in a lifetime hunt requirement for big game hunters.

[f] Big game archery hunters only, includes test and shooting.

[g] Mandatory for first-time archery hunters only.

hunters often are the majority. New Hampshire and northeastern Minnesota developed mandatory moose hunter prehunt orientation sessions. Ontario also developed course material and an instructor's manual for a proposed mandatory course for first-time moose hunters, but legislation confirming the requirement was not forthcoming as of 1996.

SHOOTING PROFICIENCY

Practical shooting exercises were provided in less than half (9 of 22) of jurisdictions conducting moose hunts in North America in 1990–1991. Most agencies that include practical shooting and marksmanship as components did so only within the context of their basic hunter education courses, and most were voluntary. Only Newfoundland and New Brunswick required shooting proficiency tests for big game hunters, and qualification was a once in a lifetime requirement. Also, all archery and muzzleloader hunters in Alaska and first-time archers in Quebec needed to take a mandatory shooting test (Table 59).

Shooting skills of North American hunters may be considered generally poor. Results obtained (1976–1988) from voluntary testing of 5,500 Ontario hunters (one of the only published evaluations made under standardized conditions) using life-size targets indicated that as many as 30 percent of shots taken could have resulted in wounding, with subsequent retrieval of the moose questionable (Timmermann 1977, Buss et al. 1989). Similarly, Newfoundland reported a 32 percent failure rate among 5,531 individuals who took part in the once-only mandatory big game hunter capability test, which included a shooting proficiency test (D. Minty personal communication: 1990). The test required participants to place two of three shots inside the vital areas (heart and lung) on a life-size target. The vital area was not outlined visibly to the shooter in these tests. In New Brunswick, where first-time moose hunters were required to hit a 16.5 by 16.5-inch (42 X 42 cm) target with two of three shots at 44 yards (40 m), the failure rate was approximately 15 percent (G. Redmond personal communication: 1990).

Both Ontario and Newfoundland developed written and illustrated material specific to moose hunting (Ontario Ministry of Natural Resources 1988a, 1988b, Interesting Services Inc. 1989, Newfoundland and Labrador Wildlife Division no date). Information includes location of the circulatory system, internal organs, vital target area, shots to be avoided, target area relative to moose position and the shooter, firearm sighting in procedures and field care and dressing (Figure 224).

BASIC INFORMATION ABOUT MOOSE HUNTING

Basic information, free of charge, is published annually in summaries of hunting regulations (Table 60). These summaries vary greatly in size, page length and quality, but list the game laws and regulations, management areas, maps, season dates, and license quotas and fees. The amount of information about moose varies from just a few words (Washington) to 24 pages (Quebec), and rarely provides hunters with specific biological- or management-related data. Only Newfoundland, in its regulations summary, provided specific educational material to moose hunters. That material included information on selective harvest rationale, identification features, shot placement, field dressing and answers to commonly asked questions (Newfoundland and Labrador Wildlife Division 1991).

SUPPLEMENTAL INFORMATION

Twelve of 22 jurisdictions provided supplemental information to moose hunters (Table 60). The number of publications (one in Newfoundland to eight in New Hampshire) and page lengths (5 in Alaska to 90 in Ontario) vary among

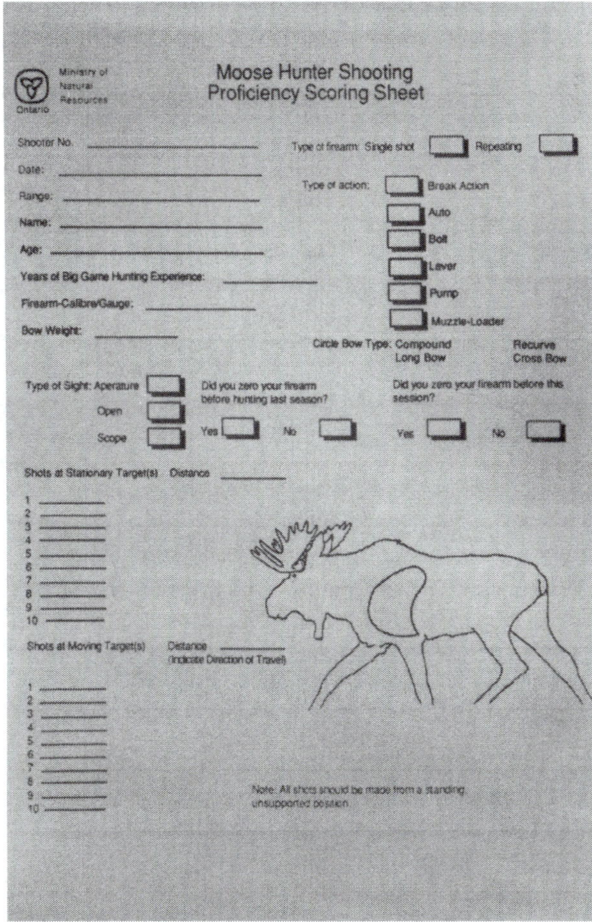

Ministry of Natural Resources
Ontario

Moose Hunter Shooting
Proficiency Scoring Sheet

Shooter No. ____________ Type of firearm: Single shot ☐ Repeating ☐

Date: ____________

Range: ____________ Type of action: ☐ Break Action ☐ Auto ☐ Bolt ☐ Lever ☐ Pump ☐ Muzzle-Loader

Name: ____________

Age: ____________

Years of Big Game Hunting Experience: ____________

Firearm-Calibre/Gauge: ____________

Bow Weight: ____________

Circle Bow Type: Compound Recurve Long Bow Cross Bow

Type of Sight: Aperature ☐ Open ☐ Scope ☐

Did you zero your firearm before hunting last season? Yes ☐ No ☐

Did you zero your firearm before this session? Yes ☐ No ☐

Shots at Stationary Target(s) Distance ____________

1 ____ 2 ____ 3 ____ 4 ____ 5 ____ 6 ____ 7 ____ 8 ____ 9 ____ 10 ____

Shots at Moving Target(s) Distance ____________ (Indicate Direction of Travel)

1 ____ 2 ____ 3 ____ 4 ____ 5 ____ 6 ____ 7 ____ 8 ____ 9 ____ 10 ____

Note: All shots should be made from a standing unsupported position.

Figure 224. Standardized scoring sheet used in Ontario to record moose hunter background, firearm type, target condition (stationary or moving) and performance (courtesy of the Ontario Ministry of Natural Resources 1992).

Table 60. Moose hunter information sources in 22 North American jurisdictions, 1990–1991 (after Timmermann 1992a)

	Printed information distributed by wildlife agencies									
	Hunt summary			Supplemental		Information from hunter harvest[c]			Attitude surveys	
State or province	Dimensions in inches (cm)	Total pages	Pages on moose	Number of publications	Total pages	Kill	Age	Reproductive tracts	Regular	Periodic
Alaska	8 x 10 (21 x 26)	50	7	3	5	M[c]	V[c]			
Colorado	4 x 7 (10 x 19)	6	0.2	3	3	V	V			✓
Idaho	7 x 10 (19 x 26)	32	4			V				✓
Utah	11 x 15 (29 x 37)	24	20			V	V		✓	
Wyoming	8 x 11 (21 x 28)	40	2			V	V			✓
Montana	11 x 8 (28 x 21)	38	2			V				✓
North Dakota	7 x 11 (21 x 28)	4	0.5			V	V		✓	
Minnesota[a]	8 x 11 (21 x 28)	14	14	2	43	M	M			
Maine[a]	4 x 8 (10 x 21)	39	1	5	34	M	V	V		✓
New Hampshire[a]	5 x 8 (13 x 21)	5		8	73	M	M	M		
Washington	8 x 11 (21 x 28)	40	0.2			M	M			
Yukon	8 x 11 (21 x 28)	16	3	1		M[d]				✓
Northwest Territories	4 x 9 (10 x 23)	8	0.5	1	8		V			
British Columbia	8 x 11 (21 x 28)	71	13	2	20	V[e]	M	V		✓
Alberta	8 x 11 (21 x 28)	41	2	1	31	V	V			✓
Saskatchewan	8 x 9 (21 x 23)	38	2			V	V			
Manitoba	11 x 8 (28 x 21)	24	2	2	20	V	V	V		
Ontario[b]	8 x 11 (21 x 28)	23	6	3	90	V	V			✓
Quebec	5 x 7 (13 x 19)	138	24	4	30	M	M		✓	
New Brunswick	4 x 9 (10 x 23)	9	0.5	3	22	M	M			
Nova Scotia	4 x 7 (10 x 19)	33	1			M	M			
Newfoundland[a]	7 x 11 (19 x 28)	56	12	1	23	M	V	V		✓

[a] Special moose hunt information booklet with application form provided.

[b] Ontario Ministry of Natural Resources (1990) included in supplemental information.

[c] M = mandatory; V = voluntary.

[d] Requirements enforced for nonresidents only.

[e] Tooth and reproductive tract required from late season, antlerless hunt harvests.

jurisdictions. Four jurisdictions (Minnesota, Maine, New Hampshire and Newfoundland) published a separate moose hunter application guide or informational booklet to assist hunters applying for a limited number or area-specific moose licenses. A number of agencies (Alaska, Minnesota, Maine, New Hampshire, Manitoba, Ontario and Newfoundland) distributed either a pamphlet handout or formal booklet to educate moose hunters about proper field care and handling procedures (Rutske 1989, Maine Department of Inland Fisheries and Wildlife no date, New Hampshire Fish and Game Department no date).

General information on moose life history is available at book stores and libraries (e.g., Van Wormer 1972, Franzmann 1978, Coady 1982). Specialized subject material includes field care and handling (National Meat Institute 1970), calling and hunting techniques (Labelle 1983, Mongrain 1986, Grenier no date), trophy antler scoring (Boone and Crockett Club 1988) and photography (e.g., Rue 1985, Hoshino 1988, Runtz 1991). And a number of books for children feature moose (e.g., Seuss 1948, Berry 1965, Fair 1992).

Also, three comprehensive booklets were compiled specifically for moose hunters (Ruel and Le Boeuf 1987, Buss and Truman 1990a, 1990b). Information includes material on basic moose biology, habitat needs, reproduction, parasites and disease, management techniques including forest management, regulations, hunter ethics, equipment, firearm sighting-in procedures, shot placement, identification features, hunting basics, and field care and handling.

Hunters have the opportunity to provide voluntary information to moose managers. In some jurisdictions, hunters are required to provide details regarding their kill (Table 61). Information collected includes age, sex, date of kill, location and in some areas, reproductive organs. Hunter attitude surveys are conducted regularly in 3 jurisdictions, periodically in 10 and not at all in 9.

CASE STUDY—ONTARIO

Voluntary moose hunter education seminars, including opportunities to shoot at life-sized moose silhouettes, were introduced in Ontario in 1976 (Timmermann 1977, 1987, Buss 1978). These sessions were supplemental to the mandatory hunter education program since 1960 for first-time hunters. Development of moose hunter seminars was motivated by A.B. Bubenik, based on his experiences with hunter education in Europe. The seminars offered a wide range of educational material, including basic moose biology, sex/age identification, proper field care and handling, hunting techniques, ethics, hunting laws and the opportunity for shooting to allow marksmanship self-evaluation.

In 1979, Ontario's moose management program faced the problems of declining moose populations and losses of related recreational and economic benefits (Ontario Ministry of Natural Resources 1980a). Public meetings involving more than 7,000 attendees were held to discuss management options. Partly as a result of these meetings, a provincial moose management policy (Ontario Ministry of Natural Resources 1980a) was approved in December 1980. It included implementation of the harvest strategy requiring two hunters to share one moose (Timmermann and Gollat 1984). This strategy was replaced by a province-wide selective harvest program in 1983 (Euler 1983a, Timmermann and Gollat 1986).

Changes in regulations were communicated to hunters by a number of ways. Pamphlets with a question-and-answer format were mailed to all previous season license holders. A 22-page moose hunter handbook (Ontario Ministry of Natural Resources 1984) was circulated, detailing 19 specific items, including moose identification features, rational and benefits of selective harvest, as well as license application rules. Subsequently, an annual moose hunter fact sheet was published and distributed at all license vendors and government offices.

Response of hunters to these new regulations and their knowledge of moose biology were measured in surveys conducted 3 and 4 years after the program was introduced (Rollins 1987, Rollins and Romano 1989). Results suggested that hunters had a high level of knowledge of harvest regulations, but less comprehension of moose management and biology. An expanded hunter education effort was recommended to strengthen hunter understanding in these weak areas. Support for the new selective harvest system was tempered by a significant amount of dissatisfaction, particularly relating to the overall enjoyment of the hunt (Ontario Ministry of Natural Resources 1987). A proportion of older hunters felt that the new regulations, which required quarry sex and age identification and eliminated party hunting, reduced their success rate.

A program review was initiated with an internal working group in January 1987 to identify specific problem areas and present alternative solutions (Wedeles et al. 1989). Assistance was provided from eight regional advisory committees, each consisting of 10 to 18 interested members of the public and representatives of special interest groups. Eventually, five major program improvements were recommended, including legalizing party hunting for adult moose, introducing a sportsman's identification card, implementing a system for group applications, maintaining a two-level preference pool system, and initiating a mandatory moose hunter education program. Of these, introduction of a mandatory moose hunter education program for first-time moose hunters was suggested as an effective way to communicate hunting ethics and skills to novices. As of

Table 61. Moose harvest assessment strategies used in North America, 1991 (after Timmermann and Buss 1995)

Agency	Hunter activity report[a]	Harvest registration[a]	Noncompliance penalty[f]
Alaska	M[b]	M[bd]	None
Washington	M	M	N/A
Idaho	M	M	Fine, loss of license, jail
Utah	V	V	N/A
Wyoming	V	V	None
Montana	V[c]	V	N/A
North Dakota	V[c]	M	None
Colorado	V	M	None
Minnesota	V	M	Fine, loss of license
Maine	M	M	Fine, jail
New Hampshire	M	M	Fine, loss of license, jail
Yukon	M	M[de]	N/A
Northwest Territories	V	M[de]	Fine
British Columbia	V	M[bg]	Fine
Alberta	V	M[d]	None
Saskatchewan	V	V	N/A
Manitoba	V	V	N/A
Ontario	V	M[de]	N/A
Quebec	V	M	Fine, loss of license
New Brunswick	M[c]	M	Fine, loss of license
Nova Scotia	M[c]	M	Fine, jail
Newfoundland	V[c]	M[d]	None

[a] M = mandatory; V = voluntary.

[b] Limited draw hunts only (British Columbia—only late season antlerless hunts).

[c] Unsuccessful hunters only.

[d] Export permit/trophy fee.

[e] Nonresident hunters only.

[f] Variable enforcement.

[g] Late season, antlerless, limited draw hunt only.

Shooting practice and proficiency testing on life-sized moose targets enable hunters to improve and evaluate their marksmanship. Some jurisdictions require hunters to meet a qualification standard before they can obtain a hunting license. Given the results of studies of hunter shooting ability, more jurisdictions should consider instituting such a standard, in the best interest of the resource as well as the hunters themselves. *Photo by M.E. Buss.*

1995, four of the five recommendations have been implemented, including printing of two booklets (Buss and Truman 1990a, 1990b) for the core curriculum for a future mandatory moose hunter education course. In addition, a draft moose hunter education instruction manual (Ontario Ministry of Natural Resources 1989a) and two videos—*Moose Hunt, a Guide to Success* (Interesting Services Inc. 1989) and *Firearms for the Moose Hunter* (Ontario Ministry of Natural Resources 1988b)—were circulated to hunter education instructors.

FUTURE DIRECTION OF MOOSE HUNTER EDUCATION

The future of hunting in North America undoubtedly depends on knowledgeable, informed and responsible hunters, characteristics that are not mutually exclusive. Education and public involvement are important for blending the biological and social context of hunting in a growing urban society (Minty 1989). Most respondents to a survey in Newfoundland (92 percent of 6,264) and Ontario (71 percent of 2,841), for example, agreed that a specific hunter education program or training course should be mandatory for all first-time moose hunters (Newfoundland and Labrador Wildlife Division 1986, Wedeles et al. 1989).

In an article entitled "Quality Hunting" Clarke (1977: 5) wrote: "The most important by far of all things brought to the hunt is knowledge and understanding, and respect, all which, so far as responsibility goes, are the things which the hunter has equipped himself. It is these that contribute most to his enjoyment."

Active Control

MANAGEMENT AREAS

All of the jurisdictions with huntable moose populations have game management areas to enable area-specific harvest control. In some jurisdictions, management areas were established on the basis of their variable differences in land capability to produce and sustain moose, as well as ease of administration of a variety of harvest management strategies. As of 1991, moose management areas varied in size from 23 to 770,000 square miles (60–2 million km^2), and numbered from 4 to 451 per jurisdiction (Table 58).

Some areas become popular because of high moose productivity, harvest success or desirable access. Such popularity can result in local overharvest of moose populations and, by intensive pressure, reduce the quality of the recreational experience. When passive methods are no longer effective to control or distribute harvest, then more direct control is required. This can be accomplished by limiting the total number or type of licenses available, limiting the number of animals that may be taken, specifying quarry age and/or sex or by a combination of these techniques.

All jurisdictions, except the Northwest Territories, used either a selective or nonselective limited hunter participation strategy, or a combination of both in 1991 (Figure 222). Most jurisdictions limited the number of licenses available for each management area, but Alaska used registration hunts to control harvests in some areas. Hunters were required to report their kill at the end of a hunt and, when a prescribed number of moose was killed, the season was closed for the area.

Limited license techniques usually require application on official forms sent to a central office. Applicants generally are allowed to indicate a choice of several areas. The licenses are awarded by random draw. In some jurisdictions where demand exceeds supply, hunters successful in such a lottery are ineligible to reapply for 1 or more years: North Dakota—a 1-year wait; Idaho and New Hampshire—3 years; Wyoming, New Brunswick and Nova Scotia—5 years; Montana—7 years or, if a bull moose is claimed, as in Idaho, Utah, Washington, Colorado and North Dakota, no future application can be made; Utah—10 years if a cow is shot and claimed. In other jurisdictions, a preference pooling system gives unsuccessful applicants priority in the following year's lottery.

LIMITED NONSELECTIVE

With this strategy, a limited number of nonselective (of sex or age) area-specific licenses are issued based on population estimates and past hunter success rates. This strategy was used by 14 agencies in 1991 (Figure 222). Seven agencies used this method exclusively, whereas four combined it with limited selective control, and four used it in conjunction with two or more forms of unlimited entry licensing. British Columbia had the widest range of harvest strategies (8), followed by Alberta, Manitoba and Alaska (4).

LIMITED SELECTIVE

Moose harvest management in Scandinavia, based on the removal of animals with currently low reproductive potential, was adopted by several North American jurisdictions. In the early 1990s, 13 agencies offered a limited number of selective licenses (Figure 222), designed to control the harvest of specific age or sex classes, maximize recreation and increase herd productivity. Saskatchewan introduced a sex and age selective hunt in 1977, and several jurisdictions, including British Columbia, Ontario and Newfoundland, followed after 1980. Each province developed a system to fit its own program objectives, which can be categorized as follows:

- To protect and increase prime breeding animals, especially cow moose, so that calf production is enhanced.
- To direct hunting pressure toward animals in the herd with the lowest reproductive potential, such as bulls and/or calves.
- To maximize recreational hunting opportunities and improve hunting quality, (MacLennan 1978, Child 1983b, Timmermann 1983).
- To maintain optimal bull/cow ratios (Figures 225 and 226).

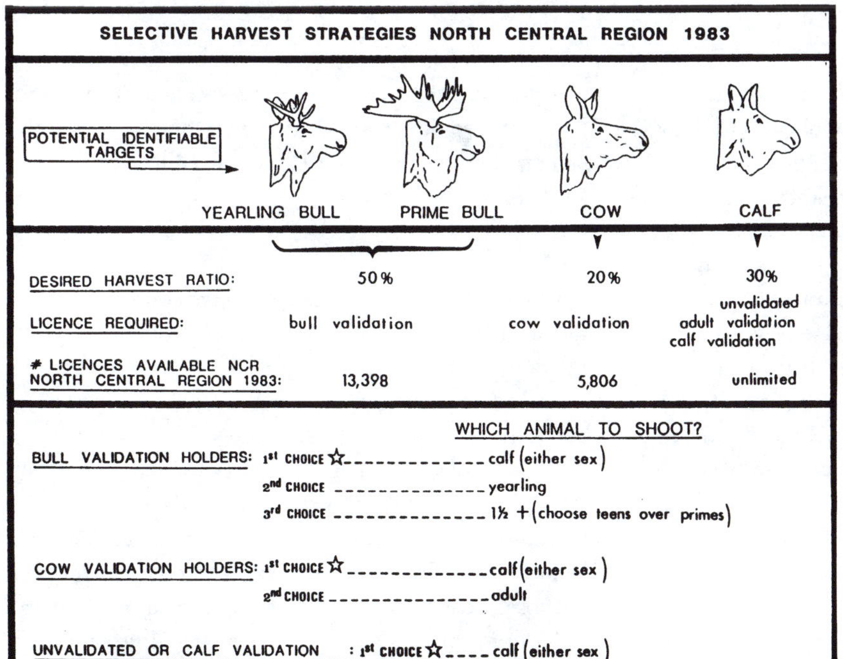

Figure 225. In 1983, Ontario introduced a province-wide selective moose harvest program. The key elements were harvest quotas, adult validation tag quotas and the allocation of these tags to hunters through a computerized draw. Using the most recent and reliable moose population estimates, harvest limits were set for almost 70 wildlife management units.

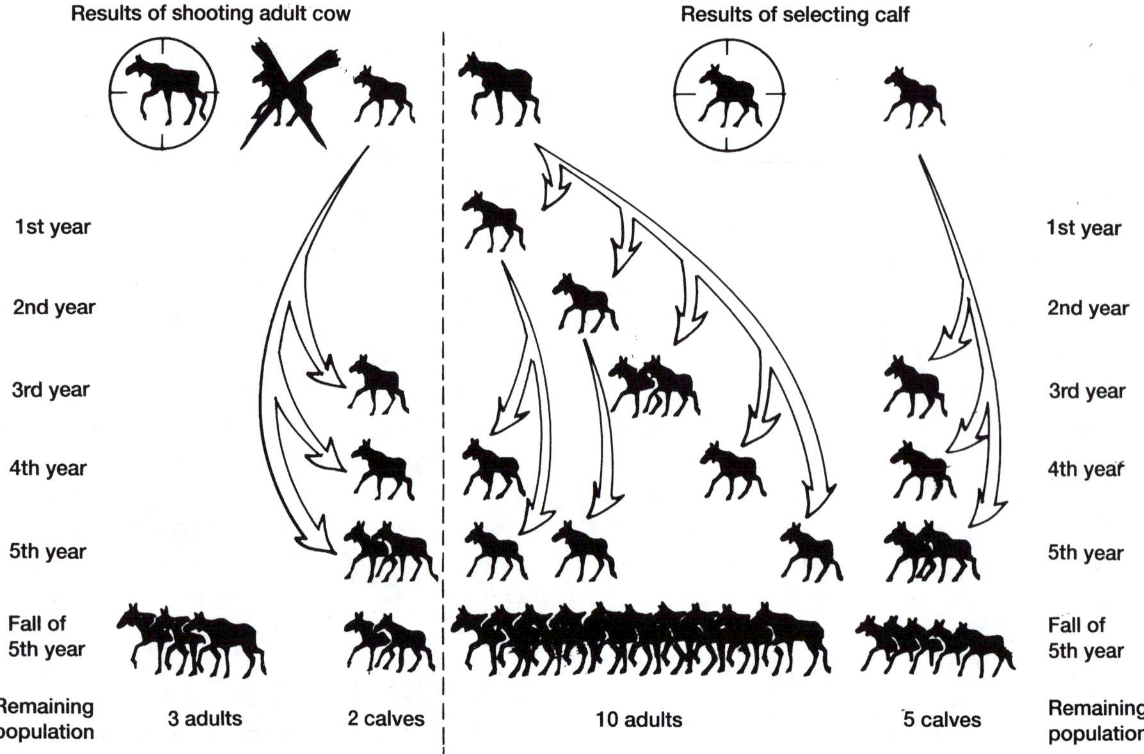

Figure 226. Any and all animals harvested by hunters influences the future herd. The model above demonstrates the long-term impact of a hunter selecting (harvesting) a calf rather than a cow (from Grenier 1979). The model assumes that when a cow with twin female calves (above left) is taken, one of the calves will die during the winter and the other will survive. The surviving calf will not produce its first young for at least another year and a half, and there is little chance that twins will be produced until its fourth or fifth year. Five years after the death of the cow, potentially just five animals—three adults and two calves—will have been produced. If hypothetically, the hunter instead selected one of the two calves (upper right), the cow would continue to produce offspring. In addition the female calf, considered by the model to survive with the protected cow, would make its own contribution to the herd. After 5 years, 10 adults and 5 calves would potentially result—three times the number than if a cow were shot.

In Ontario, the key elements of a selective harvest program for moose were harvest quotas, adult validation tag quotas and the allocation of these tags to hunters. Adult validation tags are distributed by lottery for either an adult bull or cow (Smith 1990). Harvest limits were set annually for nearly 70 wildlife management units (WMU) with moose seasons. Harvest quotas—the planned number of bulls and cows to be harvested—are set by using the most recent reliable moose population estimate for each WMU (Figure 227). Allowable harvests are calculated on the basis of a percentage of adult cows in the herd or a percentage of the total population. Ten percent of the annual province-wide planned harvest is set aside for clients of outfitters to help stimulate the economy in remote northern communities. Adult validation tag quotas combine the harvest quota for the year with the average hunter harvest success rate (a 3-year running average) in a WMU. For example, if the harvest quota for bulls in a WMU was 100 animals and the regular gun hunter success rate was 25 percent, then the bull tag quota would be 100 divided by 0.25. This means that 400 tags could be issued for an expected harvest of about 100 bulls.

HARVEST SEX RATIO

North American harvest strategies often concentrate on removal of males (Baker 1975) and, where trophy hunting is a major motivating factor, the pressure may be further concentrated on prime bulls (Cowan 1974). Concern has been expressed that low adult bull-to-cow ratios could influence conception dates and newborn sex ratios (Bishop and Rausch 1974, Crête et al. 1981a; see also Chapter 4).

Intuitively, a balanced social structure should include a sufficient proportion of prime bulls, whether for breeding purposes, retention of genetic diversity, maintaining social order or for human viewing enjoyment. Close scrutiny of age structure of the harvested component is necessary if

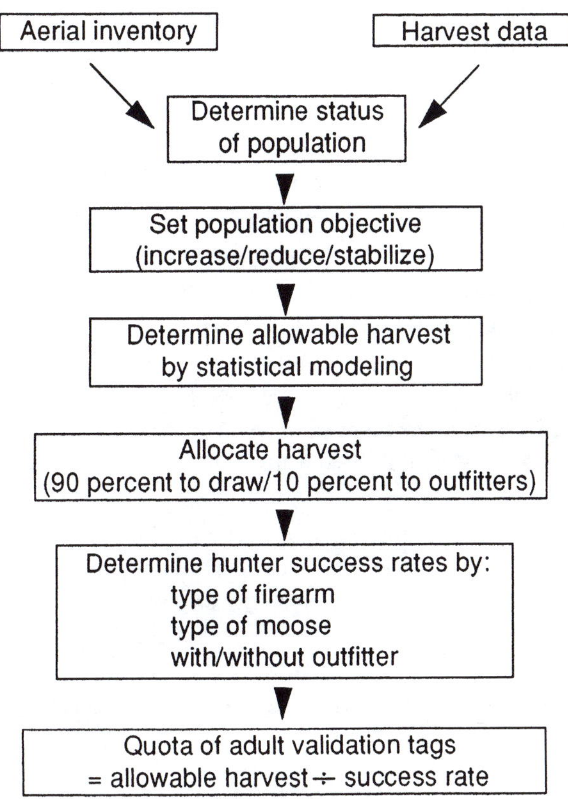

Figure 227. Diagram of the process of establishing adult validation tag quotas for a selective moose harvest in Ontario (from Smith et al. 1992).

managers hope to adjust the structure to ensure that adequate numbers of animals of prime breeding age are maintained. Herds should be socially balanced to attain maximum in utero productivity. When the mean age of harvested animals is consistently younger than 4 years of age, concern is warranted (Crichton 1992a).

Traditionally, moose harvest strategies have been directed toward males. Twelve of 22 agencies allowed a limited antlered harvest in 1991, but only Utah and Colorado used it exclusively, and the others did not specify age (Figure 222). Concern for declining bull-to-cow ratios and its potential effect on productivity led British Columbia and Alaska to experiment with a selective bull harvest strategy based on antler architecture (Child 1983b, Schwartz et al. 1992). Regulations developed by both jurisdictions in the 1980s allowed the legal harvest of bulls with spike or forked antlers (one or two points on one side only).

In addition, Alaska included bulls with antlers 50 inches (127 cm) in spread or larger or with three brow tines on one antler in some areas and with four brow tines in others. All smaller, palmated bulls are protected (Figure 228). This regulation facilitated recruitment into the prime breeding and trophy age class and increased the bull-to-cow ratio. Such a strategy appeared to offer a viable alternative to more restrictive seasons and allowed unlimited hunter participation. Because the harvest was targeted to a small segment of the male population, overharvest was no longer a major concern. Results in Alaska were positive and, after 5 years, the number of bulls in the population increased. Yearlings

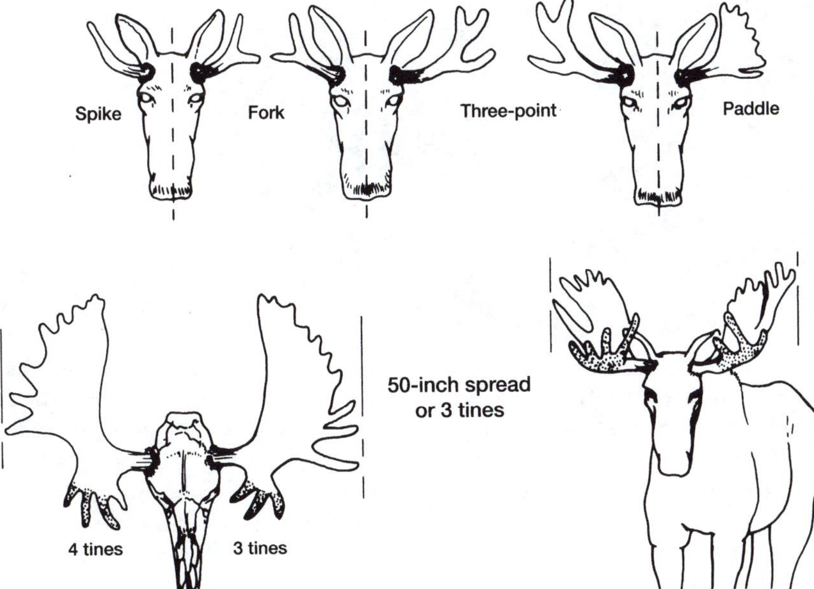

Figure 228. Moose with a spike or fork antler, with antlers greater than or equal to 50 inches (127 cm) in spread, or with three tines on at least one brow palm were legal under a selective harvest system on the Kenai Peninsula of Alaska (from Schwartz et al. 1992).

with large antlers were protected, as were about 80 percent of the 2- to 3-year-olds, and almost 50 percent of 4- to 5-year-olds. The increasing bull number generated a high public support for the program, and viewable bulls were common in many areas where before they were a rare sight after the hunting season (Schwartz et al. 1992).

Future adjustments to this strategy could include replacing the 50-inch (127-cm) regulation with a limited number of permits to harvest males surplus to the prescribed bull-to-cow ratio. Schwartz et al. (1992) suggested that such a system could provide both a general hunting season in which everyone qualifying could participate, and a special permit season in which hunting quality (success rate) could be enhanced and bull harvest controlled.

In British Columbia, prerut and postrut hunting seasons were held to protect mature bulls. After the rut, unlimited hunting was permitted for immature bulls with two or fewer points on one antler. Mature bull and cow harvests were controlled by lottery drawings.

British Columbia, Montana, Idaho, Utah, Newfoundland, North Dakota, Alberta and Wyoming all used limited antlerless permits to harvest cows and calves in selected management areas (Figure 222). The objective was to maintain balanced sex ratios and provide additional hunting/harvest opportunities.

HARVEST RATES

Gross productivity was defined by Simkin (1974) as the percentage of a population that can be removed annually on a sustained yield basis if all of the young conceived survive to the hunting season. For example, if a precalving herd numbers 100 moose and 33 calves are born, then the gross productivity is 33 divided by 133, or 24.8 percent. The rate of increase in this example is 33 percent (33/100). Estimates for North America range from 24 to 25 percent (Simkin 1974).

Predation by brown and black bears and wolves appears to be the major limiting factor for moose in northern ecosystems when human influences are minimal (Gasaway et al. 1983, Bergerud and Snider 1988, Ballard et al. 1991, Van Ballenberghe and Ballard 1994; see also Chapter 7). "Predation on moose may not be strongly limiting if, for ex-

Selective harvest of bulls was initiated on the Kenai Peninsula in the late 1980s. The objective was to facilitate recruitment into the prime breeding and trophy-age class bulls and increase bull/cow ratios. Regulations were developed to allow harvest of moose with spike or forked antlers or bulls with a minimum antler spread of 50 inches (127 cm) or at least three brow tines on one antler in some areas and four brow tines in others. Initially, the program was a hard-sell, because hunters had to learn to identify legal moose. The photo above, featuring a tame animal at the Moose Research Center, shows how to measure a moose's antler spread. The approach obviously is not possible in the wild. Local biologists used the scene for comic relief when discussing with hunters the means to identify a 50-inch (127-cm) bull. An assessment after 5 years showed a significant increase in the number of bulls, larger bulls, moose seen by hunters, and hunter acceptance of the program. In 1993, the program was extended to other areas of Alaska. *Photo by Charles C. Schwartz.*

ample, predators are much reduced in number or selectively eliminated by humans, or if human influence on other components of the ecosystem is extreme" (Van Ballenberghe and Ballard 1994: 2,076). Consequently, managers must consider predator management objectives as an integral component of any moose management plan.

Knowledge of moose productivity, adult sex ratios and estimates of annual losses is required to determine the existence of a harvestable surplus. North American moose harvest rates range from 2 to 16 percent of estimated prehunt populations (Crête 1987). If population growth is the objective, harvest rates should be sex and age specific and generally fall at less than 10 to 12 percent. Strategies could include protection of prime breeding animals (especially of cow moose so that calf production is enhanced); direction of hunting pressure toward animals in the herd with lowest reproductive potential (i.e., young bulls and calves); and reduction of nonhunting losses.

Losses to crippling/wounding, poaching and subsistence use can be significant and need to be considered in any harvest management program. Wounding losses are not well-documented and generally are considered only superficially in calculating allowable harvests (Buss et al. 1989; see also Chapter 8). In Ontario, a provincial mail survey, which sampled approximately 10 percent of nearly 100,000 hunters, suggested a 4 percent reported wounding loss (Ontario Ministry of Natural Resources 1989c). This is considerably lower than the 30 percent maximum calculated in Ontario by Buss et al. (1989) based on subjective assessment of 3,926 shots taken by 1,378 shooters on life-sized moose silhouette targets under standard conditions. Similarly, in 1990, Newfoundland reported a 32 percent failure rate among 5,531 individuals who participated in a one-time mandatory big game capability test that involved a shooting proficiency component (Newfoundland and Labrador Wildlife Division 1992). In New Brunswick, where first-time moose hunters were required to hit a 16.5- by 16.5-inch (42 x 42 cm) target two of three shots at 44 yards (40 m), the failure rate was about 15 percent (G. Redmond personal communication: 1990). Managers, therefore, should be aware of possible wounding/crippling losses and consider these in setting harvest objectives.

Poaching losses may vary in importance with respect to time and location, but are difficult to quantify (Wolfe 1987). Such illegal kill has a socioeconomic impact by reducing opportunities for legitimate hunters to harvest an animal and it deprives the managing agency of revenue from hunting license or permit sales. Estimates for moose reported by Wolfe (1987: 661) "spanned a range as low as 5–10 percent to as high as 100 percent of the legal harvest, with an unweighted mean of 30 percent." Based on data provided by Bisset (1987), Wolfe suggested that each illegally killed moose could sustain either six additional resident or three additional nonresident hunters. Most management agencies use reports of poaching losses by law enforcement personnel as a minimum estimate of this parameter.

Subsistence take by Native people "vary widely and stem from a veritable welter of geographic, historical, legislative and judicial preconditions" (Wolfe 1987: 663). For example, specific hunting rights were granted in treaties struck between Natives and the respective federal governments during the 1800s and early 1900s. Crichton (1981) believed moose harvest by treaty Indians in Manitoba was an important factor contributing to the species' population decline documented in the 1970s. Similarly, Stewart (1983) reported that hunting by treaty Indians in Saskatchewan was believed to have intensified in conjunction with road access provided by an ever-expanding timber harvesting industry in the 1970s and 1980s.

Published reports based on questionnaire–interview techniques have documented the importance of moose to Native economies (Feit 1987). Wolfe (1987: 666) summarized estimates of Native and subsistence use of moose in northwestern Ontario, northwestern Quebec and Alaska (Hamilton 1981, Behanke 1982, Drolet et al. 1982, Charnley 1983, Fall et al. 1983, Andrews and Stokes 1984). Some of these studies confirmed that moose comprise a major component (up to 45 percent of all wild foods harvested in a given year) of the diet in certain Native communities.

In Alaska, both federal and state statutes give priority to nonwasteful subsistence uses over recreational consumptive uses of wildlife (Wolfe 1987). Subsistence use is defined by Chapter 99 of the Alaska Fish and Game Code as "customary and traditional uses by rural Alaska residents for food, shelter, fuel, clothing, tools or transportation, making handcraft articles, customary trade, barter and sharing" (Wolfe 1987: 665).

In the future, managers must strive to obtain more reliable estimates of annual losses through wounding, poaching and Native and subsistence take in relation to the size of the moose populations. This information is prerequisite to sound population management in areas where these losses and harvests collectively constitute a major source of moose mortality. Such knowledge will reduce the risk of overallocating the resource in areas subject to both subsistence and non-Native recreational harvests.

Effectiveness of Harvest Strategies

To be successful, any selective harvest strategy will require compliance and increased hunter education (Timmermann 1992a).

HARVEST ASSESSMENT

Monitoring the effectiveness of a harvest strategy should be an ongoing process so that timely adjustments can be made to meet stated goals. The benefit of a harvest system can only be assessed fully if the hunter kill can be determined with reasonable accuracy. Integration of such information with population data (Larsen and Kale 1982, Gollat and Timmermann 1983) is essential in evaluating management techniques. As of 1991, hunters in 8 of 22 jurisdictions were required to report their hunting activity (Table 61). Registration of kill was compulsory in 17 of 22, with 9 agencies applying a noncompliance penalty to hunters who failed to report. Enforcement of these requirements varied among agencies.

Compliance levels are considered high in such jurisdictions as Minnesota and Maine, which have short seasons and small numbers of hunters (Timmermann 1987). Jurisdictions with longer seasons and more hunters, such as Quebec (Lacasse et al. 1984) and Newfoundland (Mercer and Strapp 1978), have measured compliance levels of 60 and 82 percent, respectively, determined from random mail surveys.

Voluntary hunter responses to mail surveys are used by 14 agencies to help calculate the estimated hunting mortality. Most of the mail surveys are random sample. Questionnaires may be followed up by one or two reminders to achieve a compliance level that varies from 70 percent in the Yukon (Kale 1982) to 70 to 85 percent in Ontario (Timmermann 1987). Results usually provide reliable information to assess trends on a jurisdictional or regional basis, but they often do not supply precise data on a management area or unit basis. This usually is the result of low sampling from hunters who hunted on small management areas (Ontario Ministry of Natural Resources 1980b). Nonresident kill is monitored by an export permit or trophy fee in eight jurisdictions. Field check stations and incentive programs frequently are used to gather additional harvest and biological data.

POPULATION STATUS RELATIVE TO HARVEST

Success or failure of a harvest strategy generally is measured against moose population status when all mortality factors are considered. Continental moose populations are estimated to have increased by as much as 16 percent between 1982 and 1991 (Table 57). Population densities in the early 1990s are believed to be relatively stable or increasing in 17 jurisdictions, stable to decreasing in 3 and decreasing in only 2 jurisdictions (Figure 203; see also Chapter 3). Both

Hunter registration of moose kill is mandatory in many jurisdictions and voluntary in others. Both sources of information are statistically tabulated and computed to assess moose population characteristics and calculate harvest impact and potential. Field check stations used by about half of the wildlife agencies in states and provinces that allow moose hunting commonly gather additional information (age, sex, weight, reproductive tracts, and body and antler measurements) to assess moose population reproductive performance, recruitment, condition, and growth. Hunters and hunting parties are encouraged to submit the lower jaws of moose harvested so that biologists can obtain additional information, primarily age data, about the harvest. A number of agencies award contributors a patch or crest in appreciation. *Left photo by H.R. Timmermann. Right photo courtesy of the Ontario Ministry of Natural Resources.*

Alberta and Manitoba reported that moose were declining 1985 to 1990. Alberta hunting seasons generally are long, with liberal bull license allocations and restrictive cow harvest regulations. Losses associated with the effects of winter ticks, predation and high harvest rates combined with increasing hunter access and use of off-road vehicles are believed to be responsible for lower populations in Alberta (A. Todd personal communication: 1992).

Crichton (1981) and Stewart (1983) suggested that extensive, unregulated hunting by treaty Indians in Manitoba and Saskatchewan was a principal cause of reduced moose populations in those provinces in the 1980s. Both provinces chose to retain an annual recreational hunt by using a mix of limited and unlimited permit (license) harvest strategies. Bull-only seasons were used extensively in Manitoba (Timmermann 1987) to protect a declining herd estimated at 28,000 in 1982 and at 27,000 in 1991 (Table 57). Stewart (1983) reported that positive results were realized in Saskatchewan during the first 3 years after introduction of a selective harvest in 1977. By 1979, the population had increased 38 percent above the 1976 level. However, between 1979 and 1981 the benefits of selective harvest were nullified by other mortality factors and in 1983, populations were 32 percent below the 1979 level. The population in 1991 was estimated at 50,000, up from 45,000 a decade earlier.

The moose hunting season in Nova Scotia reopened experimentally in 1964 after 27 years of closure. The 1964 results were analyzed in 1965 and hunting resumed in 1966; the season was closed again in 1975–1976, reopened in 1977 and closed in 1982 because of apparent population fluctuations (Pulsifer and Nette 1995). Both illegal kill and the effects of brain worm may have contributed to the pre-1982 decline. However, the legal harvest was believed to have had a minimal impact on overall populations (Timmermann 1987). Between 1986 and 1992, 200 licenses were issued annually by lottery for a 1-week season. In 1993, season length was increased to 2 weeks, and hunter success averaged nearly 80 percent.

Yukon and the Northwest Territories reported stable moose populations in the early 1990s, although there were no comprehensive population estimates over this vast area. The Yukon meets the hunting demand by directing most of the harvest at adult males. High calf losses attributable to predation were thought to affect overall productivity. Timmermann (1987) reported no adverse effects on fecundity rates at a bull-to-cow ratio of 30:100 in the southern Yukon.

Numerous jurisdictions—Washington, Montana, Idaho, Wyoming, Utah, Colorado, North Dakota, Minnesota, New Hampshire, Maine and Nova Scotia—limit the number of permits to regulate harvests and sustain moose populations where hunter demand exceeds supply by ratios of up to about 500 to 1 (Figure 222). The management objectives in both Colorado and North Dakota have been to sustain or increase populations and offer a high-quality success averaging 80 to 90 percent. Since 1975, Wyoming's moose population "increased from 6,300 to a high of 13,645 in 1991, surpassing the statewide population objective for 1993 of 12,225 moose" (Hnilicka and Zornes 1994: 105). Moose populations in Minnesota increased 33 percent between 1972 and 1982, declined slightly in the late 1980s and were considered stable to increasing in 1992 (Lenarz 1992). Hunting seasons were limited to every other year, and four resident hunters in a party were permitted to take one moose of either sex (Karns 1972). British Columbia has used the widest range of harvest strategies in an attempt to sustain or increase populations. The results of a selective harvest system initiated in the early 1980s were encouraging according to Child (1983b). The harvest of calves was nearly six times that reported before 1980, whereas the prime bull (older than 5.5 years) kill was reduced by 50 percent. However, reintroduction of postrut bull seasons in some areas in 1986 resulted in more adult bulls than calves killed, as hunters preferred taking adults when given a choice (Child and Aitken 1989).

Alaska and Ontario reported stable to increasing populations in 1991. In Alaska, past attempts to shift recreational hunting pressure from bulls and introduce antlerless seasons met considerable public resistance (Rausch et al. 1974). In Ontario, high bull harvests during seasons that spanned the rut (Timmermann 1987) reduced the proportion of mature bulls and lowered the bull-to-cow ratio to below 20:100 in some areas. Skewed sex ratios resulting from bull-only harvests were a common concern (Bubenik 1972, Baker 1975, Crête et al. 1981a). Alaska has experimented with limited antlerless seasons in some areas where recruitment exceeded preset population objectives to demonstrate that cows can be taken without precipitating a population decline. However, Van Ballenberghe and Dart (1982) argued that bull-only hunting was the only viable option in areas where moose were heavily killed by natural predators. Page (1983) countered with a caution against shooting too many bulls, and suggested limiting total harvests.

Quebec has preferred to adjust harvest regulations to monitored population trends. Limited permits were used only in specific wildlife reserves in Quebec (Desmeules 1966, Bouchard and Moisan 1974, Goudreault 1980) and not elsewhere because of imprecision of census and harvest assessment. Hunting season manipulation and multiple licenses per moose were the principal strategies used to control harvest by the continent's largest moose hunter population (Figure 222). Between 1982 and 1991, hunter numbers in Quebec increased nearly 20 percent, and overall moose populations were thought to have declined about 10 percent (Table 57). In 1994, Quebec introduced a limited quota for antlerless moose in an attempt to increase productivity.

Ontario introduced a province-wide area-specific selective harvest system in 1983 after 3 years of harvest control, using shorter seasons and requiring two hunters to share one moose (Euler 1983a, Timmermann and Gollat 1984). These latter strategies were effective in reducing overall harvests and stabilizing populations. Moose increased from approximately 80,000 in 1982 to 120,000 by 1991. Survey estimates in 1991 suggested that, in a total of 69 wildlife management units, moose were stable in 30, increasing in 17, decreasing in 8 and their status was unknown in 14 (A. Bisset personal communication: 1992).

Newfoundland moose reportedly increased 40 percent from 1973 to 1982 and doubled again by 1991 to a prehunt population estimated at 140,000. That dramatic increase occurred despite an apparent decline as recently as 1981 (Timmermann 1981, S. Ferguson personal communication: 1992). A province-wide harvest quota system initiated in the late 1970s, favorable habitat created by extensive logging and the absence of wolves were considered to be largely responsible for the increase (Figure 229).

THE ECONOMIC VALUE AND IMPACT OF MOOSE

Moose are a renewable resource of significant value. Measuring their net worth is a formidable challenge and has been the subject of considerable misunderstanding. The economic value of moose is the net benefit derived from use, which is considered the difference between what is paid to use the resource and the amount that would be paid willingly for the right to use the resource. The economic impact of moose is measured as the economic activity generated in local economies from use. Thus, economic value and economic impact are separate and distinct measures. Economic value measures the creation of wealth, whereas economic impact measures distributional effects on income flows. Traditionally, the economic value of moose has been confused with economic impact. This misunderstanding is unfortunate in that it has perpetuated the commonly held misconception that economics has little to offer in terms of appraising the relative merits of resource production and allocation of tradeoffs. The economic value of moose can be used in a cost/benefit analysis framework to appraise objectively the welfare implications of alternative resource production and allocation strategies.

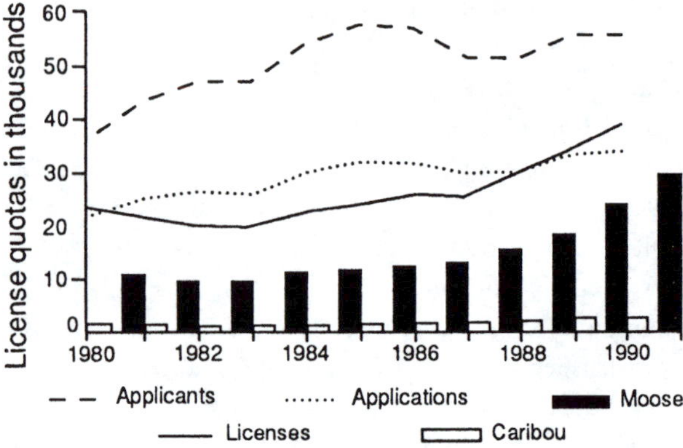

Figure 229. Moose and caribou license quotas for Newfoundland, 1980–1991 (Newfoundland and Labrador Wildlife Division 1991). The significant increase in moose in the 1980s was attributed to the province-wide quota system initiated in the late 1970s, an absence of gray wolves and favorable habitat created by extensive logging.

Although little effort can be attributed to measuring the economic value of moose, significant effort has been directed toward estimating the economic impact of moose hunting. In 1982 and based on extrapolations of average expenditures from several studies, Bisset (1987) estimated the total annual expenditures for moose hunting in North America at $464 million (Canadian) from 430,000 hunters who harvested 71,000 moose. Included were direct expenditures by moose hunters of $148 million, indirect expenditures of $241 million using a 1.63 multiplier and direct value of $75 million from the moose resource itself.

Another hunter survey conducted in eastern North America in 1989 reported an average expenditure per moose of $4,900 Canadian (G. Eason personal communication: 1992). By extrapolation, $4,900 times 88,630—the estimated 1991 North American moose harvest (Table 57)—total economic impact of moose hunting that year was approximately $434 million Canadian. (As a point of clarification, the monies spent on or otherwise associated with moose hunting were not created as a result. Had moose hunting opportunities not been available, the money likely would have been spent on alternative recreational activities elsewhere. The creation of economic welfare is not measured by such impact estimates, which only estimate the diversion of dollars from one area [largely urban] to another [largely rural].)

Techniques exist to quantify the value of moose hunting opportunities and further isolate the contribution generated by attributes (aesthetics, success rate, accessibility, etc.) of the resource. To date, little has been done to quantify economic values. However, the value of meat gained from harvesting moose is one means of illustrating economic benefit. Assuming an average yield of 352 pounds (160 kg) of processed meat per moose carcass (Hamilton 1981) with a value of $4 per pound, the recreational harvest of meat in 1991 represented about $123 million. To assume that the sole or even primary reason for hunting moose is for the meat is unrealistic, so the $123 million amount underestimates the total value of moose. Legg (1995: 4) prepared an analysis of the impact of hunting for large mammals in Ontario: "In 1993, more than 93,400 Ontario residents and nonresidents hunted for moose. In doing so, they spent 713,200 days and more than $57.2 million on activities, supplies, and services directly connected with the activity. Of the total expenditures, 95 percent were spent by resident hunters. Food and lodging accounted for 37.7 percent of the total; transportation was 27.3 percent; new equipment represented 23.8 percent of the expenditure sum. On a per-hunter basis, resident moose hunters spent an average of $598, and nonresident moose hunters spent $1,077 [Canadian]."

Nonharvest Management

Parks, Refuges and Special Areas

Intuitively, many factors appear to contribute to the increase in demand for wildlife-based nonconsumptive recreational opportunities. They include greater public knowledge of and concern for the environment, reduction of opportunities to encounter native fauna in the urban landscapes and the increased mobility of recreational users. In 1984, the total gross value of all moose-related recreation was estimated at $1.6 billion (Canadian) or $1.3 billion (United States) (Bisset 1987), which represents clear justification for commensurate management investment.

The significance of moose as an aesthetic and cultural resource has been cited by numerous writers, including Lime (1974), Bisset (1987) and Wolfe (1987). The symbolic value of moose as the largest member of the deer family and its association with wilderness areas elicit special attention. However, attributing economic value to moose, specifically nonconsumptive expenditures, is difficult.

Nearly every jurisdiction with moose in North America provides for areas where recreational hunting is not a primary management objective. These areas are found in wildlife reserves, national, state and provincial parks, and other designated nature preserves. Although few agencies have policy statements regarding moose management in protected areas, the assumed common objective is the preservation of native fauna in representative natural habitats, with a minimum of human disturbance, for the education and enjoyment of future human generations. This objective is consistent with the broader ecological goal of maintaining biodiversity and ecosystem function. However, the often-stated benefit resulting from harvest protection is provision or improvement of viewing opportunities for the public, although neither is, in itself, a consequence or product of protection in modern management context.

Only 8 of 22 jurisdictions have given special management consideration or developed special objectives for proactive management of moose populations in protected areas. In 9 of the 22, provision of viewing opportunities and natural history interpretation are within the purview of their park programs.

Not all jurisdictions prohibit hunting in parks and reserves. Manitoba and Saskatchewan, for example, conduct controlled annual moose hunts in many of their provincial parks. Recreational hunting is regulated in Ontario provincial parks by the process of master planning and thus, is specific to the park's classification and wildlife management objectives. In the case of recently regulated parks, hunting may be allowed in accordance with local historic use until such time as the master plan is formalized. The planning process evaluates park features, conservation needs and public expectations to establish activities that are consistent with the park's role. Currently, only two Ontario parks provide moose hunting opportunities that might be considered significant contributions to the hunting total. Hunting also is allowed in numerous waterway parks that vary in width up to 200 yards (183 m) either side of designated waterways. British Columbia allows moose hunting in provincial parks, but has protected zones within some to promote and accommodate viewing opportunities. The rationale in such cases has been to manage the moose population to achieve broader objectives, including nonharvest.

Lime (1974) suggested six techniques to improve opportunities for park visitors to encounter moose: (1) enhancement of the moose population through habitat manipulation; (2) provision of multimedia interpretive information that aids in locating moose; (3) development of self-guiding trails and/or auto tours in areas with a high probability of sighting moose; (4) encouragement of entrepreneurship related to moose viewing; (5) development of artificial attractants, such as salt licks, to localize moose at advantageous view sites; and (6) erection of viewing platforms.

A number of these techniques have been applied to the management of moose-viewing opportunities in protected areas. Many agencies actively promote viewing through interpretive programs and educational material available within their park systems. Much of the promotion of viewing is done in relation to summer aquatic feeding areas and peak visitor periods (Ritcey and Verbeek 1969, Cobus 1972, Fraser and Hristienko 1983). Some agencies, as in Manitoba, have actively managed sites to enhance viewing opportunities by creating artificial salt licks and constructing viewing towers near use areas.

In the past several decades, park managers have begun to recognize the actual and potential impacts of large herbi-

Wildlife photography and viewing are major uses and values of wildlife resources. These "nonconsumptive" amenities of wildlife translate into tourist dollars and more so, public interest in management that avails viewing and photographic opportunities. Clearly, the best public relations a wildlife agency can have is the presence of healthy animals. In parks, preserves and reserves where viewing of wildlife is a primary visitor incentive and a management objective, people must be reminded constantly of the significant threat wild animals pose to those who approach too closely or in a manner the animal may deem aggressive. Even tame bull moose, as top left, should be given a wide berth during the rut, as should any cow with a calf anytime. With technological advances in telephoto optics, now available at generally affordable prices, there is almost no reason for amateur photographers to approach wildlife subjects to a closeness that invites aggression or at least harasses the animals. Professional wildlife photographers, who are intimately familiar with the behavior of wildlife and intimately know the capabilities and limitations of their photo equipment, maintain a safe, respectful and reasonably "natural" distance from their subjects. Shown in the bottom scene is the late Bill Ruth of Denali, Alaska, a professional wildlife photographer, "shooting" moose during the autumn rut. *Top left photo by Charles C. Schwartz; courtesy of the Alaska Department of Fish and Game. Top right photo by Luray Parker; courtesy of the Wyoming Game and Fish Department. Bottom photo by Tom Walker.*

The art of calling moose during the rut by vocal calls, simulated thrashing of brush and antler clashing (left) can provide a unique opportunity to photograph and view moose. However, caution is advised that bulls are very aggressive at this time and such luring may not be legal in some localities. The same tactics are used by some hunters. Because of the particular susceptibility of bulls to harvest during the rut, because these animals move about extensively and are preoccupied with mating, most jurisdictions at least occasionally adjust their hunting seasons to protect bulls before the rut, to prevent overharvest. *Photo courtesy of the Ontario Ministry of Natural Resources.*

Moose viewing opportunities can be enhanced by establishing artificial salt licks (left) and constructing viewing towers, such as on Hecla Island, Lake Winnipeg, Manitoba (right). Also, an area on the Kenai Peninsula of Alaska has been managed by the Kenai National Wildlife Refuge for moose viewing. Parts of the area have been mechanically rehabilitated and others burned to create habitat, attract moose and provide better viewing. *Left photo by G. Lunney; courtesy of the National Archives of Canada. Right photo by Vince Crichton.*

vores on other natural features, such as vegetation and the abundance of large predators within parks (Crichton 1977, Diamond 1992, U.S. Department of Interior 1992). Park managers also must deal with restoration or maintenance of natural communities or even ecosystems in which fluctuations of moose numbers are affected by human-related impositions, such as forest fire suppression, logging, interspecific competition and extirpation of natural predators (Blood 1974), as well as inevitable effects of forest and range succession. Park rangers face a considerable challenge—biologically, socially and sometimes politically—to manipulate large herbivores and their habitat where regulated harvest or culling is prohibited. Such matters are exacerbated when managers must deal with animal rightist issues that focus on individual animals rather than populations or ecological systems.

Large state, provincial and federal parks, reserves and preserves contribute much to understanding of moose ecology, by functioning as ecological laboratories. Both the United States and Canadian federal governments recognize the research opportunities offered by unhunted populations in parks where large-scale habitat alterations have not taken place and "natural" ecosystem functions are allowed to proceed. Isle Royale National Park in Michigan, with a long history (beginning in the late 1950s) of basic ecological research on moose/wolf/habitat interactions, is a prime example (Jordan et al. 1973, Peterson and Page 1983). Denali National Park and Reserve and the Kenai National Wildlife Refuge in Alaska have been sites for extensive monitoring of moose numbers and related moose research (Bailey 1978, Bailey and Bangs 1980, Van Ballenberghe et al. 1989, Miquelle 1990b, Van Ballenberghe 1992). Larger provincial parks, such as Algonquin in Ontario, also have served as an outdoor laboratory for moose research (Ontario Ministry of Natural Resources 1989b, Addison et al. 1990, Wilton and Garner 1991, Forbes and Theberge 1992, Garner and Wilton 1993). In addition, Algonquin has served as a source of moose for reintroductions to former habitats (Aho and Hendrickson 1989). Although ecological research and monitoring do not constitute management per se, their results provide comparative benchmarks for management strategies applied to exploited populations and are indispensable

Table 62. Estimated moose populations in certain United States and Canadian national parks (after Timmermann and Buss 1995)

State or province	Park	Population	Year of survey	Estimation method
Alaska	Denali	2,000	1990	Aerial survey
Alberta	Beringland Bridge Preserve	200–400	1980s	
	Jasper	100–150	1992	Ground survey
	Prince Albert	975	1990	Aerial survey
	Watertown Lakes	50	1988	Aerial and ground surveys
	Wood Buffalo	1,300	1989	
British Columbia	Kootenay	<75	1985	Canadian Wildlife Service biologists
	Mount Revelstoke and Glacier	15–20	1991	Sightings and tracks
Manitoba	Riding Mountain	3,066	1992	Aerial survey
Michigan	Isle Royale	2,400	1994–1995	Aerial survey
Minnesota	Voyageurs	25	1987	
Montana	Glacier	100	1985	
New Brunswick	Fundy	123	1993	Aerial survey[a]
	Kouchibouguac	110	1995	Aerial survey[a]
Newfoundland	Gros Morne	7,738	1995	Aerial survey[a]
	Terra Nova	170	1993	Aerial survey[a]
Nova Scotia	Cape Breton Highlands	2,052	1993–1994	Aerial survey[a]
	Kejimkujik	10–15	1992	Guesstimate
Ontario	Pukaskwa	379	1990	Aerial survey
Quebec	Forillion	75	1990	Aerial survey
	La Mauricie	212 (3.9/10 km)	1989	
Wyoming	Grand Teton	120	1988	
	Yellowstone	200	1990	Aerial survey
Yukon	Kluane	316	1991	Aerial survey

[a] See Corbett 1995.

in gauging management results with naturally regulated populations. Furthermore, basic ecological research conducted on moose in protected areas contributes to overall knowledge about ecosystem functions in environments relatively unaffected by human disturbance.

Both Canadian and U.S. national parks services indicate that "protection" and "preservation" of native fauna and flora and natural processes of population regulation are paramount to their management policies (Masyk 1975, Canadian Parks Service 1979, U.S. Department of Interior 1988, Ontario Ministry of Natural Resources 1989b). To deal with maintenance of genetic diversity, policies have been developed that consider culling and/or introductions of new genetic types to restore such diversity (U.S. Department of the Interior 1988). For example, the Canadian Parks Service culled 1,130 moose from the totally enclosed and predator-free Elk Island National Park in Alberta between 1959 and 1973 (Blood 1974).

In response to statutory mandates to ensure the maintenance of wildlife populations within national parks, moose populations are monitored by a variety of estimation techniques (Table 62). Also, research being conducted in large national parks may necessitate population estimates.

Although the role of protected areas and populations still is being explored and defined, there is little doubt of their value in the maintenance and understanding of ecosystems. This value becomes more significant as development encroaches on remaining wilderness areas. Given that management of protected areas is just over 100 years old in North America, the opportunities and complexities offered by them have just begun to be recognized and appreciated fully.

ANTLER SHAPE
(LIFE)
BROWSING ON WILLOWS-
(MEMORY) - TAIL APPEARED TO BE
BURIED IN FAT OR FUR - HINDQUARTERS &
PELVIS RELATIVELY NARROW, IN CONTRAST TO
BULGING BELLY - NOTE UNDER PLANE OF
BICEPS FEMORIS, ALMOST PART OF PLANE
OF LOWER BELLY OR FLANK -
MOOSE LYING DOWN - HIND
LEGS BENT, THEN FORELEGS
DOUBLED UNDER, ONE AT A
TIME - HINDQUARTERS LAST
PART TO SETTLE
3 BULLS BROWSING
OR RESTING IN WILLOW
CLUMPS & SCATTERED
SPRUCE ON
SLOPES ABOVE
IGLOO
CREEK -
JULY 28, '56
McKINLEY
PARK,
ALASKA
HEAD TILTED
SLIGHTLY WHEN
TURNED
(MEMORY)
DEEP PURPLE-BROWN OR BLACK
ON UNDERPARTS, ETC. -
UPPERPARTS LIGHT, DULL
OCHRE-BROWN WITH
DARK FADING OUT TO
OVERLAY OF DARK-TIPPED
HAIRS -
(LIFE & MEMORY)
(DRAWN IN FROM LIFE -
CUD-CHEWING MOOSE -
COMPLETED FROM MEMORY)

18

VINCENT F.J. CRICHTON

Hunting

Much that has been recorded in the popular media about moose hunting has been an introduction of the novice hunter to the recreation and a limited source of pointers to the experienced moose hunters. The presentations provide some entertaining reading or viewing, but most offer little in terms of information about the species or ways to improve hunting ability, capability and image. Notable exceptions are two publications produced jointly by the Ontario Ministry of Natural Resources and the Ontario Federation of Anglers and Hunters (Buss and Truman 1990a, 1990b).

The intent of this chapter is twofold. The first objective is to provide readers a better understanding of the total moose hunting experience. The second is to encourage the reader to become a knowledgeable and active participant in local, provincial or state moose management programs, regardless of whether he or she chooses to hunt moose, and to appreciate the many uses and values this renewable resource represents.

Much of what is presented and discussed throughout this book deals with moose biology, ecology, habitat and management. A crucial element that seldom receives adequate attention because of its sensitive and highly dynamic nature is biopolitics—not only moose management but in all aspects of environmental management. Sooner or later, most natural resource users and enthusiasts will be faced with the realities of the biopolitical spectrum.

In particular, moose hunters must attempt to understand some of the subtleties of the relationship between their recreation and the intricate blend of art and science that moose management has become. Hunters and other conservationists, like moose biologists, will not necessarily agree on what is done to manage moose best in particular instances or places, but to assure that favorable management results are achieved, they should be aware of and familiar with the decision-making process and the factors that influence that process.

Hunter and Public Education and Communication

Manager Responsibility

All too frequently, wildlife managers have failed to communicate with the public about moose biology, management and habitat issues, and to espouse the values of the moose resource. Nearly as much as actual management of the resource, such communication is a role and responsibility of the manager.

McKenna and Lynort (1984) observed that only within the past decade or two have wildlife professionals begun to deal with private conservation groups regarding environmentally damaging projects and sensitive issues. Beyond the responsibility of communicating openly with the public, to inform them of circumstances, outlooks and management objectives, managers need to communicate often and effectively to enlist informed alliances for management programs and initiatives in times of resource difficulty as well as times of plenty. Also, sharing information and

A moose hunting camp on a remote northern Manitoba river. Proper planning can make a tent a comfortable home for an extended period. It is important to have an ethic about the camp. Once the tent is down, all evidence of human presence should be eliminated. All combustible material should be burned; biodegradables should be disposed of away from the campsite; and nonbiodegradables carried out for recycling. The area should be left so that someone to follow will likely select the same site. *Photo by Vince Crichton.*

expertise with the public has an unrealized potential in that they can ask, do and say what civil servants cannot (Crichton 1987a). The importance of this issue is illustrated by the fact that the 1984 Federal-Provincial Wildlife Conference in Canada adopted "Communicating about Wildlife" as its 1985 theme.

Mahoney (1983) suggested that biologists/managers must become better readers of the public mind and better registers of politics of the wildlife profession. Moose management programs will only be as effective as the public support they receive. Mahoney added that biologists must become better communicators and more articulate politicians. The benefits include enhanced personal and professional credibility and heightened profile and security of management programs.

Politically motivated decisions that diminish the capability of biologists to communicate reflect lack of recognition or acknowledgment of the public as the rightful owners of the resource (Crichton 1988a). Wildlife managers should view themselves as businessmen or businesswomen with a product and market it effectively. There is public demand for the product, especially charismatic big game. The fact that 86.2 percent of Canadians indicated that maintaining abundant wildlife was important to them suggests there is a broad base of support for wildlife conservation in Canada (Filion et al. 1991). In the United States, 48 percent of 225.5 million citizens 6 years of age or older spent more than 1 billion days and expended nearly $60 billion to participate in wildlife-associated recreation in 1991 (U.S. Fish and Wildlife Service 1993). Those figures indicate a strong public interest in and enthusiasm for wildlife in the other nation within North American moose range.

Hunter Responsibility

Hunters are a minority group in North America and the antihunting movement is well-established and well-funded. To protect against criticism of hunters, hence hunting itself, hunters need to be proficient, safe and socially responsible. Failure to adopt these traits eventually but invariably provides fodder for those who object to hunting, regardless of the biological, ecological, economic and recreational ramifications. Two papers by Causey (1989, 1993: 82) are recommended reading for all hunters and wildlife managers. The essence of those papers is a defense of hunters, based on their minimum environmental impact compared with that of other human activities. Causey commended hunters' intimate knowledge of natural systems and of nature's rhythms, and their rightful claim to understanding and practicing "bioregionalism long before it became a buzzword of the environmental movement." However, having had extensive experience with hunters and hunting groups, I suggest that some hunters tend to forget about or ignore other resource users, future users and the legitimacy of aboriginal use. They do not recognize the wilderness experience and the satisfaction derived from observing moose and other wildlife interacting with their environment.

In any philosophical discussion of hunting, the notion and practice of fair chase must be included. Moose have not changed for hundreds of years. But the technology to pursue them has changed, and dramatically so, within only the last three decades, with the popularization of such introductions as snow machines, all-terrain vehicles, high-powered firearms, improved ammunition and scopes. These advances and more disposable income make human hunters highly effective "predators." To protect hunting as an ethical and socially acceptable recreation, this technology must not be used in a manner that reduces hunting to an efficient exercise in killing. To do so would be to sow seeds of disrespect for the resource, for fellow hunters and for the legitimacy of other resource uses. For most hunters, the primary pleasures of and motivations for modern recreational moose hunting come not from the kill, but rather from appreciation of nature, comradeship, and the use of hunting skills (Stankey et al. 1973). Success or failure of the harvest influences the type and degree of additional satisfactions experienced by the hunter. For example, as probability of success declines other satisfactions such as general outdoor enjoyment and environmental amenities may be heightened.

Aldo Leopold (1949), widely considered the "father of modern wildlife management," suggested five components or levels of hunting satisfaction, in the following order:

1. A sense of husbandry through application of land management;
2. A perception and understanding of principles of ecology;
3. Simple pleasure of breathing fresh air and having a change of scenery;
4. Feeling close to nature; and
5. Pursuit of game with their associated symbols of achievement.

Every moose hunter knows instances where the actions of fellow hunters do not live up to the opinions or expectations about hunting as espoused by Leopold. But those five components of hunting, I believe, are embraced by a majority of hunters, if not a vast majority. That the misguided and irresponsible actions of a few hunters all too frequently are reported in the press, to the exclusion of mention of ethical hunting and hunters, is unfortunate but predictable. The negative, often sensationalized coverage has tended to implicate all hunters and hunting and caters to those who have political and philosophical agendas opposed to recreational hunting and/or contemporary moose management programs. In large measure because North American society is growing, becoming more urbanized, vocationally specialized, many people are distanced from and unassociated with the rhythms of nature (Wildlife Management Institute 1992). For them, predation in any form is unwarranted violence—an act of inhumanity. Such a view is at odds with the environment, with wildlife science and with hunting, recreational or otherwise. It is anthropocentric naivete, but it also is increasingly prevalent. Heberlein (1991) suggested that the net effect of all forces acting on hunting is that hunter numbers will decline and sport hunting eventually will be viewed as an antisocial act among the most numerous groups in society. He suggested that there is good news in this in that more is not always better. A smaller number of dedicated and knowledgeable hunters may help to improve the recreation's public image and hunting quality and somehow continue to provide the funding base that serves nearly all wildlife management programs.

In the short term at least, hunters need to be cognizant that hunting is an increasingly sensitive social issue, and that its continuation as recreation, management tool for game populations and primary funding base for game and nongame programs alike is and will be predicated foremost on the legal and ethical behavior of hunters themselves.

Ethical hunting behavior is conduct within the parameters of law and in accordance with widespread views of what is collectively respectful of the wildlife, the landscape, the landowners (public or private), the recreation and other hunters, other users of the landscape and the general public.

Ethics change as people's attitudes and societal expectations change in all areas of life (Horwath 1990). Those who hunt usually go through several dispositional phases. The first is the "shooting stage," in which the hunter's primary interest is maximum shooting. Next is the acquisition stage, in which a maximum limit or take of game is of paramount concern. Then there is the trophy stage, in which satisfaction is maximized by a particular type of hunting method, species or animal. Finally, there is the experiential stage, in which one or more aspects of the occasion—camaraderie, tradition, nature appreciation, etc.—supersede the importance of the chase or kill. Most hunters progress through one or more of those phases, as age, experience, physical conditioning, self-confidence, peer pressure and attitudes dictate. At each stage, ethical standards—usually self-imposed—are increasingly refined and strict.

Although mainly universally applicable, hunting ethics can be specific to the type of hunting (e.g., rifle versus muzzleloader, waterfowl versus moose, wilderness versus shooting preserve). They also tend to be even less flexible than hunting laws and regulations. The balance of this section addresses ethical considerations in moose hunting. The ethical behavior of the moose hunter may not be substantially different from that of the deer hunter or upland gamebird hunter, but it must be adapted to the uniqueness of the species and the species' habitat.

In the establishment of a moose hunting camp for an extended hunt, live trees should not be destroyed. Dry, dead trees make better firewood, tent poles and meat poles. The campsite should be kept free of debris and ultimately left clean with all evidence of occupation eliminated except for tent poles and a replenished woodpile. Tent poles should be leaned against a tree or rock where they will remain comparatively dry and available for future use. All combustible material should be burned. Biodegradable items can be placed in the bush some distance from camp. Tin cans should be washed out, compressed and carried away for recycling; burying is an option, but a poor one.

For moose hunting on private land, permission should be sought and secured well in advance of the season and in full compliance with provincial, state and local access/trespass statutes and landowner wishes. Reporting to the landowner upon departure will go a long way to ensuring his or her accommodation toward hunters and hunting.

There is nothing more upsetting to a hunter seated at a chosen site or more disruptive to a hunt than to have another hunter drive through the bush on an all-terrain vehicle or snow machine. All motorized vehicles should be confined to roads and designated trails. Similarly, hunters traveling by water should paddle or row, rather than motoring past places where others are hunting. Also a motorized vehicle or craft should never be used to haze or stalk moose. Roads and trails can be used to reach a hunting area, but not the specific location for a hunt. In addition, a vehicle should not be parked such that it impedes other traffic.

Gut piles in highly visible places are an eyesore and offensive to others, particularly nonhunters. Wolves, bears, ravens, eagles and other animals eventually will eliminate the remains, but that fact should not preclude sensitivity in making sure those remains are not left where they can invite criticism.

Firearms should be legally and safely cased and kept inconspicuous. Prominent display of firearms is offensive or threatening to some people, and it can invite theft. In Canada, display of a firearm in an unattended vehicle is illegal, and a vehicle or any part thereof containing a firearm must be securely locked.

Respect absolutely must be shown for others' equipment and property. Hunting must not be done in posted logging areas. Also report sign damage to the authorities or landowners.

The sighting-in of rifles should be done away from the hunting area, preferably at a certified range, and on an annual basis before the hunt.

While hunting, animals and other objects at a distance should be viewed with binoculars or spotting scopes rather than with rifle-mounted scopes, because the object might be human.

Most important, all rules of safe firearm handling and hunting must be observed by everyone. A good policy is to have each party appoint a firearm safety person. This individual can remind each hunter to double-check firearms upon return to camp or to vehicles. Hunters must not be timid about informing others (albeit politely) when their behavior is out of line and unacceptable. This applies as well to the use of alcohol before or during a hunt. A hunter must know his or her physical limitations, and conduct themselves in a fashion that will not burden others.

Hunters should notify others of when, where and how long they will be hunting a particular area. A note left readable inside a vehicle is a good practice. Most hunters also should know how to use a compass and orient themselves by a topographical relief map kept in the hunter's possession at all times. Of course, the latest technology—namely geographical or global positioning systems (GPS)—is a useful tool for finding one's way back to vehicles or camps, but hunters must become familiar with these expensive units before using them in the field, where a mistake could result in serious consequences. A knowledge of first aid is highly desirable.

Open display of carcasses during transport impresses no one, and it is disrespectful of the animal, other people and the sport. Few other ethical violations draw as much criticism from nonhunters. Persons who, by action or conversation in public, need to validate their self-professed hunting prowess should be ignored and ostracized. Hunters invariably enjoy the recounting of their experience, but such "sharing" must be done with propriety. Machismo displays and vocalizations merely serve to prove the contrary.

Moose hunters should learn all they can about the habits and life history of their quarry. A knowledge of current moose management objectives and the reasons behind

In the not too distant past, moose hunters commonly displayed their kills openly when returning from the hunt. Today, however, such conspicuous display is offensive to many people, particularly those opposed to hunting. Hunters need to refrain from actions that can reflect negatively on the sport and on hunters themselves, regardless of how innocent the motivations. *Photos by Rhys Beaulieu.*

changes in harvest regulations will enhance the total hunting experience and, through conversation, help others to appreciate the sport. Hunters should not hesitate to ask questions of conservation agency officials and register their thoughts and ideas.

Shots must not be taken at moose under any conditions or circumstances that would prevent a quick, humane kill. Long shots, running shots and obscured shots are not to be taken. Hunters should know, from practice, the capability and limitations of their firearms and ammunition. They also need to know where to aim for killing effect. Even so, wounding is possible. Therefore, moose hunters must know and be fully willing to track a wounded animal. They also should be willing to help others locate such an animal. And in all aspects of the hunt, experienced hunters should give advice, assistance and encouragement to novice hunters.

Hunters who are fortunate enough to down a moose must be prepared to dress out the carcass efficiently and expediently, to avoid waste of meat or other usable parts. This entails knowledge of field-dressing techniques, use of proper equipment and a great deal of energy.

Moose hunters should cooperate fully with wildlife management authorities in completing questionnaires, answering questions in the field or submitting biological specimens, even voluntarily. Management of moose, to the benefit of hunting, is a dynamic science that depends to an important extent on information and cooperation from hunters. In the absence of data or cooperation, management is complicated, and the first casualty is quality hunting.

Ethical hunters have no reason to be ashamed of or apologetic for pursuing their chosen outdoor activity. They should, however, assume a greater role in moose management by showing high regard for the law, respect for wildlife, the environment, fellow hunters and the general public, and by keeping abreast of wildlife programs and policies.

Moose Hunting and the Law

Recreational hunting in North America is a privilege, not a right. All jurisdictions annually promulgate laws and regulations that specify the parameters within which moose hunters must conduct themselves. The intent of regulations is to ensure that the moose resource is used and conserved wisely. The laws enacted are designed to help ensure an equitable distribution of hunting opportunity, prevent inhumane treatment of wildlife, protect private property and provide for safety of hunters and other outdoor enthusiasts (see Chapter 17).

Penalties for breaking wildlife laws and regulations range from a "slap on the wrist" in some jurisdictions to heavy fines, confiscation of hunting equipment and game, termination of hunting privileges for extended periods of time and, in some jurisdictions, restitution fees. Some violations are deliberate; others result from lack of knowledge about provincial or state laws. It is imperative that hunters take the time to read hunting brochures and understand the pertinent rules, especially changes from the previous year. If necessary, hunters should phone the management authority to clarify policies and laws that may be vague or confusing. For example, in designated route areas in Manitoba, hunters cannot take a vehicle off of such routes until a kill has been made. However, for hunters who have licensed camps in these areas, permits can be issued to travel by vehicle to and from the camp during the hunt even if it is off the route.

Hunt Preparation

General

Preparation for a moose hunt involves five different types of equipment—firearms and ammunition, clothing and personal supplies, carcass-handling devices, camp gear, and food. Table 63 lists equipment for a four-person fly-in moose hunt for 6 days.

Firearms and Ammunition

Proper preparation for a moose hunt requires the right type of firearm and ammunition. In most jurisdictions, firearms for moose hunting refer to most centerfire rifles, muzzleloaders, and even bows and certain handguns. For moose hunting, minimum calibers or pull weights are prescribed (Table 64). Also, most jurisdictions regulate projectile specifications to ensure adequate killing power and to minimize wounding losses. Most prohibit rimfire cartridges, which lack size, muzzle velocity and impact energy to produce a clean kill.

The firearm projectile kills by doing one or a combination of actions, including tissue (bone) destruction, hydrostatic shock and hemorrhage. A lethal shot is one that damages an organ essential to maintaining life (heart, lung, liver, brain, etc.), severely fractures a portion of the skeleton inhibiting locomotion, or produces massive hemorrhage resulting in major blood loss externally or internally. To bring down an animal the size of a moose, a firearm must fire a

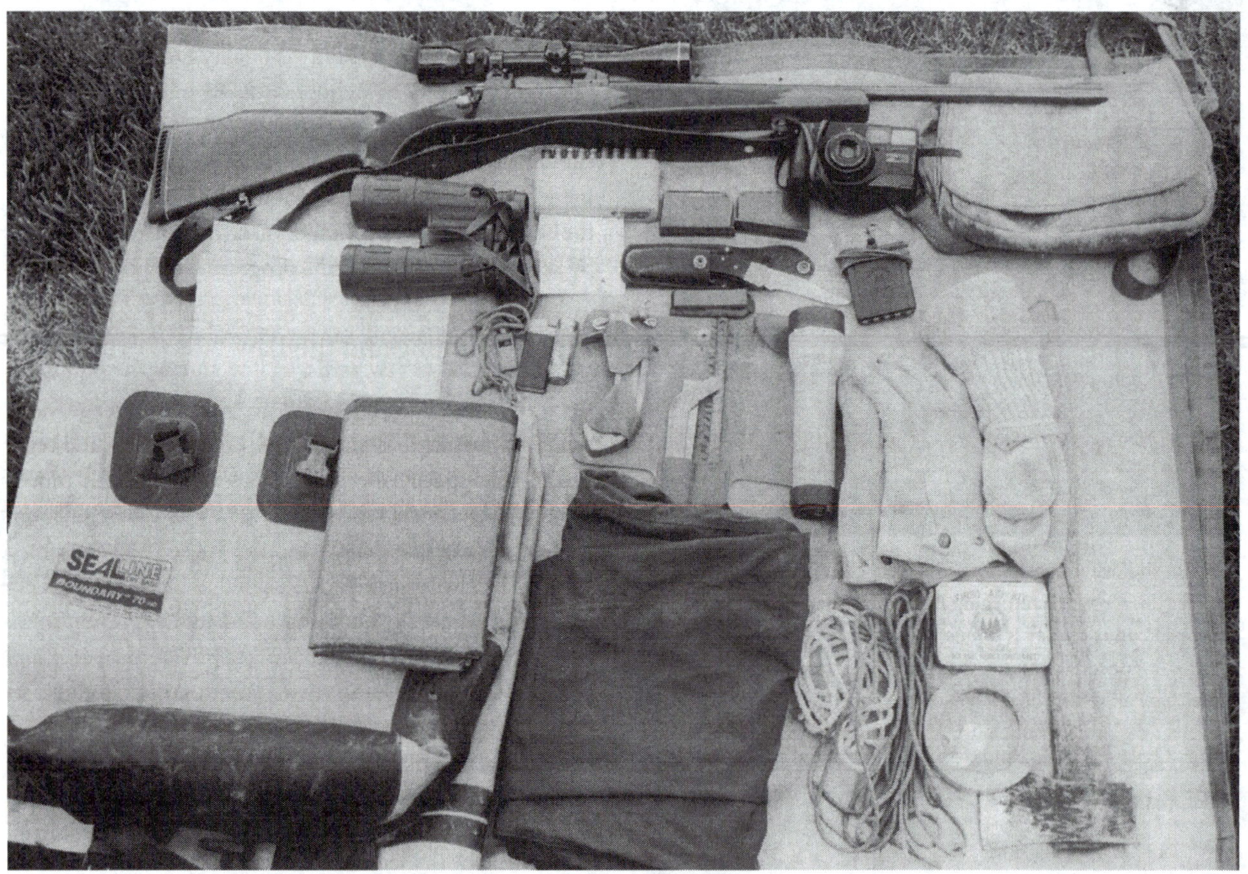

Essential gear for a day-hunt for moose includes a firearm, gloves, knife, binoculars and a waterproof knapsack containing ammunition, spare ammunition clip, matches in waterproof container and/or lighter, flagging tape, rope, small axe, sharpening stone, first aid kit, flashlight, camera and film, topographic map, compass, extra socks, hunting license/tag, survival (space) blanket, snacks, and a container of drinkable water. *Photo by Vince Crichton.*

Table 63. Recommended logistics for a 6-day, fly-in, four-person moose hunt

General
- Preparations should begin *at least* 6 weeks before the hunt
- Secure all licenses and permits
- Review all pertinent regulations
- Obtain topographic maps (one for each hunter) of the area to be hunted
- Sight-in firearm with ammunition to be used on hunt
- Clean and waterproof boots, tents and other gear
- Improve physical conditioning
- Make and confirm prehunt travel arrangements, including check on firearm / ammunition casing requirements
- Sharpen knives, saws and axes
- Identify butcher, tanner and taxidermist, as desired, for price list and special handling recommendations
- Acquire traveler's checks / cash
- Replenish propane tanks and other fuel tanks
- Clean and check camp stove and lanterns
- Preplan daily menus

Personal
- Firearm and ammunition
- License / permit
- Sleeping bag, air mattress (or similar type material) and pillow
- Tent and ground cloth
- Repair kit for air mattress and rubber boots
- Toothbrush, toothpaste, biodegradeable soap and other toiletries (no aftershave or scented deodorant)
- Two towels and face cloth
- Hip waders
- Bush boots (leather boots not recommended unless waterproof)
- Camp shoes or slippers
- Two pair of long underwear and suitable summer underwear for 8 days
- Three heavyweight shirts
- One sweater or hooded sweatshirt
- Two pair of insulated mittens or gloves
- Cap or hat (preferably with ear covering)
- Insulated coat, parka or jacket of wool or other "soft" material (not nylon)
- Two pair of pants for bush use (wool or "soft" material) and one pair for camp
- Three pair of wool socks and three pair of undersocks (silk, polypropylene or cotton)
- Warm vest with a soft outer covering (not nylon)
- Sunglasses
- Life preserver
- Raingear
- High-quality compass with lanyard
- Day bag (knapsack)
- Matches, match container and lighter
- Binoculars
- Camera equipment and film
- Pocket first aid kit (Band-Aids, bandages, tape, antiseptic, aspirin, etc.)
- Pocket sewing kit
- Flashlight and extra batteries
- Moose call(s)
- Breakdown fishing gear (no more than two units for a party of four)
- Pocketknife
- Shotgun and shells (one per party)
- Firearm cleaning kit (one per party)
- Blaze orange (or other color) vest, hat or outer garment to minimum specification
- Wind-up or battery-powered alarm clock
- Water purification kit (one per party)
- Small radio (optional) and batteries
- Space blanket
- One medium and one large dry sack
- Pint or quart plastic water bottle
- One bottle of fly dope
- Ballpoint pen and notepad
- Toilet paper, paper towels and facial tissue
- Flagging tape

Field dressing
- Medium-sized hunting knife and sharpening stone
- Light, strong rope (25–50 feet [7.6–15.2 m])
- Nylon string (5–10 feet [1.5–3.0 m])
- Meat saw, with extra blade (collapsible Wyoming saw recommended)
- Small block and tackle and come-along
- Lightweight external frame backpack or stretcher (for carrying meat and gear)
- Cheesecloth or T-shirt tubing sacks
- Large burlap bags (for transporting meat to camp)
- Nylon bags with 1 / 16-inch (0.16 cm) diameter holes (to place around hanging meat)
- Small axe (hatchet)
- Heavyweight plastic sheeting or tarp (to keep rain / snow off meat)
- 3 pounds (1.4 kg) of salt (for hide preservation)

Food
- Eight loaves of bread
- Pancake mix
- One bottle of syrup
- Five dozen eggs
- 4 pounds (1.8 kg) bacon wrapped in 1-pound (0.45-kg) packages, and an equal amount of breakfast sausages wrapped similarly
- 3 pounds (1.4 kg) butter or margarine (margarine remains soft in cold weather)
- Jam or jelly
- Sugar
- Tea bags
- Coffee
- Canned milk / Coffee-Mate or hermetically sealed whole milk for tea and coffee
- Four packs of orange juice crystals
- Luncheon meat—about five tins of packaged meat (luncheon foodstuffs can be minimized if the party opts for brunches rather than breakfasts on some days)
- Mustard and ketchup
- Salt and pepper
- 4 steaks (one meal)
- 8 pork chops (one meal)
- Premade chili (one meal)
- Premade stew (one meal)
- 4 pounds (1.8 kg) of burger in 2-pound packages (two meals)
- Premade spaghetti sauce and spaghetti (one meal)

Continued on next page

Table 63 continued

- Food for one other major meal—optional—perhaps plan on catching some fish
- Salad makings (two heads of lettuce, eight tomatoes, three cucumbers, green onions, etc.) and dressing
- Two dozen chocolate bars
- 12 oranges
- 15 apples
- Large sack of dried fruit (raisins, apricots, prunes, etc.)
- 15 small commercially prepared puddings
- 2 or 3 pounds (0.9–1.4 kg) of cookies
- One box of Minute Rice
- 16 potatoes
- Frozen vegetables for eight suppers
- One container of cooking oil
- Premixed ingredients for bannock
- 30 cans or eight plastic containers of soft drinks
- Powdered fruit drink (e.g., Tang)
- 4 pounds (1.4 kg) of munchies (chips/peanuts)
- 4 pounds (1.4 kg) of trail mix or gorp
- 2 pounds (0.9 kg) of cheese (in a block)
- Two boxes of crackers
- Two or three large onions
- Eight packets of soup mix
- One 10- by 12-foot (3 X 3.6 m) or 12- by 14-foot (3.6 X 4.2 m) dining, storage tent, plus two smaller lightweight tents each to sleep two (larger tent should have a fire-retardant ring for stovepipes and a pad and ring to hang a lantern on; rings and rope on ceiling useful for drying wet clothing)
- One "fly tarpaulin" to cover the tent (lightweight nylon tents usually come with attached fly)
- One 10- by 12-foot (3 X 3.6 m) lightweight tarp (not canvas) for tent floor
- Three 10- by 12-foot (3 X 3.6 m) lightweight tarps for construction of cook "shack" and covering meat
- Two boxes of wooden matches in waterproof containers
- Five plates, cups, knives, forks, small spoons, large spoons
- Two fry pans
- Five cereal-type bowls
- Four bowls for keeping food warm
- Tea and coffee pot
- Spatula and butcher knife
- Three large cooking pots (lightweight, aluminum cooking kit with nesting of various sized pots/pans recommended)
- Plastic wash basin for dishes
- Plastic wash basin for personal hygiene
- 1-pound (0.45-kg) portable solar shower• Large pot for heating water on open fire
- Collapsible 3- to 5-gallon (11–19 L) plastic bucket-bag, with handle, for drinking water
- Water pail
- Shovel
- Grill for use on open fire
- Two-burner camp stove
- Folding camp stove—folded size = 12.5 by 27.5 by 3.5 inches (31.3 X 68.8 X 8.8 cm), weight = 12 pounds (5.6 kg)—with oven and enough stove pipe for use in large tent
- Two lanterns
- 2 or 3 gallons (7.6–11.4 L) of lantern/stove fuel
- Small funnel
- Four spare lantern mantles
- Two axes with protective shields
- File
- Small chainsaw fully fueled and oiled, or bucksaw
- 0.5 gallon (2 L) of gas for chainsaw
- 1 quart (1 L) of chain oil
- Two aluminum square-stern canoes
- Two small outboard engines (5 horsepower maximum), spare plug(s) and shear pins
- 10-gallon (40 L) drum of mixed gas for engines (0.5-gallon [2-L] plastic drink containers are ideal for carrying mixed outboard gas in the canoe and putting in engine)
- Five canoe paddles (one is spare)
- Small tool kit (wrench, screwdriver, pliers, etc.)
- Roll of heavy-gauge vapor-barrier plastic (light, cheap and storable in bush)
- Five garbage bags—burn all garbage that can be treated this way; that which is biodegradeable should be placed in the bush away from camp and tins and bottles are to be brought home for recycling or proper disposal
- Two dish towels
- One dish cloth
- Five scouring pads
- 20 3-inch (7.5-cm) nails and spikes (for hangers and game pole and table construction)
- Roll snare wire
- Roll of duct tape
- First aid kit

projectile heavy enough to break bones and penetrate vital areas, large enough to cause maximum tissue damage, achieve velocity to impart maximum shocking power, expand and not disintegrate on impact.

The potential killing power of any firearm is determined by the impact energy of the bullet. The minimum projectile energy recommended for moose is 2,100 foot-pounds, with 3,500 foot-pounds preferred. To achieve this energy, bullets with 180 grains of powder are recommended. In recent years the .30-06, .308 and .300 Winchester magnum have replaced the once popular .303 and .30-30. Given the aforementioned energy range, calibers from .270 and higher are acceptable for conditions that most moose hunters will experience. Magnum cartridges extend one's range by about 100 yards (91 m), but in the boreal forest where most moose hunting occurs, shooting distances rarely exceed 100 to 150 yards (91–137 m). Shots exceeding 250 yards (229 m) are ill-advised and generally unnecessary.

Bullet weight and construction are extremely important and frequently overlooked considerations of inexperienced hunters. High-velocity, small caliber bullets, such as .243, often will disintegrate on impact and are less effective on moose because of their inability to penetrate, especially when used by hunters without knowledge of proper shot

Table 64. Minimum specifications for firearms, including bows, and projectiles for hunting moose in North America, 1995

	Rifle	Muzzleloader	Shotgun	Handgun	Bow
Province or state	Caliber (projectile)	Caliber (projectile)	Gauge (projectile)	Caliber (projectile)	Pullweight (projectile head)
Alaska	> .22 rimfire	.54, .45 (with 250 grains)			Cast arrow 175 yards (7/8 inch width)
Alberta	.23 (case length 1.75 inches, no full jacket)	.44	28	Prohibited	40 pounds (1 inch width, barbless, 2+ blades)
British Columbia	(no rimfire)		Prohibited	Prohibited	40 pounds (7/8 inch width)
Colorado	.24 (no full jacket)	.50	20 (slugs only)		(7/8 inch width, two cutting edges)
Idaho	(no rimfire)	.50			40 pounds (7/8 inch width)
Maine	.24 (no full jacket)	.50	20 (slugs only)		(7/8 inch width, two cutting edges)
Manitoba	center fire only (no full jacket)	.45	12 (slugs only)	Prohibited	40 pounds (7/8 inch width)
Minnesota	.230 (no full jacket)	.45	(slugs only)	.23 (1.285-inch casing)	40 pounds (7/8 inch width, barbless, two cutting edges)
Montana		.45			(no chemicals, no explosives)
New Brunswick	.23		(slug or BB)	Prohibited	44 pounds (0.8 inch width, barbless)
Newfoundland	.23 (100 grains, no full jacket)		20 (slugs only)	Prohibited	44 pounds (two cutting edges)
New Hampshire	.230 (no full jacket)	.40 single-shot			60 pounds (7/8 inch wide by 1½ inches long)
Northwest Territories	.23 (no full jacket)			Prohibited	44 pounds (1 inch width, barbless, no explosives)
Nova Scotia	.23	.45	28 (slugs only)	Prohibited	50 pounds
Ontario	(no rimfires)			Prohibited	40 pounds (7/8 inch width)
Quebec	.243 (no full jacket)	.50 (ball)	Prohibited	Prohibited	40 pounds (7/8 inch width)
Saskatchewan	.23 (no full jacket)	.45	20 (slugs only)	Prohibited	40 pounds (7/8 inch width)
Utah	.240 (no full jacket)	single-barrel (210 grains, no sabots)	20 (slugs only)		40 pounds (7/8 inch width, barbless, two cutting edges, no explosives, no chemicals)
Washington	.240 (85 grains, no full jacket)	.50 single-barrel (170 grains, no jackets, single projectile)	Prohibited	.24 (750 foot-pounds at 100 yards)	40 pounds (7/8 inch width)
Wyoming	.230 (no full jacket)	.40 (50+ grains black powder)		.41 (500 foot-pounds at 100 yards)	50 pounds, cast arrow 160 yards (1.0 inch width)
Yukon	.24 (no full jacket)	.45 (ball)	20 (slugs only)	Prohibited	45 pounds

Essential equipment for handling a downed moose includes a knife and sharpening stone, saw, axe, camera, tape measure, hipwaders (if moose falls in water), rope, cloth bags for covering meat, and a backpack for transporting the meat and desired parts to camp. Even with all this gear, the best thing to have for field dressing a moose is one or two willing assistants. *Photo by Vince Crichton.*

placement or ability to achieve such shots. Effective bullets are partitioned, expand on impact but do not disintegrate and are capable of penetrating into and through vital organs. Bullet construction is a significant contributor to the effectiveness of any shot. Controlled expansion of the bullet, which transforms the energy of the bullet into tissue damaging and shocking power, is recognized as necessary to produce a quick kill. This is acknowledged by most jurisdictions, and some prohibit the use of fully jacketed bullets, which do not expand and are capable of passing through an animal without doing sufficient damage to kill immediately.

Before a choice of caliber is made, careful consideration should be given the rifle's action—bolt, lever, pump and semiautomatic. The choice is a matter of personal preference, based on handling comfort and efficiency, balanced against such factors as type(s) of use and maintenance (e.g., hunting conditions, durability, repairability). Bolt-action rifles offer the widest range of calibers, but shooting and hunting experience *and* trial with various actions should dictate the decision. The quicker a firearm can be shouldered and accurately aimed, the more confidence the hunter will have in hitting the target. Style and bulk of clothing to be worn while hunting should be factored into determinations of a gun's "fit." An experienced gunsmith can assist with fitting for a gun and firearm adjustments accordingly.

I do not recommend the use of shotguns and muzzleloading firearms for moose hunting primarily because of the significant drop off in energy after 82 yards (75 m). However, hunting with these firearms, especially muzzleloaders, is becoming increasingly popular. Those who choose to hunt in this fashion must be experienced, prepared to work to get close for killing shots and willing to pass up shots at moving targets and animals that are at or near the range limits for these weapons. Regardless of more liberal allowances by some provinces and states, the advised minimum caliber for muzzleloader hunting of moose is .50, and at least 90 grains of powder should be used.

Archery equipment is only effective in the hands of individuals with excellent knowledge of the quarry, the hunting area and limitations of the equipment. The choice of bow type—long, recurve or compound—is a matter of the hunter's personal preference, physical capabilities and experience. Regardless of bow type, the minimum advisable draw weight for moose hunting is 45 pounds. Because the heavy bones and thick hide of moose can slow, impede or deflect arrow flight, hunters should use the heaviest draw weight they can handle and at which they can be consistently accurate. And regardless of bow type *and* draw weight, shots should not be taken at moose more than 25 to 30 yards (22.9–27.4 m) distant and unobstructed. The foremost consideration for shot distance and placement is that the animal must not be wounded and lost. Moose seldom are killed on the spot by an arrow shot, so the stalking, shooting and trailing skills of the bow hunter must be highly refined.

Arrows do not provide the hydrostatic shock associated with firearms and killing is dependent on arrow placement, speed and the cutting effect of the blade(s). Blade sharpness is critical. It is imperative that blades be kept razor sharp—anything less would be irresponsible and unethical. At minimum, arrowheads should have at least two sharp, unserrated, barbless, straight-cutting, steel-edge blades, and be at least 7 / 8 inch (22 mm) at their widest points.

There are no Canadian jurisdictions that allow hunting with handguns. Handguns are legal in only four states and their requirements vary as to caliber and energy. Handgun hunting, particularly for moose, is not for novices with such weaponry, regardless of experience with other firearms.

Regardless of the type of firearm and projectile a moose hunter chooses, he or she must gain familiarity with both, by practice, well before the hunting season. Sighting-in a firearm seldom constitutes adequate practice. Even the most experienced moose hunters are inclined to shoot upwards of 100 rounds on at least two occasions to get the feel of the gun, adjust to its recoil, regain reloading efficiency and improve marksmanship. The firearm and exact ammunition type and load to be used while hunting moose should be used in practice. Also, as much as possible, the same clothing as on the hunt should be worn.

With modern rifles, practice should be done at stationary targets 50, 100, 150 and 250 yards (46, 91, 137 and 229 m) away, to ascertain accuracy, possible mechanical adjustments and gain distance perspective through the scope or open sights to be used in the field. For shotguns and muzzleloaders, recommended practice target distances are 50 and 100 yards (46 and 91 m). For handguns and bows, target distances of 20 and 40 yards (18 and 37 m) are recommended. Whether or not sights need to be adjusted to get a tight, centered shot-group, hunters must know how to adjust them should the firearm scope or sights require alignment in the field.

Also, after the firearm is sighted in, practice shooting should be from prone, kneeling, seated or braced positions, except for archery. One or more of these will simulate shots taken on a hunt, inasmuch as offhand shooting for moose is generally considered bad form and rarely necessary (see Buss 1990, Buss and Richard 1990).

Other firearms equipment a moose hunter should consider includes a sling, scope caps, quality travel and carrying cases, a waterproof ammunition container, a spare clip and a compact cleaning kit, including oil cloth.

Clothing and Personal Supplies

Whether moose hunting is done during the autumn rut period or during winter, as legal seasons permit, hunters should prepare for weather extremes. Throughout the North American moose range, autumn / winter weather is unpredictably variable. The type of hunting to be under-

Before shooting at a moose, a hunter should use a natural rest or assume an often-practiced shooting position. A well-placed shot is likely to put the moose down quickly enough to avoid lengthy tracking and potential loss of the animal. Selection of a firearm and ammunition of adequate killing power and practice with both are essential. Firepower will not compensate for a poorly placed shot. *Photo by Vince Crichton.*

taken will dictate what to wear. Still-hunting dictates a different type of clothing than does tracking or stalking. In situations where a hunter is liable to sweat, some form of fabric worn next to the skin to absorb or "wick" perspiration is recommended. Conversely, still-hunting necessitates clothing with especially good insulative qualities. In any case, layering of clothing is advisable.

The outer shell or jacket should be constructed of wool or other similar fabric that makes little noise when the hunter moves. Outer clothing made of nylon and certain other synthetic material can alert moose to a hunter's presence; moose have extremely good hearing. Wool pants are an excellent choice, because they trap warmth, are quiet and do not readily freeze.

Several pairs of mittens or gloves and boots are essential. For suitable hand warmth, I prefer mitts with wool liners. Leather boots, unless they are absolutely waterproof (and few are), are not recommended as a moose hunter usually (and often) is in wet areas. I prefer insulated rubber-bottomed boots with leather tops. Hip (preferred) or chest waders are a distinct advantage and invariably come in handy if animals are pursued and taken in wet areas.

Many moose hunting seasons are long, and hunters should be prepared for weather extremes. Accordingly, clothing is advisably worn in layers. The technique(s) of hunting will dictate how a hunter should dress. Still-hunting for extended periods necessitates well-insulated clothing. On the other hand, stalking dictates that clothing worn next to the skin have absorbent features. In terms of color, hunter orange is best for safety, and there are some jurisdictions that require outerwear of this color. In Manitoba, for example, one must wear a minimum of 400 square inches (2,580 cm^2) of hunter orange or blaze orange visible on all sides above the waist, and also a blaze orange hat. If uncertain about clothing color requirements, hunters are well-advised to consult with the local wildlife management authority. Bow hunters may choose to hunt in camouflage clothing but when traveling to and from hunting locations, blaze orange should be worn for safety purposes. Raingear is an important element of a moose hunter's apparel. I prefer a two-piece suit made of material that is durable and allows perspiration to escape. There are many different materials that are suitable, and I suggest that economy not preempt quality when it comes to the purchase of raingear.

Other important items to be carried in the field are a high-quality compass, matches and/or a lighter, an axe, extra socks, a map of the area, first aid kit, water bottle (filled), flashlight, lightweight binoculars, and a survival (space) blanket. All of this should fit in a waterproof backpack or fanny pack, along with food for the day. Some hunters choose to carry a camera. Some carry a flare or two, and some tote pepper spray in case of chance encounter with a bear. And, of course, hunters must not forget to carry a properly signed hunting license.

Carcass-handling Devices

To field dress a moose, the minimum amount and type of equipment needed includes a medium-sized (4- to 6-inch [10–15 cm] blade) hunting knife, sharpening stone (knives should be sharpened before going afield), 16 to 20 feet (4.9–6.1 m) of strong, light rope, a meat saw, a compact block and tackle, flagging tape to mark the trail to a downed animal and a pack frame. Field dressing technique is discussed later.

Various devices, ranging from lightweight wheeled vehicles to stretchers, have been constructed to ease removal of carcasses from the bush.

Cheesecloth is useful for protecting meat, as are paper shrouds (specifically designed for covering meat). Instead of cheesecloth, which frequently tears and through which flies can penetrate, I prefer T-shirt tubing that stretches. This can be made into bags 1.5 yards (1.4 m) in length to cover each quarter totally.

For hanging moose quarters, I construct loose-fitting 1/16th-inch (2.5 cm) nylon mesh bags measuring 2 yards (1.83 m) wide by 1.5 yards (1.37 m) deep. The mesh is sown along both sides, with Velcro™ on the top. Then, this "sock" can be readily pulled over quarters hanging on a meat pole, and the Velcro™ fastened around meat hooks or ropes holding the quarters. This has the important advantages of letting air circulate but keeping flies away from the meat.

Camp Gear

If camping is part of the experience, a small air-tight heater or folding cook stove is useful for warmth, drying clothes and cooking. Tents should be modified to accommodate such a heater.

Tents should be high quality. Cheaper, same-size models are available, and many a hunter has come to rue the day he economized on this equipment. High-quality tents are those designed and tested for worst-case weather conditions. They should include a fly, ground cloth and, preferably a vestibule or awning.

Like tents, sleeping bags should be weather-rated for as cold as the hunter is likely to encounter. And in the absence of cots or other bedding platform, a thermal mattress is necessary.

A second cook stove, one lantern per three hunters in camp, a camp shovel and wood saw, metal grill, waterproof containers for extra gear, foodstuffs and cookware are ad-

A lightweight stretcher (left) constructed by the author is of great value in removing moose meat from the bush, especially in rough terrain (right). The stretcher weighs only 8 pounds (3.6 kg) and is 2 feet (0.6 m) wide by 6 feet (1.8 m) long, fastened together with carriage bolts. Careful and clean handling of the meat in the field will provide a quality product with excellent taste and nutrition. *Left photo by Vince Crichton. Right photo by Dana Slusar.*

visable equipment. Also, many experienced moose hunters carry and store extra gear and clothing in rubberized, waterproof tote bags designed specially for cold and wet weather.

A separate container at camp should contain extra matches, an emergency flahshlight, flares, area map, first aid kit and emergency rations. Other items that should be on hand include water purification device or tablets, extra flashlight batteries, gallon-size plastic ziplock bags, toilet paper and heavy-duty trash bags.

A small, collapsible stove for heating and cooking is a valuable asset on extended moose hunting trips. In addition to the firebox, the stove above has an oven. Fully erected, it measures 14 by 24 by 16 inches (36 X 61 X 41 cm). Any stove needs proper venting with sturdy pipe. *Photo by Vince Crichton.*

Food

Many moose hunters seek remote areas to get away from crowds and to enjoy the wilderness or pristine aspects of the hunting experience. Consequently, many hunting trips are for extended periods of a week or more. Logistically, this means minimum carry-in weight and bulk of provisions. To illustrate such economy, I offer my own experience with a four-man hunting group.

With the moose hunting season opening on a Monday, our group's desire is to fly to our chosen site on a Saturday morning and come out the following Sunday. Planning begins with a tabulation of the number of breakfasts, lunches and suppers required. Without proper planning, extra food would be taken and never used, representing unnecessary expense, energy and inconvenience. Breakfasts, for example, are simple and sufficient fare—pancakes, bacon and eggs, sausages and coffee or tea. For other meals, and in the interest of reducing costs and saving space, foods can be prepared and frozen at home, and packaged by meal clearly noted on the containers. Fresh meat can be vacuum-sealed and subsequently frozen. If it thaws, spoilage is not likely to occur during the course of our hunt if kept properly cool. Frozen foods are well wrapped in newspapers and placed in a cooler with ice packs or dry ice.

At the camp site, a hole for the cooler is dug in the ground in a dry, shaded site. This is covered with the surface layer of material from the hole. Wrapping the cooler in plastic assists in keeping it relatively clean. In this manner, meat usually remains frozen. Frozen foods are better than canned goods because they are lighter and reduce the amount of trash to be carried out. Frozen vegetables are packed in reusable ziplock bags in meal-sized portions. Items, such as sugar, salt and pepper, are placed in small containers. Another container has an assortment of apples, oranges and chocolate bars to carry into the field.

By careful planning and advance preparation, an adequate amount of food can be taken without burden or excess. Meals also may be supplemented with fresh fish or gamebirds, for which separate licenses are needed. Other food suggestions are given in Table 63. And, if a member of the party is successful in taking a moose, a meal of fresh liver or boiled tongue is a delicacy. Finally, extra lightweight rations should be taken for emergency purposes, in case of being weathered in.

The food list will determine what is required to prepare and eat it and clean up afterward, including knives, forks, spoons, plates, cups, cooking pots, tea/coffee pot, water pail, gas stove, fry pan, wash basin, dish cloths, drying towels and biodegradable detergent.

Vehicles

The use of vehicles and aircraft for hunting purposes is unethical and illegal in most jurisdictions. The use of vehicles for transportation to a hunting area, by way of designated roads or trails, is acceptable. But they then should be parked and left. The use of all-terrain vehicles should be restricted to transporting hunters from the campsite or the location where cars or trucks are parked to the hunting area and removing animals from the bush by the most direct and expeditious route possible.

It is illegal in most jurisdictions to shoot swimming animals and to shoot from motorized watercraft unless they are anchored or not under power. Hunters should realize that it is a formidable task to remove a downed moose from the water. Aluminum or fiberglass canoes offer the greatest durability but have the disadvantage of being noisier than wood and canvas canoes. If an aluminum canoe is the craft of choice it ought to be painted a dull camouflage color to prevent reflections. When a motorized boat is not in use, a cover over the engine also prevents reflections.

Under no circumstances should motorized equipment be used to locate or pursue a moose (or other game). This is illegal, and more so, it violates the spirit and fact of fair chase.

The Hunt

Calling

Calling moose during the rut period is considered by many to be an ultimate challenge. This technique appears to be unique to North America; in northern Europe, the drive technique is principally used. The purpose of calling moose is to lure an animal within range, rather than search for one.

There is no single "right way" to call moose, but a number of techniques work. They can be learned by listening to commercially available records or viewing videotapes, and practicing.

It is important to remember that bulls come to the "point of call" and have an uncanny ability to detect the precise location. Calling in the evening while preparing or eating supper can result in a "moose in camp" within an hour. It also can result in a midnight visit. Another reminder is that rutting moose are very aggressive.

Hunters also should be aware that moose do not vocalize when and where wolves are howling, and are not likely to be enticed by calls.

Instruments used to call moose include (from left) a birch bark horn, canoe paddle, shoulder blade from a moose and tin can call. Experience and practice will determine which device is most useful in various circumstances. Novices can learn how to call moose by watching commercial videotapes of the art or by studying with an experienced caller. *Photo by Vince Crichton.*

A moose shoulder blade rubbed against a tree or log simulates the sound of a moose raking its antlers on bushes, and can lure confrontational bulls during the rut. The shoulder blade can be conveniently carried in the field by attaching a piece of string to it through a small hole drilled through the narrow end and tied to the hunter's belt. A good source for moose scapulas is a butcher who processes wild game in moose country. Note that the hunter carries a birch bark horn by a strap over the shoulder. *Photo by Scott Crichton.*

There are a number of devices that can be used to call moose. The most traditional is the birch bark horn, but other devices such as a tin can (preferably, a quart or liter juice or coffee can), the shoulder blade from a moose and even a paddle are used with success.

Birch bark horns can be purchased commercially or made from a moderately heavy fiber material (a piece of rigid vacuum cleaner hose or similar type tubing and even plastic baseball bats with both ends cut out also will work). To fashion a birch bark horn, bark should be selected from a tree that is removed from highly visible areas. It is easily removed in spring, and can be sown together with spruce root or a piece of leather lace or glued with a glue gun. Construction begins with a piece of bark that is folded so that the inside is on the outside of the horn and folded into a conical shape 12 to 30 inches (30.5–76.2 cm) long. I prefer one about 24 inches (61 cm) long. The diameter of the opening for blowing into is about 1 inch (2.54 cm), with the opposite end opening about 8 inches (20.3 cm) in diameter. Once folded properly and glued at two or three points to ensure the shape is held, the horn can be sewn at the seams with the aforementioned material or glued along the edge of the bark. Adjustment of the calling end may be made with a sharp knife. I use a horn made in this fashion. It has lasted for 12 years and, with care, will last for many more. I carry a small roll of electrician's tape in my day bag for emergency horn repairs in the field. Posthunt repairs are made with a glue gun.

A tin can call is constructed by removing one end and placing a small hole in the center of the other. A thin piece of rope (sideline) about 0.375 inch (0.95 cm) in diameter, leather thong or skate lace about 30 inches (76.2 cm) long is knotted and threaded through the hole from the inside out. The can should be wrapped with tape to prevent reflections and to muffle the "tinny" sound. It is preferable to wet the line or apply spruce pitch to it before pulling. Wearing a glove ensures that fingers are not abraded by the friction. By pulling the line between the thumb and forefinger, the moaning sound of a cow can be produced. It is amplified by the can chamber. Short, jerky pulls can simulate the grunt of a bull.

A moose scapula (shoulder blade) is another device I like to use. For carrying, a short piece of string is attached to the blade by means of a small hole bored through the narrow end. Shoulder blades from cattle do not work well because they are smaller and of heavier bone, which does not resonate when rubbed on trees. The blade of a canoe paddle also can be used in place of a moose shoulder blade. Some hunters rattle antlers together to simulate the clashes of sparring or fighting bulls.

Calling moose is most effective during the rutting period, toward the end of September. In Manitoba, about 84 percent of the cow moose are bred between September 20 and October 7 (Crichton 1992c), with the mean breeding date being September 29. In most moose populations, breeding occurs at about this time (see Chapter 4). However, it is possible to call moose through October, but the response is not as intense. I have called moose as late as October 25, and have had reports of bulls still engaging in combat in mid- to late October.

Moose call during the winter hunting period (early December), but my experience suggests that this call (of the bull) is entirely different from vocalizations made during the rut period. It more closely resembles a cough or bark than the deep guttural grunt heard earlier, and probably is an alarm in response to the hunter's presence. Calling for moose at this time is essentially useless—tracking is a better technique. I have been informed by some hunters that they have heard cows calling in the winter hunting period, but have been unable to corroborate these assertions.

Moose reply to a call at any time of the day, but early morning and late afternoon are best. Compared with warm, windy or rainy weather, still and cool or frosty weather produces the best results. In windy weather, the call does not carry. Also, moose have a tendency to retreat into heavy cover during windy conditions. Another disadvantage is that the hunter may not hear approaching moose, and unless he or she is situated in an ideal location, the moose likely will circle and pick up his or her scent.

Many hunters believe that when moose do not respond to calls during poor weather conditions, they are not in the rut. Nothing could be farther from the truth. Females usually are receptive for less than 24 hours and if not bred at this time they will ovulate again in about 24 days (Schwartz 1992c; see Chapter 4).

Calling moose should be done from a site that offers concealment and good visibility. The general area first should be searched for evidence of recent moose activity. The calling site should allow the hunter to relocate with concealment according to the direction from which a moose approaches. Before the commencement of the rut, calling is not very effective. But during the rut moose reply to calls at any time of day, but particularly in early morning or late afternoon. Windy conditions are not ideal for calling, but will produce results if the wind is favorable. A birch bark horn merely amplifies the sound made by the caller, so it is important that one practice and gain proficiency of the appropriate vocalizations before going to the field. *Photo by Scott Crichton.*

Choosing a calling location is important to success. First, it should be in an area that represents good moose habitat and may have moose sign, such as tracks, thrashed bushes, wallows or recent browsing. However, the absence of visible sign is not proof that moose are not in the area or cannot be called. The site should offer good cover and visibility, and be near open areas where the call will carry. Trees that are easy to climb offer excellent visibility. But caution should be used in climbing trees and pulling up a firearm, which certainly must be unloaded. I always spend time before and during a moose hunt acquainting myself with the hunting area. I learn where well-used trails are and find potentially new and alternative calling sites.

Beaver ponds (dried or active), open grassy meadows, and points jutting out from the shore of lakes and rivers are particularly good calling sites. The site also should offer some opportunity to move about. Generally, moose move when calling. However, I have observed cows calling from the same location for 3 to 4 hours, at intervals varying from 5 to 10 minutes. At some of my chosen calling sites, I can move quietly about 50 yards (46 m) in several directions, call and then move back to my waiting post until the next calling time. This mobility enhances the range and direction of calls, as well as simulating reality.

A horn does not "make" the call, it merely amplifies the sound the hunter makes. And the hunter has a choice of two calls to imitate—that of a rutting bull or that of an estrous cow. I prefer a cow call because it is louder and, more important, I believe, entices bulls more readily than do bull calls to combat.

A calling area always should be entered and departed as quietly as possible. After calling, the hunter should stay well hidden, because bulls have an uncanny ability to come to the point of call quietly and with no vocalizing. Smoking or talking should be discouraged, and shiny items should be out of sight.

In the morning, I prefer to be in my calling location at least a half hour before legal shooting time. The cow call should begin with an amorous call—a soft, low, "pleading" sound. It should start low, reach a high pitch and then abruptly cease. I start the call with the horn pointing down, about 3 feet (0.9 m) from the ground or water, gradually move it upward, and finish with a downward turn. Cows know when a bull is nearby; they moderate the volume of calling accordingly. I call three or four times in succession and generally to the four cardinal points of the compass. If no answer is heard, I try again in 8 to 15 minutes. I frequently hear cows calling, and time my calls to the intervals between theirs. Sometimes, cows will vocalize only once or twice and stop; other times, they call at precise intervals. I recall listening to a cow vocalize every 8 minutes from 7:30

A.M. until 10:30 A.M., when I left the area. On another occasion, I witnessed a cow at about 100 yards (91 m) call with a loud, drawn-out moan every 5 minutes for 3 hours. Frequent calling (every few minutes) is not natural and may discourage a wary bull. In addition, calling too loud in the initial two or three calls can have the same effect. If there is no response in the first hour, I attempt to get more distance on the calls by making it louder.

If a moose answers to a call, the hunter must be patient and try to ascertain the bull's location and intention. If the animal is close and apparently moving toward the point of call, it is advisable not to call again. At close range, bulls are increasingly critical of calls, and a wrong sound may spook him. Hunters are advised to carry three or four dried sticks. If a bull is reluctant to approach, a snap or two of the sticks, to imitate a cow moving, may serve to overcome his apprehension.

If the hunter is adjacent to water, splashing water or pushing the horn up and down in the water imitates the sound of a moose (cow) walking in water. Filling a horn with water and letting it run out the narrow end mimics the sound of a cow urinating.

If an approaching bull stops nearby and out of sight, and remains motionless, the hunter should call softly with the horn pointed down and away from the bull's location. This may deceive him into thinking the cow is moving off, and he may move in pursuit. When next to water, the hunter can make a grunting noise directly down at the water, which produces a slightly different sound and may entice the bull. The secret to successful calling is patience. Hunters should avoid the urge to change location. The better alternative is a different pattern (spacing) of calls. Some bulls will make considerable effort and time to get downwind of the call source, and may stay hidden for an hour or two before exposing themselves. Bulls accompanied by cows are extremely difficult to call.

On one occasion, a large bull responded to my call at about 9:00 A.M., but stopped behind a small bluff about 300 yards (274 m) away. I knew he was there because of the sound of antlers being rubbed on trees and an occasional grunt. After a while, all was quiet. When he had not appeared after an hour or so, at about 11:00 A.M., I decided to head back to camp, convinced that he was accompanied by a cow and not about to leave even to the challenge of another bull or a seductive cow. When I returned to the area about 3:00 P.M., the bull was in a meadow and, as anticipated, with a cow. Apparently, when the moose first approached as far as the bluff, and in the company of the cow, the cow likely bedded down and the bull was reluctant to leave her to investigate my calls.

That episode illustrated the point that, if no response is heard after about a half hour to an hour, chances are that a bull has lost interest. But the hunter should not assume that because one known bull has lost interest, there are no other bulls in the vicinity. On many occasions, I have had more than one bull respond to my cow calls.

The "blade" technique—rubbing a moose scapula against a tree—can be very effective in some situations. Ideally, the rubbing is against trembling aspen or dead trees from which the bark has fallen off. Live trees, such as birch or conifers, do not give the desired echo. The sound produced is that of a bull striking his antlers against a tree. If a bull answers but is reluctant to reveal himself, the blade can be raked against small shrubs.

I prefer not to use the blade initially because too loud a noise may scare off young (subdominant) bulls. Rather, I use it when a bull is present but will not reveal himself despite my vocalizations. I have experienced occasions when raking a blade has enticed reluctant bulls to come on the run. A horn, canoe paddle or moose antler also can be rubbed against low shrubs to produce the sound of a thrashing bull.

When a bull seems to have lost all interest in a closer approach, almost any trick can be tried to lure him into suitable range. At that point, the hunter has little to lose. I tend to rely, when all else has failed, on the low, plaintive, amorous call. Once, a bull responded to my call and first grunted at about 75 yards (69 m) distance. He heard but did not see me. After stalling, he proceeded to abandon his search for the cow I represented and moved off. With a low "seductive" call, I was able to stop him at about 200 yards (183 m) and turn his attention back toward my call. Within 5 minutes, he came to within 25 yards (23 m) of my position, grunting continuously as he approached and not being particularly cautious. This bull was 3.5 years old. Young bulls frequently vocalize more often than prime males. On occasion, the latter will enter an area quietly and only advertise their presence when quite close to the point of the call.

If calling at a site in the evening brings no results, the hunter ought to visit that location early the next morning, moving into position as quietly as possible and vocalizing like a bull. One or more bulls may have moved into the area during the night.

The tin can call can be effective but is somewhat "out of context" in a pristine setting. The can should be held up at about a 45-degree angle when the call is broadcast. It should be pointed downward and away and the call softened in answer to an oncoming bull. This technique is a bit more awkward than use of a birch bark horn, but it can be mastered with practice.

Bull calls are useful when paddling quietly along waterways, because moose frequently bed down in open areas in the riparian zone. At "moosey" sites, a soft bull grunt may elicit response from a bedded animal that otherwise might not be detected.

Even when calling under ideal conditions, there will be no results. This suggests that a move to a different site is warranted. Another option is for a companion to call. The change of tone, pitch, volume, tempo and resonance may trigger a bull's interest. On one occasion I called a yearling bull to about 300 yards (274 m), but he would not come closer and eventually began to move away. My son asked for the horn, began calling in his own fashion and, within 10 minutes, the bull was within 20 (18 m) yards from us. This was my son's first calling experience.

The sounds moose make are difficult to describe, and are best learned from tapes, records and experienced callers. Video recordings of vocalizing bulls are especially useful for practicing calls.

During the rut, cows vocalize to advertise their presence and mating receptiveness to potential mates. In calm conditions, this call—the drawn-out moan—can be heard for several miles. If a bull responds, a cow will change from the moan to a plaintive or seductive call. I have heard seductive cows make myriad calls, all of which are difficult to describe or characterize adequately. Some are in response to overly amorous bulls when a cow is not receptive. Others are made to intimidate other cows trying to attract the same bull. An experienced hunter will recognize these calls as different from those of a female seeking a mate.

The cow call is imitated by moaning like a domestic cow, which is a good place to begin the learning process. The sound resembles a short *"mwar"* or a longer *"oo-oo-oo- aw."* The call emanates from the chest and is best accomplished by forcing the air up from the lungs and into the horn. Some callers suggest pinching the nose closed as the horn is blown, but this is not essential.

Some videos on calling bulls can make the art more complicated and confusing than it needs be. The best advice to a novice caller is to pick one or two cow calls and try to perfect them, and not worry whether the bull being called is a teenager, prime or postprime animal. Also, a cow call is the better call to attract bulls.

The bull call is more abrupt, lasting only about a second. It is best done by saying *"o-oh-Ah,"* with more emphasis or force at the end. It has been described as a grunt, burp or snort. Bulls repeat this grunt every few steps as they get closer, but it can be periodic. Older bulls have a lower voice, with a deeper pitch. Being more experienced, older bulls tend to answer less frequently as they approach. Younger bulls tend to be more vocal and have a higher pitch. A hunter with a deep voice has a chance to be an excellent caller.

The bottom line for successful calling is patience, alertness, practice and field experience.

Tracking

The most obvious moose sign in the bush is the track, which most hunters can recognize. Other signs also are important, and being able to read them will help improve the chances to harvest a moose. Those other signs include: fecal deposits—summer feces resemble an amorphous mass similar to those of domestic cows, whereas autumn and winter feces are oblong pellets; wallows or rutting pits in which bulls have urinated and rolled to coat themselves with the odiferous scent; and trees or small bushes beaten (thrashed) by a bull during the process of removing antler velvet or advertising his presence to other bulls and cows. Pieces of velvet on the bushes and fresh breakage of browsed trees and shrubs are indication of moose in the vicinity.

To track moose by their hoof prints, it is helpful to know whether the impressions were made by a bull, cow or calf, their freshness, what direction the animal(s) was going and how purposeful the movement (i.e., running, feeding, or milling). In most cases, direction usually is easy to detect—the tip of the long drag usually faces the direction the animal is going. Adult moose have larger tracks than calves do. Calf tracks generally are less than 3 inches (7.6 cm) in length, whereas cow tracks vary from 3.5 to 4.5 inches (8.9–11.4 cm) and those of bulls as much as 5.5 inches (14.0 cm) in width (Timmermann 1990). Cow tracks are more elongated and pointed than those of bulls, which are more rounded, especially those of rutting bulls. To ascertain track freshness, such clues as water (clear indicates old, cloudy indicates fresh) and vegetation (bent indicates fresh, upright indicates old) are helpful. Running moose usually leave indication of dew claws in the track.

Determining the gender of a trailed moose by its urine splash is not easy, contrary to widespread belief. Bull moose squat slightly to urinate and the urine is deflected about 25 degrees by the prepuce or penal opening, which faces toward the rear. Thus, the urine hits the ground near the center of an imaginary line drawn between the two rear hooves, similar to location of urine discharge of a cow. Nevertheless, urine in snow from bulls generally is more concentrated than the wider splash of cows. Some hunters are adept at picking up the subtle differences.

Fecal matter that is warm, moist and glassy is only minutes old. That which is cold, dry and not shiny is at least a few hours old.

Tracking moose in snow is a challenge, and even more so during the snow-free periods. When tracking, it pays to go

There are many signs in the bush that can indicate the presence of moose to an observant hunter. The most obvious is tracks (top left). The tip of the long drag indicates the animal's direction of travel. Scat also is convincing sign, and a hunter can learn to read the freshness of fecal deposits. In summer and early autumn, they may be an amorphous mass, similar to that of domestic cows. Later in autumn and winter, they are oblong pellets. Wallows or rutting pits (center left) are associated with bulls in the rut. Bark and leaves removed from trees or small bushes (bottom left) is evidence of thrashing by bulls to shed antler velvet. Pieces of moose antler velvet found on the bushes (top right) indicate fairly recent activity by a bull, as scavengers usually are quick to find and make off with it. Breakage of small trees indicates moose browsing activity (bottom right), and the freshness can be determined by examining the break for moisture or dryness. *Top left photo by Vince Crichton. Top and bottom right photos by Charles C. Schwartz; courtesy of the Alaska Department of Fish and Game. Center and bottom left photos by Albert W. Franzmann; courtesy of the Alaska Department of Fish and Game, Soldotna.*

slowly and quietly and be alert for sign. The hunter should make every attempt to stay downwind of the moose (Figure 230). This may require following away from the track. In such cases, the hunter should move back toward the track every couple of hundred yards to reorient to the animal's direction. This technique often is used when hunting in snow as the track is much easier to locate. It is a traditional ploy and further described in Chapter 1.

Again, the key to a moose's survival is its sight and smell, and both senses are acutely developed. I have observed moose walk downwind with a moderate breeze blowing human scent toward it and the animal has not bolted but continued on its way. Either the winds were blowing the scent above the animal or it simply chose to ignore the odor. I believe that the former was true. Swirling winds, common in the boreal region, have cost many a hunter—including me—many a shooting opportunity after a long and otherwise successful stalk.

Wet conditions are best for tracking during snow-free periods because they allow the hunter to move relatively quietly, and hoof impressions are fairly easy to see in leaves and soil.

For all tracking circumstances, the hunter should stop frequently to look and listen. A continual pace is unnatural and more noticeable. A good stalking technique, if it can be mastered, is to place the front part of the foot down first followed by the heel. It minimizes crunching underfoot. If a moose appears to be running or following a straight course,

Figure 230. When tracking moose in snow, the hunter should move on the downwind side of the track, and periodically move to the track to make certain of the animal's direction of travel (from Monk and Buss 1990).

the hunter can be assured of tracking a fair distance before catching up with it. On the other hand, if tracks are meandering, the animal likely is feeding and may be overtaken sooner than later. The hunter should pick every location for foot placement and listen for feeding noises that moose make as they clip browse and move through vegetation. If the hunter intends to stand in one location for an extended period, the snow in that spot should be removed down to bare ground to minimize the noise of boots on crunchy snow.

Frequently after feeding bouts, moose rest adjacent to the feeding area. On occasion they move past the feeding area and double back a short distance, bedding down in sight of the trail they just came along. Some refer to this as the "buttonhook" maneuver (Figure 231; see also Chapter 1). Some hunters suggest that moose usually turn to the left when they backtrack, but this tendency has never been proven.

If spooked, a moose may move a short distance and slow down, or it may move a considerable distance. Pursuit may be possible, but usually the hunter is best advised to search for a different animal that has not been alerted. Hunters who do follow a startled moose often tend to proceed too quickly and almost inevitably spook the animal again. Time of day and familiarity with the area also dictate whether to follow. Instead of following tracks late in the day, hunters should leave the track and get an early start the following morning.

In soft snow, tracking sounds are muffled, to the hunter's advantage. Hunters who locate a moose after tracking through crusted snow are simply lucky. Moose that have not been disturbed can be found by following 1- or 2-day-old tracks. Once on a track, a hunter should note by compass its direction, and subsequently recheck the compass course frequently. Tracks in snow can be checked with the hand. Fresh tracks, although packed on the bottom, will not be frozen there or on the sides. In cold, frosty weather, the track will freeze within a few hours, but this is not the case in milder conditions. The track can be checked for the presence of fresh fallen snow or for ice crystals.

Days with a moderate wind are best for tracking. If two hunters are involved, one may choose to follow the track, while the other stays downwind and slightly ahead. Windy weather will muffle many of the sounds a hunter makes, but as windy conditions increase, so does the moose's alertness. Shallow, fresh snow on a day with a continuous wind is ideal for tracking.

A tracker frequently will observe moose "beds." A large bed with one or two smaller ones signifies a cow/calf group, whereas larger beds are made by adults. During the rut, large beds together may be from a bull and a cow. In

Figure 231. If undisturbed, moose often rest after feeding. They commonly circle back on their track and bed down where they can view their back trail. This is called a "buttonhook" maneuver (from Monk and Buss 1990).

snow, antler prints made by sleeping bulls resting their antlers on the ground may be visible.

Antlers of mature bulls will reflect light, much like a mirror flash in sunlight. In addition, hunters should be cautious to differentiate yearling bulls from cows, because the formers' small antlers blend in with the ears and are difficult to see.

Bulls tend to run through branches, but cows and calves normally duck underneath obstacles. All moose normally walk around obstacles. If a moose goes straight through heavy bush, the hunter likely has been detected and the animal spooked. Cows with calves are more difficult to track than bulls, because they are especially alert to danger that may imperil their calves. Also, lone moose usually are more alert than groups. In the latter case, a loosely associated group may attribute noises a hunter makes to one of their own kind.

Moose are extremely adept at differentiating "moosey" sounds from human sounds. Hunters should avoid carrying coins, keys or other metallic, noisy things, including ammunition, in pockets, or wrap them in a handkerchief or paper towel. Also, hunters should not wear clothing that smells of camp odors. This can be overcome if the hunter changes to outer clothing in camp, keeping the hunting apparel in an airtight bag. The compass and a pocketknife should be kept in separate pockets.

Shot Placement

A well-placed shot will ensure that the animal drops immediately or within a short distance.

Although there are a number of vital shot-placement areas, preference should be for the heart/lung area (Figures 232 and 233). The brain and spinal cord also are vital areas, but shots aimed there must be especially precise. Shots at moose in questionable circumstances, such as long distance, running, behind obstructions or at bad angles, must not be taken.

Once a moose is down, the hunter should move to the animal immediately and, if it is still alive, administer a final shot in the neck region just behind the skull. Head shots are not recommended, because the skull plate to which antlers are attached may be broken, resulting in unscoreable antlers.

The reaction of a moose to being shot and its behavior afterward often are clues to where it is hit. With a heart shot, for example, the animal may go down immediately or bolt forward for a short distance and then collapse as blood drains from the heart into the thoracic cavity. With a lung shot, the moose may walk off apparently unaffected. Within 50 yards (46 m) or so, it generally will stand or lie down, often coughing or wheezing, with blood running from the nose. With most spinal shots, the animal will drop immediately. In some cases, especially if the spine is broken, the moose will die immediately. If the spine is not broken, the animal may be paralyzed, necessitating a final shot. In yet other instances, the shot may produce spinal shock rather than spinal damage, and this warrants hunter cau-

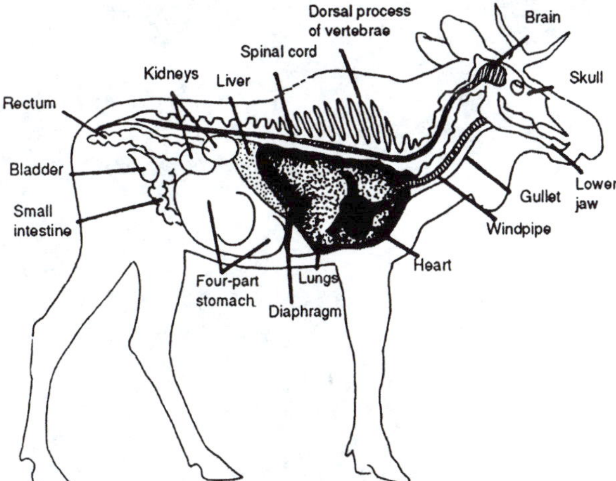

Figure 232. Internal organs of a moose. The best shot placement is in the heart/lung area. *Illustration by A.B. Bubenik.*

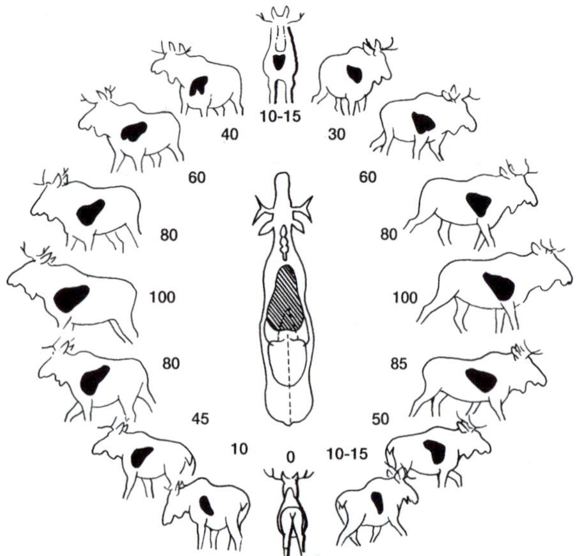

Figure 233. Proportion and percentage of the vital heart/lung area of a moose from different angles. A quick, humane kill from a well-placed shot should be every hunter's goal. Shots should not be taken at a moose that is not clearly visible or in a stance or posture that prevents proper shot placement. *Illustration produced by the Swedish Sportsmen's Association.*

tion. The animal may only be stunned, and may rise up suddenly and bolt as the hunter approaches. I have experienced this, but fortunately was in a position to make a final killing shot. With a liver shot, the animal often hunches up, moves off slowly and beds down within 500 yards (457 m) if not pursued immediately. With a stomach shot, a moose likely will run off as if unaffected, but it, too, likely will lie down within 500 yards (457 m) or so if not immediately pursued.

A quick, humane kill resulting from a well-placed shot maximizes edible meat. Meat from animals that were stressed for a prolonged period by running or wounding may taste gamey or be tough from a build-up of lactic acid in the muscles, which inhibits fiber breakdown during the aging process.

Tracking Wounded Moose

If shot placement is not perfect, the hunter likely will have to track the moose. Flagging tape to mark the animal's escape path is invaluable, and especially so during snow-free periods.

Once a wounded animal is out of sight, the hunter should not rush to get on the track. Staying still and listen-

Unless a legal moose is well within range, fairly stationary and clearly visible, shots should not be taken. A misplaced, hurried or deflected shot almost invariably is worse than no shot at all. One of the best characteristics of a moose hunter is patience. *Photo by Vince Crichton.*

ing to what the moose is doing and where it is going are important procedures. They give the hunter opportunity to orient the follow-up, get collected from the excitement of the shot and avoid pressing the animal into unnecessarily farther flight. If the moose is heard going down or perhaps "wheezing," indicating a well-placed shot, the hunter should wait at least half an hour before approaching the area cautiously. If the hunter is reasonably certain of a hit in a vital area, but the moose has left his view, he should remain in place for at least an hour before looking for the animal. If the hunter is not quite certain that the shot hit a vital area or knows the moose was "gut shot" (hit in the abdominal cavity), a wait of 2 hours before search is recommended. The only reason not to wait before tracking is if rain, snow or impending darkness may eliminate or obscure sign.

Once a search begins, the hunter should mark his shooting location with flagging tape, then move to the location where the animal was when hit and mark that site preferably in a manner that will enable it to be located from a distance. This assists in determining the precise direction the moose went. At the "hit" site, blood, hair, stomach contents, bone or other tissue may indicate where and how seriously the moose was shot. *The absence of sign at the hit site does not necessarily indicate that the animal was not hit.* If sign is not immediately found or in the absence of a clear blood trail, the hunter should walk a progressively increasing spiral course outward from the hit site in the direction the moose was last seen, until sign is found.

If possible, a companion should be enlisted to help with tracking. That person should be fully advised of the shooting circumstances and the animal's subsequent behavior and movement. This information and a plan of search should be conveyed before tracking gets underway, so the search is coordinated and without conversation. Communication should be by hand signals. Ideally, one person will stay on the sign trail and the other will be off 30 to 50 yards (27–48 m) to the downwind side. The person on the trail should move forward about 50 yards, while the other person waits and remains alert to sound or sight of the animal. Then the other person moves up, parallel to the tracker, while the tracker alertly waits. The procedure continues until the animal is located.

Well-hit animals bleed internally and usually expel blood through the wound especially when running. On some shots, particularly high ones, skin closes over the wound, preventing much blood loss. Arterial blood from the heart and lung area is bright red. Blood from a lung shot usually is frothy and pinkish. Blood from the abdominal cavity is dark red; if the stomach was hit, blood invariably contains small pieces of food. Trackers may not see blood on vegetation until it is brushed onto clothing.

A wounded moose that walks around obstacles may not be as hard hit as the one that stumbles or does not avoid the thick brush. My experience is that wounded moose tend to travel into the wind. If, in the course of tracking, the wounded moose is "started up," but cannot be dispatched, the tracker(s) should again wait half an hour at least before renewing the pursuit.

Every moose hunter has a legal and ethical obligation to make every reasonable effort to recover wounded animals. Management jurisdictions anticipate that some animals are wounded and lost, but this does not justify in any way less than a fully concerted effort to find moose that are wounded. The objective must be a zero loss.

After the Kill

Proper care and handling of moose meat is an essential aspect of the hunt. The hunter has the responsibility to utilize harvested moose as fully as possible. All jurisdictions have wanton waste statutes. In any case, any person who is unable or unwilling to process a moose fully should not get a license or exercise the hunting privilege.

Some basic knowledge is needed to process a moose carcass in the field, transport the meat and other desired parts, and prepare them for use or storage. On numerous occasions, I have heard hunters complain of bad moose meat. When I had opportunity to examine the meat and/or question those individuals, invariably I learned that meat was improperly dressed, cut, cleaned, transported, aged, packaged, frozen or cooked. Rarely has "bad meat" come from a "bad moose."

All jurisdictions require that a license or tag be notched and/or attached to the carcass once a hunter takes possession of the animal he killed. This is the hunter's first and foremost task when the animal is found dead.

Bleeding a moose by severing the jugular vein generally is unnecessary, because modern ammunition causes enough damage to drain most blood into the abdominal or thoracic cavity.

Field Dressing

Dressing a moose is possible with one person, but additional help is preferable and highly recommended. If a moose ends up in water, a block and tackle should be used to bring it to shore.

Preferably atop a lightweight plastic tarpaulin that can serve as a "work table," the moose carcass should be positioned on its back or right side because the rumen or paunch is on the left side. This facilitates relatively easy removal of the entrails. If the animal can be placed on an incline (head higher) the visceral organs will slide toward the back and further assist in dressing. Keeping the carcass on its back is difficult, but by attaching rope to the middle portion of the right front and rear legs and then to adjacent trees strong enough to support the carcass, that position can be stabilized. In the case of bulls, the head can be positioned upright, with the antler tines embedded into the ground.

If the hunter intends to have the animal's head mounted, this decision must be made before skinning, because special cutting instructions are involved. These are discussed later in the chapter.

Before skinning, the hunter may wish to remove the tarsal glands located inside of the back legs near the hocks. Some hunters suggest that secretions of these glands, if transferred to meat, can produce strong odors. However, I have never found the tarsal glands to be overly active organs, even on bulls taken during the rut, so I seldom remove them. In any case, the tarsal gland odors are not as strong as those of a white-tailed deer.

I prefer to begin the skinning process in the brisket area. The initial cut should go forward to the upper neck area and then back to the anus, passing on either side of the penis and scrotum in bulls or the vulva in females. The next cuts are around the anus (and vulva in the case of females) and large intestine to free those parts from the surrounding tissues and bone. A string tied around the anus prevents deposition of fecal material in the body cavity when it is later pulled through the pelvis. Incisions then are made from the central cut line along each leg toward the hoof. Care must be exercised to keep hair off the meat. I recommend that once the skin has been incised to start the skinning process, all cuts (sliding the cutting blade under the skin and cutting up through the skin) should be made so as not to cut hair shafts. I prefer to skin one side entirely to the middle of the back, and then the other side. This requires rolling the carcass to access the back area and other side. Once the hide is loosened totally, it can be stretched out, hair down, and used as a "table" to continue with the dressing. Then, removal of the entrails begins.

With the animal on its back, a small incision should be made into the abdominal cavity at the posterior end of the

The task in field dressing a moose is to position and stabilize the carcass for efficient skinning, cutting and gutting. The carcass above is kept on its back by means of a rope secured around the left legs and anchored by a nearby tree. Note the proper cut lines for skinning, *except* in the case of a cape mount (see Figure 236). Similarly, and regardless of whether a moose carcass is skinned in the field or quartered with hide left on, gutting is accomplished by a shallow incision along the underside center line, from at least the base of the sternum to the anus. In all instances, the gutting incision is after the skin has been split (mostly from beneath) and pulled back from the center line to minimize cutting or loosening hair. *Photo by Vince Crichton.*

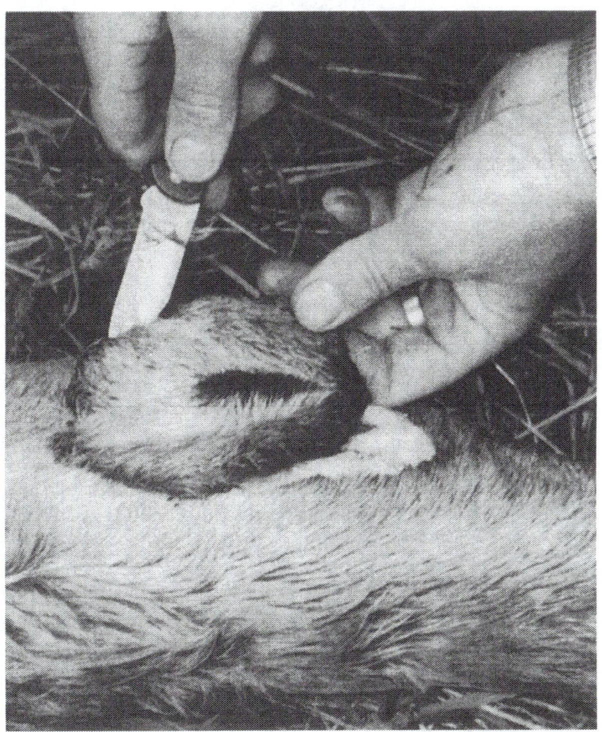

Some hunters prefer to remove the tarsal glands from the rear limb of a moose carcass to ensure that meat is not contaminated by residue of gland secretions. *Photo by Scott Crichton.*

brisket, taking care not to puncture the stomach. Two fingers are placed in the incision and, with pressure down on the stomach and the knife pointed upward, an incision is made to the anal area. Then, back at the starting point, a cut is made to the base of the neck, freeing the esophagus and trachea (windpipe). A length of strong string, about 36 inches (91 cm) long, is tied tightly around the upper end of these structures to prevent stomach contents from escaping and contaminating the meat. The string then can be used to pull the entrails backward, facilitating removal of the digestive tract. The sternum can be severed with an axe, but preferably cut with a saw, again with care not to puncture the stomach. Removal of ribs from the sternum on one side may be accomplished with a knife. Whichever option is used, the thoracic cavity will be exposed. The diaphragm is then cut from its point of attachment to the rib cage and sternum, once again carefully so as not to pierce the stomach. The string attached to the trachea and esophagus can be pulled backward on the entrails for removal with little additional cutting. Care must be taken when cutting along the backbone to ensure that the tenderloins are not cut. The colon (posterior part of the large intestine) and anus (and vulva in females), which earlier were cut free, can then be pulled through the pelvis. The entire digestive tract, along with heart and lungs, are then pulled off to the side.

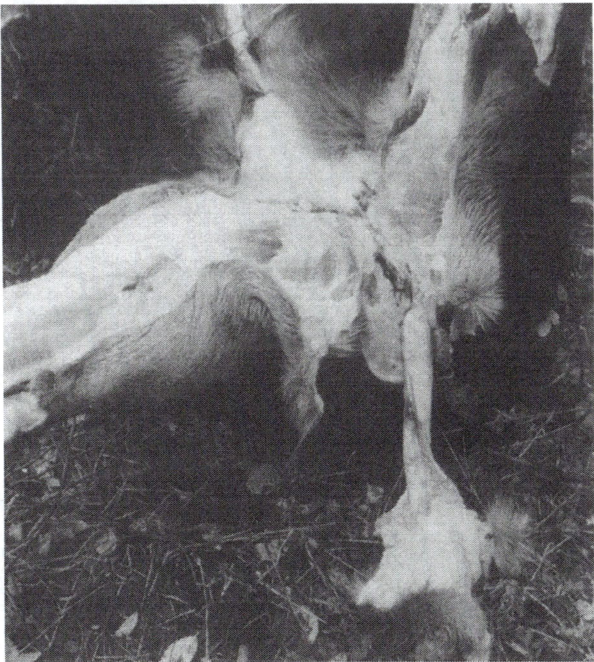

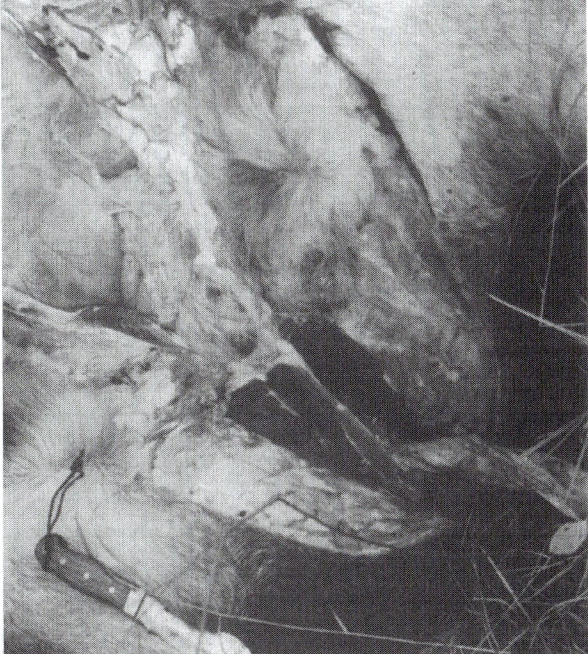

Preliminary to opening the body cavity of a moose carcass, cuts are made around the anus (and vulva of cow moose) to free it and the attached large intestine from surrounding tissue and bone (left). A string tied around the anus will prevent fecal deposition into the body cavity before the detached organs can be severed and pulled from the pelvis. Hide in the anal area then is skinned back, the penis and scrotum pulled to the posterior and cuts made into the muscle to the pelvic girdle (right). The girdle then can be sawed or carefully chopped through. *Photos by Vince Crichton.*

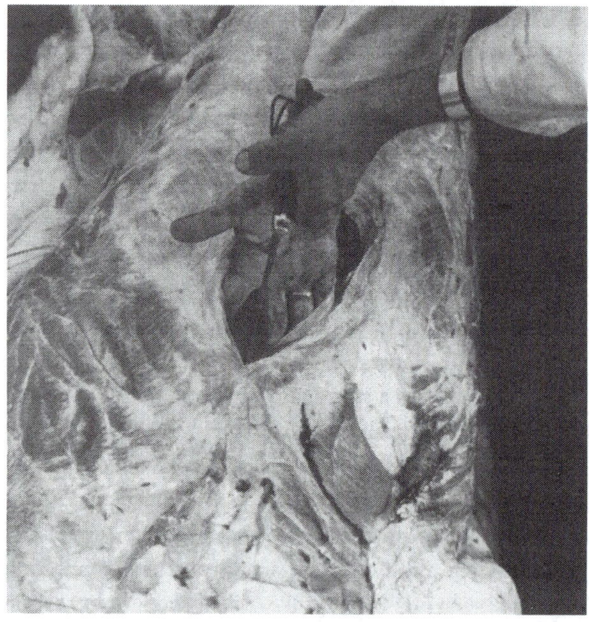

For the abdominal incision in a moose carcass, the field dresser's fingers or part of the hand should be placed through the initial cut and followed behind the inside-out incision (like most skinning cuts), to prevent accidental cutting of the rumen or paunch. The moose paunch may be very distended with both food material and gas depending on the feeding pattern of the animal before shooting. However, during peak of the rut, bull moose eat little or nothing. *Photo by Vince Crichton.*

Edible organs can be removed. The heart and liver, if saved for food, should be drained of blood and set in a clean, cool place.

If the hide is kept on to keep the quarters clean while being dragged or otherwise moved, care must be taken to keep exposed flesh clean. In any case, the hide must be removed within a few hours to facilitate cooling.

Quartering

Cutting a carcass into four or more pieces facilitates handling. A moose carcass, even quartered, is equivalent at least to half a dozen deer or several elk and presents a formidable task to move it.

Moose carcasses usually are quartered after skinning. But, when quartering is done with the hide on, care is needed to keep hair off of meat.

The first step in quartering is to locate the fourth and third last ribs and cut the muscles between them on both sides up to the backbone. Then, the backbone is cut through with a saw. Next, the back half is sawed into two pieces along the middle of the back (an axe does a messy job and results in bone splinters). If done properly, the spinal column should be visible along your cut. Leaving the lower leg on up to this point makes it easier to handle, but it now can be removed with a saw just below the hock area

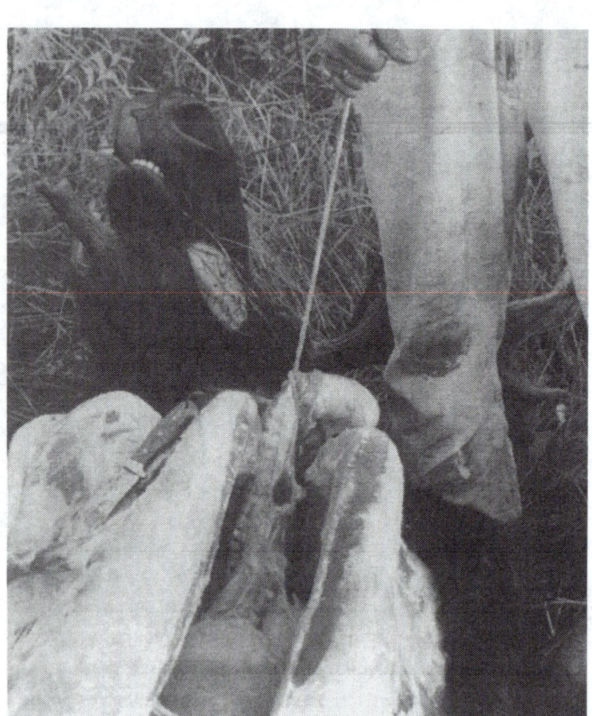

A rope attached tightly to the esophagus and trachea (left) prevents expulsion of stomach contents of a moose carcass before removal of internal organs. For removal of those organs (right), the connected rope then provides force and leverage. *Photos by Vince Crichton.*

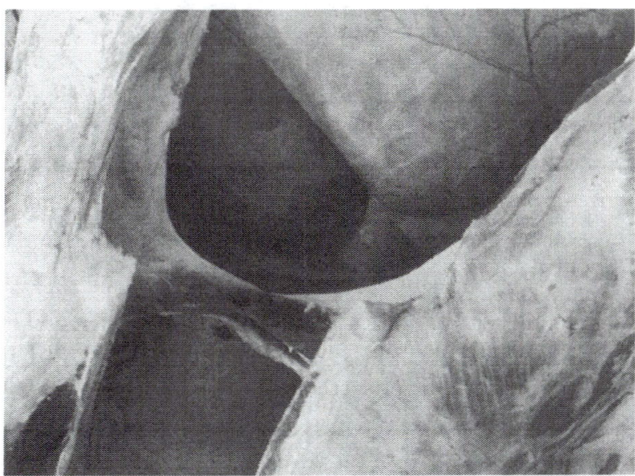

The diaphragm (arrows) of a moose is freed by cutting it along its attachment to the rib cage. *Photo by Vince Crichton.*

Use of a bone saw to divide a moose carcass down the spinal column (backbone) saves time and meat. Trying to field dress and quarter a moose carcass without the proper equipment is a wasteful exercise in futility. The correct, sharp blades for cutting also will make the eventual processing of meat much easier. *Photo by Vince Crichton.*

(leaving the Achilles tendon intact to facilitate hanging). The front half can be cut into two pieces in a similar manner, but first the head is removed in the upper neck region. By cutting slightly to one side of the backbone, the job of splitting the front half will be much easier. The most difficult part of the entire dressing procedure is cutting the front and rear halves into quarters. This is a job for two or more people. A small chain saw, with cooking oil to lubricate the chain, makes quartering the carcass relatively fast and easy.

Additional cuts can be made to lighten the load by removing the front legs from the rib cage. To do this, the muscles holding the leg and shoulder blade in place need to be cut. Cutting an animal into more than the four quarters will make transportation easier, especially if they have to be carried. Heavy loads in typical moose country can be hazardous and exhausting. Better to make an extra trip than to risk injury.

Deboning meat is a laborious job in the field, and proper equipment is needed to carry this meat out of the bush. I recommend quartering.

Hunters should be certain to collect the biological specimens requested or required by management authorities, and submit them where designated as soon as possible.

Cooling

Once quartered, the meat or quarters should be elevated on clean logs where blood can congeal and cool air is able

Cutting a moose carcass into four or more pieces eases handling. The front quarters, which are the heaviest, can be lightened by removal of the foreleg and shoulders from the rib cage. *Photo by Vince Crichton.*

to circulate around the meat. Congealing takes about 30 minutes, about the time necessary to pack up the field-dressing equipment. Thereafter, the quarters should be wrapped in game bags or cheese cloth to keep them clean.

If, for some reason, the eviscerated carcass must be left unattended until the next day, cooling can be facilitated by propping the sternum open with a stick placed on each side of the cut. Also, the hide along the inside of the legs should be cut to permit body heat to escape from the heavy leg muscles. Blood can be wiped off the meat or drained from the body cavity, but under no circumstances should water be splashed on the carcass to wash it. Water accumulates in the small tissue pockets and produces an ideal medium for bacterial growth, eventually resulting in spoiled meat. Stomach contents that may have spilled on the meat should be removed with a damp cloth. Fresh meat must never be wrapped in plastic.

Also, the gut pile should be pulled away from the immediate area of the carcass and an article of clothing "flagged" over or next to it to discourage scavenging. If possible, the carcass can be covered with brush and exposed muscle bundles covered with pepper to keep off flies.

Meat can spoil from improper cooling even in cold conditions. I once shot a moose near dark in an open bog. After dressing the carcass and forced by darkness to leave it until the next morning, I turned the carcass over and "spread-eagled" it on the snow to prevent ravens from getting at the meat. Despite freezing temperatures that night, heat was unable to escape from the hump region and about 10 pounds (4.5 kg) of meat spoiled.

Properly dressing and cooling an animal will prevent meat from tainting, which is caused by bacteria that multiply and spread. These bacteria may be present in the moose before death and develop in the absence of oxygenated blood, or they can be picked up from the ground, air or hands contaminated during field dressing.

Transportation

For transporting a moose carcass, quarters or deboned meat to a base camp or home, three cautions are given. First, a moose carcass should not be dragged a long distance or over rough terrain without putting it on a sleigh or some other raised platform. Otherwise, the continual pounding on the ground causes heavy bruising and severe damage to the meat. Second, two moose carcasses should not be placed on top of each other. This again prevents proper cooling. A useful technique is to put poles in the back of the vehicle onto which two quarters are placed. Additional poles are placed on these quarters and the remaining two quarters placed on top of them. In this manner, air can circulate around the meat (Figure 234). And third, an unskinned animal should never be left in the back of a vehicle overnight. This, too, is an invitation for spoilage due to improper cooling.

Hanging

When the meat is back at camp or home and hung, the cheesecloth is removed and the meat cleaned as best as possible by wiping with a moist clean cloth. Areas damaged by the bullet plus hair, grass, dirt and leaves should be removed (some butchers charge extra for cutting dirty meat; others will refuse it). The cheesecloth then is replaced or the meat put in other bags that do not retard air circulation. I then place a loose-fitting fly-proof sock around the meat and close it off with Velcro™. If the meat is in a sunny area, a tarpaulin should be stretched over the meat pole for shade but not in a manner that inhibits free air movement. Plastic can be placed on top of the tarpaulin to protect the meat from moisture.

Flies can be a nemesis, and every precaution must be taken before the hunt to ensure that the problem can be dealt with

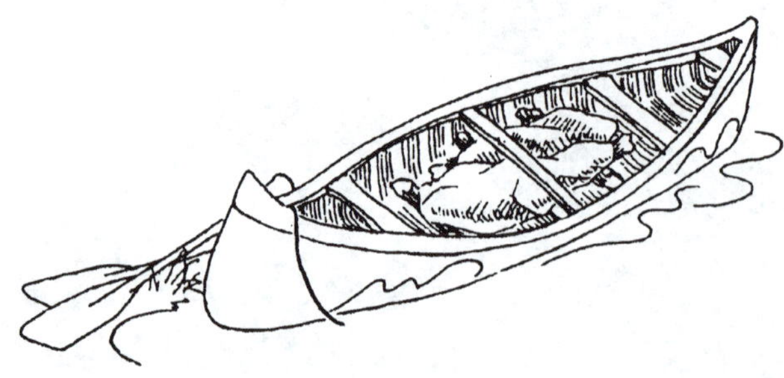

Figure 234. For transporting moose meat, every precaution should be taken to keep the meat clean and to ensure air circulation around all parts (from Monk and Buss 1990).

Plastic should only be used to keep rain off of moose and other game meat. Fresh meat should never be wrapped in plastic, which does not allow air circulation for essential, proper cooling. *Photo by Vince Crichton.*

if the weather is warm. If the meat is not protected, flies lay eggs on cut surfaces and in blood vessels. Hunters should never assume that flies will not be a problem.

Grizzly bears and/or black bears live throughout North America's moose range. They are readily drawn to gut piles and may be attracted to camps where moose meat is stored. Accordingly, if a carcass is left overnight before it can be moved to camp, returning hunters should approach the area noisily, alert and well-armed. Bears "on a gut pile" (after scavenging, they often cover the remainder with brush and dirt, before resting nearby) can be defensive of their "cache." Moose meat at camp should be hung or otherwise stored at least 40 to 50 yards (37–46 m) distant, but visible from camp.

As indicated, moose meat can spoil if improperly cooled. It can also be less tasty if cooled too quickly or allowed to freeze immediately after the animal is killed. For those hunting during freezing weather and camping in the bush for several days, the carcass should be skinned and the quarters placed on clean snow. The quarters then are covered with the hide, which itself is covered totally with a thick layer of snow. The meat will cool but not freeze under most circumstances until it can be taken to a suitable hanging facility.

Once a carcass has properly cooled, it can be hung for several days (up to a week), provided that the air temperature does not go above 45°F (7.2°C). At higher temperatures, the meat needs to be refrigerated.

The Hide

Moose hide can be used to make a variety of clothing such as mitts, vests, moccasins, slippers and jackets. With this in mind, consideration should be given before the hunt to taking salt to place on the raw hide (which, from large bulls, can weigh as much as 100 pounds [45.4 kg]), thus preventing spoilage. If salt is not available, all excess flesh and fat should be removed and the hide placed over shrubbery for drying. It should not be placed in plastic bags. If the hunter

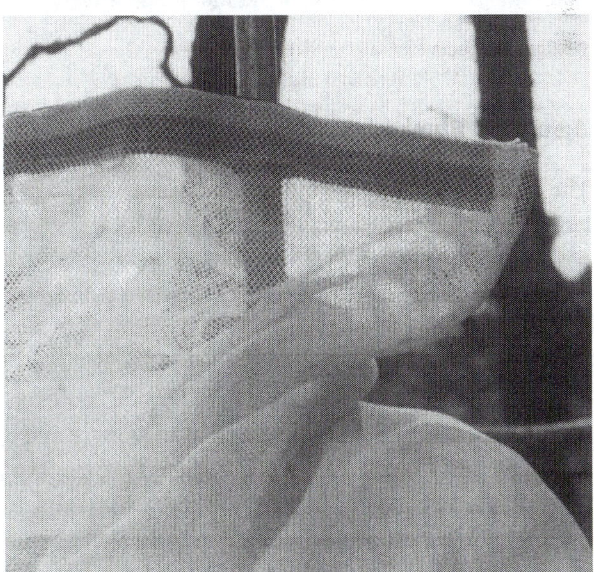

To cover moose meat or quarters hung near camp, fly-proof netting can be sewn as a shroud (left) and made to fasten at the top with Velcro (right). *Photos by Vince Crichton.*

Ontario hunters are encouraged to contribute their moose hides to First Nations people who produce smoke-tanned hides or manufacture leather goods, such as mittens and slippers, from commercially tanned hides that they purchase. In turn, the hunters receive blaze-orange caps, which feature a different emblem each year. In recent years, First Nation peoples have assumed the responsibility for collection and distribution of donated hides. Although not part of that cottage industry, moose antlers are valued by artists that carve them into a variety of forms and scenes. They also are used to make buttons, picture frames, furniture, lamps, etc. *Photo by H.R. Timmermann.*

is not intent on making use of the hide, consideration should be given to donating it to someone who will. The cost of commercial tanning in 1996 will be about $10.00 per square foot ($100/m^2) Canadian for hair-on hide and $95.00 Canadian for a dehaired hide. In many jurisdictions, there are organizations who collect the hides (and antlers) and sell them, and the funds generated are used for assorted wildlife projects (see also Chapter 17).

Aging and Butchering

The final process is to age meat properly and then have it butchered and wrapped for freezing. Persons inexperienced with butchering techniques should seek help or have it done by a professional. Most butchers also have facilities for hanging and aging the meat.

Aging meat at a few degrees above freezing for 10 to 14 days allows the muscle fibers to cure and become more tender. This process is not as critical with young moose as it is with older ones. Butchers should be advised of the length of time the animal hung in the bush and under what temperatures. This will dictate how long the meat should hang before cutting. The butcher also should be informed if the meat got wet, which will alert him to cut and freeze the meat before spoilage occurs.

Most moose meat is butchered similar to beef, with many muscle bundles cut into a single piece. An alternative to this, although more time consuming, is to separate the larger muscles by groupings on the carcass and make meal-sized portions. The advantages are that the meat is not as likely to accumulate freezer burn and the unique flavor of different muscles is not diminished (Figure 235).

Whatever technique is used to package moose meat, the wrapping must be air-tight to prevent freezer burn. Each package should be marked, dated, and the cut identified.

Nutritional Value

Moose meat that has been properly handled from kill to freezer is very palatable and nutritious. In contrast to domestic beef cattle, which have about 30 percent fat, moose have about 1.1 percent fat in the muscle. Some moose can have up to 20 percent fat, but most of this is contained in the omentum and on the muscles. This does not compromise the protein content of moose meat, which amounts to 25 percent, compared with about 26 percent for cattle (Rowland 1989).

Lean roast beef has about eight times more saturated fat than does an equivalent amount of moose meat. Saturated fats result in the human body making more cholesterol.

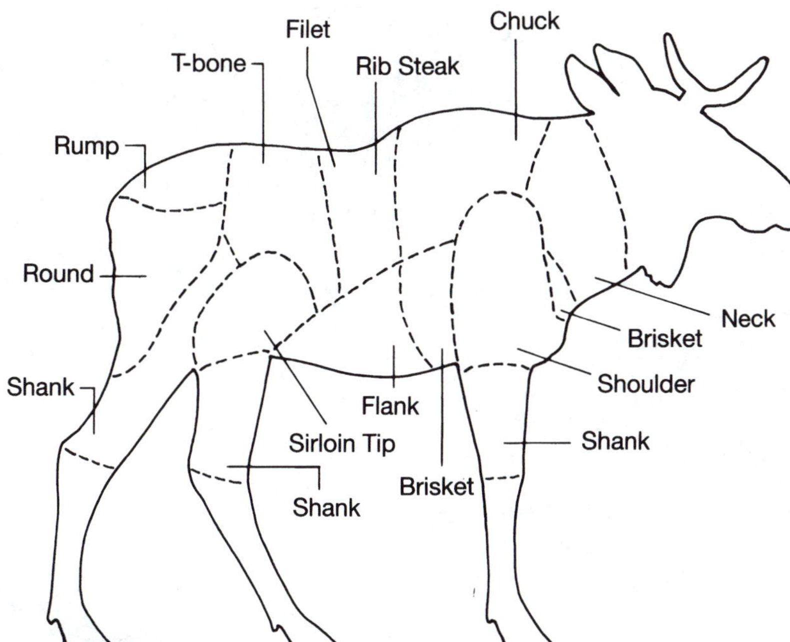

Figure 235. Primary meat cuts from a moose carcass. Roasts, steaks or both can be secured from the round, sirloin and chuck. Rib meat generally is converted to mooseburger, and shank and neck meat usually becomes mooseburger or stew meat.

Substituting moose meat for beef in the human diet can lower cholesterol levels, assuming other high-level cholesterol foods are not added.

Wild game contains relatively high levels of eicosapentaenoic acid (epa). This protective fatty acid, not found in domestic animals, improves the flow characteristics of blood. It equates to a natural antifreeze to ensure the fluids and organs of well-nourished wild animals do not stiffen in even the most frigid weather. Among others, Rowland (1989) suggested that epa in the human diet can protect against heart attack, atherosclerosis (hardening of the arteries) and certain forms of arthritis.

Moose meat also is an excellent source of calcium, vitamins B-1 and B-2, at levels 16, 0.02 and 0.37 milligrams (mg) per 100 grams (g), respectively. Comparable values in beef are 11, 0.07 and 0.19 mg per 100 g, respectively.

Unusual Conditions and Anomalies

During the process of field dressing a moose, hunters invariably will discover various anomalies, ranging from broken bones to disease conditions. The intent of this section is to alert hunters to them although the presence of parasites, diseases or broken bones, while perhaps aesthetically unpleasant, rarely makes the meat inedible. It may be necessary, however, to discard portions of meat immediately around an affected area. For further discussion of parasites and diseases with which hunters should be familiar, see Chapter 15.

Of particular importance is the fact that moose lungs, liver and muscle should *not* be fed to dogs unless those parts are well cooked, because parasite larvae may develop into adult tapeworms in pets.

Broken Bones

On occasion, moose limbs are broken and healed, leaving a large hard lump around the broken bone as a result of ossification. If the break makes the limb unusable, the associated muscles become atrophied. This meat is still edible. The lump itself and any abnormal-looking tissue in the vicinity should be trimmed away.

Rib bones frequently are found broken but healed over. Such injuries can occur from fighting or falls. This is particularly true for bulls, because they pull calcium from rib bones during antlerogenesis (see Chapter 2).

Fat Deposits

Rutting bull moose stop eating during the rut and have been reported to go without food for up to 3 weeks at this time (Schwartz et al. 1985, Miquelle 1991). During this time, they lose up to 20 percent of their body weight (Schwartz et al. 1987a); most of this loss is due to the metabolism of fat deposits. Adipose tissue (fat) is a form of connective tissue that forms when body cells take up fat for storage within the cytoplasm. The tissue associated with these fat cells is known as "areolar connective tissue" and provides both cushion and flexibility as needed. Within the subcutaneous tissues, a layer of loosely arranged areolar connective tissue fibers attaches the skin to the adjacent muscles (Frandson 1965). Moose store large quantities of fat

High-quality meat is a primary motivation for some moose hunters, but for most other moose hunters, meat is one of many desirable by-products of the hunting experience. The value of moose meat—which is high in protein, calcium and iron, and low in fat content compared with beef, pork and most other domestic meats—can easily offset the cost of most hunting trips. For example, at a conservative value of $4 per pound ($8.81/kg), an average yield of 352 pounds (160 kg) of edible meat from a single moose (Hamilton 1981) represents at least $1,408. For moose meat to have maximum value, it must be properly prepared and frozen, and consumed within a reasonable period of time. Before packaging, meat should be cleaned of all hair, clotted blood and fatty tissue, and dried (ideally, with absorbent towels). Meat then should be well-chilled, wrapped or double-wrapped in air-tight moisture/vapor-proof freezer paper or bags, and placed in a freezer away from its walls, and frozen as rapidly as possible at the freezer's lowest setting (to prevent formation of ice crystals in the meat fibers). Preferably, meat can be frozen with air space around each piece (stacking causes differential freezing). Once frozen completely, the meat should be kept frozen at 0°F (–18°C). At lower temperatures, there is greater risk of freezer burn (dehydration that discolors and changes the fiber consistency, but does not ruin meat); at higher temperatures, storage life (i.e., palatability) is reduced. As a rule, properly packaged, frozen and stored moose meat can be kept up to 16 months before meat quality begins to decline gradually. The same applies to ground moose (burger), unless beef or pork fat has been added, in which case, the mixture should be consumed within 6 to 8 months. *Photo by H.R. Timmermann.*

at these sites and these deposits frequently are associated in moose with the rump, loin, ribs and brisket. When fat is burned rapidly, as seen during the rut, the areolar tissue remains as a yellowish sometimes slimy-appearing tissue. Eventually, it shrinks and disappears. But it is burned so rapidly in rutting bulls that the tissue remains, and the yellow-brown slimy appearance is common. Hunters may attribute this to injury, and tend to discard the meat as being unfit for human consumption. However, muscles associated with this anomaly are not negatively affected and are suitable for human use.

Cadmium

The heavy metal cadmium has been found in the liver, kidneys and muscle of moose in Manitoba, Newfoundland, Ontario, Quebec and the northeastern states (Crête et al. 1986b, Scanlon et al. 1986, Glooschenko et al. 1988, Brazil and Ferguson 1989). Highest concentrations generally are in the kidney, followed by liver and muscle. Regardless of gender, older animals have higher concentrations of the metal. In some instances, the source is industrial pollution; in other cases, cadmium is naturally occurring. Concentrations in the liver and kidneys may exceed daily intake levels recommended for humans, and some jurisdictions have warned hunters not to eat the kidneys of moose or to consume no more than one meal of liver per week. Most hunters only have one or two meals of moose liver per season, and this amount does not pose a health concern. The safest course, however, is not to eat moose liver or kidneys at all.

Pus Pockets

Pus-filled cysts may be encountered in the butchering of a moose carcass. These result from infections of cuts, punctures and other wounds. In some cases, these "pockets" are totally enclosed and can be easily cut out. In other cases, where sinus tracts lead into adjacent tissue, hunters are advised to remove all the infected meat and some of the adjacent healthy meat.

Hunters and butchers may encounter large cysts filled with a clear, amber fluid. This is a seroma, and the fluid is blood plasma, also the result of an old injury, and can be readily removed.

Fatty Liver

Hunters sometimes note that the livers of bull moose taken during the rut period look unhealthy, therefore discard this organ. The condition is known as "fatty liver," and is seen in most rutting bulls (see Chapter 2). The liver appears somewhat jaundiced, friable (easily broken), and has structures resembling small glass balls on the broken surface. The latter are liver cells containing fat. During the rut, bulls reduce their food intake and metabolize fat reserves accumulated over the summer period. Contrary to what most hunters believe, such livers are edible.

Fibrinous Liver

On occasion, a moose liver will appear whitish with small plaques of fibrin on the surface, especially the diaphragm (convex surface) side. This can be peeled off, or the liver can be eaten as is.

Liver Fluke

Moose that cohabit range with white-tailed deer or elk harboring liver flukes frequently become infected with these parasites, which cause major damage to the liver. Infected livers are discolored and have large cysts containing pussy, dark-colored material. Such livers often are enlarged and as much as twice as heavy as a normal adult moose liver (about 10 pounds [4.5 kg]). Moose livers so infected are unfit for human consumption and should be discarded, but the edibility of other meat is not affected. For a more detailed discussion of this condition, hunters should consult Chapter 15, especially those who hunt in an area where the condition is endemic in moose.

Liver Tapeworm Larva

The thin-necked bladderworm is found frequently in the liver of moose taken in areas inhabited by gray wolves. The latter functions as the definitive host in the tapeworm life cycle. The cyst—ranging in number per liver from 1 to 12, and averaging 4—resembles a water blister on the surface of the liver and on occasion internally. The white head of the larva is visible inside each cyst. These larvae do not develop in humans, and can be readily removed from moose liver with a knife and destroyed. Like wolves, domestic dogs can be hosts in the life history of this parasite, therefore should not be fed liver with the cysts or suspected of possible infection, as noted earlier.

Muscle Measles

The term "muscle measles" is used to describe the presence of the larval form of the tapeworm *Taenia krabbei* in muscle (both skeletal and heart) and connective tissue. These larvae are about pea size and do not pose a risk to those eating moose meat, but they are not aesthetically pleasing. Everyone who has eaten moose meat from areas inhabited by gray wolves likely has consumed these parasites, as wolves are the definitive host in this tapeworm's life cycle. All members of the canid family function as hosts for the parasite, so again it is recommended that uncooked moose meat not be fed to dogs. The larvae commonly are found on the outer surface of the heart ventricles, and the presence of the parasite here is a sure sign of infection in other muscles. However, absence in the heart does not signify absence in the skeletal muscles. The larva also may occur in the tongue. In heavily infected moose, there can be as many as two to three larva per square inch (0.31–0.47/cm^2) of muscle.

Winter Tick

The winter tick is an external parasite that generally attaches to moose during late September and early October. By November and December, they are well established on moose. Small rust-colored larval ticks, about the size of the head of a pin, can be seen at the base of hairs. Some develop to about the size of a pea.

The presence of this parasite does not make moose flesh inedible. As the moose carcass cools, the larvae drop off. On occasion, hunters brush against vegetation where larvae have concentrated, and these larvae will attach to clothing. Larval ticks are rust-colored and can be brushed off readily. They do not live on humans.

Damaged Antlers

Damaged moose antlers are not uncommon. Damage can occur during the growing period or the rut. In the former instance, the antler may cease to grow; a bull can have one fully developed antler and a stub (see Chapter 2).

Antlers damaged during the rut have broken tines or portions of various size broken off the main beam. In some cases, the beam may break a short distance from the pedicle. Such occurrences are commonplace and should cause no concern to the hunter in terms of meat edibility. Furthermore, the presence of peruke antlers also does not affect meat edibility.

Weight

Most hunters have heard stories of 2,000-pound (907-kg) moose. But the stories are just that. The only 2,000-pound moose on record is every one that has been carried, quartered or whole, more than mile (1.6 km) to camp.

Crichton (1979, 1980) found that the average live weight of adult bulls from Hecla Island, in Lake Winnipeg, Manitoba, in early December, was 950 pounds (431 kg), with a maximum of 1,200 pounds (544 kg). Cows averaged 830 pounds (376 kg), and calves averaged about 400 pounds (181 kg) at that time of year. The weights of cow moose in eastern Manitoba were similar to those from Hecla Island, but bulls from eastern Manitoba were significantly larger than Hecla Island bulls, with some exceeding 1,600 pounds (726 kg) in early December.

Quinn and Aho (1989) reported that the mean weight of adult bulls in Ontario was 999 pounds (453 kg); that of cows was 959 pounds (435 kg). It should be noted that these animals were captured in winter, after antler shedding.

The largest subspecies of moose is the Alaskan/Yukon moose (see Chapter 2). For Alaskan moose at the Alaska Department of Fish and Game's Moose Research Center, the heaviest bull (6 years old) weighed 1,697 pounds (770 kg) (D. Johnson personal communication: 1989). Several mature bulls at the Center have weighed more than 1,650 pounds (748 kg). The heaviest female weighed 1,323 pounds (600 kg), at age 5 years (C. Schwartz personal communication: 1993).

Hunters frequently speculate about how much of a moose's whole weight is lost to dressing. Crichton (1979, 1980) reported a 30 percent reduction from live weight to dressed weight for adult bulls and cows and a 38 percent difference for calves. By knowing the dressed weight (minus liver, heart, lungs, stomach, intestines and blood), the live weight of moose can be calculated by increasing the dressed weight of adult bulls and cows by 46 percent and by 61 percent for calves.

I suggest that the four quarters (including the ribs) without the hide, feet (removed at the knee) and head represent approximately 50 percent of live weight in early December.

Using the 50 percent rule of thumb, I have taken adult moose in Manitoba in late September that figured to weigh more than 1,500 pounds (680 kg) and a 2.5-year-old bull that was calculated to weigh 1,100 pounds (499 kg).

Trophies

Antler Scoring

Every moose taken legally and ethically is a trophy in many respects. In addition, some hunters—if not most—are fascinated with the prospect of taking a bull with large antlers. Supposedly, large antlers represent an animal of superior hunting challenge, because of its genetic make-up, elusiveness and scarcity. So intense is the fascination that the term "trophy" usually has referred to moose rated by antler size and conformation as the biggest and best ever since a standardized method of measurement was established. For North American big game, including Canada moose (*Alces alces americana* and *A. a. andersoni*), Alaskan/Yukon moose (*A. a. gigas*) and Wyoming or Shira's moose (*A. a. shirasi*), the trophy standards (official scoring system) were initiated by the Boone and Crockett Club, currently headquartered in Missoula, Montana, in the early 1930s. The Club has continued and refined the system and publishes revised lists periodically. Since 1958, the Pope and Young Club, in Chatfield, Minnesota, has certified trophy big game taken by bow and arrow. And since 1988, the Longhunter Association of Friendship, Indiana, has recorded trophy big game

Table 65. Minimum trophy antler scores for North American moose, 1995[a]

Source	Moose subspecies		
	Alaska/Yukon	Canada	Wyoming
Pope and Young Club[b]	170	135	115
Longhunter Association[c]	180	145	125
Boone and Crockett Club[d]			
Record book	210	185	140
All-time	224	195	155

[a] Based on the Boone and Crockett Club's Official Scoring System.

[b] Moose taken with bow and arrow.

[c] Moose taken with muzzleloader.

[d] Moose taken during current (latest) scoring period.

taken by muzzleloaders. Both Pope and Young and Longhunter use the Boone and Crockett scoring system (Table 65).

An antler score is cumulative, using the maximum width, number of points on each side, length and width of the palms, and circumference around the base of the antler. From this score, the difference between each palm is subtracted, and that result is the final score. All scoring is done to the nearest eighth inch. Because of anomalies and decisions as to what is or is not a point, all hunters interested in having the antlers scored are advised to contact the Boone and Crockett Club for a list of scorers in a particular area.

Mount Preparation

As noted earlier, a decision to have the animal's head mounted must be made before any skinning occurs. Also before skinning, photographs should be taken of the head from different angles, and circumference measurements taken of the neck immediately behind the ears, half way to the base of the neck and at the base. Obviously, the hunter who plans on mounting a moose head, if a trophy animal is bagged, should carry a camera that is not cumbersome and can be kept weather-proofed. Measurement also should be made of the length of the head starting at the hairless spot on the front of the nose to a point midway between each antler. These reference items will greatly aid the taxidermist. If a tape measure is not available, a piece of string will do, by tying knots in it at the appropriate spot and remembering which length is for what measurement. The very best advice is to consult with a taxidermist before the hunt about mounting and his recommendations.

Figure 236 illustrates the skinning cut lines to be used if a full head mount is desired. Skinning the neck and front shoulders is not complicated, but more caution is required with the ears, eyes and lips. If the hunter is unfamiliar with the procedure and technique, consultation with the taxidermist or simply taking it to the taxidermist as soon as possible for this delicate work is recommended. The head and hide should be kept cool, and the hide should not be wrapped around the head until it has cooled sufficiently, otherwise the hair may "slip" during tanning.

When incisions are made, the throat area must not be cut. All cuts should be made on top of the neck, with care

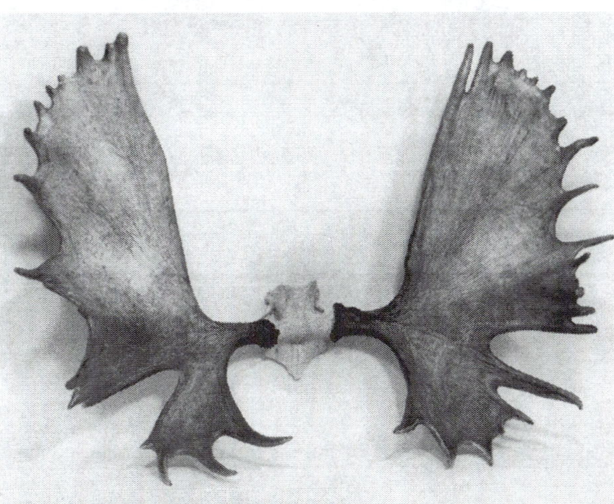

World record moose antlers (as of 1996) in the three Boone and Crockett Club categories include: (top left) Alaskan/Yukon moose (score 261 5/8), taken by John A. Crouse, September 1994, near Fortymile River in Alaska; (bottom left) Canada moose (score 242), taken by Michael E. Laub in 1980 near Grayling River, British Columbia; and (above) Wyoming or Shira's moose (score 205 4/8), taken by John M. Oakley in 1952 near Green River Lake, Wyoming. *Photos courtesy of the Boone and Crockett Club.*

Figure 236. Skinning cut lines (dashed line) for a moose head/cape mount and cut/saw line (dotted line) for antler mount (from Monk and Buss 1990). Such cuts should be made from the inside out to avoid cutting the hair shafts.

given to cutting outward, as discussed earlier, to avoid cutting the hair. Once the skin is free, any flesh adhering to it should be scraped off carefully. The head hide should then be salted thoroughly; if the hide can be frozen immediately, there is no need for salting. Finally, for transport, the hide should be wrapped with burlap or other cloth bag, *but not plastic*.

When removing antlers, enough of the top of the skull must be sawed off for the taxidermist to use as a guide when placing them on a form for mounting. Cuts should be from the back of the skull toward the front, exiting the skull in the eye sockets. This will ensure that enough bone remains with the antlers. If a bullet has caused damage, the antler can be repaired by a taxidermist.

Many hunters prefer simply to bleach the skull plate and mount the antlers on a plaque without any padding. Before bleaching, the skull plate should be boiled to remove any adhering meat, fat and connective tissue. Once this is done, bleaching can be accomplished with a commercial peroxide.

Cooking

Moose meat is lean without much fat in the muscle bundles. Consequently, well-cooked moose meat tends to be dry. To overcome this, extra fluids or a source of fat should be added during cooking. Marinades add moisture and tenderize roasts and steaks. Incisions can be made in roasts and small strips of bacon inserted. But the best safeguard is not to overcook the meat.

For preparing ground moose (burger), some pork or beef fat may be added (15 percent is recommended for lean meat) to provide additional fat. Beef fat supplement usually is preferred because it preserves longer when frozen. Another alternative is to package ground moose without fat added, but mix it with ground beef when thawed for cooking. However, moose burger, by itself, is lean and tasty.

The key to freezer storage of moose and other meats is judicious wrapping. Also, meat to be frozen should be carefully packaged in high-quality freezer wrap, *not* freezer bags. The wrap can avoid freezer burning for 2 years or more; freezer burn is likely in less than a year with freezer bags.

Conclusion

Concern exists for the future of hunting in North America. Hunters and other advocates of this recreational pursuit must become more vocal in pressing the cause for main-

A properly planned moose hunt, safely and ethically undertaken with valued companions is an unparalleled adventure, regardless of moose taken. Few experiences can match the enjoyment and bond of parent and offspring (author and son above) sharing the challenges and rewards of a wilderness moose hunt. *Photo by Ross Singleton.*

taining hunting's values and traditions, for they mirror the best qualities of the environment. Proponents should be urged to address hunting as an integral aspect of wildlife management, and as an important economic, recreational and cultural contribution. Toward this end, hunters should encourage wildlife management agencies to provide short courses on wildlife biology and management.

Hunting is a pursuit, but it is also the mornings, the nights, the seasons, the lands, the waters, the animals and the sensory grandeur of the outdoors that is poorly articulated but wonderfully experienced. Hunting is a unique and incomparable kinship with North America's wildlife heritage and a foremost link to natural resource conservation. It is a privilege to be enjoyed and protected.

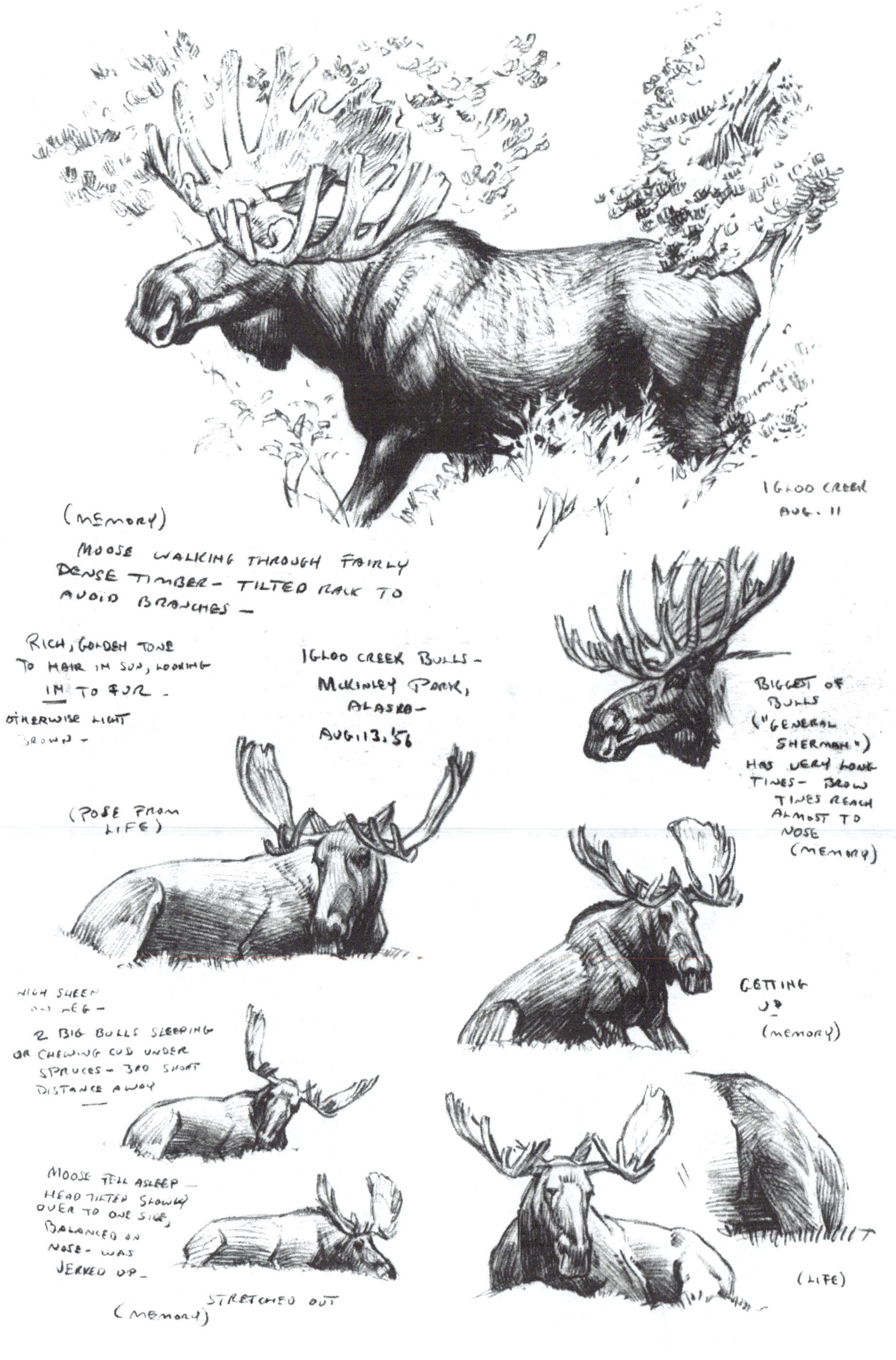
IGLOO CREEK
AUG. 11
(MEMORY)
MOOSE WALKING THROUGH FAIRLY DENSE TIMBER - TILTED RACK TO AVOID BRANCHES -
RICH, GOLDEN TONE TO HAIR IN SUN, LOOKING IN TO FUR.
OTHERWISE LIGHT BROWN -
IGLOO CREEK BULLS - McKINLEY PARK, ALASKA - AUG. 13, '56
BIGGEST OF BULLS ("GENERAL SHERMAN") HAS VERY LONG TINES - BROW TINES REACH ALMOST TO NOSE (MEMORY)
(POSE FROM LIFE)
HIGH SHEEN ON LEG -
GETTING UP (MEMORY)
2 BIG BULLS SLEEPING OR CHEWING CUD UNDER SPRUCES - 3RD SHORT DISTANCE AWAY
MOOSE FELL ASLEEP - HEAD TILTED SLOWLY OVER TO ONE SIDE, BALANCED ON NOSE - WAS JERKED UP -
STRETCHED OUT
(MEMORY)
(LIFE)

19

VINCENT F.J. CRICHTON, WAYNE L. REGELIN, ALBERT W. FRANZMANN AND CHARLES C. SCHWARTZ

The Future of Moose Management and Research

Tremendous advances have been made in the biological and scientific management of moose since the establishment of the wildlife profession in the 1930s. As the chapters of this book have demonstrated, a great deal is known about moose biology and the factors that affect populations. Continued progress will be determined by the future direction of research and management, predicated as much on social and political issues as on biology and ecology.

Wildlife in North America is a public resource. Traditionally, federal, provincial and state wildlife management agencies have emphasized management of game species to maximize sustained harvest levels. This focus and approach continue, but in recent years most agencies have expanded their program objectives to provide greater opportunities for nonconsumptive uses and enjoyment of both hunted and unhunted wildlife. Public information and education programs also have been expanded in response to public demands.

Interest in and concern about wildlife management is greater today than at anytime in the past, despite hunters representing a gradually decreasing percentage of the population (Filion et al. 1988). There are more people, and the variety and intensity of demands on wildlife resources—along a spectrum from elimination to preservation—are growing. A large portion of the demands is for viewing opportunities, especially of big game animals such as moose. However, with a growing and sprawling human population, the inevitable conflict with productive wildlife populations perhaps inevitably has given rise to situations where people are inconvenienced, frustrated or angered by the presence of wildlife.

Nevertheless, many people have become actively involved in volunteering assistance to wildlife management programs on public lands and taking greater interest and action to enhance wildlife numbers and diversity on private land. And more and more citizens are becoming involved in decision-making processes that allocate wildlife resources among various user groups.

The political systems in North America are driven in large measure by prevailing public demand. This determines which public goods or services will be supported by government. Presently, public concerns have particular focus on environmental issues and animal welfare. As societal values and circumstances change, demand for new programs or loss of support for existing ones invariably will occur. In the interim, wildlife managers in general and moose managers in particular are well-advised to avail themselves of the opportunities for and challenges of public involvement and to seek this support proactively. Similarly, the public should be made to recognize the importance of their involvement as a full partner in wildlife management programs.

Biopolitics

To conserve the moose resource for the future many management decisions will be made with political overtones, but they must be formulated with a basic understanding of the fundamentals of moose biology and the accompanying management constraints. Moose management is and will remain social and political as well as biological. The advent of First Nation claims to the resource in Canada and the obligation of governments to accommodate these peoples separately will keep wildlife management in the political spotlight for some time. Similarly, the issue of a priority for subsistence use in the United States, particularly in Alaska, will continue to focus public attention on equity issues of wildlife management and allocation.

The words "biology" and "politics" probably represent contrary views of how wildlife should be managed (Peek et al. 1982). Biology involves the gathering, analysis, interpretation, and application of data in a methodical scientific and peer-approved process to achieve goals. Politics may be defined as the processes of government administration of public affairs. Biopolitics is the correlation and interaction of biology and politics. Greenly (1971: 505) suggested that it is "the art of resolving biological game management problems in a biologically sound and politically acceptable manner." All too frequently biological/ecological issues are resolved in a politically acceptable way, but at the compromise of astute or practicable wildlife management. Responsible moose management should be based on sound scientific knowledge, but it often is clouded or misdirected by ephemeral political considerations (Crichton 1988a).

In natural resource management, biology is seldom pure science, and politics is seldom corrupting. With the former, data are collected, analyzed and conclusions drawn by individuals with a point of view drawn from biological training and predisposition. With the latter, decisions are made within the context of laws, social circumstances and financial resources. It is unrealistic to think that biology and politics are regularly compatible.

There is nothing inherently wrong with biopolitics. In fact, it is the essence of natural resource management in North America, internally and in relationship with the public. Unfortunately, most natural resource managers are not well-versed in the compromising nature of biopolitics. A good biopolitician combines biological and political savvy to achieve reasonable objectives within the prevailing constraints of legal, social and economic circumstances. The success of moose management programs and ultimately, conservation of the moose resource depend on managers who are skilled biopoliticians.

The annual promulgation of hunting seasons probably is one of the more visible biopolitical processes of wildlife management because it involves the interest of participation by a directly affected and sensitive constituency, namely hunters (Regelin 1991). Season-setting impacts a large number of individual users, and many special interest groups are focused on the deliberations.

In matters of resource use allocation, such as hunting seasons, all segments of the public should be encouraged to get involved, listen to opposing viewpoints and participate in meaningful discussion to ensure that the resource in question is wisely and equitably used, without compromising its availability for future generations. At present, energy is expended all too often in defense of special interest and posturing for particular philosophies, rather than in finding truth, reality and common ground. Inevitably lost in such opportunities is the well-being of the resource in question. Meaningful discussion and reasonable compromise are keys to resource allocation. They are achieved by understanding of biological principles, management objectives and social sensitivity.

Some Human Dimensions of Moose Management

A number of reasons account for the decline in the population proportion of hunters in North America at a time when general interest in wildlife management is increasing. First, society is increasingly urban. The majority of people in North America live in large cities, and most have had limited exposure to the rural or natural environment and the rhythms of nature. Second, television programs on nature and wildlife are the primary source of wildlife information for millions of North Americans. Some of these programs anthropomorphize animals and contain an antihunting slant, and viewers of such information can form idealistic perspectives about wildlife and nature that are remote from biological reality. Thus, at least some people are willing to devote time, energy and money to challenge wildlife management to conform to their ideals and preconceptions. Urban-based antihunting and/or animal rightist organizations have been established to influence biopolitical processes. Berryman (1987) and Thomson (1991) characterized the animal rightist movement as one of the most ominous threats facing wildlife managers. Duda et al. (1995) reported that 11 percent of adults in the United States strongly disapproved of legal hunting and another 11 percent moderately disapproved.

Another significant societal change is the proliferation of litigation. If management agencies or wildlife commissions make decisions that an individual or a group opposes, that person or contingent can sue in provincial, state or federal courts. Presently, this activity is more common in the United States than Canada. The National Environmental Policy Act (NEPA) in the United States provides opportunity for court action through the environmental impact statement process. Some states now have state laws similar to NEPA. In several provinces and states, the priority for subsistence use has resulted in numerous legal challenges to the system of allocating wildlife among various user groups.

A public opinion survey in Canada revealed more than 85 percent of Canadians stated that maintaining abundant wildlife was important to them (Filion et al. 1981). In 1991, nearly 13.9 million people in the United States, 16 years old and older, hunted an average of 17 days and spent $12.3 billion on hunting activities, or $900 each (U.S. Fish and Wildlife Service 1993). More than 76 million adults participated in nonconsumptive wildlife-related activities, including feeding, observing and photographing wildlife. These statistics clearly illustrate that people in the United States and Canada place a high value on wildlife resources and are willing to spend time and money to protect, manage and utilize them.

Wildlife viewing and study are rapidly expanding pastimes. "Watchable Wildlife" is an idea whose time has come. Viewing programs will increase public awareness, understanding and commitment to sharing wildlife through ethical and sustainable uses. Reciprocally, they serve to remind hunters that all citizens have a vested interest in wildlife resources, regardless of who pays the lion's share of the costs. The "fair chase" concept must be extended to viewing as well as to hunting. In terms of moose management, wildlife viewing accomplishes three things: (1) it puts people of all ages in contact with moose interacting with the environment; (2) it encourages participation by a cross-section of the public in conservation of the species and its habitat; and (3) it offers sustainable economic opportunities.

Moose viewing should be recognized as an ideal mechanism to inform and educate the public about all aspects of moose biology, management and research. Managers have a unique opportunity to provide leadership in promoting viewing of moose. McKenna and Lynort (1984) stated that wildlife professionals have only begun to deal with private conservation groups regarding environmentally damaging projects. Sharing information and expertise with the public can only advance sound and progressive moose management.

Subsistence Issues

CANADA

First Nation peoples have long used moose and other wildlife resources for cultural, subsistence and economic purposes throughout Canada. Presently there is an increased desire among them to ensure that the resources are used and managed in ways that will ensure ongoing use for present and future Native generations. First Nation peoples are prepared to accept reasonable regulations designed to protect resources for the future. They also are becoming involved in protecting wildlife and habitats in their traditional ways. They believe that treaty rights include the right to hunt, but realize that wildlife must be managed and include reasonable regulations to limit harvest and hunting methods. Native people, instead of merely taking advantage of their rights to hunt and fish, now are becoming directly involved in developing regulations and management plans to perpetuate wildlife populations and their values. The benefits of this cooperative approach to natural resource management are discussed in detail by Feit (1984), Usher (1987), Pinkerton (1989), Working Margins Consulting Group (1991) and Berkes et al. (1991).

An important decision by the Supreme Court of Canada occurred in the 1990 case of *Regina* v. *Sparrow*. The court ruled that the burden of proof for tests justifying limitations on harvest by First Nation peoples in Canada falls on the Crown. These tests are:

- Is there a valid legislative objective?
- Is the legislation consistent with the special trust relationship and the responsibility of the government vis-à-vis aboriginals?
- Has there been as little infringement as possible to affect the desired result?
- In a situation of expropriation, is fair compensation available?
- Has the aboriginal group in question been consulted with respect to the conservation measures being contemplated?

Provincial governments must justify regulations restricting harvests by First Nation peoples according to the tests set out in *Regina* v. *Sparrow*. This will expose contemporary policies and practices in fish and wildlife management to more rigorous testing, evaluation and greater public scrutiny and debate. It has the advantage of improving the scientific credibility of moose management programs and enhancing acceptability of these programs by the public.

Management agencies should look for opportunities to initiate new cooperative management programs. The *Regina* v. *Sparrow* decision hopefully will lead to establishment of co-management of wildlife resources in many parts of Canada. *Co-management* refers to a shared natural resource management investment and responsibility by government agencies and local communities. In the past, co-management has come about in response to critical decline of wildlife populations. Its potential benefits include:

- Sustainable harvests
- Reduced overexploitation of wildlife populations
- Termination of conflicts between users and conservation officers
- More data for management and planning
- Improved governmental/tribal relationships
- Conflict resolution among different user groups
- Maintenance of traditional values and lifestyles
- Establishment of a long-term economic base

ALASKA

An example of how the courts and state and federal legislatures dealing with the subsistence priority issue can confound sound wildlife management is exemplified by an Alaska experience.

In 1978 the Alaska legislature passed a law that provided priority, based on customary and traditional uses of fish and game for personal or family consumption, for subsistence uses of fish and wildlife. At the time, the state argued that a federal subsistence law was duplicative and unnecessary. Nevertheless, in 1980, the United States Congress passed the Alaska National Interest Lands Conservation Act (ANILCA). This act set aside more than 100 million acres (40.5 million ha) of federal land in Alaska for inclusion into the national park and national wildlife refuge systems. It ensured that *rural* residents of Alaska would have priority for nonwasteful subsistence use of fish, wildlife and other renewable resources on federal public lands. It also required the state to provide a rural subsistence priority or the federal government would take over management of fish and wildlife for subsistence purposes on federal lands.

The Alaska Board of Game passed regulations in 1982 to implement the Alaskan law and conform to the federal mandate. These regulations provided a subsistence priority based on customary and traditional uses of fish and game by rural Alaskan residents. The Board of Game also developed a process to determine which communities and areas of the state were rural and which communities had customary and traditional uses of wildlife. Regulations indicated that when demand for wildlife resources exceeded the sustained yield, use by nonresidents would be eliminated first. If harvest by subsistence users had to be restricted to maintain sustained yield, a complex system based on an individual's reliance on the resource would be established.

Administration of the subsistence regulations was complex, cumbersome and expensive. Months in advance of a hunt, hunters had to fill out applications that were subsequently evaluated and scored by the Fish and Game Department, and permits issued accordingly. The process was not popular—some people stopped hunting, and others ignored the regulation and hunted without proper permits. Numerous suits were filed in federal and state courts by people who thought the entire system or portions of it were unfair. A referendum on the priority for subsistence use was held in 1982, and 58 percent of Alaskans voted to retain the subsistence priority.

The Alaska Supreme Court ruled in February 1985 that the state regulations were invalid because eligibility for subsistence priority was based on community characteristics, not individual characteristics as required by statute. The Board of Game regulations based on rural residency became void, and the state was found not in compliance with ANILCA. The U.S. Department of the Interior then threatened to take over management of wildlife on federal lands if the state did not modify their laws to comply with the federal law.

Wildlife management was in limbo. The 1985 Alaska legislature was in session when the Supreme Court made its ruling. The Alaska Department of Fish and Game and the Board of Game waited for legislative action, but it was not forthcoming. The legislature adjourned in May without addressing the issue. This inaction placed wildlife management in turmoil and seriously jeopardized the 1985 hunting seasons. In May, the Board of Game held an emergency meeting and implemented a temporary regulatory system. Most hunters were confused. Many nonresident hunters became frustrated with Alaska and hunted elsewhere in 1985. Fewer residents hunted.

In 1986, the Alaska legislature amended the state's subsistence law to conform with the federal law. The amendments added the word "rural" to the subsistence priority and authorized the Board of Game to make community-based customary and traditional determinations. This law was immediately challenged in what is commonly known as the "McDowell case." That case challenged the constitutionality of the law, arguing that the state could not discriminate against a person based on where they lived.

While this court case moved slowly through the legal system, the Board of Game continued to regulate hunting as it had done from 1982 through 1984. There were four years (1985–1989) of relative stability in the overall subsistence regulatory framework. But there were numerous court cases challenging the authority of the state to regulate subsistence hunting. In several instances, decisions made by the Board to regulate harvests were overturned by state or federal judges. Some reversals of Board decisions were based on law, but in several cases judges simply substituted their determinations for that of the Board of Game. Wildlife management was being dictated by the courts.

On December 22, 1989, the U.S. Supreme Court finally ruled in the McDowell case and found that the portion of the law related to rural residency violated the state's constitution. Again, the state was not in compliance with ANILCA, and the federal government once more threatened to assume authority for wildlife management on federal lands.

The obvious solution to the conflict between the state constitution and ANILCA was to amend one or the other. The Alaska legislature refused to place a constitutional amendment on the ballot in 1990. And the Alaska congressional delegation refused to endorse amendments to ANILCA. A stalemate resulted. Consequently, on July 1, 1991, the federal government assumed authority for management of subsistence uses of wildlife on federal lands in Alaska. This meant a dual system of management, with duplication and sometimes conflicting state and federal regulations. The federal system now in place is administered by a large staff with a budget considerably larger than that available for such work by the Alaska Department of Fish and Game.

No solutions are apparent. Alaska's governor formed a Subsistence Advisory Committee to recommend new legislation to address the problem. After months of meetings and reasonable recommendations, the legislature initiated minor amendments to the state's subsistence law, but neither resolved the problem nor affected dual federal/state management. The Board of Game initiated implementation of the amended subsistence law, which allows for creation of nonsubsistence areas where the socioeconomic structure of the areas is not subsistence based. This law is certain to be challenged in court and the controversy will continue.

Wildlife resources and users are the ultimate losers. Dual, duplicative management is more expensive, more complex and forces expenditure of resources and personnel time on unnecessary meetings, court cases and increased bureaucracy. These resources could be much better spent on wildlife management programs. Furthermore, users are faced with a complicated system that has little biological foundation, makes no biological sense and is perceived by many as grossly unfair.

The Alaskan biopolitical scenario can potentially repeat itself in other jurisdictions. Efforts to prevent development of such occurrences fall to all who are concerned about sound and effective wildlife management, not just to wildlife biologists, not just to politicians and not just to jurists.

The Management Challenge

Many challenges face wildlife managers. They must be capable communicators and mediators as well as skilled biologists. Today, "people skills" are nearly as important as biological knowledge in the management of moose. The involved public expects moose populations to be managed professionally and to play a key role in developing moose population and human use objectives.

Organization, synthesis, analysis and interpretation of biological information, then presentation of it to the public in an understandable and useful form are difficult propositions. Nevertheless, wildlife managers must be forthright and objective in the presentation of facts, clearly identify assumptions, acknowledge any deficiencies in the information base when they occur, and avoid too many assumptions about what the public does not need to know. Public credibility is a prerequisite of effective wildlife management, and that is not achieved if the public finds or senses that matters are presented sparingly, grudgingly, patronizingly and unilaterally.

Management agencies face a major challenge in devising an efficient process to allow meaningful public input into management decisions and plans that are essential for effective moose management. Management plans in particular should provide population goals and human use objectives for all species of wildlife on an area-specific basis (Nolan et al. 1993). Other essential elements of any plan are:

- Accurate biological information on population status and trends, mortality factors, and habitat limitations
- Practical management options to achieve population and human-use goals
- Recognition and accommodation of diverse uses and values
- Time frame for evaluating and updating the plan

A proper planning process will provide several opportunities for public input during development of the plan. Planning processes often are difficult and time consuming because all segments of the public are invited to participate. Groups with widely contrasting values and interests hopefully will contribute to the planning process, although communication with and among those having fundamentally different philosophies can be difficult. However, as Decker and Brown (1987) stated, wildlife professionals must understand the views of all so that they can help unify and institute a management philosophy that is in the best interests of society. Moose managers must recognize that science is not the dominating facet in motivating, shaping and directing moose and other wildlife management programs. If a professional wildlife manager suggests that a decision is based solely on scientifically derived biological considerations, the manager either misunderstands the nature of science (by confusing scientific judgments with ethical judgments [Underwood and Porter 1991]) or is deliberately trying to dis-

←

Primary nonconsumptive use of wildlife has great economic, cultural and recreational importance to North Americans. In the United States in 1991, 76.1 million people 16 years of age or older purposefully participated in observing, feeding and/or photographing wildlife, and spent $18.1 billion on those activities. That same year, more than 18.9 million Canadians spent 1.3 billion days and $5.6 billion in purposeful wildlife-related activities. Both the activity, participation and expenditure trends show significant and continuing increase. Although the economic boon of nonconsumptive wildlife-associated activities does not contribute as much to national or local economies or nearly as much directly to resource management programs as do consumptive user dollars, there are many more nonconsumptive users than there are hunters, trappers and anglers. The nonconsumptive users, regardless of direct investment in wildlife programs, represent a majority constituency of wildlife agencies, and their demands for resource allocation must be heeded. Servicing those demands should be viewed by the agencies as no more an accommodation than providing services and opportunities for consumptive users. In the long term, this perspective of professional responsibility should engender broader support for all programs that benefit wildlife and its various user groups. *Top photo by Holly Kuper. Center photo by Michael H. Francis. Bottom photo by Charles H. Willey.*

guise or complement a value judgment under the veil of the legitimacy of science. The wildlife management profession should guard against committing such errors and should not condone their perpetuation (Decker et al. 1991).

Again, all groups wishing to assist in the development of management plans should be allowed to participate. However, some animal rightist groups try to use the planning process to stall management decisions or as political platforms to promote their viewpoints and agenda to the exclusion of all others. Many such groups have publicly avowed to subvert all wildlife management and end hunting. Members of these groups are motivated by a highly charged emotional ideology, which holds that killing of animals for any purpose is anathema (Thomson 1992). Some such groups are often well-funded and able to hire effective propagandists to promote their cause. And propaganda, which is purposeful distortion of facts, is difficult to counter. Regardless, wildlife managers have an obligation to integrity of data and to communicate that data understandably to the public.

A strong argument supporting the concepts and principles of wildlife management is in "The World Conservation Strategy" (International Union for Conservation of Nature and Natural Resources 1980). This policy statement, representing a blueprint for the survival of humans and nature, has three objectives: (1) to maintain essential ecological processes and life support systems; (2) to preserve genetic diversity; and (3) to ensure the sustainable utilization of species and ecosystems.

The last objective is especially significant inasmuch as sustainable utilization of wildlife species cannot be accomplished without sound wildlife management.

Sound wildlife management plans also make the task of allocating the wildlife resources among the various user groups much easier. An effective planning process and sound biologically based, socially responsive plans help ensure that allocation decisions are logical and consistent. They reduce litigation and help ensure that wildlife management decisions are not made by judiciary.

Research Gaps, Needs and Opportunities

Attendees at the first North American Moose Conference and Workshop in 1968 produced a list of moose research needs. Although many of these research projects were undertaken, Bubenik (1981) suggested that few of the results were incorporated into the management process. Bubenik further noted that, in fact, many research results could not be applied to management because certain general biological data on moose are incomplete or missing.

Although considerable moose research has been completed in the past 15 years, Bubenik's latter observation at least still has validity. Much about moose biology and ecology remains to be answered.

Most notable of needed additional research is that which can address the influences of sex and age structure on reproductive potential and the influence of various management strategies on the genetic diversity of moose populations. Also, there is need for continued research and evaluation of the impacts on moose—population by population and cumulatively—from human developmental activities across the species' North American range. Most research to date has focused on single biological issues or the impact of single developments, e.g., logging, hydroelectric development, mining. Cumulative impact assessments have not been adequately addressed. A major limitation has been the lack of baseline information. Agencies should initiate steps to ensure that appropriate monitoring occurs before development, and that more resources and priority are committed to data collection and analysis.

Despite substantial progress in understanding the evolutionary history and associations of moose, some important data gaps persist. The classic basis for subspeciation of moose lacks a rigorous scientific foundation. This can be corrected with developing technologies, especially in genetics. However, the pursuit of this information has been stymied by cost, perceived lack of correlation to management priorities and a general inability to coordinate globally among projects investigating evolutionary events. Recent political circumstances, particularly in Asia, now afford opportunity for communication and sharing on taxonomic classification studies.

Anatomical and morphological descriptions of moose have focused on the obvious comparisons with other species, particularly domestic ungulates. A complete anatomical description, similar to those existing for domestic species, would help explain more of the functional, behavioral and activity nuances that biologists observe.

Also lacking is knowledge that links behavioral characteristics of moose to the physiological mechanisms that regulate them. The relationships between social well-being and moose population composition need to be quantified. For example, how does social structure influence reproductive timing? How does a changing adult sex ratio influence timing and success of reproduction within a moose herd? How important is maintenance of a component of "prime bulls" in the population? Is physiological age equivalent to chronological age, or do large bulls inhibit both behavioral and physiological maturation of young bulls? Similarly, better information is required about the role of chemical signals and moose reproduction. Does a bull synchronize estrus in the female moose through biochemical means (e.g., urination in wallows or rutting pits)? How do chemical signals of moose change seasonally—with age or stage of maturation? Bubenik (1987) suggested that, in exploring these and other questions, biologists cannot simply treat moose as numbers without regard to population structure and social balance. Individual moose and moose populations are discrete biological entities.

Significant to the distribution of moose is ambient temperature. Subtle changes of temperature over time undoubtedly will influence moose distribution, population size, interspecific competition, exposures to new and/or more mortality factors, and both positive and negative human interactions. Monitoring temperature shifts and moose responses, particularly at the southern edge of the species' range, will help forecast future physiological, social and competitive vulnerabilities of moose on a broader geographic scale.

Still incomplete is understanding of all the important factors that regulate and limit moose populations. Long-term case histories are few, especially of populations in naturally regulated ecosystems, and experimental studies are limited. A combination of case history study, long-term population monitoring and large-scale experimentation is needed to enhance understanding of limiting and regulating factors. Such information would be especially useful in unraveling complexities of predator/prey systems. For example, when, where and how does predation limit moose populations? When does predation shift from compensatory to additive mortality? How do human influences on the ecosystem alter predation and other mortality factors?

Incidental mortality factors for moose have not been adequately quantified, particularly for populations where mortality may be additive. High winter mortality rates reported in recent years from moose collisions with trains and automobiles amplify the concern and research need. Not incidental to the matter are human injury, death and property damage associated with such collisions, the public reaction to such events and the consequent biopolitical influence on management.

If moose are to be treated as a featured species in a particular area, better knowledge is needed about their use of that area, the habitat conditions and the potential for impact on either by human activity. For example, myriad development activities in a forested area have the potential to affect the reproductive success of a moose population, either by displacing it or interfering with its movements and migrations. There are many examples of how these external factors have negatively impacted moose populations across North America, but the predictability and preventability of negative impacts certainly can be improved.

More research is needed on interactions and competition between moose and domestic animals, especially in terms of parasite and disease associations. Such transmissions have bearing on livestock management and implications for translocations of moose to new areas or of other wildlife into moose habitat.

One of the greatest challenges in moose management is understanding the degree (and range) of adaptability of moose to a naturally changing environment and the rate and extent of effect of human activity on that adaptability. Better methods to assess the quality and quantity of vegetation in moose habitats is needed to address the elusive and often misused concept of "carrying capacity." Refinement of techniques using moose as an indicator/emphasis species to assess habitat also needs to be pursued.

Habitat managers must identify their most important research need as better assessment of habitat management program effects on moose at the population level. This will require large-scale experimentation carefully designed to control the many variables that affect moose populations. Also, such investigation likely must have at least a 30-year term to enable sufficient information for thorough and optimally meaningful evaluation.

Better understanding is needed of how moose use time,

space and food to maintain fitness, avoid heat and cold stresses, minimize predation and achieve reproductive potential. More specifically, increased knowledge of habitat use patterns is essential to learn how and why moose select food that may be abundant but of poor quality. Likewise, more information is needed on nutritional energetics on an individual moose basis, because of the role energetics plays in individual survival in relation to population dynamics. In the future, moose will be managed near vegetative carrying capacity where food drives population dynamics, so better understanding will be needed of how moose use and process food under a variety of conditions.

Resource management entails at least three fundamental questions. First, what can the resource provide? Second, what do people want the resource to provide? And third, what information and management are required to attain and sustain a balance between biological capability and societal desirability, both of which are dynamic and mutually exclusive? More difficult is determination of best management in cases of informational gaps and/or management constraints. Most difficult is best management under those limitations when multiple user interest is high.

Rarely can any moose population meet all the demands placed on it. Moose resource allocation requires very critical planning focused on time and space. Even with maximum utilization of a particular wildlife resource, such as moose, not all users will be satisfied. And with increasing human populations, the allocation dilemma will become more and more acute over time.

Changing Perspectives in Wildlife Management

Wildlife management has changed dramatically in recent years. Today, there is increasing focus on ecosystem and habitat management, with an emphasis on biodiversity. The advent of "conservation biology" has deemphasized single species management. With this shifting, the organization, authorities and responsibilities of wildlife agencies also have changed. Few agencies can now afford the luxury of species specialists. The trend is toward landscape-level ecologists. Perhaps inevitable and likely progressive in most cases, the change has significantly altered the biologist's role, particularly in the area of communication.

For example, "moose biologists" developed a core and network of professional communication through formation of the North American Moose Conference and Workshops and its printed proceedings that spawned the journal *Alces*. In the late 1970s and early 1980s, at least one delegate from nearly every moose management province and state in North America attended the meetings and contributed to the written record. In the late 1980s, with the shift toward ecosystem management, the events began to experience a gradual attrition of attendance and participation and a correlative effect of communications and data exchange. By the mid-1990s, with special outreach efforts, the trend of attendance and moose jurisdiction representation has been improving. Recently, moose biologists began a newsletter, *The Moose Call,* which serves not only as a means of informal communication among their members but also conveys up-to-date information on moose biology and management to other interested readers.

With the prevailing direction of wildlife management toward conservation biology and its emphasis on ecosystems, with the retirement—literally or figuratively—of moose biologists, and in the absence of an encompassing communication vehicle among professionals charged with moose management and research, is continued or improved momentum of moose management a realistic expectation? Moose management is not counter to conservation biology or most other administrative program reorientations. But it can be only that which it is allowed. And a danger exists to the resource—moose—if management attempts to be too inclusive and if readjustment occurs at the sacrifice or compromise of programs that have been the mainstay of professional management all along.

Big Bull entering pond near
Wonder Lake Ranger Station –
August 20, '56 McKinley Park

Short tines —
antlers relatively simple –

APPENDIX A

Common and Scientific Names of Plants Cited

Plants are listed alphabetically by common name(s). A plant may be listed under more than one common name used in the text. Scientific names have been standardized according to the sources referenced or otherwise indicated. If a scientific name for a particular plant was not located in the source literature, other taxonomy references were consulted, including Britton and Brown (1943), Subcommittee on Standardization of Common and Botanical Names of Weeds (1966), Fernald (1970), Hotchkiss (1970), and Scott and Wasser (1980), S.G. Shetler (personal communication: 1996).

Alder, green *Alnus viridis crispa*
Alder, speckled *A. incana rugosa*
Alder, tall or Sitka *A. viridis sinuata*
Alder, thin-leaf *A. incana tenuifolia*
Arrowgrass, common seaside *Triglochin maritimum*
Arrowhead *Sagittaria* spp.
Ash, black *Fraxinus nigra*
Aspen, bigtooth or large-toothed *Populus grandidentata*
Aspen, quaking or trembling *P. tremuloides*
Aster, large-leaved *Aster macrophyllus*
Basswood *Tilia americana*
Beargrass *Xerophyllum tenax*
Beech *Fagus grandifolia*
Birch, bog *Betula nana glandulosa*
Birch, dwarf *B. nana*
Birch, Kenai *B. papyrifera kenaica*
Birch, paper or white *B. papyrifera*
Birch, silver *B. pendula*
Birch, water or river *B. occidentalis*
Birch, yellow *B. alleghaniensis*
Bitterbrush *Purshia tridentata*
Blackberry, Allegheny *Rubus allegheniensis*
Blackhead (coneflower, western) *Rudbeckia occidentalis*
Bladderwort, common *Utricularia vulgaris*
Bluegrass *Poa* spp.
Box, false *Paxistima myrsinites*
Bromegrass *Bromus* spp.
Buckbrush *Ceanothus* spp.
Buckthorn *Rhamnus* spp.
Bulrush *Scirpus* spp.
Bulrush, water *S. subterminalis*
Bunchberry, cornel, dwarf *Cornus canadensis*
Burnet, Canadian *Sanguisorba canadensis*
Burrweed *Sparganium* spp.
Calla, wild *Calla palustris*
Cattail, common *Typha latifolia*
Cedar, northern white *Thuja occidentalis*
Cedar, western red *T. plicata*
Cherry, choke *Prunus virginiana*
Cherry, pin *P. pensylvanica*
Clematis *Clematis* spp.
Cloudberry *Rubus chamaemorus*
Clover, alsike *Trifolium hybridum*
Cottonwood, black *Populus trichocarpa*
Cowlily (spatterdock) *Nuphar variegata*
Cranberry, American highbush *Viburnum trilobum*
Cranberry, European highbush *V. opulus*
Cranberry, lowbush (lingonberry) *Vaccinium vitis-idaea*
Cranberry, mountain (mooseberry, moosewood viburnum, squashberry) *Viburnum edule*

Cretaria (lichen) *Cretaria* spp.
Crowfoot, water *Ranunculus aquatilis*
Currant, black *Ribes americanum*
Currant, red *R. triste*
Dogwood, alternate-leaved *Cornus alternifolia*
Dogwood, red-osier *C. servicea*
Douglas-fir *Pseudotsuga menziesii*
Eelgrass *Zostera marina*
Elderberry, red-berried *Sambucus racemosa*
Elm, slippery *Ulnus rubra*
Fir, balsam *Abies balsamea*
Fir, subalpine or alpine *Abies lasiocarpa*
Fir, white or silver *A. grandis*
Fireweed, narrow-leaved *Epilobium angustifolium*
Fireweed, broad-leaved or dwarf *E. latifolium*
Fleabane *Erigeron* spp.
Geranium, sticky purple *Geranium viscosissimum*
Gooseberry, northern *Ribes oxyacanthoides*
Grass Poaceae
Grouseberry *Vaccinium scoparium*
Hawthorn *Crataegus* spp.
Hazel, beaked *Corylus cornuta*
Hazelnut (filbert) *C. c. californica*
Hemlock, eastern *Tsuga canadensis*
Hemlock, western *T. heterophylla*
Hobblebush (viburnum) *Viburnum alnifolium*
Holly, mountain *Nemopanthus mucronata*
Honeysuckle, American fly *Lonicera canadensis*
Honeysuckle, bracted *L. involucrata*
Honeysuckle, glaucous or wild *L. dioica*
Honeysuckle, Utah *L. utahensis*
Hornwort *Ceratophyllum* spp.
Horsetail, water *Equisetum fluviatile*
Ivy, poison *Toxicodendron radicans*
Jewelweed, touch-me-not *Impatiens* spp.
Kelp, bull *Nereocyctis marcrophyllus*
Labrador-tea *Ledum groenlandicum*
Larch, American (tamarack) *Larix laricina*
Larch, western *L. occidentalis*
Leatherwood *Dirca palustris*
Cretaria (lichen) *Cetraria* spp.
Lupine, Nootka *Lupinus nootkatensis*
Maple, boxelder *Acer negundo*
Maple, Douglas *A. glabrum*
Maple, mountain *A. spicatum*
Maple, red *A. rubrum*
Maple, striped (moosewood) *A. pensylvanicum*
Maple, silver *A. saccharinum*
Maple, sugar *A. saccharum*
Marigold, marsh *Caltha palustris*
Mayflower, Canada *Maianthemum canadense*
Meadow-sweet *Spiraea latifolia*
Menziesia *Menziesia ferruginea*
Moosebush *Viburnum lantanoides*
Mountain-ash, American *Sorbus americana*
Mountain-ash, cascade *S. scopulina*
Mountain-ash, European *S. aucuparia*
Mountain-ash, Sitka *S. sitkensis*
Mountain-balm *Ceanothus velutinus*
Mountain-laurel *Kalmia latifolia*
Mushroom, bolete *Boletus* spp.
Nannyberry *Viburnum lentago*
Nettle *Urtica dioica*
Ninebark *Physocarpus opulifolius*
Oak, northern red *Quercus rubra*
Paintbrush, Indian *Castilleja* spp.
Palm Arecaceae
Pasqueflower *Anemone patens*
Peavine *Lathyrus* spp.
Peltigera (lichen, foliose) *Petigera* spp.
Pickerelweed *Pontederia cordata*
Pine, lodgepole *Pinus contorta*
Pine, eastern white *P. strobus*
Plantain *Plantago* spp.
Plum *Prunus* sp.
Pondweed, fennelleaf or Sago *Potamogeton pectinatus*
Pondweed, fineleaf or threadleaf *P. filiformis*
Pondweed, floatingleaf or floating *P. natans*
Pondweed, grassleaf or variable *P. gramineus*
Pondweed, largeleaf or bigleaf *P. amplifolius*
Pondweed, leafy *P. foliosus*
Pondweed, northern or red *P. alpinus*
Pondweed, ribbonleaf *P. epihydrus*
Pondweed, Richardson *P. richardsonii*
Pondweed, fern or Robbins *P. robbinsii*
Poplar, balsam *Populus balsamifera*
Raisin, wild *Viburnum cassinoides*
Raspberry, red *Rubus idaeus*
Raspberry, wild red *R. i. strigosus*
Reindeer moss *Cladina* spp.
Rice, wild *Zizania aquatica*
Rhododendron (laurel) *Rhododendron* spp.
Rhododendron (or azalea, cascade or white) *Rhododendron albiflorum*
Rose, prickly or bristly *Rosa acicularis*
Rush *Juncus* spp.
Saskatoon (serviceberry, western) *Amelanchier alnifolia*
Sedge *Carex* spp.
Shadbush (Juneberry or serviceberry) *Amelanchier* spp.
Silverberry *Elaeagnus commutata*
Skunkcabbage *Symplocarpus foetidus*
Skunkcabbage, American yellow *Lysichitum americanum*
Snowberry, western *Symphoricarpos occidentalis*
Soapberry *Shepherdia canadensis*
Spikerush *Eleocharis* spp.
Spiraea *Spiraea* spp.
Spruce, black *Picea mariana*
Spruce, Englemann's *P. engelmannii*
Spruce, red *P. rubens*
Spruce, Sitka or woodland *P. sitchensis*

Spruce, white *P. glauca*
Star-grass, water *Heteranthera dubia*
Sumac *Rhus* spp.
Sundew, round-leaved *Drosera rotundifolia*
Sweetgale, low *Myrica gale*
Thimbleberry *Rubus parviflorus*
Thistle, bull *Cirsium vulgare*
Thistle, Canada *C. arvense*
Thistle, elk *C. foliosum*
Trillium *Trillium* spp.
Twisted-stalk *Steptopus amplexifolius*
Waterlily *Nymphaea* spp.
Waterlily, large yellow *Nuphar lutea advena*
Waterlily, sweet-scented *N. odorata*
Waterlily, yellow *N. lutea variegata*
Water-milfoil *Myriophyllum* spp.
Watershield *Brasenia schreberi*
Wheatgrass *Agropyron* sp.
Willow, Barclay's *Salix barclayi*
Willow, barren-ground *S. niphoclada*
Willow, basket *S. purpurea*
Willow, beaked *S. bebbiana depressa*
Willow, Bebb's *S. bebbiana*
Willow, blueberry *S. myrtillifolia*
Willow, diamond-leaf *S. planifolia*
Willow, Drummond's *S. drummondiana*
Willow, feltleaf *S. alaxensis*
Willow, grayleaf *S. glauca*
Willow, great, goat's or florista *S. caprea*
Willow, Geyer's *S. geyeriana*
Willow, Halberd's *S. hastata*
Willow, interior, sandbar *S. interior*
Willow, Lemmon's *S. lemmonii*
Willow, littletree *S. arbusculoides*
Willow, McCalla's *S. maccalliana*
Willow, meadow, slender *S. petiolaris*
Willow, park *S. padophylla*
Willow, pussy *S. discolor*
Willow, Richardson's *S. lanata*
Willow, Scouler's *S. scouleriana*
Willow, shining *S. lucida*
Willow, short-fruited (barren-ground) *S. brachycarpa*
Willow, silver *S. candida*
Willow, tall blueberry *S. boothii*
Willow, undergreen *S. commutata*
Willow, Wolf's *S. wolfii*
Willow-herb *Epilobium* spp.
Witchhazel *Hamamelis* spp.
Woodfern, toothed *Dryopteris carthusiana*
Yarrow, common *Achillea millefolium*
Yew, American (ground-hemlock) *Taxus canadensis*
Yew, Pacific or western *T. brevifolia*

(LIFE + MEMORY)

APPENDIX B

Common and Scientific Names of Animals Cited

Animals are listed alphabetically by common name. The scientific names have been standardized according to the sources from which they were cited or other taxonomy references, including Banks et al. (1987), Gill (1990), American Ornithologists' Union (1983), Hall (1981), Chapman and Feldhamer (1982), Scott (1983), Borror and White (1970), Behler and King (1979) and Thomas R. McCabe (personal communication: 1996).

In Chapter 15, some organisms are identified only by scientific name because that is how they are generally or exclusively recognized.

Ant, carpenter *Camponotus herculeonus*
Antelopes Antilopinae
Bear, black *Ursus americanus*
Bear, brown or grizzly *U. arctos*
Beaver *Castor canadensis*
Beetle, carrion *Necrophila americana*
Beetle, darkling *Upis ceramboides*
Beetle, longhorn *Anoplodera canadensis*
Beetle, scarab *Geotrupes balyi*
Beetle, tiger *Cincindela repanda*
Bison, American *Bison bison*
Bison, plains *B. b. bison*
Bison, wood *B. b. athabascae*
Blackfly *Simulium venustrum*
Boar, European *Sus scrofa*
Buffalo, African *Sycerus caffer*
Bug, ripple *Rhagovelia*
Capercaille *Tetrao urogallus*
Caribou *Rangifer tarandus*
Caribou, barren-ground *R. t. groenlandicus*
Caribou, Grant's or Alaskan barren-ground *R. t. granti*
Caribou, woodland *R. t. caribou*
Cat, domestic *Felis domestica*
Catfish Ameiuridae
Cattle, domestic *Bos taurus*
Chamois *Rupicapra rupicapra*
Chicken *Gallus gallus*
Cow, domestic *Bos taurus*
Coyote *Canis latrans*
Cricket, field *Gryllus pensylvanicus*
Deer, black-tailed, Columbian *Odocoileus hemionus columbianus*
Deer, black-tailed, sitka *O. h. sitkensis*
Deer, chital *Axis axis*
Deer, fallow *Dama dama*
Deer, giant Irish *Megalacerus giganteus*
Deer, mule *Odocoileus hemionus*
Deer, Pere David's *Elaphurus davidianus*
Deer, maral *Cervus elaphus maral*
Deer, pudu *Pudu pudu*
Deer, red *Cervus elaphus elaphus*
Deer, roe *Capreolus capreolus*
Deer, water *Hydropotes*
Deer, white-tailed *Odocoileus virginianus*
Deerfly *Chrysops, Tabanus*
Dik-dik *Madoqua kirki*
Dog, domestic *Canis familiaris*
Duck Anatinae
Duiker *Cephalophus* sp.
Eagle, bald *Haliaeetus leucocephalus*
Eagle, golden *Aquila chrysaetos*

Elephant, African *Loxodonta africana*
Elk (wapiti) *Cervus elaphus*
Elk, Rocky Mountain *C. e. nelsoni*
Eulachon *Thaleichthys pacifica*
Firefly *Lucidota corrusca*
Fisher *Martes pennanti*
Fly *Diptera*
Fly, black *Simulium* sp.
Fly, deer *Chrysops excitans*
Fly, horn *Haematobia irritans*
Fly, horse *Hybomitra sodalis*
Fly, moose *Haematobosca alcis*
Fly, snipe *Symphoromyia hirta*
Fly, stable *Stomoxys calcitrans*
Fox, red *Vulpes vulpes*
Fox, Arctic *Alopex lagopus*
Gemsbok *Oryx gazella*
Giraffe *Giraffa* sp.
Goat, domestic *Capra hircus*
Goat, Rocky Mountain *Oreamnos americanus*
Goose *Anserini*
Grouse Tetraoninae
Grouse, ruffed *Bonasa umbellus*
Hare, snowshoe *Lepus americanus*
Hare, Arctic *L. arcticus*
Hartebeest *Alcelaphus buselaphaus*
Horse, domestic *Equus caballus*
Horsefly *Hybomitra*
Human *Homo sapiens*
Ibex *Capra ibex*
Impala *Aepyceros melampus*
Jackrabbit, white-tailed *Lepus townsendii*
Jay, gray or Canada *Perisoreus canadensis*
Kudu *Tragelaphus* spp.
Leafhopper *Idioceris stellaris*
Lion, African *Panthera leo*
Lion, mountain *Felis concolor*
Lynx *F. lynx*
Mammoth *Mammothus* sp.
Marten *Martes americana*
Midge, biting *Culicoides obsoletus*
Mite *Macrocheles muscaedomesticae*
Moose *Alces alces*
Moose, Alaska/Yukon *A.a. gigas*
Moose, eastern *A. a. americana*
Moose, European or Fennoscandian *A. a. alces*
Moose, northwestern *A. a. andersoni*
Moose, Yellowstone (Shiras) *A. a. shirasi*
Mosquito *Aedes canadensis*
Mule *Equus* sp.
Musk-oxen *Ovibos moschatus*
Muskrat *Ondatra zibethicus*
Muntjac or barking deer *Muntiacus muntjak*
Oryx *Oryx beisa*
Otter, river *Lutra canadensis*
Owl Strigiformes
Ox *Bos taurus*
Panonnian maral *Cervus elaphus maral*
Pig, domestic *Sus scrofa*
Pike, northern or jackfish *Esox lucius*
Porcupine *Erithizon dorsatum*
Pronghorn *Antilocapra americana*
Raccoon *Procyon lotor*
Raven, common *Corvus covax*
Reindeer *Rangifer tarandus tarandus*
Salmon Salmonidae
Seal Phocidae
Sheep, Dall's *Ovis dalli*
Sheep, domestic *O. aries*
Sheep, mouflon *O. orientalis*
Sheep, Rocky Mountain bighorn *O. canadensis*
Snipe, common *Gallinago gallinago*
Squirrel, ground *Citellus* sp.
Tabanid, horse and deer fly *Chrysops, Tabanus, Hybromitra*
Tapeworm, hydatid cyst *Echinoccocus granulosus*
Tick, wood (winter or moose) *Dermacentor albipictus*
Tick, wood, Rocky Mountain *D. andersoni*
Tiger *Panthera tigris*
Trout Salmonidae
Tupaia *Tupaia belangeri*
Turkey, wild *Meleagris gallopavo*
Turtle Testudines
Vole, meadow *Microtus pennsylvanicus*
Walrus *Odobenus rosmarus*
Wasp, yellow-jacket *Vespula vulgaris*
Whale *Cetacea*
Whale, killer (orca) *Orcinus orca*
Whitefish *Coregonus* sp.
Wildebeest *Connochaetes* sp.
Wildebeast, blue *C. taurinus*
Wolf, grey *Canis lupus*
Wolverine *Gulo luscus*
Woodpecker Picinae
Worm, arterial or blood *Elaeophora schneideri*
Worm, brain or meningeal *Parelaphostrongylus tenuis*

References

Abercrombie, W.R. 1900. Report of a supplementary expedition into the Copper River Valley, Alaska 1884. Pages 381–408 *in* Compilation of narratives of explorations in Alaska. U.S. Govt. Print. Off., Washington, D.C.

Acharjyo, L.N. and A.B. Bubenik. 1983. The structural peculiarities of antler bone in genera *Axis, Rusa,* and *Rucervus* L. Pages 195–209 *in* R.D. Brown, ed., Antler development in Cervidae. Caesar Kleberg Wildl. Res. Inst., Kingsville, Texas.

Adair, W., P. Jordan and J. Tillma. 1991. Aquatic forage ratings according to wetland type: Modifications for the Lake Superior HSI. Alces 27: 140–149.

Adam, C.L. 1992. Impact of melatonin on time of breeding in farmed red deer. Pages 300–312 *in* R.D. Brown, ed., The biology of deer. Springer-Verlag, New York.

Adam, C.L., I. McDonald, C.E. Moir and K. Pennie. 1988. Foetal development in red deer (*Cervus elaphus* L.). I. Growth of the foetus and associated tissue. Anim. Prod. 46: 131–138.

Adamatz, L. 1928. Über angebliche Hirsch-Rinder-Bastarde mit spezieller Berucksichtigung des sogenannten "Hirschochsen von Suur." Biol. Gen. 4: 191–216.

Adams, A.T., ed. 1961. The explorations of Pierre Esprit Radisson. Ross and Haines, Inc. Minneapolis. 258 pp.

Adams, J.T. 1940. Dictionary of American history. Vol. 1. Charles Scribner's Sons, New York. 444 pp.

Adams, L. and S. Hane. 1972. Adrenal gland size as an index of adrenocortical secretion rate in the California ground squirrel. J. Wildl. Dis. 8: 19–23.

Adams, L.G., B.W. Dale and B. Shults. 1989. Population status and calf mortality of the Denali caribou herd, Denali National Park and Preserve, Alaska—1984–1988. Nat. Resour. Prog. Rep. U.S. Natl. Park Serv., Anchorage. 131 pp.

Addison, E.M. and R.F. McLaughlin. 1993. Seasonal variation and effects of winter ticks *(Dermacentor albipictus)* on consumption of food by captive moose *(Alces alces)* calves. Alces 29: 219–224.

Addison, E.M. and L.M. Smith. 1981. Productivity of winter ticks *(Dermacentor albipictus)* collected from moose killed on Ontario roads. Alces 17: 136–146.

Addison, E.M., A. Fyvie and F.J. Johnson. 1979. Metacestode of moose, *Alces alces,* of the Chapleau Crown Preserve, Ontario. Can. J. Zool. 57: 1,619–1,623.

Addison, E.M., R.F. McLaughlin and D.J.H. Fraser. 1982. Raising moose calves in Ontario. Alces 18: 246–270.

Addison, E.M., M.L. Wilton, R.F. McLaughlin and M.E. Buss. 1985. Trends in natality and calf mortality of moose in south central Ontario. Alces 21: 1–16.

Addison, E.M., J.D. Smith, R.F. McLaughlin D.J.H. Fraser and D.G. Loachim. 1990. Calving sites of moose in central Ontario. Alces 26: 142–153.

Addison, R.B. 1970. The development of a system of big game inventory utilizing an infrared scanner. Res. Branch Rept. Ontario Dept. Lands and For., Toronto. 40 pp.

Addison, R.B. 1971. Moose and caribou surveys, winter 1968–69 and 1969–70. Pages 29–43 *in* Part II, Chap. 5. Indian development study in northwestern Ontario— The Round Lake Ojibwa, the people, the land, the resources. Ontario Dept. Lands and For., Toronto.

———. 1972. The possible use of infrared thermal imagery for wildlife census. Proc. N. Am. Moose Conf. Workshop 8: 301–325.

———. 1974. Reproduction of moose measured by pigmented ovarian scars. Proc. N. Am. Moose Conf. Workshop 11: 369–390.

Addison, R.B., J.C. Williamson, B.P. Saunders and D. Fraser. 1980. Radiotracking of moose in the boreal forest of northwestern Ontario. Can. Field-Nat. 94: 269–276.

Adney, E.T. and H.I. Chapelle. 1964. The bark canoes and skin boats of North America. Smithsonian Inst., Washington, D.C. 242 pp.

Adney, T. 1900. Moose hunting with the Tro-chu-tin. Harper's New Month. Mag. C: 494–507.

Aho, R.W. and J. Hendrickson. 1989. Reproduction and mortality of moose translocated from Ontario to Michigan. Alces 25: 75–80.

Aho, R.W. and P.A. Jordan. 1979. Production of aquatic macrophytesan and its utilization by moose on Isle Royale National Park. Proc. Conf. Sci. Res. in Natl. Parks 1: 341–348.

Aitken, D.A. and K.N. Child. 1993. Relationships between in utero productivity of moose and population sex ratios: An exploratory analysis. Alces 28: 175–187.

Alados, C.L. and J.M. Escos. 1992. The determinants of social status and the effect of female rank on reproductive success in Dama and Cuvier's gazelles. Ethol. Ecol. and Evol. 50: 151–164.

Alaska Department of Fish and Game. 1990. Strategic plan for management of moose in Region 1, Southeast Alaska 1990–94. Alaska Dept. Fish and Game, Juneau. 10 pp.

Alberch, P. 1990. Natural selection and developmental constraints—External versus internal determinants of order in nature. Primate Life Hist. and Evol. 14: 15–35.

Alberta Energy and Natural Resources. 1980. Alberta conservation and hunter education. Fish and Wildl. Div. Alberta Energy and Nat. Resour., Edmonton. 243 pp.

Albon, S.D., T.H. Clutton-Brock and R. Langvatn. 1992. Cohort variation in reproduction and survival: Implications for population demography. Pages 15–21 *in* R.D. Brown, ed., The biology of deer. Springer-Verlag, New York.

Albright, C.A. and L.B. Keith 1987. Population dynamics of moose, *Alces alces,* on the south-coast barrens of Newfoundland. Can. Field-Nat. 101: 373–387.

Aldous, S.E. 1944. A deer browse survey method. J. Mammal. 25: 130–136.

Aldous, S.E. and L.W. Krefting. 1946. The present status of moose on Isle Royale. Trans. N. Am. Wildl. Conf. 11: 296–308.

Alexander, C.E. 1993. The status and management of moose in Vermont. Alces 29: 187–195.

Alexander, C.E., L.E. Garland, R.J. Regan and C.H. Willey. 1992. Moose management plan for the State of Vermont, 1992–1996. Vermont Dept. Fish and Wildl., St. Johnsbury.

Alexander, G. and D. Stevens. 1979. Discrimination of colours and grey shades by merino ewes—Tests using coloured lambs. Appl. Anim. Ethol. 5: 215–231.

Allen, A.W., P.A. Jordan and J.W. Terrell. 1987. Habitat suitability index models: Moose, Lake Superior region. Biol. Rept. 82, U.S. Fish and Wildl. Serv., Fort Collins, Colorado. 47 pp.

Allen, A.W., J.W. Terrell, W.L. Mangus and E.L. Lindquist. 1991. Application and partial validation of a habitat model for moose in the Lake Superior region. Alces 27: 50–64.

Allen, D.L. 1979. Wolves of Minong. Houghton Mifflin Co., Boston. 499 pp.

Allen, H.T. 1887. Report on an expedition to the Copper, Tananá and Koyukuk rivers, in the territory of Alaska, in the year 1885. U.S. Govt. Print. Off., Washington, D.C.

Allen, J.A. 1902. Mammals collected in Alaska and northern British Columbia by the Andrew J. Stone Expedition of 1902. Bull. Am. Mus. Nat. Hist. 16: 215–230.

Almkvist, B., T. Andre, S. Ekblom and S-A Repler. 1980. Viltolycksprojektet (Viol.) Slutrapport. (Game accidents on roads). Statens vagverk TU 146. Stockholm, Sweden. 117 pp.

Alston, E.R. 1879. On female deer with antlers. Proc. Zool. Soc. London: 296–297.

Altmann, M. 1957. A study of group dynamics in moose during the rutting season. Anat. Rec. 128: 516.

———. 1958. Social integration of the moose calf. Anim. Behav. 6: 155–159.

———. 1959. Group dynamics of Wyoming moose during the rutting season. J. Mammal. 40: 420–424.

———. 1961a. "Teenage" problems in the wilderness. Anim. Kingdom 64: 41–44.

———. 1961b. Sex dynamics within kinships of free-ranging wild ungulates. Proc. Am. Assoc. Adv. Sci. Symp. on Incest. Denver, Colorado. 189 pp.

———. 1963a. Naturalistic studies of maternal care in moose and elk. Pages 233–253 *in* H.L. Reingold, ed., Maternal behavior in mammals. John Wiley and Sons, New York. 349 pp.

———. 1963b. Seniors of the wilderness. Anim. Kingdom 66: 181–183.

Amann, R.P., J.F. Kavanaugh, L.C. Griel, Jr. and J.G. Voglmayr. 1974. Sperm production of Holstein bulls determined from testicular spermatid reserves; after cannulation of rete testes or vas deferens, and by daily ejaculation. J. Dairy Sci. 57: 93–99.

Ambrose, S.E. 1996. Undaunted courage: Meriwether Lewis, Thomas Jefferson, and the opening of the American West. Simon and Schuster, New York. 511 pp.

American Ornithologists' Union. 1983. Check-list of North American birds. 6th ed. Allen Press, Lawrence, Kansas. 877 pp.

Ammann, A.P., R.L. Cowan, C.L. Mothershead and B.R. Baumgardt. 1973. Dry matter and energy intake in relation to digestibility in white-tailed deer. J. Wildl. Manage. 37: 195–201.

Anchorage Daily News. 1992. Hungry orcas make rare meal of moose. July 21.

Andersen, R. 1991a. Habitat changes in moose ranges: Effects on migratory behavior, site fidelity and size of summer home range. Alces 27: 85–92.

———. 1991b. Habitat deterioration and the migratory behavior of moose (*Alces alces* L.) in Norway. J. Appl. Ecol. 28: 102–108.

Andersen, R., B. Wiseth, P.H. Pedersen and V. Jaren. 1991. Moose-train collisions: Effects of environmental conditions. Alces 27: 79–84.

Anderson, A.E. 1981. Morphological and physiological characteristics. Pages 27–99 *in* O.C. Walmo, ed., Mule and black-tailed deer of North America. Univ. Nebraska Press, Lincoln.

Anderson, D.J. 1982. The home range: A new nonparametric estimation technique. Ecol. 63: 103–112.

Anderson, J.R. 1975. The behavior of nose bot flies *(Cephenemyia apicata* and *C. jellisoni)* when attacking black-tailed deer *(Odocoileus hemionus columbianus)* and the resulting reactions of the deer. Can. J. Zool. 53: 977–972.

Anderson, M. 1982. Female choice selects for extreme tail length in a widowbird. Nature 299: 818–820.

Anderson, M., J. Deely, M. Krampen, J. Ransdell, T.A. Sebeok and T. Von Uexkull. 1984. A semiotic perspective on the sciences: Steps toward a new paradigm. Semiotic Circle Monogr. No. 5. Toronto.

Anderson, R.C. 1963. The incidence, development, and experimental transmission of *Pneumostrongylus tenuis* Dougherty (Metastrongyloidea: Protostrongylidae) of the meninges of the white-tailed deer *(Odocoileus virginianus borealis)* in Ontario. Can. J. Zool. 41: 755–792.

———. 1964. Neurological disease in moose infected experimentally with *Pneumostrongylus tenuis* from white-tailed deer. Path. Vet. 1: 289–322.

———. 1965a. An examination of wild moose exhibiting neurological signs, in Ontario. Can. J. Zool. 43: 635–639.

———. 1965b. The development of *Pneumostrongylus tenuis* in the central nervous system of white-tailed deer. Pathol. Vet. 2: 360–379.

———. 1970. The ecological relationships of meningeal worm (*Pneumostrongylus tenuis* Dougherty), and native cervids in North America. J. Parasitol. 56: 6–7.

———. 1971. Neurologic disease in reindeer *(Rangifer tarandus tarandus)* introduced into Ontario. Can. J. Zool. 49: 159–166.

———. 1972. The ecological relationships of meningeal worm and native cervids in North America. J. Wildl. Dis. 7: 304–309.

———. 1992. Nematode parasites of vertebrates, their development and transmission. C.A.B. International, Univ. Press, Cambridge. 578 pp.

Anderson, R.C. and M.W. Lankester. 1974. Infections and parasitic diseases and arthropod pests of moose in North America. Nat. Can. (Que.) 101: 23–50.

Anderson, R.C. and A.K. Prestwood. 1981. Lungworms. Pages 266–317 *in* W.R. Davidson, F.A. Hayes, V.F. Nettles and F.E. Kellogg, eds., Diseases and parasites of white-tailed deer. Misc. Publ. No. 7, Tall Timbers Res. Stn., Tallahasee, Florida. 458 pp.

Anderson, R.C. and U.R. Strelive. 1967. The penetration of *Pneumostrongylus tenuis* into the tissues of white-tailed deer. Can. J. Zool. 45: 285–289.

Andreska, J. 1988. Origins and development of the elk *(Alces alces* L. 1758) in southern Bohemia, Czechoslovakia. Lynx (Prague) 24: 73–77.

Andrews, E. and J. Stokes. 1984. An overview of the Upper Kuskakwim controlled use area and the use of moose by area residents, 1981–1984. Tech. Pap. No. 99. Div. of Subsistence, Alaska Dept. Fish and Game, Juneau. 6 pp.

Anonymous. 1956. The grand seigneur. The Beaver 4.

———. 1959. Her Majesty, the Queen, graciously consents to accept payment of rent from the Governor and Company of Adventurers of England Trading into Hudson's Bay. The Beaver 290: 30–31.

———. 1970. The Queen accepts. The Beaver 301: 46–47.

———. 1973. The White House, an historic guide. White House Hist. Assoc. and Natl. Geogr. Soc., Washington, D.C. 159 pp.

———. 1979. First case of rabies in moose reported in the United States. Commun. Dis. Let., Utah Dept. Soc. Serv., Salt Lake City.

———. 1988. Loyal Order of the Moose, one hundred years of progress. Loyal Order of the Moose, Mooseheart, Illinois. 20 pp.

———. 1992. Moose management objectives and development strategies proposed for 1993–1997. Losir, Chasse et Peche, Quebec. 6 pp.

Archibald, R.M. 1960. *Listeria moncytogenes* from a Nova Scotia moose. Can. Vet. J. 1: 225–226.

Arman, P., D. Hopcraft and I. McDonald. 1975. Nutritional studies on East African herbivores. 2. Losses of nitrogen in faeces. Brit. J. Nutr. 33: 265–267.

Armstrong, T., W. Craven, N. Feder, B. Haskell, R.E. Kraus, D. Robbins and M. Tucker. 1976. 200 years of American sculpture. David R. Godine, Publ., New York. 336 pp.

Arnold, G.W. 1985. Regulation of forage intake. Pages 82-101 *in* R.J. Hudson and R.G. White, eds., Bioenergetics of wild herbivores. CRC Press, Boca Raton, Florida.

Asche, M.I. 1981. Slavey. Pages 338–349 *in* J. Helm, ed., Subarctic. Handbook of North American Indians. Vol. 6. Smithsonian Inst., Washington, D.C. 837 pp.

Aschoff, J. 1954. Zeitgeber der tierischen Tagesperiodik. Naturwiss. 41: 49–56.

———. 1958. Tierische Periodik unter dem Einfluss von Zeitgebern. Z. Tierpsychol. 15: 1–30.

———. 1965a. Response curves in circadian periodicity. Pages 95–111 *in* J. Aschoff, ed., Circadian clocks. North-Holland Publ. Co., Amsterdam.

———. 1965b. The phase-angle difference in circadian periodicity. Pages 262–276 *in* J. Aschoff, ed., Circadian clocks. North-Jolland Publ. Co., Amsterdam.

———. 1969. Phasenlage der Tagesperiodik in Abhängigkeit von Jahreszeit und Breitegrad. Oekologia 3: 125–165.

———. 1979. Circadian rhythms—Influences of internal and external factors on the period measured in constant conditions. Z. Tierpsychol. 49: 225–249.

Asher, G.W. 1985. Oestrous cycle and breeding season of farmed fallow deer *(Dama dama)*. J. Reprod. Fert. 75: 521–529.

Association of Official Agricultural Chemists. 1965. Official methods of analysis. 10th ed. Assoc. Off. Agric. Chemists, Washington, D.C. 957 pp.

Astrom, M., P. Lundberg and K. Danell. 1990. Partial prey consumption by browsers: Trees as patches. J. Anim. Ecol. 59: 287–300.

Atkeson, T.D., V.F. Nettles, R.L. Marchinton and W. Branan. 1988. Nasal glands in the Cervidae. J. Mammal. 69: 153–156.

Audet, R. and P. Grenier. 1976. Habitat hivernal de l'orignal dans la Baie James, étude préliminaire. Minstere du Tourisme de la Chasse et de la Peche, Quebec. 38 pp.

Audubon, J.J. 1831–1839. Ornithological biography. 5 vols. Edinburgh, Scotland.

Audubon, J.J. and J. Bachman. 1845–1848. The viviparous quadrupeds of North America. 3 vols. Published by John J. Audubon, New York. 440 pp.

Audubon, J.J. and J.B. Chevalier. 1840–1844. The birds of America. 7 vols. John J. Audubon and J.B. Chevalier, New York. (Reprinted in 1967 by Dover Press, New York.)

Austin, P.J., L.A. Suchar, C.T. Robbins and A.E. Hagerman. 1989. Tannin-binding proteins in saliva of deer and their absence in saliva of sheep and cattle. J. Chem. Ecol. 15: 1,335–1,347.

Azzaroli, A. 1950. L'Alce di Seneneze. Paleontol. Ital. 47: 133–141.

———. 1981. On the Quaternary and recent cervid genera; *Alces, Cervalces, Libralces*. Boll. Soc. Paleontol. Ital. 20: 147–154.

———. 1985. Taxonomy of the Quaternary Alcini (Cervidae, Mammalia). Acta Zool. Fennica 170: 179–180.

Baber, D.W. 1987. Gross antler anomaly in a California mule deer: The "cactus" buck. Southwest. Nat. 32: 404–406.

Babicka, J., J. Komarek and B. Nemec. 1944. Gold in animals bodies. Bull. Int. de l'Acad. tcheque des Sci. LIV: 1–7.

Backhaus, D. 1959. Experimentelle Untersuchungen über die Sehschärfe und das Farbsehen einiger Huftiere. Z. Tierpsychol. 16: 445–467.

Backstrom, A.K. 1952. Avslutade studier over august brunsten. Svenska Jagareforbundets Tidskrift 90: 70–71.

Badger, R. 1979. The great American fair. Nelson Hall, Chicago. 177 pp.

Bailey, T.N. 1978. Moose populations on the Kenai National Moose Range. Proc. N. Am. Moose Conf. Workshop 14: 1–20.

Bailey, T.N. and E.E. Bangs. 1980. Moose calving areas and use on the Kenai National Wildlife Refuge, Alaska. Proc. N. Am. Moose Conf. Workshop 16: 289–313.

Bailey, T.N. and A.W. Franzmann. 1983. Mortality of resident versus introduced moose in confined populations. J. Wildl. Manage. 47: 520–523.

Bailey, T.N., A.W. Franzmann, P.D. Arneson and J.L. Davis. 1983. An evaluation of visual location data from neck-collared moose. J. Wildl. Manage. 47: 25–30.

Baird, S.F. 1859. Mammals of North America: The descriptions of species based chiefly on the collections in the museum of the Smithsonian Institution. J.B. Lippincott and Co., Philadelphia. 764 pp.

Baker, R.A. 1975. Biological implications of a bull moose-only moose hunting regulation in Ontario. Proc. N. Am. Moose Conf. Workshop 11: 464–476.

Baker, R.H. 1983. Michigan mammals. Michigan St. Univ. Press, East Lansing. 642 pp.

Baker, R.R. 1978. The evolutionary ecology of animal migration. Holmes and Meier, New York. 1,012 pp.

Bakker, T.C.M. and M. Milinski. 1991. Sequential female choice and previous male effect in sticklebacks. Behav. Ecol. and Sociobiol. 29: 205–210.

Ballard, W.B. 1991. Management of predators and their prey: The Alaska experience. Trans. N. Am. Wildl. and Nat. Resour. Conf. 56: 527–538.

———. 1992a. Bear predation on moose: A review of recent North American studies and their management implications. Alces (Suppl.) 1: 1–15.

———. 1992b. Modeled impacts of wolf and bear predation on moose calf survival. Alces 28: 79–88.

Ballard, W.B. and D.G. Larsen. 1987. Implications of predator–prey relationships to moose management. Swed. Wildl. Res. (Suppl.) 1: 581–602.

Ballard, W.B. and S.M. Miller. 1990. Effects of reducing brown bear density on moose calf survival in southcentral Alaska. Alces 26: 9–13.

Ballard, W.B. and K.P. Taylor. 1980. Upper Susitna Valley moose population study. Fed. Aid in Wildl. Restor. Prog. Rept. Alaska Dept. Fish and Game, Juneau. 102 pp.

Ballard, W.B. and R.W. Tobey. 1981. Decreased calf production of moose immobilized with anectine administered from a helicopter. Wildl. Soc. Bull. 9: 207–209.

Ballard, W.B., A.F. Cunning and J.S. Whitman. 1988. Hypothesis of impacts on moose due to hydroelectric projects. Alces 24: 4–47.

Ballard, W.B., C.L. Gardner and S.D. Miller. 1980. Influence of predators on summer movements of moose in southcentral Alaska. Proc. N. Am. Moose Conf. Workshop 16: 338–359.

Ballard, W.B., S.M. Miller and J.S. Whitman. 1986. Modelling a southcentral Alaska moose population. Alces 22: 201–244.

Ballard, W.B., S.D. Miller and J.S. Whitman. 1990. Brown and black bear predation on moose in southcentral Alaska. Alces 26: 1–8.

Ballard, W.B., J.S. Whitman and C.L. Gardner. 1987. Ecology of an exploited wolf population in south-central Alaska. Wildl. Monogr. 98. 54 pp.

Ballard, W.B., J.S. Whitman and D.J. Reed. 1991. Population dynamics of moose in south-central Alaska. Wildl. Monogr. 114. 49 pp.

Ballard, W.B., T.H. Spraker and K.P. Taylor. 1981a. Causes of neonatal moose calf mortality in southcentral Alaska. J. Wildl. Manage. 45: 335–342.

Ballard, W.B., C.L. Gardner, J.H Weslund and S.M. Miller. 1981b. Use of mandible versus longbone to evaluate percent marrow fat in moose and caribou. Alces 17: 147–164.

Ballard, W.B., C.L. Gardner, J.H Weslund and J.R. Dau. 1982. Susitna hydroelectric project, Phase I final report. Big game studies. Vol. 5—Wolf. Alaska Dept. Fish and Game, Juneau. 220 pp.

Ballard, W.B., A.W. Franzmann, K.P. Taylor, T. Spraker, C.C. Schwartz and R.O. Peterson. 1979. Comparison of techniques utilized to determine moose calf mortality in Alaska. Proc. N. Am. Moose Conf. Workshop 15: 362–387.

Baltzer, B.E. 1933. Swimming power of moose. Rod and Gun in Canada 85: 192–193.

Banfield, A.W.F. 1974. The mammals of Canada. Univ. Toronto Press, Toronto, Ontario. 438 pp.

Banfield, A.W.F., D.R. Flook, J.P. Kelsall and A.G. Loughrey. 1955. An aerial survey technique for northern big game. Trans. N. Am. Wildl. Conf. 20: 519–530.

Bangs, E.E. and T.N. Bailey. 1980a. Interrelationships of weather, fire, and moose on the Kenai National Moose Range, Alaska. Proc. N. Am. Moose Conf. Workshop 16: 255–274.

Bangs, E.E. and T.N. Bailey. 1980b. Moose movements and distribution in response to winter seismological exploration on the Kenai National Wildlife Refuge, Alaska. Kenai Natl. Wildl. Refuge Rept. U.S. Fish and Wildl. Serv., Kenai, Alaska. 47 pp.

Bangs, E.E., T.N. Bailey and M.F. Portner. 1984. Bull moose behavior and movements in relation to harvest on the Kenai National Moose Range, Alaska. Alces 20: 187–208.

Bangs, E.E., T.N. Bailey and M.F. Portner. 1989. Survival rates of adult female moose on the Kenai Peninsula, Alaska. J. Wildl. Manage. 53: 557–563.

Bangs, E.E., S.A. Duff and T.N. Bailey. 1985. Habitat differences and moose use of two large burns on the Kenai Peninsula, Alaska. Alces 21: 17–35.

Banks, R.C., R.W. McDiarmid and A.L. Gardner. 1987. Checklist of vertebrates of the United States, the U.S. territories, and Canada. Resour. Public. 166. U.S. Fish and Wildl. Serv., Washington, D.C. 79 pp.

Banks, W.J., Jr. and J.W. Newbrey. 1982. Antler development as a unique modification of mammalian endochondral ossification. Pages 279-306 *in* R.D. Brown, ed., Antler development in Cervidae. Caesar Kleberg Wildl. Res. Inst., Kingsville, Texas.

Banks, W.J., Jr., G.P. Epling, R.A. Kainer and R.W. Davis. 1968a. Antler growth and osteoporosis. I: Morpological and morphometric changes in the costal compacta during the antler growth cycle. Anat. Rec. 162: 387–397.

Banks, W.J., Jr., G.P. Epling, R.A. Kainer and R.W. Davis. 1968b. Antler growth and osteoporosis. II: Gravimetric and chemical changes in the costal compacta during the antler growth cycle. Anat. Rec. 162: 399–405.

Barette, C. 1977. Fighting behavior of muntjacs and the evolution of antlers. Evolution 31: 169–176.

Barker, R.W., M. Nagardeolekar, K.W. Stricker and R.E. Wright. 1990. Fecundity of partially engorged female *Dermacentor albipictus* (Acari: Ixodidae) removed by cattle grooming. J. Med. Entomol. 127: 51–56.

Barlow, P.W. 1992. A constant of temporal structure in the human heirarchy and other systems. Acta Biotheoretica 31: 321–328.

Barnhart, R.K. 1988. The Barnhart dictionary of etymology. The H.W. Wilson Co., New York. 1,284 pp.

Barraclough, G. and N. Stone, eds. 1989. The Times atlas of world history. 3rd ed. Little Brown and Co., Boston. 524 pp.

Barrell, G.K. and L.D. Staples. 1991. Melatonin implants alter calving season in farmed red deer. Pages 181-184 *in* B. Bobek, K. Perzanowski and W.L. Regelin, eds., Global trends in wildlife management. Vol. 1. Trans. 18th Cong. Int. Union Game Biol. Swiat Press, Krakow-Warszawa.

Barrett, M.W. 1972. A review of the diet, condition, diseases, and parasites of the Cypress Hills moose. Proc. N. Am. Moose Conf. Workshop 8: 60–79.

Bartholomew, G.A. 1958. The role of physiology in the distribution of terrestrial vertebrates. Pages 81–96 *in* C.L. Hubbs, ed., Zoogeography. Publ. No. 51. Am. Assoc. Adv. Sci., Washington, D.C. 509 pp.

Bartholow, J. 1985. POP-II system documentation. Fossil Creek Software. IBM-PC Ver. 4.0. Colorado St. Univ., Fort Collins. 64 pp.

Bartlett, J.R. 1859. A glossary of words and phrases usually regarded as peculiar to the United States. 2nd ed. Little, Brown and Co., Boston. 524 pp.

Bartmann, R.M., G.C. White and L.H. Carpenter. 1992. Compensatory mortality in a Colorado mule deer population. Wildl. Monogr. 121. 39 pp.

Barton, N.H. and M. Turelli. 1991. Natural and sexual selection on many loci. Genetics 127: 229–255.

Barton, R. and G.L. Nute, eds. 1948. A winter in the St. Croix Valley: George Nelson's reminiscences, 1802–03. Minnesota Hist. Soc., St. Paul. 46 pp.

Bartos, L. and V. Perner. 1991. Asynchronous antler casting in red deer. Pages 291–293 *in* B. Bobek, K. Perzanowski and W.L. Regelin, eds., Global trends in wildlife management. Vol. 1. Trans. 18th Cong. Int. Union Game Biol. Swiat Press, Krakow-Warszawa.

Bartosiewicz, L. Metacarpal measurements and carcass weight of moose in central Sweden. J. Wildl. Manage. 51: 358–359.

Bashline, L.J. and D. Saults, eds. 1976. America's great outdoors. J.G. Ferguson Co., Chicago. 367 pp.

Baskerville, G. 1985. Adaptive management: Wood availability and habitat availability. For. Chron. 61: 171–175.

Bateman, R. and R. Derry. 1981. The art of Robert Bateman. The Viking Press, New York. 178 pp.

Bateson, P. 1983. Mate choice. Cambridge Univ. Press, New York. 462 pp.

Baxter, D., D. Harmel, W.E. Armstrong and G. Butts. 1977. Spikes versus forked-antlered bucks. Texas Parks and Wildl. (March): 6–7.

Bayford, S. 1975. Redirecting deer movements by the use of flagging behavior models. M.A. thesis, Millersville St. Coll., Millersville, Pennsylvania.

Beach, F.A. 1976. Sexual attractivity, proceptivity and receptivity in female mammals. Hormones and Behav. 7: 105–138.

Beach, T.D. 1992. Transmission of meningeal worm: An analysis of sympatric use of habitat by white-tailed deer, moose and gastropods. M.S. thesis, Univ. New Brunswick, Fredericton. 54 pp.

Bearden, H.J. and J.W. Fuquay. 1984. Applied animal reproduction, 2nd ed. Reston Publ. Co., Inc., Reston, Virginia. 382 pp.

Beaulieu, R. 1984. Moose calf mortality study. Wildl. Pop. Manage. Info. Base, 84-WPM-8. Saskatchewan Parks and Renew. Resour., Saskatoon. 5 pp.

———. 1992. Moose. Pages 35–54 *in* Big game status report, hunting season proposed for 1992. Saskatchewan Dept. Parks and Renew. Resour., Regina.

Bedard, J., M. Crête and E. Audy. 1978. Short-term influence of moose upon woody plants of an early seral wintering site in Gaspé Peninsula, Quebec. Can J. For. Res. 8: 407–415.

Becker, E.A. 1969. A treasury of Alaskana. Superior Publ. Co., Seattle. 183 pp.

Becker, E.F. 1988. Susitna hydroelectric project, big game studies. Vol. V. Moose carrying capacity estimate. Final Rept. Alaska Dept. Fish and Game, Anchorage. 113 pp.

Becker, E.F. and C.L. Gardner. 1991. Wolf and wolverine density estimation techniques. Fed. Aid in Wildl. Restor. Rept. Alaska Dept. Fish and Game, Juneau. 17 pp.

Becker, E.F. and C.A. Grauvogel. 1991. Relationship of reduced train speed on moose–train collisions in Alaska. Alces 27: 161–168.

Becker, E.F. and W.D. Steigers, Jr. 1987. Susitna hydroelecric project. Big game studies. Vol. III—Moose forage biomass in the middle Susitna River basin, Alaska. Final Rept. Alaska Dept. Fish and Game, Anchorage. 113 pp.

Behler, J.L. and F.W. King. 1979. The Audubon Society field guide to North American reptiles and amphibians. Alfred A. Knopf, New York. 719 pp.

Behnke, S.R. 1982. Wildlife utilization and the economy of Nondalton. Tech. Pap. No. 47. Div. of Subsistence, Alaska Dept. Fish and Game, Juneau. 74 pp.

Bejšovec, J. 1955. Špatny vyvoj parůžků u srnčí zvěře. (Kümmerliche Geweihentwicklung beim Rehwild). Acta Soc. Zool. Bohemosloven 19: 119–137.

Belknap, J. 1812. The history of New-Hampshire. 3 vols. Printed for O. Crosby and J. Varney, by J. Mann and J.K. Remick, Dover, New Hampshire.

Bell, J.F. 1979. The pictographs of Slocan Lake. Biennial Conf. of Can. Rock Art Assoc. 4: 23–47.

Bell, J.M. 1903. The fireside stories of the Chippwyans. J. Am. Folk-Lore XVI: 73–84.

Bell, R.H.V. 1970. The use of the herb layer by grazing ungulates in the Serengeti. Pages 111–123 *in* A. Watson, ed., Animal populations in relation to their food resource. Blackwell Scientific Publ., Oxford.

———. 1971. A grazing system in the Serengeti. Sci. Am. 225: 86–93.

Bell, W.J. 1991. Searching behavior—The behavioral ecology of finding resources. Chapman and Hall, New York. 372 pp.

Belovsky, G.E. 1978. Diet optimization in a generalist herbivore: The moose. Theor. Pop. Biol. 14: 105–134.

———. 1981. A possible population response of moose to sodium availability. J. Mammal. 62: 631–633.

———. 1984. Moose and snowshoe hare competition and a mechanistic explanation from foraging theory. Oecologia 61: 150–159.

———. 1990. A reply to Hobbs. Pages 415–422 *in* R.N. Hughes, ed., Behavioral mechanisms of food selection. Springer-Verlag, New York.

Belovsky, G.E. and P.A. Jordan. 1978. The time-energy budget of a moose. Theor. Pop. Biol. 14: 76–104.

Bendell, J.F. 1974. Effects of fire on birds and mammals. Pages 73–138 *in* T.T. Kozlowski and C.E. Ahlgren, eds., Fire and ecosystems. Academic Press, New York.

Bennett, G.F. 1962. On the biology of *Cephenomyia phobifera* (Dipter: Oestridae), the pharyngeal bot of the white-tailed deer, *Odocoileus virginianus*. Can. J. Zool. 40: 1,195–1,210.

Bennett, G.F. and C.W. Sabrosky. 1962. The nearctic species of the genus *Cephenemyia* (Diptera, Oestridae). Can. J. Zool. 40: 431–448.

Bennett, L.J., P.F. English and R. McCain. 1940. A study of deer populations by use of pellet-group counts. J. Wildl. Manage. 4: 398–403.

Benson, D.A. 1958. "Moose sickness" in Nova Scotia. Can. J. Comp. Med. 22: 244–248.

Benson, D.A. and D.G. Dodds. 1977. Deer of Nova Scotia. Nova Scotia Dept. Lands and For., Halifax. 92 pp.

Bent, A.C. 1946. Life histories of North American jays, crows, and titmice. U.S. Natl. Mus. Bull. 191. U.S. Govt. Print. Off., Washington. (Reprinted in 1964 in 2 vols. by Dover Publ., New York.)

Bentley, W.W. 1961. Moose population inventories: Interior and arctic Alaska. Pages 161–171 *in* Fed. Aid in Wildl. Restor. Rept. Alaska Dept. Fish and Game, Juneau.

Berg, W.E. 1971. Habitat use, movements, and activity patterns of moose in northwestern Minnesota. M.S. thesis, Univ. Minnesota, St. Paul. 98 pp.

———. 1975. Management implications of natural mortality of moose in northwestern Minnesota. Proc. N. Am. Moose Conf. Workshop 9: 332–342.

Berg, W.E. and R.L. Phillips. 1972. Winter spacing of moose in northeastern Minnesota. Proc. N. Am. Moose Conf. Workshop 8: 166-176.

Berg, W.E. and R.L. Phillips. 1974. Habitat use by moose in northwestern Minnesota with reference to other heavily willowed areas. Nat. Can. (Que.) 101: 101–116.

Bergerud, A.T. 1962. Moose management report 1961–62. Newfoundland and Labrador Wildl. Div., St. John's. 33 pp.

———. 1968. Numbers and densities. Pages 21–42 *in* F.B. Golley and H.K. Brechner, eds., A practical guide to the study of the productivity of large herbivores. Handbook No. 7, Int. Biol. Prog. Blackwell Scientific Publ., London. 300 pp.

———. 1972. Food habits of Newfoundland caribou. J. Wildl. Manage. 36: 913–923.

———. 1974. Relative abundance of food in winter for Newfoundland caribou. Oikos 25: 379–387.

———. 1981. The decline of moose in Ontario—A different view. Alces 17: 30–43.

———. 1992. Rareness as an antipredator strategy to reduce predation risk. Trans. Int. Union Game Biol. Cong. 19: 15–25.

Bergerud, A.T. and W.B. Ballard. 1989. Wolf predation on Nelchina caribou herd: A reply. J. Wildl. Manage. 53: 344–357.

Bergerud, A.T. and F. Manuel. 1968. Moose damage to balsam fir-white birch forests in central Newfoundland. J. Wildl. Manage. 32: 729–746.

Bergerud, A.T. and F. Manuel. 1969. Aerial census of moose in central Newfoundland. J. Wildl. Manage. 33: 910–916.

Bergerud, A.T. and J.B. Snider. 1988. Predation in the dynamics of moose populations: A reply. J. Wildl. Manage. 52: 559–564.

Bergerud, A.T., F. Manuel and H. Whalen. 1968. The harvest reduction of a moose population in Newfoundland. J. Wildl. Manage. 32: 722–728.

Bergerud, A.T., W. Wyett and J.B. Snider. 1983. The role of wolf predation in limiting a moose population. J. Wildl. Manage. 47: 977–988.

Bergstrom, R. and K. Danell. 1986. Moose winter feeding in relation to morphology and chemistry of six tree species. Alces 22: 91–112.

Berkes, F., P. George and R. Preston. 1991 Co-management: The evolution of the theory and practice of joint administration of living resources. Technology Assessment in Subarctic Ontario (TASO) Research Rept. 2nd ser., No. 1. McMaster Univ., Hamilton, Ontario. 37 pp.

Bernhard, K., G. Brubacher, H. Hediger and H. Bruhin. 1953. Untersuchungen über chemische Zusammenzetzung und Aufbau des Hirschgeweihes. Experientia 9: 138–140.

Berry, K.H. 1986. Introduction: Development, testing, and application of wildlife habitat models. Pages 3–4 *in* J. Verner, M.L. Morrrison and C.J. Ralph, eds., Wildlife 2000: Modelling habitat relationships of terrestrial vertebrates. Univ. Wisconsin Press, Madison.

Berry, W.D. 1965. Deneki, an Alaskan moose. Macmillan Publ. Co., New York.

———. 1967. Deneki, an Alaskan moose, 2nd. ed. Macmillan Publ.Co., New York. 47 pp.

Berryman, J.H. 1987. Socioeconomic values of the wildlife resource: Are we really serious? Pages 5–11 *in* D.J. Decker and G.R. Goff, eds., Valuing wildlife and economic and social perspectives. Westview Press, Boulder, Colorado.

Berson, H. 1975. A moose is not a mouse. Crown Publishers, New York. 32 pp.

Bertalanffy, L. von. 1960. Principles and theory of growth. Pages 137–259 *in* W.W. Nowinski, ed., Fundamental aspects of normal and malignant growth. Elsevier, Amsterdam.

Best, D.A., G.M. Lynch and O. Rongstad. 1977. Annual spring movements of moose to mineral licks in Swan Hills, Alberta. Proc. N. Am. Moose Conf. Workshop 13: 215–228.

Best, D.A., G.M. Lynch and O. Rongstad. 1978. Seasonal activity patterns of moose in the Swan Hills, Alberta. Proc. N. Am. Moose Conf. Workshop 14: 109–125.

Bethune, W.C. 1937. Canada's western northland. Dept. Mines and Resour., Ottawa. 162 pp.

Bevins, J.S., C.C. Schwartz and A.W. Franzmann. 1990. Seasonal activity patterns of moose on the Kenai Peninsula, Alaska. Alces 26: 14–23.

Bewick, T. 1807. A general history of quadrupeds. 5th ed. Printed by Edward Walker for T. Bewick and S. Hodgson, Newcastle upon Tyne, England. 525 pp.

Bider, J.R. and D.H. Pimlott. 1973. Access to hunting areas from major urban centers and big game kills in Quebec. Proc. N. Am. Moose Conf. Workshop 9: 59–80.

Biggar, H.P., ed. 1924. The voyages of Jacques Cartier. Public. of the Public Archives of Canada No. 11. F.A. Acland, King's Printer, Ottawa. 330 pp.

Birch, L.C. 1957. The meaning of competition. Am. Nat. 91: 5–18.

Bishop, C.A. 1969. The Northern Chippewa: An ethnohistorical study. Ph.D diss., St. Univ. of New York, Buffalo.

———. 1981. Northeastern Indian concepts of conservation and the fur trade: A critique of Calvin Martin's thesis. Pages 39–58 *in* S. Krech III, ed., Indians, animals and the fur trade. Univ. Georgia Press, Athens. 207 pp.

Bishop, J.B. 1920. Theodore Roosevelt and his time, shown in his own letters. 2 vols. Charles Scribner's Sons, New York.

Bishop, R.H. and R.A. Rausch. 1974. Moose population fluctuations in Alaska. 1950–1972. Nat. Can. (Que.) 101: 559–593.

Bisset, A.R. 1987. The economic importance of moose *(Alces alces)* in North America. Swed. Wildl. Res. (Suppl.) 1: 677–698.

———. 1991. The moose population of Ontario revisited: A review of survey data 1975-1991. Ontario Min. Nat. Resour., Toronto. 30 pp.

Bisset, A.R. and R.S. Rempel. 1991. Linear analysis of factors affecting the accuracy of moose aerial inventories. Alces 27: 127–139.

Bittman, E.L. and F.J. Karsch. 1984. Nightly duration of pineal melatonin secretion determines the reproductive response to inhibitory day lengths in the ewe. Biol Reprod. 30: 585–593.

Bittman, E.L., R.J. Dempsey and F.J. Karsch. 1983. Pineal melatonin secretion drives the reproductive response to day length in ewe. Endocrin. 113: 2,276–2,283.

Blair, E.H., ed. 1911. The Indian tribes of the Upper Mississippi Valley and the region of the Great Lakes. 2 vols. The Arthur H. Clark Co., Cleveland, Ohio.

Blauel, G. 1935. Beobachtungen über die Entstehung der Perücke beim Rehbock. Endokrin. 15: 321–329.

———. 1936. Beobachtungen über die Entstehung der Perücke beim Rehbock. 2. Endokrin. 17: 369–372.

Blaxter, K.L. 1962. The energy metabolism of ruminants. C.C. Thomas, Springfield, Illinois. 332 pp.

———. 1965. Prediction of the amount of methane produced by ruminants. Brit. J. Nutr. 19: 511–522.

Bliss, W.L. 1939. Early man in western and northwestern Canada. Sci. 89(2,312): 365–366.

Blissitt, M.J., K.P. Bland and D.F. Cottrell. 1990. Discrimination between the odours of fresh oestrus and monoestrus ewe urine by rams. Appl. Anim. Behav. Sci. 25: 51–59.

Blood, D.A. 1974. Variation in reproduction and productivity of an enclosed herd of moose *(Alces alces)*. Trans. Int. Congr. Game Biol. 11: 59–66.

Blood, D.A., J.R. McGillis and A.L. Lovass. 1967. Weights and measurements of moose in Elk Island National Park. Can. Field-Nat. 81: 263–269.

Blumenthal, L. 1952. Die Milz des Elches. Z. Mikrosk. Anal. Forsch. 58: 230–255.

Blyth, C.B. and R.J. Hudson. 1987. A plan for the management of vegetation and ungulates, Elk Island National Park. Elk Island Natl. Park and Dept. of Anim. Sci. Rept. Univ. Alberta, Edmonton. 398 pp.

Bo, S. and O. Hjeljord. 1991. Do continental moose ranges improve during cloudy summers? Can. J. Zool. 69: 1,875–1,879.

Bock, P.K. 1978. Micmac. Pages 109–122 *in* B.G. Trigger, ed., Northeast. Handbook of North American Indians. Vol. 15. Smithsonian Inst., Washington, D.C. 924 pp.

Boer, A.H. 1978. Management of deer wintering areas in New Brunswick. Wildl. Soc. Bull. 6: 200–205.

———. 1987a. Reproductive productivity of moose in New Brunswick. Alces 23: 49–60.

———. 1987b. Hunting and the population dynamics of moose in New Brunswick. Ph.D. thesis, Univ. New Brunswick, Fredericton. 95 pp.

———. 1988. Mortality rates of moose in New Brunswick: A life table analysis. J. Wildl. Manage. 52: 21–25.

———. 1990. Spatial distribution of moose kills in New Brunswick. Wildl. Soc. Bull. 18: 431–434.

———. 1991. Hunting: A product or a tool for wildlife managers? Alces 27: 74–78.

———. 1992a. Fecundity of North American moose *(Alces alces)*: A review. Alces (Suppl.) 1: 1–10.

———. 1992b. History of moose in New Brunswick. Alces (Suppl.) 1: 16–21.

Boer, A.H. and D.M. Keppie. 1988. Modelling a hunted moose population in New Brunswick. Alces 24: 201–217.

Boertje, R.D., D.G. Kelleyhouse and R.D. Hayes. In press. Methods for reducing natural predation on moose in Alaska and the Yukon: An evaluation. *In* L.N. Carbyn, S.H. Fritts and D.R. Seip, eds., Ecology and conservation of wolves in a changing world. Can. Circumpolar Inst., Univ. Alberta, Edmonton.

Boertje, R.D., W.C. Gasaway, D.V. Grangaard and D.G. Kelleyhouse. 1988. Predation on moose and caribou by radio-collared grizzly bears in eastcentral Alaska. Can. J. Zool. 66: 492–499.

Boertje, R.D., W.C. Gasaway, D.V. Grangaard, D.G. Kelleyhouse and R.O. Stephenson. 1987. Factors limiting moose population growth in subunit 20E. Fed. Aid in Wildl. Restor. Rept. Alaska Dept. Fish and Game, Juneau. 86 pp.

Boertje, R.D., W.C. Gasaway, D.J. Reed, J.L. Davis, D.F. Holleman, R.O. Stephenson and W.B. Ballard. 1992. Testing socially acceptable methods of managing predators—reducing wolf predation on moose through increased caribou abundance. Fed. Aid in Wildl. Restor. Proj. W-23-4 Final Rept. Alaska Dept. Fish and Game, Juneau. 21 pp.

Bogaczyk, B.A. 1990. A survey of metastrongyloid parasites in Maine cervids. M.S. thesis, Univ. Maine, Orono. 60 pp.

Bogomolova, E.M., J.A. Kurochkin and A.A. Nikolskii. 1982. Akusticheskaya signalizatsiya zhivotnykh. Zvukovie reaktsii losya *(Alces alces)* [The sound reactions of the moose]. Nauchny biologich. senter issledovaniya AN SSSR, Pushchino. 28 pp.

Bogomolova, E.M., J.A. Kurochkin and P.K. Anokhin. 1992. Observations of moose behavior on a moose farm. Alces (Suppl.) 1: 216.

Bolotin, N. 1980. Klondike lost. Alaska Northwest Publ. Co., Anchorage. 128 pp.

Bonar, R.L. 1975. Wildlife resources and habitat in the Williston reservoir area. British Columbia Hydro and E.L.U.C. Secretariat, Victoria. 96 pp.

Bond, J.H. 1959. Moose-skin boat. The Beaver 290: 44–45.

Bontaites, K.M. and K. Guftason. 1993. This history and status of moose management in New Hampshire. Alces 29: 163–168.

Boone and Crockett Club. 1988. Records of North American big game. 9th ed. Boone and Crockett Club, Dumfries, Virginia. 498 pp.

Boonstra, R. and A.R.E. Sinclair. 1984. Distribution and habitat use of caribou, *Rangifer tarandus caribou*, and moose, *Alces alces andersoni*, in the Spatsizi Plateau Wilderness Area, British Columbia. Can. Field-Nat. 98: 12–21.

Borror, D.J. and R.E. White. 1970. A field guide to the insects of America north of Mexico. Houghton Mifflin Co., Boston. 404 pp.

Botkin, D.B., P.A. Jordan, A.S. Dominski, H.S. Lowendorf and G.E. Hutchinson. 1973. Sodium dynamics in a northern forest ecosystem. Proc. Natl. Acad. Sci. 70: 2,745–2,748.

Bouchard, R. and G. Moisan. 1974. Chasse contrôlée a l'orignal dans les parcs et reserves du Quebec (1962–1972). Nat. Can. (Que.) 101: 689–704.

Boucher, P. 1664. Histoire véritable et naturelle des moeures & productions du pays de la Nouvelle France, vulgairement dite Le Canada. Chez Florentin Lambert, Paris. 168 pp.

Boudreau, F. and D. Bisson. 1983. Observations sur la distribution et l'habitat de l'orignal en moyenne et basse Côté-Nord. Ministère de l'environment. Pap. SIEN-2, Quebec. 40 pp.

Boulanger, J.G. and G.C. White. 1990. A comparison of home range estimators using Monte Carlo simulation. J. Wildl. Manage. 54: 310–315.

Bourque, C. 1982. Les variations de l'habitat et du régime alimentaire de l'orignal *(Alces alces)* avec la progression de l'hiver en Abitibi-est. M.S. thesis, Univ. Quebec, Montreal. 153 pp.

Bourque, C. and R. Higgins. 1984. Serologic studies on brucellosis, leptospirosis and tularemia in moose *(Alces alces)* in Quebec. J. Wildl. Dis. 29: 95–99.

Boutin, S. 1992. Predation and moose population dynamics: A critique. J. Wildl. Manage. 56: 116–127.

Bowman, R.I. 1955. Aerial reconnaissance of moose in summer. J. Wildl. Manage. 19: 383–387.

Bowyer, R.T. 1986. Antler characteristics as related to social status of male southern mule deer. Southwest Nat. 31: 289–298.

Bowyer, R.T. and D.W. Kitchen. 1987. Sex and age class differences in vocalization of Roosevelt elk during rut. Am. Midl. Nat. 118: 225–235.

Bowyer, R.T., J.L. Rachlow, V. VanBallenberghe and R.D. Guthrie. 1991. Evolution of a rump patch in Alaskan moose: An hypothesis. Alces 27: 12–23.

Boyce, M.S. 1991. Natural regulation or the control of nature? Pages 183–208 *in* R.B. Reiter and M.S. Boyce, eds., The greater Yellowstone ecosystem, redefining America's wilderness heritage. Yale Univ. Press, New Haven, Connecticut.

Braefield, A.E. and M.J. Llewellyn. 1982. Animal bioenergetics. (tertiary level biology). Blackie and Son Ltd., Glasgow. 168 pp.

Bragdon, K.J. 1996. Native people of southern New England, 1500–1650. Univ. Oklahoma Press, Norman. 301 pp.

Brandt, K. 1901. Das Gehorn und die Entstehung monströser Formen. P. Parey Verlag, Berlin. 212 pp.

Brassard, J.M., E. Audy, M. Crête and P. Grenier. 1974. Distribution and winter habitat of moose in Quebec. Nat. Can. 101: 67–80.

Brasser, T.J. 1978. Early Indian-European contacts. Pages 78–88 *in* B.J. Trigger, ed., Northeast. Handbook of North American Indians. Vol. 15. Smithsonian Inst., Washington, D.C. 924 pp.

Brazil, J. and S. Ferguson. 1989. Cadmium concentrations in Newfoundland moose. Alces 25: 52–57.

Bretzlaff, K., J. Edwards, D. Forrest and L. Nuti. 1993. Ultrasonographic determination of pregnancy in small ruminants. Vet. Med. (Jan.): 12–24.

Briedermann, L. 1968. Die Westvorstösse des Elchwildes (*Alces alces* L.) in Mitteleuropa in ihrer populationsdynamischen Bedeutung. Zool. Garten 35: 224–229.

———. 1971. Ermittlungen zur Aktivitätsperiodik des mitteleuropäischen Wildschweines. Zool. Garten 40: 302–327.

———. 1993. Unser Muffelwild. Neumann-Neudam, Neudam. 175 pp.

Brindle, Y., R. Letarte and J. Gagnon. 1979. Investigation of antibodies against *Fasciola hepatica* in moose *Alces americana*. Can. J. Microbiol. 25: 788–789.

Britton, N.L. and A. Brown. 1943. An illustrated flora of the United States, Canada and the British possessions. 3 vols. Lancaster Press, Inc., Lancaster, Pennsylvania.

Brody, S. 1945. Bioenergetics and growth. Reinhold Publ. Co., New York. 1,023 pp.

Bronson, F.H. 1989. Mammalian reproductive biology. Univ. Chicago Press, Chicago, Illinois. 325 pp.

Brooke, V. 1878. On the classification of the Cervidae with a synopsis of the extinct species. Proc. Zool. Soc. L. 1878: 883–928.

Brown, F.A., Jr. 1965. A unified theory for biological rhythms—Rhythmic duplicity and the genesis of "circa" periodirism. Pages 231–261 *in* J. Aschoff, ed., Circadian clocks. North-Holland Publ. Co., Amsterdam.

Brown, F.A., Jr., J.W. Hastings and J.D. Palmer. 1970. The biological clock: Two views. Academic Press, New York. 94 pp.

Brown, R.D. 1990. Nutrition and antler development. Pages 426–441 *in* G.A. Bubenik and A.B. Bubenik, eds., Horns, pronghorns and antlers. Springer-Verlag, New York.

Brusewitz, G., trans. by W. Wheeler. 1969. Hunting; hunters, game, weapons and hunting methods from the remote past to the present day. Stein and Day Publ., New York. 251 pp.

Brush, G. De F. 1892. Killing the moose. Century XLII: 600, 638.

Brusnyk, L.M. and F.F. Gilbert. 1983. Use of shoreline timber reserves by moose. J. Wildl. Manage. 47: 673–685.

Brutt, E.H., Jr., ed. 1979. The behavioral significance of color. Garland STPM Press, New York. 486 pp.

Bryant, J.P., P.J. Kuropat, B.P. Reichardt and T.P. Clausen. 1991. Controls over the allocation of resources by woody plants to chemical anti-herbivore defenses. Pages 84–102 *in* R.T. Palo and C.T. Robbins, eds., Plant defenses against mammalian herbivory. CRC Press, Boca Raton, Florida.

Bryson, R.A. and W.M. Wendland. 1967. Radiocarbon isochromes of the retreat of the Laurentide ice sheet. Tech. Rept. No. 35. Dept. Meteor., Univ. Wisconsin, Madison. 25 pp.

Bubenik, A.B. 1952. Kurs jeleního troubeni na revnicí [Kurs des Hirschrohrens in Wort und auf der Gramophon-Platte]. Gramofonové Závody, N.P. Prague. (Recorded on disk.)

———. 1956. O parožních atavismech srnců a jelenů [On antler-atavisms in roe and red deer]. Vestník Čsl. Zool. Spol. 3: 249–266.

———. 1959. Grundlagen der Wildernährung. Deutscher Bauernverlag, Berlin. 300 pp.

———. 1960. Le rhythme nycthemeral et le régime journalier des ongules sauvages. Problemes théoretiques. Rhythme d'activité du chevreuil. Mammal. 24: 1–59.

———. 1963. Prečo niektore jelenie shody krvacaju? [Why do some cast antlers of red deer bleed?] Polovnictvo a Rybarstvo (Bratislava-SR) 15: 8.

———. 1965a. Osobennosti sna u kopytnych. Okhota i Okhotnichie Khozyaystvo 128: 516.

———. 1965b. Beitrag zur Gegurtskunde und zu den Mutter-Kind-Beziehungen des Reh- (*Capreolus capreolus*), und Rotwildes (*Cervus elaphus* L.). Z. Säugetierkunde 30: 65–128.

———. 1966a. Die Sprache des Hirsches—Handhabung und Anwendung des Hirschrufes. Die Pirsch 18: 982–989.

———. 1966b. Das Geweih. P. Parey Verlag, Hamburg. 214 pp.

———. 1968a. The significance of the antlers in the social life of Cervidae. Deer 6: 208–214.

———. 1968b. Trophy scoring from formulas in view of behavioral research. Trans. C.I.C. Congr. Mammalia: 1–9.

———. 1971. Rehwildhege und Rehwildbiologie. F.C. Mayer Verlag, Munich. 59 pp.

———. 1972. North American moose management in light of European experiences. Proc. N. Am. Moose Conf. Workshop 8: 279–295.

———. 1973. Antlers as releaser and Gestalt in the social life of animals. Int. Ethologic Conf., XIII: 12.

———. 1975a. Taxonomic value of antlers in genus *Rangifer*. Pages 9–11 *in* J.R. Luick, P.C. Lent, D.R. Klein and R.B. White, eds., Proc. First Int. Reindeer and Caribou Symp. Biol. Pap., Spec. Rept. No. 1. Univ. Alaska-Fairbanks.

———. 1975b. Significance of antlers in the social life of barren-ground caribou. Pages 436–461 *in* J.R. Luick, P.C. Lent, D.R. Klein and R.B. White, eds., Proc. First Int. Reindeer and Caribou Symp. Biol. Pap., Spec. Rept. No. 1. Univ. Alaska-Fairbanks.

———. 1981. Moose research and sociobiological management. Alces 17: 79–94.

———. 1982a. Physiology. Pages 125–179 *in* J.W. Thomas and D.E. Toweill, eds., Elk of North America: Ecology and management. Stackpole Books, Harrisburg, Pennsylvania. 698 pp.

———. 1982b. Proposals for standard nomenclature for bony appendices. Pages 187–194 *in* R.D. Brown, ed., Antler development in Cervidae. Caesar Kleberg Wildl. Res. Inst., Kingsville, Texas.

———. 1982c. The behavioral aspects of antlerogenesis. Pages 389–449 *in* R.D. Brown, ed., Antler development in Cervidae. Caesar Kleberg Wildl. Res. Inst., Kingsville, Texas.

———. 1984. Ernährung, Verhalten und Umwelt des Schalenwildes. BLV Verlag, Munich. 272 pp.

———. 1985. Reproductive strategies in Cervids. Pages 367–373 *in* P.F. Fennessy and K.R. Drew, eds., Biology of deer production. Bull. 22. R. Soc. New Zealand, Aukland.

———. 1986. Taxonomic position of *Alcinae* Jerdon, 1974 and the history of *Alces* Gray, 1821. Alces 22: 1–67.

———. 1987. Behaviour of moose *(Alces alces)* of North America. Swed. Wildl. Res. (Suppl.) 1: 333–366.

———. 1989a. Sport hunting in continental Europe. Pages 115–133 *in* R.J. Hudson, K.K. Drew and L.M. Baskin, eds., Wildlife production systems—Economic utilization of wildlife ungulates. Cambridge Univ. Press, New York.

———. 1989b. Coevolution of forest ecosystems as a basis for coexistence. Pages 19–39 *in* Algonquin Region forest/wildlife workshop. Ontario Min. Nat. Resour., Toronto.

———. 1989c. Socio-biological versus hunter's viewpoints of antlers and horns. Pages 355–385 *in* W. Trense, ed., The big game of the world. Verlag Paul Parey, Hamburg.

———. 1990. Epigenetical, morphological, physiological and behavioral aspects of evolution of horns, pronghorns and antlers. Pages 3–113 *in* G.A. Bubenik and A.B. Bubenik, eds., Horns, pronghorns and antlers. Springer-Verlag, New York.

———. 1991. The wildlife resource—Mine, ours or whose? Pages 143–151 *in* Proc. 2nd Int. Wildl. Ranching Symp. Univ. Alaska-Fairbanks.

———. 1992. Who is arrogant in Banff—Man or wapiti? Paper presented at the Elk Conflicts Workshop, Banff Natl. Park, Alberta. 17 pp.

———. 1993. Evolution of cranial protuberances of Cervoids from velericorn stage into annually deciduous antlers. Pages 106–109 *in* N. Ohtaishi and H.-I. Sheng, eds., Int. Symp. on Deer in China. Elsevier Science Publ., Amsterdam.

Bubenik, A.B. and T.J. Bellhouse. 1980. Brain volume of taiga moose (*Alces alces* sp.) in relation to skull parameters—A pilot study. Proc. N. Am. Moose Conf. Workshop. 16: 11–64.

Bubenik, A.B. and T.J. Bellhouse. 1985. Volumetric measurement of brain case cavity. Mammalia 49: 415–420.

Bubenik, A.B. and J. Brna. 1968. The kinetics of stag *(Cervus elaphus maral)* roaring in the riparian forests of Belje. Bull. Jelen Beograd 6: 83–95.

Bubenik, A.B. and J. Bubenikova. 1967. 24-hour periodicity in red deer. Trans. Cong. Int. Union Game Biol. 7: 344–349.

Bubenik, A.B. and P. Časnocha. 1965. Die Nutzung der Umwelt durch das Rotwild (Cervus elaphus L.). Trans. Cong. Int. Union Game Biol. 6: 175–199.

Bubenik, A.B. and V. Munkačevič. 1967. Karakteristika parozne trofike jelena (*Cervus elaphus* L.). [Peculiarities of antler trophic in red deer]. Trans. Int. Union Game Biol. 7: 255–259.

Bubenik, A.B. and R. Pavlansky. 1965. Trophic responses to trauma in growing antlers. J. Exp. Zool. 159: 289–302.

Bubenik, A.B. and B.A. Pond. 1992. Principles of sociobiological modelling of moose *(Alces alces andersoni)* of the North American taiga. Alces (Suppl.) 1: 29–51.

Bubenik, A.B. and R.H. Timmermann. 1982. Spermatogenesis in the taiga-moose of north central Ontario. Alces 18: 54–93.

Bubenik, A.B., G.A. Bubenik and D.G. Larsen. 1990. Velericorn antlers on a mature male moose *(Alces alces gigas)*. Alces 26: 115–128.

Bubenik, A.B., R. Tachezy and G.A. Bubenik. 1976. The role of the pituitary-adrenal axis in the regulation of antler growth processes. Saugetierkundl. Mitt. 24: 12–15.

Bubenik, A.B., R.H. Timmermann and B. Saunders. 1975. Simulation of population structure and size in moose on behalf of age structure of harvested animals. Proc. N. Am. Moose Conf. Workshop 11: 391–463.

Bubenik, A.B., O. Williams and H.R. Timmermann. 1977. Visual estimation of sex and social class in moose *(Alces alces)* from the ground and the plane. A preliminary study. Proc. N. Am. Moose Conf. Workshop 13: 157–176.

Bubenik, A.B., O. Williams and H.R. Timmermann. 1978a. Some characteristics of antlerogenesis in moose. A preliminary report. Proc. N. Am. Moose Conf. Workshop. 14: 157–177.

Bubenik, A.B., O. Williams and H.R. Timmermann. 1978b. The significance of hooves in moose management. A preliminary report. Proc. N. Am. Moose Conf. Workshop 14: 209–226.

Bubenik, A.B., M. Dombalagian, J.W. Wheeler and O. Williams. 1979. The role of the tarsal glands in the olfactory communication of Ontario moose. A preliminary report. Proc. N. Am. Moose Conf. Workshop 15: 119–147.

Bubenik, G.A. 1972. Seasonal variations of nuclear size of the hypothalmic cells in roe buck. J. Anim. Sci. 35: 967–973.

———. 1982. The endocrine regulation of the antler cycle. Pages 73–106 *in* R.D. Brown, ed., Antler development in Cervidae. Caesar Kleberg Wildl. Res. Inst., Kingsville, Texas.

———. 1990a. Neuroendocrine regulation of the antler cycle. Pages 265–297 *in* G.A. Bubenik and A.B. Bubenik, eds., Horns, pronghorns and antlers. Springer-Verlag, New York.

———. 1990b. The role of the nervous system in the growth of antlers. Pages 339–370 *in* G.A.Bubenik and A.B. Bubenik, eds., Horns, pronghorns and antlers. Springer-Verlag, New York.

———. 1990c. The antler as a model in biomedical research. Pages 474–487 *in* G.A. Bubenik and A.B. Bubenik, eds., Horns, pronghorns, and antlers. Springer-Verlag, New York.

———. 1991. Regulatory mechanisms of the antler cycle and the selection of deer breeding stock by endocrine tests. Pages 521–529 *in* L.A. Renecker and R.J. Hudson, eds., Wildlife production: Conservation and sustainable development. AFES Misc. Pub. 91-6. Univ. Alaska-Fairbanks.

———. 1993a. Morphological differences in the antler velvet of Cervidae. Pages 56–64 *in* N. Ohtaishi and H.-I. Sheng, eds., Proc. Int. Symp. on Deer in China. Elsevier Science Publ., Amsterdam.

———. 1993b. The neuronal and endocrine regulation of antler development and growth. Pages 110–112 *in* N. Ohtaishi and H.-I. Sheng, eds., Proc. Int. Symp. on Deer in China. Elsevier Science Publ., Amsterdam.

Bubenik, G.A. and A.B. Bubenik. 1978. The role of sex hormones in the growth of antler bone tissue: Influence of an antiandrogen therapy. Saugetierkdl. Mitt. 26: 284–291.

Bubenik, G.A. and A.B. Bubenik. 1985. Seasonal variations in hair pigmentation of white-tailed deer and their relationship to sexual activity and plasma testosterone. J. Exp. Zool. 235: 387–395.

Bubenik, G.A. and A.B. Bubenik. 1986. Phylogeny and ontogeny of antlers and neuroendocrine regulation of the antler cycle—A review. Saugetierkdl. Mitt. 33: 97–123.

Bubenik, G.A. and D. Schams. 1986. Relationship of age to seasonal levels of LH, FSH, prolactin and testosterone in male white-tailed deer. Comp. Biochem. Physiol. 83A: 179–183.

Bubenik, G.A. and H.J. Smith. 1986. The effect of thyroxine (T-4) administration on plasma levels of Tri-Iodothyronine (T-3) and (T-4) in male white-tailed deer. Comp. Biochem. Physiol. 83A: 185–187.

Bubenik, G.A. and P.S. Smith. 1987. Circadian and circannual rhythms of melatonin in plasma of male white-tailed deer and the effect of oral administration of melatonin. J. Exper. Zool. 241: 81–89.

Bubenik, G.A., A.B. Bubenik and A. Frank. 1985. Aussersaisonale Brunft beim Edelhirsch *(Cervus elaphus maral)*. Z. Jagdwiss. 31: 129–133.

Bubenik, G.A., D. Schams and G. Coenen. 1987a. The effect of artificial periodicity and antiandrogen treatment on the antler growth and plasma levels of LH, FSH, testosterone, prolactin and alkaline phosphotase in the male white-tailed deer. Comp. Biochem. Physiol. 86A: 551–559.

Bubenik, G.A., A.J. Sempere and J. Hamr. 1987b. Developing antler, a model for endocrine regulation of bone growth: Concentration gradient of T_3, T_4, and alkaline phosphotase in the antler, jugular, and saphenous vein. Calcified Tissue Intern. 41: 38–43.

Bubenik, G.A., J.H. Smith and A. Flynn. 1988. Plasma levels of b-endorphine in white-tailed deer. Seasonal variation and the effects of thyroxine, GnRH, dexamethasone and ACTH administration. Comp. Biochem. Physiol. 90A: 309–313.

Bubenik, G.A., R.D. Brown and D. Schams. 1990. The effect of latitude on the seasonal pattern of reproductive hormones in male white-tailed deer. Comp. Biochem. Physiol. 98A: 253–257.

Bubenik, G.A., A.B. Bubenik, G.M. Brown, A. Trenkle and D.A. Wilson. 1975. Growth hormone and cortisol levels in the annual cycle of white-tailed deer *(Odocoileus virginianus)*. Can. J. Physiol. Pharmacol. 53: 787–792.

Bubenik, G.A., A.B. Bubenik, D. Stevens and A. Binnington. 1982. The effects of neurogenic stimulation on the development and growth of bony tissue. J. Exp. Zool. 219: 205–216.

Budyko, M.I. 1974. Climate and life. Academic Press, New York. 508 pp.

Buel, J.W. 1894. The magic city, a massive portfolio of original photographic views of the Great World's Fair. Historical Publ. Co., St. Louis, Missouri.

Buffon, G.L.L. de. 1749–1769. Histoire naturelle générale et particulière. 13 vols. Planteaux, Paris.

Bunnell, F.L. and M.P. Gillingham. 1985. Foraging behavior: Dynamics of dining out. Pages 53–59 *in* R.J. Hudson and R.G. White, eds., Bioenergetics of wild herbivores. CRC Press, Boca Raton, Florida. 314 pp.

Burch, E.S., Jr. 1979. Indians and Eskimos in north Alaska, 1816–1977. Arctic Anthropol. 16(2): 123–151.

Burger, J.F. and J.R. Anderson. 1974. Taxonomy and life history of the moose fly, *Haematobosca alcis,* and its association with the moose, *Alces alces shirasi* in Yellowstone National Park. Ann. Entomol. Soc. Am. 67: 204–214.

Burr, D.B. and R.B. Martin. 1989. Errors in bone remodelling—Toward a unified theory of metabolic bone disease. Am. J. Anat. 186: 186–216.

Burris, O.E. and D.E. McKnight. 1973. Game transplants in Alaska. Wildl. Tech. Bull. No. 4. Alaska Dept. Fish and Game, Juneau. 57 pp.

Burt, W.H. 1940. Territorial behavior and populations of some small mammals in southern Michigan. Misc. Publ. 45. Univ. Mich. Mus. Zool., Ann Arbor. 58 pp.

———. 1943. Territory and home range as applied to mammals. J. Mammal. 24: 346–352.

Burton, A.C. and O.G. Edholm. 1955. Man in a cold environment; physiological and pathological affects of exposure to low temperatures. Monogr. of the Physiol. Soc. II. Arnold: London, England.

Bush, M. 1992. Remote drug delivery systems. J. Zoo Wildl. Med. 23: 159–180.

Buss, M.E. 1978. Post graduate education for moose hunters. Can. Wildl. Admin. 2: 36–37.

———. 1990. Using the right tools. Pages 42–47 *in* M.Buss and R. Truman, eds.,The moose in Ontario. Book II: Moose hunting techniques, hunting ethics and the law. Ontario Min. Nat. Resour. and Ontario Fed. Anglers and Hunters, Toronto.

Buss, M.E. and J. Richard. 1990. Sighting in and making the shot. Pages 48–54 *in* M.E. Buss and R. Truman, eds., The moose in Ontario. Book II: Moose hunting techniques, hunting ethics and the law. Ontario Min. Nat. Resour. and Ontario Fed. Anglers and Hunters, Toronto.

Buss, M.E. and R. Truman, eds. 1990a. The moose in Ontario. Book I. Moose biology, ecology and management. Ontario Min. Nat. Resour. and Ontario Fed. of Anglers and Hunters, Toronto. 33 pp.

Buss, M.E. and R. Truman, eds. 1990b. The moose in Ontario. Book II. Moose hunting techniques, hunting ethics and the law. Ontario Min. Nat. Resour. and Ontario Fed. Anglers and Hunters, Toronto. 78 pp.

Buss, M.E., R. Gollat and H.R. Timmermann. 1989. Moose hunter shooting proficiency in Ontario. Alces 25: 98–103.

Butler, C.E. 1986. Summer food utilization and observations of a tame moose *(Alces alces)*. Can. Field-Nat. 100: 85–88.

Cahalane, V.H. 1942. Wildlife vistas of the eastern highlands. Audubon Mag. 44: 101–111.

———. 1947. Mammals of North America. The Macmillan Co., New York. 682 pp.

Cahalane, V.H. and W.A. Weber. 1946. Meeting the mammals. The Macmillan Co., New York. 133 pp.

Cairns, A.L. 1976. Distribution and food habits of moose, wapiti, deer, bison, and snowshoe hare in Elk Island National Park, Alberta. M.S. thesis, Univ. Calgary, Calgary, Alberta. 167 pp.

Cairns, A.L. and E.S. Telfer. 1980. Habitat use by four sympatric ungulates in boreal mixedwood forest. J. Wildl. Manage. 44: 849–857.

Cameron, A.W. 1949. Report on biological investigations of game and fur-bearing animals in Nova Scotia—1948. Nova Scotia Dept. Lands and For., Halifax. 28 pp.

Campbell, G.D., E.M. Addison, I.K. Barker and S. Rosendal. 1994. *Erysipelothrix rhusiopathiae,* serotype 17, septicemia in moose *(Alces alces)* from Algonquin Park, Ontario. J. Wildl. Dis. 30: 436–438.

Camsell, C. 1954. Son of the North. Ryerson Press, Toronto, Ontario.

Canada Surveys and Mapping Branch. 1974. The national atlas of Canada. 4th ed. Depart. Energy, Mines and Resour., Ottawa.

Canadian Parks Service. 1979. Parks Canada policy. Min. Environ., Ottawa.

Cantin, M. and J. Genest. 1986. The heart as an endocrine gland. Sci. Am. 254: 76–81.

Carbyn, L. 1983. Wolf predation on elk in Riding Mountain National Park, Manitoba. J. Wildl. Manage. 47: 963–976.

Carlstrom, N.W. 1990. Moose in the garden. Harper Collins Publishers., New York. 32 pp.

Carpenter, L.H. 1996. Helicopter moose capture opens new doors. The Moose Call 3: 3.

Carpenter L.H. and J.I. Innes. 1995. Helicopter netgunning: A successful moose capture technique. Alces 31: 181–184.

Carr, D.E. 1972. The forgotten senses. Doubleday and Co., Inc., Garden City, New York. 377 pp.

Carver, J. 1838. Carver's travels in Wisconsin. 3rd London ed., Harper and Brothers, New York. 376 pp.

Catesby, M. 1743. The natural history of Carolina, Florida and the Bahama Islands. 2 vols. Printed for the author, London, England.

Caton, J.D. 1877. The antelope and deer of America. Hurd and Houghton, New York. 426 pp.

Catto, T.M. and C. Mandryk. 1990. Geology of the postulated ice-free corridor. Pages 80–85 *in* L.D. Agenbroad, J.I. Mead and L.W. Nelson, eds., Megafauna and man: Discovery of America's heartland. Univ. N. Arizona Press, Flagstaff.

Caughley, G. 1966. Mortality patterns in mammals. Ecology 47: 906–917.

———. 1974. Bias in aerial survey. J. Wildl. Manage. 38: 921–933.

———. 1976. Wildlife management and the dynamics of ungulate populations. Adv. Appl. Biol. 1: 183–247.

———. 1977. Analysis of vertebrate populations. John Wiley and Sons, New York. 234 pp.

———. 1979. What is this thing called carrying capacity? Pages 2–8 *in* M.S. Boyce and L.D. Hayden-Wing, eds., North American elk: Ecology, behavior, and management. Univ. Wyoming Press, Laramie.

———. 1981. Comments on natural regulation of ungulates (What constitutes a real wilderness?). Wildl. Soc. Bull. 9: 232–234.

Caughley, G. and C.J. Krebs. 1983. Are big mammals simply little mammals writ large? Oecologia 59: 7–17.

Causey, A.S. 1989. On the morality of hunting. Environ. Ethics 11: 327–343.

———. 1993. A reverence for life: Is hunting moral? Bugle 10: 82–88.

Cederlund, B-M. 1987. Parturition and early development of moose (*Alces alces* L.) calves. Swed. Wildl. Res. (Suppl.) 1: 399–422.

Cederlund, G. and G. Markgren. 1987. The development of the Swedish moose population, 1970-1983. Swed. Wildl. Res. (Suppl.) 1: 55–62.

Cederlund, G. and H. Okarma. 1988. Home range and habitat use of adult female moose. J. Wildl. Manage. 52: 336–343.

Cederlund, G. and H.K.G. Sand. 1991. Population dynamics and yield of a moose population without predators. Alces 27: 31–40.

Cederlund, G. and H.K.G. Sand. 1992. Dispersal of subadult moose *(Alces alces)* in a migratory population. Can. J. Zool. 70: 1,309–1,314.

Cederlund, G. and H.K.G. Sand. 1994. Home-range size in relation to age and sex in moose. J. Mammal. 75: 1,005–1,012.

Cederlund, G., F. Sandegren and K. Larsson. 1987. Summer movements of female moose and dispersal of their offspring. J. Wildl. Manage. 15: 342–352.

Chaddock, T.T. 1940. Chemical analysis of deer antlers. Wisconsin Conserv. Bull. 5: 42.

Chadwick, H.W. 1096. Plant succession and big game winter movements and feeding habits in the sand dune area of Fremont County, Idaho. M.S. thesis, Univ. Idaho, Moscow. 121 pp.

Chamberlain, A.F. 1906. Indians of the eastern provinces of Canada. Pages 122–136 *in* Ann. archeol. rept. 1905, Report of the Minister of Education. L.K. Cameron, Toronto, Ontario. 294 pp.

Chamberlain, L.C. 1977. Cursed pond of the big noise. Ontario Fish and Wildl. Rev. 16: 19–23.

Chapman, D.I. 1975. Antlers, bones of contention. Mammal. Rev. 5: 121–172.

Chapman, D.M. 1985. Histology of moose *(Alces alces)* interdigital glands and associated green hairs. Can. J. Zool. 63: 899–911.

Chapman, J.A. and G.A. Feldhammer, eds. 1982. Wild mammals of North America. The Johns Hopkins Univ. Press, Baltimore, Maryland. 1,147 pp.

Chappel, R.W. and R.J. Hudson. 1978. Winter bioenergetics of Rocky Mountain bighorn sheep. Can. J. Zool. 56: 2,388–2,393.

Chappel, R.W. and R.J. Hudson. 1979. Energy cost of standing in Rocky Mountain bighorn sheep. J. Wildl. Manage. 43: 261–263.

Charlevoix, P. de. 1761. Journal of a voyage to North-America undertaken by order of the French King, containing the geographical description and natural history of that country, particularly Canada. 2 vols. Printed for R. and J. Dodsley, London, England.

Charnley, S. 1983. Moose hunting in two central Kuskakwin communities: Chutathbaluk and Steelmute. Tech. Pap. No. 76. Div. of Subsistence, Alaska Dept. Fish and Game, Bethel. 59 pp.

Charnov, E.L. 1976. Optimal foraging, the marginal value theorem. Theor. Pop. Biol. 9: 129–136.

Cheatum, E.L. 1949. The use of corpora lutea for determining ovulation incidence and variations in the fertility of white-tailed deer. Cornell Vet. 39: 282–291.

———. 1951. Disease in relation to winter mortality of deer in New York. J. Wildl. Manage. 15: 216–220.

Chermnykh, N.A. 1987. Advances in physiology studies on moose in the USSR. Swed. Wildl. Res. (Suppl.) 1: 327–332.

Child, K.N. 1983a. Railways and moose in the central interior of British Columbia: A recurrent management problem. Alces 19: 118–135.

———. 1983b. Selective harvest of moose in the Omineca: Some preliminary results. Alces 19: 162–177.

Child, K.N. and D.A. Aitken. 1989. Selective harvests, hunters and moose in British Columbia. Alces 25: 81–97.

Child, K.N. and L.J. Foubister. 1986. A preliminary study of the effectiveness of ultrasonic warning devices in reducing wildlife-vehicle collisions. British Columbia Min. Environ., Prince George. 12 pp.

Child, K.N. and K.M. Stuart. 1967. Vehicle and train collision fatalities of moose: Some management and socio-economic considerations. Swed. Wildl. Res. (Suppl.) 1: 699–703.

Child, K.N., S.P. Barry and D.A. Aitken. 1991. Moose mortality on highways and railways of British Columbia. Alces 27: 41–49.

Chittendon, H.R. 1935. The American fur trade of the far West. 2 vols. Press of the Pioneers, Inc. New York. 1,014 pp. (Reprinted in 1986 by Univ. Nebraska Press.)

Choquette, L.P.E. 1970. Anthrax. Pages 256–260 *in* J.W. Davis, L.H. Karstad and D.O. Trainer, eds., Infectious diseases of wild mammals. Iowa State Univ. Press, Ames.

Christopherson, R.J., R.J. Hudson and R.J. Richmond. 1978. Comparative winter bioenergetics of American bison, yak, Scottish Highland and hereford calves. Acta. Theriol. 23: 49–54.

Church, D.C. 1971. Digestive physiology and nutrition of ruminants. Vol. 2., Nutrition. D.C. Church, Publ. Oregon St. Univ. Book Stores, Inc., Corvallis. 801 pp.

Churcher, C.S. 1983. Faunal correlations of Pleistocene deposits in western Canada. Pages 145–158 *in* W.C. Mahaney, ed., Correlation of Quaternary chronologies. Proc. Symp. Correlation of Quaternary Chronologies. GEO Books, New York.

Churcher, C.S. and J.D. Pinsof. 1987. Variation in the antlers of North American *Cervalces* (Mammalia; Cervidae): Review of new and previously recorded specimens. J. Vert. Paleontol. 7: 373–397.

Clark, A.M. 1981. Koyukon. Pages 582–601 *in* J. Helm, ed., Subarctic. Handbook of North American Indians. Vol. 6. Smithsonian Inst., Washington, D.C. 837 pp.

Clark, R.A. and R.T. Bowyer. 1987. Occurrence of protostrongylid nematodes in sympatric populations of moose and white-tailed deer in Maine. Alces 23: 313–321.

Clark, W.P. 1885. The Indian sign language. L.R. Hamersly and Co., Philadelphia. 443 pp.

Clarke, C.H.D. 1977. Quality hunting. Excerts from a keynote address entitled "Moose hunter ethics" given at the moose hunter seminar, Confederation Coll., Thunder Bay, Ontario. 10 pp.

Clayton, J.L. 1967. The growth and economic significance of the American fur trade. Pages 62–72 *in* Aspects of the fur trade: Selected papers of the 65th North American Fur Trade Conference. Minnesota Hist. Soc., St. Paul.

Clayton Resources, Ltd. 1989. Wildlife fencing and control on the Okanagan connector highway. A benefit cost analysis. Rept. for the Envir. Serv. Sect. and the Highway Engineer. Branch, British Columbia Min. of Transp. and Highways, Victoria. 32 pp.

Clinton, De Witt. 1822. Letter XLIV. Pages 189–193 *in* Letters on the natural history and internal resources of the State of New York by Hibernicus (a pseudonym). Bliss and White, New York.

Clutton-Brock, T.H. 1982. The function of antlers. Behav. 79: 108–125.

Clutton-Brock, T.H. and G.R. Iason. 1986. Sex ratio variation in mammals. The Quart. Rev. of Biol. 61: 339–374.

Clutton-Brock, T.H. and K. McComb. 1993. Experimental tests of copying and choice in fallow deer *(Dama dama)*. Behav. Ecol. 4.3: 191–193.

Clutton-Brock, T.H., S.D. Albon and F.E. Guiness. 1981. Parental investment in male and female offspring in polygamous animals. Nature 289: 486–489.

Clutton-Brock, T.H., S.D. Albon and F.E. Guiness. 1987. Interactions between population density and maternal characteristics affecting fecundity and juvenile survival in red deer. J. Anim. Ecol. 56: 857–871.

Clutton-Brock, T.H., F.E. Guiness and S.D. Albon. 1982. Red deer behavior and ecology of two sexes. Univ. Chicago Press, Chicago, Illinois. 378 pp.

Coady, J.W. 1974a. Influence of snow on behavior of moose. Nat. Can. (Que.) 101: 417–436.

———. 1974b. Late pregnancy of moose in Alaska. J. Wildl. Manage. 38: 571–572.

———. 1980. History of moose in northern Alaska and adjacent regions. Can. Field-Nat. 94: 61–68.

———. 1982. Moose. Pages 902–922 *in* J.A. Chapman and G.A. Feldhamer, eds., Wild mammals of North America. The Johns Hopkins Univ. Press, Baltimore.

———. 1989. Analyse du systeme de suive de l'original. Ministere du Loisir, la Chasse et la Peche, Quebec. 48 pp.

Cobus, M. 1972. Moose as an aesthetic resource and their summer feeding behaviour. Proc. N. Am. Moose Conf. Workshop 8: 244–275.

Cochran, W.C. 1980. Wildlife telemetry. Pages 507–520 *in* S.D. Schemnitz, ed., Wildlife management techniques manual, 4th . ed. The Wildl. Soc., Washington, D.C.

Cogho, D.R. 1986. Über die Veränderungen der Rosenstöcke beim Geweihwechsel der Edelhirsche. P. Wolf Verlag, Leipzig. 50 pp.

Cogley, T.P. 1987. Effects of Cephenemyia spp. (Diptera: Oestridae) on the nasopharynx of black-tailed deer *(Odocoileus hemionus columbianus)*. J. Wildl. Dis. 23: 596–605.

Cole, G. F. 1981. Alternative hypotheses on ecological effects of meningeal parasite *(Paralaphostrongylus tenuis)*. Minn. Acad. Sci. 47: 8–10.

Coles, E.H. 1967. Veterinary clinical pathology. W.B. Saunders Company. Philadelphia. 455 pp.

Compton, B.B. and L. E. Oldenburg. 1994. The status and management of moose in Idaho. Alces 30: 57–62.

Connolly, G.E. 1978. Predators and predator control. Pages 369–394 *in* J.L. Schmidt and D.L. Gilbert, eds., Big game of North America. Stackpole Books, Harrisburg, Pennsylvania. 494 pp.

———. 1981. Limiting factors and population regulation. Pages 245–285 *in* O.C. Wallmo, ed., Mule and black-tailed deer of North America. Univ. Nebraska Press, Lincoln.

Cook, H.W., R.A. Rausch and B.E. Baker. 1970. Moose milk. 1. Gross composition, fatty acid and mineral constitution. Can. J. Zool. 48: 213–215.

Cook, R. and J.O. Jacobson. 1978. A visibility biased model for aerial surveys of moose on the Aoserp study area. Alberta Oil Sands Environ. Res. Prog. Rept., Edmonton. 25 pp.

Cook, R.D. and F.B. Martin. 1974. A model for quadrat sampling with visibility bias. J. Am. Stat. Assoc. 69: 345–349.

Cooper, J.M. 1938. Snares, deadfalls, and other traps of the Northern Algonquians and Northern Athapaskans. Anthropol. Series 5. Catholic Univ. of America, Washington, D.C.

Cooper. S.F. 1968. Rural hours. Syracuse Univ. Press, Syracuse, New York. 337 pp.

Cooper, W.E. and G.M. Burghardt. 1990. Vomerolfaction and vomodor. J. Chem. Ecol. 16: 103–105.

Cope, L. 1919. Calendars of the Indians north of Mexico. Univ. California Publs. American Archeol. and Ethnol. 16: 119–176.

Corbett, G.N. 1995. Review of the history and present status of moose in the national parks of the Atlantic region: Management implications. Alces 31: 255–267.

Corner, A.H. and R. Connell. 1958. Brucellosis in bison, elk, and moose in Elk Island National Park, Alberta, Canada. Can. J. Comp. Med. 22: 9–20.

Cosgrove , G.E. and L.D. Fay. 1981. Viral tumors. Pages 424–428 *in* J.W. Davis, L.H. Karstad and D.O. Trainer, eds., Infectious diseases of wild mammals. Iowa St. Univ. Press, Ames.

Cottam, C and C.S. Williams. 1943. Speed of some animals. J. Mammal. 24: 262–263.

Coues, E., ed. 1897. The manuscript journals of Alexander Henry and David Thompson. 2 vols. Ross and Haines, Inc., Minneapolis, Minnesota.

Courtois, R. and M. Crête. 1993. Predicting moose population parameters from hunting statistics. Alces 29: 75–90.

Courtois, R. and H. Jolicoeur. 1993. The use of Schaefer's and Fox's surplus-yield models to estimate optimal moose harvest and hunting effort. Alces 29: 149–162.

Courtois, R. And G. Lamontagne. 1997. Management system and current status of moose in Quebec. Alces 33: 97–114.

Covington, W.W., D.B. Wood, D.L. Young, D.P. Dykstra and L.D. Garrett. 1988. TEAMS: A decision support system for multiresource management. J. For. 86: 25–33.

Cowan, I. McT. 1946a. Report of wildlife studies in Jasper, Banff and Yoho National Parks, 1944, and parasites, diseases and injuries of game animals in the Rocky Mountain national parks, 1942–1944. Can. Wildl. Serv., Ottawa. 84 pp.

———. 1946b. Parasites, diseases, injuries, and anomalies of the Columbian black-tailed deer, *Odocoileus hemionus columbianus* (Richardson), in British Columbia. Can. J. Res. Sect. D, 24: 71–103.

———. 1950. Some vital statistics of big game on overstocked mountain range. Trans. N. Am. Wildl. Conf. 15: 581–588.

———. 1951. The diseases and parasites of big game mammals of western Canada. Proc. Ann. British Columbia Game Conf. 5: 37–64.

———. 1956. Life and times of the coastal black-tailed deer. Pages 523–617 *in* W.P. Taylor, ed., The deer of North America. The Stackpole Company, Harrisburg, Pennsylvania.

———. 1974. Management implications of behavior in the large herbivorous mammals. Pages 921–934 *in* V. Geist and F. Walters, eds., The behaviour of ungulates and its relation to management, Vol. 1. New Ser. Publ. 24. Int. Union Conserv. Nature and Nat. Resour., Morges, Switzerland.

Cowan, I. McT. and V.C. Brink. 1949. Natural game licks in the Rocky Mountain national parks of Canada. J. Mammal. 30: 379–387.

Cowan, I. McT., W.S. Hoar and J. Hatter. 1950. The effect of forest succession upon the quantity and upon the nutritive values of woody plants used as food by moose. Can. J. Zool. 28: 249–271.

Cowlishaw, G., R.I. Dunbar and M. June. 1991. Dominance rank and mating success in male primates. Anim. Behav. 41: 1,045–1,056.

Craig, W. 1918. Appetites and aversions as constituents of instincts. Biol. Bull. 34: 91–108.

Crampton, E.W. and L.E. Lloyd. 1959. Fundamentals of nutrition. W.H. Freeman and Co., San Francisco. 494 pp.

Crawford, H.S., Jr. 1982. Seasonal food selection and digestibility by tame white-tailed deer in central Maine. J. Wildl. Manage. 46: 974–982.

Crawford, R.C. and F.D. Johnson. 1985. Pacific yew dominance in tall forests, a classification dilemma. Can. J. Bot. 63: 592–602.

Crête, M. 1977. Importance de la coupe forestière sur l'habitat hivernal de l'orignal dans le sud-uest de Quebec. Can. J. For. Res. 7: 241–257.

———. 1982. Évaluation des modifications apportes a la reglementation de la chasse a l'orignal au Quebec de 1971 a 1981 et perspectives d'avenir. Alces 18: 329–356.

———. 1987. The impact of sport hunting on North American moose. Swed. Wildl. Res. (Suppl.) 1: 553–563.

———. 1989. Approximation of K carrying capacity for moose in eastern Quebec. Can. J. Zool. 67: 373–380.

Crête, M. and E. Audy. 1974. Individual and seasonal variation in the diameter of browsed twigs by moose. Proc. N. Am. Moose Conf. Workshop 10: 145–159.

Crête, M. and A. Beaumont. 1986. Fécondité de l'original au Quebec d'après l'examen macroscopique d'ovaries récoltes au début de l'automne. Alces 22: 277–302.

Crête, M. and J. Bedard. 1975. Daily browse consumption of moose in the Gaspe Peninsula, Quebec. J. Wildl. Manage. 39: 368–373.

Crête, M. and C. Dussault. 1987. Using hunting statistics to estimate density, cow-calf ratio and harvest rate of moose in Quebec. Alces 23: 227–242.

Crête, M. and F. Goudreault. 1980. Les bois, la tache vulvaire et la couleur du museau pour déterminer le sexe des orignaux *(Alces alces americana)* en Janvier dans le sud- ouest du Quebec. Proc. N. Am. Moose Conf. Workshop 16: 275–288.

Crête, M. and H. Jolicoeur. 1987. Impact of wolf and black bear removal on cow-calf ratio and moose density in southwestern Quebec. Alces 23: 61–87.

Crête, M. and P.A. Jordan. 1981. Régime alimentaire des orignaux du sud-ouest quebecois pour les mois d'avril à octobre. Can. Field-Nat. 95: 50–56.

Crête, M. and P.A. Jordan. 1982a. Production and quality of forage available to moose in southwestern Quebec. Can. J. For. Res. 12: 151–159.

Crête, M. and P.A. Jordan. 1982b. Population consequences of winter forage resources for moose, *Alces alces,* in southwestern Quebec. Can. Field-Nat. 96: 467–475.

Crête, M. and F. Messier. 1984. Response of moose to wolf removal in southwestern Quebec. Alces 20: 107–128.

Crête, M., R.J. Taylor and P.A. Jordan. 1981a. Optimization of moose harvest in southwestern Quebec. J. Wildl. Manage. 45: 598–611.

Crête, M., R.J. Taylor and P.A. Jordan. 1981b. Simulating conditions for the regulation of a moose population by wolves. Ecol. Model. 12: 245–252.

Crête, M., L. Rivest, H. Jolicoeur, J. Brassard and F. Messier. 1986a. Predicting and correcting helicopter counts of moose with observations made from fixed-wing aircraft in southern Quebec. J. Appl. Ecol. 23: 751–761.

Crête, M., M. Potvin, P. Walsh, J.L. Bendetti, J.P. Weber, G. Paillard and J. Gagnon. 1986b. Présence de cadmium dans le foie et les reins d'originaux et de cerfs de Virginie au Quebec. Direction générale de la fauna Rept. Ministère du Loisir de la Chasse et de la Pêche, Quebec City. 49 pp.

Crichton, V.F.J. 1977. Hecla Island—Manitoba's answer to Isle Royale. Proc. N. Am. Moose Conf. Workshop 13: 191–199.

———. 1979. An experimental moose hunt on Hecla Island, Manitoba. Proc. N. Am. Moose Conf. and Workshop 15: 245–279.

———. 1980. Manitoba's second experimental moose hunt on Hecla Island. Proc. N. Am. Moose Conf. Workshop 16: 489–526.

———. 1981. The impact of treaty Indian harvest on a Manitoba moose herd. Alces 17: 56–63.

———. 1987a. Moose management in North America. Swed. Wildl. Res. (Suppl.) 1: 541–551.

———. 1987b. Procedure for standardized moose sex and age surveys in Manitoba. Manitoba Dept. Nat. Resour., Winnipeg. 12 pp.

———. 1988a. Toward century 21—A moose management perspective. Alces 24: 1–6.

———. 1988b. In utero productivity of moose in Manitoba. Alces 24: 143–149.

———. 1992a. Management of moose populations: Which parameters are used? Alces (Suppl.)1: 11–15.

———. 1992b. A plan for the management of Manitoba's moose herd. Manitoba Dept. Natur. Resour., Winnipeg. 81 pp.

———. 1992c. Six year (1986/87–1991/92) summary of in utero productivity of moose in Manitoba, Canada. Alces 28: 203–214.

———. 1993. Hunter effort and observations—The potential for monitoring trends of moose populations: A review. Alces 29: 181–186.

Crockford, J.A., F.A. Hayes, J.H. Jenkins and S.D. Feurt. 1958. An automatic projectile type syringe. J. Vet. Med. 53: 115–119.

Cronin, M.A. 1989. Molecular evolutionary genetics and phylogeny of cervids. Ph.D. thesis, Yale Univ., New Haven, Connecticut.

———. 1992. Intraspecific variation in mitochondrial DNA of North American cervids. J. Mammal. 73(1): 70–82.

Cronon, W. 1983. Changes in the land: Indians, colonists and ecology of New England. Hill and Wang, New York. 241 pp.

Croon, G.W., D.R. McCullough, C.E. Olson, Jr. and L.M. Queal. 1968. Infrared scanning techniques for big game censusing. J. Wildl. Manage. 32: 751–759.

Crossley, A. and J.R. Gilbert. 1983. Home range and habitat use of female moose in northern Maine—A preliminary look. Trans. Northeast Sect. The Wildl. Soc. 40: 67–75.

Crow, J.R. and P.R. Obley. 1981. Han. Pages 506—513 *in* J. Helm., ed., Handbook of North American Indians. Vol. 6. Smithsonian Inst., Washington, D.C. 837 pp.

Cruikshank, J. 1986. Understanding Yukon history: Contributions from oral traditions. Pages 286–316 *in* R.B. Morrison and C.R. Wilson, eds., Native peoples: The Canadian experience. McClelland and Stewart Inc., Toronto. 639 pp.

Culin, S. 1907. Games of the North American Indians. Twenty-fourth Ann. Rept. Bur. Amer. Ethnol. Smithsonian Inst., Washington, D.C. 846 pp.

Cullen, C.T., ed. 1983. The papers of Thomas Jefferson. 21 vols. Princeton Univ. Press, Princeton, New Jersey.

Cumbie, P.M. 1975. Mercury in hair of bobcats and raccoons. J. Wildl. Manage. 39: 419–425.

Cumming, H.G. 1974a. Moose management in Ontario from 1948 to 1973. Nat. Can. (Que.) 101: 673–687.

———. 1974b. Annual yield, sex and age of moose in Ontario as indices to the effects of hunting. Nat. Can. (Que.) 101: 539–558.

Cumming, H.G. and E. Evans. 1978. Repeatability of incisor cementum age determinations for moose. Proc. N. Am. Moose Conf. Workshop 14: 68–88.

Curtis, N. 1923. The Indians' book. Harper and Brothers Publ., New York. 583 pp.

Cutright, P.R. 1989. Lewis and Clark: Pioneering naturalists. Univ. Nebraska Press, Lincoln. 506 pp.

Dahl, E. 1968. Varfor skall vi skjuta kalv? [Why do we shoot calf moose?]. Svensk Jakt 106: 304–309.

Dale, M.L. 1736. A letter . . . to Sir Hans Sloane, Bart., president of the Royal Society, containing the descriptions of the moose-deer of New-England and a sort of stag in Virginia; with some remarks relating to Mr. Ray's description of the flying squirrel of America. Philos. Trans. Roy. Soc. 39(444): 384–389.

Dall, W.H. 1870. Alaska and its resources. Lee and Shepard, Boston. 627 pp.

Dalton, W.J. 1990. Moose density estimation with line transect survey. Alces 26: 129–141.

Dalton, W.J. and G.D. Francis. 1988. The status of applied moose aging technology. Alces 24: 69–77.

Damas and Smith Company. 1983. Wildlife mortality in transportation corridors in Canada's national parks. Impact and mitigation. Consultants rept. to Parks Can., Ottawa. 397 pp.

Damuth, J. 1981. Home range, home range overlap, and species energy use among herbivorous animals. Biol. J. Linnean Soc. 15: 185–193.

Dance, P.S. 1978. The art of natural history: Animal illustrators and their work. The Overlook Press, Woodstock, New York. 224 pp.

Danell, K. 1983. Shoot growth of *Betula pendula* and *B. pubescens* in relation to moose browsing. Alces 18: 197–209.

Danell, K. and L. Ericson. 1986. Foraging by moose on two species of birch when these occur in different proportions. Holarctic Ecol. 9: 79–84.

Danell, K., L. Edenius and P. Lunberg. 1991. Herbivory and tree stand composition: Moose patch use in winter. Ecol. 72: 1,350–1,357.

Darby, W.R. and W.O. Pruitt. 1984. Habitat use, movements, and grouping behavior of woodland caribou, *Rangifer tarandus caribou*, in southeastern Manitoba. Can. Field- Nat. 98: 184–190.

Darwin, C. 1859. On the origin of species by means of natural selection, or the preservation of favoured races in the struggle for life. J. Murray, London. 479 pp.

———. 1872. The expression of the emotions in man and animals. D. Appleton and Co., New York. 374 pp.

Dasmann, R.F. 1964. Wildlife biology. John Wiley and Sons. New York. 212 pp.

Dauphine, T.C. 1976. Biology of the Kaminuriak population of barren-ground caribou. Part 4: Growth, reproduction and energy reserves. Rept. No. 38. Can. Wildl. Serv., Ottawa. 71 pp.

Davidson, P.A. and R. Dawson. 1990. Williston wildlife compensation program management plan. British Columbia Min. Environ., Prince George. 444 pp.

Davies, A.M., C. Brantlow, R. Heumann, R. Korsching, H. Rohrere and H. Thonen. 1987. Timing and site of nerve growth factor synthesis in developing skin in relation to innervation and expression of the receptor. Nature 326: 353–358.

Davies, K.G. and A.M. Johnson, eds. 1963. Northern Quebec and Labrador journals and correspondence, 1819–35. Vol. XXIV. The Hudson's Bay Rec. Soc., London, England. 415 pp.

Davies, K.G. and A.M. Johnson. 1965. Letters from Hudson Bay, 1703–40. Vol XXV. The Hudson's Bay Rec. Soc., London, England. 455 pp.

Davis, D.E. and R.L. Winstead. 1980. Estimating the numbers of wildlife populations. Pages 221–246 *in* S.D. Schemnitz, ed., Wildlife management techniques manual. 4th ed. The Wildl. Soc., Washington, D.C.

Davis, J.L. and A.W. Franzmann. 1979. Fire-moose-caribou interrelationships: A review and assessment. Proc. N. Am. Moose Conf. Workshop 15: 80–118.

Davis, M.B. 1986. Climate instability, time lags and community disequilibrium. Pages 269–284 *in* J. Diamond and T.J. Case, eds., Community ecology. Harper and Row, New York. 665 pp.

Day, G.M. 1978. Western Abenaki. Pages 148–159 *in* B.G. Trigger, ed., Northeast. Handbook of North American Indians. Vol. 15. Smithsonian Inst., Washington, D.C. 924 pp.

Dean, F.C. 1976. Aspects of grizzly bear population ecology in Mount McKinley National Park. Pages 111–119 *in* M.R. Pelton, J.W. Lentfer and G.E. Folk, Jr., eds., Bears—Their biology and management. Publ. New Ser. No. 40. Int. Union Conserv. Nature and Nat. Resour., Morges, Switzerland.

de Bry, T. 1590. Admiranda narratio, fida tamen, de commodis et incolarum ritibus Virginae [America. pt. 1. Latin] . . . Anglico scripta sermone à Thomas Hariot...Nunc autem primum Latio donata à C.C.A. Francoforti ad Moenum, Typus I. Wechli, sumtibus T. de Bry, venales reperiùtur in officina S. Feierabendii. 91 pp.

Decker, D.J. and T.L. Brown. 1987. How animal rightists view the "wildlife management-hunting system." Wildl. Soc. Bull. 15: 599–602.

Decker, D.J., R.E. Shanks, L.A. Nielsen and G.R. Parsons. 1991. Ethical and scientific judgements in management: Beware of blurred distinctions. Wildl. Soc. Bull. 19: 523–527.

DeGayner, E.J. and P.A. Jordan. 1987. Skewed fetal sex ratios in white-tailed deer: Evidence and evolutionary speculation. Pages 178–188 *in* C.M. Wemmer, ed., Biology and management of Cervidae. Smithsonian Inst. Press, Washington, D.C. 577 pp.

DeKay, J.E. 1842. Mammalia. Part 1. Pages 1–146 *in* Zoology of New York. W. and A. White and J. Visshner, Albany, New York.

De Laguna, F. 1990. Tlingit. Pages 203–228 *in* W. Suttles, ed., Northwest coast. Handbook of North American Indians. Vol. 7. Smithsonian Inst., Washington, D.C. 777 pp.

De Laguna, F. and C. McClellan. 1981. Ahtna. Pages 641–663 *in* J. Helm, ed., Subarctic. Handbook of North American Indians. Vol. 6. Smithsonian Inst., Washington, D.C. 837 pp.

DelFrate, G.G. and T.H. Spraker. 1992. Moose-vehicle interactions and an associated public awareness program on the Kenai Peninsula, Alaska. Alces 27: 1–7.

DelGiudice, G.D., L.D. Mech and U.S. Seal. 1989. Physiological assessment of deer populations by analysis of urine in snow. J. Wildl. Manage. 53: 284–291.

DelGiudice, G.D., R.O. Peterson and U.S. Seal. 1991. Differences in urinary chemistry profiles of moose on Isle Royale during winter. J. Wildl. Dis. 27: 407–416.

DelGiudice, G.D., R.O. Peterson and W.M. Samuel. In press. Long-term trends of winter nutritional restrictions, ticks, and numbers of moose on Isle Royale. J. Wildl. Manage.

DelGiudice, G.D., L.D. Mech, W.J. Paul and P.D. Karns. 1986. Effects of fawn survival of multiple immobilization of captive pregnant white-tailed deer. J. Wildl. Dis. 22: 245–248.

DelGiudice, G.D., L.D. Mech, U.S. Seal and P.D. Karns. 1987. Winter fasting and refeeding effects on urine characteristics in white-tailed deer. J. Wildl. Manage. 51: 860–864.

Denney, R.N. 1973. The hunter as seen by the nonhunter. Pages 20–22 *in* Big Game Hunter and Fisherman's Conf., Game Conserv. Intl., San Antonio, Texas.

Denniston, G. 1981. Sekani. Pages 433–441 *in* J. Helm, ed., Subarctic. Handbook of North American Indians. Vol. 6. Smithsonian Inst., Washington, D.C. 837 pp.

Denniston, R.H., II. 1956. Ecology , behaviour and population dynamics of Wyoming or Rocky Mountain moose *(Alces alces shirasi)*. Zool. 41: 105–115.

Denys, N. 1672. The description and natural history of the coasts of North America (Acadia). 2 vols. Publ. II. The Champlain Soc., Toronto, Ontario. 625 pp.

Deschamp, J.A., P.J. Urness and D.D. Austin. 1979. Summer diets of mule deer from lodgepole pine habitats. J. Wildl. Manage. 42: 154–161.

De Smet, P.J. 1847. Oregon missions and travels over the Rocky Mountains, in 1845–46. Edward Dunnigan, New York.

Des Meules. P. 1962. Intensive study of an early spring habitat of moose (*Alces alces americana* cl.) in Laurentides Park, Quebec. Proc. Northeast Wildl. Conf. 12 pp.

———. 1964. The influence of snow on the behavior of moose. Ministère du tourisme, de la chasse et de la pêche, Quebec. Rapport No. 3: 51–73.

———. 1965. Hyemal food and shelter of moose (*Alces alces americana* Cl.) in Laurentide Park, Quebec. M.S. thesis, Univ. Guelph, Guelph, Ontario. 138 pp.

———. 1966. Controlled moose hunts in Quebec's provincial parks. Proc. Northeast Sect. Wildl. Soc., Boston, Massachusetts. 15 pp.

———. 1968. Détermination du nombre de tas crottins rejectes et du nombre de reposées établies, par jour, par l'orignal *(Alces alces)*, en hiver. Nat. Can. (Que.) 95: 1,153–1,157.

de Vos, A. 1958. Summer studies of moose in Ontario. Trans. N. Am. Wildl. Conf. 21: 510–525.

———. 1964. Range changes of mammals in the Great Lakes region. Am. Midland Natur. 71(1): 210–231.

Dewdney, S. and K.E. Kidd. 1973. Indian rock paintings of the Great Lakes. Univ. Toronto Press, Toronto, Ontario. 191 pp.

Diamond, J. 1990a. Playing dice with megadeath. Discover 11: 55–59.

———. 1990b. War babies. Discover 11: 70–75.

———. 1992. Must we shoot deer to save nature? Nat. Hist. (Aug.): 3-4, 6, 8.

Dickson, W.M. 1970. Endocrine glands. Pages 1,189–1,252 *in* M.J. Swenson, ed., Duke's physiology of domestic animals. Cornell Univ. Press, Ithaca, New York.

Dièreville, Sieur de. 1708. Relation of the voyage to Port Royal in Acadia or New France. J.C. Webster, ed. Publ. No. XX. The Champlain Soc., Toronto, Ontario. 324 pp.

Dieterich, R.A., J.K. Morton and R.L. Zarnke. 1991. Experimental *Brucella suis* biovan 4 infection in moose. J. Wildl. Dis. 27: 470–472.

Dietrich, H. 1979. Die Saisonveränderungen der Hoden und Neben hoden von Gamsboecken verschiedener Altersstufen in der Vor- , Haupt- und Nachbrunft. Ph.D. thesis, Vet. Med. Univ., Wien. 54 pp.

Dixon, B. 1978. Beyond the majic bullet. Harper and Row, New York. 249 pp.

Dixon, J.S. 1938. Birds and mammals of Mount McKinley National Park. Fauna of the national parks of the U.S. Department of Interior. U.S. Natl. Park Serv. 3: 1–236.

Dixon, K.R. and J.A. Chapman. 1980. Harmonic mean measure of animal activity. Ecol. 61: 1,040–1,044.

Dobyns, H.F. 1976. Native American historical demography: A critical bibliography. The Newberry Library Center for the History of the American Indian, Bibliograph. Ser. Indiana Univ. Press, Bloomington.

D'Occhio, M.J. and J.M. Suttie. 1992. The role of the pineal gland and melatonin in reproduction in male domestic ruminants. Anim. Reprod. Sci. 30: 135–155.

Dodds, D.G. 1955. A contribution to the ecology of the moose in Newfoundland. M.S. thesis, Cornell Univ., Ithaca, New York. 116 pp.

———. 1958. Observations of pre-rutting behavior of Newfoundland moose. J. Mammal. 39: 412–416.

———. 1959. Feeding and growth of a captive moose calf. J. Wildl. Manage. 23: 231–232.

———. 1960. Food competition and range relationships of moose and snowshoe hare in Newfoundland. J. Wildl. Manage. 24: 52–60.

———. 1963. The present status of moose *(Alces alces americana)* in Nova Scotia. Proc. Northeast Wildl. Conf., Portland, Maine. 40 pp.

———. 1974. Distribution, habitat and status of moose in the Atlantic provinces of Canada and northeastern United States. Nat. Can. (Que.) 101: 51–65.

Doerr, J.G. 1983. Home range size, movements and habitat use in two moose populations in southeastern Alaska. Can. Field-Nat. 97: 79–88.

Domblagian, M.J. 1979. Analysis of the tarsal gland secretion and its biological role in Ontario moose. Ph.D. thesis, Howard Univ., Washington, D.C.

Donne, T.E. 1924. The game animals of New Zealand. John Murray, London. 322 pp.

Donnelly, J.P., ed. 1967. Wilderness kingdom; Indian life in the Rocky Mountains 1840–1847: The journal and paintings of Nicolas Point. S.J. Holt, Rinehart and Winston, New York. 274 pp.

Dorn, R.D. 1970. Moose and cattle food habits in southwest Montana. J. Wildl. Manage. 34: 559–564.

Doty, R.L., ed. 1976. Mammalian olfaction, reproductive processes and behavior. Academic Press, New York. 344 pp.

Douglas-Lithgow, R.A. 1909. Dictionary of American Indian place and proper names in New England. Salem Press, Salem, Massachusetts.

Draper, L.C. and R.G. Thwaites, eds. 1903-1915. Collections of the State Historical Society of Wisconsin. 21 vols. St. Hist. Soc. of Wisconsin, Madison.

Dratch P.A. and J.M. Pemberton. 1992. Application of biochemical genetics to deer management. Pages 367–383 *in* R.D. Brown, ed., The biology of deer. Springer- Verlag, New York.

Dreesen, D.W. 1990. *Toxoplasma gondii* infections in wildlife. J. Amer. Vet. Med. Assoc. 196: 274–276.

Drescher-Kadren, U. and K. Walser-Kärst. 1970. Untersuchungen ueber den Vitamin-A-Gehalt in der Leber von wildlebenden Ruminantia im Vergleich zu domestizierten Wiederkaeuern. Trans. Intl. Union Game Biol. Congr.: IX: 823–829.

Drew, M.L. and W.M. Samuel. 1986. Reproduction in winter tick, *Dermacentor albipictus,* under field conditions in Alberta, Canada. Can. J. Zool. 64: 714–721.

Driver, H.E. 1969. Indians of North America. 2nd ed. Univ. Chicago Press, Chicago, Illinois. 632 pp.

Driver, H.E. and W.C. Massey. 1957. Comparative studies of North American Indians. Trans. Am. Philos. Soc. 47(2): 165–456.

Drolet, C.A., H.A. Feit, I. Juniper and B. Sinard. 1982. The wealth of the land: Wildlife harvests by the James Bay Cree, 1972–73 to 1978–79. James Bay and Northern Quebec Native Harvest. Res. Comm., Quebec City.

Drozdz, A. and A. Osiecki. 1973. Intake and digestibility of natural foods by roe-deer. Acta Theriol. 13: 81–91.

Dubey, J.P. 1980. *Sarcocystis* species in moose *(Alces alces),* bison *(Bison bison),* and pronghorn *(Antilocapra americana)* in Montana. Am. J. Vet. Res. 41: 2,063–2,065.

———. 1981. Isolation of encysted *Toxoplasma gondii* from musculature of moose and pronghorn in Montana. Am. J. Vet. Res. 42: 126–127.

———. 1986. Toxoplasmosis. J. Amer. Vet. Med. Assoc. 189: 166–170.

Dubey, J.P. and C.S.F. Williams. 1980. *Hammondia heydorni* infection in sheep, goats,moose, dogs, and coyotes. Parasitol. 77: 123–127.

Duda, M. D., S. J. Bissell and K. C. Young. 1995. Factors related to hunting and fishing participation in the United States. Phase V: Final Report. West. Assoc. Fish and Wildl. Agencies, Respons. Manage., and U.S. Fish and Wildl. Serv., Harrisonburg, Virginia. 174 pp.

Dudin, V.A. and A.P. Mikhailov. 1990. Domestication and commercial exploitation of moose on the Kostroma Moose Farm. Alces (Suppl.) 1: 220.

Dudley, P. 1721. A description of the moose in America. Philos. Trans. Roy. Soc. 31(368): 165–168.

Dugatkin, L.A. 1992. Sexual selection and imitation: Females copy the mate choice of others. Am. Nat. 139: 1,384–1,389.

Dukes, H.H. 1947. The physiology of domestic animals. Comstock Publishing Co., Ithaca, New York. 817 pp.

Dunn, F.D. 1966. Moose census. Fed. Aid Wildl. Restor. Rept. W-37-R-15. Maine Dept. Inland Fish. and Game, Augusta.

———. 1969. Moose investigations. Fed. Aid Wildl. Restor. Rept. W-37-R-18. Maine Dept. Inland Fish. and Game, Augusta.

———. 1971. Aerial moose census. Fed. Aid Wildl. Restor. Rept. W-66-R-2. Maine Dept. Inland Fish. and Game, Augusta.

Dunn, J.E. and P.S. Gipson. 1977. Analysis of radio telemetry data in studies of home range. Biometrics 33: 85–101.

Dunraven, Fourth Earl of. 1876. The great divide. Republished in 1927 as Hunting in Yellowstone. H. Kephart, ed. Outing Publishing Co., New York. 333 pp.

Duvall, A.C. and G.S. Schoonveld. 1988. Colorado moose: Reintroduction and management. Alces 24: 188–194.

Duvall, A.C. and S.H. Porter. 1987. Moose management plan: North Park data analysis unit (M-1). Colorado Div. Fish and Wildl., Denver. 18 pp.

Dyer, N.J. 1948. Preliminary plan for wildlife management on Baxter State Park. M.S. thesis, Univ. Maine, Orono. 79 pp.

Eason, G. 1985. Overharvest and recovery of moose in a recently logged area. Alces 21: 55–75.

———. 1989. Moose response to hunting and 1 km^2 block cutting. Alces 25: 63–74.

Eason, G., E.R. Thomas, R. Jerrard and K. Oswald. 1981. Moose hunting closure in a recently logged area. Alces 17: 111–125.

Eastman, C.A. 1916. From the deep woods to civilization: Chapters in the autobiography of an Indian. Little, Brown and Co., Boston. 206 pp.

Eastman, D.S. 1974. Habitat use by moose of burns, cutovers and forests in northcentral British Columbia. Proc. N. Am. Moose Conf. Workshop 10: 238–256.

Eastman, D.S. and I. Hatter. 1983. Compensatory and additive mechanisms of hunting mortality in moose: The Canadian perspective. Fish and Wildl. Branch, British Columbia Min. Environ., Victoria. 43 pp.

Eastman, D.S. and R. Ritcey. 1987. Moose habitat relationships and management in British Columbia. Swed. Wildl. Res. (Suppl.) 1: 101–118.

Ebedes, H. 1991. Reducing translocation mortalities with tranquilizers. Pages 378–380 *in* L.A. Reneker and R.J. Hudson, eds., Wildlife production: Conservation and sustained development. AFES Misc. Publ. 91-6. Univ. Alaska-Fairbanks.

Eberhardt, L.L. 1960. Estimation of vital characteristics of Michigan deer herds. Rept. 2282. Michigan Dept. Conserv., Lansing. 192 pp.

Eberhardt, L.L. 1988. Using age structure data from changing populations. J. Appl. Ecol. 25: 373–378.

Eberhardt, L.L., A.K. Majorowicz and J.A. Wilcox. 1982. Apparent rates of increase for two feral horse herds. J. Wildl. Manage. 46: 367–374.

Ebling, F.J.P. and M.H. Hastings. 1992. The neural basis of seasonal reproduction. Ann. Zootech. 41: 239–246.

Ecoregions Working Group. 1989. Ecoclimatic regions of Canada, first approximation. Ecological Land Classif. Ser. No. 23. Sustainable

Develop. Branch, Canadian Wildl. Serv., Conserv. and Protect., Environ. Canada, Ottawa. 119 pp.

Edenius, L. 1991. The effects of resource depletion on the winter feeding behavior of a browser: Winter foraging by moose on Scots pine. J. Appl. Ecol. 28: 318–328.

Edwards, J. 1983. Diet shifts in moose due to predator avoidance. Oecologia 60: 185–189.

Edwards, R.Y. 1952. An aerial moose census. Res. Note No. 23. British Columbia For. Serv., Victoria. 9 pp.

———. 1954. Comparison of an aerial and ground census of moose. J. Wildl. Manage. 18: 403–404.

———. 1956a. Snow depths and ungulate abundance in the mountains of western Canada. J. Wildl. Manage. 20: 159–168.

———. 1956b. Moose pellet group data, Wells Gray Park, British Columbia. British Columbia For. Serv., Victoria. 3 pp.

———. 1957. Damned waters in a moose range. Murrelet 38: 1–3.

Edwards, R.Y. and R.W. Ritcey. 1956. The migrations of a moose herd. J. Mammal. 37: 486–494.

Edwards, R.Y. and R.W. Ritcey. 1958. Reproduction in a moose population. J. Wildl. Manage. 22: 261–268.

Eiben, B. and K. Fischer. 1983. Untersuchungen über die Beziehungen der Aktivität der alkalischen Phosphotase und dem Geweihbildungszyklus beim Damhirsch *(Dama dama L.)*. Z. Jagdwiss. 29: 244–247.

Eiben, B., S. Scharla, K. Fischer and H. Schmidt-Gayak. 1984. Seasonal variations of serum 1,25-dihydroxyvitamin D3 and alkaline phosphotase in relation to the antler formation in fallow deer (*Dama dama* L.). Acta Endocrinol. 107: 141–144.

Eibl-Eibesfeld, I. 1970. Ethology, the biology of behavior. Holt, Rinehart and Winston, New York. 530 pp.

Eide, S.H., S.D. Miller and M.A. Chihuly. 1986. Oil pipeline crossing sites utilized in winter by moose, *Alces alces,* and caribou, *Rangifer tarandus,* in southcentral Alaska. Can. Field-Nat. 100: 197–207.

Eisenberg, J.F. 1983. The mammalian radiations. Univ. Chicago Press, Chicago, Illinois. 610 pp.

Eley, A., H.E. Anthony and R.R. M. Carpenter. 1939. North American big game. Boone and Crockett Club. Charles Scribner's Sons, New York. 533 pp.

Elliot, D.C.M. 1988. Large area moose census in northern Manitoba. Alces 24: 48–55.

Ellis, J.E., J.A. Weins, C.F. Rodell and J.C Anways. 1976. A conceptual model of diet selection as an ecosystem process. J. Theor. Biol. 60: 93–108.

Elman, R. 1972. The great American shooting prints. Alfred A. Knopf, New York.

Erhardova-Kotrla, B. 1971. The occurrence of *Fascioloides magna* (Bassi, 1875) in Czechoslovakia. Academia. Czechoslovakian Acad. Sci., Prague. 155 pp.

Erikson, V.O. 1978. Maliseet-Passamaquoddy. Pages 123–136 *in* B.G. Trigger, ed., Northeast. Handbook of North American Indians. Vol. 15. Smithsonian Inst., Washington, D.C. 924 pp.

Essen, L. and G. Sparre. 1975. Palme bjod franska topp-politiker till jakten. Svengk Jakt 113: 14–16.

Estes, R.D. 1972. The role of vomeronasal organs in mammalian reproduction. Mammalia 36: 315–340.

Euler, D. 1983a. Selective harvest, compensatory mortality and moose in Ontario. Alces 19: 148–161.

———. 1983b. Selective harvest for Ontario moose. Ontario Angler and Hunter 7: 26–28.

Evans, C.D., W.A. Troyer and C.J. Lensink. 1966. Aerial census of moose by quadrat sampling units. J. Wildl. Manage. 30: 767–776.

Eve, J.H. and F.E. Kellogg. 1977. Management implications of abomassal parasites in southeastern white-tailed deer. J. Wildl. Manage. 41: 169–177.

Fair, J. 1992. Moose for kids. Northword Press, Inc., Minocqua, Wisconsin. 48 pp.

Fall, J.A., J. Foster and R.T. Stanek. 1983. The use of moose and other wild resources in the Tyonek and Upper Yentna areas: A background report. Tech. Pap. No. 74. Alaska Dept. Fish and Game, Div. of Subsistence, Anchorage. 44 pp.

Farnham, C.H. 1884. A winter in Canada. Harper's New Monthly Mag. LXVIII: 392–408.

Faro, J.B. and A.W. Franzmann. 1978. Alaska Peninsula moose productivity and physiology study. Fed. Aid Wildl. Restor. Final Rept. Alaska Dept. Fish and Game, Juneau. 29 pp.

Fees, P. and E. Boehme. 1988. Frontier America. Harry N. Abrams, Inc., New York. 128 pp.

Feit, H.A. 1973. The ethno-ecology of the Waswanipi Cree: Or, how hunters can manage their resouces. Pages 115–125 *in* B.A. Cox, ed., Cultural ecology: Readings on the Canadian Indians and Eskimos. McClelland and Stewart, Toronto.

———. 1984. Conflict areas in the management of renewable resources in the Canadian north. Pages 435–458 *in* National and regional interests in the north. Canadian Arctic Resour. Comm., Ottawa.

———. 1986. Hunting and the quest for power: The James Bay Cree and whitemen in the twentieth century. Pages 181–224 *in* R.B. Morrison and C.R. Wilson, eds., Native peoples: The Canadian experience. McClelland and Stewart Inc., Toronto. 639 pp.

———. 1987. North American native hunting and management of moose populations. Swedish Wildl. Res. (Suppl.) 1: 25–42.

Fennessy, P.F., J.M. Suttie, S.F. Crosbie, I.D. Corson, H.J. Elgar and K.R. Lapwood. 1988. Plasma LH and testosterone responses to gonadotrophin-releasing hormone in adult red deer *(Cervus elaphus)* stags during annual antler cycle. J. Endocrinl. 117: 35–41.

Fennessy, P.F., I.D. Corson, J.M. Suttie and R.P. Littlejohn. 1992. Antler growth patterns in young red deer stags. Pages 487–492 *in* R.D. Brown, ed., The biology of deer. Springer- Verlag, New York.

Fenstermacher, R. and W.L. Jellison. 1933. Diseases affecting moose. Agric. Exp. Stn. Bull. No. 294. Univ. Minnesota, St. Paul. 20 pp.

Fenstermacher, R. and O.W. Olsen. 1942. Further studies of diseases affecting moose, III. Cornell Vet. 32: 241–254.

Fenton, W.N. 1978. Northern Iroquoian culture patterns. Pages 296–321 *in* B.G. Trigger, ed., Northeast. Handbook of North American Indians. Vol. 15. Smithsonian Inst., Washington, D.C. 924 pp.

Fenzel, L.D. 1974. Occurrence of moose in food of wolves as revealed by scat analysis: A review of North American studies. Nat. Can. (Que.) 101: 467–479.

Ferguson, S.H. 1992. Newfoundland moose model 1: Harvest management based on relationship between productivity and vegetation. Draft rept., Newfoundland and Labrador. Wildl. Div., Dept. Environ. and Lands, St. John's. 39 pp.

———. 1993. Use of cohort analysis to estimate abundance, recruitment and survivorship for Newfoundland moose. Alces 29: 99–114.

Fernald, M.L. 1970. Gray's manual of botany. 8th ed. D. Van Nostrand Co., New York. 1,632 pp.

Fernández-Armesto, F., ed. 1991. The Times atlas of world exploration. Harper Collins Publ., New York. 286 pp.

Ferris, G.T. 1897. Wild animals in a New England game-park. The Century Mag. LIV: 924–937.

Ferris, R.G., ed. 1968. Explorers and settlers. U.S. Dept. Inter., U.S. Govt. Print. Off., Washington, D.C. 506 pp.

Filion, F.L., S. Parker and E. DuWors. 1988. The importance of wildlife to Canadians: Demand for wildlife to 2001. Can. Wildl. Serv., Ottawa. 29 pp.

Filion, F.L., S.W. James, J.L. Ducharme, W. Pepper, R. Reid, P. Boxall and D. Teillet. 1981. The importance of wildlife to Canadians. 47th Fed.-Prov. Wildl. Conf., Can. Wildl. Serv., Ottawa. 40 pp.

Filion, F.L., E. DuWors, P. Boxall, P. Bouchard, R. Reid, P.A. Gray, A. Bath, A. Jacquemot and G. Legare. 1991. The importance of wildlife to Canadians: Highlights of the 1991 survey. Can. Wildl. Serv., Environ. Canada, Ottawa. 60 pp.

Filonov, C.P. 1974. Geographical variation of moose predation. Proc. N. Am. Moose Conf. Workshop 10: 299–313.

———. 1983. Los i lesnaya promyshlenost. Moskva. 248 pp.

Finger, S.E., I.L. Brisbin, Jr., M.H. Smith and F.D. Urbston. 1981. Kidney fat as a predictor of body condition in white-tailed deer. J. Wildl. Manage. 45: 964–968.

Fischer, K. 1987. The fallow deer's year. Seasonality of reproduction and of morphogenical and physiological parameters. Page 62 *in* Abstr. 18th Cong. Int. Union Game Biol. Jagiellonian Univ., Krakow, Poland.

Fisher, R.A. 1958. Genetical theory of natural selection. 2nd rev. ed. Dover, New York. 291 pp.

Fitzhugh, W., ed. 1985. Cultures in contact: The European impact on native cultural institutions in eastern North America, A.D. 1000–1800. Smithsonian Inst. Press, Washington, D.C.

Flerov, C. 1931. A review of the elks or moose (*Alces* Gray) in the Old World. Doklady AN USSR 6: 71–74.

Flerov, K.K. 1950. Morphologiya i ecologiya olenjeobraznykh v processe ikh evolutsii. Pages 50–69 *in* Materialy pro chetvertychnom periodu SSSR. Vol. 2. Nauk SSSR, Moskva-Leningrad.

———. 1952. Kabargi i oleni. Fauna SSSR. Mlekopitayishchie, Vol. I. Izdat. AN SSSR, Moskva-Leningrad. 256 pp.

Flint, R.F. 1948. Glacial geology of the Pleistocene epoch. J. Wiley and Sons, Ltd., New York. 589 pp.

Flook, D.R. 1964. Range relationships of some ungulates native to Banff and Jasper national parks, Alberta. Pages 119–128 *in* D.J. Crisp, ed., Grazing in terrestrial and marine environments. Blackwell Scientific Publ., Oxford, England.

———. 1970. Causes and implications of an observed sex differential in the survival of wapiti. Rept. Ser. 11. Can. Wildl. Serv., Queen's Printer for Canada, Ottawa. 71 pp.

Flynn, A. and A.W. Franzmann. 1987. Mineral element studies in North American moose. Swed. Wildl. Res. (Suppl.) 1: 289–300.

Flynn, A., A.W. Franzmann and P.D. Arneson. 1975. Sequential hair shaft as an indicator of prior mineralization in the Alaskan moose. J. Anim. Sci. 41: 906–910.

Flynn, A., A.W. Franzmann and C.C. Schwartz. 1980. Seasonal calcium flux in moose. Proc. N. Am. Moose Conf. Workshop 16: 69–81.

Flynn, A., A.W. Franzmann, P.D. Arneson and O.A. Hill, Jr. 1974. Determination of past trace element uptake in a wild animals by longitudinal analysis of the hair shaft. Die Naturwiss. 61: 362–363.

Flynn, A., A.W. Franzmann, P.D. Arneson and J.L. Oldemeyer. 1977. Indications of a copper deficiency in a subpopulation of Alaskan moose. J. Nutr. 107: 1,182–1,189.

Foley, J.P., ed. 1900. The Jeffersonian cyclopedia. 2 vols. Russell and Russell, New York.

Folk, G.E. 1974. Textbook of environmental physiology. 2nd ed. Lea and Febinger, Philadelphia. 465 pp.

Foose, T.J. 1982. Trophic strategies of ruminants versus nonruminant ungulates. Ph.D. thesis, Univ. Chicago, Chicago, Illinois. 337 pp.

Forbes, L.B. and S.V. Tessaro. 1993. Transmission of brucellosis from reindeer to cattle. J. Am. Vet. Med. Assoc. 203: 289–294.

Forbes, G.F. and J.B. Theberge. 1992. Importance of scavenging on moose by wolves in Algonquin Park, Ontario. Alces 28: 235–241.

Forbes, L.B., S.V. Tessaro and W. Lees. 1996. Experimental studies of *Brucella abortus* in moose (*Alces alces*). J. Wildl. Dis. 32: 94–104.

Ford, M.J. The changing climate: Responses of the natural flora and fauns. George Allen, London. 190 pp.

Foreyt, W.J. 1981. Trematodes and cestodes. Pages 237–265 *in* W.R. Davidson, F.R. Hayes, V.F. Nettles and F.E. Kellogg, eds., Diseases and parasites of white-tailed deer. Misc.Pub. No. 7. Tall Timbers Res. Sta., Tallahasee, Florida.

Fowler, M.E. 1978. Restraint and handling of wild and domestic animals. Iowa St. Univ. Press, Ames. 332 pp.

———, ed. 1986. Zoo and wild animal medicine, 2nd ed. W.B. Saunders Co., Philadelphia. 1,127 pp.

Frandson, R.D. 1965. Anatomy and physiology of farm animals. Lea and Febinger, Philadelphia. 501 pp.

Frankenberger, Z. 1951. The first stages of the development of the antlers in the Cervidae. Biol. Listy (Suppl.) II: 127–141.

Frankenberger, Z. 1954. Pučnice a paroh. [The pedicle and the antler]. Čsl. Morfologie 2: 89–95.

———. 1957. Cirkumanální a circumgenitální žlázy našich cervidů. [Circumanal and circumgenital glands of our cervidae]. Čsl. Morfologie 5: 255–265.

———. 1961. K otázce mechanismu shozů parohů u našich jelenů. [Some remarks on the mechanism of the shedding of the antlers in the deer]. Čsl. Morfologie 9: 41–45.

Franklin, J. 1823. Narrative of a journey to the shores of the Polar Sea, in the years 1819, 20, 21, and 22. John Murray, London, England.

Franzmann, A.W. 1977. Condition assessment of Alaskan moose. Proc. N. Am. Moose Conf. Workshop. 13: 119–127.

———. 1978. Moose. Pages 67–88 *in* J.L. Schmidt and D.L. Gilbert, eds., Big game of North America: Ecology and management. Stackpole Books, Harrisburg, Pennsylvania. 494 pp.

———. 1981. *Alces alces*. Mammalian Species No. 154. Am. Soc. Mammal. 7 pp.

———. 1982. An assessment of chemical immobilization of North American moose. Pages 393–407 *in* L. Nielson, J.C. Haigh and M.E. Fowler, eds., Chemical immobilization of North American wildlife. Wisconsin Hum. Soc., Inc., Milwaukee. 447 pp.

———. 1985. Assessment of nutritional status. Pages 239–259 *in* R.J. Hudson and R.G. White, eds., Bioenergetics of wild herbivores. CRC Press, Boca Raton. 314 pp.

———. 1988. A review of Alaskan translocations. Pages 210–229 *in* L. Nielsen and R.D. Brown, eds., Translocation of wild animals. Wisconsin Hum. Soc., Inc., Milwaukee. 333 pp.

Franzmann, A.W. and P.D. Arneson. 1974. Immobilization of Alaskan moose. J. Zoo Anim. Med. 5: 26–32.

Franzmann, A.W. and P.D. Arneson. 1975. A winch/tripod device for weighing moose. J. Zoo Anim. Med. 6: 10–12.

Franzmann, A.W. and P.D. Arneson. 1976. Marrow fat in Alaskan moose femurs in relation to mortality factors. J. Wildl. Manage. 40: 336–339.

Franzmann, A.W. and T.N. Bailey. 1977. Moose Research Center report. Fed. Aid Wildl. Restor. Prog. Rept. Alaska Dept. Fish and Game, Juneau. 114 pp.

Franzmann, A.W. and W.B. Ballard. 1993. Use of physical and physiological indices for monitoring moose population status—A review. Alces 29: 125–133.

Franzmann, A.W. and R.E. LeResche. 1978. Alaskan moose blood studies with emphasis on condition evaluation. J. Wildl. Manage. 42: 334–351.

Franzmann, A.W. and R.O. Peterson. 1978. Moose calf mortality assessment. Proc. N. Am. Moose Conf. Workshop 14: 247–269.

Franzmann, A.W. and C.C. Schwartz. 1978. Moose Research Center report. Fed. Aid Wildl. Restor. Final Rept. Alaska Dept. Fish and Game, Juneau. 129 pp.

Franzmann, A.W. and C.C. Schwartz. 1982. Evaluating and testing of techniques for moose management. Fed. Aid Wildl. Restor. Final Rept. Alaska Dept. Fish and Game, Juneau. 45 pp.

Franzmann, A.W. and C.C. Schwartz. 1983a. Moose productivity and physiology. Fed. Aid Wildl. Restor. Final Rept. Alaska Dept. Fish and Game, Juneau. 129 pp.

Franzmann, A.W. and C.C. Schwartz. 1983b. Management of North American moose populations. Pages 517–522 *in* C.M. Wemmer, ed., Biology and management of the Cervidae. Smithsonian Inst. Press, Washington, D.C.

Franzmann, A.W. and C.C. Schwartz. 1985. Moose twinning rates: A possible population condition assessment. J. Wildl. Manage. 49: 394–396.

Franzmann, A.W. and C.C. Schwartz. 1986. Black bear predation on moose calves in highly productive versus marginal moose habitats on the Kenai Peninsula, Alaska. Alces 22: 139–154.

Franzmann, A.W., P.D. Arneson and D.E. Ullrey. 1975a. Composition of milk from Alaskan moose in relation to other ruminants. J. Zoo Anim. Med. 6: 12–14.

Franzmann, A.W., A. Flynn and P.D. Arneson. 1975b. Serum corticoid levels relative to handling stress in Alaskan moose. Can. J. Zool. 53: 1,424–1,426.

Franzmann, A.W., A. Flynn and P.D. Arneson. 1975c. Levels of some mineral elements in Alaskan moose hair. J. Wildl. Manage. 39: 374–378.

Franzmann, A.W., J.L. Oldemeyer and A. Flynn. 1975d. Minerals and moose. Proc. N. Am. Moose Conf. Workshop 11: 114–139.

Franzmann, A.W., P.D. Arneson and J.L. Oldemeyer. 1976a. Daily winter pellet groups and beds of Alaskan moose. J. Wildl. Manage. 40: 374–375.

Franzmann, A.W., A. Flynn and P.D. Arneson. 1976b. Moose milk and hair element levels and relationships. J. Wildl. Dis. 12: 202–207.

Franzmann, A.W., J.L. Oldemeyer, P.D. Arneson and R.K. Seemel. 1976c. Pellet-group count evaluation for census and habitat use of Alaskan moose. Proc. N. Am. Moose Conf. Workshop 12: 127–142.

Franzmann, A.W., R.E. LeResche, P.D. Arneson and J.L. Davis. 1976d. Moose productivity and physiology. Fed. Aid Wildl. Restor. Final Rept. Alaska Dept. Fish and Game, Juneau. 87 pp.

Franzmann, A.W., A. Flynn and P.D. Arneson. 1977. Alaskan moose hair element values and variability. Comp. Biochem. Physiol. 57A: 299–306.

Franzmann, A.W., R.E. LeResche, R.A. Rausch and J.D. Oldemeyer. 1978. Alaskan moose measurements and weights and measurement-weight relationships. Can. J. Zool. 56: 298–306.

Franzmann, A.W., C.C. Schwartz and R.O. Peterson. 1980a. Moose calf mortality in summer on the Kenai Peninsula, Alaska. J. Wildl. Manage. 44: 764–768.

Franzmann, A.W., W.B. Ballard, C.C. Schwartz and T.H. Spraker. 1980b. Physiologic and morphometric measurements in neonatal moose and their cows in Alaska. Proc. N. Am. Moose Conf. Workshop 16: 106–123.

Franzmann, A.W., A. Flynn, C.C. Schwartz, D.G. Calkins and L. Nichols. 1981. Beta-endorphine levels in blood from selected mammals. J. Wildl. Dis.. 17: 593–596.

Franzmann, A.W., C.C. Schwartz and D.C. Johnson. 1982. Chemical immobilization of moose at the Moose Research Center, Alaska (1968–1981). Alces 18: 94–115.

Franzmann, A.W., C.C. Schwartz and D.C. Johnson. 1984a. Baseline body temperature, heart rates, and respiratory rates of moose. J. Wildl. Dis. 20: 333–337.

Franzmann, A.W., C.C. Schwartz and D.C. Johnson. 1984b. Kenai Peninsula moose calf mortality study. Fed. Aid Wildl. Restor. Prog. Rept. Alaska Dept. Fish and Game, Juneau. 30 pp.

Franzmann, A.W., C.C. Schwartz and D.C. Johnson. 1987. Monitoring status (condition, nutrition, health) of moose via blood. Swed. Wildl. Res. (Suppl.) 1: 281–288.

Franzmann, A.W., C.C. Schwartz, D.C. Johnson, J.B. Faro and W.B. Ballard. 1984c. Immobilization of moose with carfentanil. Alces 20: 259–282.

Fraser, C.M., ed. 1991. The Merck veterinary manual. 7th ed. Merck and Co., Rahway, New Jersey. 1,832 pp.

Fraser, D.J. 1976. An estimate of hunting mortality based on the age and sex structure of the harvest. Proc. N. Am. Moose Conf. Workshop. 12: 236–273.

———. 1980. Moose and salt: A review of recent research in Ontario. Proc. N. Am. Moose Conf. Workshop 16: 51–68.

Fraser, D.J. and H. Hristienko. 1983. Effects of moose, *Alces alces*, on aquatic vegetation in Sibley Provincial Park, Ontario. Can. Field-Nat. 7: 57–61.

Fraser, D.J. and E. Reardon. 1980. Attraction of wild ungulates to mineral-rich springs in central Canada. Holarctic Ecol. 3: 36–40.

Fraser, D.J. and E.R. Thomas. 1962. Moose-vehicle accidents in Ontario: Relation to highway salt. Wildl. Soc. Bull. 10: 261–265.

Fraser, D.J., B.K. Thompson and D. Arthur. 1982. Aquatic feeding by moose: Seasonal variation in relation to plant chemical composition and use of mineral licks. Can. J. Zool. 60: 3,121–3,126.

Fraser, D.J., E.R. Chavez and J.E. Paloheimo. 1984. Aquatic feeding by moose: Selection of plant species and feeding areas in relation to plant chemical composition and characteristics of lakes. Can. J. Zool. 62: 80–87.

Fraser, D.J., D. Arthur, J.K. Morton and B.K. Thompson. 1980. Aquatic feeding by moose in a Canadian lake. Holarctic Ecol. 3: 218–223.

Fraser, J. 1881. Three months among the moose. Lowell, Montreal.

Frechette, J.L., M. Beauregard, A.L. Giroux and D. Clairmont. 1973. Infection des jeunes bovins par Trichuris discolor. Can. Vet. J. 14: 243–246.

Freedman, D.H. 1993. In the realm of the chemical. Discover 14: 69–76.

Freeman, L.C., S.C. Freeman and A.K. Romney. 1992. The implications of social structure for dominance heirarchies in red deer (*Cervus elaphus* L.). Anim. Behav. 44: 239–245.

French, C.W., L.C. McEwan, N.D. Magruder, R.H. Ingram and R.W. Swift. 1956. Nutrient requirements for growth and antler development in white-tailed deer. J. Wildl. Manage. 20: 221–232.

Frenzel, B. 1967. Klimaschwankungen des Eiszeitalters. Fr. Viewig und Sohn, Braunschweig. 296 pp.

Fretwell, S.D. and H.L. Lucas. 1970. On territorial behaviour and other factors influencing habitat distribution in birds. Acta Biotheoretica 19: 16–36.

Frick, C. 1937. Horned ruminants of North America. Bull. Am. Mus. Nat. Hist., New York. 669 pp.

Frodelius, R.B. 1973. Determination of anti-hunting organizations by content analysis of their literature. Ph.D. thesis, St. Univ. New York, Syracuse. 259 pp.

Frontenac, L. de Baude de. 1867. Letter to M. De Seignalay, dated November 12, 1690, relating to the exploits of Sieur de Louvigney. Wisconsin Hist. Coll. V: 65–67.

Fronval, G. and D. Dubois. 1978. Indian signs and sign language. Wings Books, Avenel, New Jersey. 81 pp.

Fruetel, M. and M.W. Lankester. 1988. *Nematodirella alcidis* (Nematoda: Trichostronglyoidea) in moose on northwestern Ontario. Alces 24: 159–163.

Frund, J.L. and A.B. Bubenik. 1982. The high season of moose. OOPIK, Ltd., Montreal. (16mm film and JVC video.)

Frund, J.L. and A.B. Bubenik. 1988. Avoir du Panache. OOPIK, Ltd., Montreal. (JVC video.)

Fryxell, J.M., W.E. Mercer and R.B. Gellately. 1988. Population dynamics of Newfoundland moose using cohort analysis. J. Wildl. Manage. 52: 14–21.

Fuller, T.K. 1989. Population dynamics of wolves in northcentral Minnesota. Wildl. Monogr. 105. 41 pp.

Fuller, T.K. and L.B. Keith. 1980. Wolf population dynamics and prey relationships in norteastern Alberta. J. Wildl. Manage. 44: 583–602.

Fuller, T.K. and L.B. Keith. 1981. Woodland caribou population dynamics in northeastern Alberta. J. Wildl. Manage. 45: 197–213.

Gaab, J.E. 1948. Field autopsy report. Fed. Aid Wildl. Restor. Rept. Montana Fish and Game Dept., Helena. 80 pp.

Gall, C.M. and P.J. Isackson. 1989. Limbic seizures increase neuronal production of messenger RNA for nerve growth factor. Science 245: 758–761.

Gandini, G., H. Ebedes and R. Burroughs. 1989. The use of long-acting neuroleptics in impala. S. Afr. Vet. J. 16: 206–207.

Ganong, W.F., ed. 1910. New relation of Gaspesia. Publ. V. The Champlain Soc., Toronto, Ontario. 452 pp. (Reprinted in 1968 by Greenwood Press, New York.)

Garner, F.M. and L.W. Schwartz. 1969. Spontaneous hematopoietic neoplasms of free living and captive wild mammals. Natl. Cancer Inst. Monogr. 32: 153–156.

Garner, D.L. and W.F. Porter. 1990. Movements and seasonal home ranges of bull moose in a pioneering Adirondack population. Alces 26: 80–85.

Garner, D.L. and W.F. Porter. 1991. Prevalence of *Parelaphostrongylus tenuis* in white- tailed deer in northern New York. J. Wildl. Dis. 27: 594–598.

Garner, D.L. and M.L. Wilton. 1993. The potential role of winter tick *(Dermacentor albipictus)* in the dynamics of a south central Ontario moose population. Alces 29: 169–173.

Garrett, W.E., ed. 1988. Historical atlas of the United States. Natl. Geogr. Soc., Washington, D.C. 289 pp.

Garrod, A.H. 1877. Notes on the visceral anatomy and osteology of ruminants. Proc. Zool. Soc. London 145: 2–18.

Garton, E.O., M.D. Samuel and J.M. Peek. 1985. Analysis of moose home ranges. Alces 21: 77–89.

Garvin, J.W., ed. 1925. Wanderings of an artist among the Indians of North America. The Raddison Soc. of Canada, Ltd., Toronto, Ontario. 323 pages.

Gasaway, W.C. and J.W. Coady. 1974. Review of energy requirements and rumen fermentation in moose and other ruminants. Nat. Can. (Que.) 101: 227–262.

Gasaway, W.C. and S.D. Dubois. 1985. Initial response of moose *(Alces alces)* to a wildfire in interior Alaska. Can. Field-Nat. 99: 135–140.

Gasaway, W.C. and S.D. Dubois. 1987. Estimating moose population parameters. Swed. Wildl. Res. (Suppl.) 1: 603–617.

Gasaway, W.C., D. Haggstrom and O.E. Burris. 1977. Preliminary observations on the timing and causes of calf mortality in an interior Alaskan moose population. Proc. N. Am. Moose Conf. Workshop 13: 54–70.

Gasaway, W.C., D.B. Harkness and R.A. Rausch. 1978a. Accuracy of moose age determinations from incisor cementum layers. J. Wildl. Manage. 42: 558–563.

Gasaway, W.C., A.W. Franzmann and J.B. Faro. 1978b. Immobilizing moose with a mixture of etorphine and xylazine. J. Wildl. Manage. 42: 687–690.

Gasaway, W.C., S.D. Dubois and K.L. Brink. 1980. Dispersal of subadult moose from a low density population in interior Alaska. Proc. N. Am. Moose Conf. Workshop 16: 314–337.

Gasaway, W.C., R.O. Stephenson, J.L. Davis, P.E.K. Shepherd and O.E. Burris. 1983. Interrelationships of wolves, prey, and man in interior Alaska. Wildl. Monogr. 84. 50 pp.

Gasaway, W.C., S.D. Dubois and S.J. Harbo. 1985a. Biases in aerial transect surveys for moose during May and June. J. Wildl. Manage. 49: 777–784.

Gasaway, W.C., S.D. Dubois, D.J. Preston and D.J. Reed. 1985b. Home range formation and dispersal of subadult moose in interior Alaska. Fed. Aid Wildl. Restor. Final Rept. Alaska Dept. Fish and Game, Juneau. 26 pp.

Gasaway, W.C., S.D. Dubois, D.J. Reed and S.J. Harbo. 1986. Estimating moose population parameters from aerial surveys. Biol. Pap. 22. Univ. Alaska-Fairbanks. 108 pp.

Gasaway, W.C., D.J. Preston, D.J. Reed and D.D. Roby. 1987. Comparative antler morphology and size of North American moose. Swed. Wildl. Res. (Suppl.) 1: 311–326.

Gasaway, W.C., S.D. Dubois, R.D. Boertje, D.J. Reed and D.T. Simpson. 1989. Response of radio-collared moose to a large burn in central Alaska. Can. J. Zool. 67: 325–329.

Gasaway, W.C., R.D. Boertje, D.V. Grandgard, K.G. Kellyhouse, R.O. Stephenson and D.G. Larsen. 1990. Factors limiting moose population growth in subunit 20E. Fed. Aid Wildl. Restor. Final Rept. Alaska Dept. Fish and Game, Juneau. 106 pp.

Gasaway, W.C., R.D. Boertje, D.V. Grandgard, K.G. Kellyhouse, R.O. Stephenson and D.G. Larsen. 1992. The role of predation in limiting moose at low densities in Alaska and Yukon and implications for conservation. Wildl. Monogr. 120. 59 pp.

Gaskoin, J.S. 1856. On some defects in the growth of antlers and some results of castration in Cervidae. Proc. Zool. Soc. L. 24: 151–159.

Gates, C.C. and R.J. Hudson. 1979. Effects of posture and activity on metabolic responses of wapiti to cold. J. Wildl. Manage. 43: 564–567.

Gates, D.M. 1972. Man and his environmental climate. Harper and Row, New York. 175 pp.

———. 1985. Energy and ecology. Sinauer Assoc., Sunderland, Massachusetts. 377 pp.

Gauthier, D.A. and D.G. Larsen. 1985. Geographic variation in moose antler characteristics, Yukon. Alces 21: 91–101.

Gawley, D.J. and J.B. Dawson. 1965. An evaluation of the track density system of estimating relative moose population densities in Gagoma District, January 1964. Ontario Dept. Lands and For. Resour. Manage. Rept. 80: 13–28.

Geist, V. 1960. Diurnal activity of moose. Soc. pro Fauna at Flora Fennica 35: 95–100.

———. 1963. On the behaviour of North American moose (*Alces alces andersoni* Peterson, 1950) in British Columbia. Behavior. 20: 377–416.

———. 1966. Ethological observations on some North American Cervidae. Zool. Beitrage 2: 219–250.

———. 1968. Moose (Genus *Alces*). Pages 229–242 *in* H.C.B. Grzimek, ed., Grzimek's animal life encyclopedia. English ed. Vol. 13, Mammals IV. Van Nostrand Reinhold Co., New York.

———. 1971. Mountain sheep. A study in behaviour and evolution. Univ. Chicago Press, Chicago, Illinois. 383 pp.

———. 1974a. On the evolution of reproductive potential in moose. Nat. Can. (Que.) 101: 527–537.

———. 1974b. On the relationship of social evolution and ecology in ungulates. Am. Zool. 14: 205–220.

———. 1986. On specialization in ice age mammals with special reference to cervids and caprids. Can. J. Zool. 65: 1,061–1,084.

———. 1987. On the evolution and adaptations of *Alces*. Swed. Wildl. Res. (Suppl.) 1: 11–24.

———. 1991. Bones of contention revisited: Did antlers enlarge sexual selection as a consequence of neonatal security strategies? Appl. Anim. Behav. Sci. 29: 453–469.

Gentilcore, R.L., ed. 1993. Historical atlas of Canada. Vol. II. The land transformed, 1800–1891. Univ. Toronto Press, Toronto, Ontario.

Georgii, B. 1981. Activity patterns of female red deer (*Cervus elaphus* L.) in the Alps. Oecologia 49: 127–136.

Gerlach, A.C., ed. 1970. The national atlas of the United States. U.S. Geol. Surv., Washington, D.C. 417 pp.

Gianni, M., M. Studer, G. Carpani, G. Terao and E. Garattini. 1991. Retinoic acid induces liver bone kidney-type alkaline phosphotase gene expression in F9 teratocarcinoma cells. Biochem. J. 274: 673–678.

Gibbs, E.P.J. 1982. Bluetongue—An analysis of current problems, with particular reference to importation of ruminants to the United States. J. Am. Vet. Med. Assoc. 182: 1,190–1,194.

Gibson, B.H. 1971. A study of moose browsing and population decline on two islands of Lake Nipigon. Pages 21–39 *in* Resour. Manage. Rept. 106. Ontario Dept. Lands and For., Toronto.

Gibson, D. 1957. The ovary as an indicator of reproductive history in the white-tailed deer *Odocoileus virginianus borealis* (Miller). M.S. thesis, Univ. Toronto, Toronto, Ontario. 61 pp.

Gibson, R.M. and P.A. Jewell. 1982. Semen quality, female choice and multiple mating in domestic sheep: A test of Triver's sexual competence hypothesis. Behav. 80: 9–31.

Gilbert, B. 1973. The trailblazers. Time-Life Books, New York. 236 pp.

Gilbert, F.F. 1974. *Parelaphostrongylus tenuis* in Maine: II—Prevalence in moose. J. Wildl. Manage. 38: 42–46.

———. 1992. Retroductive logic and the effects of meningeal worms: A comment. J. Wildl. Manage. 56: 614–616.

Gilchrist, L.M. 1986. The Canadian national railway's effects on larger ungulates in northern British Columbia. B.S. thesis, Univ. British Columbia, Victoria. 65 pp.

Gill, F.B. 1990. Ornithology. W.H. Freeman and Co., New York. 660 pp.

Gill, R. 1990. Monitoring the status of European and North American cervids. GEMS Information Ser. No. 8, Global Environ. Monitor. Syst. United Nations Environ. Programme. Nairobi. 277 pp.

Gillard, B.D.E. 1965. Comparison of the hemicelluloses from plants belonging to two different plant families. Phytochemistry. 4: 631–634.

Gillespie, B.C. 1968–1971. Unpublished ethno-graphic notes from field work among Dogrib, Bear Lake Indians and Mountain Indians, Northwest Territories, Canada. (Manuscripts in author's possession.)

———. 1981a. Major fauna in the traditional economy. Pages 15–18 *in* J. Helm, ed., Subarctic. Handbook of North American Indians. Vol. 6. Smithsonian Inst., Washington, D.C. 837 pp.

———. 1981b. Mountain Indians. Pages 326–337 *in* J. Helm, ed., Subarctic. Handbook of North American Indians. Vol. 6. Smithsonian Inst., Washington, D.C. 837 pp.

Gillmore, P. No date. Prairie and forest: A guide to the field sports of North America. W.H. Allen and Co., Ltd., London. 396 pp.

Girard, F. and R. Joyal. 1984. L'effet des coupes à blanc sur les populations d'orignaux du nord ouest du Quebec. Alces 20: 40–53.

Glines, M.V. and W.M. Samuel. 1989. The effect of *Dermacentor albipictus* (Acarina: Ixodidae) on blood composition, weight gain and hair coat of moose, *Alces alces*. Exper. and Appl. Acarology 6: 197–213.

Globe and Mail. 1986. Lonely bull moose seeks romance with dairy cows. Toronto, Ontario. September 22.

Glooschenko, V., C. Downes, R. Frank, H.E. Braun, E.M. Addison and J. Hickie. 1988. Cadmium levels in Ontario moose and deer in relation to soil sensitivity to acid precipitation. Sci. Total Environ. 71: 173–186.

Glover, R., ed. 1962. David Thompson's narrative 1784–1812. The Champlain Soc., Toronto, Ontario. 410 pp.

Goddard, I. 1978. Delaware. Pages 213–239 *in* B.G. Trigger, ed., Northeast. Handbook of North American Indians. Vol. 15. Smithsonian Inst., Washington, D.C. 924 pp.

Goddard, J. 1970. Movements of moose in a heavily hunted area in Ontario. J. Wildl. Manage. 34: 439–445.

Godman, J.D. 1826. American natural history. 3 vols. H.C. Carey and I. Lea, Philadelphia.

Goering, H.K. and P.J. Van Soest. 1970. Forage fiber analyses (apparatus, reagents, procedures, and some applications). Handbook 379. U.S. Dept. Agric.,Washington, D.C. 20 pp.

Goetzmann, W.H. 1959. Army exploration in the American West 1803–1863. Univ. Nebraska Press, Lincoln. 489 pp.

———. 1966. Exploration and empire. W.W. Norton and Co., New York. 656 pp.

Goetzmann, W.H. and G. Williams. 1992. The atlas of North American exploration. Prentice Hall Gen. Ref., New York. 224 pp.

Gollat, R. and H.R. Timmermann. 1983. Determining quotas for a moose selective harvest in northcentral Ontario. Alces 19: 191–203.

Golley, F.B. 1961. Energy valuse of ecological materials. Ecology 42: 581–584.

Goodchild, P. 1984. Survival skills of North American Indians. Chicago Review Press, Chicago, Illinois. 234 pp.

Goode, G.B. 1883. The fisheries and fishery industries of the United States. U.S. Govt. Print Off. Washington, D.C.

Gordon, M.S. 1972. Animal physiology: Principles and adaptations. 2nd ed. McMillan Publ. Co., New York. 592 pp.

Gosling, L.M. 1986. Biased sex ratios in stressed animals. Am. Nat. 127: 893–896.

Goss, R.J. 1980. Is antler asymmetry in reindeer and caribou genetically determined? Pages 364–372 *in* E. Reimers, E. Gaare and S. Skjenneberg, eds., Proc. 2nd Intern. Reindeer/Caribou Symp. Direktoatet for vilt og ferskvannsfisk. Trondheim, Norway.

———. 1983. Deer antlers: Regeneration, function, and evolution. Academic Press, New York. 316 pp.

———. 1985. Tissue differentiation in regenerating antlers. Pages 229–238 *in* P.F. Fennessy and K.R. Drew, eds., Biology of deer production. Bull. 22. R. Soc. New Zealand, Wellington.

Goss, R.J., A. Vanpraagh and P. Brewer. 1992. The mechanism of antler casting in the fallow deer. J. Exp. Zool. 264: 429–436.

Goudreault, F. 1980. L'influence d'un parc de conservation et d'un réserve sur la récolte des orignaux dans les territoires adjacents intensement chasses dans le centre-sud du Québec. Proc. N. Am. Moose Conf. workshop 16: 527–548.

Gould, S.J. and R.C.Levontin. 1979. The spandrels of San Marco and the Panglossian paradigm: A critique of the adaptationist program. Proc. R. Soc. L. 205: 581–598. (Ser. B)

Graf, R.P. 1992. Status and management of moose in the Northwest Territories, Canada. Alces Suppl. 1: 22–28.

Graham, N. McC., T.W. Searle and D.A. Griffiths. 1974. Basal metabolic rate in lambs and young sheep. Australian J. Agric. Res. 25: 957–971.

Grant, M. 1894. The vanishing moose, and their extermination in the Adirondacks. The Century Mag. XLVII: 345–356.

———. 1902. Moose. Ann. Rept. New York St. For., Fish and Game Commiss. 7: 225–238.

Grant, W.L., ed. 1907. Voyages of Samuel de Champlain, 1604–1618. Charles Scribner's Sons, New York. 374 pp.

———. 1911–1914. The history of New France. Vols. II and III; Publ. VII and XI. The Champlain Soc., Toronto, Ontario.

Grasse, J.E. 1950. Trapping the moose. Wyoming Wildl. 14: 12–18, 36, 38.

Grauvogel, C.A. 1984. Seward Peninsula moose population identity study. Fed. Aid Wildl. Restor. Final Rept. Alaska Dept. Fish and Game, Juneau. 93 pp.

Graves, H.B., E.D. Bellis and W.M. Knuth. 1972. Censusing white-tailed deer by air-borne thermal infrared imagery. J. Wildl. Manage. 36: 875–884.

Gray, J.E. 1821. On the natural arrangement of vertebrose animals. L. Med. Reposit. 15: 296–310.

Greatrex, C.B. 1854. Whittlings from the West. J. Hogg, London. 442 pp.

Greenleaf, M. 1829. A survey of the State of Maine. [A map exhibiting the principal, original grants and sales of lands, etc.] Shirley and Hyde, Portland, Maine.

Greenly, J.C. 1971. The effects of biopolitics on proper game management. Proc. West. Assoc. St. Game and Fish Commiss. 51: 505–509.

Greenwood, P.J. 1980. Mating systems, philopatry and dispersal in birds and mammals. Anim. Behav. 28: 1,140–1,162.

———. 1983. Mating systems and the evolutionary consequences of dispersal. Pages 116–131 *in* I.R. Swingland and P.J. Greenwood, eds., The ecology of animal movement. Clarendon Press, Oxford.

Greenwood, P.J. and I.R. Swingland. 1983. Animal movement: Aproaches, adaptations, and constraints. Pages 1–6 *in* I.R. Swingland and P.J. Greenwood eds., The ecology of animal movement. Clarendon Press, Oxford.

Greif, G. 1992. Saskatchewan forest habitat project discussion paper. Saskatchewan For. Hab. Proj., Prince Albert. 41 pp.

Grenier, M.P. 1979. New regulations on moose hunting in Quebec. Can. Wildl. Admin. 2: 28–31.

———. n.d. Moose calling and hunting techniques. Mont-laurier, Quebec. 108 pp.

Grenier, P.A. 1973. Moose killed on the highway in the Laurentides Park, Quebec, 1962–1972. Proc. N. Am. Moose Conf. Workshop 9: 155–194.

Griffith, E. 1827. The animal kingdom arranged in conformity with its organization, by the Baron Cuvier. 5 vols. Printed for Geo. B. Whittaker, London, England.

Gross, J.E., J.E. Rolle and G.L. Williams. 1973. Program ONRPOP and information processor: A systems modelling and communications project. Colorado Coop. Wildl. Res. Unit Prog. Rept. Colorado St. Univ., Fort Collins. 327 pp.

Gross, R.J. 1983. Deer antlers: Regeneration, function, and evolution. Academic Press, New York. 316 pp.

Groves, C.P. and P. Grubb. 1987. Relationship of living deer. Pages 21–59 *in* C.M. Wemmer, ed., Biology and management of the Cervidae. Smithsonian Inst. Press, Washington, D.C.

Groves, C.P. and P. Grubb. 1990. Muntiacidae. Pages 134–168 *in* G.A. Bubenik and A.B. Bubenik, eds., Horns, pronghorns, and antlers. Springer-Verlag, New York.

Guédon, M. 1981. Upper Tanana River potlach. Pages 577–581 *in* J. Helm, ed., Subarctic. Handbook of North American Indians. Vol. 6. Smithsonian Inst., Washington, D.C. 837 pp.

Guiness, F., G.A. Lincoln and R.V. Short. 1971. The reproductive cycle of the female red deer, *Cervus elaphus* L. J. Reprod. Fert. 27: 427–438.

Gupta, R.P. and H.C. Gibbs. 1971. Infectivity of *Dictycaulus viviparus* (moose strain) to calves. Can. Vet. J. 56: 56.

Gustafson, K.A. 1993. Infrared thermal imagery. *In* H.R. Timmermann, ed., Methods of population estimation and monitoring. Alces 29: 281–284.

Gustin, R.E. 1973. Population estimates of moose and deer in the Kenora District. Ontario Min. Nat. Resour., Toronto. 9 pp.

Guthrie, R.D. 1966. Bison horns, cores-character choice and systematics. J. Paleontol. 3: 738–740.

———. 1971. A new theory of mammalian rump patch evolution. Behav. 38: 132–145.

———. 1990a. Frozen fauna of the mammoth steppe: The story of blue babe. Univ. Chicago Press, Chicago, Illinois. 323 pp.

———. 1990b. New dates in Alaskan Quaternary moose, *Cervalces-Alces*. Archeological, evolutionary and ecological implications. Curr. Pleistocene Res. 7: 111–112.

Gutin, J-A. C. 1993. Good vibrations. Discover 14: 45–54.

Gyles, J. 1851. Memoirs of odd adventures, strange deliverances, etc., in the captivity of John Gyles, Esq., written by himself [1736]. Pages 73–109 *in* S. Drake, Indian captivities. Derby and Miller, Auburn, New York.

Haber, G.C. 1977. Socio-ecological dynamics of wolves and prey on a subarctic ecosystem. Ph.D. thesis, Univ. British Columbia, Vancouver. 786 pp.

Hagerman, A.E. and C.T. Robbins. 1992. Specificity of tannin-binding salivary proteins relative to diet selection by mammals. Can. J. Zool. 71: 628–633.

Haggstrom, D. 1994. Some ramifications of fire and forest management policies on the boreal forest and wildlife of interior Alaska. Fire Effects in Alaska Workshop, Wainwright, Alaska.

Hahn, M.E., C. Jensen and B.C. Dudek. 1979. The role of development in the brain-behavior relationship. Pages 372–380 *in* M.E. Hahn, C. Jensen and B.C. Dudek, eds., Development and evolution of brain size. Academic Press, New York.

Haigh, J.C. 1979. Hyaluronidase as an adjunct in an immobilizing mixture for moose. J. Am. Vet. Med. Assn. 175: 916–917.

———. 1982. Reproductive seasonality of male wapiti. M.S. thesis, Univ. Saskatchewan, Saskatoon. 131 pp.

Haigh, J.C., E.W. Kowal, W. Runge and G. Wobeser. 1982. Pregnancy diagnosis as a management tool for moose. Alces 18: 45–53.

Haigh, J.C., R.R. Stewart and W. Mytton. 1980. Relationships among linear measurements and weights for moose. Proc. N. Am. Moose Conf. Workshop. 16: 1–10.

Haigh, J.C., R.R. Stewart, G. Wobeser and P.S. MacWilliams. 1977. Capture myopathy in a moose. J. Am. Vet. Med. Assoc. 171: 924–926.

Hairston, N.G., F.E. Smith and L.B. Slobodkin. 1960. Community structure, populaion control, and competition. Am. Nat. 94: 421–424.

Halfpenny, J.C. and R.D. Douglas. 1989. Winter: An ecological handbook. Johnson Publ. Co., Boulder, Colorado. 273 pp.

Hall, E.R. 1981. The mammals of North America. 2 vols. John Wiley and Sons, New York.

Hall, P. 1991. Early glimpses of New France. Gilcrease Mag. Am. Hist. 13(4): 28–32.

Hall, E.R. and K.R. Kelson. 1959. The mammals of North America. The Ronald Press Co., New York. 1,083 pp.

Hall, E.S., Jr. 1973. Archeological and recent evidence for expansion of moose range in northern Alaska. J. Mammal. 54: 294–295.

Hall, E.T. 1959. The silent language. Doubleday and Co., Inc., Garden City, New York. 240 pp.

———. 1967. The hidden dimension. Doubleday and Co., Inc. Garden City, New York. 202 pp.

Hall, S.J.G., M.A. Vince, E. Shillito-Walser and P.J. Garson. 1988. Vocalizations of the Chillingham cattle. Behavior 104: 78–104.

Hall, T.C., E.B. Taft, W.H. Baker and J.C. Aub. 1953. A preliminary report on the use of flaxedil to produce paralysis in the white-tailed deer *(Odocoileus virginianus borealis)*. J. Wildl. Manage. 17: 517–520.

Halliday, T.R. 1981. Sexual selection and mate choice. Pages 180–313 *in* J.R. Krebs and N.B. Davies, eds., Behavioral ecology. An evolutionary approach. Sinauer Assoc., Inc., Sunderland, Massachusetts.

Hallowell, A.I. 1960. Ojibwa ontology, behavior, and world view. Pages 19–52 *in* S. Diamond, ed. Culture in history; essays in honor of Paul Radin. Columbia Univ. Press., New York.

Halls, L.K., ed. 1984. White-tailed deer: Ecology and management. Stackpole Books, Harrisburg, Pennsylvania. 870 pp.

Haltenorth, T. 1963. Klassifikation der Säugetiere: Artiodactyla. Vol. I. Walter De Gruyter and Co., Berlin. 167 pp.

Hamell, G.R. 1987. Mythical realities and European contact in the Northeast during the sixteenth and seventeenth centuries. Man in the Northeast 33: 63–87.

Hamilton, G.D. 1981. Practical importance of moose and other foods to natives in a remote northern Ontario community. Alces 17: 44–55.

Hamilton, G.D. and P.D. Drysdale. 1975. Effects of cutover use width on browse utilization by moose. Proc. N. Am. Moose Conf. Workshop 11: 5–26.

Hamilton, G.D., P.D. Drysdale and D.L. Euler. 1980. Moose winter browsing patterns on clearcuts in northern Ontario. Can. J. Zool. 58: 1,412–1,416.

Hamilton-Smith, C. 1827. The Ruminantia. Vol. 4. The Class Mammalia. Pages 1–496 *in* E. Griffiths, ed., The animal kingdom by Baron Cuvier with additional descriptions of all species hitherto named and many not before noticed. George Whittaker, London.

Hammel, H.T., T.R. Houpt, K.L. Andersen and S. Skenneberg. 1962. Thermal and metabolic measurements on a reindeer at rest and in exercise. Arctic Aeromed. Lab. Tech. Doc. Aaltdr-61-54. 34 pp.

Hammond, S.H. 1854. Hills, lakes, and forest streams. J.C. Derby, New York. 340 pp.

Hanley, T.A. 1982. The nutritional basis for food selection by ungulates. J. Range Manage. 35: 146–151.

Hansel, W. 1959. The estrous cycle of the cow. Pages 223–265 *in* H.H. Cole and P.T. Cupps, eds., Reproduction in domestic animals. Academic Press, New York.

Hansen, H.L., L.W. Krefting and V. Kurmis. 1973. The forest of Isle Royale in relation to fire history and wildlife. Tech. Bull. 294. Univ. Minnesota, St. Paul. 43 pp.

Hansen, P.J. 1985. Photoperiodic regulation of reproduction in mammals breeding under long days versus mammals breeding under short days. Anim. Reprod. Sci. 9: 301–315.

Hanson, H.M. 1971. Effects of fetal undernourishment on experimental anxiety. Nutr. Rept. Intern. 5: 307–314.

Harbo, S.J. and F.C. Dean. 1983. Historical and current perspectives on wolf management in Alaska. Pages 51–65 *in* L.N. Carbyn, ed., Wolves in Canada and Alaska. Rept. Ser. 45. Can. Wildl. Serv., Ottawa.

Hardin, G. 1960. The competitive exclusion principle. Science 131: 1,292–1,297.

Hardy, C. 1855. Sporting adventures in the New World; or, days and nights of moose-hunting in the pine forests of Acadia. Vol. 1. Hurst and Blackett, London. 304 pp.

Harlan, R. 1825. Fauna Americana: Being a description of the mammiferous animals inhabiting North America. J. Harding, Philadelphia. 318 pp.

Harnel, R.A., J.D. Williams and W.E. Armstrong. 1989. Effects of genetics and nutrition on antler development and body size in white-tailed deer. Fed. Aid Wildl. Restor. Rept. Texas Parks and Wildl. Dept., Austin. 55 pp.

Harper, T.A., R.A. Ruttan and W.A. Benson. 1955. Hydatid disease *(Echinococcus granulosus)* in Saskatchewan big game. Trans. N. Am. Wildl. Conf. 20: 198–208.

Harris, D. 1945. Symptoms of malnutrition in deer. J. Wildl. Manage. 9(4): 319–322.

Harris, R.C. and G.J. Matthews. 1987. Historical atlas of Canada. Vol. I. From the beginning to 1800. Univ. Toronto Press, Toronto, Ontario. 198 pp.

Harrison, J.L. 1958. Range of movements of some Malayan rats. J. Mammal. 39: 190–206.

Harry, G.B. 1957. Winter food habits of moose in Jackson Hole, Wyoming. J. Wildl. Manage. 21: 53–57.

Harthoorn, A.M. 1982. Physical aspects of both mechanical and chemical capture. Pages 62–71 *in* L. Nielsen, J.C. Haigh and M.E. Fowler, eds., Chemical immobilization of North American wildlife. Wisconsin Hum. Soc., Inc., Milwaukee. 447 pp.

Hartl, G.B., F. Reimoser, R. Willing and J. Koller. 1991. Genetic variability and differentiation in roe deer (*Capreolus capreolus* L.) of Central Europe. Genet. Selection Evol. 23: 281–299.

Hartl, K. 1965. Pes a vlčice - Některé poznatky z křížení psa å vlčicí. Pes přítel člověka (Dog—man's best friend): 86-90. Ročenka Kynologie (Cynegetic yearbook), Prague.

Hartwig, H. and H.G. Hartwig. 1975. Über die Milz der Cervidae (Gray, 1821). Quantitative= morphologische Untersuchungen an Milzen des Rothirsches (*Cervus elaphus* L., 1758) und des Rehes (*Capreolus capreolus* L., 1758). Gegenbauers Morph. Jahrb. 121: 669–697.

Hasling, C., E.F. Eriksen, P. Charles and L. Mosekilde. 1987. Exogenous triidothyronine activates bone remodelling. Bone 8: 65–69.

Hatler, D.F. 1983. Concerns for ungulate collision mortality along surface routes. MacLaren Plansearch Corp., Vancouver. 47 pp.

Hatt, G. 1916. Moccasins and their relation to Arctic footwear. Memoirs of the Am. Anthropol. Assoc. 3: 151–250.

Hatter, I.W. 1988. Effects of wolf predation on recruitment of black-tailed deer on northern Vancouver Island. Wildl. Rept. R-23. British Columbia Min. Environ., Victoria. 82 pp.

Hatter, I.W. and W.A. Bergerud. 1991. Moose recruitment, adult mortality, and rate of change. Alces 27: 65–73.

Hatter, I.W., K.N. Child and H. Langin. 1990. Draft provincial moose statement for B.C. 1990–1995. British Columbia Min. Environ., Victoria. 52 pp.

Hatter, J. 1948. Summarized interim report on a study of moose of central British Columbia, 1946. Pages 42–52 *in* British Columbia Game Comm. Rept., 1946. Victoria.

———. 1949. The status of moose in North America. Trans. N. Am. Wildl. Conf. 14: 492–501.

Hauge, T.M. and L.B Keith. 1981. Dynamics of moose populations in northeastern Alberta. J. Wildl. Manage. 45: 573–597.

Haukioja, E. and K. Lehtila. 1992. Moose and birch: How to live on low quality diets. Tree 7: 12–22.

Hayne, D.W. 1949. Calculation of size of home range. J. Mammal. 30: 1–18.

———. 1984. Population dynamics and analysis. Pages 203–210 *in* L.K. Halls, ed., White-tailed deer: Ecology and management. Stackpole Books, Harrisburg, Pennsylvania. 870 pp.

Hearne, S. 1795, 1911. A journey from Prince of Wale's Fort in Hudson's Bay to the Northern Ocean in the years 1769, 1770, 1771, and 1772. Publ. VI. The Champlain Soc., Toronto, Ontario. 437 pp. (Reprinted in 1968 by Greenwood Press, New York.)

Heberlein, T.A. 1991. Changing attitudes and funding for wildlife preserving the sport hunter. Wildl. Soc. Bull. 19: 528–533.

Hebert, D.M. and R.J. McFetridge. 1979. Chemical immobilization of North American game animals. 2nd ed. Fish and Wildl. Div., Alberta Energy and Nat. Resour., Edmonton. 250 pp.

Hedden, M. 1996. 3,500 years of shamanism in Maine rock art. Pages 7–24 *in* C.H. Faulkner, ed., Rock art of the eastern woodlands. Occas. Paper 2. Am. Rock Art Res. Assoc., Lemon Grove, California.

Hediger, H. 1946. Zur psychologischen Bedeutung des Hirschgeweihs. Verhandl. d. Schweiz. Naturforsch. Gesellsch., Basel 126: 162–163.

———. 1969. Comparative observations on sleep. Proc. R. Soc. Med. 62: 153–156.

———. 1976. Proper names in the animal kingdom. Expermentia 32: 1,357–1,364.

Hegerová, E. and O. Štěrba. 1959. Epaxialní a hypaxialní svalovina tělního kmene jelena (*Cervus elaphus* L.). [Epaxial and hypaxial muscles of red deer trunk]. Acta Univ. Agric. Silvicult. Vol. 7 (Part B) 368: 429–436.

Heidenreich, C.E. 1978. Huron. Pages 368–388 *in* B.G. Trigger, ed., Northeast. Handbook of North American Indians. Vol. 15. Smithsonian Inst., Washington, D.C. 924 pp.

Heidenreich, C.E. and A.J. Ray. 1976. The early fur trades: A study in cultural interaction. McClelland and Stewart, Ltd., Toronto. 95 pp.

Heinselman, M.L. 1973. Fire in the virgin forests of the Boundary Waters Canoe Area, Minnesota. Quat. Res. 3: 329–382.

Heintz, E. and F. Poplin. 1981. *Alces carnutorum* (Laugel 1862) du Pleistocene de Saint-Prest (France). Syste'matique et evolution des Alcines (Cervidae, Mammalia). Quat. Paleontol. 4: 105–122.

Heisey, D.M. and T.K. Fuller. 1985. Evaluation of survival and cause-specific mortality rates using telemetry data. J. Wildl. Manage. 49: 668–674.

Helm, J. 1961. The Links Point people: The dynamics of a northern Athapaskan band. Anthropol. Ser. 53, Bull. 176. Natl. Mus. Canada, Ottawa.

———. 1981. Subarctic. Vol. 6. Handbook of North American Indians. Smithsonian Inst., Washington, D.C. 837 pp.

Helm, J. and N.O. Lurie. 1961. The subsistence economy of the Dogrib Indians of Lac La Martre in the Mackenzie District of the Northwest Territories. Dept. North. Affairs and Nat. Resources, North. Coord. and Res. Centre, Ottawa.

Helm, J., E.S. Rogers and J.G.E. Smith. 1981. Intercultural relations and cultural change in the Shield and Mackenzie borderlands. Pages 146–157 *in* J. Helm, ed., Subarctic. Handbook of North American Indians. Vol. 6. Smithsonian Inst., Washington, D.C. 837 pp.

Henshaw, J. 1971. Antlers—The unbrittle bones of contention. Nature 231: 469.

Heptner, W.G. and A.A. Nasimowitsch. 1967. Der Elch—*Alces alces*. A. Ziemsen Verlag, Wittenberg-Lutherstadt, Germany. 269 pp.

Heptner, W.G. and A.A. Nasimowitsch. 1974. Der Elch—*Alces alces*. 2nd ed. A. Ziemsen Verlag, Wittenberg-Lutherstadt, Germany. 269 pp.

Hermansson, N. and J. Boethius 1975. Algen naring och miljo-jakt och vard. Svenska Jagare for bundet, Stockholm. 219 pp.

Hershkovitz, P. 1969. The evolution of mammals on southern continents. VI. Recent mammals of the neotropic region: A zoogeographic and ecological review. Quart. Rev. Biol. 44: 1–70.

Hettier de Boislambert, A.J. 1992. Bent-nosed phenomenon. Deer 8: 558.

Heusser, C.J. 1960. Late Pleistocene environments of the North Pacific North America. Spec. Pap. 35. Am. Geogr. Soc., New York. 308 pp.

Heydon, C., D. Euler, H. Smith and A. Bisset. 1992. Modelling the selective moose harvest program in Ontario. Alces 28: 111–121.

Hibbard, C.W., D.E. Ray, D.E. Savage, D.W. Taylor and J.E. Guilday. 1965. Quaternary mammals of North America. Pages 509–525 *in* H.E. Wright, Jr. and D. G. Frey, eds., The Quaternary of the United States. Princeton Univ. Press, Princeton, New Jersey. 922 pp.

Hibbs, N. 1890. Moose-hunting in the Rocky Mountains. Pages 17–44 *in* G.O. Shields, ed., The big game of North America. Rand, McNally and Co., New York. 581 pp.

Hibler, C.P. and J.L. Adcock. 1971. Elaeophorosis. Pages 263–278 *in* J.W. Davis and R.C. Anderson, eds., Parasitic diseases of wild mammals. Iowa St. Univ. Press, Ames.

Hibler, C.P. and A.K. Prestwood. 1981. Filarial nematodes of white-tailed deer. Pages 351–362 *in* W.R. Davidson, F.A. Hayes, V.F. Nettles and F.E. Kellogg, eds., Diseases and parasites of white-tailed deer. Misc. Pub. No. 7. Tall Timbers Res. Sta., Tallahasee, Florida.

Hickie, P.F. n.d. Michigan moose. Game Div., Michigan Dept. Conserv. Lansing. 57 pp.

Hicks, A.C. 1986. The history and current status of moose in New York. Alces 22: 245–252.

———. 1993. Using road-kills as an index to moose population change. Alces 29: 243–247.

Higginson, F. 1929. New-England plantation. Massachusetts Hist. Soc. Proc. 62.

Hill, H.F. and C.R. Chapman. 1989. Brain activity measures in assessment of pain and analgesia. Issues in Pain Measurement 12: 231–247.

Hindelang, M., R.O. Peterson and A.L. MacLean. 1992. Osteoporosis in moose on Isle Royale: A pilot study of bone mineral density using CT scans. Alces 28: 35–39.

Hirotani, A. 1989. Social relationships of reindeer *(Rangifer tarandus)* during rut: Implications for female choice. Appl. Anim. Behav. Sci. 24: 183–202.

Hjeljord, O., F. Sundstol and H. Haagenrud. 1982. The nutritional value of browse to moose. J. Wildl. Manage. 46: 333–343.

Hjeljord, O., N. Hovik and H.B. Pedersen. 1990. Choice of feeding sites by moose during summer: The influence of forest structure and plant phenology. Holarctic Ecol. 13: 281–292.

Hnilicka, P. and M. Zornes. 1994. Status and management of moose in Wyoming. Alces 30: 101–107.

Hobbs, N.T. 1987. Fecal indices to dietary quality: A critique. J. Wildl. Manage. 51: 317–320.

———. 1990. Diet selection by generalist herbivores: A test of the linear programming model. Pages 395–414 *in* R.N. Hughes, ed., Behavioral mechanisms of food selection. Springer-Verlag, New York.

Hobbs, N.T. and T.A. Hanley. 1990. Habitat evaluation: Do use/availability data reflect carrying capacity? J. Wildl. Manage. 54: 515–522.

Hobbs, N.T. and D.M. Swift. 1985. Estimates of habitat carrying capacity incorporating explicit nutritional constraints. J. Wildl. Manage. 49: 814–822.

Hobbs, N.T., D.L. Baker and R.B. Gill. 1983. Comparative nutritional ecology of montane ungulates during winter. J. Wildl. Manage. 47: 1–17.

Hobbs, N.T., D.L. Baker, J.E. Ellis and D.M. Swift. 1981. Composition and quality of elk winter diets in Colorado. J. Wildl. Manage. 45: 156–171.

Hobbs, N.T., D.L. Baker, J.E. Ellis, D.M. Swift and R.A. Green. 1982. Energy and nitrogen-based estimates of elk winter range carrying capacity. J. Wildl. Manage. 46: 12–21.

Hobusch, E. 1980. Fair game: A history of hunting, shooting and animal conservation. Arco Publ. Co., New York. 280 pp.

Hodge, F.W., ed. 1907. Handbook of American Indians north of Mexico. Part 1. Bur. Am. Ethnol. Bull. 30. U.S. Govt. Print. Off., Washington, D.C. 972 pp.

———, ed. 1910. Handbook of American Indians north of Mexico. Part 2. Bur. Am. Ethnol. Bull. 30. U.S. Govt. Print. Off., Washington, D.C. 1,221 pp.

Hoeve, J. 1982. Factors affecting the transmission of the rumen fluke, *Paramphistomum liorchis*, Fischoeder, 1901, from aquatic snails in moose, *Alces alces* L. Hons. thesis, Lakehead Univ., Thunder Bay, Ontario. 79 pp.

Hoeve, J., D.G. Joachim and E.M. Addison. 1988. Parasites of moose *(Alces alces)* from an agricultural area of eastern Ontario. J. Wildl. Dis. 24: 371–374.

Hoff, G.L. and D.O. Trainer. 1978. Blue tongue and epizootic hemorrhagic disease viruses: Their relationship to wildlife species. Adv. Vet. Sci. Comp. Med. 22: 111–132.

Hoff, G.L. and D.O. Trainer. 1981. Hemorrhagic disease of wild ruminants. Pages 45–53 *in* J.W. Davis, L.H. Karstad and D.O. Trainer, eds., Infectious diseases of wild mammals. Iowa St. Univ. Press, Ames.

Hoffman, C.F. 1839. Wild scenes in the forest and prairie. 2 vols. Richard Bentley, London.

Hoffman, D.G. 1952. Paul Bunyan, last of the frontier demigods. Univ. Pennsylvania Press for Temple Univ. Publ., Philadelphia. 213 pp.

Hoffman, W.J. 1888. Pictographs and shamanistic rites of the Ojibway. Am. Anthropol. 1: 209–229.

———. 1891. The Midewiwin or "Grand Medicine Society" of the Ojibway. Seventh Ann. Rept. 1885–86, Bur. Ethnol. Smithsonian Inst., Washington, D.C.

———. 1896. The Menomini Indians. Pages 1–328 *in* J.W. Powell, ed., Fourteenth Ann. Rept. Bur. Ethnol., U.S. Govt. Print. Off., Washington, D.C.

Hofmann, R.R. 1973. The ruminant stomach. E. African Monogr. in Biol. East African Lit. Bur., Nairobi. 54 pp.

———. 1985. Digestive physiology of deer—Their morphophysiological specialization and adaptation. Pages 393–407 *in* K.R. Drews and P.F. Fennessy, eds., Biology of deer production. Bull. 22. R. Soc. New Zealand, Wellington.

Hofmann, R.R. 1987. Sertolizelltumor—Tubulares Androblastom—des Eierstocks bei einer Gehörnten Ricke. Z. Jagdwiss. 33: 62–65.

Hofmann, R.R. and K. Nygren. 1992. Morphological specialization and adaptation of the moose digestive system. Alces (Suppl.) 1: 91–100.

Hofmann, R.R. and H. Thome. 1986. Zur funktionellen Morphologie und Mikroarchitektur des Zirkumkaudalorgans von *Cervus elaphus* und *C. nippon*. Berlin Munich Tierärztl. Wschr. 99: 418–424.

Hogg, D. 1990. Moose management: The forest habitat. Pages 30–33 *in* M. Buss and R. Truman, eds., The moose of Ontario. Book I: Moose biology, ecology and management. Ontario Min.Nat. Res. and Ontario Fed. Anglers and Hunters, Toronto.

Hohle, P. and J. Lykke. 1986. Elg og elgjakt i Norge Gyldendal Norsk Forlag. Oslo, Norway. 550 pp.

Holechek, J.L., M. Vavra and R.D. Pieper. 1982. Botanical composition determination of range herbivore diets: A review. J. Range Manage. 35: 309–315.

Holling, C.S. 1978. Adaptive environmental assessment and management. John Wiley and Sons, New York. 337 pp.

Hollingsworth, S. 1787. The present state of Nova Scotia with a brief account of Canada and the British Islands on the coast of North America. Creech, Edinburgh, Scotland. 221 pp.

Holsworth, W.N. 1960. Interactions between moose, elk and buffalo in Elk Island National Park, Alberta. M.S. thesis, Univ. British Columbia, Vancouver. 66 pp.

Holter, J.B. and H.H. Hayes. 1977. Growth in white-tailed deer fed varying energy and constant protein. J. Wildl. Manage. 49: 814–822.

Holter, J.B., H.H. Hayes and S.H. Smith. 1979. Protein requirements of yearling white-tailed deer. J. Wildl. Manage. 43: 872–879.

Holter, J.B., W.E. Urban, H.H. Hayes, H. Silver and H.R. Skutt. 1975. Ambient temperatiure effect on the physiological traits of white-tailed deer. Can. J. Zool. 53: 679–685.

Hölzenbein, S. and R. L. Marchington. 1992. Spatial integration of maturing male white-tailed deer into the adult population. J. Mammal. 73: 326–344.

Honigmann, J. J. 1946. Ethnography and acculturation of the Fort Nelson Slave. Yale Univ. Public. in Anthropol. 33. New Haven, Connecticut. 169 pp.

———. 1956. The Attawapiskat Swampy Cree: An ethnographic reconstruction. Anthropol. Papers of the Univ. of Alaska 5(1): 23–82.

———. 1981. Kaska. Pages 442–450 *in* J. Helm, ed., Subarctic. Handbook of North American Indians. Vol. 6. Smithsonian Inst., Washington, D.C. 837 pp.

Hope, R.D. 1972. Aerial survey of big game in Saskatchewan. Tech. Rept. Saskatchewan Dept. Nat. Resour., Regina. 7 pp.

Hopkins, D.M., ed. 1967. The Bering land bridge. Stanford Univ. Press, Palo Alto, California. 495 pp.

Horn, H.S. 1984. Some theories about dispersal. Pages 54–62 *in* I.R. Swingland and P.J. Greenwood, eds., The ecology of animal movement. Clarendon Press, Oxford.

Hornaday, W.T. 1900. The New York zoological park. The Century Mag. LXI: 85–102.

———. 1910. The American natural history. Charles Scribner's Sons, New York. 449 pp.

Horwath, C. 1990. An ethic for moose hunting. Pages 39–41 *in* M. Buss and R. Truman, eds., The moose in Ontario. Book II: Moose hunting techniques, hunting ethics and the law. Ontario Min. Nat. Resour. and Ontario Fed. Anglers and Hunters, Toronto. 78 pp.

Hoshino, M. 1988. Moose. Chronicle Books, San Francisco. 80 pp.

Hosley, E.H. 1981a. Environment and culture in the Alaska plateau. Pages 533–545 *in* J. Helm, ed., Subarctic. Handbook of North American Indians. Vol. 6. Smithsonian Inst., Washington, D.C. 837 pp.

———. 1981b. Intercultural relations and cultural change in the Alaska plateau. Pages 546–555 *in* J. Helm, ed., Subarctic. Handbook of North American Indians. Vol. 6. Smithsonian Inst., Washington, D.C. 837 pp.

Hosley, N.W. 1949. The moose and its ecology. Wildl. Leafl. 312. U.S. Fish and Wildl. Serv., Washington, D.C. 51 pp.

Hosley, N.W. and F.S. Graser. 1952. Triplet Alaskan moose calves. J. Mammal. 33: 247.

Hotchkiss, N. 1970. Common marsh plants of the United States and Canada. Resour. Public. 93. U.S. Bur. Sport Fish. and Wildl., U.S. Govt. Print. Off., Washington, D.C. 99 pp.

Houser, A.M. 1972. Basis for assessing land capability and degree of effort for wildlife in Ontario. Proc. N. Am. Moose Conf. Workshop 8: 209–222.

Houston, C.S., ed. 1974. To the Arctic by canoe, 1819–1821; the journal and paintings of Robert Hood, midshipman with Franklin. Arctic Inst. N. Am. McGill-Queen's Univ. Press, Montreal. 217 pp.

———. 1984. Arctic ordeal: The journal of John Richardson, surgeon-naturalist with Franklin, 1820–1822. McGill-Queen's Univ. Press, Montreal. 349 pp.

Houston, D.B. 1968. The Shiras moose in Jackson Hole, Wyoming. Tech. Bull. 1. Grand Teton Nat. Hist. Assoc. 110 pp.

———. 1970. Immobilization of moose with M-99 etorphine. J. Mammal. 51: 396–399.

———. 1974. Aspects of the social organization of moose. Pages 690–696 *in* V. Geist and F. Walther, eds., The behaviour of ungulates and its relation to management. Vol. II. Publ. 24. Int. Union Conserv. Nature and Nat. Resour., Morges, Switzerland.

———. 1982. The northern Yellowstone elk: Ecology and management. Macmillan Publ. Co., New York. 474 pp.

Howard, W.E. 1960. Innate and environmental dispersal of individual vertebrates. Am. Midl. Nat. 63: 152–161.

Howell, B.W. 1930. Doctor George Fordyce and his times. Ann. Med. Hist. NS2: 281–296.

Howells, W.D. 1873. A chance acquaintance. Houghton, Mifflin and Co., Boston. 271 pp.

Hubbert, M.E. 1987. The effect of diet on energy partitioning in moose. Ph. D. thesis, Univ. Alaska-Fairbanks. 158 pp.

Huckabee, J.W., F. Cartan, G. Kennington and F. Camenzind. 1973. Mercury concentration in the hair of coyotes and rodents in Jackson Hole, Wyoming. Bull. Environ. Contam. Toxicol. 9: 38–43.

Hudson, R.J. 1976. Resource division within a community of large herbivores. Nat. Can. (Que.) 103: 153–167.

———. 1985. Body size, energetics, and adaptive radiation. Pages 3–24 *in* R.J. Hudson and R.G. White, eds., Bioenergetics of wild herbivores. CRC Press, Boca Raton, Florida.

Hudson, R.J. and R.J. Christopherson. 1985. Maintenance metabolism. Pages 121–159 *in* R.J. Hudson and R.G. White, eds., Bioenergetica of wild herbivores. CRC Press, Boca Raton, Florida.

Hudson, R.J. and S. Frank. 1987. Foraging ecology of bison in aspen boreal habitats. J. Range Manage. 40: 71–75.

Hudson, R.J. and M.T. Nietfeld. 1985. Effect of forage depletion on the feeding rate of wapiti. J. Range Manage. 38: 80–82.

Hudson, R.J. and W.G. Watkins. 1986. Foraging rates of wapiti on green and cured pastures. Can. J. Zool. 64: 1,705–1,708.

Hughes, J., ed. 1988. The world atlas of archaeology. Portland House, New York. 423 pp.

Hughes, J.D. 1983. American Indian ecology. Texas West. Press, El Paso. 174 pp.

Hume, C., ed. 1987. From the wild. NorthWord Press, Ashland, Wisconsin. 192 pp.

Hume, I.D. and A.C.I. Warner. 1980. Evolution of microbial digestion in mammals. Pages 665–684 *in* Y. Ruckelbusch and P. Thivend, eds., Digestive physiology and metabolism in ruminants. AVI Publ. Co., Westport, Connecticut.

Hundertmark, K.J., W.L. Eberhardt and R.E. Ball. 1983. Winter habitat utilization by moose and mountain goats in the Chilkat Valley. Haines-Klukwan Coop. Resour. Study Final Rept. Alaska Dept. Fish and Game, Juneau. 47 pp.

Hundertmark, K.J., C.C. Schwartz and D.C. Johnson. 1989. Evaluation and testing of techniques for moose management. Fed. Aid Wildl. Restor. Prog. Rept. Alaska Dept. Fish and Game, Juneau. 30 pp.

Hundertmark, K.J., W.L. Eberhardt and R.E. Ball. 1990. Winter habitat use by moose in southeastern Alaska: Implications for forest management. Alces 26: 108–114.

Hundertmark, K.J., C.C. Schwartz, C.C. Shuey and D.C. Johnson. 1992a. Evaluation and testing of new techniques for moose management. Fed. Aid Wildl. Restor. Prog. Rept. Alaska Dept. Fish and Game, Juneau. 22 pp.

Hundertmark, K.J., P.E. Johns and M.H. Smith. 1992b. Genetic diversity of moose from the Kenai Peninsula, Alaska. Alces 28: 15–20.

Hungate, R.E. 1966. The rumen and its microbes. Academic Press, New York. 533 pp.

Hunt, C.B. 1967. Physiography of the United States. W.H. Freeman, San Francisco.

Hunter, M.D. and P.W. Price. 1992. Plating chutes and ladders: Bottom-up top-down forces in natural communities. Ecol. 73: 724–732.

Huot, J. 1988. Review of methods for evaluating the physical condition of wild ungulates in northern environments. Centre d'etudes nordiques, Univ. Laval, Quebec City. 30 pp.

Hurnik, J. 1987. Sexual behaviour of domestic mammals. Pages 423–461 *in* E.O. Price, ed., Farm animal behavior: Veterinary clinics of North America. W.B. Saunders, Philadelphia.

Huston, J.E. 1978. Forage utilization and nutrient requirements of the goat. J. Dairy Sci. 61: 988–993.

Hutchinson, T. 1764. The history of the colony of Massachusetts Bay from the first settlement thereof in 1628 until its incorporation with the colony of Plymouth, province of Main, by the charter of King William and Queen Mary, in 1691. Thomas and John Fleet, Boston. 566 pp.

Huxley, J.S. 1931. The relative size of antlers in deer. Proc. Zool. Soc. L. 72: 819–864.

Iason, G.R., D.A. Elston and D.A. Sim. 1993. Ultrasonic scanning for determination of stage of pregnancy in the llama *(Lama dama)*: A critical comparison of calibration techniques. J. Agric. Sci. 120: 371–377.

Iglesias, R.M.R., N.H. Ciccioli, H. Irazoqui and B.T. Rodrigues. 1991. Importance of behavioral stimuli in ram induced ovulation in seasonally anovular Corriedale ewes. Appl. Anim. Behav. Sci. 30: 323–332.

Interesting Services. 1989. Moose hunt: A guide to success. Interesting Serv., Inc., Emsdale, Ontario. (Video.)

International Union for the Conservation of Nature and Natural Resources. 1980. World conservation strategy. Int. Union Conserv. Nature and Nat. Resour., Gland, Switzerland.

International Wildlife Veterinary Service. 1991. Wildlife restraint series. Int. Wildl. Vet. Serv., Inc., Salinas, California. 266 pp.

Irvin, A.D. 1970. The epidemiology of wildlife rabies. Vet. Rec. 87: 333–348.

Irwin, L.L. 1975. Deer-moose relationships on a burn in northeastern Minnesota. J. Wildl. Manage. 39: 653–662.

Irwin, R.S. 1984. The providers: Hunting and fishing methods of the North American natives. Hancock House Publ., Inc., Blaine, Washington. 294 pp.

Ivankina, N.F., S.V. Isay and N.G. Busarova. 1990. Adaptive alterations of lipids of the reindeer *(Rangifer tarandus)* caused by low temperatures. Page 139 *in* N.E. Kochanov, ed., Abstract of papers and posters. Third Int. Moose Symp., Syktyvar, KOMI USSR.

Jackson, C.S. 1947. Picture maker of the old West: William H. Jackson. Charles Scribner's Sons, New York. 308 pp.

Jackson, D. 1966. The journals of Zebulon Montgomery Pike. 2 vols. Univ. Oklahoma Press, Norman.

Jackson, G.L., G.D. Racey, J.G. McNicol and L.A. Godwin. 1991. Moose habitat interpretation in Ontario. Tech Rept. 52. Northwest. Ontario For. Tech. and Wildl. Develop. Unit, Ontario Min. Nat. Resour., Toronto. 74 pp.

Jackson, V.W. 1926. Fur and game resources of Manitoba. Indust. Develop. Bd. of Manitoba, Winnipeg. 55 pp.

Jacobson, H.A. and R.N. Griffen. 1982. Antler cycle of white-tailed deer in Mississippi. Pages 15–22 *in* R.D. Brown, ed., Antler development in Cervidae. Caesar Kleberg Wildl. Res. Inst., Kingsville, Texas.

Jacobson, M. 1811. Description anatomique d'un organe observe dans les Mammiferes. Ann. du Mus. d'Historie Nat. 18: 412–431.

Jacoby, G.C. and R. D'Arrigo. 1989. Reconstructed northern hemisphere annual temperature since 1671 based on high altitude tree-ring data from North America. Climatic Change 14: 39–59.

Jaczewski, Z. 1954. The effect of change of the daylight on the growth of antlers in red deer (*Cervus elaohus* L.). Folia Biol. 2: 133–134.

———. 1980. Geweih eines europäischen Elches (*Alces alces alces* L. 1758) besonder Starke. Z. Jagdwiss. 29: 72–75.

———. 1990. Experimental induction of antler growth. Pages 371–395 *in* G.A. Bubenik and A.B. Bubenik, eds., Horns, pronghorns and antlers. Springer-Verlag, New York.

Jaczewski, Z., J. Zaniewski and W. Zurowski. 1965. Observation on the circulation in the pedicle arteries of red deer (*Cervus elaphus L.*). Pages 145–155 *in* Trans. 6th Cong. Int. Union Game Biol. The Nature Conservancy, London.

Jalanka, H.H. and B.O. Roeken. 1990. The use of medetomidine, medetomidine-ketamine combinations, and antipamezole in non-domestic mammals: A review. J. Zoo Wildl. Med. 21: 259–285.

James, D.D. 1983. Seasonal movements, summer food habits, and summer predation rates on wolves in northwest Alaska. M.S. thesis, Univ. Alaska-Fairbanks. 105 pp.

James, E. 1956. A narrative and adventures of John Tanner, during 30 years residence among the Indians. Ross and Haines, Minneapolis.

Janssen, D.L., J.L. Allen, J.P. Raath, G.E. Swan, D. Jessup and T.H. Stanley. 1991. Field studies with the narcotic immobilizing agent A3080. Pages 340–342 *in* Proc. Am. Assoc. Zoo Vet. Ann. Meeting, Calgary, Alberta.

Jaques, F.P. 1944. Snowshoe country. Univ. Minnesota Press, Minneapolis. 110 pp.

———. 1973. Francis Lee Jacques, artist of the wilderness world. Doubleday and Co., New York. 370 pp.

Jaques, F.P. and F.L. Jaques. 1938. Canoe country. Univ. Minnesota Press, Minneapolis. 79 pp.

Jaren, V., R. Andersen, M. Ulleberg, P.H. Pedersen and B. Wiseth. 1991. Moose-train collisions: The effects of vegetation removal with a cost-benefit analysis. Alces 27: 93–99.

Jarman, P.J. 1974. The social organization of antelope in relation to their ecology. Behav. 48: 215–266.

Jarman, P.J. and A.R.E. Sinclair. 1979. Feeding strategy and the pattern of resource partitioning in ungulates. Pages 130–163 *in* A.R.E. Sinclair and N. Norton-Griffiths, eds., Serengeti-dynamics of an ecosystem. Univ. Chicago Press, Chicago, Illinois.

Jarman, P.J. and M.V. Jarman. 1974. A review of impala behavior and its relevance to management. Pages 871–881 *in* V. Geist and F. Walthers, eds., The behavior of ungulates and its relation to management. Publ. New Ser. 24. Int. Union Conserv. Nature and Nat. Resour., Morges, Switzerland.

J.C.B. 1941. Travels in New France. Pennsylvania Hist. Commiss., Harrisburg. 167 pp.

Jefferson, T. 1787. Notes on the State of Virginia. John Stockdale, London.

Jelinski, D.E. and L.J. Fisher. 1991. Spatial variability in the nutrient composition of *Populus tremuloides:* Clone-to-clone differences and implications for cervids. Oecologia 88: 116–124.

Jellison, W.L., C.W. Fishel and E.L. Cheatum. 1953. Brucellosis in a moose, *Alces americanus*. J. Wildl. Manage. 17: 217–218.

Jenkins, K.J. and R.G. Wright. 1987. Dietary niche relationships among cervids relative to winter snowpack in northwestern Montana. Can. J. Zool. 65: 1,397–1,401.

Jenness, D. 1932. The Indians of Canada. Bull. 65. Natl. Mus. Canada, Ottawa. 446 pp.

———. 1937. The Sekani Indians of British Columbia. Anthropol. Series 20, Bull. 84. Natl. Mus. Canada, Ottawa.

Jennrich, R.L. and F.B. Turner. 1969. Measurement of non-circular home range. J. Theoretical Biol. 22: 227–237.

Jerdon, T.C. 1874. The mammals of India; A natural history of all the animals known to inhabit continental India. 2nd ed. J. Weldon, London. 323 pp.

Jerison, H.J. 1982. Allometry, brain size, cortical surface, and convolutedness. Pages 77–84 *in* E. Armstrong and D. Falk, eds., Primate brain evolution. Plenum Press, New York.

Jerison, H.J. and I. Jerison, eds. 1988. Intelligence and evolutionary biology. Springer-Verlag, Berlin. 483 pp.

Jewell, P.H. 1966. The concept of home range in mammals. Symp. Zool. Soc. London 18: 85–109.

Jezierski, W. 1987. Bark feeding by the elk and productivity of stands of trees: The problem of deciduous species. Page 19 *in* C.G.J. Richards and T.Y. Ku, eds., Control of mammal pests. Taylor and Francis, London.

John, P.D., B.W. Ritchie and L. Kuck. 1985. An evaluation of unregulated harvest of Shiras moose in Idaho. Alces 21: 231–252.

Johnson, A.M., ed. 1967. Saskatchewan journals and correspondence. Vol. XXVI. The Hudson's Bay Rec. Soc., London, England. 368 pp.

Johnson, B. 1990. Analogy 1.0—A wildlife population model. Tech. Rept. 90-2. Manitoba Dept. Nat. Resour., Winnipeg.

Johnson, D.H. 1980. The comparison of usage and availability measurements for evaluating resource preference. Ecol. 61: 65–71.

Johnson, G.L. 1901. Contributions to the comparative anatomy of the mammalian eye, chiefly based on ophthalmoscopic examination. Philos. Trans. R. Soc. Lond., Ser. B 194: 1–82.

Johnson, R. 1874. Notice of a new species of deer from the Norfolk Forest Bed. Annals and Mag. Nat. Hist., Ser. 4, 13: 1.

Johnson, S. 1755. A dictionary of the English language. 2 vols. W. Strahan, London, England.

Johnston, P.C. 1995. An artist apart. Sporting Classics XIV(5): 66–71, 95.

Johnston, W.A. 1933. Quaternary geology of North America in relation to the migration of man. Pages 11–45 *in* American aborigines, their origin and antiquity: a collection of papers. . . . Published for presentation at the 5th Pacific Sci. Congr., Canada. Univ. Toronto Press, Toronto, Ontario.

Jolicouer, H. and M. Crête. 1988. Winter survival and habitat use of orphaned and nonorphaned moose calves in southern Quebec. Can. J. Zool. 66: 919–924.

Jones, J.S. 1987. An asymmetrical view of fitness. Nature 325: 298–299.

Jones, S. 1872. The Kutchin tribes. Pages 320–327 *in* Notes on the Tinnich or Chepewyan Indians of British and Russian America. Ann. Rept. of the Smithsonian Inst. for the Year 1866. Washington, D.C.

Jones, W. 1906. Central Algonkin. Pages 136–146 *in* Ann. Archaeol. Rept. of 1905; Rept. of Min. of Educ. L.K. Cameron, Toronto. 249 pp.

Jordan, P.A. 1987. Aquatic foraging and the sodium ecology of moose: A review. Swed. Wildl. Res. (Suppl.) 1: 119–137.

Jordan, P.A. and M.L. Wolfe. 1980. Aerial and pellet-count inventory of moose on Isle Royale. Proc. 2nd Conf. on Sci. Res. in Natl. Parks 12: 363–393.

Jordan, P.A., D.B. Botkin and M.L. Wolfe. 1970. Biomass dynamics in a moose population. Ecol. 52: 147–152.

Jordan, P.A., M.L. Wolfe and D.B. Botkin. 1972. Numbres of moose at Isle Royale and a report of current ecosystem studies there. Proc. N.Am. Moose Conf. Workshop 8: 208.

Jordan, P.A., R.O. Peterson, P. Campbell and B. McLaren. 1993. Comparison of pellet counts and aerial counts for estimating density of moose at Isle Royale: A progress report. Alces 29: 267–278.

Jordan, P.A., D.B. Botkin, A.S. Dominicki, H.S. Lowendorf and G.E. Belovsky. 1973. Sodium as a critical nutrient for the moose of Isle Royale. Proc. N. Am. Moose Conf. Workshop 9: 13–42.

Joyal, R. 1976. Winter foods of moose in La Verendrye Park, Quebec: An evaluation of two browse survey methods. Can. J. Zool. 54: 1,765–1,770.

———. 1987. Moose habitat investigations in Quebec and management implications. Swed. Wildl. Res. (Suppl.) 1: 139–152.

Joyal, R. and B. Scherrer. 1978. Summer movements and feeding by moose in western Quebec. Can. Field-Nat. 92: 252–258.

Jubb, K.V.F., P.C. Kennedy and N. Palmer. 1985. Pathology of domestic animals. 3rd. ed., Vol. 2. Academic Press, Orlando, Florida. 582 pp.

Judson, K.B., ed. 1914. Myths and legends of the Mississippi Valley and the Great Lakes. A.C. McClurg and Co., Chicago. 215 pp.

Kahlke, H.D. 1956. Die Cervidenreste aus den altpleistozänen Sänden von Mosbach (Biebrich-Wiesbaden). Tiel I. Die Geweihe, Gehorne und Gebisse. Abhandl. d.D. Akad. Wiss. Berlin. 62 pp.

Kale, W. 1982. Estimation of moose harvest for smaller management units in the Yukon. Alces 18: 116–141.

Kaletskyi, A.A. 1965. Yuvelnaya linka losyat, yeye svyaz se sutochnym potrebleniyem kormov i izmerniyem veza. [The juvenile molt of calf moose and its relation to photoperiodically controlled food intake and gain of body weight]. Pages 281–284 *in* A.G. Bannikov, ed., Biologiya i promysel losya. Kolos, Moscow.

———. 1967. Sutochnyi cykl aktivnosti losey v pervyi god ikh zhizni. Pages 247–257 *in* A.G. Bannikov, ed., Biologija i promysel losya. Kolos, Moscow.

Kalm, P. 1770. Travels into North America. William Eyres, Warrington, England.

Kane, P. 1859. Wanderings of an artist among the Indians of North America. Longman, Brown, Green, Longmans, and Roberts London. (Reprinted in 1925 by the Radisson Soc. of Canada, Toronto. 323 pp.)

Kaneko, J.J. 1980. Clinical biochemistry of domestic animals. 3rd ed. Academic Press, New York. 832 pp.

Kapherr, E., von. 1924. Das Hirschgeweih. Die Entwicklung des Kopfschmuckes der Cerviden nach heutigem Stande der Wissenschaft. J. Neumann Verlag, Neudamm, Germany. 128 pp.

Kaplan, E.L. and P. Meier. 1958. Nonparametric estimation from incomplete observations. J. Am. Stat. Assoc. 53: 457–481.

Karapelou, J. 1993. Phantom in the brain. Discover 14: 6.

Karns, P.D. 1967a. *Pneumostrongylus tenuis* in deer in Minnesota and implications for moose. J. Wildl. Manage. 31: 299–303.

———. 1972. Minnesota's 1971 moose hunt: A preliminary report on the biological collections. Proc. N. Am. Moose Conf. Workshop 8: 115–123.

———. 1976. Relationships of age and body measurements to moose weight in Minnesota. Proc. N. Am. Moose Conf. Workshop 12: 274–284.

———. 1979. Winter—The grim reaper. Pages 47–53 *in* R.L. Hine and S. Nehls, eds., White-tailed deer population management in the north central states. Proc. Symp. Northcentral Sect. Wildl. Soc. The Graphic Print. Co., Eau Claire, Wisconsin. 116 pp.

———. 1982. Twenty-plus years of aerial moose census in Minnesota. Alces 18: 186–196.

Karns, P.D., H. Haswell, F.F. Gilbert and A.E. Patton. 1974. Moose management in the coniferous-deciduous ecotone of North America. Nat. Can. (Que.) 101: 643–656.

Kastner, J. 1991. The animal illustrated 1550–1900. Harry N. Abrams, Inc., New York. 128 pp.

Kaufmann, J. H. 1962. Ecology and social behavior of the coati, *Nasua nirica,* on Barro Colorado Island Panama. Univ. California Publ. Zool. 60: 95–222.

Kautz, M.A., G.M. Van Dyne and L.H. Carpenter. 1982. Energy cost for activities of mule deer fawns. J. Wildl. Manage. 46: 704–710.

Kay, R.N.B., W.V. Engelhardt and R.G. White. 1980. The digestive physiology of wild ruminants. Pages 743–761 *in* Y. Ruckebusch and P. Thivend, eds., Digestive physiology and metabolism in ruminants. AVI Publ. Co., Westport, Connecticut.

Kay, R.N.B., M. Phillippo, J.M. Suttie and G. Wenham. 1981. The growth and mineralization of antlers. J. Physiol. 322: 1–4.

Kearney, S.R. and F.F. Gilbert. 1976. Habitat use by white-tailed deer and moose on sympatric range. J. Wildl. Manage. 40: 645–657.

Kearney, S.R. and F.F. Gilbert. 1978. Terrestrial gastropods from the Himsworth Game Preserve, Ontario, and their significance in *Parelaphostrongylus tenuis* transmission. Can. J. Zool. 56: 688–694.

Keiser, M., ed. 1987a. Montana hunter education manual. Rev. ed. Outdoor Empire Publ., Inc., Seattle, Washington. 87 pp.

———. 1987b. Wyoming hunter education students handbook. Outdoor Empire Publ., Inc., Seattle, Washington. 87 pp.

Keith, L.B. 1974. Some features of population dynamics in mammals. Trans. Int. Congr. Game Biol. 11: 17–58.

———. 1983. Population dynamics of wolves. Pages 66–77 *in* L.N. Carbyn, ed., Wolves in Canada and Alaska: Their status, biology, and management. Rept. Ser. 45. Can. Wildl. Serv., Ottawa.

Kellogg, L.P., ed. 1917. Early narratives of the Northwest, 1634–1699. Charles Scribner's Sons, New York. 382 pp.

———, ed. 1923. Journal of a voyage to North America, translated from the French of Pierre François Xavier de Charlevoix. 2 vols. R.R. Donnelley and Sons, Chicago.

Kellum, F. 1941. Cusino's captive moose. Michigan Conserv. 10: 4–5.

Kelsall, J.P. 1969. Structural adaptations of moose and deer for snow. J. Mammal. 50: 302–310.

———. 1972. The northern limits of moose *(Alces alces)* in western Canada. J. Mammal. 53: 129–138.

———. 1987. The distribution and status of moose *(Alces alces)* in North America. Swed. Wildl. Res. (Suppl.) 1: 1–10.

Kelsall, J.P. and W. Prescott. 1971. Moose and deer behaviour in snow. Rept. Ser. 15. Can. Wildl. Serv., Ottawa. 27 pp.

Kelsall, J.P. and E.S. Telfer. 1974. Biogeography of moose with particular reference to western North America. Nat. Can. (Que.) 101: 117–130.

Kelsall, J.P., E.S. Telfer and T.D. Wright. 1977. The effects of fire on the ecology of the boreal forest, with particular reference to the Canadian north: A review and selected bibliography. Occas. pap. 323. Can. Wildl. Serv., Ottawa. 58 pp.

Kendrick, K. 1990. Through a sheep's eye. New Scientist 6: 60–65.

Kennedy, M.J., M.W. Lankester and J.B. Snider. 1985. *Paramphistomum cervi* and *Paramphistomum liorchis* (Digenea: Paramphistomatidae) in moose, *Alces alces,* from Ontario. Can. J. Zool. 63: 1,207–1,210.

Kennelly, J.J., F.X. Aherne and M.J. Apps. 1980. Dysprosium as an inert marker for swine digestibility studies. Can. J. Anim. Sci. 60: 441–446.

Kennaugh, J.H., D.I. Chapman and N.G. Chapman. 1977. Seasonal changes in the prepuce of adult fallow deer *(Dama dama)* and its probable function as a scent organ. J. Zool. Soc. London. 183: 301–310.

Kenton, E., ed. 1956. Black gown and redskins. Lowe and Brydone Ltd., London. 527 pp.

Kenward, R. 1987. Wildlife radio tagging. Academic Press, San Diego, California. 222 pp.

Khan, E. 1970. Biostratigraphy and paleontology of the Sangamon deposit at Fort d'Apelle, Saskatchewan. Paleontol. 5: 52–66.

Kheruvimov, V.D. 1969. Los-Sravnitelnye issledovaniya na primere tambovskoi populatsii. [Moose-comparative study sampled from the Tambovsk population]. Tcentral. Chernozem. Knizhoe Izdat, Voroniezh. 432 pp.

Kidd, W. 1903. Direction of hair in animals and man. Adam and Charles Black, London. 154 pp.

Kie, J.Q., J.A. Baldwin and C.J. Evans. 1996. CALHOME: A program for estimating animal home ranges. Wildl. Soc. Bull. 24: 342–344.

Kierdorf, H. 1993. Effects of an antiandrogen treatment on the antler cycle of male fallow deer (*Dama dama* L.). J. Exper. Zool. 266: 195–205.

Kierdorf, U. 1985. Gehornte Ricken. [Antlered females]. WJSC-Blatter 61: 1–12.

King, L.S. 1939. Moose encephalitis. Am. J. Path. 15: 445–454.

Kingsley, J.S., ed. 1884. The Riverside natural history. Vol. V. Mammals. Houghton, Mifflin and Co., Boston, Massachusetts. 451 pp.

Kingston, N. 1981a. Protozoan parasites. Pages 193–236 *in* W.R. Davidson, F.A. Hayes, V.F. Nettles and F.E. Kellogg, eds., Diseases and parasites of white-tailed deer. Misc. Publ. No. 7. Tall Timbers Res. Sta., Tallahasee, Florida. 458 pp.

———. 1981b. Trypanosoma. Pages 166–169. *in* R.A. Dieterich, ed., Alaskan wildlife diseases. Univ. Alaska-Fairbanks.

Kipp, R. 1991. Currier's price guide to Currier and Ives prints. 2nd ed. Currier Public., Brockton, Massachusetts. 270 pp.

Kirkpatrick, M. 1987. Sexual selection by female choice in polygynous animals. Ann. Rev. Ecol. Syst. 18: 43–70.

Kirkpatrick, R.L. 1980. Physiological indices in wildlife management. Pages 99–112 *in* S.D. Schemnitz, ed., Wildlife Management techniques manual. 4th ed. The Wildl. Soc., Washington, D.C. 686 pp.

Kistner, T.P., C.E. Trainer and N.A. Hartmann. 1980. A field technique for evaluating physical condition of deer. Wildl. Soc. Bull. 8: 11–17.

Kitchener, A. 1992. The functional design of the antlers of the Irish elk. Pages 311–313 *in* B. Bobek, K. Perzanowski and W.L. Regelin, eds., Global trends in wildlife management. Vol. 1. Trans. Cong. Int. Union Game Biol. Swiat Press, Krakow-Warszawa, Poland.

Kleiber, M. 1961. The fire of life: An introduction to animal energetics. John Wiley and Sons, New York. 454 pp. (Reprinted in 1975 by Robert E. Kreiger Publ. Co., New York.)

Klein, D.R. 1964. Range related differences in growth of deer reflected in skeletal ratios. J. Mammal. 45: 226–235.

———. 1981. The problems of overpopulation of deer in North America. Pages 119–127 *in* P.A. Jewell and S. Holt, eds., Problems in management of locally abundant wild mammals. Academic Press, New York.

Knight, R.R. 1970. The Sun River elk herd. Wildl. Monogr. 23. 66 pp.

Knorre, E.P. 1949. Ekologiya losya v svyazi s ego odomashnivaniyem. Nauchno-metod. zapiski Glavn. upravlenyia po zapovednikam. Vol. 13.

———. 1959. Ekologiya losya. [Moose ecology]. Trudy Pechoro-Ilych. Gos. Zapov. 7: 5–167.

———. 1961. The results and perspectives of domestication of moose. Trudy Pechora-Ilych. Gos. Zapov. 9: 5–113.

———. 1974. Changes in the behavior of moose with age and during the process of
domestication. Nat. Can. (Que.) 101: 371–377.

Knowles, W.C. 1984. The ethological analysis of the use of antlers as social organs by rutting bull moose (*Alces alces gigas* Miller). M.S. thesis, Univ. Alaska-Fairbanks. 93 pp.

Knowlton, F.F. 1960. Food habits, movements, and populations of moose in the Gravelly Mountains, Montana. J. Wildl. Manage. 24: 162–170.

Knox, W.M., K.V. Miller and R.L. Marchinton. 1988. Recurrent estrous cycles in white-tailed deer. J. Mammal. 69: 384–386.

Knue, J. 1991. Big game in North Dakota. North Dakota Game and Fish Dept., Bismarck. 343 pp.

Ko, K.M., T.T. Yip, S.W. Tsao, Y.C. Kong, P. Fennessy, M.C. Belew and J. Porath. 1986. Epidermal growth factor from deer *(Cervus elaphus)*. Submaxillary gland and velvet antler. Gen. Comp. Endocrinol. 63: 431–440.

Kocan, A.A., S.J. Barron, J.C. Fox and A.W. Franzmann. 1986a. Antibodies to *Toxoplasma gondii* in moose *(Alces alces)* from Alaska. J. Wildl. Dis. 22: 432.

Kocan, A.A., A.W. Franzmann, K.A. Waldrup and G.J. Kubat. 1986b. Serological studies of select infectious diseases of moose *(Alces alces)* from Alaska. J. Wildl. Dis. 22: 418–420.

Kodrick-Brown, A. and J.H. Brown. 1985. Why the fittest are prettiest: Peacock's plumes and elk's antlers advertise good genes. The Sciences 25: 26–33.

Koeppl, J.W., N.A. Slade and R.S. Hoffmann. 1975. A bivariate home range model with possible application to ethological data analysis. J. Mammal. 59: 870–871.

Kojola, I. and E. Eloranta. 1989. Influences of maternal body weight, age, and parity on sex ratio in semidomesticated reindeer *(Rangifer t. tarandus)* Evolution 43: 1,331–1,336.

Kolda, J. 1951. Osteologický atlas. Zdravot. Naklad., Prague. 194 pp.

Komárek, V.I. 1958. Některé poznámky k svalovine hrudní končetiny jelena. Sborník VSZL, Brno, 6: 147–159.

Koong, L.J., J.A. Nienaber, J. Pekas and J.C. Yen. 1982. Effect of plane of nutrition on organ size and fasting heat production in pigs. J. Nutr. 112: 1,638–1,642.

Korablev, P.N. 1989. Pathological changes in the maxilla of the moose. E'kologiya. 20: 294–298.

Korablev, P.N. and R.I. Likhotov. 1990. On symmetry of mammal skulls. Viestnik Zool. 5: 52–58.

Kowal, E.H. and W. Runge. 1981. Experimental pregnancy testing of moose in game management zone 27. Resour. Branch, Dept. N. Saskatchewan, Prince Albert. 15 pp.

Kozhukhov, M.V. 1959. Intravitam traumas of organs and tissues in moose. Pages 179–181 *in* G.A. Novikov, ed., Trudy Pechoro-Ilych. Vol. 7. Gos. Zapov, Moscow.

———. 1990. A rare case of the birth of moose with fangs. Page 35 *in* N.E. Kochanov, ed., Abst. Pap. and posters. Third Int. Moose Symp. AN USSR, Ural Div., Syktyvkar, USSR.

Kozlowski, T.T. 1971. Growth and development of trees. Vol.1. Academic Press, New York. 423 pp.

Krausman, P.D., B.D. Leopold, R.F. Seegmiller and S.G. Torres. 1989. Relationships between desert bighorn sheep and habitat in western Arizona. Wildl. Monogr. 102. The Wildl. Soc., Bethesda, Maryland. 66 pp.

Krebill, R.G. 1972. Mortality of aspen on the Gros Vente elk winter range. Res. Pap. INT-129. USDA For. Serv., Washington, D.C. 16 pp.

Krebs, C.J. 1978. Ecology: The experimental analysis of distribution and abundance. Harper and Row, New York. 678 pp.

Krebs, C.J., B. Keller and R. Tamarin. 1969. Microtus population dynamics. Ecol. 50: 587–607.

Krech, S. III. 1981. "Throwing bad medicine": Sorcery, disease, and the fur trade among the Kutchin and other Northern Athapaskans. Pages 73–108 *in* S. Krech III, ed., Indians, animals and the fur trade. Univ. Georgia Press, Athens. 207 pp.

Krefting, L.W. 1951. What is the future of the Isle Royale moose herd? Trans. N. Am. Wildl. Conf. 16: 461–472.

———. 1974a. The ecology of the Isle Royale moose with special reference to the habitat. Agr. Expt. Sta. Tech. Bull. 297. Univ. Minn., Minneapolis. 75 pp.

———. 1974b. Moose distribution and habitat selection in North America. Nat. Can. (Que.) 101: 81–100.

Krefting, L.W. and C.J. Shine. 1960. Counting deer pellet groups with a multiple-random-start-systematic sample. Sch. For., For. Note No. 89. Univ. Minnesota, St. Paul. 2 pp.

Kristchinski, A.A. 1974. The moose in northeast Siberia. Nat. Can. (Que.) 101: 179–184.

Kroeber, A.L. 1939. Cultural and natural areas of North America. Univ. California Public. in Am. Archaeol. and Ethnol. 38. Berkeley.

Krog, L.J. and M. Monson. 1954. Notes on the metabolism of a mountain goat. Am. J. Physiol. 178: 515.

Krog, J., O.B. Reite and P. Fjellheim. 1969. Vasomotor responses to the growing antlers of reindeer *(Rangifer tarandus)*. Nature 223: 99–100.

Kruuk, H. 1972. The spotted hyena: A study of predation and social behavior. Univ. Chicago Press.

Krysl, L.J., F.T. McCollum and M.L. Galyean. 1984. Estimation of fecal output and particulate passage rate with a pulse dose of ytterbium-labeled forage. J. Range. Manage. 38: 180–182.

Kubota, J. 1974. Mineral composition of browse plants for moose. Nat. Can. (Que.) 101: 291–305.

Kucera, T.E. 1991. Adaptive variation in sex ratios of offspring in nutritionally stressed mule deer. J. Mammal. 72: 745–749.

Kufield, R.C. 1992. Development of census methods and determination of movements, habitat selection, and degree of calf mortality of moose in northcentral Colorado. Colorado Div. Wildl., Fort Collins. 7 pp.

———. 1994. Status and management of moose in Colorado. Alces 30: 41–44.

Kuhlmann, R.E., R. Rainey and R. O'Neil. 1962. Biochemical investigations of deer antler growth. Part II. Quantitative microchemical changes associated with antler and bone formation. J. Bone Joint Surg. 45A: 345–350.

Kuhme, W. 1974. Klauenpflege beim Nord-Elch *(Alces alces alces)*. Zeit. Kölner Zoo 17: 23–28.

Kurtén, B. 1968. Pleistocene mammals of Europe. Weidenfeld and Nicholson, London. 317 pp.

Kurtén, B. and E. Anderson. 1980. Pleistocene mammals of North America. Columbia Univ. Press, New York. 442 pp.

Labelle, E. 1983. Moose hunters guide. Imprimrvie Carriere, Inc., Maniwaki, Quebec. 88 pp.

Labetskaya, N.V., N.F. Ivankina and S.V. Isay. 1990. Fatty acid composition phosphatidylinositol antlers of a sika deer. Page 142 *in* N.E. Kochanov, ed., Abstr. of pap. and posters. Third Int. Moose Symp., AN USSR, Ural Div., Syktyvar, USSR.

Lacasse, M. J.L. Ducharme, J. Pelletier and G. Pamontaque. 1984. Wildlife management and use in Quebec, development strategies. Losir, Chasse et Peche, Quebec City. 43 pp.

Lack, D. 1954. The natural regulation of animal numbers. Clarendon Press, Oxford, England. 343 pp.

Lahontan, Baron. 1703. New voyages to North America. 2 vols. Printed for H. Bonwicke, T. Goodwin, M. Wottan, B. Tooke and S. Manship, London.

Lamb, M.J. 1977. Biology of ageing. John Wiley and Sons, New York. 184 pp.

Lamson, A.L. 1941. Maine moose disease studies. M.S. thesis, Univ. Maine, Orono. 61 pp.

Lance, W.R. 1991. New pharmaceutical tools for the 1990's. Proc. Am. Zoo Vet. Assoc.: 354–359.

Lankester, M.W. 1974. *Parelaphostrongylus tenuis* (Nematoda) and *Fascioloides magna* (Trematoda) in moose in southeastern Manitoba. Can. J. Zool. 52: 235–239.

———. 1977. Neurologic disease in moose caused by *Elaphostrongylus cervi* Cameron 1931 from caribou. Proc. N. Am. Moose Conf. Workshop 13: 177–190.

———. 1987. Pests, parasites, and diseases of moose *(Alces alces)* in North America. Swed. Wildl. Res. (Suppl.) 1: 461–490.

Lankester, M.W. and R.C. Anderson. 1968. Gastropods as intermediate hosts of *Pneumostrongylus tenuis* Dougherty. Can. J. Zool. 57: 1,384–1,392.

Lankester, M.W. and T.J. Bellhouse. 1982. Pathological anomalies in moose of northwestern Ontario. Alces 18: 17–24.

Lankester, M.W. and D. Fong. 1989. Distribution of Elaphostrongyline nematodes (Metastrongyloidea: Protostrongylidae) in Cervidae and possible effects of moving *Rangifer* spp. into and within North America. Alces 25: 133–145.

Lankester, M.W. and P.L. Hauta. 1989. *Parelaphostrongylus andersoni* (Nematoda: Protostrongylidae) in caribou *Rangifer tarandus* of northern and central Canada. Can. J. Zool. 67: 1,966–1,975.

Lankester, M.W. and S. Luttich. 1988. *Fascioloides magna* (Trematoda) in woodland caribou *(Rangifer tarandus caribou)* of the George River herd, Labrador. Can. J. Zool. 66: 475–479.

Lankester, M.W. and T.H. Northcott. 1979. *Elaphostrongylus cervi* Cameron 1931 (Nematoda: Metastrongyloidea) in caribou *(Rangifer tarandus caribou)* of Newfoundland. Can. J. Zool. 57: 1,384–1,392.

Lankester, M.W. and W.J. Peterson. 1996. The possible importance of wintering yards in the transmission of *Parelaphostrongylus tenuis* to white-tailed deer and moose. J. Wildl. Dis. 32: 31–38.

Lankester, M.W. and R.D. Sein. 1986. The moose fly, *Haematobosca alcis*, and skin lesions on *Alces alces*. Alces 22: 361–375.

Lankester, M.W. and J.B. Snider. 1982. *Rumenfilaria andersoni* (Nematoda: Filarioedea) in moose, *Alces alces* (L), from northwestern Ontario, Canada. Can. J. Zool. 60: 2,455–2,458.

Lankester, M.W., J.E.G. Smits, M.J. Pybus, D. Fong and J.C. Haigh. 1990. Experimental infection of fallow deer *(Dama dama)* with Elaphostrongyline nematodes (Nematoda: Potostrongylidae) from caribou *(Rangifer tarandus caribou)* in Newfoundland. Alces 26: 154–162.

Lantz, D.E. 1910. Raising deer and other large game animals in the United States. Bull. No. 36. U.S. Bur. Biol. Surv. U.S. Govt. Print. Off., Washington, D.C. 62 pp.

Larkin, R.P. And D. Halkin. 1994. A review of software packages for estimating animal home ranges. Wildl. Soc. Bull. 22: 274–287.

Larsen, D.G. and D.A. Gauthier. 1989. Effects of capturing pregnant moose and calves on calf survivorship. J. Wildl. Manage. 53: 564–567.

Larsen, D.G. and W. Kale. 1982. Integration of moose aerial survey and hunter harvest information to establish future harvests in the southwest Yukon. Alces 18: 168–185.

Larsen, D.G., D.A. Gauthier and R.L. Markel. 1989a. Causes and rate of moose mortality in the southwest Yukon. J. Wildl. Manage. 53: 548–557.

Larsen, D.G., D.A. Gauthier, R.L. Markel and R.D. Hayes. 1989b. Limiting factors on moose population growth in the southwest Yukon. Yukon Dept. Renew. Resour., Whitehorse. 105 pp.

Lautenschlager, R.A. and H.S. Crawford. 1983. Halter-training moose. Wildl. Soc. Bull. 11: 187–189.

Lautenschlager, R.A. and P.A. Jordan. 1993. Potential use of track-pellet group counts for moose censusing. Alces 29: 175–180.

Lavsund, S. and F. Sandegren. 1991. Moose-vehicle relations in Sweden: A review. Alces: 118–126.

Leacock, E. 1969. The Montagnais-Naskapi. Pages 1–17 *in* D. Damas, ed. Contributions to anthropology: Band societies. Anthropol. Ser. 84, Bull. 228. Nat. Mus. Can., Ottawa.

———. 1995. The Montagnais-Naskapi of the Labrador Peninsula. Pages 150–180 *in* R.B. Morrison and C.R. Wilson, eds., Native peoples: The Canadian experience. McClelland and Stewart Inc., Toronto, Ontario. 639 pp.

Leechman, D. 1974. Comodityes besides furres. The Beaver 304: 46–52.

Legg, D. 1995. The economic impact of hunting for large mammals in Ontario, 1993. Soc. and Econ. Res. and Anal. Sect., Ontario Min. Nat. Resour., Toronto. 7 pp.

Leege, T.A. 1990. Moose management plan 1991–1995. Idaho Dept. Fish and Game, Boise. 22 pp.

Lenarz, M.S. 1992. 1991–92 aerial moose survey results. Wildl. Res. Rept. Minnesota Dept. Nat. Resour. Grand Rapids. 8 pp.

Lent, P.C. 1974. A review of rutting behavior in moose. Nat. Can. (Que.) 101: 307–323.

Lenz, M.F., A.W. English and A. Dradjat. 1993. Real-time ultrasonograph for pregnancy diagnosis and foetal aging in fallow deer. Aust. Agric. Sci. 70: 373–375.

Leo, J. and R. Agemian 1986. The hidden power of body odors. Time (Dec.) 1: 70.

Leopold, A. 1933. Game management. Charles Scribner and Sons, New York. 481 pp.

———. 1949. A sand county almanac. Oxford Univ. Press, New York. 226 pp.

Leopold, A.S. and F.F. Darling. 1953. Effects of land use on moose and caribou in Alaska. Trans. N. Am. Wildl. Conf. 18: 553–562.

Leptich, D.J. and J.R. Gilbert. 1989. Summer home range and habitat use by moose in northern Maine. J. Wildl. Manage. 53: 880–884.

LeResche, R.E. 1966. Behaviour and calf survival in the Alaska moose. M.S. thesis, Univ. Alaska-Fairbanks. 85 pp.

———. 1970. Moose report. Fed. Aid Wildl. Restor. Rept. Alaska Dept. Fish and Game, Juneau. 13 pp.

———. 1972. Migrations and population mixing of moose on the Kenai Peninsula (Alaska). Proc. N. Am. Moose Conf. Workshop 8: 185–207.

———. 1974. Moose migrations in North America. Nat. Can. (Que.) 101: 393–415.

LeResche, R.E. and J.L. Davis. 1971. Moose research report. Fed. Aid Wildl. Restor. Prog. Rept. Alaska Dept. Fish and Game, Juneau. 88 pp.

LeResche, R.E. and J.L. Davis. 1973. Importance of nonbrowse foods to moose on the Kenai Peninsula, Alaska. J. Wildl. Manage. 37: 279–287.

LeResche, R.E. and G.M. Lynch. 1973. A trap for free-ranging moose. J. Wildl. Manage. 37: 87–89.

LeResche, R.E. and R.A. Rausch. 1974. Accuracy and precision of aerial moose censusing. J. Wildl. Manage. 38: 175–182.

LeResche, R.E., R.H. Bishop and J.W. Coady. 1974a. Distribution and habitats of moose in Alaska. Nat. Can (Que.) 101: 143–178.

LeResche, R.E., U.S. Seal, P.D. Karns and A.W. Franzmann. 1974b. A review of blood chemistry of moose and other Cervidae, with emphasis on nutritional status. Nat. Can. (Que.) 101: 263–290.

Lewin, V. and J.G. Stelfox. 1967. Functional anatomy of the tail and associated behavior in woodland caribou. Can. Field-Nat. 81: 63–66.

Li, C., Z. Liu and S. Zhao. 1988. Variation of testosterone and estradiol levels in plasma during every stage of growth of sika deer antler. Acta Theriol. Sinica 8: 224–231.

Lidicker, W.Z., Jr. 1975. The role of dispersal in the demography of small mammals. Pages 103–128 *in* K. Petrusewicz, E.B. Golley and L. Ryzkowski, eds., Small mammals: Productivity and dynamics of populations. Cambridge Univ. Press, London.

Likhachev, A.I. 1955. Razmonozhenie i embrionalnye razvitie losya. Pages 269–282 *in* Trudy Tomoskogo Gos. Univ., Band 131, Tomsk, USSR.

———. 1956. Prisposobitelnye morfo-funktsionalnye osobennosti v sisteme organov dvizheniya losey. Zool. Zh. 35: 445–458.

Lime, D.W. 1974. Moose as a nongame recreational resource. Proc. N. Am. Moose Conf. Workshop 10: 110–134.

Lincoln, G.A. 1992. Biology of antlers. J. Zool. Soc. L. 226: 517–528.

Lincoln, G.H. 1971. Puberty in a seasonally breeding male, the red deer stag (*Cervus elaphus* L.). J. Repr. Fert. 25: 41–54.

Lindholdt, P.J., ed. 1988. A critical edition of "Two voyages to New England." Univ. Press of New England, Hanover, New Hampshire. 221 pp.

Lindstedt, S.L. and W.A. Calder III. 1981. Body size, physiological time, and longevity of homeothermic animals. Quart. Rev. Biol. 56: 1–15.

Lindstedt, S.L., B.J. Miller and S.W. Buskirk. 1986. Home range, time, and body size in mammals. Ecol. 67: 413–418.

Linnaeus, C. 1758. Systema neturae per regna tria naturae, secundum classes, ordines, genera, differentriis, synonymus, locic. Edition decima, reformata. Laurentii Salvii, Stockholm. 824 pp.

Little, W. 1870. The history of Warren: A mountain hamlet located among the White Hills of New Hampshire. Heritage Books, Inc., Bowie, Maryland.

Lloyd, L.E., B.E. McDonald and E.W. Crampton. 1978. Fundamentals of nutrition. 2nd ed. W.H. Freeman and Co., San Francisco. 466 pp.

Lojda, Z. 1956. Histogenesis of antlers of our Cervidae and its histochemical picture. Csl. Morphol. 4: 43–62.

Loken, K.I., J.C. Schlotthauer, H.J. Kurtz and P.D. Karns. 1965. *Pneumostrongylus tenuis* in Minnesota moose *(Alces alces)*. Wildl. Dis. Assoc. Bull. 1: 7.

Lomnicki, A. 1988. Population ecology of individuals. Princeton Univ. Press, Princeton, New Jersey. 264 pp.

Long, J. 1791. Voyages and travels of an Indian interpreter and trader . . . . Printed for the author, London. (Reprinted in 1904 *in* R.G. Thwaites, ed. Early western travels . . . . Vol. II. The Arthur H. Clark Co., Cleveland.)

Longfellow, H.W. 1855. The song of Hiawatha. Ticknor and Fields, New York.

Loranger, A.J., T.N. Bailey and W.W. Larned. 1991. Effects of forest succession after fire in moose wintering habitats on the Kenai Peninsula, Alaska. Alces 27: 100–109.

Lorenz, K. 1954. Das angeborene Erkennen. Natur und Volk 84: 285–295.

———. 1959. Gestaltwahrnehmung als Quelle wissenschaftlichter Erkenntnis. Z. Exper. Angew. Psychol. 6: 118–165.

———. 1965. Darwin hat recht gesehen. Opuscula aus der Wissensch. u. Dichtung. 75 pp.

———. 1966. Evolution and modification of behavior. 2nd ed. Methuen and Co., London. 121 pp.

———. 1974. Analogy as a source of knowledge. Science 185: 229–234.

———. 1982. Die acht Todsünden der zivilisierten Menschheit. R Piper and Co. Verlag, Munchen. 112 pp.

Lorenz, K. and P. Leyhausen. 1968. Antriebe tierischen und menschlichen Verhaltens. Gesammelte Abhandlungen PC R. Piper and Co.Verlag, Münich. 472 pp.

Lorenz, K. and P. Leyhausen. 1973. Motivation of human and animal behavior. An ethological view. Van Nostrand Reinhold Co., New York. 423 pp.

Loudon, A.S.I. and J.D.Curlwiss. 1988. Cycles of antler and testicular growth in an aseasonal tropical deer *(Axis axis)*. J. Reprod. Fertil. 83: 729–738.

Ludewig, H.A. and R.T. Bowyer. 1985. Overlap in winter diets of sympatric moose and white-tailed deer in Maine. J. Mammal. 66: 390–392.

Lugger, O. 1896. Gad-flies. Breeze-flies (Tabanidae). Rept. St. Entomol. Minn. No. 2: 166–170.

Lukaski, H.C. 1987. Methods for the assessment of human body condition: Traditional and new. Am. J. Clin. Nutr. 46: 537–556.

Lumsden, H.G. 1959. Ontario moose inventory, winter 1958-59. Ontario Dept. Lands and For., Toronto. 11 pp.

Lunberg, K. 1956. Algens forstlingshorn. Svensk Jakt 94: 417, 423.

Lundberg, P. and M. Astron. 1990. Functional response of optimally foraging herbivores. J. Theor. Biol. 144: 367–377.

Lundberg, P. and K. Danell. 1990. Functional response of browsers: Tree exploitation by moose. Oikos 58: 378–384.

Lundberg, P., M. Astrom and K. Danell. 1990. An experimental test of frequency-dependent food selection: Winter browsing of moose. Holarctic Ecol. 13: 177–182.

Lutz, H.J. 1960. History of the early occurence of moose on the Kenai Peninsula and in other sections of Alaska. Res. Center Misc. Publ. 1. USDA For. Ser., Juneau. 25 pp.

Lydekker. 1898. The deer of all lands: A history of the family Cervidae living and extinct. R. Ward, London. 329 pp.

Lykke, J. and I. M. Cowan. 1968. Moose management and population dynamics on the Scandanavian Peninsula, with special reference to Norway. Proc. N. Am. Moose Conf. Workshop 5: 1–18.

Lynch, G.M. 1971. Ungulate population surveys conducted in the Edison region. Alberta Dept. Lands and For., Edmonton. 37 pp.

———. 1973. Influence of hunting on an Alberta moose herd. Proc. N. Am. Moose Conf. Workshop 9: 123–135.

———. 1976. Some long-range movements of radio-tagged moose in Alberta. Proc. N. Am. Moose Conf. Workshop 12: 220–235.

———. 1986. A practical population model for big game managers. Fish and Wildl. Div. Rept., Alberta For., Lands and Wildl., Edmonton. 23 pp.

Lynch, G.M. and L.E. Morgantini. 1984. Sex and age differential in seasonal home range of moose in northwestern Alberta. Alces 20: 61–78.

Lynch, G.M. and G.E. Shumaker. 1995. GPS and GIS assisted moose surveys. Alces 31: 145–151.

MacArthur, R.H. and O.E. Wilson. 1967. The theory of island biogeography. Monogr. in population biology. Princeton Univ. Press, Princeton, New Jersey. 268 pp.

MacCracken, J.G. 1992. Ecology of moose on the Copper River Delta, Alaska. Ph.D. thesis, Univ. Idaho, Moscow. 338 pp.

Mace, G.M., P.H. Harvey and T.H. Clutton-Brock. 1983. Vertebrate home-range size and energetic requirements. Pages 32–53 *in* I.R. Swingland and P.J. Greenwood, eds., The ecology of animals movement. Clarendon Press, Oxford.

MacGregor, W.C. 1987. Moose *(Alces alces)* management and wolf control in British Columbia. Swed. Wildl. Res. (Suppl.) 1: 767–769.

Machida, S. 1979. Differential use of willow species by moose in Alaska. M.S. thesis, Univ. Alaska-Fairbanks. 96 pp.

Mackenzie, A. 1801. Voyages from Montreal, on the river St. Laurence, through the continent of North America to the frozen and Pacific Oceans; in the years 1789 and 1793. T. Cadell, London, England. (Reprinted in 1970 as The journals and letters of Sir Alexander Mackenzie, W.K. Lamb, ed. Cambridge Univ. Press, Cambridge, England. 551 pp.)

Mackie, R.J. 1981. Interspecific relationships. Pages 487–508 *in* O.C. Wallmo, ed., Mule and black-tailed deer of North America. Univ. Nebraska Press, Lincoln. 605 pp.

MacLennan, R.R. 1974. Reasons for the decline of Saskatchewan's moose population. Proc. N. Am. Moose Conf. Workshop 10: 63–75.

———. 1978. Preliminary evaluation of Saskatchewan's new moose management program. Can. Wildl. Admin. 2: 23–24.

MacNally, L. 1992. Bent-nosed roebuck. Deer 8: 538.

MacNeish, R.S. 1959a. Men out of Asia: As seen from the northwest Yukon. Anthropol. Pap. of Univ. Alaska 7(2): 41–70.

———. 1959b. A speculative framework of northern North America prehistory as of April 1959. Anthropol. 1(1–2): 7–23.

———. 1963. The early peopling of the New World—As seen from the southwestern Yukon. Anthropol. Pap. of Univ. of Alaska. 10(2): 93–106.

Macoun, J. 1882. Manitoba and the great Northwest. The World Publ. Co., Guelph, Ontario. 687 pp.

Madden, D.J., T.R. Spraker and W.J. Adrian. 1991. *Elaeophora schneideri* in moose *(Alces alces)* from Colorado. J. Wildl. Dis. 27: 340–341.

Maehlum, J. 1981. Elgkalv in Januar. Fauna 34: 131.

Mahoney, S. 1983. The trend toward bio-politics. A Newfoundland case study. Pages 28–35 *in* Wildlife management today and tomorrow. Can. Wildl. Admin., Ottawa.

Maine Department of Inland Fisheries and Wildlife. n.d. Care of your moose, from field to freezer. Maine Dept. Inland Fish.and Wildl., Augusta. 4 pp.

———. 1990. Final report, moose census testing 1989–90. Maine Dept. Inland Fish. and Wildl., Augusta. 24 pp.

———. 1991. 1991 update, moose assessment (1986–90) and strategic plan. Maine Dept. Inland Fish. and Wildl., Augusta. Pages 55–62.

Malende, H.J. 1988. Beobachtungen an markiertem Damwild I. Teil. Wild und Hund 91: 46–49.

Malkmus, B. 1901. Outlines of the clinical diagnosis of the internal diseases of domestic animals. Alexander Eger, Inc., Chicago. 205 pp.

Mallery, G. 1893. Picture-writing of the American Indians. Tenth Ann. Rept. of Bur. Ethnol. to Sec. of the Smithsonian Inst., 1888–1889. U.S. Govt. Print. Off., Washington, D.C. 822 pp.

Mangel, M. 1991. Adaptive walks on behavioural landscapes and the evolution of optimum behaviour by natural selection. Ecology 72: 30–39.

Manning, J.T. 1089. Age advertisement and the evolution of the peacock's tail. J. Evol. Biol. 2: 379–384.

Manweiler, J. 1941. The future of Minnesota's moose. Conserv. Volunteer 3: 38–45.

Marchinton, R.L., J.R. Fudge, J.C. Fortson, K.W. Miller and D.A. Dobie. 1991. Genetic and environment as factors in production of record class antlers. Pages 325–318 *in* B. Bobek, K. Perzanowski and W.L. Regelin, eds., Global trends in wildlife management. Vol. 1. Trans. Cong. Int. Union Game Biol. Swiat Press, Krakow-Warszawa.

Marcy, R.B. 1866. Thirty years of Army life on the border. J.B. Lippincott Co., Philadelphia. 266 pp.

Marest, Fr. G. 1694. Letter from Father Marest, missionary of the Company of Jesus, to Father De Lamberville of the Company of Jesus, overseer of the missions of Canada. Pages 105–142 *in* J.B. Tyrrell, ed., Documents relating to the early history of Hudson Bay. 1931. The Champlain Soc., Toronto, Ontario. 419 pp.

Markgren, G. 1964. Puberty, dentition and weight of yearling moose in a Swedish county. Viltrevy 2: 409–416.

———. 1966. A study of hand reared moose calves. Viltrevy 4: 1–35.

———. 1969. Reproduction of moose in Sweden. Viltrevy 6: 127–299.

———. 1975. Winter studies on orphaned moose calves in Sweden. Viltrevy 9: 193–218.

Marma, B. 1972. K voprosu nenormalnoga rosta kopyt i rogov v oslovi-akh zooparkov. Pages 287–289 *in* 14th Int. Symp. Erkrank. Zootier, Wroclaw.

Marshall, F.H.A. 1936. Sexual periodicity and the causes which determine it. Philos. Trans. R. Soc. Lond., Ser.B. 226: 423–456.

———. 1937. On the change over in the oestrous cycle in animals after transference across the equator, with further observations on the incidence of the breeding seasons and the factors controlling sexual periodicity. Proc. R. Soc. Lond., Ser. B. 122: 413–428.

Marston, H.R. 1948. Energy transactions in the sheep. I. The basal heat production and heat increment. Aust. J. Res. Bull. 1: 93–129.

Martin, C. 1978a. Keepers of the game: Indian-animal relationships and the fur trade. Univ. California Press, Berkeley.

———. 1978b. The war between Indians and the animals. Nat. Hist. 87(6): 92–96.

Martinka, C.J. 1969. Population ecology of summer resident elk in Jackson Hole, Wyoming. J. Wildl. Manage. 33: 465–481.

Mason, J.A. 1946. Notes on the Indians of the Great Slave Lake area. Yale Univ. Public. in Anthropol. 34. New Haven, Connecticut. 43 pp.

Mason, O.T. 1896. Primitive travel and transportation. Rept. of the U.S. Natl. Mus. for 1894: 239–593.

Mason, W. 1979. Ontogent of social behavior. Pages 1–28 *in* P. Marler and

J.C. Vandenbergh, eds., Handbook of behavioral neurobiology, Vol. 3: Social behavior and communication. Plenum Press, New York.

Mastenbrook, B. and H.G. Cumming. 1989. Use of residual strips of timber by moose within cutovers in northwestern Ontario. Alces 25: 146–155.

Masyk, J. 1975. Wildlife management in national parks, past and future. Proc. N. Am. Moose Conf. Workshop 11: 264–276.

Matchett, M.R. 1985. Habitat selection by moose in the Yak River drainage, northwestern Montana. Alces 21: 161–190.

Matsell, G.W. 1859. Vocabulum; or the rogue's lexicon. G.W. Matsell and Co., New York. 130 pp.

Matson, P.A. and M.D. Hunter. 1992. The relative contributions of top-down and bottom-up forces in population and community ecology. Ecol. 73: 723.

Mautz, W.W. 1971. Confinement effects on dry matter digestion coefficients displayed by deer. J. Wildl. Manage. 35: 366–368.

———. 1978. Nutrition and carrying capacity. Pages 321–348 *in* J.L. Schmidt and D.L. Gilbert, eds., Big game of North America: Ecology and management. Stackpole Books, Harrisburg, Pennsylvania. 494 pp.

Mautz, W.W. and J. Fair. 1980. Energy expenditure and heart rate for activities of white-tailed deer. J. Wildl. Manage. 44: 333–342.

Mautz, W.W., P.J. Perkins and J.A. Warren. 1985. Cold temperature effects on metabolic rate of white-tailed deer, mule deer and black-tailed deer in winter. Pages 453–457 *in* P.F. Fennessy and K.R. Drew, eds., Biology of deer production. Bull. No. 22. R. Soc. of New Zealand, Wellington.

Maximilian, Prince of Wied. 1843. Travels in the interior of North America. Ackermann and Co., London, England. 520 pp.

Maxwell, J.A., ed. 1978. America's fascinating Indian heritage. The Reader's Digest Assoc., Inc. Pleasantville, New York. 416 pp.

Mayer, M.J. 1989. Klondike women: True tales of the 1897–98 gold rush. Ohio Univ. Press, Athens. 267 pp.

Mayr, E. 1970. Populations, species and evolution. Harvard Univ. Press, Cambridge, Massachusetts. 453 pp.

Maze, R.J. and C. Johnstone. 1986. Gastropod intermediate hosts of the meningeal worm *Parelaphostrongylus tenuis* in Pennsylvania: Observations on their ecology. Can. J. Zool. 64: 185–188.

McArthur, C., A. Hagerman and C.T. Robbins. 1991. Physiological strategies of mammalian herbivores against plant defenses. Pages 103–114 *in* R.T. Palo and C.T. Robbins, eds., Plant defenses against mammalian herbivory. CRC Press, Boca Raton, Florida.

McArthur, P.H. and E.R. Pianka. 1966. An optimal use of a patchy environment. Am. Nat. 100: 603–609.

McCabe, R.E. 1982. Elk and Indians: Historical values and perspectives. Pages 60–123 *in* J. W. Thomas and D.E. Toweill, eds., Elk of North America: Ecology and management. Stackpole Books, Mechanicsburg, Pennsylvania. 698 pp.

McClellan, C. 1981a. Inland Tlingit. Pages 469–480 *in* J. Helm, ed., Subarctic. Handbook of North American Indians. Vol. 6. Smithsonian Inst., Washington, D.C. 837 pp.

———. 1981b. Intercultural relations and cultural change in the Cordillera. Pages 387–401 *in* J. Helm, ed., Subarctic. Handbook of North American Indians. Vol. 6. Smithsonian Inst., Washington, D.C. 837 pp.

———. 1981c. Tagish. Pages 481–492 *in* J. Helm, ed., Subarctic. Handbook of North American Indians. Vol. 6. Smithsonian Inst., Washington, D.C. 837 pp.

———. 1981d. Tutchone. Pages 493–505 *in* J. Helm, ed., Subarctic. Handbook of North American Indians. Vol. 6. Smithsonian Inst., Washington, D.C. 837 pp.

McClellan, C. and G. Denniston. 1981. Environment and culture in the Cordillera. Pages 372–386 *in* J. Helm, ed., Subarctic. Handbook of North American Indians. Vol. 6. Smithsonian Inst., Washington, D.C. 837 pp.

McCollough, M.A. and K.A. Pollard. 1993. *Parelaphostrongylus tenuis* in Maine moose and the possible influence of faulty Baermann procedures. J. Wildl. Dis. 29: 156–158.

McCorquodale, S.M. and R.F. DiGiacoma. 1985. The role of wild North American ungulates in epidemiology of bovine brucellosis: A review. J. Wildl. Dis. 21: 351–357.

McCullough, D.R. 1979. The George Reserve deer herd. Univ. Michigan Press, Ann Arbor. 271 pp.

———. 1984. Lessons from the George Reserve, Michigan. Pages 211–242 *in* L.K. Halls, ed., White-tailed deer: Ecology and management. Stackpole Books, Harrisburg, Pennsylvania. 870 pp.

———. 1990. Detecting density dependence: Filtering the baby from the bathwater. Trans. N. Am. Wildl. and Nat. Resour. Conf. 55: 534–543.

McCullough, D.R., C.E. Olsen, Jr. and L.M. Queal. 1969. Progress in large animal census by thermal mapping. Pages 138–147 *in* P.L. Johnson, ed., Remote sensing in ecology. Univ. Georgia Press, Athens. 244 pp.

McDonald, M.G. 1991. Moose movement and mortality associated with the Glenn Highway expansion, Anchorage, Alaska. Alces 27: 208–219.

McEwan, B.S., P.G. Davis, B. Parsons and D.W. Pfaff. 1979. The brain as a target for steroid hormone action. Ann. Rev. Neurosci. 2: 65–112.

McEwan, E.H. and P.E. Whitehead. 1970. Seasonal changes in the energy and nitrogen intake in reindeer and caribou. Can. J. Zool. 48: 905–913.

McFarland, W.N., F.H. Pough, T.J. Cade and J.B. Heiser. 1985. Vertebrate life. Macmillan Publ. Co., New York. 636 pp.

McGillis, J.R. 1972. The kidney fat index as an indicator of condition in various age and sex classes of moose. N. Am. Moose Conf. Workshop 8: 105–114.

McGinley, M.A. 1984. The adaptive value of male-biased sex ratios among stressed animals. Am. Nat. 124: 597–599.

McIntosh, M. 1993. Lord of the lakes. Wildl. Art News XII(6): 120–126.

McIntyre, E.G. 1972. Barkstripping—A natural phenomenon. J. Scottish For. Soc. 26: 43–50.

McKenna, M.G. and B. Lynort. 1984. Taking the offensive in wildlife management. Wildl. Soc. Bull. 12: 79–81.

McKennan, R.A. 1959. The Upper Tanana Indians. Yale Univ. Publ. in Anthropol. 55. New Haven, Connecticut.

———. 1981. Tanana. Pages 562–576 *in* J. Helm, ed., Subarctic. Handbook of North American Indians. Vol. 6. Smithsonian Inst., Washington, D.C. 837 pp.

McLaren, B.E. and R.O. Peterson. 1994. Wolves, moose, and tree rings on Isle Royale. Science 266(5190): 1,555–1,558.

McLaughlin, R.F. and E.M. Addison. 1986. Tick *(Dermacentor albipictus)*—Induced winter hairloss in captive moose *(Alces alces)*. J. Wildl. Dis. 22: 502–510.

McMillan, J.F. 1953a. Measures of association between moose and elk on feeding grounds. J. Wildl. Manage. 17: 162–166.

———. 1953b. Some feeding habits of moose in Yellowstone National Park. Ecol. 34: 102–110.

McNab, B.K. 1963. Bioenergetics and the determination of home range. Am. Nat. 97: 130–140.

———. 1983. Ecological and behavioral consequences of adaptation to various food resources. Pages 664–697 *in* J.F. Eisenberg and D.G. Kleiman, eds., Advances in the study of mammalian behavior. Spec. Publ. 7. Am. Soc. Mammal. Dept. Zoology, Univ. Florida, Gainesville.

McNay, M. 1989a. Moose survey-inventory progress report 1987-1988, Game Management Unit 20A. Pages 212–224 *in* S.O. Morgan, ed.,

Annual report of survey- inventory activities. Moose. Fed. Aid in Wildl. Restor. Rept. Alaska Dept. Fish and Game, Juneau.

———. 1989b. Wolf survey-inventory progress report 1987–1988, Game Management Units 20A, 20B, 20C, 20F, and 25C. Pages 88–102 *in* S.O. Morgan, ed. Annual report of survey-inventory activities. Wolf. Fed. Aid in Wildl. Restor. Rept. Alaska Dept. Fish and Game, Juneau.

McNeill, M.A. and M.E. Rau. 1987. *Echinicoccus granulosus* (Cestoda: Taeniidae) infections in moose *(Alces alces)* from southwestern Quebec. J. Wildl. Dis. 23: 418–421.

McNicol, J. 1990. Moose and their environment. Pages 11–18 *in* Ontario. Moose biology, ecology and management. Book 1. Ontario Min. Natur. Resour., Toronto.

McNicol, J.G., H.R. Timmermann and R. Gollat. 1980. The effects of heavy browsing pressure over eight years on a cutover in Quetico Park. Proc. N. Am. Moose Conf. Workshop 16: 360–373.

McRoberts, R.E., L.D. Mech and R.O. Peterson. 1995. The cumulative effect of consecutive winters' snow depth on moose and deer populations. J. Anim. Ecol. 64: 131–135.

Mech, L. D. 1966. The wolves of Isle Royale. Fauna Ser. No. 7. U.S. Natl. Park Serv. U.S. Govt. Printing Off., Washington, D.C. 210 pp.

———. 1970. The wolf: The ecology and behavior of an endangered species. Nat. Hist. Press, Garden City, New York. 384 pp.

———. 1983. Handbook of animal radio-tracking. Univ. Minnesota Press, Minneapolis. 107 pp.

———. 1987. Age, season, distance, direction, and social aspects of wolf dispersal from a Minnesota pack. Pages 55–74 *in* B.D. Chepko-Sade and Z. Halpin, eds., Mammalian dispersal patterns. Univ. Chicago Press, Chicago, Illinois. 342 pp.

Mech, L.D. and P.D. Karns. 1977. Role of wolf in a deer decline in Superior National Forest. Res. Pap. NC-148. USDA For. Serv., St. Paul, Minnesota. 23 pp.

Mech, L.D., R.E. McRoberts, R.O. Peterson and R.E. Page. 1987. Relationships of deer and moose populations to previous winters snow. J. Anim. Ecol. 56: 615–627.

Mech, L.D., T.J. Meier, J.W. Burch and L.G. Adams. 1995. Patterns of prey selection by wolves in Denali National Park, Alaska. Pages 231–243 *in* L.N. Carbyn, S.H. Fritts and D. Seip, eds., Ecology and conservation of wolves in a changing world. Proc. 2nd Int. Wolf Symp. Canadian Circumpolar Inst., Univ. Alberta, Edmonton.

Meile, P. and A.B. Bubenik. 1979. Zur Bedeutung sozialer Auslöser für das Sozialverhalten der Gemse (*Rupicapra rupicapra* L.). Säugetierkdl. Mitt. (Sonderheft: Gemsen). 42 pp.

Meister, W. 1956. Changes in the histological structure of the long bones of white-tailed deer *(Odocoileus virginianus)* during the growth of antlers. Anat. Rec. 116: 701–729.

Mercer, W.E. and D.A. Kitchen. 1968. A preliminary report on the extension of moose range in the Labrador Peninsula. Proc. N. Am. Moose Conf. Workshop 5: 62–81.

Mercer, W.E. and F. Manuel. 1974. Some aspects of moose management in Newfoundland. Nat. Can. (Que.) 101: 657–671.

Mercer, W.E. and M. Strapp. 1978. Moose management in Newfoundland 1972–1977. Proc. N. Am. Moose Conf. Workshop 14: 227–246.

Meredith, M. 1980. The vomeronasal organ and the accessory olfactory system in the hamster. Pages 303–326 *in* D. Müller-Schwarze and R.M. Silverstein, eds., Chemical signals. Vertebrates and invertebrates. Plenum Press, New York.

Merrill, S. 1916. The moose book, facts and stories from northern forests. E.P. Dutton and Co., New York. 399 pp.

———. 1920. The moose book, facts and stories from the northern forests. Rev. ed. E.P. Dutton and Co., New York. 399 pp.

Mertens, D.R. 1973. Application of theoretical mathematical models to cell wall digestion and forage intake in ruminants. Ph.D. thesis, Cornell Univ., Ithaca, New York. 187 pp.

Messier, F. 1991. The significance of limiting and regulating factors on the demography of moose and white-tailed deer. J. Anim. Ecol. 60: 377–393.

———. 1994. Ungulate population models with predation: A case study with the North American moose. Ecol. 75: 478–488.

Messier, F. and M. Crête. 1984. Body condition and population by food resources in moose. Oecologia 65: 44–50.

Messier, F. and M. Crête. 1985. Moose–wolf dynamics and the natural regulation of moose populations. Oecologia 65: 503–512.

Messier, F., M.E. Rau and M.A. McNeill. 1989. *Echinococcus granulosus* (Cestoda:Taeniidae) infections and moose–wolf population dynamics in southwestern Quebec. Can. J. Zool. 67: 216–219.

Metcalfe, J., M.K. Stock and D.H. Barron. 1988. Maternal physiology during gestation. Pages 2,145–2,176 *in* E. Knobil and J.D. Neill, eds., The physiology of reproduction. Raven Publishing Co., New York.

Michael, H.N., ed. 1967. Lieutenant Zagoskin's travels in Russian America, 1842–1844. Arctic Inst. North America. Univ. Toronto Press, Toronto, Ontario. 358 pp.

Milchunas, D.G. and D.L. Baker. 1982. In vitro digestion—Sources of within and between trial variability. J. Range Manage. 35: 199–203.

Milchunas, D.G., M.I. Dyer, O.C. Wallmo and D.E. Johnson. 1978. In-vivo/in-vitro relationships of Colorado mule deer forages. Spec. Rept. No. 43. Colorado Div. Wildl., Fort Collins. 44 pp.

Milke, G.C. 1969. Some moose–willow relationships in the interior of Alaska. M.S. thesis, Univ. Alaska-Fairbanks. 79 pp.

Millais, J.G. 1915. Moose, caribou, wapiti, white-tailed and mule deer. Pages 235–340 *in* D. Carruthers, P.B. Van der Byl, R.L. Kennion, J.G. Millais, H.F. Wallace and F.G. Barclay, eds., The gun, home and abroad: Big game of Asia and North America. The London and Counties Press Assoc. Ltd., Covent Garden, England.

Millar, J.B. 1953. An ecological study of the moose in the Rock Creek area of Alberta. M.S. thesis, Univ. Alberta, Edmonton. 123 pp.

Miller, G.S., Jr. 1899. A new moose from Alaska. Proc. Biol. Soc. Wash. 13: 57–59.

Miller, B.K. and J.A. Litvaitis. 1992. Habitat segregation in a boreal forest ecotone. Acta Ther. 37: 41–50.

Miller, F.L., E. Broughton and E.M. Land. 1972. Moose fatality from over-extension of range. J. Wildl. Dis. 8: 95–98.

Miller, K.V., R.L. Marchinton and P.B. Bush. 1985. Variations in density and chemical composition of white-tailed deer antlers. J. Mammal. 66: 693–701.

Miller, K.V., R.L. Marchinton and P.B. Bush. 1991. Signpost communication by white-tailed deer: Research since Calgary. Appl. Anim. Behav. Sci. 29: 195–204.

Miller, N.L., J.K. Frenkel and J.P. Dubey. 1972. Oral infections with toxoplasma cysts and oocysts in felines, other mammals, and in birds. J. Parasitol. 58: 928–937.

Miller, S.D. and W.B. Ballard. 1992. Analysis of efforts to increase moose calf survival in south-central Alaska by increasing sport harvests of brown bear. Wildl. Soc. Bull. 20: 445–454.

Miller, V.P. 1986. The Micmac: A maritime Woodland group. Pages 347–374 *in* R.B. Morrison and C.R. Wilson, eds., Native peoples: The Canadian experience. McClelland and Stewart Inc., Toronto. 639 pp.

Milne, A. 1957. The natural control of insect populations. Can. Ent. 89: 193–213.

Milner, R. 1996. Crossing guard. Nat. Hist. 7: 68–69.

Milton, V. and W.B. Cheadle. 1865. The north-west passage by land: Being the narrative of an expedition from the Atlantic to the Pacific. Cassell, Petter, and Galpin, London.

Minnesota Department of Natural Resources. 1990. Moose. Pages17.1–17.17 *in* Public review draft. Chap. 17. Minnesota Dept. Nat. Resour., St. Paul.

Minty, D. 1989. Education and public participation. Workshop session. Alces 25: 167.

Miquelle, D.G. 1990a. Why don't bull moose eat during the rut? Behav. Ecol. Sociobiol. 27: 145–151.

———. 1990b. Behavioral ecology of moose in Denali National Park and Preserve, Alaska. Ph.D. thesis, Univ. Idaho, Moscow. 152 pp.

———. 1991. Are moose mice? The function of scent urination in moose. Am. Nat. 138: 460–477.

Miquelle, D.G. and P.A. Jordan. 1979. The importance of the diversity of the diet of moose. Proc. N. Am. Moose Conf. Workshop 15: 54–79.

Miquelle, D.G. and V. Van Ballenberghe. 1985. The moose bell: A visual or olfactory communicator? Alces 21: 191–214.

Miquelle, D.G. and V. Van Ballenberghe. 1989. Impact of bark stripping by moose on aspen–spruce communities. J. Wildl. Manage. 53: 577–586.

Miquelle, D.G., J.M. Peek and V. Van Ballenberghe. 1992. Sexual segregation in Alaskan moose. Wildl. Monogr. 122. The Wildl. Soc., Bethesda, Maryland. 57 pp.

Mitchell, H.B. 1970. Rapid aerial sexing of antlerless moose in British Columbia. J. Wildl. Manage. 34: 645–646.

Modaferri, R.D. 1982. Susitna hydroelectric project, big game studies. Vol. II. Moose—Downstream. Phase I Final Rept. Alaska Dept. Fish and Game, Juneau. 114 pp.

———. 1991. Train–moose kill in Alaska: Characteristics and relationship with snowpack depth and moose distribution in Lower Susitna Valley. Alces 27: 193–207.

———. 1992. In utero pregnancy rate, twinning rate, and fetus production for age-groups of cow moose in southcentral Alaska. Alces 28: 233–234.

Moen, A. 1973. Wildlife ecology: An analytical approach. W.H. Freeman and Co., San Francisco, California. 458 pp.

Moen, A. and F. Ausenda. 1987. Sensitive population parameters in modelling long-lived species such as moose. Alces 23: 33–47.

Mohr, C.O. 1947. Table of equivalent populations of North American small mammals. Am. Midl. Nat. 37: 223–249.

Mohr, C.O. and W.A. Stumpf. 1966. Comparison of methods for calculating areas of animal activity. J. Wildl. Manage. 30: 293–303.

Moisan, G. 1952. Investigations on moose populations. Pages 25–43 *in* General report of the Minister of Game and Fisheries for the fiscal years 1950–51 and 1951–52. Quebec Dept. Fish and Game, Quebec City.

Moldenhauer, J.J. ed. 1974. The illustrated Maine woods. Princeton Univ. Press, Princeton, New Jersey. 346 pp.

Moll, D. and B.K. Moll. 1976. Moose triplets on Isle Royale. Trans. Illinois St. Acad. Sci. 69:151–152.

Molvar, E.M. and R.T. Bowyer. 1994. Costs and benefits of group living in a recently social ungulate: The Alaskan moose. J. Mammal. 73: 621–630.

Monaghan, J. 1963. The book of the American West. Simon and Schuster, New York. 608 pp.

Monfort, S.L., C.C. Schwartz and S.K. Wasser. 1993. Monitoring reproduction in moose using urinary and fecal steroid metabolites. J. Wildl. Manage. 57: 400–407.

Mongrain, R. 1986. The great hunt, a handbook for moose hunters. St.Tite, Quebec. 115 pp.

Monk, C. and M. Buss. 1990. Moose hunting basics. Pages 60–67 *in* M. Buss and R. Truman, eds., The moose in Ontario. Book II: Moose hunting techniques, hunting ethics and the law. Ontario Min. Nat. Resour. and Ontario Fed. of Anglers and Hunters, Toronto. Ontario. 78 pp.

Montana Department of Fish, Wildlife and Parks. 1992. End of year project report/job progress report FY 92. Montana Dept. Fish, Wildl. and Parks, Helena. 21 pp.

Mooney, J. 1928. The aboriginal population of America north of Mexico. Smithsonian Misc. Coll. 80(7). Washington, D.C.

Mooring, M.S. and B.L. Hart. 1955. Costs of allogrooming in impala: Distraction from vigilance. Anim. Behav. 49: 1,414–1,416.

Mordosov, I.I. 1990. To the ecology of the *Alces alces* L. in Yakutia. Page 114 *in* N.E. Kochanov, ed., Abstracts of papers and posters. 3rd Int. Moose Symp. Syktyvar, USSR.

Morgan, L.H. 1954. League of the Ho-De-No Sau-Nee or Iroquois. 2 vols. Reprinted for Hum. Relat. Area Files, New Haven, Connecticut.

Morice, A.G. 1895. Notes archaeological, industrial, and sociological on the western Dènès with an ethnological sketch of the same. Trans. Canadian Inst. 4:1–222.

Morris, E.J. 1984. Degredation of the intact plant cell wall of subtropical and tropical herbage by rumen bacteria. Pages 378–398 *in* F.M.C. Gilchrist and R.I. Mackie, eds., Herbivore nutrition in the tropics and subtropics. The Science Press, Craighill, England.

Morris, K. and K. Elowe. 1993. The status of moose and their management in Maine. Alces 29: 91–98.

Morrison, J.A. 1960. Ovarian characteristics in elk of known breeding history. J. Wildl. Manage. 24: 297–307.

Morrow, M. 1975. Indian rawhide: An American folk art. Univ. Oklahoma Press, Norman. 243 pp.

Morton, T. 1632. New English Canaan or New Canaan, containing an abstract of New England. 3 books, 2nd book. Jacob Frederick Stam, Amsterdam.

Mould, E.D. 1977. Habitat relationships of moose in northern Alaska. Proc. N. Am. Moose Conf. Workshop 13: 144–156.

———. 1979. Seasonal movements related to habitat of moose along the Colville River, Alaska. Murrelet 60: 6–11.

Mould, E.D. and C.T. Robbins. 1981. Nitrogen metabolism in elk. J. Wildl. Manage. 45: 323–334.

Mould, E.D. and C.T. Robbins. 1982. Digestive capabilities in elk compared to white-tailed deer. J. Wildl. Manage. 46: 22–29.

Moulton, G.E., ed. 1986–1990. The journals of Lewis and Clark expedition. 6 vols. Univ. Nebraska Press, Lincoln.

Mrosovsky, N. and T.L. Posley. 1977. Set points for body weight and fat. Behav. Biol. 20: 205–223.

Mueller, R.J. 1964. A short illustrated topical dictionary of Western Kutchin. Summer Inst. Linguist., Fairbanks, Alaska.

Muir, P.D. and A.R. Sykes. 1988. Effect of winter nutrition on antler development in red deer *(Cervus elaphus)*: A field study. New Zealand J. Agric. Res. 31: 145–150.

Muir, P.D., A.R. Sykes and G.K. Barrell. 1987a. Calcium metabolism in red deer *(Cervus elaphus)* offered herbages during antlerogenesis: Kinetic and stable balance studies. J. Agric. Sci. 109: 357–364.

Muir, P.D., A.R. Sykes and G.K. Barrell. 1987b. Growth and mineralisation of antlers in red deer *(Cervus elaphus)*. New Zealand J. Agric. Res. 30: 305–315.

Müller-Schwartze, D. 1991. The chemical ecology of ungulates. Appl. Anim. Behav. Sci. 29: 389–402.

Müller-Schwartze, D., N.J. Volkman and K.F. Zemanek. 1977. Osmetrichia: Specialized scent hair in black-tailed deer. J. Ultrastr. Res. 59: 223–230.

Müller-Using, D. and R. Schloeth. 1967. Das Verhalten der Hirsche (Cervidae). Pages 1–60 *in* J.G. Helmcke, H.V. Lengerken, D. Starck and H. Wermuth, eds., Handbuch der Zoologie. Vol. 8. W. de Gruyter and Co., Berlin.

Munro, J.A. 1947. Observations of birds and mammals in central British Columbia. Occas. Pap. No. 6. British Columbia Prov. Mus., Victoria. 165 pp.

Murdock, G.P. and T.J. O'Leary. 1975. Ethnographic bibliography of North America. 5 vols., 4th ed. Hum. Relat. Area Files Press, New Haven, Connecticut.

Murie, A. 1934. The moose of Isle Royale. Misc. Publ. No. 25. Univ. Michigan Mus. Zool., Ann Arbor. 44 pp.

———. 1944. The wolves of Mount McKinley. Fauna Ser. No. 5. U.S. Natl. Park Serv., U.S Govt. Printing Off., Washington, D.C. 238 pp.

———. 1961. A naturalist in Alaska. Devin-Adair, New York. 302 pp.

———. 1981. The grizzlies of Mount McKinley. Sci. Monogr. Ser. No. 14. U.S. Dept. Int., U.S. Govt. Print. Off., Washington, D.C. 251 pp.

Murie, O.J. 1928. Abnormal growth of moose antlers. J. Mammal. 9: 65.

Murray, J. 1984. The last buffalo: The story of Frederick Arthur Verner, painter of the Canadian West. Pagurian Press, Toronto, Ontario. 192 pp.

Murray, J.A.H. 1908. A new English dictionary on historical principles. 20 vols. Clarendon Press, Oxford, England.

Muzzi, P.D. and A.R. Bisset. 1990. The effectiveness of ultrasonic warning devices to reduce moose fatalities along railway corridors. Alces 26: 37–43.

Mytton, W.R. and L.B. Keith. 1981. Dynamics of moose populations near Rochester, Alberta, 1975–1978. Can. Field-Nat. 95: 39–49.

Nagpal, S.K., L.S. Sudhaker and Y. Singh. 1988. Anatomy of the vomeronasal organ in camel. Indian J. Anim. Sci. 58: 218–220.

Nasimovich, A.A. 1955. The role of the regime of snow cover in the life of ungulates in the USSR. Moskva Akad. Nauk SSR. 403 pp.

National Geographic Society. 1982. Indians of North America. Washington, D.C. (Map.)

National Meat Institute. 1970. The moose. Natl. Meat Inst., Inc., Montreal. 192 pp.

National Rifle Association. 1982. Basic hunter's guide. Natl. Rifle Assoc. of America, Washington, D.C. 300 pp.

Neff, D.J. 1968. The pellet group count technique for big game trend census and distribution: A review. J. Wildl. Manage. 32: 597–614.

Nelson, E.W. 1914. Description of a new subspecies of moose from Wyoming. Proc. Biol. Soc., Washington, D.C. 27: 71–74.

———. 1916. The larger North American mammals. Natl. Geog. Mag. 30(5): 385–472.

Nelson, L.J. and J.M. Peek. 1982. Effect of survival and fecundity on rate of increase in elk. J. Wildl. Manage. 46: 535–540.

Nelson, R.K. 1973. Hunters of the northern forest: Designs for survival among the Alaskan Kutchin. Univ. Chicago Press, Chicago, Illinois. 339 pp.

Nesbitt, W.H. and P.L. Wright, eds., 1981. Records of North American big game. 8th ed. The Boone and Crockett Club, Alexandria, Virginia. 409 pp.

Neu, C.W., C.R. Byers and J.M. Peek. 1974. A technique for analysis of utilization-availability data. J. Wildl. Manage. 38: 541–545.

Newfoundland and Labrador Wildlife Division. No date. Newfoundland and Labrador hunter capability and study guide. Infor. and Educ. Branch, Newfoundland and Labrador Wildl. Div. St. John's. 48 pp.

———. 1986. Preliminary results of a public green paper on hunting. Newfoundland and Labrador Wildl. Div., St. John's. 6 pp.

———. 1991. 1991 hunting guide. Newfoundland and Labrador Wildl. Div., St. John's. 64 pp.

———. 1992. Hunting guide. Newfoundland. AlC 5T7, Newfoundland and Labrador Wildl. Div., St. John's. 64 pp.

New Hampshire Fish and Game Department. No date. Moose field techniques and game care. New Hampshire Fish and Game Dept., Concord. 11 pp.

Newman, P.C. 1987. Caesars of the wilderness. Vol. II Company of adventurers, the story of the Hudson's Bay Company. Penguin Books Canada., Ltd., Markham, Ontario. 450 pp.

———. 1989. Empire of the Bay: An illustrated history of the Hudson's Bay Company. Madison Press Books, Toronto, Ontario. 223 pp.

Newton, J. and S.G. Hyslop, eds. 1992. The spirit world. Time-Life Books, Alexandria, Virginia. 176 pp.

Newton, J. and S.G. Hyslop, eds. 1995. Hunters of the northern forest. Time-Life Books, Alexandria, Virginia. 184 pp.

Nicholson, A.J. 1933. The balance of animal populations. J. Anim. Ecol. 2: 132–178.

Nicholson, M.C., R.T. Bowyer and J.G. Kie. 1997. Habitat selection and survival of mule deer: Tradeoffs associated with migration. J. Mammal. 78: 483–504.

Nicholoson, H.P. and C.E. Michel. 1950. The black flies of Minnesota (Simulinidae). Agric. Exp. Sta. Tech. Bull. No. 192. Univ. Minnesota, Minneapolis. 64 pp.

Noble, W.C. 1981. Prehistory of the Great Slave Lake and Great Bear Lake region. Pages 97–106 *in* J. Helm, ed. Subarctic. Handbook of North American Indians. Vol. 6. Smithsonian Inst., Washington, D.C. 837 pp.

Noji, S., T. Mohno, E. Koyana, K. Muto, K. Ohyana, Y. Aoki, K. Tamura, K. Ohsugi, H. Ide, S. Taniguchi and T. Saito. 1991. Retinoic acid induces polarizing activity but is unlikely to be a morphogen in the chicken limb bud. Nature 350: 83–86.

Nolan, T., L. Goldstein and J.W. Pfeiffer. 1993. Plan or die! 10 keys to organizational success. Pfeiffer and Co., San Diego, California. 179 pp.

Northcott, T.H. 1974. The land mammals of insular Newfoundland. Newfoundland Wildl. Div., Dept. of Tourism, St. John's. 90 pp.

Norton, T.E. 1974. The fur trade in colonial New York 1686–1776. Univ. Wisconsin Press, Madison. 243 pp.

Novak, M. 1978. Observations of moose in winter. Ontario Fish and Wildl. Rev. 14.4: 3–10.

Novak, M. and J. Gardner. 1975. Accuracy of aerial moose surveys. Proc. N. Am. Moose Conf. Workshop 11: 154–180.

Nowlin, R.A. 1978. Habitat selection and food habits of moose in northeastern Alberta. Proc. N. Am. Moose Conf. Workshop 14: 178–193.

———. 1985. Distribution of moose during occupation of vacant habitat in north central Colorado. Ph.D. thesis, Colorado St. Univ., Fort Collins. 60 pp.

Nudds, T.D. 1990. Retroductive logic in retrospect: The ecological effects of meningeal worms. J. Wildl. Manage. 54: 396–402.

———. 1992. Retroductive logic and the effects of meningeal worms: A reply. J. Wildl. Manage. 56: 617–619.

Nygren, K., S. Paatsama and B. Degritz. 1990. Structural variation in the nasal bone region of European moose (*Alces alces* L.). Acta Vet. Scand. 31: 385–391.

Nygren, K., R. Silvennoinen and M. Karna. 1992. Antler stress in the nasal bone region of the moose. Alces (Suppl.)1: 84–90.

Obergfell, F.A. 1957. Vergleichende Untersuchungen an Dentitionen und Dentale altburdigaler Cerviden von Wintershof-West in Bayern und rezenter Cerviden. Paleontographica 109: 71–166.

O'Callaghan, E.B. 1849–1851. The documentary history of the state of New York; arranged under the direction of the Hon. Christopher Monsan, Secretary of State. 4 vols. Weed, Parsons and Co., New York.

O'Connor, J. 1961. The big game animals of North America. E.P. Dutton and Co., New York. 264 pp.

Odum, E.P. 1959. Fundamentals of ecology. W.B. Saunders Co., Philadelphia. 546 pp.

Odum, E.P. 1983. Fundamentals of ecology, 3rd. ed. CBS College Publ., New York. 613 pp.

Oftedal, O.T. 1981. Milk, protein and energy intakes of suckling mammalian young: A comparative study. Ph.D. thesis, Cornell Univ., Ithaca, New York. 456 pp.

———. 1985. Pregnancy and lactation. Pages 216–238 *in* R.J. Hudson and R.G. White, eds., Bioenergetics of wild herbivores. CRC Press, Boca Raton, Florida.

O'Gara, B.W. 1990. The pronghorn *(Antilocapra americana)*. Pages 231–264 *in* G.A. Bubenik and A.B. Bubenik, eds., Horns, pronghorns and antlers. Springer-Verlag, New York.

Ohtaishi, N. and T. Too. 1974. The possible thermo-regulatory function and its character of the velvety antlers in the Japanese deer *(Cervus nippon)*. J. Mammal. Soc. Japan 6: 1–11.

Oldemeyer, J.L. 1975. Characteristics of paper birch saplings browsed by moose and snowshoe hares. Proc. N. Am. Moose Conf. Workshop 11: 53–62.

———. 1981. Estimation of paper birch production and utilization and an evaluation of its response to browsing. Ph.D. thesis, Pennsylvania St. Univ., College Park. 58 pp.

———. 1983. Browse production and its use by moose and snowshoe hares at the Kenai Moose Research Center, Alaska. J. Wildl. Manage. 47: 486–496.

Oldemeyer, J.L. and A.W. Franzmann. 1981. Estimating winter defecation rates for moose, *Alces alces*. Can. Field-Nat. 95: 208–209.

Oldemeyer, J.L. and W.L. Regelin. 1980. Response of vegetation to tree crushing in Alaska. Alces 16: 429–443.

Oldemeyer, J.L. and W.L. Regelin. 1987. Forest succession, habitat management, and moose on the Kenai National Wildlife Refuge. Swed. Wildl. Res. (Suppl.) 1: 163–179.

Oldemeyer, J.L. and R.K. Seemel. 1976. Occurrence and nutritive quality of lowbush cranberry on the Kenai Peninsula, Alaska. Can. J. Bot. 54: 966–970.

Oldemeyer, J.L., A.W. Franzmann, A.L. Brundage, P.D. Arneson and A. Flynn. 1977. Browse quality and the Kenai moose population. J. Wildl. Manage. 41: 533–542.

Oldmixon, J. 1741. The British Empire in America, containing the history of the discovery, settlement, progress, and state of the British colonies on the continent and islands of America. Vol. I. Printed for J. Brotherton, London, England. 567 pp.

Olsen, O.W. and R. Fenstermacher. 1942. Parasites of moose in northern Minnesota. Am. J. Vet. Res. 3: 403–408.

Olsen, S.J. 1964. Mammal remains from archaeological sites. Part I. Southeastern and southwestern United States. Peabody Mus., Cambridge, Massachusetts. 162 pp.

Olson, S.F. 1963. Runes of the north. Alfred A. Knopf, New York. 255 pp.

Olterman, J.H., D.W. Kenvin and R.C. Kufeld. 1994. Moose transplant to southwestern Colorado. Alces 30: 1–8.

O'Neil, L.J. and A.B. Carey. 1986. Introduction: When habitats fail as predictors. Pages 207–209 *in* J. Verner, M.L. Morrison and C.J. Ralph, eds., Wildlife 2000. Modelling habitat relationships of terrestrial vertebrates. Univ. Wisconsin Press, Madison.

Ontario Ministry of Natural Resources. 1980a. Moose management in Ontario, a report of open house public meetings. Wildl. Branch, Ontario Min. Nat. Resour., Toronto. 14 pp.

———. 1980b. Moose management policy. Wildl. Branch, Ontario Min. Nat. Resour., Toronto. 21 pp.

———. 1984. Moose hunter's handbook 1984. Wildl. Branch, Ontario Min. Nat. Resour., Toronto. 21 pp.

———. 1987. Improving the quality and enjoyment of moose hunting in Ontario: Background report. Wildl. Branch, Ontario Min. Nat. Resour.,Toronto. 25 pp.

———. 1988a. Moose anatomy for the hunter. Wildl. Branch, Ontario Min. Nat. Resour., Toronto. 7 pp.

———. 1988b. Firearms for the moose hunter. Wildl. Branch, Ontario Min. Nat. Resour., Toronto. (Video.)

———. 1988c. Timber management guidelines for the provision of moose habitat. Ontario Min. Nat. Resour., Toronto. 33 pp.

———. 1989a. Advanced moose hunter education instructor's manual. Wildl. Branch, Ontario Min. Nat. Resour., Toronto. 135 pp.

———. 1989b. Algonquin Park bibliography. Rev. ed. Ontario Min.. Nat. Resour., Whitney. 90 pp.

———. 1989c. Annual provincial mail survey. Wildl. Branch, Ontario Min. Nat. Resour., Queens Park, Toronto.

———. 1990. The moose in Ontario. Book 1—Moose biology, ecology and management. Chapts. 1–7, 34 pp. Book 2—Moose hunting techniques, hunting ethics and the law. Chapts.8–14, pages 35–78. Ontario Fed. of Anglers and Hunters, Peterborough.

Oosenbrug, S.M. and S.H. Ferguson. 1992. Moose mark-recapture survey in Newfoundland. Alces 28: 21–29.

Oosenbrug, S.M., E.W.Mercer and S.H. Ferguson. 1991. Moose-vehicle collisions in Newfoundland—Management considerations for the 1990's. Alces 27: 220–225.

Oosenbrug, S.M., R.E. McNeily, E.W. Mercer and J.F. Folinsbee. 1986. Some aspects of moose–vehicle collisions in Newfoundland, 1973–1985. Alces 23: 377–393.

Orlowski, N. and H. Kallmeyer. 1936. Zu: Ist die Geweihbildung beim Elchschaufler oder Strangenhirsch eine Rassen- oder Ernährungs- und Gesundheitsfrage? Wild und Hund 42: 679–680.

Orskov, O.R. 1982. Protein nutrition in ruminants. Academic Press, New York. 160 pp.

Osborne, T.O., T.F. Paragi, J.L. Bodkin, A.J. Loranger and W.N. Johnson. 1991. Extent, causes, and timing of moose calf mortality in western interior Alaska. Alces 27: 24–30.

Osgood, C. 1936. Contributions to the ethnology of the Kutchin. Yale Univ. Public. in Anthropol. 7: 3–23.

———. 1937. The ethnography of the Tanaina. Yale Univ. Public. in Anthropol. 16. New Haven, Connecticut.

———. 1958. Ingalik social culture. Yale Univ. Publ. in Anthropol. 53. New Haven, Connecticut.

———. 1971. The Han Indians: A compilation of ethnographic and historial data on the Alaska-Yukon boundary area. Yale Univ. Publ. in Anthropol. 74. New Haven, Connecticut.

Osipov, A.N. 1966. Life cycle of *Setaria altaica* (Rajewskaja, 1928), a parasite of the brain of Siberian deer. Doklady Akademii Nawk SSSR 168: 247–248.

Osuji, P.O. 1974. The physiology of eating and the energy expenditure of the ruminant at pasture. J. Range Manage. 27: 437–443.

Oswald, K. 1982. A manual for aerial observation of moose. Wildl. Branch, Ontario Min. Nat. Resour., Toronto. 103 pp.

Overton, W.S. and D.E. Davis. 1969. Estimating the numbers of animals in wildlife populations. Pages 403–455 *in* Wildlife management techniques. 3rd. ed. The Wildl. Soc., Washington, D.C.

Ozoga, J.J. and L.J. Verme. 1970. Winter feeding patterns of penned white-tailed deer. J. Wildl. Manage. 34: 431–439.

Ozoga, J.J. and L.J. Verme. 1975. Activity patterns of white-tailed deer during estrus. J. Wildl. Manage. 39: 679–683.

Ozoga, J.J. and L.J. Verme. 1978. The thymus gland as a nutritional status indicator in deer. J. Wildl. Manage. 42: 791–798.

Page, R.E. 1983. Population dynamics in relation to moose management. Alces 19: 83–97.

———. 1987. Integration of population dynamics for moose management—A review and synthesis of modelling approaches in North America. Swed. Wildl. Res. (Suppl.) 1: 491–501.

Page, R.E. 1989. The inverted pyramid: Ecosystem dynamics of wolves and moose on Isle Royale. Ph.D. thesis, Michigan Tech. Univ., East Lansing. 62 pp.

Papi, F. 1990. Homing phenomena—Mechanisms and classifications. Ethol. Ecol. and Evol. 2: 3–10.

Parker, A.C. 1954. The Indian how book. Dover Publ., Inc., New York. 335 pp.

Parker, G.R. 1966. Moose disease in Nova Scotia. M.S. thesis, Acadia Univ., Wolfville, Nova Scotia. 126 pp.

Parker, G.R. and L.D. Morton. 1978. The estimation of winter forage and its use by moose on clearcuts in northcentral Newfoundland. J. Range Manage. 31: 300–304.

Parker, K.L. and C.T. Robbins. 1984. Thermoregulation in mule deer and elk. Can. J. Zool. 62: 1,409–1,422.

Parker, K.L. and C.T. Robbins. 1985. Thermoregulation in ungulates. Pages 161–182 *in* R.J. Hudson and R.G. White, eds., Bioenergetics of wild herbivores. CRC Press, Inc., Boca Raton, Florida.

Parker, K.L., C.T. Robbins and T.A. Hanley. 1984. Energy expenditures for locomotion by mule deer and elk. J. Wildl. Manage. 48: 474–488.

Passarge, H. 1971. Die Gehörnlangensumme als Masstab für die Gütterklassenansprache beim Rehbock. Unsere Jagd 15: 64–66.

Passmore, R.C. 1963. Experimental use of aerial photography in censusing moose. Pages 30–33 *in* Manage. Report. 71. Ontario Dept. Lands and For. Resour. Maple.

Passmore, R.C., R.L. Peterson and A.T. Cringan. 1955. A study of mandibular tooth-wear as an index to age of moose. Pages 223–238 *in* R.L. Peterson, ed., North American moose. Univ. Toronto Press, Toronto, Ontario.

Pastor, J., R.J. Naiman, B. Dewey and P. McInnes. 1988. Moose, microbes, and the boreal forest. Biosci. 38: 770–777.

Patch, J.R., G.C. Moore and W.R. Watt. 1986. Moose and forest management in New Brunswick. Alces 22: 443–447.

Pauls, R.W., R.J. Hudson and S. Sylven. 1981. Energy metabolism of free-ranging wapiti. Pages 87–90 *in* M.A. Price and A.R. Robblee, eds., 60th Annual Feeders' Day rept. Dept. of Anim. Sci., Univ. Alberta, Edmonton.

Pearson, N.E. 1974. Optimal foraging theory. Quant. Sci. Pap. 310. Univ. Washington, Seattle. 48 pp.

Peek, J.M. 1962. Studies of moose in the Gravelly and Snowcrest mountains, Montana. J. Wildl. Manage. 26: 360–365.

———. 1971a. Moose habitat selection and relationships to forest management in northeastern Minnesota. Ph.D. thesis, Univ. Minnesota, St. Paul. 250 pp.

———. 1971b. Moose-snow relationships in northeastern Minnesota. Proc. Snow and Ice Symp.: 39–49.

———. 1974a. On the nature of winter habitats of Shiras moose. Nat. Can. (Que.) 101: 131–174.

———. 1974b. Initial response of moose to a forest fire in northeastern Minnesota. Am. Midl. Nat. 91: 435–438.

———. 1974c. A review of food habit studies in North America. Nat. Can. (Que.) 101: 195–215.

———. 1980. Natural regulation of ungulates (What constitutes a real wilderness?). Wildl. Soc. Bull. 8: 217–227.

———. 1982. Moose (Alaska). Pages 264–265 *in* D.E. Davis, ed., CRC handbook of census methods for terrestrial vertebrates. CRC Press, Boca Raton, Florida.

———. 1986. A review of wildlife management. Prentice-Hall, Englewood Cliffs, New Jersey. 486 pp.

Peek, J.M., L.W. Krefting and J.C. Tappeiner. 1971. Variation in twig diameter-weight relationships in northern Minnesota. J. Wildl. Manage. 35: 501–507.

Peek, J.M., R.E. LeResche and D.R. Stevens. 1974. Dynamics of moose aggregations in Alaska, Minnesota and Montana. J. Mammal. 55: 126–137.

Peek, J.M., D.L. Urich and R.J. Mackie. 1976. Moose habitat selection and relationships to forest management in northeastern Minnesota. Wildl. Monogr. 48. The Wildl. Soc., Washington, D.C. 65 pp.

Peek, J.M., R.J. Petersen and J.W. Thomas. 1982. The future of elk hunting. Pages 599–625 *in* J.W. Thomas and D.E. Toweill, eds., Elk of North America: Ecology and management. Stackpole Books, Harrisburg, Pennsylvania. 698 pp.

Peek, J.M., V. Van Ballenberghe and D.G. Miquelle. 1986. Intensity of interactions between rutting bull moose in central Alaska. J. Mammal. 67: 423–426.

Peek, J.M., D.J. Pierce, D.C. Graham and D.L. Davis. 1987. Moose habitat use and implications for forest management in northcentral Idaho. Swed. Wildl. Res. (Suppl.) 1: 195–199.

Peet, S.D. 1890. Emblematic mounds and animal effigies. Prehistoric America. Vol. II. Am. Antiquar. Off., Chicago, Illinois. 351 pp.

Pemberton, J., S.D. Albon, F.E. Guiness, T.H. Clutton-Brock and R.J. Berry. 1988. Genetic variation and juvenile survival in red deer. Evolution 42: 921–934.

Peninsula Clarion. 1988. Sleepy moose killed when airplane comes in. Jan. 7. Kenai, Alaska.

———. 1990. MarkAir airplane clips cow moose on Kenai runway. Dec. 24. Kenai, Alaska.

Pennant, T. 1784. Class I. Quadrupeds. Pages 1–163 *in* Arctic zoology. Vol I. Henry Hughs, London, England. 586 pp.

Penner, D.F. 1978. Some relationships between moose and willow in the Fort Providence, N.W.T. area. M.S. thesis, Univ. Alberta, Edmonton. 183 pp.

Perron, P., ed. 1981. The neurological basis of signs in communication processes. Proc. Symp. Neurol. Basis of Signs in Commun. Processes. Univ. Toronto Press, Toronto, Ontario. 280 pp.

Perrot, N. 1864. Mémoire sur les moeurs, costumes et religion des Sauvages de l'Amérique septentrionale, par Nicholas Perrot, Public pour la première fois par le R.P. Tailhan de la compagnie de Jésus. Librairies à Frank, Paris. 341 pp. (Reprinted in 2 vols. In 1911, E.H. Blair, ed., by Arthur H. Clark Co., Cleveland.)

Peteet, D.M. 1986. Modern pollen rain and vegetational history of the Malaspina Glacier District. Quat. Res. 25: 100–120.

Peterle, T.J. 1977. Hunters, hunting, anti-hunting. Wildl. Soc. Bull. 5: 151–161.

Peterson, G.R. and I.L. Withler. 1965. Effects on fish and game species of development of Mica Dam for hydro-electric purposes. Manage. Publ. No. 10. British Columbia Fish and Wildl. Branch, Victoria. 62 pp.

Peterson, R.L. 1950. A new species of moose from North America. Occ. Pap. R. Ontario Mus. 9: 1–7.

———. 1952. A review of the living representatives of the genus *Alces*. Contrib. No. 34. R. Ontario Mus. Zool. Paleont., Ottawa. 30 pp.

———. 1953. Studies on the food habits and the habitat of moose in Ontario. Contrib. No. 36. R. Ontario Mus. Zool. Paleont., Ottawa. 40 pp.

———. 1955. North American moose. Univ. Toronto Press, Toronto, Ontario. 280 pp.

———. 1974. A review of the general life history of moose. Nat. Can. (Que.) 101: 9–21.

Peterson, R.O. 1976. The role of wolf predation in a moose population decline. Proc. Conf. on Sci. and Res. in Natl. Parks 1: 329–333.

———. 1977. Wolf ecology and prey relationships on Isle Royale. U.S. Natl. Park Serv. Sci. Monogr. Ser. 11. 210 pp.

———. 1979. The wolves of Isle Royale—New developments. Pages 3–18 *in* E. Klinghammer, ed., Proc. Symp. Behav. and Ecol. of Wolves. Garland STPM Press, New York.

———. 1982. Wolf-moose investigations on the Kenai National Wildlife Refuge. Michigan Tech. Univ., Houghton. 146 pp.

———. 1988. Increased osteoarthritis in moose from Isle Royale. J. Wildl. Dis. 24: 461–466.

———. 1989. Ecological studies of wolves on Isle Royale, annual report 1988–1989. Michigan Tech. Univ., Houghton. 14 pp.

———. 1992. Ecological studies of wolves on Isle Royale, annual report 1991–1992. Michigan Tech. Univ., Houghton. 16 pp.

Peterson, R.O. and D.L. Allen. 1974. Snow conditions as a parameter in moose-wolf relationships. Nat. Can. (Que.) 101: 481–492.

Peterson, R.O. and R.E. Page. 1983. Wolf–moose fluctuations at Isle Royale National Park, Michigan, U.S.A. Acta Zool. Fennica 174: 251–253.

Peterson, R.O. and R.E. Page. 1988. The rise and fall of Isle Royale wolves, 1975-1986. J. Mammal. 69: 89–99.

Peterson, R.O. and R.E. Page. 1993. Detection of moose in midwinter from fixed-wing aircraft over dense forest cover. Wildl. Soc. Bull. 21: 80–86.

Peterson, R.O. and J.D. Woolington. 1982. The apparent extirpation and reappearance of wolves on the Kenai Peninsula, Alaska. Pages 334–344 *in* F.H. Harrington and P.C. Paquet, eds., Wolves of the world. Noyes Publ., Park Ridge, New Jersey.

Peterson, R.O., J.M. Scheidler and P.W. Stephens. 1982. Selected skeletal morphology and pathology of moose from the Kenai Peninsula, Alaska, and Isle Royale, Michigan. Can. J. Zool. 60: 2,812–2,817.

Peterson, R.O., C.C. Schwartz and W.B. Ballard. 1983. Eruption patterns of selected teeth in three North American moose populations. J. Wildl. Manage. 47: 884–888.

Peterson, R.O., J.D. Woolington and T.N. Bailey. 1984. Wolves of the Kenai Peninsula, Alaska. Wildl. Monogr. 88. 52 pp.

Peterson, W.J. and M.W. Lankester. 1991. Aspects of the epizootiology of *Parelaphostrongylus tenuis* in a white-tailed deer population. Alces 27: 183–192.

Petitot, E. 1888. En route pour la mer glaciale. 2nd ed. Letouzey et Ané, Paris.

Petrov, A.K. 1957. Zakonomyernosti rosta skeleta losya i krupnogo rogatogo skota. Trudy Mosk. Vet. Akad. 19: 423–437.

———. 1958a. Eksternye dannye losya i krupnogo rogotogo skota v ontogeneze. Trudy Mosk. Vet. Akad. 22: 3–16.

———. 1958b. Izmenenie teloslozheniya v ontogeneze losya i krupnogo skota. Sbornik nauk trudov Uralskogo selskokhozyast. Inst. 16: 193–203.

———. 1964. Okosteniya skeleta v individualnom razvitii losya [Ossification of the skeleton in the ontogenesis of elk]. Zool. Zhurn. 43: 1,837–1,847.

Petrov, A.K., N.F. Pleshakov, E.A. Isayenekov, M.D. Vishnevskaya and A.M. Smirnov. 1992. Morphological changes of organs and systems in ontogenesis of elk, cattle and small cattle as ruminants with differences in their ecology. Alces (Suppl.) 1: 234

Péwé, T.L. and D.M. Hopkins. 1967. Mammal remains of pre-Wisconsin age in Alaska. Pages 266–270 *in* D.M. Hopkins, ed., The Bering land bridge. Stanford Univ. Press, Stanford, California. 495 pp.

Phillips, J.G. 1975. Environmental physiology. Halsted Press, New York. 198 pp.

Phillips, P.C. 1961 The fur trade. 2 vols. Univ. Oklahoma Press, Norman. 1,416 pp.

Phillips, R.L. and W.E. Berg. 1972. Automatic recording of moose activity patterns. Proc. 34th Midwest Wildl. Conf. 14 pp.

Phillips, R.L., W.E. Berg and D.B. Siniff. 1973. Moose movement patterns and range use in northeastern Minnesota. J. Wildl. Manage. 37: 266–278.

Pianka, E.R. 1970. On r and K selection. Am. Nat. 104: 592–597.

———. 1976. Competition and niche theory. Pages 114–141 *in* R.M. May, ed., Theoretical ecology: Principles and applications. W.B. Saunders Co., Philadelphia.

Pierce, D.J. 1984. Shiras moose forage selection in relation to browse availability in north central Idaho. Can. J. Zool. 62: 2,404–2,409.

Pierce, D.J. and J.M. Peek. 1984. Moose habitat use and selection patterns in north central Idaho. J. Wildl. Manage. 48: 1,335–1,343.

Pilarski, W. and T. Roskosz. 1959. Beobachtungen über den Rumpfabschnitt der Wirbelsäule des Elches, *Alces alces* L. Acta Theriol. 3: 1–16.

Pimlott, D.H. 1953. Newfoundland moose. Trans. N. Am. Wildl. Conf. 18: 563–581.

———. 1959. Reproduction and productivity of Newfoundland moose. J. Wildl. Manage. 23: 381–401.

———. 1961. The ecology and management of moose in North America. Terre et Vie 2: 246–265.

———. 1963. Influence of deer and moose on boreal forest vegetation in two areas of eastern Canada. Trans. Int. Union Game. Biol. Congr. 6: 106–116.

Pimlott, D.H. and W.J. Carberry. 1958. North American moose transplantations and handling techniques. J. Wildl. Manage. 22: 51–62.

Pinkerton, E. 1989. Co-operative management of local fisheries. Univ. British Columbia Press, Vancouver. 299 pp.

Pinkwater, D. 1993. Blue moose—Return of the moose. Random Books, New York. 112 pp.

Pinus, A. 1928. Beitrag zur Entwicklungsgeschichte der Nasenregion des Elches (*Alces alces* L.). Z. Wiss. Biol. 13: 35–65.

Platt, T.R. and W.M. Samuel. 1978. *Parelaphostrongylus odocoilei*: Life cycle in experimentally infected cervids including mule deer, *Odocoileus h. hemionus*. Exp. Parisitol. 46: 330–338.

Pledger, D.J., W.M. Samuel and D.A. Craig. 1980. Black flies (Diptera: Simuliidae) as possible vectors of legworm (*Onchocerca cervipedis*) in moose of central Alberta. Proc. N. Am. Moose Conf. Workshop 16: 171–202.

Plotka, E.D., U.S. Seal and L.J. Verme. 1982. Morphologic and metabolic consequences of pinealectomy in deer. Pages 153–169 *in* R.J. Reiter, ed., The pineal gland. Vol. III: Extra-reproductive effects. CRC Press Inc., Boca Raton, Florida.

Pocock, R.I. 1923. On the external characters of *Elaphorus, Hyropotes, Pudu*, and other Cervidae. Proc. Zool. Soc. L. 1923: 181–210.

Poelker, R.J. 1972. The Shiras moose in Washington. Washington Game Dept., Olympia. 46 pp.

Poliquin, A. 1978. Mise en évidence des caracteristiques des aires hivernales de broutage de l'orignal dans le parc du Mont-Tremblant. M.S. thesis, Univ. Quebec, Montreal. 102 pp.

Poll, D.M. 1989. Wildlife mortality on the Kootenay Parkway, Kootenay National Park. Kootenay Nat. Park Final Rept., Radium Hot Springs, British Columbia. 105 pp.

Pollock, K.H., S.R. Winterstein, C.M. Bunck and P.D. Curtis. 1989. Survival analysis in telemetry studies: The staggered entry design. J. Wildl. Manage. 53: 7–15.

Poliquin, A., B. Scherrer and R. Joyal. 1977. The characteristics of moose winter yards as described by factorial discrimination analysis. Proc. N. Am. Moose Conf. Workshop 13: 128–144.

Pont, A.C. 1986. Muscidae. Pages 57–215 *in* A. Soos and L. Papp, eds., Catalogue of Palaearctic Diptera. Vol. 11. Akademiai Kiado, Budapest.

Poole, J.H. 1989. Mate guarding, reproductive success and female choice in African elephants. Anim. Behav. 37: 842–849.

Poole, T.B. 1985. Social behaviour in mammals. Blackie and Son, Ltd., Glasgow. 248 pp.

Porter, W.F. and K.E Church. 1987. Effects of environmental pattern on habitat preference analysis. J. Wildl. Manage. 51: 681–685.

Portmann, A. 1953. Das Tier als soziales Wesen. Rhein Verlag A.G. Zürich. 382 pp.

———. 1960. Die Tiergestalt. Fr. Reinhardt Verlag, Basel. 274 pp.

———. 1961. Animals as social beings. The Viking Press, New York. 249 pp.

Post, A. and G. Streveler. 1976. The tilted forest: Glaciological–geological implications of vegetated neoglacial ice at Lituya Bay, Alaska. Quat. Res. 6: 111–117.

Potter, D.R., J.C. Hendee and R.N. Clark. 1973. Hunting satisfaction: Game, guns or nature? Trans. N. Am. Wildl. and Nat. Resour. Conf. 38: 220–229.

Potvin, F. and J. Huot. 1983. Estimating carrying capacity of a white-tailed deer wintering area in Quebec. J. Wildl. Manage. 47: 463–475.

Powell, A.M. 1909. Trailing and camping in Alaska. A. Wessels, New York.

Powell, J.E. 1856. Untitled letter. Proc. Acad. Natl. Sci., Philadelphia. VII: 342–344.

Preble, E.A. 1902. A biological investigation of the Hudson Bay region. N. Am. Fauna No. 23. U.S. Bur. Biol. Surv., U.S. Govt. Print. Off., Washington, D.C. 140 pp.

———. 1908. A biological investigation of the Athabaska-Mackenzie region. N. Am. Fauna No. 27. U.S. Bur. Biol. Surv., U.S. Govt. Print. Off., Washington, D.C. 574 pp.

Prescott, W.H. 1974. Interrelationships of moose and deer of the genus *Odocoileus*. Nat. Can. (Que.) 101: 493–504.

Presidente, P.J.A., D.E. Worley and J.E. Katlin. 1972. Gross transmission experiments with *Dictyocaulus viviparus* isolates from Rocky Mountain elk and cattle. J. Wildl. Dis. 8: 57–62.

Prestwood, A.K. and S.R. Pursglove. 1977. Prevalence and distribution of *Setaria yehi* in southeastern white-tailed deer. J. Am. Vet. Med. Assoc. 171: 933–935.

Preston, R.J., III. 1964. Ritual hangings: An aboriginal "survival" in a northern North American trapping community. Man 64(180):142–144.

Price, E.O., S.J.R. Wallach and M.R. Dally. 1991. Effects of sexual stimulation on the sexual performance of rams. Appl. Anim. Behav. Sci. 30: 333–340.

Price, M.A. and R.G. White. 1985. Growth and development. Pages 183–213 *in* R.J. Hudson and R.G. White, eds., Bioenergetics of wild herbivores. CRC Press, Boca Raton, Florida. 314 pp.

Price, W.A. 1945. Nutrition and physical degeneration: A comparison of primitive and modern diets and their effects. Paul B. Hoeber, New York.

Priklonskiy, S.G. 1965. Pereschetnye koeffitsienty dlya obrabotki dannykh zimnego marshrutnogo ucheta promyslovykh zvery po sledam. (Conversion coefficients for processing data of the winter transect method of census taking of game animals by their tracks.) Byulleten Mosk. obsch. isp. prirody, otd. Biologii, LXX (6). (Unedited draft translation from Russian by Can. Wildl. Serv., Ottawa. 21 pp.)

Prior, R. 1993. The three original "Baillie Monsters." Deer 8: 684.

Proulx, G. 1983. Characteristics of moose *(Alces alces)* winter yards on different exposures and slopes in southern Quebec. Can. J. Zool. 61: 112–118.

Proulx, G. and R. Joyal. 1981. Forestry maps as an information source for description of moose winter yards. Can. J. Zool. 59: 75–80.

Pruitt, W.O. 1959. Snow as a factor in the winter ecology of the barren ground caribou. Arctic 12: 159–179.

Pulsifer, M.D. and T.L. Nette. 1995. History, status and present distribution of moose in Nova Scotia. Alces 31: 209–219.

Purchas, S. 1625. Hakluytus posthumus or Purchas his pilgrimes. 20 vols. (Reprinted in 1906 by James MacLehose and Sons, Glasgow, Scotland.)

Pybus, M.J. 1990. Survey of hepatic and pulmonary parasites of wild cervids in Alberta, Canada. J. Wildl. Dis. 26: 453–459.

Pybus, M.J. and W.M. Samuel. 1980. Pathology of the muscleworm, *Parelaphostrongylus odocoilei* (Nematoda: Metastrongyloidea) in moose. Proc. N. Am. Moose Conf. Workshop 16: 152–170.

Pyke, G.H., H.R. Pulliam and E.L. Charnov. 1977. Optimal foraging: A selective review of theory and test. Quart. Rev. Biol. 52: 137–154.

Quimby, G.I. 1960. Indian life in the upper Great Lakes 11,000 B.C. to A.D. 1800. Univ. Chicago Press, Chicago, Illinois. 182 pp.

Quinn, N.W.S. and R.W. Aho. 1989. Whole weights of moose from Algonquin Park, Ontario, Canada. Alces 25: 45–51.

Racey, G.D, T.S. Whitfield and R.A. Sims. 1989. Northwestern Ontario forest ecosystem interpretations. NWOFTDU Tech. Rept. 46. Ontario Min. Nat. Resour., Toronto. 90 pp.

Rajnovich, G. 1987. Visions in the quest for medicine: An interpretation of the Indian pictographs of the Canadian shield. Midcont. J. Archaeol. 14(2):179–225.

———. 1994. Reading rock art: Interpreting the Indian rock paintings of the Canadian shield. Nat. Herit./Nat. Hist., Inc., Toronto. 191 pp.

Ralph, J. 1890. Antoine's moose yard. Harper's New Month. Mag. LXXXI: 650–666.

———. 1892. On Canada's frontier. Harper and Brothers, New York. 325 pp.

Ramsey, P.R., J.C. Avise, M.H. Smith and D.F. Urbston. 1979. Biochemical variation and genetic heterogeneity in South Carolina deer populations. J. Wildl. Manage. 43: 136–142.

Ramusio, G.B., ed. 1556. Terzo volume della navigationi et viaggi. S tamperia de Givnti, Venice, Italy.

Raskin, E. 1974. Moose, goose, and little nobody. Parents' Mag. Press, New York. 31 pp.

Rasmussen, P. 1977. Calcium deficiency, pregnancy and lactation in rats. Calcium Tiss. Res. 23: 87–94.

Rausch, R.A. 1956. The problem of railroad moose conflicts in the Susitna Valley, 1956–57. A preliminary report. U.S. Fish and Wildl. Serv., Anchorage, Alaska. 19 pp.

———. 1959. Some aspects of population dynamics of the railbelt moose population of Alaska. M.S. thesis, Univ. Alaska-Fairbanks. 81 pp.

———. 1965. Annual assessments of moose calf production and mortality in southcentral Alaska. Ann. Conf. W. Assoc. St. Fish and Game Commiss. 45: 140–146.

———. 1967. Some aspects of the population ecology of wolves, Alaska. Am. Zool. 7: 253–265.

———. 1969. A summary of wolf studies in south-central Alaska, 1957-1969. Trans. N. Am. Wildl. and Nat. Resour. Conf. 34: 117–131.

Rausch, R.A. and A. Bratlie. 1965. Annual assessments of moose calf production and mortality in south central Alaska. Ann. Conf. West. Assoc. St. Game and Fish Commiss. 45: 140–146.

Rausch, R.A., R.J. Sommerville and R.H. Bishop. 1974. Moose management in Alaska. Nat. Can. (Que.) 101: 705–721.

Rausch, R.L. 1959. Notes on the prevalence of hydatid disease in Alaskan moose. J. Wildl. Manage. 23: 122–123.

———. 1993. The biology of *Echinococcus granulosus*. Pages 24–56 *in* F.L. Andersen, J. Chai and F. Liu, eds., Compendium on cystic echinococcosis with special reference to the Xinjiang Uygur Autonomous Region, the People's Republic of China. Brigham Young Univ. Press, Provo, Utah.

Rausch, R.L. and B. Huntley. 1978. Brucellosis in reindeer, *Rangifer tarandus* L., inoculated experimentally with *Brucella suis*, type 4. Ca. J. Microbiol. 24: 129–135.

Ray, A.J. 1974. Indians in the fur trade. Univ. Toronto Press, Toronto, Ontario. 248 pp.

———. 1978. Competition and conservation in the early subarctic fur trade. Ethnohist. 25:347–357.

———. 1987. The fur trade in North America: An overview from a historical geographical perspective. Pages 21–30 *in* M. Novak, J.A. Baker, M.E. Obbard and B. Malloch, eds., Wild furbearer management and conservation in North America. Ontario Min. Nat. Resour., Toronto. 1,150 pp.

Raymond, R.B. and R.B. Forbes. 1975. Lead in hair of urban and rural small animals. Bull. Environ. Contam. Toxicol. 13: 551–553.

Rayner, V. and S.W.B. Ewen. 1981. Do the blood vessels of the antler velvet of the red deer have an adrenergic innervation? Quat. J. Exp. Physiol. 66: 81–86.

Reed, D.J. 1989. Moosepop, program documentation and instructions. Draft. Alaska Dept. Fish and Game, Fairbanks. 15 pp.

Regelin, W.L. 1971. Deer forage quality in relation to logging in the spruce–fir and lodgepole pine type in Colorado. M.S. thesis, Colorado St. Univ., Fort Collins. 50 pp.

———. 1991. Wildlife management in Canada and the United States. Pages 55–64 *in* B. Bobek, K. Perzanowski and W. Regelin, eds., Global trends in wildlife management. Vol. 1. Trans. Int. Union Game Biol. Congr., Swiat Press, Krakow-Warszawa, Poland.

Regelin, W.L., C.C. Schwartz and A.W. Franzmann. 1981a. Respiratory chamber for study of energy expenditure of moose. Alces 17: 126–135.

Regelin, W.L., C.C. Schwartz and A.W. Franzmann. 1981b. Energy expenditure of moose on the Kenai National Wildlife Refuge. Ann. Prog. Rept. Proj. W-17-11. Kenai Alaska Field Station, Alaska Dept. Fish and Game, Juneau. 44 pp.

Regelin, W.L., C.C. Schwartz and A.W. Franzmann. 1982. Raising, training and maintaining moose *(Alces alces)* for nutritional studies. Trans. Int. Congr. Game Biol. 14: 425–428.

Regelin, W.L., C.C. Schwartz and A.W. Franzmann. 1985. Seasonal energy metabolism of adult moose. J. Wildl. Manage. 49: 394–396.

Regelin, W.L., C.C. Schwartz and A.W. Franzmann. 1986. Energy cost of standing in adult moose. Alces 22: 83–90.

Regelin, W.L., C.C. Schwartz and A.W. Franzmann. 1987a. Effects of forest succession on nutritional dynamics of moose forage. Swed. Wildl. Res. (Suppl.) 1: 247–264.

Regelin, W.L., M.E. Hubbert, C.C. Schwartz and D.J. Reed. 1987b. Field test of a moose carrying capacity model. Alces 23: 243–284.

Reid, J.T. and J. Robb. 1971. Relationship of body composition to energy intake and energy efficiency. J. Dairy Sci. 54: 553–564.

Reid, J.T., O.D. White, R. Anrique and A. Fortin. 1980. Nutritional energetics of livestock: Some present boundaries of knowledge and future research needs. J. Anim. Sci. 51: 1,393–1,415.

Reid, R.L. 1968. Rations for maintenance and production. Pages 190–200 *in* F.B. Golley and H.K. Buechner, eds., A practical guide to the study of the productivity of large herbivores. IBP Handbook No. 7. Blackwell Scientific Publ., Oxford, England.

Reiss, M.J. 1989. The allometry of growth and reproduction. Cambridge Univ. Press, Cambridge, England. 182 pp.

Remington, F. 1899. The story of the dry leaves. Harper's New Month. Mag. XCIX(DLXXXIX): 95–101.

Rempel, R.S., A.R. Rodgers and K.F. Abraham. 1995. Performance of a GPS animal location system under boreal forest canopy. J. Wildl. Manage. 59: 543–551.

Renecker, L.A. 1987a. Bioenergetics and behavior of moose *(Alces alces)* in the aspen dominated boreal forest. Ph.D. thesis, Univ. Alberta, Edmonton. 265 pp.

———. 1987b. The composition of moose milk following a late parturition. Acta Theriol. 32:129–133.

———. 1989. Seasonal nutritional cycles of ungulates in Elk Island National Park. Elk Island Natl. Park, Fort Saskatchewan. 35 pp.

Renecker, L.A. and R.J. Hudson. 1983. Winter energy budgets of free-ranging moose, using a calibrated heart rate index. Proc. Int. Conf. Wildl. Biotelemetry 4: 187–211.

Renecker, L.A. and R.J. Hudson. 1985a. Estimation of dry matter intake of free-ranging moose. J. Wildl. Manage. 49: 785–792.

Renecker, L.A. and R.J. Hudson. 1985b. Telemetered heart rate as an index of energy expenditure in moose *(Alces alces)*. Comp. Biochem. Physiol. 82A: 161–165.

Renecker, L.A. and R.J. Hudson. 1985c. A technique for estimating dry matter intake of tame free-ranging moose. Alces 21: 267–277.

Renecker, L.A. and R.J. Hudson. 1986a. Seasonal foraging rates of free-ranging moose. J. Wildl. Manage. 50: 143–147.

Renecker, L.A. and R.J. Hudson. 1986b. Seasonal energy expenditure and thermoregulatory response of moose. Can. J. Zool. 64: 322–327.

Renecker, L.A. and R.J. Hudson. 1988. Seasonal quality of forages used by moose in the aspen-dominated boreal forest, central Alberta. Holarctic Ecol. 11: 111–118.

Renecker, L.A. and R.J. Hudson. 1989a. Seasonal activity budgets of moose in aspen-dominated boreal forest. J. Wildl. Manage. 53: 296–302.

Renecker, L.A. and R.J. Hudson. 1989b. Ecological metabolism of moose in aspen-dominated boreal forests, central Alberta. Can. J. Zool. 67: 1,923–1,928.

Renecker, L.A. and R.J. Hudson. 1990a. Digestive kinetics of moose *(Alces alces)*, wapiti *(Cervus elaphus)* and cattle. Anim. Prod. 50: 51–61.

Renecker, L.A. and R.J. Hudson. 1990b. Behavioral and thermoregulatory responses of moose to high ambient temperatures and insect harassment in aspen-dominated forests. Alces 26: 66–72.

Renecker, L.A. and R.J. Hudson. 1992a. Thermoregulatory and behavioral responses of moose. Is large body size an adaptation or constraint? Alces (Suppl.) 1: 52–64.

Renecker, L.A. and R.J. Hudson. 1992b. Habitat and forage selection of moose in the aspen-dominated boreal forest, central Alberta. Alces 28: 189–201.

Renecker, L.A. and R.J. Hudson. 1993. Morphology, bioenergetic, and resource use: Patterns and processes. Pages 141–163 *in* J.B. Stelfox, ed., Hoofed mammals of Alberta. Lone Pine Press, Edmonton, Alberta. 241 pp.

Renecker, L.A. and H.M. Kozak. 1987. Game ranching in western Canada. Rangelands 9: 215–218.

Renecker, L.A. and W.M. Samuel. 1990. Growth and seasonal weight changes as they relate to spring and autumn set points in mule deer. Can. J. Zool. 69: 744–747.

Renecker, L.A., R.J. Hudson, M.K. Christophersen and C. Arelis. 1978.

Effect of posture, feeding, low temperature and wind on energy expenditure of moose calves. Proc. N. Am. Moose Conf. Workshop 14: 126–140.

Renecker, L.A., R.J. Hudson and R. Berzins. 1982. Nylon bag digestibility and rate of passage of digesta in moose, wapiti and cattle. Alces 18: 1–16.

Renecker, L.A., R.J. Hudson and G.W. Lynch. 1987. Moose husbandry in Alberta, Canada. Swed. Wildl. Res. (Suppl.) 1: 775–780.

Rerabek, J. and A. Bubenik. 1963. The metabolism of phosphorus and iodine in deer. Translation Ser. AEC-tr-5631. U.S. Atomic Energy Commiss., Washington, D.C. 51 pp.

Reuterwall, C. 1981. Temporal and spatial variability of the calf sex ratio in Scandinavian moose *Alces alces*. Oikos 37: 39–45.

Reynolds, J.K. 1953. Inventory of moose in Ontario—1953. Div. Fish and Wildl, Ontario Dept. Lands and For., Maple. 4 pp.

Reynolds, S.H. 1929. Monograph of the British Pleistocene. Mammalia: The red deer, reindeer and roe deer. Paleontol. Soc. L. 4: 1–46.

Rich, E.E. 1938. Journal of occurrences in the Athabasca department by George Simpson, 1820 and 1821, and report. Vol. I. Published by The Champlain Soc. for the Hudson's Bay Rec. Soc., Toronto, Ontario. 498 pp.

———. 1942. Minutes of the Hudson's Bay Company, 1671–1674. Vol. V. Published by The Champlain Soc. for the Hudson's Bay Rec. Soc., Toronto, Ontario. 276 pp.

———. 1945. Minutes of the Hudson's Bay Company. First part 1679–82. Vol. VIII. Published by The Champlain Soc. for the Hudson's Bay Rec. Soc., Toronto, Ontario. 378 pp.

———. 1946. Minutes of the Hudson's Bay Company 1682–1684. Vol. II, Pt. 2. Published by The Champlain Soc. for the Hudson's Bay Rec. Soc., Toronto, Ontario. 368 pp.

———. 1947. Part of a dispatch from George Simpson Esqr., Governor of Ruperts Land, to the Governor and Committee of the Hudson's Bay Company London. Vol. X. Published by The Champlain Soc. for the Hudson's Bay Rec. Soc., London, England. 277 pp.

———. 1960. Hudson's Bay Company, 1670-1870. Vol. 1. McClelland and Stewart Ltd., Toronto, Ontario. 687 pp.

Rich, E.E. and A.M. Johnson, eds. 1949. James Isham's observations on Hudsons Bay, 1743. Vol. XII. Published by The Champlain Soc. for the Hudson's Bay Rec. Soc., London, England. 352 pp.

Rich, E.E. and A.M. Johnson, eds. 1950. Peter Skene Ogden's Snake country journals, 1824-25 and 1825-26. Vol. XIII. The Hudson's Bay Rec. Soc., London, England. 283 pp.

Rich, E.E. and A.M. Johnson, eds. 1952. Cumberland House journals and inland journals 1775–82. 2nd ser. Hudson's Bay Rec. Soc., London, England. 313 pp.

Rich, E.E. and A.M. Johnson, eds. 1957. Hudson's Bay copy booke of letters, commissions, instructions outward, 1688–1696. Vol. XX. Hudson's Bay Rec. Soc., London, England. 357 pp.

Richards, G.N. 1976. Search for factors other than "lignin-shielding" in protection of cell wall polysaccharides from digestion in the rumen. Pages 129–135 *in* Carbohydrate research in plants and animals. Misc. Pap. 12, Landbouwhogeschool, Wageningen, Germany.

Richardson, D.W. 1990. The 3rd carpal bone. J. Equine Vet. Sci. 10: 258.

Richardson, J. 1829. Fauna Boreali-Americana, or the zoology of the northern parts of British America. Vol. 1. The quadrupeds. John Murray, London, England. 300 pp.

———. 1852. Arctic searching expedition: A journal of a boat-voyage through Rupert's Land and the Arctic Sea. Harper, New York.

Rickaby, F. 1926. Ballads and songs of the shanty-boy. Harvard Press, Cambridge, Massachusetts. 242 pp.

Rickard, L.G., B.B. Smith, E.J. Gentz, A.A. Frank, E.G. Pearson, L.L. Wlaker and M. Pybus. 1994. Experimentally induced meningeal worm *(Parelaphostrongylus tenuis)* infection in the llama *(Lama glama)*: Clinical evaluation and implications for parasite translocation. J. Zoo and Wildl. Medicine 25: 390–402.

Ricklefs, R.E. 1967. A graphical method of fitting equations to growth curves. Ecol. 48: 978–983.

Ridington, R. 1981. Beaver. Pages 350–360 *in* J. Helm, ed., Subarctic. Handbook of North American Indians. Vol. 6. Smithsonian Inst., Washington, D.C. 837 pp.

———. 1986. Freedom and authority: Teachings of the hunters. Pages 232–259 *in* R.B. Morrison and C.R. Wilson, eds., Native peoples: The Canadian experience. McClelland and Stewart Inc., Toronto. 639 pp.

Riese, G. and H.H. Sambraus. 1976. Untersuchungen über das Farbsehvermögen des Schafes. Säugetierkundl. Mitt. 24: 288–295.

Rinehart, J. and L. Fay. 1981. Brucellosis. Pages 148–154 *in* W.R. Davidson, F.A. Hayes, V.F. Nettles, F.E. Kellogg, eds., Diseases and parasites of white-tailed deer. Misc. Publ. No. 7. Tall Timbers Res. Sta., Tallahassee, Florida. 458 pp.

Riney, T. 1955. Evaluating condition of free-ranging red deer *(Cervus elaphus)*, with special reference to New Zealand. Part 1. Description of techniques for determination of red deer *(Cervus elaphus)*. New Zealand J. Sci. Tech. 36: 429–463.

———. 1957. The use of faeces counts in studies of several free-ranging mammals in New Zealand. New Zealand J. Sci. Tech. B38: 507–532.

———. 1982. Study and management of large mammals. John Wiley and Sons, New York. 552 pp.

Ringberg, T.M., R.G. White, D.F. Holleman and J.R. Luick. 1981. Prediction of carcass composition in reindeer (*Rangifer tarandus tarandus* L.) by use of selected bones and muscles. Prediction of carcass composition in reindeer. Can. J. Zool. 59: 583–588.

Risenhoover, K.L. 1985. Interspecific variation in moose preference for willows. Pages 58–63 *in* F. Provenza, J.T. Flinders and E.D. McArthur, comps., Proc. Symp. on Plant-Herbivore Interactions. Intermountain Res. Sta., USDA For. Serv., Ogden, Utah.

———. 1986. Winter activity patterns of moose in interior Alaska. J. Wildl. Manage. 50: 727–734.

———. 1987. Winter foraging strategies of moose in subarctic and boreal forest habitats. Ph.D. thesis, Michigan Tech. Univ., Houghton. 108 pp.

———. 1989. Composition and quality of moose winter diets in interior Alaska. J. Wildl. Manage. 53: 568–576.

Risenhoover, K.L. and S.A. Maass. 1987. The influence of moose on the composition and structure on Isle Royale forests. Can. J. For. Res. 17: 357–364.

Risenhoover, K.L. and R.O. Peterson. 1986. Mineral licks as a sodium source for Isle Royale moose. Oecologia 71: 121–126.

Ritcey, R.W. 1965. Ecology of moose winter range in Wells Gray Park, British Columbia. British Columbia Fish and Game Branch, Victoria. 15 pp.

———. 1974. Moose harvesting programs in Canada. Nat. Can. (Que.) 101: 631–642.

Ritcey, R.W. and R.Y. Edwards. 1956. Trapping and tagging moose on winter range. J. Wildl. Manage. 20: 324–325.

Ritcey, R.W. and R.Y. Edwards. 1958. Parasites and diseases of the Wells Gray moose herd. J. Mammal. 39: 139–145.

Ritcey, R.W. and N.A.M. Verbeek. 1969. Observations of moose feeding on aquatics in Bowron Lake Park, British Columbia. Can. Field-Nat. 83: 339–343.

Ritchie, B.W. 1978. Ecology of moose in Fremont County, Idaho. Wildl. Bull. 7. Idaho Dept. Fish and Game, Boise. 33 pp.

Ritchie, B.W. and D.A. Barney. 1972. Use of snowmobiles to capture moose. Proc. N. Am. Moose Conf. Workshop 8: 296–300.

Ritzenthaler, R.E. 1978. Southwestern Chippewa. Pages 743–759 *in* B.G. Trigger, ed., Northeast. Handbook of North American Indians. Vol. 15. Smithsonian Inst., Washington, D.C. 924 pp.

Rivest, L.P., H. Crepeau and M. Crête. 1990. A two-phase sampling plan for the estimation of the size of a moose population. Biometrics 46: 163–176.

Rizzo, B. and E. Wilken. 1992. Assessing the sensitivity of Canada's ecosystems to climatic change. Climatic Change. 21: 37–55.

Robbins, C.T. 1973. The biological basis for the determination of carrying capacity. Ph.D. thesis, Cornell Univ., Ithaca, New York. 239 pp.

———. 1983. Wildlife feeding and nutrition. Academic Press, New York. 343 pp.

———. 1993. Wildlife feeding and nutrition. 2nd ed. Academic Press, San Diego, California. 352 pp.

Robbins, C.T. and A.N. Moen. 1975. Uterine composition and growth in pregnant white-tailed deer. J. Wildl. Manage. 39: 684–691.

Robbins, C.T., P.J. Van Soest, W.W. Mautz and A.N. Moen. 1975. Feed analyses and digestion with reference to white-tailed deer. J. Wildl. Manage. 39: 67–79.

Robbins, R.L., D.E. Redfern and C.P Stone. 1982. Refuges and elk management. Pages 479–507 *in* J.W. Thomas and D.E. Toweill, eds., Elk of North America: Ecology and management. Stackpole Books, Harrisburg, Pennsylvania. 698 pp.

Roberts, F.H.H., Jr. 1940. Developments in the problem of the North American Paleo-Indian. Pages 51–116 *in* Essays in historical anthropology of North America. Smithsonian Misc. Coll. 100. Washington, D.C.

Robinette, W.L., D.A. Jones and H.S. Crane. 1955. Fertility of mule deer in Utah. J. Wildl. Manage. 19: 115–136.

Robinson, T.J. 1959. The estrous cycle of the ewe and doe. Pages 291–334 *in* H.H. Cole and P.T. Cupps, eds., Reproduction in domestic animals. Academic Press, New York.

Roese, J.H., K.L. Risenhoover and L.J. Folse. 1991. Habitat heterogeneity and foraging efficiency: An individual-based model. Ecol. Model. 57: 133–143.

Rogers, E.S. 1962. The Round Lake Ojibwa. Royal Ontario Mus. Art and Archeaol. Div., Occas. Pap. 5(III): 1–79.

———. 1963. The hunting group–hunting territory complex among the Mistassini Indians. Anthrol. Ser. 63, Bull. 195. Natl. Mus. Canada, Ottawa.

———. 1967. The material culture of the Mistassini. Anthrol. Series 80, Bull. 218. Natl. Mus. Canada, Ottawa.

———. 1973. The quest for food and furs: The Mistassini Cree, 1953–1954. Publ. in Ethnol. 5. Natl. Mus. Man, Ottawa.

Rogers, E.S. and E. Leacock. 1981. Montaganais-Naskapi. Pages 169–189 *in* J. Helm, ed., Subarctic. Handbook of North American Indians. Vol. 6. Smithsonian Inst., Washington, D.C. 837 pp.

Rogers, E.S. and J.G.E. Smith. 1981. Environment and culture in the shield and Mackenzie Borderlands. Pages 130–145 *in* J. Helm, ed., Subarctic. Handbook of North American Indians. Vol. 6. Smithsonian Inst., Washington, D.C. 837 pp.

Rogers, E.S. and J.G. Taylor. 1981. Northern Ojibwa. Pages 231–243 *in* J. Helm, ed., Subarctic. Handbook of North American Indians. Vol. 6. Smithsonian Inst., Washington, D.C. 837 pp.

Rolley, R.E. 1982. Moose (Alberta). Pages 262–263 *in* D.E. Davis, ed., CRC handbook of census methods for terrestrial vertebrates. CRC Press, Boca Raton, Florida.

Rolley, R.E. and L.B. Keith. 1980. Moose population dynamics and winter habitat use at Rochester, Alberta, 1965–1969. Can. Field-Nat. 94: 9–18.

Rollins, R. 1987. Hunter satisfaction with the selective harvest system for moose in northern Ontario. Alces 23: 181–194.

Rollins, R. and L. Romano. 1989. Hunter satisfacion with the selective harvest system for moose management in Ontario. Wildl. Soc. Bull. 17: 470–475.

Romer, A.S. 1956. Osteology of the reptiles. Univ. Chicago Press, Chicago, Illinois. 772 pp.

Romer, A.S. and T.S. Parsons. 1978. The vertebrate body. Saunders College, Philadelphia. 476 pp.

Rongstad, O.J. and J.R. Tester. 1969. Movements and habitat use of white-tailed deer in Minnesota. J. Wildl. Manage. 33: 366–379.

Roosevelt, T.R. 1893. The wilderness hunter. G.P. Putnam's Sons, New York. 472 pp.

———. 1908. Outdoor pastimes of an American hunter. Charles Scribner's Sons, New York. 420 pp.

———. 1909. The wilderness hunter, an account of the big game of the United States and its chase with horse, hound, and rifle. G. P. Putnam's Sons, New York. 472 pp.

———. 1916. A curious experience. Scribner's Mag. LIX: 155–167.

Rorig, A. 1900. Über Geweihentwicklung und Geweihbildung. II. Abschnitt. Die Geweihentwicklung in histologischer und histogenetischer Hinsicht. Arch. Entwickl.-mech. Organ. 17: 618–645.

———. 1906. Das Wachstum des Geweihes von—*Cervus elaphus, Cervus barbarus,* und *Cervus canadensis*. Arch. f. Enwickl. -mech. Organ. 20: 507–536.

———. 1908. Das Wachstum des Geweihes von—*Capreolus vulgaris*. Arch. F. Entwickl.-mech. Organ. 25: 423–430.

Rosenzweig, M.L. 1981. A theory of habitat selection. Ecol. 62: 327–335.

Ross, B.R. 1861. An account of the animals useful in an economic point of view to the various Chipewyan tribes. Canadian Nat. and Geol. (6):433–444.

Rostlund, E. 1952. Fresh water fish and fishing in native North America. Univ. California Publ. in Geogr. 9. Berkeley.

Rounds, R.C. 1981. First approximation of habitat selectivity on extensive winter ranges. J. Wildl. Manage. 45: 187–196.

Rousseau, J. 1952. Persistences païennes chez les indiens de la forêt boréale. Cahiers des Dix 17:183–208.

Roussel, Y.E. and R. Patenaude. 1975. Some physiological effects of M-99 etorphine on immobilized free-ranging moose. J. Wildl. Manage. 39: 634–636.

Roussel, Y.E. and C. Pichette. 1973. Review of techniques used to restrain and mark moose in Laurentides Park. Proc. N. Am. Moose Conf. Workshop 9: 43–58.

Roussel, Y.E., A. Audy and F. Potvin. 1975. Preliminary study of seasonal moose movements in Laurentides Provincial Park, Quebec. Can. Field-Nat. 88: 47–52.

Rowe, J.S. 1982. Forest regions of Canada. For. Branch Bull. 123. Can. Dept. North. Affairs and Nat. Resour., Ottawa. 71 pp.

Rowe, J.S. 1972. Forest regions of Canada. Rev. ed. Publ. 1300. Can. For. Serv., Ottawa.

Rowland, D. 1989. Outdoor health and nutrition. Eating wild game is no trivial pursuit. Courier (Spring): 18.

Roy, K.K. 1974. Early relations between the British and Indian medical systems. Proc. Int. Congr. for Hist. Medic. XXIII(1): 697–703.

Rue, L.L., III. and W. Owen. 1985. Meet the moose. Dodd, Mead and Co., New York. 78 pp.

Ruel, J. and D. Le Boeuf. 1987. Eastern moose. Sentier Chasse-Pêche, Montreal. 175 pp.

Rulcker, J. and F. Stalfelt. 1986. Das Elchwild. Verlag P. Parey, Hamburg. 285 pp.

Rumney, G.R. 1969. Climatology and the world's climates. MacMillan Co., New York. 656 pp.

Runtz, M.M.P. 1991. Moose country, saga of the woodland moose. Stoddart Publ. Co., Ldt., Toronto. 100 pp.

Russell, F. 1898. Explorations in the far North. Univ. Iowa, Iowa City. 290 pp.

Rutberg, A.T. 1986. Lactation and fetal sex ratios in American bison. Am. Nat. 127: 89–94.

Rutske, L. 1989. A guide to field care of moose meat and trophies. Wildl. Sect. Minnesota Dept. Nat. Resour., St. Paul. 28 pp.

Rutter, N. 1988. Late Pleistocene history of the western Canada ice-free corridor. Can. J. Anthro. 1: 1–8.

Ryan, M.J. 1990. Signals, species, and sexual selection. Am. Scientist 79: 46–52.

Rykovskyj, A.S. 1964. Nablyudeniya za revom losya v Kaluzhskoy oblasti [Observations during the bellowing of European elk *(Alces alces)* in Kaluzhskoi region.] Pages 143–150 *in* A.G. Bannikov, ed., Biologiya i promysel losya. Rosselkhozizd, Moscow.

Ryman, N., C. Reuterwall, K. Nygren and T. Nygren. 1980. Genetic variation and differentiation in Scandinavian moose *(Alces alces)*: Are large mammals monomorphic? Evolution 34: 1,037–1,049.

Sablina, T.B. 1970. Evolutsiya pishchevarietelnoy sistemy oleney. Nauka, Moscow. 248 pp.

Sadlier, R.M.F.S. 1982. Reproduction of female cervids. Pages 123–144 *in* C.M. Wemmer, ed., Biology and management of the Cervidae. Smithsonian Inst. Press, Washington, D.C. 577 pp.

Saether, B-E. 1987. Patterns and processes in the population dynamics of the Scandanavian moose *(Alces alces)*: Some suggestions. Swed. Wildl. Res. (Suppl.) 1: 525–537.

———. 1990. The impact of different growth patterns on the utilization of tree species by a generalist herbivore, the moose *(Alces alces)*: Implications on optimal foraging theory. Pages 323–341 *in* R.N. Hughes,ed., Behavioral mechanisms of food selection. Springer-Verlag, New York.

Saether, B-E. and R. Andersen. 1990. Resource limitation in a generalist herbivore, the moose, *(Alces alces)*: Ecological constraints on behavioural decisions. Can. J. Zool. 68: 993–999.

Saether, B-E. and H. Haagenrud. 1983. Life history of the moose *Alces alces*: Fecundity rates in relation to age and carcass weight. J. Mammal. 64: 226–232.

Saether, B-E. and H. Haagenrud. 1985. Life history of the moose *Alces alces*: Relationship between growth and reproduction. Holarctic Ecol. 8: 100–106.

Sagard, G. 1632. Le grand voyage du pays des Hurons. Chez Denys Moreau, Paris.

Salwasser, H. 1986. Modelling habitat relationships of terrestrial vertebrates. Pages 419–424 *in* J. Verner, M.L. Morrison and C.J. Ralph, eds., Wildlife 2000. Modelling habitat relationships of terrestrial vertebrates. Univ. Wisconsin Press, Madison.

———. 1993. The role of wildlife management in sustainable development. The Wildlifer 261:58, 66, 71–72.

Samuel, M.D. and E.O. Garton. 1985. Home range: A weighted normal estimate and tests of underlying assumptions. J. Wildl. Manage. 35: 338–346.

Samuel, M. D., D. J. Pierce, E. O. Garton, L. J. Nelson and K. R. Dixon. 1983. User's manual for program home range. For. Wildl. Exp. Sta. Tech. Rept. 15. Univ. Idaho, Moscow. 64 pp. (Revised 1985.)

Samuel, M. D., R.K. Steinhorst, E.O. Garton and J.W. Unsworth. 1992. Estimation of wildlife population ratios incorporating survey design and visibility bias. J. Wildl. Manage. 56: 718–725.

Samuel, W.M. 1972a. Encounters between hunters and parasites of moose in Alberta. Occas. Pap. No. 2. Alberta Fish and Wildl. Div., Edmonton. 21 pp.

———. 1972b. *Taenia krabbei* in the musculature of moose: A review. Proc. N. Am. Moose Conf. Workshop 8: 18–41.

———. 1991a. Grooming by moose *(Alces alces)* infested with the winter tick, *Dermacentor albipictus* (Acari): A mechanism for premature loss of winter hair. Can. J. Zool. 69: 1,255–1,260.

———. 1991b. A partially annotated bibliography of meningeal worm, *Parelaphostrongylus tenuis* (Nematoda), and its close relatives. *In* M.J. Kennedy, ed., Synopses of the parasites of vertebrates. Anim. Health Div., Alberta Agric., Edmonton. 36 pp.

Samuel, W.M. and M. Barker. 1979. The winter tick, *Dermacentor albipictus* (Packard, 1869) on moose, *Alces alces* (L.), of central Alberta. Proc. N. Am. Moose Conf. Workshop 15: 303–348.

Samuel, W.M. and D.A. Welch. 1991. Winter ticks on moose and other ungulates: Factors influencing their population size. Alces 27: 169–182.

Samuel, W.M., M.W. Barrett and G.M. Lynch. 1976. Helminths in moose of Alberta. Can. J. Zool. 54: 307–312.

Samuel, W.M., D.A. Welch and M.L. Drew. 1986. Shedding of the juvenile and winter hair coats of moose *(Alces alces)*. Alces 22: 345–360.

Samuel, W.M., M.J. Pybus, D.A. Welch and C.J. Wilke. 1992. Elk as a potential host for meningeal worm: Implications for translocation. J. Wildl. Manage. 56: 629–639.

Samuels, P. 1990. Remington: The complete prints. Crown Publ., Inc. New York. 167 pp.

Samuels, P. and H. Samuels. 1985. Samuel's encyclopedia of artists of the American West. Castle, Secaucas, New Jersey. 549 pp.

Sancetta, C. and S.W. Robinson. 1983. Diatom evidence on Wisconsin and Holocene events in the Bering Sea. Quat. Res. 20: 232–245.

Sandegren, F. and P.Y. Sweanor. 1988. Migration distances of moose populations in relation to river drainage length. Alces 24: 112–117.

Sandegren, F., R. Bergstron and P.Y. Sweanor. 1985. Seasonal moose migrations related to snow in Sweden. Alces 21: 321–338.

Sanderson, G.C. 1966. The study of mammal movements—A review. J. Wildl. Manage. 30: 215–235.

Saskatchewan Forest Habitat Project. 1992. Annual report (1991/92) and workplan (1992/94). Saskatchewan For. Habitat Proj. Int. Rept., Regina. 98 pp.

Saugstad, S. 1942. Aerial census of big game in North Dakota. Trans. N. Am. Wildl. Conf. 7:343–356.

Saunders, B.P. and J.C. Williamson. 1972. Moose movements from ear tag returns. Proc. N. Am. Moose Conf. Workshop 8: 177–184.

Savishinsky, J.S. 1970. Kinship and the expression of value in an Athabascan bush community. Pages 31–59 *in* R. Darnell, ed., Athabascan studies. W. Can. J. Anthropol. 2(1).

Savishinsky, J.S. and H.S. Hara. 1981. Hare. Pages 314–325 *in* J. Helm, ed., Subarctic. Handbook of North American Indians. Vol. 6. Smithsonian Inst., Washington, D.C. 837 pp.

Scanlon, P.F., K.I. Morris, A.G. Clark, N. Fimreite and S. Lierhagen. 1986. Cadmium in moose tissue: Comparison of data from Maine, U.S.A., and from Telemark, Norway. Alces 22: 303–312.

Schamberger, M.L. and L.J. O'Neil. 1986. Concepts and constraints of habitat-model testing. Pages 5-10 *in* J. Verner, M.L. Morrison and C.J. Ralph, eds., Wildlife 2000: Modelling habiatat relationships of terrestrial vertebrates. Univ. Wisconsin Press, Madison.

Schierbeck, O. 1929. Is it right to protect the female of the species at the cost of the male? Can. Field-Nat. 43: 6–9.

Schladweiler, P. and D.R. Stevens. 1973. Reproduction of Shiras moose in Montana. J. Wildl. Manage. 37: 535–544.

Schlegel, M. 1976. Factors affecting calf elk survival in northcentral

Idaho—A progress report. Proc. Ann. Conf. W. Assoc. St. Game and Fish Commiss. 56: 342–355.

Schmidt, J.L. and D.L. Gilbert, eds. 1978. Big game of North America: Ecology and management. Stackpole Books, Harrisburg, Pennsylvania. 494 pp.

Schmidt-Nielsen, L. 1975. Animal physiology: Adaptation and environment. MacMillan of Canada, Toronto. 699 pp.

Schmitt, S.M. and R.W. Aho. 1988. Reintroduction of moose from Ontario to Michigan. Pages 258–274 *in* L. Nielsen and R.D. Brown, eds., Translocation of wild animals. Wisconsin Hum. Soc., Milwaukee.

Schnare, H. and K. Fischer. 1987. Sekundäre Geschlechtsmerkmale und physiologische Parameter beim Damhirsch (*Dama dama* L.) in Abhängigkeit von der Photoperiode. Pages 43–44 *in* Abstracts: 61. Hauptversamml. Deut. Gesell. Säugetierkde. 82, 83.

Schoener, T.W. 1968. The anolis lizards of Bimini: Resource partitioning in a complex fauna. Ecol. 49: 704–726.

———. 1971. Theory of feeding strategies. Ann. Rev. Ecol. Syst. 2: 369–404.

———. 1981. An empirical based estimate of home range. Theor. Pop. Biol. 20: 281–325.

Schoolcraft, H.R. 1851. Personal memoirs of a residence of thirty years with the Indian tribes on the American frontiers. Lippencott, Grambo and Co., Philadelphia.

Schoonover, C. 1976. Frank Schoonover: Illustrator of the North American frontier. Watson-Guptill Publ., New York. 207 pp.

Schowalter, D.B., J.O. Iverson, L.C. Corner and J.R. Gunson. 1980. Prevalence of antibodies to *Toxoplasma gondii* in striped skunks from Saskatchewan and Alberta. J. Wildl. Dis. 16: 189–194.

Schuh, E. 1982. Zahn- und Zahnbetterkrankungen bei alterndem Elch *(Alces alces)*. Ein Beitrag zur Vergleichenden Paradontologie. [Disease of tooth and jaw in an ageing moose. A contribution to comparative paradontology]. Z. f. Jagdwiss. 28: 123–130.

Schuhmacher, S. 1939. Jagd und Biologie—Ein Grundriss der Wildkunde. J. Springer-Verlag, Berlin. 236 pp.

Schwab, F.E. and M.D. Pitt. 1991. Moose selection of canopy cover types related to operative temperature, forage, and snow depth. Can. J. Zool. 69: 3,071–3,077.

Schwartz, C.C. 1987. Evaluation and testing of techniques for moose management. Fed. Aid Wildl. Rest. Prog. Rept. Alaska Dept. Fish and Game, Juneau. 36 pp.

———. 1992a. Physiological and nutritional adaptations of moose to northern environments. Alces (Suppl.) 1: 139–155.

———. 1992b. Techniques of moose husbandry. Alces (Suppl.) 1: 177–192.

———. 1992c. Reproductive biology of North American moose. Alces 28: 165–173.

———. 1993. Constructing simple population models for moose management. Alces 29: 235–242.

Schwartz, C.C. and B. Bartley. 1991. Reducing incidental moose mortality: Considerations for management. Alces 27: 227–231.

Schwartz, C.C. and A.W. Franzmann. 1981. Moose Research Center report. Fed. Aid. Wildl. Restor. Prog. Rept. Alaska Dept. Fish and Game, Juneau. 49 pp.

Schwartz, C.C. and A.W. Franzmann. 1983. Effects of tree crushing on black bear predation on moose calves. Int. Conf. Bear Res. and Manage. 5: 40–44.

Schwartz, C.C. and A.W. Franzmann. 1989. Bears, wolves, moose, and forest succession, some management considerations on the Kenai Peninsula, Alaska. Alces 25: 1–10.

Schwartz, C.C. and A.W. Franzmann. 1991. Interrelationship of black bears to moose and forest succession in the northern coniferous forest. Wildl. Monogr. 113. The Wildl. Soc., Bethesda, Maryland. 58 pp.

Schwartz, C.C. and N.T. Hobbs. 1985. Forage and range evaluation. Pages 25–52 *in* R.J. Hudson and R.G. White, eds., Bioenergetics of wild herbivores. CRC Press, Boca Raton, Florida.

Schwartz, C.C. and K.J. Hundertmark. 1993. Reproductive characteristics of Alaskan moose. J. Wildl. Manage. 454–468.

Schwartz, C.C., W.L. Regelin and A.W. Franzmann. 1980. A formulated ration for captive moose. Proc. N. Am. Moose Conf. Workshop 16: 82–105.

Schwartz, C.C., A.W. Franzmann and D. Johnson. 1981. Moose Research Center report. Fed. Aid in Wildl. Rest. Proj. W-21-2. Alaska Dept. Fish and Game, Juneau. 65 pp.

Schwartz, C.C., W.L. Regelin and A.W. Franzmann. 1982. Male moose successfully breed as yearlings. J. Mammal. 63: 334–335.

Schwartz, C.C., A.W. Franzmann and D. Johnson. 1983. Moose Research Center report. Fed. Aid in Wildl. Rest. Proj. W-21-1. Alaska Dept. Fish and Game, Juneau. 65 pp.

Schwartz, C.C., W.L. Regelin and A.W. Franzmann. 1984. Seasonal dynamics of food intake in moose. Alces 20: 223–244.

Schwartz, C.C., W.L. Regelin and A.W. Franzmann. 1985. Suitability of a formulated ration for moose. J. Wildl. Manage. 49: 137–141.

Schwartz, C.C., W.L. Regelin and A.W. Franzmann. 1987a. Seasonal weight dynamics of moose. Swed. Wildl. Res. (Suppl.) 1: 301–310.

Schwartz, C.C., W.L. Regelin and A.W. Franzmann. 1987b. Protein digestion in moose. J. Wildl. Manage. 51: 352–357.

Schwartz, C.C., W.L. Regelin, A.W. Franzmann and M. Hubert. 1987c. Nutritional energetics of moose. Swed. Wildl. Res. (Suppl.) 1: 265–280.

Schwartz, C.C., M.E. Hubbert and A.W. Franzmann. 1988a. Changes in body composition of moose during winter. Alces 24: 178–187.

Schwartz, C.C., M.E. Hubbert and A.W. Franzmann. 1988b. Energy requirements of adult moose for winter maintenance. J. Wildl. Manage. 52: 26–33.

Schwartz, C.C., M.E. Hubbert and A.W. Franzmann. 1988c. Estimates of digestibility of birch, willow, and aspen mixtures in moose. J. Wildl. Manage. 52: 33–37.

Schwartz, C.C., W.L. Regelin, A.W. Franzmann, R.G. White and D.F. Holleman. 1988d. Food passage rate in moose. Alces 24: 97–101.

Schwartz, C.C., A.B. Bubenik and R. Claus. 1990. Are sex pheromones involved in moose breeding behavior? Alces 24: 178–187.

Schwartz, C.C., M.E. Hubbert and A.W. Franzmann. 1991. Energy expenditure in moose calves. J. Wildl. Manage. 55: 391–393.

Schwartz, C.C., K.J. Hundertmark and T.H. Spraker. 1992. An evaluation of selective bull moose harvest on the Kenai Peninsula, Alaska. Alces 28: 1–13.

Schwarz, K. 1974. New essential trace elements (Sn, V, F, Si): Progress report and outlook. Pages 355–380 *in* W.G. Hoekstra, J.W. Suttie, H.E. Gantes and W. Mertz, eds., Trace element metabolism in animals—2. Univ. Park Press, Baltimore, Maryland. 755 pp.

Scott, J.D. and O. Sweet. 1981. Moose. G.P. Putnam's Sons, New York. 64 pp.

Scott, J.P. 1967. The development of social motivation. Pages 111–132. Vol. 12 *in* D. Levine, ed., Nebraska Symp. on Motivation. Univ. Nebraska Press, Lincoln.

Scott, S., ed. 1983. Field guide to the birds of North America. Natl. Geog. Soc., Washington, D.C. 464 pp.

Scott, T.G. and C.H. Wasser. 1980. Checklist of North American plants for wildlife biologists. The Wildl. Soc., Washington, D.C. 58 pp.

Scott, W.B. 1885. *Cervalces americanus*, a fossil moose, or elk from the Quaternary of New Jersey. Proc. Acad. Natl. Sci. 1885: 181-202.

Scribner, K.T. and M.H. Smith. 1990. Genetic variability and antler development. Pages 460–473 *in* G.A. Bubenik and A.B. Bubenik, eds., Horns, pronghorns and antlers. Springer- Verlag, New York.

Seal, U.S., S.M. Schmitt and R.O. Peterson. 1985. Carfentanil and xylazine for immobilization of moose *(Alces alces)* on Isle Royale. J. Wildl. Dis. 21: 48–51.

Seaman, D.E. 1993. Home range and male reproductive optimization in black bears. Ph.D. thesis, North Carolina St. Univ., Raleigh. 100 pp.

Sears, H. 1898. William's moose. Harper's New Month. Mag. XCVII: 87–95.

Sebeok, T.A. 1962. Evolution of signaling behavior. Behav. Sci. 6: 430–442.

———. 1963. The informational model of language: Analog and digital coding in animal and human communication. Pages 47–64 *in* P.L. Garvin, ed., Natural language and the computer. McGraw-Hill, New York.

———. 1965. Animal communication. Science 147: 1,006–1,014.

Sein, R. 1985. Biting flies (Diptera) attracted to, and their effect on the behaviour of, captive moose *(Alces alces)* and captive woodland caribou *(Rangifer tarandus caribou)* in northwestern Ontario. Hons. thesis, Lakehead Univ., Thunder Bay, Ontario. 98 pp.

Seip, D.R. 1992. Factors limiting woodland caribou populations and their interrelationships with wolves and moose in southeastern British Columbia. Can. J. Zool. 70: 1,494–1,503.

Selous, F.C. 1907. Recent hunting trips in British North America. Charles Scribner's Sons, New York. 400 pp.

Selye, H. 1950. The physiology and pathology of exposure to stress. Acta Publ., Montreal. 150 pp.

Sempere, A.J. and J. Boissin. 1983. Neuroendocrine and endocrine control of the antler cycle in roe deer. Pages 109-122 *in* R.D. Brown, ed., Antler development in Cervidae. Caesar Kleberg Wildlife Research Institute, Kingsville, Texas.

Semyonov, B.T. 1956. Kolichestveny uchet losey po sledam samoleta. [Numerical census of European elk (moose) by aerial survey]. Voprosy biologii Pushnykh Zverey, Moscow. 72 pp. (Translated from Russian by J.M. MacLennan, Can. Wildl. Serv. 22 pp.)

Serafinski, W. 1969. Reproduction and dynamics of moose *(Alces alces)* in Kampoinos National Park. Ekologia Polska, Ser. A 17: 709–718.

Sergeant, D.E. and D.H. Pimlott. 1959. Age determination in moose from sectioned incisor teeth. J. Wildl. Manage. 23: 315–321.

Serveen, G. and L.J. Lyon. 1989. Habitat use by woodland caribou in the Selkirk Mountains. J. Wildl. Manage. 53: 230–237.

Seton, E.T. 1909. Life-histories of northern animals. Vol. I. Grass-eaters. Charles Scribner's Sons, New York. 673 pp.

———. 1929. Lives of game animals. Vol. III, Pt. I. Hoofed animals. Doubleday, Doran and Co., Garden City, New York. 412 pp.

———. 1940. Trail of an artist-naturalist. Charles Scribner's Sons, New York. 412 pp.

———. 1953. Lives of game animals. Vol. 3. Charles T. Branford Co., Boston. 411 pp.

Seuss, Dr. (pseud.) 1948. Thidwick: The big-hearted moose. Random House, New York.

Severtsov, S.A. 1951. Problemy ekologii zhivotnykh—Neopublikovanye roboty. Akad. Nauk SSR, Moscow. 172 pp.

Seyfarth, M. 1938. Pathogene Wirkung und innerer Bau von Paramphistomum cervi. Deut. Tierarz. Wochenschrift 46: 515–518.

Shank, C.C. 1985. Inter- and intrasexual segregation of chamois *(Rupicapra rupicapra)* by altitude and habitat during summer. Zeit. Saugeterkunde 50: 117–125.

Shaw, D.L. 1973. The hunting controversy: Attitudes and arguments. Ph.D. thesis. Colorado St. Univ., Fort Collins. 174 pp.

Shelford, V.E. 1963. The ecology of North America. Univ. Illinois Press, Urbana.

Shelton, P.C. and R.O Peterson. 1983. Beaver, wolf and moose interactions in Isle Royale National Park, USA. Acta Zool. Fennica 174: 265–266.

Shepp, J.W. and D.B. Shepp. 1893. Shepp's world's fair photographed. Globe Bible Publ. Co., Chicago, Illinois. 529 pp.

Sher, A.V. 1987. History and evolution of moose in USSR. Swed. Wildl. Res. (Suppl.) 1: 71–100.

Sherrod, D. 1990. Quelling controversy through public relations—implementing a controlled moose hunt. New Hampshire Fish and Game Dept., Concord. 10 pp.

Shields, G.O., ed. 1890. The big game of North America. Rand, McNally and Co., New York. 581 pp.

Shields, W.M. 1983. Optimum inbreeding and the evolution of philopatry. Pages 132–159 *in* I.R. Swingland and P.J. Greenwood, eds., The ecology of animal movement. Clarendon Press, Oxford.

———. 1987. Dispersal and mating systems: Investigating their casual connections. Pages 3–24 *in* B.D. Chepko-Sade and Z.T. Halpin, eds., Mammalian dispersal patterns. Univ. Chicago Press, Chicago, Illinois.

Shipley, L. and D.E. Spalinger. 1992. Mechanics of browsing in dense food patches: Effects of plant and animal morphology on intake rate. Can. J. Zool. 70: 1,743–1,752.

Shiras, G., III. 1906. Photographing wild game with flashlight and camera. Natl. Geog. Mag. XVII: 366–423.

———. 1908. One season's game-bag with the camera. Natl. Geog. Mag. XIX: 387–446.

———. 1912. White sheep, giant moose, and smaller game of the Kenai Peninsula, Alaska. Natl. Geog. Mag. XXIII: 423–494.

———. 1913. Wild animals that took their own pictures by day and night. Natl. Geog. Mag. XXIV: 763–834.

———. 1921. Wild life of Lake Superior. Past and present: The habits of deer, moose, wolves, beavers, muskrats, trout, and feathered woodfolk studied with camera and flashlight. Natl. Geog. Mag. XL: 113–204.

———. 1932. Wild life of the Atlantic and Gulf coasts. A field naturalist's photographic record of nearly a half-century of fruitful exploration. Natl. Geog. Mag. LXII: 261–309.

———. 1935. Hunting wild life with camera and flashlight. 2 vols. Natl. Geog. Soc., Washington, D.C.

Shively, L. 1989. Mechanics of foraging behavior of boreal herbivores. M.S. thesis, Univ. Maine, Orono. 96 pp.

Shostak, A.W. and W.M. Samuel. 1984. Moisture and temperature effects of survival and infectivity of first-stage larvae of *Parelaphostrongyles odocoilei* and *P. tenuis* (Nematoda: Metastrongyloidea). J. Parasitol. 70: 261–269.

Shreeve, J. 1993. Touching the phantom. Discover 14: 35–42.

Siepierski, S.J., C.E. Tanner and J.A. Embil. 1990. Prevalence of antibody to *Toxoplasma gondii* in moose (*Alces alces americana* Clinton) of Nova scotia, Canada. J. Parasitol. 76: 136–138.

Silver, H.H., N.F. Colovos, J.B. Holter and H.H. Hayes. 1969. Fasting metabolism of white-tailed deer. J. Wildl. Manage. 33: 490–498.

Silver, H.H., J.B. Holter, N.F. Colovos and H.H. Hayes. 1971. Effect of falling temperature on heat production in fasting white-tailed deer. J. Wildl. Manage. 35: 37–46.

Simkin, D.W. 1959. Big game browse and pellet survey in Sioux Lookout District. Pages 19–26 *in* Fish and Wildl. Manage. Rept. 46. Ontario Dept. Lands and For., Toronto.

———. 1962. Weights of Ontario moose. Ontario Fish and Wildl. Rev. 1: 10–12.

———. 1963. Tagging moose by helicopter. J. Wildl. Manage. 27: 136–139.

———. 1965. Reproduction and productivity of moose in northwestern Ontario. J. Wildl. Manage. 29: 740–750.

———. 1968. A comparison of three methods to age moose. Proc. N. Am. Moose Conf. Workshop 5: 46–61.

———. 1974. Reproduction and productivity of moose. Nat. Can. (Que.) 101: 517–525.

Simpson, G.G 1946. The principles of classification and a classification of mammals. Vol. 85. Bull. Am. Mus. Nat. Hist., New York. 350 pp.

Sims, R.A., W.D. Towill, K.A. Baldwin and G.M. Wickware. 1989. Forest ecosystems classification for northwestern Ontario. Ontario For. Res. Dev. Agreement Publ., Toronto. 191 pp.

Sinclair, A.R.E. 1977. The African buffalo—A study of resource limitation of populations. Univ. Chicago Press, Chicago, Illinois. 355 pp.

———. 1983. The function of distance movements in vertebrates. Pages 240–258 *in* I.R. Swingland and P.J. Greenwood, eds., The ecology of animal movement. Clarendon Press, Oxford.

———. 1989. Population regulation in animals. Pages 197–241 *in* J.M. Cherrett, ed., Ecological concepts: The contribution of ecology to the understanding of the natural world. Blackwell Scientific Publ., Oxford, England.

———. 1991. Science and the practice of wildlife management. J. Wildl. Manage. 55: 767–773.

Sinclair, A.R.E. and M. Norton-Griffiths. 1979. Serengeti. Dynamics of an ecosystem. Univ. Chicago Press, Chicago, Illinois. 389 pp.

Singer, F.J. and J. Dalle-Molle. 1985. The Denali ungulate-predator system. Alces 21: 339–358.

Siniff, D.B. 1986. Sex ratio variation in relation to maternal condition and parental investment in wild reindeer *(Rangifer t. tarandus)* Oikos 46: 417–419.

———. 1991. What are the effects of predators on large ungulate populations? Oikos 61: 401–411.

Siniff, D.B. and R.O. Skoog. 1964. Aerial censusing of caribou using stratified random sampling. J. Wildl. Manage. 28: 391–401.

———. 1991. What are the effects of predators on large ungulate populations? Oikos 61: 401–411.

Skinner, A.B. 1912. Notes on the Eastern Cree and Northern Saulteaux. Anthropol. Pap. of Am. Mus. Nat. Hist. 9: 1–179.

Skogland, T. 1991. What are the effects of predators on large ungulate populations? Oikos 61: 401–411.

———. 1985. The effects of density-dependent resource limitations on the demography of wild reindeer. J. Anim. Ecol. 54: 359–374.

Skogland, T. 1986. Sex ratio variation in relation to maternal condition and parental investment in wild reindeer *Rangifer t. tarandus*. Oikos 46: 417–419.

Skuncke, F. 1949. Algen-Studier, jakt ach vard. P.A. Norstedt und Soners Forlag, Stockholm. 400 pp.

Slaby, O. 1990. Morphogenesis of the nasal apparatus of the red deer (*Cervus elaphus* L.). Folia Morphologica 38: 212–223.

Slauson, W. L., B. S. Cade and J. D. Richards. 1991. Users manual for BLOSSOM statistical software. Natl. Ecol. Res. Ctr., U.S. Fish and Wildl. Serv., Fort Collins, Colorado. 61 pp.

Slijper, E.J. 1938. Vergleichende anatomische Untersuchungen uber den Penis der Säugetiere. Acta Neerlandica Morphol. 1: 375–418.

Slobodin, R. 1981. Kutchin. Pages 514–532 *in* J. Helm, ed., Subarctic. Handbook of North American Indians. Vol. 6. Smithsonian Inst., Washington, D.C. 837 pp.

Slomke, A.M., M.W. Lankester and W.J. Peterson. 1955. Infrapopulation dynamics of *Parelaphostrongylus tenuis* in white-tailed deer. J. Wildl. Dis. 31: 125–135.

Smith, A.H. 1984. Kutenai Indian subsistence and settlement patterns, northwest Montana. U.S. Army Corps of Engineers, Seattle, Washington. 291 pp.

Smith, B.L. 1982. The history, current status and management of moose on the Wind River Reservation. U.S. Fish and Wildl. Serv., Lander, Wyoming. 98 pp.

———. 1985. Moose and their management on the Wind River Indian Reservation, Wyoming. Alces 21: 359–391.

Smith, H. 1990. Managing a moose population. Pages 25–29 *in* M. Buss and R. Truman, eds., The moose in Ontario. Book 1. Wildl. Branch. Ontario Min. Nat. Resour., Toronto.

Smith, H.L. 1976. Identification of social classes by physical features amd management implications for white-tailed deer *(Odocoileus virginianus)*. M.S. thesis, Univ. Guelph, Guelph, Ontario. 103 pp.

Smith, J. 1616. A description of New England: Or the observations, and discoveries, of Captain John Smith (Admirall of that country) in the North of America, in the year of our Lord 1614. 3 vols. Humfrey Lownes, London, England.

———. 1624. The general historie of Virginia, New-England and the Summer Isles, together with the true travel adventures, and observations. . . . I.D. and I.H., London. (Reprinted as The complete works of Captain John Smith, 1986, 3 vols., by the Univ. North Carolina Press, Chapel Hill.)

Smith, J.G.E. 1981a. Western Woods Cree. Pages 256–270 *in* J. Helm, ed., Subarctic. Handbook of North American Indians. Vol. 6. Smithsonian Inst., Washington, D.C. 837 pp.

———. 1981b. Chipewyan. Pages 271–284 *in* J. Helm, ed., Subarctic. Handbook of North American Indians. Vol 6. Smithsonian Inst., Washington, D.C. 837 pp.

Smith, N.S. 1962. The fall and winter ecology of the Shiras moose in Rock Creek drainage, Granite County, Montana. M.S. thesis, Univ. Montana, Missoula. 52 pp.

Smith, W.P. 1983. A bivariate normal test for elliptical home-range models: Biological implications and recommendations. J. Wildl. Manage. 47: 613–619.

Smith, H.J. and R.M. Archibald. 1967. Moose sickness, a neurological disease os moose infected with the common cervine parasite, *Elaphostrongylus tenuis*. Can. Vet. J. 8: 173–177.

Smith, N.E. and R.L. Baldwin. 1974. Effects of breed, pregnancy and lactation on weight of organs and tissues in dairy cattle. J. Dairy Sci. 57: 1,055–1,060.

Smith, H.J., R.M. Archibald and A.H. Corner. 1964. Elaphostrongylosis in maritime moose and deer. Can. Vet. J. 5: 287–296.

Smith, M.H., K.T. Scribner, L.H. Carpenter and R.A. Garrott. 1990. Genetic characteristics of Colorado mule deer *(Odocoileus hemionus)* and comparisons with other cervids. Southwest. Nat. 35: 1–8.

Smith, S.H., J.B. Holter, H.H. Hayes and H.H. Silver. 1975. Protein requirements of white-tailed deer fawns. J. Wildl. Manage. 39: 582–589.

Smith, S.M., D.M. Davies and V.I. Golini. 1970. A contribution to the bionomics of the Tabanidae (Diptera) of Algonquin Park, Ontario: Seasonal distribution, habitat preferences, and biting records. Can. Entomol. 102: 1,461–1,473.

Snider, J.B. 1980. An evaluation of mandibular fat as an indicator of conditionin moose. Proc. N. Am. Moose Conf. Workshop. 16: 37–50.

———. 1985. The biology of rumen flukes (Tremadoda: Paramphistomatidae) in moose, *Alces alces L.*, in northwestern Ontario. M.S. thesis, Lakehead Univ., Thunder Bay, Ontario. 90 pp.

Snider, J.B. and M.W. Lankester. 1986. Rumen flukes (*Paramphistomum* spp.) in moose of northwestern Ontario. Alces 22: 323–344.

Snow, D.R. 1978. Eastern Abenaki. Pages 148–159 *in* B.G. Trigger, ed., Northeast. Handbook of North American Indians. Vol. 15. Smithsonian Inst., Washington, D.C. 924 pp.

Snow, J.H. 1981. Ingalik. Pages 602–617 *in* J. Helm, ed., Subarctic. Handbook of North American Indians. Vol. 6. Smithsonian Inst., Washington, D.C. 837 pp.

Snow, W.A. 1891. The moose fly—A new Haematobia. Can. Entomol. 23: 87–92

Sokolov, L. 1969. The biology of degenerative joint disease. Univ. Chicago Press, Chicago, Illinois. 162 pp.

Sokolov, V.E. 1964. Stroenie kozhnogo pokrova i jazyka losya. [Morphology of moose skin and tongue]. Pages 174–195 *in* A.G. Bannikov, ed., Biologiya i promysel losya. Kolos, Moscow.

Sokolov, V.E. and O.F. Chernova. 1987. Morphology of the skin of moose (*Alces alces* L.). Swed. Wildl. Res. (Suppl.) 1: 367–376.

Sokolov, V.E. and L.V. Stepanova. 1980. Ultrastructural organization of the interdigital gland of the elk. Doklady Akademii Nauk SSSR 250: 451–453.

Solow, A.R. 1989. A randomization test for independence of animal relocations. Ecol. 70: 1,546–1,549.

Soltys, M.A., C.E. Andress and A.L. Fletch. 1967. Johne's disease in a moose *(Alces alces)*. Wildl. Dis. Assoc. Bull. 3: 183–184.

Sopuck, L.G. and D.J. Vernam. 1986. Distribution and movements of moose *(Alces alces)* in relation to the Trans-Alaska oil pipeline. Arctic 39: 138–144.

Sotheby's. 1989. John James Audubon. The birds of America: The viviparous quadrupeds of North America. Auction catalog, 5889 "PEABODY," June 23–24, 1989, and sale results. Sotheby's, New York.

Spallinger, D.E. and N.T. Hobbs. 1992. Mechanisms of foraging in mammalian herbivores: New models of functional responses. Am. Nat. 140: 325–348.

Speakman, A. 1987. Place cells in the brain: Evidence for a cognitive map. Sci. Prog. 71: 511–530.

Speck, F.G. 1911. Huron moose hair embroidery. Am. Anthrop. 13(1): 1–14.

———. 1935. Naskapi: The savage hunters of the Labrador Peninsula. Univ. Oklahoma Press, Norman. 257 pp.

———. 1940. Penobscot man: The life history of a forest tribe in Maine. Univ. Pennsylvania Press, Philadelphia.

Spence, L. 1994. Myths of the North American Indians. Gramercy Books, New York. 393 pp.

Spencer, D.L. and E.F. Chatelain. 1953. Progress in the management of the moose in south central Alaska. Trans. N. Am. Wildl. Conf. 18: 539–552.

Spencer, D.L. and J.B. Hakala. 1964. Moose and fire on the Kenai. Proc. Tall Timbers Fire Ecol. Conf. 3: 11–33.

Spraker, T.R. 1982. An overview of the pathophysiology of capture myopathy and related conditions that occur at the time of capture of wild animals. Pages 81–118 *in* L. Nielsen, L., J.C. Haigh and M.E. Fowler, eds., Chemical immobilization of North American wildlife. Wisconsin Hum. Soc., Inc., Milwaukee.

Spry, I.M., ed. 1968. Papers of the Palliser Expedition, 1857–1860. Publ. XLIV. The Champlain Soc., Toronto, Ontario. 694 pp.

Stankey, G.H., R.C. Lucas and R.R. Ream. 1973. Relationships between hunting success and satisfaction. Trans. N. Am. Wildl. Nat. Resour. Conf. 38: 235–242.

Stanley, T.H., S. McJames, J. Kimball, J.D. Port and N.L. Pace. 1988. Immobilization of elk with A-3080. J. Wildl. Manage. 52: 577–581.

Steen M. 1991. Elaphostrogylosis: A clinical, pathological, and taxonomical study with special emphasis on the infection in moose. Ph.D. thesis, Swedish Univ. Agric. Sci., Uppsala. 102 pp.

Steen M. and L. Roepstorff. 1990. Neurological disorder in two moose calves (*Alces alces* L.) naturally infected with *Elaphostrongylus alces*. Rangifer (Spec. Iss. 3): 399–406.

Steen M., A.G. Chabaud and C. Rehbinder. 1989. Species of the genus *Elaphostrongylus* parasite of Swedish Cervidae. A description of *E. alces*. Ann. Parasitol. Hum. Comp. 64: 134–142.

Stefansson. V. 1944. Pemmican. Milit. Surg. 95 (2):89–98.

Steigers, W.D., Jr. and E.F. Becker. 1986. Susitna hydroelectric project moose browse inventory, food habits, and nutritional quality of forage in the Middle Susitna Basin—A progress report for FY85. LGL Alaska Res. Assoc., Anchorage. 57 pp.

Steinbring, J.H. 1981. Saulteaux of Lake Winnipeg. Pages 244–255 *in* J. Helm, ed., Subarctic. Handbook of North American Indians. Vol. 6. Smithsonian Inst., Washington, D.C. 837 pp.

Stelfox, J.G. 1981. Effects of ungulates of clearcutting in western Alberta: The first 25 years. Can. Wildl. Serv., Edmonton. 46 pp.

Stelfox, J.G. and R.D. Taber. 1969. Big game in the northern Rocky Mountains coniferous forest. Pages 197–220 *in* R.D.Taber, ed., Coniferous forests of the northern Rocky Mountains. Univ. Montana Found., Missoula.

Stelfox, J.G., G.M. Lynch and J.R. McGillis. 1976. Effects of clearcut logging on wild ungulates in the central Albertan foothills. For. Chron. 52: 65–70.

Stenhouse, G.B., P.B. Latour, L. Kutny, N. MacLean and G. Glover. 1994. Productivity, survival, and movements of female moose in a low-density population, Northwest Territories, Canada. Arctic 48: 57–62.

Stenseth, N.C. 1983. Causes and consequences of dispersal in small mammals. Pages 63–101 *in* I.R. Swingland and P.J. Greenwood, eds., The ecology of animal movement. Clarendon Press, Oxford.

Stephens, P.W. and R.O. Peterson. 1984. Wolf-avoidance strategies of moose. Holarctic Ecol. 7: 239–244.

Stephenson, R.O., W.B. Ballard, C.A. Smith and K. Richardson. 1995. Wolf biology and management in Alaska 1981–1991. Pages 43–54 *in* L.N. Carbyn, S.H. Fritts and D. Seip, eds., Ecology and conservation of wolves in a changing world. Proc. 2nd Int.. Wolf Symp. Canadian Circumpolar Inst., Univ. Alberta, Edmonton.

Stephenson, T.R., K.J. Hundertmark, C.C. Schwartz and V. VanBallenberghe. 1994. Ultrasonic fat measurement of captive yearling bull moose. Alces 29: 115–123.

Sterling, K.B., ed. 1974. Notes on animals of North America. Arno Press, New York.

Stevens, D.R. 1970. Winter ecology of moose in the Gallatin Mountains, Montana. J. Wildl. Manage. 34: 37–46.

———. 1971. Shiras moose. Pages 89–95 *in* T.W. Mussehl and F.W. Howell, eds., Game management in Montana. Fed. Aid in Wildl. Restor. Proj. Rept. Montana Dept. Fish, Wildl. and Parks, Helena.

———. 1974. Rocky Mountain elk—Shiras moose range relationships. Nat. Can. (Que.) 101:505–516.

Stevens, J. 1948. Paul Bunyan. Alfred A. Knopf, New York. 245 pp.

Stewart, A.J. 1970. A preliminary report relating ARDA wildlife capability ratings to present moose populations. Ontario Dept. Lands and For., Toronto. 28 pp.

Stewart, R.R. 1978. Introduction of sex and age specific hunting licenses for the moose harvest in Saskatchewan. Proc. N. Am. Moose Conf. Workshop 14: 194–208.

———. 1983. A sex and age-selective harvest strategy for moose management in Saskatchewan. Pages 229–238 *in* Proc. Symp. Game Harvest Manage. Texas A&I Univ., Kingsville.

Stewart, R.R. and A. Flynn. 1978. Mineral element levels in Saskatchewan moose hair. Proc. N. Am. Moose Conf. Workshop. 14: 141–156.

Stewart, R.R., R.R. MacLennan and J.D. Kinnear. 1977. The relationship of plant phenology to moose. Tech. Bull. No. 3. Saskatchewan Dept. Tourism and Renew. Resour., Regina. 20 pp.

Stewart, R.R., L.M. Comishen-Stewart and J.C. Haigh. 1985a. Levels of some reproductive hormones in relation to pregnancy in moose: A preliminary report. Alces 21: 393–402.

Stewart, R.R., E.H. Kowal, R. Beaulieu and T.W. Rock. 1985b. The impact of black bear removal on moose calf survival in east-central Saskatchewan. Alces 21: 403–418.

Stewart, R.R., L.M. Comishen-Stewart and J.C. Haigh. 1987. Gestation periods in two yearling captive moose, *Alces alces,* in Saskatchewan. Can. Field-Nat. 101: 103–104.

Stewart-Scott, I.A., P.D. Pearce, G.H. Moors and P.F. Fennessy. 1990. Freemartinism in the red deer (*Cervus elaphus* L.). Cytogenet. Cell Genet. 54: 58–59.

Stock, T.M. and M.W. Barrett. 1983. Helminth parasites of the gastrointestinal tracts and lungs of moose *(Alces alces)* and wapiti *(Cervus elaphus)* from Cypress Hills, Alberta, Canada. Proc. Helminthol. Soc. Wash. 50: 246–251.

Stocker, M. 1981. Optimization model for a wolf ungulate system. Ecol. Model. 12: 151–172.

Stone, A., C.W. Sabrosky, W.W. Wirth, R.H. Foote and J.R. Coulson. 1965. Catalogue of the Diptera of America north of Mexico. U.S. Dept. Agric. Res. Serv., Washington, D.C. 1,696 pp.

Stone, A.J. 1924. The moose: Where it lives and how it lives. Pages 291–325 *in* T.R. Roosevelt, T.S. Van Dyke, D.G. Eliott and A.J. Stone, and et al., eds. The deer book. Macmillan Co., New York.

Stone, J.L. 1971. Winter movements and distribution of moose in Upper Rock Creek drainage, Granite County, Montana. M.S. thesis, Univ. Montana, Missoula. 80 pp.

Strickland, D. 1989. Michael and the moose. Raven—Algonquin Park News 30: 1–4.

Stringham, S.F. 1974. Mother-infant relations in moose. Nat. Can. (Que.) 101: 325–369.

Stuart, K.M. 1984. Wildlife–vehicle collisions in British Columbia. British Columbia Min. Environ., Victoria. 37 pp.

Stubbs, M. 1977. Density dependence in the life-cycles of animals and its importance in K- and r- selected strategies. J. Anim. Ecol. 46: 677–688.

Stumer, H.M. 1978. This was Klondike fever. Superior Publ. Co., Seattle, Washington. 159 pp.

Stuve, G. 1986. The prevalence of *Elaphostrongylus cervi* infection in moose *(Alces alces)* in southern Norway. Acta Vet. Scandinavica 27: 397–409.

Stuve, G. and A. Skorping. 1987. Experimental *Elaphostrongylus cervi* infection in moose *(Alces alces)*. Acta Vet. Scandinavica 28: 165–171.

Suber, T. 1940. Minnesota moose—Then and now. Conserv. Volunteer 1: 41–44.

Subcommittee on Standardization of Common and Botanical Names of Weeds. 1966. Standardized names of weeds. Weeds 14(4): 347–386.

Sugden, L.G. 1964. An antlered calf moose. J. Mammal. 45: 490.

Sumanik, K. and A.S. Harrison. 1968. Williston report. British Columbia Min. Environ., Prince George. 76 pp.

Surrendi, C. 1983. Preliminary report on wildlife mortality along C.N. rights-of-way, and observation on habitat quality in Jasper National Park. Rept. to Can. Natl. Railways by Corsal Enterprises Ltd., Edmonton, Alberta. 46 pp.

Suttie, J.M. 1990. Experimental manipulation of the neural control of antler growth. Pages 359–370 *in* G.A. Bubenik and A.B. Bubenik, eds., Horns, pronghorns and antlers. Springer- Verlag, New York.

Suttie, J.M. and P.F. Fennessy. 1990. Antler regeneration: Studies with antler removal, axial tomography, and angiography. Pages 313–338 *in* G.A. Bubenik and A.B. Bubenik, eds., Horns, pronghorns and antlers. Springer-Verlag, New York.

Suttie, J.M. and P.F. Fennessy. 1991. Growth promoting hormones and antler development. Pages 283–286 *in* B. Bobek, K. Perzanowski and W.L. Regelin, eds., Global trends in wildlife management. Vol. 1. Trans. Int. Union Game Biol. Congr., Swiat Press, Krakow-Warszawa, Poland.

Suttie, J.M. and P.F. Fennessy. 1992. Recent advances in the physiological control of velvet antler. Pages 471–478 *in* R.D. Brown, ed., Biology of deer. Springer-Verlag, New York.

Suttie, J.M., R.G. White, B.H. Brier and P.D. Gluckman. 1991. Photoperiod associated changes in insulin-like growth factor-I in reindeer. Endocrin. 129: 679–682.

Sutton, J.B. 1890. Evolution and disease. Walter Scott, London. 285 pp.

Svendsen, C.R. 1990. Use of riparian willow stands during summer by moose, muskoxen, and reindeer on the Seward Peninsula, Alaska. Pages 47–54 *in* Role of willow communities in the ecology of herbivores on the Seward Peninsula, Alaska. Alaska Coop. Wildl. Res. Unit, Univ. Alaska-Fairbanks.

Swales, W.E. 1935. The life cycle of *Fascioloides magna* (Bassi, 1875), the large liver fluke of ruminants in Canada. Can. J. Res., D, 12: 177–215.

Sweanor, P.Y. and F. Sandegren. 1988. Migratory behavior of related moose. Holarctic Ecol. 11: 190–193.

Sweanor, P.Y. and F. Sandegren. 1989. Winter range philopatry of seasonally migratory moose. J. Appl. Ecol. 26: 25–33.

Sweanor, P. Y. and F. Sandegren. 1991. Migration patterns of moose population in relation to calf recruitment. Pages 631–635 *in* B. Bobek, K. Perzanowski and W. Regelin, eds., Global trends in wildlife management. Swiat Press, Krakow, Poland.

Swihart, R.K. and N.A. Slade. 1985a. Influence of sampling interval on estimates of home range size. J. Wildl. Manage. 49: 1,019–1,025.

Swihart, R.K. and N.A. Slade. 1985b. Testing for independence of observations in animal movements. Ecol. 66: 1,176–1,184.

Swihart, R.K., N.A. Slade and B.J. Bergstrom. 1988. Relating body size to the rate of home range use in mammals. Ecol. 69: 393–399.

Sylven, S. 1980. Study of the reproductive organs of female moose in Sweden. Proc. N. Am. Moose Conf. Workshop 16: 124–136.

Tankersley, N.G. and W.C. Gasaway. 1983. Mineral lick use by moose in Alaska. Can. J. Zool. 61: 2,241–2,249.

Taylor, C.R. 1978. Why change gaits? Recruitment of muscles and muscle fibers as a function of speed and gait. Am. Zool. 18: 153–161.

Taylor, C.R., K. Schmidt-Nielsen and J.L. Raab. 1970. Scaling of energetic costs of running to body size in mammals. Am. J. Physiol. 291: 1,104–1,107.

Taylor, K.P. and W.B. Ballard. 1979. Moose movements and habitat use along the Susitna River near Devil's Canyon. Proc. N. Am. Moose Conf. Workshop 15: 169–186.

Taylor, K.P. and R.A.J. Taylor. 1977. Aggregation, migration and population mechanics. Nature 265: 415–421.

Taylor, W.P. 1956. The deer of North America. Stackpole Books, Harrisburg, Pennsylvania. 209 pp.

Telfer, E.S. 1967a. Comparison of moose and deer winter range in Nova Scotia. J. Wildl. Manage. 31: 418–490.

———. 1967b. Comparison of a deer yard and a moose yard in Nova Scotia. Can. J. Zool. 45: 485–490.

———. 1969. Twig weight-diameter relationships for browse species. J. Wildl. Manage. 33: 917–921.

———. 1970. Winter habitat selection by moose and white-tailed deer. J. Wildl. Manage. 34: 553–559.

———. 1972. Forage yield and browse utilization on logged areas in New Brunswick. Can. J. For. Res. 2: 346–350.

———. 1974. A trend survey for browse ranges using the Shafer twig count technique. Proc. N. Am. Moose Conf. Workshop 10: 160–171.

———. 1978a. Cervid distibution, browse and snow cover in Alberta. J. Wildl. Manage. 42: 352–361.

———. 1978b. Habitat requirements of moose—the principal taiga range animal. Proc. Int. Rangeland Cong.: 462–465.

———. 1984. Circumpolar distribution and habitat requirements of moose *(Alces alces)*. Pages 145–182 *in* R. Olson, R. Hastings and F. Geddes, eds., Northern ecology and resource management. Univ. Alberta Press, Edmonton.

———. 1988. Habitat use by moose in southwestern Alberta. Alces 24: 14–21.

———. 1995. Moose range under presettlement fire cycles and forest management regimes in the boreal forest of western Canada. Alces 31: 153–166.

Telfer, E.S. and A.L. Cairns. 1986. Resource use by moose versus sympatric deer, wapati and bison. Alces 22: 113–137.

Telfer, E.S. and J.P. Kelsall. 1979. Studies of morphological parameters affecting ungulate locomotion in snow. Can. J. Zool. 57: 2,153–2,159.

Telfer, E.S. and J.P. Kelsall. 1984. Adaptations of some large North American mammals for survival in snow. Ecol. 65: 1,828–1,834.

Tessaro, S.V. and L.B. Forbes. 1986. Brucella suis biotype 4: A case of granulomatous nephritis in a barren-ground caribou (*Rangifer tarandus groenlandicus* L.), with a review of the distribution of rangiferine brucellosis in Canada. J. Wildl. Dis. 22: 479–483.

Thaller, C. and G. Eichele. 1987. Identification and spatial distribution of retinoids in the developing chick limb bud. Nature 327: 625–628.

Thelenius, J. 1990. Plant succession on earthquake uplifted coastal wetlands, Copper River Delta, Alaska. Northwest Sci. 64: 259–262.

Thomas, D.C. 1990. Moose diet and use of successional forests in the Canadian taiga. Alces 26: 24–29.

Thomas, D.C., S.C. Barry and H.P. Kiliaan. 1989. Fetal sex ratios in caribou: Maternal age and condition effects. J. Wildl. Manage. 53: 885–890.

Thomas, D.S. and I. McT. Cowan. 1975. The pattern of reproduction in female Columbian black-tailed deer, *Odocoileus hemionus columbianus*. J. Reprod. Fert. 44: 261–272.

Thomas, J.E. and D.G. Dodds. 1988. Brainworm, *Parelaphostrongylus tenuis*, in moose, *Alces alces*, and white-tailed deer, *Odocoileus virginianus*, of Nova Scotia. Can. Field-Nat. 102: 639–642.

Thomas, J.W. and D.E. Toweill, ed. 1982. Elk of North America: Ecology and management. Stackpole Books, Harrisburg, Pennsylvania. 736 pp.

Thomas, L.J. and A.R. Cahn. 1932. A new disease in moose. I: Preliminary report. J. Parasitol. 18: 219–231.

Thomas, P.A. 1985. Cultural change on the southern New England frontier, 1630–1665. Pages 131-162 *in* W. Fitzhugh, ed., Cultures in contact: The European impact on native cultural institutions in eastern North America, A.D. 1000–1800. Smithsonian Inst. Press, Washington, D.C.

Thomasson, R.D. 1973. Ontario land inventory series: Wildlife. Ontario Dept. Lands and For., Toronto. 71 pp.

Thompson, D.V. 1942. On growth and form. Univ. Cambridge Press, Cambridge, England. 1,116 pp.

Thompson, I.D. 1979. A method of correcting population and sex and age estimates from aerial transect surveys for moose. Proc. N. Am. Moose Conf. Workshop 15: 148–168.

———. 1988. Moose damage to pre-commercially thinned balsam fir stands in Newfoundland. Alces 24: 56–61.

Thompson, I.D. and D.L. Euler. 1987. Moose habitat in Ontario: A decade of change in perception. Swed. Wildl. Res. (Suppl.) 1: 181–193.

Thompson, I.D. and R.O. Peterson. 1988. Does wolf predation alone limit the moose populationin Pukaskwa Park? A comment. J. Wildl. Manage. 52: 556–559.

Thompson, I.D. and M.F. Vukelich. 1981. Use of logged habitats in winter by moose cows with calves in northeastern Ontario. Can. J. Zool. 59: 2,103–2,114.

Thompson, I.D., R.E. McQueen, P.B. Reichardt, D.G. Trenholm and W.G. Curran. 1989. Factors influencing choice of balsam fir twigs from thinned and unthinned stands by moose. Oecologia 81: 506–509.

Thompson, I.D., W.J. Curran, J.A. Hancock and C.E. Butler. 1992. Influence of moose browsing on successional forest growth on black spruce sites in Newfoundland. For. Ecol. and Manage. 47: 29–37.

Thompson, J. 1994. From the land: Two hundred years of Dene clothing. Can. Mus. Civil., Hull, Quebec. 134 pp.

Thompson, M.E. 1987. Seasonal home range and habitat use by moose in northern Maine. M.S. thesis, Univ. Maine, Orono. 57 pp.

Thompson, M.H. 1938. High trails of Glacier National Park. The Caxton Printers, Caldwell, Idaho. 137 pp.

Thompson, W.R. 1929. On natural control. Parasitol. 21: 269–281.

Thompson, Z. 1853. Natural history of Vermont. Charles E. Tuttle Company, Rutland, Vermont, 1972. 286 pp.

Thomson, R. 1990. Crown land wildlife habitat protection and development project, northwest portion of central region. Alberta Dept. For. Lands and Wildl., Edmonton. 48 pp.

Thomson, R. 1992. The wildlife game. Nyala Wildlife Publication Trust, Westville, South Africa. 292 pp.

Thomson, R.N. 1991. An analysis of the influence of male age and sex ratios on reproduction in British Columbia moose (*Alces alces* L.) populations. M.S. thesis, Univ. British Columbia, Victoria. 45 pp.

Thoreau, H.D. 1864. The Maine woods. Houghton Mifflin, Boston, Massachusetts. 328 pp.

Thorne, E.T. 1979. Diagnosis of diseases in wildlife. Fed. Aid Wildl. Restor. Prog. Rept. Wyoming Dept. Fish and Game, Cheyenne. 13 pp.

Thorne, E.T., R.E. Dean and W.G. Hepworth. 1976. Nutrition during gestation in relation to successful reproduction in elk. J. Wildl. Manage. 40: 330–335.

Thorne, E.T., N. Kingston, W. Jolley and R. Bergstrom. 1982. Diseases of wildlife in Wyoming. 2nd. ed. Wyoming Game and Fish Dept., Cheyenne. 353 pp.

Threlfall, W. 1967. Parasites of moose *(Alces alces)* in Newfoundland. J. Mammal. 48: 668–669.

———. 1969. Further records of helminths from Newfoundland mammals. Can. J. Zool. 47: 197–201.

Thwaites, R.G., ed. 1896. Acadia 1610–1613. Vol. 1. The Jesuit relations and allied documents: Travel and exploration of the Jesuit missionaries in New France 1610–1791. Burrows Bros., Cleveland, Ohio. 319 pp.

———. 1896–1901. The Jesuit relations and allied documents, travels and explorations of the Jesuit missionaries in New France 1610–1791; the original French, Latin, and Italian texts, with English translations and notes. 73 vols. Burrows Bros., Cleveland, Ohio.

———. 1897a. Acadia 1611–1616. Vol. 3. The Jesuit relations and allied documents: Travel and exploration of the Jesuit missionaries in New France 1610–1791. Burrows Bros., Cleveland, Ohio. 301 pp.

———. 1897b. Acadia, Quebec 1616–1629. Vol. 4. The Jesuit relations and allied documents: Travel and exploration of the Jesuit missionaries in New France 1610–1791. Burrows Bros., Cleveland, Ohio. 272 pp.

———. 1897c. Quebec 1633–1634. Vol. 6. The Jesuit relations and allied documents: Travel and exploration of the Jesuit missionaries in New France 1610–1791. Burrows Bros., Cleveland, Ohio. 330 pp.

———. 1897d. Quebec, Hurons, Cape Breton 1634–1636. Vol. 8. The Jesuit relations and allied documents: Travel and exploration of the Jesuit missionaries in New France 1610–1791. Burrows Bros., Cleveland, Ohio. 314 pp.

———. 1897e. Quebec 1636. Vol. 9. The Jesuit relations and allied documents: Travel and exploration of the Jesuit missionaries in New France 1610–1791. Burrows Bros., Cleveland, Ohio. 315 pp.

———. 1898a. Quebec 1637. Vol. 12. The Jesuit relations and allied documents: Travel and exploration of the Jesuit missionaries in New France 1610–1791. Burrows Bros., Cleveland, Ohio. 277 pp.

———. 1898b. Quebec, Hurons 1639. Vol. 16. The Jesuit relations and allied documents: Travel and exploration of the Jesuit missionaries in New France 1610–1791. Burrows Bros., Cleveland, Ohio. 259 pp.

———. 1898c. Lower Canada, Iroquois 1642–1643. Vol. 24. The Jesuit relations and allied documents: Travel and exploration of the Jesuit missionaries in New France 1610–1791. Burrows Bros., Cleveland, Ohio. 312 pp.

———. 1898d. Hurons, Iroquois, Lower Canada 1645–1646. Vol. 28. The Jesuit relations and allied documents: Travel and exploration of the Jesuit missionaries in New France 1610–1791. Burrows Bros., Cleveland, Ohio. 320 pp.

———. 1898e. Gaspé, Hurons, Lower Canada 1647–1648. Vol. 32. The Jesuit relations and allied documents: Travel and exploration of the Jesuit missionaries in New France 1610–1791. Burrows Bros., Cleveland, Ohio. 313 pp.

———. 1899a. Lower Canada, Abenakis 1651–1652. Vol. 37. The Jesuit relations and allied documents: Travel and exploration of the Jesuit missionaries in New France 1610–1791. Burrows Bros., Cleveland, Ohio. 267 pp.

———. 1899b. Hurons, 1653. Vol. 39. The Jesuit relations and allied documents: Travel and exploration of the Jesuit missionaries in New France 1610–1791. Burrows Bros., Cleveland, Ohio. 267 pp.

———. 1899c. Lower Canada, Acadia, Iroquois 1659–1660. Vol. 45. The Jesuit relations and allied documents: Travel and exploration of the Jesuit missionaries in New France 1610–1791. Burrows Bros., Cleveland, Ohio. 272 pp.

———. 1899d. Lower Canada, Ottawas, Canadian Interior 1659–1661. Vol. 46. The Jesuit relations and allied documents: Travel and exploration of the Jesuit missionaries in New France 1610–1791. Burrows Bros., Cleveland, Ohio. 304 pp.

———. 1899e. Lower Canada, Iroquois 1663–1665. Vol. 49. The Jesuit relations and allied documents: Travel and exploration of the Jesuit missionaries in New France 1610–1791. Burrows Bros., Cleveland, Ohio. 278 pp.

———. 1899f. Lower Canada, Iroquois, Ottawas 1664–1667. Vol. 50. The Jesuit relations and allied documents: Travel and exploration of the Jesuit missionaries in New France 1610–1791. Burrows Bros., Cleveland, Ohio. 328 pp.

———. 1899g. Ottawas, Lower Canada, Iroquois 1666–1668. Vol. 51. The Jesuit relations and allied documents: Travel and exploration of the Jesuit missionaries in New France 1610–1791. Burrows Bros., Cleveland, Ohio. 295 pp.

———. 1899h. Lower Canada, Iroquois, Ottawas 1670–1672. Vol. 55. The Jesuit relations and allied documents: Travel and exploration of the Jesuit missionaries in New France 1610–1791. Burrows Bros., Cleveland, Ohio. 322 pp.

———. 1899i. Lower Canada, Iroquois, Ottawas, Hudson Bay 1671–1672. Vol. 56. The Jesuit relations and allied documents: Travel and exploration of the Jesuit missionaries in New France 1610–1791. Burrows Bros., Cleveland, Ohio. 304 pp.

———. 1900a. Lower Canada, Iroquois 1667–1687. Vol. 63. The Jesuit relations and allied documents: Travel and exploration of the Jesuit missionaries in New France 1610–1791. Burrows Bros., Cleveland, Ohio. 307 pp.

———. 1900b. Lower Canada, Abenakis, Louisiana 1716–1727. Vol. 67. The Jesuit relations and allied documents: Travel and exploration of the Jesuit missionaries in New France 1610–1791. Burrows Bros., Cleveland, Ohio. 344 pp.

———. 1904. Voyages and travels of an Indian interpreter and trader. The Arthur H. Clark Co., Cleveland, Ohio. 329 pp.

———. 1904–1905. Original journals of the Lewis and Clark Expedition, 1804–1806. 7 vols. Dodd, Mead and Co., New York.

———. 1904-1907. Early western travels, 1748–1846. 32 vols. The Arthur H. Clark Co., Cleveland, Ohio.

Tilley, J.M.A. and R.A. Terry. 1963. A two stage technique for in vitro digestion of forage crops. J. Brit. Grassland Soc. 18: 104–111.

Timmermann, H.R. 1971. The antlers of the moose development related to age. Ontario Fish and Wildl. Rev. 10: 11–18.

———. 1974. Moose inventory methods: A review. Nat. Can. (Que.) 101: 615–629.

———. 1972. Some observations of the moose hunt in the Black Sturgeon area of northwestern Ontario. Proc. N. Am. Moose Conf. Workshop. 8: 223–239.

———. 1977. The killing proficiency of moose hunters. Proc. N. Am. Moose Conf. Workshop 13: 13–25.

———. 1979. Morphology and anatomy of the moose (*Alces alces* L.) bell and its possible function. M.S. thesis, Lakehead Univ., Thunder Bay, Ontario. 90 pp.

———. 1981. Managing moose in the 1980's. Alces 17: 21–45.

———. 1983. A selective harvest for moose in Ontario. Ontario Angler and Hunter 8: 25–29.

———. 1987. Moose harvest strategies in North America. Swed. Wildl. Res. (Suppl.) 1: 565–579.

———. 1990. Basic moose biology. Pages 1-10 *in* M. Buss and R. Truman, eds., The moose of Ontario. Book I: Moose biology, ecology and management. Ontario Min. Nat. Resour. and Ontario Fed. of Anglers and Hunters, Toronto. 78 pp.

———. 1991. A look at moose in the Soviet Union. Angler and Hunter (May): 23–24, 31.

———. 1992a. Moose hunter education in North America—A review. Alces (Suppl.) 1: 65–76.

———. 1992b. Moose sociobiology and implications for harvest. Alces 28: 59–77.

———. 1993. Use of aerial surveys for estimating and monitoring moose populations—A review. Alces 29: 35–46.

Timmermann, H.R. and M.E. Buss. 1995. The status and management of moose in North America—Early 1990's. Alces 31: 1–14.

Timmermann, H.R. and R. Gollat. 1982. Age and sex structure of harvested moose related to season manipulation and access. Alces 18: 301–328.

Timmermann, H.R. and R. Gollat. 1984. Sharing a moose in north central Ontario. Alces 20:161–186.

Timmermann, H.R. and R. Gollat. 1986. Selective moose harvest in north central Ontario—A progress report. Alces 22: 392–417.

Timmermann, H.R. and R. Gollat. 1994. Early winter social structure of hunted versus unhunted moose populations in north central Ontario. Alces 30: 117–126.

Timmermann, H.R. and M.W. Lankester. 1978. Joint disease in ungulates with special reference to moose. Proc. N. Am. Moose Conf. Workshop. 14: 89–108.

Timmermann, H.R. and G.D. Racey. 1989. Moose access routes to aquatic feeding site. Alces 25: 104–111.

Timmermann, H.R. and H. Whitlaw. 1992. Selective moose harvest in north central Ontario—A progress report. Alces 28: 137–163.

Timmermann, H.R., M.W. Lankester and A.B. Bubenik. 1985. Morphology of the bell in relation to sex and age of moose *(Alces alces)*. Alces 21: 419–446.

Timmermann, H.R., M.W. Lankester and A.B. Bubenik. 1988. Vascularization of the moose bell. Alces 24: 90–96.

Timmermann, H.R., H. Whitlaw and A. Rodgers. 1993. Testing moose harvest data sensitivity to changes in aerial population estimates. Alces 29: 47–53.

Tinbergen, N. 1955. Tiere untereunander. Soziales Verhalten bei Tieren insbesondere Wirbeltieren. Paul Parey, Berlin. 150 pp.

———. 1990. Social behavior of animals—With special reference to vertebrates. Chapman and Hall, New York. 184 pp.

Tinsley, R. 1983. Call of the moose and other Fiordland hunting adventures. Reed Books, Aukland, New Zealand. 168 pp.

Tobey, M.L. 1981. Carrier. Pages 413–432 *in* J. Helm, ed., Subarctic. Handbook of North American Indians. Vol. 6. Smithsonian Inst., Washington, D.C. 837 pp.

Tobey, R.W. 1989. Wolf survey–inventory progress report, GMU 13. Pages 57–66 *in* S.O. Morgan, ed., Annual report of survey-inventory activities. Wolf, 1 Jan. 1988–31 Dec. 1988. Fed. Aid in Wildl. Restor. Rept. Alaska Dept. Fish and Game, Juneau.

Todd, W.A. 1991. Management plan for moose in Alberta. Fish and Wildl. Div. Alberta For., Lands and Wildl., Edmonton, Alberta. 168 pp.

Topinski, P. 1975. Abnormal antler cycles in deer as a result of stress inducing factors. Acta Theriol. 20: 267–279.

Torrey, B. 1906. Friends on the shelf. 2 vols. Houghton, Mifflin and Co., Boston.

Toufexis, A. 1993. The right chemistry—Evolutionary roots, brain imprints, biological secretions. Time 141: 39–41.

Townsend, J.B. 1981. Tanaina. Pages 623-640 *in* J. Helm, ed. Subarctic. Handbook of North American Indians. Vol. 6. Smithsonian Inst., Washington, D.C. 837 pp.

Trainer, D.O. and M.M. Jochim. 1969. Serologic evidence of bluetongue in wild ruminants of North America. Am. J. Vet. Res. 30: 2,007–2,011.

Traquair, R. 1945. The coat of arms. Beaver 276: 42–46.

Treichel, B. 1979. Cypress Hills elk and moose rumen analysis. Alberta Fish and Wildl. Div., Edmonton. 24 pp.

Trigger, B.G., ed. 1978. Northeast. Handbook of the North American Indians. Vol. 15. Smithsonian Inst., Washington, D.C. 924 pp.

Trivers, R.L. and D.E. Willard. 1973. Natural selection of parental ability to vary the sex ratio of offspring. Science 179: 90–92.

Trnka, R. 1990. Návrat losa evropského [Return of European moose]. Polovn. a Rybarstvo. 42: 10–11.

Trotter, R.H. 1958. An aerial census technique for moose. Abstract. Proc. Northeast Sect. Wildl. Soc., Montreal. 10 pp.

Trottier, G.C., S.R. Rollins and R.C. Hutchison. 1983. Range, habitat and foraging relationships of ungulates in Riding Mountain National Park. Large Mammal System Stud. Rept. No. 24. Can. Wildl. Serv., Edmonton. 224 pp.

Trumbull, J.H., ed. 1963. The complete writings of Roger Williams. Vol. 1. Russell and Russell Inc., New York. 112 pp.

Turner, C.D. 1948. General endocrinology. W.B. Saunders Co., Philadelphia. 604 pp.

Tyrrell, J.B., ed. 1911. A journey from Prince of Wales' Fort in Hudson's Bay to the northern ocean in the years 1769, 1770, 1771, and 1772. Publ. VI. The Champlain Soc., Toronto. 437 pp.

———. 1916. David Thompson's narrative of his explorations in western America, 1784–1812. Publ. XII. The Champlain Soc., Toronto, Ontario. 582 pp.

———. 1931. Letters of La Potherie. Pages 144–370 *in* Documents relating to the early history of Hudson Bay. The Champlain Soc., Toronto, Ontario. 419 pp.

Ullrey, D.E., W.G. Youatt, H.E. Johnson, L.D. Fay and B.L. Bradley. 1967. Protein requirements of white-tailed deer fawns. J. Wildl. Manage. 31: 679–685.

Ullrey, D.E., W.G. Youatt, H.E. Johnson, L.D. Fay, B.L. Schoepke and W.T. Magee. 1969. Digestible energy requirements for winter maintenance of Michigan white-tailed does. J. Wildl. Manage. 33: 482–490.

Ullrey, D.E., W.G. Youatt, H.E. Johnson, L.D. Fay, B.L. Schoepke and W.T. Magee. 1970. Digestible and metabolizable energy requirements for winter maintenance of Michigan white-tailed does. J. Wildl. Manage. 34: 863–869.

Underwood, E.J. 1971. Trace elements in human and animal nutrition. 3rd ed. Academic Press, New York. 543 pp.

Underwood, H.B. and W.F. Porter. 1991. Values and science: White-tailed deer management in eastern national parks. Trans. N. Am. Wildl. and Nat. Resour. Conf. 56: 67–73.

Upshall, S.M., M.D.B. Burt and T.G. Dilworth. 1986. *Parelaphostrongylus tenuis* in New Brunswick: A parasite in terrestrial gastropods. J. Wildl. Dis. 22: 582–585.

Upshall, S.M., M.D.B. Burt and T.G. Dilworth. 1987. *Parelaphostrongylus tenuis* in New Brunswick: The parasite in white-tailed deer *(Odocoileus virginianus)* and moose *(Alces alces)*. J. Wildl. Dis. 23: 683-685.

U.S. Department of Interior. 1988. Management policies of the U.S. Department of the Interior National Parks Service. Chapts. 4: 5, 4: 6, 4: 7 and 4: 10. U.S. Dept. Inter., Washington, D.C.

———. 1992. Voyageurs National Park draft. Sect II (5)21: 1–20 in Natural resource management plan. U.S. Natl. Park Serv., International Falls, Minnesota.

U.S. Fish and Wildlife Service. 1993. 1991 national survey of fishing, hunting and wildlife-associated recreation. U.S. Dept. Inter., Dept. of Comm. and Bur. of Census. U.S. Govt. Print Off., Washington, D.C. 124 pp.

U.S. Geological Survey. 1979. Alaska geographical names. 2 books. USGS, Reston, Virginia.

———. 1985. Maine geographic names information system. USGS, Reston, Virginia. 406 pp.

U.S. Navy Department. 1969. Dictionary of American Naval fighting ships. Vol. IV. Naval Hist. Div., Washington, D.C. 745 pp.

U.S. Senate. 1855-1860. Reports of explorations and surveys to ascertain the most practicable and economical route for a railroad from the Mississippi River to the Pacific Ocean. Exec. Doc. 78, 33rd Congress, 2nd Sess.; Exec. Doc. 36th Congress, 1st Sess. 12 vols. Beverley Tucker, Printer and Thomas H. Ford, Printer, Washington, D.C.

———. 1900. Compilation of narratives of explorations in Alaska. Report No. 1023, 56th Congress, 1st Sess. U.S. Govt. Print. Off., Washington, D.C. 856 pp.

Usher, P.J. 1987. Indigenous management systems and the conservation of wildlife in the Canadian north. Alternatives 14: 3–9.

Utah Division of Wildlife Resources. 1988. Moose. Pages 67–90 *in* A strategic plan for the comprehensive management of Utah's wildlife resources. Utah Div. Wildl. Resour., Salt Lake City.

Vacek, Z. 1955. The innervation of the velvet of growing antlers of the Cervidae. Csl. Morfol. 3: 249–264.

Vachon, A. 1982. Dreams of empire: Canada before 1700. Pub. Arch. Can., Ottawa. 387 pp.

VanStone, J.W. 1974. Athapaskan adaptations: hunters and fishermen of the subarctic forests. Aldine, Chicago, Illinois.

Vaughan, A., ed. 1977. New England's prospect. Univ. Massachusetts Press, Amherst. 162 pp.

Valkenburg, P., R.O. Boertje and J.L. Davis. 1983. The effects of darting and netting of caribou in Alaska. J. Wildl. Manage. 47: 1,233–1,237.

Van Ballenberghe, V. 1977. Migratory behavior of moose in southcentral Alaska. Intl. Congr. Game Biol. 13: 103–109.

———. 1978. Final report on the effects of the trans-Alaskan pipeline on moose movements. Spec. Rept. 23. Joint St. / Fed. Fish and Wildl. Advis. Team, Anchorage, Alaska. 41 pp.

———. 1979. Productivity estimates of moose populations: A review and re-evaluation. Proc. N. Am. Moose Conf. Workshop 15: 1–18.

———. 1980. Utility of multiple equilibrium concepts applied to population dynamics of moose. Proc. N. Am. Moose Conf. Workshop 16: 571–586.

———. 1981. Population dynamics of wolves in the Nelchina Basin, southcentral Alaska. Pages 1,246–1,258 *in* J.A. Chapman and D. Pursley, eds., Proc. Worldwide Furbearer Conf., Frostburg, Maryland.

———. 1982. Growth and development of moose antlers in Alaska. Pages 37–48 *in* R.D. Brown, ed., Antler development in Cervidae. Caesar Kleberg Wildl. Res. Inst., Kingsville, Texas.

———. 1983. The rate of increase in moose populations. Alces 19: 98–117.

———. 1987. Effects of predation on moose numbers: A review of recent North American studies. Swed. Wildl. Res. (Suppl.) 1: 431–460.

———. 1989. Twenty years of moose immobilization with succinycholine chloride. Alces 25: 25–30.

———. 1992. Behavioral adaptations of moose to treeline habitats in subarctic Alaska. Alces (Suppl.) 1: 193–206.

Van Ballenberghe, V. and W.B. Ballard. 1994. Limitation and regulation of moose populations: The role of predation. Can. J. Zool. 72: 2,071–2,077.

Van Ballenberghe, V. and J. Dart. 1982. Harvest yields from moose populations subject to wolf and bear predation. Alces 18: 258–275.

Van Ballenberghe, V. and D.G. Miquelle. 1990. Activity of moose during spring and summer in interior Alaska. J. Wildl. Manage. 54: 391–396.

Van Ballenberghe, V. and J.M. Peek. 1971. Radiotelemetry studies of moose in northeastern Minnesota. J. Wildl. Manage. 35: 63–71.

Van Ballenberghe, V., D.G. Miquelle. and J.D. MacCracken. 1989. Heavy utilization of woody plants by moose during summer at Denali National Park, Alaska. Alces 25: 31–35.

Vandenbergh, J.G., ed. 1983. Pheromones and reproduction in mammals. Academic Press, New York. 298 pp.

———. 1987. Regulation of puberty and its consequences on populaton dynamics of mice. Am. Zool. 27: 891–898.

Vandenberghe, E.P., F. Wernerus and R.R. Warner. 1989. Female choice and the mating cost of peripheral males. Anim. Behav. 38: 875–884.

Van der Eems, K., R.D. Brown and C.M. Grundberg. 1988. Circulating level of 1,25-dihydroxyvitamin D, alkaline phosphotase, and osreocalcin associated with antler growth in white-tailed deer. Acta Endocrinol. 118: 407–414.

Van der Eems, K., R.D. Brown and C.M. Grundberg. 1991. Urinary cyclic-AMP and gamma-carboxyglutamic acid associated with antler growth in the white-tailed deer. Pages 287–290 *in* B. Bobek, K. Perzanowski and W.L. Regelin, eds., Global trends in wildlife management. Vol. 1. Trans. Cong. Int. Union Game Biol. Swiat Press, Krakow-Warszawa, Poland.

Van Dyne, G.M., N.R. Brockington, Z. Szocs, J. Duek and C.A. Ribic. 1980. Large herbivore subsystem. Pages 269–538 *in* A.J. Breymeyer and G.M. Van Dyne, eds., Grasslands, systems analysis, and man. Cambridge Univ. Press, Cambridge, England.

Van Horne, B. 1983. Density is a misleading indicator of habitat quality. J. Wildl. Manage. 47: 893–901.

Van Rooijen, N. 1977. Immune complexes in the spleen: Three concentric follicular areas of immune complex trapping, their interrelationships and possible function. J. Reticuloendothelial Soc. 21: 143–151.

Van Soest, P.J. 1963a. Use of detergents in the analysis of fibrous feeds. I. Preparation of fiber residues of low nitrogen content. J. Assoc. Off. Agric. Chem. 46: 825–829.

———. 1963b. Use of detergents in the analysis of fibrous feeds. II. A rapid method for the determination of fiber and lignin. J. Assoc. Off. Agric. Chem. 46: 829–835.

———. 1965. Non-nutritive residues: A system of analysis for the replacement of crude fiber. J. Assoc. Off. Agric. Chem. 49: 546–551.

———. 1966. Forage intake in relation to chemical composition and digestibility: Some concepts. Proc. South. Pasture Forage Crop Improve. Conf. 24 pp.

———. 1967. Development of a comprehensive system of feed analysis and its application to forages. J. Anim. Sci. 26: 119–128.

———. 1973. The uniformity and nutritive availability of cellulose. Fed. Proc. 32: 1,804–1,808.

———. 1982. Nutritional ecology of the ruminant. O & M Books, Inc. Corvallis, Oregon. 374 pp.

Van Wormer, J. 1972. The world of the moose. J.B. Lippincott Co., Philadelphia, Pennsylvania. 160 pp.

Varley, G.C. and G.R. Gradwell. 1968. Population models for the winter moth. Pages 132-141 *in* T.R.E. Southwood, ed., Insect abundance. Blackwell Scientific Publ., Oxford, England.

Vaughan, A.T., ed. 1977. New England's prospect. Univ. Massachusetts Press, Amherst. 132 pp.

Vecellio, G.M., R.D. Deblinger and J.E. Cordoza. 1993. Status and management of moose in Massachusetts. Alces 29: 1–7.

Vercoe, J.E. 1973. The energy cost of standing and lying in adult cattle. Brit. J. Nutr. 30: 207–210.

Vereshchagin, N.K. 1967. Geologicheskaya istoriya losya i jevo osvoyeniya pervobytnom chelovekom. Pages 3–37 *in* A.G. Bannikov, ed., Biologiya i promysel losya. Rozselkhozizdat, Moskva, USSR.

Verme, L.J. 1965. Swamp conifer deer yards in northern Michigan: Their ecology and management. J. For. 63: 523–529.

———. 1969. Reproductive patterns of white-tailed deer relative to nutritional plane. J. Wildl. Manage. 33: 881–887.

———. 1970. Some characteristics of captive Michigan moose. J. Mammal. 51: 403–405.

———. 1974. Problems in appraising reproduction in Cervidae. Proc. N. Am. Moose Conf. Workshop 10: 22–36.

———. 1983. Sex ratio variation in *Odocoileus:* A critical review. J. Wildl. Manage. 47: 573–582.

———. 1989. Maternal investment in white-tailed deer. J. Mammal. 70: 438–442.

———. 1991. Decline in doe fawn fertility in southern Michigan deer. Can. J. Zool. 69: 25–28.

——— and J.J. Ozoga. 1980. Effects of diet on growth and lipogenesis in deer fawns. J. Wildl. Manage. 44: 315–324.

Viereck, L.A. 1970. Forest succession and soil development adjacent to the Chena River in interior Alaska. Arct. and Alpine Res. 2: 1–26.

Viereck, L.A. and C.T. Dyrness. 1980. A preliminary classification system for vegetation in Alaska. Gen. Tech. Rept. PNW-106. U.S. For. Serv., Washington, D.C. 38 pp.

Vislobokova, I.A. 1986. Elks in the Pliocene of the USSR. Quatarpal. 6: 239–242.

Vislobokova, I.A. 1990. Iskopaenye Oleni Evrazii [The fossil deer of Eurasia]. Trudy Paleontol. Inst. Vol. 240. NAUKA, Moskva. 208 pp.

Vivas, H.J. and B.-E. Saether. 1987. Interactions between a generalist herbivore, the moose *(Alces alces)* and its food resources: An experimental study of winter foraging behavior in relation to browse availability. J. Anim. Ecol. 56: 509–520.

Voland, E. and C. Engel. 1990. Female choice in humans—A conditional mate selection strategy of the Krummhorn women (Germany, 1720–1874). Ethol. 84: 144–154.

Volkman, N.J., K.F. Zemanek and D. Muller-Schwarze. 1978. Antorbital and forehead secretions of black-tailed deer *(Odocoileus hemionus columbianus)*—Their role in age- class recognition. Anim. Behav. 26: 1,098–1,160.

Von Holst, D. 1969. Sozialer Stress bei Tupajas. Z. Vergl. Physiol. 63: 1–58.

Von Uexkull, J. 1921. Umwelt und Innenwelt der Tiere. 2nd ed. Springer Verlag, Berlin. 224 pp.

———. 1937. Umweltforschung. Z. Tierpsychol. 1: 33–34.

Von Uexkull, T. 1982. Introduction: Meaning and science in Jakob von Uexkull's concept of biology. Semiotica 42: 1–24.

Vozeh, G.E. and H.G. Cumming. 1960. A moose population census and winter browse survey in Gogama District, Ontario. Ontario Dept. Lands and For., Toronto. 31 pp.

Wake, D.B. 1991. Homoplasty—The result of natural selection, or evidence of design limitations? Am. Nat. 138: 543–567.

Wakefield, P. and L. Carrarra. 1992. A moose for Jessica. Puffin Books, New York. 64 pp.

Walcott, F.C. 1939. Hunting the Canada moose. Pages 266–272 *in* North American big game. Boone and Crockett Club Comm. on Records of N. Am. Big Game. Charles Scribner's Sons, New York. 533 pp.

Waldman, C. 1985. Atlas of the North American Indian. Facts on File Publ., New York. 276 pp.

Waldo, C.M. and G.B. Wislocki. 1951. Observations on the shedding of the antlers of Virginia deer *(Odocoileus virginianus borealis)*. Am. J. Anat. 88: 351–395.

Waldo, C.M., G.B. Wislocki and D.W. Fawcett. 1949. Observations on the blood supply of growing antlers. Am. J. Anat. 84: 27–61.

Wallace, B. 1975. Hard and soft selection revisited. Evolution 29: 465–473.

———. 1977. Automatic culling and population fitness. Evol. Biol. 10: 265–276.

Wallace, F.G. 1934. Parasites collected from moose, *Alces americanus,* in northern Minnesota. J. Am. Vet. Med. Assoc. 84: 770–775.

Wallace, G.I., A.R. Cahn and L.J. Thomas. 1933. *Klebsiella paralytica:* A new pathogenic bacterium from "moose disease." J. Inf. Dis. 53: 386–414.

Wallace, W.S., ed. 1932. John McLean's notes of a 25 years' service in the Hudson's Bay Territory [1849]. The Champlain Soc., Toronto, Ontario.

Wallis, W.D. and R.S. Wallis. 1955. The Micmac Indians of eastern Canada. Univ. Minnesota Press, Minneapolis.

Wallmo, O.C., ed. 1981. Mule and black-tailed deer of North America: Ecology and management. Univ. Nebraska Press, Lincoln. 624 pp.

Wallmo, O.C., L.C. Carpenter, W.L. Regelin, R.B. Gill and D. L. Baker. 1977. Evaluation of deer habitat on a nutritional basis. J. Range Manage. 30: 122–127.

Walters, C.J. and C.S. Holling. 1990. Large-scale management experiments and learning by doing. Ecol. 71: 2,060–2,068.

Walters, C.J., M. Stocker and G.C. Haber. 1981. Simulation and optimization models for a wolf-ungulate system. Pages 317–337 *in* C.W. Fowler and T.D. Smith eds., Dynamics of large mammal populations. John Wiley and Sons, New York.

Walther, F. 1966. Mitt Horn and Huf. P. Parey Verlag, Hamburg. 171 pp.

Walther, F.R. 1984. Communication and expression in hoofed animals. Indiana Univ. Press., Bloomington. 423 pp.

Walton, A. 1960. The male genital tract. Pages 133–160 *in* A.S. Parks, ed., Marshall's physiology of reproduction. Longmans, London.

Ward, C.C. 1878. Moose-hunting. Scribner's Month. Mag. XV: 449–465.

Warren, W. W. 1884. History of the Ojibway people. Coll. of Minnesota Hist. Soc. 5: 21–394.

Waser, P.M. and W.T. Jones. 1983. Natal philopatry among solitary mammals. Quart. Rev. Biol. 58: 355–390.

Waterman, C.F. 1973. Hunting in America. Holt, Rinehart and Winston, New York. 250 pp.

Watkins, W.G., R.J. Hudson and P.L.J. Fargey. 1991. Compensatory growth of wapiti *(Cervus elaphus)* on aspen parkland ranges. Can. J. Zool. 69: 1,682–1,688.

Webb, J.S. 1898. The river trip to the Klondike. Century Mag. LV: 672–691.

Webb, S.B. 1984. Early history of the Boone and Crockett records keeping. Pages 3–6 *in* H. Nesbitt and P.L. Wright, eds., Records of North American big game. Boone and Crockett Club, Alexandria, Virginia.

Webb, S.D. and B.E. Taylor. 1980. The phylogeny of hornless ruminants and a description of the cranium of Archaeomeryx. Am. Mus. Nat. Hist. Bull. 167: 117–158.

Weber, R.J. and A. Pert. 1989. The periaqueducal gray matter mediates opiate-induced immunosuppression. Science 245: 188–189.

Weber, W.A. 1960. An artist's glimpses of our roadside wildlife. Natl. Geogr. 93: 16–32.

Webster, A.J.F. 1971. Prediction of heat losses from cattle exposed to cold outdoor environments. J. Appl. Physiol. 30: 684–690.

Webster, A.J.F. and D. Valks. 1966. The energy cost of standing in sheep. Proc. Nutr. Soc. 25: 19.

Weckerly, F.W. 1992. Territoriality in North American deer: A call for a common definition. Wildl. Soc. Bull. 20: 228–231.

Wedeles, C.H.R., H. Smith and R. Rollins. 1989. Opinions of Ontario moose hunters on changes to the selective harvest system. Alces 25: 15–25.

Wein, R.W. and J.M. Moore. 1977. Fire history and rotations in the New Brunswick Acadian Forest. Can. J. For. Res. 7: 285–294.

Weiner, J. 1977. Energy metabolism of roe deer. Acta Theriol. 22: 3–24.

Weinmann, C.J., J.R. Anderson, W.M. Longhurst and G. Connolly. 1973. Filarial worms of Columbian black-tailed deer in California. I. Observations in vertebrate host. J. Wildl. Dis. 9: 213–220.

Weins, J.A. 1977. On competition and variable environments. Am. Sci. 65: 590–597.

Weiss, J. 1981. Deer do see color. Outdoor Life 167: 64, 108–117.

Weiss, R. 1990. Hints of another signaling system in brain. Sci. News 137: 54.

Welch, D.A., M.L. Drew and W.M. Samuel. 1985. Techniques for rearing moose calves with resulting weight gains and survival. Alces 21: 475–491.

Welch, D.A., W.M. Samuel and C.J. Wilke. 1991. Suitability of moose, elk, mule deer, and white-tailed deer as hosts for winter ticks *(Dermacentor albipictus)*. Can. J. Zool. 69: 2,300–2,305.

Wells, H.P. 1887. Moose hunting. Harper's New Month. Mag. LXXIV: 332, 448–461.

Welsh, D.A., K.P. Morrison, K. Oswald and E.R. Thomas. 1980. Winter habitat utilization by moose in relation to forest harvesting. Proc. N. Am. Moose Conf. Workshop 16: 398–428.

Wentzel, W. 1889–1890. Letters to the honorable Roderic McKenzie, 1807–1824. Pages 67–153 *in* L.R. Masson, ed., Les Bourgeois de la Compagnie du nord-ouest. Vol. 1. A. Coté, Quebec.

Wernick, R. 1993. At Monticello, a big birthday for the former owner. Smithsonian 24(2): 80–92.

Wesley, D.E., K.L. Knox and J.G. Nagy. 1973. Energy metabolism of pronghorn antelopes. J. Wildl. Manage. 37: 563–573.

Westoby, M. 1974. An analysis of diet selection by large generalist herbivores. Am. Nat. 108: 290–304.

———. 1980. Black-tailed jack rabbit diets in Curlew Valley, northern Utah. J. Wildl. Manage. 44: 942–948.

Westworth, D., L. Brusnyk, J. Roberts and H. Veldhuzien. 1989. Winter habitat use by moose in the vicinity of an open pit copper mine in north-central British Columbia. Alces 25: 156–166.

Wharton, D. 1972. The Alaska gold rush. Indiana Univ. Press, Bloomington. 302 pp.

Wheather, P.R., H.G. Burkitt and V.G. Daniels. 1979. Functional histology—Text and coloration atlas. Churchill Livingstone, Edinburgh. 278 pp.

White, G.C. 1983. Numerical estimation of survival rates from band recovery and biotelemetry data. J. Wildl. Manage. 47: 506–511.

White, G.C. and R.A. Garrott. 1990. Analysis of wildlife radio-tracking data. Academic Press, San Diego. 383 pp.

White, R.G. and M.K. Yousef. 1978. Energy expenditure of reindeer walking on roads and on tundra. Can. J. Zool. 56: 215–223.

Whitaker, G. A., E.R. Roach and R.H. McCuen. 1978. Inventorying habitats and rating their value for wildlife species. Proc. Ann. Conf. Southeast. Assoc. Fish and Wild. Agencies 30: 590–601.

Whitlaw, H.A. and M.W. Lankester. 1994a. A retrospective evaluation of the effects of parelaphostrogylosis on moose populations. Can. J. Zool. 72: 1–7.

Whitlaw, H.A. and M.W. Lankester. 1994b. The co-occurrence of moose, white-tailed deer and *Parelaphostrongylus tenuis* in Ontario. Can. J. Zool. 72: 819–825.

Whittaker, R.H., S.A. Levin and R.B. Root. 1973. Niche, habitat and ecotype. Am. Nat. 107: 321–338.

Whitten, K.R., G.R. Garner, F.J. Mauer and R.B. Harris. 1992. Productivity and early calf survival in the Porcupine caribou herd. J. Wildl. Manage. 56: 201–212.

Whymper, F. 1868. Travel and adventure in the territory of Alaska . . . . John Murray, London. 331 pp. (Reprinted in 1966 by Univ. Microfilms, Inc., Ann Arbor, Michigan.)

Whymper, F. 1869. Travel and adventure in the territory of Alaska, formerly Russian America—now ceded to the United States—and in other parts of the North Pacific. Harper and Bros., New York. 331 pp.

Whyte, J. and E.J. Hart. 1985. Carl Rungius: Painter of the western wilderness. Salem House, Salem, New Hampshire. 184 pp.

Wickler, W. 1961. Ökologie und Stammesgeschichte von Verhaltensweisen. Fortschr. Zool. 13: 303–365.

———. 1972. Verhalten und Umwelt. Hoffmann and Campe Verlag, Hamburg. 193 pp.

Wickler, W. and U. Steibt. 1983. Monogamy, an ambitious concept. Pages 33–50 *in* P. Bareson, ed., Mate choice. Cambridge Univ. Press, Canbridge, England.

Wickstrom, M.L., C.T. Robbins, T.A. Hanley, D.E. Spalinger and S.M. Parish. 1984. Food intake and foraging energetics of elk and mule deer. J. Wildl. Manage. 48: 1,285–1,301.

Wiens, J.A. 1977. On competition and variable environments. Am. Sci. 65: 590–597.

Wika, M. and L. Edvinsson. 1978. In vitro studies of antler blood vessels. Acta Physiol. Scand. 70A: 102–103.

Wika, M. and J. Krog. 1971. Surface temperature and avenues of heat loss during summer conditions in reindeer with growing antlers. Norw. J. Zool. 19: 89–91.

Wika, M. and J. Krog. 1980. Antler "disposable vascular bed." Pages 422–424 *in* E. Reimers, E. Gaare and S. Skjenneberg, eds., Proc. 2nd Int. Reindeer/Caribou Symp. Direktoratet for Vilt og Ferskvannsfisk, Trondheim, Norway.

Wika, M., J. Krog, P. Fjellheim, A. Blix and U. Rasmussen. 1975. Heat loss from growing antlers of reindeer (*Rangifer tarandus* L.) during heat and cold stress. Norw. J. Zool. 23: 93–95.

Wilbur, C.K. 1978. The New England Indians. The Globe Pequot Press, Chester, Connecticut. 108 pp.

Williams, E.S., E.T. Thorne and H.A. Dawson. 1984. Malignant catarrhal fever in a Shiras moose. J. Wildl. Dis. 20: 230–232.

Williams, G., ed. 1969. Andrew Graham's observations on Hudson's Bay, 1767–91. Vol. XXVII. Hudson's Bay Rec. Soc., London, England. 423 pp.

Williams, G.C. 1979. The question of adaptive sex ratio in outcrossed vertebrates. Proc. R. Soc. Lond., Ser. B, 205: 567–580.

Williams, M.L., ed. 1956. Schoolcraft's Indian legends. Michigan St. Univ. Press, East Lansing. 322 pp.

Williamson, V.H.H. 1961. Determination of hairs by impressions. J. Mammal. 32: 80–84.

Williamson, W.D. 1832. The history of the State of Maine: From its discovery A.D. 1602, to the separation, A.D. 1820. 2 vols. Glazier, Masters. Hollowell, Maine.

Wilson, D.E. 1971. Carrying capacity of key browse species for moose on the north slopes of the Uinta Mountains, Utah. Resour. Publ. No. 71-9, Utah Div. Wildl., Salt Lake City. 57 pp.

Wilson, M.J. and E. Spaziani. 1976. The melogenic response to testosterone in scrotal epidermis: Effects on tyrosinase activity and protein synthesis. Acta Endocrinol. 81: 435–448.

Wilton, M.L. 1983. Black bear predation on young cervids—A summary. Alces 19: 136–147.

———. 1992. Implications of hunting moose *(Alces alces)* during the period of pre-rut and rut activity. Alces 28: 31–34.

Wilton, M.L. and D.L. Garner. 1991. Calving site selection in south central Ontario, Canada. Alces 27: 111–117.

Wilton, M.L. and D.L. Garner. 1993. Preliminary observations regarding mean April temperatures as a possible predictor of tick-induced hair loss on moose in south central Ontario, Canada. Alces 29: 197–200.

Winick, M. 1976. Malnutrition and brain development. Oxford Univ. Press, Oxford, England. 169 pp.

Winthrop, T. 1861. Life in the open air and other papers. Ticknor and Fields, Boston. 374 pp.

Wirz, K. 1950. Studien über die Cerebralisation. Zur quantitativen Bestimmung der Rangordnung bei Säugetieren. Acta Anat. 8: 134–196.

Wishart, W.D. 1980. Perukes in wild moose. Can. Field-Nat. 94: 458–459.

———. 1990. Velvet antlered female moose. Alces 26: 64–65.

Wislocki, G.B. and M. Waldo. 1953. Further observations on the histological changes associated with the shedding of the antlers of the white-tailed deer *(Odocoileus virginianus borealis)*. Anat. Rec. 117: 353–375.

Wislocki, G.B., J.C. Aub and M. Waldo. 1947. The effects of gonadectomy and the administration of testosterone proprionate on the growth of antlers in male and female deer. Endocrin. 40: 202–224.

Witzel, D.A., M.D. Springer and H.H. Mollenhauer. 1978. Cone and rod photoreceptors in the white-tailed deer *(Odocoileus virgianus)*. Am. J. Vet. Res. 39: 699–701.

Wobeser, G. 1989. Atlas of important diseases of wild ungulates. Saskatchewan Park and Renew. Resour., Regina. 65 pp.

Wobeser, G. and W. Runge. 1975. Arthropathy in white-tailed deer and a moose. J. Wildl. Dis. 11: 116–121.

Wobeser, G. and T.R. Spraker. 1980. Post-mortem examinations. Pages 89–98 *in* S.D. Schemnitz, ed., Wildlife management techniques manual. 4th ed. The Wildl. Soc., Washington, D.C. 696 pp.

Wobeser, G., T.R. Spraker and V.L. Harms. 1980. Collection and field preservation of biological materials. Pages 537–551 *in* S.D. Schemnitz, ed., Wildlife management techniques manual. 4th ed. The Wildl. Soc., Washington, D.C. 686 pp.

Wobeser, G., A.A. Gajadhar and H.M. Hunt. 1985. *Fascioloides magna:* Occurrence in Saskatchewan and distribution in Canada. Can. Vet. J. 26: 241–244.

Wolfe, M.L. 1974. An overview of moose coactions with other animals. Nat. Can. (Que.) 101: 437–456.

———. 1977. Mortality patterns in the Isle Royale moose population. Am. Midland Nat. 97: 267–279.

———. 1982. Moose (Michigan). Pages 266–267 *in* D.E. Davis, ed., CRC handbook of census methods for terrestrial vertebrates. CRC Press, Boca Raton, Florida.

———. 1987. An overview of the socioeconomics of moose in North America. Swed. Wildl. Res. (Suppl.) 1: 659–675.

Wolff, J.O. 1980. Moose snowshoe hare competition during peak hare densities. Proc. N. Am. Moose Conf. Workshop 16: 238–254.

Wolff, J.O. and J.C. Zasada. 1979. Moose habitat and forest succession on the Tanana River floodplain and Yukon-Tanana upland. Proc. N. Am. Moose Conf. Workshop 15: 213–244.

Wong, B. and K.L. Parker. 1988. Estrus in black-tailed deer. J. Mammal. 69: 168–171.

Wood, T.J. 1974. Competition between arctic hares and moose in Gros Morne National Park, Newfoundland. Proc, N. Am. Moose Conf. Workshop 10: 215–237.

Working Margins Consulting Group. 1991. Self-government and co-management of natural resources for the Swampy Cree First Nations. Report prepared for the Swampy Cree Tribal Council, Winnipeg, Manitoba. 78 pp.

Worley, D.E. 1975. Observations on epizootiology and distribution of *Elaeophora schneideri* in Montana ruminants. J. Wildl. Dis. 11: 486–488.

Worley, D.E., C.K. Anderson and K.R. Greer. 1972. Elaeophorosis in moose from Montana. J. Wildl. Dis. 8: 242–244.

Wormer, J.V. 1972. The world of the moose. J.B. Lippincott Co., New York. 160 pp.

Worton, B.J. 1987. A review of models of home range for animal movement. Ecol. Model. 38: 277–298.

———. 1989. Kernel methods for estimating the utilization distribution in home-range studies. Ecol. 70: 164–168.

Wright, B.S. 1956. The moose of New Brunswick. A report to the New Brunswick Min. Lands and Mines, St. Johns. 63 pp.

Wright, G.F. 1892. Man and the glacial period. D. Appleton and Co., New York. 385 pp.

Wright, J.V. 1981. Prehistory of the Canadian shield. Pages 86–96 *in* J. Helm, ed., Subarctic. Handbook of North American Indians. Vol. 6. Smithsonian Inst., Washington, D.C. 827 pp.

Wurzinger, H. 1961. Die gegenwärtige Situation der Schnüffelkrankeit (SK) in der Tschechoslovakischen Sozialistischen Republik. Sbornik CSL. Akad. Zemed. Vet. Med. 6: 173–182.

Wuthier, R.E. 1984. Calcification of vertebrate hard tissues. Pages 411–471 *in* H. Sigel and A. Sigel, eds., Metal ions in biological systems. 17: Calcium and its role in biology. Marcel Decker, Inc., New York.

Wyatt, C.L., M. Trivedi and D.R. Anderson. 1980. Statistical evaluation of remotely sensed thermal data for deer census. J. Wildl. Manage. 44: 397–402.

Wyatt, C.L., D.R. Anderson, R. Harshbarger and M.M. Trivedi. 1984. Deer census using a multispectral linear array instrument. Proc. Int. Symp. Remote Sensing Environ. 18: 1,475–1,488.

Wyeth, N.C. 1912. The moose hunter, a moonlight night. Scribner's Mag. LII: facing 468.

Yazan, Y.P. 1961. Vliyanie temperatury vozdukha na aktivnost losy vo vremya ossenie-zimnoy migratsii. Zool. Zh. 40: 469–471.

Yazan, Y.P. and T. Knorre. 1964. Domesticating elk in a Russian national park. Oryx 7: 301–304.

Yellon, S.M., D.L. Foster, L.D. Lango and J.M. Suttie. 1992. Ontogeny of the pineal melatonin rhythm and implications for reproductive development in domestic ruminants. Anim. Reprod. Sci. 30: 91–112.

Youngman, P.M. 1975. Mammals of the Yukon Territoty. Publ. Zool., Mus. Nat. Sci. Canada, Toronto. 192 pp.

Yuncheung, K., K. Kamming, Y. Taitung and T. Saiwah. 1987. Epidermal growth factor of the cervine velvet antler. Acta Zool. Sinica 33: 301–308.

Zannier-Tanner, E. 1965. Vergleichende Untersuchungen über das Hinlegen und Aufstehen bei Huftieren. Z. Tierpsychol. 22: 696–723.

Zarnke, R.L. 1991. Serological survey of Alaskan wildlife for microbial pathogens. Fed. Aid Wildl. Restor. Final Rept. Alaska Dept. Fish and Game, Juneau. 58 pp.

Zarnke, R.L., R. Dieterich, K. Neiland and G. Ranglack. 1983. Serologic and experimental investigations of contagious ecthyma in Alaska. J. Wildl. Dis. 19: 170–174.

Zarnke, R.L., W.M. Samuel, A.W. Franzmann and R. Barrett. 1990. Factors influencing the potential establishment of the winter tick *(Dermacentor albipictus)* in Alaska. J. Wildl. Dis. 26: 412–415.

Zender, S. 1991. 1991 moose status report. Washington Dept. Wildl., Olympia. 7 pp.

Zheleznov, N.K. 1993. Historic and current distribution of moose in the northeast USSR. Alces 29: 213–218.

Index

Page numbers in bold refer to photographs; page numbers followed by *t* refer to tables; page numbers followed by *f* refer to figures